The hunger for salt. Two Merino sheep avidly ingesting a mixture of sodium chloride and sodium bicarbonate. The sheep on the left is salt deplete as a result of loss of alkaline parotid saliva from a unilateral parotid fistula. The teat of the fistula is visible on the animal's cheek. The sheep on the right is sodium replete but the concentration of sodium in cerebrospinal fluid (CSF) has been reduced 10 mmol/l by infusion of isotonic mannitol in artificial CSF into the lateral ventricle at 1 ml/h. The second infusion line connected to the head probes is available for episodic sampling of CSF composition from the contralateral ventricle. The reduction of the sodium concentration in the CSF is a powerful specific stimulus of salt appetite. The animal also has an autotransplant of the left adrenal gland into a combined carotid artery–jugular vein skin loop on the left side.

Derek Denton

The Hunger for Salt

An Anthropological, Physiological and Medical Analysis

With 346 Figures

Springer-Verlag
Berlin Heidelberg New York Tokyo 1984

Derek Denton (MB, BS, FRACP, FAA. Foreign Medical Member, Royal Swedish Academy of Sciences)

Director and Research Professor, Howard Florey Institute of Experimental Physiology and Medicine, University of Melbourne, Parkville 3052, Australia

Second printing of first edition 1984

ISBN 3-540-13480-8 Springer-Verlag Berlin Heidelberg New York Tokyo
ISBN 0-387-13480-8 Springer-Verlag New York Heidelberg Berlin Tokyo

Library of Congress Cataloging in Publication Data
Denton, Derek A. The hunger for salt. Bibliography: p. Includes index.
1. Salt in the body. I. Title. [DNLM: 1. Sodium chloride. 2. Appetite. QU 130 D415h]
QP535.N2D46 612'.01524 82-3336
ISBN 0-387-11286-3 (U.S.) AACR2

Filmset by Willmer Brothers Limited, Birkenhead, Merseyside.
Printed and bound in Great Britain by William Clowes Limited, Beccles and London.

2128/3916-54321

For Maggie, Matthew and Angus

Preface

This book, as described in the introductory chapter, aims to set the hunger for salt in its broad biomedical and anthropological context and therefore synthesize knowledge from many fields.

I am deeply indebted to many colleagues for the privilege of discussion, and the provision of source material and experimental data. This is acknowledged in the body of the text and also below. In the case of figures and photographs, the source (authors and publication) is noted in each instance by way of reference to the bibliography of each chapter. The permission to reproduce illustrations which has been generously given by individual authors and publishers of journals and books instanced is much appreciated in terms of allowing a wide representation of research from many laboratories.

By virtue of its scope, the book has turned out to be rather large. To facilitate access to the material, most chapters have an introductory summary which indicates the line of argument and some salient material covered in the chapter. Taken with the detailed Table of Contents, each summary should allow the reader to consider any particular aspects of interest. Sequential reading of the summaries alone will give an overview of the book, though, of course substantial themes which embody material from many sources may in some cases be omitted or receive scant mention in these summaries. The chapters differ greatly in their character. Some, like Chapter 7 on the physiological effects of sodium deficiency, represent in the main a repository of data for reference, whereas others, such as those on cannibalism, the genesis of salt appetite, salt and hypertension, electrical stimulation of the brain, salt appetite in reproduction, and the renin–angiotensin system and salt appetite, involve matter of considerable controversy. It seemed possible that many readers might be interested in particular fields as distinct from all areas in the book. Accordingly, in relation to experimental methodology and some concepts, there has been deliberately some brief repetition in the endeavour to make some chapters self-contained.

The author's interest and experimental work in the field of salt appetite covers some thirty years. Much of this time has involved concurrent investigation of the mode of control of aldosterone secretion and comparative physiological aspects of both questions. I am enormously indebted particularly to Professor R. Douglas Wright and Dr John Coghlan for ideas, suggestions, discussion, experimental planning and warm collaboration, and similarly to Dr Bruce Scoggins, Dr John Blair-West, Dr Marelyn Wintour-Coghlan, Dr James Munro and the late Dr James Goding. In the earliest stages of the medical work which led eventually to experimentation on sheep I had the privilege of working with Professor Victor Wynn, now of London University, Dr Ian MacDonald, now of Monash University, and Professor Henry Harris, now of Oxford University, as well as with Professor Wright. In the investigation of salt appetite in our laboratory and in experiments in the field, a major contribution was made in the early stages by Dr Elspeth Orchard, Dr Sigrid Weller, Dr Jack Nelson and Dr John Sabine, and later by Dr Richard Weisinger, Dr Michael McKinley, Dr Suzanne Abraham, Dr John McKenzie and Mrs Margaret Smith, and by our overseas visitors including Dr Edward Blaine, Dr and Mrs Leon Krainz, Dr David Mouw, Dr Robin Kay, Professor Alex Muller, Dr Eva Tarjan and Dr Lars Leksell, and many other colleagues as described in the text.

For the collection of reference material I am greatly in the debt of my research collaborators referred to above and also to Dr Judith Whitworth, Miss Theya Mollison of the British Museum (Natural History) in London, Dr Hugh Niall, Dr Sylvia Tait of the Middlesex Hospital in London, Ms Judith Pugh, the late Sir Russell Drysdale and Lady Drysdale, Dr Ken Myers, Dr Victor Benavides of Barcelona, Dr Carl Pfaffmann of New York, Professor Etienne Emile Baulieu and Professor Philippe Meyer of Paris, Ms Aldona Butkus, Professor Bengt Andersson of Stockholm, Dr Maria New of New York, Dr Curt Richter of Baltimore, Mrs Charles W. Engelhard of New York, Professor Stanley Peart of London, Dr William Oliver of Minnesota and Dr Lotte Schenkel of Basel.

I am extremely appreciative of the great help given me by Mrs Jane Farrell, copy editor for Springer-Verlag, Dr John Blair-West and Dr Margaret Sumner who have assisted in many ways with the editing of the text, and by Mrs Farrell, Dr Blair-West, Dr Eva Tarjan and Miss Angela Gibson in the proof-reading. Dr Blair-West has prepared the index of the book.

Many colleagues have generously read particular chapters or several of them in draft form and have made many suggestions, though they are not responsible for any shortcomings in the final material. My thanks are due to Miss Theya Mollison, Dr Bruce Scoggins, Professor Alex Muller, Dr Eva Tarjan, Dr John Coghlan, Dr Hugh Niall, Dr Carl Pfaffmann, Professor R. Douglas Wright, Dr Richard Weisinger, Dr Trefor Morgan, Dr Michael McKinley, Ms Judith Pugh, Dr David Mouw and Dr Jack Nelson. Miss Jane Ansett, Miss Margret Cooke, Mr John Hartley Clayton and Mr Sam Critchley prepared figures and photographs. In preparing and editing the references, much help was given me by Miss Ann Compton, Mrs Elsa Newman, Miss Jannette Tresham and Mrs Elizabeth Vorrath.

The large task over several years of typing the manuscript through many drafts was carried out by Mrs Kaye Thorpe and I am very grateful to her, and also to Miss Anne Fairbrace of London who typed the early stages of the work, and to Miss Janine Copland who helped at the end. Dr Kingsley Allen gave valuable assistance in organizing the text onto the University word processor.

The preparation of the book serves as an opportunity to express warm thanks to many generous benefactors who have supported this particular aspect of the work of the Howard Florey Institute of Experimental Physiology and Medicine. These include the National Health and Medical Research Council of Australia, the National Institutes of Health, Washington, the Myer Family Trusts, The Ian Potter Foundation, the Trustees of the Robert J. Jr and Helen C. Kleberg Foundation of Texas, The Sunshine Foundation, the Rural Credits Fund of the Reserve Bank and the Wool Industry Research Fund of the Commonwealth.

Finally, the task of writing, involving space to set out hundreds of reprints, books and other records, was made possible over the last few years by some periods of many months of seclusion away from the day to day pleasures of a busy research institute, and I am most deeply appreciative of the hospitality at Lodsbridge Mill with the care of Mrs Joyce Miles, and also of Sir Ian and Lady Potter at Lake Eucembene, of Dr Frank Tait in London, of Mrs Neilma Gantner at Sorrento, and of Lady Drysdale and the late Sir Russell Drysdale at Pittaedie in New South Wales.

Lodsbridge Mill,
Selham via Petworth,
United Kingdom
October 1981

Derek Denton

Acknowledgements

Where figures and tables from other publications have been reproduced in the book, beneath each there is acknowledgement of the authors, and the bibliographic references at the end of each chapter identify the journal, or book and publisher, from which permission was also obtained for reproduction. In addition, I wish to thank the following for kind permission for reproduction:

Fig. 3-4 was reproduced by permission of UNESCO; Figs. 3-5, 3-11, 3-16, 10-2, 16-4 and Table 16-4 were reproduced by permission of Pergamon Press Inc., New York and Oxford; Figs. 3-20, 10-10, 10-11, 11-10, 27-24 were reproduced by permission of the Rockefeller University Press, New York; Fig. 4-1 was reproduced by permission of the Royal Anthropological Institute of Great Britain and Ireland; Figs. 4-4, 4-5, 22-8, 22-9 and Table 22-3 are copyright of and were reproduced by permission of the American Association for the Advancement of Science; Figs. 5-2, 5-3 were reproduced by permission of H. Roger Viollet, Paris; quotations from the *Journals of Captain James Cook* edited by J.C. Beaglehole were reproduced by permission of the Hakluyt Society, The British Library, London; quotations from the *Endeavour Journal* of Sir Joseph Banks edited by J.C. Beaglehole were reproduced by permission of the Trustees of the Mitchell Library, Sydney; Figs. 7-3, 7-4, 13-15, 13-16, 15-1, 15-2 were reproduced by permission of the *Australian Journal of Experimental Biology and Medicine*, Dept. of Microbiology, Adelaide; Fig. 7-6 was reproduced by permission of Blackwell Scientific, Oxford, UK; Figs. 11-9, 12-14, 14-3, 14-22, 20-9, 20-10 and Table 16-3 are copyright of and were reproduced by permission of the American Psychological Association; Figs. 11-1, 13-11, 13-19, 14-4, 24-3, 26-5, 24-6, 24-7, 24-8, 24-9, 24-13, 24-15 and Tables 13-1, 20-4, 20-5, 20-6, 24-1 were reproduced by permission of Academic Press, New York; Fig. 24-4 was reproduced by permission of the New York Academy of Science; Figs. 19-5, 19-6 were reproduced by permission of the Williams and Wilkins Company, Baltimore; Fig. 20-14 and Table 20-10 were reproduced by permission of Elsevier/North-Holland Biomedical Press; Fig. 21-12 was reproduced by permission of Plenum Press, New York, from *Control of Renin Secretion* ed. T.A. Assaykeen (1972); Figs. 22-12, 22-13, 22-14, 24-36, 25-1 and Table 22-2 were reproduced by permission of Macmillan, London and Basingstoke; Fig. 27-16 was reproduced by permission of the *New England Journal of Medicine*, Massachusetts Medical Society; Fig. 27-1 and Table 27-1 were reproduced by permission of Grune and Stratton, New York; Figs. 27-6, 27-7, 27-8, 27-9, 27-29 and Tables 27-4, 27-5, 27-6, 27-7, 27-8, 27-25 were reproduced by permission of the American Heart Association Inc.; Fig. 27-2 was reproduced by permission of Pergamon Press, New York; Fig. 27-11 was reproduced by permission of the American Medical Association (copyright 1970).

Contents

1 Introduction: The scope of the analysis in this book

The elements sodium and potassium were discovered by Humphry Davy in 1807. This was done by passing an electric current through moist caustic potash or caustic soda. He was successful first with potash, the current causing minute globules of potassium to burst through the crust of potash and take fire. At the sight of this, Davy 'bounded about the room in ecstatic delight' (John Davy 1839).

In the century which followed, researchers in biomedical science defined the role of sodium as the principal cation of the circulating blood and tissue fluids of animals. It predominates over the other constituents of the tissue fluids, such as potassium, calcium, magnesium, phosphate, bicarbonate, sulphate and chloride. Normal sodium content is determinant of normal osmotic pressure, acid–base balance, and volume of these extracellular or tissue fluids—and therefore the integrity of the blood circulation, and the chemical milieu in which the various specialized and excitable tissues function. The regulation of sodium is central to the concept of the constancy of the *milieu intérieur*, as enunciated by Bernard and elaborated by Pfluger, Fredericq, Richet and Cannon. Curt Richter of Johns Hopkins University has pioneered study of the behavioural regulators of this constancy, in relation not only to sodium but to many other components of the body.

This book will consider the elective intake of sodium by the organism—sodium appetite, more commonly called salt appetite, or salt hunger. This will be done against a broad anthropological, physiological and medical background.

In large areas of the planet, as a result of meteorological conditions, there is very little sodium. Accordingly, for a wide diversity of animal species in different ecological niches, there is great survival value in the possession of effective brain mechanisms for acquisition of salt, and of endocrine mechanisms for its retention in the body. This will be considered at various levels of the evolutionary scale, and evidence presented showing sodium deficiency of animals is a quite common event in nature. In particular, the primates and feral man will be discussed because of the insight this may give into our contemporary civilized circumstances. The important role of salt in the history of civilization will be recounted. The thesis is that this is no accident when considered against the metabolic phylogeny of feral man and the selection pressures which must have operated—particularly during the reproductive process. Put another way, it is not fortuitous that salt is one of the four primary modalities of taste throughout the vertebrate kingdom. As with sweet and bitter for other reasons, evident advantage has accrued to animals in being able to detect salt and ingest it.

Because of the wealth of evidence to be presented in the course of physiological analysis that salt appetite is innate, I will consider eventually some general concepts and problems which have emerged in the study of instinctive behaviour. This involves the manner in which a genetically programmed neuronal organization subserving a particular behaviour is excited and a drive state generated. This, of course, draws heavily in discussion upon the distinguished studies of the ethologists. They have shown how fixed motor patterns are transmitted in the genome and inherited like physical traits. Distance receptor stimuli, often complex in nature and perceived as 'gestalts', act as the innate releasing mechanism of both simple and complex behaviour patterns. But we will be more concerned here with the manner in which chemical and hormonal changes within the *milieu intérieur* can induce genetically programmed behavioural drives. Along with salt appetite, these include the crucial vegetative functions of thirst, hunger, sexual and maternal behaviour, and appetites for other minerals. Comparisons of salt appetite with these functions, particularly thirst, will be made.

Following a short review of the general organization of control of sodium homeostasis in the higher mammals (Fig. 1-1), and the physiological effects of sodium deficiency, the methodology we and other scientists have used in the study of salt appetite will be surveyed. This will include an account of why our studies have centred on the sheep. I will explain the advantages in relation to the understanding of salt appetite in its broad

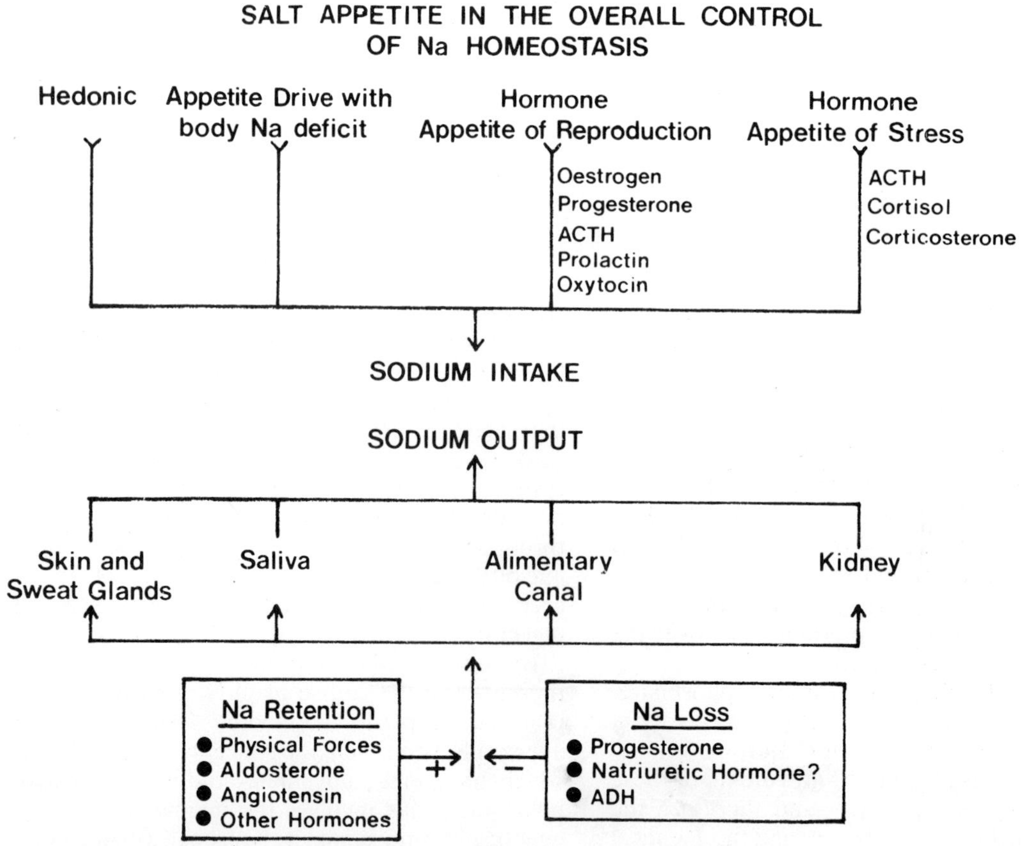

Fig. 1-1. Salt appetite in the organization of sodium homeostasis. Diagrammatic representation of the four mechanisms which may generate elective intake of sodium salts.In relation to control of sodium output from the body, aldosterone secretion reduces loss from the kidney and skin and reduces salivary sodium and increases sodium reabsorption from the gut. Prolactin also facilitates sodium retention. Sodium output is facilitated by antidiuretic hormone and progesterone, and, as well as physical intrarenal processes, a natriuretic hormone which stimulates sodium excretion is postulated.

biological context which I think have accrued from the choice of sheep as experimental animals.

The present knowledge of salt taste, electrophysiological and behavioural, and the influence of hormones on it will be reviewed.

Thereafter, some dozen chapters will be devoted to the formal physiological analysis of salt appetite. A central observation is that a sheep depleted of 500–700 mmol of sodium (twice the salt content of the blood plasma), when presented, for example, with a 300 mmol/l sodium bicarbonate solution, will rapidly drink in 2–5 min an amount closely related to the extent of body deficit. A precipitate decline of motivation follows. Two major questions which emerge are: How is an appetitive drive commensurate with the amount of body salt deficit developed in the brain? How does the appetite become satiated by drinking the correct amount, and drinking thereafter cease, long before the salt drunk could be absorbed from the gut and correct any deviation of body chemistry postulated as causing the appetitive drive?

These basic problems of motivation and satiation, as they present in the several species studied by us and others, have been dissected by physiological, surgical, electrophysiological, endocrinological and chemical techniques. The influence on the mechanisms revealed of other physiological states, in particular pregnancy and lactation, will be recounted. There are relatively recently discovered and remarkable influences of the hormones of the reproductive process on salt appetite. Reproduction requires extra sodium for the tissues of the developing young *in utero* and for lactation. Much of the earth is sodium deficient. Thus I will propose that our finding, that a spectrum of steroid and peptide hormones involved in reproduction can generate a powerful salt hunger in the sodium-replete animal, reflects a most important survival mechanism for a wide diversity of species in nature. We will consider, in the context of these findings, the appetitive aberrations of pregnancy in the human, and the historically fascinating subject of 'pica', aberrant eating of strange and nutritionally useless materials. To place the issue of salt appetite in a more general biological context, the data will be compared with those on appetite for phosphate, calcium, magnesium and potassium and also for the vitamins of the B complex. This opens the issue of learnt as distinct from innate appetite. It highlights the remarkably facilitated capacities of animals for learnt association of alimentary-related sensation with the taste and smell sensations which accompanied previous ingestion. This is so even if the ingestion has taken place many hours beforehand. This would appear to involve an extensive neurophysiological organization in which the solitary nucleus of the vagus is pivotal. Also it is necessary to appreciate and discuss the role of hedonic or palatability factors in determining intake of nutritious substances.

In the last part of the physiological analysis, we come to the theories of genesis of salt appetite. Here the neurochemical and neurophysiological changes which might be envisaged in the light of the known facts are discussed. In particular, this involves recognition of the need for any theory to encompass the fact that as well as response to sodium deficiency, the neural organization subserving sodium appetite may be powerfully activated in the sodium-replete animal by a number of steroid and peptide hormones. It is postulated that activation of salt appetite drive involves genetic transcription within the special neuronal systems which subserve this innate behaviour. The substantial body of data bearing upon the slow genesis of salt appetite, and its slow offset in the case of hormonal induction, is considered. The special neuronal system subserving salt appetite is evidently an endocrine target organ, and is thus analogous to the systems subserving sexual behaviour. Indeed the synthesis of the behaviour of the whole animal as it is manifest in the early stages of lactation necessitates the jointly sufficient and severally necessary action of four, perhaps five, steroid and peptide hormones in appropriate time sequence. In general, the contrast with the fast physiological reaction to water depletion and thirst is highlighted. However, recent work from our Institute has shown that acute alterations in ionic composition of cerebrospinal fluid and brain extracellular fluid may influence salt appetite much more rapidly than previously thought. Thus the possibility of direct influence at the level of translation or at synaptic membranes also arises—at least under experimentally contrived conditions. The possibility that intracerebrally generated angiotensin has a major role in the genesis of salt appetite in response to body sodium deficit is also extensively discussed.

Finally, the role of dietary salt in the genesis of the high incidence of hypertension in the modern affluent industrialized civilization is examined. No scientific meeting or symposium dealing with the experimental analysis of high blood pressure proceeds far without the role of sodium becoming crucial in the discussion. Thus the epidemiology of hypertension and its relation to salt content of diet and other possible factors is outlined. Animal experimental studies of hypertension are compared and contrasted with human data. With human communities in the past, as outlined in the early part of the book, salt was a very rare commodity and highly prized to the point of warfare. With the advent of technological civilization and commerce salt has become abundant and cheap. The demonstrably high intake from infancy onwards in modern communities raises the issue of the subtle aggregate effects of long-term exposure to great excess of a substance which is nonetheless a normal, indeed essential, component of the diet. The exposure over the growing period of life may be particularly important. There are, in my view, good grounds for proposing excessive salt intake, concurrent with other factors, is a major factor in the causation of the large incidence of high blood pressure in modern societies. Though some aspects of the epidemiological evidence are strong the case is not proven, and there are great difficulties in proving it beyond cavil. However, if high salt intake is not directly and commensurately causal, the high intake from infancy to middle life of a large segment of the population may be a necessary condition for the expression of other factors, including genetic ones, which contrive the hypertensive state.

2 Sodium: The main cation of body fluids

Summary

1. Sodium is the predominant cation of the circulating blood plasma and tissue fluids. Relative to the wide diversity of changes in morphology and function which have occurred during animal evolution, the remarkable constancy of the ionic pattern of *milieu intérieur* throughout the vertebrates suggests this pattern is a very basic character of animal life reflecting the phylogenetically determined set of a variety of regulatory functions.

2. This pattern possibly represents the composition of the late Cambrian ocean at the time when metazoan organisms first evolved a closed system of circulating fluids. The conditions under which cell life is possible are very restricted indeed, and have not changed substantially since life began (Baldwin 1948). Regulation of sodium is paramount to constancy of osmotic pressure, acid–base balance and volume of this circulating milieu. This involves brain mechanisms controlling elective intake of salt, and endocrine secretions modulating loss from the body. Evidence on the phylogenetic emergence of these systems at or near the stage of bony fishes is discussed.

Palaeochemistry of body fluids

In a discussion in *Physiological Reviews* in 1926, Macallum compared the blood sera of a range of invertebrate and vertebrate species. Whilst there is a wide range in the percentage concentrations of the electrolytes, there is a remarkable constancy in the ratios of the electrolytes. We have confirmed this finding with relation to ionic pattern in the Australian monotremes and marsupials (see Table 2-1). Macallum believed that this similarity in ratios of electrolytes throughout the animal kingdom points unmistakably to a common origin in the remote past, and holds that this dates 'from the time when Metazoan organisms, in which the internal medium was the then ocean water, acquired a closed circulatory system and developed organs which regulated and kept constant, within certain limits, the inorganic composition of the internal medium whatever changes in composition the external medium, the ocean, underwent in subsequent time—this endowment may be held to be exemplified in the Vertebrate.'

He observed that there still remained a striking resemblance in the ionic pattern of blood serum of mammals and the composition of the ocean today. This does not hold, of course, for concentration since the salinity of the ocean is approximately three times that of blood serum. He speculated that this pattern of composition arose in the late Cambrian when protovertebrates first appeared, and emphasizes the predominant role of the kidney in maintaining the set of these particular values. The anomaly in relation to magnesium, which is in much lower proportion in plasma than in sea water, where it is three- to fourfold greater than potassium or calcium, Macallum attributes to changes in ocean composition since the protovertebrates evolved.

Table 2-1. Plasma analyses of some Australian animals: a reptile, two monotremes and six marsupials. Analyses of Merino sheep and man are included for comparison. Findings confirm that the extracellular electrolyte patterns of phylogenetically widely separated animals are remarkably similar (Macallum 1926).

	Plasma (mmol/l)						
Animal	Na^+	K^+	Ca^{2+}	Mg^{2+}	Cl^-	PO_4^{2-}	Cl^-/Na^+ ratio
Blue-tongued lizard (*Teliqua scinoides*)	145	1.9	7.0	1.3	97	2.3	0.67
Platypus (*Ornithorhynchus*)	165	3.5	5.0	1.7	128	2.5	0.77
Spiny anteater: echidna (*Tachyglossus aculeatus*)	139	5.0	5.6	0.9	104	3.8	0.75
Short-nosed bandicoot (*Isoodon obesulus*)	153	6.0	4.4	1.7	123	4.2	0.80
Rat kangaroo (*Aepyprymnus rufesceus*)	149	4.6	4.4	2.0	107	3.8	0.72
Brush-tail possum (*Trichosurus vulpecular*)	154	4.8			117	4.5	0.76
Koala (*Phascolarctos cinereus*)	144	3.6	4.8	3.0	105	3.2	0.73
Wombat (*Vombatus hirsutus*)	138	5.8	5.2	2.1	97	2.9	0.70
Black-tail wallaby (*Wallabia bicolor*)	143	4.8	5.2	1.7	99	5.4	0.69
Merino sheep	145	4.4	4.2	0.7	106	4.4	0.73
Man	143	4.5	4.5	2.0	105	1.8	0.73
Platypus urine	69	46			35	66	

From Denton et al. (1951), with permission.

Of the nine or more inorganic elements which have been found to be constantly associated with living matter, the proportions are quite different in the cell to the medium which surrounds it. He sees the conditions which prevail within the cell, with the dominance of potassium, as reflecting a more ancient origin than that of the surrounding medium—indeed harking back to the time when all life on earth was unicellular. A gradual adjustment occurred in the far past between the cytoplasm and its internal inorganic content. This adjustment became stabilized when the superficial layer of the cytoplasm acquired the power of regulating the diffusion of inorganic salts into or from it. Were it not for this regulating function there would be a correspondence in inorganic composition between the cell and its external medium.

In putting forward this theory, Macallum notes that there are forms in the ocean today, such as *Limulus polyphemus* (the horseshoe crab), which have ancestral prototypes as far back as the Cambrian geological era and which have a blood plasma composition very similar to that of the contemporary ocean. He contrasts these, and like organisms, which have what he terms neochemical inorganic composition, with those, including ocean-dwellers such as whales, which he sees as having a palaeochemical composition of blood plasma.

Macallum's interpretation of the analytical facts has been criticized by Pantin (1931) and Baldwin (1939, 1948), who place much more emphasis on the dynamic aspects of the problem. They point out that physicochemical factors such as the diffusion of ions at different velocities may be important in determining the differences between the ionic pattern of the *milieu intérieur* and the ocean water: for example the different potassium to magnesium ratios could be due to the fact that potassium is a faster-moving ion than magnesium, rather than being a reflection of the ocean composition at the geological era when the extracellular fluid closed off. The comprehensive analyses of teleostean and elasmobranch plasma made by Homer Smith are in this regard consistent with the Macallum hypothesis, as are the analyses of echinoderm coelomic fluid and of jellyfish quoted by Prosser (1950). Further, the main stress on ionic pattern of these animals could arise from the ingestion of sea water swallowed with food. The work of Smith (1932) suggests that this is the case. If this water were totally absorbed from the gut, then much more magnesium would enter the extracellular fluid than potassium, despite ionic mobility differences. Thus the prevailing pattern would be much more a reflection of renal activity than differential absorption. The urine analyses of sea fishes made by Pitts (1934) and Grollman (1929) showed very high magnesium and sulphate excretion.

Relative to the wide diversity of changes in morphology and function which have occurred during animal evolution, the comparative constancy of the chemical pattern of the extracellular fluid suggests that this pattern reflects a very basic genetic character of animal life. The mechanism which determines and controls this pattern is the kidney. Hence the selective permeability of the membranes of the tubular cells, and the functional reserve of the enzyme systems responsible for the active reabsorption and secretion, are phylogenetically determined, and the 'set' of the integrated function is so as to produce the characteristic extracellular ionic pattern. This is exemplified by consideration of the very similar extracellular patterns of, for example, man, sheep, whale, platypus, marsupials, marine and terrestrial birds, fresh-water fish, reptiles, ganoids and goosefish (Pantin 1931; Baldwin 1948; Florkin 1949; Denton et al. 1951; Prosser and Brown 1961; Holmes et al. 1963), despite the fact that they are continually submitted to large distorting ionic stresses of quite different type as a consequence of marked differences in feeding habits. This dynamic situation is reflected in the marked difference of urine composition of man, cellulose-digesting ruminants, the fresh-water mud-sifting monotreme, and the sea-dwelling teleost. Clearly control of sodium balance is a primary condition in the maintenance of this constancy of the *milieu intérieur*.

Baldwin (1948) summarizes by emphasizing that the remarkable similarity in ionic composition of the blood of animals suggests that the conditions under which cell life is possible are very restricted indeed, and have not changed substantially since life began, and he concludes that evolution of the different forms of life has necessarily been attended by development of mechanisms to maintain these conditions.

Earliest vertebrates in the fossil record are the heavily armoured Ostracodermi, which were jawless and are grouped with the living cyclostomes in the superclass Agnatha. Homer Smith suggests that this heavy armour plating represented an adaption favouring survival in the face of the osmotic stress of the fresh-water environment. There are many theories as to the origin of the vertebrates from invertebrate forms. The Bateson–Garstang speculation, as termed by J. Z. Young, proposes that the chordates originated from a sessile creature feeding by means of a lophophore and rather like any annulate, and that initially only the larvae were fish-like with muscles, notochord and nerve tube. Another possible consideration is that the Grand Canyon uplift in the Cambrian period provided the condition of rivers flowing into the ocean. Protovertebrate types, with spindle body and segmentally arranged muscles adapted to rhythmic contractions, evolved in response to the motion of the rivers (Homer Smith 1953). I have discussed briefly some aspects of these theories elsewhere (see Denton 1965).

Evolutionary emergence of the renin–angiotensin system and aldosterone secretion

In the evolution of the vertebrates, the emergence of mechanisms which specifically maintained a normal body content of sodium, and thus a normal circulation of the blood and tissue fluids, was of cardinal importance. The evolutionary process involved the transition from marine, brackish and fresh-water habitats to dry land and many subsequent re-invasions of fresh water and ocean. The resulting stresses on sodium status have been correspondingly diverse. At its simplest, regulation involves intake and output; in the case of sodium, this regulation is both endocrine and neural. Sodium loss from the integument and from specialized cell surfaces, and in reabsorptive systems of excretory and secretory organs, has been modulated by development of special hormone systems with secretion in response to some consequence of deviation of the sodium content of the organism from normal. In this regard, renin seems to have appeared first during early evolution of the bony fishes (Sokabe et al. 1980). It has not been identified in the elasmobranchs or cyclostomes but incubation of kidneys with homologous plasma in holocephalians resulted in a pressor substance of low activity (Nishimura 1980). (The holocephalians are cartilaginous fish but differ from elasmobranchs.) The juxtaglomerular (JG) apparatus in non-mammalian vertebrates is incomplete. Primitive bony fishes, teleosts, lungfishes, amphibians and reptiles have granulated epithelioid cells resembling mammalian JG cells in the media of small arteries and arterioles of the kidney. Kidneys from aglomerular teleosts have granulated cells. These have not been found in cyclostomes or elasmobranchs but holocephalians have granulated cells differing somewhat in appearance and distribution from those in teleosts (Nishimura and Ogawa 1973; Nishimura 1980).

The possibility could exist that elasmobranchs and cyclostomes produce an angiotensin which is not pressor in mammalian systems—i.e. in the rat assay preparation (Ogawa and Oguri 1978;

Nishimura 1978). A teleostean angiotensin derived from the aglomerular kidney of Japanese goosefish has been identified as [Asn[1], Val[5]] angiotensin I (Sokabe and Watanabe 1977). Sokabe et al. (1978) speculate that with teleostean emergence into fresh water, it was necessary to excrete excess water, and the renin–angiotensin system appeared to increase glomerular filtration rate by virtue of its role as a blood pressure regulating system. Macula densa cells are first clearly apparent in avian species (Nishimura 1980). Fig. 2-1 depicts this phylogeny. Selected

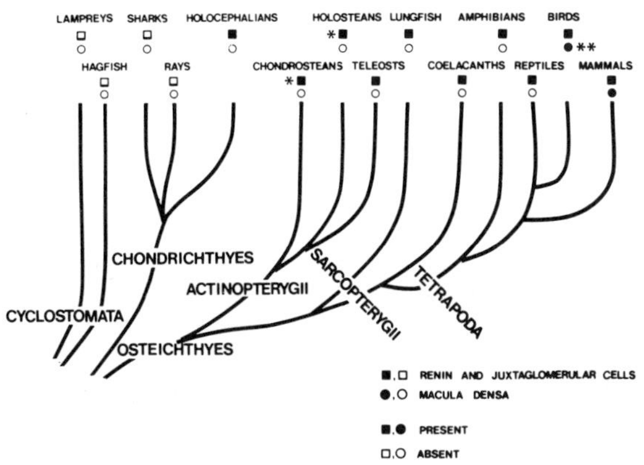

Fig. 2-1. Diagrammatic representation of the vertebrate phylogenetic tree and the presence or absence of the renin–angiotensin system among various classes of the vertebrates. *Renin activity was noted, but granules were not stained or showed different properties with Bowie's method. **A primitive form of a macula densa. (After Nishimura 1980.)

species of all classes from elasmobranchs to mammals give blood pressure increase with angiotensin, so presumptively the angiotensin receptor appeared early in vertebrate phylogeny. In relation to extrarenal renin in primitive forms, the Corpuscles of Stannius (CS) from the European eel (Chester Jones et al. 1966), carp, goldfish and aglomerular Japanese goosefish (Sokabe et al. 1970) form an angiotensin-like substance when incubated with homologous plasma. The granules in the CS stain with Bowie's method but differ slightly from JG granules of the kidney. The CS contain a substance termed 'hypocalcin' that decreases plasma calcium (Pang et al. 1974) and it remains to be seen whether CS angiotensin or renin is chemically the same substance (Nishimura 1980).

With relation to emergence of aldosterone secretion, we were unable to show significant production of the hormone following incubation of bronze whaler shark and coachwhip ray interrenals with [^{3}H]progesterone (J. McDougall, J. Coghlan, E. Blaine and D. Denton, unpublished). Blood aldosterone level in six elasmobranchs was low (49±11 pmol/l) and corticosterone and cortisol were 3.2±0.8 nmol/l and 0.83±0.3 nmol/l respectively. Multiple derivative formation and chromatography were performed to check specificity of these determinations. Bern et al. (1962), however, found aldosterone present in incubates of interrenal tissue from three species of chondrichthyean fishes (dogfish, skate and ratfish). The precursor was [^{3}H]progesterone. This is presumptive evidence that aldosterone is formed in vivo in elasmobranchs. Aldosterone is present in the blood of the bony fishes (Phillips et al. 1959; Idler et al. 1959; Denton, 1965). In sodium-deplete teleosts, in contrast to mammals, plasma renin either decreases or does not change. Plasma renin increases during sea-water adaptation. Nishimura (1978) suggests these findings may be understandable on the basis that angiotensin stimulates cortisol secretion in teleosts. Cortisol has an important role in sea-water osmoregulation in teleosts by promoting extrusion of sodium across the gill and increasing water absorption from the gut. Thus both renin activity and cortisol increase in plasma during sea-water adaptation of fresh-water eels. She also notes that there is not a correlation between plasma renin activity and mineralocorticoid secretion in the toad adapted to fresh water (Garland and Henderson 1975) although administration of angiotensin increased aldosterone and corticosterone secretion in the frog (Johnston et al. 1967).

The osmoregulatory role of prolactin was reported first by Pickford and Phillips (1959) who showed the hormone would keep the hypophysectomized killifish *Fundulus heteroclitus* alive in fresh water. In teleosts, it is the principal fresh-water adaptation hormone, and high prolactin secretion occurs in fresh water or in face of dilution of plasma osmolality. Cortisol is the main sea-water adaptation hormone and acts synergistically with prolactin also in fresh water. Prolactin acts by reducing ion permeability in the gills preventing passive loss of sodium. It also inhibits osmotic inflow of water. In the kidney in the fresh-water fish it has a diuretic effect and may increase sodium absorption in some species (Hirano 1977, 1980). In hypophysectomized *Fundulus*, prolactin increases renal Na,K-ATPase (Bern 1975; Hirano 1977). Prolactin receptors are found in mammalian kidneys and it influences renal formation of cyclic AMP. It modulates the renal effect of other hormones rather than being a primary controlling factor (Horrobin 1980).

Evolutionary emergence of salt appetite

In parallel though not contemporaneously, in the course of evolution of the brain there has emerged the capacity to react to depletion of specific body components such as sodium by genesis of a specific appetitive drive. The ability to detect the substance in the environment and to repair the deficit quantitatively by a consummatory act of ingestion has conferred a great survival advantage in allowing free-moving species to range over more diverse environments.

The exact period of emergence of the salt appetite neurophysiological system is unknown. The experiments on kangaroos in the Snowy Mountains of Australia reported in Chapter 3 indicate clearly a specific sodium appetite organized in the brain of the marsupial; this takes the phylogenetic emergence at least back to the Cretaceous period. It was suggested (Denton 1965) that the manifestation of saline preference in the migratory euryhaline teleosts may be the first emergence of systems eventually subserving salt appetite. Stevens (1973) states that the olfactory epithelium of fish is sensitive to salinity and to temperature. Thyroxine treatment, which has been shown in mammals to cause increased usage of ATP by the sodium pump (Ismail-Beigi and Edelman 1970) alters the sensing of sodium, and also alters the adaptive responses to changes in the salinity of the environment. Thyroxine increases the sensitivity of olfactory epithelium to the detection of salinity (Oshima and Gorbman 1966) and causes a preference for increased salinity, while thiourea favours fresh-water preference (Baggerman 1963). Thus there is some suggestion of neural–endocrine interaction in behaviour towards salt emergent at this phylogenetic level, but the subject is more or less unexplored below the Metatheria, though some data exist on insects, as discussed in Chapter 10.

The way in which the central nervous and the endocrine systems have become functionally interlinked in the organization of sodium control in higher mammals, as a consequence of their parallel evolution, is a most interesting field of enquiry. This question will be dealt with in Chapters 18, 21 and 23.

References

Baggerman B (1963) Effect of TSH and antithyroid substances on salinity preference and thyroid activity in juvenile Pacific salmon. Can J Zool 41: 307

Baldwin E (1939) Perspectives in biochemistry (ed J Needham). Cambridge University Press, p 99

Baldwin E (1948) An introduction to comparative biochemistry, 2nd edn. Cambridge University Press

Bern HA (1975) Prolactin and osmoregulation. Am Zoologist 15: 937

Bern HA, deRoos D, Biglieri EG (1962) Aldosterone and other corticosteroids from chondrichthyean interrenal glands. Gen Comp Endocrinol 2: 490

Chester Jones I, Henderson IW, Chan DK, Rankin JC, Mosely W, Brown JJ, Lever AF, Robertson JIS, Tree M (1966) Pressor activity in extracts of the corpuscles of Stannius from the European eel (*Anguilla anguilla* L.). J Endocrinol 34: 393

Denton DA (1965) Evolutionary aspects of the emergence of aldosterone secretion and salt appetite. Physiol Rev 45: 245

Denton DA, Wynn V, McDonald IR, Simon S (1951) Renal regulation of the extracellular fluid. II. Renal physiology in electrolyte subtraction. Acta Med Scand 140 [Suppl 261]: 1

Florkin M (1949) Biochemical evolution (Translated by S Morgulis). Academic Press, New York

Garland HO, Henderson IW (1975) Influence of environmental salinity on renal and adrenocortical function in the toad *Bufo marinus*. Gen Comp Endocrinol 27: 136

Grollman A (1929) The urine of the Goosefish (*Lophius piscatorius*): Its nitrogenous constituents with special reference to the presence in it of trimethylamine oxide. J Biol Chem 81: 267

Hirano T (1977) Prolactin and hydromineral metabolism in the vertebrates. Gunma Symp Endocrinol 14: 45

Hirano T (1980) Prolactin and osmoregulation. In: Proceedings of the Sixth International Congress of Endocrinology, Melbourne. Australian Academy of Sciences, Canberra. p 186

Holmes WN, Phillips JG, Chester Jones I. (1963) Adrenocortical factors associated with adaptation of vertebrates to marine environments. Recent Prog Horm Res 19: 619

Horrobin DF (1980) Prolactin as a regulator of fluid and electrolyte metabolism in mammals. Fed Proc 39: 2567

Idler DR, Ronald AP, Schmidt PJ (1959) Isolation of cortisone and cortisol from the plasma of Pacific salmon (*Onchorynchus nerka*). J Am Chem Soc 81: 1260

Ismail-Beigi F, Edelman IS (1970) Mechanism of thyroid calorigenesis: Role of active sodium transport. Proc Natl Acad Sci USA 67: 1071

Johnston CI, Davis JO, Wright FS, Howards FS (1967) Effects of renin and ACTH on adrenal steroid secretion in the American bullfrog. Am J Physiol 213: 393

Macallum AB (1926) Paleochemistry of the body fluids and tissues. Physiol Rev 6: 316

Nishimura H (1978) Physiological evolution of the renin–angiotensin system. Jpn Heart J 19: 806

Nishimura H (1980) Comparative endocrinology of renin and angiotensin. In: Johnson A, Anderson R (eds) The renin angiotensin system 1980. Plenum, New York

Nishimura H, Ogawa M (1973) The renin angiotensin system in fishes. Am Zoologist 13: 823

Ogawa M, Oguri M (1978) Occurrence of renin-angiotensin system in the vertebrates. Jpn Heart J 19: 791

Oshima K, Gorbman A (1966) Influence of thyroxine and steroid hormones on spontaneous and evoked unitary activity in the olfactory bulb of goldfish. J Comp Endocrinol 7: 482

Pang PKT, Pang RK, Sawyer WH (1974) Environmental calcium and the sensitivity of the killifish (*Fundulus heroclitus*) in bioassays for the hypocalcemic response to Stannius corpuscles from killifish and cod (*Gadus morhua*). Endocrinology 94: 548

Pantin CFA (1931) The origin of the composition of the body

fluids in animals. Cambridge Philosophical Society, Biological Reviews and Proceedings VI: 459
Phillips JG, Holmes WN, Bondy PK (1959) Adrenocorticosteroids in salmon plasma (*Onchorhynchus nerka*). Endocrinology 65: 811
Pickford GE, Phillips JG (1959) Prolactin, a factor in promoting survival of hypophysectomized killifish in fresh water. Science 130: 454
Pitts RF (1934) Urinary composition in marine fish. J Cell Comp Physiol 4: 389
Prosser CL (1950) In: Prosser CL, Brown FA, Jr, Bishop DW, Jahn TL, Wuulf VJ (eds) Comparative animal physiology. Saunders, Philadelphia, pp 6, 75
Prosser CL, Brown FA Jr (1961) Comparative animal physiology. Saunders, Philadelphia, p 57
Smith HW (1932) Evolution of vertebrate kidney. Q Rev Biol 7: 1
Smith HW (1953) From fish to philosopher. Little Brown, Boston
Sokabe H, Watanabe TX (1977) Evolution of the chemical structure of angiotensins and their physiological roles. Gunma Symp Endocrinol 14: 83
Sokabe H, Nishimura H, Ogawa M, Oguri M (1970) Determination of renin in the corpuscles of Stannius of the teleost. Gen Comp Endocrinol 14: 510
Sokabe H, Pang PKT, Gorbman A (1978) Some frontiers in comparative studies of the renin angiotensin system. Jpn Heart J 19: 783
Sokabe H, Nakajima T, Ogawa M, Watanabe TX (1980) Evolution of the renin angiotensin mechanism. In: Proceedings of the Sixth International Congress of Endocrinology, Melbourne. Australian Academy of Sciences, Canberra
Stevens ED (1973) The evolution of endothermy. J Theor Biol 38: 597

3 The natural history of sodium deficiency and salt appetite in wild animals

Summary

1. The influence of marine aerosols does not extend far from the coast. As a result of meteorological conditions, large areas of mountains and the interior of continents are deficient in sodium.

2. Sodium content in plants is accordingly very low. Many animal species show unequivocal evidence—behaviourally, metabolically and endocrinologically—of sodium deficiency. This involves both wild animals and domesticated grazing herbivores on several continents where such studies have been made.

3. Primitive human communities may exist with sodium status marginally above deficiency.

4. Many stresses may aggravate the environmentally determined deficiency. These include pregnancy and lactation, temperature control, infectious disease and parasitism, trace metal imbalance and increased population density with its contingent endocrine disruption of sodium conservation.

5. Over a geological time scale sodium deficiency has been a powerful selection pressure on vertebrates occupying diverse ecological niches. Considerable survival advantage accrued with the phylogenetic emergence of central nervous mechanisms for elective acquisition of salt—including the development of salt taste as one of the primary modalities of chemoreception. Parallel to these central nervous mechanisms, a similar advantage followed the emergence of endocrine systems facilitating salt conservation within the body.

6. The ruminant species represent a large proportion of the domesticated pastoral and the game animals of the world. In their rumen or forestomach at any one time they have sequestrated a large volume of sodium-rich salivary secretion as part of the dynamics of the digestive process. In stringent environmental conditions where sodium is deficient in food, or is lost in abnormal amounts, the animal can draw progressively upon these sodium reserves as a result of the influence of aldosterone hypersecretion by the adrenal glands. This lowers the Na/K ratio of the large salivary flow. The increased potassium for salivary secretion is readily available from the animal's food. A similar process may operate without large reduction of extracellular sodium content if environmental changes alter diet so that increased volumes of saliva are sequestrated in the rumen pool. It is proposed that the spectacular range of the reciprocal change in sodium and potassium concentration in ruminant saliva discovered to occur under the influence of aldosterone has been a major factor in the evolutionary success of the Ruminantia.

Introduction

Given the paramount role of sodium in animal metabolism, the issue arises as to whether or not it is readily available in all parts of the planet. It is abundant in the oceans, and is readily available in some volcanic regions such as the Rift Valley of Africa, and in sea coast areas. It may also be in rich supply under arid conditions where evaporation in inland drainage areas is high and accumulation occurs in soil. However, there are large areas of the planet where salt is very scarce. Accordingly it is present in only trace amounts in the plants. The reason is that, in the absence of geological sources, the source of salt is from rainwater. It has been established by the Swedish meteorologist, Eriksson (1952), and also by the Australian, Hutton (1958), that the sodium content of rainwater decreases rapidly with distance from the ocean so that 150 km or more from the shore, rain may be virtually distilled water. In the absence of these marine aerosols, alpine areas and interiors of continents may be severely deficient in sodium. In the alpine regions this may be accentuated as the result of the freezing and thawing breaking up the soil. The melting snow leaches out any soluble sodium present. Major monsoonal inundation in the interior of continents, such as, for example, in Central Australia, may also effectively leach sodium out of large areas, and concentrate it in restricted saline lakes and deposits. Our studies and those of Botkin, Jordan and colleagues, which will be instanced in detail below, have shown these environmental effects. The content of sodium is very low in plants in alpine areas, and in those of the mountain and plain regions moderately or far removed from the sea coast. (An exception may be halophytic plants growing in shallow lake water, which concentrate sodium.) Thus there is the possibility of sodium deficiency becoming a severe stress in these regions. In particular, it could jeopardize the all-important biological processes of reproduction. The mammal requires quite large amounts of sodium for the tissues of the developing young and for milk during the nursing period.

There is no reason to suppose that the basic conditions which have determined the incidence of environmental sodium deficiency in substantial areas of the planet have not operated over much of geological time. Given the phenomenon of continental drift and thus the much closer packing of the land masses in, for example, the Jurassic and Permian, it would follow that conditions in relation to sodium availability may have been more stringent in the very early periods of evolution of land-dwelling vertebrates.

In this chapter, I record first the evidence which has accumulated from historical and natural history records, and specific ecological studies on the salt-seeking habits of wild and domestic animals. This will lead up to studies conducted over the last decade or so which have shown, beyond cavil, that this well-attested behaviour has a metabolic basis.

Salt appetite in the wild

Herbivores

Zeuner (1954) suggests that at the outset of the process of domestication of wild animals, possibly in Magdalenian times, man appears to have availed himself of the eagerness of reindeer for salty matter, notably human urine. This substance attracts and binds reindeer to human camps. The craving arises from lack of salt in the available water, which is derived from melting snow. Even today the human nomads take full advantage of this weakness of the reindeer so that supply of this delicacy provides a meeting ground on which the social media of the two species overlap. This is a most remarkable parallel to the conditions under which societies of social insects are invaded. Some scavengers which have become adopted by ant colonies feed on the larvae of their hosts who nevertheless suffer their presence since they enjoy the flavour of the exudates.

The reindeer is not the only animal domesticated through salt licks. Another example is the mithan of Assam, a cross of the gaur with domesticated humped cattle. These humped animals are of considerable age since the Indus civilization of about 2500 B.C. was familiar with them.

In one of the tales in the Pentameron, Giambattista Basile (1575–1635) remarks on the gluttony of sheep for salt. Bunge (1873) commented on the fact that it was the herbivores rather than the carnivores that showed appetite for salt in nature. He stated that stags, deer, chamois and other plant eaters seek out salty pools and rocks to lick salt, and that hunters lie in wait for them in such places, or put out salt to attract them. There are places in the Altai Mountains where animals have licked clean entire caves in the salty soft argillaceous slates. In this, he was quoting Ledebour's journey through the Altai Mountains, published in Berlin in 1830. He stated that this sort of thing has never been observed in beasts of prey

Table 3-1. Analyses of samples of grasses from the regions of continental Australia in which the animals were studied.

			Concentration (mmol/kg dry wt)		
Location	Season	Sample type	Na	K	Cl
Snowy Mountains	Summer	Mixed	1.7	648	
		Mixed	1.3	170	
		Mixed	0.9	160	
		Mixed	1.9	117	
		Mixed	1.9	446	
	Autumn	Mixed	1.6	170	
		Mixed	2.0	160	
		Rumex	1.1	226	
		Poa	3.6	141	
		Restio	2.2	163	
		Stylidium			
		Seeds	20	103	
		Stalks	83	90	
		Seed husks	175	110	
	Winter	Mixed	3.8	50	
		Mixed	3.6	106	
		Mixed	0.6	52	
	Spring	Mixed	<0.2	428	
		Mixed	<0.1	485	
		Mixed	8.5	165	
Grasslands (Canberra)		Mixed	5	320	
		Mixed	10	280	
		Mixed	10	105	
		Mixed	5	450	
Central desert		*Atriplex*	2200	310	
		Mixed	200	110	
		Mixed (young shoots)	96	65	
Victorian sea coast		Mixed	309	349	
		Mixed	68	135	
		Mixed	90	173	
		Mixed	208	395	
Victoria (temperate coastal)					
		Oaten chaff (15 specimens)	136 (20–350)	233 (170–315)	172 (80–265)
		Lucerne chaff (14 specimens)	80 (20–200)	487 (360–650)	184 (150–210)
Gilruth Plains (Central Queensland)					
		Mitchell grass	3.5	70	31
		Blue grass	3	75	46
		Mulga–Mitchell	2.5	68	23
		Mulga	2.5	177	85
		Rosewood	3	380	103
		Wilga	1.4	395	150
		Gidgyea	2.2	195	230
		Leopard-wood	4.8	260	135
		Supplejack	1.4	220	37
Argadargarda (Central Northern Territory)					
		Windgrass	19	95	68
		Mulga	11	200	92
		Gidgyea	7	235	218
		Supplejack	16	270	58
		Whitewood	9	350	126
		Gidgyea (ground)	1.5	265	150
		Lucerne (pasture irrigated with bore water (Na^+ content, 5 mmol/l)) (1)	62	565	295
		(2)	67	570	280

and proposed that this was all the more striking because the amount of salt taken in the organic foodstuffs of herbivores was not at all smaller than that of the carnivores. He attributed this difference in behaviour to the different potassium content of the herbivores' diet. Bunge appears to have been unaware of the enormous variation that can occur in the sodium content of herbivores' food (see Table 3-1 as an example) and that sodium content may be exceedingly low. Also, any effect of high potassium intake on sodium excretion and balance is of short duration, and Richter (1956) has shown that substantial increase in the potassium content of an animal's diet does not increase its salt appetite. Various other data against this hypothesis bearing upon the great variation of Na/K ratio of herbage are cited by Russell and Duncan (1956).

Cambodia is one of the great gamelands of the world, the number of big game animals in northern and eastern Cambodia being second only to Africa. Here mineral or 'salt' licks are a conspicuous part in the life of most large hoofed animals, and Wharton (1957) observed herds of water buffalo, banting, kouprey, Eld's deer, wild cattle, elephant and wild pigs visiting them. A group of kouprey were observed to visit a particular lick about once a week. The average Cambodian lick is a stinking quagmire of mud pocketed with thousands of cattle tracks (Wharton 1957), to one side of which is the freshly exposed stratum with large holes several feet in diameter eaten out of it. Teeth marks are very much in evidence; elephants, though, dislodge the clay with their toenails and transport it in lumps to their mouth with the trunk. After heavy rains the licks fill with muddy water. Animals come still and drink. The lick was not salty to the human taste and distribution of several hundred pounds of rock salt at one lick did not increase the frequency of visits by wild cattle, but Wharton does not record whether it was taken. Termite mounds are one of the commonest form of licks. Either termites bring up soil from a lower mineralized layer or the termite-treated soil is attractive.

Wharton suggests that any animal whose diet contains a large amount of vegetation may seek mineral supplements. Thus earth eating is a natural phenomenon and should be distinguished from instances of depraved appetite caused by severe deficiencies—such as osteophagia (bone chewing). He notes porcupines will eat bacon but also devour axe handles impregnated with salts and oils of human perspiration. Dr R. L. Noble of British Columbia (personal communication) has described how porcupines may eat the verandah posts of cottages in the mountains where these posts have been impregnated with human urine.

The fact that the herbivores regularly visit the licks and have worn deep trails to and from them is taken advantage of by leopards, tigers and man for predatory purposes. Whilst the licks are generally in open terrain, giving the prey animals good surveillance, some studied were in heavy cover.

In relation to kouprey habitat in Cambodia, Wharton (1957) notes that there were plentiful licks where kouprey were, and no kouprey in areas where there were no licks. This fact could hardly be ignored in establishing refuges for the animals.

The material of the licks is usually gummy white or yellow clay in beds suggesting, by depth, that it is a hardpan with a concentration of minerals probably leached from the topsoil. There are also the termite mounds with a pH of usually around 8.

Theories for the use of salt licks noted by Wharton are: a hunger for a specific mineral such as phosphorus; a craving for sodium chloride; the need for some vital trace element; the alteration of intestinal flora and fauna; the elimination of intestinal parasites, and the chemical alteration of the acidity of the digestive tract. The consequences include important factors such as: the social value of having regular meeting places; the predation by carnivores and man due to continued visits along specific routes; the over-grazing of nearby range, and the passing or exchange of diseases and internal and external parasites by way of the lick itself or the nearby vegetation.

In India, Caius and Chhapgar (1933) analysed some licks and compared them with surrounding earths. They found that sodium and phosphorus were present but that there was no distinction between the licks and the surrounding regions. Cows, deer, chital, sambhar, barking deer, four-horned deer, and sometimes pigs, porcupines and black faced monkeys visit the licks by night. They propose the animals take the earth because it acts as a purgative.

Leek (1917) described how in the United States hunters would make a salt lick to lure deer in to be shot, or would lie in wait in the vicinity of salt licks. He noted how wild animals that range in the region of Jacksons Hole in the Western mountains must in many cases travel for miles to reach the nearest lick and that great trails are worn in the process, some of which have been in use for a considerable time. The earth at the licks is eaten by the animals in quantity for the salt or alkali which it contains, and without it they become sluggish, their hair has a dead look, their eyes lose their fire, their spirit and energy are diminished, and sometimes their actions denote debility. In the western United States in these mountain areas, all herbivorous animals—elk, deer, moose, antelope, mountain sheep, horses and cattle—seem to require either salt or alkali water. In the Jacksons Hole area, the few elk that range outside the limits of the natural alkali springs will travel for miles to the nearest lick and, on drawing near it, will come in on the run. They begin immediately to sip the water or eat the earth. They will then hang around the springs for several days before returning to their range. In general, where licks were in this terrain the elk would be found during the summer. Leek drew attention to the fact that salt grass and sage grass, which were a natural source of sodium for these animals, had become restricted and this had altered their behaviour in terms of winter range.

Herds of reindeer in the Arctic have been reported to drink sea water, and snowshoe hares have been seen gnawing wood which has been immersed in salt water (Bell 1963).

Discussing conditions in Africa, Orr (1929) states that a pica for earth and salt licks is reputed to occur in both east and west Africa, Somalia (Somaliland), Malawi (Nyasaland) and Uganda. The presence of salt licks has for generations been recognized by natives as a valuable asset to any grazing and they drive flocks and herds suffering from pica to these licks. European stock-owners have adopted this practice and found that when cattle were allowed occasional access to the salt licks they kept in better condition. French (1945) reviewed this question of geophagia in African animals and dealt with the analysis of many licks. In a considerable number of them, sodium could well have been the material which the animals found attractive, but this was not so clearly the case at others. Iron was sometimes a significant component but phosphate was usually low. French raised the question as to whether the attraction of some of these licks might be the presence of trace metals. He theorizes that pica may be associated with cobalt and copper deficiencies. This would seem surprising with cobalt deficiency, where anorexia is a characteristic feature.

Orr does note, with lick behaviour, the fact that animals may learn by experience. He quotes evidence of Rigg and Askew (1934) on this point. When sheep on 'bush sickness' pastures were given access to soil which would cure the disease certain sheep were still suffering from the disease after some months, and it was recorded that some animals were not eating the earth. French wondered whether sheep were less capable of profiting by 'flock experience' than cattle. Imitation appears to play an important part in the feeding habits of cattle. When a completely new food such as banana leaves or cottonseed is made available to a herd, they usually do not take it straight away and it is only consumed in any quantity when the herd has watched one or two of its bolder members eating it with relish.

Henshaw and Ayeni (1971) studied ten major mineral licks along the Gaji River in Nigeria that showed varying degrees of use and had craters of up to 1200 m^3. They made a total of 385 hours' observations at one lick. The use by different species was variable. Hartebeest, waterbuck and warthog made frequent visits (Western hartebeest travelling 10 km or more) and baboon were also seen, but other species did not appear. Peak use was at 14.00 hours in the day. Analysis of soil samples did not indicate that any particular element was attracting the wildlife, although the licks were richer in most minerals than were soil samples from surrounding areas.

Cowan and Brink (1949) discussed the natural game licks in the Rocky Mountain national parks of Canada. In British Columbia and Alberta, mineral springs attract game and natural mineral licks are abundant and widely distributed. In this area there are large populations of bighorn sheep, mountain goat, moose, elk, caribou, mule deer and white-tailed deer, all of which species visited the licks to a greater or lesser degree, though usually in the summer months only. Animals in their first summer were seldom observed to eat the soil of the licks or drink at the mineral springs. There was a difference between species in their preference for dry or wet licks but all made use of salt blocks when they were made available to them. Cowan and Brink record the amusing incident of being able to induce three bighorn sheep to consume an entire 3 pound trout by sprinkling the fish with salt. Bone chewing, suggestive of phosphate deficiency, is also commonly seen with the elk in the area. However, while sodium was present in the licks analysed by Cowan and Brink it could not be established as the source of attraction, and they drew attention to studies of Honness and Frost (1942) on bighorn sheep in Wyoming, where phosphorus was present in all five licks analysed but sodium in measurable quantities was present in only two. They record another instance in Alaska where calcium phosphate appeared to be the important element of attraction. They also noted that other possibilities, including trace elements, could not be excluded.

Stockstad et al. (1953) experimented on the chemical characteristics of natural licks used by big game animals in western Montana. Following Richter's approach they put out 16 mineral 'cafeterias', each consisting of a rack holding clay flower-pots containing mixtures of chemical compounds made up in non-lick soils. The compounds used were water soluble, contained elements known to be essential in animal nutrition, and were used in the amounts necessary to make them

Table 3-2. Comparative use by big game animals of mixtures offered in 16 mineral cafeterias in Western Montana for a 2-year period (1951–1952).

Compounds used	No. of cafeterias in which mixture was offered	No. of cafeterias in which mixture was used	Frequency of use (%)	Total amount of mixture offered (lb)	Amount of mixture used (lb)	Mixture consumed (%)	Relative use index
$NaHCO_3$	9	9	100.0	21	19.1	91.0	1.91
NaI	16	16	100.0	56	46.0	82.4	1.82
NaCl	16	16	100.0	56	40.0	71.5	1.72
NaH_2PO_4	16	16	100.0	56	41.1	62.3	1.62
$CoCl_2{\cdot}6H_2O$	16	5	31.2	56	2.8	5.0	0.36
KCl	16	4	25.0	56	2.0	3.6	0.29
$MgCl_2{\cdot}6H_2O$	16	3	18.7	56	1.7	3.0	0.22
$(NH_4)_2HPO_4$	6	1	16.7	26	0.7	3.0	0.20
$CaCl_2{\cdot}6H_2O$	16	2	12.5	56	1.5	2.7	0.15
H_3PO_4	5	0	0	8	0	0	0
KH_2PO_4	5	0	0	14	0	0	0
$Mg_3(PO_4)_2{\cdot}4H_2O$	1	0	0	2	0	0	0
NH_4Cl	16	0	0	12	0	0	0
HCl	4	0	0	9	0	0	0
$CuSO_4$	16	0	0	56	0	0	0
$FeSO_4{\cdot}7H_2O$	14	0	0	56	0	0	0
H_2SO_4	15	0	0	16	0	0	0
$(NH_4)_2SO_4$	8	0	0	28	0	0	0
CaI_2	5	0	0	11	0	0	0
MgI_2	4	0	0	6	0	0	0
NH_4I	5	0	0	11	0	0	0
$KHCO_3$	4	0	0	6	0	0	0
Control	16	0	0	56	0	0	0

From Stockstad et al. (1953), with permission.

equivalent to the number of sodium ions in 5 g of sodium chloride. The mixtures were weighed at intervals to determine their use by the animals over a period of 2 years (Table 3-2). To obtain a comparative use rating for each compound, an indexed number was calculated by adding the percentage frequency with which it was taken and the percentage of mixture consumed. On the basis of this rating all sodium compounds received a much greater use than did any other compound. Chloride compounds other than sodium chloride received only a minor amount of use. Phosphorus compounds other than sodium phosphate and ammonium phosphate received no use. A particularly interesting feature of the data, upon which the authors have not commented, was the use of cobalt by the animals. This occurred in five out of the 16 cafeterias, while in contrast copper and iron were not touched.

These workers also made an extensive study of the results of soil impregnation with various salts. Impregnation was done on a grid pattern, each compound being repeated three to five times in each grid. The type and amount of compounds used for treatment were the same as in the mineral cafeterias. Again, by far the predominant intake was of the sodium-treated soil. Soil treated with potassium chloride or with magnesium chloride was also licked extensively, and soil impregnated with cobalt chloride was used once in substantial quantity. Again copper and iron received no attention.

When 18 natural licks were analysed and compared with the non-lick areas nearby it was found that the average pH value of the lick samples was 8.37 while that of the non-lick samples was 6.72. The average salt content of all the lick samples combined was about five times that of the non-lick samples and sodium content on their analysis was as high as 100 mmol/kg in some instances. Calcium, magnesium and potassium were found in fairly large amounts in all lick areas but phosphorus was present in only very small amounts.

Dalke et al. (1965) made an interesting series of behaviour experiments on elk, deer and cattle in Idaho. These involved the distribution of salt blocks and observation of animal behaviour and attraction to them relative to the natural licks. They found in this mountainous area of Idaho, near the Selway River, that salt hunger developed in elk during May after they had been feeding on the succulent forage of spring and early summer for two to three weeks. The maximum desire for salt was in the period of late May/early June, which, coincidentally, was the time of calving amongst the elk. No actual attraction to salt was noticeable until about the third week in April, and before the middle of May the appetite of the animals for new vegetation was considerably greater than their desire for salt, for the elk were seen scattered widely throughout the area as distinct from being concentrated on artificial salt grounds. At most sites the salt blocks were preferred by both elk and deer, and were consumed before there was noticeable use of salt-impregnated soil. The dominant animals used the salt blocks while the subordinate animals competed for the salty soil. Competition between elk and livestock was also observed. The number of elk and cattle accommodated before a saturation point was reached depended on the size of the lick. Groups as small as five cattle and five elk or fewer used the lick without demonstrating animosity, but when ten or more domestic cows were on a lick, elk were usually driven away without resistance. Elk therefore normally used the cattle salt grounds early in the mornings or late in the evenings when the cattle were either feeding or bedded down on distant slopes. When elk were present on a lick, deer were immediately driven from it. All age classes of elk, except calves, were involved in driving deer away. All elk of the herd had the choice of lick cavities and salt blocks. The composition of the herds on artificial licks varied from spring to summer, but cows and yearlings were invariably the most numerous. This is an interesting observation in relation to the data that will be reported later on the influence of the hormones of reproduction on salt appetite. In relation to 22 natural licks that Dalke et al. analysed, the amount of sodium was relatively constant in all lick waters. It was not exceeded by any other element, except at one where there was a higher concentration of calcium than sodium. However, the amount of sodium was actually very low, 1–4 mmol/l. In some of the natural soil licks, phosphorus was found to be as high as sodium (6 mmol/l).

Knight and Mudge (1967) examined the characteristics of some natural licks in the Sun River area of Montana which is in the Sawtooth Range of mountains. They found that sodium bicarbonate and sodium sulphate are the only compounds that occur in much greater quantities in the licks than elsewhere, though boron and lithium occur in slightly larger quantities in the licks. Use of the licks was greatest during the spring and autumn. The groups of animals observed adjacent to the licks varied in size with the seasons, being 30–150 during spring and as many as 300 in late autumn. Knight and Mudge note that the results of their analyses substantiate those of most other investigators but that the reasons why ruminants prefer the sodium compounds are too little understood to permit more than conjecture on the subject.

Hebert and Cowan (1971b) studied the attraction of mountain goats to salt licks in southeast British Columbia. During early spring/summer the goats range high and descend to the licks, which are in the Douglas fir zone, making journeys of from 3 to 24 km to do so. In no instance did an animal make more than one trip from the uplands to the lick in a summer. The journey caused increased exposure to predation, but it was shown in an instance where a good new lick was opened up that the goats would cover the extra distance despite increased risk. The licks appeared to be at accumulation sites from higher runoffs, and were a dry brownish to whitish clay with a sodium content of 5–240 mmol/kg. Analysis of plants in the various regions (levels) did not indicate much seasonal variation in sodium content from a mean of 2 mmol/kg dry wt, but three or four sodium-accumulators were identified such as *Poa alpine* with a sodium content of ca 20 mmol/kg dry wt. Serum sodium levels gave no evidence of severe sodium deficiency (mean 150 mmol/l) but Hebert and Cowan did observe that with the succulent new growth forage of spring, goats coming to the licks had diarrhoea. Further, faecal sodium content was over twice as high in goats fed on succulent diet as in those on dry forage. Thus the spring seasonal change of diet could engender sodium loss and this was the time the animals searched out the licks. The females came later to the licks but this was attributed to their staying on the alpine areas until after the young were born. The authors note that Frens (1958) has reported in cattle that a diet of new-growth grass can increase faecal loss of sodium to the point where sodium deficiency is detected. Blaxter and Wood (1953) reported large losses of sodium in faeces of young calves with diarrhoea, as did Fayet (1968), and the transfer of large amounts of sodium into the intestinal lumen of calves with diarrhoea was shown by Whitten and Phillips (1971). However, Rook and Balch (1959) were unable to confirm that the more liquid faeces produced on spring pasture involved increased sodium loss. Reduction of dry matter was involved. It is of interest that working in the Rocky Mountain trench area of British Columbia, Hebert and Cowan (1971a) found a significant incidence of white muscle disease and high blood serum glutamic-oxaloacetic transaminase in wild goats associated with low selenium content of vegetation.

Shortridge's (1934) extensive review of the

mammals of South West Africa includes accounts of attraction of animals to salt. Eland, he notes, are browsers by inclination, favouring grass only when green and fresh, and sometimes cropping the top of young river reeds. They are harmless and gentle creatures, easily tamed even when quite large. As a rule 'they take a bullet silently'. They drink water regularly when it is at hand but in its absence may go for long periods without it, probably obtaining the liquid necessary for their subsistence from wild melons and various roots which grow in the desert parts of South Africa. Eland visit salt pans regularly and Wilhelm found as much as 1 kg of brak soil in the stomachs of some animals. The eland, which are described as particularly fond of salt licks, also drink the brackish water the licks contain during the rains. The licks are visited at night, whether moonlit or otherwise, and in undisturbed areas by day. He records in one set of observations how little antelope spoor of any kind is seen around water holes, in comparison with the trampled condition of salt licks. Kudu are also browsers, but occasionally, like other bush antelope, they venture into the open to feed upon young grasses. They are fond of wild fruit. They too visit salt licks, on moonlit nights. Wildebeest are reported as drinking water so salty that bream and barbel were dying in it (Cooke 1964).

V. Benavides (personal communication), who is very familiar with the Amazonian rain forest and the eastern slopes of the Andes, reports there is a very great scarcity of salt in the leached soils. He learned that the hunters of deer and jaguar, wild pigs and other smaller animals have specially preferred spots for finding these animals, usually in the mornings. These spots are either water holes or what they call *collpas*, that is, places where evidently there are salty waters or salty soils. All the jungle animals congregate there and eat or lick the salty soils.

Anthony Sutcliffe of the Palaeontology Department of the British Museum (Natural History) has made pioneering observations on the bone-chewing behaviour of phosphate-deficient wild animals, to be referred to in Chapter 25. He has also given a most interesting description of salt-seeking behaviour of wild animals in Africa (1973). The observations were made on Mount Elgon in the East African Rift Valley. Mount Elgon, with a basal breadth of over 70 km and rising to a height of 4321 m, is a volcano of the Mio-Pliocene age that contains innumerable caves. Since many of these caves are situated in thick forest, the number has never been actually determined, but Sutcliffe thinks on the published evidence that it must go to several hundred. They have been explored at altitudes ranging from 1800 to 3600 m. Geology of Mount Elgon, described by Davies (1952), is that of a soda-calcium-rich volcano made up predominantly of pyroclastic materials (tuffs and agglomerates) with a smaller proportion of lavas. Many animals (including elephants, buffaloes, antelopes and leopards) live in the forest zone of the mountain. The elephants occur mainly between 2500 and 3000 m, sometimes emerging at night on to the lower part of the moorland for grazing. The lowering of their territory is determined by the presence of man.

The caves of Mount Elgon show many remarkable features. These include the kinds of secondary mineral salts on the cave walls, former mining of the caves for these salts, and the fact that they are occasionally the habitation of man. They have a rich remaining fauna, composed partly of cave-dwelling species and partly of animals not usually attracted underground, but which have been drawn to the caves by the salt deposits there. The kinds of secondary mineral salts in the caves have been described by many writers, who have noted crystals of mirabilite (Glauber's salt, i.e. sodium sulphate). Many of the caves have been excavated and rock has been taken out by the local people and placed in troughs outside the caves for consumption by their cattle. Sutcliffe (1973) describes visiting a number of these caves with Anthony Perkins who had earlier described their habitation by man and wild animals (Perkins 1965). The most spectacular cave they visited was Kitum cave at 2360 m, which had a gigantic opening (70 m wide with a stream falling across it) leading to an extensive chamber with a level floor beyond which a rock-fall piled up to a height of about 16 m. There were secondary deposits of a salty substance on the cave walls just beyond the limit of daylight. The cave showed evidence of being frequented by many animals, in particular elephants, antelopes and a leopard. The level area at the front of the cave was strewn with elephant droppings. Some of the droppings were also lying on the lower part of the roof-fall, showing that the elephants had been climbing on the boulders at a place about 150 m from the cave entrance, where to human eyes at least, it was almost completely dark. At another place near the rock-fall, in complete darkness, was the mummified carcass of an antelope, which had apparently fallen into a gap between some large boulders while looking for mineral salt and had been unable to climb out.

Another cave, Makingnen cave, about a kilometre from Kitum cave, was noteworthy for the diversity of animals living there. Again, there was evidence of elephants in the chambers: trampled and fresh droppings where the cave was dry, extensive areas of water-filled footprints where the cave was wet, and tusk marks on the wall. Evidently, the elephants had also found their way into a small, wet chamber on the side of the cave and droppings were present there. Again, there was evidence that elephants were prepared to climb over a large area of fallen rocks, part of it 8 m above the general floor level at a place where it was almost dark. Local residents described how the elephants go from one cave to another. A big herd may remain around a cave entrance for up to two nights, going inside four to ten at a time. It was clear that they went inside for the salts, as did antelopes, and a Mr Barton who lived in the area showed Sutcliffe several sets of Elgon elephant tusks which had the tips worn down and abraded as a consequence of being struck against the cave walls. The elephants were at times observed to have accidents as a result of endeavouring to enter these caves. The local residents were of the view that the leopards that lived in the caves had their hunting facilitated as antelopes came there as a result of being attracted by the salt.

Sutcliffe raises the issue of the exact nature of the attraction of the caves to the animals. On the face of it, it would appear that the sodium in the sodium sulphate is the specific attraction, although he speculates whether there may be trace elements also attracting the animals. He notes that Watson (1969) has described the mining of Glauber's salt by pre-Columbian Indians in the salt caves in Kentucky and suggested that the salty seasoning may have been attractive to these people. The role of Glauber's salt as a purgative is also noted. Sutcliffe also remarks on the possible role of elephants in the excavation of some of these caves and records that a substantial overhang at a salt lake in the Ngorongoro Crater, Tanzania, has been made by elephants.

There are many other accounts of wild and domestic animals being attracted to salt. Hutyra and Marek (1922) reported the craving for saline material by mountain sheep who licked the perspiration-soaked clothing of mountain climbers. Loeb records how cattle will seek and lick the saddle cloths of horses. Salt is used to muster herds, for example in the highlands of southern Australia. Hemingway has described how, in the African game reserves, animals are attracted to salt pans and salt is distributed to attract animals to observation posts. Hunters lie in wait at the salt pans.

Primates

Jane Goodall (1963) has reported observations on a population of chimpanzee numbering some 60–100 animals in the Gombe Stream reserve in Tanzania. Over 800 hours were spent in direct observation of the chimpanzees, of which about 300 hours were concerned with actual feeding behaviour. She notes that in March and April, during the rainy season, the chimpanzees visit a certain cliff near the southern boundary of the reserve, where on five occasions they were observed scraping soil by hand and eating it. The animals concerned were all females, including four with infants between 2 and 3 years of age. In each instance the individuals remained by the cliff for about 30 minutes. Analysis of a sample of the soil showed that a small amount of sodium chloride was present.

Towards the end of the field study a 1 kg red block of mineral salts, as used for cattle, was put out. The following day it was discovered that the block had been removed. During the afternoon a group of chimpanzees, one of which was carrying the block of salt, were seen coming over the mountain ridge on the other side of the valley. Four other chimpanzees were following, stretching out their hands in a begging gesture. By that time the block had been carried for at least 3 km.

George Schaller (1963) in his studies on the mountain gorilla in Central Africa found three places at which gorillas had eaten soil. In two cases the feeding areas were located on the slopes of Mount Mikeno. The gorillas ate volcanic soil from small bare patches at the base of bluffs where the substratum was quite dry, crumbly and of a light brown to grey colour. The soil tasted chalky to Schaller. On one occasion he observed a black-back male, a silver-back male and a female as they stood on a patch of bare soil about 5 m long and 2 m wide. They fed intently, touching their mouths directly on the ground or picking up small objects between the tip of the thumb and the side of the index finger. Examination later showed that they had scraped along the soil with their upper incisors leaving grooves. On another occasion a group of animals appeared in a rainstorm with soil-smeared faces resembling grey masks. Inspection of the area the following day revealed the animals had scraped the soil with their incisors, and older tooth marks indicated that the soil had been eaten on previous occasions. Schaller collected soil samples from the three sites. Sample 3 from Miya, Utu region, showed little sodium (7 mmol/kg). Sample 1 from Mount Mikeno was very high in potassium (217 mmol/kg), calcium and magnesium and quite high in sodium (150 mmol/kg). Sample 2 had a sodium content of 477 mmol/kg and also had a high content of calcium, magnesium and potassium. He notes that the gorillas ate the soil occasionally but they did not seek out the sites habitually. At least three times groups were observed to pass within 30 m of the locality without deviating from their course.

Harrison (1960) records that earth was regularly eaten by all young orang-utans from an early age—as soon as they left the nest area (ca 14 months) and began coming down to the ground.

Baboons in the Cape region of Africa are recorded by Hall as eating mussels and crabs, which coincidentally would have a high salt content. Baboons have also been seen to eat clay—a fact confirmed by de Vore (1964).

Sweet potato washing (SPW) is an example of preculture characteristic of the troop of monkeys (*Macaca fuscata*) in Koshima (a small islet in Majazaka Prefecture, Kyushu, Japan). It was observed initially in 1953 by Kawamura and Kawai (Kawamura 1954; Kawai 1965). The behaviour began with a single monkey of a wild troop—a 1.5-year-old female. This inventive behaviour was propagated to others and by August 1962, 73% of 49 monkeys above 2 years old did this. There were two periods of propagation. In the first stage (1953–1957) propagation was by lineage and playmate relationship. The behaviour was acquired at the age of 1 or 2 years without any difference due to sex. Monkeys older than 5 years who acquired the behaviour were all females. Adult monkeys who did not acquire the behaviour could not acquire it even after 5 years. In the second period of study (1958–1964) propagation was from mother to young. Almost all young born after 1958 acquired SPW behaviour.

In SPW behaviour (Fig. 3-1), fresh water was used at first, then gradually salt water was used. Following the progress of the behaviour, it was apparent that some monkeys washed potatoes in fresh or salt water. Kawai's (1965) observations led to the view that the monkeys much preferred salt water but subordinate monkeys often avoided coming to the shore for fear of the dominant ones. Kawai thinks they prefer the taste with salt water. In some of the monkeys that acquired the behaviour in the second period a third develop-

Fig. 3-1. Sweet potato washing (SPW) in salt water by a monkey (*Macaca fuscata*) in Kyushu, Japan. (Photograph kindly supplied by Professor Kawai, The Japan Monkey Centre, Aichi.)

ment was seen, which consisted in dipping the potato into the water every time after gnawing it once or twice. Kawai regards the monkeys to be seasoning the potato with salt water—essentially a hedonic taste effect. Parallel to this it was noted that when a potato was covered in sand, the sand was removed by a complete brushing-wash behaviour. The monkey has the technique of a brushing-wash but uses it, it seems, only when it feels it necessary. This would not appear to be the explanation of the seasoning behaviour which is, on the face of it, due to salt taste. Other practices observed to develop included wheat washing and swimming during which seaweed was collected as food.

The fact that monkeys indulge in sham louse-picking or grooming is well known to students of primate behaviour and to zoo keepers. As Ewing (1935) records, Knottnerus-Meyer, Director of the Zoological Gardens of Rome, states:

> In this connection it should be noted monkeys are *not* hunting for fleas, since healthy monkeys are never or only very rarely afflicted with these tormentors. The toilet consists rather of combing out the fur; and what the monkeys are constantly searching for and eating is the crystallized secretions of the skin, whose salty flavour tickles their palates. So long as monkeys are able to bathe frequently, as in the Roman monkey paradise, it will be found that they very rarely indulge in these so-called flea hunts.

Ewing notes sham flea-hunting is only one of the general behaviours of monkeys coming under the heading of grooming. On this same general theme Zuckerman (1932) writes,

> Monkeys and to a lesser extent apes spend a great part of the day grooming one another. An animal will carefully examine a fellow's coat with its fingers, eating many of the odds and ends that it finds. The fruits of the search generally turn out to be small, loose, scaly fragments of skin, particles of skin secretion, thorns and other foreign matter.

Ewing, however, in examining a variety of monkeys at the National Zoological Park in Washington, was unable to see any crystallized secretions, or any scaly fragments of the same, even with a hand lens. From an experiment with human sweating, he draws attention to the minute nature of the crystals which are formed. However, he showed that if, unbeknown to the monkey concerned, he shook some salt crystals into its fur, a second monkey on arriving at the area would continuously and enthusiastically groom and lick the first. Ewing thinks that the sweaty secretions of the skin impart a salty flavour to any particles adhering to the surface and also, in particular, to the hairs. Thus grooming behaviour involves removal and eating of loose hairs. Ewing notes in his zoo garden studies that the 'louse-hunting type' of grooming was seen in nearly all species of Old World monkeys, but has been observed in only a few New World species.

The metabolic basis of salt hunger in wild and range domestic animals

Australia

Knowledge of some aspects of the behaviour of wild animals cited above, together with a particular body of experimental data arising from a laboratory study on sodium balance in ruminants, led us into this work at the Howard Florey Institute.

Laboratory data on sheep

Relative to that in other mammals, the volume of digestive secretions in ruminants is large. This is mainly because of the large continuous alkaline salivary secretion (Denton 1957; Hyden 1958; Kay 1960) which buffers the acid products of microbial fermentation in the rumen. In sheep and goats, the daily salivary secretion of 10–15 litres contains more than five times the sodium content of the blood plasma. In cattle the daily volume of salivary secretion of similar sodium concentration may be ten times this amount (Bailey 1961a).

It is clear that particular ecological conditions could involve considerable stress on sodium balance and circulatory homeostasis. In arid continental or mountain pastures, sodium intake may be much reduced. The type of food may increase the rate of digestive secretion and thus the volumes of sodium-rich fluid physiologically sequestrated in the gut, or alternatively may cause mild diarrhoea and sodium loss.

The normal cation content of ruminant parotid saliva is: sodium 160–175 mmol/l, potassium 4–6 mmol/l, Na/K ratio 25–40 (McDougall 1948; Denton 1957; Kay 1960; Bailey 1961a, b; Bailey and Balch 1961a, b) (Fig. 3-2). In a sheep with a permanent unilateral parotid fistula, the sodium deficit resulting from the loss of salivary secretion causes the salivary sodium concentration to fall and potassium to rise reciprocally so that the Na/K ratio of the continuous parotid secretion may decrease to 1.0 or less (Denton 1957; Denton et al. 1959; Kay 1960) (see Fig. 3-3 and Table 3-3). Sheep may survive for weeks with a large residual sodium deficit, and with potassium as the predominant cation of the salivary secretions (Denton et al. 1961) The same finding was made in

Table 3-3A. Parotid fistula no. 1: The effect of sodium withdrawal on electrolyte balance (21 May–28 June 1954).

	Na (mmol)			K (mmol)			Cl (mmol)			Weight at end of period (kg)
	Total intake	Total output	Balance	Total intake	Total output	Balance	Total intake	Total output	Balance	
Control period, 5 days	3448	3392	+56	2013	2104	−91	878	917	−39	31.9
Withdrawal period, 14 days	1223	2193	−970	5635	5378	+257	2275	2516	−241	29.7
Post-withdrawal period: A, 7 days	4232	3570	+652	2698	3209	−511	1103	878	+225	32.7
Post-withdrawal period: B, 12 days	8274	7988	+286	4968	4998	−30	1890	1850	+40	32.0

Reprinted from Denton (1957), with permission.

Table 3-3B. Parotid fistula no. 1: The relation of parotid salivary sodium and potassium concentrations to the external sodium balance.

Period	Na intake (mmol)	Na output (mmol)	Daily Na balance (mmol)	Cumulative Na balance for period (mmol)	Saliva Volume (l/day)	Saliva Na (mmol/l)	Saliva K (mmol/l)
Control period, 5 days	3448	3392	+11	+56	3.12	178	21
Na withdrawal							
1st day	87	415	−328	−328	2.65	154	35
2nd day	87	325	−238	−566	2.28	140	56
3rd day	87	218	−131	−697	2.02	105	86
4th day	87	113	−26	−723	1.47	73	119
5th day	87	150	−63	−786	1.89	76	113
6th day	87	161	−74	−860	2.00	79	120
7th day	87	119	−32	−892	1.73	67	130
8th day	87	134	−47	−939	1.90	68	126
9th day	87	98	−11	−950	1.72	55	125
10th day	87	77	+10	−940	1.42	52	130
11th day	87	146	−59	−999	2.56	56	130
12th day	87	74	+13	−986	1.31	54	134
13th day	87	70	+17	−969	1.47	46	140
14th day	87	85	+2	−967	1.60	50	136
Na replacement							
1st day	689	261	+428	−539	1.64	122	62
2nd day	689	407	+282	−257	1.70	164	33
3rd day	689	532	+157	−100	1.80	185	19
4th day	689	648	+41	−59	2.21	189	14
5th day	689	599	+90	+31	1.47	203	11

Reprinted from Denton (1957), with permission.

cattle and goats with a permanent unilateral parotid fistula (Denton et al. 1961) (see Fig. 3-4), and with sheep and cattle on a low sodium intake (Bailey and Balch 1961a, b; Dobson 1961, 1965). The change in parotid salivary Na/K ratio is caused by a five- to twentyfold increase in aldosterone secretion (Figs. 3-5 and 3-6). Also, concurrently, the parotid gland becomes more sensitive to the effect of this adrenal hormone during sodium deficiency (Goding and Denton 1959; Blair-West et al. 1963, 1964) (Fig. 3-7). These changes in sodium balance, salivary Na/K ratio and aldosterone and also renin secretion were demonstrated to be accompanied by the development of an avid specific appetite for sodium salts that was commensurate with the degree of sodium deficit (Denton and Sabine 1960).

As noted above, the sodium content of rainwater decreases with increasing distance from the sea coast. Thus sodium content of plants from areas of Central Australia, with seasonal monsoonal rains, may be as low as 1–10 mmol/kg dry wt (Denton et al. 1961) (Table 3-1). In 1957 the theory was advanced that the laboratory experiments on parotid fistulae have revealed a quantitatively spectacular vector of endocrine

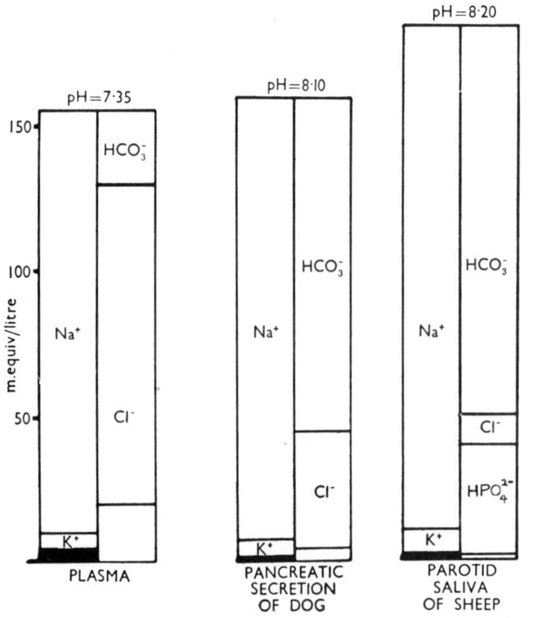

Fig. 3-2. The composition of the parotid saliva of sheep compared with plasma, and pancreatic juice of the dog. (Reprinted from Denton 1957, with permission.)

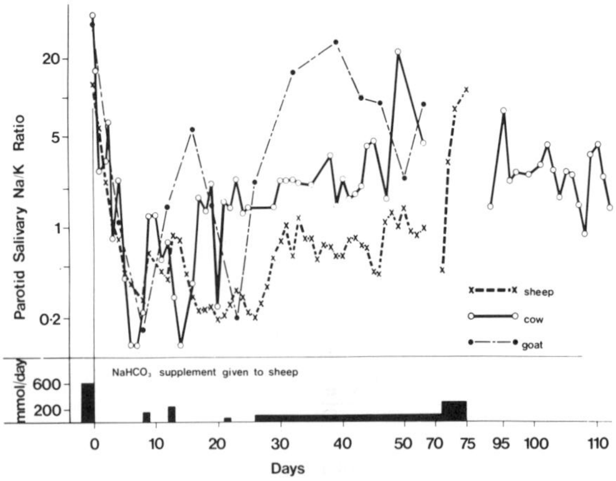

Fig. 3-4. The effect on salivary Na/K ratio of making a parotid fistula in a sheep. During the first 26 days little sodium bicarbonate supplement was given. Between day 28 and day 72 100 mmol $NaHCO_3$/day were given by rumen tube, and saliva composition changed to a new equilibrium. It returned to normal on day 72 when 300 mmol of sodium bicarbonate were given each day. On the same graph the influence of making a parotid fistula on parotid saliva composition of a cow and also a goat is shown. (After Denton et al. 1961.)

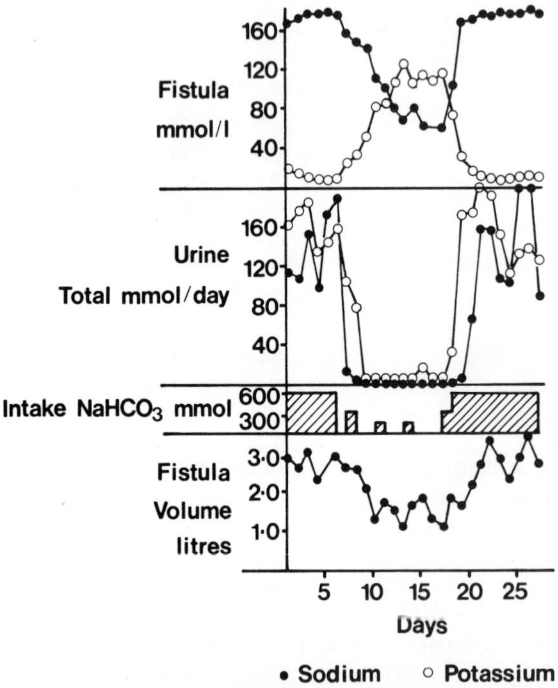

Fig. 3-3. The effect of withholding daily intake of 595 mmol of sodium bicarbonate on the volume and sodium and potassium concentrations of the parotid saliva of the sheep. The sodium concentration fell and the potassium concentration rose. The change in daily excretion of these ions by the kidney is also shown. Sheep: parotid fistula 1. (After Denton, 1956.)

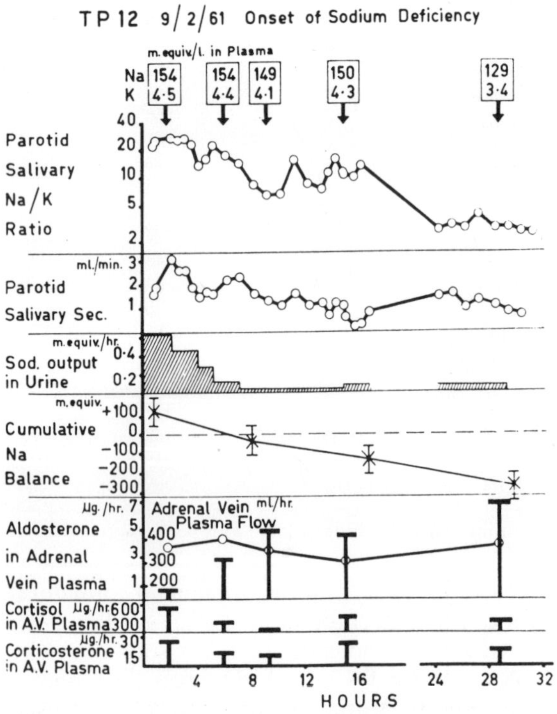

Fig. 3-5. Adrenal Transplant 12. The effect of onset of sodium deficiency on the secretion of aldosterone, cortisol and corticosterone, as measured in adrenal venous plasma. The effects on parotid salivary Na/K ratio, parotid secretion rate, urinary sodium excretion, adrenal plasma flow and sodium and potassium concentrations in adrenal venous plasma are also shown. The cumulative sodium balance is shown and the degree of uncertainty as to the sodium status at the commencement of the external balance is designated as ±75 mmol. (Reprinted from Blair-West et al. 1964, with permission.)

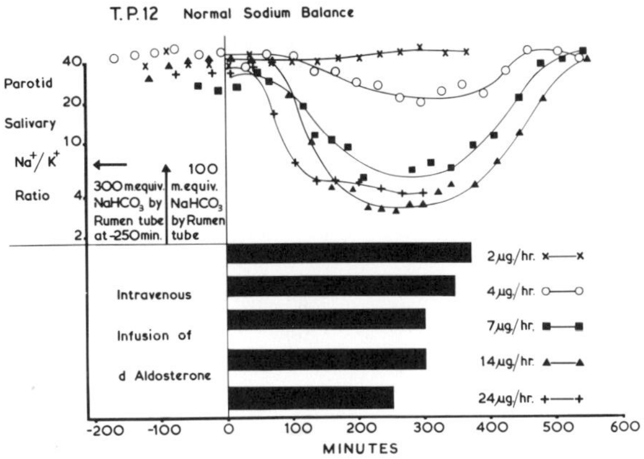

Fig. 3-6. Adrenal Transplant 12 (normal sodium balance). The effect on parotid salivary Na/K ratio of intravenous infusion of D-aldosterone at the rates designated over the time intervals shown on the abscissa. (Reprinted from Blair-West et al. 1964, with permission.)

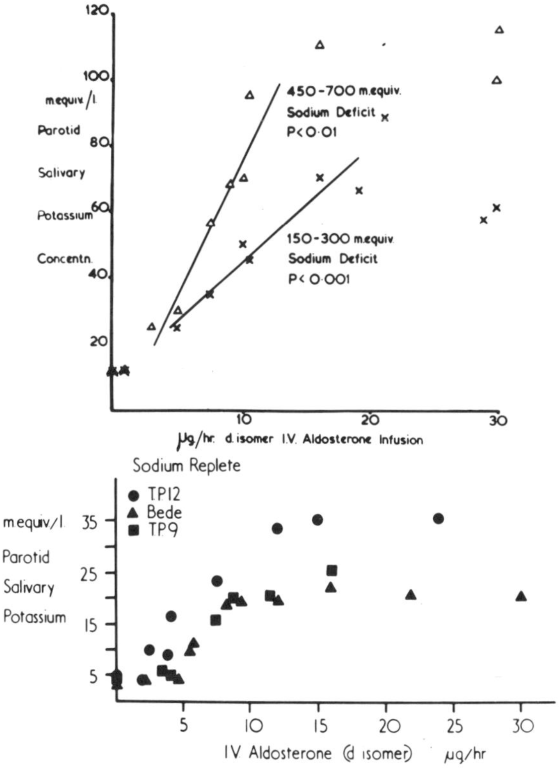

Fig. 3-7. The effect on parotid salivary potassium concentration of intravenous infusion of D-aldosterone at the rates indicated on the abscissa. The lower panel shows results on three sheep which were sodium-replete. The upper panel shows results on a group of adrenalectomized sheep at two different levels of sodium deficit. (Reprinted from Blair-West et al. 1964, with permission.)

effect on salivary Na/K ratio which represents an adaptation of vital importance in the evolution of the ruminant-type animal (Denton 1957; Denton et al. 1961). This adaptation has conferred survival advantage under the stringent ecological conditions of regions where seasonal pasture change involves need of increased volumes of saliva in the rumen, and the sodium content of plants available is very low. Here the obligatory needs of digestion place a stress on sodium balance of animals.

Cattle

As part of the investigation of this theory, field studies were conducted on two herds of cattle in circumstances where it seemed possible that sodium deficiency might occur (Bott et al. 1964). The first herd consisted of Aberdeen Angus cattle which were grazing on river-plain pastures irrigated at fortnightly intervals by water (sodium content <1 mmol/l) from the Murray River in inland Victoria, Australia. Analysis of the pasture the animals were eating showed: sodium 30–44 mmol/kg dry wt, potassium 590–860 mmol/kg dry wt, which represents a sodium intake of approximately 10–12 mmol/kg wet wt of food. One group (1A) had access to a daily supplement of crushed oats and sodium chloride (about 1500 mmol Na/day). The Na/K ratio of the parotid saliva of this group was normal (Table 3-4). All animals of the other group (1B), which ate only the irrigated pastures, showed moderate to large reduction in salivary Na/K ratio and the sodium concentration of the urine was very low, though the specific gravity was high in some instances.

The second herd comprised Hereford cattle that had had summer grazing at 1300 m on mountain grasslands of Victoria for three months. The grasslands are covered in winter by snow, which melts each spring. They are over 100 km from the ocean. Prevailing winds from the south west travel a great deal further overland. The control group (2A, ten cows and one steer: Table 3-5), which were grazing on pastures in the foothills at 400 m, had normal parotid salivary Na/K ratio and a high urinary sodium excretion. Analysis of the pasture eaten by the animals showed: sodium 283–290 mmol/kg dry wt, potassium 583–566 mmol/kg dry wt. This represents a sodium intake of approximately 84 mmol/kg wet wt. Group 2B (seven cows and five calves; Table 3-5), that had been grazing on the mountain pastures, showed a large reduction of salivary Na/K ratio, and, in most instances, a low urinary sodium excretion. Analysis of the pasture eaten by these animals showed: sodium 6 mmol/kg dry wt, potassium 220 mmol/kg

Table 3-4. Analyses of parotid saliva and urine of two groups of Aberdeen Angus cattle grazing on irrigated pasture (10–12 mmol Na/kg wet wt).

	Saliva			Urine		
Group	Na (mmol/l)	K (mmol/l)	Na/K	Na (mmol/l)	K (mmol/l)	Specific gravity
Group A[a]						
G.6	141	4.2	33.6		no sample	
G.16	148	6.3	23.5	3.6	100	—
G.20	117	15.4	7.6	53	67	1.008
G.30	131	8.0	16.4	29	100	1.009
G.33	140	6.2	22.6	2.3	174	1.018
Group B						
G.17	131	26	5.0	0.9	103	1.007
G.28	122	23	5.3	1.7	350	1.030
G.37	106	37	2.9	1.7	410	1.037
F.47	61	80	0.8	1.0	67	1.005
X	113	30	3.8	<1.0	210	1.019
Y	96	23	4.2	2.5	200	—

Reprinted from Bott et al. (1964), with permission.
[a]This group had access to a supplement of crushed oats and sodium chloride (about 1500 mmol Na/day).

dry wt, with a sodium intake of approximately 2 mmol/kg wet wt.

These data showed that in cattle grazing on pastures of low sodium content, sodium was conserved by the kidney and that there was a low salivary Na/K ratio indicative of a significant sodium deficit and hypersecretion of aldosterone. This field result substantially supported the hypothesis that this endocrine mechanism, involving reciprocal substitution of potassium for sodium in the large volumes of fluid obligatorily involved in the digestive cycles of the ruminant, is an important factor in ecological adaptation by these animals. The stress on sodium balance caused by low-sodium pasture would be greater in young growing animals, and would be aggravated by sodium loss in milk in the case of cows with calves.

These mountain cattle were shown to have a strong appetite for salt blocks: film records were made of herds of them running out of the bush when salt was put out from a car. And cattlemen in the region appear to have conditioned animals so that they can be mustered from out of the thick scrub with the cry of 'salt'. Dobson and McDonald (1963) showed reduced Na/K ratio in the mixed saliva of sheep grazing a pasture of low sodium content.

In considering the mechanisms producing sodium deficiency in animals eating a pasture with a sodium content of 10 mmol/kg wet wt, a very important factor is faecal sodium loss. Brouwer (1961) and van Weerden (1961) have shown that cattle may excrete 15–45 kg of faeces per day, with 18 mmol Na/kg of faecal water. Change of diet from moderate to low sodium content reduced loss of sodium in faeces from 752 to 196 mmol/day (Renkema et al. 1962). These investigations demonstrate the large sodium loss in cattle faeces, the likelihood of sodium deficiency when pasture has a low sodium content, and the importance of sodium reabsorption from the lower intestine in control of sodium balance. Bailey and Balch (1961b) found that a daily sodium intake of 600–700 mmol was required to prevent salivary changes indicative of sodium deficiency in cows. Reduction of sodium content of faeces in sodium deficiency and with administration of electrolyte-active corticosteroid has been shown in sheep, dogs and man, and it seems probable that faecal sodium conservation in cattle, like the conservation of sodium in ruminant saliva, is determined by increased secretion of aldosterone.

Study in Snowy Mountains of south-eastern Australia

Following these observations, Myers (1967) of the Council of the Scientific and Industrial Research Organization's Wildlife Division in Canberra published most interesting data from a general study of the rabbit in a variety of ecological situations, and the influence of population density, mineral balance and stress upon its adrenal gland.

As general historical background to his studies, Myers (1967, 1970) notes that on Christmas Day 1859 the brig *Lightning* arrived in Melbourne with about a dozen 'wild-type' rabbits, bound for Barwon Park, a property in Western Victoria. In 1863 a bush fire destroyed the fences enclosing one colony of rabbits, and initiated events which

Table 3-5. Analyses of parotid saliva and urine of two groups of Hereford cattle.

	Saliva			Urine		
Group	Na (mmol/l)	K (mmol/l)	Na/K	Na (mmol/l)	K (mmol/l)	Specific gravity
Group A[a]						
1	159	7	23	158	249	1.032
2	147	5	30	126	90	1.016
3	130	9	14	22	103	1.024
4	141	8	18	46	175	1.020
5	148	6	25	33	57	1.008
6	115	14	8	20	30	—
7	128	10	13	74	200	—
8	134	9	15	96	304	1.040
9	138	10	14	140	260	1.033
10	117	12	10		no sample	
11	140	8	18	23	138	—
Group B[b]						
1	74	63	1.2	5	57	
2		no sample		4	113	
3	37	74	0.5	1	91	
4	44	40	1.1	15	52	
5	53	14	3.8	2	41	
6	21	11	1.9	1	49	
7	32	7	4.6	1	42	
8	17	59	0.3	1	90	
9	70	32	2.2	22	151	
10	33	105	0.3	2	61	
11	96	17	5.6	2	125	
12	93	60	1.6	1	30	

Reprinted from Bott et al. (1964), with permission.
[a]Grazing on pasture in the foothills (84 mmol Na/kg wet wt).
[b]Grazing on mountain grasslands (2 mmol Na/kg wet wt).

changed the economy of Australia and became the despair of governments and pastoralists. Today the rabbit inhabits the southern portion of the continent up to more or less the level of the Tropic of Capricorn, with more directions of spread above this—an area of about 3.84 million km^2 (1 500 000 square miles).

Myers found that the mean weights of the adrenal glands taken in the sub-alpine Snowy Mountain regions and in the sub-tropical continental interior regions were 30% higher than those in hot arid regions. He described dramatic histological changes accompanying reproduction, stress, high population density and mineral deficiencies. With the stress of high population density it was found the fasciculata zone replaces the glomerular zone, which may be reduced to 5% of the cortex. However, in the alpine areas, and also in the sub-tropical regions the converse may occur. There was apparently increased cell division along the interface of the fasciculata and glomerulosa regions as indicated by mitotic figures. In the alps, glomerular zones forming half the cortex were common (see Fig. 3-8). Myers also observed instances where concurrent rapid enlargement of fasciculata and glomerulosa occurred to the extent that distinct nodules were formed which eventually budded off as small adrenals (composed of glomerulosa and fasciculata) covered in fine vessels and lying free on the mesentery (see Fig. 3-9). These disappeared in the population after a period of time and Myers suggested the effect resulted from concurrent operation of mineral deficiency and stress, for he found that if rabbits were maintained on a low-sodium diet and adrenocorticotrophic hormone (ACTH) given, incipient nodulation was produced.

These great enlargements of glomerulosa which Myers presumed to be related to mineral deficiency were associated with an avid appetite for soft wooden pegs soaked in salt—particularly sodium chloride but also potassium and magnesium chloride, though not calcium or ferric chloride (see Fig. 3-10). Previously, in 1932, Cook had reported from a study of the effect of wood preservatives on soft pegs in preventing termite attack below ground that, in fact, the pegs were attacked above ground by rabbits. Only pegs containing sodium salts were attacked and he suggested that the rabbits were suffering from sodium deficiency.

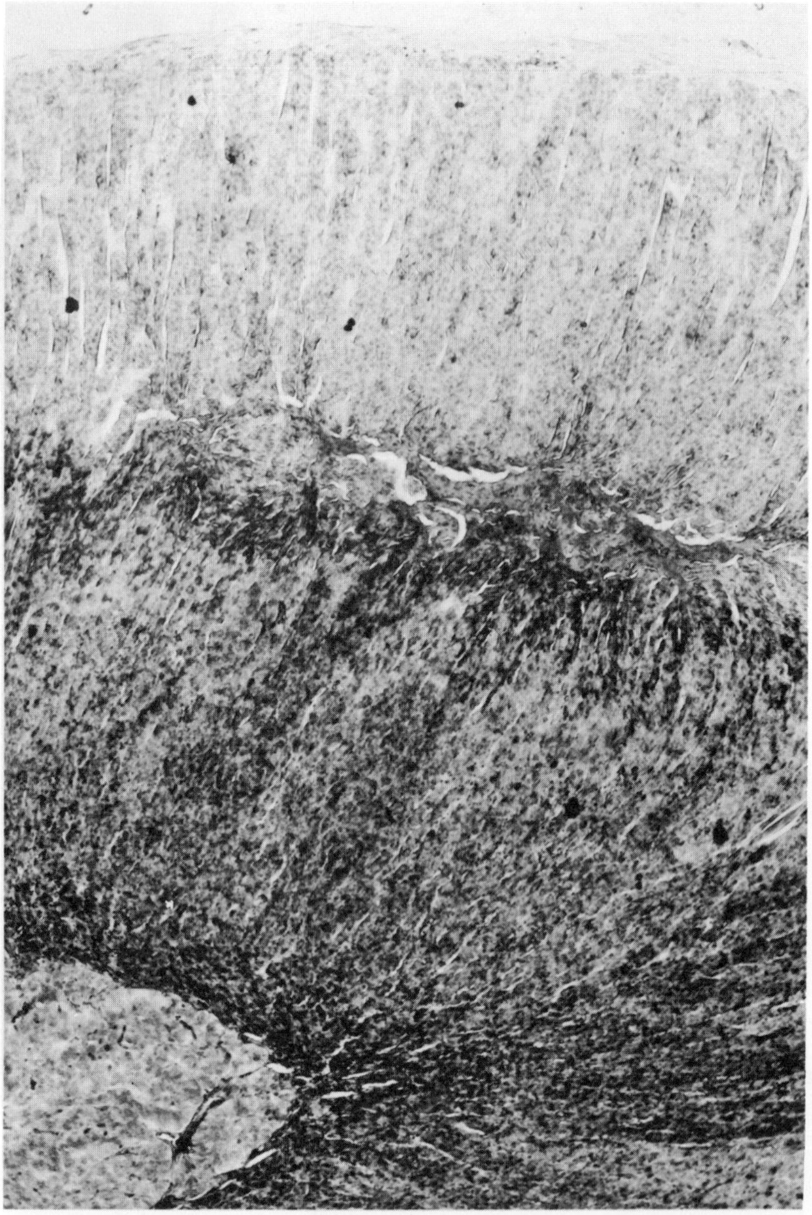

Fig. 3-8. Section of adrenal gland from adult female rabbit from sub-alpine zone in Snowy Mountains showing large glomerular zone occupying about half the cortex. ×40. (Reprinted from Myers 1967, with permission.)

Cook's work was done in the Canberra region, inland near the Snowy Mountains.

At this point in time, our group and Myers and his collaborators decided upon a collaborative in-depth study of the native and introduced animals in the Snowy Mountains regions in relation to their sodium balance, biochemical and endocrinological status, and behaviour (Blair-West et al. 1968). Control observations were made on coastal areas where marine aerosols ensured a high sodium content of soil and grass, and on particular continental areas where drainage conditions had produced the same salinity effect. Carlos Luiz Junqueira of São Paulo, Brazil, also collaborated in the investigation; he was most interested in the salivary gland changes.

First an analysis was made of the soil and grass of the regions (Tables 3-1 and 3-6). The locations to be compared with the Snowy Mountains (a region, incidentally, with winter snow-cover over an area larger than Switzerland) were deserts near Broken Hill, Central Australia, an inland temperate grassland area near the national capital Canberra, and the Southern Australian (Victoria) ocean coast (Fig. 3-11).

Observations on the sodium status of the

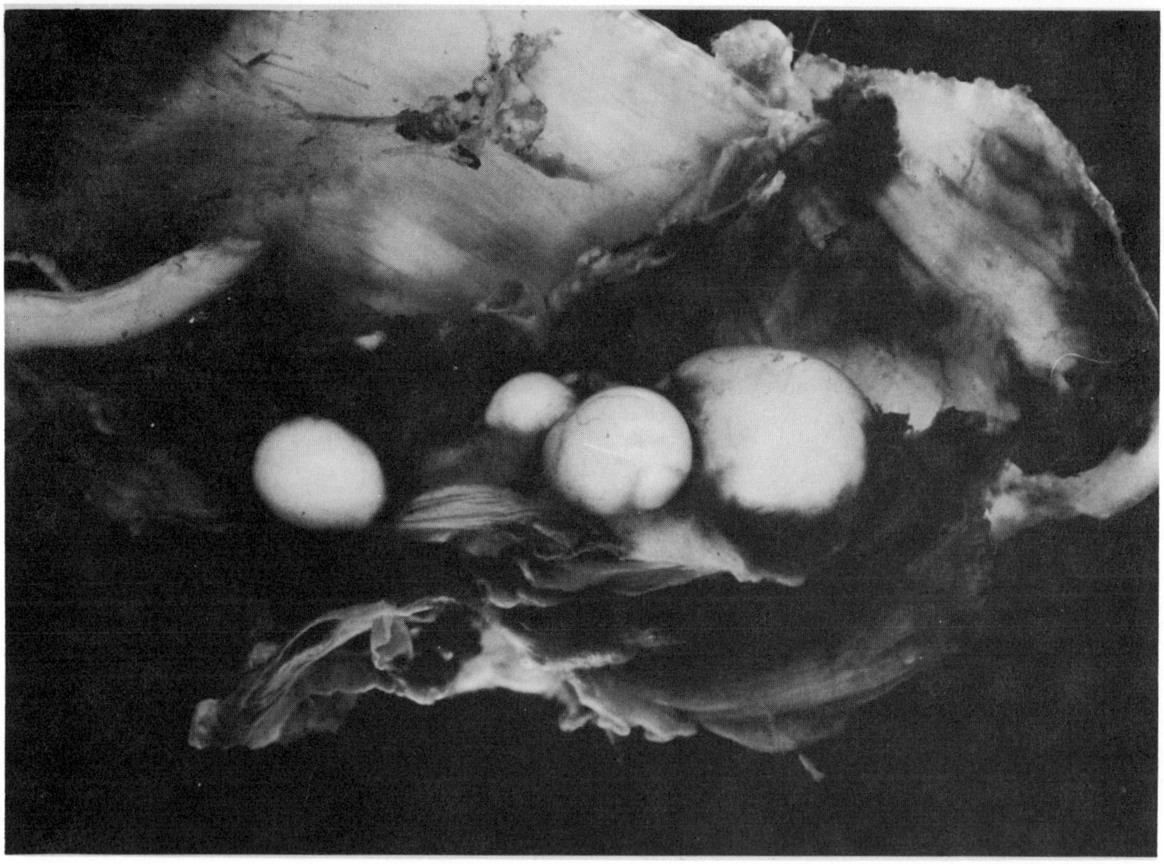

Fig. 3-9. Adrenal gland from adult male rabbit from sub-tropical region showing three large accessory adrenals lying free in the body cavity. Each accessory adrenal consists of fasciculata and glomerular tissue only. ×5. (Reprinted from Myers 1967, with permission.)

Fig. 3-10. Wild rabbits in the sodium-deficient alpine areas attacking pegs impregnated with sodium bicarbonate and sodium chloride. Photograph was taken by Mr E. Slater, CSIRO Wildlife Division, Canberra.

Table 3-6. Analysis of soil samples from the regions of continental Australia in which the animals were studied.

Location	No. of samples	Concentration (mmol/kg dry wt)					
		Na	K	Mg	Ca	P	Cl
Snowy Mountains	7	0.5	11	28	108	1.7	0.56
Grasslands (Canberra)	3	0.8	4.3	27	45	0.3	0.60
Central desert	3	3.6	11	25	52	1.2	0.51
Victorian sea coast	7	3.3	6.0	29	59	0.5	1.19

Reprinted from Blair-West et al. (1968), with permission.

animals were made on two introduced species, wild rabbits *Oryctolagus cuniculus* (L.) and sheep, and on two native animals, wombats *Vombatus hirsutus* (Perry) and kangaroos *Macropus giganteus* (Shaw).

One aim was to measure the concentration of steroids in peripheral blood by using highly sensitive methods of double-isotope dilution derivative assay (Coghlan and Scoggins 1967). Laboratory experiments (Blair-West et al. 1969) on the time course of action of ACTH following direct adrenal arterial infusion had shown that 2–3 min elapse before corticosteroid secretion rises in the adrenal venous effluent. In order to obtain a measurement on wild animals in conditions where the concentrations were not modified by probable pituitary stimulation associated with pursuit and capture, it was considered that the most practical approximation to the normal would be achieved by location of the nocturnal grazing animals by spotlight from a travelling car, and instant killing by shooting through the brain. Cardiac blood was immediately withdrawn and frozen. Under these conditions the steroid levels in the peripheral blood should reflect those in the grazing animal before disturbance.

Fig. 3-11. Map of Australia to show localities under investigation. A, alpine area in Snowy Mountains; B, grassland area at Canberra; C, sea coast at Welshpool; D, desert at Calindary. (Reprinted from Scoggins et al. 1970, with permission.)

Table 3-1 shows that a very low sodium content of grass was found in the Snowy Mountains in spring with the lush growth, and in summer. The exceptional observation in the spring was on an area where a bush fire had burnt 2 years previously. The sodium content of grasses was greater in autumn and winter. A sodium-accumulating plant (*Stylidium*) was identified in the alpine region in autumn, attention being drawn to it by the fact of the rabbits selectively eating the tops off the plants. In relation to the alpine figures (ca 1 mmol/kg wet wt) approximately a week's intake of food by a sheep or rabbit is required to provide up to 5% of the sodium content of the circulating plasma whereas, on normal pasture, sodium equivalent to 10%–25% of the plasma content is ingested *each day*. At birth, a rabbit weighs about 40 g and total body sodium is about 5 mmol. Subsequent to birth a litter of four rabbits each grows about 9 g/day. If the doe provides 100 g of milk of normal composition each day, of which the ash content is 2.5% and sodium is 5% of the ash (Blaxter 1961), it may be calculated that 20% of the sodium content of circulating plasma would be required each day if lactation were normal. The profound stress on sodium homeostasis is evident from the following data.

Table 3-7 and Fig. 3-12 show that in spring and summer the mountain rabbits excreted urine which was virtually free of sodium, whereas in autumn and winter urine contained 2–6 mmol/l. Correspondingly, the sodium content of hard and soft faeces was much lower than in the desert region. The sodium content of urine of kangaroos and wombats in the alps was virtually nil compared with that of sea coast animals, where kangaroos had a mean of 268 mmol/l and wombats of 23 mmol/l (Table 3-8). The lower sodium content of the urine of the sea coast wombat possibly reflects its different habit as a root eater. A large unexplained variation was found in urinary sodium content of sheep grazing in the mountain areas in

Table 3-7. Measures (mean±s.d.) of physiological parameters in wild rabbits in the regions designated.

Region	Urine Na (mmol/l)	Urine K (mmol/l)	Peripheral blood aldosterone (ng/100 ml)	Kidney renin (units/g cortex)	Faeces[a] Hard (mmol/kg dry wt) Na	Faeces[a] Hard (mmol/kg dry wt) K	Faeces[a] Soft (mmol/kg dry wt) Na	Faeces[a] Soft (mmol/kg dry wt) K
Snowy Mountains								
Spring	0.59± 0.36 (19)[b]	208 (19)	130±104 (15)	162±184 (10)	2	155	5	518
Summer	0.53± 0.12 (18)	269 (18)	69±66 (12)	103±80 (10)				
Autumn	2.6 ± 2.9 (12)	466 (6)		18±7 (6)				
Winter	6.4 ± 1.2 (5)	390 (5)	74±58 (11)					
Grasslands (Canberra)	18.0 ±14.6 (12)	219 (12)	21±10 (14)	39±17 (8)				
Desert	139 ±97 (15)	319 (15)	9±6 (11)	30±8 (7)	28	90	42	343

Reprinted from Blair-West et al. (1968), with permission.
[a]Hard pellets are normal faeces voided directly onto the ground. Soft pellets, resulting from caecal fermentation, are eaten from the anus.
[b]Numbers in parentheses are numbers of animals in each group.

early summer (from 0.3 to 130 mmol/l, $n=14$); this suggests the possibility of selective grazing of sodium-accumulating plants like *Poa* grass or access to an undetected source of salt.

Only limited observations of blood electrolytes were made. The mean sodium content of plasma of

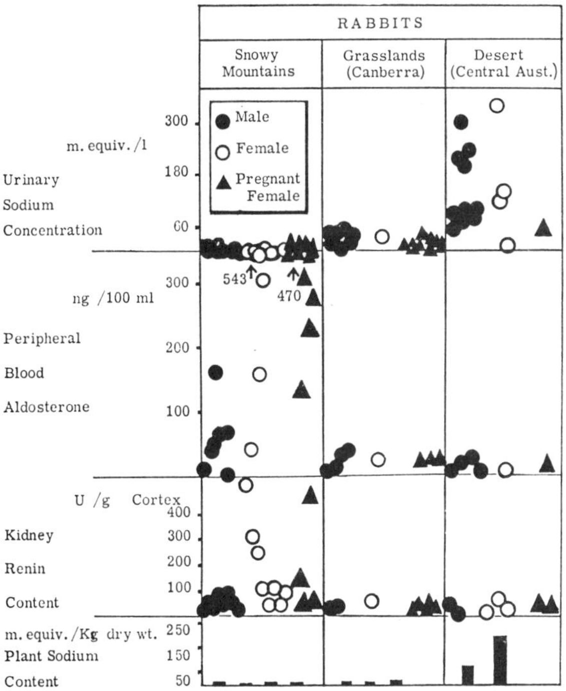

Fig. 3-12. The sodium status of rabbits in different ecosystems in Australia. The urinary sodium concentration, peripheral blood aldosterone concentration and kidney renin concentration are shown. The lowest panel records the sodium concentration in grass samples collected in the region. See Fig. 3-11 for location of regions.

Table 3-8. Mean plasma and urinary sodium and potassium concentrations (mmol/l), range, and number of samples for wombats and kangaroos from alpine area and sea coast.

	Alpine (spring)	Sea coast (summer–winter)
Kangaroo		
Plasma Na	138 (136–140) (n=7)	140 (131–150)(n=5)
Plasma K	8.8 (7.7–10.0)	8.7 (5.9–10.0)
Urinary Na	0.4 (0.2–0.8) (n=8)	268 (79–496) (n=5)
Urinary K	336 (48–665)	130 (45–275)
Wombat		
Plasma Na	134 (128–141) (n=9)	131 (127–134) (n=2)
Plasma K	8.5 (5.1–9.5)	6.4 (6.2–6.5)
Urinary Na	0.3 (0.1–0.8) (n=13)	23 (2–50) (n=4)
Urinary K	134 (17–262)	108 (47–165)

From Scoggins et al. (1970), with permission.

the alpine kangaroos (138 mmol/l) was little different from that of the sea coast animals (140 mmol/l) (Table 3-8). Plasma potassium levels of both kangaroos and wombats shot while grazing were surprisingly high (ca 8 mmol/l) despite immediate centrifugation and absence of haemolysis (see also Scoggins et al. 1970).

Table 3-7 shows that the concentration of peripheral blood aldosterone was very high in the alpine rabbits relative to that of the desert and Canberra grassland animals. The highest aldosterone levels were found in pregnant animals during the spring in the alps. Blood corticosteroid levels of kangaroos and wombats from the alpine and coast areas are shown in Fig. 3-13. Aldosterone levels were elevated in sodium-deplete kangaroos of both sexes. The high level of 13.6 ng/100 ml seen

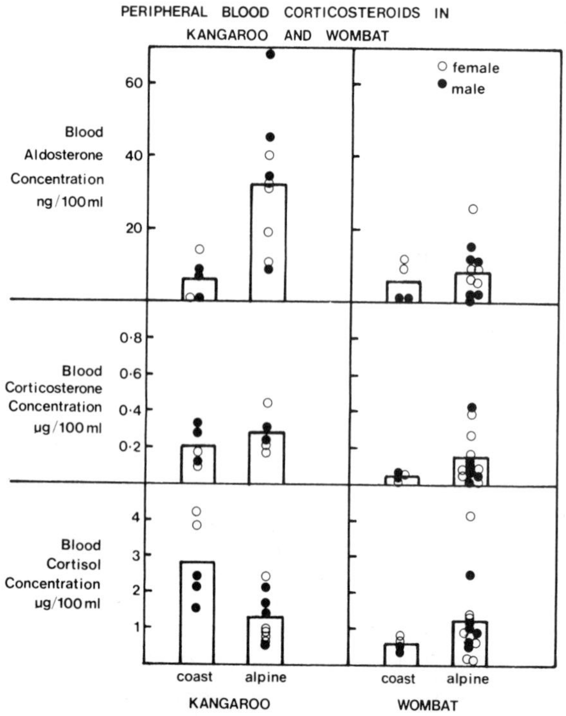

Fig. 3-13. Blood aldosterone, cortisol and corticosterone levels in wombats and kangaroos from the alpine and sea coast areas. (Reprinted from Scoggins et al. 1970, with permission.)

in one sodium-replete female was in an animal with a 60-day-old pouch young. One of the females (11.0 ng/100 ml) had a foetus in the pouch. With the exception of two female sodium-replete kangaroos, cortisol and corticosterone levels were similar in both groups of animals. In the case of wombats from the same two areas, however, a number of animals of both sexes from the sodium-deficient region with low urinary sodium had blood aldosterone levels similar to those seen in the replete animals; the means of the two groups are not significantly different. The two female wombats from the Victorian sea coast (Welshpool) area had pouch young, 3 and 12 weeks old respectively. Blood cortisol and corticosterone levels were similar in animals of both sexes and from the two areas.

The peripheral blood aldosterone levels in three sheep (wethers) from the Snowy Mountains (36, 77 and 108 ng/100 ml) were also very high compared with the normal sodium-replete level of about 1 ng/100 ml. Aldosterone secretion rates were measured by cannulation of the adrenal vein under anaesthesia in other sheep from this area. In three animals in which urine was virtually free of sodium and a low salivary Na/K ratio was found, high aldosterone secretion rates (14.0, 15.6 and 19.8 μg/h) were observed. In four other animals which were apparently sodium-replete and had sodium in urine, aldosterone secretion rates were about 2 μg/h, as seen in normal sodium-replete animals in the laboratory.

The concentration of renin was measured in the kidney cortex (Blair-West et al. 1968). In alpine rabbits (Table 3-7) the mean concentration (units/g) was greatly elevated in the alps in spring and summer. Highest values were recorded in females, as with aldosterone values, possibly suggesting the additional stress on sodium balance of the reproductive process. The mean value for animals from the Canberra grasslands area was slightly higher than for desert animals. As with rabbits, kangaroos and wombat kidneys were incubated with prepared sheep substrate (Blair-West et al. 1968). The concentrations were less than 1 unit/g, which may be caused by species specificity of the renin/renin substrate reaction. The kidneys of sheep in the alps had renin levels in the high-normal range but no values were found as high as those in laboratory animals which were sodium-depleted by a parotid fistula.

The adrenal glands of the alpine rabbits in spring were heavier than those of rabbits in other regions, and seasonal variation occurred in the alpine animals (Table 3-9). The area of zona glomerulosa represented a much greater percentage of the total area of the adrenal in the alpine animals in spring

Table 3-9. The weight and area of adrenal glands and the percentage of total area of adrenals represented by zona glomerulosa in wild rabbits in different regions of Australia.

Region	Adrenal weight (g)	n	Total area of adrenals (mm^2)	Area of zona glomerulosa (% total)
Snowy Mountains				
Spring	0.18	(24)	22.6	34.4
Summer	0.12	(21)	18.2	22.6
Autumn	0.14	(12)	21.2	17.3
Winter	0.14	(13)	22.2	17.8
Central desert				
Summer	0.11	(14)	18.4	15.6
Canberra grasslands				
Spring	0.13	(12)	20.6	15.3

From Blair-West et al. (1968), with permission.

and summer than in the desert and grassland animals. The adrenal glands of alpine kangaroos (1.1 g) and wombats (0.7 g) were approximately double the weight of the control animals from the sea coast (0.6 and 0.4 g, respectively) and the zona glomerulosa represented twice the percentage of total area in kangaroos, but was only 20% greater in mountain wombats (Fig. 3-14).

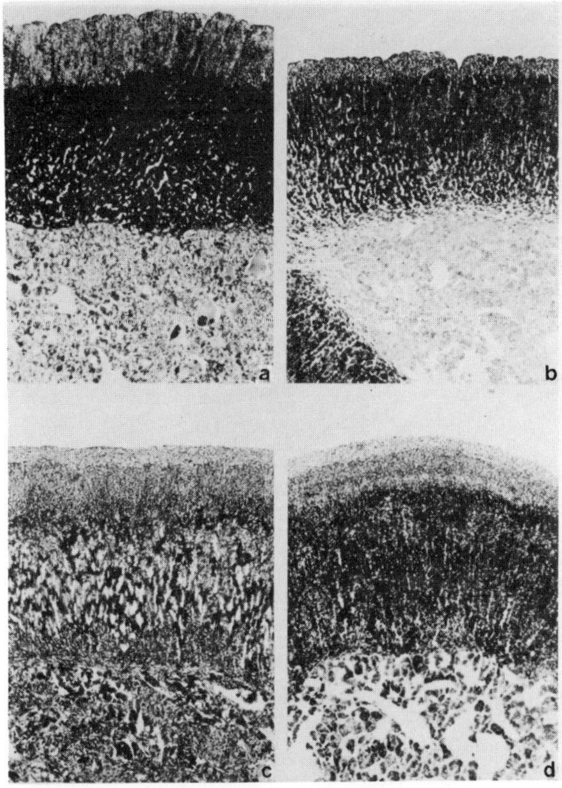

Fig. 3-14. Morphology of the adrenal cortex of the sodium-deficient (a) and sodium-replete (b) grey kangaroo and of the sodium-deficient (c) and sodium-replete (d) wombat. The zona glomerulosa is wider in the sodium-deficient animals. ×27. (Reprinted from Scoggins et al. 1970, with permission.)

There were striking structural changes in the salivary glands of the alpine animals when compared with those of the same species living in the sodium-rich sea coast areas. Table 3-10 records the much more extensive duct system of the parotid gland of sodium-deplete specimens of the grey kangaroo. A similar finding was made in the submandibular glands of the same species, where the percentage of total gland of sodium-deplete animals represented by the striated ducts was four times that of the replete animals. The height of parotid duct cells (n=100) in sodium-deficient animals was 19.51±3.20 μm, and in replete animals 14.04±3.05 μm (P<0.01). The vascular pattern around the glands in the sodium-deficient animals was considerably different from that usually observed. Blood vessels were extraordinarily abundant around striated ducts and Fig. 3-15 shows that loops of capillaries penetrate the outer wall of the striated ducts and run in the epithelial zone. The wombat material showed similar changes in the epithelium though not to the same degree.

Table 3-10. Constitution of grey kangaroo parotid gland expressed as a percentage of total gland volume.

Condition	Serous cell (%)	Striated duct (%)	Excretory duct (%)
Sodium-deficient	86.3	5.3	3.1
Sodium-deficient	79.6	12.4	3.6
Sodium-deficient	83.2	9.5	4.8
Sodium-replete	95.4	0.0	1.5
Sodium-replete	94.3	0.3	0.6
Sodium-replete	92.7	0.2	1.3

From Blair-West et al. (1968), with permission.

In the parotids of rabbits deficient in sodium the volume of striated ducts was 5.9% of the total volume compared with 1.5% in the sodium-replete animals. The number of mitochondria in duct cells was considerably greater than in the sodium-replete animals and the epithelium was approximately twice as high (18.3 μm and 10.9 μm respectively). Striated ducts were present in the submandibular glands whereas they were virtually absent in the sodium-replete rabbits. Electron micrographic study of the rabbit parotids showed increased density of acinar granules in sodium deficiency. The striated ducts in sodium-replete animals (Fig. 3-16a) had low prismatic or cubic cells with very few or no infoldings of the basal cell membrane. Mitochondria were sparsely distributed and separated by cytoplasm, their cristae were not very closely packed and no intracellular channels were visible. In the sodium-deficient animals (Fig. 3-16b), not only were the cells higher but there were profuse infoldings of the basal cell membranes. These infoldings are limiting portions of cell cytoplasm that are packed with mitochondria with very little cytoplasm between them. Furthermore, the cristae of the mitochondria were also very closely arranged suggesting a higher content of cristae per unit volume. The conspicuous intercellular clear channels are suggestive of involvement in ion and fluid transport. The morphological changes in the duct system of salivary glands were an important new finding presumably indicative of chronic hyperactivity in active sodium reabsorption (Fig. 3-15). Such structural changes might be attributable to either chronic sodium deficiency and/or the action of aldosterone. Our previous observations on the influence of desoxycorticosterone acetate (DOCA) in producing a great increase in the height of duct epithelium in sheep suggest a hormonal trophic influence.

The kidneys of the sodium-deficient wombats showed hypertrophy of the juxtaglomerular region and much greater height of proximal tubular cells. Changes in the kangaroo kidneys were less marked.

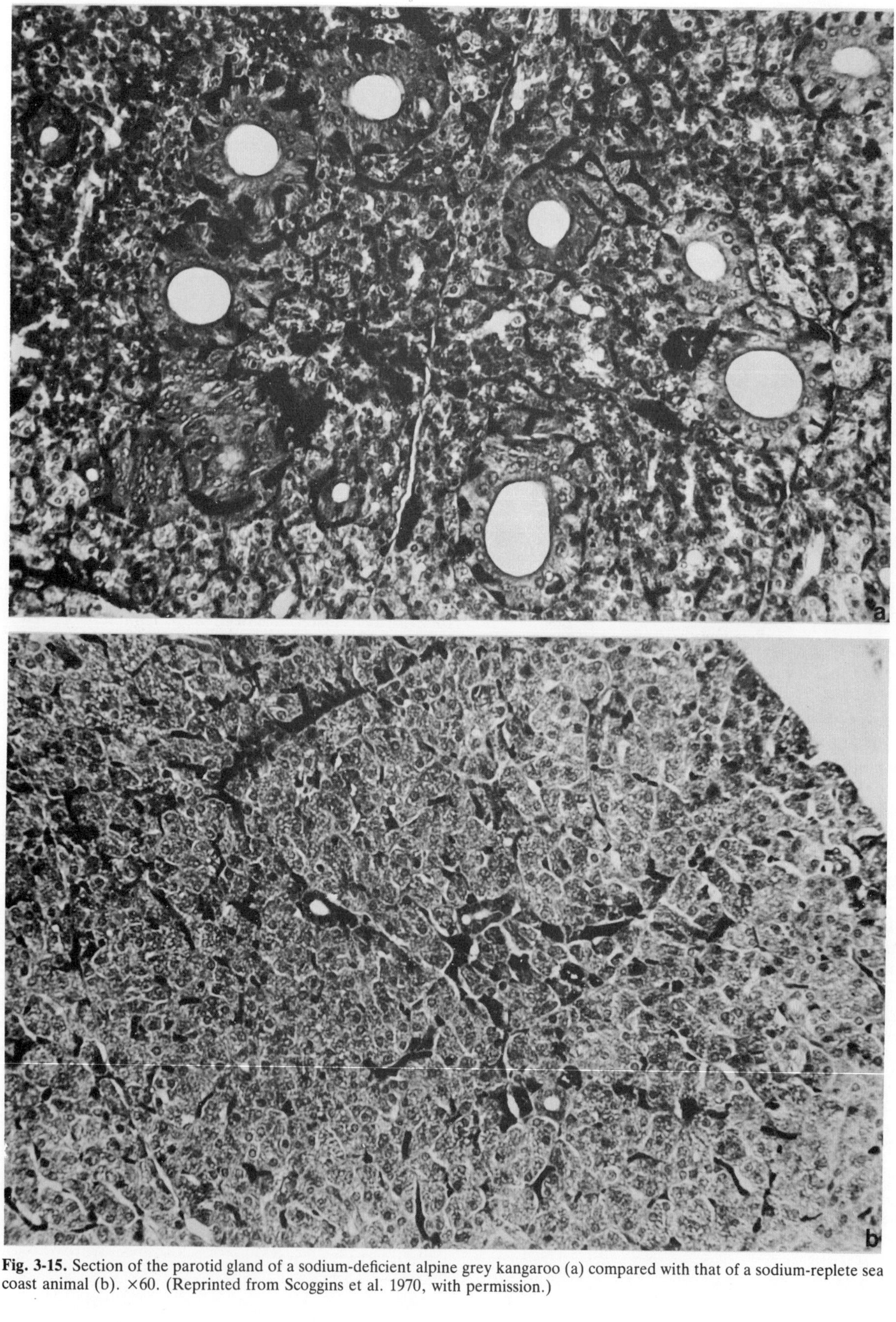

Fig. 3-15. Section of the parotid gland of a sodium-deficient alpine grey kangaroo (a) compared with that of a sodium-replete sea coast animal (b). ×60. (Reprinted from Scoggins et al. 1970, with permission.)

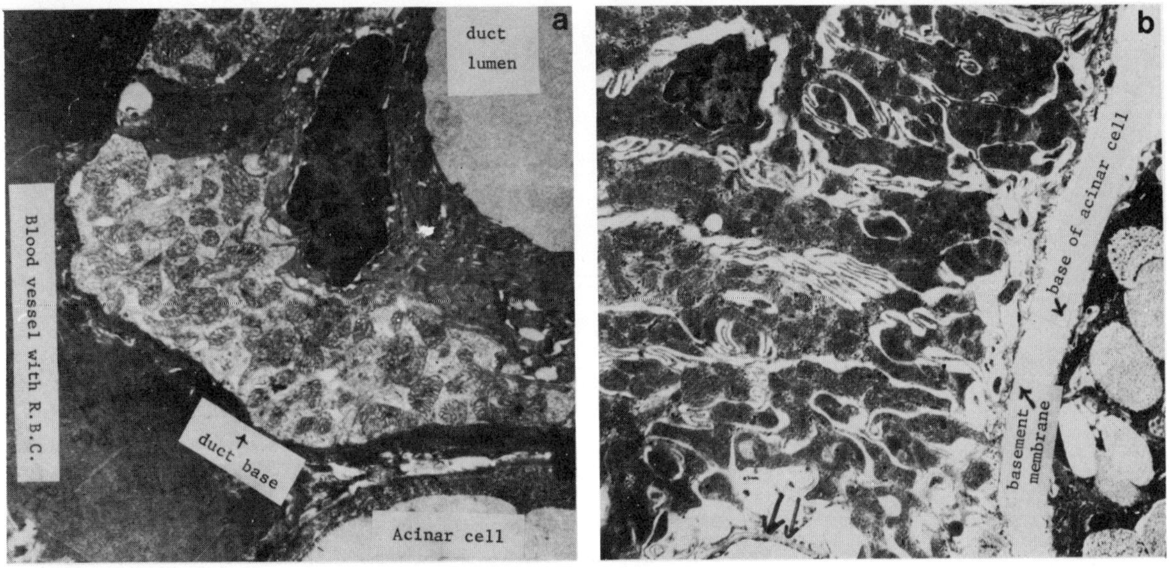

Fig. 3-16. Electron micrograph of the parotid gland of a sodium-replete rabbit (a) compared with that of a sodium-deficient alpine wild rabbit (b). ×10 000. (Reprinted from Scoggins et al. 1970, with permission.)

Experiments were undertaken in which groups of softwood sticks impregnated with different inorganic salts were firmly driven into the ground in the region of rabbit warrens. Analysis showed that after impregnation the outer wood contained up to 1800 mmol Na/kg whereas the core contained 20 mmol/kg. The content varied with different salts. Figs. 3-17 and 3-10 show that in the spring and early summer in the Snowy Mountains (top three panels of Fig. 3-17) the wild rabbits avidly devoured the pegs impregnated with sodium chloride and sodium bicarbonate but largely ignored those with magnesium chloride and distilled water. Some gnawing of sticks impregnated with potassium chloride occurred. In some instances the pegs impregnated with sodium were dug out of the ground and completely eaten. In the late autumn the rabbits did not show significant salt appetite. This corresponded with the urine analyses (mean sodium content, 6 mmol/l), zona glomerulosa width (Table 3-9) and analyses of grasses (Table 3-1) including *Stylidium*. Pegs put out near warrens at Canberra (mean urinary sodium, 18 mmol/l) were untouched, as were pegs in the desert area. Some instances of intake of magnesium have been seen in pegs left out in the alps for a long time in late summer (Myers 1967). This requires further investigation but might be related to the influence of a sustained very high blood aldosterone in spring and early summer in causing magnesium depletion (Care and Ross 1963).

In 1973 we (Abraham et al.) also reported further studies on the mountain kangaroos which we had shown to be severely sodium-deplete. Blocks of filter-paper pulp which were heavily impregnated with sodium, potassium, magnesium or calcium chloride (or soaked in water, as controls) were put out in alpine forest glades known to be traversed and grazed upon by grey mountain kangaroos (*Macropus robustus*). The animals behaviour was filmed by Dr Edward Blaine from a hide elevated in a gum tree. Visits were usually at twilight. The kangaroos showed an unequivocal preference for the blocks impregnated with sodium chloride (see Fig. 3-18) and this was particularly so with females with a joey (young kangaroo) in their pouch. Often an entire block was broken up and devoured during the night (Abraham et al. 1973). Horses in this region also showed an appetite for paper pulp blocks impregnated with sodium. Reid and McDonald (1969) show also that a salt appetite developed with diuretic-induced sodium depletion in another marsupial, the possum *Trichosurus vulpecula*.

On the question of detrimental effects of sodium deficiency, experiments with wild rabbits (K. Myers, unpublished) studied in enclosures with 35, 50 or 200 animals to the acre (86, 124 or 494 per hectare) showed a highly significant reduction of glomerulosa and an increase in fasciculata areas in animals at high density (Table 3-11) (see also Myers and Bults 1979). This is consistent with the proposal that density is an additional aggravating stress in sodium-deficient zones. The extensive data from our Institute on the degeneration of the zona glomerulosa and impaired response to sodium deficiency caused by ACTH will be

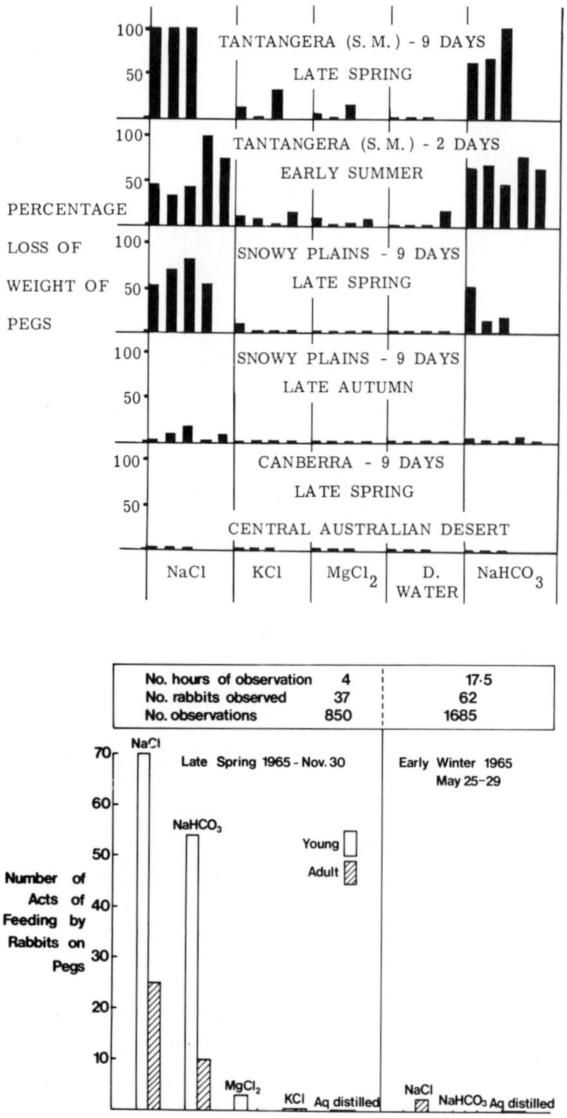

Fig. 3-17. Upper section: the percentage loss of weight of soft wooden pegs impregnated with various salt solutions. Pegs were placed in the vicinity of warrens for 9 days. The upper four panels record seasonal observations from two different sites in the alpine area and the bottom panel results from the grassland and desert regions. S.M., Snowy Mountains. (Reprinted from Scoggins et al. 1970, with permission.) Lower section: the number of acts of gnawing pegs in warren area by rabbits during periods of observation at two different seasons in the Snowy Mountains. (Data kindly provided by Dr K. Myers and colleagues, CSIRO Wildlife Division, Canberra.)

detailed later. In preliminary experiments examining the nodule-producing effects of ACTH, Myers and Bults found that, of itself, the stress of simply handling the sodium-deficient mountain rabbits resulted in weight loss and morbidity which were not seen in animals from non-deficient regions.

In 1970, Myers published further data on the dynamics of rabbit populations in various parts of Australia. The diminished reproductive performance of rabbits in the alpine area was significant relative to other areas. However, the possibility of effect of other factors such as worm infestation in the alpine animals was not able to be excluded. Animals in a sea coastal region with adequate salt also had impaired reproductive performance, apparently as a result of parasite infestation. The quantitative reduction in young produced per year was greatest, however, in the alpine region. The sub-tropical Queensland region (Mitchell), which also showed impairment of reproductive function relative to the temperate Mediterranean-type areas, was sufficiently inland for sodium deficiency to be a contributory factor (see Murphy and Plasto 1973) but the appropriate determinations were not included—except that Myers (1967) reported zona glomerulosa enlargement of rabbits from the area. In a further analysis of his large body of data on rabbit population dynamics, K. Myers (personal communication) has shown a much greater incidence of adrenal nodulation in rabbits from the sodium-deficient regions of the mountains and of tropical Queensland; this is true in both males and females, but is greatest in females during lactation.

African game

In 1969 Weir published most interesting data on the behaviour of African elephants on the north-eastern portion of the Wankie National Park of Zimbabwe. It is an area of flat aeolian Kalahari sand deposits and analyses of soil samples in these areas showed that salt licks created and used by elephants have high concentrations of water-soluble sodium. The licks may be from 3 to 25 m from side to side and are dug out. They are used mainly in the hot dry season when elephants can frequently be seen pounding the soil in the licks with their feet, and shoving it into their mouths with their trunks (which appear to hold the soil by twisting the tip of the trunk into a loop). In the wet season the salt licks fill with water and elephants are occasionally seen drinking there. Buffaloes are often to be seen in the wet season using these licks as wallows, and have occasionally been seen drinking from the pools in the licks.

Soils with concentrations of calcium, magnesium and potassium, as in, for instance, termite mounds, are used as licks only in regions where soluble sodium is not present in quantity in the soil. Weir (1969) also noted that the odour given to the region by the elephant dung may serve to attract the animals back at later seasons. Giraffe were also seen at the licks biting the soil, and tracks indicate other ungulates come to them. Elephants also dig

Fig. 3-18. Photograph taken with a telescopic lens at twilight in the Snowy Mountains. A cafeteria of filter-paper pulp blocks impregnated heavily with sodium, potassium, magnesium or calcium chloride (or soaked in water, as controls) has been placed out in a forest clearing. The wild kangaroos showed a specific appetite for sodium and often the entire block (2 kg) was devoured in a night.

Table 3-11. Effects of density on zonation of adrenal glands of wild rabbits in experimental populations.

					Median cross-section of right adrenal			
	Population density (rabbits/acre)	n	Total adrenal weight (g)	Total adrenal area (mm^2)	Area zona glomerulosa (mm^2)	Area zona glomerulosa (% total)	Area zona fasciculata-reticularis (% total)	Area medulla (% total)
Males	200	20	0.38	26.27	* 2.50	10.29	82.49	7.23
	50	10	0.40	28.53	3.18 *	11.43 *	80.84 *	7.62
	35	5	0.30	25.59	3.78	16.07	76.71	7.27
Females	200	19	0.29	21.94	* 2.47	11.92 **	80.54 **	7.54
	50	10	0.25	19.10	3.35 **	17.83	73.25	8.93
	35	10	0.24	20.65	* 3.56	17.92	73.94	8.12

From Blair-West et al. (1968), with permission.
*$P<0.01$; **$P<0.001$.

up areas where ash from burnt trees is abundant, and this may represent another sodium source for the elephants.

In a subsequent more detailed behaviour study, Weir (1972) examined the relation of sodium content of bore waters provided for game in the park to the density of population of elephants. The attraction to high-sodium sources was clear-cut, the density of elephant population in any region being directly proportional to the sodium content of the bore water (Fig. 3-19). This was independent of the availability of browse food to the animals. Further, excavations for licks occurred in regions where sodium content of the bore water was low and not in the regions where it was high. Sodium was the only cation likely to have been determinant. Calcium and magnesium showed no relation to elephant numbers, and while chloride concentration was high in some bores, again there was no correlation with attraction for elephants. Some levels of sodium in water were very high (70–160 mmol/l). Once more the behaviour of elephants in digging for ash was recorded. Sodium

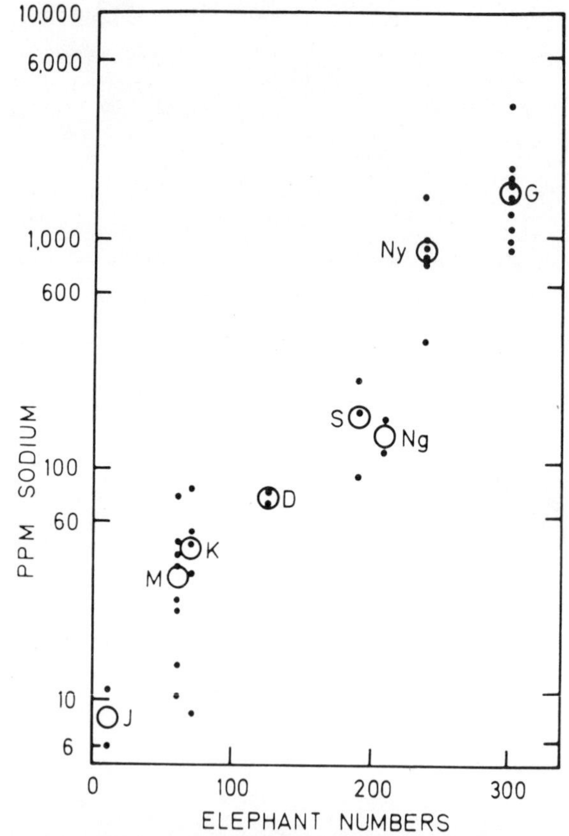

Fig. 3-19. Numbers of elephants censused at eight pans are shown as mean values plotted against all available sodium analyses for water at these pans irrespective of season. Mean values of sodium are also shown, the plot being log-normal. (Reprinted from Weir 1972, with permission.)

was higher in the analyses of lick soils than in other soils not used as licks but it was interesting that phosphate content was also higher.

Using such data as were available to him, Weir approximated the sodium requirements of a 3175 kg elephant. He cited Benedict's (1936) figures for the Indian elephant suggesting that 70 g Na/day would not maintain the animal in equilibrium. A large faecal loss occurred in Benedict's study but a figure of nearly 3000 mmol sodium loss in the urine does not seem to me compatible with sodium deprivation in view of the sodium conservation capacities of a mammalian kidney. However, the magnitudes of turnover were highlighted by the discussion. Data were cited on the necessity of providing salt for zoo and working elephants but amounts of only 600–1000 mmol Na/day were suggested. An estimate of faecal loss (21 kg/day) on a Kenyan elephant gave a sodium loss of 1300 mmol/day. Analysis of browse food eaten by elephants suggests a sodium intake of from 16 to 80 mmol/kg. The daily water intake of the elephants cited was 160–320 litres. Given the sodium content of the various bore waters (see Fig. 3-19) it is clear that at many regions sodium intake could be 10–25 mol/day whereas at other regions it would be only 300 mmol, and these latter animals may have definite physiological need of additional sources. Weir noted that grass was often a substantial element of elephant diet in some regions, and with sodium shortage a movement to browse feed would help rectify deficiency since its sodium content is higher than grass. He points out that an elephant is not a ruminant but that its digestive processes produce fatty acids as do the ruminants, and it may well be that the need for sodium in the digestive processes is similar to that demonstrated in the sheep and other cud chewers by Denton (1957). The amount of sodium in the saliva in the 150 kg of stomach content of an elephant may be considerable and this need may fluctuate. It is clear that blood aldosterone and other determinations on the elephants in these regions would be of great interest.

Central Asia

Findings of sodium deficiency in wild animals have been confirmed in the Pamirs of Central Asia by scientists from the Physiological Institute of Science City, Novosibirsk (Bazhenova and Kolpakov 1969). They found that field mice living at 3800 m altitude in the Eastern Pamirs concentrated in areas where there was sharply increased sodium content of soil, possibly from a geological source. Analysis of grass in the region showed very high sodium and potassium (320 and 575 mmol/kg dry wt) and analysis of secretions of the animals confirmed this high intake. The authors suggest the distribution confirms the findings of Aumann and Emlen (1965). Sheep transferred from a habitat of 900 m to 3800 m showed, associated with a sharp fall in sodium content of soil, a fall in plasma sodium and rise in plasma potassium. After adaptation to 3800 m the plasma sodium concentration rose again.

North America

Smith (1977) has studied the snowshoe hares in Southern Ontario near Lake Huron in summer and winter and shown the occurrence of sodium deficiency. The vegetation of the peninsula contained 10 mmol Na/kg dry wt, the soil 3 mmol/kg and the water <0.04 mmol/l. Mean urinary sodium of the animals was 1.0 mmol/l. This was consistent with daily intake of about 100 g vegetation, mainly coniferous plants (Smith et al. 1978). Zona glomerulosa was enlarged, particularly in the females, and nodulation of the zona glomerulosa was seen as described by Myers (1967). Blood aldosterone was raised though considerable variation in levels was observed. The seasonal effect on zona glomerulosa and presumably aldosterone was consistent with data (Blair-West et al. 1968) on maximal stress being in spring/early summer (Smith et al. 1978). The evidence was of greater stress on the females on this sodium-deficient diet.

The most comprehensive single study of the sodium dynamics in a specific ecosystem has been made in North America by a group from Yale University (Botkin et al. 1973; Jordan et al. 1973) on the moose population on Isle Royale National Park, Lake Superior. This is a northern temperate wilderness archipelago, essentially undisturbed by man, where wolves (*Canis lupus*) and moose (*Alces alces*) exist in apparent equilibrium, though both are relatively new to the island. The moose were first observed there 65 years ago and the wolves about 25 years ago. There are 1000–1200 moose on the 550 km^2 archipelago and this is one of the most dense populations of moose known. Before the appearance of the wolves, the forage demands of this population exceeded the productive capacity of available vegetation, and mortality in the moose was widespread.

The Yale group's studies on the flow of minerals through the soil, vegetation, moose–wolf system at Isle Royale have indicated that sodium is not easily available to the moose. For mammals, sodium is a major constituent of the body comprising about 0.1% of the live weight. In contrast, few terrestrial green plants are known to require sodium, and such plants need only trace amounts. Except for halophytes, terrestrial green plants do not concentrate sodium. Four other elements of key nutritional importance (nitrogen, potassium, calcium and magnesium) appear to be present in adequate concentration in the vegetation eaten by the moose, though the availability of phosphorus may be marginal. Their inventory of the sodium available in the ecosystem is set out in Table 3-12.

Moose have a year-round diet consisting mainly (around 90%) of browse—current leaves and twigs of woody plants with occasional bark—supplemented by some herbaceous plants in spring and some aquatic plants in mid-summer. The browse has an average sodium concentration of 10 ppm at Isle Royale, which is only 1% of the level recommended for domestic ruminants (0.1% or more of diet). The animals nevertheless appear healthy. The moose remove about 10%–20% of the annual browse production and this provides about 7%–14% of their estimated annual sodium requirement. The small but unknown proportion of herbaceous plants they also take would not significantly alter their sodium intake because these plants have a very low sodium content compared with the woody plants. The estimated annual sodium requirement against which this figure of browse intake has been set was derived from the amount of sodium required for all new tissues grown during the year, plus the amount needed to balance urine and faeces losses. For the whole moose population, this rough calculation gave a figure of 243 kg Na/year. The calculated total sodium in the available browse on the island was 170 kg/year of which the animals eat only 10%–20%: this leaves an overall deficit of over 200 kg for the population.

It was also noted that some animals use natural mud licks on the island. However, the sodium content in the mud and water at the licks averaged only 24 ppm (1 mmol/kg), so this was not a significant source of sodium.

During 8 weeks or more in the summer, moose of the region fed extensively on aquatic macrophytes. The analyses made on the common plants of this type in the area (see Table 3-12) show the sodium content of submerged and floating aquatics to be about 500-fold, and that of emergent aquatics and those rooted along the shore to be 50-fold greater than that of the terrestrial plants of the region. This concentration of sodium by the submerged and floating aquatics is spectacular.

Measurements of the production of aquatic macrophytes at Isle Royale are said by the authors to be inadequate. They have used a figure of 0.5%

Table 3-12. Sodium values (mmol/kg) in various components of the Isle Royale ecosystem (calculated from Jordan et al. 1973).

Material	Leaves mean	n	Twigs mean	n
A. Terrestrial woody plants				
Abies balsamea[a]	0.1	26		
Acer spicatum[b]	0.4	22	0.4	31
Alnus rucosa	0.6	5	1.2	2
Betula alleghaniensis[b]	0.8	5	1.0	5
Betula papyrifera	0.7	31	0.5	45
Corylus cornuta[b]	0.2	8	0.4	7
Populus tremuloides	0.3	14	0.3	20
Myrica gale	1.2	2	0.5	1
Sorbus americana[b]	0.3	35	0.2	49

Material	Mean	n
B. Aquatic vegetation		
1. Typically emergent or rooted along shore		
Calamagrostis canadensis	0.1	1
Calla palustris	74.5	1
Carex aquatilis[c]	16.4	2
Carex rostrata[c]	10.7	4
Carex scabrata[c]	9.7	1
Carex sp.[c]	31.0	1
Eleocharis smallii	62.6	1
Iris versicolor	1.2	3
Juncus gerardii	1.9	1
Menyanthes trifoliata	38.8	2
Polygonum sp.	25.0	2
Potentilla palustris	13.0	2
2. Typically submerged or floating		
Callitriche sp.	153	2
Chara sp.[c]	44	5
Equisetum fluviatile[c]	68	1
Myriophyllum tenellum	206	1
Nuphar sp.	408	1
Potamogeton gramineus[c]	270	6
Potamogeton richardsonii[c]	314	3
Utricularia vulgaris[c]	350	3
C. Moose tissues		
Bone[d]	130	10
Liver[d]	30	1
Muscle[d]	34	1
Skin and hair[d]	68	1
Winter faeces (from snow)[e]	4.9	5
Winter faeces (in rectum)[e]	44	1
Summer faeces[f]	3.9	23
D. Substrate and water		
Rain[g]	0.01	7
Surface water[h]	0.07	15
Alluvium	0.4	3
Pond bottom, mainly inorganic	1.8	7
Pond bottom, mainly organic	13.3	4
Pond bottom, inorganic and organic	5.9	11
Mud lick water	1.0	9
Mud lick soil	1.0	4

[a]Sample includes leaves and twigs; this species is taken extensively by the moose only in winter.

[b]Important summer and winter food for moose.

[c]Observed to be taken by moose at Isle Royale.

[d]Live-weight basis; all tissues from one adult male autopsied in winter.

[e]Dry weight basis.

[f]Dry weight basis; collected within 24 h, no precipitation intervening.

[g]Six of seven were equivalent to that of glass-distilled water; one value was 1.1 ppm, which was suspected to be due to contamination.

[h]Collected from ponds, harbours and streams.

of the annual incoming solar radiation, or half the value reasonably expected for agricultural crops or forests, and on this basis estimated aquatic production to be 5×10^5 kg/km^2 per year. From map readings they calculate the area of shallow water which could support aquatic plants to be 30 km^2. Assuming 10 km^2 of this to be colonized, and the production per square kilometre to be as above, this gives an annual production of 5×10^6 kg. They take 500 ppm as a mean figure for the content of sodium in the aquatics, which gives an amount of 2500 kg of sodium, some ten times the calculated requirement of the moose herd. All this production is available to the animals, but field observations of moose feeding indicate the animals consume primarily submerged plants, which have a higher than average sodium concentration. However, the central point for consideration is that even if the herd's *annual* sodium requirement can be met in this way, the feeding period is only approximately 2 months, so that there would need to be some mechanism for retaining sodium during the 10 months when it is available in only minimal quantities.

Jordan and colleagues (1973) estimate the factors in the annual sodium balance of an individual moose to be as set out in Table 3-13. The

Table 3-13. Estimated annual sodium budget for a hypothetical adult moose based on data from Isle Royale.

		Sodium (g)
Input		
Food		
Terrestrial plants		13
Aquatic plants		199
Others: water from licks		20
	Total	232
Requirements		
Replacements		
Summer faeces		10
Non-summer faeces		96
Urine		19
New tissues		107
	Total	232
	Balance	0

After Jordan et al. (1973).

subject moose is a hypothetical adult animal assigned a weight of 358 kg but not in a reproductive condition. If this moose is to avoid a negative balance, long-term sodium outflow cannot exceed input. The data are derived from measurements and from the estimates given above. Sodium demand was of two types: that lost in excrement, which has to be replaced on a regular basis, and that required for growth of new tissues. The excremental losses are mainly through faeces. The faecal sodium level remains relatively constant through the year (see Table 3-12), outflow of the element being proportional to the passage of undigested food. New tissues include those annually replaced such as hair, epithelium and antlers, and those accumulated as body growth. Sodium in growth includes that accumulated in foetus, amniotic fluid and milk. The amount of growth assigned to the hypothetical individual is equal to the sum of all growth in the population divided by the number of adult equivalents. The latter is determined by dividing the population biomass by the weight of the adult equivalent. Since this budget estimates only an average need, minimum requirements for some classes, such as the lactating cow, are underestimated.

It was shown that a moose eats 7 kg of browse per day during the 241 day non-summer period and 4.8 kg of forage per day including 0.6 kg of aquatic vegetation during the summer. Each moose took an estimated 2.25 litres of mud-lick water per day. Although these computations are rough, they indicate (Table 3-12) that the moose may keep itself in satisfactory sodium balance over the period of the year, though it would be compelled to expend considerable extra time and energy on obtaining adequate amounts of sodium. The lactating cow would be the first to experience the effects of sodium deficiency. Urine was collected as it was passed into the snow and the estimates from its analysis indicated that the animals were probably extremely efficient at sodium conservation, for the sodium content of the urine was less than 0.2 mmol/l. Faecal sodium was 4–5 mmol/kg which, allowing that extraction of sodium from browse and other food may not be 100% efficient, probably reflects the action of adrenal cortical hormones in facilitating the reabsorption of sodium from the secretions in the large intestine.

As regards the way in which the animal is able to utilize the large sodium intake over 2 months to facilitate its survival and well-being over the whole year, the authors considered two possibilities. First, that there is mineralization of extra sodium in hard tissues for release on demand, as, for example, is seen with calcium and phosphates. Secondly, that there is a substitution of potassium for sodium in the saliva by the parotid gland, so that the sodium in the rumen fluid can be expended as from a reserve.

They note that if there were seasonal storage and release of sodium in hard tissues, then this would be reflected in bone and they are carrying out an analysis accordingly. A study by McDougall et al. (1974) on the effect of sodium deficiency in sheep has indicated that in the ruminant, as in other species, bone sodium may provide some store for the animal, but it is not a major source. The sheep studied were in three groups: normal animals, animals with parotid fistulae which were allowed to become sodium-deficient and animals which were normal but given only sodium chloride solution (300 mmol/l) to drink, so that they were heavily sodium-loaded. Those which were on a very high sodium intake did not have a higher level of bone sodium than the two other groups: paradoxically, it was lower.

The effect of short-term sodium depletion on bone sodium in sheep was determined. Sodium depletion for 72 h caused a mean external sodium loss of 540 ± 110 mmol and a small but significant decrease in bone sodium from 333 ± 14 to 318 ± 10 mmol/kg fat-free dry wt ($P<0.01$, *t*-test). As bone mass comprises 16%–18% of total body mass, the mean decrease in bone sodium was 15 mmol/kg fat-free dry wt, and as water- and acetone-extractable material makes up 17.5% of wet bone mass, the authors calculated that 16% of the total amount of sodium lost from the body was from bone. There was no correlation between the calculated sodium deficit and the decrease in bone sodium. Their experiments also showed that short-term sodium depletion over 24–48 h decreased the total exchangeable sodium of the sheep as measured by ^{24}Na injection. Long-term sodium depletion over a period of weeks decreased bone sodium but not to a substantially greater extent than did short-term loss. Further, 3 weeks after returning to a normal sodium intake, the bone sodium still had not recovered to its pre-depletion value. These data therefore indicate that whereas bone sodium could contribute as a reservoir, it was not the major source of the sodium which the animals lost. There was, from these studies, no evidence that the bone acted as a significant store in the instance of high sodium intake.

Jordan and colleagues' (1973) second suggestion as to how the moose maintains its sodium balance, i.e. that it accommodates physiologically to sodium deficiency by substituting potassium for sodium in the saliva and thus drawing upon sodium in the rumen pool, is put quantitatively by them as follows:

If moose are similar to sheep with respect to substituting K, then, extrapolating from Denton's sheep on a live weight basis, we calculate that our adult equivalent moose has 274 grams of Na to contribute annually from its rumen pool. This 274 grams exceeds what we estimate is our moose's total twelve month requirement [Table 3-13], thus suggesting that Isle Royale moose would maintain year-round balance using this strategy. We analysed rumen contents from the adult male autopsied in winter. Adjusting for Na and K contributed by forage, we calculated that the animal's salivary Na/K ratio had been 0.45. This value is intermediate between Denton's 18 for those sheep which were on adequate intake and 0.055 for those on low Na for 14 days. It is consistent with the substitution hypothesis and has prompted us to further study.

It is presumed that when sodium intake rises in the summer the composition of saliva changes accordingly, and the composition of the fluid in the rumen is restored to normal. In effect, the potential reservoir is reconstituted. Jordan et al. go on to note that if the moose were not physiologically adapted to withstand such sodium shortages, then the species would not exist in such habitats. The fact of their presence serves to corroborate the calculation that they are adapted to withstand prolonged sodium shortage (cf. Fig. 3-4 and Table 3-3), so long as their deficits can be made up during some season of the year. Obviously, a downward trend in the available sodium would lead to a severe stress on the population, and a population decline. Some of the observations on the Isle Royale suggest decrease in aquatic plants and a reduction in the pond sediments that are a source of sodium, and the authors warn that any undue impact on the productivity of the aquatic habitats or an over-exploitation of them could cause this resource to fail and therefore create a nutritional crisis for the moose population. It appears that the growth of an animal population is limited solely by the supply of a single chemical element.

These data represent an important ratification of the thesis (Denton 1957) that the capacity of ruminant animals to change their spectacular volume of salivary secretion from a predominantly sodium to a predominantly potassium fluid represents a major mechanism in the evolutionary emergence of ability to colonize environments where conditions involve severe stress on sodium balance.

Weeks and Kirkpatrick (1976) studied salt appetite occurring seasonally in animals of southern Indiana, USA. They reported that fox squirrels and woodchucks showed increased seasonal salt drive but their main work was on the deer. On a 251 km^2 area of land, largely forested, they collected water from non-lick and lick sources, and also analysed monthly the ionic content of the ten most important plant species. Deer drank from standing water in licks, and also ate the soil. Lick soil was shown to have a mean of 27 mmol Na/kg compared with ca 1 mmol Na/kg in other areas. The sodium content of water in licks ranged from 1 to 7 mmol/l. No significant differences were observed in lick use between different sexes and ages; does were pregnant for several months before initiation of salt drive. Intensity of use was highly seasonal, though, being greatest in spring; March and April were characterized dietetically by deer shifting to succulent green grass and forbs. Most species eaten by deer varied little in seasonal sodium content, which was generally low (2 mmol/kg). Potassium content was usually below 250 mmol/kg but increased dramatically with spring growth when peak levels reached 1 mol/kg, and this very high potassium intake with the lush spring vegetation coincided with the increased salt appetite.

In terms of sodium needs for reproduction, the authors calculate that a twin conceptus including amniotic fluid would need about 800 mmol of sodium during development and about 50 mmol Na/day in milk. Also, antler growth in the male would sequestrate a variable amount of sodium ranging up to 280 mmol. In spring there was an increase in the area of zona glomerulosa of the deer adrenal glands corresponding in time exactly to the greatest frequency of lick usage. In October–December zona glomerulosa decreased in males. As most does become pregnant in November, it is interesting to consider whether social stress of the mating season for the male causes this reduction via ACTH release (see later part of this chapter). At no stage was urinary sodium concentration high in the animals (mean, 2–7 mmol/l). Even allowing that urine volume doubled or trebled on the lush diet in spring, a great loss by this route did not occur. The authors stress the possibility that the very high potassium intake could have increased sodium loss, but though urinary sodium concentration was highest in April–May (7 mmol/l) this was also the time the animals used licks most intensively. Analysis showed there was some increase in faecal loss of sodium during the spring.

This study reflects well the capacity of selective appetite to maintain a wild ungulate in good condition in the face of minimal sodium in the forage. The fact that urinary sodium was always low suggests that sodium status was verging on deficiency. But in the absence of data on salivary Na/K ratios, aldosterone and renin levels a clear picture is lacking.

In discussing the data of Botkin et al. (1973) on survival and reproduction of the Isle Royale moose Weeks and Kirkpatrick raise the question of whether drawing on rumen sodium will be effective over an extended period of deficiency, noting that the rumen pool is not a static amount of fluid but is being constantly added to by salivation and reduced by reabsorption. Because the animals feed episodically (i.e. with days between meals) they suggest there is a high probability of rapid disappearance of the sodium pool in the rumen. Whereas a markedly episodic pattern of feeding may cause some movement of sodium from rumen to extracellular fluid and possibly also sodium excretion (Denton 1957), the role of the rumen as a major sodium reservoir under chronic conditions is quantitatively a different matter. Bailey (1961b) has shown a close relationship between Na/K ratio of mixed saliva and the contemporaneous sodium and potassium content of rumen fluids of cattle. Different saliva compositions over a wide range were produced by feeding diets of different composition and sodium content. Further it has been shown (Fig. 3-4) that a sodium-deficient animal may live for months with a consistently low salivary Na/K ratio, i.e. in sodium equilibrium but with an existing deficit. In fistula experiments this situation develops fairly rapidly. However, in the wild, when there is a clear seasonal change in dietary sodium, it is clear that with the dynamics of salivary secretion of 100–200 litres of saliva per day in cattle (Bailey 1961a) and reabsorption of comparable volume from the rumen, the onset could well be very gradual. With the gradual fall of rumen fluid Na/K ratio following the aldosterone-determined salivary change, *the additional sodium would be available to extracellular fluid, the tissues of the developing young* in utero *etc*. Thus dynamic equilibrium may be perturbed to the disadvantage of the animal by episodic feeding, but once mild sodium deficiency develops, it is unlikely, with a high blood aldosterone level, that any significant amount of sodium would be lost in the urine. Weeks and Kirkpatrick (1976) do raise the interesting question of the mode of survival of non-ruminant herbivores in sodium-deficient regions.

Further observations on grazing cattle

Parallel to these various findings indicative or demonstrative of sodium deficiency in nature, it is worthwhile to add some additional evidence on the circumstances of sodium deficiency in grazing animals.

Investigations of the requirements of cattle for common salt were made by Babcock and Carlyle in 1905 (see Orr 1929). They showed that when store-fed cows were not given access to salt they developed a craving for it within 2 to 3 weeks. Later, there was a complete breakdown in health with loss of appetite, haggard appearance, lustreless eyes, rough coat, and rapid decline of live-weight and milk yield. When this occurred in a high-producing cow after calving, death might follow suddenly. Sodium chloride cured or prevented the disorder. South African workers investigated the quantitative requirements for sodium by cows and reached the conclusion from a series of experiments that 1.5 g Na/day was adequate for growth, but that if the cow were lactating of the order of 11 g/day was required. Given the amount of dry matter a cow consumes the requirements would be met for a lactating cow by diet with 0.15% sodium in it. These data and those of the US National Research Council have indicated that in a number of grazing conditions the amount of sodium available to cattle may not be adequate for normal lactation. The literature has been surveyed by Russell and Duncan (1956).

Sodium deficiency was shown to impair growth severely in beef cattle fed on an all-*Sorghum* ration in southern Queensland, Australia. Sodium needed to be 0.05% of dry matter intake to give optimal growth (Morris and Gartner 1971).

Murphy and Plasto (1973) examined a herd of cattle on the Eastern Darling Downs in Queensland, which is a district with a subtropical climate that is about 150 km from the sea. The elevation was 500 m and annual rainfall was 900 mm, mainly as summer rain. They studied two groups of 40 Hereford cows with suckling calves ranging from 3 to 13 weeks of age. One group was given access to sodium chloride ad libitum. The saliva composition of both groups at the outset (mean Na, 60 mmol/l; K, 71 mmol/l) was consistent with moderately severe sodium deficiency causing high blood aldosterone secretion. After 21 days of sodium supplementation, the saliva composition of the treated group was normal (mean Na, 139 mmol/l; K, 7.4 mmol/l), and it remained normal throughout the 12-week study. Analyses of four predominant pasture species eaten by the cattle showed a mean sodium content of 3–7 mmol/kg dry wt. The cattle offered access to sodium chloride ingested 151 g/day per head during the first 21 days and the mean intake over the period was 95 g/day per head. Intake exceeded 800 g per head during the first 2 days of access.

Pica, which had also been manifest in the herd, ceased on access to the sodium chloride supplement. Cows receiving the supplement had significantly better live-weight gain (0.35 vs 0.14 kg/day per head, $P<0.01$) than unsupplemented cows, and their calves grew significantly faster (0.77 vs 0.62 kg/day per head, $P<0.01$). Blood and milk concentrations of sodium and potassium were not affected by the supplementary intake. It is interesting in this regard that Yagil et al. (1973) have reported that aldosterone causes a significant (ca 30%) decrease in sodium content of milk in rats. After 12 weeks, the calves of the unsupplemented group had significantly lower sodium concentrations in saliva (79 vs 119 mmol/l) than the supplemented group. The authors suggest that, in relation to calf growth, a predominant influence of sodium is via the increased milk output of the cow. Aines and Smith (1957) reported a 100% increase in milk yields when sodium-deficient dairy cows were fed a sodium supplement.

These results indicate that sodium deficiency occurred in the cattle grazing these pastures and the authors propose the same conditions may hold in other pastures of low sodium content, such as cover a wide region of Australia (Denton et al. 1961). In this regard Arnold's (1964) experiments on sheep are noteworthy. By the use of an oesophageal fistula he examined the selective grazing capacities under field conditions of normal animals and those made sodium deficient by a parotid fistula. He found that the sodium-deficient animals biased their selection of plants towards those with a higher sodium content. This suggests that if there are better sodium-accumulators among plant species available, selective grazing may help obviate the impact of deficiency. In the Snowy Mountains rabbits remove the high-sodium-content seed husks of *Stylidium* plants. A good review of this question and its relation to sodium deficiency in grazing animals, and the current evidence on physiological benefit from sodium supplementation has been provided by Morris (1980). He points out that sodium has not been shown to be an essential element for plant growth. He points out that while some forage plants in low rainfall areas, such as *Atriplex*, may accumulate sodium up to 8% of dry matter (Wilson 1966), temperate region plants may show large and consistent differences in capacity to take up sodium. Analysis of tropical species also suggests the existence of high-sodium and low-sodium plants, with a possible predominance of those with low-sodium potential (Jones 1963; Playne 1970; Gartner and Murphy 1974). Herbage with less than 0.001% sodium has been recorded in Highland New Guinea (Leche 1977), and dramatic response of cattle to salt feeding in this area has been reported. In pastures in Mediterranean climates marked seasonal variation in sodium content occurs. Morris also notes that speculation on the benefits of feeding salt to cattle is ancient; Columella (about 40 A.D.) in *De rustica* says that salt contributes to good bodily health in cattle and that they gladly have recourse to it.

Frens (1958), working in The Netherlands, noted that the large quantities of water combined with small amounts of structure-giving components that have to pass the intestinal tract of cattle living only on young grass put such high demands on reabsorptive capacity that water reabsorption from the intestines is less effective than it physiologically should be. This causes soft faeces with which extra amounts of mineral are lost. Sodium and copper requirements of cattle pastured on intensively managed grassland are especially increased as a result of this laxative effect, and the amounts of the minerals ingested in grass are often inadequate to balance this. Farmers see animals to be in poor condition and in areas where there is pasture haemoglobinuria the animals are very susceptible. Frens calculates that cows on such pastures may require 100 g sodium daily as a result of faecal loss and milk production. A pasture content of 0.6% dry matter of sodium would meet this, but, as the usual content is only about 0.3%, supplemental sodium (as 2.5% NaCl) is required.

Jones et al. (1967) found that feeding sheep a low-sodium food over 4 weeks caused salivary changes but there were no deleterious effects on the animals detected over this time period.

General consideration of ruminants and their adaptive capacities with change of diet

For the ruminant, and perhaps other herbivorous animals, there are particular problems of sodium status contingent on the prodigious salivary flow (Denton 1957, 1969). This, in effect, is a second circulation, since up to twice the sodium content of the extracellular fluid passes down the oesophagus each day and is reabsorbed. The dynamics of this equilibrium involve at any one time a pool of fluid in the rumen or forestomach. In the case of the sheep this is 2–5 litres of fluid containing 250–600 mmol sodium. Rumen analyses on 12 camels by Schmidt-Nielsen et al. (1956) showed that the semi-solid part of rumen contents contained on average 17% solid matter and 83% water, while the rumen fluid contained 2% solids and 98% water. The amount of rumen content (27–66 kg in a 350–500 kg animal) was the same order as reported by Ritzman and Benedict (1938) for cattle—10%–15% of body weight. The analyses of the rumen fluids showed a mean sodium concentration of 106 mmol/l, potassium of 17.6 mmol/l and chloride of

17.6 mmol/l. As the authors pointed out, these figures approximate the rumen fluid of other animals and there is a conspicuous similarity to the composition of mixed saliva as described in sheep with oesophageal fistulae (Denton et al. 1952a, b). Thus they discounted the old legend of the camel storing drinking water in its forestomach, though they pointed out there could well be substance to the tale of the desperate desert traveller saving his life by killing his camel and drinking the rumen fluid. In relation to our primary point, their data show that a camel may have 2500–8500 mmol of sodium sequestrated in its rumen at any one time. It is of interest that Paque (1963) reports that sources of drinking water for camels in the Sahara contain 170 mol/l of sodium.

Given these data, and the dynamics of secretion, several considerations present. Changes in amount or quality of diet of ruminant animals as, for example, from lush green grass with high water content to very dry material may alter substantially the amount of salivary secretion, and also the volume of fluid in the rumen at any one time. This type of change in feeding conditions may occur over weeks to months in the sub-tropical monsoon-watered interior of continents where summer rains may be followed by heat and dry winds. The metabolic problem of providing large additional amounts of sodium-rich digestive fluid in the rumen may be aggravated if the grass has a very low sodium content as, for example, in Central Australia (Table 3-1). In these circumstances the animal may draw on its extracellular sodium reserves and increased aldosterone secretion with a fall in salivary Na/K ratio will result. If a sheep's increased salivary secretion results in an extra 1–2 litres of saliva being sequestrated in the rumen, the provision of its sodium content of 150–300 mmol may require many weeks of food intake if the animal eats 2 kg/day and the sodium content is 1–10 mmol/kg (Table 3-1). Another factor relevant to electrolyte dynamics is the effect of starvation or for that matter episodic feeding. If food intake is suddenly reduced, absorption of rumen fluid progresses. We have shown that parotid saliva secretion decreases in starvation. Thus a quantity of sodium from the rumen will enter the extracellular compartment. Will this sodium be excreted, so that when feeding resumes at the same level some time later, the animal will need again to draw on extracellular content and deplete itself of sodium in order to resume full digestive activity? In relation to this, a preliminary impression we have from studies made some years ago is that with starvation in sheep no sudden increase in urinary sodium occurs. Further, A. W. Turner and B. Hodgetts (personal communication) found gradual starvation of sheep caused no fall in plasma bicarbonate concentration, as it does in carnivores. In contrast to the well-known 'alkaline tide' phenomenon in man, where gastric secretion results in compensatory sodium excretion in the urine, the evidence suggests that sheep and other ruminants may have adaptive mechanisms to conserve sodium in the face of fluctuations in rumen sodium content.

Data which also show change in sodium status and rumen composition in the face of changing dietary pattern have come from the studies of Dobson (1965).

Man

Some evidence of interest in salt by primates was cited above, but there are no data available on their mineral status in the wild. However, there is some interesting metabolic material on feral man, indicating that the level of sodium intake may habitually be very low by virtue of diet and region of habitation. Consistent with this sodium status, the 'set' of a number of functions is strikingly different from parameters as measured in urbanized man. The evidence, as will be recounted in other sections, indicates that this low sodium status is no physiological handicap, and may indeed be advantageous. However, it does indicate that any of a number of recognizable metabolic stresses which may occur naturally may rapidly move individuals in such populations into severe sodium deficit.

Dutch anthropologists working in New Guinea noted that the Highlanders living on a predominantly sweet potato diet had a K/Na ratio of intake of the order of 200 to 1. Daily sodium intake was as low as 1–3 mmol (Oomen et al. 1961). In Papuan communities food intake is often dependent on a single staple food be it sago, sweet potato or taro—or rarely yam or banana. Not uncommonly this accounts for 90% of the intake.

Macfarlane et al. (1964) studied the Chimbu tribe in the tropical New Guinea Highlands, whose diet consists of sweet potato, breadfruit, yams, taro, banana and beans. The urine of these vegetarians was alkaline, with a high potassium concentration (200–400 mmol/l) and very low sodium concentration. Consistent with the low sodium intake (ca 10 mmol/day), the mean plasma sodium concentration was low—128 mmol/l. Mean plasma potassium was 5.1 mmol/l. The sodium contents of sweat, saliva and milk of the Chimbu were much lower than those of Europeans living in the same region, and total body water and extracellular space were higher than in the Europeans. In collaboration with Macfarlane and colleagues, Drs J. Coghlan and B. Scoggins determined peripheral blood aldosterone levels on a group of these Highlanders and Dr S.L. Skinner the renin levels (Denton et al. 1969). Fig. 3-20 records these results along with determinations made on other New Guineans with access to European food. Mean blood aldosterone level of the Highlanders from remote regions was four times higher than those of urbanized New Guineans or urban Australians (i.e. 28 ng/100 ml vs 6 ng/100 ml in the latter two categories). Renin showed no difference, and did not correlate with high aldosterone.

Oliver et al. (1975) studied the Yanomamö Indians living in the tropical rain forests of Venezuela and Brazil. About 12 000 to 15 000 people are scattered in 150 villages over an area of 256 000 km^2 (100 000 square miles), and apparently have lived in isolation for thousands of years. They eat bananas, vegetables and occasional wild game, insects and fish. Urinary sodium excretion was about 1 mmol/day and urinary potassium 150 mmol/day. Several individuals excreted as little as 0.1 mmol of sodium a day—a remarkable physiological feat. Urinary aldosterone (74.5 ± 44.9 (s.d.) μg/day) was high compared with urbanized man (17 μg/day). All plasma renin activity measurements were within or exceeded the range seen in ambulatory Americans receiving a diet with 10 mmol/day of sodium.

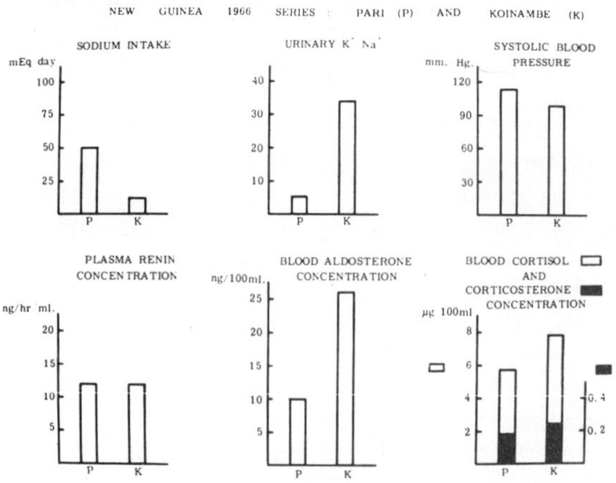

Fig. 3-20. The peripheral blood aldosterone concentrations of New Guineans in remote areas of highlands (K) in relation to dietary sodium and Na/K ratio in the urine compared with New Guineans with some access to European diet (P). Collaborative study of Howard Florey Institute (Drs J. Coghlan and B. Scoggins) with Prof. Victor Macfarlane's group and Dr S. L. Skinner. (Reprinted from Denton et al. 1969, with permission.)

In the Highlands of New Guinea at Tukisenta, Sinnett and Whyte (1973) have studied a population living in valleys between 1850 and 2600 m. Diet consisted of sweet potato, breadfruit and yams and bananas. Pig herding is practised so occasional meat-eating occurs. Urinary sodium averaged 13 mmol/day in females ($n=135$) and 6 mmol/day in males ($n=138$).

These selected examples, which will be elaborated in certain respects in Chapter 27 that deals with the epidemiology of high blood pressure and diet, are indicative that feral man may have a very low sodium intake.

Aggravation of low sodium status by superimposed natural causes

Depletion of body sodium reduces the blood and extracellular volume, which in turn impairs the functional capacity of circulatory adjustment. Thus, in the wild animal, speed and endurance during flight or pursuit are jeopardized. It will also affect the distance ranged and thus the grazing capacity of animals. Sodium deficiency is aggravated by a series of natural phenomena which include reproduction, temperature regulation and increased population density. During reproduction additional sodium is required for the foetal tissues *in utero*, the changes in adnexa, and also during lactation. For example, Bactrian camels produce 5–15 litres of milk per day in a lactation of up to 2500 litres (Kneraskov 1939). Lactation stress will vary between species. Whereas human milk contains 5–6 mmol/l of sodium, cows' milk may contain 25-40 mmol/l. The sodium concentration does not reduce with sodium depletion though volume does. I have referred above to the manner in which change in physical character of food with alteration of salivary secretion (Wilson and Tribe 1963; Wilson 1963) may influence sodium requirements and thus have impact during sodium deficiency.

Additional stress on sodium homeostasis may occur in primates and other species as a result of temperature regulation. Anthropological data support the contention that feral man is a tropical mammal having a natural water turnover of up to 19 litres/day (Martin 1930; Macfarlane 1963). Cattle have been observed to drivel large volumes of saliva in extreme heat and thus a large sodium loss may occur. Sweating may have a similar effect. Macfarlane (1964) quotes sweat loss in camels of up to 8–10 litres/day with sodium loss of 80–100 mmol. However, both sheep and camels are relatively effective in avoiding sodium loss in this way as their sweat has a much higher potassium content.

McCance (1936) notes that a man may lose 150–200 mmol of sodium per day in sweat, and Collins (1963) records losses of up to 400 mmol/day in unacclimatized man. 'Salt frost' may be seen on the skin (Black 1953). Acclimatization, in which the adrenal hormones play an important role (Conn 1949), effectively reduces the sodium content of sweat, and Conn showed subjects were able to approach sodium equilibrium on very low sodium intake (ca 40 mmol/day) though losing 5–9 litres of sweat per day. It is clear, however, that in tropical conditions, or with sudden heatwaves in temperate areas, a continual loss of variable degree may occur in this way. If diet were vegetarian under these conditions, McCance (1936) points out he would, for example, have needed to eat about twice his body weight daily in potatoes to get his sodium requirement.

In both sheep and cattle, plasma volume increases with exposure to hot environment, and with sheep, increase of extracellular volume has also been demonstrated (Macfarlane et al. 1959; Dale et al. 1956). Macfarlane et al. (1959) give measurements on sheep that indicate extracellular volume increase of 5 litres in tropical heat with a resultant increase in sodium requirement of 700 mmol. Bazett et al. (1940) and Scott et al. (1940) have shown that men working effectively in hot environments maintain an increased blood and interstitial fluid volume and often show an increased peripheral circulation. Myers and Bults (1979) emphasize the additional stress on animals, particularly small animals, when a sodium-deficient environment is combined with severe cold and the additional metabolic demand this involves.

Many species suffer infectious diseases which involve pyrexia with additional sweating. They may also have diarrhoea or scouring because of intestinal infection or infestation with parasites. Sodium loss ensues. These latter effects are more likely in gregarious species where transfer of the infection or parasite from host to host is more effective. For example, in Bovidae, *Salmonella* bacterial, or various viral diseases such as infectious bovine diarrhoea (mucosal disease) occur. Diarrhoea is evidently common in human societies, and Schaller (1963) describes death from gastroenteritis among the wild mountain gorillas. The extreme example in man is cholera, where stool loss may range from 40 to 600 ml/h, with sodium and potassium concentrations ranging from 90 and 30 mmol/l at the slow rate to 140 and 10 mmol/l at the fast rate (Nalin and Cash 1976). It is worth noting that with infectious disease involving electrolyte loss, the initial defence for survival will be in the area of biochemical and endocrinological regulation since the animal needs to withstand the impact for some days before immunological mechanisms become effective (Denton et al. 1951). As well as having infectious causes, severe diarrhoea can arise in ecological conditions involving trace metal imbalance: for example, copper deficiency, molybdenum intoxication or selenium deficiency. As indicated above, eating a lush green pasture can cause it in cattle, and wild game. Man loses sodium too as a result of the vomiting that occurs in pregnancy and cyclically in children (McCance 1936). There appears little information as to whether these latter phenomena occur in primates, or for that matter in primitive human communities.

Another cause of stress on sodium balance appears to arise from increased population density. Christian and Davis (1964) reviewed a substantial body of evidence indicating that hormone changes may be operative in some of the major population-regulating processes recorded. Several types of density-dependent mechanisms have evolved for regulation of many populations of mammals within the limits enforced by the environment, including food. Thus these species avoid the hazard of destroying their own environment, and the hazard of their own extinction. One effect of increased numbers and aggressive behaviour involves stimulation of the pituitary–adrenal axis and inhibition of reproductive function. The increased adrenal cortical function, with its effects such as thymic involution, would also reduce resistance to disease. Correlations of these indices with population density have been made in laboratory experiments on mice, which showed the diverse influence of increased numbers on sexual maturation, spermatogenesis and lactation. The possibility was considered of increased adrenal cortical secretion during pregnancy affecting the brain maturation of the young with subsequent behavioural results (Keeley 1962; Howard 1963). The question also arises of whether ACTH secretion, through stimulation of adrenal androgens, has a suppressive effect on gonadotrophin secretion, and thus on sexual maturation and reproductive function. Emotional changes, themselves contingent on density, may also have a direct effect on the pituitary. Social rank has a large influence on endocrine changes in high population density, the dominant animals being much less affected than subordinate ones. In this regard, the most important consideration emerges that the young are subordinate animals and thus bear the full brunt of these density-induced endocrine changes. A particular aspect of this is stressed by Christian and Davis. Immature mice secrete appreciable amounts of cortisol. When grouped at high density this amount increases relative to corticosterone. With sexual maturation in the male mouse, this capacity to produce cortisol diminishes. If there is delayed maturation because of stress, this capacity to secrete cortisol, which is a much more potent glucocorticoid than corticosterone, persists. Thus similar order of stimulation of the adrenals in immature mice results in much more profound effects than in the mature mouse—even if there were no difference in social rank. This has been ratified by comparison of the effects of ACTH, such as thymic involution, in immature and adult mice.

In the light of these general considerations there is very great interest in the studies of Aumann and Emlen (1965), who examined the influence of population density on sodium selection by microtine rodents. They note that a collection of 55 published records of population densities of rodents of the genus *Microtus* reveals a strong correlation between numbers reached at population peaks and local soil sodium levels, as reported in soil surveys or deduced from indirect evidence. In areas of low soil sodium, the highest density reported was 230 animals per acre (568 per hectare). Populations exceeding 30 individuals per acre (74 per hectare) were seen or noted in only 12 of 63 reports. In regions of intermediate soil sodium, *Microtus* populations ranged up to 400 per acre (998 per hectare), with 16 of 22 reports listing more than 30 per acre (74 per hectare). In regions of high soil sodium there were 14 records in excess of 1000 animals per acre (2470 per hectare), and 30 of 33 records were more than 30 (74 per hectare). Thus although cyclic population fluctuations occur in animals regardless of sodium levels, the densities reached at population peaks were characteristically many times higher in regions of high soil sodium than in regions of low soil sodium.

The authors recognize that other factors may have been involved, including other ions, but nevertheless the data appeared to them to warrant a number of laboratory experiments on meadow voles (*Microtus pennsylvanicus*) from a laboratory colony established with wild animals trapped at Madison, Wisconsin. Two experiments were conducted to test the effect of restricted and unrestricted sodium chloride diets on reproduction and population growth. Populations supplied with the unlimited sodium chloride solution produced 60 young in the two male/two female groups to one cage, and 104 young in the one male/three female groups to one cage, compared with 43 and 64 young in the restricted sodium diet groups respectively ($P<0.05$, paired *t*-test).

In a second group of experiments, segregated-sex and mixed-sex groups of 5- to 8-week-old voles were tested for self-selection of sodium chloride at various levels of crowding. They were provided with a choice of distilled water and sodium chloride solution in paired tubes, which were calibrated for direct measurements of daily fluid consumption. It was found that animals at high density selected about 70% more of the sodium chloride solution in the case of males, and 90% more in the case of females, than did the animals at low density animals ($P<0.01$, paired *t*-test). In a second test series, sodium chloride selection varied directly with crowding when a group of 24 animals was subjected to four 14-day tests at different densities. This was most striking in males, where the ratio of sodium chloride solution to distilled water selected increased from about 2:1 to 6:1 as density was doubled from four to eight per cage, rose again to 10:1 when all animals were in one cage, and then fell to 4:1 as density was decreased to four per cage again.

In a further experiment, groups consisting of two males and two females selected a significantly higher ratio of sodium chloride solution to distilled water (6.2:1) than did groups of one male and three females to a cage (2.1:1).

Aumann and Emlen raised the question of whether the higher selection of sodium chloride by the animals under conditions of continued crowding was a manifestation of sodium deficiency, perhaps due to inadequate adrenal cortical regulation of sodium metabolism. The field data on the relation of population to sodium imply that sodium is the critical factor limiting vole populations in many areas. In regions with high sodium levels in the soil enough sodium may be ingested with vegetation to satisfy the increased requirements of crowded animals, thereby permitting unrestricted reproduction and population growth to the point where some other food or factor becomes critical. Christian also proposes an explanation of limitation of population in this context. That is, loss of sodium due to increased glucocorticoid secretion, as a result of the increased population density, may reach levels that cannot be compensated by dietary intake in a low sodium environment. He also proposes that potassium loss stimulated by steroids may be limiting when dietary sodium is adequate. Another facet of the overall biology resides in the powerful stimulating effects ACTH has on salt appetite (see Chapter 23). For example, Weisinger et al. (1978) have shown specific induction of sodium appetite by ACTH in rats, involving turnover of three to four times total body sodium content per day.

Considerable new light on the pattern of

hormone secretion at sustained high ACTH levels, and the metabolic consequences that ensue, has come from a series of studies conducted on sheep in our Institute. These studies were originally designed to examine the influence of ACTH on adrenal biosynthesis of aldosterone. However, because of the percipient observation by a student (Dr J. Fan) that the blood pressure of the sheep so treated was raised, the programme developed into an extensive investigation of the metabolic effects of ACTH and its influence in consistently raising blood pressure. In relation to the considerations raised in the previous paragraphs, a number of important facts have emerged, which are described in detail below.

ACTH was administered to the sheep for periods of 5 to 12 days, either intramuscularly (40 international units b.d. of long-acting ACTH) or by intravenous infusion. During the first 12–24 h the aldosterone secretion rate is raised, together with that of cortisol, corticosterone, desoxycorticosterone and desoxycortisol. However, whereas the levels of these other hormones are still substantially elevated at 5 days, the aldosterone concentration in peripheral blood has fallen to below the control pre-injection level by this time (see Fig. 3-21). It is characteristic that during the first 2 days of adminstration there is a statistically significant retention of sodium by the kidney (Fig. 3-22). This is reversed at day 3 in that output is similar to thc control level. By days 4 to 5 the urinary excretion of sodium exceeds input and this holds over a period of some days of ACTH administration. When ACTH is stopped, there is a significant natriuresis, so that at the end of the episode the animal is in negative sodium balance. Increased water turnover is a characteristic feature of the experiment. The adrenal glands of these animals undergo changes as a result of the ACTH. These involve substantial reduction in size of the zona glomerulosa and general degenerative changes. Eventually, there appears to be little glomerulosa remaining (see Fig. 3-23). Coincident with these observations, the response of the adrenal glands to sodium deficiency and a number of other stimuli has been tested following 5 days of ACTH administration (Fig. 3-24). It is evident that

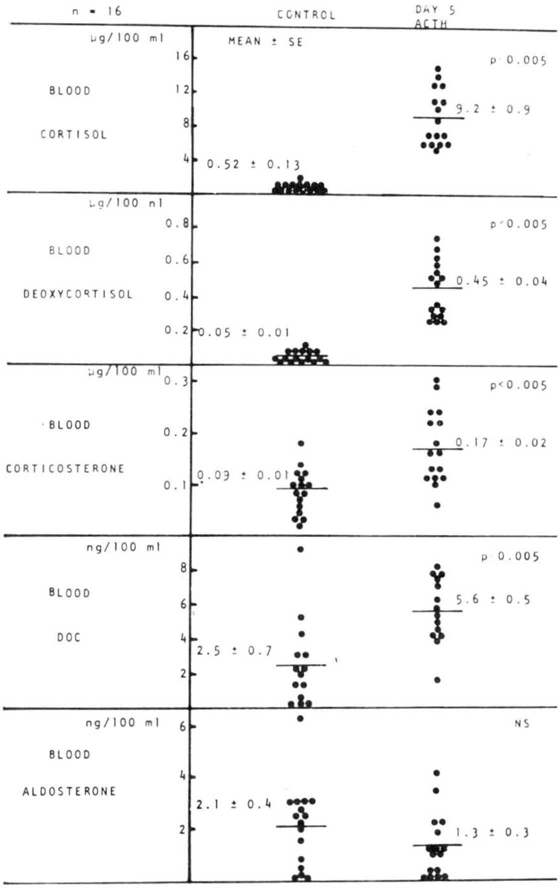

Fig. 3-21. Effect of ACTH administration (80 IU/day for 5 days) on peripheral blood cortisol, desoxycortisol, corticosterone, desoxycorticosterone (DOC) and aldosterone levels in 16 experiments on sheep. Individual values are shown with mean ±s.e. Statistic for paired *t*-test analysis is also shown. (From Scoggins et al. 1974, with permission.)

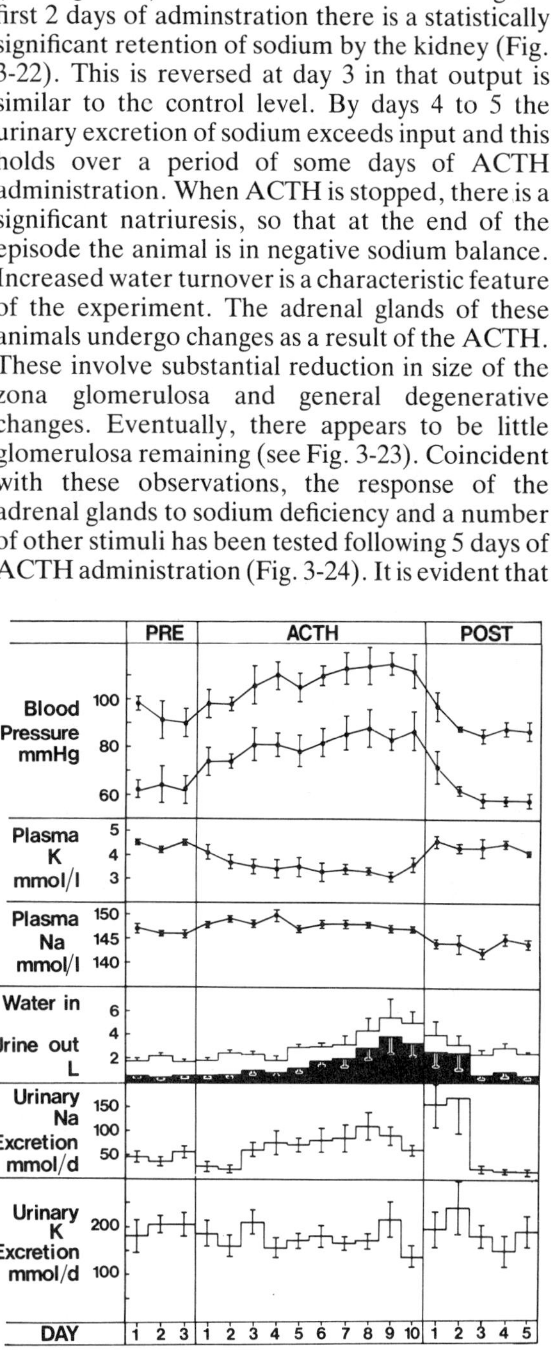

Fig. 3-22. The effect of ACTH administration (80 IU/day for 10 days) on systolic and diastolic blood pressure and other parameters in five sheep. (From Scoggins et al. 1974, with permission.)

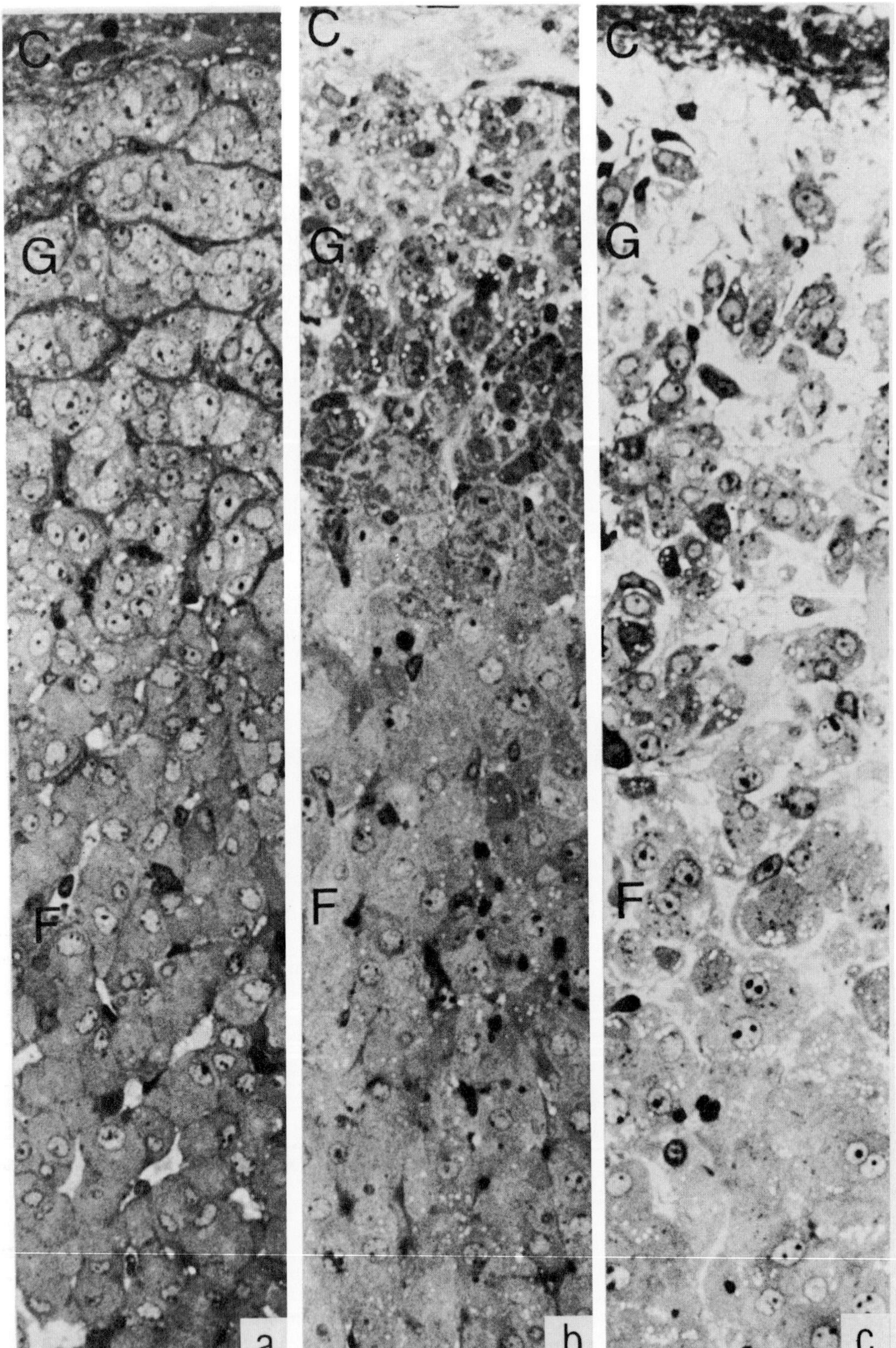

Fig. 3-23. Semi-thin plastic sections of sheep adrenal gland stained with toluidine blue for light microscopy. Each micrograph shows part of the capsule (C), the zona glomerulosa (G) and part of the zona fasciculata (F). (a) Normal adrenal. The glomerulosa cells are found in small groups surrounded by fine connective tissue elements. They fill the space between the capsule and the fasciculata. (b) After 1 day of ACTH treatment. The glomerulosa cells show lipid accumulation and slight separation from the capsule and from one another. (c) After 5 days of ACTH treatment. The organization of the glomerulosa is quite disrupted. Cells in this zone appear fewer in number and the separation of cells extends deep into the fasciculata. ×400. (From McDougall et al. 1980, with permission.)

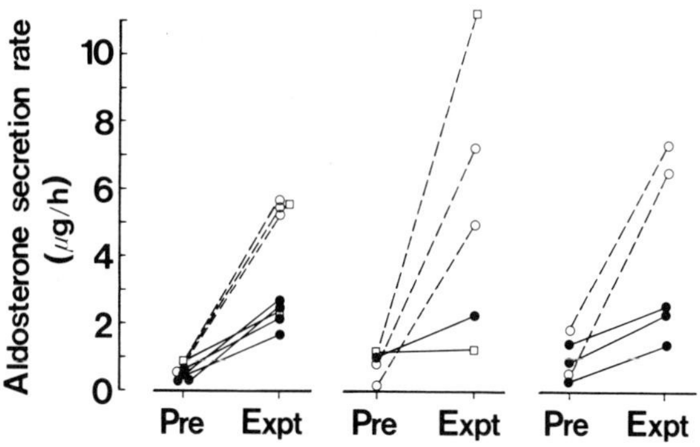

Fig. 3-24. Effect of 5 days of treatment with ACTH (80 IU/day) on the aldosterone secretion rate in response to angiotensin II infusion and 48 h of sodium depletion in three sheep. Results can be compared with control experiments without ACTH. ○---○, response to 50 μg angiotensin II/h; □---□, response to sodium depletion; ●——●, response to 50 μg angiotensin II/h after ACTH; □——□, response to sodium depletion after ACTH. (From Coghlan et al. 1979, with permission.)

relative to the control response to angiotensin, potassium or to sodium deficiency itself, the animals which had received ACTH showed a clear reduction of their response in terms of aldosterone secretion. These data, together with other studies in the laboratory showing an impairment of 18-hydroxylation as a result of sustained ACTH administration, strongly support the notion that a protracted high level of ACTH secretion may have deleterious effects on the zona glomerulosa and thus on the capacity to adapt to a sodium-deficient environment. If, as indeed appears to be the case, the increased hierarchical competition and aggression that result from crowding cause a marked increase in ACTH secretion, with fasciculata hypertrophy and glomerulosa atrophy of the adrenals, the effects of a sodium-deficient environment could be greatly aggravated in high-density populations. This would have deleterious consequences on the population of animals (see Scoggins et al. 1974).

References

Abraham SF, Blaine EH, Blair-West JR, Coghlan JP, Denton DA, Mouw DR, Scoggins BA, Wright RD (1973) New factors in control of aldosterone secretion. In Scow RO (ed) International Conference Series 273, Proceedings of the Fourth International Congress of Endocrinology. Excerpta Medica, Amsterdam, p 733

Aines TD, Smith SE (1957) Sodium versus chlorine for the therapy of salt deficient dairy cattle. J Dairy Sci 40: 682

Arnold GW (1964) Some principles in the investigation of selective grazing. Proc Soc Animal Prod 5: 258

Aumann GD, Emlen JT (1965) Relation of population density to sodium availability and sodium selection by microtine rodents. Nature 208: 198

Bailey CP (1961a) Saliva secretion and its relation to feeding in cattle. The rate of secretion of mixed saliva in the cow during eating, with an estimate of the magnitude of the total daily secretion of mixed saliva. Br J Nutr 15: 443

Bailey CP (1961b) Saliva secretion and its relation to feeding in cattle. The relationship between the concentrations of sodium, potassium, chloride and inorganic phosphate in mixed saliva and rumen fluid. Br J Nutr 15: 489

Bailey CP, Balch CC (1961a) Saliva secretion and its relation to feeding in cattle. The composition and rate of secretion of parotid saliva in a small steer. Br J Nutr 15: 371

Bailey CP, Balch CC (1961b) Saliva secretion and its relation to feeding in cattle. The composition and rate of secretion of mixed saliva in the cow during rest. Br J Nutr 15: 383

Bazett HC, Sunderman FW, Doupe J, Scott JC (1940) Climatic effects on the volume and composition of the blood in man. Am J Physiol 129: 69

Bazhenova AG, Kolpakov MG (1969) The state of sodium homeostasis in herbivores adapting to high altitude conditions. In: Proceedings of the USSR Academy of Sciences Fourth Physiol Conference of Central Asia and Kasakhstan, Novosibirsk, p 9

Bell FR (1963) The variations of taste thresholds of ruminants associated with sodium depletion. In: Zotterman Y (ed) Proceedings of the First International Symposium on Olfaction and Taste. Pergamon Press, Oxford, p 299

Benedict FJ (1936) The physiology of the elephant. Carnegie Institution, Washington

Black DAK (1953) Body fluid depletion. Lancet I: 305

Blair-West JR, Coghlan JP, Denton DA, Goding JR, Wright RD (1963) The effect of aldosterone, cortisol and corticosterone upon the sodium and potassium content of sheep's parotid saliva. J Clin Invest 42: 484

Blair-West JR, Coghlan JP, Denton DA, Goding JR, Wright RD (1964) The effect of adrenal cortical steroids on parotid salivary secretion. In: Sreebny LM, Meyer J (eds) Salivary glands and their secretions. Pergamon Press, Oxford, p 253

Blair-West JR, Coghlan JP, Denton DA, Nelson JF, Orchard E, Scoggins BA, Wright RD, Myers K, Junquiera CL (1968) Physiological, morphological and behavioural adaptation to a sodium deficient environment by wild native Australian and introduced species of animals. Nature 217: 922

Blair-West JR, Coghlan JP, Denton DA, Scoggins BA, Wintour EM, Wright RD (1969) The onset of effect of ACTH, angiotensin II and raised plasma potassium concentration on the adrenal cortex. Steroids 15: 433

Blaxter KL (1961) In: Kon SK, Cowie AT (eds) Milk: Lactation and growth of the young. The mammary gland and its secretion, vol 2. Academic Press, New York London, p 305

Blaxter KL, Wood WA (1953) Some observations on the biochemical and physiological events associated with diarrhoea in calves. Vet Rec 65: 889

Botkin DB, Jordan PA, Dominski AS, Lowendorf HS, Hutchinson GE (1973) Sodium dynamics in a northern ecosystem. Proc Natl Acad Sci USA 70: 2745

Bott E, Denton DA, Goding JR, Sabine JR (1964) Sodium deficiency and corticosteroid secretion in cattle. Nature 202: 461

Brouwer E (1961) In: Lewis D (ed) Digestive physiology and nutrition of the ruminant. Butterworth, London, p 154

Bunge G (1873) On the significance of common salt and the behaviour of potassium salts in the human organism. Z Biol 9: 104

Caius JF, Chhapgar SK (1933) Earth eating and salt licking in India. J Bombay Nat Hist Soc 37: 455

Care AD, Ross DB (1963) The role of the adrenal cortex in magnesium homeostasis and in the aetiology of hypomagnesaemia. Res Vet Sci 4: 24

Christian JJ, Davis DE (1964) Endocrines, behaviour and

population. Science 146: 1550
Coghlan JP, Scoggins BA (1967) The measurement of aldosterone in peripheral blood of man and sheep. J Clin Endocrinol Metab 27: 1470
Coghlan JP, Blair-West JR, Denton DA, Fei DT, Fernley RT, Hardy KJ, McDougall JG, Puy R, Robinson PM, Scoggins BA, Wright RD (1979) Control of aldosterone secretion. J Endocrinol 81: 55P
Collins KJ (1963) Endocrine control of salt and water in hot conditions. Fed Proc 22: 716
Conn JW (1949) The mechanism of acclimatization to heat. Adv Intern Med 3: 373
Cook GA (1932) Rabbits and sodium deficiency. J CSIRO 5: 196
Cooke HBS (1964) In: Howell F, Bourlière F (eds) Discussion of African ecology and human evolution. Methuen, London, p 579
Cowan IM, Brink VC (1949) Natural game licks in the Rocky Mountain national parks of Canada. J Mammal 30: 379
Dale HE, Brodie S, Burge GJ (1956) Effects of environmental temperature rhythms on blood and serum volumes and body water in dairy cattle. Fed Proc 15: 43
Dalke PD, Beeman RD, Kindel FJ, Robel RJ, Williams PR (1965) Use of salt by elk in Idaho. J Wildl Management 29: 319
Davies KA (1952) The building of Mt Elgon (East Africa). Mem Geol Serv Uganda 7: 1
Denton DA (1956) The effect of Na depletion on the Na:K ratio of the parotid saliva of the sheep. J Physiol (Lond) 131: 516
Denton DA (1957) The study of sheep with permanent unilateral parotid fistulae. Q J Exp Physiol 42: 72
Denton DA (1969) Salt appetite. Nutritional Abstracts and Reviews 39: 1043
Denton DA, Sabine JR (1960) The selective appetite for Na shown by Na deficient sheep. J Physiol (Lond) 157: 97
Denton DA, Wynn V, McDonald IR, Simon S (1951) Renal regulation of the extracellular fluid. II. Renal physiology and electrolyte subtraction. Acta Med Scand 140 [Suppl 261]
Denton DA, McDonald IR, Munro J, Williams W (1952a) Excess sodium subtraction in the sheep. Aust J Exp Biol Med Sci 30: 213
Denton DA, Maxwell M, McDonald IR, Munro J, Williams W (1952b) Renal regulation of the extracellular fluid in acute respiratory acidemia. Aust J Exp Biol Med Sci 30: 489
Denton DA, Goding JR, Wright RD (1959) Control of adrenal secretion of electrolyte-active steroids. Br Med J ii: 447, 522
Denton DA, Goding JR, Sabine JR, Wright RD (1961) Adaptation of ruminant animals to variations of salt intake. In: Salinity problems in the arid zones. Proceedings of the Teheran Symposium. UNESCO, Paris, p 193
Denton DA, Nelson JF, Orchard Elspeth, Weller Sigrid (1969) The role of adrenocortical hormone secretion in salt appetite. In: Pfaffmann C (ed) Proceedings of the Third International Symposium on Olfaction and Taste. Rockefeller University Press, New York, p 535
Dobson A (1961) Discussion of E Brouwer. Mineral relationships of the ruminant. In: Lewis D (ed) Digestive physiology and nutrition of the ruminant. Butterworth, London, p 165
Dobson A (1965) In: Dougherty RW et al. (eds) Physiology of digestion in the ruminant. Butterworth, London, p 88
Dobson A, McDonald I (1963) Changes in composition of the saliva of sheep on feeding heavily fertilized grass. Res Vet Sci 4: 247
Eriksson B (1952) Composition of atmospheric precipitation. II. Sulphur chloride iodine compounds bibliography. Tellus 4: 280
Ewing HE (1935) Sham louse picking, or grooming, among monkeys. J Mammal 16: 303
Fayet JC (1968) Water and mineral balances in normal calves and calves with diarrhoea. Recherches vétérinaires 99: 109
French MH (1945) Geophagia in animals. East Afr Med J [Apr]: 103
Frens AM (1958) Physiological aspects of the nutrition of grazing cattle. Eur Assoc Animal Prod Publ 6: 93
Gartner RJW, Murphy GM (1974) Evidence of low sodium status in beef cattle grazing Coloniao guinea grass pasture. Proc Aust Soc Anim Prod 10: 95
Goding JR, Denton DA (1959) The response to sodium depletion in adrenalectomized sheep with parotid fistulae. Aust J Exp Biol Med Sci 37: 211
Goodall J (1963) Feeding behaviour of wild chimpanzees. A preliminary report. Symp Zool Soc Lond 10: 39
Harrison RJ (1960) A study of orang-utan behaviour in semi-wild state. Sarawak Museum Journal 9: 15
Hebert DM, Cowan IM (1971a) White muscle disease in the mountain goat. J Wildl Management 35: 752
Hebert D, Cowan IM (1971b) Natural salt licks as a part of the ecology of the mountain goat. Can J Zool 49: 605
Henshaw J, Ayeni J (1971) Some aspects of big-game utilization of mineral licks in Yankari Game Reserve, Nigeria. East Afr Wildl J 9: 73
Honness RF, Frost NM (1942) A Wyoming bighorn sheep study. Wyoming Game Fish Dept Bull 1: 127
Howard E (1963) Effects of corticosterone on the developing brain. Fed Proc 22: 270 (abstr 657)
Hutton JT (1958) The chemistry of rain water with particular reference to conditions in south east Australia. In: Climatology and microclimatology. UNESCO, Paris, p 285
Hutyra F, Marek J (1922) Spezielle Pathologie und Therapie der Haustiere, 6th edn. G. Fischer, Jena
Hyden S (1958) Description of two methods for the collection of saliva in sheep and goats. Kungl Lantbrukshogskolans Annaler 24: 55
Jones DIH (1963) Mineral content of some cultivated grasses grown in Northern Rhodesia. Rhod J Agri Res 2: 57
Jones DIH, Miles DG, Sinclair KB (1967) Some effects of feeding sheep on low sodium hay with and without sodium supplement. Br J Nutr 21: 391
Jordan PA, Botkin DB, Dominiski AS, Lowendorf HS, Belovsky GE (1973) Sodium as a critical nutrient for the moose of Isle Royale. In: Proceedings of the North American Moose Conference Workshop, vol 9, p 13
Kawai M (1965) Newly acquired pre-cultural behaviour of the natural troop of Japanese monkeys on Koshima Islet. Primates 6: 1
Kawamura S (1954) A new type of action expressed in feeding behaviour of Japanese monkeys in the wild. Seibutsu Shinka 2: 11
Kay RNB (1960) The rate of flow and composition of various salivary secretions in sheep and calves. J Physiol (Lond) 150: 515
Keeley K (1962) Prenatal influence on behaviour of offspring of crowded mice. Science 135: 44
Kneraskov S (1939) The manufacture of cheese from camels' milk. Molochn.-Maslodel'anay Prom 6: 14
Knight RR, Mudge MR (1967) Characteristics of some natural licks in the Sun River area, Montana. J Wildl Management 31: 293
Leche TF (1977) Effect of a sodium supplement on lactating cows and their calves on tropical pastures. Papua New Guinea Agr J 28: 11
Leek SN (1917) Salt licks and alkali springs for elk. Sci Am [Suppl] 2179
Macfarlane WV (1963) Endocrine functions in hot environments. In: Environmental physiology and psychology in arid conditions. UNESCO, Paris, p 153
Macfarlane WV (1964) Terrestrial animals in dry heat—

ungulates. In: Dill DB, Adolph EF, Wilber CJ (eds) Handbook of physiology: Adaptations to the environment. American Physiological Society, Washington, p 509
Macfarlane WV, Morris RJH, Howard B, Budtz-Olsen OE (1959) Extracellular fluid distribution in tropical Merino sheep. Aust J Agr Res 10:269
Macfarlane WV, Howard B, Hipsley E (1964) Water and salt metabolism of the Chimbu. Proc Aust Physiol Soc 26
Martin CJ (1930) Thermal adjustments of man and animals to external conditions. Lancet II: 561, 617, 673
McCance RA (1936) Medical problems in mineral metabolism. Lancet I: 643
McDougall EI (1948) The composition and output of sheep's saliva. Biochem J 43: 99
McDougall JG, Coghlan JP, Scoggins BA, Wright RD (1974) Effect of sodium depletion on bone sodium and total exchangeable sodium in sheep. Am J Vet Res 35: 923
McDougall JG, Butkus A, Coghlan JP, Denton DA, Muller J, Oddie CJ, Robinson PM, Scoggins BA (1980) Biosynthetic and morphological evidence for inhibition of aldosterone production following administration of ACTH to sheep. Acta Endocrinol (Copenh) 94: 559
Morris JG (1980) Assessment of sodium requirements of grazing beef cattle: A review. J Anim Sci 50: 145
Morris JG, Gartner RJW (1971) The sodium requirements of growing steers given an all-sorghum grain ration. Br J Nutr 25: 191
Murphy GM, Plasto AW (1973) Live weight response following sodium chloride supplementation of beef cows and their calves grazing native pastures. Aust J Exp Agr Anim Husb 13: 369
Myers K (1967) Morphological changes in the adrenal glands of wild rabbits. Nature 213: 147
Myers K (1970) The rabbit in Australia. Dynamics of populations. In: den Boer PJ, Gradwell J (eds) Proceedings of the Advanced Study Institute on dynamics of numbers in populations. Oosterbork, The Netherlands, p 478
Myers K, Bults HG (1979) Stress in the rabbit. In: Proceedings of the World Lagomorph Conference, Guelph University. Guelph University Press, p 1
Nalin DR, Cash RA (1976) Sodium content in oral therapy for diarrhoea. Lancet II: 957
Oliver WJ, Cohen EL, Neel JV (1975) Blood pressure, sodium intake and sodium related hormones in the Yanomamo Indians. A 'no salt' culture. Circulation 2: 146
Oomen HAPC, Spoon W, Heesterman JE, Ruinard J, Luyken R, Slump P (1961) The sweet potato as the staff of life of the highland Papuan. Trop Geogr Med 13: 55
Orr JB (1929) Minerals in pasture and their relation to animal nutrition. Lewis, London
Paque C (1963) Un puits saharien à chameaux: Hassi Zehar. Teneurs de l'eau en sodium et autres éléments minéraux. Mammalia 27: 310
Perkins AJ (1965) Some notes on the caves of Mt Elgon. Newsletter Cave Exploration Group of East Africa 3: 21
Playne MJ (1970) The sodium concentration of some tropical pasture species with reference to animal requirements. Aust J Agr Anim Husb 10: 32
Reid IA, McDonald IR (1969) The renin angiotensin system in a marsupial, *Trichosurus vulpecula*. J Endocrinol 44: 231
Renkema JA, Senshu T, Gaillard B, Brouwer E (1962) Regulation of sodium excretion and retention by the intestine in cows. Nature 195: 389
Richter CP (1956) Salt appetite of mammals. Its dependence on instinct and metabolism. In: L'instinct dans le comportment des animaux et de l'homme. Masson et Cie, Paris, p 577
Rigg T, Askew HO (1934) Soil and mineral supplements in the treatment of Bush Sickness. Emp J Exp Agr 2: 1
Ritzman EG, Benedict FG (1938) Nutritional physiology of the adult ruminant. Carnegie Institution, Washington (Publication no 494)
Rook JAF, Balch CC (1959) The physiological significance of the fluid consistence of faeces from cattle grazing spring grass. Proc Nutr Soc 18: xxxv
Russell FCM, Duncan D (eds) (1956) Minerals and pasture: Deficiencies and excesses in relation to animal health. Commonwealth Bureau of Animal Nutrition, p 113
Schaller GB (1963) The mountain gorilla. Ecology and behaviour. University of Chicago Press, Chicago, p 149
Schmidt-Nielsen K, Schmidt-Nielsen B, Houpt TR, Jarnum SA (1956) The question of water storage in the stomach of the camel. Mammalia 20: 1
Scoggins BA, Blair-West JR, Coghlan JP, Denton DA, Myers K, Nelson JF, Orchard E, Wright RD (1970) Physiological and morphological responses of mammals to changes in their sodium status. Cambridge University Press, London, p 577 (Memoirs of the Society for Endocrinology 18)
Scoggins BA, Coghlan JP, Denton DA, Fan SK, McDougall JG, Oddie CJ, Shulkes AA (1974) The metabolic effects of ACTH in the sheep. Am J Physiol 226: 198
Scott JC, Bazett HC, Mackie GC (1940) Climatic effects on cardiac output and circulation in man. Am J Physiol 129: 102
Shortridge GC (1934) The mammals of South West Africa. Heinemann, London
Sinnett PF, Whyte HM (1973) Epidemiological studies in a total highland population, Tukisenta, New Guinea. Cardiovascular disease and relevant clinical, electrocardiographic, radiological, and biochemical findings. J Chronic Dis 26: 265
Smith MC (1977) Studies of seasonal availability of electrolytes on adrenal physiology of a wild population of snowshoe hares. PhD dissertation, University of Guelph, Ontario, Canada
Smith MC, Leatherland JF, Myers K (1978) Effects of seasonal availability of sodium and potassium on the adrenal cortical morphology of a wild population of snowshoe hares, *Lepus americanus*. Can J Zool 56: 1869
Stockstad DS, Morris MS, Lory EC (1953) Chemical characteristics of natural licks used by big game animals in western Montana. North American Wildlife and Natural Resources Conference: Transactions 18: 247
Sutcliffe AJ (1973) Caves of the East African Rift Valley. Transactions of the Cave Research Group of Great Britain 15: 41
Vore I de (1964) In: Howell F, Bourlière F (eds) Discussion of African ecology and human evolution. Methuen, London, p 588
Watson PJ (1969) The pre-history of salt cave Kentucky. Rep Invest Illinois State Mus 16: 1
Weeks HP, Kirkpatrick CM (1976) Adaptions of white-tailed deer to naturally occurring sodium deficiencies J Wildl Management 40: 610
Weerden GJ van (1961) The osmotic pressure and the concentration of some solutes of the intestinal contents and the faeces of the cow, in relation to the absorption of the minerals. J Agr Sci 56: 317
Weir JS (1969) Chemical properties and occurrence on Kalahari sand of salt licks created by elephants. J Zool (Lond) 158: 293
Weir JS (1972) Spatial distribution of elephants in an African national park in relation to environmental sodium. Oikos 23: 1
Weisinger RS, Denton DA, McKinley MJ, Nelson JF (1978) ACTH-induced sodium appetite in the rat. Pharmacol Biochem Behav 8: 339
Wharton GW (1957) An ecological study of the Kouprey. Institute of Science and Technology, Manila (Monograph no. 5)

Whitten EH, Phillips RW (1971) In vitro intestinal exchanges of Na^+, K^+, Cl^-, and H_2O in experimental bovine neonatal enteritis. Am J Digest Dis 16: 891
Wilson AD (1963) The effect of diet on secretion of parotid saliva by sheep. Aust J Agr Res 14: 808
Wilson AD (1966) The intake and excretion of sodium by sheep fed on species of *Atriplex* (saltbush) and *Kochia* (bluebush). Aust J Agr Res 17: 155
Wilson AD, Tribe DE (1963) The effect of diet on the secretion of parotid saliva by sheep. I. The daily secretion of saliva by caged sheep. Aust J Agr Res 14: 670
Yagil R, Etzion Z, Berlyne GM (1973) The effect of D-aldosterone and spironolactone on the concentration of sodium and potassium in the milk of rats. J Endocrinol 59: 633
Zeuner FE (1954) Domestication of animals. In: Singer C et al. (eds) A history of technology. OUP, London, p 327
Zuckerman S (1932) The social life of monkeys and apes. Harcourt Brace, New York

4 Hominoid evolution and the influences on sodium homeostasis

Summary

1. The impact of a sodium-deficient environment falls directly on the herbivores and vegetarians. The carnivores receive adequate sodium by virtue of the sodium content of the muscles and viscera of their prey. Studies of omnivores in sodium-deficient alpine areas showed variable sodium status with instances of substantial deficiency.

2. The dietetic history of hominoids over the 30 million years from the late Oligocene onwards to the present has probably determined the selection pressures operating to influence level of development of salt appetite and endocrine mechanisms of sodium conservation. There are data suggesting that from prosimians onwards through *Dryopithecus*, *Ramapithecus* and *Australopithecus* up to the late Pliocene time, the diet was predominantly, if not exclusively, vegetarian. *Ramapithecus* may have been graminiferous like the contemporary *Theropithecus*, incisal and canine reduction with molar predominance coming from a diet of small tough objects associated with open grassland habitat. Evidence of significant hunting occurs in the early Pleistocene though meat-eating may well have preceded this in the late Pliocene. Early *Homo*, from about 1.7 million years ago, was a hunter of large animals. Contemporary evidence on hunter-gatherer societies shows that great variation occurs in the meat component of the diet. Africa, as representative of tropical and warm-temperate regions, provides data suggesting that hunting has seldom if ever been in any exclusive sense the staff of hominid life. Usually meat is less than 50% of the diet in hunter-gatherer societies and may be much less. Some contemporary feral societies are almost entirely vegetarian, with very low sodium status.

3. The great apes have a diet which is predominantly, and in some instances exclusively, vegetarian. Calculations on meat-eating in chimpanzees from Goodall's detailed records suggest that over the course of a year meat provides an additional sodium intake of ca 1 mmol/day. This would not preclude severe stress on sodium status in such animals living in tropical jungle in the interior of a continent. During pregnancy 500 mmol of sodium may be sequestrated, and additional sodium will be required for milk. On the grounds of the relatively small differences between the amino acid composition of protein in the great apes and man, the African apes may have diverged from hominids later (ca 5 million years ago) than some, but not all, interpreters of the morphological and palaeontological evidence believe.

4. The question of the influence of the discovery of fire on food choice and intellectual development is discussed, including its possible role in the migration to cold-temperate areas and thus an increase in the protein content of the diet.

5. The behaviour towards salt of a wide variety of species of wild

herbivorous animals including primates has been recorded in Chapter 3. The environmental, dietetic and metabolic conditions which may cause sodium deficit could have affected hominoids over a large part of their evolution. It follows that selection pressures would favour retention of the salt appetite mechanisms—both hedonic liking and the hunger with deficiency—which developed at earlier stages of phylogeny. There could also be elaboration of the physiological system of salt retention by aldosterone secretion. This has a multifactorial system of control that includes special adaptations to the assumption of the upright posture.

Dietary sodium and salt appetite in herbivores, carnivores and omnivores

It has been documented in the preceding chapter that sodium is in very short supply in large areas of the planet, and that animals become sodium deficient. A major fact emerging is that the primary impact of a sodium-deficient environment is directly upon the herbivores and vegetarians, not the carnivores. This is reflected in the field data. There are no accounts of carnivores seeking out salt sources and being observed at licks. They do, however, lie in wait for herbivorous prey at such places or on trails leading to them.

Herbivores and vegetarians have the grasses, browse and fruits of their native region as their sole source of food. As analyses show, sodium may be virtually a trace element in this diet, or at least in very short supply relative to the capacities of the animals to conserve it and their metabolic needs (including those of reproduction). By contrast the carnivore will always receive adequate sodium because of the obligatory sodium content of the viscera, blood and muscles of its prey. This varies within relatively narrow limits irrespective of whether sodium is abundant or scarce in the environment, and will provide the carnivore with from 30 to 80 mmol Na/kg of food according to whether it is eating predominantly muscle or viscera (McCance 1936). In this regard it is interesting to note that the literature on laboratory experimental investigation of salt appetite has been remarkably devoid of studies on the carnivore. One exception is Chernigovsky's (1963) investigation of the salt preference of dogs given salt solutions of various concentrations in a pyloric pouch, where an aversion to salt was seen when high-sodium solutions were placed in the pouch. This, however, does not really bear upon the matter of sodium deficiency in carnivores.

There is a recent study of salt appetite in dogs by Fregly (1980). Seven female pure-bred beagle dogs were tested with choice of distilled water and increasing concentrations of saline to drink. None showed a clear-cut preference for any saline solution, and intake decreased as concentration rose. Hydrochlorthiazide given over 4 days to four beagles failed to increase sodium intake where the dogs were offered either 0.08 M or 0.15 M sodium chloride and water. In four other dogs, ethacrynic acid had no effect on intake of saline, nor did 4 days of desoxycorticosterone acetate (DOCA) at 25 mg/day. The diuretics are known to stimulate salt appetite in rat or sheep. The data are suggestive, but as apparently no records of the animals' sodium balance were made, there is no audit that significant sodium deficiency did actually occur in the face of the sodium content of the food given. The DOCA results are perhaps more suggestive in relation to the broad issue. In the absence of any selection pressure because of the perennially adequate sodium content of their diet, carnivores may not have evolved the repertoire of hedonic preference, and response to sodium deficiency or hormones, which have developed in the herbivores.

However, evidence contrary to this general line of reasoning has come from studies by Ramsay and colleagues at San Francisco. Eight dogs had sodium restricted in the diet for 5 days, and received frusemide (20 mg intramuscularly) on the first 3 days. Plasma sodium concentration fell from 142±1 to 132±2 mmol/l, and plasma renin activity increased from 3.5±0.6 to 24.5±1.9 ng ml^{-1} 3 h^{-1}. When offered water and 0.3 M saline (which is normally aversive in the dog) on the fifth day, the dogs drank 406±110 ml of saline and 261±73 ml of water in a 4-h period. Five days of treatment with

30 mg DOCA per day did not induce sodium appetite. Four hours of infusion of angiotensin II caused water drinking but no salt appetite. Giving the angiotensin II infusion after 5 days of DOCA did not cause salt appetite, and nor did giving intravenous renin after DOCA (Ramsay and Reid 1979). Dogs were prepared with chronic cannulae in the third ventricle and angiotensin II (1 μg kg^{-1} h^{-1} at 1 μl/h) infused for 7 days using osmotic minipumps (Brown et al. 1980). A maintained very large water intake was seen over the 7 days, the dogs drinking 50%–100% their body weight per day. The treatment reduced plasma sodium concentration 20–25 mmol/l, and lowered plasma renin and antidiuretic hormone. There was no evidence of salt appetite compared with vehicle-infused controls. The question clearly requires further investigation, the initial data being suggestive of a salt appetite mechanism existing in the dog but organized differently to that in other species. The large stimulating effect of intraventricular renin and angiotensin on salt appetite in rats and sheep will be discussed in Chapter 22, including the issue of whether this is a pharmacological or physiological effect.

There is a similar dearth of information regarding the sodium status of wild omnivores. With Dr A. Newsome and Mr L. Corbett from CSIRO Wildlife Division we studied foxes in the Snowy Mountains of Australia. We presumed the foxes would probably have adequate sodium intake in this sodium-deficient region by virtue of rabbits and other small rodents in the diet. However, blood and bladder urine analyses of these animals shot in the wild showed (Table 4-1) that some were definitely sodium deficient: blood aldosterone was high and urinary sodium concentration low when compared with animals shot on the coast. This conclusion was supported by histological examination of the adrenal glands, where it was found the zona glomerulosa region was enlarged.

Studies of the stomach contents of foxes shot in these mountains in spring and summer showed a variety of insects, lizards and blackberries, rather than always predominantly mammals (Table 4-2).

Table 4-1. Adrenal steroids in the blood of foxes.

	Aldosterone (ng/100 ml)	Corticosterone (μg/100 ml)	Cortisol (μg/100 ml)	Urinary Na (mmol/l)	Sex	Remarks
	25.7	0.50	1.28	6	M	Foxes from the Snowy Mountains
	107.9	0.58	3.84	10	M	
	43.4	0.26	1.09	2.5	M	
	21.5	0.28	1.42	3.5	M	
	61.1	0.96	0.22	2	M	
	17.1	0.09	0.91	47	M	
	7.7	0.33	0.80	23	F	
Mean	40.9	0.43	1.37	13		
s.d.	±34.5	±0.26	±1.16	±17		
	11.9	0.77	3.38	12.9	M	Foxes from Victorian coastal areas
	17.6	0.55	2.37	71	M	
	26.0	0.37	1.01	29	F(lact.)	
	22.2	0.45	1.09	66	M	
Mean	19.4	0.54	1.96	74		
s.d.	± 6.1	±0.36	±1.13	±41		

The differences in blood aldosterone and urinary sodium are significant (P<0.01) between the two groups. Differences in corticosterone and cortisol are not significant.

Table 4-2. Gut contents of foxes

No.	Sex	Locality	Gut contents
5	M	alps	Grass 98%; lizard tail 2%
6	M	alps	Lepidoptera 95%; bone fragments and grass 5%
7	M	alps	Lepidoptera 60%; horse faeces 23%; rabbit 17%
11	F(lact.)	coast	Rabbit 75%; Lepidoptera (larvae) 17%; Orthoptera 4%; lizards 2%; Coleoptera 2%
13	M	alps	Lepidoptera (larvae) 100%
14	M	alps	Bird 70%; freshwater crayfish 30%
15	F	alps	Rabbit 85%; Lepidoptera (larvae) 15%
16	M	alps	Blackberries 90%; Coleoptera 10%
17	F	alps	Blackberries 100%

McCance (1936) has pointed out that just as plants may grow with little or no sodium, insects may thrive similarly. *Drosophila* can flourish when 95% of its sodium is removed and it may be able to do without any at all—an extraordinarily interesting observation. We know nothing of the sodium content of insects in the alpine areas, but in Lepidoptera potassium is the predominant cation. As various moth species appear to make up a sizeable portion of the foxes' diet at seasons, their very low sodium content may explain the sodium-deficient status found in many foxes. Overall the data on this wild omnivore in the alps indicate that the end result of the balance of foods in a sodium-deficient environment may be sodium deficiency, at least episodically.

MacIntosh (1963) describes the fox in the Canberra region, a relatively high inland plain area, as an opportunist predator and scavenger eating a wide range of foods. The predominant items are of vertebrate origin, supplemented by invertebrates and fruits according to availability and abundance. Sheep and rabbit are the most important foods, most of the sheep being obtained as carrion. Similar data on preponderance of vertebrates in the diet were obtained in the state of Victoria (Coman 1973) and in the arid interior, where the predominant food was kangaroo carcass (Martens 1971). However, the opportunistic nature of this omnivore was indicated by a report showing a large intake of cabbage by foxes in a region of Victoria (Coman et al. 1973). Stomach and faecal analyses suggested a predominantly vegetarian diet.

Hominoid evolution

As Washburn (1964) points out, the interest in human classification and evolution has been so great that it has produced a bewildering quantity and variety of terms and theories. As classifications are made by people with varied training and interests there is no reason why a particular classification should seem equally reasonable to all of them. Louis Leakey, writing in the same monograph, stated that the situation was so confused that there were scarcely any two workers in the field who employed even the simplest terms with identical meaning and value. In general (Simons 1977), the superfamily Hominoidea (hominoids) is taken to include all greater and lesser apes, and all humans and prehumans. It has two divisions: the family Pongidae, which includes all great apes, and the family Hominidae (hominids), which includes all humans and prehumans. The genera within the Pongidae are the living great apes—orang-utan, chimpanzee and gorilla—and the extinct *Dryopithecus*, *Sivapithecus* and *Gigantopithecus*. Within the Hominidae are *Homo sapiens* (man) and the extinct *Ramapithecus*, *Australopithecus* and *Homo erectus*.

The primary question about hominid evolution from the viewpoint of this analysis is the dietetic history over the course of phylogenetic development from the Oligocene/early Miocene to present times. Whereas it is clear that meat-eating is a fact of hominid development in the Pleistocene era and probably earlier, and that meat is a major feature of the diet of many communities of feral man, the question remains of how significant this has been quantitatively over the 20–30 million years involved. If for over 90% of this time the progenitors of man ate fruit, plants and seeds, it would seem likely, for the reasons already elaborated, that selection pressures on sodium homeostasis would have operated at a high intensity. If they had an omnivorous diet as hunter-gatherers over the remaining time, low sodium intake and episodic deficiency could have occurred, which would have favoured the phylogenetic emergence of appetite systems geared to the liking and seeking of salt and endocrine capacities for highly effective sodium conservation. When meat-eating did develop, there is also the issue of whether meat was the major part of the diet or whether it was ancillary to the substantial vegetable, nut, seed and plant intake characteristic of the many contemporary hunter-gatherers (see Clark (1970) for a detailed analysis).

While precise answers to these questions are not available, the body of data to be reviewed is suggestive that over the bulk of the evolutionary period involved, the dietetic habits of our hominoid and hominid progenitors made likely a high level of stress on sodium homeostasis. Therefore, contemporary man may carry a greater legacy of these selection pressures in his physiological organization and hedonic salt taste mechanisms than would have been the case had he been the descendant of a line of carnivores extending back unbroken over 20–30 million years.

Tool-making and the definition of man

Kenneth Oakley in his essay 'Skill as a human possession' (1954) refers to Benjamin Franklin's (1778) remark that 'Man is a tool-making animal'. The pivotal issue in the palaeontological analysis of the emergence of man is the time at which tool-making appeared. Oakley goes on to emphasize man's uniqueness in this respect by considering other animals which use tools but which, unlike man, do not make them. He cites many instances. The invertebrate North American solitary burrowing wasp has been observed to seize a pebble in its mandibles and use it to pound down the sand grains at the neck of its burrow. One of Darwin's Galapagos Islands finches, which feeds on insects embedded in the branches or trunks of trees, uses a cactus spine or twig held lengthwise in its beak to poke them out. The southern sea otter breaks hard shells on a stone 'anvil' which it carries in the water. It may lie on its back in the water with the stone on its chest and pound a shell held in its paws on it.

Oakley points out that it is a common fallacy to suppose that monkeys cannot oppose the thumb to the other digits: most Old World monkeys can do this when catching insects. But it is true that man and the apes have developed a greater power of rotating the thumb, which facilitates its opposition to other digits. He quotes Wood Jones: 'We look in vain if we seek for movements that a man can do and a monkey cannot, but we will find much if we look for the purposive actions that a man does do, and a monkey does not.' Clearly manual skill reflects fine central nervous mechanism rather than specially delicate distal muscular apparatus.

In comparing the use of an object immediately at hand for the incentive of visible reward with actually making a tool, Oakley reflects that the power of abstraction—conceptual thought— is basic to the regular manufacture of tools. A man seeking to make a tool for a particular purpose will visualize it, for example, in a formless lump of stone which he will then chip at until that which is imagined is actualized. That is, man is basically an artist in the sense of Aristotle's definition of art—the conception of the result to be produced before its realization in the material.

There is the possibility of gradation between the extreme of perceptual thought in apes and conceptual thought in man, but there is a large gap between the forethought involved in making even the crudest Palaeolithic artefacts and the occasional manufacture of tools by apes. In this latter regard, a chimpanzee under study in captivity by Kohler was seen to fit one stick into another, having sharpened the point with its teeth, in order to reach a bunch of bananas. Jane Goodall (1963) describes the use by chimpanzees in the wild of grass stalks or sticks as tools for extracting termites from their nests, and recounts an instance where a male in the Gombe Stream area of Tanzania was observed apparently carefully selecting four grass stalks for this purpose. Once at the termite nest he broke them off to a length of about nine inches before use, the spares being tucked in the groin or laid on the ground. Goodall remarks that in this area chimpanzees do not only use as tools objects which happen to be lying close by, but will actually seek out material for a tool, modify it if necessary (e.g. stripping leaves from a branch) and carry it for several yards to use it for a specific purpose. The chimpanzee differs in this from other apes and monkeys, though Oakley observes that baboons will sometimes use pebbles to kill a scorpion—a favourite food—and if pursued will sometimes scamper up a hillside and dislodge stones or roll boulders down the slope to deter their pursuers. Kohler held that the time in which a chimpanzee lives is limited in past and future. This is the chief difference to be found between anthropoids and the most primitive human beings. In summary, the activities of anthropoids direct to the incentive of a visible reward—as distinct from the usefulness of shaping an object for an imagined eventuality.

Washburn characterizes the primates by the adaptation of climbing by grasping (as opposed to climbing using claws), which is a complex specialization distinguishing them from other living mammals. It was fully developed in the prosimians and was present in the Eocene. It involves elongated digits, flattened terminal phalanges with nails, specialized palms and soles, and brachiation (emphasis on long highly mobile arms with powerful flexor muscles and the reduction in importance of the lower back in locomotion).

Man's earliest ancestors

The Prosimii, the original ancestors of the monkeys and apes, evolved from tiny insectivores similar to those found in Cretaceous rocks in Mongolia (Oakley 1954), and by the middle Eocene, when subtropical conditions were widespread, there were prosimians in almost all parts of the world. They resembled, in many ways, their modern survivors the lemurs, tarsiers and tree shrews, and like them probably lived mainly on insects and fruit. Life in the trees made demands on vision. In contrast to the earth-bound animal with its emphasis on sniffing and olfactory examination of objects, the arboreal creature fingered objects while examining them. There was enlargement of the cortical elements associated with co-ordination of impressions other than smell. Prosimians at the end of the Oligocene evolved in three directions. One line, the New World monkeys, acquired extreme agility in the trees by devloping prehensile tails. A second line, the Old World monkeys, mostly restricted to forests, retained the tail but only as a balancer. A few species became ground-dwellers. The earliest Old World monkey known is *Oligopithecus*, whose fossilized remains were found south-west of Cairo on what used to be a forested sea coast. The skull has 32 teeth like most Old World monkeys, apes and men; prosimians have 34 or more teeth.

A third line included the ancestors of apes and men and were perhaps not very different from *Proconsul*. In 1856 Lartet, a French lawyer and palaeontologist, reported on a primate jaw found in clay of the Miocene era and named it *Dryopithecus fontani*, literally 'oak ape', reflecting his belief that it had lived in forest. Animal and plant remains found in conjunction with other *Dryopithecus* fossils since Lartet's time strengthen his conjecture (Simons 1977). Simons suggests that these cosmopolitan apes evidently preferred wooded tropical and subtropical environments where they lived by browsing on leaves and fruit. The Leakeys and collaborators discovered *Proconsul* when working in the fossil-rich Miocene deposits on the islands in Lake Victoria (now Lake Nyanza). Eventually a wide variety of material was unearthed from these sites, varying in size from gibbon to gorilla. Extensive comparison with other fossils has led to one present view that *Proconsul* is not a unique genus but an African member of the cosmopolitan genus *Dryopithecus* (Simons 1977). According to this view these apes flourished over a period of some 20 million years, not only in France and Africa but in other regions of Europe (e.g. Barcelona, Hungary), Asia Minor, the Indian subcontinent and China. The teeth and jaw structure point to *Proconsul* being ape-like rather than hominid. (The human palate is arched and teeth run back on each side in a broad curve, while apes have no arching palate and the teeth are parallel to one another in a U-shape.)

On the basis of fossils found in Europe and Asia since 1970 it is suggested that between 10 and 15 million years ago *Dryopithecus* gave rise to at least three other genera. Two of them, *Sivapithecus* and *Gigantopithecus*, were primates with a large face whereas the other, *Ramapithecus*, had a small face. *Ramapithecus* evidently a crucial find, was discovered by Lewis of Yale University in the Siwalik Hills about 150 km north of New Delhi. Lewis was impressed by the short snout of the creature and also the arched palate and the wide curving jaws. He placed the jaw fragment in the Hominidae. The fossils referred to *Ramapithecus* were discovered by Leakey near Fort Tenan in Kenya. Following this, other discoveries described as *Ramapithecus* were made, especially in Greece, Hungary (Kretzoi 1975) and Pakistan. The evidence from the two dozen or so *Ramapithecus* jaws which have now been discovered leads to the conclusion that it had adapted to a way of life quite different from that followed by most of the forest-dwelling *Dryopithecus* group. Its jaws and teeth reflect adaptation resembling the later African hominid *Australopithecus*.

Jolly's hypothesis

The mode of existence of *Ramapithecus* has been a matter of close consideration by Jolly (1970) from which he develops the lines of argument set out below. Direct fossil evidence for the use of 'raw' tools or weapons is necessarily tenuous, and the use of fabricated stone artefacts appears relatively late, actually in the late Pliocene (R. Leakey 1970; M. Leakey 1970), or at about 2.61±0.26 million years on potassium/argon dating (Fitch and Miller 1970)—now remeasured as early Pleistocene or ca 1.8 million years ago (Drake et al. 1980). Nevertheless, as Holloway (1967) stated, the current orthodox theory regarded these elements as pivotal in the evolution of the hominoid adaptive complex, probably antedating and determining the evolution of the upright posture, and certainly in some ways determining the reduction of the anterior teeth, the loss of sexual dimorphism in the canines, and the expansion of the cerebral cortex.

Jolly (1970) challenges the proposition that it was artefact use which made the canines redundant as weapons, and the incisors as tools. He says hominoids with front teeth smaller than living or fossil Pongidae were widespread at the close of the Miocene period: *Oreopithecus* in southern Europe and Africa, and *Ramapithecus* (which probably includes *Kenyapithecus*) in Africa, India, southern Europe and China (Simons and Pilbeam 1965). Thus the theory to be consistent requires that regular tool and weapon making be extended well back into the Miocene and be attributed to Hominoidea whether one considers this to be *Ramapithecus*, *Oreopithecus* or neither. Simons regarded *Ramapithecus* as too early to be a tool-maker but he and Pilbeam suggest it was a regular tool-user like the savannah chimpanzee (Goodall 1964). Jolly thinks this may well be likely but is not an explanation for anterior dental reduction since the chimpanzee has relatively the largest canines and incisors of any pongid, much larger than those of the gorilla which has never been seen to use artefacts in the wild. In his view, to explain dental reduction on these grounds it would be necessary to postulate that the basal hominids were much more dependent on the use of artefacts than the chimpanzee without any explanation of why this was so. One might expect associated evidence of tool-making in the fossil records as early as the first signs of dental reduction. Further, the more artefactually sophisticated the wild chimpanzee is shown to be, the weaker the logic of the tool/weapon determinant theory of dental reduction.

Jolly also considers the issue of the postulated increase of meat-eating beyond that usual in primates which would follow 'open country' adaptation, and the view that the peculiarities of hominids ultimately represent adaptations to hunting. Chimpanzees living in open savannah woodland have been seen catching and eating mammals (Goodall 1971) while those living in rain forest have not. However, in Jolly's eyes, by a similar logic to that applied to the artefact determinant theory, the more proficient a hunter the non-bipedal large-canined chimpanzee is found to be, the less plausible it becomes to attribute the origin of converse hominid traits to hunting. Hunting by chimpanzees does not involve features which could lead to evolution of hominid characteristics: neither weapon use nor bipedalism is prominent. Prey is captured and killed with the bare hands, and is dismembered like other fleshy food with the incisors. Thus given a population of chimpanzees adapting to a hunting life in the savannah, there is no reason to predict incisor reduction, weapon use or bipedalism. In fact incisal reduction would make for less efficient processing of all fleshy food including meat. It is therefore difficult to justify the hominid characteristics of australopithecines as functional adaptations to life as a 'carnivorous chimpanzee'.

Jolly emphasizes that there is absence of fossil evidence of efficient hunting before the lower Pleistocene, and suggests that it is legitimate to put aside the 'carnivorous chimpanzee' model as an explanation for the anatomical differences between basal hominids and pongids. His alternative explanation centres on his observation that many of the characters that distinguish basal hominids from pongids also distinguish the grassland baboon *Theropithecus* from its woodland savannah and forest relatives *Papio* and *Mandrillus*. These differences are correlated with different, but no less vegetarian, dietetic habits.

Theropithecus gelada in the wild uses a precision grip for most of its food collecting. Its food consists mainly of grass blades, seeds and rhizomes which are picked up singly with thumb and index finger and collected in the fist until a mouthful is accumulated. The feeding method is facilitated by a well-developed thumb and very short index finger, which gives *T. gelada* the highest opposability index of any animal including *Homo sapiens*. Whereas it has this high precision grip, outclassing the chimpanzee on this score, it has not been seen making or using artefacts in the wild. It spends most of its day sitting in an upright position, as its Pleistocene relatives probably did. When foraging, it even moves in its truncally erect position, shuffling slowly on its haunches. This truncal erectness is habitual and explains the large mastoid process which in Hominidae can be related to the erect posture (Krantz 1963). Also, with this mode of existence the forelimbs are more liberated from locomotor function than in any other non-biped because it rarely locomotes. Sitting upright allows both hands to be used simultaneously for rapid gathering of small food objects. This pattern is seen more rarely with *Papio* where tripedal stance, leaving one hand free, is associated with a diet mainly of larger items (Crook and Aldrich-Blake 1968).

The temporal muscles of the jaw in *Theropithecus* are set well forward, efficiency of incisal action for nibbling being sacrificed to adding power to the masseter and pterygoid muscles for cheek-tooth chewing. This anatomical evidence for molar dominance in *Theropithecus* agrees well with the data on diet in the natural habitat. Jolly suggests that in the few areas in the Ethiopian Highlands where Pleistocene-type sympatry of *Theropithecus* and *Papio* still exists, the former eats small food objects requiring little incisor preparation but prolonged chewing, while *Papio* (which elsewhere in its wide range is a catholic feeder) concentrates on fleshy fruits and other tree products most of which require peeling or nibbling with the incisors (Crook and Aldrich-Blake 1968). He sees no reason against attributing the *Theropithecus*-like incisal proportions and jaw characters of the early hominids to a similar adaptation to a diet of small tough objects. There is no need to postulate a compensatory use of cutting tools for food preparation unless it can be shown archaeologically that such tools were being made.

In relation to the selection mechanisms that led to incisal reduction in molar-dominant forms, he suggests that teeth not needed for processing food may be a liability in the sense of being a site of injury or infection and of requiring raw materials for their formation and maintenance. Natural selection would then favour the genotype producing a jaw structure of the size and complexity such as to confer greatest net advantage. Thus in a monkey or hominid adapting to a diet similar to that of *Theropithecus*, a unit of tooth material allotted genetically to a molar would bring a greater return in food processed than a unit allotted to an incisor, and selection should favour incisors of the smallest size consistent with their residual function. This embraces a positive advantage in reduction. The concomitant canine reduction is considered in terms of it favouring increased efficiency in rotary chewing by avoiding canine 'locking' and producing more even molar wear. Alternatively it might be considered a secondary character dependent on incisal reduction, possibly at a genetic level with canine reduction being a simple pleiotropic effect of a genotype which primarily determined incisal reduction. Jolly cites evidence for this.

Given this working hypothesis of adaptation to a terrestrial life and a diet of small food items as the initial hominid divergence from the Pongidae, Jolly (1970) speculates about the characters of soft tissue and social organization which might follow, though admitting these can never be tested upon the fossil record. Initially he notes that female epigamic characters pectorally and ventrally rather than perineally are a feature which can be correlated, in *Theropithecus*, with a way of life in which the majority of time is spent sitting down. This arrangement is unique to *Theropithecus* among non-primates, but also occurs in *Homo sapiens*. Fatty pads on the buttocks, adjacent to true ischial callosities, are another *Theropithecus* peculiarity (Pocock 1925) which again can be related to sitting while feeding, and they occur uniquely in *Homo sapiens* amongst the Hominoidea.

Experiments with grain chewing in man suggest that food items not crushed in one masticatory stroke tend to be rolled in the oral cavity from where they are guided back to the teeth by the tongue. This demands constant agile lingual action—more than in eating a fibrous bolus of fruit. Thus chewing physiology of the hominid type might be expected to include a thick muscular mobile tongue accommodated in a large oral cavity. A highly arched palate, capacious interramal space and absence of symphyseal shelf are elements of such a large oral cavity which, significantly, provide preadaptations to articulate speech. This aggregate of hominid characters suggests that the small food items of the hominid diet were solid, spherical and hard. Many potential foods fit the description but only one is widespread enough in open country to be a likely staple. This is seeds of grasses and annual herbs, which still substantially provide for man. Other resources were probably exploited when available, but the diet of basal hominids is postulated to centre upon cereal grains, as that of the chimpanzee is upon fruit, and of the gorilla upon herbage.

Thus Jolly sees that populations of dryopithecines, destined to become hominids, began to exploit more and more exclusively a habitat in which grass and seeds constituted most of the available resources while trees were scarce or absent. This he postulates as phase I of hominid development. The putative sequelae are set out in Fig. 4-1. They derive from the change from a fruit- or herbage-centred diet to one based upon cereals. The ability to exploit grass seeds as a staple requires agile hand and hand–eye coordination in the higher primate to pick up and eat such small food items fast enough to support the metabolism of a large body. Other grassland resources obtained by individual foraging or cooperation would also be utilized. These might include small animals, invertebrates and vertebrates, leafy parts of herbs and shrubs, occasional fruits and tubers, which may be important sources of vitamins such as B_{12} and C, as well as various minerals. On such a stable adaptive plateau dryopithecines could have persisted for millions of years and gradually accumulated adaptations of an 'open country' species. Jolly (personal communication) has also stated that such hominids may also have lived in temperate woodlands where nuts and acorns could have represented a major component of diet.

Jolly sees a medium-sized *Dryopithecus* of the Miocene as a starting point for such differentia-

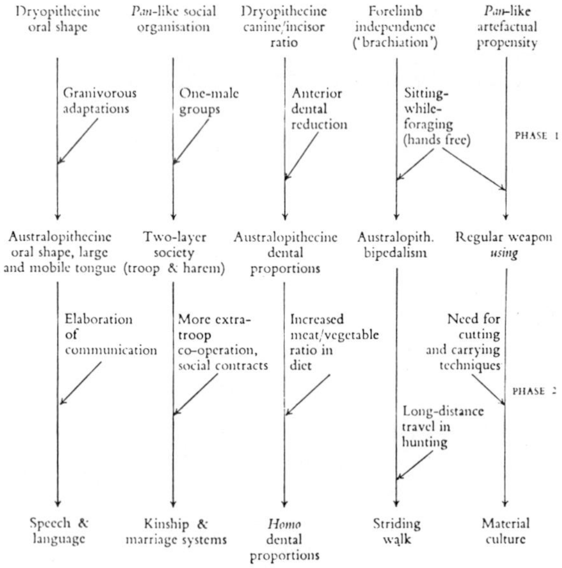

Fig. 4-1. A model of the development of some of the major hominid characters. During Phase I a series of Dryopithecine heritage characters (top line) is modified by the functionally determined requirements of the sedentary seed-eating complex (second line), producing the characters of evolved Phase I hominids (line three). These are preadaptive heritage characters of Phase 2 which determine the fact that adaptation to the demands of a hunting way of life (fourth line) takes the form of the human traits listed in the bottom line. The illogicalities of previous models tend to arise from omitting the vital second line, inserting the elements of the hunting complex in its place and invoking feedback. (From Jolly 1970, with permission; from the Royal Anthropological Institute of Great Britain and Ireland.)

tion. The increased emphasis on seasons of the middle to upper Miocene and lower Pliocene make it likely his phase I differentiation could have occurred at that time. The jaw and teeth of *Ramapithecus* are precisely of the form to be expected in a hominid of this type. In Jolly's eyes, amongst the early Pleistocene hominids, the robust australopithecines *Australopithecus robustus* and subspecies show the combination of characters to be expected in the long-term of his phase I (Fig. 4-1) adaptation. In fact, his model puts in perspective the apparent paradox of these hominids. He says their specializations such as incisal and canine reduction would be related to their seed-eating habits while their apparent primitiveness (relatively small cranial capacity and inefficient bipedalism: Napier 1964; Tobias 1970) reflects failure to develop what Jolly has termed phase II specialization. Robinson (1962, 1964) also has viewed *Australopithecus robustus* as a primitive stage of hominid evolution rather than an aberrant and late line, and has held that the basal hominids are unlikely to have been more carnivorous than pongids. On the basis of recent discoveries, *Australopithecus robustus* had a history of millions of years in Africa, leading to an adaptive plateau based on a vegetarian diet of small food items.

Jolly sees phase I hominids as uniquely preadapted to develop more human characteristics following further ecological shift. This may have involved the increasing assumption by adult males of a role as providers of meat, with the corollary that females and juveniles would have had to collect enough vegetable food for themselves and the hunters. The environmental change which provoked the institution of hunting may have been a slight change of seasonality in a marginal tropical area which placed a premium on exploiting meat as an additional staple. This dietary change may have been relatively small but reflected dentally in a moderate reversal of the phase I molar dominance in favour of incisal breadth needed to tear meat. These changes may have given great advantage to the development of cutting tools to prepare the kill and to transport it. The skilful hands and upright posture of phase I would predispose the hominid to solve these problems of adaptation by the development of artefactual propensity into a material culture. Cooperation between local bands in hunting activity could develop also, and symbolic communication by speech, instead of gesture, may be facilitated by and a consequence of the seed-eating mouth. Language, intellect and ritual develop. Thus the second distinctly human phase is more comprehensible in Jolly's view when superimposed on the base of preadaptations to the type of vegetarianism postulated rather than the 'carnivorous chimpanzee' type. There was no reason why some populations of phase I should not have continued to exist sympatrically with those in the phase II cycle.

Some discoveries of the Leakeys

In 1959, L.S.B. Leakey reported the discovery of a skull in Bed I of Olduvai Gorge in Tanzania. The skull, whose position in Bed I indicated it was almost 2 million years old, was almost complete and beautifully preserved. This find by Mary Leakey was in the course of a search for the makers of the stone tools Leakey had found in Bed I and named as an Oldowan culture after the Gorge. The skull lay on a well-defined surface which at that point had been used as a camp site by the makers of the Oldowan culture. From the floor not only the skull was recovered but also Oldowan stone tools. There was a hammerstone and many waste flakes as well as the broken-up bones of the animals upon which the hominids who occupied this camp site had fed. The animal bones included those of birds, frogs, lizards, fish, rats and mice, and the young of several of the giant mammals of the period. The animal bones, unlike the human skull, had all been deliberately broken up. This was in contrast to the skull, from which it was inferred he was the owner not the victim. Leakey notes that the fauna of Bed I contained many giant forms: giant ostrich, antlered giraffe, pigs, wild sheep, wild cattle, hippopotamus and much else. The adults of these creatures were hunted during Bed II time by later hominids, but the makers of the Oldowan culture were not good hunters. They had in Leakey's view only just emerged from a purely vegetarian diet and caught only small mammals, birds and reptiles and a few young creatures. Isaac (1971) suggests that early hominids may also have tapped the rich protein resources of the microfauna, as more recent peoples on arid terrains do.

Whereas Leakey (1959a, b) emphasized that the skull, which he called *Zinjanthropus bosei* (now known as *Australopithecus bosei*), showed many significant differences from the australopithecines, he included it in the subfamily Australopithecinae together with the monumental discoveries of Raymond Dart and Broom and Robinson: *Australopithecus africanus* and *Australopithecus robustus* found at Tuang, Makapan, Sterkfontein and Swartkrans. It had reduced canines and incisors relative to molars, very large mastoid processes and a long face. It was the oldest stone-tool-maker yet found and as a tool-maker was ranked as man. Oakley (1959), commenting on the find, suggested that this demonstration of the physical nature of the earliest tool-makers may prove to rank with the importance of the acceptance at the Royal Society in 1859 of evidence of the geological antiquity of man. That the discovery of the 'first men' should have been made in Africa was remarkably appropriate, he remarked, to the year of Darwin's centenary, for Darwin had said that it was probable that Africa was formerly inhabited by now-extinct apes closely allied to the gorilla and chimpanzee, and that as these two species are now man's nearest allies it is somewhat more probable that our early progenitors lived on the African continent than elsewhere. Oakley states that the australopithecines lived in open country, walked upright and had teeth of a hominid pattern. The question of whether they should be counted as 'man' or 'near-man' had been controversial but the Leakey's discovery of these tools, including standardized types representing a well-defined tradition, meant their makers were counted as 'human', however bestial they may have looked.

Though Leakey's interpretation has had to be modified, in one important respect at least, in the light of later data, the contemporary records reflect strongly the excitement and enthusiasm which is endemic to the subject.

In 1960 remains of another hominid were found at Olduvai, and at a level slightly below that of *Zinjanthropus*. They included part of the side and back bones of the skull of a juvenile, a clavicle and fifteen hand bones of two individuals, and most of the foot bones of an adult. With them were ten worked stone tools and evidence of carnivore activity. The cranial fragments have been shown to belong to a somewhat larger-brained hominid with dental pattern different from *Zinjanthropus* and closer to *Australopithecus africanus* (the gracile hominid). Features of the foot suggested the owner was adapted to running not striding, and study of the leg bones also found suggest a bipedal gait. Leakey and his associates have termed these remains a new species of genus *Homo—Homo habilis*—which means 'man having the ability to manipulate tools'. Stone artefacts were found associated with these remains from the bottom of Bed I into the lower part of Bed II. Later (1964) Leakey stated that it was not as likely as it once seemed that *Zinjanthropus* made the industry which is concentrated on the living floor of the site on Bed I described above. The pattern of distribution of bones and stone tools and waste flakes on the site makes it clear that the *Zinjanthropus* skull like other larger specimens was on the outskirts of the site. It is therefore possible that it was, like these other specimens the remains of a meal. Leakey particularly emphasizes the parietal bones of the juvenile found at another site, also on a living floor in association with stone tools of the Oldowan culture. The parietals and the thumbs show a size comparable to a *Homo erectus* and not an australopithecine. Geologically *Zinjanthropus* and this child found in an older bed than *Zinjanthropus* must have overlapped. The type of hominid represented by the big-brained child may have been the maker of the Oldowan culture. This line may have led to *Homo*, whereas *Zinjanthropus* was heading for extinction. The coexistence of the two types may be likened to the coexistence of Pygmies, Bantu, Europeans, gorillas and chimpanzees in the Eastern Congo today.

In 1970 Richard Leakey and colleagues discovered in the Koobi Fora deposits near Lake Rudolf (now called Lake Turkana) a rich collection of vertebrates which included five hominid specimens. The two hominid skulls were nearly complete. One resembled *Zinjanthropus bosei* from Olduvai, the other was probably *Australopithecus africanus* or a very early representative of *Homo*. With the bones were early artefacts, the total of 51 specimens including five choppers, twenty-four flakes, two refresher flakes and two roughly flaked and battered cobbles. Mary Leakey (1970) who described the material suggests the finding was not unexpected since the multiple 'tool kit' found at Olduvai predicted that tool-making must have been in practice for a considerable period before this time. Richard Leakey (1970) remarks that as it is generally accepted opinion that *Zinjanthropus* was not a tool-maker, the discovery of artefacts in the Koobi Fora deposits adds support to the evidence from the fossil material that there were two different representatives of hominids in the area investigated. Work by French and American teams on the evidence of potassium/argon dates has shown that both *A. africanus* and *A. robustus* were living in the Omo Basin north of Lake Rudolf in the late Pliocene/early Pleistocene, i.e. between 3.5 and 1.8 million years ago (Clark 1970; Coppens 1970, 1963, 1975a, b; Bonnefille et al. 1973; Howell and Coppens 1973).

From the viewpoint of the general theme I am considering, it is interesting that meat-eating, with its implications for sodium homeostasis, extended back this far into the Plio-Pleistocene, and possibly back further as an occasional activity of *Ramapithecus*. Leakey (1968) reports that bones found in upper Miocene fossil beds at Fort Ternan, Kenya, show evidence of having been broken up by some form of blunt instrument, possibly a peculiar lump of lava exhibiting several battered edges. The possibility exists that the upper

Miocene hominid *Kenyapithecus wickeri* was already making use of stones to break open animal skulls to get at the brain, and bones to get at the marrow. Ardrey (1976) sees an aggregate of data as pointing to systematic hunting going on long before the middle Pleistocene, and quotes Oakley and Washburn as supporting this view. The presence of very large prey animals at butchering sites at Olduvai almost 2 million years old—before evidence of the modern human brain—is taken as proof that hunting may have been practiced for unknown earlier periods of time. However, as Isaac (1971) notes, there is no way of telling whether these butchered large animals (elephant and *Deinotherium*) were killed or found dead.

Hunters and gatherers: Evidence on diet of early man

Isaac (1971) in his review of the diet of early man, describes the substantial element of controversy that has been evident in the subject and the way in which various points with regard to human evolution and human nature have been backed by citation of only selected items of archaeological evidence. He thinks that despite its manifest imperfections the implications of the whole corpus of data ought to have been considered. He notes the importance of hunting in creating the selection pressures that have directed the human evolution towards brain expansion and effective linguistic communication. And Tiger (1969) has argued that the widespread phenomenon of male bonding in human societies has its origin in social arrangements that were adapted for early humans who lived by hunting. But the vegetable component of human diet has received much less dramatic attention in discussion of human nature and evolution. Isaac suggests that the aggregate of evidence from a number of studies suggests that the terms such as 'hunters' or 'carnivores' should probably be applied to recent representatives of the human species only with qualification. A small sample of studies from Africa and Australia suggests that in the tropics and warm-temperate regions the meat component of diet may often have been appreciably less than 50%. This matter has been extensively reviewed by Blainey (1975) in his history of ancient Australia. For example, amongst a wide variety of studies, he quotes the work of the anthropologist Mervyn Meggitt to the effect that perhaps 70%–80% by bulk of the food eaten by aborigines in tropical Australia is vegetable. There could also be a latitudinal gradation in which the importance of meat and fish increases from Equator to Pole. Thus in Arctic and sub-Arctic peoples as much as 80%–90% of the diet is protein foods.

Isaac (1971) suggests that the occurrence of opportunistic hunting among baboons, chimpanzees and other anthropoids can perhaps be taken to indicate that this is a generalized anthropoid trait shared by ancestral stock, even before the evolutionary divergence of the hominids. As this tendency is evidently intensified in man it is important to determine the history of changes leading to increased carnivorous proclivity. But it would be wise to bear in mind the great variation found in ethnography. The evolution of human behaviour has involved not simply increasing intensity of food production but the unusual development of a flexible system of joint dependence on plant and animal food. In many recent human communities the retention of food-gathering as a major dietary component represented an insurance that rendered hunting possible under marginal conditions. This appears to have depended on differentiation in the subsistence activities of the two sexes—probably partly genetically determined—and was associated with the distinctive human practices of food sharing and occupation of home bases.

Food transport back to specific localities results in the localized accumulation of refuse which is what has made the study of prehistoric life possible. But because bone is more durable than vegetable matter, it is more likely to be preserved and survive to be studied as a palaeoanthropological document. Thus the archaeological record is liable to exaggerate the carnivorous proclivities of early men unless deliberate steps are taken to counter the bias. Microscopic plant food refuse survives in small quantities only at Choukoutien and Kalambo Falls. Some sites where there were high intensities of occupation and low relative abundance of bone refuse may be those where diet consisted principally of gathered foods. Isaac thinks the evidence for use of animal bones for tools is similarly inconclusive. He goes on to say that the data set out by Dart and others prove conclusively the existence of an assemblage showing a high degree of selectivity of body part representation. This is consistent with a tool and weapon hypothesis. Subsequent research, however, has demonstrated that natural and midden bone assemblages involve highly selective patterns of preservation and destruction of different body parts in circumstances where tool use can be ruled out as an explanation. On the other hand certain depressed fracture, breakage and damage patterns and jamming of bones into one another are potential evidence of hominoid activity. Overall the supposition that *Australopithecus africanus* was partially carnivorous and a tool-user remains eminently reasonable.

In summary, the available evidence shows clearly that a wide range of meat foods figured in the diet of lower Pleistocene hominids and that the quantities were probably substantial. But there is no real way of assessing the relative proportions of the total diet that accrued from hunting, gathering and scavenging. All three were probably important activities. With relation to the middle Pleistocene, however, even when allowance has been made for distortion of the bone remains by scavengers and poor preservation, it does appear that most sites contain comparatively modest amounts of bone. Some, such as Latamme, Olorgesailie H/6 and Cave of Hearths contain only minor amounts of bones in spite of apparently favourable conditions for preservation (Isaac 1971). If the spectrum of variation in bone abundance in the middle Pleistocene is amplified by further data, it could indicate that a great variation in the meat component of diet of recent non-agricultural peoples has, in fact, been a consistent pattern through much of human evolution.

There are a number of sites where large accumulations of bones from ungulates and baboons suggest drives in the course of hunting. In Africa, known lower and early middle Pleistocene sites are confined to relatively dry regions where subsistence by hunting and/or gathering is feasible to judge by contemporary examples of Bushmen and Hadza. Distribution of the Acheulian industries in the late middle Pleistocene covers all of the African grasslands, savannahs and much of woodland, but not the heavily forested areas. Again Isaac (1971) sees this as consistent with the variable generalized hunting and gathering pattern envisaged for this timespan in the tropics and warm-temperate areas. In general, Africa, as representative of tropical and warm-temperate regions, provides data suggesting that hunting has seldom if ever been in any exclusive sense the staff of hominid life (see also Lewin 1976). The pattern suggested is more one of broadly based subsistence than intensive and voracious predation. Division between male hunters and female gatherers is unique to man and should be stressed more as a feature of human evolution than mere predation. Invasion of cold-temperate and sub-Arctic regions led to hominids having a new ecological adjustment where proteins were the dietary staples.

Some more recent findings

Considerable controversy surrounds the status of *Ramapithecus* in hominid evolution. Two possibilities are illustrated in Fig. 4-2. *Ramapithecus* will be

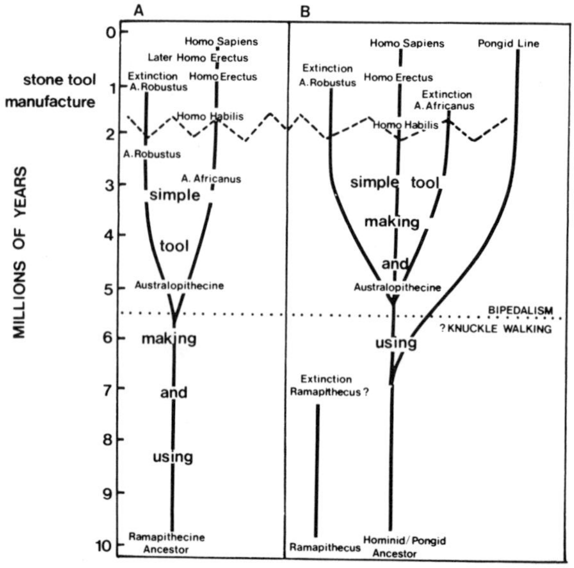

Fig. 4-2. Two possible interpretations of hominid evolution in relation to *Ramapithecus*. (After Clark 1970.)

considered in detail here because of the implication that for perhaps 10–15 million years since it possibly emerged from a *Proconsul*-type of dryopithecine, this ancestor of the hominids was largely, if not exclusively, vegetarian. Simons (1977), however, notes that there is a large gap in the fossil record between *Ramapithecus*, with its jaw structure resembling the African hominid *Australopithecus*, and which is not known by fossil specimens in Eurasia since 8 million years ago, and the earliest fossils of *Australopithecus* and *Homo* found by Johanson and colleagues in Ethiopia and by Mary Leakey in Tanzania which are less than 4 million years old (Leakey et al. 1976; Johanson and Coppens 1976; Taieb et al. 1975). Leakey et al. (1976) report the discovery of 13 early hominids in the Laetoli Beds of Northern Tanzania. The potassium/argon dating of the fossiliferous deposits gives an upper limit averaging 3.59 million years and a lower limit of 3.77 million years. These specimens are regarded as very similar to fossil material from the Hadar region of Ethiopia. They indicate a strong resemblance between these two groups of hominids and the later radiometrically dated specimens assigned to the genus *Homo* in East Africa.

Johanson and White (1979), reviewing the data on the Hadar and Laetoli finds (Fig. 4-3) as the

Fig. 4-3. Some of the main sites in Africa where important skeletal finds have been made.

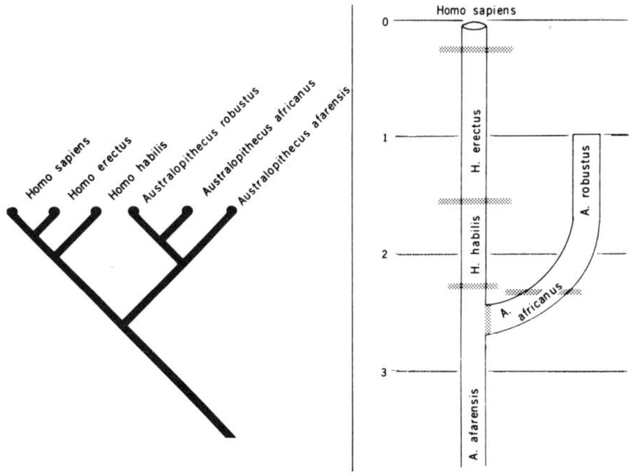

Fig. 4-4. Cladogram of the family Hominidae (left) and phylogenetic tree of the family Hominidae, with a scale in millions of years. (From Johanson and White 1979, with permission; copyright 1979 by American Association for the Advancement of Science.)

earliest substantial record of the family Hominidae, propose a new taxon *Australopithecus afarensis* for these Pliocene hominid fossils. They are the most primitive group of demonstrable hominids recovered so far, and are characterized by a substantial sexual dimorphism. They state that the ancestry of these hominids is not understood, and that the primitive appearance of the Laetoli and Hadar material suggests that a late divergence from hominoids, such as *Ramapithecus*, must remain a possibility. The ultimate resolution must await collection and analysis of further hominid remains dating between 5 and 15 million years. It will be critical to this resolution to recover specimens of lineages of the existing pongids. They note that as well as the biochemical evidence discussed further on in this chapter, some interpretations of affinities of post-cranial anatomy between extant African apes and modern man suggest pongid–hominid divergence was late in time (Washburn 1971; Lewis 1973). Others place the divergence much earlier, in the middle Miocene or even the Oligocene (e.g. Simons 1977). Johanson and White's (1979) representation of *Australopithecus afarensis*' position in the phylogenetic tree is shown in Fig. 4-4.

In a review, Coppens (1975a) notes tooth discoveries at N. Gorora in the basin of Lake Baringo and at Lukeino, in Kenya, which date at approximately 9.3–12.0 and 6.5 million years ago respectively. At Lothagam in the basin of the River Kerio at the south-east of Lake Rudolf, a half-mandible still carrying a first molar resembling an Australopithecine type has been found in a bed dated at 6 million years. Simons suggests there was a hominid stock, ancestral to primitive *Homo* and to *Australopithecus*, which resembled *Ramapithecus* more closely than later representatives of *Australopithecus* did. He notes that Leakey favoured a view that *Australopithecus* and *Homo* branched directly and independently from *Ramapithecus*.

The middle Pleistocene saw great advances in the physical evolution of *Homo* and his tool-making capacity. At the beginning of this era *Homo erectus* appeared, with a greater brain size and new techniques of stone tool manufacture as, for example, in the bifacial hand axes. Finds of *Homo erectus* have been made at Olduvai, at Choukoutien near Peking in China, in Algeria, Swartkrans and in Java. In Olduvai the finding of the skull cap of a large-brained *Homo erectus* form in Bed II was associated with Acheulian artefacts, and the same relationship is seen at the sites in north-west Africa. In China there is evidence of meat-eating and use of fire and of pointed sticks for hunting and stone tools for stripping kills. Clark (1970) notes that the evidence is clear that *Homo erectus* possessed a much more extended range of abilities and indulged in a greater variety of activities than did *Homo habilis*, including hunting large and medium-sized animals. Clark suggests that what is termed the 'Acheulian industrial complex' lasted for about a million years. Thus over the course of this time *Australopithecus robustus*, which had coexisted from late Pliocene with *A. africanus*, and through the development of *Homo habilis* and *Homo erectus* up to about a million years ago (as judged from finds at Lake Natron in Tanzania), had not advanced. The currently accepted view is that *A. robustus* failed to become a tool-maker, though, possibly, may have been a tool-user, and eventually succumbed to the competition of the tool-maker.

With the first half of the late Pleistocene, Neanderthal man began to emerge with ever increasing cultural complexity, with more sophisticated tools, and, apparently, ritual burial of the dead.

Chronology of divergence of Hominidae and apes and dietetic implications

Diet of pongids

Before considering the further development of *Homo* and his diet, the data on the diet of pongids will be considered. This becomes a matter of increased importance, if, as the immunological evidence suggests, the separation of hominids and pongids was very much later than most physical anthropologists suppose. That question will be considered in the following section of this chapter.

Chimpanzees

Jane Goodall (1971), as a result of her landmark studies on the chimpanzees of the Gombe Stream Reserve in Tanzania, records that, like man, the chimpanzee is an omnivore, feeding on vegetables, insects and meat.

Over 90 different species of tree and plant were used by the chimpanzees as food. Fifty types of fruit and over 30 types of leaf and leaf bud were eaten at different times, and the animals also ate some blossoms, seeds, barks and piths. Insect foods were eaten in large quantities throughout the year. In particular, three species of ant, two species of termite and one species of caterpillar were consumed. When the chimps raided bees' hives, larvae were eaten as well as the honey. Birds' eggs and fledglings were also eaten and these were occasionally taken out of the nests.

Chimpanzees are relatively efficient hunters. Prey is captured and killed with bare hands and is dismembered like other fleshy foods with the incisors. Goodall records that a group of about 40 individuals may catch over 20 prey animals per year. The most common prey species are young bush buck, bush pigs and baboons, and young or adult Colobus monkeys. Occasionally a red-tailed monkey or a blue monkey is killed. In relation to the implications of meat-eating for sodium homeostasis, a rough approximation from Goodall's figures is as follows. If 20 animals are killed per year each with a meat and organ weight of approximately 15 kg (which is probably a liberal estimate for the creatures detailed), then allowing an average sodium content for the mixture of meat and viscera of 50 mmol/kg wet wt, the total intake of sodium over the year would be 15 000 mmol. This spread over 40 chimpanzees would give approximately 1 mmol sodium/day per animal. This may represent 10%–20% of the daily requirements of a chimpanzee, though there are few data upon which to base an estimate. As recorded in Chapter 3 the chimpanzees sometimes eat small quantities of soil containing salt. If the composition of chimpanzee milk resembles that of human milk it is possible that requirements for a female during lactation could reach 10–20 mmol/day. Teleki (1973) reviewing the data from Gombe Stream suggests an estimate of 12 kills per year by the troop, which would give a smaller sodium intake than as above. Furthermore, if we take into account that the individual human sequestrates about 500 mmol of sodium during pregnancy (Chapter 23), and we accept the chimpanzee's requirement is half this, it could well be that the reproducing chimpanzee enter the lactation phase already under considerable stress on sodium homeostasis.

In relation to the meat-eating, between each bite of meat Goodall observed that a mouthful of leaves was invariably eaten. Another interesting feature was that in the instance of the meat being in the possession of one individual, it was seen that other chimpanzees came and sat around and begged for pieces by stretching out their hands to touch his lips with their fingers. Another begging gesture was that in which the palm was held uppermost. Both mature males and females were observed to beg in this fashion. Teleki (1973) terms the chimpanzee an omnivorous forager-predator which supplements a basic vegetarian diet in various ways, including predation on fellow primates.

Gorillas

Schaller (1963) in his book on the mountain gorilla has summarized published data and his own observations on the diet of these animals. He states that nearly all authors agree that the gorillas of West Africa favour afromomum and cultivated bananas. Autopsies on five animals by Sabater Pi and de Lassaletta showed that the marrow of banana stems provided about 80% of the bulk eaten, followed by manihot at 10% with the remainder consisting of afromomum, sugar cane and fruits. A hundred different species of plants are used as food, of which herbs, vines and trees predominate numerically. Many of these plants Schaller found bitter or otherwise unpleasant to taste. The animals usually forage leisurely, alternately sitting and walking. Characteristically each gorilla sits and reaches for food in all directions. Apparently gorillas do not share food, for Schaller did not observe a single instance in which one animal offered vegetable matter to another in the wild. They showed great dexterity in combining the use of hands and mouth to expose and eat the palatable parts of each plant.

Schaller states there is no reliable evidence that gorillas eat animal matter in the wild. There are some records but these are not well documented, and Yerkes and Yerkes (1929) in their summary of great ape literature did not list a single unquestionable instance of meat-eating in free-living gorillas. Visual examination of several thousand sections of gorilla dung revealed not a single instance of hair, chitinous material, bone, skin or other evidence that animal matter had been ingested. He saw cases of groups passing recently dead animals and ignoring them. One group rested within 4 m of an incubating olive pigeon without destroying the obvious nest. However, as Yerkes and Yerkes (1929) pointed out, in captivity gorillas readily eat meat. During his visit to Columbus Zoo, Schaller saw a keeper feed a slab of boiled beef nightly to two adult gorillas.

Baboons

De Vore and Washburn (1964) have attempted to reconstruct elements of the evolution of human behaviour by comparing the social behaviour and ecology of baboons with that of contemporary hunter-gatherer groups and applying these comparisons to the archaeological evidence. They studied baboon behaviour and ecology, including troop structure, in the game reserves of southern Zimbabwe (Rhodesia) and Kenya.

They found that baboons, like the macaques of Asia, eat a wide variety of foods. Although the bulk of their diet is vegetable food, they will also eat insects, eggs, and an occasional small mammal. Grass is the baboons' single most important food. During 10 months of observations not a day passed when de Vore and Washburn did not observe the baboons eating grass, and for many weeks during the dry season grass constituted an estimated 90% of their diet. When the tassels of grass contained seeds these were harvested by pulling the tassel through the closed palm or clenched teeth. Most often, however, baboons pulled the grass shoots in order to eat the thick lower stem at the base of the culm. In the middle of the dry season, when grass shoots are rare, baboons concentrate on digging up rhizomes, the thick root-like runners of the grasses which lie 4–8 cm beneath the surface. This ability to find food below-ground in the dry season is one of their important adaptations to the dry conditions of the grasslands. No baboon was ever observed using a digging stick or sharp stone to help extract the rhizomes, though this would have greatly increased their efficiency.

The baboons' diet is extended by eating berries, buds, blossoms and seed pods of various bushes, flowering plants and shrubs. The bean-like seed pods of acacias were eaten and also those of the fever trees. They eat many types of insect, the most common being the ant living in the galls of the acacia trees. Though insect food is minor in the overall baboon diet, a heavy infestation of army worm caterpillars at certain periods caused them to be the baboons' most important food.

On six, perhaps seven, occasions during the months of their study the authors saw baboons eating freshly killed animals. Twice they caught and ate half-grown African hares. Two or three times baboons were seen eating birds of some ground-nesting species, probably the plover. Two very young Thompson's gazelles were caught and eaten by adult males of the troop. In all the cases the flesh-eating had one thing in common: it involved the eating of immature animals whose defence was to hide 'frozen in the grass', and in each case their discovery by the baboons seemed fortuitous. Nothing resembling the systematic search of an area or the stalking of prey was observed nor was fresh meat eaten except when it was found alive and caught immediately by a waiting baboon.

In summary, baboons may be described as very inefficient predators. It would appear that meat-eating is learned by each generation and meat never becomes an important source of food for the whole troop. De Vore and Washburn feel that the importance of meat in the baboon diet has been over-stressed on the basis of the reports on carnivorous baboons in South Africa (see below). They also discuss the issue of scavenging. They consider this an unlikely method of obtaining food because the chances of a kill within the range of any particular baboon troop are small, and as most kills are at night little meat is likely to be left in the morning. Also, the vicinity of the kill is dangerous and most baboons seem uninterested in dead animals. They remark in relation to scavenging that once man became a skilled tool-user he could extend tool-use to the hunting of large animals, to defence and to driving carnivores from their kills. Scavenging may thus have become a source of meat when man had become sufficiently skilled to take meat away from carnivores, but the hunting of small animals and defenceless young is much more likely to lie at the root of the human hunting habit. Leakey (1964), however, has expressed a divergent opinion. He considers scavenging to be quite feasible since lions often leave their kill for several hours, and a lion is not dangerous when it has just fed. In the Olduvai area a small hominid family could have scavenged at least twice a week from kills, and it is also possible to catch and kill adult small antelopes with one's bare hands. Therefore, a creature which desired meat, unlike the predominantly vegetarian baboons, could definitely obtain it. Monod (1964) noted that villagers in East Chad Republic take part of the kills of lions.

Dart (1963) records an impressive catalogue of instances of baboons attacking and devouring the young of Thompson's gazelle, impala and other buck in the South African game reserves. He also includes accounts of baboon troops marauding and killing large numbers of lambs, and also mature sheep. He regards these observations as indicative that the baboon's need of animal protein is perpetual and not a seasonal one, and as supporting the belief that the carnivorous needs of our australopithecine ancestors were also perpetual.

The account by Strum (1975) of a study over 16 months of a baboon troop in Kenya also emphasizes meat-eating as a component of the diet, though insects even more than meat help fill out the primates' mainly vegetarian diet. The killers and consumers were almost exclusively adult males. But in the course of the study it was found that certain female baboons, juveniles and even infants showed an interest in meat. As the study proceeded it emerged that cooperation occurred between baboons in the hunt for, for example, Thompson's gazelle, where males ran down the prey in relays. Animals captured included Cape hares, steenbok, dik-dik and klipspringer, together with an occasional bird. In relation to the context of these observations on a ranching area, Strum (1975) remarks that the animals killed more prey than baboons observed anywhere else. The logic, without proof, is that ranching operations extended cattle pasture and introduced irrigation, while the leopards, lions and hyenas which are the baboons' natural enemy, diminished as a result of shooting. So more game was available to the baboons, and their own risk of being attacked diminished.

Orang-utans

George Schaller (1961) in reporting his observations on the orang-utan in the forests of Sarawak stated that they are primarily frugivorous. Hornaday (1885) recorded that they eat the leaves of certain trees, but Schaller did not observe this directly. The shoots and fruits of palms are eaten. One instance of an animal gnawing the inner lining of bark was noted. At times, the animals raid cultivated fruits at the edge of the forest. Possibly they sometimes eat insects and one observer maintains they eat the eggs of birds. Many of the fruits of the forest the orang-utans ate were also eaten by man and various monkeys.

Chronology of divergence

In relation to the chronology of divergence of the Hominidae and the apes, Washburn (1964) notes that the point of divergence appears to be based on the new adaptive complex of the Hominidae in which tool-making and erect bipedalism are basic. Large brains and small faces followed long after the Hominidae were distinct from the apes. It has become traditional to date the separation of the Hominidae from the Pongidae to the Miocene. But the distinctive size of the human brain evolved at the end of the lower or beginning of the middle Pleistocene. Most of the human characteristics, as for example pelvic, seem to have evolved well within the Pleistocene. In Washburn's view, there is no need to postulate an early separation of man and the ape.

Goodman (1973) points out that the amino acid sequences of primate proteins are evolutionary documents containing a chronicle of phylogeny. Each change in an amino acid sequence of a protein can be related to a genic mutation. The resulting change may affect the biological functioning of a protein and thereby the survival chances of the organism carrying the mutant gene.

According to Goodman (1964) the phylogenetic stage of the placental mammals introduced a new restrictive factor in the divergence of genetic codes: maternal immunizations to foetal antigens, particularly with the haemochorial type of placenta. Proteins such as albumen synthesized early in foetal life thus showed a decreasing rate of evolution, but proteins such as gamma globulin not synthesized until after birth still evolved at a relatively rapid rate because divergence was not selected against by maternal immunizations. He proposed that this immunological process operated with particular force in the phyletic line leading to man.

Here there was a necessity for both cerebral evolution and placental evolution. Placental evolution increased foetal–maternal vascular intimacy, allowing the developing foetus to obtain a rich supply of nutrients and oxygen. Therefore as gestation lengthened and the barrier between embyronic tissues and maternal blood decreased, maternal immunizations selected for a state of genetic homozygosity. But as cerebral evolution progressed, and the ancestors of man invaded broader ecological domains, the more varied external environment selected for a state of genetic heterozygosity. This contradiction was resolved by selecting for genetic codes which increasingly shifted to the post-natal phase of ontogeny the maturation of somatic systems concerned with adaptive responses to the external environment. Maternal immunizations would not select against gene determined polymorphisms in such late-maturing systems or for proteins within the central nervous system. This delay in appearance of adult proteins increased the helplessness of the young and thereby induced development of the protective care of the young by the mother and other adults. Goodman states it is not by chance that man, who shows the greatest cerebral development and occupies the broadest environmental range, is also the most retarded in his maturation.* He cites data to show marked deceleration in primates of albumen evolution but not of gamma globulin evolution. Various serological tests also showed chimpanzee has more recent common ancestry with man and gorilla than with the orang-utan or gibbon.

Wilson and Sarich (1969) cite their own and other published data from immunological techniques on serum albumens, transferrins, haemoglobins and DNA that show man is genetically much more similar to the African apes than to Old World monkeys. For example, according to immunological criteria the albumens of chimpanzee, gorilla and man are only about one-sixth as different from one another as they are from that of the rhesus monkey. Given a regularity in protein evolution, it would follow that the African apes and man diverged about six times more recently than did rhesus monkey and man. Assuming a divergence of Hominoidea and Cercopithecoidea (Old World monkeys) is unlikely to be greater than 30 million years, it would follow that man and ape lineage divergence could have occurred approximately 5 million years ago. The haemoglobin data (Table 4-3) do not allow such a clear conclusion, as

Table 4-3. Comparison of amino acid sequences of haemoglobins in different species

Species compared	No. of amino acid differences	Mutational distance[a]
Man vs chimpanzee	0	0
Man vs gorilla	2	2
Monkey vs man	12	15
Monkey vs chimpanzee	12	15
Monkey vs gorilla	14	17
Horse vs man	43	52
Horse vs chimpanzee	43	52
Horse vs gorilla	45	54
Horse vs monkey	43	52

From Wilson and Sarich (1969), with permission.
[a]The minimum number of base substitutions required to account for the observed amino acid substitutions, calculated by the method of Jukes, and Fitch and Margoliash.

*Parenthetically, it is of interest to draw attention to another aspect of accelerated cerebral evolution in relation to the birth process. The hormone relaxin, secreted by the ovary, is closely related structurally to insulin and insulin-like growth factor. It is composed of two chains of amino acids, 22 and 31 units respectively, linked by two interchain disulphide bonds. It is involved in the birth process, being responsible for the loosening of the ligaments of the pelvis and, in particular, the pubic symphysis. It also inhibits the spontaneous contractions of the uterus, and softens and increases the distensibility of the cervix.

Following the determination of the structure of the hormone (James R., Niall H., Kwok S. and Bryant Greenwood G. (1977) *Nature* 267: 544; Scwabe C., McDonald J. K. and Steinetz B. G. (1977) *Biochem. Biophys. Res. Comm.* 75: 503), the entire sequence of the gene coding for it in the rat has been resolved (Hudson P., Haley J., Cronk M., Shine J. and Niall H. (1981) *Nature* 291: 127). In relation to the potential medical importance of the hormone, it is noteworthy that studies have shown that approximately 50% of cases of spasticity and cerebral palsy in children are attributable to injuries in the birth process. These are in the context of delayed second stage of labour with anoxia of the foetus, application of forceps, etc., as well as with prematurity where conceivably deficiency of relaxin might also be involved. Cerebral haemorrhage occurs in up to 10% of premature infants.

These considerations highlight and perhaps reflect the latter stages of hominid evolution. That is, Le Gros Clarke (*History of Primates* (1970), Trustees of the British Museum) records the cranial capacity of australopithecines to be in the range of 700 cc, *Homo erectus* to be an average of 860 cc in the Javanese specimens and an average of 1075 cc in the Chinese specimens, as compared with 1350 cc for modern human races. This very large change has occurred over 2–3 million years, and accordingly an intense selection pressure may well have operated favouring the hormonal mechanisms controlling delivery of the progressively larger brain and head through the bony ligamentous pelvis. The incidence of brain damage in the birth process may reflect the continuing selection on this basis, though clearly many other factors may be involved. Similarly, undetermined factors such as the extent to which post-natal brain growth in *Homo sapiens* vs *Australopithecus* is reflected in the ultimate difference in cranial size will evidently be relevant to such an hypothesis, as will eventual determination of the quantitative role of relaxin in the human birth process.

a statistically reliable number of amino acid differences between ape and human haemoglobins have not yet accumulated. Citing Zuckerkandl and Pauling and other groups, Wilson and Sarich propose that the average rate of evolutionary change amongst mammalian haemoglobins is only one amino acid replacement per approximately 3.5 million years. Thus, species which diverged 3–4 million years ago would be expected to differ in haemoglobin sequences by two amino acid replacements and those which diverged 30 million years by about 17 amino acid replacements. Table 4-3 shows the ape and Old World monkey data are consistent with this, and suggest a phylogenetic relationship (Fig. 4-5) different from the data of

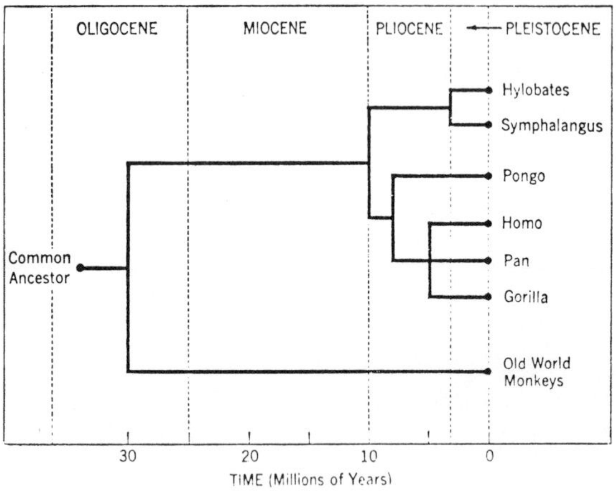

Fig. 4-5. Times of divergence between the various hominids as estimated from immunological data. The time of divergence of hominids and Old World monkeys is assumed to be 30 million years. (From Sarich and Wilson 1967, with permission; copyright 1967 by American Association for the Advancement of Science.)

physical anthropology. The concordance between haemoglobin and albumen results suggests the human lineage diverged from that leading to the African apes far more recently than generally supposed, most probably 4–5 million years ago (cf. Washburn and Moore 1974).

This interpretation of the timing of the divergence of man and apes has been the subject of considerable controversy. Discussion of the issue has been focused in the symposium *Molecular Anthropology* edited by Goodman and Tashian (1976). Reconciliation of the molecular data with the extensive work of morphology and palaeontology, has been suggested on the basis that the 'clock' of protein evolution has slowed considerably in the Catarrhini (Hominoidea and Cercopithecoidea) as compared with non-anthropoid primates or non-primate mammals (Dene et al. 1976; Lasker 1976).

Diet of hominids

The principal consideration in this discussion of the 30 million years of evolution of the Hominoidea, is that diet was probably almost exclusively vegetarian during the first 25 million years. If the deductions on hunter-gatherer societies based on the Kalahari Bushmen and the Hadza are valid in relation to the hominids, then diet may also have been predominantly vegetarian over the last five million years, though there was considerable variation in diet according to the prevailing climate and conditions. This is perhaps reflected by contemporary feral communities where diet may be predominantly vegetarian, as in some regions of Highland New Guinea or in the Yanomamö Indians of the Amazon Basin (see Chapter 6), or predominantly of animal products as, for example, with the Eskimos and the Maasai. As will be detailed in Chapter 27, these vegetarian communities have very low rates of urinary sodium excretion.

The behaviour towards salt of a wide variety of species of wild herbivorous mammals has been recorded in Chapter 3. The environmental, dietetic and metabolic conditions which determined it could have operated on hominoids over a large part of their evolution. It follows that selection pressure would favour retention and, perhaps, elaboration of salt appetite mechanisms—both hedonic liking and the hunger with deficiency—which were developed at earlier stages of phylogeny. This may also have held for the elaboration of the physiological system of salt retention by aldosterone secretion, which has a multifactorial mode of control that includes special adaptations to the assumption of the upright posture. This selection pressure would have been ameliorated by meat-eating with the consequential obligatory sodium supply during the latter 2–3 million years of hominid evolution. Whether this resulted in regression of these mechanisms to any extent is a matter of speculation (see Chapter 26).

Considered against the background of the dietetic behaviour of the pongids, the possible divergence of the line leading to *Homo sapiens* later than generally conceived would imply a persistence of largely vegetarian habit, with its implications for sodium status, longer in hominid evolution than otherwise envisaged.

Oakley (1954) remarks that dentally and from the alimentary point of view we should be vegetarians. We lack the teeth of true carnivores, and we have the long gut associated with herbivorous diet. Furthermore, our nearest living relations the anthropoid apes are herbivores and consume only small quantities of animal protein.

Man's change of habit from herbivore to semi-carnivore gave a new potential. It is easy to see how tool-making might arise out of the adoption of carnivorous habits. Though they may have killed game easily enough, proto-humans must often have had difficulty in removing skin and fur and in dividing flesh. Without strong canines, sharp pieces of stone would provide the solution. As Oakley suggests, this may have been the origin of the tradition of tool-making. Clark (1970), however, emphasizes that vegetable foods form about 60%–80% of the food of hunters and gatherers in warm and temperate climates today. This proportion decreases, however, as one approaches the Poles, with Arctic and sub-Arctic populations being up to 80%–90% dependent on meat and fish (Isaac 1971).

The discovery that animal skins were a means of conserving heat in cold weather may well have been another important factor in the ability of early humans to move into the temperate and cooler areas of the Old World, for those that originated in the tropical or semi-tropical areas were covered, at most, with a relatively thin growth of hair that provided little protection against the cold.

Clark (1964) has detailed observation of the diet of Hukive Bushmen, who are contemporary hunter-gatherers in the Zambesi–Mashi area. The group of about 100 lived in open woodland and moved between five waterholes about 30 km apart over an arc of some 250 km^2. The main foods were oil-rich nuts which were available all the year round either on the tree or fallen to the ground, roots, caterpillars, lizards, snakes, mice, honey (which was particularly prized) and some game animals such as eland, roan, zebra, kudu, waterbuck, duiker and giraffe. They were not seen to eat salt. Clark shot a large bull eland and it lasted a group of 30 Bushmen for 2–3 days. Everything was eaten, including the skin.

Clark (1970) notes vegetable foods, however, leave very little evidence in the archaeological record. It is, nevertheless, possible to regard the numerous rough polyhedral spheroids and hand axes not only as tools for breaking animal bones and dealing with tough parts of skin also eaten by present-day hunters, but also for breaking open nuts and preparing the otherwise unpalatable parts of plants by the breaking down of the fibrous portions. In his eyes the Olduwan artefacts thus comprise the basic equipment necessary for preparing a varied and unselected supply of plants and animal foods which were carried back to home base. He sees the home base as one of the most significant features of early hominid sites since it represents a place of continuous temporary occupation. It reflects the prolongation of pre-adult life and the greater dependence of the young upon adults. Thus a group became mutually dependent and shared the results of their hunting and foraging activities. Clark again refers to gathering as a major source of food, and notes that at Kalambo Falls various edible foods, seeds and nuts are found associated with the Acheulian living sites.

Isaac (1971) remarks that when allowance is made for distortions due to scavengers and poor preservation, it does appear that most Acheulian sites contain comparatively modest amounts of bone. He summarizes his conclusions as follows. Hominids, although their ancestral stock was largely vegetarian, have evolved so as to include more protein in their diet, its importance increasing from Equator to Pole. Flexible joint dependence on animal and plant foods, establishment of home bases, food sharing, and differentiation of the subsistence activities of the sexes together constitute an integrated behavioural complex already partly established some 2 million years ago. Large animals are certainly represented in the lower Pleistocene food refuse; they become commoner in the middle Pleistocene, when evidence for effective co-operative large-scale hunting also first occurs.

Discovery of fire

Evidence of fire has not been found on Acheulian sites in Africa except on the latest, which are dated at 40 000 to 60 000 years ago. Its use is, however, confirmed from several sites in Europe and Asia during the early and middle Pleistocene and in the colder high-latitude regions from Choukoutien, Torralba, Ambrona, Hoxne and other places. Clark is of the opinion that this fundamentally important source of heat and roasting ability was similarly employed in the tropics but the resulting charcoals have probably been broken down and dispersed by soil fauna and so are not preserved. Peking man presumably had considerable skill as a hunter since his own remains were associated with quantities of bones of butchered animals, mainly deer, but also bison, horse, rhinoceros, elephant, bear, hyaena and tiger. As Oakley points out (1954) the killing of some of these must have involved the use of pits or traps. During their long period of occupation of the caves, these Choukoutien hunters regularly used fire. The discovery of fire was mankind's greatest step forward in gaining freedom from the dominance of the environment. Leopold and Ardrey (1972) emphasize that cooking was a major advantage in extending spectacularly the range of food, since cooking destroyed many toxic substances in plant foods. They note that because of cooking, present day hunter-gatherers may utilize vegetables more extensively than early hominids. Importance of vegetable component of diet in the hunter-gatherer may have been over-estimated relative to meat. However, in this discussion consideration should be given also to taste faculties and ability to distinguish nutritious substances, and reject poisonous (often bitter) materials, and also of the capacities to learn to avoid foods after only one experience of deleterious effects (Garcia and Hankins 1975; Pfaffmann 1977). Clearly pongids survive and flourish on this basis, and there seems little reason to suppose hominids could not do likewise, though the added advantage of fire and cooking is evident.

This hypothesis on the effect of the discovery of fire has been given a wholly new level of physiological credibility and range of implication by the experimental work and deductions of Richter (1977). Richter has made a detailed study of the 24-h biological clock of a primate, the squirrel monkey. These data, together with those which he has gathered on the 24-h clock in other mammals such as rats and hamsters, he feels are sufficient to be able to predict with some degree of assurance the sequences and interrelations of events before and after the discovery of fire by man. They bring out the great effects the use of fire must have had on the intellectual and cultural evolution of *Homo erectus*. The chief tools for a study of the 24-h clock of the squirrel monkeys, as well as for rats, were gross bodily activity and eating and drinking times. These he called the behavioural hands of the clock, that tell us the clock is running and at what rate.

Fig. 4-6 shows a 40-day excerpt of the activity distribution record from a normal squirrel monkey. The record extends from noon to noon and Richter kept the laboratory completely dark from 6 p.m. to 6 a.m. and well-lit from 6 a.m. to 6 p.m. Black bars and lines show activity. This monkey was continually active during the 12-h light period and in deep sleep in the 12-h dark period. The onset of the daily active period was entrained to the start of the light period and the animal was thus strictly a light-active animal. When the entraining effects of light were eliminated by keeping the monkeys in constant darkness or by blinding (cutting optic nerves) there was direct evidence of existence of the 24-h clock. Fig. 4-7 shows that before blinding the experimental monkey became active each day, as in Fig. 4-6, at the start of the light period. After blinding, onset of its daily active period occurred 11 min earlier each day. Its clock thus had a period of 23 h 49 min. Monkey B on the right of the figure has onset of active phase a few minutes later each day. Richter observes that, once freed from the entraining effects of light, clocks of blinded animals function with a very constant length of period, either a little longer or shorter than 24 h. The clocks are remarkably accurate as confirmed by 6-month observations. In effect the clock also measures 12-h periods with the same order of accuracy.

A diversity of experimental procedures ranging from environmental events to endocrine ablations and production of stupor by phenobarbital had essentially no influence on the clock. Nor did the production of severe convulsions by electroshock. Fig. 4-8 shows that under a regime of 16 h of light and 8 h of darkness the monkeys tended to sleep well into the light period by 2–3 h. This was presumably because of the clock's tendency to keep the length of active and inactive phases the same, namely 12 h. Richter notes it is possible that had these animals been in the temperate zone during long days and short nights of summer and exposed to daytime enemies they would have laid themselves open to attack while sleeping on into daylight for 2–3 h. If there were no enemy it would not have made any difference whether the monkeys overslept or failed to retreat in time, and as a consequence they might have migrated well out of the tropics. He notes, however, that the distribution of the squirrel monkey shows a strict limitation to tropical areas. It can also be noted that the three large apes—chimpanzee, gorilla and orang-utan—are similarly limited to the tropics.

He postulates that australopithecines and hominids throughout the millions of years of their existence likewise must have had a 24-h clock that made them retire at dusk and awaken 12 h later at dawn. On this basis he suggests that early man, *Homo erectus*, could not leave the tropics until a definite change in his 24-h clock made this possible. The proposed situation contrasts with modern man who does not show the sharp onsets and terminations of active and inactive phases. Man does not retire and fall asleep at or soon after dusk not to awaken until 12 h later. He falls asleep more or less at will, not forced to by his clock. Not all animals studied possessed this remarkable 24-h clock of the squirrel monkey. Some, for example the gerbil, have very unstable clocks.

Considering the use of fire—ranging from ready-made fire caused by spontaneous combustion, lightning, volcanoes and other causes, to its direct production—Richter notes that light at night was undoubtedly fire's greatest contribution to the evolution of early man. It changed the functioning of his 24-h clock and he now had the opportunity to

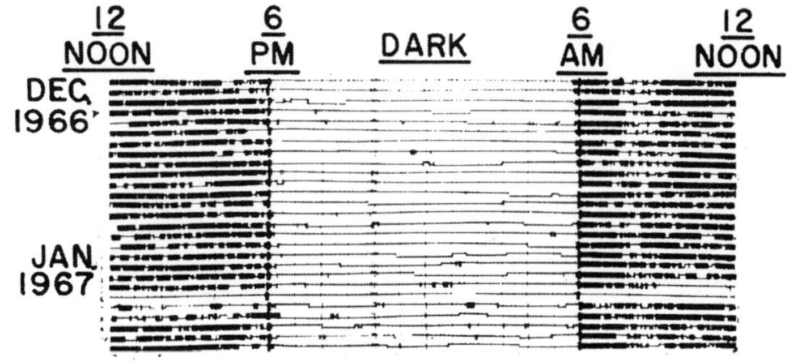

Fig. 4-6. Activity distribution chart of normal squirrel monkey kept in alternating periods of 12-h light and 12-h darkness. (Figure kindly provided by Dr Curt Richter.)

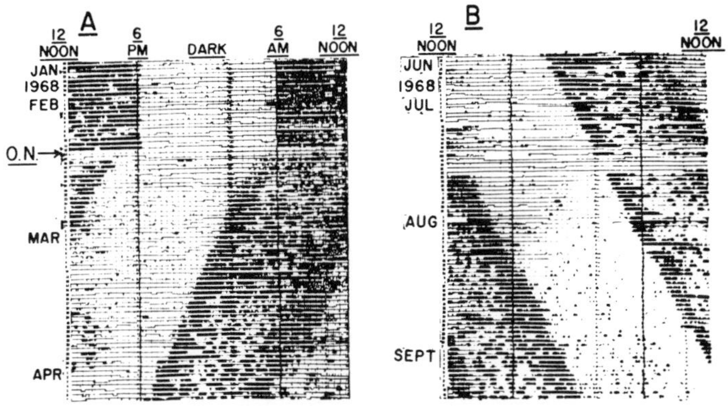

Fig. 4-7. Activity distribution chart of normal squirrel monkey (A) before and after section of the optic nerves (O.N.), and (B) a record of another monkey subsequent to cutting nerves. (Figure kindly provided by Dr Curt Richter.)

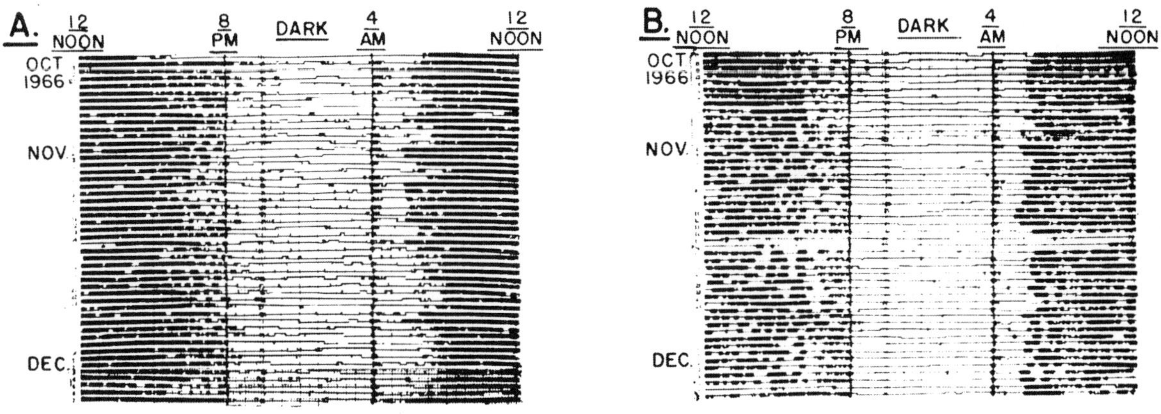

Fig. 4-8. Activity distribution charts of squirrel monkeys kept under a regime of 16 h of light and 8 h of complete darkness. (Figure kindly provided by Dr Curt Richter.)

stay awake to make use of the extra hours of night. His 24-h clock tended to put him out of action soon after dusk, but individuals with the least rigid clocks had a better chance of staying awake longer than others and so had better chances of development. These extra hours of light and the techniques for achieving them could be used for cultural and intellectual development. The hearth fire became a gathering place of all members of the family. It offered opportunity for communication and development of language and planning hunts. It gave opportunity to develop tools in quiet, undisturbed circumstances, and for observing the effects of fire on wood and stone and food. It was a stimulus to the imagination.

After *Homo erectus* became independent of the alternating 12-h shifts of activity and inactivity and had full use of fire, he was, according to this prediction, ready to leave the tropics for temperate zones where he could function independently of the seasonal differences in lengths of day and night. Early man migrated to Europe, Asia and China, making full use of the warmth of the fire and using it for cooking. About three-quarters of a million years ago he reached Escale in southern France, and China by four hundred thousand years ago—even perhaps much earlier. Richter suggests there is evidence from experiments involving isolation and constant light to suggest the 24-h clock still remains in humans. Considerable data from psychiatric observations, and from medical conditions affecting the central nervous system such as blows on the head, encephalitis, or brain tumour point to the same conclusion.

References

Ardrey R (1976) The hunting hypothesis. Collins, London

Blainey G (1975) Triumph of the nomads—A history of ancient Australia. Macmillan, Adelaide

Bonnefille R, Brown FH, Chavaillon J, Coppens Y, Haesaerts P, de Heinzelin J, Howell FC (1973) Situation stratigraphique des localités à hominides des gisements Plio-Pleistocenes de L'Omo en Ethiopie. CR Acad Sci [D] (Paris) 276: 2781

Brown CJ, Thrasher TN, Keil LC, Ramsay DJ (1980) Chronic cerebroventricular infusions of angiotensin II stimulate excessive drinking but not salt appetite in dogs. Neurosci Abstr 6: 529

Chernigovsky B (1963) Neurophysiology of visceral afferent systems and the behaviour of animals towards food (feeding behaviour). Nova Acta Leopoldina 28: 317

Clark JD (1964) In: Howell FC, Bourlière F (eds) African ecology and human evolution. Methuen, London, p 635

Clark JD (1970) The pre-history of Africa. Praeger, New York, Washington

Coman BJ (1973) The diet of red foxes (*Vulpes vulpes* (L.)) in Victoria. Aust J Zool 21: 391

Coman BJ, Stevens PL, Stark RW (1973) An unusual food item in the diet of red foxes (*Vulpes vulpes* (L.)). Victorian Naturalist 90: 42

Coppens MY (1963) Les restes d'hominides des formations Plio-Villafranchiennes de L'Omo en Ethiopie. CR Acad Sci [D] (Paris) 276: 1823

Coppens MY (1970) Localization dans le temps et dans l'éspace des restes d'hominides des formations Plio-Pleistocenes de L'Omo (Ethiopie). CR Acad Sci [D] (Paris) 271: 1968

Coppens MY (1975a) La grande aventure paleontologique est-africaine. Le Courrier du CNRS, April: 30

Coppens MY (1975b) Evolution des Mammifères de leurs fréquences et de leurs associations, au cours du Plio-Pleistocene dans la basse vallée de L'Omo en Ethiopie. CR Acad Sci [D] (Paris) 281: 1571

Crook JH, Aldrich-Blake P (1968) Ecological and behavioural contrasts between sympatric ground dwelling primates in Ethiopia. Folia Primatol 8: 192

Dart RA (1963) The carnivorous propensity of baboons. Symp Zool Soc Lond 10: 49

Dene HT, Goodman M, Prychodko W (1976) Immunodiffusion evidence on the phylogeny of the primates. In: Goodman M, Tashian RE (eds) Molecular anthropology. Plenum, New York, p 173

de Vore I, Washburn SL (1964) Baboon ecology and human evolution. In: Howell FC, Bourlière F (eds) African ecology and human evolution. Methuen, London, p 335

Drake RE, Curtis GH, Cerling TE, Cerling BW, Hampel J (1980) KBS Tuff dating and geochronology of tuffaceous sediments in the Koobi Fora and Shungura formations, East Africa. Nature 283: 368

Fitch FJ, Miller JA (1970) Radioisotopic age determinations of Lake Rudolf artifact site. Nature 226: 226

Fregly MJ (1980) On the spontaneous NaCl intake by dogs. In: Kare M, Bernard R, Fregly MJ (eds) Biological and behavioural aspects of NaCl intake. Academic Press, New York, p 55

Garcia J, Hankins WG (1975) The evolution of bitter and the acquisition of toxiphobia. In: Denton DA, Coghlan JP (eds) Proceedings of the Fifth International Symposium on Olfaction and Taste. Academic Press, New York, p 39

Goodall J (1963) Feeding behaviour of wild chimpanzees. A preliminary report. Symp Zool Soc Lond 10: 39

Goodall J (1964) Tool using and aim throwing in a community of free-living chimpanzees. Nature 201: 1264

Goodall J van Lawick (1971) In the shadow of man. Collins, London

Goodman M (1964) Man's place in the phylogeny of the primates as reflected in serum proteins. In: Washburn SL (ed) Classification and human evolution. Methuen, London

Goodman M (1973) The chronicle of primate phylogeny contained in proteins. Symp Zool Soc Lond 33: 339

Goodman M, Tashian RE (1976) Molecular anthropology. Plenum Press, New York

Holloway RL (Jr) (1967) Tools and teeth: Some speculations regarding canine reduction. Am Anthrop 69: 63

Howell FC, Coppens Y (1973) Deciduous teeth of hominidae from the Pliocene/Pleistocene of the lower Omo Basin, Ethiopia. J Hum Evol 2: 461

Isaac G (1971) The diet of early man. World Archaeol, Feb: 278

Johanson CD, Coppens Y (1976) A preliminary anatomical diagnosis of the first plio-pleistocene hominid discoveries in the central Afar, Ethiopia. Am J Phys Anthropol 35: 217

Johanson CD, White TO (1979) A systematic assessment of early African hominids. Science 203; 321

Jolly CJ (1970) The seed eaters: A new model of hominid differentiation based on a baboon analogy. Man 5: 5

Krantz GS (1963) The functional significance of the mastoid process in man. Am J Phys Anthropol 21: 591

Kretzoi M (1975) New ramapithecines and *Pliopithecus* from the Lower Pliocene of Rudabánya in north-eastern Hungary. Nature 257: 578

Lasker GW (1976) What is molecular anthropology? In: Goodman M, Tashian RE (eds) Molecular anthropology. Plenum Press, New York, p 3

Leakey LSB (1959a) A new fossil skull from Olduvai. Nature 184: 491

Leakey LSB (1959b) The first men: Recent discovery in East Africa. Antiquity 33: 285

Leakey LSB (1964) Very early East African hominidae and their ecological setting. In: Howell FC, Bourlière F (eds) African ecology and human evolution. Methuen, London, p 587

Leakey LSB (1968) Bone smashing by Late Miocene Hominidae. Nature 218: 528

Leakey M (1970) Early artifacts from the Koobi Fora area. Nature 226: 228

Leakey MD, Hay RL, Curtis GH, Drake RE, Jackes MK, White TD (1976) Fossil hominids from the Laetotil bed. Nature 262: 460

Leakey REF (1970) New hominid remains and early artifacts from northern Kenya. Nature 226: 223

Leopold AC, Ardrey R (1972) Toxic substances in plants and the food habits of early man. Science 176: 513

Lewin R (1976) Ancient hunting in an ecological perspective. New Sci 70: 298

Lewis OJ (1973) The hominid os capitatum, with special reference to the fossil bones from Sterkfontein and Olduvai Gorge. J Hum Evol 2: 1

MacIntosh D (1963) Food of the fox in the Canberra district. CSIRO Wildl Res 8: 1

Martens Z (1971) Observations on the food of the fox *Vulpes*

vulpes (L.) in an arid environment. CSIRO Wildl Res 16: 73
McCance RA (1936) Medical problems in mineral metabolism. Lancet I: 643
Monod T (1964) In: Howell FC, Bourlière F (eds) African ecology and human evolution. Methuen, London, p 587
Napier JR (1964) The evolution of bipedal walking in the hominids. Arch Biol (Liege) 75 [Suppl]: 673
Oakley KP (1954) Skill as a human possession. In: Singer C (ed) History of technology, vol. 1. Oxford University Press, London
Oakley K (1959) Commentary on the first men: Recent discovery in East Africa of LSB Leakey. Antiquity 33: 285
Pfaffman C (1977) Biological and behavioural substrates of the sweet tooth. In: Weiffenbach JM (ed) Taste and development. US Dept Health Education and Welfare, Bethesda (DHEW Publication NIH 77-1068), p 3
Pocock RI (1925) External characters of catarrhine monkeys and apes. Proc Zool Soc Lond 2: 1479
Ramsay DJ, Reid IA (1979) Salt appetite in dogs. Fed Proc 38: 3906
Richter CP (1977) Discovery of fire by man—its effects on his 24 hour clock and intellectual and cultural evolution. Johns Hopkins Med J 141: 47
Robinson JT (1962) The origins and adaptive radiation of the Australopithecines. In: Kurth J (ed) Evolution and hominization. Gustav Fischer, Stuttgart
Robinson JT (1964) Adaptive radiation in the Australopithecine and the origin of man. In: Howell FC, Bourlière F (eds) African ecology and human evolution. Methuen, London
Sarich VM, Wilson AC (1967) Immunological time scale for hominid evolution. Science 158: 1200
Schaller G (1961) The orang-utan in Sarawak. Zoologica, New York Zool Soc 46: 2
Schaller GB (1963) The mountain gorilla—ecology and behaviour. University of Chicago Press, Chicago
Simons EL (1977) *Ramapithecus*. Sci Am 236: 28
Simons EL, Pilbeam D (1965) Preliminary revision of the Dryopithecinae. Folia Primatol (Basel) 3: 81
Strum SC (1975) Life with the 'Pumphouse Gang'. Natn Geogr Mag 147: 672
Taieb M, Johanson CD, Coppens Y (1975) Expedition internationale de L'Afar, Ethiopie (3e campagne 1974); Decouverte de hominides Plio-Pleistocenes à Hadar. CR Acad Sci [D] (Paris) 281: 1297
Teleki G (1973) The omnivorous chimpanzee. Sci Am 228: 33
Tiger L (1969) Men in groups. Nelson, London
Tobias PB (1970) Cranial capacity in fossil hominidae. Lecture, American Museum of Natural History. Cited in Jolly CJ (1970) Man 5: 5
Washburn SL (1964) Behaviour and human evolution. In: Washburn SL (ed) Classification and human evolution. Methuen, London, p 190
Washburn SL (1971) The study of human evolution. In: Dolhinow P and Sarich V (eds) Background for man. Little Brown, Boston, p 82
Washburn SL, Moore R (1974) Ape into man: A study of human evolution. Little Brown, Boston
Wilson AC, Sarich VM (1969) A molecular time scale for human evolution. Proc Natl Acad Sci USA 63: 1088
Yerkes RM, Yerkes AW (1929) The great apes. New Haven

5 Salt in history: Symbolic, social and physiological aspects

This is a vast subject which has provoked the writing of many books during the past century. Here I will touch only on some selected aspects reflecting the human preoccupation with salt since early times. In some instances, it is suggestive that the motivation of behaviour may have a physiological basis, but in others, custom and superstition seem predominant. Overall, from ancient times the history is one of great liking and of effort to obtain salt. It rises, in some instances, to a craving. But some important anthropological records of distaste for salt are recounted. The sources of salt and the influence on history is a fascinating field also, but again we can only mention the subject here. The most recent major review is Multhauf's *Neptune's Gift* (1978).

Symbolism

In relation to the symbolic significance of salt over the course of recorded history, Ernest Jones (1951) conceives it in the following terms. Salt is pure, white, immaculate and incorruptible. It seems irreducible into further constituents and is indispensable to living creatures. Correspondingly it has been regarded as the essence of things in general, the quintessence of life, and the very soul of the body. It has been invested with the highest general significance, more than any other article of diet. It was the equivalent of money in other forms of wealth. Its presence was indispensable for undertaking any enterprise, particularly a new one. In religion, it was one of the most sacred objects. Magical powers were ascribed to it. Pungent, stimulating flavour of salt was applied metaphorically to a telling wit or discourse. This contributed to the concept of it as an essential element. To be without salt was to be insipid or to have something essential lacking. Its durability and its immunity against decay made it an emblem of immortality. It was believed to be important in favouring fertility and fecundity and in preventing barrenness. This idea is connected with other attributes. The word 'salacious' as well as 'salary' comes from this source. The idea still survives in the joke that the cook who has put too much salt into the soup must be in love. Shakespeare uses salt in the same sense in the *Merry Wives of Windsor* 'though we are justices . . . we have some salt of our youth in us'.

Jones suggests that the permanence of salt helped create the idea that for one person to partake of the salt of another formed a bond of lasting friendship and loyalty between the two. The substance played an important part in the rites of hospitality, in the confirming of oaths, ratifying of compacts and sealing of solemn covenants. Conception of a bond was related to the capacity salt has for combining with the second substance and imparting to this its peculiar properties. These include the power to preserve against decay. For one important substance—water—it had a natural and curious affinity.

The idea of salt as the essence of things, particularly of life itself, underlies the biblical phrase 'ye are the salt of the earth'. In ancient Egypt, salt and the burning candle represented life and were placed over a dead body to express the ardent desire of prolonging the life of the deceased.

History

Mediterranean

In Plutarch's *Moralia VIII: Table Talk IV*, Symmachus discourses:

> If a relish [*opson*] is something that makes a dish palatable, then the best relish would be the one that does most to attract our appetite . . . first there is salt without which practically nothing is eatable. Salt is added even to bread and enriches its flavour: this explains why Poseidon shares a temple with Demeter. Salt is also the best relish to season other relishes . . . beyond that salty food aids digestion of any other: it makes any food tender and more susceptible to concoction: the salt contributes at once the savour of a relish and the good effect of a medicine. Moreover, the other delicacies of the sea, in addition to being most gratifying to the taste, are also the safest to eat: for they have the character of meat without its heaviness, and are easily digested and assimilated.

Diogenes Laertius says: 'Of salt he said it should be brought to table to remind us of what is right: for salt preserves whatever it finds, and it arises from the purest sources, sun and sea. This is what Alexander says that he found in the Pythagorean memoirs.'

Pliny in his *Natural History* (Book XXXI) writes on salt sources in the ancient world:

Salt is dried out of the Tarentine lake by summer sun when the whole pool turns into salt. Also in Sicily from a lake called Cocanicus and from another near Gela . . . all salt from pools is fine powder and not in blocks. Another kind is produced from sea water spontaneously as foam left on the edge of the shore and on rocks. All this is condensation from drift, and that found on rocks has the sharper taste. There are also three kinds of native salt: for in Bactra are two vast lakes which exude salt, while in Memphis salt is taken out of the lake and dried in the sun. But the surface too of rivers may condense into salt, the rest of the stream flowing as it were under ice, as near the Caspian Gates are what are called 'rivers of salt', and also around the Mardy and the Armenians. There are mountains of natural salt such as Oromenus in India where it is cut out like blocks of stone from a quarry and even replaces itself bringing greater revenues to the rajas than those from gold and pearls. At Gerra, a town of Arabia, the walls and houses are made of blocks of salt cemented with water . . . In Hither Spain too, at Egelesta, salt is cut into almost transparent blocks: to this for some time past most physicians have given the first place among all kinds of salt. Every region in which salt is found is barren and nothing will grow there. In Chaonia, there is a spring from which they boil water and on cooling it obtain a salt that is insipid and not white. In the provinces of Gaul and Germany they pour salt water on burning logs. The wood used also makes a difference. The best is oak, for its pure ash by itself has the properties of salt: in some places hazel finds favour. Sheep, cattle and draught animals are encouraged to pasture in particular by salt: supply of milk is much more copious and there is even a far more pleasing quality in the cheese. Therefore, heaven knows, a civilized life is impossible without salt and so necessary is this basic substance that its name is applied metaphorically even to intense mental pleasures. We call them '*sales*' (wit): all the humour of life, its supreme joyousness, and relaxation after toil, are expressed by this word more than any other. It has a place in magistracies also and on service abroad from which comes the term salary—'salt money': it had great importance among the men of old as is clear from the Salarian Way, since by it, according to agreements, salt was imported to the Sabines.

The importance of salt in the Middle East is exemplified by the quotation 'Ought ye not to know that the Lord God of Israel gave the kingdom over Israel to David forever, even to him and to his sons by a covenant of salt?' (II Chronicles xiii, 5). This use of salt as an emblem of perpetuity or in the eyes of others as a token of confirmation is widespread. As Manley (1884) points out, engagements among eastern tribes are ratified with salt and oaths taken on salt are considered the most binding. The Hindus swear by their salt, and during the great Indian mutiny the Sepoys were often held in restraint by being reminded they had sworn by their salt to serve faithfully the Queen of England. Arabs offer salt as a sign that their guest is safe in their hands and even the Bedouin robber will not violate the laws of hospitality to one who has once tasted of his salt. As an illustration of the strength of this bond, Price in his *Mahommedan History* gives the following incident:

Yaakoob, the son of Eb-Leys Es-Suffar, having adopted a predatory life excavated a passage one night into the palace of Dirhem, the Governor of Seestan. After he had made up a convenient bale of gold and jewels, and the most costly stuffs he was proceeding to carry it off when he happened in the dark to strike his foot against something hard on the floor. Thinking it might be a jewel of some sort or other he picked it up and put it to his tongue. To his equal mortification and astonishment he found it to be a piece of rock salt. Having thus tasted the salt of the owner his avarice gave way to his respect for laws of hospitality, and throwing down his precious booty left it behind him and withdrew empty-handed to his habitation. The treasurer of Dirhem, reappearing on the next day according to custom to inspect his charge was equally surprised and alarmed. He observed that a great part of the treasure had been removed. But on examining the packages that lay on the floor his astonishment was not less to find that not a single article had been taken away. He reported it immediately to his master. The latter caused it to be proclaimed through the city to the effect that the author of this proceeding had his free pardon. He further announced that if he repaired to the palace he would be distinguished by the most encouraging marks of favour. It is further recounted that Yaakoob availed himself of this invitation relying upon the promise. It was fulfilled to him and from this period he gradually rose in power until he became the founder of a dynasty.

From the studies in ancient technology, it seems that salt monopolies were commonly found in most of the Hellenistic states. In Egypt the salt was extracted in the Nile delta from sea water and also from lakes near Memphis. Every inhabitant of Egypt (except a few privileged groups) paid the salt tax, as well as paying for the salt consumed. Women paid less than men, slaves paid half the tariff, children were exempt. Freedom from this tax was sometimes allotted to prominent officials and victors in the athletic games, but even priests had to pay it. The money from this hated tax was spent in the province where it was levied. Salt production was organized on similar lines in Hellenistic Palestine and the salt tax there, the *halike*, was equally unpopular. As in other Hellenistic countries the salt pans were private possessions of the king and an important source of revenue for the crown.

In relation to influence on animals, Aristotle (*Historia animalium*: viii, 10) states sheep are much improved in condition by drinking, and accordingly the flocks are given salt every five days in summer, to the extent of 1 *medimnus* (about 10 gallons) to the 100 sheep, and this is found to render a flock healthier and fatter. In fact salt is mixed with the greater part of their food; a large amount of salt is mixed into their bran for the reason that they drink more when thirsty, and in autumn they get cucumbers with a sprinkling of salt on them; this mixture of salt in their food tends also to increase the quantity of milk in the ewes. If sheep be kept on the move at midday they will drink more copiously towards evening: and if the ewes be fed with salted food as the lambing season draws near they will get larger udders. Sheep are fattened by twigs of the olive or of the oleaster, by vetch, and bran of every kind: and these articles of food fatten all the more if they be first sprinkled with brine.

In the *Georgics* (3.394–397), Virgil says 'that the man who is out for milk should bring plenty of clover, trefoil and salted grass to the goat pens, and do it in person. A thirst for water is wetted thus, and their udders swell, and they will impart to the milk a faint sub-flavour of salt.'

Salt finds its way into poetic mythology, as in a conversation at Papkos in Robert Graves' *White Goddess* where Theophilus and Paulus, a Roman governor-general of Cyprus, discuss the Vestal Virgins of King Tarquin and the use of spring water mixed with powdered and purified brine in their sacrifices. The leaping priests of Mamurius or Pallas performed their orgiastic dance for hour after hour around the blazing bonfires and must have sweated terribly and come to fainting. As in harvest fields where countrymen always refresh themselves in brine-water in preference to plain, brine fetched by the Vestals at the midsummer orgy must have restored the vigour of the shepherds like a charm!

The Bible is replete with references to salt, as in Leviticus (ii, 13): 'And every oblation of thy meat offering shalt thou season with salt; neither shall thou suffer the salt of the covenant of thy God to be lacking from thy meat offering: with all thine offerings thou shall offer salt.'

Tacitus (*Annals*, Book XIII, Chapter 57) records the importance of salt to inland races:

> in the course of the same summer a battle was fought with great rage and slaughter between the Hermundurians and the Cattians. The exclusive property of a river which flowed between both nations impregnating the stalls of salt was the cause of their mutual animosity. To the natural fierceness of barbarians who know no decision but that of the sword, they added the gloomy motives of superstition. According to the creed of these savage nations, that part of the world lay in the vicinity of the heaven and thence the prayers of man were wafted to the ears of the gods. The whole region was, by consequence, peculiarly favoured: and to that circumstance it was to be ascribed that the river and the adjacent woods teemed with quantities of salt, not, as in other places, a concretion on the seashore formed by the foaming of waves but produced by the simple act of throwing the water from the stream on a pile of burning wood where, by the conflicting elements, the substance was engendered. For this salt a bloody battle was fought. Victory declared in favour of the Hermundurians and the event was the more destructive to the Cattians as both armies with their usual ferocity had devoured the vanquished as a sacrifice to Mars and Mercury.

Grimm refers to the fact that the Burgundians and Allamens also fought for salt springs.

China

In the *Encyclopaedia Sinica* (Gouling 1973) it is recorded that taxation of salt in China dates from about 2200 B.C., tribute salt being introduced during the reign of Yu, first emperor of the Hsia dynasty. The Ch'i kingdom (seventh century B.C.), in which the manufacture of salt was encouraged under government control, supplied salt from Shantung to all the neighbouring inland kingdoms: under the administration of Kuan Tzu it derived a very large revenue from salt taxes and was reputed to be 'the richest kingdom in the world'. During the reign of the emperor Wu Ti (Han dynasty) special officials were appointed in charge of salt taxes throughout China. At the beginning of the T'ang dynasty (seventh century A.D.) there were said to be 18 salt lakes and 640 salt wells under the control of the Board of Revenue. Taxes were also levied on sea salt. During the Sung dynasty a system of transportation of salt by merchants was introduced. Permits called *yen* came from the central government on payment of tax, and the country was divided into *yen* areas each of which established a merchants' monopoly. Towards the end of the Ming dynasty (early seventeenth century) the salt administration decayed and revenue fell. In 1910 a central salt office was established at Peking under the control of the minister of finance. The Chinese Government Reorganization Loan Agreement in 1913 involved the Chinese government, with the assistance of foreigners, taking immediate steps towards the reorganization of the system of collection of the salt revenues of China.

Marco Polo noted the great importance of salt in Central Asia, and its value in the thirteenth century. Because of the distance from the sea, salt was obtained from springs. Cakes of it were officially stamped with the Great Khan's head and used as currency. Very large revenue was obtained from salt. Parsons (1951) notes that even recently salt money could be found in Africa, Borneo, China and other parts of the East. In Yunnan the custom is said to have been retained, and in Ethiopia (Abysinnia) salt bars were legal tender as recently as 1933. Multhauf (1978) suggests from his reading of Chinese literature that by the eighteenth century salt merchants operating under the government monopoly held the largest aggregate wealth of any commercial or industrial group in the empire, and were responsible to a degree for the flourishing culture along the lower Yangtse River in that era.

In terms of obtaining salt, the Chinese were a millennium ahead of the West, and their methods were sophisticated in antiquity. Apart from solar evaporation, drilling for salt was practised in the province of Szechwan. Records show drilling of wells in 67 B.C. and during Han times (206 B.C.–220 A.D.) natural gas was used for evaporating the brine. The wells were drilled with derricks and the brine hauled up with the aid of a windlass if it did not itself flow. It was poured in a hopper and led through bamboo pipes to evaporation pans. An oven held five such pans and was heated by natural gas, also brought by bamboo pipes from wells. The reliefs on Han tombs depict this industry. A comprehensive account can be found in Li Ch'iao-Ping's writing (1948) on the chemical arts of old China.

Europe

The rock salt mines of the eastern Alps—at Salzburg, Hallstatt and Durrenberg—played an important role in the culture of prehistoric Europe. They appeared to be worked from an early period on a scale that far exceeded local needs. Galleries were driven into the mountains for as far as 350 m. Stones for grinding the picks used in the process have been found in the galleries, as have torches, which were stuck in staves about a metre long. Some timber work was used to prevent the galleries from caving in and some bronze carpenter's tools have been found too. The lumps of rock salt produced were broken with the help of wedges and hammers. The tools used suggest that the mines were operated during the later stages of the Bronze Age (from 1400 B.C. onwards), and may have been worked earlier.

Transylvania is richer in rock salt than any other portion of Europe. The area consists of a central basin, that of the Maros River, and the basins of the upper courses of the Szamos and Olt rivers. The whole territory is more or less mountainous and the deposits of rock salt are frequently found along the banks of the small rivers amongst the hills. The supply of salt is almost inexhaustible. There are centres of salt mining on the Maros River which are favourably situated for water communication.

Schleiden (1875) remarks that the great importance attached to salt is seen from the fact that there was hardly a place that produced salt where this fact was not expressed in its name: from the Indian Lavanapura and the Austrian Salzburg (both meaning salt town) to the Prussian Salzkotten and the Scottish Saltcoats. Bloch (1963) notes that early settlements grew up around salty springs which hunting tribes located by following animals to them. It was difficult to concentrate salt from these dilute sources but the hunters kept wood fires going and boiled away the water. These salt-boiling civilizations go back to Neolithic times. There were centres in the Tyrol region of the Alps, the Moselle and the Franche-Comte areas in France, the Saale and Lüneburg areas in Germany and Droitwich in England. Whole forests were burned up in this industry and the salt-makers had to haul their wood from farther and farther away, often on river rafts. Before the Norman conquest of England, salt pans of Cheshire and Worcestershire supplied not only Britain but northern Gaul. The trade route went south to the Thames which was forded at Westminster, where the river was shallower and broader. This trade had much to do with the establishment of the village which became London. Bloch also describes the practice, prevalent in Europe from the ninth to the sixteenth century, of making salt from peat. The peat would be soaked in sea water, dried, burnt and the salt extracted from the ashes with sea water. The extract was filtered and evaporated in cauldrons over peat fires. Over the period from Neolithic times until about 400 years ago when the peat gave out in many regions, millions of tons were harvested for salt-making.

In Poland the famous Wieliczka salt mine was a major source of wealth. Parsons (1951) records that the early workings there were made without adequate shoring and entire streets of the old town subsided with heavy loss of life. They were worked by convict labour and were infamous throughout Europe for hundreds of years. The underground vaults and chambers now extend over more than 31 km^2 (12 square miles), each a maximum depth of 3700 m (12 000 feet), and include sights which are unique. There is a vast ballroom, with pillars, statues and huge hanging candelabra all made from crystal salt where festivities were held up until the period between the world wars. In addition there are great underground lakes, and a sumptuous chapel where services were held.

In the United States the pre-Columbian salt river settlement of Arizona had a salt mine at Camp Verde. Mesopotamian cities appear to have transported salt up the Tigris and Euphrates rivers from centres near the river mouth where it was evaporated. Jericho, which dates back to 8000 B.C. and is one of the oldest known agricultural communities, was near the Dead Sea and the salt mountain of Mount Sodom.

Multhauf (1978) describes an elaborate history of salt trade by Venice. In the sixth century salt was the chief article of commerce and was produced from the lagoons around the town. In 1184 Venice asserted control over the export of Chioggia salt and entered agreements to supply it to various Italian cities. By the fourteenth century she was supplying salt from as far afield as Alexandria, Cyprus and the Balearic Islands as part of her trade agreements. Later, other Italian cities such as Genoa and Pisa entered the trade.

An important era in the history of European salt production began in the fourteenth century with the development of the salted herring trade, the process being attributed to a Dutchman. Fish had been preserved in ancient times by dry salting, but the large-scale preservation of fresh fish by salting allowed its shipment in barrels to inland places for consumption. Salt counteracts putrefaction by giving the tissue fluid such a high osmolarity that bacteria cannot grow in it. Multhauf quotes Schleiden as recording that when the trade was near its peak it amounted to three billion herring requiring 123 million kilograms of salt a year, and caused an enormous increase in demand from the North Sea herring fisheries. The salt was obtained from the nearest sources: Lüneburg, England and the Low Countries. For many centuries the salt made in England was considered inferior, and soon after its constitution the Royal Society of London directed its attention to improving the art of manufacturing white salt (Manley 1884).

Josephson (1971) notes that at the beginning of this century salt pork and salt herring were the two main sources of protein for the majority of the population of Scandinavia. According to Alwall (1958) the average consumption of salt in Sweden during the middle of the sixteenth century could have been about 100 g/day. Meneely (1954) suggests that widespread use of salted foods may well have played an important role in dissociating the human taste for salt from the need for it. He also notes that it is not possible to trace a common word for salt farther back than primitive Greek; there is no word for it in ancient Sanskrit, the oldest of the Indo-European languages. He takes this as oblique evidence that primitive man may have developed his craving for salt concurrently with progress from a nomadic to an agricultural existence, and that soon thereafter the preservative properties of salt were observed and exploited.

In the Highlands of Scotland it was a common custom to place on the breast of the deceased a wooden platter containing a small quantity of salt and earth, separate and unmixed. The earth was an emblem of the corruptible body, the salt an emblem of the immortal spirit. The expressions 'above the salt' and 'below the salt' are well known. These arose because it was formerly the custom in great families—in France as well as in England and Scotland—to place a large piece of plate called the salt vat in the middle of the table, the guests being seated above or below the vessel according to their several ranks. Above was the mark of a gentleman while below indicated a humble station in society. This distinction also extended to the fare, the wine frequently circulating only above the salt vat or cellar, and the dishes below it being of a coarser kind than those near the head of the table. As Manley points out, the custom is one which must be considered as debasing salt from its high symbolism of hospitality to being used as a means of making invidious distinctions.

Africa

One of the most famous aspects of the transport of salt has been the great caravan which twice each year carried salt from the Taoudeni salt swamp in the Sahara Desert to Timbuktu in Mali. To maintain the celebrated market at Timbuktu a caravan of 2000 camels would leave the town on a 720 km journey to the swamp for the purpose of bringing back about 300 tons of salt. It was an important factor in the slave trade in Africa, for in the interior, children were sold into slavery in return for salt. The salt from Timbuktu was distributed southwards to only a limited extent, and in vast areas of equatorial Africa the source of salt, predominantly potassium, was from plant ashes. Over other areas to the east and in the south of Africa it was obtained from consumption of animal blood or of cattle urine. Jarcho (1958), writing on the salt of the Sahara, notes the records of salt caravans in the writing of Herodotus and describes a report by Cadamosto in the fifteenth century. Cadamosto was a Venetian who made two voyages to Africa in Portuguese ships during 1454–1456 or thereabouts. A full translation of his work has been published by the Hakluyt Society in London. He describes how the caravans from Timbuktu go to Melli (on the Niger River), the empire of the blacks, where so rapidly is the salt sold that it is disposed of within 8 days of its arrival for 200 or 300 duckets a load. The country of Melli is very hot and the pasturage cannot support any four-footed animals, so only about 25 out of 100 camels return. Many Arabs also sicken and die on account of the great heat. Cadamosto enquired as to what the merchants of Melli did with this salt. He was told that the equatorial climate, where the day is constantly about as long as the night and where it is extremely hot at certain seasons of the year, causes the blood to putrefy, so that were it not for this salt they would die. The remedy is to take a small piece of the salt, mix it in a jar with a little water, and drink it every day. The remainder of the salt they take with them on their long journeys, in pieces as large as a man can with a certain knack bear on his head.

In relation to the indigenous salt sources of the Sahara there are west, central and eastern groups. Taoudeni is in the central group, and its salt caravans—*les azalai*—are in the realms of both history and legend. The route from Taoudeni to Timbuktu via Araouan is the 'axis of symmetry' of the salt routes in the Sahara (Fig. 5.1). The wealth

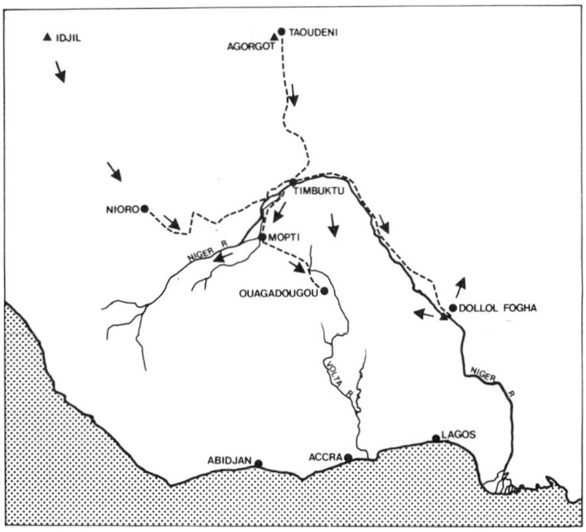

Fig. 5-1. North Africa: some main routes of distribution of salt from Taoudeni via Timbuktu, and also from Idjil. (After Pales 1950.)

of Timbuktu was founded upon salt, and the city renowned throughout Europe as a market for bartering this commodity. As Carver (1946) recounts, at the height of its glory Timbuktu was the resort of wealthy and learned classes, many of the houses boasted large libraries and education reached a high standard. Salt is still of outstanding importance in the economy of Africa.

Pales (1950), who was Chief of the Anthropological Mission of French West Africa, gives a comprehensive description of the salt trade of the Sahara in the first part of this century. He recalls how the grand salt routes of the Sahara have been the scene of great drama over the ages; countless browns, blacks and whites have died along them. From earliest times there were two great salt caravans a year from Araouan to the mines of Agorgot near Taoudeni (Figs. 5-2 and 5-3). The winter caravan, which left in November and entered Timbuktu at the end of December, was the more important; the summer one left in February and returned in April or May. In 1909, which was the *vrai apogée* of the old era of salt production, 25 000 camels went to Taoudeni. In 1932 the two caravans together would have comprised only 10 000 animals. The tribes of Kel Araouan made up a large proportion of the caravan but all the nomads from Gao to Timbuktu who possessed camels participated. Timbuktu was the grand market of the area. As indicated on Fig. 5-1, it was a short route to take the salt of Taoudeni from Timbuktu to the greatest river route of West Africa, the Niger, from where it went as far as Niamey to the east and Segou to the west. In the route to the west, the large river port of Mopti was the centre for exchange of all the products of the interior. If Timbuktu was the great market for salt, Mopti was the crossroads, the economic key to the route. Dastre (1901) recounts how the history of the salt sources at Taoudeni is the story of the struggles of the Moroccans and the Moors against the sultans and masters of Timbuktu. The old adage 'who has land has war' could be formulated 'who has salt has war'. These salt deposits north of Timbuktu are residues of evaporation from salt lakes. There is evidence that the deposits which feed them are from the Triassic era.

On the eastern side of Africa at the 72-km wide Karum salt lake in the Great Rift Valley, the Afar people would work in very high temperatures cutting blocks of salt out of the lake; these were taken on camel trains to the market at Makale in the Ethiopian highlands where they were traded for feeding to cattle—salt being very short on the high plateau.

Dastre (1901) has given a comprehensive and romantic account of salt in the Sahara. He remarks that men have not flinched from any effort, any sacrifice or any danger to procure this precious substance. They have sought it by war and fraud, and for it have suffered the exhaustion of long journeys. Primitive people have displayed remarkable ingenuity to procure it. For example, the natives of Sunda Islands created a kind of rudimentary chemical industry to extract it from the mangrove swamps. Mungo Park saw the inhabitants of the Sierra Leone coast give all they possessed, even their wives and children, to get it. This is also true in Central Africa. He remarks that the hunger for salt is not limited to man, and that many animals search for this substance avidly. Buffon writes: 'Nothing stimulates the appetite of the sheep more than salt.' Barral and others have

Fig. 5-2. Salt caravan of 5000 camels preparing to leave Colomb-Bechar in Central Africa. (Photograph by Harlingue-Viollet, Paris.)

Fig. 5-3. Salt blocks prepared for loading onto camels in a salt caravan to the Sahara. (Photograph by Harlingue-Viollet, Paris.)

taught that an animal suffers cruelly from the scarcity of salt but will prosper with its usual ration. In agricultural practice it is desirable to give 2–5 g of salt a day to a sheep, 30–50 g to a horse and 60–100 g to a cow. After citing a number of societies where salt is not used, which will be referred to in the next section, Dastre notes the account of Mungo Park who at the beginning of the nineteenth century explored the loop of the Niger and was struck by the avidity of the Negro agricultural population for salt. Park reported that in the interior of the country salt is the most precious gift. It is a strange sight for Europeans to see a child sucking a stick of salt as though it were sugar. When one says that someone eats salt at his meals, this indicates that he is a rich man.

France

Mollat (1968), writing on the role of salt in history, underlines its double nature: it has both economic and fiscal importance. He draws attention to Italian studies showing that it is necessary to understand the fiscal system of Venice in relation to salt in order to understand the emergence of Venetian capitalism and the way in which the first capitalists amassed their considerable fortunes. This is also true of France. The French politics of salt emerged in the fourteenth century and two symbolic events emphasize the economic and fiscal factors in the history of salt. These were the acquisition in 1291 by Philippe le Bel of the *salines* of Languedoc, especially those of Peccais, and the institution in 1341 of the *gabelle*, the tax on salt. This had a dramatic economic effect both on states that exported salt and on those that imported salt either because they were non-producers or because, like the Low Countries and northern Italy, they used salt a great deal for the preservation of fish, meat and other foods.

Stocker (1949) has written a major review on salt, particularly in the context of French history. He explains that the word *gabelle* was derived from the Latin *gablum*, which in turn came from the Hebrew *gab*, meaning tribute. After the fifteenth century, however, *gabelle* was designated especially to the salt tax which forced citizens to buy salt from the king's storehouses at a price fixed by him. Before this, salt was liable to taxes which the nobility collected through the domains by way of toll charges for crossing bridges, fording rivers, etc. The toll collector often took the tax in kind in the form of a handful of salt. Initially the *gabelle* was a temporary measure appearing, for example, in the reign of Jean II in the fourteenth century for the purposes of raising an army, and upon the king's return from captivity to ensure the payment of his ransom. In 1680 Colbert formulated the great ordinance which was the code of legislation until the Revolution. The administration and tariffs of the *gabelle* were very diverse but never extended to the whole of France. Some provinces did not submit to it as a result of agreement at the time when they were annexed; others enjoyed exemption or a lower rate because of their proximity to saline marshes or to a foreign border. Yet others paid a ransom to be exempt. In fact France was divided into six parts as regards the *gabelle*. These were:

1. Provinces of the 'grande gabelle'. They comprised the areas round the Ile de France and it was here that the fiscal exploitation was greatest. Though only comprising one-third of France they furnished two-thirds of the tax. The cost of salt was about 20 times its cost price and tax was calculated to have absorbed about an eighth of the annual income of an agricultural family. A certain quantity of salt had to be bought from a specified storehouse.

2. Areas of the 'petite gabelle'. These were the south-east provinces supplied by the salt marshes of the Midi. Tax reached five to ten times the value of salt. The salt duty did not exist and the inhabitants could buy the quantity they wished in the storehouse of their choice.

3. The ransomed areas. In these areas, including Perigord, there was a major revolt in 1548, with killing of the officials and pillaging of the salt storehouses. Following a punitive expedition of great ferocity, an indemnity was paid to Henry II but the provinces were then declared free of the *gabelle* in perpetuity.

4. Free areas. In these, which included Brittany, the kings had sought to win the fidelity and attachment to the throne of the population by the privilege of exemption. The consumption of salt was, however, regulated in order to prevent smuggling to the other areas.

5. Area of salt pans. These included Alsace and Lorraine, and here tax was about five times the value of salt. A certain amount had to be bought each year.

6. Area of 'quart-de-bouillon'. This was an area of lower Normandy which enjoyed the right to manufacture salt by boiling the salt sands gathered from the coast. Formerly the salt manufacturers gave the king a quarter of their yield. The laws concerning salt were moderate and so the temptation to traffic in salt with the provinces of the *grande gabelle* was very strong. There was surveillance along the border zone and also of the amount of salt used.

The collection of these taxes was farmed out to

the highest bidder, which augmented the severity of the tax. To see that the two great tax monopolies—salt and tobacco—were respected, the authorities had a veritable army of detectives with its own hierarchy. There were also companies of infantry and cavalry which were available to strengthen and reinforce the actions of the tax officers. The punishments for smuggling and fraud were very rigorous. Initially there was a fine but if this was not paid then the men were put in the galleys and there was whipping or banishment for women. If offences were repeated it was the galleys for life or hanging. In spite of these punishments fraud continued. Stocker writes that because the *gabelle* was such a heavy tax on an irreplaceable commodity it was particularly hard on the underprivileged.

Bloch (1963) recounts a story told by a French cleric, J. Bion, in 1708. In France there are poor peasant families who eat no soup for a whole week, though it is their common food, because they have no salt. A man distressed to see his wife and children in a starving languishing condition goes to buy salt in the provinces where it is only a quarter of the price. If discovered he is sent to the galleys, and it is a melancholy sight to see wife and children lamenting their father laden with chains and irrevocably lost—and this for no other crime than trying to procure subsistence for those whom he had procreated. In the last years of the tax before the Revolution, a total of 3500 people were condemned to death or to forced labour or prison for offences against the salt laws alone.

Because of the *gabelle* the consumption of salt was reduced to a bare minimum; it was 4.5 kg per head per year in the *grand gabelle* areas compared with twice that amount in the ransomed areas. There was no question of it being given to animals and thus it was said the tax did more harm to agriculture than did frost or hail. By the eve of the Revolution, after more than five centuries of fiscal oppression, the *gabelle* had reached it maximum of harassment and iniquities. Stocker (1949) quotes Necker as saying in 1781 that 'of all the taxes it is the most intolerable; a universal cry is raised against it and one must regard it with horror'. In the face of unrest, the tax was modified in 1789 and some of the worst features removed. In 1790 there was complete suppression of the last remnants of the *gabelle*. The Empire believed itself powerful enough to re-establish the tax and under Napoleon a tax was levied on salts leaving their place of production. Thereafter, as Stocker points out, the history of the tax mirrored the political history of France. During 1812–1813, the time of the retreat from Russia, the salt tax was doubled. In 1814, when Louis XVIII came to the throne, he lowered the tax again to gain popularity. After the revolution of February 1848, the Second Republic abolished every salt tax. In later years little by little the salt tax was re-imposed, until it was finally abolished in 1946.

Salt taxes have caused anger and revolts in other countries. In Hungary there was a state monopoly on salt from the Middle Ages. At the beginning of the sixteenth century conditions were very bad: the country was divided and there were civil wars and great misery. Transylvania became independent after the battle of Mohacs in 1526 when the Turks laid waste to the country. In 1597 a revolt broke out because of the country-wide salt shortage. The salt offices were stormed and the officials thrown out of the windows. The salt stocks were then freely distributed (Josephson 1971).

India

Gandhi (1929) in his *Monograph on Common Salt* describes how from the earliest period of British rule India had suffered from the peculiar government policy towards this important commodity. The supply of salt was monopolized at one time. It was made open to British manufacturers and traders from abroad at another. Furthermore the increase or decrease in salt consumption from year to year and from decade to decade followed the level of salt duty. The system of salt duty was resented by the whole nation and it was even more intolerable that foreign salt accounted for a third of the total consumption in India and 90% of the salt consumed in Bengal proper. This was despite the fact that India, with 8000 km of ideal seaboard together with its inexhaustible stores of crystalline salt in the Punjab and Sind, could not only be self-supporting in salt but could export it.

Gandhi emphasized the essential nature of common salt for human existence, especially in a tropical country like India, and also the need of animals for it. The cereals and green foods were particularly low in sodium and this exacerbated the situation in India where, like other tropical people, they are mostly vegetarians and rice-eaters. Thus the consumption of salt per head for maintaining a decent standard of health and hygiene is much greater than in Europe and North America. In general a poor country needs more salt per head than a rich one, and in the same way a poor man needs more salt for maintaining his health than a rich man. Gandhi goes on to state that the policy of salt taxation was therefore to be condemned as much on general principle as on the ground of its relation to the special conditions of India. The contention that the salt duty was the only levy upon the poor classes of India is open to the telling retort of the late Dr Naoroji, namely:

> What a humiliating confession to say that after the length of the British rule, the people are in such wretched plight that they have nothing that the Government can tax and that the Government must therefore tax an absolute necessity of life to an inordinate extent ... and how can anything be a greater condemnation of the results of British line of policy than that the people have nothing to spend and enjoy and pay tax on but that they must be pinched and starved in the necessity of life.

In 1930 Gandhi led 78 supporters on a 'salt march' to the sea. He walked 300 km from Ahmedabad to the sea. There he swam and took a lump of salt from the beach and returned with it. For so breaking the monopoly he was arrested and jailed. A revolt against the salt monopoly then occurred and vast numbers of people went to the beach to fetch salt in the same way. The authorities arrested 100 000 Indians but action against them was not feasible and so they were released and the law changed to allow people to produce salt for their own use. The taxes were abolished in 1946. J.B.S. Haldane (1956), in discussing the imperious need for salt in a country such as India, said that it was evident that Gandhi knew a great deal more physiology than the Viceroy. Haldane showed that miner's or glassworker's cramp, caused by work in hot mines or standing over glass-melting furnaces, was due to excessive perspiration—which contains about 0.5% sodium chloride. In the case of the glassworkers, as much as 36 litres of water a day was required to make good the loss from the skin, but cramp could be prevented by drinking a weak salt solution.

The deaths from salt loss which may occur during military activity under tropical conditions were described in Marriott's (1950) classic monograph on water and salt depletion. During the Second World War, studies by American scientists also showed that soldiers in a desert climate could lose up to 24 litres of sweat per day (Smith 1951). This is equivalent to a sodium chloride loss of 70–100 g/day. Decline would be very rapid if the salt were not replaced. Similarly it was well known that English coalminers, who did hard physical work in high temperatures, had a strong preference for salt food such as bacon and pickled herrings, and would salt their beer. The situation changed when the mine management provided lightly salted drinking water at the mines.

Gandhi's monograph refers to the findings of the 1836 Select Committee on Salt in British India and the 1830 Select Committee of Lords on the East India Company. The evidence brings out the limitation on consumption that resulted from the salt levies. As a result of the monopoly the price was greatly increased. In the areas where it was examined, the consumption of salt per head ranged between 4 and 5 kg/year. Evidence before the committees suggested that consumption in England was about 11 kg per head annually. Other evidence suggested that the cost of salt to the rural labourer, that is to the main mass of people in Bengal, for a family was equal to about one-sixth of his total annual earnings. By contrast a Bengal or Madras Sepoy on foreign service received up to 20 kg per year with salt fish in addition. The enquiry also revealed the practice of adulterating salt by mixing it with earth.

Condiments, and some early views on salt in this context

Manley (1884) writing on salt and other condiments at the time of the International Health Exhibition in London noted that the word condiment signified to compound or concoct something to give seasoning or flavouring to food. Whilst this may include every article, liquid or solid, which is not primarily food itself, attention is mainly confined to the common elements found on the table, namely salt and mustard, pepper and vinegar. These are usually used without combination with others and applied directly to the food by the eater. Manley muses as to whether or not man might be regarded as an animal that uses condiments with its food. Of course it is not known whether he used condiments from the beginning. It might be argued that he only developed a taste for them, as he has for the various forms of narcotics and alcohol, as his life gradually became more artificial. Notwithstanding, we find different races in different parts of the world which have long selected and used certain condiments with their food, and this was not based on any chemical knowledge but apparently some natural instinct.

Manley goes on to say that salt is the most widely accepted of all condiments. So universal has been its use that it may be called the cosmopolitan condiment. So great is the craving for it, and relish of it, that we are led to consider that a love of it is one of the most potent of our natural instincts and that salt itself is necessary to the health and even the life of man. There is a natural craving for it when it does not exist in sufficient quantity in food. This universal existence of an appetite for salt surely indicates that the substance serves more important functions than that of merely gratifying the palate, a conclusion which the most elementary considerations of human physiology fully substantiate. Further, the same craving is seen in animals. Manley quotes instances known in his time of the experimental consequences of deprivation of salt, including the failure to thrive of cattle which were deliberately fed low-salt diets. There was also the fact that in the old criminal law of Holland, a particularly stringent punishment was to confine criminals to a diet of water and bread which did not contain any salt.

He notes that there are writings which show an entirely opposite attitude to salt, suggesting it as the source of all the ills the flesh is heir to and proposing that abstention from it is their cure. An example is Dr Howard's (1851) *Treatise on Salt*. Manley considers the writer a painful instance of the lengths people will go to in perverting historical and other facts in the texts of scripture when it comes to a craze and they are apostles of a movement. He does cite the many instances of native tribes and populations known not to use salt and points out that this is associated with the fact either that the water supply is brackish or saline or alternatively that there is plenty of salt in the meat foods which are eaten. He reflects that it is possible to eat salt in such enormous quantities as to produce very injurious results. This is also the case for many other substances perfectly harmless and wholesome if taken in moderation.

As a brief digression on the general context of condiments, it is of interest to note that several botanically different species of plants are similar in possessing powerful and peculiar smells and tastes arising from essential oils containing sulphur. Instances are garlic, onions, horseradish and mustard. Mustard was probably first used as a condiment in Asia and found its way to Europe through Egypt in classical times. It is mentioned by Pythagoras and was used in medicine by Hippocrates in B.C. 480. Pliny mentions there were three kinds of mustard cultivated in his day. The first record of use of mustard known to Manley in England is in the household book of the Duke of Northumberland in the reign of Henry VII where it is stated that 160 gallons of mustard seed was the allowance per annum for his retainers and servants. French mustard is chiefly made at Dijon and, as is the case with other continental mustards, is mixed with tarragon and other vinegars flavoured with a variety of herbs, spices and other substances such as walnut or mushroom ketchup or the liqueurs of the richer pickles. Hence they have not the same pungency as the English table condiment.

In his review Manley (1884) stated that there are three kinds of pepper in domestic use: black, white, and red or cayenne. Black and white are produced from the one and the same plant and cayenne is not from a pepper plant. The pepper of commerce is furnished by *Piper nigrum*. It is indigenous to the forests of Travancore and Malabar and is cultivated in both East and West Indies and Java. It is a shrubby climbing plant which attains the height of 2.5–3 m. The active properties of pepper depend upon an acrid resin, a volatile oil and crystallizable substance called piperine. Pepper is mentioned in Shakespeare in the metaphorical sense of irritability, as is vinegar.

Vinegar in one or other form is one of the oldest condiments of the world. The ancients knew that the juices of fruits after becoming vinous, that is alcoholic, from fermentation were subject to another change by which they were rendered sour. Vinegar was used by the Greeks, and Roman soldiery when campaigning mixed it with water as their ordinary drink. All liquids capable of vinous fermentation may be made to produce vinegar but in all cases the sugar is first converted into alcohol, and the alcohol by oxidation into acetic acid—the acid of pure vinegar. It is this volatile principle that gives the aroma and pungency. In the manufacture of malt vinegar a mixture of malt and unmalted grain is mashed with hot water and the resulting wort fermented as in the process of brewing. Wine vinegar is prepared in wine-producing countries from grape juice and inferior new wines by processes similar to those adopted for malt vinegar. As with other condiments vinegar is mentioned in the old dramatists. Sir Andrew Ague-Cheek in Shakespeare's *Twelfth Night* says 'Here is the challenge, read it: I warrant there is vinegar and pepper in it'.

Ethnographic data on use and non-use of salt

As both Lapicque (1896) and Bunge (1874) propose, it is important to examine ethnographic evidence on salt habits in relation to understanding basis of intake. Several authors have summarized extant knowledge in the field, and to some extent provide mutually confirmatory accounts as indicated.

As Dastre (1901) records, there are ethnic groups who have never adopted the use of salt. The Egyptian priests never salted their foods. Plutarch was astonished at this strange disregard. All the nomadic tribes of northern Russia and Siberia refrained from salting their food, not because it was lacking—salt beds, outcrops, and salt lakes abound in these regions—but because these people live by hunting and fishing and show a marked aversion to salt. An explorer who lived for a long time with the 'Kamtschadales' and the

'Tungouses' peoples, the well-known mineralogist C. von Ditmar, enjoyed making them taste the salted foods which he used, so that he could watch the grimaces of repulsion which this simple seasoning provoked. These people did not, however, show any great delicacy of taste. Normally they ate an unnamable mixture made up from fish heaped in huge silos which decayed at leisure waiting for the moment of consumption. The Russian government, wishing to improve these repugnant and unhealthy feeding habits, taught the people the art of preserving fish by salting it, and established for this purpose salt works near the tribal encampments and delivered salt to them at a very low price. The obedient population complied. They salted the fish but did not eat it.

Examples of indifference or antipathy to salt are found in other latitudes. The Kirghiz of Turkestan, who feed on milk and meat in the salty steppes, use no salt. Neither do the Numidians, who have a similar diet. The Bedouins of Arabia find the use of salt ridiculous.

In Mexico, a sedentary people devoted to cultivation used salt regularly (Dastre 1901). Another facet of this behaviour is shown by the Negro tribes over a wide part of central Africa on either side of the Congo River who eat *sel de cendres*. These vegetable salts contain predominantly potassium chloride and are made by burning particular plants and extracting the ashes. The salt has a flavour which resembles sodium chloride but leaves the bitter after-taste of potassium salts. The natives are said to prefer it to ordinary sodium chloride. Dastre, in referring to Bunge's (1874) hypothesis that a high-potassium diet contingent on vegetarian status creates a need for sodium because of increased sodium loss, remarks that clearly a potassium salt would aggravate the situation.

Perhaps vegetable substances create a special appetite which can be satisfied by the taste of salt—chloride of either sodium or potassium—that is, there is a need for a salty flavour, a particular kind of taste sensation. Lapicque (1896) is cited as proposing that the taste for salt is a case of a very special penchant, the taste for condiments, common to all people who are vegetarian. The Indians and Malaysians disguise the flat taste of rice, their basic food, with curry. The Ethiopians enhance the insipidness of their maize with a piquant sauce. Finally, Dastre refers to the work of Pavlov just appearing at that time and speculates whether the role of a condiment, be it sodium chloride or any other, is that it sets in motion the secretion of gastric juice and thus has beneficial effects for digestion. It is interesting that Pales (1950) in his book on *Les sels alimentaires* reports that in French West Africa, the areas with the dominant use of vegetable salts and those with endemic goitre are superimposable. In fact 700 000 people out of the 16 million inhabitants appear to be affected.

Manley (1884) records that the Damaras in south-western Africa never take salt and Europeans travelling there never feel the want of it. But the well water in Damara land is nearly always brackish or saline and appears to taste of carbonate of soda rather than chloride. Their neighbours the Namaquas set no store by salt and the Hottentots of Walfisch Bay hardly ever take the trouble to collect it. The wild game in the Swakop do not frequent salt rocks to lick them as they do in America. In the colds of Siberia also, as in the heats of Africa, a similar disregard of salt sometimes prevails. Most of the Russians at Berezov eat their food without any salt though the condiment can easily be obtained at a trifling cost. Soups, vegetables and roast meats are prepared and eaten without salt. Manley suggests the explanation of these cases is that when food contains a sufficiency of salt there is no occasion or craving to add more.

Lapicque (1896) refers to the basic proposition of Bunge that the use of salt will be largely unknown to people who are hunters and pastoralists and live exclusively or mostly on meat or milk whereas it should be used constantly by people who are agriculturists and who live entirely as vegetarians. He cites several examples that support the hypothesis, including that of Finland. The Finnish language hasn't a word for salt. Finns of the west who now practice agriculture consume salt and designate it by the German word. The eastern Finns who are still hunters and nomads, however, still do not consume salt.

The Indian tribes of North America at the time of their first encounter with whites lived by hunting and fishing but did not consume salt though the prairies have plenty of this mineral. Kroeber (1941) discusses in detail the behaviour of the Indians in this regard. The outstanding fact regarding the salt in western North America is that it was used in half of that area and not used in the other half. The line of demarcation was sinuous and there were virtually no exceptions to the rule that salt was eaten everywhere to the south and not eaten everywhere to the north of this line. The boundary begins on the coast at the mouth of the Columbia, follows the Cascades south, cuts across the north-east corner of California to Nevada and then turns south and then swings back into south-eastern Oregon. South of the boundary seaweed is eaten mainly as seasoning or relish, a piece of seaweed cake being broken off and occasionally nibbled at between spoonfuls of food. In British Columbia salt as such is universally denied, and the seaweed is not only spoken of as a food but treated as such. Kroeber found it difficult to attribute this distinction to a physiological basis in terms of animal against plant food or temperature. It appeared to be the social custom. In parts of California and Nevada, a certain salt grass that grows on salt marshes and alkaline soil is roasted or burned in a pit from the bottom of which salt would then be collected.

The shepherds of the South American pampas, who don't take vegetable food and declare it fit only for beasts, do not take salt though the pampas is covered with saline lakes and deposits. Quite close to them the Araucans who were already agriculturists at the time of the discovery of America utilize both sea salt and rock salt of the mountains. However, in relation to the validity of generalization along these lines, Lapicque referred to his investigations in Africa involving the consuming of plant salt. The area between Chad and Congo is very large and was one of the least explored areas of Africa at that time. The Blacks of the region like most of the African Blacks were agriculturists. They deployed particular ingenuity in extracting salt from plants to satisfy their taste for salt. However, analysis by colleagues of Lapicque and others of plant salt obtained under authentic conditions showed that the salts are almost exclusively of potassium. Furthermore, at that time, a large part of the region where the plant salt was consumed had been opened to commerce and the marine salt had penetrated there.

Lapicque remarks that if it were true that the vegetable regime acts on the organism by way of potassium salts to create a deficit and need for sodium chloride, the usage of the plant salt by the Africans would only increase this need. Notwithstanding, the population appears to manage quite well. Lapicque suggests that the explanation of the usage of salt is that it should be considered as a condiment and not a food. It is a substance agreeable and at the same time useful because of its action on the senses but not a substance necessary for the continual supplement of the body. With the relation which has been recognized between the vegetable regime and the taste for salt, it is necessary to search the cause in the relative insipidness of vegetable diet. In this regard it is not only salt which is sought by agricultural people but also other condiments such as pepper and capsicum.

Alexander Caldcleugh (1825), who travelled in South America on horse to Las Chilchitas, records arriving at the banks of a salt lake called Beberero which was seven leagues in circumference. Round the edges the most beautiful white salt was encrusted in cubes. It was carried to Mendoza and other adjacent parts. But in the locality the consumption was trifling from a strong prejudice against its use. The people said that it produced a premature appearance of age, and the women, in particular, carefully abstained from using it.

Eskimos

Stefansson (1921) wrote on his experiences on living with the Eskimo. He remarked that most people thought of articles of diet like salt as necessities whereas he had not found them so. The longer you go without grain foods and vegetables the less you long for them. He felt salt was like a narcotic poison in that it was hard to break off its use, similar to tobacco. But after you have been a month or so without salt you cease to long for it, and after six months you find the taste of meat boiled in salt water distinctly disagreeable. But in the case of fat, the longer you are without it the more you long for it, until the craving becomes much more intense than is the hunger of a man who fasts. With the uncivilized Eskimos, the dislike for salt is very strong—so much so that a saltiness imperceptible to Stefansson would prevent them from eating at all. He found this circumstance useful in his travels about Coronation Gulf. Whenever his Eskimo visitors threatened to eat him out of house and home, he could put a little pinch of salt in the food, and thus husband resources without seeming inhospitable. The man who tasted anything salty at the table would quickly think that he had plenty of more palatable fare in his own house. In the meat diet of the Eskimos there would be ample salt. Ehrstrom (1934) notes that the Eskimos drank blood from the game they killed and as Josephson (1971) points out, blood contains more salt than muscle tissues and this could contribute to covering their need of salt.

Bolivian Indians

Holmberg (1950) made an extensive study of a primitive society, the Siriono of Eastern Bolivia. They were of particular interest because they were semi-nomadic and were reported to suffer from lack of food. Holmberg found in the jungle that supply of food is rarely abundant and always insecure. Game was not plentiful and the techniques of hunting, fishing and agriculture are very limited. Patterns of food storage did not exist.

Consequently eating habits depended largely upon quantities of food available for consumption at the moment. Estimates led him to the view that the average Indian probably ate less than a pound of meat per day.

The society is also a gathering society and the edible heart of the palm is an important food. There are quite a variety of fruit-bearing trees in the area. Fish also play a small part in the diet. A limited degree of agriculture was carried out, including maize, manioc, papaya. In this tropical area fresh meat has to be cooked within 8 h after it is killed in order to prevent spoilage. The natives have no salt with which to preserve meat nor have they developed any techniques for drying or smoking meat to render it edible for more than 2–3 days. Meats, fish, maize, manioc and camote are never eaten raw but are always cooked. No condiment of any kind is used in cooking. Salt is unknown to the Siriono living under aboriginal conditions. Holmberg presumes that the foods contain enough salt. He introduced salt to some Indians for the first time and they expressed a distaste for eating it. By using small quantities in cooking, however, they soon developed a craving for it. In some instances this craving, once an Indian had become accustomed to using salt, became so great as to become an important factor in establishing and maintaining friendly relations with the whites.

In relation to Holmberg's account of the Indians' first encounter with salt, it is not clear whether or not something like part of a handful of salt was given to them for tasting. In such concentrated impact it could well be distasteful and it is interesting that he records that by using small quantities in cooking a craving soon developed. Holmberg notes that he had travelled with primitive groups when all went without salt for as long as 43 days without suffering any ill effects from such a diet. With intake of an average of a quarter to half a kilogram of meat per day adequate sodium would be obtained.

Sahara nomads

Paque (1963, 1974) has summarized his extensive studies on the nomads of the Sahara. In spite of their environment, the consumption of water appeared to be remarkably low and constant. Very strict dietary practices, the salinity of the water and the special behavioural customs appear to be the basis of this strict economy of fluid intake. He wonders whether genetic factors might also be partly responsible. Salt waters in the Sahara, as in Morocco, are a basic factor of the environment. Paque remarks that for thousands of years its population seems to have consumed them without ill effect. Investigation with regard to sodium has revealed that amongst the peoples of the Sahara the consumption of heavily salted waters (up to ca 50 mmol/l of sodium) results in total exclusion of salt from the diet. No foodstuff is salted, the water alone supplies the sodium. The urinary sodium levels found with usage of the Tarfaya well are within the normal range. The use of less salty water, e.g. ca 10 mmol/l of sodium equally results in habitual exclusion of salt from the diet. It would appear the consumption of salt among the peoples of the Sahara is maintained at a low level. With waters notoriously saline this level remains physiologically tolerable. Paque reports that in spite of the salt intake imposed by the environment there is no frequency of cardiovascular troubles or of hypertension.

Pattern and seeming paradox: A summary

There is an emphasis in relevant historical records on a human preoccupation with salt. Sometimes this may amount to a craving and people may endure great hardships and take risks to obtain it. The situation is conspicuously different from that with other condiments, and this would appear to be clearly related to the fact that sodium is a critical element in the body. Deficiency of it leads to a progression of deleterious physiological consequences. Fairly comprehensive ethnographic data support the view that there is a general, if imprecise, relation of salt-seeking behaviour to bodily need. Thus, the mainly nomadic herdsmen for whom meat is the major item of diet or who live on animal products such as milk, and who therefore have an adequate supply of sodium in their diet, commonly take little or no salt with their food. Peoples who are agriculturalist and have high cereal or vegetable intake, which usually involves a diet low in sodium, mostly add salt to the diet. The peoples with high meat intake may, in some instances, change their traditional attitude to salt if it is gradually added in increasing amounts to their cooking; then, eventually, the taste becomes desired. In general, the data could be consistent with the idea that the habit of adding salt to food largely emerged with the change in human society from a hunting-gathering to an agricultural existence.

Within the overall pattern, there are, however, some apparent inconsistencies. The use of salt from plant ash—essentially a potassium compound—by the natives in the area between Congo and Chad, allegedly in preference to sodium chloride when that became available (Lapicque 1896), seems paradoxical. But while one might presume in vegetarians in a tropical area a significant need for sodium, a true evaluation of this behaviour is not feasible without knowledge of the actual sodium status of the people. At the time Lapicque was writing this area between the Congo and Lake Chad was perhaps the least explored in Africa. Therefore it could be that they had other sources of sodium; for example, cannibalism was alleged to occur (Chapter 6). If their sodium status were adequate, albeit borderline, there would be little or no retroactive benefit or sense of well-being from sodium ingested. In these circumstances the basic liking for salt taste might have been satisfied by the potassium salts with their bitter aftertaste, and cultural habituation could have resulted in the persistence of the preference. Whether this preference for potassium chloride would have continued in the face of frank and severe sodium deficiency is an interesting question. In terms of inbuilt mechanisms, it remains noteworthy that what the natives in this region sought, and went to considerable trouble to get, was a salty taste.

References

Alwall N (1958) Med Forum 11: 65

Bloch MR (1963) The social influence of salt. Sci Am 209: 89

Bunge G (1874) Ethnologischer Nachtrag zur Abhandlung über die Bedeutung des Kochsalzes. Z Biol 10: 111

Caldcleugh A (1825) Travels in South America during 1819–21. John Murray, London

Carver JE (1946) Salt. Contemporary Review 169: 112

Dastre A (1901) Le sel. Revue des Deux Mondes 1: 1

Ehrstrom R (1934) Die Diät- und Kostführung der Nordischen Länder in historischer Beleuchtung. Acta Med Scand 81: 583

Gandhi M (1929) Monograph on common salt. Federation of Indian Chambers of Commerce and Industry, Calcutta

Gouling S (ed) (1973) Encyclopaedia Sinica. Ch'eng Wen, Taiwan

Graves R (1961) The White Goddess. Faber & Faber, London

Grimm JLK (1883–1900) Teutonic Mythology vols 2 and 4 (Translated by JS Stallybrass). G Bell, London

Haldane JBS (1956) Les aspects physico-chimiques des instincts. In: L'instinct dans le comportement des animaux et de l'homme. Masson et Cie, Paris

Holmberg A (1950) Nomads of the long bow. The Siriono of Eastern Bolivia. US Government Printing Office, Washington (Smithsonian Institute of Social Anthropology publication no 10)

Howard R (1851) A treatise on salt. W & T Piper, London

Jarcho S (1958) A note on the medical use of salt in fifteenth century Africa. Med Hist 2: 226

Jones E (ed) (1951) The symbolic significance of salt in folklore and superstition. In: Essays in folklore, anthropology and religion, vol 2. Hogarth Press, London, p 22

Josephson B (1971) Salt. Sartryck ur nordisk medicine 85: 3

Kroeber AL (1941) Culture elements distribution. XV. Salt, dogs, tobacco. Anthropol Rec, Univ Calif 6: 1

Lapicque L (1896) Documents ethnographiques sur l'alimentation minérale. L. Anthropologie 7: 35

Li Ch'iao Ping (1948) The chemical arts of old China. Easton

Manley JJ (1884) Salt and other condiments. Clowes, London (International health exhibition handbooks)

Marriott HL (1950) Water and salt depletion. Thomas, Springfield, Ill

Meneely GR (1954) Salt—an editorial. Am J Med 106: 1

Mollat M (1968) Le rôle du sel dans l'histoire. Presses Universitaires de France

Multhauf RP (1978) Neptune's gift. Johns Hopkins Press, Baltimore

Pales L (1950) Les sels alimentaires—sels minéraux. Director General of Public Health, Dakar (Publication of General Government of French West Africa)

Paque C (1963) Alimentation sans sel dans une population nomade saharienne utilisant des eaux exclusivement salées: Approche de l'excrétion urinaire. CR Acad Sci [D] (Paris) 256: 331

Paque C (1974) Sahariens et eaux salées Sahariennes. From Contribution à l'étude scientifique de la Province de Tarfaya. Institut Scientifique Cherifien, Faculté des Sciences, Rabat, p 227

Parsons JA (1951) Sodium chloride: The salt of life. Discovery, Nov and Dec: 360, 385

Schleiden MJ (1875) Das Salz, Leipzig. Cited in Multhauf RP (1978)

Smith HW (1951) From fish to philosopher. Little Brown, Boston

Stefansson V (1921) My life with the Eskimo. Macmillan, New York

Stocker J (1949) Le sel. Presses Universitaires de France, Paris

6 Cannibalism

Summary

1. The incidence of cannibalism is an issue of some importance in relation to the question of alleviation of stress on mineral homeostasis which may arise in feral man inhabiting equatorial jungle and other environments where critical elements may be in short supply. The Amazon Basin with its enormous rainfall is of particular interest by virtue of the established fact of very low sodium intake marginal to deficit in natives in a substantial part of the region, and the possibility, for similar ecological reasons, of marginal intake of calcium and phosphorus. The practice of endocannibalism, particularly in relation to the ingestion of ground-up human bones added to caxiri, an alcoholic beverage, and other foods may represent a ritualized method of recycling critical mineral nutrients in tropical rain forest conditions. It is possible that a practice initiated under circumstances of extreme deprivation produced a substantial sense of well-being at the time for direct metabolic reasons, and that behaviour derived from experience of retroactive benefit bccamc ritualized with appropriate general overlay of myth. There is a large body of data showing remarkable capacities of animals to learn rapidly benefit or toxicity effects from ingestion of specific foods (Chapters 11, 23 and 25), and there seems no a priori reason to deny the possibility of some similar capacities in feral man. There are some clinical experimental data to support such a notion—particularly in relation to children. The total body calcium of a 70-kg man is about 50 000 mequiv (1 kg), of which 99% is in the skeleton, and total body phosphorus is 500–600 g of which 88% is in the skeleton. Sodium content of skeleton could amount to 40 g. Thus a group of 10–20 natives could get a significant intake of minerals in the course of such a ritual, giving rationale to the myth of the virtues of the deceased being transmitted to the drinkers. However, in the absence of metabolic data before and after the practice, which will be extremely difficult to ever obtain, the idea remains a speculative indulgence.

2. There are some data on cannibalism in animals including primates but generally the incidence of the behaviour would seem low in mammals.

3. The general tenor of historical records has been that cannibalism has been widespread and there is evidence of it in anthropological discoveries. Many motivations have been proposed. Apart from survival instances, desire for flesh in the face of boredom with a farinaceous or vegetarian diet, and revenge and cruelty have been proposed in the case of exocannibalism. Endocannibalism, the eating of kinsmen or dead of the same tribe, is proposed to involve ritualization of respect and affection for the departed, and assimilation of desirable qualities.

4. In relation to the attribution of any validity to the hypothesis set out above in (1) and, for that matter, the actual historical existence of the

behaviour, this chapter must needs give detailed consideration to the recent monograph by Arens, *The Man Eating Myth*. Arens states that despite extensive review of highly recommended and generally accepted sources he could not isolate a single reliable complete first-hand account by an anthropologist of this purported conventional way of disposing of the dead. He is dubious about the actual existence of this act as an accepted practice for any time and place. In the course of dissecting a large amount of material to illustrate his point, Arens has made an important scholastic contribution in focusing on standards requisite in this field where sensationalism and hearsay abound. However, in arriving at a sweeping conclusion he has overlooked the objective and mutually confirmatory accounts of a number of distinguished scientists involved in the voyages of Captain James Cook. There is also an important record by Alfred Russell Wallace, and a contemporary eye witness account of the same endocannibalism behaviour in the Amazon by Dr Napoleon Chagnon. There are other first-hand accounts from Fiji and Australia.

Introduction. Cannibalism—fact or largely fiction?

Historical records of cannibalism suggest it has been widespread throughout human history. Many motives have been proposed. Survival during famine is an obvious one. Revenge and cruelty, and desire for flesh in the face of boredom with a farinaceous or vegetarian diet have been suggested. On the other hand, as regards cannibalism of kinsmen, ritualization of respect and affection for the departed, involving passage of desirable qualities to the members of the tribe, has been advanced as an explanation. In relation to the theme of this book, there are interesting records equally suggestive that need of minerals might sometimes have been operative as a cause, particularly with in-group cannibalism (endocannibalism). The occurrence of particular cannibalistic behaviours—such as ingestion of ground-up bones—within tribal groups in equatorial rain forests where dietetic availability of sodium, calcium and phosphorus may range from marginal to deficit is one basis for putting forward this view.

From the remotest periods of history there is evidence of *Homo sapiens* having eaten his fellows. With the remains of Pekin man, broken-up skulls have been found which are suggestive of brains having been extracted and cooked. Similar findings point to the same conclusion for Neanderthal and Cro-Magnon man. The bases of most of the eleven Neanderthal skulls from N'gandorg in Java had been opened, presumably for extraction of the brain (von Koenigswald 1958). Evidence of unusual circumstances was strengthened by the fact that only two tibiae were found with the skulls, although the rich accompanying fauna yielded many complete skulls and jaws, and even larger joined pieces of backbone. Remains of cannibalistic meals appear evident among early Neanderthals from Weimar-Ehringsdorf and also at La Quina (Geiseler, 1952; Behm-Blancke 1959–1960). Other reports deal with cuts on cervical vertebrae implying decapitation, openings in the base of the skull, and of the skull and mutilations of the skeleton and effects of fire (Geiseler 1953; Behm-Blancke 1956, 1958).

The Greek historians Herodotus and Strabo give accounts of cannibalism among the Scythian Massagetae and Issedones, including deliberate killing and eating of old people of their tribes. Similar practices were alleged in Ireland. Instances of cannibalism were reported in Europe during the Middle Ages, including that of the women of Elvira in Spain cutting up and eating the body of the Arab chief, Sauwar, who had massacred their menfolk. Marco Polo and other travellers report Chinese and Tibetan tribes eating the flesh of fellows.

It is necessary to say at the outset that the whole basis of assumption that cannibalism has been a fairly widespread element of human behaviour has been challenged by Arens (1979) in *The Man-Eating Myth*. He states that as a result of directed research, conversations with colleagues, and some deliberation, he is dubious about the actual existence of this act as an accepted practice for any time and place. He does not deny recourse to cannibalism under survival conditions or as a rare instance of antisocial behaviour for any culture. His position, he says, of course, flies in the face of conventional wisdom and numerous reports, but little of genuine interest would result from attempting to confirm the acceptable and obvious. In the words of Montaigne, who gave some thought to cannibalism, 'We are to judge by the eye of reason, and not from common report'.

Arens says if the custom of eating the dead was well documented, and confirmed independently by others, then such an approach as the one guiding his study would be unnecessarily tedious. But instead of demonstrated fact, we constantly encounter assumption and rumour, and the fear that these might be true. Further, in the summary of his book, Arens states that despite extensive review of highly recommended and generally accepted sources, he could not isolate a single reliable complete first-hand account by an anthropologist of this purported conventional way of disposing of the dead. Instead he found a legion of reports from non-specialists that ranged from highly suspect to entirely groundless when viewed from the perspective of objective scholarship and commonsense. He adds that 'those members of the anthropological fraternity who would readily admit that the custom has been blown out of proportion to its actual occurrence by the laity are merely retreating behind the transparent claim of possessing arcane knowledge not available to the uninitiated. However this assertion is nothing more than an academic ideological strategy based upon similarly vague preconceptions lacking a solid historical or ethnographic foundation.'

From the foregoing, it could be inferred that Arens' criterion of acceptance as truth in this field is observation by the professional anthropologist. He would appear to reject accounts of first-hand observation, spaced over a century or more by several individuals, of, for example, cannibalism in Fiji. Retrospectively, it would be difficult to provide proof that one or two of them may have been telling the truth as to what they personally witnessed though, as Arens correctly emphasizes, there is no difficulty in seeing that much of the material in the literature is written in the past tense and quotes the observations of others. First-hand accounts involve missionaries, traders, masters of vessels and, though, in some instances they are in considerable detail, it is not possible to say that they are not fiction for a gullible or susceptible audience excited by stories of a distant land.

Against this background I will briefly refer to some animal data, and will then cite for illustrative purposes some of the material which is representative of general accounts from historical records of various types which receive a lambasting from Arens. Then a more detailed account of some of Arens' basic challenge in specific instances will be given including the kuru story in New Guinea.

Notwithstanding his polemical style, it seems clear that Arens has made a substantial scholastic contribution in focusing on standards to be met in acceptance of material in this field. However, he himself, in arriving at a viewpoint, seems to have neglected some critical sources which invalidate the sweeping nature of his conclusions. These include the five volumes of accounts written during Cook's voyages which have been edited by Dr Beaglehole. They contain the observations of many distinguished scientists who began with a very active cynicism about the whole matter, stating clearly many of the criteria of evidence and reasons for misgiving that Arens cites two centuries later. Each finishes independently with the same clear conclusion on the general occurrence of cannibalism in New Zealand, and the accounts corroborate one another on particular events. There are also accounts by some of the ship's company who had no 'academic investment' in anything in this area. I will quote these accounts in considerable detail and also draw on other material from the Pacific. Not only is the detail of great intrinsic interest, but the establishment of cannibalism as an incontrovertible fact, rather than fiction, is crucial to the last part of the chapter, which bears on motivation in certain classes of endocannibalism. On this theme evidence from the Amazon will then be discussed, with its important relation also to Chapter 25. The South American evidence starts with the records of Alfred Russell Wallace (1853), who spent three years in the upper Amazon Basin before travelling to the East Indies and eventually writing the paper 'On the tendency of varieties to depart indefinitely from the original type—in which principles of Natural Selection were explained'.

Animal data

Wilson (1975) states that the colonies of all termite species so far investigated promptly eat their own dead and injured. The behaviour is so pervasive that it can be said to be a way of life in these insects. Cook and Scott (1933) found cannibalism became intense in colonies of *Zootermopsis angusticollis* when they were kept on a diet of pure cellulose and thus deprived of protein. When sufficient quantities of casein were added to the diet, cannibalism dropped almost to zero. Richter (1955) described how bodies of dead rats were used to poison other rats, and how rats distinguished between dead and live animals. So long as a rat shows any sign of life its conspecifics will not touch it. But as soon as the last breath has left the animal they eat the entire body, often leaving neither teeth nor bones.

Schaller (1972) reports that lion cubs are sometimes killed and eaten during territorial disputes when nomadic male lions invade territories of prides. Hyenas are also said to be habitual cannibals. Mothers must stand guard while their cubs are feeding on a carcass in order to prevent them being eaten by other members of the clan (Kruuk 1972).

Cannibalism has also been observed in the course of studies of the wild chimpanzee. Bygott (1972) states that the behaviour is rarely seen among wild mammals and was not recorded in wild primates until recently, when in two instances a group of adult males acquired an infant chimpanzee and began eating it alive. Suzuki encountered a dominant male in Budongo Forest, Uganda, eating the flesh of a mutilated newborn infant. Other males showed great interest in touching the infant. Bygott made his observation in Gombe National Park, Tanzania, when a group of chimpanzees encountered two adult females. The elder female was immediately attacked by five males, and after the fighting the five males were observed alone clustered under a tree. They held an infant, 1.5 years old, which was bleeding from the nose. The dominant male beat its head against a branch and 3 minutes later began to eat the flesh of the legs. Another male tore off a foot of the infant, which had ceased to call. The dominant male remained in possession of the carcass for one and a half hours. Eventually he began playing with the carcass, grooming or punching it. Other chimpanzees handled or nibbled it. The oldest chimpanzee, who had taken the foot, now ate the carcass for about an hour and a half. Other chimpanzees had it but after 6 hours only the legs, genitals and one hand had been eaten. Bygott suggests the prey was less attractive as a food object and more an object of curiosity than other primates, such as colobus monkeys, which chimpanzees have been seen to eat. Wild chimpanzees are reported to ignore meat they themselves have not killed, and adult males eat much more meat than females do. Adult males are also more aggressive and in charging displays often flail sticks; rarely they may pick up and flail an unguarded infant, usually without injuring it. Cannibalism might occur in an instance of extreme aggressive arousal with an unfamiliar infant not recognized as a member of the group.

Recently, Goodall (1979) has given an account of a sequence of cannibalistic acts involving recently born infants which have followed attacks by a particular group of three chimpanzees in the Gombe region. It involved a mother and her two children. They appeared to treat chimpanzee babies as normal prey and after killing there was begging and sharing of meat. Other chimpanzees appeared to become aware of their behaviour and the potential danger to their infants. It is not yet clear to what extent this has been an aberrant event emerging in a particular group. However, other observations during the second decade of study at Gombe have shown considerable violence in chimpanzee society with one group making vicious attacks on members of another group occupying a different territory, leading to the gradual extermination of the second group.

Historical background

Hogg (1958) proposes in his book *Cannibalism and Human Sacrifice* that the researches in the eighteenth and nineteenth century revealed that eating of human flesh was practised in almost every part of the world. He notes, however, the preponderance of the behaviour in tropical regions. The book deals with the more sensational accounts and is in contrast to the more sober material from, for example, Cook's voyages.

Diverse motives for cannibalism are cited in the course of Hogg's consideration of a vast literature, mainly derived from the records of the various missionary societies, the British Museum, and the Royal Anthropological Society. They include revenge as a dominant one. There is also the transfer of desirable qualities from the dead to the living, the prevention of any form of after-life for the victim (including the possibility of him haunting his killers) and sometimes, to the exclusion or subordination of all other motives, the desire for human flesh coupled with a passion for sheer cruelty. He also suggests that boredom with an unwelcome farinaceous diet has led to experimenting with human flesh as food.

Thus, for example, it is asserted by Hogg that certain African and Australian tribes devoured dead kinsfolk in the belief that this was the most flattering way of burying them. Indian tribes of the north-west Pacific coast ate human flesh as part of an elaborate ceremonial aimed at good relations with their gods. The Bagesu of the Uganda Protectorate held cannibal feasts in order to honour their recent dead and regularly ate corpses of their fellow tribesmen. The act of eating an enemy produced the maximum degree of satisfaction among victors. In Africa, South America and in much of cannibal Melanesia, as, for example, among the Hallenga or the Fangs of Gaboon, part or all of the body of an enemy would be devoured as a final gesture of contempt. In some instances the live victim would be methodically dismembered, an arm or a leg at a time. The limbs would be cooked and eaten in front of him as a supreme gesture of scorn. The Fangs of Gaboon exemplify the widespread custom of cannibals of filing the teeth to better deal with their food.

Arens emphasizes the lack of first-hand reports, the predominance of the past tense, and the frequency in the instance of anthropological accounts of the people having been cannibals 'long ago', or 'until recently', or 'before contact', or 'before the researcher took up residence'. He stresses that it is not possible to divorce entirely the way in which the cannibal notion has been manipulated, by scholar and layman alike, from its use as ideological justification for some very real forms of human exploitation.

Fiji

Few missionaries in any part of the world are alleged to have faced greater horrors than in Fiji in Melanesia. In 1838, Rev. David Cargill of the Methodist Society wrote back to London:

> Some of the circumstances connected with the immolation of human victims are most revolting and diabolical. The passions of the people during the performance of these horrible rites seem inflamed by a fiendish ferocity which is not exceeded by anything we have ever heard of in the annals of human depravity.
>
> When about to offer a human sacrifice, the victim is selected from among the inhabitants of a distant territory, or is procured by negotiation from a tribe which is not related to the persons about to sacrifice. The victim is kept for some time, and supplied with abundance of food, that he may become fat.
>
> When about to be immolated, he is made to sit on the ground with his feet under his thighs and his hands placed before him. He is then bound so that he cannot move a limb or a joint. In this posture he is placed on stones heated for the occasion (and some of them are red-hot), and then covered with leaves and earth, to be roasted alive. When cooked, he is taken out of the oven and, his face and other parts being painted black, that he may resemble a living man ornamented for a feast or for war, he is carried to the temple of the gods and, being still retained in a sitting posture, is offered as a propitiatory sacrifice.
>
> These ceremonies being concluded, the body is carried beyond the precincts of the consecrated ground, cut into quarters, and distributed among the people; and they who were the cruel sacrificers of its life are also the beastly devourers of its flesh . . .
>
> The unnatural propensity to eat human flesh exists among them in its most savage forms. The Feegeeans eat human flesh, not merely from a principle of revenge, nor from necessity, but from choice. Captives and strangers are frequently killed and eaten. The natives of Thakanndrove kidnap men, women and children to glut their appetite for human flesh; it is said that, as if they were human hyenas, they disinter dead bodies, even after they have been two or three days beneath the ground; and that, having washed them in the sea, they roast and devour them. The flesh of women is preferred to that of men, and when they have a plentiful supply the head is not eaten. In some cases the heart is preserved for months. The bones of those persons whose bodies have been eaten are never buried, but are thrown about as the bones of beasts, and the smaller ones are formed into needles. Recently a boat's crew from the vessel *Active* was attacked by the natives in the expectation of obtaining their clothing and belongings. The four unhappy men were cooked and eaten, and *their* bones have now been formed into needles for making sails.

There are accounts from other writers of feasts of dozens of bodies following warfare and raids, and wrecks upon the shores. Early observers stated that Fijians loved human flesh for its own sake and possibly the absence of any animal they could eat gave rise to the custom. The pig was introduced only in the eighteenth century.

In seeking first-hand source material on cannibalism in the Pacific, I am indebted to the librarians at the Mitchell Library, Sydney and Ms Judith Pugh for valuable unpublished material located in the archives.

The diary for 1836–1844 of Dr Richard Burdsall Lyth, a surgeon missionary working in Somosomo, Fiji, includes many entries concerning cannibalism, and though much punctuated with the matters of the soul which took him and his family there, the accounts are explicit enough and at times the descriptions reflect his professional training. The following excerpt from 14 September 1840, is representative:

> Tuiilaila and a good part of the people returned yesterday at noon bringing with them two baolas or dead men—who with a third had been slain by a party of the king's people on the preceding days . . . One was, we are informed, eaten immediately. The other two were baked yesterday afternoon. One of them Reverend Hunt and I saw cut up but we came away before the other was proceeded with. In the former a piece of a spear 12 inches long was found in his insides having entered about the fifth rib on the left side. A piece of spear was also extracted from the loins of the other victim 3 inches and a half long.
>
> The one we saw (after it was nearly cut up but whilst the process was still going on) was divided as follows. The arms were cut off at the shoulders. The legs at the hip joints. The trunk was flayed and the viscera taken out. The liver, heart, kidneys were washed. The lungs and the entrails thrown into the sea. The head was cut off and put on one side to be dressed. The pectoral muscles were separated from the ribs and the chest being cleared of its viscera was added to the other parts that were ready for baking—the body thus cut up being removed to the oven. The limbs were then one after another laid on the heated stones and when they had been on the fire a short time were scraped. I observed a boy scraping a leg in this manner. Another a man was scraping another extremity. The remainder, I suppose, underwent the same process after we came away. By then we understood the different parts would be wrapt up in leaves and baked having been further cut up and divided into a great number of small parcels. The instrument

employed was a bamboo which was used with great dexterity, every now and again taking a fresh piece of bamboo or tearing a piece off with the teeth to have a new edge.

The whole was done with the utmost coolness and self-complacency with no more appearance of revenge than in the ordinary preparing of food.

The women were not silent spectators at the events of the day. As soon as the news reached Somosomo that the warriors were returning with baola the [...] which before wore the appearance of a forsaken village was all alive. The women turned out of their houses dressed in their better [...] with joy sparkling in their faces and hastened to the sea beach dancing as they went. As the canoes approached the shore the women greeted the returning warriors with loud singing and dancing with all their might using at the same time the most violent gestures and I understand, for I did not see them, gestures of indecent signification. This appears to be the custom of the women on such occasions when the men return successful from their wars. Before the return of the canoes the sabbath had passed very agreeably and peacefully.

On Tuesday 21 February 1843 he writes:

Canoes arrive from Horoivonu soon after sunset with nine bodies of Netewa people who with six others had been killed in the morning in a fight ...

Wednesday was a day of cannibalism and an additional body was brought making 10 in all brought to Somosomo 8 or 9 of which were cooked and eaten in Somosomo. A human head was wantonly thrown on the beach in front of our premises, and in returning from the town on Friday a part of a human pelvis was picked up by my dog. I had both the head and the bone buried.

The results of my observations this week are my full confirmation in the opinion that human flesh is very much relished by this people and that they have almost a passion for it. They are at the same time ashamed to acknowledge it, one blames the other, but few will speak out the truth and especially when the abominable food is in their possession. But I believe that most of them eat it both chiefs and people old and young and even many of the women and children. The Chiefs consider the eating of human flesh as their special prerogative and from them it passes into the hands of the common people. I have heard that in some parts of Feejee the priests do not touch it but it is not so here. They have a goodly portion and are distinguished cannibals. That parents give it to their children is a notorious fact ... I asked Raivalita one of Tuiilaila's sons perhaps 11 years of age if he had eaten baola. He answered yes. I asked him if it was good. Yes he said it is good. I asked him is it better than turtle and pork and fowls. No he replied they are all alike good and it is no better than the rest. All this was said with the greatest simplicity. He had been taught to eat it. He did not see the evil of it. He therefore spoke without any hesitation or reserve ...

Another fact, for such I believe it to be, is that they not only eat it when fresh but salt it and thus are enabled to keep it for several days. This is what the Somosomo people themselves have told us.

The entry for Sunday 28 May 1843 reads:

Learning that the oven in which the baola was cooked on Friday was not open yesterday as I expected it would be, I was curious to see whether they would eat it after being dead so long this being the 4th day. So I walked as far as Nasima. I learned from a young chief [...] that the oven was about to be opened. [...] into the king's house Seru informing Tuittakau at the same time as I was desirous of seeing the baola. The king told me to sit down. In a few moments four baskets containing the cooked baola were brought in and portioned out to the heads of the houses. One old man present turned his head on one side and commenced eating his portion given him from Tuittakau's basket. Several passing remarks were made by the parties on the qualities of the baola—one old man said that it looked very sick. Another (a priest) observed that it smelt well and so on. Some younger men had portions from Tuittakau's basket which as they received they walked off. Having seen the whole divided out I followed one young man. When I overtook him he was eating it and turning to me said it was sweet. I used one word of remonstrance and then hastened home. This was the first time of my seeing baola in its cooked state and people in the act of eating it. It appears the flesh was coarse and dark coloured.

Mexico

Hogg (1958) recounts that with the Aztecs, human sacrifice was an essential element of the culture.

A regular supply of victims for the priests was necessary for ensuring the beneficial motion of the sun and the fertility of crops and men. The procedures of sacrifice were elaborated constantly. Each had a deep symbolism. Following the ritual removal of the heart by the High Priest, the breast of the victim having been opened with a knife of obsidian, the heart was held up to the sun that each might give strength to the other. The body was thrown down a flight of stone steps where it was cut up and eaten by the warriors and their close associates.

Toxcatl, the great Aztec feast, involved sacrificing a carefully selected victim who was groomed for the year beforehand as something between a king and a god. For the three weeks of April preceding the ceremony he was attended day and night by four young and beautiful maidens. They were the Goddess of Flowers, the Goddess of Young Corn, the Goddess of our Mother among-the-Water and the Goddess of Salt. They were beautiful, ardent and dedicated and he lived a life of supreme voluptuousness. On the day, he said farewell to the four goddesses on a hilltop near the great lake. He walked on alone to the temple where the priests awaited, and as his shadow crossed the threshold, the priests seized him and the High Priest sacrificed him. Immediately his heart ceased to palpate his successor was announced and the ritual recommenced. The powerful Aztec influence resulted in similar customs being transferred to other tribes of Central America, the sacrificial and ritual element being dominant and any cannibalism minimal.

Harris (1978) has proposed this ritual of the Aztecs as a basis for provision of animal protein to the community in the form of human flesh. It was ritual slaughter in a state-sponsored system. This viewpoint has been strongly attacked by Sahlins (1978) who proposes the records show no shortage of animal protein, and that the cost of such rituals would have made them a poor return for effort from a nutritional viewpoint. He emphasizes the point made above, that cannibal consumption took only a small part in the ritual. He quotes, for example, old Spanish chroniclers such as Sahugan (Book IX, p. 67): 'They cooked him in an olla. Separately, in an olla, they cooked the grains of maize. They served [his flesh] on it. They placed only a little on top of it. No chili did they add to it; they only sprinkled salt on it.'

Arens describes Sahugan's 13-volume record of Aztec history, entitled *The Florentine Codex:*

General History of the Things of New Spain, as an admirable scholarly feat, wherein the author took pains to learn Aztec and through the use of informants systematically recorded numerous aspects of the traditional native culture. Sahugan was born in Spain and arrived in Mexico after about ten years of Spanish rule. Only two volumes, *Merchants* and *Ceremonies* contain fragmentary references to the Aztecs eating human flesh. They are not informants' admissions to having participated nor are they eye witnesses' accounts. In reviewing this and the large amount of material from various sources Arens says that the sound scholarly foundation which is so often assumed or alluded to evaporates upon investigation. This does not prove the Aztecs did not consume human flesh in private rituals nor demonstrates they did not engage in it on a large scale as a result of protein deficiency. What can be said with some assurance is that the evidence is too sparse, too ephemeral and too suspect on various grounds to allow a positive assertion.

Africa

Hinde (1897), who was involved in the war between the Zanzibar Arabs and the Belgians, wrote:

> Nearly all the tribes in the Congo Basin either are, or have been, cannibals; and among some of them the practice is on the increase. Races who until lately do not seem to have been cannibals, though situated in a country surrounded by cannibal races, have, from increased intercourse with their neighbours, learned to eat human flesh.
>
> Soon after the Station of Equator was established, the residents discovered that wholesale human traffic was being carried on by the natives of the district between this station and Lake M'Zumba. The captains of the steamers have often assured me that whenever they try to buy goats from the natives, slaves are demanded in exchange; the natives often come aboard with tusks of ivory with the intention of buying a slave, complaining that meat is now scarce in their neighbourhood.
>
> There is not the slightest doubt in my mind that they prefer human flesh to any other. During all the time I lived among cannibal races I never came across a single case of their eating any kind of flesh raw; they invariably either boil, roast or smoke it. This custom of smoking flesh to make it keep would have been very useful to us, as we were often without meat for long periods. We could, however, never buy smoked meat in the markets, it being impossible to be sure that it was not human flesh.
>
> The preference of different tribes for various parts of the human body is interesting. Some cut long steaks from the flesh of the thighs, legs or arms; others prefer the hands and feet; and though the great majority do not eat the head, I have come across more than one tribe which prefers this to any other part. Almost all use some part of the intestines on account of the fat they contain.
>
> A young Basongo chief came to our Commandant while at dinner in his tent and asked for the loan of his knife, which, without thinking, the Commandant gave him. He immediately disappeared behind the tent and cut the throat of a little slave-girl belonging to him, and was in the act of cooking her when one of our soldiers saw him. This cannibal was immediately put in irons, but almost immediately after his liberation he was brought in by some of our soldiers who said he was eating children in and about our cantonment. He had a bag slung round his neck which, on examining it, we found contained an arm and a leg of a young child.

Members of the Baptist Missionary Society such as the Reverends Bentley and Grenfell gave accounts of the Congo Basin. Bentley (1900), in *Pioneering on the Congo*, wrote:

> The whole wide country seemed to be given up to cannibalism, from the Mobangi [a major tributary of the Congo] to Stanley Falls, for six hundred miles on both sides of the main river, and the Mobangi as well. Often did the natives beg Grenfell to sell some of his steamer hands, especially his coast people; coming from the shore of the great salt sea, they must be very 'sweet'—salt is spoken of as sweet, in the same way as sugar. They offered two or three of their women for one of those coast men. They could not understand the objections raised to the practice. 'You eat fowls and goats, and we eat men; why not? What is the difference?'

And he adds that bad as cannibalism was on the Congo, it was worse on the Mobangi. Tribes kept and fattened slaves for butchery as we do cattle and poultry. They conducted raids along the river.

Arens regards accounts of this type by 'Congo hands' as trifling or absurd and notes that David Livingstone, who spent a good portion of his adult life in East and Central Africa, considered that the case for cannibalism among an African group with which he was familiar was not proven.

Kalous (1974) in *Cannibals and Tongo Players of Sierra Leone* notes it may be fair to question the objectivity of the source documents of his book—colonial papers— but they were not written with any particular purpose in mind apart from the recording of certain facts, and probably do give a historically true picture of this aspect of Sierra Leone's past. The documents give first-hand accounts of cannibalism that were given as evidence in the course of judicial processes, where testimony, in some instances, led to the death sentence for murder.

New Guinea

Cannibalism has been alleged to be widespread in New Guinea. Murray (1912), who was Lieutenant Governor and Chief Judicial Officer, stated that certain tribes like human flesh and do not see why they should not eat it. A government anthropologist studying the Orakaiva society gave the reason as the simple desire for good food.

Murray records that many tribesmen regard human flesh as civilized men regard beef and mutton. A witness at a trial stated: 'We boil the bodies. We cut them up and boil them in a pot. We boil babies too. We cut them up like a pig. We eat them cold or hot. We eat the legs first. We eat them because they are like fish. We have fish in the creeks, and kangaroos in the grass. But men are our real food.' Seligmann (1910) described human flesh as usually being boiled though it is also cooked in native ovens. He noted the brain, extracted from the foramen magnum of the boiled skull, was broken up and considered a very special delicacy. The tongue, solid viscera, testes, penis and vulva, hands, feet and mammae were all eaten—some also regarded as special delicacies.

Seligmann (1910) stressed the powerful motive of revenge in New Guinea cannibalism as well as the liking for flesh. Walker (1909: quoted by Hogg 1958) described how he joined a punitive expedition by the Resident Magistrate against the Dobodura tribe:

> We decided to rush the village of Kanau, but when we got there we found it deserted. In the centre of the village was a kind of small, raised platform, on which were rows of human skulls and quantities of bones, the remnants of many a gruesome cannibal feast. Many of the skulls were quite fresh, with small bits of meat still sticking to them, but for all that, they had been picked very clean. Every skull had a large hole punched in the side, varying in size but uniform as regards position. The explanation for this we soon learnt from the Notus, and later it was confirmed by our prisoners.
>
> When the Doboduras capture an enemy, they slowly torture him to death, practically eating him alive. When he is almost dead, they make a hole in the side of the head and scoop out his brains with a kind of wooden spoon. These brains, which are eaten warm and fresh, were regarded as a great delicacy. No doubt the Notus recognized some of their relatives amid the ghastly relics.

The alleged practices with the brain are relevant to the story of kuru. This is a chronic fatal disorder of the nervous system, the main symptoms being a shivering-like tremor, lack of muscular coordination and speech difficulties. Burnet (1971) writes that Gajdusek must be given the lion's share of the credit both for defining the disease and its distribution and carrying out the first successful transfer of the disease to chimpanzees. Kuru is found only in the Fore and other tribes of the Okapa area of New Guinea, where it first appeared in 1910. Initially it affected adult women, but 15 years later it was seen in children of both sexes. It was associated with the entry of cannibalism into the area. This started as a surreptitious practice which South Fore women adopted from their neighbours, but then generalized and involved eating incompletely cooked brains from kinswomen who had died from any natural cause not considered dangerous. Burnet (1971) says that it must be assumed that at about the same time the kuru virus entered the region and caused a woman's death. Presumably several women were infected from eating the brain or other viscera of the initial source and eventually became victims of the disease. At death their brains would have contained large amounts of virus and thus propagation continued. Young children were stated to be present at the cannibalistic feasts and undoubtedly received morsels from the corpse. Men were disinclined to join the women in ritualistic cannibalism, particularly since it involved eating women, and it is asserted they had primitive fear of female magic. Glasse (1967) gives a detailed account of cannibal practices, as he learnt of them from the South Fore, including methods of cooking the body and kinship rights to it.

As Arens remarks, in New Guinea it is the professional anthropologist rather than the bygone traveller who gives life to the notion of cannibalism, and almost every anthropological book on the area contains a series of references to it. Thus Margaret Mead (1950), in a chapter on 'The Pace of Life in a Cannibal Tribe', includes a footnote saying that the former practice of cannibalism was outlawed by the Australian authorities three years before her arrival and that she did not directly encounter it. She learned of the Mundugumor passion for human flesh from their neighbours the 'gentle Arafesh'. Arens goes on to say that as we close in on the kuru area of the highlands, academic standards seem to function as an almost-forgotten ideal, rather than as a standard operating procedure. Anthropologists with well-deserved reputations become the victims of their own sensationalism and poor scholarship. The convergence of both 'medical and anthropological hypothesis on the existence of cannibalism prepares the way for outstanding leaps of fantasy and demand for the reader to follow along'.

Kuru was first mentioned in the routine reports of the Australian colonial patrol officers who administered the area. Berndt (1952, 1954) suggested kuru was a psychosomatic disease brought about by the shock of contact with the European world. Zigas, who set about investigating it with Gajdusek, was aware that virtually everyone afflicted died. Arens traces the history of the investigation, the search for different vectors, and emphasizes the early reports of Gajdusek which, though noting the Fore were New Guinea cannibals, regarded any connection between cannibalism and kuru as outlandish.

In dealing with the developments parallel to Gajdusek's (Gajdusek and Gibbs 1966) showing that the disease was caused by a virus as shown by transmission from infected kuru brain to a chimpanzee, Arens notes that the increasing notoriety and potential scientific interest drew social anthropologists to investigate social and cultural effects of kuru on the Fore. He states that the now prevailing notion that cannibalism rather than some other form of close contact is the principal means of transmitting the virus among the Fore (Glasse and Lindenbaum 1976; Gajdusek 1977) is based on circumstantial evidence and the act has never been seen by an outsider. He notes that the anthropological groups concerned in proposing cannibalism state that the practice ceased three or four years before their arrival. Thus, in his eyes, Glasse and Lindenbaum relied upon Berndt's idiosyncratic discussion of the material, the fact the Fore had a reputation among the surrounding people for eating their dead, the odd report someone had eaten someone else, and the belief among the males that the great majority of women were cannibals. Further, it was argued the Fore became cannibals at the turn of the century, which would conform to their statements of the initial appearance of the disease. The proclivity of the women for eating human flesh would conform to the distribution in adult and the distribution over both sexes in young children since the mother would pass on the infected material to children. The gradual decline in kuru deaths since the height of the disease in 1961 conforms to effective European stamping out of remaining vestiges of cannibalism. Arens says this all amounts to an association, not a demonstrated cause and effect relation. He also raises critical objections to the idea. For example, the native view that only women and children are cannibals may be a symbolic statement about females in a culture area renowned for sexual antagonism. Further, specialists are called upon to prepare the deceased for burial, since contact with a corpse is thought to be dangerous—a paradoxical situation if a corpse is thought of as something to be eaten and also as dangerous. And after each death there is a mortuary feast involving slaughter of pigs and distribution of vegetables (Glasse and Lindenbaum 1976) and such a period of abundance of animal protein would seem the least likely time to resort to cannibalism. Arens states that Gajdusek's current position (in 1978) on cannibalism as the vector of kuru is more cautious, involving the statement that there is so far no convincing evidence that the infection can be acquired by eating or drinking infected material.

Payer (1979) in a review of Arens' book remarks there are many more things on earth than can be witnessed and recorded by anthropologists. The review involves quoted discussion with Lindenbaum who did the field research on the Fore in collaboration with Glasse. She is quoted as saying Arens' thesis is important but 'I think cannibalism has been all over the place as a fact, and it has been all over the place as a slander with no basis in fact'. She thinks it regrettable that Arens will accept only evidence from Western anthropologists and not from 'black people' who were actual participants, as were her informants. Arens ridicules the idea that a widely accepted custom could be abolished suddenly just before anthropologists can arrive to observe it yet, she says, it does not seem to have occurred to him to search the records of the colonial officers who did the outlawing and enforced it. Lindenbaum adds that her informants date the end of the custom precisely to 'the building of the road' which brought the Fore into contact with missionaries and a culture of infinitely superior technology which promised and threatened much.

New Zealand

There are first-hand accounts of cannibalism in the South Pacific by persons of scientific eminence whose objectivity is evidently as stringent as that of any contemporary specialist. These are in the volumes recording Captain Cook's three voyages of exploration in the Pacific Ocean. The accounts are written independently. In the light of the issues raised by Arens it is desirable to go into some considerable detail, including the sequence of acquisition of data, and the writers' own reflections on criteria of validity.

Cook's first voyage: 1768–1771

Banks

Sir Joseph Banks, who was the botanist on Cook's first voyage, in the *Endeavour*, was one of the great Englishmen of the latter part of the eighteenth century. For over 40 years he was President of the Royal Society and the scientific advisor of the British government. He was one of the founders of the Kew Gardens and also of the colony of New South Wales and had diverse scientific interests. Banks was accompanied on the voyage by Dr Daniel Carl Solander, a Swedish naturalist who was a pupil of Linnaeus. He subsequently became Banks's librarian and keeper of the Natural History Department of the British Museum in 1773. He was elected a Fellow of the Royal Society in 1764.

Banks' *Endeavour Journal* (1768–1771) was published by the Trustees of the Public Library of New South Wales and edited by Dr J.C. Beaglehole, one of the world's foremost authorities on Cook and Banks.

The first description of cannibalism in New Zealand comes on 16 January 1770. Banks gives the following account:

After dinner we went in the boat towards a cove about a mile from the ship. As we rowd along something was seen floating upon the water which we took to be a dead seal; we rowd up to it and it provd to our great surprize to be the body of a Woman who seemd to have been dead some time. We left it and proceeded to our cove where we found a small family of Indians [Maoris] who were a little afraid of us as they all ran away but one; they soon however returnd except an old man and child who staid in the woods but not out of sight of us; of these people we inquird about the body we had seen. They told Tupia [the Tahitian who accompanied Cook and Banks; he was well understood by the Maoris and they conversed with him] that the woman was a relation of theirs and that instead of Burying their dead their custom was to tie a stone to them and throw them into the sea, which stone they supposed to have been unloosed by some accident.

The family were employd when we came ashore in dressing their provisions, which were a dog who was at that time buried in their oven and near it were many provision baskets. Looking carelessly upon one of these we by accident observd 2 bones, pretty clean pickd, which as apeard upon examination were undoubtedly human bones. Tho we had from the first of our arrival upon the coast constantly heard the Indians acknowledge the custom of eating their enemies we had never before had a proof of it, but this amounted almost to demonstration: the bones were clearly human, upon them were evident marks of their having been dressd on the fire, the meat was not intirely pickd off from them and on the grisly ends which gnawd were evident marks of teeth, and these were accidentaly found in a provision basket. On asking the people what bones are these? they answerd, The bones of a man.—And have you eat the flesh?—Yes.—Have you none of it left?—No.—Why did not you eat the woman who we saw today in the water?—She was our relation.—Who then is it that you do eat?—Those who are killd in war.—And who was the man whose bones these are?—5 days ago a boat of our enemies came into this bay and of them we killd 7, of whoom the owner of these bones was one.—The horrour that apeard in the countenances of the seamen on hearing this discourse which was immediately translated for the good of the company is better conceivd than describd. For ourselves and myself in particular we were before too well convincd of the existence of such a custom to be surprizd, tho we were pleasd at having so strong a proof of a custom which human nature holds in too great abhorrence to give easy credit to.

On the following day Banks states:

A small canoe came this morn from the Indian town. As soon as they came alongside Tupia began to enquire into the truth of what we had heard yesterday and was told over again the same story. But where are the skulls said Tupia—do you eat them? Bring them and we shall then be convinced that these are men whose bones we have seen. We do not eat the heads answered the old man who had first come on board the ship but we do the brains and tomorrow I will bring one and show you. Much of this conversation passed after which the old man went home.

On 20 January Banks records:

The old man came this morn according to his promise with the heads of 4 people which were preserved with the flesh and hair on and kept I suppose as trophies, as possibly scalps whereby the North Americans before the Europeans came among them: the brains were however taken out as we had been told, maybe they are a delicacy here. The flesh and skin upon these heads were soft but they were somehow preserved so as not to stink at all.

Banks records later:

We made another excursion to the Bay this morning. The bay everywhere where we have yet been is very hilly but we have hardly seen a flat large enough for a potato garden. Our friends here do not seem to feel the want of such places as we have not yet seen the least appearance of cultivation, I suppose they live entirely upon fish, dogs and enemies.

Banks' entry of 4 February suggests that he is well aware of the pitfalls which Arens (1979) emphasized two centuries later:

Prevented from sailing by our hay which had been so thoroughly soked by the late rains that it was too wet to put on board. Some conversation passd today concerning a report we heard yesterday. Two of our boats went out different ways and returnd at different times; the people of one said that they had met a double canoe who told them that they had a few days ago lost a female child who they suspected had been stole and eat by some of their neighbours; the other said that they had also met a double canoe whose people told them that they had yesterday eat a child, some of whose bones they sold them. From hence many of our gentlemen were led to conclude that thefts of this kind are frequent among these Indians. This story in my opinion throws very little light upon the subject as I am inclind to beleive that our two boats who went out at very different times in the morn both in the same direction, one only farther than the other, saw one and the same canoe and only differently interpreted the conversation of the people, as they know only a few words of the language, and eating people is now always the uppermost Idea in their heads. This however I must say, that when such families have come off to the ship even with an intention to fight with us they have very often brought Women and young children in arms as if they were afraid to leave them behind.

Again on the next day Banks notes: 'Tupia all along warned us not to believe too much anything these people told us: for says he they are given to lying, they told you that one of their people was killed by a musket and buried which was absolutely false.'

In March 1770 Banks remarks: 'as we intend to leave this place tomorrow morn I shall spend a few sheets in drawing together what I have observed of this country and its inhabitants'. He records that it was discovered by Abel Jansen Tasman on 13 December 1642. Tasman, however, 'never went ashore probably for fear of the natives who when he had come to anchor set upon one of his boats and killed three or four out of seven people that were in her'. Banks notes that 'the disposition of both sexes seems mild gentle and very affectionate to each other but implacable towards their enemies who after having killed they eat, probably out of principle of revenge.' Dr Beaglehole in a footnote here remarks that there seemed to have been a number of motives for cannibalism—revenge or exultation at the end of a battle or seige; acquisition of *mana*, or prestige; ritual; lack of flesh food; simple hunger.

Speaking of their food, in the use of which they seem to be moderate, Banks states that it consists of dogs, birds, especially sea fowl such as penguins, albatrosses etc., fish, sweet potatoes, yams, taro, some few wild plants, palm cabbage and above all the roots of a species of fern very common upon the hills.

As for the flesh of men although they certainly do eat it I cannot in my opinion debase human nature so much as to imagine that they relish it as a dainty or even look upon it as a part of common food. Though thirst of revenge may drive men to great lengths when the passions are allowed to take their full swing yet nature through all the superior part of creation shows how much she recoils at the thought of any species preying upon itself: dogs and cats show visible signs of disgust at the very sight of the dead carcass of their species even wolves or bears were never said to eat one another except in case of absolute necessity when the stings of hunger have overcome the precepts of nature in which case the same has been done by the inhabitants of the most civilized nations.

He adds that among fish and insects indeed there are many instances which prove that those who live by prey regard little whether what they take is of their own or any other species, but returns finally to remarking that whoever considers this will easily see that no conclusion in favour of such a practice can be drawn from the actions of a race of beings placed so infinitely below us in the order of nature. Further on in this summary Banks records:

That they eat the bodies of such of their enemies as are killd in war is a fact which, tho universaly acknowledg'd by them from our first landing at every place we came into, I confess I was very loth to give credit to till I by accident found the bones of men well pick'd in the very baskets where these people keep their provision: so convincing a proof I could not withstand, so I proceeded to inquire as well as I could with the small knowledge of their language which I had and the Assistance of Tupia what were their customs upon this occasion. They told us that a few days before a canoe of their enemies had been surprizd by them and that out of her they killd 7 persons, to one of whoom the bones in the basket had belongd, that now all the flesh of these people was eat up and most of the bones thrown away, which we found to be true for in almost every cove where we landed fresh bones of Men were found near the places where fires had been made. The whole was still more confirmed by the old man who we supposd to be the cheif of an Indian town which was very near us, coming a few days afterwards and at our desire bringing with him in his Canoe 6 or 7 heads of men preservd with the flesh on. These it seems these people keep after having eat the brains as trophies of their victories in the same manner as the Indians of North America do scalps; they had their ornament in their ears as when alive and some seemd to have false eyes. He was very jealous of shewing them. One I bought tho much against the inclinations of its owner, for tho he likd the price I offerd he hesitated much to send it up, yet having taken the price I insisted either to have that returnd or the head given, but could not prevail untill I enforc'd my threats by shewing Him a musquet on which he chose to part with the head rather than the price he had got, which was a pair of old Drawers of very white linnen. It appeard to have belongd to a person of about 14 or 15 years of age, and evidently shewd by the contusions on one side of it that it had receivd many violent blows which had chippd of a part of the scull near the eye: from hence and many more circumstances I am inclind to beleive that these Indians give no quarter, or ever take prisoners to eat upon a future occasion as is said to have been practisd by the Floridan Indians; for had they done so this young creature who could not make much resistance would have been a very proper subject.

Dr Beaglehole notes in relation to these observations of Banks that the heads of friends as well as enemies might be preserved, the former to be wept over the latter to be reviled. They were first steamed to soften and dispose of interior matter, the eyes taken out and the eyelids sewn down and then were smoke dried and oiled. The 'false eyes' referred to by Banks were pieces of haliotis shell as used for the eyes of carved figures. It was these 'smoked heads' that were a popular article of Maori–European trade in the 1820s and 1830s especially when the tattooing was good. The Maori, however, was not a 'head-hunter'. He remarks in relation to the question of quarter that quarter was given and prisoners became slaves of the victors. Battles were not infrequently followed by the slaying and consumption of prisoners, and slaves were killed and eaten on ritual occasions, but it does not seem that the sort of stocking of a larder to which Banks refers was ever a matter of general practice.

Cook

Cook's journal, *The Voyage of the Endeavour 1768–1771*, was published by the Hakluyt Society and also edited by Dr Beaglehole (1955). Captain James Cook, FRS, has been called the greatest British maritime explorer ever and the greatest explorer of his age. He led three expeditions to the Pacific between 1768 and 1779 when he was killed in Hawaii. He managed by appropriate procedures to eliminate scurvy from his ships, which had hitherto made long sea voyages a disaster. On his second voyage of over 3 years he lost not one man from scurvy. He was a great navigator and practical scientist. He was the first European seaman to map the east coast of Australia, to cross the Antarctic Circle, to circumnavigate New Zealand and to chart the north-west coast of America.

In the early part of the exploration of New Zealand Cook records in his journal a statement about the eating of enemies. On Wednesday 17 January 1770 he writes:

Light airs Calms and pleasent weath^{r} PM righted the Ship and got ready for heeling out the other side, and in the evening hauled the Saine and caught a few fish, while this was doing some of us went in the Pinnace into a nother Cove not far from where the Ship lays; in going thether we met with a Woman floating upon the water who to all appearence had not been dead many days. Soon after we landed we met with two or three of the Natives who not long before must have been regailing themselves upon human flesh, for I got from one of them the bone of the fore arm of a Man or a Woman which was quite fresh and the flesh had been but lately pick'd off which they told us they had eat, they gave us to understand that but a few days ago they had taken Kill'd and eat a Boats crew of their enemies or strangers, for I beleive that they look upon all strangers as enemies; from what we could learn the Woman we had seen floating upon the water was in this boat and had been drowned in the fray. There was not one of us that had the least doubt but what this people were Canabals but the finding this Bone with part of the sinews fresh upon it was a stronger proof than any we had yet met with, and in order to be fully satisfied of the truth of what they had told us, we told one of them that it was not the bone of a man but that of a Dog, but he with great fervency took hold of his fore-arm and told us again that it was that bone and to convence us that they had eat the flesh he took hold of the flesh of his own arm with his teeth and made shew of eating. — AM Careen'd scrubed and pay'd the Starboard side of the Ship: While this was doing some of the natives came along side seemingly only to look at us, there was a Woman among them who had her Arms, thighs and legs cut in several places, this was done by way of Mourning for her husband who had very lately been kill'd and eat by some of their enimies as they told us and pointed towards the place where it was done which lay some where to the Eastward. Mr Banks got from one of them a bone of the fore arm much in the same state as the one before mention'd and to shew us that they had eat the flesh they bit a[nd] naw'd the bone and draw'd in thro' their mouth and this in such a manner as plainly shew'd that the flesh to them was a dainty bit.

Beaglehole's footnotes to this entry are of interest since they cover entries from other journals of the voyage.

'... the Capt bought one of their bones half pick'd to Convince the World that there are Cannibals. Nam'd the Bay Cannibal Bay' (Anon). Cannibal Bay is the name commonly given in the logs. Cook's bare statement is not enough for Wilkinson, who likes his horrors: 'they Saw Several Indians round an Oven made of Stones where the[y] Bake their fish or Flesh ... even the Intrels was Laying on a Bank of Grass by them'. Nor for Pickersgill: '... we saw one of the bodys and two arms with flesh upon them which we saw them eat this is the first Proof Possitive we have had of the Inhabitants being CANNIBALS and I belive these are the only People who kill their fellow creatuers Puerly for the meat which we are well Assured they do by their laying in wait one for another as a sportsman would for his game and they carry this detestable crime so far as to glory in carrieing in their ears the Thumbs of those unhappy sufferrs who fell in their way ... their was a young girl seized upon by some People in the same bay & eat one morning whilst we were here and about two hours afterwards they brought the Bones to sell to some of our People these with several other Instances of Barbarrous cruelty these savages is guilty of which ought to make them be abhord by all who may have occaison to tutch at these Islands.' This was evidently written up later.

Cook's entry of two days later reads

Winds and weather as yesterday and the employment of the people the same. In the PM some of our people found in the skirts of the wood three hip bones of Men, they lay near to a hole or hoven, that is a place where the natives dress their Victuals, this circumstance trifleing as it is, is still a farther proff that this people eat human flesh.

Sidney Parkinson, the natural history artist with Banks and Solander, who was described as a good journal-keeper and ardent collector of miscellaneous knowledge, also records the incident: 'We also found human bones in the woods, near the ovens, where they used to partake of their horrid midnight repasts.' Beaglehole remarks Mr Parkinson had read too deeply in romance; the Maori epicure would have been shocked at the idea of waiting till midnight.

On 20 January Cook recounts the instance which Banks covered in detail

Winds Southerly, fair and pleasant weather. Employ'd Wooding Watering &c[a] and in the AM sent part of the powder a Shore to be air'd. Some of the Natives brought along side in one of their Canoes four of the heads of the men they had lately kill'd, both the Hairy scalps and skin of the faces were on: Mr Banks bought one of the four, but they would not part with any of the other on any account whatever, the one Mr Banks got had received a blow in the Temple that had broke the skull.

On Wednesday 24 January Cook notes that

This forenoon some of us visited the Hippa. The inhabitants of this place showed not the least dislike at our coming but on the contrary with a great deal of seeming good nature showed us all over the place. We found among them some human bones the flesh of which they told us they had eaten. They likewise informed us that there was no passage into the sea through this inlet as I had imagined there might.

On Saturday 31 March 1770 Cook gives a general account of New Zealand prior to quitting the land. He writes in relation to our subject:

It is hard to account for what we have every w[h]ere been told of eating their enimies kill'd in battle which they most certainly do, circumstance enough we have seen to convince of the truth of this. Tupia who holds this custom in great aversion hath very often argued with them against it but they always as strenuously support it and never would own that it was wrong. It is reasonable to suppose that men with whome this Custom is found seldom or never give quarter to those they overcome in battle and if so they must fight desperatly to the very last. A strong proff of this supposision we had from the people of Queen Charlottes Sound who told us but a few days before we arrived that they had kill'd and eat a whole boats crew; surely a single boats crew or at least a part of them when they found themselves beset and over powerd by number would have surrender'd themselves prisioners was such a thing practised among them. The heads of these unfortunate people they preserved as trophies; four or five of them they brought off to shew to us, one of which Mr Banks bought or rather forced them to sell for they parted with it with the utmost reluctancy and afterwards would not so much as let us see one more for anything we could offer them.

Endeavour Journal

A further source of information in this field comes from the anonymous journal of the voyage of the *Endeavour*. It was published two months after the ship's return and nearly 2 years before Hawkesworth's account. It was entitled *Journal of a Voyage Round the World in His Majesty's Ship Endeavour in the Years 1768–1769–1770–1771*, D. Becket and P.A. De Hondt, London 1771. It was the first publication containing a narrative of the voyage. At one stage it was attributed to Banks and Solander or thought to have been the journal of one of the civilians who died at Batavia. Professor G.A. Wood in his *Discovery of Australia* (1922) suggested that the author was James Magra, the American midshipman who 12 years later, with name changed to Matra, brought forward after consultation with Banks a noteworthy proposal for the foundation of a colony in New South Wales. The account includes many detailed instances of confrontation with the Maoris. The entry for Tuesday 16 January 1770 reads:

Tuesday the 16th as we prepared to careen our ship several Indians and canoes came alongside with fish for sale which they offered to the deputy purser: but on his giving them the stipulated price they suddenly withdrew their fish and would have killed him with their hepatoos had he not precipitately escaped. This insidious attempt being represented to Captain Cook he seized a fouling piece, ready loaded with duckshot, and fired at the aggressor who being almost directly under him received the charge in his knee, which was thereby shattered in pieces, a few scattering shot likewise passed through his great toe. His wounds produced a plentiful haemorrhage, he bathed them in salt water, and the pain being acute he angrily threw the fish which he had sold and for which he had been paid into the sea. The Indians who were in the other canoes did not appear surprised either at the report of the gun or the wounds it had made though they all paddled round and examined them: nor did the wounded Indian retire, but wrapping himself up in mats he continued about the ship several hours. A little before this transaction, two of these Indians being prevented from coming on board by the master, who thought there were more on the deck than could be prudently admitted, they immediately drew their spears to assault him, and force submittance, and nothing but actual violence could drive them back to their canoes.

The same afternoon the captain, with several gentlemen, went in the pinnace to the other side of the bay, where they met several Indians who were employed in fishing. They had several baskets in their canoes, which we examined: and to our great surprise, found in them several limbs, and other parts of human bodies, which had been roasted and which it was evident they had lately eaten by the marks of their teeth, which we discovered in the flesh, and which appeared to have been recently gnawed and torn. We had been assured, that the inhabitants of New Zealand were cannibals from their own concurrent testimonies in many different places, but had never occular demonstration of the fact until this time.

When we enquired of these people how this human flesh came into their possession, they told us that 5 or 6 days before a canoe containing 10 men with 2 women had been driven into their bay from a different district, and they had attacked and killed them excepting one woman, who in attempting to swim, had been drowned: and that their bodies were afterwards divided among them, of which the flesh we had seen was a part. Perhaps they thought, like a celebrated philosopher, that it was as well to feed on the bodies of their enemies (for by their own account they eat no other) as to leave them to be devoured by crows. It is however certain they had no belief of any interpitude in this practice, because they were not ashamed of it: but, on the contrary, when we took up an arm for examination, they imagined us to be desirous of the same kind of food, and with great good nature promised that they would the next day spare a human head ready roasted if we would come or send to fetch it. Some gentlemen, who never left their own homes, have ventured, on the strength of speculative reasoning, to question the veracity of those travellers who have published accounts of cannibals in Africa and America; treating as falsehood every relation, which, from their ignorance of human nature, appears to them improbable: but let them not indulge the same freedom on this occasion; the fact will be too well attested to be rendered doubtful by their visionary impertinent objections.

While we [were] conversing with our cannibal, we observed something on shore roasting after the manner practised by the inhabitants of Georges Island, which they told us was a young dog: but suspecting it to be human flesh, we were going to open the oven, when we saw the hair and the entrails of a dog, which satisfied us concerning the truth of their account.

Wednesday, having finished careening our ship we began to wood and water, but in going to that part of the bay where we had discovered the baskets of human flesh, we found the body of a woman floating on the water, which was supposed to be the same that had been drowned in attempting to escape by swimming, as before related: but she was soon after claimed by an Indian, who told us she was his sister and having died had been sunk in the sea, according to the custom of their tribe: a custom which however is peculiar to the inhabitants of this bay.

In this part of New Zealand we saw many towns whose inhabitants had either fled or been exterminated: some of whom appeared to have been deserted or uninhabited 4 or 5 years being overgrown with shrubs and high grass.

Cook's second voyage: 1772–1775

Cook and his officers

The volume of Cook's journals covering his second voyage, with the *Resolution* and the *Adventure* from 1772 to 1775, have an introduction by Dr Beaglehole in which he summarizes how the two ships became separated and how Cook sailed with the *Resolution* on 25 November 1773, four days before the *Adventure* at last got into Queen Charlotte Sound. Furneaux, the commander of the *Adventure*, had had his fill of bad weather. They found a bottle with Cook's message and prepared to set out to sea as soon as possible. They were ready for sea on 17 December when a tragedy occurred. The ship's cutter went out to gather greens and failed to return and Burney, sent next day to investigate, made a report that struck the world as much as most other things in Cook's journals. Ten men had been killed and their bodies gone into the Maori *hangi* or cooking pit. This was cannibalism indeed, by the side of which the spectacle on board the *Resolution* to be recounted below was almost an innocent joke. These events are covered in the journals of several members of the ship.

However, first Cook's account of events prior to this massacre is recorded. On 23 November 1773 he writes:

Calm or light airs from the Northward so that we could not get to sea as I intended, some of the officers went on shore to amuse themselves among the Natives where they saw the head and bowels of a youth who had lately been killed, the heart was stuck upon a forked stick and fixed to the head of their largest Canoe, the gentlemen brought the head on board with them, I was on shore at this time but soon after returned on board when

I was informed of the above circumstances and found the quarter deck crowded with the Natives. I now saw the mangled head or rather the remains of it for the under jaw, lip &c[a] were wanting, the scul was broke on the left side just above the temple, the face had all the appearence of a youth about fourteen or fifteen, a peice of the flesh had been broiled and eat by one of the Natives in the presince of most of the officers. The sight of the head and the relation of the circumstances just mentioned struck me with horor and filled my mind with indignation against these Canibals, but when I considered that any resentment I could shew would avail but little and being desireous of being an eye wittness to a fact which many people had their doubts about, I concealed my indignation and ordered a piece of the flesh to be broiled and brought on the quarter deck where one of these Canibals eat it with a seeming good relish before the whole ships Company which had such effect on some of them as to cause them to vomit. [Oediddee] was [so] struck with horor at the sight that [he] wept and scolded by turns, before this happened he was very intimate with these people but now he neither would come near them or suffer them to touch him, told them to their faces that they were vile men and that he was no longer their friend, he used the same language to one of the officers who cut of the flesh and refused to except, or even touch the knife with which it was cut, such was this Islanders aversion to this vile custom. I could not find out the reason of their undertaking this expedition, all I could understand for certain was that they had gone from hence into Admiralty Bay and there fought with their enemies many of whom they killed, they counted to me fifty number which exceeded all probabillity by reason of the smallness of their own number, I think I understood them for certain that this youth was killed there and not brought away a prisoner, nor could I learn that they had brought away any more which increased the improbabillity of their having killd so many. We had reason to beleive that they did not escape without some loss, a young woman was seen, more than one, to cut and scar herself as is the custom when they loose a friend or relation.

That the New Zealanders are Canibals can now no longer be doubted, the account I gave of it in my former Voyage was partly founded on circumstances and was, as I afterwards found, discredited by many people. I have often been asked, after relateing all the circumstance, if I had actually seen them eat human flesh my self, such a question was sufficient to convence me that they either disbelieved all I had said or formed a very different opinion from it, few considers what a savage man is in his original state and even after he is in some degree civilized; the New Zealanders are certainly in a state of civilization, their behaviour to us has been Manly and Mild, shewing allways a readiness to oblige us; they have some arts a mong them which they execute with great judgement and unweared patience; they are far less addicted to thieving than the other Islanders and are I believe strictly honist among them-selves. This custom of eating their enimies slain in battle (for I firmly believe they eat the flesh of no others) has undoubtedly been handed down to them from the earliest times and we know that it is not an easy matter to break a nation of its ancient customs let them be ever so inhuman and savage, especially if that nation is void of all religious principles as I believe the new zealanders in general arê and like them without any settled form of government; as they become more united they will of concequence have fewer Enemies and become more civilized and then and not till then this custom may be forgot, at present they seem to have but little idea of treating other men as they think they should be treated under the same circumstances. If I remember right one of the arguments they made use on against Tupia who frequently expostulated with them against this custom, was that there could be no harm in killing and eating the man who would do the same by you if it was in his power, for said they 'can there be any harm in eating our Enimies whom we have killed in battle, would not those very enimies have done the same to us?' I have often seen them listen to Tupia with great attention, but I never found that his arguments had any weight with them or that they ever once owned that this custom was wrong and when [Oediddee] shewed his resentment against them they only laughed at him, indeed it could not be supposed that they would pay much attention to a youth like him. I must here observe that [Oediddee] soon learnt to convirse with these people tolerable well as I am perswaided he would have done with those of Amsterdam had he been the same time with them.

In relation to the presence of most of the officers at this incident, the journal of Charles Clark the Second Lieutenant records (21 November):

I ask'd him if he'd eat a peice there directly to which he very chearfully gave his assent. I then cut a peice off carry'd [it] to the fire by his desre and gave it a little broil upon the Grid Iron then deliver'd it to him—he not only eat it but devour'd it most ravenously, and suck'd his fingers ½ a dozen times over in raptures: the Captain was at this time absent, he soon after came on board, when I cut & dress'd my friend the other steak which he Eat upon the Quarter Deck before Capt. Cook and both were before the Ships Crew.

Captain Cook makes a further note on this incident on a separate slip of paper, which Beaglehole thinks may have been written as an afterthought, probably in England when Cook was preparing the journal for publication. It reads as follows:

Among many reasons which I have heard assigned for the practice of this horrid custom, the want of animal food has been one; but how far this is deducible from either facts or circumstances, I shall leave those to find out who advanced it, as [in] every part of New Zealand which I have been in, Fish have been found in such plenty that the Natives have generally caught as much as served both themselves and us; they have also plenty of Dogs, nor is there any want of wild fowl, which they know very well how to kill. So that neither this nor the want of food of any kind, can in my opinion, be the reason, but whatever m[a]y be it, I think it was but too evident that they have a great liking for this kind of food.

There is an account of the events relating to the *Adventure* in the log of James Burney, the Second Lieutenant. He was the son of Dr Charles Burney the musician and musical historian, and also accompanied Cook on his third voyage. He wrote a great five-volume work on the chronological history of the discoveries in the South Sea or Pacific Ocean, and also papers on defence, scientific and geographical matters. He was elected a Fellow of the Royal Society in 1809. He was highly regarded by a circle of friends that included Dr John Charles Lamb (Beaglehole 1961). Burney described how on Saturday 18 December 1773 he was ordered in the launch, which was well manned and armed, to go in quest of the missing cutter.

I now kept close to the East Shore and came to another Settlement where the Indians invited us ashore. I enquired of them about the Boat, to which they pretended ignorance—they appeared very friendly here & sold us some fish—within an hour after we left this place, in a small beach adjoining to Grass Cove we saw a very large double canoe just hauld up, with 2 men & a Dog—the men on seeing us left the Canoe & ran up into the woods—this gave me reason to Suspect I should here get some tidings of our Cutter—we went ashore & Searchd the Canoe where we found one of the Rullock ports of the Cutter & some Shoes one of which was known to belong to Mr Woodhouse, one of our Midshipmen, who went with Mr Rowe—one of the people at the same time brought me a piece of meat, which he took to be some of the Salt Meat belonging to the Cutter's Crew—on examining this & smelling to it I found it was fresh meat—Mr Fannin (the Master) who was with me, supos'd it was Dog's flesh & I was of the same opinion, for I still doubted their being Cannibals: but we were Soon convinced by most horrid & undeniable proofs—a great many baskets (about 20) laying on the beach tied up, we cut them open, some were full of roasted flesh & some of fern root which serves them for bread—on further search we found more shoes & a hand which we immediately knew to have belong'd to Thos Hill one of our Forecastlemen, it being markd T.H. which he had got done at Otaheite with a tattow instrument—I went with some of the people a little way up the woods, but saw nothing else—coming down again was a round spot cover'd with fresh earth, about 4 feet diameter, where Something had been buried [presumably the *hangi* or *umu*, the 'earth-oven' where the bodies had been cooked: Beaglehole 1961)]: having no spade we began to dig with a Cutlass—in the mean time I launchd the Canoe with an intention to destroy her—but seeing a great smoke ascending over the nearest hill, I got all the people in the boat & made what haste I could to be with them before Sunsett—on opening the next bay, which was Grass Cove, we saw 4 Canoes—a Single, & 3 double ones—a great many People on the beach—a large fire was on the top of the High Land beyond the woods, from whence all the way down the Hill the place was throngd like a Fair—those who were near the Shore had retreated to a small hill within a Ships length of the Water side, where they stood talking to us—as we came in I order'd a Musquetoon to be fired through one of the Canoes, as we suspected they might be full of men laying down in the bottom, but nobody was in them—the Savages on the little hill still kept hollowing & making Signs for us to come ashore—however as soon as we had got close in we all fired the first Volley did not seem to affect them much—but on the 2d they began to scramble away as fast as they could, some of them howling—we continued firing as long as we could see the leasst glimpse of a man through the bushes—amongst the Indians were 2 very stout men who never offer'd to move till they found themselves forsaken by their companions & then they walkd away with great composure & deliberation—their pride not Suffering them to run—one of them however stumbled, &, just made Shift to crawl off on all fours—the other got clear without any apparent hurt—I then landed with the Marines & left Mr Fannin to guard the boat—on the beach were 2 bundles of Cellery which had been gather'd for loading the Cutter—a plain proof that the attack was made here—a broken piece of an Oar was stuck upright in the Ground to which they had tied their Canoes—I then searchd all along at the back of the beach to see if the Cutter was there—we found no boat—but instead of her—Such a shocking scene of Carnage & Barbarity as can never be mentioned or thought of, but with horror.—whilst we remained almost stupified on this spot Mr Fannin call'd to us that he heard the Savages gathering together in the Valley, on which I returned to the Boat & hauld alongside the Canoes, 3 of which we demolished—whilst this was transacting, the fire on the top of the High Land disappeard & the Indians had gatherd together in the wood, where we heard them at very high words, doubtless quarelling whether or no they should come to attack us & try to save their Canoes—it now grew dark. I therefore just stept out & lookd once more along the back of the beach to see if the Cutter had been hauld up in the bushes—but seeing nothing of her returned & put off—our whole force would have been but barely sufficient to have gone up the Hill, & to have ventured with half (for one half must have been left to guard the Boat) would have been madness—As we open'd the upper part of the Sound we saw a very large fire about 3 or 4 miles higher up—this fire formd a complete Oval, reaching from the top of a hill down to the water Side—the middle space being inclosed all around by the fire, like a hedge—I consulted with Mr Fannin & we were both of the Opinion that we could expect to reap no other advantage than the poor Satisfaction of killing some of the Savages—at leaving Grass Cove we had fired a general Volley towards where we heard the Indians talking—but by going in & out of the boat the Arms had got wet & some 4 of the pieces mist fire—what was still worse it began to rain—ammunition was more than half expended & we left 6 Large Canoes behind us in one place—I therefore did not think it worth while to proceed where nothing could be hoped for but revenge.

Coming between 2 round Islands that lay to the Southward of East Bay we imagined we heard somebody calling—we lay on our Oars & listened but heard no more of it—we hollowd several times but to little purpose the poor Souls were far enough out of hearing—& indeed I think it some comfort to reflect that in all probability every man of them must have been killd on the Spot. We got on board between 11 & 12—

The people lost in the Cutter were Mr Rowe, Mr Woodhouse, Francis Murphy Quartermaster, Wm Facey, Thos Hill, Edwd. Jones, Michael Bell, Jno. Cavenaugh Thos. Milton & James Swilley the Captns Man—4 of them belonged to the Forecastle & 2 to the After guard—being 10 in all —most of these were of our very best Seamen—the stoutest & most healthy people in the Ship—We brought on board 2 Hands—one belonging to Mr Rowe, known by a hurt he had received in it the other to Thomas Hill as beforementiond, & the head of the Captns Servant—these with more of the remains were tied in a Hammock & thrown overboard with ballast & shot sufficient to sink it—we found none of their Arms or Cloaths except part of a pair of Trowsers, a Frock & 6 shoes—no 2 of them being fellows—

I am not inclined to think this was any premeditated plan of these Savages, as the morning Mr Rowe left the Ship he met 2 Canoes who came down & staid all the forenoon in Ship Cove. It might probably happen from Some quarrel, or the fairness of the Opportunity tempted them; our people being so very incautious & thinking themselves to Secure—another thing which encouraged the Zealanders was, they were sensible a Gun was not infallible, that they sometimes mist & that when discharged they must be loaded again, which time they knew how to take advantage of. After their success I imagine was a general meeting on the East Side of the Sound—the Indians of Shag Cove were there—this we knew by a Cock which was in one of the Canoes, & by a long Single Canoe which I had seen 4 days before in Shag Cove where I had been with Mr Rowe in the Cutter.

Wales

William Wales was the astronomer who accompanied Cook on his second voyage. Wales had observed the transit of Venus for the Royal Society at Hudson's Bay in 1769–1770. After his return with Cook in 1775, he edited the record of his and Bayly's observations and was elected a Fellow of the Royal Society in 1776. He was Secretary of the Board of Longitude from 1795 to 1798. He was characterized by Charles Lamb in his recollections of Christ's Hospital as a hardy sailor as well as an excellent mathematician and co-navigator with Captain Cook (Beaglehole 1971)

The journals of Wales are pertinent to one of the concerns of Dr Arens. Cook evidently had a high opinion of Wales and writes 'The Mr Wales, whose abilities is equal to his assiduity lost no one observation that could possibly be obtained.' Writing at Queen Charlotte Sound on 6 June 1773 Wales says:

Being going to leave this land of Canibals, as it is now generally thought to be, it may be expected that I should record what bloody Massacres I have been a witness of; how many human Carcases I have seen roasted and eaten; or at least relate such Facts as have fallen within the Compass of my Observation tending to confirm the Opinion, now almost universally believed, that the New Zeelanders are guilty of this most detestable Practice. Truth, notwithstanding, obliges me to declare, however unpopular it may be, that I have not seen the least signs of any such custom being amongst them, either in Dusky Bay or Charlotte sound; although the latter place is that where the only Instance of it was *seen* in the *Endeavour*'s Voyage. I know it is urged as a proof positive against them, that in the representation of their War-Exercise, which they were very fond of shewing us, they confessed the Fact. The real state of the Case is this. They first began with shewing us how they handled their Weapons, how they defyed the Enemy to Battle, how they killed him: they then proceeded to cut of his head, legs & arms; they afterwards took out his Bowels & threw them away, and lastly shewed us that they went to eating. But it ought by all means to be remarked that all this was shewn by signs which every one will allow are easily misunderstood, and for any thing that I know to the contrary they might mean they Eat the Man they had just killed; but is it not as likely that after the Engagement they refreshed themselves with some other Victuals which they might have with them? It ought farther to be remarked that I did not see one out of the many who went through those Massacres, who did not stop before he made the sign of eating; or that did it before some of us made the sign, as if to remind him that he had forgot that part of the Ceremony. This circumstance is brought as a proof both that they are, and that they are not Caniballs. One says, it is plain they Eat their Enemy after they have killed him, but are ashamed to acknowledge it, because they know we disapprove of it. The others say no; it is plain they looked on the Action as complete & stoped, untill you reminded them that it was necessary to take some refreshment after their Labour.

No stronger proofs, than the above have been seen by any person on board the *Resolution* this Voyage: but more substantial ones are said to have been seen on board the *Adventure*. One of the Gentlemen found amongst the Baggage in a Canoe, a raw human head, cut off close to the shoulders. This discovery is said to have produced much consternation amongst the People in the canoe, who put it out of sight again with all possible expedition, nor could it be found afterwards. But hard as this circumstance is made to bear against them, it does by no means follow with certainty that this Head was preserved to be eaten: Saul, had he been a stranger to the Affair, might with equal justice have concluded that David was a Caniball when he was brought before him with the Head of Goliath in his hand. On the other side it must be admitted that after what is reported to have been seen here in the *Endeavour*'s Voyage it was most natural to conclude that this head was intended for a very different purpose than that which David carried his up to Jerusalem for.

I cannot help relating a circumstance or two before I quit the subject entirely, as they tend to shew how far we are liable to be misled by Signs, report, & prejudice. Two Canoes of Natives came alongside one of the Ships & after they had traded such little matters as they had, went away out of the Sound. Sometime after One Canoe only returned & the people of the ship enquired ernestly what was become of the other. They made the same sign they do at the conclusion of their Exercise and it was immediately concluded that they had met with Enemies who had killed & eaten them; and it would certainly have passed so, had not the other Canoe returned a day or two after with every soul in it alive & well. Another circumstance is the following. A person on board the *Adventure* not only asserted that he saw one Indian killed by the others, but also related the particulars and manner in which it was done: nay it had gone so far on board our Ship with some as to suppose that a fire which [he] saw on shore, had been kindled to dress him at, & it was expected that Capt. Cook who was gone on shore at the place would surprize them in the very fact & probably bring on board some part of the unhappy Victim, to silence all unbelievers of this Custom for the future. The Capt notwithstanding returned without having seen any thing of the sort; & it was afterwards certainly known that the man was not only un-roasted, but even alive and well.

On return to Queen Charlotte Sound in New Zealand in November 1773 Wales records on the third of the month an incident in a shore expedition with Cook undertaken to set up an observatory:

We had not anchored before great numbers of the Natives came along side in their Boats, amongst whom we soon recognized several of those who came into the sound a day or two before we left it in June last: Capt. Cook also remembered one old Man who was chief at this place, and resident in it all the time that the *Endeavour* lay here but there was not one of the family amongst them who was here all the time the *Resolution* lay in this place in June last, and were so much frightened at the arrival of the newcomers. Our People soon learned *with certainty* that they had all been killed and eaten.

In his diary entry for 12–16 November, Wales notes

Moderate Wind at South-East; South, & South-West and very fine Weather. In the course of these five Days The Man and his family who were here in May & June last made their appearance amongst us, having had (I suppose) a joyful Resurrection from the Bowels of their Brother Cannibals; for nothing, it seems could be more certain than our Information that they were eaten! This is the third time, however, that I have been witness to our intellegence of this nature being false, or misunderstood.

On 23 November Wales writes:

I have this day been convinced beyond the possibility of a doubt that the New-Zeelanders are Cannibals; but as it is possible others may be as unbelieving, as I have been in this matter, I will, to give all the satisfaction I possibly can, relate the whole affair just as it happened. After dinner some of the Officers went on shore at a place where many of the Natives generally dwelt to purchase Curiosities, and found them just risen from feasting on the Carcase of one of their own species. It was not immediately perceived what they had been about; but one of the Boats Crew happening to see the head of a Man lying near one of their Canoes, they began to look round them more narrowly, and in another place found the Intestines Liver Lungs &c lying on the ground, as fresh as if but just taken out of the Body, and the Heart stuck on the points of a two pronged spear & tied to the Head of their largest Canoe. One of the Natives with great gayety struck his spear into one lobe of the Lungs, and holding it close to the Mouth of one of the Officers made signs for him to eat it; but he beged to be excused, at the same time taking up y[e] Head & making signs that he would Accept of that which was given to him, and he presented them with two Nails in return. These Gentlemen saw no part of the Carcase nor even any of the Bones; but understood that the unhappy Victim had been brought from Admiralty Bay, where these natives had lately been on a *hunting Party*; and one of them took great pains to inform them that he was the person who killed him.

When the Head was brought on board, there happened to be there several of the Natives who resided in another part of the sound, and who although in friendship with were not of the Party of whom the Head was purchased. These were, it seems very desirous of it; but that could not be granted: However one of them who was a great favorite was indulged with a piece of the flesh, which was cut off carried forward to the Gally, broiled and eaten, by him before all the officers & ship's Company then on board. Thus far I speak from report: the Witnesses are however too credible & numerous to be disputed if I had had no better authority; but coming just now on board with the Capt. and Mr. Forster, to convince us also, another Steake was cut off from the lower part of the head, behind, which *I saw* carried forward, broiled, and eaten by one of them with an avidity which amazed me, licking his lips and fingers after it as if affraid to lose the least part, either grease or gravy, of so delicious a morsel.

The Head as well as I could judge had been that of a Youth under twenty, and he appeared to have been killed by two blows on the Temple, with one of their *Pattoos*, one crossing the other; but some were of opinion that the whole might have been done at one blow, and that what appeared to have been caused by the other, was only a cross fracture of the Scull arrising from the first.

My Account of this matter would be very defective, was I to omitt taking notice of the Behavior of the young Man whom we brought with us from Uliateah & who came on board with the Captain, &c. in the Pinnace. Terror took possession of him the moment he saw the Head standing on the Tafferal of the Ship; but when he saw the piece cut off, and the Man eat it, he became perfectly motionless, and seemed as if Metamorphosed into the Statue of Horror: it is, I believe, utterly impossible for Art to depict that passion with half the force that it appeared in his Countenance. He continued in this situation untill some of us roused him out of it by talking to him, and then burst into Tears nor could refrain himself the whole Evening afterwards.

From this Transaction the following Corollaries are evidently deducible, viz—1st) They do not, as I supposed might be the Case, eat them only on the spot whilst under the Impulse of that wild Frenzy into which they have shewn us they can & do work themselves in their Engagements; but in cool Blood: For it was now many Days since the Battle could have happened.

2[d])That it is not their Enemies only whom they may chance to kill in War; but even any whom they meet with who are not known Friends: since those who eat the part of the head on board, could not know whether it belonged to a friend or Enemy.

3[d])It cannot be through want of Annimal food; because they every day caught as much Fish as served both themselves and us: they have moreover plenty of fine Dogs which they were at the same time selling us for mere trifles; nor is there any want of various sorts of fowl, which they can readily kill if they please.

4th) It seems therefore to follow of course, that their practice of this horrid Action is from Choice, and the liking which they have for this kind of Food; and this was but too visibly shewn in their eagerness for, and the satisfaction which they testified in eating, those inconsiderable scrapts, of the worst part on board the Ship: It is farther evident what esteem they have for it by the risks which they run to obtain it; for although our neighbours feasted so luxuriously, we had abundant reasons to conclude that they came off no gainers in the Action, since almost all of them had their foreheads & Arms scarrified, which is, it seems, their usual custom, when they lose any near Relation in War.

The following day Wales records going on shore with some officers to the Indian village where he saw the man's heart sticking on a spear and part of the head on the canoe as described. Intestines were lying dried out in the sun but the liver and lungs were not found and he supposes that they had now found an appetite for these also, the carcass being eaten.

On Wednesday 10 August 1774, when the *Resolution* was in the New Hebrides, Cook notes:

Yesterday Mr Forster obtained from these people the name of the island (Tanna) and today I got from them the names of those in the neighbourhood. They gave us to understand in such a manner which admitted of no doubt that they eat human flesh, they began the subject themselves by asking us if we did . . . otherwise we should never have asked them such a question. Have heard people argue that no nation would be cannibals if they had of other flesh to eat, or did not want victuals, and so lay the Custom to necessity; the people of this island can be under no such necessity, they have fine Pork and Fowls and plenty of roots and fruit; but sence we have not actually seen them eat human flesh, it will admit of some doubt that they are Cannibals.

Forster, however, got pretty explicit information that the Tanese and other New Hebrideans were, and enjoyed being, cannibals.

Cook's third voyage: 1776–1780

In the diaries where Cook describes his third and last voyage—that of the *Resolution* and *Discovery* 1776–1780—he records his presence at a human sacrifice in the Society Islands, and that at the Morais where this occurred there were 49 skulls every one of which was of a man who had been sacrificed there. He notes here that 'this is not the only barbarous custom we find amongst these people. We have great reason to believe there was a time when they were cannibals. However I will not insist upon this but confine myself [to that] as we have unquestionable authority for.' When in Fiji, Cook again discusses cannibalism and remarks that 'now I am again with cannibals let me ask those men who maintain that the want of food first brings men to this custom. What it is that induces them to keep it up in midst of plenty?'

On 24 February 1777 Cook recorded his conversation with Kahourah, the chief who headed the party that had cut off Captain Furneaux's men. He notes that it was the third visit that the chief had paid to them without showing the least mark of fear. He remarks that many of the Maoris had said he was a very bad man and urged him to kill the chief. Cook believed that they were not a little surprised that he had not, for according to their ideas of equity this ought to have been done. But he remarks that if he had followed the advice of all pretended friends he might have exterminated the whole race for the people of each hamlet or village by turns applied to destroy the other, a very striking proof of the divided state in which they lived. He could not misunderstand them as Omai who understood their language perfectly well was his interpreter. Cook gives a very interesting account of his attitude to these people in the light of what had occurred at this point of his journal. He remarks that from his own observations New Zealanders must live under perpetual apprehension of being destroyed by each other. There are few tribes that have not received some injury or another from some other which they are continually upon the watch to revenge and perhaps the idea of a good meal may be no small incitement. He was told that many years will sometimes elapse before a favourable opportunity happens and that the son never loses sight of an injury that has been done his father. Their method of executing these horrible designs is by stealing upon the party in the night. If they find them unguarded which however he believes to be very seldom the case they kill every soul that falls in their way not even sparing the women and children and then either feast and gorge themselves on the spot or carry off as many of the dead as they can and do it at home with acts of brutality horrible to relate.

The journals of William Anderson the surgeon on the *Resolution* and David Stamwell the surgeon's second mate also deal with the history of the cutter of the *Adventure* and other accounts of the outcome of tribal warfare between the New Zealanders.

The aggregate of these accounts involves first-hand observation and collaborative verification by a number of distinguished scientists. The inconsistencies, e.g. in the number of preserved heads seen, seem minor. The scientific quality, obvious from the overall achievements of the expeditions and the men, is all the more forcibly conveyed when these records are read in the context of the several thousand pages from which they are abstracted. Further, at the outset the scepticism of the writers on the issue of cannibalism is clear. They did not accept that which they could not verify. In any event there would seem no reason to discount the Maoris' own consistent statements of what they did, since Cook's evidence and initial accounts of this indicate in the main little or no desire at all to hide it. This material refutes Arens' main contention. It was, of course, available to him, as Tayler (1980) has also pointed out.

Australia

The distinguished Australian historian Geoffrey Blainey (1975) states the episodes of fighting in aboriginal Australia run counter to the beliefs of most anthropologists. Thus Moni Nag (1973), surveying the attitudes of living anthropologists, concluded that they did not see war as a major cause of death amongst hunters and food gatherers. 'Whatever evidence we have of armed conflicts among the contemporary hunters and gatherers tends to support this view.' In fact, Blainey states, evidence from the large part of Arnhem Land and Western Victoria defies this view. Blainey cites records of intertribal fighting among groups of aboriginals in relation to the fact that violent death—by spearing or clubbing—was a restraint, along with other forces, on the growth of population. In raids and ambushes, women and children were killed, as well as the men in pitched battles. In making computations from the accounts and the population densities he concludes that, in the Second World War, probably only the Germans and the Russians who lost 1 person in 150 of the population and 1 in 130 respectively, had higher losses than could be surmised from the aboriginal data, where the comparison was minimized since the conservative aboriginal figures, 1 in 270–300, involved comparing their normal life over years with only the war years of the industrialized nations. Thus the real effect on population over a comparable period of time might have been as great or greater than the combined effect of the Russo-Japanese, the First World War, the War of Intervention and the Second World War on the Russian population. As regards cannibalism, Blainey's view, from the records of the anthropologists and others who lived among the aborigines, is that cannibalism probably was more often the ritual aftermath of a death than a motive for murder. On occasions, however, aboriginals were deliberately killed so their flesh could be eaten.

Blainey remarks

> Every generation seems to writhe under a different nightmare, and in the nineteenth century—that heroic era of inland exploration—many Europeans in their nightmare imagined themselves falling into the hands of cannibals. In that century cannibalism was often regarded as the greatest depravity, the antithesis of civilization, and was so viewed even by many who regularly took Holy Communion and believed they were thereby eating the body and drinking the blood of Christ. In fact many aboriginals ate human flesh in the same spirit, believing that they thus acquired some of the strength of those who had

died. In many tribal areas from the De Grey River near the Indian Ocean to headlands on the Pacific Ocean, pieces of flesh were cut from a dead enemy and eaten. There are many records—some of them false or exaggerated—of aboriginals eating flesh which they had cut from those who had died peacefully.

Baldwin Spencer and Gillen (1927) reported that among the Loriitja-speaking people of Central Australia a very young child would occasionally be killed so that its flesh could be eaten by an older but weaker child. The Herbert River in North Queensland, the western district of Victoria and several other localities yield convincing evidence that certain aboriginals committed murder in order to eat flesh.

Howitt (1904) reports that the Mukjarawaint killed a man who eloped with a woman of the same totem. They roasted the flesh of the thighs and upper arms and ate them but the rest of the body was chopped up and left lying on a log to be eaten by fellow totemites including his brothers. He reports also a first-hand account by Hugh Murray of the Witaurong tribe who, having murdered an old man and child of the Colac tribe, brought on their spears portions of the flesh, which Murray saw them eat.

Spencer (1914) reports on the eating of the dead in northern Australian tribes, discussing the practice among the Mara and other tribes of Arnhem Land and near the Roper River. The corpse, he says, is cooked by men. Those who remove it from the oven and cut it up are either mother, brother or son to the deceased. There are rules governing who may eat whom. Later the bones are wrapped in bark and left for some time on a tree platform. Lumholtz (1889) records that aboriginals along the Herbert River practised cannibalism and during his time amongst them they killed two aboriginal men and ate parts of the flesh, especially the fat around the kidneys. The flesh of dead relatives is ritually consumed on the Daintree and Mollsman Rivers on the eastern coast of Cape York Peninsula. Those eligible to eat are the deceased's immediate family and members of the same clan or moiety (McConnel 1937).

Berndt and Berndt (1964) give a critical account of the literature on cannibalism in Australia, and amongst other strictures note that it was not unusual to hear aborigines of one tribe speak of the people in the next as cannibals, without any real evidence. Evidence becomes more difficult to assess as the traditional aboriginal world recedes into the past. Burial cannibalism was apparently fairly widespread, and in most cases only parts of the body were eaten. The expressed aims included the desire to absorb some of the dead person's qualities or attributes.

An eye-witness account of cannibalism of the body of a young woman killed in intertribal warfare in the Portland Bay region is included in the despatch from the Governor of New South Wales, Sir George Gipps, to Lord Stanley that was published in the British Parliamentary Papers of 1844. After her tribe had indulged in most violent grief at her loss she was devoured by her own friends and relations. The full details are given, including the emotional reactions of the eye-witness.

The Amazon Basin: The journals of Alfred Russell Wallace

Alfred Russell Wallace, the distinguished naturalist and scientist, spent three years in the Amazon before going to Borneo and eventually proposing, contemporaneously with Darwin, the theory of evolution by natural selection. In his journal he writes:

In my communications and inquiries among the Indians on various matters, I have always found the greatest caution necessary, to prevent one's arriving at wrong conclusions. They are always apt to affirm that which they see you wish to believe, and, when they do not at all comprehend your question, will unhesitatingly answer, 'Yes'. I have often in this manner obtained, as I thought, information, which persons better acquainted with the facts have assured me was quite erroneous. These observations, however, must only be taken to apply to those almost uncivilized nations who do not understand, at all clearly, any language in which you can communicate with them. I have always been able to rely on what is obtained from Indians speaking Portuguese readily, and I believe that much trustworthy information can be obtained from them. Such, however, is not the case with the wild tribes, who are totally incapable of understanding any connected sentence of the language in which they are addressed.

He goes on: 'As I am best acquainted with the Indians of the River Uaupes, I shall first state all I know of them, and then point out the particulars in which other nations differ from them.' He lists some thirty tribes which distinguish themselves from the inhabitants of other rivers, and gives detailed descriptions of their physical appearance and way of life. They had permanent homes and were basically an agricultural people cultivating a number of crops such as sugar-cane, sweet potatoes, maize, pineapples and bananas. They also ate meat and game occasionally, but their favourite food was fish, which they would catch daily. He notes that they are very fond of hot peppers, 'preferring the small red ones, which are of excessive pungency: when they have no fish, they boil several pounds of these peppers in a little water, and dip their bread into the fiery soup'.

Wallace states:

From the Fruits of the Baccaba, Patawa, and Assai palms (*Oenocarpus Baccaba*, *Oe. Batawa*, *Euterp oleracea* and allied species), they produce wholesome and nourishing drinks.

Besides these they make much use of sweet potatoes, yams, roasted corn, and many forest fruits, from all of which, and from mandiocca cakes, they make fermented drinks, which go under the general name of 'caxiri'. That made from the mandiocca is the most agreeable, and much resembles good table-beer. At their feasts and dances they consume immense quantities of it, and it does not seem to produce any bad effects. They also use, on these occasions, an intensely exciting preparation of the root of a climber—it is called capi ...

All the fish [they catch that are] not used at the time are placed on a little platform of sticks over the fire, till they are so thoroughly dried and imbued with smoke, as to keep good any length of time. They are then used for voyages, and to sell to travellers, but, having no salt, they are a very tasteless kind of food.

Salt is not so much sought after by these Indians as by many other tribes; for they will generally prefer fish-hooks and beads in payment for any articles you may purchase of them. Peppers seem to serve them in place of salt. They do, however, extract from the fruits of the Inaja palm (*Maximiliana regia*) and the Jara palm (*Leopoldinia major*), and also from the Caruru (a species of *Lacis* very common on the rocks in the falls), a kind of flour which has a saline taste, and with which they season their food. The caruru, indeed, has quite the smell of salt water, and is excellent eating, both boiled as a vegetable, or with oil and vinegar as a salad.

In relation to endocannibalism he has this to say:

The dead are almost always buried in the houses, with their bracelets, tobacco-bag, and other trinkets upon them: they are buried the same day they die, the parents and relations keeping up a continual mourning and lamentation over the body, from the death to the time of interment; a few days afterwards, a great quantity of caxiri is made, and all friends and relations invited to attend, to mourn for the dead, and to dance, sing, and cry to his memory. Some of the large houses have more than a hundred graves in them, but when the houses are small, and very full, the graves are made outside.

The Tarianas and Tucanos, and some other tribes, about a month after the funeral, disinter the corpse, which is then much decomposed, and put it in a great pan, or oven, over the fire, till all the volatile parts are driven off with a most horrible odour, leaving only a black carbonaceous mass, which is pounded into a fine powder, and mixed in several large couches (vats made of hollowed trees) of caxiri: this is drunk by the assembled company till all is finished; they believe that thus the virtues of the deceased will be transmitted to the drinkers.

He labels only one tribe in the area, the Cobeus, as 'real' cannibals:

They eat those of other tribes whom they kill in battle, and even make war for the express purpose of procuring human flesh for food. When they have more than they can consume at once, they smoke-dry the flesh over the fire, and preserve it for food a long time. They burn their dead, and drink the ashes in caxiri, in the same manner as described above.

They scarcely seem to think that death can occur naturally, always imputing it either to direct poisoning or the charms of some enemy, and, on this supposition, will proceed to revenge it. This they generally do by poisons, of which they have many which are most deadly in their effects: they are given at some festival in a bowl of caxiri, which it is good manners always to empty, so that the whole dose is sure to be taken. One of the poisons often used is most terrible in its effects, causing the tongue and throat, as well as the intestines to putrefy and rot away, so that the sufferer lingers some days in the greatest agony: this is of course again retaliated, on perhaps the wrong party, and thus a long succession of murders may result from a mere groundless suspicion in the first instance.

These extracts, as do the rest of the journals, give a clear impression that Wallace saw these things first-hand—even if he does not always explicitly state that he did. His evident caution in his communications with the Indians lends more weight to the view that the accounts are strict scientific reportage of what was intimately known to him.

Further support for endocannibalism among the Amazonian tribes has come from Dr Napoleon Chagnon (personal communication), author of *Yanomamö: the Fierce People* (1977):

I have in fact witnessed Yanomamö endocannibalism a number of times during the some 40 months I lived among them between 1964 and 1975. At death, the body is placed on a pyre of wood and completely covered with additional wood. The pyre is ignited and allowed to burn down to ashes which, when cool, are meticulously sifted and all pieces of bone and teeth carefully extracted from the ashes. The bones are then placed, a few at a time, into a hollowed-out log about 1.5 metres long and pulverized into ashes by using a long wooden pestle. The ashes are then placed into small, hollow gourds (if adult) or consumed entirely at that time (infants). In either case, a thick soup made from bananas (plantains, ripe) is poured into the hollow trough and consumed by drinking—usually by close kin. Men killed in warfare are never consumed by other men; both sexes consume the ashes of infants and adults if death was not caused by war. The ashes of a slain kinsman are never completely consumed until the death is avenged; I have seen ashes of men who were killed in 1965 still being used in endocannibalistic pre-raid rituals 8 years later ... their deaths not having yet been completely avenged.

Fig. 6-1 is a photo taken by Dr Chagnon of an endocannibalism ceremony.

Other aspects are described in Chagnon (1977), and he states there that the ashes of the dead are consumed so that the living will see their departed friends and relatives in *hedu*. Members of allied villages show their friendship to each other by sharing the ashes of particularly important people as a very intimate act. Endocannibalism, to the Yanomamö, is the supreme way of displaying friendship and solidarity.

Helmuth (1973) also says that endocannibalism was common in South America. In the belief that the soul resided in the bones, the Amahuaca, Tucana, Jumano, Waika, Surara, Pakidai and other tribes used to mix the ashes of the dead with their drinks in order to preserve for the tribe the life located in the bones. Helmuth states that according to Steward's *Handbook of South American Indians* (1946–1950) 16 tribes or sub-tribes practise or practised endocannibalism. The number of tribes reported to have practised exocannibalism (38) is much higher, and compari-

Fig. 6-1. Yanomamö Indians engaged in drinking the ashes of a deceased kinsman, mixed in banana (plantain) soup. (Photograph taken and kindly supplied by Dr Napoleon Chagnon, The Pennsylvania State University.)

son of these two groups of tribes in terms of their economic and social organization gives a clear-cut and statistically significant difference. Thus, of those with a hunter-gatherer economy with agriculture missing or unimportant, 14 practised endocannibalism and 6 exocannibalism. Of those with plantations and farming as the basic form of subsistence, 2 practised endocannibalism and 32 exocannibalism. If classification were on the basis of social structure, 13 of the tribes that were nomadic and had minimal social differentiation were endocannibals and 4 exocannibals, whereas of those with a permanent village-type organization 3 were endocannibals and 34 exocannibals. Possibly, just as hunting-gathering communities are more ancient and primitive than planting and farming communities, endocannibalism is older than and primary to exocannibalism.

Motivation, including nutritional status

Arens, discussing the motivation of cannibalism remarks that sociobiology, which attempts to integrate the accumulated wisdom of the ages from various intellectual branches, is now becoming fashionable. The new synthesis deals with the matter of why man eats man which is seen as the ultimate in human nastiness. Rampant cannibalism is assumed. Only the shared pessimism on the human condition provides a common meeting ground, as both the civilized and primitive come in for a vigorous lambasting in the debate—the issue being in the main whether the aggressiveness of man is at least partially genetically programmed or whether these tendencies are learned cultural responses. Freeman (1964)—'The history of primitive peoples with their bizarre expressions of human cruelty and aggression, in sacrificial rites, ceremonies of initiation, ritual mutilations, head hunting and cannibalistic cults and murderous societies'—is quoted as exemplifying this pessimistic view. Arens also quotes Raymond Dart's views along these lines, and Dart's incautious acceptance of all reports on cannibalism from Strabo onwards, and comments that even Wilson (1975), the grand synthesizer, with a major academic investment in the notion of adapted aggressive instincts, regards Dart's fulminations as dubious anthropology.

On the other side, the rationale is that cultural

responses to various situations have produced a potential cannibalistic nature in man. Montagu (1976), the standard bearer of these forces, states that the ban on cannibalism is often only a civilized prejudice. Therefore, if civilized man has been able to shed the anthropophagic cloak in making moral progress, cannibalism or its denial is no more than a cultural feat. In this respect, Arens notes that Helmuth (1973) prophetically states that the assumption of cannibalism as an aggressive act is only an example of the application of contemporary moral notions on the human past. He suggests primitive South American endocannibalism could imply a long history of friendly loving feelings and affection. Arens adds that with such scholarly apologists as defenders, the contemporary South American Indians need no enemies in the struggle to prevent extermination of their way of life. In the closing part of this discussion Arens records that Montagu (1976) who has maintained a cautious scholarly position on cannibalism has seriously undermined the entire evolutionary perspective. He points out that reports of cannibalism for more recent times are almost non-existent on hunting-gathering peoples, who have the simplest known economic system of adaptation. This means that it is not possible to point a relationship between primitive man and cannibalism as an atavistic aggressive trait, as many have done.

It is interesting that the accounts of cannibalism in the literature are predominantly from the tropical equatorial areas of the world—Melanesia, Indonesia, Borneo, equatorial Africa (French Gaboon, the Congo, Uganda, Kenya) and the Amazon area. When we consider nutritional status in relation to hypertension (see Chapter 27), the verified low sodium intake in the unacculturated societies in these areas and also the poor protein intake in some of them, but not in others, will be central to the analysis. In the tropical regions there is additional sodium loss contingent on temperature regulation.

The additional requirement of salt for reproduction in women, possibly coupled with central effects of hormones on appetites, may all conspire to make the desire for salt an element in the desire for meat. The high incidence of pica in women, particularly pregnant women, is analysed in Chapter 23, and in the ancient and medieval literature on this subject there are accounts of desire for human flesh—though the likely occasional co-existence of metabolically determined pica and psychosis could obviously produce any bizarre result. However, the predominance of kuru in women is interesting in this regard if indeed it be true that it was caused by cannibalism.

Wagner (1966), in discussing Sumner's review of the subject, states that cases are recorded in which primitive people finding themselves confined to a vegetarian diet without salt for a considerable time were driven to killing one of their own to satisfy the craving not necessarily for human flesh, but just for flesh. Sumner (1906) says the passion for meat among people who have to live on a heavy starch diet is especially strong. Hence they eat worms, insects and offal. It is asserted that the appetite for human flesh, once the eating of it is habitual, becomes a passion. When salt is not to be had the passion for meat reaches its highest intensity (Sumner 1906).

While there might be instances where there is substance in such generalities, in many societies where there are reliable accounts of cannibalism there is no evidence that it could have been related to any shortage of protein. This was clearly recognized in the accounts from Cook's voyages: the oceanic islanders had ready access to fish, and thus a source of some sodium. However, investigation of a population on the coral atoll of Ontong, Java, by Lot Page (personal communication) showed that despite the availability of seafood they seemed to have a low salt intake, and blood renin levels similar to those in a US population on a diet with 10 mmol Na/day. So some caution on this issue is merited. But in the equatorial jungles and mountains, very low sodium status, or deficiency, does occur and could conceivably underly other overt and seemingly immediate motives for cannibalism.

In the initial account of the tragic air crash in the Andes in the early 1970s, when the young survivors existed until rescue by eating the bodies of the dead from the crash, it was described how the blood clots around the heart were eaten and how texture and taste were different from the flesh and fat, of which staple they were sick. It was suggested that 'it was not just that senses clamoured for different tastes. Their bodies cried out for those minerals of which they had been deprived for so long—above all for salt. It was for this reason in obedience to these cravings that the less fastidious survivors began to eat those parts of the body that had started to rot' (Read 1974).

In relation to cannibalism in Africa, the Zulu, and the Maasai, who are cattle-breeders and live on a diet of blood and milk, are believed never to have been cannibals. With the Maasai the drinking of blood would, of course, supply adequate salt. In the accounts by Ehrstrom (1934) of the Eskimos and their diet, this habit of drinking blood from game was also noted. Josephson (1971) suggests this is because it has a higher salt content than muscle tissue. Motley (1959), in describing cannibal behaviour in the Belgian Congo, cites the view that lack of salt was causal, and that the reason there was no cannibalism in the north was not only because the people were Moslem but also because the great salt mines in the region of Timbuktu provided a source of salt. Allegedly, anthrophagy was commonest in the saltless regions.

Sumner (1906) in discussing endocannibalism suggests that its motive is affection, rather than the enmity which characterizes the out-group type. Thus, mothers ate their babies, if the latter died, in order to recover strength lost in bearing them. Sumner records many accounts of the eating of parents and the old from the literature. Tartars killed their fathers when old, burned the corpses and mingled the ashes with their daily food. The Taure of Brazil burn their dead, preserving the ashes in reeds and mixing them with their daily meals. The Jumanas on the head-waters of the Amazon regard the bones as the seat of the soul. They burn the bones of the dead, grind them to a powder, mix the powder with intoxicating liquor and drink it that the dead may live again in them. The Kobena drink in their *caxiri* the powdered bones of their dead relatives. In Australian aborigine cannibalism there was no desire to gratify or appease the appetite but the idea of saving the strength which would be lost, or of acquiring the qualities of the dead. There were no great feasts. Some tribes of Western Australia explained their practice of eating every tenth child born, as necessary to keep the tribe from increasing beyond the carrying capacity of the territory (Sumner 1906).

With regard to the virtues of the deceased being transmitted to the drinkers, it is plausible that there is a rationale. The vast tropical jungle of the 8 million km^2 (about 3 million square miles) of the Amazon Basin, with its rainfall contributing a fifth of all river water entering the oceans, is an area where certain minerals are very likely to be scarce in vegetation. The metabolic studies on the Yanomamö Indians recounted in full in Chapter 27, and the accounts in Chapter 3 of salt appetite in wild animals, give pertinent data on the likelihood or fact of sodium, and possibly phosphate and calcium, being in very short supply. The analysis of the water of the Amazon shows it to be conspicuously lower in total salts than the St Lawrence, Nile, Rhine and Thames, and particularly in those of calcium, sodium and phosphorus (the main salts in it being of iron and silicon: Kobayashi 1957). The inhabitants may live near marginal status for these three elements, with episodic body deficiency as a result of reproduction or incidence of disease. As far as sodium status is concerned, existence near the edge of deficit is clear-cut for the Yanomamö.

The behaviour of animals under these conditions, for example of hens in the Amazon eating the shells of their own eggs, is noted in Chapter 25 in relation to calcium deficiency. It is speculative but conceivable that customs with bones such as those recounted above are a ritualized method of recycling critical mineral nutrients in tropical rain forest conditions. It is possible that what was initiated at times of extreme deprivation produced a substantial sense of well-being for direct metabolic reasons by processes discussed in Chapters 11, 23 and 25, and that the behaviour, derived from experience of retroactive benefit, became ritualized with appropriate general overlay of myth. There seems no reason to deny feral man capacities that are clearly demonstrable in experimental animals. Metabolic studies on Yanomamö or similar groups in relation to calcium and phosphorus, to parallel the data on sodium would be of very great interest in this context. Mitchell et al. (1945) and other authors quoted by Fourman and Royer (1960) estimate total body calcium of a 70-kg man at about 1 kg (50 000 mequiv) of which 99% is in the skeleton. Total phosphorus is 500–600 g of which 88% is in the skeleton, and there would be approximately 1700 mmol of sodium in the skeleton. During a lactation period of 6 months a mother may lose as much as 50 g of calcium. The division of the mineral contents of a skeleton among 10–50 people could thus represent a highly significant addition to mineral intake. On the other hand, a study by Chagnon and Haines (1979) suggests protein intake of the Yanomamö is quite adequate, which could argue against significant calcium and phosphorus deficiency amongst them (see Chapter 27).

References

Arens W (1979) The man-eating myth. Anthropology and anthropophagy. Oxford University Press, New York

Banks J (1962) The endeavour journal of Joseph Banks, 2 vols, Beaglehole JC (ed). Trustees of the Public Library of New South Wales in association with Angus & Robertson, Sydney

Behm-Blancke G (1956) Bronze- und hallstattzeitliche Kulthöhlen im Gips-Gebirge bei Bad Frankenhausen. Ausgrabungen und Funde 1: 176

Behm-Blancke G (1958) Höhlen, Heiligtümer, Kannibalen. Archaologische Forschungen im Kyffhäuser, Leipzig

Behm-Blancke G (1959/60) Die altsteinzeitlichen Rastplätze im Travertingebiet von Taubach, Weimer, Ehringsdorf (Der Kannibalismus in Ehringsdorf). Alt-Thürignen 4: 131

Bentley WH (1900) Pioneering on the Congo (2 vols). Religious Tract Society, London

Berndt RM (1952) A cargo movement in the Eastern Central Highlands of New Guinea. Oceania 23: 40, 137

Berndt RM (1954) Reaction to contact in the Eastern Highlands of New Guinea. Oceania 24: 190, 255
Berndt RM, Berndt CH (1964) The world of the first Australians. Ure Smith, Sydney
Blainey G (1975) Triumph of the nomads: A history of ancient Australia. Macmillan, Melbourne
Burnet FM (1971) Reflections on kuru. Hum Biol in Oceania 1: 3
Bygott JD (1972) Cannibalism among wild chimpanzees. Nature 239: 410
Cargill (1838) Cited in Hogg (1958)
Chagnon NA (1977) Yanomamö: The fierce people. Holt, Rinehart & Winston, New York
Chagnon NA, Haines RB (1979) Protein deficiency and tribal warfare in Amazonia: New data. Science 203:910
Cook J (1955–1967) The Journals of Captain James Cook, 4 vols, Beaglehole JC (ed). Cambridge University Press for the Hakluyt Society. *The Voyage of the Endeavour 1768–1771* (1955), *The Voyage of the Resolution and Adventure 1772–1775* (1961), *The Voyage of the Resolution and Discovery 1776–1780* (1967)
Cook SF, Scott KG (1933) The nutritional requirements of *Zootermopsis (Termopsis) angusticollis*. J Cell Comp Physiol 4: 95
Dole G (1962) Endocannibalism among the Amahuaca Indians. Trans NY Acad Sci [II] 24: 567
Ehrstrom R (1934) Die Diät- und Kostführung der nordischen Länder in historischer Beleuchtung. Ein Beitrag zur Geschichte der Kostführung in vorhistorischer Zeit, des Kochsalzers und der Avitaminosen. Acta Med Scand 81: 583
Fourman P, Royer P (1960) Calcium metabolism and the bone. Blackwell Scientific, Oxford
Freeman D (1964) Human aggression in an anthropological perspective. In: Carthy JD, Ebling FJ (eds) The natural history of aggression. Academic Press, London, p 109
Gajdusek DC (1977) Unconventional viruses and the origin and disappearance of kuru. Science 197: 943
Gajdusek DC, Gibbs CJ (1966) Attempts to demonstrate a transmissible agent in Kuru, amyotrophic lateral sclerosis, and other sub-acute and chronic nervous system degenerations of man. Nature 204: 257
Geiseler W (1952) Schädelverletzungen, Kannibalismus und Bestattungen im europäischen Paläolithikum. Aus der Heimat 59: H12
Geiseler W (1953) Das jungpaläolithische Skelett von Neuessing. Aus der Heimat 61: H7/8
Glasse R (1967) Cannibalism in the kuru region of New Guinea. Trans NY Acad Sci 29: 748
Glasse R, Lindenbaum S (1976) Kuru at Wanitabe. In: Hornabrook RW (ed) Essays on Kuru. Classey, Farringdon, p 28
Goodall J (1979) Life and death at Gombe. Nat Geogr Mag 155: 592
Harris M (1978) Cannibals and kings: The origins of cultures. Random House, New York
Helmuth A (1973) Cannibalism in paleoanthropology and ethnology. In: Montagu A (ed) Man and aggression. Oxford University Press, London, p 229
Hinde SL (1897) The fall of the Congo Arabs. Methuen, London
Hogg G (1958) Cannibalism and human sacrifice. Robert Hale, London
Howitt AW (1904) The native tribes of south-east Australia. Macmillan, London
Josephson B (1971) Salt. Sartryck ur Nordisk Medicin 85: 3
Kalous M (1974) Cannibals and Tongo players of Sierra Leone. Auckland. Distributed by Kegan Paul, Trench, Trubner
Kobayashi J (1957) On geographical relationship between chemical nature of river water and death rate from apoplexy. Berichte des OHara i.f. Landwirtschaflichte Biologie, Okayama Universität 11: 12
von Koenigswald GHR (1958) Der Solo-Mensch von Java: Ein Tropischer Neanderthaler. In: Hundert Jahre Neanderthaler. Kemink, Utrecht, p 21
Kruuk H (1972) The spotted hyena: A study of predation and social behaviour. University of Chicago Press, Chicago, p 335
Lumholtz C (1889) Among cannibals. Murray, London
Lyth RB (Dr) Journal of Richard Burdsall Lyth, vol I 1836–1842, vol II 1842–44. B33, B534, in Mitchell Library, Sydney
McConnel UH (1937) Mourning ritual among the tribes of Cape York Peninsula. Oceania 7/3
Mead M (1950) Sex and temperament in three primitive societies. Mentor Books, New York
Mitchell HM, Hamilton TS, Steggerda FR, Bean HW (1945) The chemical composition of the adult human body and its bearing on the biochemistry of growth. J Biol Chem 158: 627
Montagu A (1976) The nature of human aggression. Oxford University Press, New York
Moni Nag (1973) Anthropology and population: Problems and perspectives. Population Studies 27: 61
Motley Mary (1959) Devils in waiting. Longman, Harlow, Essex p 99
Murray JHP (1912) Papua, or British New Guinea. Fisher Unwin, London
Payer C (1979) A question of cannibals. Nation, 16 June: 730
Read PP (1974) Alive: The story of the Andes survivors. Secker & Warburg, London (Reprinted in The Australian, 18 July 1974)
Richter CP (1955) Self-regulatory functions during gestation and lactation. In: Villee CA (ed) Macy Foundation Conference on Gestation. Josiah Macy Foundation, New York, p 11
Sahlins M (1978) Culture as protein and profit. The New York Review, Nov, p 45
Schaller GB (1972) The Serengeti lion: A study of predator/prey relations. University of Chicago Press, Chicago, p 480
Seligmann CG (1910) The Melanesians of British New Guinea. Cambridge University Press
Spencer B (1914) Native tribes of the Northern Territory of Australia. Macmillan, London
Spencer B, Gillen FJ (1927) The Arunta. A study of a Stone Age people, 2 vols. Macmillan, London
Steward JH (1946–50) Handbook of South American Indians (vols I–VI). Smithsonian Institute Bulletin, p 143
Sumner WG (1906) Folkways. Ginn, London
Tayler D (1980) Cannibals. Times Literary Supplement, 15 Feb
Wagner PM (1966) Food as a ritual. In: Farber SM, Wilson NL, Wilson RHL (eds) Food and civilization, a Symposium. Thomas, Springfield, Ill, p 60
Wallace AR (1853) Travels on the Amazon. Ward Lock, London
Wilson EO (1975) Sociobiology. Harvard University Press, Cambridge, Mass.
Wood GA (1922) Discovery of Australia. Macmillan, London

7 Physiological effects of sodium deficiency; the putative brain renin–angiotensin system

Summary

1. The body may be in sodium equilibrium, with daily output and input equal, over a wide range of sodium status. That is, turnover may involve 500–600 mmol Na/day as in inhabitants in Northern Honshu, Japan, or 2–10 mmol/day as in inhabitants of Highland New Guinea or the Amazon jungle. In other words, there is continuum of positive sodium status, and the 'set' of a variety of parameters as, for example, blood aldosterone, or the half-life of ^{22}Na in the body, will vary accordingly. The neutral status or 'set point' of the system has been defined as the amount of sodium in the body when the individual is in balance with no sodium intake or with intake which covers the minimal obligatory loss in faeces and urine and from skin (ca 1–2 mmol/day). Negative sodium status, sodium deficiency, denotes actual loss from the body. That is, sodium administered is retained until the 'set point' of neutral sodium status is reached. A striking example of negative sodium status is a ruminant animal, such as a sheep, with a parotid fistula. Here a situation can be reached where loss of sodium in saliva of low Na/K ratio may be exactly balanced by sodium intake in food, and the animal may remain in equilibrium for weeks with a negative sodium status of 500–700 mmol. It is noteworthy that it is a misnomer to call the clinical procedure whereby a subject is changed from a ward diet with 100–200 mmol Na/day to a diet with 5–10 mmol/day the production of sodium deficiency. It is a simple change of the equilibrium or steady state, both being on the positive side of neutral sodium status.

2. The onset of negative sodium status and its progression to severe deficiency entrains a very wide range of physiological effects any one of which, or any combination of several, might be responsible for the genesis of salt appetite.

3. The extensive catalogue of effects includes weight loss, and, in the case of loss of intestinal secretions, acid–base balance changes according to whether gastric or pancreatic or parotid secretion is lost. Languor, weakness, cramps and changed taste sensations are seen. Extracellular fluid and blood volume and eventually blood pressure decrease as does pressor response to infused angiotensin. Urinary sodium excretion, glomerular filtration rate and effective renal plasma flow decrease. Intestinal mobility decreases. Total exchangeable sodium, bone sodium and plasma sodium decrease. Adrenal response to angiotensin II and parotid response to aldosterone are increased. Muscle intracellular [Na] (sodium concentration) decreases.

4. Brain intracellular [Na] decreases as does [Na] in the cerebrospinal fluid (CSF). Brain intracellular [K] also decreases coincident with hyponatraemia.

5. Blood levels of renin, angiotensin and aldosterone are increased, and a variety of morphological changes occur in adrenal and salivary glands

(Chapter 3). [Na] in sweat and saliva decrease.

6. Angiotensin II and aldosterone concentration rise in CSF.

7. The evidence on the existence of a brain renin–angiotensin system is discussed in terms of the source of the increased angiotensin II found in CSF in sodium deficiency, and as a preliminary to consideration of any role of such a putative system in the genesis of salt appetite in sodium deficiency.

If salt intake were purely incidental to food intake, the appeal of which was governed by other factors, the operation of renal mechanisms which excrete salt in the presence of excess and can almost completely conserve it when intake is low could give relatively effective sodium regulation. This is probably the basis of control in many natural situations, though Arnold's (1964) data on sheep with oesophageal fistula, showing selective grazing by sodium-deficient animals of sodium-accumulating varieties of plants, indicated that ingestive regulation may operate unsuspected unless a very painstaking class of study is made. This is shown also in striking fashion by the selective grazing of aquatic halophytes by the moose of Isle Royale, Canada, as described in Chapter 3.

Sodium status

Sodium steady state in an organism may be established at different levels of status (Fig. 7-1). In the case of man, for example, equilibrium may involve an intake of 500–600 mmol/day as in the population in districts of Northern Honshu in Japan, 50–400 mmol/day as in urbanized American, European or Australian society, or 2–10 mmol/day as in New Guinea Highlanders or Yanomamö Indians of Brazil. In each instance balance is maintained but the setting of various physiological parameters is different. For example, the aldosterone concentration in peripheral blood (Denton et al. 1969) of New Guinea Highlanders is 28 ng/100 ml compared with 7 ng/100 ml in urbanized New Guineans. In this regard it should be emphasized that the low-sodium diet used in clinical experiments, where the change to this from a previous ward diet is characterized as 'sodium depletion', really amounts to a change to low turnover but still at or near neutral or positive status. A new equilibrium is established with turnover at 5–10 mmol/day, as in healthy unacculturated societies, rather than the previous 100–200 mmol/day. Negative sodium status can be characterized as that state where a substantial extra intake of sodium does not increase the total loss in urine, faeces and from skin by the same amount. As with positive status, there will be a gradation of effect, and attention has been drawn in Chapter 3 to the external balance studies on ruminants such as the sheep where there may, in fact, be equilibrium (i.e. daily balance) where sodium intake in food equals daily sodium loss in parotid saliva of low Na/K ratio in circumstances where a residual body sodium deficit of e.g. 500–600 mmol persists over weeks (Denton 1957b). More severe deficit, even in the

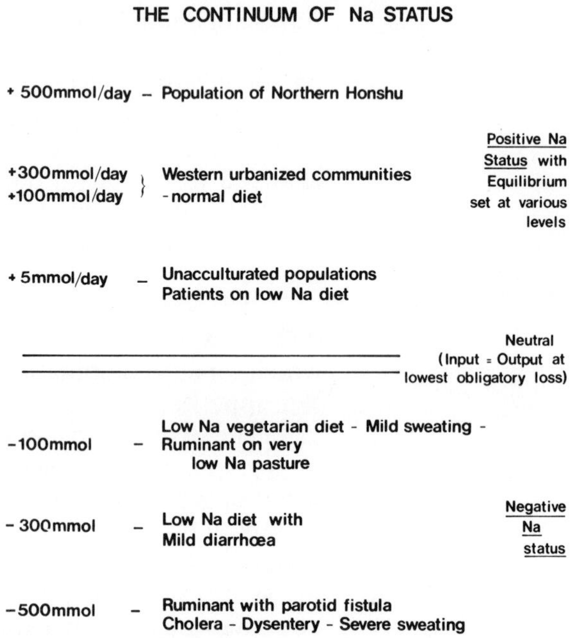

Fig. 7-1. Diagrammatic representation of the continuum of sodium status reflecting that steady state may hold at various levels of intake (positive sodium status) and that a neutral point or 'set point' exists. In the instance of negative sodium status the situation may hold where in an animal with a parotid fistula and low salivary Na/K ratio as a result of aldosterone secretion, salivary sodium loss equals sodium intake in food, and the animal is in steady state with residual body deficit of e.g. 500 mmol sodium.

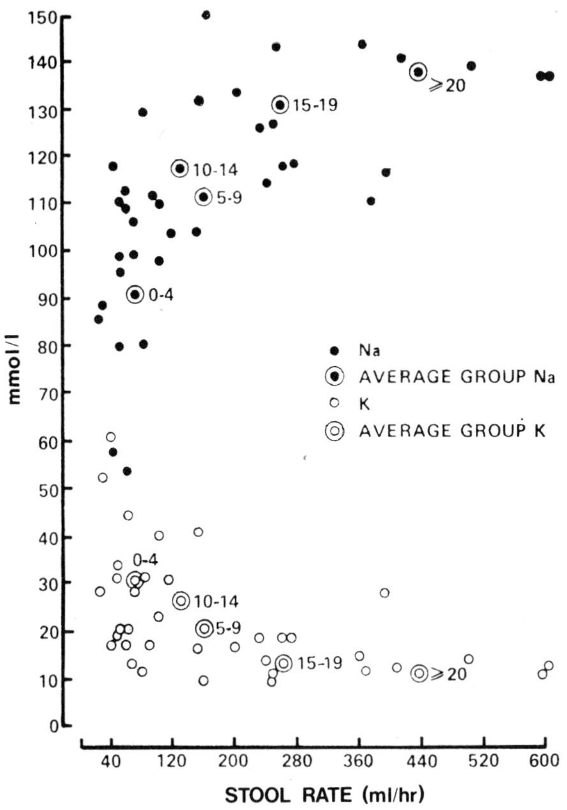

Fig. 7-2. Stool sodium and potassium analyses (mmol/l) from 36 cholera patients during period of maximum diarrhoea (12–24 h after admission). At all ages sodium tends to rise and potassium to fall at higher diarrhoea rates. The number of patients were 12 (0–4 yr), 10 (5–9 yr), 5 (10–14 yr), 2 (15–19 yr), 7 (20 yr and over). (After Nalin and Cash 1976.)

ruminant does, of course, cause physiological deterioration, as is the case in man when rapid intestinal or skin loss of sodium occurs as a result of dysentery, cholera or severe sweating. Fig. 7-2 indicates the rate of sodium loss in cholera in children.

A searching analysis of the 'set point' of sodium homeostasis has been made by Hollenberg (1980). He emphasizes that sodium homeostasis, especially control of extracellular volume, is an excellent example of a feedback control system. The afferent and efferent limbs of the system have received considerable attention but the 'set point' has received remarkably little emphasis. He notes a personal clinical observation of a normal man, in sodium balance on a 10 mmol/day intake, inadvertently being given a 30 mmol sodium load intravenously and promptly excreting it. This was in the face of the fact that the negative sodium balance which had accrued during the interval he came into balance on his new reduced sodium intake was over 300 mmol. Similar findings have been documented in several studies (Braunwald et al. 1965; Hollenberg et al. 1972), including the outstanding work of Strauss and colleagues (1958). They showed that a person in balance on a low-salt diet responded with a prompt natriuresis to a 30 mmol sodium load. If a diuretic was administered beforehand and 100 mmol of sodium lost, then the 30 mmol load did not cause natriuresis. Indeed this did not occur until the 100 mmol lost with the diuretic was replaced. The data suggest the 'set point' around which sodium balance cycles in the normal person is that amount of sodium chloride in the body when the person is in balance on no-salt intake (Hollenberg 1980). This 'set point' might be viewed also as the situation where small obligatory loss in urine and faeces and from skin is balanced by sodium in food: 1–2 mmol/day.

Normally we are in a state of sodium excess, but when excretion of sodium is reduced on restriction of sodium intake, it does not reflect activation of mechanisms designed to retain sodium. Rather these mechanisms are suppressed by an expanded sodium space, and reduction of sodium intake causes reduction of the stimulus to suppression. Thus the system is running downhill, and the exponential reduction observed in the amount of sodium excreted is appropriate, as it reflects a progressive reduction of the stimulus to excretion of sodium (Hollenberg 1980).

It is noteworthy here that a 30 mmol change in sodium balance produces a clear-cut physiological reaction. Data discussed in Chapter 13, including the 'error' in body sodium necessary to evoke salt appetite and other physiological changes in large animals like sheep, suggest that effects are produced by a change in sodium status not much greater than this. For example, the natriuretic effect of intraventricular microinjection of hypertonic sodium chloride is blunted by loss of 50 mmol sodium from a parotid fistula and abolished by loss of 100 mmol (Abraham et al. 1975). It is also interesting to note the individual variations in behaviour of sheep with constant salivary loss from a parotid fistula and 24-h access to a bar-press delivering sodium bicarbonate solution. The behaviour records show episodes of bar pressing at 2–5-h intervals, but while some animals bring their sodium status up to or near the neutral 'set point', where there is little or no sodium in the urine, others evidently always continue ingesting sodium so that their status is much on the positive side of the 'set point', and they are excreting a large amount of sodium in the urine. (For further details see Chapter 13.)

Hollenberg (1980) raises interesting questions about the possible role of a change of 'set point' in the aetiology of disease states such as Conn's syndrome, and states of oedema including the puzzling cyclic oedema.

Positive sodium status

Studies on change of sodium status with a high level of intake (300–400 mmol/day), in some instances on a short-term basis, have shown weight increase (Lyons et al. 1944; Grant and Reischsman 1946), increase in exchangeable body sodium (Jagger et al. 1963), increased half-life in the body of injected radiosodium (^{22}Na) (Dahl 1960; Smilay et al. 1961) and increase in plasma volume, extracellular volume and venous pressure. Nichols and Nichols (1956b, 1957) report that in the rat high sodium intake may cause an 8.6% increase in bone mineral sodium—equivalent to sequestration of 213 mmol of sodium in the adult human—but Forbes et al. (1959) did not find a comparable effect in the cat and nor did McDougall et al. (1974) in the sheep. Other aspects of positive sodium status will be discussed in Chapter 27, which reviews salt intake and hypertension.

Negative sodium status and its effects

The effects of negative sodium status ranging from equilibrium with small deficit to very severe body deficiency involve a wide diversity of physiological changes. It is appropriate to recount them in some detail, as, on the face of it, any one alone or in combination with others could be the initiating stimulus which generates the salt appetite drive. Self-evidently, some changes appear more likely than others to have such a role, but at present the answer is not known—though there is suggestive evidence on certain factors.

I have chosen also to include in this chapter some detailed discussion of the putative brain renin–angiotensin system since consideration of it arises logically from data on change of angiotensin II concentration in the cerebrospinal fluid (CSF) in sodium deficiency.

Under natural conditions, sodium deficiency will result when any gland excretes more sodium than the organism's intake or when any intestinal secretion which is normally reabsorbed is lost from the gut, or sequestrated in it. In disorders of body chemistry, e.g. diabetic acidosis, or with drug administration, loss by urine can also deplete the body. Severe depletion is more likely in any of these circumstances if sodium intake is low. It is also evident that whereas with sweat loss there will be little distortion of ionic pattern and pH, loss of alkaline body fluids such as pancreatic juice or ruminant saliva will produce, as well as sodium deficit, an acidaemia. Similarly loss by vomiting of gastric secretions, which contain chloride in excess of sodium relative to extracellular proportions, will produce an alkalaemia as well as sodium deficit. These concurrent stresses on acid–base balance induce urinary reactions and other ionic changes which will be discussed.

Historically, as I have aimed to illustrate in the references cited, major advance in knowledge of effects of salt depletion on body fluids was made in the period 1930–1960, with consequent great advance in clinical medicine. The paediatricians played a major role.

Growth

The chronic effects of a very low sodium diet on the growth and development of weaned rats was investigated by Orent-Keiles, Robinson and McCollom (1937), using a diet which eliminated the possibility of concurrent vitamin deficiency. The experimental animals grew fairly rapidly for a few weeks and then began to lose weight. No noticeable signs were apparent until the sixth to eighth week, when the animals began to show striking eye changes: some rats began to exhibit a bluish-grey cornea, some had perforated ulcers and all showed a thin sanguine secretion covering both eyes. After another 2 weeks nearly all rats had eye changes and histological section showed the whole cornea infiltrated with wandering cells. The authors note these changes were markedly different from those produced by vitamin A deficiency, and the diet used had adequate vitamin A. There was a 100% mortality by 18 to 21 weeks. At post-mortem the body was lean, there was marked atrophy of muscle tissue, the lungs were badly infected and the bones were softer and cut more easily than those of the controls. There was very little cartilage and the testes were smaller than those of the controls. The suprarenals were stated to be orange rather than the pale pink colour of controls. Sexual maturity of female rats was somewhat delayed and the ovulation rhythm was gravely affected, there being a prolongation of the oestrous desquamation stage that resembled that in vitamin A deficiency. The sodium-deplete females rarely copulated when mated with normal males. The sodium-deplete males remained fertile and mated with normal females for 75 to 80 days. In control experiments where similar rats were given the same diet but deficient in chloride instead of sodium, growth was retarded but no other signs developed over 90 days. On diets deficient in both sodium and chloride, the animals were not so retarded in growth as those on low-sodium alone. They developed severe alopecia of the anterio-dorsal part of the body but the eyes and bones were stated to be normal. The authors consider that the decreased resistance to infection may not be due directly to sodium depletion but rather be an indirect effect.

Body weight and physical condition

The body weight usually falls when sodium intake is significantly restricted. In Crabbé et al.'s (1958) study of human volunteers on a diet with less than 10 mmol Na/day, the body weight fell 1.0–1.5 kg in 5–6 days; this is less than in McCance's (1936a) study, where sweating occurred also. Weight loss of 4.6 kg in 4 days occurring during the osmotic diuresis induced in diabetic patients by withdrawals of insulin therapy was partly attributable to sodium deficiency (Atchley et al. 1933). Leard and Freis (1949) observed weight loss of 2.1–2.9 kg in normal subjects made sodium-deficient by low-sodium diet and mercurhydrin injections.

In dogs made severely sodium-deficient by loss of pancreatic secretion from a permanent fistula (Gamble and McIver 1928) body weight declined even though food intake was maintained for the first 10 days. A similar weight decline despite constant food intake has been observed in sheep made sodium-deficient as a result of loss of parotid secretion from a permanent fistula (Denton 1957b). Observations parallel to the apathy and lassitude of sodium-deficient volunteers recounted by McCance, and by workers studying sodium deprivation in the tropics (McEwen 1935; Marriott 1950) have been made regularly on sodium-deficient animals. Following Pavlov's (1902) description of dogs with pancreatic fistulae, Gamble and McIver (1928) and Elman and Hartmann (1930) also reported on the eventual asthenia and apathy of dogs with pancreatic fistulae, and the latter authors also noted marked gastric irritability and vomiting. Darrow and Yannet (1935) reported languor in monkeys, rabbits and dogs made sodium-deficient by intraperitoneal glucose injection. In the monkeys and rabbits a greyish pallor similar to that seen in sodium-deficient humans developed. When sheep become grossly sodium-deficient, apathy and loss of appetite have been observed, and this may eventually be associated with decline of the characteristic appetitive behaviour exhibited when sodium solutions are prepared for presentation to sodium-deficient sheep (Denton and Sabine 1963).

Reported symptoms of sodium deficiency in man include mild to severe cramps (Black 1953; Marriott 1950), peculiar sensation in mouth, headache, nausea, breathlessness, tetany (Hartmann and Elman 1929) or hyperventilation (Gamble and McIver 1928; Denton 1949; Denton et al. 1951). The particular symptoms are evidently dependent on the mode of production of sodium deficiency, with its resultant alkalosis or acidosis, and clearly concurrent potassium depletion may contribute to some of the deterioration cited above.

Marriott (1950) in his monograph stresses the absence of thirst, and notes that for the production of pure salt depletion there must be liberal intake of water. This clinical observation, which serves to contrast for differential diagnosis the state of sodium depletion from that of pure water depletion, would not, however, seem to prejudice the fact that precise balance studies on such patients might show some increase of water intake. The data from studies on dogs by Cizek et al. (1951) and our studies on sheep with parotid fistulae (Abraham et al. 1976) show a clear increase in water intake as a result of sodium deficiency. Fitzsimons (1979) has analysed the data fully.

McCance (1936a, b) in his classical study of human volunteers who underwent salt deficiency reported that he and his subjects noticed a peculiar sensation in the mouth. He states:

> Their sense of flavour and taste was affected. E interpreted this aberration or lack of sensation as thirst. She complained of it constantly and drank freely but without obtaining any relief. RAM recognized the feeling as distinct from thirst. His mouth was not unusually dry but food was tasteless, even highly flavoured food, and this was the more noticeable because such foods were eagerly sought to make the meals more appetizing. RBN was not so much troubled by this symptom but felt it from time to time. He noted once that he was 'thirsty all morning—drank a lot but water seems to make little difference' and on another day reported that he had a funny feeling in the mouth. (McCance 1936a)

Body fluid compartments and blood pressure

Of major significance in the symptoms and signs of sodium deficiency is the shrinkage of the extracellular fluid (Black 1953; Elkinton and Danowski 1955).

Haemoconcentration, with rise in plasma protein concentration (Gamble and McIver 1928; Elkinton et al. 1946, 1947; Denton 1949; Marriott 1950) and decline in plasma volume (Darrow and

Yannet 1935; Leard and Freis 1949; Elkinton and Danowski 1955), has been reported by most workers studying sodium deficiency caused by external loss of sodium, or by intraperitoneal injection of 5% glucose or hypodermoclysis with 5% glucose. Elkinton et al.'s (1946) comprehensive studies involved comparison of blood and haemodynamic changes in acute water deficiency produced by osmotic diuresis with 10% urea and 5% glucose infusion, with those in salt deficiency produced by intraperitoneal glucose injection and withdrawal 4–5 h later. Though some electrolyte was lost in the osmotic diuresis, the difference in haemodynamic change in the two conditions in unanaesthetized dogs was striking, plasma volume declining much more in salt deficiency than water deficiency for an equivalent change in extracellular volume. In the salt deficiency large amounts of circulating plasma protein were lost which were not recovered in the peritoneal fluid, whereas in water depletion protein loss was small. The authors considered that the circulatory deterioration in this special type of salt deficiency may well be related to this loss of protein from the circulation. Loss of total circulating protein has also been recorded in sodium deficiency in man (Abbott et al. 1952). Marriott (1950) has emphasized the importance of increased viscosity of the blood in circulatory changes, as shown experimentally by Seligman and colleagues (1946).

Associated with the large decrease in plasma volume instanced above, these early studies of acute salt depletion showed that the mean arterial blood pressure fell by 20–80 mmHg, cardiac output decreased by 60%, the arteriovenous oxygen difference increased and the peripheral resistance rose 60%–200% (Elkinton et al. 1946). Blood pressure decrease in dogs with sodium deficiency produced in a similar manner was recorded by Gilman (1934). In dogs with acute sodium deficiency, decreased oxygen consumption and impaired cardiorespiratory function have been reported (Khoyi 1967; Fulton and Ridolpho 1971). Atrial pressure and pressure in large venous capacitance vessels decline, and blood flow through organs decreases. Black (1953), discussing clinical features of sodium deficiency, records that hypovolaemia with reduced cardiac output leads to inadequate circulation; reflex vasoconstriction may maintain the central blood pressure for a time but in severe sodium depletion blood pressure falls.

Marriott (1950) notes that in 1942 in India there were 1959 admissions of troops to hospital with heat effects, with 136 deaths. He describes salt-depleted patients as exhausted and pallid with a clammy sweat and cold extremities. Eyes are sunken and in later stages the appearance resembles the Hippocratic facies. He states that blood pressure is at first maintained and the diastolic pressure may be raised (Ladell et al. 1944) but later falls. The pulse pressure is diminished and cardiac rate increased, progressing towards oligaemic circulatory failure. Other marked clinical features are the giddiness and liability to orthostatic fainting and incidence of cramps in 50%–70% of cases. In Leard and Freis's patients (1949) no fall in blood pressure was observed despite a mean decline of plasma volume of 0.5 litres, whereas Abbott and colleagues (1952) record blood pressure decrease with sodium deficiency and plasma volume reduction produced by hypodermoclysis of 5% glucose. Clearly this is a matter of degree and efficacy of several compensatory mechanisms. In sheep made severely sodium-deficient by several days loss of saliva from a parotid fistula the systolic blood pressure fell (Denton 1957a).

More detailed haemodynamic studies on sheep by Dr B. Scoggins and colleagues (unpublished) have shown that unilateral parotid cannulation with mean sodium loss over 48 h of 630 mmol (n=7) resulted in:

a) No change in mean arterial pressure (MAP), which remained at 68–70 mmHg.
b) A significant fall in cardiac output, from 4.7 to 3.8 l/min, due to fall in stroke volume. The heart rate rose significantly.
c) An increase in total peripheral resistance from 14.2 to 17.1 mmHg $l^{-1}min^{-1}$.
d) A rise in plasma renin concentration (PRC) from 1.4±0.2 to 10.8±2.3 pmol h^{-1} ml^{-1}.
e) A significant fall in plasma [Na] from 147 to 143 mmol/l.

Sodium restriction for 14 days in sheep (n=9) produced:

a) A fall in urinary sodium excretion from 82±5 to 8±2 mmol/day with calculated mean deficit of 130 mmol.
b) No change in MAP, heart rate, cardiac output or total peripheral resistance or plasma sodium.
c) A rise in PRC from 1.9±0.4 to 7.3±2.8 pmol h^{-1} ml^{-1}.

In both situations intravenous infusion of converting enzyme inhibitor (SQ20881) produced substantial falls in MAP (15–20 mmHg) compared with those seen in sodium-replete animals (2–5 mmHg). Coghlan et al. (1977) showed parotid cannulation of sheep with loss of 250–670 mmol of sodium caused reduction of extracellular fluid volume but this only accounted for ca 20% of sodium lost, the remainder presumably coming from the rumen and, to a small extent, bone.

Sullivan et al. (1980) in an important study of haemodynamic changes with sodium status in normal and borderline hypertensive humans showed MAP was 82±1.9 mmHg on an ad libitum sodium diet (mean daily sodium excretion, 167±24.1 mmol) and 78.5±2.1 mmHg on a 10 mmol/day diet (mean daily sodium excretion, 24.2±3.5 mmol).

Lohmeier et al. (1980) studied intact and adrenalectomized dogs when sodium intake was reduced from 50 to 5 mmol/day. No change occurred in MAP in intact dogs, but a small fall (8 mmHg) was seen in adrenalectomized dogs sustained on basal steroid infusion. Glomerular filtration rate was unchanged in both groups, but, in both, effective renal plasma flow fell 7%–9%. Conway et al. (1979) reduced sodium intake of dogs from 70 to 5 mmol/day for 8 days and also administered diuretics. Sodium depletion increased plasma renin activity (PRA) from 1.11±0.15 to 26.48±3.77 ng ml^{-1} h^{-1}. There was little change in mean arterial pressure. Cardiac output reduced progressively from 2.6 to 2.0 l/min, and central venous pressure was reduced. Heart rate and total peripheral resistance increased progressively. Administration of converting enzyme inhibitor caused a 20–35 mmHg fall in MAP which was related to the PRA level extant.

Coleman et al. (1975) showed that the blood pressure of rats given a low-salt diet for one month was no different from normal, as was cardiac output and total peripheral resistance. Under anaesthesia, the low-salt rats showed a large fall in blood pressure (−47 mmHg) with converting enzyme inhibitor, comparison with the normal controls indicating that an increase in total peripheral resistance was supported by angiotensin. Agabiti-Rosei et al. (1979) showed in the human given frusemide and maintained on a low-sodium diet that saralasin caused a small fall in blood pressure in three of five subjects.

Decrease of extracellular volume, cited as 16.4 litres or 23% of body weight in a 70-kg man (Moore et al. 1956), is a principal result of sodium deficiency in man (McCance 1938; Gamble 1949; Wolf 1958), dog (Elkinton and Danowski 1955) and sheep (Denton et al. 1952; Coghlan et al. 1977). There is evidence that in the initial stages of sodium deficiency decrease in volume of interstitial fluid precedes decrease in plasma volume (Gamble and McIver 1928) though in the investigations by Mellors et al. (1942), decrease in plasma volume began in the early stages of deficiency. As shown clearly by McCance (1936a), if free access to water is permitted, initially loss of sodium is followed *pari passu* by loss of water, the tonicity of extracellular fluid being maintained, but as sodium deficiency progresses the volume of extracellular fluid is preserved to some extent at the expense of a fall in osmotic pressure, a physiological process which involves some measure of accommodation of the osmoreceptors (Verney 1947). Schoelkens et al. (1980) report no change in plasma antidiuretic hormone (ADH) following reduction of sodium content of diet in dogs; the values were 3.97±0.66 pg/ml in animals on normal diet and 3.62±0.38 pg/ml in those on low-sodium diet (n=5). The process may also involve reduction of tonic inhibition of thirst arising from stretch receptors in the walls of the low pressure system—the thoracic vessels and atria (Fitzsimons 1979).

Organ blood flow

Consistent with the experimental evidence recorded above of reduced cardiac output and increased peripheral resistance in dogs with severe salt deficiency (Elkinton et al. 1946) there are some observations on reduced blood flow through organs in salt deficiency. This is an important consideration since such circulatory changes may have effects on the specific functions of the organ. Decline in renal blood flow and glomerular filtration rate with rise in blood urea and the characteristic circulatory changes in the skin with, presumably, attendant decline in the temperature-regulatory capacities of this organ, have been cited above and later. A comprehensive study has been made of the blood flow of both the abdominal and the transplanted adrenal gland in sodium deficiency and decline of rate by 25%–50% is seen in moderate to severe sodium deficit (Blair-West et al. 1963b, 1964a). In the gradation of sodium status from positive to neutral or very small deficit, a rise in aldosterone secretion may occur without significant change in adrenal blood flow or despite the fact that, because of other factors influencing cardiac output such as, for example, digestive processes, there may coincidentally be a small rise in adrenal blood flow. Black (1953) states that salt deficiency is associated with decline of splanchnic as well as renal blood flow. Davis (1962) and Davis et al. (1965) reported there was no change in the metabolic clearance rate of aldosterone in sodium-deficient dogs, a somewhat surprising finding if, as found by the Taits and their collaborators (1961), the metabolic clearance rate of aldosterone is directly dependent on liver blood flow.

Plasma sodium concentration

Changes in plasma sodium concentration clearly depend on whether there is simply an alteration in intake, and thus in the relation to neutral 'set point', or whether there is frank negative status and, if so, how it is produced (for example whether water drinking is feasible).

Luetscher and Axelrad (1954) studied two subjects on a low-salt diet for 6 days. A small decrease (2–3 mmol/l) in plasma [Na], and increased plasma [K] occurred during the deficiency period, when aldosterone excretion increased fivefold. In the experiments of Crabbé et al. (1958) on normal humans on a low-sodium diet, plasma sodium fell in all five subjects. Leaf and Couter (1949) also record small falls in plasma sodium and chloride under the same circumstances in normal individuals. Renwick et al. (1955) and Romero et al. (1968) did not observe a fall in plasma [Na] in humans on a low-salt diet.

In clinical deficiency involving large loss of sodium-containing fluid from intestines, decrease in plasma [Na] often occurs (Marriott 1950; Black 1953), as, for example, with post-gastrectomy duodenal fistula in man (Denton et al. 1951), where there may be a plasma [Na] decrease of 10–20 mmol/l.

In normal dogs, induction of sodium deficiency by intraperitoneal administration of 5.5% glucose resulted in a fall in plasma sodium concentration and a significant movement of water into cells, which was reflected in a rise in plasma specific gravity (Gilman 1934). In similar experiments on dogs, rats and monkeys where 100 cc of 5% glucose/kg of body weight were injected intraperitoneally, plasma sodium decreased 13–20 mmol in 3.5–6.0 h (Elkinton et al. 1947). In dialysis experiments on dogs (Nichols and Nichols 1956a, b) plasma sodium concentration decreased over 3–6 h by 7–18 mmol/l and total base concentration decreased in dogs with pancreatic fistulae (Gamble and McIver 1928).

In sheep made acutely sodium-deficient by loss of parotid saliva, there is a small change in plasma sodium concentration of 3–5 mmol/l with negative external balance of 400–600 mmol of sodium (Abraham et al. 1976). Plasma sodium concentration is reduced in sheep with gross deficit by 10–20 mmol/l (Denton 1957b).

Exchangeable sodium and bone sodium

Total exchangeable sodium of the body has been studied by a number of workers by estimation of the sodium space (Moore et al. 1956; Adesman et al. 1960; Jagger et al. 1963). Values ranged from 36.8 to 43.8 mmol/kg body wt (Forbes and Perley 1951; Ikkos et al. 1954; Edelman et al. 1954a). The method of neutron activation has been used to assess total body sodium and in two patients values of 45–46 mmol/kg were derived (Anderson et al. 1964), but the values of exchangeable sodium were also a little higher. A higher value of 54.7 mmol/kg was found in sheep, and this may be accounted for by the large amount of exchangeable sodium in the rumen (McDougall et al. 1974). The total exchangeable sodium is considerably less than the total body sodium; 31%–55% of the skeletal sodium is non-exchangeable (Davies and Kornberg 1952; Edelman et al. 1954a, b; Moore et al. 1956; Nichols and Nichols 1956b) with possibly some variation being attributable to species difference and age (Forbes and McCoord 1963). Moore and colleagues cite the total exchangeable sodium of a 70-kg mature adult man as 2870 mmol (41 mmol/kg body wt) with 2350 mmol present in the extracellular water. Total exchangeable sodium is observed to decrease in the human (Hine et al. 1962; Jagger et al. 1963) and rat (Forbes and Perley 1951) on a low-salt diet.

In 1894, Gabriel demonstrated the presence of considerable quantities of sodium in bone. Harrison et al. (1936) showed that more than half of this was not present in the extracellular space. Comparing the structure and distribution of sodium in muscle and bone, Nichols and Nichols (1956a) point out that muscle consists of some 20% organic solids and 80% water. Of this water, 20% is extracellular containing sodium at a concentration of 150 mmol/l. The intracellular water of muscle probably contains about 10 mmol/l. The total amount of sodium contained in a kilogram of muscle is about 28 mmol, of which 79% is in the extracellular phase. Bone water, comprising 20% of bone by weight, includes three fractions: intracellular water in two phases (kinetically free and water of crystallization) and extracellular water. In the adult animal the extracellular fluid constitutes some 80% of the total bone water and contains 24 mmol Na/kg of bone, the same concentration as the extracellular water of muscle. The amount of intracellular water of bone is so small that the total amount of sodium in it is less than 1 mmol/kg bone. But by far the greatest fraction of the 200 mmol of sodium found in a kilogram of wet bone—176 mmol, or 88%—is contained in the crystal lattice. Bone mineral is composed of small hexagonal flat crystals, so that the surface area per unit weight is enormous: it has been estimated at 40 ha in a 70-kg man (Neuman and Neuman 1953). It is presumed that in normal bone a dynamic equilibrium exists between the ions at the crystal surface and the ions of the hydration shell, and that, in turn, the ions of the hydration shell are in equilibrium with the ions of the surrounding fluids. This is an electrochemical equilibrium and does not imply equality of concentration. Actually, the concentration of both calcium and sodium at the crystal surface far exceeds their concentration in extracellular fluid. In general, conditions which cause a loss of body sodium lower bone sodium. In adult animals it would appear that 6%–13% of bone mineral sodium can be removed from adult bone by various hyponatraemic conditions such as prolonged low-sodium diet, mercurial diuresis, adrenalectomy and hyperchloraemic acidosis and sodium depletion (Edelman et al. 1954a, b; Nichols and Nichols 1955, 1956b; Munro et al. 1957; Norman 1963; McDougall et al. 1974). The data suggest the loss is possibly greater in juvenile animals (Bergstrom and Wallace 1954) and when sodium deficiency is associated with acidosis (Norman 1963; Nichols and Nichols 1956a, b). It is clear, however, that these figures are considerably smaller than the figures for total exchangeable sodium in bone, indicating that exchangeable bone sodium is not identical with available bone sodium. After long-term sodium depletion it was found the reduced bone sodium content was not repaired after 3 weeks of sodium repletion (McDougall et al. 1974).

The bone reservoir may be one factor involved in the fact that loss of body sodium may not affect plasma concentration of the ion, and that it is often found that more sodium is required to correct hyponatraemia than calculations based on extracellular volume and plasma concentration would indicate.

McCance's observations on the effect of sodium deficiency in producing negative nitrogen balance and elevated blood urea have been confirmed by other experimental studies on normal man (Leaf and Couter 1949) and by balance studies on patients with sodium deficiency as a result of loss of intestinal secretions (Darrow 1946; Denton et al. 1951; Elkinton and Danowski 1955).

Renal function

Wiggins et al. (1951) observed that glomerular filtration rate was reduced by an average of only 4 cc/min and renal plasma flow by 44 cc/min in patients depleted by low-salt diet and injections of 2 cc of Mercuzanthin per day for 3 days. After loading by administration of 3 litres of saline intravenously, glomerular filtration increased by an average of 6.1 cc/min, and renal plasma flow by 26 cc/min. During depletion sodium and chloride excretion in the urine was negligible. In the study by Black et al. (1950) of subjects on a low-sodium diet for 5–6 days, decrease of glomerular filtration rate was observed, in one instance, from 150 to 90 ml/min. Rapid infusion of hypertonic sodium chloride (60–300 mmol) did not cause an immediate large increase in the very low rate of sodium excretion though a large increase in glomerular filtration rate and an increase in plasma sodium and thus the tubular load of sodium occurred. In Leaf and Couter's studies (1949) one subject showed reduction in glomerular filtration rate but the other did not. Urinary sodium excretion was reduced to a very low level (1–2 mmol/day) by the regime of 9 mmol Na/day intake. Inulin and creatinine clearance did not change significantly in Luetscher and Axelrad's experiments (1954) though urinary sodium excretion was minimal. In sheep becoming progressively depleted of sodium as a result of salivary loss from an oesophageal fistula, there was a progressive decline in inulin clearance rate to as low as 10% of normal in the very severely depleted animal (Denton et al. 1952). In McCance's (1938) study, salt depletion altered the response to respiratory alkalosis in that the urine did not become alkaline and there was no increase in volume or in sodium or potassium content.

Plasma potassium

In man when sodium deficiency occurs as a result of loss of alkaline intestinal secretions, plasma potassium concentration rises even though the total potassium content of the body may be reduced (Denton et al. 1951).

With sodium deficiency and alkalaemia resulting from loss of gastric secretions, the plasma potassium concentration is reduced (Elkinton and Danowski 1955; Scribner and Burnell 1956). Serum potassium increased in rats which were made sodium-deficient and acidotic (Cotlove et al. 1951), as it did in dogs (Nichols and Nichols 1956a, b). There was little change in plasma potassium concentration in sheep with onset of sodium deficiency as a result of salivary loss from a parotid fistula (Blair-West et al. 1963b). However, the arterial plasma potassium concentration of sheep is increased in severe sodium deficiency (Coghlan et al. 1960).

Pressor response

Sodium-depleted normal humans and dogs show a reduced pressor response to angiotensin relative to the sodium-replete state (Carpenter et al. 1961; Davis 1962; Laragh 1962). This reduced response was not observed with the pressor effect of noradrenaline. Sodium-deficient sheep show a large reduction of pressor response to angiotensin (Fig. 7–3), which is proportional to the degree of sodium deficit (Blair-West et al. 1963a). This was seen also in adrenalectomized animals. A reduction of response to tyramine but not noradrenaline also occurred.

Sensitivity of tissue response to hormones

In sodium-replete animals the sodium status—moderate intake to loaded as reflected in the plasma [Na]—significantly affects aldosterone response to adrenal arterial infusion of angiotensin II (Fig. 7-4) (Blair-West et al. 1970). Sodium deprivation results in an increased sensitivity of the adrenal aldosterone response to stimulation by

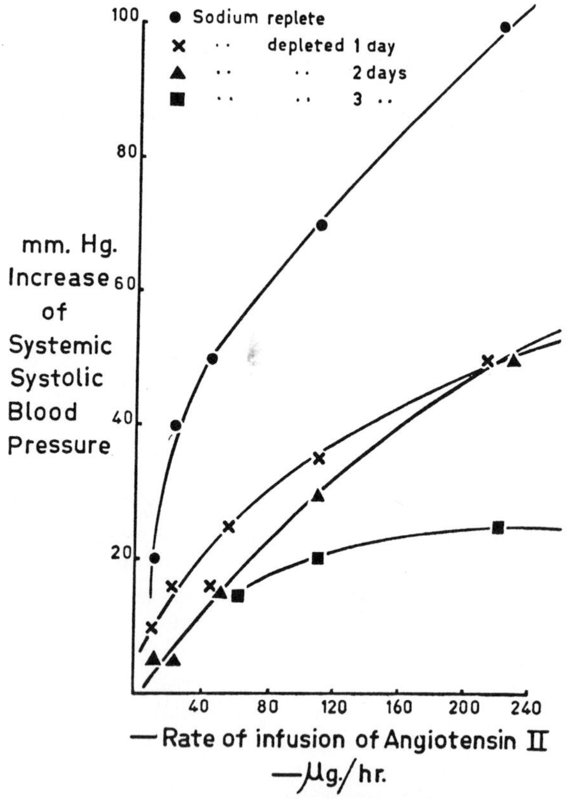

Fig. 7-3. The effect of intravenous infusion of Val^5-angiotension II amide upon the systemic systolic blood pressure when sodium-replete and on consecutive days following withdrawal of dietary sodium supplement. The direct relationship between rate of infusion of angiotensin II and increase of systolic blood pressure is indicated for each day by a trend-line. (From Blair-West et al. 1963a, with permission.)

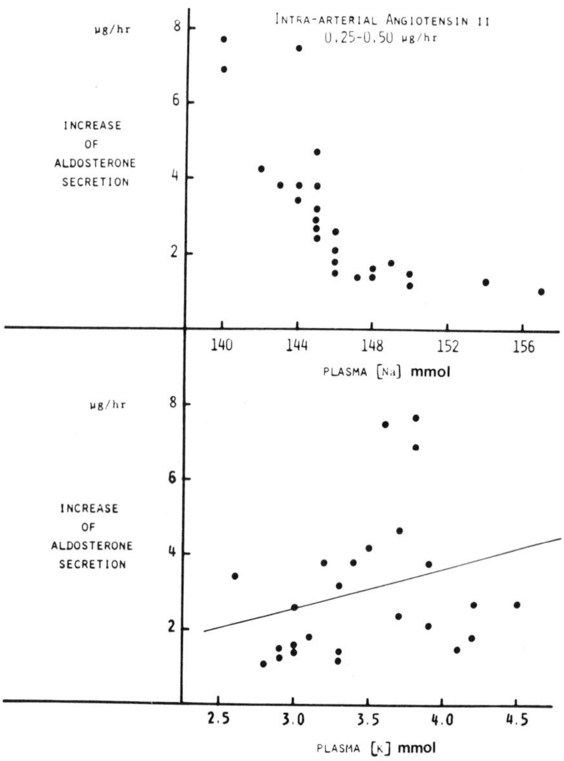

Fig. 7-4. Relation between increment of aldosterone secretion rate due to angiotension II-amide infusion and plasma sodium concentration (upper panel) and plasma potassium concentration (lower panel). (From Blair-West et al. 1970, with permission.)

angiotensin II (see Chapter 9). However, in moderate sodium deficiency (i.e. negative status) in the sheep, adrenal arterial infusion of angiotensin II does not cause a further increase of aldosterone secretion whereas further loss of sodium does.

Sodium depletion increases the sensitivity and response of the parotid gland to intravenous infusion of aldosterone, as described in Chapter 3. Sodium depletion also increases the sensitivity of aldosterone response to the adrenal arterial infusion of ACTH in the sheep (Blair-West et al. 1963b) and the dog (Gordon et al. 1980) (see also Chapter 9). The possibility that sodium deficiency may alter the response of brain receptors to humoral stimuli also formally arises.

Intestinal mobility

Sodium deficiency is associated with lower intestinal mobility (Marriott, 1950; Streeten, 1950; Black 1953), and possibly the delayed water diuresis associated with water ingestion which is seen in salt deficiency is attributable to a considerable extent to delayed water absorption. The absence of thirst in salt deficiency has been the subject of comment (Marriott 1950), though animal experimental studies show water intake rises. McCance's (1936a, b) experiments emphasized the changed taste sensations.

Tissue composition

Experimental studies of tissue composition during acute sodium deficiency produced by vivodialysis (Nichols and Nichols 1956a, b) and peritoneal dialysis (Woodbury 1956) have been made using chloride space as the index of extracellular fluid. When 23% of total body sodium was removed over 4–6 h by vivodialysis of dogs, 70% of the sodium removed came from the extracellular phase, 26% from the mineral phase of bone, and intracellular sodium contributed only 4%. Of the sodium lost from extracellular phase, more than 60% came from skin and extracellular water of muscle, and the loss from these highly vascular tissues was rapid initially and declined with time whereas avascular tissues such as tendon showed an increasing rate of sodium loss with time. There was significant sodium and chloride loss from the heart, but the liver was exceptional in that it gained sodium. Under these conditions of vivodialysis which resulted in decreased plasma pH the tissue analysis showed that bone, skin and liver lost water, whereas the water contents of heart and tendon were virtually unchanged. During the first 3 h muscle lost water, but subsequently muscle water rose above its initial value. In rats with acute hyponatraemia induced by intraperitoneal injection of isosmolar glucose solution (Woodbury 1956) muscle sodium, potassium and chloride concentrations decreased but cellular hydration occurred. In the heart, sodium and chloride spaces decreased without a change in potassium concentration; cellular hydration occurred here also. Bone sodium decreased. In the skin, water content decreased, and sodium and chloride fell moderately.

Studies on sodium-deficient male rats (Cotlove et al. 1951) showed that when sodium deficiency was associated with acidosis there was a decrease in the sodium in intracellular water of muscle from 9 to 1 mmol/kg, whereas when it was associated with chloride deficiency and alkalosis the sodium content was 22 mmol/kg and with potassium deficiency 39 mmol/kg. The characteristic decrease in the potassium content of muscle water with alkalosis (Darrow et al. 1948) was observed. Intracellular magnesium concentration was measured also, and there was little or no change in the conditions cited above. Leonard (1962) studied rats on a sodium-deficient diet for 14 days and, using an estimate of extracellular fluid based on chloride space, showed that deficiency was associated with increases in the intracellular sodium concentrations of heart, liver and adrenals, but no change in that of abdominal muscle.

Meyer and colleagues (1950) made a similar finding in relation to heart and liver in rats fed a low-sodium diet for a long period. Cox et al. (1962) reported a rise in intracellular sodium following sodium restriction in the human and use of ion-exchange resin. They found that whereas extracellular sodium as deduced from ^{82}Br space fell by a mean of 528 mmol, total exchangeable sodium fell only 182 mmol, weight declined by a mean of 2.7 kg, total body water decreased by 2.7 litres, extracellular space by 3.5 litres, and intracellular space increased by 0.80 litre. The authors regard it highly unlikely that this sodium entered the exchangeable fraction of bone sodium and regard it probable that it entered the intracellular compartment of soft tissue. A rise in intracellular sodium of heart muscle has also been observed in adrenalectomized rats, suggesting that aldosterone secretion was not the factor responsible for the rise in intracellular sodium concentration (Leonard 1963). It seems possible from the data that the rate of change of sodium balance as well as pH change may be influential on the intracellular changes which result.

Sodium deficiency in pregnancy

Observations on the effects of sodium deficiency in pregnancy in the rat have been made by Kirksey and Pike (1962) and Kirksey et al. (1962). Pregnant animals fed a low-sodium diet showed general langour and debility and ate less food and gained less weight than those receiving the control diet, but the level of dietary sodium did not have a significant effect on reproductive performance as evaluated by the number of live young, number of resorptions, foetal weight and placental–foetal ratios. The decreases in haematocrit and total protein characteristic of the pregnant state and haemodilution were not as great in the animals that were receiving low-sodium diet. Maternal plasma sodium concentration decreased slightly as a result of pregnancy, but on the low-sodium diet a significant decrease in sodium and an increase in potassium levels in plasma occurred. The low-sodium intake also caused an approximately 35% decrease in sodium and 4% increase in potassium concentrations in muscle, and a significant increase in muscle water in the pregnant animals. Bone sodium, and brain sodium and potassium, were significantly decreased in the pregnant groups receiving the low-sodium intake. Potassium concentration and water were not affected by the dietary sodium or by pregnancy. Foetal plasma levels of sodium and potassium were not significantly influenced by the level of sodium in the maternal diet. In general, the evidence was of marked maternal changes at the expense of maintenance of normal levels of sodium in the foetus. Data in the sheep have shown very high levels of aldosterone secretion when pregnancy is associated with sodium deficiency and also a fall in amniotic fluid sodium concentration.

Brain electrolytes

Katzman and Pappius (1973) open their discussion of the effect of sodium depletion on brain electrolytes by noting that loss of sodium without abnormal retention of water will eventually lead to hyponatraemia and appearance of neurological symptoms similar to those seen with mild water intoxication. This has been described in healthy individuals exposed to sustained heat in the tropics (Saphir 1945) and in a surprisingly high proportion of cardiac patients treated with mercurial diuretics and a low-sodium diet (Schroeder 1949). The symptoms are relieved rapidly by giving salt. Hyponatraemia as a result of excessive sodium loss in sweat in children with cystic fibrosis can also cause heat prostration (Barbero and Sibinga 1964).

The effects of hyponatraemia on brain water and electrolyte content have been studied both in salt depletion and in water intoxication. Yannet (1940) used isosmolar glucose placed in the peritoneal cavity to deplete cats of 10% body weight. When measured 24 h after dialysis, there was a decrease in extracellular volume and increase in intracellular volume of kidney and liver but not of brain. Here, no change in total water or its distribution was found but the systemic sodium depletion caused a decrease in sodium and potassium content which was proportional to the decrease in serum sodium. Swinyard (1949) and Woodbury (1956) confirmed this effect in the rat, with measurement 4 h after the dialysis (The technique of dialysis avoided any water-loading.) Yannet (1940) suggested that release of intracellular potassium is a mechanism whereby the brain adjusts to extracellular electrolyte depletion, in contrast to other tissues where shifts of water into cells maintain isosmolarity between the two compartments. As Katzman and Pappius (1973) remark, later results from shorter term studies by several workers showed a movement of water into brain cells and that potassium movement begins 2–3 h after the extracellular change.

Pappius et al. (1967) and Wakim (1969) used haemodialysis in the dog, with a low-sodium bath fluid. Increase in cerebral tissue water occurred after 30 min, and decreases in sodium and potassium content of brain were seen after 60 min and by 90 min respectively. The changes in muscle sodium involved a fall of the same order as that in the serum, and swelling of cells was considerable. A rise in CSF pressure was observed which was not correlated with presence or absence of cerebral swelling, and appeared to be due to movement of water into CSF space (Pappius 1975). Davenport (1949), however, found in adrenalectomized rats with free access to drinking water and with mild hyponatraemia (130 mmol Na/l) that there was no change in sodium, potassium or water in the brain. Arieff et al. (1976), following the time course of dialysis in rabbits over 17 days, showed a clear fall in sodium and potassium content of brain with only small rise in brain water; these changes were associated with a fall in serum sodium from 122 mmol/l to, eventually, 99 mmol/l. Pappius et al. (1967) and Pappius (1975) discuss the question of whether, in the chronic situation, the decrease in cerebral potassium is a response to hypo-osmolarity or to hyponatraemia. On the basis of data obtained from uraemic dogs where hyponatraemia occurred in the presence of hyperosmolarity, and where significant loss of potassium from cerebral cortex was demonstrated, it was concluded hyponatraemia was the important factor. They suggest that several other studies support this conclusion. The mechanism by which hyponatraemia causes depletion of cerebral potassium and diminishes chronic osmotically induced cerebral swelling is unknown. In the face of the otherwise rigid control of potassium content of brain it is most intriguing (Pappius 1975).

In discussing the implications of the fall in brain sodium with sodium depletion, Arieff and Guisado (1976) note that sodium is essential for the excitability properties of both nerve and muscle tissue. Without sodium ions in the extracellular fluid, both nerve and muscle become inexcitable (Overton 1904). Synaptic transmission and generation of action potentials are both dependent on presence of sodium, and the sodium equilibrium potential influences the membrane potential at the height of the action potential and the rate of rise of the action potential itself. The release of neurotransmitters at the presynaptic terminal is also influenced by Na (Gage and Quastel 1965), as is the re-uptake of transmitter by the nerve endings (Bogdanski and Brodie 1966; Bogdanski et al.

1970). Arieff and Guisado (1976) also draw particular attention to the fact that transport systems for both release and re-uptake of neurotransmitter amino acids have been shown to exist in synaptosomes of brain and spinal cord (Bennett et al. 1973) as well as in peripheral nerve (Yamaguchi et al. 1970; Baker and Potashner 1973), and a sodium gradient is assumed to be one of the major ways of applying energy for transport of metabolites against a concentration gradient (Schultz and Currant 1970). In brain and spinal cord synaptosomes, the neurotransmitter amino acid transport systems are almost totally sodium-dependent (Bennett et al. 1973; Lajtha and Sershen 1975) and amino acid uptake in brain slices is decreased or completely inhibited by absence of sodium (Lahiri and Lajtha 1964).

In terms of other biochemical effects of low osmolality, reduction of brain phosphocreatine has been observed, but not of ATP, ADP or AMP (Fishman 1974). Arieff and Guisado (1976) remark that hyponatraemia may affect brain function by any of several mechanisms. As hyponatraemia becomes prolonged, and brain becomes depleted of sodium and potassium, there may be inhibition of brain metabolism and interference with neurotransmitter amino acid release at the synaptic level.

McCance (1938) showed that sodium deficiency in man caused a reduction in CSF [Na] and [Cl]. And in sheep with a parotid fistula, a significant decrease in CSF [Na] associated with sodium deficiency and a fall in plasma [Na] has been shown (Fig. 7-5); there is also a significant fall in CSF

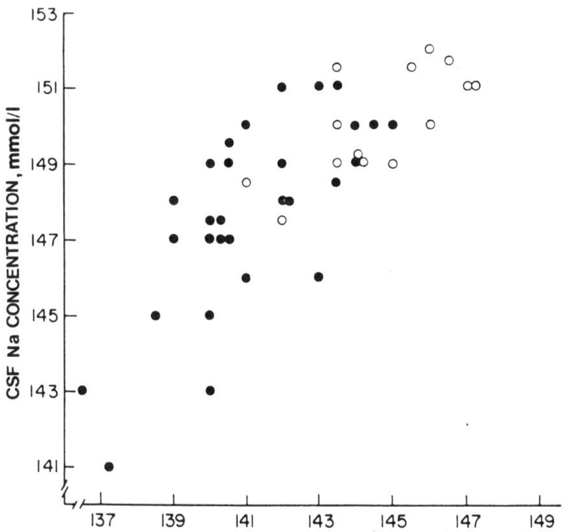

Fig. 7-5. Relation of CSF and plasma sodium concentration in normal (sodium-replete: open circles) and sodium-deplete (filled circles) sheep. (After Mouw et al. 1974.)

[HCO_3] but no change in potassium, calcium or magnesium (Mouw et al. 1974).

Aldosterone secretion and action

Deane (1962) has reviewed the evidence that sodium deficiency or sodium excess causes morphological and cytochemical changes in the adrenal glomerulosa, and Blair-West et al. (1968) have reported large morphological changes in secretory tissues with sodium deficiency in wild animals.

Sodium deficiency causes increased secretion of aldosterone. For example in an early study using isotope dilution techniques, Jones et al. (1959) found secretion rates in normal man between 72 and 315 μg/day; with a low-sodium diet the rate rose to 780 μg/day. There was no change in cortisol secretion, a finding frequently ratified by direct sampling of adrenal venous blood with onset of sodium deficiency in the conscious sheep (Fig. 7-6). The outcome of a number of studies at this

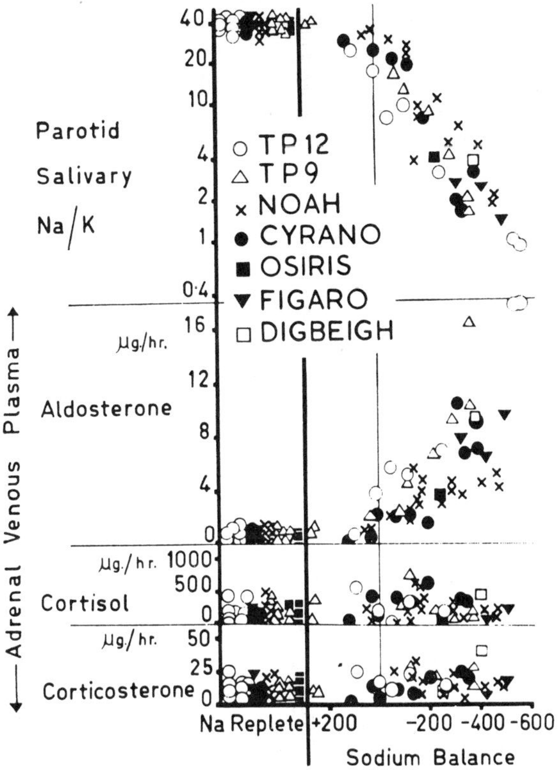

Fig. 7-6. The results of experiments on seven sheep showing the relationship between sodium balance (abscissa) and (1) the parotid salivary Na/K ratio, and the secretion of (2) aldosterone, (3) cortisol, (4) corticosterone as measured in adrenal vein plasma. Additional observations made on the animals when sodium-replete but not under precise balance conditions are shown on the left. (From Blair-West et al. 1964b, with permission.)

time showed that mild to moderate sodium deficiency increased aldosterone secretion 20–25-fold (Muller et al. 1959; Peterson 1959; Laragh et al. 1960; Tait et al. 1961; Bougas et al. 1964). Mills (1962) reported an unpublished study of Gowenlock and Hartley in which a patient addicted to purgatives and grossly depleted of sodium and potassium had an aldosterone secretion rate of 5700 μg/day. When repleted of potassium the rate rose to 6750 μg/day. The levels in mild sodium deficiency are of the order seen in Conn's tumour and normal pregnancy, but they are very much higher in pregnant women on a low-sodium diet, in nephrosis or malignant hypertension, where values of 2000–10 000 μg/day have been reported (Ulick et al. 1958; Laragh et al. 1960). Singer and Stack-Dunne (1955) also found that sodium deficiency in the rat which elevated aldosterone secretion did not increase corticosterone output.

As discussed in Chapter 3, and reviewed by Gaunt and Chart (1962), the increased aldosterone secretion affects sodium transport in target organs. The salivary Na/K ratio falls after a latent period which reflects the binding of aldosterone to its receptor and the intranuclear transport of the complex (Edelman 1968). Similar effects on Na/K ratio occur in sweat gland secretion and in the secretion–reabsorption processes in the large intestine as reflected by faecal composition in sodium deficiency (Denton 1957b). Sodium reabsorption in the distal tubule is stimulated.

Sodium deficiency causes a rise in plasma renin, and in angiotensin I, II and III in blood. An early study showing the changes in plasma renin with sodium deficiency and sodium-loading was made by Brown et al. (1964). Precise correlations established in sheep between sodium balance, plasma renin, angiotensin II and aldosterone are given in Chapter 9.

Angiotensin and steroids in cerebrospinal fluid

In examining the influence of sodium deficiency on CSF angiotensin concentration, studies were made on 31 conscious sheep with permanent indwelling guide-tubes to the lateral ventricles. Sodium deficiency was produced as a result of saliva loss from a permanent parotid fistula or by parotid cannulation. Paired arterial blood and ventricular CSF specimens were taken from sodium-deplete animals, extracted, and subjected to chromatography and radioimmunoassay (Fei et al. 1979). As reported initially by Abraham et al. (1975) sodium

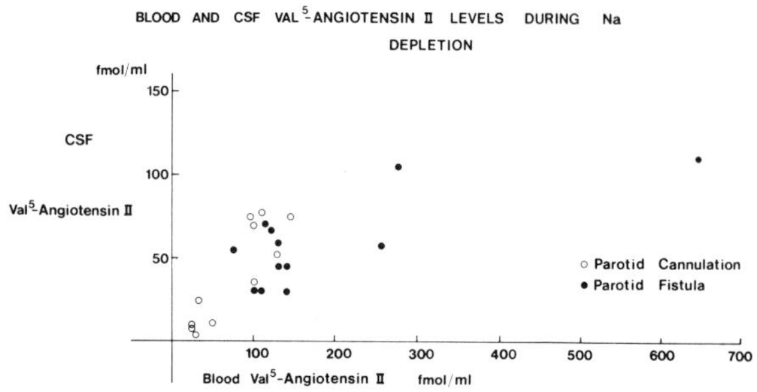

Fig. 7-7. The relation between blood Val^5-angiotensin II concentration and CSF Val^5-angiotension II concentration in sodium-replete and sodium-deficient sheep. Sodium deficiency was produced by parotid fistula or parotid cannulation. (S. F. Abraham, J. P. Coghlan, D. Denton, D. T. W. Fei, M. J. McKinley and B. Scoggins, unpublished.)

depletion resulted in elevation of both blood and CSF angiotensin II concentrations; values ranged from 7.5 to 27.5 ng/100 ml for blood and 2.8 to 10.5 ng/100 ml for CSF (Fig. 7-7). There is a significant linear relation between CSF and blood concentration over the range of blood angiotensin II concentration 0–25 ng/100 ml (r=0.62, P<0.05). The mean ratio of CSF to blood angiotensin II concentration was 0.47±0.06 (n=14). In a single experiment with a very high blood angiotensin II concentration of 64.5 ng/100 ml the corresponding CSF level was 11.2 ng/100 ml (Fig. 7-7). In sodium-replete animals CSF angiotensin was 2.4±0.5 ng/100 ml (n=7) with a corresponding blood concentration of 3.8±1 ng/100 ml (n=7). Chromatography of the CSF showed that the main immunoreactive material migrates like angiotensin II, though there were some smaller immunoreactive fragments, one corresponding to angiotensin III (Fig. 7-8). An additional slower moving peak was identified, but direct radioimmunoassay of CSF extract using an angiotensin I antibody which had the sensitivity of detecting 5 fmol of angiotensin I with confidence did not show any significant immunoreactivity above the background. This question of the presence of angiotensin I in CSF requires further investigation. The first reports of the presence of angiotensin in brain tissue by Fischer-Ferraro et al. (1971) and Ganten et al. (1971a, b) indicated that most of the angiotensin was angiotensin I but that angiotensin II was also present.

Plasma renin concentration of sodium-replete sheep was 1.1±0.2 ng h^{-1} ml^{-1} (n=6). In sodium-deplete sheep equivalent values were 16±4.4 ng h^{-1} ml^{-1} (n=6). Renin was not detected in any of the CSF samples (n=6).

In another series of experiments (Abraham et al. 1980), where angiotensin II was infused intravenously at 20–50 μg/h for 6 h, giving blood

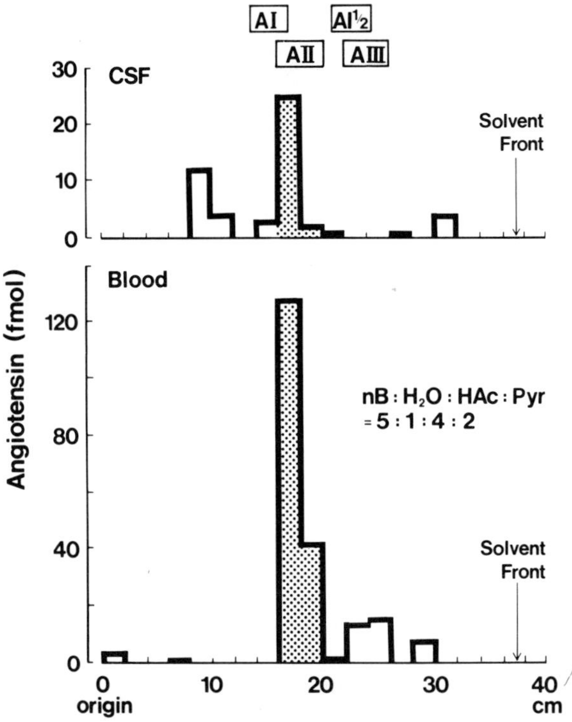

Fig. 7-8. Chromatogram of arterial blood extract (lower panel) and CSF extract (upper panel) obtained simultaneously from a sodium-deficient sheep. The positions of Val[5]-angiotensin II (AII), [des-Asp[1]]angiotensin II (AIII), angiotensin I (AI) and [des-Asp[1]]angiotensin I (A1 1/2) run in this solvent system (n-butanol:acetic acid:water:pyridine=5:1:4:2) are shown. (S. F. Abraham, J. P. Coghlan, D. Denton, D. T. W. Fei, M. J. McKinley and B. Scoggins, unpublished.)

concentrations up to 70 ng/100 ml, all measurable CSF angiotensin II concentrations in the sodium-replete sheep were below 5 ng/100 ml (Fig. 7-9). In six sodium-deplete animals with similar infusions at 40 μg/h for 6 h there was no significant elevation of CSF angiotensin II concentrations above the pre-infusion levels.

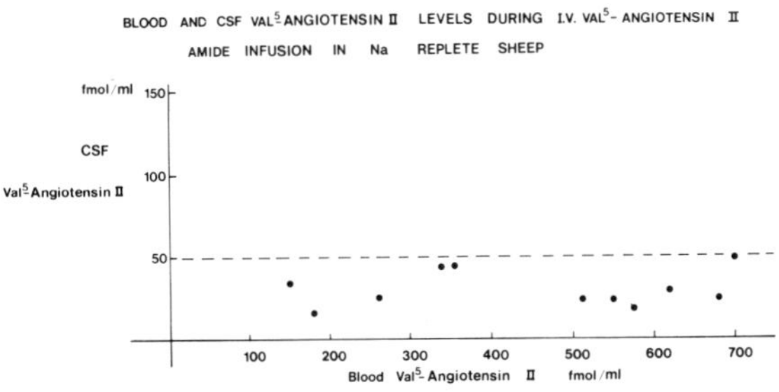

Fig. 7-9. Effect of intravenous infusion of Val[5]-angiotensin II on CSF angiotensin II concentration in sodium-replete sheep. CSF and blood samples were taken simultaneously 6 h after commencement of infusion at rates ranging from 10 to 60 μg/h. (S. F. Abraham, J. P. Coghlan, D. Denton, D. T. W. Fei, M. J. McKinley and B. Scoggins, unpublished.)

The increase in CSF angiotensin II concentrations associated with sodium depletion could be attributable either to passage of angiotensin II from the blood across the blood–brain barrier into the CSF, or to production of angiotensin within the brain. The failure of angiotensin II concentration in CSF (in contrast to that in arterial blood) to fall following nephrectomy in the sodium-deficient sheep (Fig. 7-10) favours the latter proposition.

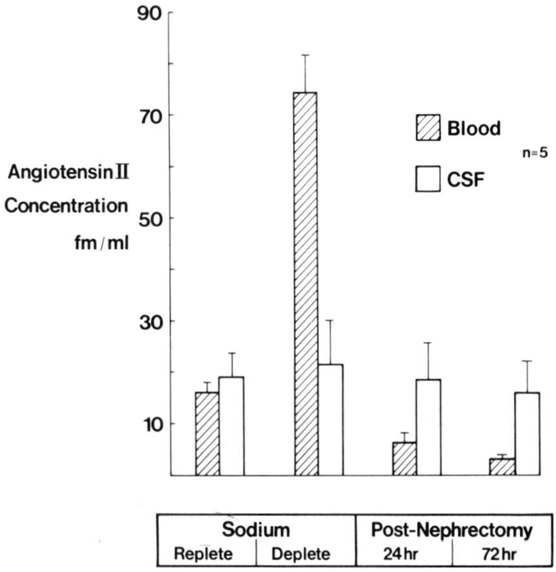

Fig. 7-10. Angiotensin II concentration in arterial blood and CSF in conscious sheep when sodium-replete, mildly sodium-deplete and after bilateral nephrectomy. (D. T. W. Fei, J. P. Coghlan, D. Denton, M. J. McKinley and B. Scoggins, unpublished.)

In the course of these experiments it was also shown that there was a correlation between arterial blood aldosterone concentration and CSF aldosterone concentration (r=0.94, P<0.001) (Fig. 7-11). The relation of CSF to blood

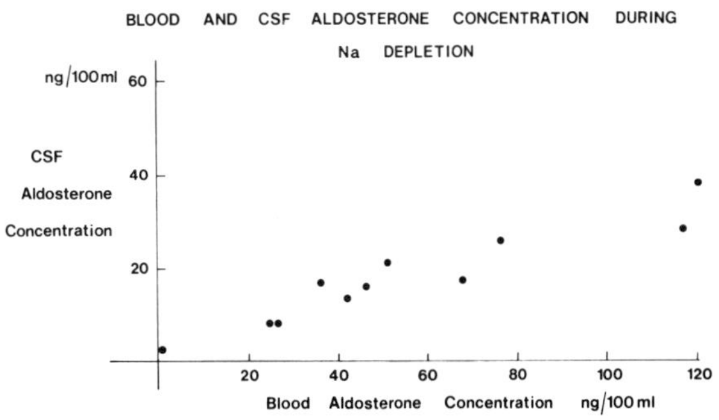

Fig. 7-11. Correlation of CSF and arterial blood aldosterone concentration during sodium depletion. (S. F. Abraham, J. P. Coghlan, D. Denton, D. T. W. Fei, M. J. McKinley and B. Scoggins, unpublished.)

aldosterone concentration for 11 paired samples was 0.38±0.03. In this series of experiments there was no significant relationship between blood and CSF concentrations of cortisol and corticosterone, though levels of cortisol in CSF were always less than those in blood.

Ganten and colleagues (1978) Ganten and Speck (1978) also conclude that systemic angiotensin II does not penetrate the blood–brain barrier freely. They interpret experiments where it has been shown that intraventricular saralasin will block dipsogenic effects of systemic angiotensin II, and where systemic saralasin will block dipsogenic effects of intraventricular angiotensin II as reflecting the opening of the blood–brain barrier by an abrupt rise of blood pressure, as has been shown with horseradish peroxidase (see also Chapter 24). Schelling et al. (1976) have also measured angiotensin II in CSF and angiotensin III has been detected by Hutchinson et al. (1978). The levels are claimed to be high in spontaneously hypertensive rats (Ganten et al. 1976).

The postulated brain renin–angiotensin system: Its possible role in sodium deficiency

Ganten and Speck (1978) give the following evidence derived from a wide variety of experiments to support the notion of a specific activity of a brain renin–angiotensin system in vivo.

1) Brain renin converts naturally occurring homologous and heterologous angiotensinogen into angiotensin I.
2) Specific active-site-directed peptide enzyme inhibitors inhibit brain renin in a competitive manner. Purified human brain renin does not hydrolyse the following substrates: angiotensin I, angiotensin II, denatured haemoglobin (which cathepsin D does), and beta endorphin.
3) Injection of purified human brain renin into the brain ventricles of conscious rats entails angiotensin I formation from CSF angiotensinogen, and blood pressure increases.
4) Injection of angiotensinogen into brain tissue elicits drinking behaviour. This effect is reversed by inhibition of brain angiotensin I converting enzyme, and by angiotensin II blockade.
5) Angiotensin I and angiotensin II can be extracted from brain tissue and have been localized by immunohistochemical techniques in neural tissues and specific brain areas (see Chapter 22).
6) Blockade of endogenous brain angiotensin II at the receptor level results in altered biological function: decrease of electrical firing of single neurones, decrease of blood pressure in spontaneously hypertensive rats, and reduction of water drinking.

The question of activation of this system by a deficit of body sodium becomes an important issue when the genesis of sodium appetite is discussed in Chapters 22 and 24.

The proposal that there is a brain renin–angiotensin system is not universally accepted. The arguments against simple interpretation of the data have been set out by Reid (1979) and Ramsay (1979). Reid points out that the initial studies of brain renin have revealed marked differences between the properties of this enzyme and those of renal renin. The activity of the brain enzyme was found to decrease markedly above pH 5 so that at the pH of plasma, pH 7.4, little or no measurable activity could be demonstrated. Renal renin is, of course, active at pH 7.4. There is now considerable evidence that most of the renin-like activity in the brain is due to the lysosomal acid protease cathepsin D. The evidence may be summarized as follows:

1) Angiotensin I is formed when cathepsin D is incubated with natural and synthetic renin substrates.
2) Brain-renin-like and cathepsin D activities co-purify during several procedures including ammonium sulphate fractionation.
3) Purified brain renin closely resembles cathepsin D with respect to substrate specificity, inhibition of angiotensin formation from tetradecapeptide renin substrate by proteins, acid protease activity, sensitivity to inhibition by pepstatin, and pH dependence of angiotensin formation.
4) The sub-cellular distribution of the renin-like activity in the brain closely parallels that of cathepsin D.

As pointed out by Hackenthal et al. (1978) cathepsin D is not an isoenzyme of renin so that the term isorenin, commonly used to describe the brain enzyme, should no longer be used. Reid notes that the identification of brain 'renin' as cathepsin D does not exclude the possibility that the enzyme may function as an angiotensin-forming enzyme. However, cathepsin D has very low renin-like activity compared with renal renin and the enzymatic activity of cathepsin D is very low at about pH 6. For this reason significant activity could be expressed only in some intracellular compartment, presumably in lysosomes where pH is lower than in the extracellular compartment

(Reijnoud and Tager 1973). In this regard it is noteworthy that measurement of intracellular pH of skeletal muscle with pH-sensitive glass microelectrodes showed that the normal muscle cell had a pH of 5.99 (Carter et al. 1967). Ramsay remarks there is little concrete information on the intracellular fluid pH within organelles in brain cells. He notes, however, that the lysosomes contain many proteases including angiotensinase. If angiotensinogen were taken up by lysosomes and acted upon by cathepsin D to produce angiotensin II, the angiotensin would immediately be degraded. For angiotensin to have an action in the brain either at a peptidergic synapse or by more widespread diffusion through the CSF, presumably it has to be extruded from the cell. Ramsay finds it difficult to see how, if angiotensin II were formed in the lysosomes, it could reach the outside of the cell without being degraded.

In relation to these arguments, Phillips et al. (1979) place particular emphasis on the findings of Inagami and colleagues (Hirose et al. 1978) who used an affinity column capable of separating renin from general acid proteases. Using whole brain homogenates from nephrectomized rats they purified brain renin and found it to be similar to renal renin (and distinct from cathepsin D) in its response to anti-renin antibodies, lack of proteolytic activity and neutral pH optimum (between 6 and 7.6). Reid (1979) regards this as the first convincing evidence that renin is present in the brain. However, he notes that the amount of renin measured by these investigators was very low—about a tenth of the concentration of renin present in the plasma of normal rats. Furthermore the distribution of the enzyme within the brain was not determined. Since blood vessels from other parts of the body are known to contain renin (Ganten et al. 1976) it is possible the activity measured by Hirose et al. (1978) represented renin contained in the walls of blood vessels. Additional studies are required to evaluate this (Reid 1979).

The presence of renin substrate and converting enzyme in the brain has been demonstrated in vivo. It has been shown that renin produces a variety of effects when administered centrally. Effects are prevented by angiotensin antagonists or converting enzyme inhibitor and are accompanied by a marked increase in the concentration of angiotensin II in CSF. Reid and Ramsay (1975) assayed samples of CSF for angiotensinogen and found concentrations of 205 ng/ml. This was one-fifth the concentration in plasma but because of the lower protein concentration in CSF the specific activity was much higher than in plasma. Ramsay (1979) remarks, though, that the source of the substrate remains to be determined. It is unlikely that this substance, with molecular weight 60 000, could cross the blood–brain barrier passively. The material therefore is either made within the central nervous system or must cross by some form of transport system. Reid (1979) states that experiments involving central administration of renin substrate have not produced convincing evidence for the existence of renin activity in the brain in vivo. The approach involves a number of assumptions one of which is that the concentration of endogenous renin substrate in the brain is rate-limiting. The early experiments of Epstein et al. (1974) involving synthetic substrate and the elicitation of drinking allowed no conclusion on the issue of the presence of renin since it has been shown (Dorer et al. 1975) that the converting enzyme can form angiotensin II from tetradecapeptide in the absence of renin by successively removing dipeptides from the C terminal of the molecule. It was shown that the dipsogenic action of the tetradecapeptide in rats was unaffected by administration of the renin inhibitor pepstatin but was abolished by the converting enzyme inhibitor SQ20881. Similar observations were made in dogs.

Reid (1979) proposes that experiments utilizing natural renin substrate have been unimpressive. Hoffman et al. (1976) have shown that partially purified natural renin substrate stimulated drinking when administered in rats. Reid suggests the responses were small and inconsistent. He notes that the investigators reported that a highly purified substrate preparation was less effective in eliciting drinking than the less pure preparation, suggesting that the drinking was caused by a non-specific factor. In Reid's laboratory, natural renin substrate prepared from plasma or from CSF has consistently failed to elicit drinking when administered centrally in rats or dogs. Fitzsimons and Kucharczyk (1978) have also reported that natural renin substrate fails to produce significant stimulation of drinking in the dog. To approach the problem directly Reid and Moffat (1978) measured the concentration of angiotensin II in CSF after injecting purified natural renin substrate into the ventricular system of dogs. CSF renin substrate concentration increases three-fold but, despite this, CSF angiotensin II concentration was undetectable before or after the injection. Blood pressure was also unchanged by the injection.

As Reid (1979) points out, the most direct test of the hypothesis that angiotensin is generated by the brain renin–angiotensin system is to determine whether angiotensin is present in the brain. His overview of the attempts at detection and assay of angiotensin I and II in CSF and brain by a number of laboratories is that the results have been extremely variable. In their laboratory, angiotensin II has not been detected in rat CSF and he notes that early estimates of its presence at 9–17 ng/100 ml (Ganten et al. 1975; Simpson et al. 1976) have had to be revised downward to 0–1.5 ng/100 ml by Epstein and Ganten (1977). In Reid's laboratory, angiotensin II has not been detected in the CSF of dog by either bioassay or radioimmunoassay. Attention was drawn above to the report of Abraham et al. (1980) bearing upon the persistence of an elevated angiotensin II concentration measured by radioimmunoassay following chromatography in the 72 h nephrectomized sheep. No renin or angiotensin I were detected in CSF under these conditions.

Coming to a consideration of direct measurement of angiotensin in the brain, Reid notes that the demonstration of angiotensin-like immunoreactive material in rat brain is not without problems. There are marked discrepancies concerning the distribution of the immunoreactive material. He states that immunohistochemical studies with several high-titre antibodies in his laboratory have failed to reveal angiotensin I or angiotensin II in the brain and suggests these techniques are subject to problems of nonspecificity and need to be supported by direct measurements of angiotensin; this has not been achieved. Biological and immunological activity resembling angiotensin have been measured in extracts of brain from various animal species, the reported values varying over a wide range. There is the possibility as pointed out by Horvath et al. (1977) that pressor activity from brain is attributable to antidiuretic hormone. It was not neutralized by saralasin, or antibodies to angiotensin I or angiotensin II. Studies by Reid et al. (1977) have shown that much of the apparent angiotensin immunoreactivity measured in brain extracts is actually an artefact caused by angiotensinase activity. The peptidases present in brain extracts destroy the ^{125}I-labelled angiotensin used in the radioimmunoassay for angiotensin thus reducing the amount of labelled angiotensin available for binding to the antibody. The resulting reduction in binding could be erroneously interpreted as displacement of labelled angiotensin by authentic angiotensin. If extraction procedures are used which eliminate angiotensinase activity the apparent angiotensin immunoreactivity is reduced to low or undetectable levels.

A line of evidence which may point to a central origin of angiotensin is the data showing that angiotensin antagonists lower blood pressure when given centrally to strains of spontaneously hypertensive rats (Schoelkens et al. 1976; Phillips et al. 1977; Mann et al. 1978). The result was not confirmed by Elghozi et al. (1976). To minimize any influence of circulating angiotensin II, Mann et al. (1978) showed that saralasin given intraventricularly did lower the blood pressure of nephrectomized spontaneously hypertensive rats, but some plasma renin activity remained after nephrectomy. Overall, Reid is of the view that the major argument presented for the brain renin–angiotensin system is that the components are there. There are serious doubts that the different components interact to form a functional system. Most renin-like activity found in extracts of brain is due to cathepsin D. Central administration of renin substrate fails to demonstrate a renin–substrate interaction in vivo, and the evidence that angiotensin is actually present is not convincing. Whereas the possibility of a renin–angiotensin system being present in the brain is not ruled out, he feels it is premature to conclude that such a system exists without further evidence.

These unresolved matters become of great relevance when we come to consider in Chapter 22 the issue of possible activation of a brain renin–angiotensin system with sodium deficiency, and a postulated role of angiotensin II so generated in sodium appetite.

References

Abbott WE, Levey S, Foreman RC, Krieger H, Holden WD (1952) The danger of administering parenteral fluids by hypodermoclysis. Surgery 32: 305

Abraham SF, Baker RM, Blaine EH, Denton DA, McKinley MJ (1975) Water drinking induced in sheep by angiotensin—a physiological or pharmacological effect? J Comp Physiol Psychol 88: 503

Abraham SF, Coghlan, JP, Denton DA, McDougall JG, Mouw DR, Scoggins BA (1976) Increased water drinking induced by sodium depletion in sheep. Q J Exp Physiol 61: 185

Abraham SF, Coghlan JP, Denton DA, Fei DTW, McKinley MJ, Scoggins BA (1980) Correlation of cerebrospinal fluid and blood angiotensin II in sheep. In: Proceedings of the Sixth International Congress of Endocrinology, Melbourne, abstr 725

Adesman J, Goldberg M, Castleman L, Friedman IS (1960) Simultaneous measurement of body sodium and potassium using Na^{22} and K^{42}. Metabolism 9: 561

Agabiti-Rosei E, Brown JJ, Brown WCB, Fraser R, Trust PM, Lever AF, Morton JJ, Robertson JIS (1979) Effect of the angiotensin II antagonist saralasin on plasma aldosterone concentration and on blood pressure before and during sodium depletion in normal subjects. Clin Endocrinol (Oxf) 10: 227

Anderson J, Osborn SP, Tomlinson RWS, Newton D, Rondo J, Salmon L, Smith JW (1964) Neutron-activation analysis in man in vivo. A new technique in medical investigation. Lancet II: 1201

Arieff AI, Guisado R (1976) Effects on the central nervous system of hypernatremic and hyponatremic states. Kidney Int 10: 104

Arieff A, Lacc HF, Massry SG (1976) Neurological manifestations and morbidity of hyponatraemia: Correlation with brain water and electrolytes. Medicine 55: 121

Arnold GW (1964) Some principles in the investigation of selective grazing. Proc Soc Anim Prod 5: 258
Atchley DW, Loeb RF, Richards DW Jr, Benedict EM, Driscoll ME (1933) On diabetic acidosis. A detailed study of electrolyte balances following the withdrawal and reestablishment of insulin therapy. J Clin Invest 12: 297
Baker PF, Potashner SJ (1973) The role of metabolic energy in the transport of glutamate by invertebrate nerve. Biochim Biophys Acta 318: 123
Barbero GJ, Sibinga FS (1964) The electrolyte abnormality in cystic fibrosis. Pediat Clin North Am 11: 983
Bennett JP, Logan WJ, Snyder SH (1973) Amino acids as central nervous transmitters: The influence of ions, amino acid analogues, and ontogeny on transport systems for L-glutamic and L-aspartic acids and glycine into central nervous synaptosomes of the rat. J Neurochem 21: 1533
Bergstrom WH, Wallace WM (1954) Bone as a sodium and potassium reservoir. J Clin Invest 33: 867
Black DAK (1953) Body-fluid depletion. Lancet I: 305
Black DAK, Platt R, Stanbury RW (1950) Regulation of sodium excretion in normal and salt depleted subjects. Clin Sci 9: 205
Blair-West JR, Coghlan JP, Denton DA, Goding JR, Munro JA, Wright RD (1963a) The reduction of the pressor action of angiotensin II in sodium-deficient conscious sheep. Aust J Exp Biol Med Sci 41: 369
Blair-West JR, Coghlan JP, Denton DA, Goding JR, Wintour M, Wright RD (1963b) The control of aldosterone secretion. Recent Prog Horm Res 19: 311
Blair-West J, Boyd GW, Coghlan JP, Denton DA, Goding JR, Wintour M, Wright RD (1964a) The functional capacity of the abdominal and transplanted left adrenal glands of sheep. J Physiol (Lond) 175: 78P
Blair-West JR, Boyd GW, Coghlan JP, Denton DA, Goding JR, Wintour M, Wright RD (1964b) Experimental observations on aldosterone secretion in the sheep. In: Baulieu EE, Robel P (eds) Aldosterone. Blackwell Scientific, Oxford, p 203
Blair-West JR, Coghlan JP, Denton DA, Nelson JF, Orchard E, Scoggins BA, Wright RD, Myers K, Junqueira CL (1968) Physiological, morphological and behavioural adaptation to a sodium deficient environment by wild native Australian and introduced species of animals. Nature 217: 922
Blair-West JR, Coghlan JP, Denton DA, Scoggins BA, Wintour EM, Wright RD (1970) Effect of change of sodium balance on the corticosteroid response to angiotensin II. Aust J Exp Biol Med Sci 48: 253
Bogdanski DF, Brodie BB (1966) Role of sodium and potassium ions in storage of norepinephrine by sympathetic nerve endings. Life Sci 5: 1563
Bogdanski DF, Blaszkowske PP, Tissari AH (1970) Mechanisms of biogenic amine transport and storage. IV. Relationship between K and the Na requirement for transport and storage of 5-hydroxytryptamine and norepinephrine in synaptosomes. Biochim Biophys Acta 211: 521
Bougas J, Flood C, Little B, Tait JF, Tait SAS, Underwood R (1964) Dynamic aspects of aldosterone metabolism. In: Baulieu EE, Robel P (eds) Aldosterone. Blackwell Scientific, Oxford, p 25
Braunwald E, Plauth WH Jr, Morrow AG (1965) A method for the detection and quantification of impaired sodium excretion. Circulation 32: 223
Brown JJ, Davies DL, Lever AF, Robertson JIS (1964) Influence of sodium deprivation and loading on the plasma-renin in man. J Physiol (Lond) 173: 408
Carpenter CCJ, Davis JO, Ayers CR (1961) Relation of renin, angiotensin II, and experimental renal hypertension to aldosterone secretion. J Clin Invest 40: 2026
Carter NW, Rector FC, Campion DS, Seldin DW (1967) Measurement of intracellular pH of skeletal muscle with pH-sensitive glass microelectrodes. J Clin Invest 46: 920
Cizek LJ, Semple RE, Huang KC, Gregersen MI (1951) Effect of extracellular electrolyte depletion on water intake in dogs. Am J Physiol 164: 415
Coghlan JP, Denton DA, Goding JR, Wright RD (1960) The control of aldosterone secretion. Postgrad Med J 36: 76
Coghlan JP, Fan JSK, Scoggins BA, Shulkes AA (1977) Measurement of extracellular fluid volume and blood volume in sheep. Aust J Biol Sci 30: 71
Coleman TG, Cowley AW, Guyton AC (1975) Angiotensin and the hemodynamics of chronic salt deprivation. Am J Physiol 229: 167
Conway J, Hatton R, Keddie J, Dawes P (1979) The role of angiotensin in the control of blood pressure during sodium depletion. Hypertension 1: 402
Cotlove E, Holliday MA, Schwartz R, Wallace WM (1951) Effects of electrolyte depletion and acid–base disturbance on muscle cations. Am J Physiol 167: 665
Cox JR, Leonard PJ, Singer B (1962) Changes in sodium distribution in body fluid compartments and urinary aldosterone excretion. Clin Sci 23: 13
Crabbé J, Ross EJ, Thorn GW (1958) The significance of the secretion of aldosterone during dietary sodium deprivation in normal subjects. J Clin Endocrinol Metab 18: 1159
Dahl LK (1960) Possible role of salt intake in the development of essential hypertension. In: Bock KD, Cottier PT (eds) Essential hypertension. Springer, Berlin, p 53
Darrow DC (1946) Retention of electrolytes during recovery from severe dehydration due to diarrhoea. J Pediatr 28: 515
Darrow DC, Yannet H (1935) The changes in the distribution of body water accompanying increase and decrease in extracellular electrolyte. J Clin Invest 14: 266
Darrow DC, Schwartz R, Iannucci JF, Coville F (1948) The relation of serum bicarbonate concentration to muscle composition. J Clin Invest 27: 198
Davenport VD (1949) Relation between brain and plasma electrolytes and electroshock seizure thresholds in adrenalectomized rats. Am J Physiol 156: 322
Davies RE, Kornberg HL (1952) Relation between total and exchangeable sodium in the body. Nature 170: 979
Davis JO (1962) Adrenocortical and renal hormonal function in experimental cardiac failure. Circulation 25: 1002
Davis JO, Olichney MJ, Brown TC, Binnion PF, Casper A (1965) Metabolism of aldosterone in several experimental situations with altered aldosterone secretion. J Clin Invest 44: 1433
Deane HW (1962) The anatomy, chemistry, and physiology of adrenocortical tissue. In: Deane HW (ed) Handbook of experimental pharmacology. The adrenocortical hormones, Part 1. Springer, Berlin, p 1
Denton DA (1949) Renal regulation of the extracellular fluid: A study of homeostasis in a patient with duodenal fistula. Med J Aust 2: 521
Denton DA (1957a) The effect of variations in blood supply on the secretion rate and composition of parotid saliva in Na-depleted sheep. J Physiol (Lond) 135: 227
Denton DA (1957b) The study of sheep with permanent unilateral parotid fistulae. Q J Exp Physiol 42: 72
Denton DA, Sabine JR (1963) The behaviour of Na deficient sheep. Behaviour 20: 364
Denton DA, Wynn V, McDonald IR, Simon S (1951) Renal regulation of the extracellular fluid. II. Renal physiology in electrolyte subtraction. Acta Med Scand 140 [Suppl 261]: 1
Denton DA, McDonald IR, Munro J, Williams W (1952) Excess Na subtraction in the sheep. Aust J Exp Biol Med Sci 30: 213
Denton DA, Nelson JF, Orchard E, Weller S (1969) The role of adrenocortical hormone secretion in salt appetite. In:

Pfaffmann C (ed) Proceedings of the Third International Symposium on Olfaction and Taste. Rockefeller University Press, New York, p 535

Dorer FE, Kahn JE, Lentz KE, Levine M, Skeggs LT (1975) Formation of angiotensin II from tetradecapeptide renin substrate by angiotensin-converting enzyme. Biochem Pharmacol 24: 1137

Edelman IS (1968) Aldosterone and sodium transport. In: McKerns KW (ed) Functions of the adrenal cortex, vol 1. Appleton-Century-Crofts, New York, p 79

Edelman IS, James AH, Baden H, Moore FD (1954a) Electrolyte composition of bone and the penetration of radiosodium and deuterium oxide into dog and human bone. J Clin Invest 33: 122

Edelman IS, James AH, Brooks L, Moore FD (1954b) Body sodium and potassium. IV. The normal total exchangeable sodium, its measurement and magnitude. Metabolism 3: 530

Elghozi JL, Altman J, Devynck MA, Liard JF, Grunfeld JP, Meyer P (1976) Lack of hypotensive effect of central injection of angiotensin inhibitors in spontaneously hypertensive (SH) and normotensive rats. Clin Sci Mol Med 51: 385S

Elkinton JR, Danowski TS (1955) The body fluids. Williams & Wilkins, Baltimore

Elkinton JR, Danowski TS, Winkler AW (1946) Hemodynamic changes in salt depletion and in dehydration. J Clin Invest 25: 120

Elkinton JR, Winkler AW, Danowski TS (1947) The importance of volume and of tonicity of body fluids in salt depletion shock. J Clin Invest 26: 1002

Elman R, Hartmann AF (1930) The cause of death following rapidly the total loss of pancreatic juice. Arch Surg 20: 333

Epstein AN, Ganten D (1977) Reciprocity of plasma and CSF angiotensin: Failure to confirm (Abstract). Fed Proc 36: 481

Epstein AN, Fitzsimons JT, Johnson AK (1974) Peptide antagonists of the renin–angiotensin system and the elucidation of the receptors for angiotensin-induced drinking. J Physiol (Lond) 238: 34P

Fei DTW, Coghlan JP, Fernley RT, Scoggins BA (1979) Blood clearance rates of angiotensin II and its metabolism in sheep: Presence of immunoreactive fragments in arterial blood. Clin Exp Pharmacol Physiol 6: 129

Fischer-Ferraro C, Nahmod VE, Goldstein DJ, Finkielman S (1971) Angiotensin and renin in rat and dog brain. J Exp Med 133: 353

Fishman RA (1974) Cell volume, pumps, and neurologic function. Brain's adaptation to osmotic stress. In: Plum F (ed) Brain dysfunction in metabolic disorders, p 159. Raven Press, New York (Research publications of the Association for Research into Nervous and Mental Diseases, no. 53)

Fitzsimons JT (1979) The physiology of thirst and sodium appetite. Cambridge University Press

Fitzsimons JT, Kucharczyk J (1978) Drinking and haemodynamic changes induced in the dog by intracranial injection of components of the renin–angiotensin system. J Physiol (Lond) 276: 419

Forbes GB, McCoord A (1963) Changes in bone sodium during growth in the rat. Growth 27: 285

Forbes GB, Perley A (1951) Estimation of total body sodium by isotopic dilution. I. Studies on young adults. J Clin Invest 30: 558

Forbes GB, Tobin RB, Lutz A (1959) Response of bone sodium to acute changes in extracellular fluid composition (cat). Am J Physiol 192: 69

Fulton RL, Ridolpho P (1971) Physiologic effects of acute sodium depletion. Ann Surg 173: 344

Gabriel S (1894) Chemische Untersuchungen über die Mineralstoffe der Knocken und Zähne. Z Physiol Chem 18: 257

Gage PW, Quastel DMJ (1965) Influence of sodium ions on transmitter release. Nature 206: 1047

Gamble JL (1949) Chemical anatomy, physiology and pathology of extracellular fluids. Harvard University Press, Cambridge, Mass.

Gamble JL, McIver MA (1928) Body fluid changes due to continued loss of the external secretion of the pancreas. J Exp Med 48: 859

Ganten D, Speck G (1978) The brain renin angiotensin system: A model for the synthesis of peptides in the brain. Biochem Pharmacol 27: 2379

Ganten D, Marquez-Julio A, Granger P, Hayduk K, Karsunky KP, Boucher R, Genest J (1971a) Renin in dog brain. Am J Physiol 221: 1733

Ganten D, Minnich JL, Granger P, Hayduk K, Brecht HM, Barbeau A, Boucher R, Genest J (1971b) Angiotensin-forming enzyme in brain tissue. Science 173: 64

Ganten D, Hutchinson JS, Schelling P (1975) The intrinsic brain isorenin–angiotensin system in the rat: Its possible role in central mechanisms of blood pressure regulation. Clin Sci Mol Med 48: 265S

Ganten D, Hutchinson JS, Schelling P, Ganten U, Fischer H (1976) The isorenin–angiotensin systems in extrarenal tissue. Clin Exp Pharmacol Physiol 3: 103

Ganten D, Fuxe K, Phillips MI, Mann JFE, Ganten U (1978) The brain isorenin–angiotensin system: Biochemistry, localization, and possible role in drinking and blood pressure regulation. In: Ganong WF, Martini L (eds), Frontiers in neuroendocrinology, vol 5. Raven Press, New York, p 61

Gaunt R, Chart JJ (1962) Mineralocorticoid action of adrenocortical hormones. In: Deane HW (ed) Handbook of experimental pharmacology: The adrenocortical hormones, part 1. Springer, Berlin, p 514

Gilman A (1934) Experimental sodium-loss analogous to adrenal insufficiency and the resulting water shift and sensitivity to haemorrhage. Am J Physiol 108: 662

Gordon RD, Nicholls MG, Tree M, Fraser R, Robertson JIS (1980) Influence of sodium balance on ACTH/adrenal corticosteroid dose–response curves in the dog. Am J Physiol 238: E543

Grant H, Reischsman F (1946) The effects of the ingestion of large amounts of sodium chloride on the arterial and venous pressures of normal subjects. Am Heart J 32: 704

Hackenthal E, Hackenthal R, Hilgenfeldt U (1978) Isorenin, pseudorenin, cathepsin D and renin: A comparative enzymatic study of angiotensin-forming enzymes. Biochim Biophys Acta 522: 574

Harrison HE, Darrow DC, Yannet H (1936) The total electrolyte content of animals and its probable relation to the distribution of body water. J Biol Chem 113: 515

Hartmann AF, Elman R (1929) The effects of loss of gastric and pancreatic secretions and the methods for restoration of normal conditions in the body. J Exp Med 50: 387

Hine GJ, Jagger PI, Burrows BA (1962) Use of a clinical body counter for long-term exchangeable sodium status. In: Whole body counting. Atomic Energy Agency, Vienna, p 413

Hirose S, Yokosawa H, Inagami T (1978) Immunochemical identification of renin in rat brain and distinction from acid proteases. Nature 274: 392

Hoffman WE, Schelling P, Phillips MI, Ganten D (1976) Evidence for local angiotensin formation in brain of nephrectomized rats. Neurosci Lett 3: 299

Hollenberg NK (1980) Set point for sodium homeostasis: Surfeit, deficit, and their implications. Kidney Int 17: 423

Hollenberg NK, Adams DF, Solomon HS, Abrams HL, Merrill JP (1972) What mediates renal vascular response to a salt load in normal man? J Appl Physiol 33: 491

Horvath JS, Baxter C, Furby F, Tiller DJ (1977) Endogenous

angiotensin in brain. Prog Brain Res 47: 161
Hutchinson JS, Csicsman J, Korner PI, Johnston CI (1978) Characterization of immunoreactive angiotensin in canine cerebrospinal fluid as Des-Asp[1]-Angiotensin II. Clin Sci Mol Med 54: 147
Ikkos D, Luft R, Sjogren B (1954) Distribution of fluid and sodium in healthy adults. Metabolism 3: 400
Jagger PI, Hine GJ, Cardarelli JA, Burrows BA (1963) Influence of sodium intake on exchangeable sodium in normal human subjects. J Clin Invest 42: 1459
Jones KM, Lloyd-Jones R, Riondel A, Tait JF, Tait SAS, Bulbrook RD, Greenwood FC (1959) Aldosterone secretion and metabolism in normal men and women in pregnancy. Acta Endocrinol (Copenh) 30: 321
Katz B (1966) Nerve muscle and synapse. McGraw Hill, New York
Katzman R, Pappius HM. (1973) Brain electrolytes and fluid metabolism. Williams & Wilkins, Baltimore
Khoyi MA (1967) Electrocardiographic changes in oxygen consumption in acute salt depletion. Am Heart J 73: 217
Kirksey A, Pike RL (1962) Some effects of high and low sodium intakes during pregnancy in the rat: I. J Nutr 77: 33
Kirksey A, Pike RL, Callahan JA (1962) Some effects of high and low sodium intakes during pregnancy in the rat: II. J Nutr 77: 43
Ladell WSS, Waterlow JC, Hudson MF (1944) Desert climate. Physiological and clinical observations. Lancet II: 491, 527
Lahiri S, Lajtha A (1964) Cerebral amino acid transport in vitro. I. Some requirements and properties of uptake. J Neurochem 11: 77
Lajtha A, Sershen H (1975) Inhibition of amino acid uptake by the absence of sodium in slices of brain. J Neurochem 24: 667
Laragh JH (1962) Hormones and the pathogenesis of congestive heart failure: Vasopressin, aldosterone, and angiotensin II. Further evidence for renal–adrenal interaction from studies in hypertension and in cirrhosis. Circulation 25: 1015
Laragh JH, Ulick S, Januszewicz V, Deming QB, Kelly WG, Lieberman S (1960) Aldosterone secretion and primary and malignant hypertension. J Clin Invest 39: 1091
Leaf A, Couter WT (1949) Evidence that renal sodium excretion by normal human subjects is regulated by adrenal cortical activity. J Clin Invest 28: 1067
Leard SE, Freis ED (1949) Changes in the volume of the plasma, interstitial and intracellular fluid spaces during hydration and dehydration in normal and oedematous subjects. Am J Med 7: 647
Leonard PJ (1962) Effect of sodium restriction and experimental nephrosis on tissue spaces and tissue sodium distribution in the rat. J Endocrinol 35: 323
Leonard PJ (1963) Aldosterone and water and sodium distribution in normal and adrenalectomized rats. J Endocrinol 26: 525
Lohmeier TE, Kastner PR, Smith MJ, Guyton AC (1980). Is aldosteronism important in the maintenance of arterial blood pressure and electrolyte balance during sodium depletion? Hypertension 2: 497
Luetscher JA Jr, Axelrad BJ (1954) Increased aldosterone output during sodium deprivation in normal men. Proc Soc Exp Biol Med 87: 650
Lyons RH, Jacobson SD, Avery NL (1944) Increases in the plasma volume following the administration of sodium salts. Am J Med Sci 208: 148
Mann JFE, Phillips MI, Dietz R, Haebara H, Ganten D (1978) Effects of central and peripheral angiotensin blockade in hypertensive rats. Am J Physiol 234: H629
Marriott HL (1950) Water and salt depletion. Thomas, Springfield, p 40
McCance RA (1936a) Experimental sodium chloride deficiency in man. Proc R Soc Lond [Biol] 119B: 245
McCance RA (1936b) Medical problems in mineral metabolism. III. Experimental human salt deficiency. Lancet I: 823
McCance RA (1938) The effect of salt deficiency in man on the volume of the extracellular fluids, and on the composition of sweat, saliva, gastric juice, and cerebrospinal fluid. J Physiol (Lond) 92: 208
McDougall JG, Coghlan JP, Scoggins BA, Wright RD (1974) Effect of sodium depletion on bone sodium and total exchangeable sodium in sheep. Am J Vet Res 35: 923
McEwen OR (1935) Salt loss as a common cause of ill-health in hot climates. Lancet I: 1015
Mellors RC, Muntwyler E, Mautz FC, Abbott WE (1942) Changes of the plasma volume and 'available (thiocyanate) fluid' in experimental dehydration. J Biol Chem 144: 785
Meyer HH, Grunert RR, Zepplin MT, Crummer RH, Bohstedt G, Phillips PH (1950) Effect of dietary levels of sodium and potassium on growth and on concentrations in blood plasma and tissue of white rat. Am J Physiol 162: 182
Mills JN (1962) Aldosterone secretion in man. Br Med Bull 18: 170
Moore FD, McMurrey JD, Parker HV, Magnus IC (1956) Body composition. Total body water and electrolytes: Intravascular and extravascular phase volumes. Metabolism 5: 447
Mouw DR, Abraham SF, Blair-West JR, Coghlan JP, Denton DA, McKenzie JS, McKinley MJ, Scoggins BA (1974) Brain receptors, renin secretion, and renal sodium retention in conscious sheep. Am J Physiol 226: 56
Muller AF, Veyrat R, Manning EL (1959) Study of aldosterone secretion by aldosterone marked with tritium. Helv Chir Acta 26: 714
Munro DS, Satoskar RS, Wilson GM (1957) The exchange of bone sodium with isotopes in rats. J Physiol (Lond) 139: 474
Nalin DR, Cash RA (1976) Sodium content in oral therapy for diarrhoea. Lancet II: 957
Neuman WF, Neuman MW, (1953) The nature of the mineral phase of bone. Chem Rev 53: 1
Nichols G Jr, Nichols N (1956a) Changes in tissue compositions during acute sodium depletion. Am J Physiol 186: 383
Nichols G Jr, Nichols N (1956b) The role of bone in sodium metabolism. Metabolism 5: 438
Nichols N, Nichols G Jr (1955) The relation of bone sodium to sodium excretion in mercurial diuresis. Clin Res Proc 3: 44
Nichols N, Nichols G Jr (1957) Effect of large loads of sodium on bone and soft tissue composition. Proc Soc Exp Biol Med 96: 835
Norman N (1963) The participation of bone in the sodium and potassium metabolism of the rat. II. The effect of variation of electrolyte intake, acidosis and alkalosis. Acta Physiol Scand 57: 373
Orent-Keiles E, Robinson A, McCollom EV (1937) The effects of sodium deprivation on the animal organism. Am J Physiol 119: 651
Overton E (1904) Beiträge zur allgemeinen Muskel- und Nerven-physiologie. II. Ueber die Unentbehrlichkeit von Natrium-(oder Lithium) Ionen für den Contractionsact des Muskels. Pflügers Arch 92: 346
Pappius HM (1975) Normal and pathological distribution of water in brain. In: Cserr HF, Fenstermacher JD, Fencl V (eds) Fluid environment of the brain. Academic Press, New York, p 183
Pappius HM, Oh JH, Dossetor JB (1967) The effects of rapid haemodialysis on brain tissues and cerebrospinal fluid of dogs. Can J Physiol Pharmacol 45: 129
Pavlov IP (1902) The work of the digestive glands. Charles Griffin, London
Peterson RE (1959) The miscible pool and turnover rate of

adrenocortical steroids in man. Recent Prog Horm Res 15: 231
Phillips MI, Mann JFE, Haebara H, Hoffman WE, Dietz R, Schelling P, Ganten D (1977) Lowering of hypertension by central saralasin in the absence of plasma renin. Nature 270: 445
Phillips MI, Weyhenmeyer J, Felix D, Ganten D, Hoffman WE (1979) Evidence for an endogenous brain renin–angiotensin system. Fed Proc 38: 2260
Ramsay DJ (1979) The brain renin–angiotensin system: A re-evaluation. Neuroscience 4: 313
Reid IA (1979) The brain renin–angiotensin system: A critical analysis. Fed Proc 38: 2255
Reid IA, Moffat B (1978) Angiotensin II concentration in cerebrospinal fluid following intraventricular injection of angiotensinogen or renin. Endocrinology 103: 1494
Reid IA, Ramsay DJ (1975) The effects of intracerebroventricular administration of renin on drinking and blood pressure. Endocrinology 97: 536
Reid IA, Day RP, Moffat B, Hughes HG (1977) Apparent angiotensin immunoreactivity in dog brain resulting from angiotensinase. J Neurochem 28: 435
Reijnoud DJ, Tager JM (1973) Measurement of intralysozymal pH. Biochim Biophys Acta 297: 174
Renwick R, Robson JS, Stewart CP (1955) Observations upon the withdrawal of sodium chloride from the diet in hypertensive and normotensive individuals. J Clin Invest 34: 1037
Romero JC, Staneloni RJ, Dufau ML, Dohmen R, Binia A, Kliman B, Fasciolo JC (1968) Changes in fluid compartments, renal hemodynamics, plasma renin and aldosterone secretion induced by low sodium intake. Metabolism 17: 10
Saphir W (1945) Chronic hypochloremia simulating psychoneurosis. J Am Med Assoc 129: 510
Schelling P, Hutchinson JS, Ganten U, Sponer G, Ganten D (1976) Impermeability of the blood–cerebrospinal fluid barrier for angiotensin II in rats. Clin Sci Mol Med 51[Suppl. 3]: 399s
Schoelkens BA, Jung W, Steinbach R (1976) Blood pressure response to central and peripheral injection of angiotensin II and 8-C-phenylglycine analogue of angiotensin II in rats with experimental hypertension. Clin Sci Mol Med 51[Suppl 3]: 403s
Schoelkens BA, Jung W, Becker R, Rascher W, Schonnig A, Unger Th, Ganten D (1980) Endogenous brain angiotensin II increases arterial blood pressure by stimulation of pituitary hormones and plasma catecholamines. Acta Endocrinol (Copenh) 94 [Suppl 234]: 132
Schroeder HA (1949) Renal failure associated with low extracellular sodium chloride. JAMA 141: 117
Schultz SJ, Currant PF (1970) Coupled transport of sodium and organic solute. Physiol Rev 50: 637
Scribner BH, Burnell JM (1956) Interpretation of serum potassium concentration. Metabolism 5: 468
Seligman AM, Frank HA, Fine J (1946) Traumatic shock. XII. Hemodynamic effects of alterations of blood viscosity in normal dogs and in dogs with shock. J Clin Invest 25: 1
Simpson JB, Saad WA, Epstein AN (1976) The subfornical organ, the cerebrospinal fluid, and the dipsogenic action of angiotensin. In: Onesti G, Fernandes M, Kim KE (eds) Regulation of blood pressure by the central nervous system. Grune & Stratton, New York, p 191
Singer B, Stack-Dunne MP (1955) The secretion of aldosterone and corticosterone by the rat adrenal. J Endocrinol 12: 130
Smilay MG, Dahl LK, Spraragen SC, Silver L (1961) Isotope sodium turnover studies in man. Evidence of a minimal sodium (Na^{22}) retention 6 to 11 months after administration. J Lab Clin Med 58: 60
Strauss MB, Lamdin E, Smith WP, Bleifer SJ (1958) Surfeit and deficit of sodium. Arch Intern Med 102: 527
Streeten DPH (1950) The effects of sodium and chloride lack on intestinal motility and their significance in paralytic ileus. Surg Gynecol Obstet 91: 421
Sullivan JM, Ratts TE, Taylor JC, Kraus DH, Barton BR, Patrick DR, Reed SW (1980) Hemodynamic effects of dietary sodium in man. A preliminary report. Hypertension 2: 506
Swinyard EA (1949) Effect of extracellular electrolyte depletion on brain electrolyte pattern and electroshock seizure threshold. Am J Physiol 156: 163
Tait JF, Tait SAS, Little B, Laumas KR (1961) The disappearance of 7-H^3-D-aldosterone in the plasma of normal subjects. J Clin Invest 40: 72
Ulick S, Laragh JR, Lieberman S (1958) The isolation of a urinary metabolite of aldosterone and use to measure the rate of secretion of aldosterone by the adrenal cortex of man. Trans Assoc Am Physicians 71: 225
Verney EB (1947) The antidiuretic hormone and the factors which determine its release. Proc R Soc Lond [Biol] 135: 25
Wakim KG (1969) Predominance of hyponatraemia over hypo-osmolality in simulation of the dialysis disequilibrium syndrome. Mayo Clin Proc 44: 433
Wiggins WS, Maury CH, Lyons RH, Pitts RF (1951) The effect of salt loading and salt depletion on renal function and electrolyte excretion in man. Circulation 3: 275
Wolf AV (1958) Thirst. Physiology of the urge to drink and problems of water lack. Thomas, Springfield, Ill
Woodbury DM (1956) Effect of acute hyponatremia on distribution of water and electrolytes in various tissues in the rat. Am J Physiol 185: 281
Yamaguchi M, Yanos T, Yamaguchi T, Lajtha A (1970) Amino acid uptake in the peripheral nerve of the rat. J Neurobiol 1: 4419
Yannet H (1940) Changes in the brain resulting from depletion of extracellular electrolytes. Am J Physiol 128: 683

8 Techniques in study of salt appetite

Summary

1. The study of salt appetite offers exceptional opportunities for the dissection of basic mechanisms of instinctual behaviour. Loss of sodium from the body and appetitive behaviour to repair the deficit can be measured precisely. But more than this, the state of body balance can be rapidly manipulated physiologically by, for example, intravenous infusion, and the time characteristics of control system reaction assessed. Large, but physiological, changes in concentration of highly relevant variables such as sodium concentration can be made locally in, for example, the cerebrospinal fluid with trivial influence on body balance generally. Thus there are real, though relative, practical advantages in the investigation of salt appetite compared to other instinctual patterns of great interest—such as sex, territoriality and maternal behaviour.

2. It is practical to contrive that a young animal enters an experimental situation with no previous experience of sodium deficiency.

3. The various ways low sodium status or frank body sodium deficit have been produced in rats are described.

4. Emphasis is given to the fact that physiological amounts of a number of hormones involved in the response to stress have a large capacity to induce salt appetite. Some techniques used to produce sodium deficit in rats may have effects due to such hormone secretion. Thus the resulting salt appetite behaviour may not be a simple response to body sodium deficit.

5. The great advantages in the use of a permanent unilateral parotid fistula in ruminants such as sheep for the study of salt appetite are described. There is continuous profuse secretion of an alkaline saliva with high sodium concentration, which contains no enzymes to damage the skin. Large deficits of sodium and also, if desired, phosphate can be induced at will. Sheep are particularly suitable experimental animals because they are docile, easily trained and may be kept indefinitely in a metabolism cage. Methods of human surgery are readily applicable to study of sheep, and the large folds of neck skin of the Merino breed make plastic surgery and organ transplantation straightforward.

Introduction

Salt hunger is a classic example of an innate neural organization generating a behavioural drive in the instance of nutritional deficiency. As described elsewhere, salt intake may occur for other reasons—that is, coincident with the character of food eaten, pleasure of taste, habituation, or for reasons as striking as psychosocial bonding ceremonies in Middle Eastern religions. But salt appetite is a royal road to insight of how instinctive behaviours are activated by deficiency and how desires are satisfied, particularly because it lends itself to precise measurement of the body deficit and the intake behaviour it causes.

Furthermore, a large variety of techniques is available for experimental manipulation of sodium balance. But the advantage of precise quantitation, in contrast to some other innate behavioural patterns of comparable interest—e.g. sex, aggression, territoriality—goes much further than exact catalogue of deficits and intake. It is open to make very rapid correction of sodium balance in the deficient animal by systemic infusion. Thus the time and·quantitative characteristics of reaction of the systems which determine appetite can be appraised. On this same theme of accessing the control system, it is possible to make relatively localized, physiologically graded and sustained changes in sodium concentration in regions suspected to be important: for example in the cerebrospinal fluid of the third ventricle. These modes of experimental analysis are, obviously, not without counterpart in examining other behaviours, but the facility of correlation with physiological and endocrinological measurement deserves remark.

As will be detailed further on, surgical construction of a permanent unilateral parotid fistula in the ruminant is tantamount to placing a sodium tap on the bloodstream. Over twice the salt content of the circulating plasma may be lost per day. Thus, at will, a large negative sodium balance can be contrived rapidly, and corrected rapidly, and the behaviour thus evoked evaluated.

There is also the advantage that Wolf et al. (1974) see in the fact that standard laboratory chow contains an excess of sodium. Thus rats raised on this in the laboratory may have no or minimal experience of sodium deficiency. Further, one can be certain that rats so reared will never have had the opportunity to taste sodium independently of other strong concomitant taste stimuli. Thus they propose it is possible in the adult animal to study a relatively virgin drive system without special manipulation of past experience. Similar consideration applies to the sheep, where the land on which birth occurred can be known exactly in terms of sodium content of pasture and of the water sources.

In terms of ease of laboratory study, the amount of salt voluntarily ingested by rats under baseline sodium-replete conditions can be varied by changing the concentration of solutions provided. With sodium chloride, a 1% solution (171 mmol/l) is ingested in relatively large quantities—more than 30 ml/day—while with a 3% solution (513 mmol/l) intake is small—1–5 ml/day (Bare 1949; Richter 1956). The palatable solutions are useful when experimental manipulations cause a decrease in sodium appetite because they provide a well-elevated baseline. The aversive concentrations facilitate observing increases in sodium appetite because they give a stable near-zero baseline. Wolf et al. (1974), in addition, suggest that sodium appetite has an important advantage in relation to studies of brain lesions in, for example, the lateral hypothalamus. They note that whereas rats may recover relatively normal food and water intake a few days after lesioning, they show little evidence of recovery of salt appetite in spite of a long period of post-operative experience. Possibly the functions which recover are mediated by other brain regions which contain programmes for food and water ingestion as a result of numerous pre-operative feeding and drinking experiences, while the failure of recovery of sodium appetite is due to the absence of such learned programmes for sodium intake.

The methods used for study of sodium appetite, principally in the rat, are instanced below. In a number of cases reference will be brief only, as more detailed description and analysis of methodology occurs in later chapters. The main discussion of methods in the sheep will be deferred until the end of the chapter.

Sodium depletion

Low-sodium diet

In the laboratory rat short-term deprivation does not readily induce sodium appetite. According to Fregly et al. (1965), after 5 days of sodium deprivation a small increase in intake is seen on offer of salt, but is significant only after 24 h of access. Jalowiec (1974) similarly reports that maintenance of rats on a low-sodium diet did not elicit appetite for 0.51 mol/l sodium chloride though ionic changes indicative of increased mineralocorticoid secretion occurred. A somewhat different emphasis emerges from Richter (1956). He records the use of the McCollum diet (graham flour, skimmed milk powder, casein, butter, calcium carbonate and sodium chloride) which contained 1% by weight of sodium chloride, the amount calculated by McCollum as necessary for the rat. When the sodium was removed from this diet and the rats had access to a 3% sodium chloride solution (0.51 mol/l) and water, without a single exception they drank salt solution freely on the first day of the experiment and continued to drink essentially constant amounts which closely approximated the 1% of total diet. A similar finding on elective salt intake was made in the more complicated circumstance of a cafeteria of some 17 different substances.

Chronic exposure to a diet with very low sodium content can induce a powerful sodium appetite as attested by behaviour of native rats, rabbits and marsupials in the Australian alps. However, for laboratory studies, methods involving actual sodium loss or sodium sequestration have been more useful. Recently, in studies at the Howard Florey Institute, it has been shown sheep can be maintained in good health on a *Sorghum* grain diet which provides ca 1 mmol of sodium and 6 mmol of potassium per day. Mineral status can be manipulated by intrarumen infusion of electrolytes. To date this regime has been used in studies of hypertension induced by adrenocorticotrophic hormone (ACTH) (Scoggins et al. 1978), but it should prove a useful addition to methodology of the study of salt appetite.

Adrenalectomy

This was the classic method used by Richter (1936, 1942–1943, 1956) to produce external sodium loss in rats. It has subsequently been used to study salt appetite in sheep and rabbits. When withdrawal of supportive hormones has been protracted, deterioration and weakness from the variety of metabolic consequences in addition to sodium deficiency may modify ingestive behaviour and introduce ambiguity into experimental interpretation. Compromise of adrenal secretion by methods such as reduction of thyroid function by propylthiouracil (PTU) (Fregly 1969) or metabolic inhibitors is also used.

Peritoneal dialysis

This method, usually involving 5% glucose solution, is effective in rapidly producing sodium deficit. However, as will be recounted, questions arise of it being a stressful procedure with physical effects causing deterioration of animals, particularly if done sequentially. This could compromise interpretation.

Polyethylene glycol

In the rat subcutaneous injection of 5 ml of a 10%, 20% or 30% solution of polyethylene glycol (PEG) (Carbowax, Compound 20-M; Union Carbide), a biochemically inert colloid, decreases intravascular volume by raising the oncotic pressure in the region of the injection and drawing protein-free plasma from the circulation. Thus sequestration and functionally effective withdrawal of fluid isosmotic with plasma occurs. This method (Stricker 1966; Stricker and Wolf 1969) has given valuable insight into both thirst and sodium appetite. Sodium appetite is stimulated after 12–14 h. Fitzsimons used the method, but with intraperitoneal route of injection, to study thirst in 1961. However, Stricker and MacArthur (1974) showed that intraperitoneal injection did not stimulate sodium appetite because PEG enters the plasma within 6–12 h and the distortion of body fluid is not sustained as it is with subcutaneous injection.

We have examined the influence of PEG in the wild rabbit. The data would suggest some caution before assuming that this is a satisfactory procedure for study of behavioural effects of water and sodium deficiency uncomplicated by other factors—at least in this species. Eight rabbits were given 4 g/kg body wt of PEG as a 30% aqueous solution injected subcutaneously. This resulted in a fairly rapid accumulation of fluid which had become quite large 12 h later. A small increase in volume continued until day 2 after injection. The volume had decreased to about two-thirds on day 3 when it was sampled, was quite small on day 4 and fluid had disappeared on day 5. The haematocrit (mean) increased from a pre-injection value of 43% to 45% in 2 h, was 48% at 12 h, 47% at 24 h, 38% on day 2, 32% on day 3 and 25% on day 4 after injection. High erythrocyte sedimentation rates were noted on days 2–4. The rabbits were all distressed for several days, remaining hunched up in a corner of the cage for most of the time. Their general condition appeared to be poor for 5–6 days with loss of smooth glossy coat. This correlated with decreased food consumption. The fluids aspirated on day 3 averaged 128 mmol Na/l (range 115–139 mmol/l) and were almost free of red cells. Table 8-1 shows the effect of the treatment on voluntary daily intake of water, 500 mmol/l sodium chloride solution, and food. Balance data are shown also.

In general it would seem that evaluation of the use of PEG in the study of salt appetite requires

Table 8-1. Effect of subcutaneous PEG (12 ml 30% aq. solution/kg body wt) on food, sodium chloride and total fluid intake, urinary sodium excretion and sodium and water balance.

A. Mean control values for previous 7 days

Food (g)	Voluntary intake of 500 mM NaCl (mmol)	Total Na intake (mmol)	Urine Na (mmol)	Difference (mmol)	Total fluid intake (ml)	Urine volume (ml)	Difference (ml)
62	1.1	5.44	5.20	0.24	92	57	35

B. Mean daily values after PEG

Day	Food (g)	Voluntary intake of 500 mM NaCl (mmol)	Total Na intake (mmol)	Urine Na (mmol)	Difference (mmol)		Total fluid intake (ml)	Urine volume (ml)	Difference (ml)		Cumulative	
					Uncorrected	Corrected			Uncorrected	Corrected	Na balance (mmol)	H_2O balance (ml)
1	18	6.0	7.3	1.9	+5.4	+5.3	168	46	122	87	+5.3	+87
2	17	10.6	11.9	5.5	+6.4	+6.3	89	39	50	15	+11.6	+102
3	33	0.8	3.1	1.4	+1.7	+1.6	82	49	33	−2	+13.2	+100
4	38	2.0	4.7	9.1	−4.4	−4.6	27	55	−28	−63	+8.6	+37
5	32	0.5	2.7	7.2	−4.5	−4.6	66	65	1	−34	+4.0	+3
6	46	1.4	4.6	7.1	−2.5	−2.7	82	58	24	−11	+1.3	−8
7	50	3.2	6.7	7.5	−0.8	−1.0	107	64	43	8	+0.3	0
8	53	0.7	4.4	4.6	−0.2	−0.4	97	55	42	7	−0.1	+7
9	46	1.5	4.7	5.2	−0.5	−0.7	94	60	34	−1	−0.8	+6

Ratio of sodium to water accumulated: day 1, 61 mmol/l; day 2, 114 mmol/l; day 3, 132 mmol/l; day 4, 233 mmol/l.
Daily water balance after PEG was calculated from total fluid intake and urine volume. It was corrected for the observed difference (35 ml), which presumably represents evaporative and faecal loss, in the control period. Daily sodium balance after PEG was calculated from the total intake and urinary excretion. It was corrected for the observed difference (presumably faecal) between total sodium intake and urine sodium output, which was 0.4 mmol/100 g of food eaten under control conditions.

much more knowledge of its influence upon ACTH and adrenal secretion in rat and rabbit. The clear-cut and specific effect of ACTH on salt appetite in ten wild rabbits offered free access to 500 mmol/l solutions of sodium, potassium, magnesium and calcium chloride and water is recounted in Chapter 23 (Fig. 23-9). There is also a very large effect of ACTH on salt appetite in the rat (Weisinger et al. 1978) and sheep (Weisinger et al. 1980).

In relation to the rat, Stricker et al. (1979) have published data on the influence of plasma volume deficit produced by 5 ml of 30% PEG on appetites and also plasma renin, aldosterone and corticosterone concentrations. Mean plasma aldosterone in control estimations was 9.8±0.6 ng/100 ml and corticosterone was 10.0±2.2 μg/100 ml (n=14). These values, derived with radioimmunoassay after chromatographic purification, are, in the case of corticosterone, about three times higher than found in our laboratory with use of the double-isotope dilution derivative assay. PEG treatment caused a 10-fold rise in plasma renin activity and aldosterone concentration when measured 4 h later, and mean corticosterone had increased 5-fold. Ligation of the inferior vena cava caused a similar order of rise in aldosterone and corticosterone concentrations. With the nephrectomy experiments in this series (Stricker et al. 1979) both concentrations rose about 3-fold, and when PEG treatment was superimposed in the nephrectomized animal, the steroid concentrations approximated those seen with PEG alone in the normal animal. By contrast, with uncomplicated sodium depletion there was a 5-fold increase in aldosterone concentration with no change in corticosterone concentration. With caval ligation in the sodium-deplete animal, corticosterone was no higher than with caval ligation in the sodium-replete animal, whereas aldosterone was elevated about 36-fold. The thrust of these considerations is the fact that increase in corticosterone is not a feature of sodium deficiency and the biosynthetic mechanisms involved in the elevation of aldosterone secretion, but is a result of ACTH stimulation. Stricker et al. (1979) suggest that ACTH may be contributing to aldosterone hypersecretion. Also the rise in plasma potassium concentration with PEG may do this. But the data emphasize that a possible role of ACTH in the stimulation of salt appetite by PEG also has to be considered. The delay in onset of salt appetite for 6–8 h and peak activity by 16 h is consistent with the time delay of effect of steroid hormones, itself a result of their mode of action that involves transcription and translation processes.

Abdominal ligature of the inferior vena cava

This procedure effectively evokes thirst and water intake. Stricker (1971, 1973) reports that it does not elicit sodium appetite when rats are given both water and either 0.15 or 0.51 mol/l sodium chloride for the 24 h following operation. Fitzsimons (1969, 1972), however, states that with this procedure there is marked oliguria, and water intake causes a fall in serum sodium and osmolality. Later quite a marked preference for sodium chloride develops. Stricker et al. (1979) found no increase in 0.5 mol/l sodium chloride intake 8 and 24 h after caval ligation despite a 6- and 10-fold increase in plasma renin activity and aldosterone concentration respectively.

At all events the method has been little used in the study of salt appetite. In limited experience with sheep where prior surgical placement of a suprahepatic polythene choker round the inferior vena cava permitted its graded constriction in the conscious animal with verified haemodynamic effects, there was no evocation of salt appetite over 24 h. Ligature of hemi-azygous veins had precluded anastomotic channels circumventing the constriction, but further investigation is probably warranted.

Acute hyponatraemia

Acute reduction of plasma sodium concentration has been produced by intragastric loading of water equal to 5% body weight (Stricker 1966). No sodium appetite was produced. In man, however, De Wardener and Herxheimer (1957) showed that drinking 8–10 litres of water per day for 11 days induced an intense desire for salt. But this in one subject was associated with negative sodium balance. Taste threshold for salt decreased. The method of excess water intake unless combined with other factors appears to have little value in the study of salt appetite.

Frusemide diuresis

Frusemide (Lasix; Hoechst) has been used to produce acute water and electrolyte loss in rats (Jalowiec 1974; Wolf et al. 1974) and sheep (Zimmerman et al. 1978; M. J. McKinley, E. H. Blaine, D. Denton and R. S. Weisinger, unpublished). The drug inhibits renal sodium reabsorption in distal and proximal tubules and also in the ascending loop of Henle (Dirks and Seely 1969; Morgan et al. 1970). It has been given by intraperitoneal or subcutaneous injection or by intravenous infusion and a second dose given 4 h after the first can augment the effect. Jalowiec

(1974) states there are no adverse behavioural effects but Wolf et al. (1974) record anorexia in their frusemide studies, as also seen with subcutaneous formalin. Cruz, Perelle and Wolf (1977) report on the use of frusemide in combination with desoxycorticosterone acetate (DOCA) to produce a large abrupt increase in sodium intake without need to exceed physiological levels of mineralocorticoids by heroic amounts. Wilson and Wong (1975) have used Aldactazide (a combination of aldactone and hydrochlorathiazide) to induce salt appetite in rats.

Formalin

The technique of injection of 1.25–2.5 ml of 1.5% formalin (15 parts of 10% formalin phosphate in 85 parts of distilled water) under the thick skin of the back was introduced into the study of sodium appetite by Wolf and Steinbaum (1965). The hypovolaemia that is produced is due to sequestration of protein-rich plasma in the local interstitium, probably as a result of damage and increased permeability of the capillary walls. Furthermore, extracellular sodium accumulates in the damaged cells in the injection area reducing osmotic pressure and the sodium concentration of the extracellular fluid. This causes movement of water into the intracellular fluid. Urine sodium concentration falls but there is little change in its potassium concentration (Wolf and Steinbaum 1965). The method has been valuable in studies of thirst in terms of comparison of the effects of formalin with those of PEG. Both induce water drinking. Since formalin elicits cellular volume changes and PEG does not, the results point to an important role of decreased intravascular volume in thirst. However, the situation may be much more complicated in relation to sodium deficiency, as we will recount.

Initially Wolf and Steinbaum (1965) suggested that the technique had simplicity and was well tolerated. It therefore had advantages over techniques such as adrenalectomy, peritoneal dialysis, prolonged sodium deprivation or fistulation of the salivary glands which in their view were complicated and time-consuming. However, Wolf et al. (1974) draw attention in their experimental protocols to anorexic effects of formalin which may result from general stress or debilitation. They also state that formalin is an aversive stimulus (Woods et al. 1971) and can cause local skin ulceration (Porter and Relinger 1972), in addition to having well-known general stress effects. Wolf and Steinbaum (1965) noted that Braun-Menendez had used formalin to produce stress. In this regard Selye (1937), using 0.5 cc of a 4% solution (approximately the same amount as Wolf and Steinbaum (1965) but higher concentration), describes effects such as peritoneal and pleural exudates, retroperitoneal oedema, pancreatic, lymphatic, thymic, hypophyseal and adrenal changes, etc., as part of the 'alarm reaction'. Makara et al. (1969) showed that with 3% formalin injection in the hind paw there was a large rise in ACTH as evidenced by corticosterone increase, and that this effect was greatly attenuated by contralateral hemisection of the spinal cord but not by homolateral section. Wolf et al. (1974) go on to add that in order to obtain further evidence on possible stressful side effects of formalin they made close observation of spontaneous behaviour of four rats during the 8 h period following injection of formalin (1.25 ml), frusemide (10 mg) and fludrocortisol (5 mg). Whereas the latter two drugs had no influence on spontaneous behaviour, the small dose of formalin caused the animals to adopt abnormal prone postures with eyes partially closed and to drag their hind legs when locomoting. They seemed sensitive to touch at the injection site and squealed frequently when handled. Porter and Relinger (1972) have suggested that with 0.5% formalin the external skin damage effects are much reduced. While Wolf et al. (1974) raise the question of whether a long-acting anaesthetic would abolish these effects, they recommend the use of formalin in sodium appetite studies be discontinued. Morrison (1971) has pointed to anomalies in usual potassium chloride preference over calcium chloride which occur in formalin-treated animals and these data could reflect hormone action. Notwithstanding this viewpoint the technique is still being employed to produce sodium deficiency in studies involving interpretation of effects of precise brain lesioning (Walsh and Grossman 1977).

In preliminary studies in the rabbit, injection of 1.5% formalin in water (6 ml/kg body wt) did not produce the necrosis or local oedema reported in the rat and no increased water or saline intake occurred.

As will recur through the text wherever interpretation of data with formalin and PEG studies is involved, there is another major problem not yet adequately considered in the rat studies. That is, ACTH of itself (Blaine et al. 1975) and the adrenal glucocorticoid secretion it provokes (Blaine et al. 1975) have powerful salt-appetite-inducing effects in rabbit, rat and sheep. In the case of the rabbit, ACTH action appears to be directly on the brain, apart from its adrenal effects (Blaine et al. 1975; see also Chapter 23). The problem emerging is that apart from formalin inducing sodium appetite through fluid shifts, hyponatraemia, and also mineralocorticoid secretion, its effects probably could include major and variable stimulation of ACTH and glucocorticoid secretion. These may operate on appetite through different mechanisms. Interpretation becomes questionable if no measurement of these variables is made and probably is highly complicated even with such data available. The same essential objections apply to production of hypovolaemia and salt appetite by application and release of a tourniquet on a limb.

In a study of 18 rats by Nelson and Weisinger, 12 were injected subcutaneously with 2.5 ml of 0.6% formaldehyde (1.5 ml formalin in 0.02 mol/l sodium phosphate buffer) while six were given a control injection of buffer only. Of the experimental group six had access to 500 mmol/l sodium chloride. They were killed 6 h later by decapitation and steroids measured by double-isotope dilution derivative assay by Coghlan and Scoggins. Peripheral blood corticosterone of the control group was 3.4±1.5 μg/100 ml (mean±s.e.), of the formalin-injected rats with no access to sodium 10.0±1.7 μg/100 ml, and of the experimental group which had access to 500 mmol/l sodium chloride 12.2±1.0 μg/100 ml. At sacrifice the mean volume of blood obtained from the formalin-injected animals was 4.1 ml and from the controls 6.3 ml. There were 5–7 ml of fluid at the injection site in the formalin-treated animals. However, no increase in voluntary sodium intake was seen during the 6 h of access in the formalin-treated group. Blood aldosterone rose approximately 3-fold. Plasma desoxycorticosterone measurements were not made in these studies. This mineralocorticoid is elevated by ACTH (Scoggins et al. 1974) and thus there is the possibility of it also, in addition to corticosterone, having an action in inducing salt appetite during stress. Important data directly relevant to a role of ACTH come from the study of Jalowiec, Stricker and Wolf (1970) on salt appetite induced by formalin in hypophysectomized rats. They found the sodium appetite was slower in onset and only half as great in the hypophysectomized animals though plasma volume change was as large. They raise the question of whether less aldosterone was secreted in the hypophysectomized animal, but it may be that the reduction in desoxycorticosterone and glucocorticoids was more important.

In relation to the salt-appetite-inducing effects of stress, and this in the absence of any changes in fluid distribution or loss, some remarkable data have emerged during a study of wild rabbits (Denton and Nelson 1980). In the course of aiming to make intravenous infusions in these creatures, jackets were prepared in anticipation of fixing infusion apparatus to the backs of the free-moving but caged animals. The attachment caused enormous agitation and the animals usually managed to become free of it at night. However, with more efficient preparation a jacket was eventually contrived which remained fixed. Fig. 8-1 shows the profound effect on salt appetite of

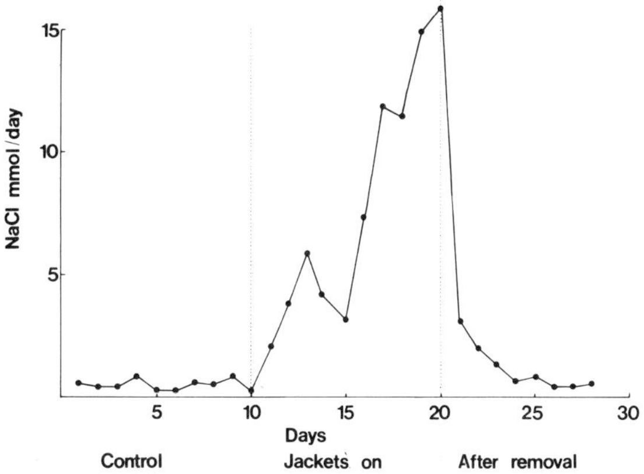

Fig. 8-1. Mean daily intake of 500 mmol/l sodium chloride solution by wild rabbits before, during and after attachment of jackets to their backs (n=6). (After Denton and Nelson 1980.)

these manoeuvres. The peaking of effect occurred when the jackets were permanently attached, following the days of intermittent attachment. The effect reversed when the jackets were removed in the face of the evident distress of the animals. Whether this influence on salt appetite was attributable solely to ACTH release or to influence of emotional state on the neurophysiological systems subserving appetite has not been determined.

Sweating

This has been used in association with a low-sodium diet to produce substantial sodium deficit in man (McCance 1938; Denton et al. 1951).

Hormones

Reproduction

The effect of pregnancy and lactation in inducing salt appetite was first recognized by Richter and Barelare (1938) in rat studies. The phenomenon has since been extensively explored and elucidated with metabolic balance studies in the rabbit (Denton and Nelson 1971). Sheep also show greater salt appetite in pregnancy. The outcome of the rabbit studies has been to show that a number of steroid and peptide hormones involved in reproduction have powerful effects on appetite and this involves interdependent relations (Denton et al. 1977).

Mineralocorticoids

Rice and Richter (1943) and Braun-Menendez (1952) made the somewhat unexpected finding that injection of DOCA produced salt appetite in rats. Increased water intake which also occurred was shown to be secondary to the increased salt intake. DOCA, on a low-salt diet, did not cause thirst. This hormonic effect with either DOCA or aldosterone has been used extensively in the analysis of the mechanism of sodium appetite. Aldosterone has been assigned a physiological causal role in sodium appetite in sodium deficiency by some workers. The issue is debatable.

Permanent unilateral parotid fistula in the ruminant

Early experimental studies of body fluid balance and the results of sodium deficiency were made on dogs with a pancreatic fistula (Gamble and McIver 1928; Gamble 1947). The large effects on plasma sodium (fixed base), acid–base balance and the circulation caused by loss of alkaline fluid were described.

Of course, with the pancreatic fistula in the dog, the fluid loss causes severe disruption of digestion as well as the problems of abdominal skin excoriation due to enzymes. To date there has been little study of salt appetite in the dog though there are issues of great theoretical importance at stake. Ramsay et al. (1980) report that injection into the third ventricle of Nerve Growth Factor can stimulate salt appetite in dogs by virtue of production of angiotensin II. The published accounts of salt appetite in the wild do not include any observation of carnivores going to salt sources, and enquiries I have made through wardens of game parks, though not extensive, confirm this. Experimental knowledge on the carnivore would be highly desirable though techniques other than the pancreatic fistula will be preferable. Adrenalectomy or adrenal inhibitors may be the best approach.

The method of the permanent unilateral parotid fistula in the ruminant, initially applied to the sheep, has six great advantages.

1. There are no enzymes which cause damage to the skin.

2. There are two parotid glands. The secretion from one, together with those of inferior molar, sublingual and other alkali-secreting oral glands (Kay 1960), is wholly adequate for normal digestive processes.

3. In contrast to the carnivore, where secretion is in response to oral influences or conditioned stimuli, flow from the ruminant parotid is continuous. But it may be augmented or diminished by a variety of influences including eating, rumination, apprehension, and conditioned stimuli including gregarious factors (Denton 1957). The composition of the secretion (see Fig. 3-2) is more alkaline than pancreatic juice. The daily loss is 1–4 litres. If adequate supplementary sodium bicarbonate is given to an animal it may live in good health with a fistula for years, and we have studied in our Institute dozens of animals which have had fistulae for as long as 2–10 years. Over such time intervals it can be calculated the animals may have lost up to 1500 mol of sodium. They also lose significant phosphate in the saliva: 10–40 mmol/l. Thus if the phosphate content of the diet is marginal, it is desirable to include sodium phosphate in the maintenance regime. Otherwise, over a long period, phosphate deficiency and bone deterioration occur.

4. Sheep are docile, are easily handled and trained and, if adequately fed, will live quietly in a metabolism cage for years. Use of a metabolism cage with an anterior compartment for saliva collection and a posterior compartment which separates hard pelleted faeces from urine is straightforward, and precise external electrolyte balance studies are feasible (Denton 1956).

5. The animal's size, 20–50 kg, allows direct application of methods of human surgery and clinical measurement with invasive techniques (Denton et al. 1959; Borrie and Mitchell 1960). Ample volumes of biological fluids are available for analysis. Conversely, however, if special extracts or hormones in short supply are under scrutiny for appetite-inducing, endocrine or other effects, the animal is large from the bioassay viewpoint. This can be circumvented to a considerable extent in the instance where the arterial supply of the effector organ is isolated and freely accessible, as with the adrenal autotransplant preparation noted below. This is also the case with infusions into the ventricular system of the brain.

6. In the case of the Merino sheep with its large folds of neck skin, it is possible to autotransplant the left adrenal gland to a combined carotid artery/jugular vein skin loop in the neck. The right adrenal gland is removed. This allows free access to arterial supply and venous drainage of the sole

adrenal gland in the conscious animal. Thus, in the overall analysis of sodium homeostasis and the role of salt appetite, the function and response to experimental situations of the other major sodium regulatory system, the adrenal secretion of aldosterone, can be contemporaneously directly studied. Indeed, quite novel data have emerged. This bears upon the functional interrelationship of the two systems.

Sodium depletion has also been produced in sheep and goats by unilateral parotid cannulation (Hyden 1958). This is a convenient method for short-term studies and can be done without anaesthesia by use of intravenous Valium (Abraham et al. 1976). The duct is cannulated via the papilla in the mouth and the cannula passed through a needle in the cheek to the exterior. After an experiment lasting e.g. 2–4 days the cannula can be removed, and, after an interval, the procedure repeated. In five animals (ten experiments) the loss of saliva following cannulation of one parotid gland ranged from 1700 to 4950 ml, containing 263–767 mmol sodium, in the first 24 h. Stewart and Stewart (1961) have made bilateral cannulation of the parotid ducts. The ducts were divided without interference with nerve supply and special cannulae with ends prepared to facilitate a tight connective tissue seal were inserted into each end of the cut duct. The two ends were connected so that saliva normally re-entered the mouth and, under given conditions, flow rate on both sides could be measured concurrently. The preparations lasted up to 3 months. Though not used by the present author for this purpose, the method, or similar principle, might be useful for very rapid production of severe sodium deficiency.

With very severe sodium deficiency in the sheep caused by 3–4 days of uncompensated parotid salivary loss there is often, as with other techniques in the rat, a decrease in food appetite and evidence of lassitude in the animal. This decline in general condition as a result of sodium deficiency could jeopardize drinking behaviour and quantitative response to deficit. However, as will be recounted in succeeding chapters, this does not often appear to be the case, though when severe sodium deficiency is concurrent with adrenal insufficiency, appetite is compromised.

Parotid fistulation has also been used in calves (Bell 1963) and in goats (Baldwin 1976) to study salt appetite. As described later (Chapter 25) we have used the method also to produce phosphate deficiency and study phosphate appetite in cattle.

References

Abraham S, Coghlan JP, Denton DA, McDougall JG, Mouw DR, Scoggins BA (1976) Increased water drinking induced by sodium depletion in sheep. Q J Exp Physiol 62: 185

Bare JK (1949) The specific hunger for sodium chloride in normal and adrenalectomized white rats. J Comp Physiol Psychol 42: 242

Baldwin BA (1976) Effects of intracarotid or intraruminal injections of NaCl or $NaHCO_3$ on sodium appetite in goats. Physiol Behav 16: 59

Bell FR (1963) The variation in taste thresholds of ruminants associated with sodium depletion. In: Zotterman Y (ed) Proceedings of the First International Symposium on Olfaction and Taste. Pergamon Press, Oxford, p 299

Blaine EH, Covelli MD, Denton DA, Nelson JF, Shulkes AA (1975) The role of ACTH and adrenal glucocorticoids in the salt appetite of wild rabbits (*Oryctolagus cuniculus* (L)). Endocrinology 97: 793

Borrie J, Mitchell RM (1960) The sheep as an experimental animal in surgical science. Br J Surg 204: 435

Braun-Menendez E (1952) Aumento del apetito especifico para la sal provocado por la desoxicorticosterona. II. Sustancias que potencian o inhiben esta accion. Rev Soc Arg Biol 29: 23

Cruz CE, Perelle IB, Wolf G (1977) Methodologic aspects of sodium appetite: An addendum. Behav Neural Biol 20: 96

Denton DA (1956) The effect of Na depletion on the Na:K ratio of the parotid saliva of sheep. J Physiol (Lond) 131: 516

Denton DA (1957) A gregarious factor in the natural conditioned salivary reflexes of sheep. Nature 197: 341

Denton DA, Nelson JF (1971) Effects of pregnancy and lactation on the mineral appetites of wild rabbits (*Oryctolagus cuniculus*(L.)). Endocrinology 88: 31

Denton DA, Nelson JF (1980) The influence of reproductive processes on salt appetite. In: Kare M, Fregly M, Bernard R (eds) Proceedings of the International Conference on Biological and Behavioural Aspects of Salt Intake. Academic Press, New York

Denton DA, Wynn V, McDonald IR, Simon S (1951) Renal regulation of the extracellular fluid. II. Renal physiology and electrolyte subtraction. Acta Med Scand 140 [Suppl 261]: 1

Denton DA, Goding JR, Wright RD (1959) The control of adrenal secretion of electrolyte active steroids. Br Med J ii: 447

Denton DA, Blair-West JR, Coghlan JP, Scoggins BA, Wright RD (1977) Gustatory-alimentary reflex inhibition of aldosterone hypersecretion by the autotransplanted adrenal gland of sodium deficient sheep. Acta Endocrinol (Copenh) 84: 119

Dirks JH, Seely JF (1969) Micropuncture and diuretics. Annu Rev Pharmacol Toxicol 9: 73

Fitzsimons JT (1969) The role of a renal thirst factor in drinking induced by extracellular stimuli. J Physiol (Lond) 201: 349

Fitzsimons JT (1972) Thirst. Physiol Rev 52: 468

Fregly MJ (1969) Preference threshold and appetite for NaCl solution as affected by propylthiouracil and desoxycorticosterone acetate in rats. In: Pfaffmann C (ed) Proceedings of the Third International Symposium on Olfaction and Taste. Rockefeller University Press, New York, p 554

Fregly MJ, Harper JM, Radford EP (1965) Regulation of sodium chloride intake by rats. Am J Physiol 209: 287

Gamble JL (1947) Chemical anatomy, physiology and pathology of the extracellular fluid. A lecture syllabus. Harvard University Press, Cambridge, Mass.

Gamble JL, McIver M. (1928) Gastric secretions. J Exp Med 48: 837
Hyden S (1958) Description of two methods for the collection of saliva in sheep and goats. Kgl Lantbruks-Hogskol An 24: 55
Jalowiec J (1974) Sodium appetite elicited by furosemide: Effects of differential dietary maintenance. Behav Neurol Biol 10: 313
Jalowiec JE, Stricker EM, Wolf G (1970) Restoration of sodium balance in hypophysectomized rats after acute sodium deficiency. Physiol Behav 5: 1145
Kay RNB (1960) The rate of flow and composition of various salivary secretions in sheep and calves. J Physiol (Lond) 150: 515
Makara GB, Stark E, Mihaly K (1969) Corticotrophin release induced by injection of formalin in rats with hemisection of the spinal cord. Acta Physiol Acad Sci Hung 35: 331
McCance RA (1938) The effect of salt deficiency in man on the volume of extracellular fluids, and on the composition of sweat, saliva, gastric juice and cerebrospinal fluid. J Physiol (Lond) 92: 208
Morgan T, Tadokoro M, Martin D, Berliner RW (1970) Effect of furosemide on Na and K transport studied by microperfusion of the rat nephron. Am J Physiol 218: 292
Morrison GR (1971) Effects of formalin-induced Na deficiency on $CaCl_2$ and KCl acceptability. Psychonomic Sci 25: 167
Porter J, Relinger H (1972) Daily sodium intake as a function of time of measurement and formalin injection volume and concentration. Psychonomic Sci 26: 276
Ramsay DJ, Thrasher TN, Keil LC (1980) Nerve growth factor stimulates angiotensin II dependent drinking and vasopressin secretion in dogs. In: Proceedings of the Twenty-eighth Congress of Physiological Sciences, Budapest. Hungarian Academy of Sciences, p2879
Rice KK, Richter CP (1943) Increased sodium chloride and water intake of normal rats treated with desoxycorticosterone acetate. Endocrinology 33: 106
Richter CP (1936) Increased salt appetite in adrenalectomized rats. Am J Physiol 115: 155
Richter CP (1942–43) Total self-regulatory function in animals and human beings. Harvey Lect 38: 63
Richter CP (1949) Domestication of the Norway rat and its implications for the problem of stress. Proc Assoc Res Nerv Ment Dis 29: 19
Richter CP (1956) Salt appetite of mammals: Its dependence on instinct and metabolism. In: L'instinct dans le comportement des animaux et de l'homme. Masson et Cie, Paris, p 576
Richter CP, Barelare B Jr (1938) Nutritional requirements of pregnant and lactating rats studied by the self-selection method. Endocrinology 23: 15
Scoggins BA, Coghlan JP, Denton DA, Fan JSK, McDougall JG, Oddie CJ, Shulkes AA (1974) Metabolic effects of ACTH in sheep. Am J Physiol 226: 198
Scoggins BA, Butkus A, Coghlan JP, Denton DA, Fan JSK, Humphery TJ, Whitworth JA (1978) Adrenocorticotropic hormone-induced hypertension in sheep: A model for the study of the effect of steroid hormones on blood pressure. Circ Res [Suppl] 43: 76
Selye H (1937) Studies on adaption. Endocrinology 21: 169
Stewart WE, Stewart DG (1961) Technique for cannulation of parotid salivary duct of sheep. J Appl Physiol 16: 203
Stricker EM (1966) Extracellular fluid volume in thirst. Am J Physiol 211: 232
Stricker EM (1971) Effects of hypovolaemia and/or caval constriction in water and NaCl solution drinking by rats. Physiol Behav 6: 299
Stricker EM (1973) Thirst, sodium appetite, and complementary physiological contributions to the regulation of intravascular fluid volume. In: Epstein AN, Kissileff HR, Stellar E (eds) Neuropsychology of thirst. New findings and advances in concepts. Winston, Washington, p 357
Stricker EM, MacArthur JP (1974) Physiological bases for different effects of extravascular colloid treatments on water and NaCl solution drinking by rats. Physiol Behav 13: 389
Stricker EM, Wolf G (1969) Behavioural control of intravascular fluid volume: Thirst and sodium appetite. Ann NY Acad Sci 157: 553
Stricker EM, Vagnucci RH, Leanen FH (1979) Renin and aldosterone secretions during hypovolemia in rats: Relation to NaCl intake. Am J Physiol 237: R45
Walsh LL, Grossman SP (1977) Electrolytic lesions and knife cuts in the region of the zona incerta impair sodium appetite. Physiol Behav 18: 587
De Wardener HE, Herxheimer A (1957) The effect of high water intake on salt consumption, taste thresholds and salivary secretion in man. J Physiol (Lond) 139: 53
Weisinger RS, Denton DA, McKinley MJ, Nelson JF (1978) ACTH-induced sodium appetite in the rat. Pharmacol Biochem Behav 8: 339
Weisinger RS, Coghlan JP, Denton DA, Fan JSK, Hatzikostas S, McKinley MJ, Nelson JF, Scoggins BA (1980) ACTH-elicited Na appetite in sheep. Am J Physiol 239: E45
Wilson CS, Wong R (1975) Aldactazide-induced consummatory and operant responding to sodium by rats. Am J Psychol 88: 377
Wolf G, Steinbaum EA (1965) Sodium appetite elicited by subcutaneous formalin. J Comp Physiol Psychol 59: 335
Wolf G, McGovern JF, DiCara LV (1974) Sodium appetite: Some conceptual and methodologic aspects of a model drive system. Behav Biol 10: 27
Woods J, Weisinger R, Wall B (1971) Conditioned aversions produced by subcutaneous injections of formalin in rats. J Comp Physiol Psychol 77: 410
Zimmerman MB, Stricker EM, Blaine EH (1978) Water and NaCl intake after furosemide treatment in sheep. J Comp Physiol Psychol 92: 501

9 The sheep (a ruminant) as a felicitous creature for research in experimental endocrinology and body fluid regulation: Control of aldosterone secretion

Summary

1. The digestive organization of ruminant animals involves continuous secretion of large volumes of alkaline parotid saliva, rich in sodium and phosphate, to buffer the volatile fatty acids produced by fermentation in the rumen.

2. The preparation of a permanent unilateral parotid fistula in the sheep allows subtraction of large volumes of saliva from the body. Thus large deficits of sodium can be produced at will.

3. The initial impetus for development of methodology of electrolyte subtraction from the body in sheep came from the need for an experimental model of clinical surgical and medical conditions such as, for example, post-gastrectomy duodenal fistula where large volumes of alkaline pancreatic juice are lost from the body with resulting progressive circulatory deterioration. Study of renal mechanisms of electrolyte homeostasis was greatly facilitated.

4. The docile easily trained sheep with large flaps of neck skin is an ideal animal for direct application of clinical plastic and vascular surgical techniques to cervical autotransplantation of the adrenal gland, the kidney, and several other endocrine glands. The aim is to elucidate the role of direct action of local changes in composition of arterial blood on the secretory function of the organ. In addition, episodic sampling of the venous effluent of the organ for chemical assay is simple, and the mechanisms of physiological control can be determined in the conscious undisturbed animal.

5. Desiderata for valid experimental investigation of the causation and manner of secretion of a hormone are proposed. The central consideration in the discussion is the control of secretion of the salt-retaining hormone, aldosterone, discovered by J.F. and S.A.S. Tait in 1953.

6. In the light of the phylogenetic emergence of the two physiological systems, aldosterone secretion and salt appetite, in the face of the same selection pressure—sodium deficiency—it is feasible that the same stimuli might activate both systems. Present knowledge on the mode of control of aldosterone is briefly reviewed.

Increase in plasma potassium concentration and decrease in plasma sodium concentration, increase in ACTH, and increase in angiotensin II and angiotensin III concentration stimulate aldosterone secretion. Angiotensin II and III are equipotent, at least in the sheep. Basal levels of angiotensin II in blood are essential to aldosterone secretion in response to sodium deficiency, and the increase in aldosterone secretion in sodium deficiency may be partly or largely attributable to increase in peripheral blood angiotensin II concentration. In sodium deficiency there is an increase in sensitivity of the adrenal glomerulosa to angiotensin II and ACTH which may be attributable to ionic changes in the glomerulosa cells or, in the former case, to induction of additional receptors to angiotensin II by the hormone itself. Experimental evidence showing the antidopaminergic drug metaclopramide will increase aldosterone secretion, coupled with a large body of data showing that rapid satiation of salt appetite will cause large rapid but evanescent reduction of aldosterone secretion in the sodium-deficient animal (see Chapter 21), point to an as yet unidentified central nervous mechanism in the control of aldosterone secretion.

Introduction

Claude Bernard observed that the solution of a particular physiological or pathological problem often depends solely on the appropriate choice of an animal for experiment so as to make the result clear and searching.

The pioneering studies of some of the basic phenomena of salt appetite behaviour were made in the rat. Outstanding figures in this advance include Richter, Braun-Menendez, Stellar, Epstein, Pfaffmann, Nachman, Wolf, Mook, Fregly, Stricker, Covian, Jalowiec, Handal, Krieckhaus and others of whom I will write in some detail in relation to specific issues which have emerged in the last few years, and also in relation to the broader canvas of ingestive behaviour generally. The early salt appetite enquiry has been predominantly, indeed overwhelmingly, an American contribution, but with recent major advances coming also from Fitzsimons' group at Cambridge, UK. Richter (1949, 1956) (Fig. 9-1) has written extensively and cogently on the history and virtues of the laboratory rat as an experimental animal, and has considered the influences of domestication on its physiological and endocrinological organization and also its behaviour. He sees the species as closely related nutritionally to man, and an especial opportunity to examine the influences of domestication by comparison with the wild rat. The development of experimental approaches has been gradual, though obviously some major ideas and special circumstances determined direction at times. Thus the Second World War and issues of poisoning wild rats influenced some aspects of Richter's enquiries, and, also, his original behaviourist inclinations led him to look first for peripheral alterations (i.e. at the taste bud level) in the genesis of appetite. Some reservations about the overwhelmingly predominant use of inbred domesticated laboratory rats in behaviour studies are set out in Chapter 25.

Fig. 9-1. Curt Paul Richter, Emeritus Professor of Psychobiology, The Johns Hopkins University. Dr Richter has pioneered major studies in the field of appetite research and has been acclaimed internationally for his distinguished work.

New dimensions in salt appetite behaviour and revealing insights in the analysis have come from studies on the sheep employing methodology quite different from anything used, or for that matter feasible, in the rat. The choice of the sheep and the development of some unique methods did not come about by chance any more than it did in the rat. But the initial circumstances were not concerned with salt appetite. They involved the aim of solving an important problem in the resuscitation of medical and surgical patients. I recount the history here because it answers the question asked by a majority of visitors to the Howard Florey Institute on finding in total dozens of sheep throughout its ten floors: Why have you always chosen to use sheep in your experiments? It also fills in some history on experimentation with fistulae and the problems involved. In recounting also the primary interest of our scientific group in the mode of control of aldosterone secretion, the brief history highlights the physiological systems thought to be involved, and this, of course, parallels the enquiry in salt appetite. Since both systems have emerged phylogenetically in response to the same environmental stress of sodium deficiency, it is legitimate initially to ask whether the same humoral changes could activate both systems— as appears to be the case with thirst and antidiuretic hormone release.

Influence of digestive secretion on blood chemistry

The amount and type of electrolyte-containing fluids required in digestion vary in different animal species. Correspondingly there is great variation in the effects of the digestive process on the *fixité du milieu intérieur*. For example, at the extreme, when an animal such as an alligator is fed, the secretion of gastric juice may cause the animal's plasma chloride concentration to fall from 100 to 40 mmol/l, bicarbonate concentration to rise commensurately, and pH to rise from 7.5 to 8.1. In man, electrolyte changes in plasma following a meal are small, and a regulatory response is evidenced by the 'alkaline tide' in the urine. In ruminants such as sheep, goats and cattle, and a wide variety of game species, the initial digestion of the large amounts of grass and foliage consumed occurs in the rumen. The microbial action produces fatty acids, and copious secretion of alkaline saliva provides the buffered medium. Secretion is continuous, though variable in rate over the short term in response to particular physiological stimuli. Longer term variation in the composition of the food may vary the volume of fluid sequestrated in the rumen. In the face of this situation, in a sodium-poor environment there are highly adaptive mechanisms which involve mineralocorticoid secretion, conservation of sodium in the copious salivary flow and commensurate replacement of it by potassium (which is readily available in the herbivore's food). In this fashion the impact of digestive processes, and other physiological variation, on the *fixité du milieu intérieur* in a sodium-deficient environment is ameliorated. The role of aldosterone in offsetting the effect of salivary digestive processes on stability of blood composition in ruminants in sodium-deficient environments represents the quantitatively most spectacular effect of this hormone in nature.

Barcroft (1934), in his monograph *The Architecture of Physiological Function*, emphasized that the development of mechanisms ensuring a stable chemical composition of the blood in the face of internal and external stress was an essential condition for the evolutionary emergence of higher nervous activity—a contention well supported by the evidence of deterioration in function of the human brain with changes in pH, osmotic pressure and glucose concentration as occur, for example, in diabetic acidosis, water intoxication and hypoglycaemia. In the Ruminantia, the central nervous development can be seen to be limited by the basic circumstances of a life largely committed to munching low-calorie food. Nonetheless their nervous capacities in terms of detection and evasion of predators, including alarm behaviour, and of chemical nutritional regulation by selective ingestion are well developed. Some species have developed elaborate mating procedures with male competition and territorial defence.

The peculiarity of the ruminant in prefermenting its food in a rumen–reticular sac by microbial action carries other general physiological consequences worth noting. The ruminants, of course, include a large number of the wild grazing species as well as the domesticated animals, and, by convergent evolution, Australian marsupials, the wallaby for example, have developed a similar stomach system and blood sugar level (Moir et al. 1956). The pre-fermentation of cellulose precedes the digestion by enzymes which begins in the abomasum or stomach. The rumen–reticular sac is usually never empty and gastric and intestinal secretion are, like salivary secretion, continuous (Cuthbertson 1958). The large capacity of the rumen–reticular sac has a particular implication. It has been found, in experimental studies to be recounted, that when water-deplete or salt-deplete, the sheep has the capacity to repair deficiency in a rapid consummatory act over 1–5 min when the appropriate fluid—water or salt solution—is offered. This may involve drinking 1–6 litres in this time. The camel, also a ruminant, when water-deprived for 5–10 days can drink 50–100 litres in 5–10 min (Schmidt-Nielsen et al. 1956). At times when it is travelling to water, or salt licks or springs, and drinking there, the ruminant animal is particularly exposed to predators. The ability to satisfy a need accurately with a rapid consummatory act, and then go to safer places, has therefore most likely been a survival advantage to the numerous species involved, and the large rumen capacity has been important in this regard.

Sheep as experimental animals in the study of body fluid homeostasis

The use of sheep in experiments has some history (Lewinsohn 1954). Pasteur used them in his pioneering studies of vaccination against anthrax. The Montgolfier brothers included a sheep, a hen and duck in the first hot-air balloon ascent with passengers in Paris in 1783. Dr Guillotine used sheep in developing his apparatus for execution. Coming to physiological science, Barcroft employed them extensively as experimental animals in the study of foetal respiration, and Dawes of Oxford also used them later in work on foetal physiology. However, though agricultural institutes and scientists studied them for insight into problems particular to their pastoral management, nutrition and reproduction, it would be true to say that they did not loom large in the history of basic medical physiological investigation, where the choice has centred on the dog, cat, rat, mouse, rabbit and monkey as laboratory animals. Nowadays there is a large upsurge in their use, particularly in endocrinology and body fluid physiology. Knowledge emerging from techniques developed at the Howard Florey Institute in Melbourne has been influential in this regard. There has also been a great appreciation of the virtues of the goat for study in the field of central control of fluid regulation, as pioneered by Andersson, Olsson and colleagues (Andersson and Olsson 1973), though some experimental procedures are not as easy as in the sheep because of paucity of neck skin (e.g. endocrine gland transplantation).

Our preoccupation with the sheep developed in the following fashion. As exemplified by the work of Dr H. L. Marriott, during the Second World War the British Army in India and elsewhere made studies on a number of problems of the tropical environment. These involved the issues of heat stroke and sweating, management of battle casualties with fluid loss, and medical and surgical situations involving fluid balance disorder (burns, vomiting, diarrhoea, paralytic ileus, etc.). As a working basis, which was embodied in his Croonian Lectures and his book *Water and Salt Depletion* (1950), Marriott concluded that the best practical method of deciding the crucial issue as to whether a patient was or was not depleted of salt was to measure the chloride in the urine. The normal kidney was an effective regulator. Thus in the event of jeopardization of circulation due to salt depletion and therefore body fluid volume depletion, urinary chloride excretion would be minimal to nil. At this time there were no flame photometers for monitoring the balance of the clinical electrolytes sodium and potassium (see Domingo and Klyne 1949; Wynn et al. 1950). Marriott noted that even if rapid plasma analysis were feasible, in some situations where there were mixed losses of water and salt, interpretation of balance state could not be made from the plasma chloride concentration. Accordingly he strongly advocated a very simple ward test for urinary chloride concentration developed by Fantus (1936). This involved adding 2.9% silver nitrate drop by drop from a pipette to the patient's urine to which 20% potassium chromate had been added. If colour changed on the first drop chloride was virtually absent. Each drop before colour change equalled 1 g/l of sodium chloride in the urine (ca 20 mmol/l). This test was a great practical clinical advance in the diagnosis of patients with electrolyte depletion, and was used widely.

In 1947, as a junior resident medical officer at the Royal Melbourne Hospital, I was responsible for the care of a 17-year-old boy with a duodenal fistula (Denton 1948, 1949) following a partial gastrectomy performed for a large duodenal ulcer. The essential elements of such a clinical situation and experimental results are shown in Figs 9-2 and

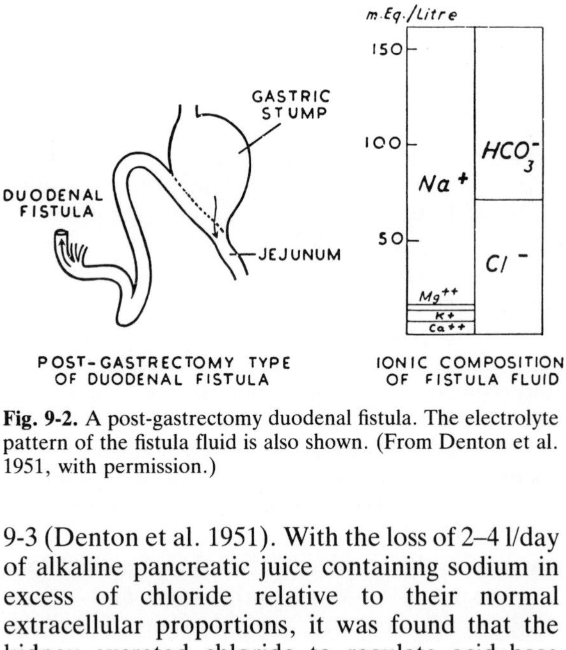

Fig. 9-2. A post-gastrectomy duodenal fistula. The electrolyte pattern of the fistula fluid is also shown. (From Denton et al. 1951, with permission.)

9-3 (Denton et al. 1951). With the loss of 2–4 l/day of alkaline pancreatic juice containing sodium in excess of chloride relative to their normal extracellular proportions, it was found that the kidney excreted chloride to regulate acid–base balance. Large chloride excretion occurred (Fig. 9-4) despite the progressive depletion of both sodium and chloride, which caused a fall in circulatory volume and the blood pressure, and signs of collapse. However, the excretion of chloride ceased when sodium in excess of chloride was given intravenously as an isotonic solution of sodium lactate (Fig. 9-4). The initial finding (Denton 1948, 1949), which was confirmed in detail in a subsequent patient (Denton et al. 1951), appeared important for two reasons. First, it involved modification of Marriott's approach to clinical management, in that it was recognized that

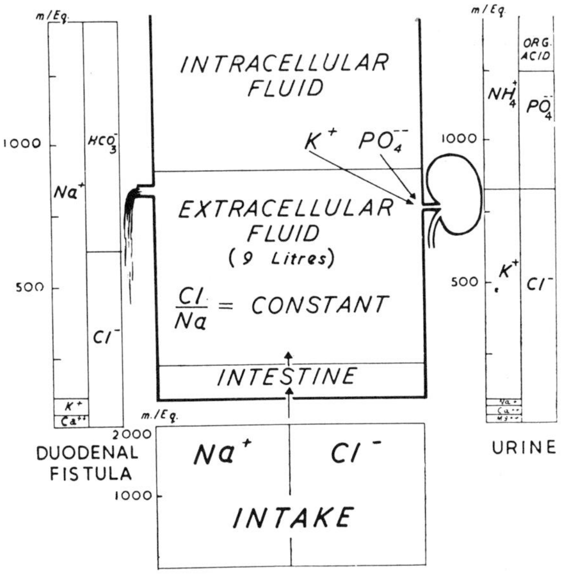

Fig. 9-3. Electrolyte homeostasis in a post-gastrectomy type of duodenal fistula. The kidney responds to excess sodium subtraction by excreting chloride in excess of sodium (From Denton et al. 1951, with permission.)

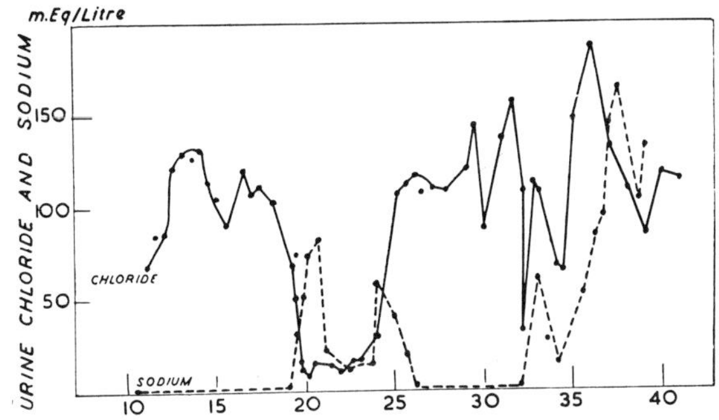

Fig. 9-4. Urinary chloride and sodium excretion in a patient with a post-gastrectomy duodenal fistula studied over 30 days. On day 19 in the face of clinical findings of rising respiratory rate and circulatory collapse despite high urinary chloride excretion, 2 litres of sodium lactate (310 mmol of sodium) were given over 6 h. Blood pressure rose and cardiac rate fell, and clinical condition improved. Urinary chloride excretion fell. (From Denton et al. 1951, with permission.)

if there were subtraction of sodium in excess of chloride from the body (i.e. alkaline fluid loss), as in some intestinal fistulae or diarrhoea, the simple application of the Fantus test for evaluation of salt balance may not be reliable (Denton 1948, 1949; Marriott 1950).

Second, and more important, a basic issue in renal physiology arose. The excretion of chloride was not determined by the plasma concentration or 'renal threshold' in these conditions of excess sodium subtraction because large excretion occurred when plasma chloride concentration was 10–20 mmol/l below normal (Fig. 9-5). Similarly it was not determined by the total body chloride content, which was clearly reduced by fistula loss and the continuing urinary chloride excretion. It was presumably not determined by any maximum tubular function type mechanism (Tm) since the load presenting to the tubules would have been greatly reduced by decrease in blood pressure and glomerular filtration as well as the fall in plasma chloride concentration. Many aspects of this were considered (Denton et al. 1951) and suffice it to say here, it was decided to develop an experimental model. Particularly this was to confirm the facts in relation to Tm, and generally to analyse the proposal that subtraction of electrolytes from the body in disproportionate amounts relative to normal extracellular proportions of sodium and chloride (presumptively a common stress in nature because of diarrhoeal disease) evoked renal regulatory capacities which differed from those seen when ions were added in excess to the body, or there was simple deprivation of salts.

In considering whether this stimulus to a large chloride excretion was pH change, or, alternatively, some other fundamental consequence of distortion of the extracellular electrolyte pattern, we were drawn to the sheep as a result of a lunchtime conversation at the Walter and Eliza Hall Institute in Melbourne with Dr A. W. Turner of the Animal Health Division of the Commonwealth Scientific and Industrial Research Organ-

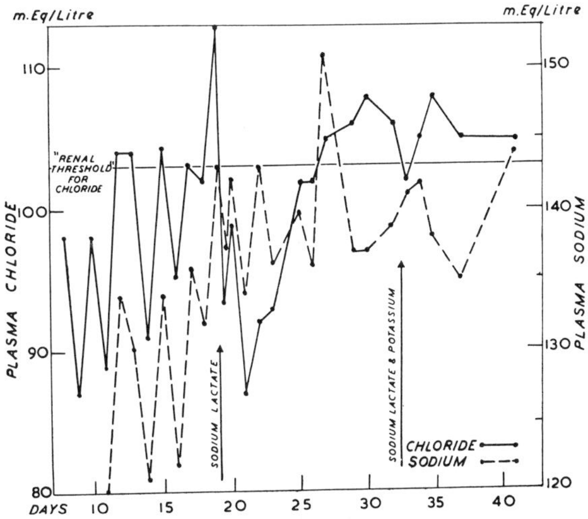

Fig. 9-5. Plasma concentration of chloride and sodium in a patient with a post-gastrectomy duodenal fistula. The horizontal line represents 'the renal threshold' for chloride. Reference to Fig. 9-4 shows large excretion of chloride even though the plasma concentration was much below the renal threshold. (From Denton et al. 1951, with permission.)

ization. He described to me the occurrence of severe acidaemia in sheep as a result of wheat feeding. With Dr Victor Wynn of our group, the blood pH fall due to accumulation of lactic acid with wheat feeding was investigated. The absence of any increase in chloride excretion, though a large pH fall due to organic anion accumulation occurred, was noted. Therefore we decided to make surgically in the sheep the equivalent of the post-gastrectomy duodenal fistula. When we anaesthetized the creature as a trial run, Wynn, Dr Ian McDonald and I became aware of the profuse alkaline parotid secretion—already, of course, well known and described chemically by McDougall (1948). The amount of sodium in excess of chloride in the parotid secretion far exceeded that in the pancreatic juice: indeed it was over 10 to 1.

It was immediately crystal clear that subtracting these secretions from the body was the way of choice for producing gross sodium deficiency. Initially this was done using an oesophageal fistula (Denton et al. 1951). This succeeded in giving an immediate large loss over 24–48 h but secretion fell away rapidly in the absence of normal eating and rumination. However, the excess chloride excretion was confirmed and shown to occur in the face of a falling load of filtered chloride (Denton et al. 1952).

Development of the permanent unilateral parotid fistula in the sheep

To carry the enquiry further it evidently needed a chronic preparation contrived so that the animal had normal digestion, and where the conditions allowed repeated experiments with the animal used as its own experimental control. Thus attention eventually turned to the parotid fistula since it was from this gland that the profuse alkaline secretion arose, and Pavlov (1910) had already made monumental contributions to biological knowledge with this method in the dog. The necessity of having a preparation which allowed long-term study in the conscious confident animal, if one were to derive knowledge of basic physiological validity, had been brought home to me as the result of the experience of working in the laboratory of Professor Ernest Basil Verney at the University of Cambridge.

Verney's (1947) studies of the regulation of the release of antidiuretic hormone (ADH) in the conscious dog were classic and, in 1953, Bengt Andersson of Stockholm and I had the privilege of working with Verney on the influence of ADH on sodium excretion (Verney 1957).

The accounts in the literature bearing on the parotid fistula approach in the ruminant were anything but encouraging. Scheunert and Trautmann (1921) made the first attempt to prepare permanent unilateral parotid fistulae in sheep by means of the Pavlov–Glinski operation for explanting the duct and papilla on to the cheek. Most of their animals developed mouth fistulae and lost much chewed food and mouth fluids. The sheep were unable to ruminate properly, ejected dry lumps of food and soon died in a cachectic state. They concluded that the lack of the secretion of even one parotid gland caused a fatal nutritional failure. Hence the long-term study of the secretion of the parotid gland in the sheep could not be successful. The collection of saliva also was difficult. The profuseness of fluid, the lively movements of the head, the excursions of the lower jaw during chewing, and the unfavourable surface of the cheek soon detached the collecting funnels. Finally, they cannulated the parotid duct and made short-term experiments (see also Scheunert, et al. 1929). Scheunert and Trautmann studied the histology of parotid glands in the animals which survived the Pavlov operation for a period; they found that the gland on the fistula side was smaller and fibrosed and the excretory ducts dilated.

From 1930, workers in Russia studied salivary secretion in sheep. However, they also described difficulty in making a Pavlov–Glinski fistula because the papilla is very close to the gum margin. They cut the parotid duct in the cheek and transplanted the cut end into the wound or into a separate stab wound (Popov 1932; Kudriavstev and Anurov 1932; Epanishnikov 1935; Blokh 1939; Serebryakov 1950; Ugolev 1953; Sineschekov and Sheremet 1953). The disadvantages (duct stenosis, occluding plugs, etc.) consequent on neglecting Pavlov's principle of preserving the papilla were described by Kvastnitsky (1955). He described a method of preserving the papilla when the duct was explanted. However, when I first attempted the Pavlov–Glinski procedure on the sheep in the Physiology Department of the University of Melbourne the procedure worked well though a long time was spent isolating the papilla by an oral approach using a retractor as Pavlov had. The procedure was then simplified to a cheek approach using a needle introduced through the mouth to mark a point immediately in front of the papilla. It was further elaborated by Professor R. D. Wright using a plastic surgical method (Fig. 9-6) (see Denton 1957a) to give a skin teat with a dependent point from which the profuse salivary

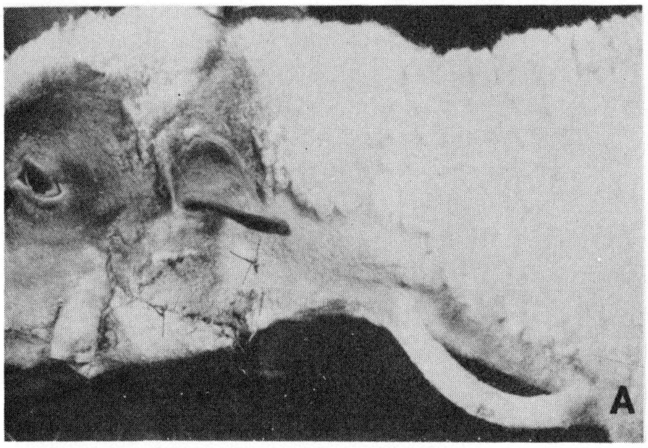

Fig. 9-6. Sheep PF7, a Suffolk ewe. This photograph, taken 4 days after operation, shows the rotation of the skin flaps used in preparation of a Wright parotid fistula. The animal also had bilateral carotid loops. (From Denton 1957, with permission.)

flow could drip without wetting the face. This also allowed direct observation of psychic influences on salivary flow without attachment of any apparatus to the animal (Fig. 12-8) (Denton 1957b). Important gregarious effects on salivary secretion were discovered. Animals lived for years with a fistula and there was no histological evidence of

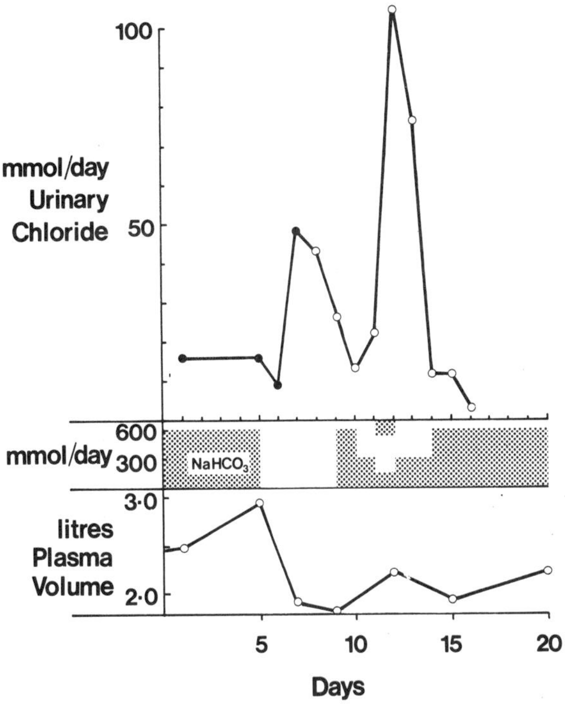

Fig. 9-7. Sheep losing 2–3 litres of saliva per day from a parotid fistula. The effect on urinary chloride excretion of 4 days of withdrawal of the daily supplement of 600 mmol of sodium bicarbonate, and the change in plasma volume produced by negative sodium balance are shown. On day 11 a supplement of 100 mmol sodium chloride was given under conditions of partial restoration of sodium bicarbonate supplement.

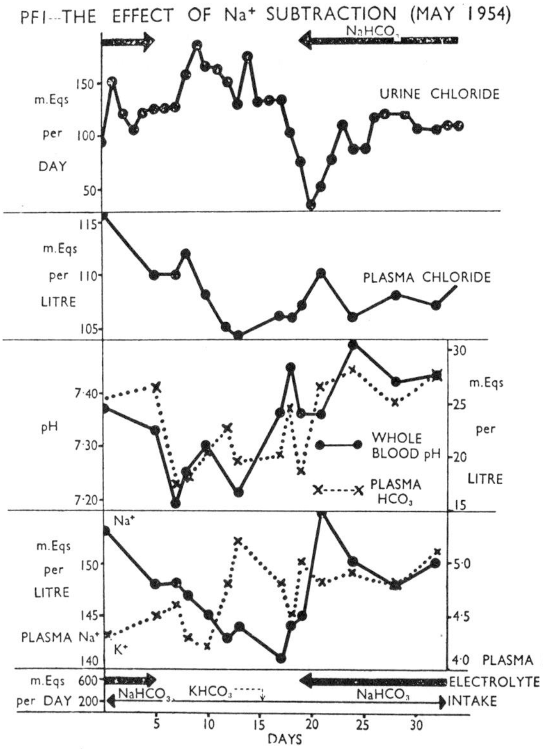

Fig. 9-8. Data from sheep PF1, showing that sodium depletion caused a significant increase in the total amount of chloride excreted in the urine each day. There was commensurate retention of chloride by the kidney when sodium bicarbonate supplement was replaced. The increased chloride excretion occurred despite a fall in plasma chloride concentration. Plasma sodium fell during sodium depletion and plasma potassium rose. Blood pH and plasma bicarbonate concentration fell. (From Denton 1957, with permission.)

complications in the gland as described by Scheunert and Trautmann (1921): see Denton (1957a). The normal physiological stimuli of secretion (eating, rumination, etc.) and the basic mechanisms of ion transfer in the parotid have been described (Denton 1957b; Coats et al. 1956, 1958).

The phenomenon seen in man of increasing urinary chloride excretion despite falling plasma chloride concentration in response to uncompensated sodium bicarbonate subtraction was confirmed (Figs. 9-7 and 9-8), as well as the concurrent fall in plasma volume and blood pressure caused by salivary loss.

Appetite for rock salt in sheep with a parotid fistula

Whereas sheep with a parotid fistula were maintained indefinitely in good condition if given

600 mmol of sodium bicarbonate in 1.5 litres of water daily by rumen tube, it soon became obvious that the animals developed an avid salt appetite if the sodium bicarbonate supplement were withdrawn. Such a sheep, if offered a block of rock salt (95% sodium chloride), licked it avidly. Sheep without a fistula licked little from a salt block; for instance, animal PF7 licked 26 mmol/day for 21 days, PF6 8.5 mmol/day for 4 days, and PF4 none during 2 days. Sheep with a parotid fistula which were not given the sodium supplement usually licked off 85–205 mmol/day. The sheep varied, but some were maintained quite well if given ordinary diet and access to a block of salt. The salivary composition indicated, however, that the animals did not maintain normal sodium balance. Most of the time they fluctuated around an equilibrium involving a moderate degree of sodium depletion. The urinary composition showed that sodium was usually completely conserved and that a very large chloride excretion was paralleled by an equivalent excretion of ammonium and potassium. The daily water intake increased after establishment of the fistula. When a sodium-deplete sheep is offered a salt block for the first time, it might lick off enough in a short time to restore its sodium balance. For example, PF16, when sodium-depleted for 6 days, licked off 958 mmol in 3 days, at the end of which time salivary composition was normal.

During this period in the mid-1950s a most striking phenomenon upon which our attention centred was the fact that the sodium tap we had contrived on the bloodstream for producing large sodium deficit changed progressively into a potassium tap as the animal became more sodium-deplete. From an initial Na/K ratio of 170/5=34, composition could change as far as 10/170=0.06. Indeed it was clear that the animals could live for months with a digestive system which functioned adequately to maintain condition with a salivary Na/K of e.g. 90/90=1 (Chapter 3).

The discovery of aldosterone

At this time, the Taits announced their landmark discovery of aldosterone (Simpson et al. 1953), the salt-retaining component of the amorphous fraction of the adrenal gland. In the sheep it was shown adrenalectomy abolished the Na/K ratio change in response to sodium deficiency (Goding and Denton 1956, 1959). It became clear with experiments involving aldosterone infusion (Fig. 3-6) that the parotid salivary Na/K ratio in the sheep was an in-built bioassay system *par excellence*, reflecting the secretion and contemporary systemic arterial blood level of aldosterone. The change in composition of both ions in sheep saliva is at least ten times as great as that occurring in human saliva in sodium deficiency (McCance 1938; Thorn et al. 1956). Moreover, in contrast to the human parotid (Thorn et al. 1956), the salivary composition varies little with secretion rate when the animal is in normal sodium balance (Denton 1957c). Thus a normal salivary Na/K ratio at any secretion rate can be taken to indicate normal sodium balance. Furthermore, it was observed in the sodium-deficient sheep that if a rapid systemic infusion of sodium chloride was given to correct substantial sodium deficiency (e.g. 65 ml of 4 M NaCl=260 mmol Na, during 60 min), after a time delay of 60–90 min the salivary Na/K ratio rose rapidly to normal. This showed that some physiological effect of the infusion had inhibited aldosterone secretion (McDonald and Denton 1956). Salivary Na/K ratio rose to normal with the same time characteristics if a sodium-deficient sheep were adrenalectomized.

The great importance of discovery of aldosterone resided, amongst many other things, in the fact that Conn (1955) had described the increase of blood pressure with small aldosterone-secreting adenoma of the adrenal gland. The issue of aldosterone involvement in oedema of heart, liver and kidney disease, as well as in hypertension came under intense study.

The question of the normal mode of control of aldosterone secretion became central to the whole field. This involved three principal questions: (1) the site of the receptor (or receptors) sensitive to the state of sodium balance, (2) the exact nature of the stimulus to it, and (3) the site of origin and the mode of transmission of the stimulus of aldosterone secretion to the adrenal.

Desiderata for valid experimental investigation of the causation and manner of secretion of a hormone

In setting out to find what it is precisely that causes secretion of a hormonic material under physiological conditions, we proposed (Denton et al. 1959) the following seven desiderata.

1) All observations should be made on conscious undisturbed animals in normal correspondence with their environment. It was shown by Verney (1947) that the normal hormonal mechanism regulating osmotic pressure is interfered with by anaesthesia or fright. This is true also in the case

of hormonal factors contributing to control of sodium balance. This places strictures on the methods of experimental approach, but solution of the problems by stage-by-stage surgery ultimately carries the advantage that the results derived are those holding for a normal animal.

2) The physiological stresses causing a response dependent upon the endocrine gland under investigation should be precisely defined and standardized, and be capable of graded variation.

3) The arterial blood supply of the gland should be exclusively accessible so that the effect of supposed active agents can be tested by direct local injection or sustained infusion.

4) The nerve supply of the gland, when relevant, should be accessible for stimulation or blocking.

5) The venous effluent exclusively from the gland should be accessible.

6) There should be one or more biological systems for quantitative assay of the activity of the venous effluent.

7) The chemical assay of the venous effluent must be developed until it detects quantitatively each substance known to be active on the biological indicator, and the overall effect of these substances must correspond quantitatively to the action of the entire venous effluent.

At that time, inferences concerning the control of aldosterone secretion had been based on urinary excretion, or upon measurement of adrenal secretion by cannulation of the adrenal vein in the anaesthetized animal or one which had been operated upon very recently (Farrell 1958). The influence of trauma, anaesthesia, and blood loss due to the large amount needed to be collected for analysis using insensitive methods clouded interpretation.

When we considered (3) and (5) above and the evident difficulty of isolating and varying independently one of the many factors recognized as changing during sodium deficiency and its rapid correction, we were led to the view that an essential and possibly decisive step towards answering both the question of the locus and that of the stimulus to the receptor regulating aldosterone secretion might be the devising of a preparation in which there was local access to the arterial supply of the adrenal in a conscious undisturbed animal (Figs. 9-9 and 9-10). Thus, in view of an adrenal blood flow of only 5–20 ml/min, it would be possible to insert a needle into the artery and impose for some hours a significant local alteration in the concentration of sodium or any other electrolyte in the blood perfusing the gland yet cause only negligible change in the body balance of the animal. This might decide whether

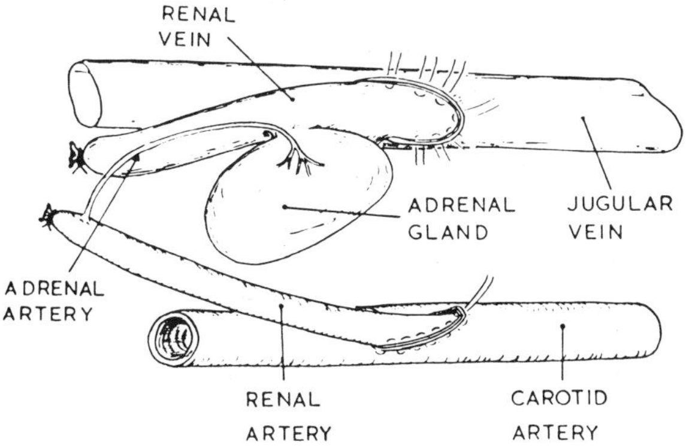

Fig. 9-9. Diagrammatic representation of the anastomosis between the renal and carotid arteries and the renal and jugular veins. The adrenal gland receives an artery from the renal artery and the adrenal vein drains into the left renal vein. (From McDonald et al. 1958, with permission.)

(a) the adrenal gland itself responded directly to changes in sodium or potassium concentration, or (b) the gland responded to a stimulus from a distance.

In relation to how this was achieved in 1956 (McDonald, Goding and Wright 1958), Professor Wright, in summarizing some 15 years of experience with adrenal transplants (Wright et al. 1972), gives the following account:

> At this stage, chromatographic separation of aldosterone from corticosterone and cortisol in extracts of adrenal venous blood was perfectly feasible, and so our sights were set on a preparation in which suitable amounts of adrenal venous blood could be collected in the relatively undisturbed sheep in contrived situations of direct control of adrenal arterial electrolyte concentrations and sodium status. The procedures of transplanting to the anterior chamber of the eye or under the renal capsule were rejected as not likely to give enough blood or enough concentration of hormone, and temporary sequestration of areas of inferior vena cava gave little hope of long-term experiments. In a group discussion of what might be done, Denton, probably reflecting his recent stay with Verney in Cambridge, expressed the hope but not high expectation that a transplant into a skin loop in the neck with access to artery and vein would be feasible—the idea being to make the carotid the sole artery of the adrenal in a fashion analogous to Verney contriving that the carotid in skin loop became the end artery of the hypothalamus by ligature of ipsilateral external carotid, anterior and middle cerebral, and posterior communicating arteries (Verney 1947; Jewell and Verney 1957). Goding and Wright, who had had some experience of vascular surgery, expressed the view that transplantation of the left adrenal by end to side anastomosis of left renal artery and vein to carotid and jugular after nephrectomy should be reasonably easy. Van Leersum's carotid loop gave the stimulus to wrapping artery, vein and transplant in a skin loop giving easy access. At the time Levy and Blalock's (1939) transplantation of kidney and adrenal to the neck without skin loop in the dog as a convenient arrangement for easy adrenalectomy was not known to us.

Levy and Blalock transplanted the kidney and adrenal to the neck using an end to end

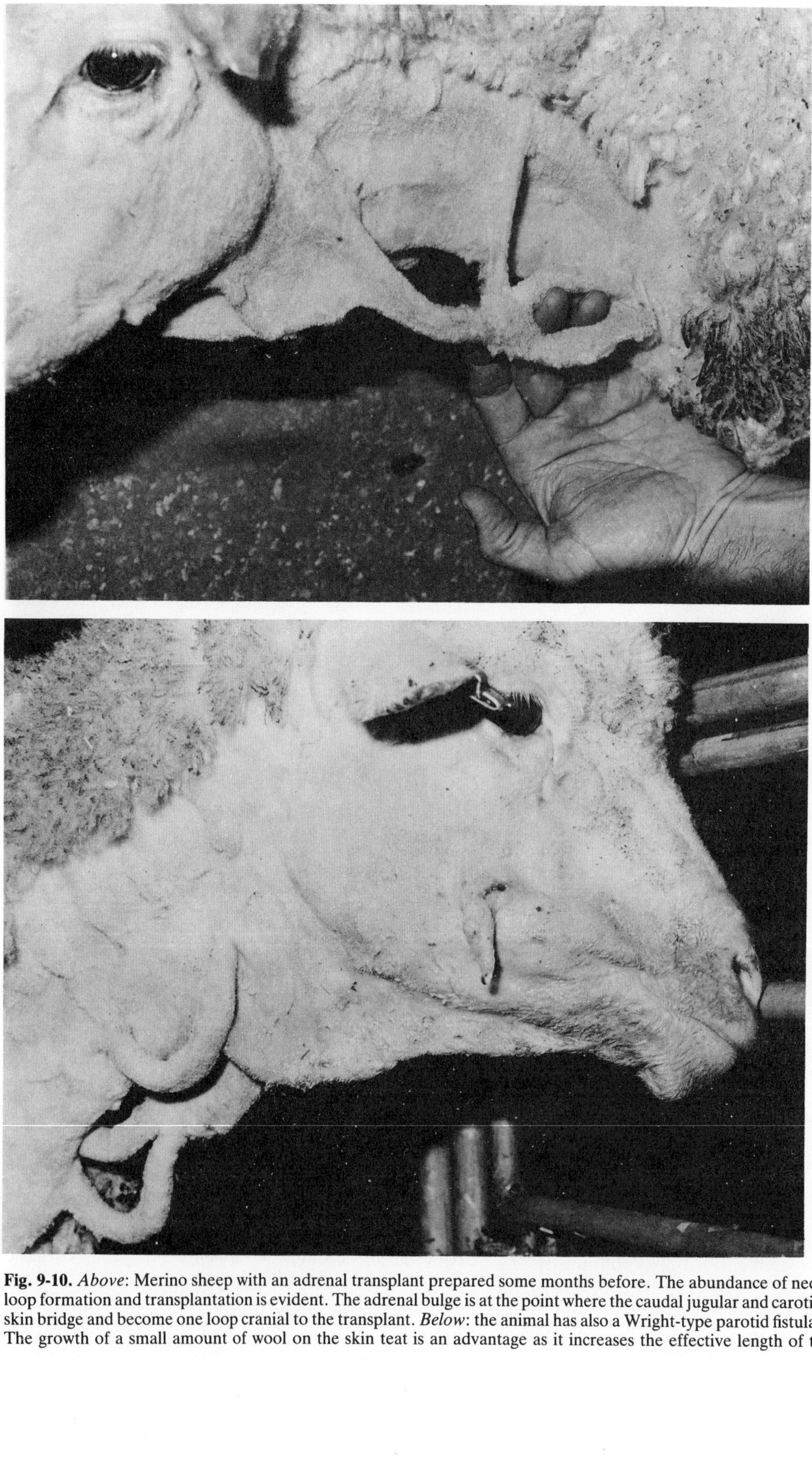

Fig. 9-10. *Above*: Merino sheep with an adrenal transplant prepared some months before. The abundance of neck skin facilitating loop formation and transplantation is evident. The adrenal bulge is at the point where the caudal jugular and carotid loops join to the skin bridge and become one loop cranial to the transplant. *Below*: the animal has also a Wright-type parotid fistula on the right side. The growth of a small amount of wool on the skin teat is an advantage as it increases the effective length of the teat.

anastomosis of the renal vessels and the carotid artery and external jugular vein. The kidney was removed subsequently (8–93 days later). The non-transplanted adrenal was removed 13–32 days later. The successful transplants, as well as allowing easy access for adrenalectomy under local anaesthesia or roentgen treatment, permitted sampling of venous effluent by puncture of the external jugular. But total collection, with rate of flow measurement, was not feasible as a survival preparation (Levy and Blalock 1939); nor was arterial infusion.

In the sheep, a skin loop of an H or Y design containing carotid artery and jugular vein was made beforehand, and also right adrenalectomy was performed some weeks previously. In relation to the skin loops, an enormous advantage resides in the Merino sheep. The large folds of skin, particularly in the neck, which the wool growers had selected for in terms of maximal surface area to grow wool provided liberally for plastic surgical procedures in the cervical region (Fig. 9-10). The left adrenal receives an artery from the left renal artery and the left adrenal vein drains into the left renal vein. Thus the kidney is removed and the left renal artery and vein, together with attached adrenal gland, are anastomosed end to side to the carotid artery and jugular vein in the skin loop in the neck (Fig. 9-9). The adrenal sits in the cross-piece of the H of the loop. On occasions, the aortic branch to the left adrenal is also used, a button of aorta with the arterial origin being removed.

In the chronic preparation, occlusion of the carotid artery cranial to the transplant by a cuff inflated to 300 mmHg contrives that the carotid artery which normally carries ca 300 ml/min becomes the adrenal artery with a flow of 5–20 ml/min, and sustained direct adrenal arterial infusion is feasible (Fig. 9-11). Occlusion of jugular vein caudal to the transplant with fingers or with a cuff at 20 mmHg allows collection of total venous effluent from a jugular vein cannula with its tip at the level of the transplant. The jugular cannula is usually inserted into the vein in the cranial part of the neck and fed down to the transplant. The cuff on the loop cranial to the transplant effectively occludes the jugular vein as well as the artery. Because of the small blood flow, infusions which make a big change in the concentration of substances in the adrenal blood have little effect systemically. At the simple practical level, consequent on the end to side anastomosis, it follows that puncture for arterial infusion and cannulation for episodic total venous collection is made with vessels that normally carry ca 300 ml/min. When this is being done for experiments at intervals of 2–3 weeks over 2–10 years, it is obviously a different proposition from working with vessels carrying 10–20 ml/min. In experiments lasting e.g. 6–8 h, six to eight collections of 10–20 ml of adrenal venous blood are made, and measurement of aldosterone, cortisol, corticosterone, desoxycorticosterone, desoxycortisol and adrenal androgens can be made by double-isotope dilution derivative assay (Kliman

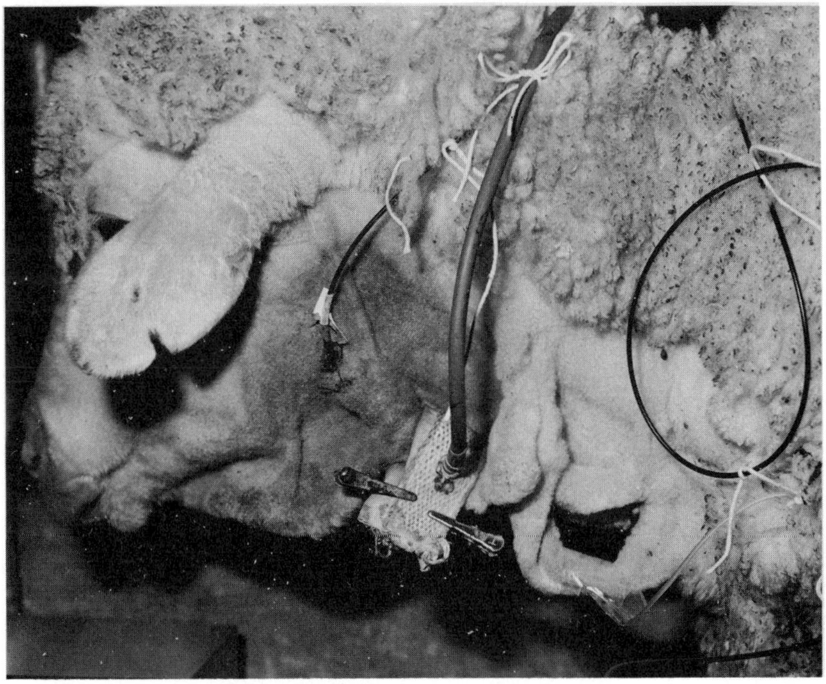

Fig. 9-11. Sheep with an adrenal transplant showing an infusion being made into the arterial supply. A pneumatic cuff inflated at 300 mmHg cranial to the transplant occludes both carotid and jugular vessels. A polyethylene cannula passed down the jugular from the angle of the jaw, with its tip near the transplant, permits episodic collection of total venous effluent of the adrenal when the jugular is temporarily occluded caudal to the transplant.

and Peterson 1960; Hudson et al. 1964; Coghlan et al. 1966; Oddie et al. 1972). Also, adrenal arterial infusion of radiolabelled precursors in the biosynthesis of aldosterone can be made. Venous effluent can be measured for percentage incorporation of label in secreted aldosterone under conditions of varying aldosterone secretion rate which can be measured concurrently as was done in collaboration with Drs J. F. and S. A. S. Tait (Baniukiewicz et al. 1968).

Development of this preparation and its use in elucidating basic aspects of the control of aldosterone secretion—particularly delineation of agents with direct action on the adrenal glomerulosa—led to the same principles being applied to other endocrine glands, as initially predicted (Denton et al. 1960). The adrenal transplant itself was used by New Zealand workers (Espiner et al. 1963, 1972) for bioassay of ACTH and study of cortisol secretion. Harrison and McDonald (1966) used it at the Institute of Animal Physiology at Babraham, Cambridge, for study of cortisol secretion. Holland (1969) and Bell et al. (1970) have succeeded in transplanting to the neck the left lobe of the sheep's pancreas with splenic vessels of supply. The pancreatic duct was ligated at a prior operation. The preparations continued to secrete insulin and in studies on the conscious animal it was shown that direct arterial infusion of glucose and butyrate were powerful stimuli of insulin secretion.

Falconer (1963) used the same principle to obtain access to the thyroid arterial supply and venous drainage in the sheep by transferring a lobe of the thyroid and its vessels into a carotid artery/jugular vein skin loop in the neck. The exteriorized lobe responded normally to thyroid stimulating hormone, and other studies, including stimulatory effects of emotions, were made.

Goding et al. (1967a, b) succeeded in transplanting the ovary to the neck in similar fashion using a plaque of aorta with the origin of the ovarian artery and the middle uterine vein (Figs. 9-12 and 9-13). Harrison et al. (1968) transplanted the ovary and uterus. Cyclic activity was disturbed when the ovary alone was transplanted to the neck but this was restored when the ovary and the uterus were transplanted together to a cervical site (Goding et al. 1971). A countercurrent mechanism by which uterine prostaglandin $F_2\alpha$ gained access to the ovarian artery and caused luteolysis was proposed.

More recently, Professors Wright and Hardy of the Howard Florey Institute have successfully carried out an ingenious new method of Wright's for kidney transplantation which allows precise physiological study of the organ. At initial operations, an ipsilateral parotid fistula with

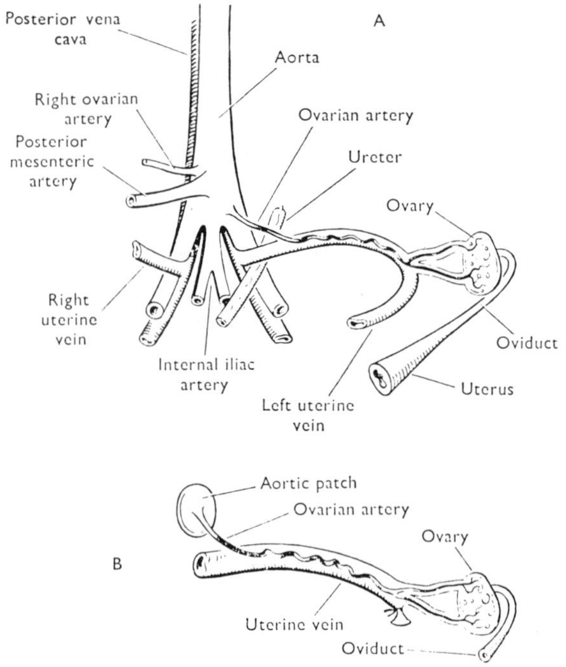

Fig. 9-12. A, Surgical anatomy of the blood supply of the sheep's ovary. B, Ovary and pedicle before transplantation. (From Goding et al. 1967, with permission.)

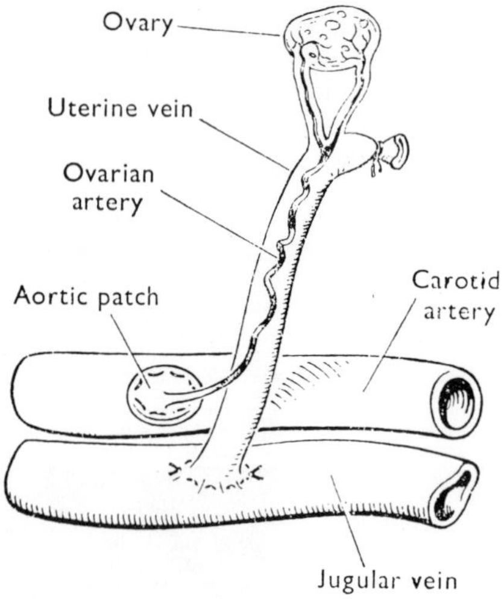

Fig. 9-13. Completed ovarian transplant. (From Goding et al. 1967, with permission.)

parotid papilla implanted in skin teat, and ipsilateral skin loops containing carotid artery and jugular vein are made. Further, a prosthesis is implanted in the neck immediately behind the angle of the jaw to make a convenient bed in which the kidney can lie when it is eventually transplanted 2–3 months later. At the time of transplantation the prosthesis is removed and the renal artery and carotid, and the renal vein and jugular are anastomosed end to end. The critical innovation is that the ipsilateral parotid is denervated and the parotid duct divided, the proximal end being ligatured and the ureter anastomosed onto the distal end. This provides a natural duct papilla for the ureter. In such preparations, as recorded on cine film 6–12 months after operation, there are regular and large ureteric contractions with expulsion of urine at times in jets. With the natural orifice and papilla, this mitigates against retrograde infection which can be common with ureteric explants to skin in cervical kidneys (Dempster and Joekes 1953) and has prevented significant long-term survival studies with this preparation.

The preparation described above (Fig. 9-14) allows contemporaneous access to renal arterial and renal venous blood and the urine. Evidently the influence of arterial infusion of specific substances upon secretion by the kidney of humoral agents, and the urine composition, can be examined repetitively in the one animal over years—using it as its own experimental control as with the adrenal transplant. To date we have four animals so prepared—one now 14 months after operation—and Dr J. McDougall has shown that the response of the transplanted kidney to water deprivation and loading and to infusion of ADH, and its haemodynamic parameters and glomerular filtration rate are normal. Moderate compression of the artery to this sole kidney causes elevation of systemic blood pressure within 30 min, and reversal occurs over the 30–60 min following cessation of renal artery constriction.

Care et al. (1965) have used the same basic principle in devising vascular arrangements for study of the secretion of the parathyroid glands in the sheep.

Mode of control of aldosterone secretion

Studies with the adrenal transplant have revealed, as will be recounted in Chapter 21, that in the course of the phylogenetic emergence of control of sodium homeostasis there has been some functional integration within the central nervous system of the mechanisms subserving satiation of salt appetite and those of aldosterone control. Thus it is most relevant to this later discussion to record here certain aspects of the history of development of ideas on the mode of control of aldosterone secretion. Since both major elements in the control of sodium homeostasis serve the same end, the possibility exists that the stimuli evoking reaction of the two systems could be the same, rather as they are with thirst and ADH release.

The outstanding achievement of the Taits in the discovery of aldosterone resulted from the de-

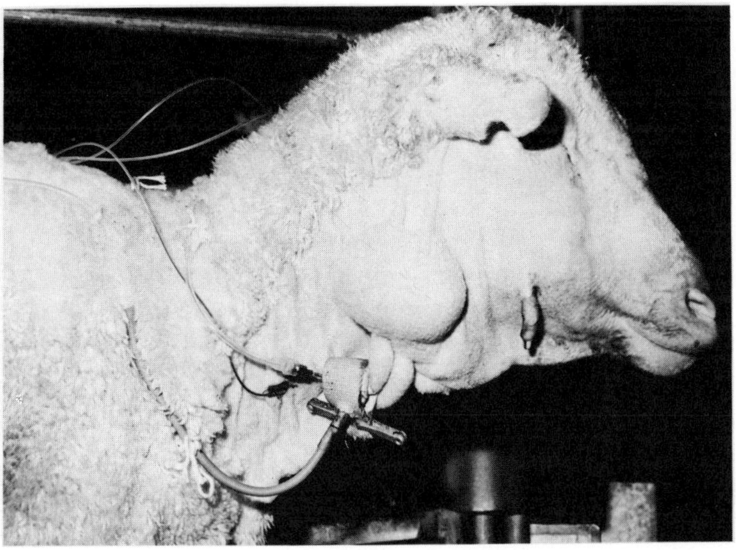

Fig. 9-14. Renal transplant. The photograph shows the kidney resting in the previously prepared bed near the angle of the jaw. The skin loops contain respectively the carotid artery which is anastomosed end to end with the renal artery, and the jugular vein which is anastomosed end to end to the renal vein. The ureter has been anastomosed end to end to the parotid duct and urine drips from or is expelled in squirts from the parotid papilla in the skin teat. The parotid gland is denervated and the duct has been ligated proximal to the anastomosis. The preparation allows contemporaneous access to urine and renal arterial and venous blood, and permits elective reversible constriction by pneumatic cuff of the renal arterial supply in the conscious animal as shown in the photograph. Concurrent infusion of agents into the renal arterial supply is also possible as shown.

velopment of a sensitive bioassay for the activity in the amorphous fraction of the adrenal gland and the application of then recently developed procedures of chromatography (Simpson and Tait 1952; Simpson et al. 1953). Twenty-one milligrams of aldosterone were isolated from 500 kg of beef adrenals. A year later the structure of the material was determined, and accordingly it was named aldosterone (Simpson et al. 1954). The first synthetic material—racemic DL-aldosterone—was produced in 1955 (Schmidlin et al. 1955). The isolation was just in advance of another group (Mattox et al. 1953). Using a different approach, Luetscher and colleagues showed that a potent steroid material, present in urine of patients with nephrotic syndrome and congestive cardiac failure, caused sodium retention in test animals (Deming and Luetscher 1950; Luetscher and Axelrad 1954a, b; Luetscher and Johnson 1954a, b). It was shown subsequently that this material was aldosterone (Luetscher et al. 1955). At this time, it was also demonstrated by Farrell that desoxycorticosterone was secreted by the dog adrenal, apparently under control of ACTH (Farrell 1954). This followed Reichstein and von Euw's (1938) isolation of the compound from adrenal tissue.

Soon after the discovery of aldosterone, Bartter et al. (1959) showed an influence of blood volume change on the rate of aldosterone secretion. Laragh and Stoerk (1957) found that increased potassium intake increased aldosterone secretion. Several groups of investigators demonstrated that ACTH had an important but restricted role in the control of secretion in different species (Liddle et al. 1956; Davis, 1961; Hume 1962; Blair-West et al. 1963; Williams et al. 1970). Aldosterone hypersecretion in response to sodium deficiency or thoracic caval constriction occurred in the hypophysectomized animal or human (Liddle et al. 1956; Blair-West et al. 1963; Davis 1975). At low physiological dosage ACTH increased only cortisol and corticosterone secretion but when the rate of secretion of these components was increased by ACTH to about 50% of the functional capacity of the adrenal glands to secrete them, increased secretion of aldosterone also occurred (Blair-West et al. 1963). In man the effect of ACTH was large and amplified by sodium deficiency (Tucci et al. 1967). The successful autotransplantation of the left adrenal gland to the neck of the Merino sheep was followed by demonstration that physiological decrease in plasma sodium or increase in plasma potassium concentration acted directly on the adrenal to stimulate aldosterone secretion (Denton et al. 1959; Blair-West et al. 1963).

However, another early finding with this preparation was that a local increase in adrenal arterial plasma sodium concentration to normal level or above in the sodium-deficient animal had only a small effect (ca 25% reduction) on the aldosterone hypersecretion, and, thus, ionic change was not a major factor in the response to sodium deficiency. Denton, Goding and Wright (1959) reported discovery of an unidentified aldosterone-stimulating hormone by cross-circulation experiments on conscious sheep (Figs. 9-15 and 9-16). A sodium-deficient adrenalecto-

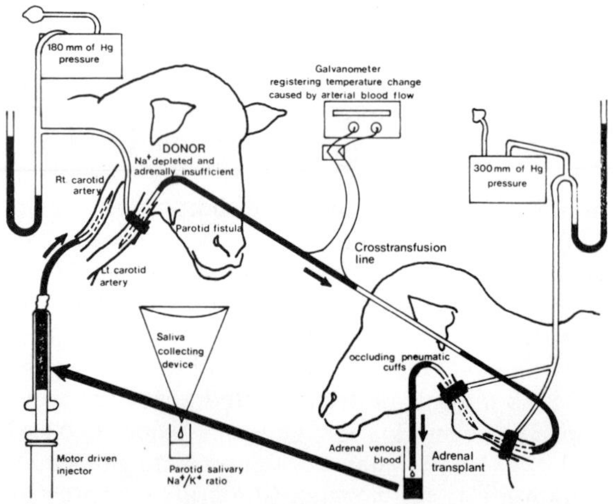

Fig. 9-15. Method of cross-circulation in conscious animals. A wide-bore needle attached to a polyethylene tube inserted into the left carotid artery of a sodium-deplete adrenalectomized donor leads blood to the carotid artery supplying the adrenal transplant of the recipient. This loop is occluded above and below the transplant by cuffs inflated to 300 mmHg. The adrenal venous effluent is collected from an indwelling jugular cannula, and blood is reinfused into right carotid artery loop of donor. Saliva from the right parotid fistula of the donor is collected and analysed at frequent intervals. (From Denton et al. 1959, with permission.)

mized animal was donor and the blood was led to the adrenal transplant of a sodium-replete recipient. Concurrently and independently, Yankopoulos, Davis, Kliman and Peterson (1959) reported the same result in the anaesthetized dog, the donor having the thoracic vena cava constricted. In a lecture at University College London in 1958 describing the cross-circulation experiments (Denton et al. 1959), we suggested that 'alternative to the usual view that the adrenal controlled the kidney in certain respects, the kidney might be the site of a receptor sensitive to haemodynamic change or supply of Na, and accordingly secrete an adrenotrophic hormone', and that the receptor mechanism, whether in the same place as the tissue secreting the hormone or not, is: '(a) akin to a Na accounting machine, in

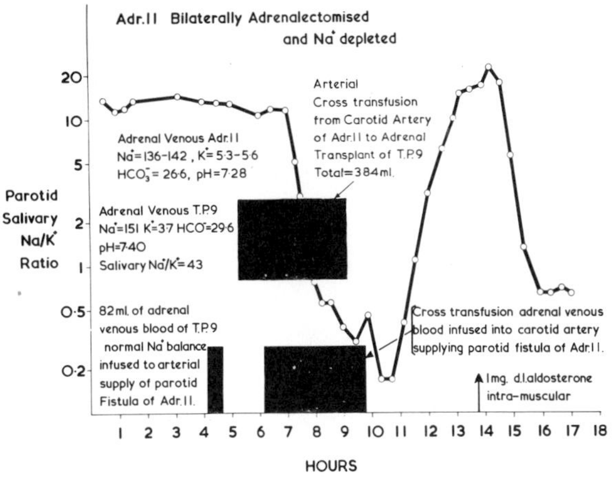

Fig. 9-16. Adr.11 (bilaterally adrenalectomized and sodium-deplete). Effect on parotid salivary Na/K ratio of : ipsilateral intracarotid infusion of 82 ml of adrenal venous blood from TP9 (normal sodium balance); ipsilateral intracarotid infusion of 384 ml of its own peripheral blood after cross-circulation through adrenal gland of TP9; intramuscular injection of 1 mg DL-aldosterone. (From Denton et al. 1959, with permission.)

that, for example, serially arranged transfer mechanisms in a specific local vascular field register the load of Na/unit time in the local blood supply; or that (b) volume change causes stimulation of a pressure or stretch receptor or determines the area of surface available for a particular transfer process'. These proposals are not unlike certain facets of current views of the pathway of aldosterone control through renin. However, further experiments of ours showed that onset of sodium deficiency caused increased aldosterone secretion in the nephrectomized sheep—a finding ostensibly against a renal control of aldosterone (Denton et al. 1959). Retrospectively, in the face of data strongly in favour of a renal origin for the stimulus, we wholly attributed this result to ionic effects associated with nephrectomy. This may now have to be re-evaluated in the light of evidence on new factors, possibly from the central nervous system, in aldosterone control (see Chapter 21).

Davis (1961), in view of our result with nephrectomy, first examined the effect of hepatectomy on aldosterone hypersecretion in response to haemorrhage, found the response unchanged, and then made the important finding that nephrectomy did abolish this response. Ganong and Mulrow (1961) independently reported the same result.

Gross in 1958 had made the inspired suggestion, on the grounds of the inverse relation between sodium balance on the one hand and renin content of kidneys and aldosterone on the other, that the renin system may play an important role in aldosterone secretion. He stated that his results indicated 'there is a close relationship between the incretory function of the kidney which is shown by the discharge of renin and similar materials, and the function of the adrenal cortex, especially the zona glomerulosa, which is the place of production of the sodium retaining hormone aldosterone'.

The discovery of a major role of the kidney in aldosterone hypersecretion was followed by the demonstration by Genest et al. (1960) and Laragh et al. (1960) that intravenous infusion of angiotensin II stimulated aldosterone secretion. Our laboratory showed that the action of angiotensin II was directly upon the adrenal transplant and not via a secondary pathway (Blair-West et al. 1962). It was shown that nephrectomy reduced aldosterone hypersecretion in sodium deficiency to basal level in dog and sheep (Davis et al. 1961; Blair-West et al. 1968a). With development of the enzyme kinetic assay for renin, and the radioimmunoassay for angiotensin II, it was shown at several centres that renin and angiotensin are increased in sodium deficiency (Brown et al. 1964; Blair-West et al. 1968a; Gocke et al. 1969). Angiotensin has been widely accredited with the dominant role in physiological control of aldosterone secretion, as stated in 1965 by Binnion et al.: 'increased activity of the renin–angiotensin system is the primary mechanism leading to hyperaldosteronism during Na depletion'.

The analysis of the mode of control of aldosterone secretion is a complex and large subject and no attempt will be made here to make a balanced appraisal of diverse data derived from many species under conditions varying from the responses to simple reduction of sodium intake to those evoked by profound body deficit caused by loss of intestinal secretions. Certainly it has been shown, perhaps in as comprehensive a fashion in the sheep as in any species, that with progressive sodium deficit, as caused by uncompensated salivary loss, there is a highly significant correlation of blood aldosterone, blood angiotensin II concentration and plasma renin concentration (Fig. 9-17). Blood angiotensin II concentration and blood aldosterone were highly significantly related, which in the light of the known action of angiotensin on aldosterone secretion in this regard makes a strong *prima facie* case for primacy of the renin–angiotensin system in causation of aldosterone secretion. Similar findings have been made in man (Williams et al. 1970; Laragh and Sealey 1973; Hollenberg et al. 1974; Fraser et al. 1979).

However, a large series of experimental findings in sheep show that dislocation of the usual parallel movement of angiotensin and aldosterone levels in

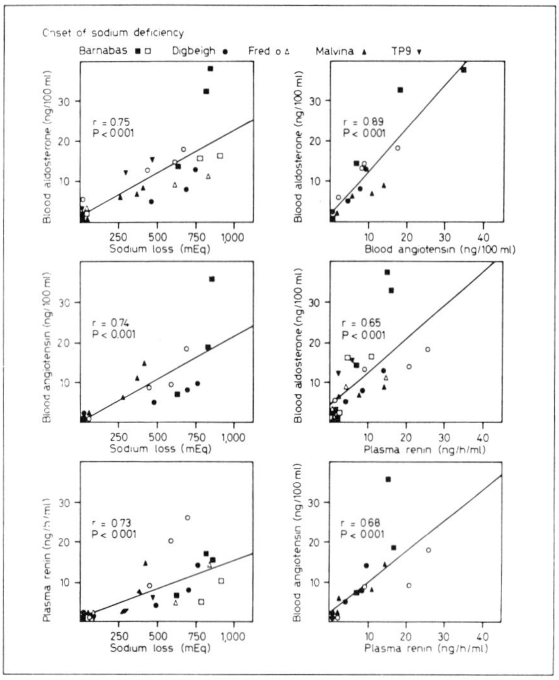

Fig. 9-17. The effect of onset of sodium deficiency on blood aldosterone concentration, arterial blood angiotensin II and plasma renin concentration in five sheep with a parotid fistula and an autotransplanted adrenal gland. The relation of the variables to one another is shown also. There was a highly significant relation between blood aldosterone and blood angiotensin II concentration, blood aldosterone and plasma renin, and blood angiotensin II and plasma renin concentration. (From Blair-West et al. 1970, with permission.)

sodium deficiency can be contrived. Further, such changes are not necessarily related to changes in the other known proximate causes of aldosterone secretion: plasma sodium and potassium concentrations and ACTH concentration (Blair-West et al. 1970a, b, 1972). Other reviews and reports have dealt extensively with the complexities of the control system, and anomalies in the notion of a simple and direct causal relation between contemporary blood level of angiotensin II and aldosterone secretion (Müller 1971; Coghlan et al. 1980). Current assessment ranges between those who see the evidence as showing angiotensin II is commensurately causal of aldosterone hypersecretion in sodium deficiency (Brown et al. 1972; Davis 1975; Hollenberg et al. 1974; Walker et al. 1976) to those who see it as essential but permissive of the action of other factors. An intermediate position is that it is an essential and also contributory causal factor. A major question is whether there is an unidentified factor involved in the control system.

For the purposes of discussion here, which is largely to identify factors under consideration in aldosterone control which might come to attention

STIMULATION	ZONA GLOMERULOSA	INHIBITION
K^+ ↑		Na^+ ↑
Na^+ ↓	Changed Sensitivity due to changed intracellular ionic composition in Na^+ deficiency	K^+ ↓
ACTH		Humoral Inhibitor
Angio II		PGE 1
Heptapeptide		
'x' factor		

Fig. 9-18. Diagrammatic representation of the factors affecting aldosterone secretion as presented by Blair-West et al. (1969) at Coranado Conference (see text).

in relation to the parallel stimulation and satiation of salt appetite, it will suffice to reproduce a diagram on the mode of control of aldosterone secretion (Fig. 9-18) we showed in 1969 at the Coronado Conference on Aldosterone Control organized by the NIH Study Group in General Medicine, and The High Blood Pressure Council of the US Heart Association (Mulrow et al. 1969). The left side shows five stimuli of aldosterone secretion known to act directly upon the adrenal glomerulosa. The inclusion of angiotensin III (heptapeptide) reflected the then recent observation by our group which we published two years later (Blair-West et al. 1971) that angiotensin III was equipotent with angiotensin II in the stimulation of aldosterone secretion (see also Blair-West et al. 1980)—though the hormone has only about 20% of the pressor activity. Today it would be necessary to add to this left-hand section a component to reflect that the antidopaminergic drug metoclopramide will increase aldosterone secretion by an unknown vector (Norbiato et al. 1977; Carey et al. 1978; Coghlan et al. 1980). It has been shown also that adrenal arterial infusion of prostaglandin A_1 in sheep has a direct action in increasing aldosterone secretion (Blair-West et al. 1968b). In man, subpressor doses increase aldosterone secretion, an effect which may be associated with increase of renin (Fichman et al. 1972; Golub et al. 1976). On the right-hand side of the diagram the ionic changes which reduce aldosterone secretion are recorded, as is the reduction produced in sodium-deplete animals by adrenal arterial infusion of prostaglandin E_1 (Blair-West et al. 1971). The question of an unidentified inhibitory factor acting upon the adrenal is examined in Chapter 21. This records the large evanescent reduction in aldosterone secretion which occurs in sodium-deficient sheep with adrenal transplants when they rapidly satiate salt appetite. No concurrent change in plasma ions, renin, angiotensin II or ACTH secretion occurs.

The centre of Fig. 9-18 records the proposal that with sodium deficiency the adrenal develops an increased sensitivity to the proximate stimuli of aldosterone secretion. This followed from data on this effect with ACTH in sodium deficiency (Thorn et al. 1957; Blair-West et al. 1963; Ganong et al. 1966; Kinson and Singer 1968; James et al. 1968) and the possibility that a similar change of response to angiotensin II would explain some apparent anomalies in aldosterone response in sodium-deficient animals (Ganong et al. 1966; Blair-West et al. 1967). That such a change of sensitivity in response to angiotensin II infusion occurred with sodium deficiency was shown by Oelkers et al. (1974), Hollenberg et al. (1974) and Nicholls et al. (1978). Possible explanations for the mechanism of this change of sensitivity include a change in ionic composition of the adrenal with sodium deficit resulting in an alteration of response characteristics in the biosynthetic pathways (Blair-West et al. 1970a, b; Müller 1971; Fraser et al. 1979) or a humoral influence mediating this effect on the adrenal.

One possibility in relation to this latter alternative arises from a large body of data from in vitro studies by Catt and colleagues (1979) emphasizing the importance of angiotensin II. This includes its ability to induce an increase in the number of its receptor sites, so that Catt et al. propose most changes seen in sodium deficiency could be mediated by increased plasma angiotensin II concentration. During sodium deficiency in the rat there was increased angiotensin binding to receptor cells. With up to 26 h of deficiency this was predominantly an increase in receptor affinity with a minor increase in receptor numbers. Longer periods of sodium deficiency led to a rise in angiotensin receptors while binding affinity returned to normal. The opposite changes were seen with sodium loading. Infusion of angiotensin at 50–100 ng/min to conscious rats for 1–6 days caused an increase in angiotensin II receptor concentration and in aldosterone response to angiotensin II in vitro (Hauger et al. 1978). This finding contrasts with the failure of infusion of physiological amounts of angiotensin for 2–3 days to sustain aldosterone secretion in sheep for more than 12 h. Cowley and McCaa (1976) found angiotensin II would not sustain aldosterone secretion in the sodium-replete dog, as did Bean et al. (1979). Catt et al. (1979) see the increased receptors and affinity as the basis of the increased sensitivity of the adrenal to angiotensin which develops in sodium deficiency (Ganong et al. 1966; Blair-West et al. 1967; Kinson and Singer 1968; Boyd et al. 1972; Oelkers et al. 1974; Aguilera et al. 1978; McCaa 1978; Carey et al. 1978). The similar change in sensitivity of the adrenal glomerulosa to ACTH with sodium deficiency was noted above (Blair-West et al. 1963; Ganong et al. 1966; Kinson and Singer 1968; Fraser et al. 1979).

In incubation studies with isolated rat and canine adrenal glomerulosa cells (Douglas et al. 1978; Aguilera et al. 1979) it was shown that angiotensin II and angiotensin III were equipotent in stimulation of aldosterone secretion though the effect declined with angiotensin III due to its more rapid metabolism. In sheep (Blair-West et al. 1971, 1980), angiotensins II and III are equipotent in stimulation of aldosterone secretion (Fig. 9-19).

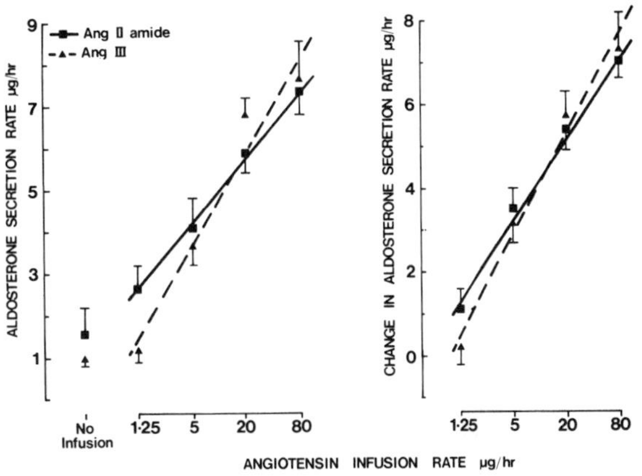

Fig. 9-19. Relation between intravenous infusion rate (μg/h) of [Val^5]-angiotensin II amide (squares) and [des-Asp^1Val^5]-angiotensin II (triangles), and aldosterone secretion rate (left) and change in aldosterone secretion rate (right) in sodium-replete sheep. Values are mean $\pm$s.e.m. (From Blair-West et al. 1980, with permission.)

In an overview of the complex issue of control of aldosterone secretion, one fact evident is that set out in the introduction to Chapter 7. This draws attention to the quite different changes in sodium status resulting from, on the one hand, a low-sodium diet coupled with, for example, a diuretic, and, on the other, that produced by the uncompensated drainage from a salivary fistula in a ruminant or, for that matter dysentery or cholera in a human, who might also be an inhabitant of a tropical region and, perhaps, initially on a low-sodium diet. It is possible that mechanisms operate in the latter circumstances of negative sodium status which do not have influence in the instance of a change in sodium status contrived simply by a low-sodium diet. This will be a relevant approach in future investigation, as will be the question of whether chronic exposure to episodes of sodium deficiency over months may result in development, at the biochemical level, of mechanisms which are not immediately evident during initial response to sodium deficiency; this circumstance is analogous in a way to the influence of frequent episodes of sodium deficiency on salt appetite, when learning factors become superimposed on the innate. There is also the question of the implications of the local formation of angiotensin II in peripheral tissues, presumably from sequestrated renal renin, as demonstrated by contemporaneous arterial and venous assay of angiotensin II and III by Coghlan et al. (1979), a finding recently confirmed by demonstration of high circulating angiotensin II concentration in the presence of low plasma renin activity in nephrectomized rats (Aguilera et al. 1981). Aguilera et al. (1981) also showed the presence of immunoreactive angiotensin II in high concentration relative to other tissues in the adrenal capsules of nephrectomized rats.

References

Aguilera G, Hauger RL, Catt KJ (1978) Control of aldosterone secretion during sodium restriction: Adrenal receptor regulation and increased adrenal sensitivity to angiotensin II. Proc Natl Acad Sci USA 75: 975

Aguilera G, Capponi A, Baukal A, Fujita K, Hauger R, Catt KJ (1979) Metabolism and biological activities of angiotensin II and des-Asp^1-angiotensin II in isolated adrenal glomerulosa cells. Endocrinology 104: 1279

Aguilera G, Schirar A, Baukal A, Catt KJ (1981) Circulating angiotensin II and adrenal receptors after nephrectomy. Nature 289: 507

Andersson B, Olsson K (1973) On central control of body fluid homeostasis. Conditional Reflex 8: 147

Baniukiewicz S, Brodie A, Flood C, Motta M, Okamoto M, Tait JF, Tait SAS, Blair-West JR, Coghlan JP, Denton DA, Goding JR, Scoggins BA, Wintour M, Wright RD (1968) Adrenal biosynthesis of steroids in vivo and in vitro using continuous superfusion and infusion procedures. In: McKerns KW (ed) Functions of the adrenal cortex, vol 1. Appleton-Century-Crofts, New York p 153

Barcroft J (1934) The architecture of physiological function. Cambridge University Press.

Bartter FC, Mills I, Biglieri EG, Delea C (1959) Studies on the control and physiologic action of aldosterone. Recent Prog Horm Res 15: 311

Bean BL, Brown JJ, Casals-Stenzel J, Fraser R, Lever AF, Millar JA, Morton JJ, Petch B, Riegger AJG, Robertson JIS, Tree M (1979) The relation of arterial pressure and plasma angiotensin II concentration: A change produced by prolonged infusion of angiotensin II in the conscious dog. Circ Res 44: 452

Bell JP, Salamonsen LA, Holland GW, Espiner EA, Beaven DW, Hart DS (1970) Autotransplantation of the pancreas in sheep: Insulin secretion from the transplant. J Endocrinol 48: 411

Binnion PR, Davis JV, Brown TC, Olichney MJ (1965) Mechanisms regulating aldosterone secretion during sodium depletion. Am J Physiol 208: 655

Blair-West JR, Coghlan JP, Denton DA, Goding JR, Munro JA, Peterson RE, Wintour M (1962) Humoral stimulation of adrenal cortical secretion. J Clin Invest 41: 1606

Blair-West JR, Coghlan JP, Denton DA, Goding JR, Wintour EM, Wright RD (1963) The control of aldosterone secretion. Recent Prog Horm Res 19: 311

Blair-West JR, Coghlan JP, Denton DA, Goding JR, Orchard E, Scoggins BA, Wintour M, Wright RD (1967) Mechanisms regulating aldosterone secretion during sodium deficiency. In: Schreiner GE (ed) Proceedings of the Third International Congress on Nephrology. Karger, Basel, p 201

Blair-West JR, Coghlan JP, Denton DA, Goding JR, Wintour M, Wright RD (1968a) The effect of nephrectomy on aldosterone secretion in the conscious sodium-depleted hypophysectomized sheep. Aust J Exp Biol Med Sci 46: 295

Blair-West JR, Coghlan JP, Denton DA, Funder JW, Scoggins BA, Wright RD (1968b) The effect of prostaglandins on adrenocortical steroid secretion. Proc Aust Soc Med Res 2: 184

Blair-West JR, Cain MD, Catt KJ, Coghlan JP, Denton DA, Funder JW, Scoggins BA, Stockigt JR, Wright RD (1970a) Further facets of aldosterone regulation. In: Hormonal steroids. Excerpta Medica, Amsterdam, p 572 (International congress series no. 219)

Blair-West JR, Cain MD, Catt KJ, Coghlan JP, Denton DA, Funder JW, Scoggins BA, Wintour EM, Wright RD (1970b) The mode of control of aldosterone secretion. In: Proceedings of the Fourth International Congress on Nephrology, vol 2. Karger, Basel, p 33

Blair-West JR, Coghlan JP, Denton DA, Funder JW, Scoggins BA, Wright RD (1971) The effect of the heptapeptide (2–8) and hexapeptide (3–8) fragments of angiotensin II on aldosterone secretion. J Clin Endocrinol Metab 32: 575

Blair-West JR, Coghlan JP, Denton DA, Funder JW, Scoggins BA (1972) The role of the renin–angiotensin system in control of aldosterone secretion. In: Assaykeen TA (ed) Control of renin secretion. Plenum, New York, p 167

Blair-West JR, Coghlan JP, Denton DA, Fei DTW, Hardy KJ, Scoggins BA, Wright RD (1980) A dose–response comparison of the actions of angiotensin II and angiotensin III in sheep. J Endocrinol 87: 409

Blokh EL (1939) The work of the parotid and submaxillary glands in cattle. Fiziol Zh SSSR 27: 200

Boyd GW, Adamson AR, Arnold M, James VHT, Peart WS (1972) The role of angiotensin II in the control of aldosterone secretion in man. Clin Sci 49: 91

Brown JJ, Davies DL, Lever AF, Robertson JIS, Peart WS (1964) The estimation of renin in plasma. In: Baulieu EE, Robel P (eds) Aldosterone. Blackwell Scientific, Oxford p 417

Brown JJ, Fraser R, Lever AF, Love DR, Morton JJ, Robertson JIS (1972) Raised plasma angiotensin II and aldosterone during dietary sodium restriction in man. Lancet II: 1106

Care AD, Copp DH, Henze KG (1965) A study of factors affecting calcitonin secretion in the conscious sheep. J Physiol (Lond) 176: 31P

Carey RM, Vaughan ED, Peach MJ, Ayers CR (1978) Activity of [des-aspartyl^1]-angiotensin II and angiotensin II in man. J Clin Invest 61: 20

Catt KJ, Aguilera G, Capponi A, Fujita K, Schirar A, Fakunding J (1979) Angiotensin II receptors and aldosterone secretion. J Endocrinol 81: 37P

Coats DA, Denton DA, Goding JR, Wright RD (1956) Secretion by the parotid gland of the sheep. J Physiol (Lond) 131: 13

Coats DA, Denton DA, Wright RD (1958) The ionic balances and transferences of the sheep's parotid gland during maximal stimulation. J Physiol (Lond) 144: 108

Coghlan JP, Wintour M, Scoggins BA (1966) The measurement of corticosteroids in the adrenal vein blood of sheep. Aust J Exp Biol Med Sci 44: 639

Coghlan JP, Blair-West JR, Denton DA, Fei DT, Fernley RT, Hardy KJ, McDougall JG, Puy R, Robinson P, Scoggins BA, Wright RD (1979) Control of aldosterone secretion. J Endocrinol 81: 55
Coghlan JP, Blair-West JR, Butkus A, Denton DA, Hardy KJ, Leksell LG, McDougall JG, McKinley MJ, Scoggins BA, Tarjan E, Weisinger RS, Wright RD (1980) Factors regulating aldosterone secretion. In: Cummings IM, Funder JW, Mendelsohn FAO (eds) Proceedings of the Sixth International Congress on Endocrinology. Australian Academy of Science, Canberra, p 385
Conn JW (1955) Primary aldosteronism, a new clinical syndrome. J Lab Clin Med 45: 6
Cowley AW Jr, McCaa RE (1976) Acute and chronic dose response relationships of angiotensin, aldosterone, and arterial pressure at varying levels of sodium intake. Circ Res 39: 788
Cuthbertson DP (1958) Digestion in the ruminant. Advancement of Science 58: 1
Davis JO (1961) Mechanisms regulating the secretion and metabolism of aldosterone in experimental secondary hyperaldosteronism. Recent Prog Horm Res 17: 293
Davis JO (1975) Regulation of aldosterone secretion. In: Handbook of physiology, Sect 7: Endocrinology, vol VI: Adrenal gland. American Physiological Society, Washington, p 77
Davis JO, Ayers CR, Carpenter CCJ (1961) Renal origin of an aldosterone-stimulating hormone in dogs with thoracic caval constriction and in sodium-depleted dogs. J Clin Invest 40: 1466
Deming QB, Luetscher JA (1950) Bioassay of deoxycorticosterone-like material in urine. Proc Soc Exp Biol Med 73: 171
Dempster WJ, Joekes AM (1953) Functional studies of the kidney autotransplanted to the neck of dogs. Acta Med Scand 147: 99
Denton DA (1948) Renal regulation of the extracellular fluid. Nature 162: 618
Denton DA (1949) Renal regulation of the extracellular fluid. I. A study of homeostasis in a patient with duodenal fistula. Med J Aust 2: 521
Denton DA (1957a) A gregarious factor in the natural conditioned salivary reflexes of sheep. Nature 179: 341
Denton DA (1957b) The study of sheep with permanent unilateral parotid fistulae. Q J Exp Physiol 42: 72
Denton DA (1957c) The effect of variations in blood supply on secretion rate and composition of parotid saliva in sodium depleted sheep. J Physiol (Lond) 135: 227
Denton DA, Wynn V, McDonald IR, Simon S (1951) Renal regulation of the extracellular fluid. II. Renal physiology in electrolyte subtraction. Acta Med Scand 140 [Suppl 261]: 1
Denton DA, McDonald IR, Munro J, Williams W (1952) Excess Na subtraction in the sheep. Aust J Exp Biol Med Sci 30: 213
Denton DA, Goding JR, Wright RD (1959) Control of adrenal secretion of electrolyte-active steroids. Br Med J ii: 447, 522
Denton DA, Goding JR, Wright RD (1960) The control of aldosterone secretion. In: Astwood EB (ed) Clinical endocrinology, vol 1. Grune & Stratton, New York, p 373
Domingo WR, Klyne W (1949) Photoelectric flame photometer. Biochem J 45: 400
Douglas J, Condo T, Bartley P, Catt KJ (1978) Evidence that [des-Asp[1]]-angiotensin II is not an obligatory intermediate in the steroidogenic action of angiotensin II. Endocrinology 102: 1921
Epanishnikov DA (1935) Mechanism of ruminant parotid secretion. In: Physiology of farm animal digestion, p 11. Quoted by Serebryakov PN in IP Pavlov's Theories and farm animal physiology, p 85 Selkhozgiz, Moscow, 1950
Espiner EA, Hart DS, Beaven DW (1963) Bioassay of corticotrophin using the sheep's transplanted adrenal gland. J Endocrinol 27: 267
Espiner EA, Hart DS, Beavan DW (1972) Cortisol secretion during acute stress and response to dexamethasone in sheep with adrenal transplants. Endocrinology 90: 94
Falconer IP (1963) The exteriorization of the thyroid gland and measurement of its function. J Endocrinol 26: 241
Fantus JB (1936) Fluid postoperatively (a statistical study). JAMA 107: 14
Farrell G (1954) Steroids in adrenal venous blood of dog: Venous–arterial differences across adrenal. Proc Soc Exp Biol Med 86: 338
Farrell G (1958) Regulation of aldosterone secretion. Physiol Rev 38: 709
Fichman MP, Littenberg G, Brooker G, Horton R (1972) Effect of prostaglandin A_1 on renal and adrenal function in man. Circ Res 30/31 [Suppl 2]: 19
Fraser R, Brown JJ, Lever AF, Mason PA, Robertson JIS, (1979) Control of aldosterone secretion. Clin Sci 56: 389
Ganong WF, Mulrow PJ (1961) Evidence of secretion of an aldosterone-stimulating substance by the kidney. Nature 190: 115
Ganong WF, Biglieri EG, Mulrow PJ (1966) Mechanisms regulating adrenocortical secretion of aldosterone and glucocorticoids. Recent Prog Horm Res 22: 381
Genest JE, Koiw E, Nowaczynski W, Sandor T (1960) Study of urinary adrenocortical hormones in human arterial hypertension. (Abstract) In: Proceedings of the First International Congress on Endocrinology, Denmark. Endocrine Society, p 173
Gocke DA, Gerten J, Sherwood LM, Laragh JH (1969) Physiological and pathological variations of plasma angiotensin II in man. Correlation with renin activity and sodium balance. Circ Res [Suppl] 1: 131
Goding JR, Denton DA (1956) Adrenal cortex and the parotid secretion of sodium-depleted sheep. Science 123: 986
Goding JR, Denton DA (1959) The response to sodium depletion in adrenalectomized sheep with parotid fistulae. Aust J Exp Biol Med Sci 37: 211
Goding JR, Harrison FA, Heap RB, Linzell JL (1967a) Ovarian activity in the ewe after autotransplantation of the ovary or uterus to the neck. J Physiol (Lond) 191: 129
Goding JR, McCracken JA, Baird DT (1967b) The study of ovarian function in the ewe by means of a vascular autotransplantation technique. J Endocrinol 39: 37
Goding JR, Baird DT, Cumming IA, McCracken JA (1971) Functional assessment of autotransplanted endocrine organs. Karolinska Institute, Stockholm, p 169 (Symposia on Research Methods in Reproductive Endocrinology)
Golub MS, Speckart PF, Zia PK, Horton R (1976) The effect of prostaglandin A_1 on renin and aldosterone in man. Circ Res 39: 574
Gross F (1958) Renin und Hypertensin, physiologische oder pathologische Wirkstoffe? Klin Wochenschr 36: 693
Harrison FA, McDonald IR (1966) The arterial supply to the adrenal gland of the sheep. J Anat 100: 189
Harrison FA, Heap RB, Linzell JL (1968) Ovarian function in the sheep after transplantation of the ovary and uterus to the neck. J Endocrinol 40: Proc XIII
Hauger RL, Aguilera G, Catt KJ (1978) Angiotensin II regulates its own receptor sites in the adrenal glomerulosa zone. Nature 271: 176
Holland GW (1969) Pancreatic autotransplantation in sheep. Aust NZ J Surg 30: 75
Hollenberg NK, Chenitz WR, Adams DF, Williams GH (1974) Reciprocal influence of salt intake on adrenal glomerulosa and renal vascular responses to angiotensin II in normal men. J Clin Invest 54: 34
Hudson B, Coghlan JP, Dulmanis A, Wintour M (1964) The

measurement of testosterone in biological fluids in the evaluation of androgen activity. In: Proceedings of the Second International Congress on Endocrinology. Excerpta Medica, Amsterdam, p 1127. (International Congress Series no 83)
Hume DM (1962) Direct measurement of adrenal secretion in man with observations on the mechanism of ACTH release in trauma and the role of ACTH in the secretion of aldosterone. In: Proceedings of the International Congress on Hormonal Steroids. Excerpta Medica, Amsterdam, p 20 (International Congress Series no. 51)
James VH, Landon J, Fraser R (1968) Some observations on the control of corticosteroid secretion in man. Mem Soc Endocrinol 17: 141
Jewell PA, Verney EB (1957) An experimental attempt to determine the role of the neurohypophyseal osmoreceptors in the dog. Proc R Soc Lond [Biol] 240: 197
Kinson GA, Singer B (1968) Sensitivity to angiotensin and adrenocorticotrophic hormone in the sodium deficient rat. Endocrinology 83: 1108
Kliman B, Peterson RE (1960) Double isotope derivative assay of aldosterone in biological extracts. J Biol Chem 235: 1639
Kudriavstev AA, Anurov SN (1932) Physiology of the sheep. Collection of papers, p 5. Cited in EK Krasusky et al. (1932) Fiziol Zh SSSR 28: 372
Kvastnitsky OV (1955) A new method of externalizing farm animals' parotid ducts. Ukranian Physiol J 1: 120
Laragh JH, Sealey JE (1973) The renin–angiotensin–aldosterone hormonal system and regulation of sodium, potassium, and blood pressure homeostasis. In: Orloff J, Berliner RW (eds) Handbook of physiology, Sect 8: Renal physiology. Waverly Press, Baltimore, p 831
Laragh JH, Stoerk HC (1957) A study of the mechanisms of secretion of the sodium retaining hormone (aldosterone). J Clin Invest 36: 383
Laragh JH, Angers M, Kelly WG, Lieberman S (1960) Hypotensive agents and pressor substances. JAMA 174: 234
Levy SE, Blalock A (1939) A method of transplanting the adrenal gland of the dog with re-establishment of its blood supply. Ann Surg 109: 84
Lewinsohn R ('Morus') (1954) Animals, men and myths. Victor Gollancz, London
Liddle GW, Duncan LE, Bartter FC (1956) Dual mechanism regulating adrenocortical function in man. Am J Med 21: 380
Luetscher JA, Axelrad BJ (1954a) Sodium-retaining corticoid in urine of normal children and adults and of patients with hypoadrenalism or hypopituitarism. J Clin Endocrinol Metab 14: 1086
Luetscher JA, Axelrad BJ (1954b) Increased aldosterone output during sodium deprivation in normal men. Proc Soc Exp Biol Med 87: 650
Luetscher JA, Johnson BB (1954a) Chromatographic separation of the sodium-retaining corticoid from the urine of children with nephrosis, compared with observations of normal children. J Clin Invest 33: 276
Luetscher JA, Johnson BB (1954b) Observations of the sodium-retaining corticoid (aldosterone) in the urine of children and adults in relation to sodium balance and edema. J Clin Invest 33: 1441
Luetscher JA, Dowdy A, Harvey J, Neher R, Wettstein A (1955) Isolation of crystalline aldosterone from the urine of a child with the nephrotic syndrome. J Biol Chem 217: 505
McCaa RE (1978) Aldosterone response to long-term infusion of angiotensin II and angiotensin III in conscious dogs before and after dietary sodium restriction. Endocrinology 103: 458
McCance RA (1938) Effect of salt deficiency in man on volume of extracellular fluids and on composition of sweat, saliva, gastric juice and cerebrospinal fluid. J Physiol (Lond) 92: 208
McDonald IR, Denton DA (1956) Delayed effect of ipsilateral intracarotid infusion of sodium chloride on the composition of parotid saliva of sodium depleted sheep. Nature 177: 1035
McDonald IR, Goding JR, Wright RD (1958) Transplantation of the adrenal gland of the sheep to provide access to its blood supply. Aust J Exp Biol Med Sci 36: 83
McDougall EI (1948) The composition and output of sheep's saliva. Biochem J 43: 99
Marriott HL (1950) Water and salt depletion. Thomas, Springfield, Ill
Mattox VR, Mason HL, Albert A (1953) Isolation of sodium-retaining substance from beef adrenal extract. Mayo Clin Proc 28: 569
Moir RJ, Somers M, Waring H (1956) Studies on marsupial nutrition. I. Ruminant-like digestion in a herbivorous marsupial (*Setonix brachyurus quoy gaimard*). Aust J Biol Sci 9: 293
Müller J (1971) Regulation of aldosterone biosynthesis. Monogr Endocrin 5
Mulrow PJ, Bartter FC, Kirkendall WM, Peterson RE, Tait JF (1969) Adrenal cortex, aldosterone and hypertension. Circulation 40: 739
Nicholls MG, Tree M, Brown JJ, Douglas BH, Fraser R, Hay GD, Lever AF, Morton JJ, Robertson JIS (1978) Angiotensin II/aldosterone dose response curves in the dog: Effect of changes in sodium balance. Endocrinology 102: 485
Norbiato G, Beirlacqua M, Raggi U, Micossi P, Moroni C (1977) Metoclopramide increases plasma aldosterone concentration in man. J Clin Endocrinol Metab 45: 131
Oddie CJ, Coghlan JP, Scoggins BA (1972) Plasma deoxycorticosterone levels in man with simultaneous measurement of aldosterone, corticosterone, cortisol and 11-deoxycortisol. J Clin Endocrinol Metab 34: 1039
Oelkers W, Brown JJ, Fraser R, Lever AF, Morton JJ, Robertson JIS (1974) Sensitization of the adrenal cortex to angiotensin II in sodium-deplete man. Circ Res 34: 69
Pavlov IP (1910) The work of the digestive glands, 2nd edn. Charles Griffin, London
Popov NF (1932) Physiology of the sheep. Cited in Krasusky VK et al. (1932) Fiziol Zh SSSR 28: 372
Reichstein T, von Euw JV (1938) Über Bestandteile der Nebennierenrinde. 20. Isolierung der Substanzen Q (Desoxycorticosteron) und R sowie weiterer Stoffe. Helv Chim Acta 21: 1197
Richter CP (1949) Domestication of the Norway rat and its implications for the problem of stress. Proc Assoc Res Nerv Ment Dis 29: 19
Richter CP (1956) Salt appetite of mammals: Its dependence on instinct and metabolism. In: L'Instinct dans le comportement des animaux et de l'homme. Masson et Cie, Paris, p 576
Scheunert A, Trautmann A (1921) Zum Studium der Speichelsekretion. II. Über die Sekretion der Parotis und Mandibularis des Schafes. Pflügers Arch 192: 33
Scheunert A, Krzywanek W, Zimmermann K (1929) Zum Studium der Speichelsekretion; die Dauersekretion der Parotis des Schafes und ihre Bedeutung. Pflügers Arch 223: 453
Schmidlin J, Anner G, Billeter JR, Wettstein A (1955) Über Synthesen in der Aldosteronreihe. I. Totalsynthese des racemischen Aldosterons. Experientia 11: 365
Schmidt-Nielsen K, Schmidt-Nielsen B, Houpt TR, Jarnum SA (1956) The question of water storage in the stomach of the camel. Extrait de Mammalia 20: 1
Serebryakov PN (1950) Digestion of the ruminant. In: IP. Pavlov's Theories and farm animal physiology. Selkhozgiz, Moscow, p 85
Simpson SA, Tait JR (1952) A quantitative method for the bioassay of the effect of adrenal cortical steroids on mineral metabolism. Endocrinology 50: 150
Simpson SA, Tait JF, Wettstein A, Neher R, von Euw J,

Reichstein T (1953) Isolierung eines neuen kristallisierten Hormons aus Nebennieren mit besonders hoher Wirksamkeit auf den Mineralstoffwechsel. Experientia 9: 333

Simpson SA, Tait JF, Wettstein A, Neher R, von Euw J, Schindler O, Reichstein T (1954) Konstitution des Aldosterons, des neuen Mineralcorticoids. Experientia 10: 132

Sineschekov AD, Sheremet ZI (1953) Method of complex study of the physiology of nutrition. In: Sineschekov AD (ed) Physiology of farm animal nutrition. Selkhozgiz, Moscow, p 24

Thorn NA, Schwartz IL, Thaysen JH (1956) Effect of Na restriction on secretion of sodium and potassium in human parotid saliva. J Appl Physiol 9: 477

Thorn GW, Ross EJ, Crabbe J, van't Hoff W (1957) Studies on aldosterone secretion in man. Br Med J ii: 955

Tucci JR, Espiner EA, Jagger PI, Pauk GL, Lauler DP (1967) ACTH stimulation of aldosterone secretion in normal subjects and in patients with chronic adrenocortical insufficiency. J Clin Endocrinol Metab 27: 568

Ugolev AM (1953) Methods of study of the comparative physiology of salivary secretion. In: Study of regulation of physiological functions under natural conditions in the organism's existence. Acad Sci USSR 2: 237

Verney EB (1947) The antidiuretic hormone and the factors which determine its release. Proc R Soc Lond [Biol] 135: 25

Verney EB (1957) Renal excretion of water and salt. Lancet II: 1237, 1295

Walker WG, Moore MA, Horvath JS, Whelton PK (1976) Arterial and venous angiotensin II in normal subjects. Relation to plasma renin activity and plasma aldosterone concentrations, and response to posture and volume changes. Circ Res 38: 477

Williams GH, Dluhy RG, Underwood RH (1970) The relationship of dietary potassium intake to the aldosterone stimulatory properties of ACTH. Clin Sci 39: 489

Wright RD, Blair-West JR, Coghlan JP, Denton DA, Goding JR, Nelson JF, Scoggins BA (1972) The structure and function of adrenal autotransplants. Aust J Exp Biol Med Sci 50: 873

Wynn V, Simon S, Morris R, McDonald IR, Denton DA (1950) The clinical significance of sodium and potassium analyses of biological fluids: Their estimation by flame spectrophotometry. Med J Aust 1: 821

Yankopoulos NA, Davis JO, Kliman B, Peterson RE (1959) Evidence that a humoral agent stimulates the adrenal cortex to secrete aldosterone in experimental secondary hyperaldosteronism. J Clin Invest 38: 1278

10 Salt taste and the response to sodium deficiency

Summary

1. The taste of foods bears a systematic relation to metabolic consequences. It is no phylogenetic accident that salt is one of the four primary modalities of taste throughout the vertebrate phylum.

2. Taste buds in the phylogenetically recent anterior portion of the tongue innervated by the seventh cranial nerve are sensitive to substances appropriate to ingestion. In the more ancient posterior portion innervated by the ninth nerve they are sensitive to substances evoking rejection.

3. The comparative physiology of taste is discussed from an anatomical and behavioural viewpoint. Matters of specificity of individual taste fibres as revealed by electrophysiological studies, and also the question of intravascular taste, are examined.

4. Salt depletion has been shown by Contreras and Frank (1979) to cause clear changes in the electrical behaviour of salt-specific fibres which could have significant influence on the behaviour of sodium-deficient animals towards salt solutions.

5. Adrenal insufficiency in man is reported to reduce greatly the threshold for detection of salt and many other compounds, but to elevate the recognition threshold. Normal function is restored by administration of carbohydrate-active steroids but not mineralocorticoid.

6. The ubiquitous presence of salt receptors in insects is a puzzle, since Dethier (1977) reports lack of evidence that either herbivorous or carnivorous species of insects ever suffer salt deficiency in nature. Some species of butterfly show an attraction to salt. Salt receptors in insects may be primarily involved in warding against hypersalinity.

Introduction

In the preceding chapters the occurrence of sodium deficiency in large areas of the planet has been recorded. By the nature of its causation, this situation will have held over geological time. In fact the meteorological condition of absence of marine aerosols inside continents probably operated more intensively in earlier geological eras before continental drift, though, possibly, volcanically derived sources of salt were more abundant during these times. It is thus no accident that salt is one of the four primary modalities of taste throughout the vertebrate phylum. The selection pressure favouring its emergence would have operated under diverse ecological conditions. This reflection is consonant with the view of Nachman and Cole (1971), who noted that while the odour or texture of a substance does not bear any systematic relationship to its caloric or other metabolic effects, the taste of food clearly does (Le Magnen 1967). Over a wide range of animal species there appears to be an inherent biological relationship between the taste of a substance, the acceptability or hedonic aspects of that taste and the consequences of ingestion for nutrition. In general, sweet substances are preferred by animals and are associated with positive nutritive consequences (Pfaffmann 1960, 1975) while bitter substances are aversive and are often toxic if ingested (Garcia and Hankins 1975).

Richter (1956) expressed this general consideration as regards salt, thus:

> First of all, the ability to taste salt is definitely inherited in all animals. Secondly, there is a universal liking for salt, a primitive attraction to salt. So far, we have not found any instance of a definite aversion to salt. Attention may be called here to the simple fact that all nutritive substances apparently have a characteristic taste—that is provided they are ingested in their natural form and are soluble in the mouth, and not chemically bound. It is likely the ability to taste nutritive substances was inherited on the basis of natural selection—that is animals best able to taste the nutritious substances had a better chance of surviving.

In considering the basis of the increased appetite for sodium chloride by adrenalectomized rats, and the selection of sodium chloride solutions in preference to water at much lower concentrations in these animals, Richter (1956) and Richter and Eckert (1938) proposed that the physiological change was one of increased gustatory sensitivity. In 1956 Richter expressed this as follows:

> It has always seemed likely to us that appetite changes occur at the taste bud membrane. Under conditions of salt deficiency, when all tissues of the body are in a state of salt deficiency, there is also less salt in the taste bud side of the membrane. It follows that sodium ion will be transported across the membrane from the ingested solution to the cell fluids at a lower concentration than would be the case if the latter had a higher (normal) salt content. It is conceivable that when salt solution comes into contact with the salt bud membrane, chemical changes are produced that make the salt much more recognizable (or attractive) than it would otherwise be if ample amounts of salt are present in the blood.

A survey of the literature on salt taste shows that this hypothesis has been provocative of a great amount of experimental work. It will be convenient further on to review briefly some material on the basic analysis of salt taste in relation to it. However, first some general background will be provided on taste and how the quality of salt taste arises. Then electrophysiological and behavioural studies of salt taste will be considered.

The cells responsible for the sense of taste are localized mainly in the taste buds. These receptors are found within the distinctively shaped papillae which are located on the edges and rear of the upper surface of the tongue.

Vinnikov (1965) notes that the diversity of the structure, chemical properties and function of receptor cells of the sense organs reflects their strict specialization. Each of the sense organs reacts selectively to a particular type of chemical or physical energy in the external environment: light, substances with odour or taste, sound waves, gravitation, angular acceleration and vibration. This reaction, usually described as reception, initiates and is accompanied by dynamic molecular changes in the specific and non-specific biologically active chemical substances localized in the organoids of the receptor cells of the sense organ. The presence of these specialized structures at first sight renders any comparison between, for example, rods and cones of photoreceptors and the peripheral processes of the olfactory or acoustic cells, impossible. However, Vinnikov points out that modern biochemical, cytochemical and electron microscopical methods of investigation have shown conclusively that despite certain cardinal differences, the structure and functional evolution of all receptor cells of the sense organs in vertebrates (and evidently in most invertebrates) is based on a modification of a cell supplied with motile hairs or a flagellum (Fig. 10-1). This, in the

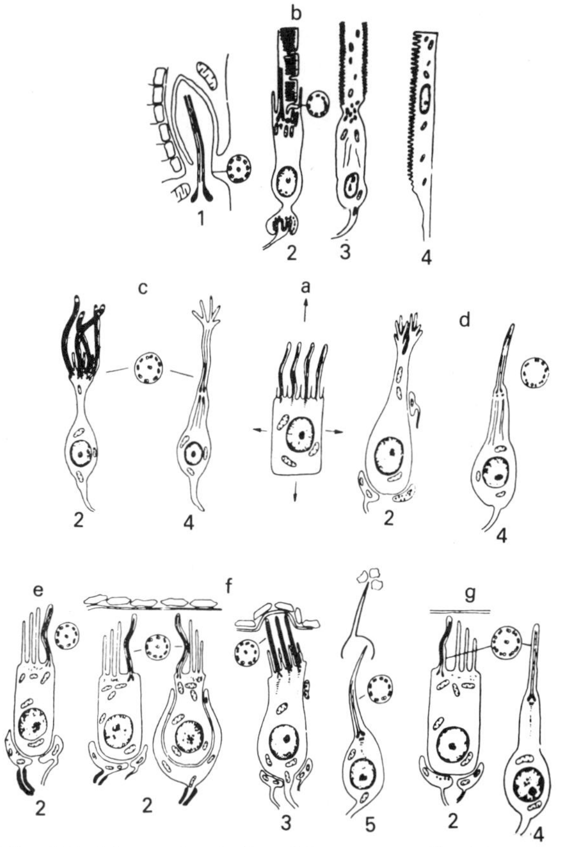

Fig. 10-1. Scheme of evolution of the receptor cells of the most highly developed representatives of phyletic lines of animals. Cells: (a) flame; (b) photoreceptors; (c) olfactory; (d) taste; (e) lateral line receptor; (f) of organ of gravitation; (g) of auditory organ. Eyes of: 1, *Euglena*; 2, vertebrates; 3, cephalopods and molluscs; 4, insects; 5, higher crustaceans. (After Vinnikov 1965, 1975.)

simplest forms of organisms, performs both a locomotor and a receptor function. All receptor cells of the sense organs, regardless of the different pathways of their functional evolution, contain nine pairs of peripheral and two central fibrils. These fibrils consist of macromolecular strands similar in nature to myosin. They contain ATPase, and contract under the influence of ATP, which is supplied from the mitochondria. Vinnikov holds that investigation of the taste buds in members of all classes of vertebrates has revealed the similarity of their substructures and cytochemical organization, possibly because of the similar character of the four basic taste modalities. However, particular importance is evidently attached in vertebrates to replacement of the flagellum in the taste cells by a 'stub' containing bound protein, mucopolysaccharides and alkaline phosphatase. This 'stub' or 'brad' probably plays the role of an ion exchange apparatus for it is situated between the taste substances and the receptor cells.

Writing on this issue of transduction Beidler (1978) remarks that most receptors respond to a stimulus with a change in sodium conductance across the plasma membrane. However, this event does not occur at the location of the stimulus–taste cell interactions but rather towards the base of the taste bud, where innervation occurs.

Some species differences in anatomy of taste reception

In man, taste buds are found in the fungiform, the foliate and the circumvallate papillae but only rarely in the filiform papillae. Different estimates have been made of the number of taste buds on a human tongue: Harper (1972) puts the number at 10 000. The number of taste buds in a single papilla may differ considerably. Circumvallate papillae may contain between 30 and 500, a typical number being about 250 (Harper 1972). Fungiform papillae contain only about four taste buds. Other sensitive parts of the oral cavity, including the pharynx and soft palate, also contain a few taste buds but these are bedded in the epithelium rather than in distinctive papillae.

There are considerable species differences in the total number and types of taste buds and these do not correlate well with the taste thresholds of the species. As Kare (1971) records, many birds, such as chicken, pigeon, bullfinch, starling, duck and parrot, have only between 24 and 350 taste buds, whereas kittens have 473, rabbits 17 000, goats 15 000, calves 25 000 and the catfish 100 000. However, although chickens have only a few dozen buds, they can perceive some solutions and concentrations imperceptible to the cow which has a thousand times as many buds.

There are also species differences in distribution. For example, the posterior part of the cow's tongue has by far the greatest number, in the circumvallate papillae. Furthermore, Bernard (1964) has reported that the chorda tympani nerve has a posterior receptive field on the tongue of the cow. This has been related to the possibility that chemoreception plays a significant role in the mechanism of rumination. In the dog, the greatest concentration of taste buds is on the anterior portion of the tongue. Domestication has reduced the number of fungiform papillae on the tongue of the rat from an average of 217 to 178 (Richter 1954).

There are 40 to 60 taste cells connected to form each taste bud. The cells have an average life span of about 250–300 h (Beidler 1971)

Innervation

Several nerve fibres enter each taste bud, and each fibre may connect with one or more taste cells. In effect, a network of peripheral nerve fibres connects the taste buds with one another. The front two-thirds of the tongue are innervated by the lingual (seventh cranial) nerve and the rear part by the glossopharyngeal (ninth cranial) nerve. Peripheral gustatory afferents reach the brain via the seventh, ninth and tenth cranial nerves. They synapse at the anterior end of the nucleus of the solitary tract.

Nowlis (1977), of the Rockefeller University, New York, in discussing the phylogeny of neural organization of taste draws particular attention to the pioneering work of C. Judson Herrick in the late nineteenth century. Examining the taste buds distributed on the outer surface of the body of many fishes, Herrick (1903) found that they were all connected to the brain stem via the seventh cranial nerve whether on lips, fins or the whisker-like barblets. This is the nerve that innervates the taste buds of the anterior two-thirds of the mammalian tongue. The taste buds on the fish's tongue, which is roughly homologous to the posterior third of the human tongue, are connected to the brain stem via the ninth cranial nerve. Herrick demonstrated in fish that all reflexes of intake had as their afferent limb the seventh cranial nerve. The primitive tongue of fishes, with taste buds innervated by the ninth nerve, is primarily used for rejection. In species adapted to bottom-feeding, mouthfuls of mud and debris are taken in along with potential food, most of which must be rejected. Taste buds on the tongue of reptiles and amphibians are usually also innervated by the ninth nerve.

In mammals there are no external taste buds so the decision on ingestion or rejection is made entirely by the superbly mobile tongue. Thus it is no accident that the anterior, phylogenetically new portion of the tongue has its taste buds innervated by the seventh nerve, with its history of sensitivity to substances appropriate for ingestion. The tip is most sensitive to sweet, and the area behind this most sensitive to salt. Still further to the rear, sour is more strongly represented and once the transition is made to the area innervated by the ninth nerve, bitterness predominates. Nowlis cites the data of Pfaffmann et al. (1967) comparing the integrated whole nerve activity from seventh and ninth cranial nerves in the rat (Fig. 10-2). With sodium chloride solutions, the activity in the ninth nerve begins to increase at a higher concentration

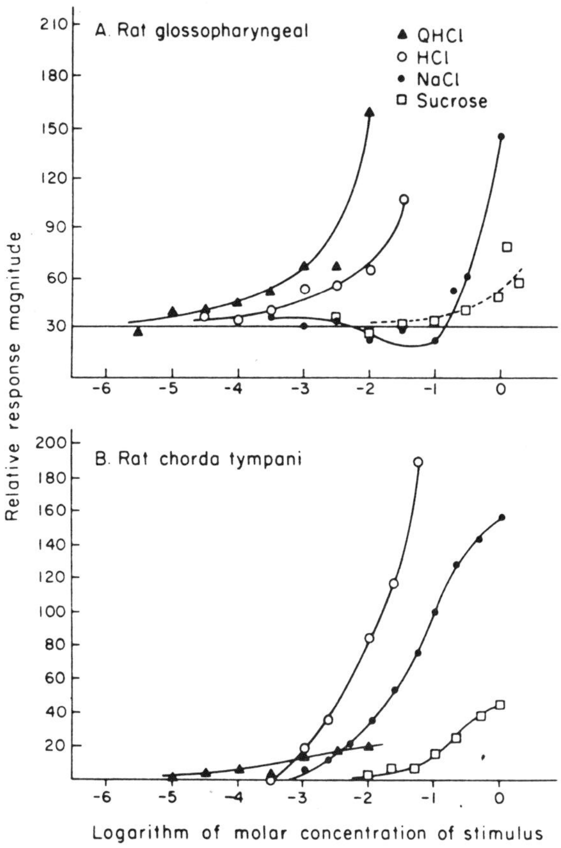

Fig. 10-2. Composite graph showing integrated whole nerve responses in arbitrary units from rat glossopharyngeal nerve (top half) and rat chorda tympani (lower half), as a function of concentration of solutions of quinine hydrochloride (bitter), hydrochloric acid (sour), sodium chloride (salty) and sucrose (sweet). (From Pfaffmann et al. 1967, with permission.)

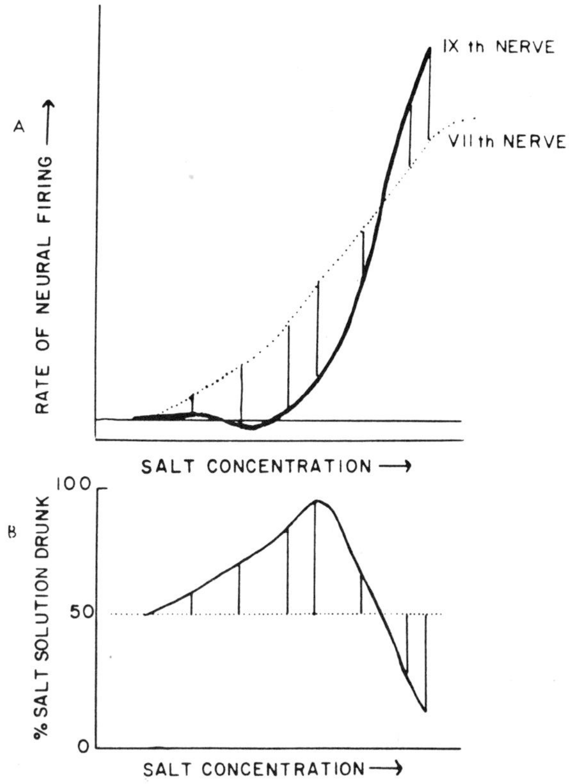

Fig. 10-3. Schematization of derivation of rat preference–aversion function for sodium chloride from neural activity functions. A, Integrated whole nerve responses in seventh and ninth nerves as a function of concentration of sodium chloride, replotted from data presented in Fig. 10-2. B, 'Preference–aversion' function for sodium chloride, derived by subtracting ninth nerve activity from seventh nerve activity at the points indicated by vertical lines. (After Nowlis 1977, with permission.)

than in the seventh but it increases at a steeper rate. At lower salt concentrations there is more seventh nerve activity and at higher more ninth nerve.

Fig. 10-3B illustrates the procedure of subtracting the activity curve of the ninth nerve from that of the seventh nerve. When the difference between the two curves is positive it is then plotted above the 50%, or neutral palatability level, as an increment in palatability. When the difference is negative, it is plotted below the 50% level, as a decrement in palatability. Such a theoretical curve bears a close resemblance to the preference–aversion curve for saline solutions determined behaviourally in the rat. However, Nowlis adds that obviously the behaviour of the intact animal is not as simple as such a construct would suggest, as cutting one or other of the nerves should then eliminate the positive or negative function, and this is not the case (Pfaffmann 1952). Most likely, learning of preferences, from experience of

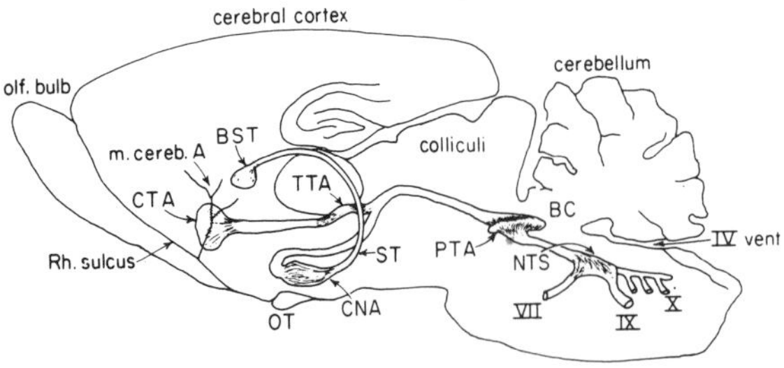

Fig. 10-4. Basic pathways of the central gustatory system superimposed on a parasagittal section of the rat brain. Abbreviations: BC, brachium conjunctivum; BST, bed nucleus of the stria terminalis; CNA, central nucleus of the amygdala; CTA, cortical taste area; IV vent, fourth ventricle; IX, glossopharyngeal nerve; LH, lateral hypothalamus; m. cereb. A, middle cerebral artery; NTS, nucleus of the solitary tract; olf. bulb, olfactory bulb; OT, optic tract; PBN, parabrachial nuclei; Rh.sulcus, rhinal sulcus; ST, stria terminalis; TTA, thalamic taste area; VII, intermediate (facial) nerve; X, vagus nerve. This figure shows the left side of a bilateral system which has decussations at the level of solitary and pontine nucleus and major crossover in the thalamus. The figure was kindly supplied by Dr Carl Pfaffmann. (From Norgren 1977, with permission.)

reflexive reactions, becomes a major element of behaviour.

Norgren and Pfaffmann (1975) and Norgren (1977) describe neurones in the rostral part of the nucleus of the solitary tract which project ipsilaterally through the lateral reticular formation into the dorsal pons (Fig. 10-4). Cells in the caudal parabrachial area, located between the principal and mesencephalic trigeminal nuclei, receive afferents from the solitary tract nucleus and respond to gustatory stimuli applied to the tongue. From the dorsal pons, axons of these third-order gustatory neurones ascend, again primarily ipsilaterally, via the central tegmental tract to synapse in the thalamic taste relay on the medial edge of the ventro-basal complex.

Norgren notes that in addition to synapsing in the thalamus, pontine gustatory axons ramify extensively in subcortical limbic structures. From the thalamus taste area they pass ventrally through subthalamus and dorsal lateral hypothalamus, possibly making functional connections, and penetrate the internal capsule. Rostral to the internal capsule the fascicles break into a maze of individual fibres which shift laterally through substantia innominata, and into the anterior amygdala. These pontine axons appear to end in the central nucleus of the amygdala. Some axons, however, communicate by the stria terminalis with another terminal field in the bed nucleus of the stria terminalis. The third-order gustatory neurones in the pons not only provide a basis for the thalamo-cortical elaboration of taste sense, but give a direct link to centres controlling ingestive behaviour. Pfaffmann et al. (1977) note that lesions involving the gustatory cortex raise taste thresholds under some circumstances and interfere with gustatory learning but do not affect preference behaviour. On the other hand injury to the amygdala disrupts feeding behaviour and taste preferences. They suggest that given the importance of the amygdala in modulating motivated behaviour, and its connections to the hypothalamus, they would tentatively assign discriminative functions to the thalamo-cortical gustatory system, hedonic and motivational aspects of gustatory stimulation to the ventral gustatory pathway.

Discrimination and psychophysics

Lord Adrian remarked that the physiology of sensation began by comparing the stimulus with the sensory experience in man (Adrian 1963). The problem is to decide how the events in the environment excite the receptors, and how the information is conveyed so as to show not only the intensity, time course and localization of the stimulus but the special characters which reveal its origin. The information must contain enough data to make the animal react differentially to stimuli which are of the same type and differ only in some specific quality.

Adrian pointed out that the taste receptors have a distribution which is half-way between that of the scattered receptors in the skin, and that of the concentrated sheets in the eye, ear and nose. In these latter three distance receptors the stimulus is focused on an extended sheet of receptors joined to the brain by thousands or millions of nerve fibres. With this sort of arrangement there are two possible ways in which the stimulus might give the necessary information about itself. One is by producing a particular spatial pattern of excitation over a receptor surface, with or without a particular temporal pattern showing the sequence and timing of excitation in different regions; or it might excite a particular class of receptor which is specially sensitive to a particular quality. In the ear, the spatial and temporal analysis seems to be enough. But with the eye there are special receptors to deal with colour vision in addition to the changing pattern of light and shade on the retina. From his own work on the olfactory sense of the rabbit, Adrian believed the part played by specific receptors in the olfactory discharge would give more information about the nature of the smell than could be gathered from the general distribution of the excitation pattern. Thus he saw a gradation across these three senses: the recognition of sound depends entirely on the general excitation pattern; with vision it depends mainly on pattern but is aided by specific colour receptors; and the recognition of smell depends mainly, perhaps almost entirely, on specific receptors for specific smells but may be aided by a characteristic pattern.

In the context of these considerations, and stimulated by the spectrum theory of taste (the evidence for which we will consider below), von Bekesy (1966) set out to determine, using a point electrode transmitting low-voltage stimuli, whether he could find any new taste qualities apart from the well-known four of sour, salt, bitter and sweet. However, the conclusion from experiments on the human tongue was that there were only these four taste qualities, which large-area stimulation showed to be distributed broadly as indicated by Fig. 10-5 (von Bekesy 1966). von Bekesy's (1964) results also indicated that monogustatory sweetness as evoked by electrical stimulation was distinct from the taste of sugar. (It was described by the subjects as 'heavenly sweet' and glycine was most like it.)

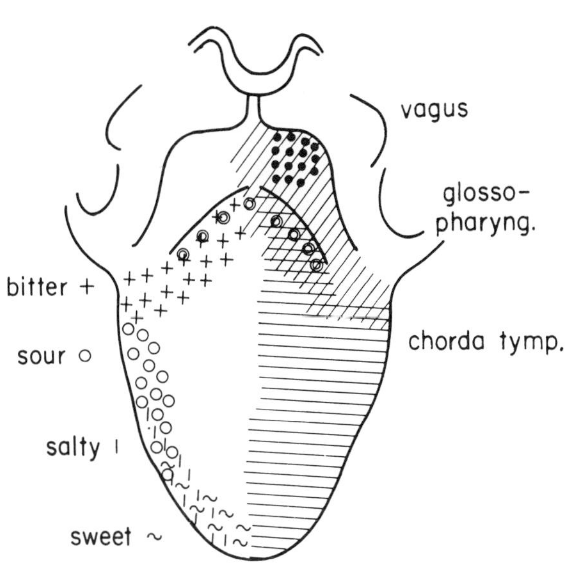

Fig. 10-5. The classical view of the distribution of taste sensitivity over the surface of the human tongue. (From von Bekesy 1966, with permission.)

Following this, he (von Bekesy 1966) went on to determine whether chemical stimulation of a single taste papilla evoked only a separate and distinct sensation of one of the four taste qualities. By means of delicate micromanipulations he managed by suction to elevate a single papilla and apply a discrete drop to it. Two thresholds were determined: the first threshold at which the subject was able to report a sensation; then the second threshold when the subject could tell the quality of the taste. Both were repeatedly tested. The quality of sensation was not changed by an increase in the concentration of the solution when a single papilla was stimulated. Most of the papillae, but not all, always showed exactly the same quality when stimulated around the whole wall. The larger papillae, especially in the soft palate, usually showed several taste qualities, in some cases as many as three. He further found that there was complete agreement in taste quality on each individual papilla for both chemical and electrical stimulation. There were a few papillae which were markedly insensitive. For example, in mapping the tongue of an experienced observer he identified 44 acid, 17 salt, 29 quinine, 13 sugar and 15 insensitive papillae. He concluded that the map produced by electrical stimulation and the map obtained by chemical stimulation of single taste papillae seemed to indicate that individual papillae have sensitivities for specific taste qualities.

Although he suggested that the pattern of nervous discharges does not play a role in the discrimination of basic monogustatory stimuli, he believed this pattern to be very important in the analysis of complex tastes. This followed from the fact that delays are produced between stimulation of the different kind of taste buds when a fluid moves about on the surface of the tongue that must result in a complex pattern of summations and inhibitions. The mobility of the tongue during tasting probably contributes to the development of such a time pattern and thus the analysis of taste (von Bekesy 1966). Taste papillae on the soft and hard palates probably play a part in the analysis, a consideration which came to von Bekesy's attention initially by the happy experience of observing the validated ability of expert wine-tasters to discriminate not only the year of vintage but the side of the hill on which a wine was grown.

Taste electrophysiology

Pfaffmann (1970) remarks that electrophysiological studies of single primary afferent taste neurones of animals uniformly agree that individual fibres very often have multiple sensitivities (Pfaffmann 1941; Cohen et al. 1955; Sato et al. 1969). That is, when one of the four basic taste stimuli—sodium chloride, quinine, hydrochloric acid or sucrose—is placed on the tongue surface, most fibres respond to several of these stimuli. For example, Sato et al. (1969) observed 48 taste units in the rat, 9 of which responded to only one kind of stimulus and 8 to two kinds. Such specific units barely responded to cooling of the tongue, and showed a low rate of spontaneous discharge. Nineteen units, on the other hand, responded to three kinds of stimuli and 12 to all four basic stimuli. Such non-specific units were sensitive to cooling and showed a high rate of spontaneous discharge. In the hamster, 5 units among 28 responded to only one kind of stimulus and 5 units to two. Fig. 10-6 shows a representative individual recording and Fig. 10-7 the profile of 50 chorda

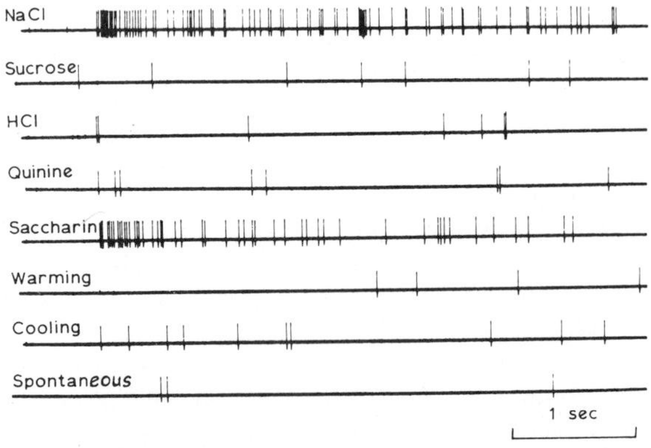

Fig. 10-6. Impulse discharges in a single chorda tympani nerve fibre of rat elicited by application to the tongue of 0.1 M sodium chloride, 0.5 M sucrose, 0.01 M hydrochloric acid, 0.02 M quinine hydrochloride, 0.02 M sodium saccharin, 40°C water and 20°C water. The bottom trace shows the spontaneous discharge. Temperature of the tongue was kept at 25°C throughout the experiment by rinsing it with 25°C water, but for warming, 40°C water was applied to the tongue at 20°C and for cooling, 20°C water was applied at 40°C. (From Ogawa et al. 1968, with permission.)

tympani fibres, both from the rat. Ogawa et al. (1968) and Sato et al. (1969) suggested that neural information for a quality depends on the pattern of finite discharges across many neurones, as proposed by Pfaffmann (1955) and Erickson (1967). Pfaffmann (1970) noted that most work on afferent fibres in mammals had used the chorda tympani, which responds well to sodium chloride or hydrochloric acid but is not very sensitive to sucrose and quinine. Therefore he and Frank studied sensitivity of single units from the glossopharyngeal (ninth cranial) nerve, which is more sensitive to quinine and sucrose. They found that fibres which responded to two of the four basic stimuli were the most common, then those which responded to only one stimulus, next those which responded to three stimuli; and those which responded to all four basic stimuli were least common.

Pfaffmann points out that such data do not indicate whether multiple sensitivity occurs because there is much branching as fibres enter the papilla on the tongue, as indicated by studies of

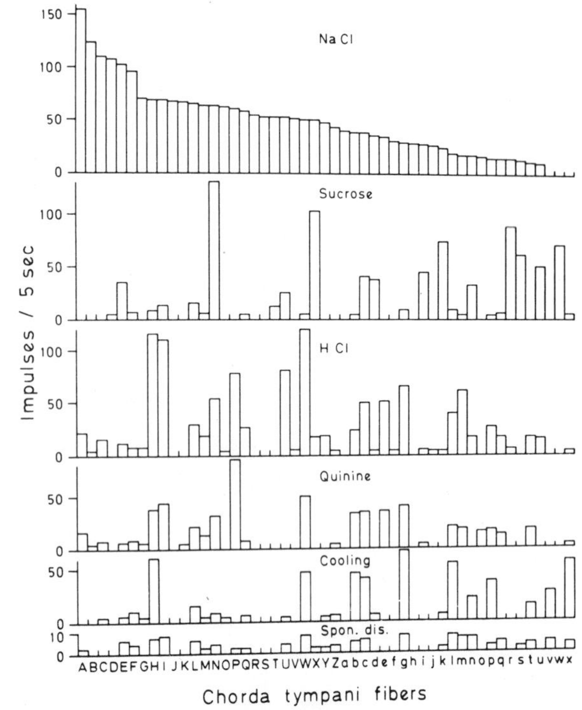

Fig. 10-7. Response profile of 50 chorda tympani fibres of rat (A,B,C, etc., to x), in which fibres were arranged in the order of responsiveness to 0.1 M sodium chloride. Stimuli were, from the top, 0.1 M sodium chloride, 0.5 M sucrose, 0.01 M hydrochloric acid, 0.02 M quinine hydrochloride, and cooling (20°C water to 40°C tongue). The bottom profile shows spontaneous discharge. (From Ogawa et al. 1968, with permission.)

neural innervation by Beidler (1969), or because single receptor cells in the papillae have multiple sensitivities, as shown by Kimura and Beidler's (1961) microelectrode recordings. Either mechanism could account for multiple sensitivity. Work by a number of investigators on the frog and the rat has indicated that several receptors are connected to a single afferent fibre with a mean of about five or six receptors per receptive field. Further, individual gustatory receptors are part of the receptor field of more than one afferent fibre. It is not yet known how these interact, or the nature of their excitatory and possible inhibitory relations. Zotterman (1970) agrees with Pfaffmann in that single-fibre work in monkey, pig and dog in his laboratory shows that 8%–12% of fibres were one quality.

A number of workers recognize that the basic problem in the field is that there is no information on single-fibre experiments in man. To what extent one can extrapolate from the animal data or compare them with those of von Bekesy in man, is a major problem.

Another important facet of taste stimulation has been illuminated by Bradley (1973). Working in

Beidler's laboratory he carried out electrophysiological investigations of intravascular taste. Responses were recorded from the chorda tympani while the tongue was perfused through an extracorporeal circuit. Sodium saccharin, sodium cyclamate, sodium chloride, sodium dehydrocholate and thiamine hydrochloride caused taste responses when added to the perfusion fluid. When stimuli of the same concentration were applied simultaneously to the tongue and intravascularly, the responses added. However, repetitive intravascular stimulation caused loss of responsiveness whereas the surface response in the same preparation was unaffected. It emerged that the stimulation route caused differences in responses attributable possibly to diffusion characteristerics. Thus glucose and sucrose were ineffectual intravascular stimuli compared with sodium chloride. The intravascular stimulus has to pass from the bloodstream to the sites of stimulation—putatively the membranes of the taste cells. Since the permeability of the capillary wall to sodium chloride is over twice that to glucose and almost four times that to sucrose (Landis and Pappenheimer 1963), the greater effect of sodium chloride may be so explained.

Bradley raises the general question of the taste bud as an interoceptor as well as an exteroceptor. He notes the tasting of drugs when injected intravenously. From this line of reasoning it might be questioned whether some 'taste in the mouth' changes experienced at times of illness may be direct sensory effects rather than due to passage of substances into the saliva. Bradley notes, in particular, that whereas increase in sodium chloride concentration relative to isotonicity led to increase in neural activity, a decrease depressed activity relative to baseline. The tongue surface responses to sodium chloride remained the same. A hypothesis is that a large amount of spontaneous activity of the chorda tympani is due to constant stimulation of the intravascular receptors by substances in the blood. Reduction of salt concentration is signalled to the brain and this acts as a stimulus to drive. However, in our experiments in sheep, cutting the lingual and glossopharyngeal nerves does not appear to diminish appetitive behaviour with sodium deficiency. Nor in the short term—i.e. over 10–15 min—does perfusion of the tongue with blood with sodium concentration increased by 15–30 mmol/l appear to alter salt appetite in the sodium-deficient sheep (Beilharz et al. 1965).

The effect of salt deprivation on taste sensitivity

This section might appropriately begin with Richter's original observations in 1939. Rats receiving the standard McCollum diet, and kept separately in cages, were given access to two graduated inverted bottles. In the beginning both bottles contained distilled water. Records were taken until the daily intake from each bottle became fairly constant. This rarely required more than 6–10 days. Then one bottle was filled with a sodium chloride solution of a salt taste concentration estimated to be subliminal. Each day thereafter, the salt concentration was increased in very small steps. It was found that with concentrations of 0.055% (1 part of salt to 2000 of water), or ca 10 mmol/l, normal rats began to drink more salted and less distilled water. This concentration was taken to be threshold. Adrenalectomized rats made this distinction with concentrations of 0.003% (1 part of salt to 33 000 of water), or about 1 mmol/l. To determine whether such a lowered preference threshold was the result of increased gustatory sensitivity, Pfaffmann and Bare (1950) compared the gustatory sensory threshold of normal and adrenalectomized rats by recording afferent nerve discharges elicited by sodium chloride applied to the tip of the rat's tongue. The weakest concentration required to produce the discharge of taste impulses was taken as a measure of the sensory threshold. It was shown that the thresholds were essentially the same in both the normal and the adrenalectomized animals.

Subsequently, Nachman and Pfaffman (1963) studied the response of the chorda tympani in sodium deficiency. In this case, both experimental and control animals were on a sodium-deficient diet but one group had sodium added to the diet. There was no difference between the afferent discharges from the chorda tympani when both groups were tested. This was so for sodium chloride at varying concentrations and also for other stimuli. The authors proposed that a central mechanism is affected by the sodium deficiency which causes the alterations in taste preferences they found. They pointed out that, in the rat, sodium chloride concentrations below electrophysiological threshold, in the sense that they do not produce an increase in discharge do, nevertheless, produce a smaller decrease in resting discharge than does pure water. They consider that this might be a basis of detection of threshold. It is interesting that Zotterman and colleagues (Cohen et al. 1955) have suggested that in some species the presence of 'water fibres' which respond with increasing intensity to sodium chloride solutions of decreasing concentration (0.03 M to 0 M) might extend saline sensitivity into the extreme hypotonic range. Pfaffmann (1963) notes, however, that the animals in all those experiments were surgically anaesthetized, so these observations do not rule out the possibility of afferent modulation by reticular influences or by other neural efferents. Pfaffmann and McBurney (Pfaffmann 1963) also showed the importance of the ambient salivary concentration in influencing the threshold of detection of sodium chloride. Rinsing the tongue in distilled water lowered the threshold.

There is an element of paradox in some implications of the theory of increased sensitivity of the peripheral receptor in sodium deficiency. As Richter has shown, and others confirmed, with the two-bottle preference–aversion test (Fig. 10-8) the

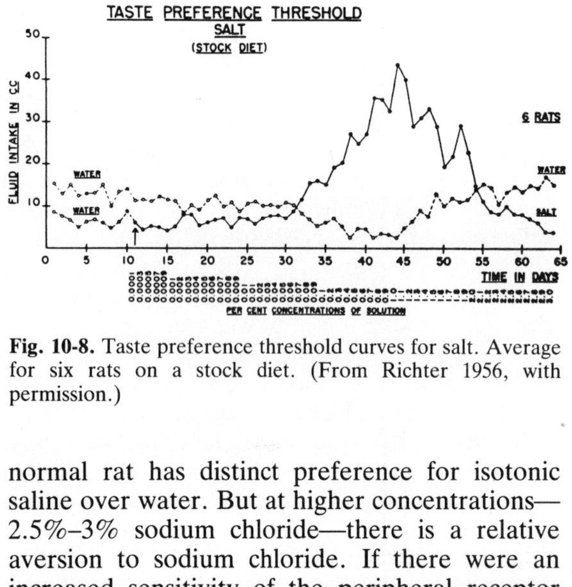

Fig. 10-8. Taste preference threshold curves for salt. Average for six rats on a stock diet. (From Richter 1956, with permission.)

normal rat has distinct preference for isotonic saline over water. But at higher concentrations—2.5%–3% sodium chloride—there is a relative aversion to sodium chloride. If there were an increased sensitivity of the peripheral receptor with sodium deficiency, one line of reasoning might propose that this might also amplify the aversive characteristics of the high concentrations. However, it is seen that the sodium-deficient rat (Nachman and Cole 1971) increases its voluntary intake of sodium over all concentrations including the high, previously aversive level.

Contreras (Contreras and Hatton 1975; Contreras 1977) notes that Bradley (1973) has shown that the concentration of sodium chloride in blood bathing the taste receptors influences the afferent gustatory input. This is the only study to substantiate the notion that the changed salt preference in sodium deficiency may be a result of changed neural response at the receptor level. Previous studies of sodium deficiency had used multifibre or whole-nerve recordings. Contreras studied the gustatory nerve discharges with single units during sodium deficiency in rats anaesthetized with urethane. He succeeded in isolating fibres which were 'salt-best' in the sense that they responded more to sodium chloride than to other stimuli. He also showed that the salt-best fibres were either broadly tuned in that they did respond to other stimuli, or narrowly tuned in that their response to salt was almost exclusive. Animals which were sodium-deprived for a period of 10 days differed from those which were sodium-replete only in their responses and response variance to 0.1 M sodium chloride, i.e. the distribution of responses to 0.5 M sucrose, 0.01 M hydrochloric acid or 0.02 M quinine hydrochloride of the two groups of fibres did not differ either in mean or variance. Analysis revealed that this difference resided in the fibres which gave a high salt response, defined as a response of over 75 impulses in 10 s. Sodium deprivation produced a reduction in chorda tympani fibre response to 0.1 M sodium chloride, primarily in salt-best fibres (see Fig. 10-9). Further, it was observed that the

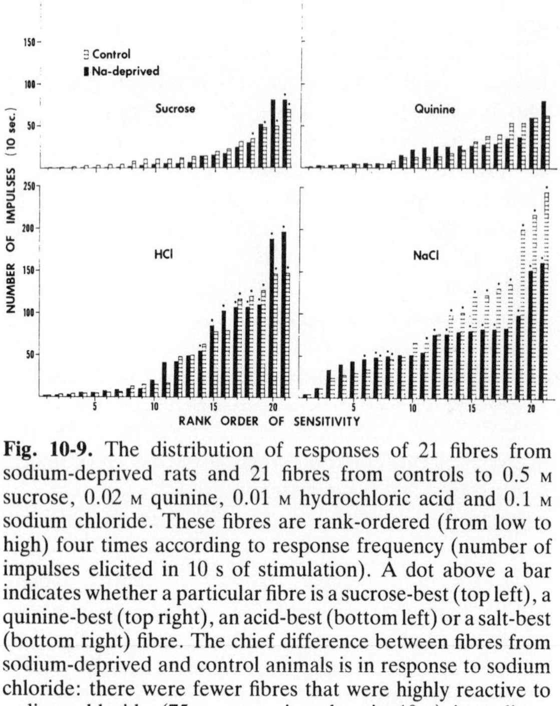

Fig. 10-9. The distribution of responses of 21 fibres from sodium-deprived rats and 21 fibres from controls to 0.5 M sucrose, 0.02 M quinine, 0.01 M hydrochloric acid and 0.1 M sodium chloride. These fibres are rank-ordered (from low to high) four times according to response frequency (number of impulses elicited in 10 s of stimulation). A dot above a bar indicates whether a particular fibre is a sucrose-best (top left), a quinine-best (top right), an acid-best (bottom left) or a salt-best (bottom right) fibre. The chief difference between fibres from sodium-deprived and control animals is in response to sodium chloride: there were fewer fibres that were highly reactive to sodium chloride (75 or more impulses in 10 s) in sodium-deprived animals. (From Contreras 1977, with permission.)

spontaneous firing rate, and also the response to water, were more variable in the fibres sampled from the sodium-deprived rats.

Contreras hypothesized that if satiety or cessation of drinking requires a critical total number of salt-driven impulses in the chorda tympani nerve, more time and more drinking would be required for the sodium-deprived rat to accumulate this critical total number of impulses (Contreras and Hatton 1975). Certainly these data, elaborated by Contreras and Frank (1977) at the Rockefeller University in New York and reported at the Sixth International Symposium on Olfaction and Taste, throw new light on the overall aspects of ingestion behaviour in the sodium-deprived animal. Whereas they would not appear to negate the essential central genesis of the sodium appetite drive, they could indicate an important component in increased intake in sodium deficiency. Salt-deficient animals would not be as likely to reject solutions at concentrations which are aversive to normal animals. In the light of the data of Pfaffmann et al. (1967) on integrated activity of seventh and ninth nerves (Fig. 10-2), detailed comparison of seventh and ninth nerves in sodium deficiency would be of great interest.

In a later paper, Contreras and Frank (1979) elaborated this study with data on whole nerves and single fibres. Two measures of whole-nerve activity were taken from their strip chart records derived when fluid was passed over the anterior portion of the tongue. One was the maximum pen deflection above an estimated pre-stimulation level; this, was designated as the initial peak neural discharge. The other was the pen deflection above the same pre-stimulation level at approximately 10 s after the beginning of the response; this was designated the tonic neural discharge. The data in Fig. 10-10 show both peak and tonic responses for ten whole nerves tested for the primary modalities. In sodium-deprived rats the whole-nerve response to sodium chloride is smaller than in control rats.

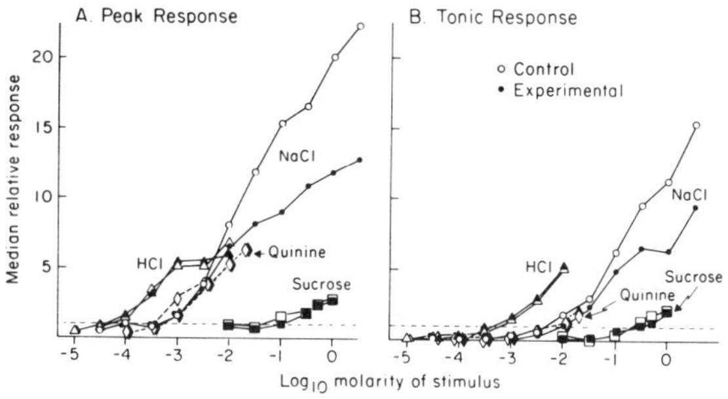

Fig. 10-10. The median responses of ten whole nerves of sodium-deprived (filled symbols) and control (open symbols) rats to a range of concentrations of sodium chloride, hydrochloric acid, sucrose and quinine. Two measures of the summed evoked activity of the nerve were taken: one (A) a measure of the peak discharge, the other (B) a measure of the tonic discharge. The numbers plotted are ratios of evoked responses to a reference level of activity during continuous water flow (indicated by a dashed line). Both peak and tonic responses to sodium chloride (0.03–3.0 M) are smaller after deprivation. (From Contreras and Frank 1979, with permission.)

The threshold concentrations of sodium chloride for peak and tonic responses do not change but the slopes and maximal responses to salt solutions are lower after sodium deprivation. The reduction is statistically reliable over a range of sodium chloride concentrations (0.03–3.0 M) for both initial peak and later tonic responses. The response functions for sucrose, hydrochloric acid or quinine did not differ between control and deprived rats in threshold, slope or maximum response generated. The differences in responses to 0.1 M potassium chloride and 0.1 M lithium chloride between control and sodium-deprived animals were not statistically significant on the basis of the amount of data collected.

In discussing this landmark study, Contreras and Frank note that it is as yet undetermined whether the same changes would occur in adrenalectomized rats with sodium loss. In a general discussion they note that, to humans, a sodium chloride solution applied to the tongue following adaptation to that stimulus is tasteless; both lower and higher concentrations have a taste, the former being bitter-sour but the latter salty. The intensity of the taste increases with deviation from the adapting concentration (Bartoshuk, McBurney and Pfaffmann 1964). They remark that one possibility why adrenalectomized rats prefer weak concentrations of saline is because they avoid the greater bitter-sour taste of plain water which could arise due to elevated salivary sodium levels. Adrenalectomy with its accompanying increase of sodium in the saliva would not reduce the recognition threshold of sodium chloride—it could only increase it. In contrast, sodium-deprived animals, with lower salivary sodium levels, would be predicted, if such levels have anything to do with salt preferences, to behave differently with solutions of lower sodium concentration. This possibility has not been formally tested. This question of salivary sodium concentration being determinant of salt appetite will be discussed in Chapter 24.

Concerning possibilities of sensory change after sodium deprivation, the authors note three. First, the change may be affected through the circulatory system changing either the immediate environment of the taste receptors themselves, their development or the transmission of activity to the afferent nerves innervating them. Secondly, the autonomic nervous system via efferent fibres in the chorda tympani could be determinant (Farbman and Hellekant 1976). This effect could be mediated by efferent impulses which influence receptors via salivary outflow or blood vessel activity. A third possibility could be related to the animal not having tasted sodium chloride for a long period of time, so that the receptor system might undergo some internal change simply through disuse. At present the explanation is not clear.

Contreras and Frank (1979) note that interpretation is set against a background of basic disagreement as to whether gustatory quality consists of only four different sensations—sweet, salty, sour and bitter—or whether there is a continuum of qualitatively different sensations. The former view proposes four different taste receptor systems each sensitive to stimuli of one quality and four quality-specific fibre types which relay information about that single taste quality to the central nervous system. The division of fibres according to the stimulus to which each unit is most sensitive is the neural basis for a labelled line system (Pfaffmann et al. 1976; Nowlis and Frank 1977). Others (Travers and Smith 1976; Erickson 1977; Scott 1977) propose that variability and sensitivity profiles of units of the taste system defy the grouping of units into four categories and that broad tuning of these units necessitates a 'population' or 'a cross neurone pattern' code for taste quality.

The large and consistent difference in taste neurones sampled from rats deprived of sodium was found in fibres highly sensitive to sodium chloride, many of which were salt-best units. According to the labelled line notion such units would code salt by the number of impulses evoked by given concentration of sodium chloride. This would be decreased by sodium deficiency, and, therefore, accordingly a change in the intensity of evoked saltiness would result. For example, if the deprived animal has 0.3 M sodium chloride on its tongue this evokes a signal in salt-best units which approximates the size of the signal evoked by 0.07 M in replete animals. If there is no change in the central neural interpretation of the intensity signal, the deprived animals would sense very strong sodium chloride solutions as relatively weak.

In correspondence which arose out of Dethier's article in relation to the fundamental basis of salty taste, Strong (1978) suggests that the simplest approach is to hypothesize that the taste of a compound is the sum of the tastes of its ions, and that there is evidence suggesting the saltiness of salt resides in the chloride ion. Dethier (1978) responds that there are many complex ionic and molecular interactions at the receptor membrane site, as for example various kinds of peripheral inhibition: mixtures of ions can cause synergism and inhibition. Peripheral neural interaction in the mammalian taste bud (cross innervation, multiple innervation) is so great that it is impossible to know what a particular ion is doing in how it tastes. Further, quality—sweet or salty—is a central phenomenon with high-order interneuronal integration of multifibre input. In insects, electrophysiological data show unambiguously that the receptor which responds to a solution of sodium chloride is responding to the sodium. What this tastes like is not known. In man nobody has succeeded in placing an electrode into a gustatory receptor cell (rather than the taste bud), and thus it is not known which ion it responds to, nor how many ions it responds to. Data suggest each receptor at a molecular level is a patchwork of molecular sites some of which respond to cations and others which respond to anions. But how all this is translated into the sensation 'salty' is unknown.

The increased acceptability of concentrated solutions to sodium-deprived adrenalectomized rats was shown in the brief access tests of Smith et al. (1969). Similar change has been shown by Bell (1963) and Bell and Williams (1959) with sodium-depleted calves. The findings of Contreras and Frank (1979) may also be indicative of the basis of the altered taste sensations which have been described as experienced in sodium deficiency in man (McCance 1936).

Insects and salt appetite

Dethier (1977) has written a most interesting and provocative review on salt taste and the reader is referred to the extensive bibliography in relation to the data on insects. He notes that the largest community of herbivores, both in terms of number of species and individuals, is the insects. Though they neither sweat nor dribble they are confronted with the problem of conserving the small amount of sodium chloride that vegetable food provides. Salt balance involves maintaining haemolymph constancy and is affected by rectal absorption. In general the haemolymph of insects is high in potassium. The concentration of sodium in fluids immediately surrounding neurones is higher than elsewhere. In the haemolymph of moths and butterflies potassium is the dominant cation while in the flies it is sodium. Herbivorous species often have a higher concentration of potassium, the main cation of plants, but its correlation with diet is not strict.

Dethier notes that the adaptive value of possession of salt receptors appears obvious, but the ubiquitous occurrence of such receptors, as well as their presence in animals that do not exhibit salt appetites, is something of a puzzle. They occur in herbivores and carnivores alike, in mammals, birds and fishes, in many freshwater and soil invertebrates as for example nematodes, and in all insects in which they have been sought. Among the insects they are well developed in herbivorous species such as caterpillars, grasshoppers and locusts, in blood-sucking flies such as mosquitoes, in nectar-feeders, and in carrion-feeders and omnivores such as houseflies, blowflies and cockroaches.

There is no evidence that insects ever suffer a salt deficiency in nature, nor do they exhibit unusual salt appetites when their food is low in sodium. The herbivorous insects do not differ behaviourally from carnivorous species in this respect. It has not been possible to induce a salt appetite in laboratory insects by rearing them on a salt-free diet, but few experiments have been attempted. The marine organisms may be an outstanding exception to the universal presence of salt receptors but few studies have been made. Further, no studies on oceanic birds and mammals that drink sea water have been reported. Some insects on normal diets do exhibit a response to dilute salt. This has been shown with cockroaches and houseflies but it has not been established whether it involves an aversion to distilled water as distinct from a preference to salt.

The anatomy of the gustatory system is less complex in insects than in mammals. All of the taste receptors are primary bipolar neurones; the dendrite is exposed to the stimulating solution and encoded information is transmitted directly to the central nervous system by the axon. Most recent evidence suggests that the tuning of insect receptors is somewhat similar to that of mammalian receptors and gustatory fibres. The water receptor exhibits the narrowest specificity; the other receptors have broader tuning. For example, the sugar receptor does respond to salt but is clearly specialized to respond to certain carbohydrates, and the salt receptor responds best to salt but also to amino acids, organic acids and glycosides. Present evidence on response patterns would suggest that, for example, the blowfly (*Phormia*) would be unable to distinguish between sodium and potassium chlorides.

Against the background of little selection pressure to distinguish between sodium and potassium salts, there is one striking behavioural exception. In the tropics some species of butterflies flock to puddles, the scenes resembling those of mammals going to waterholes. Various theories have been suggested, such as the acquisition of nutrients or salts. Though the reason is undetermined, it has been shown that the swallowtail butterfly can unerringly distinguish between sodium and other cations. Arms et al. (1973) found that when sand trays soaked with 0.1 M aqueous solutions of calcium chloride, sodium chloride, potassium chloride, magnesium chloride or sodium phosphate were put out, 'puddling' occurred specifically on the sodium salts. Dethier points to the interest in conducting an electrophysiological study of the tarsal chemoreceptors of swallowtails, particularly as Kusano (1965) found behavioural differences between the male and female in response to various salts applied to the tarsi. Females did not accept any salts, but sodium salts evoked acceptance in males, with lowest threshold for sodium chloride. The threshold for salts was decreased with starvation. Potassium chloride was ineffective.

Dethier raises the question of whether the salt receptor, rather than having evolved as part of a mechanism responsive to sodium, may be primarily involved in warding against hypersalinity and can be thought of as an evolutionary legacy from inherent irritability in all cells. Any neural element that salt can reach is likely to respond to it. Whether or not the response makes any behavioural sense could depend on the probability of salt impinging on the receptor in the normal course of events. Corneas of vertebrates have free nerve endings which respond to salt, a sensitivity advantageous as a warning device to forestall injury to the eye. In experimental situations vertebrates and invertebrates alike avoid concentrated saline solutions, and in nature animals are more likely to encounter salt in their drinking water than in their food. Since some of the cells most insensitive to salt are sugar receptors and, in insects, water receptors, Dethier suggests that the evolution of receptors with a blindness for salt may have been even more innovative than the evolution of receptors with an enhanced sensitivity to it.

Behavioural thresholds and preferences in salt deficiency

Though the analysis of the increased voluntary intake of salt which occurs in sodium deficiency and adrenalectomy will be made in Chapters 11, 12, 13 and 18, some observations, as they apply to threshold and taste sensitivity, are appropriate here.

Richter and MacLean (1939) found that the threshold at which humans first recognized salt taste was 11 mmol/l. This agreed closely with the preference threshold in normal rats. However, in the same series of experiments they showed that humans first distinguished a difference between distilled water and a salt solution at 1.5–3 mmol/l. Noting this difference between threshold of 'detection' and of 'recognition' they considered rats may well detect a difference at lower concentrations but do not prefer to drink the salt solution until the taste is recognizable. Since the electrophysiological threshold is not altered in adrenalectomized rats, it would appear that the sodium-deficient animal takes the salt solution as soon as it is detectable—a change which reflects a central nervous mechanism. That a change of motivation is involved is supported by conditioning experiments in rats. These were aimed at determining behaviourally the lowest concentration at which sodium chloride is detected from distilled water on the basis of the receipt of an electric shock when saline was chosen. Carr (1952) found this concentration to be very near the electrophysiological threshold, and not altered by adrenalectomy. Harriman and MacLeod (1953) found discrimination at a much lower level—one-tenth of the electrophysiological threshold. Attention was drawn above to the data of Nachman and Pfaffmann (1963) and Zotterman and colleagues (Cohen et al. 1955; Liljestrand and Zotterman 1956) as indicative of a possible basis of detection at lower concentrations. This is in the light of change of resting discharge, and of water fibre response in some species.

Yensen (1959) showed that salt deprivation in two normal humans reduced the taste threshold for salt to approximately 1 mmol/l, but did not alter the threshold for sour, sweet or bitter substances. No salt craving developed, but at meals the subjects felt the food would taste better with salt. Forced drinking of water, which resulted in negative sodium balance, lowered taste threshold and this was not associated with lowered salivary Na/K ratio (de Wardener and Herxheimer 1957). An intense craving for salt developed in the subjects.

In rats, the preference threshold for salt in a two-bottle preference test with water is decreased by sodium deficiency (Bare 1949; Young and Chaplin 1949). Richter (1956) did not record a clear-cut result in this regard in the course of studies on groups of rats on a low-sodium diet, a stock diet, and a high-sodium diet. He did, however, find that the concentration of maximum preference shifted under these conditions. The animals on a stock diet exhibited maximum choice of sodium chloride solution at isotonic concentration whereas those on a low-salt diet chose a maximum concentration at 1.2% sodium chloride. The rats on a high-salt diet had a maximum preference below isotonic. Fregly and Waters (1965) added the anti-thyroid drug propylthiouracil (PTU) to the food of rats. They demonstrated this to induce an increased sodium appetite reminiscent of adrenalectomy, and also to reduce the preference threshold for saline from 0.02–0.05 M to 0.003–0.001 M. Titlebaum, Falk and Mayer (1960) showed that diabetes insipidus caused the preference for saline to shift to concentrations below isotonic, while relative rejection began at regions where normal animals ingested maximally.

Kawamura et al. (1970) also found rats accept 1% sodium chloride solution in larger quantity than any other concentration and reject totally a 5% solution. After ablation of the cortical taste area bilaterally, rats accepted the latter concentration. They noted that nerve fibres which responded to 6% sodium chloride did not respond to noxious stimuli and suggested the aversion of the animal to this concentration was not dependent on pain sensation. Pfaffmann's (1952) study showed that combined removal of chorda tympani and glossopharyngeal nerves flattened the preference–aversion curve for saline. The fact that the aversion portion of the curve was flattened (i.e. more concentrated solutions were accepted) suggests that both parts are determined by taste stimulation *per se* and not by some other modality like pain (Weiner and Stellar 1951). In any event, pain sensitivity was left intact.

Specific effects of adrenal insufficiency on taste

A series of observations on sensory acuity in patients with adrenal insufficiency were made by Henkin, Gill and Bartter (1963), Henkin and Bartter (1966), Henkin et al. (1967) and Henkin (1969, 1970). They determined the median detection threshold of sodium chloride in normal and adrenally insufficient patients by asking them to distinguish between water and a solution of electrolyte. As shown in Fig. 10-11 the median detection threshold for normal volunteers was 12 mmol/l whereas for patients with untreated adrenal cortical insufficiency it was 0.1 mmol/l. That is, normal subjects required a solution approximately 100 times more concentrated in order to make the distinction. This dramatic change in acuity was seen also with other substances: other salts (potassium chloride and sodium bicarbonate) and also sucrose, urea and hydrochloric acid. Again, the threshold was dropped 100-fold by adrenal insufficiency.

Henkin and Solomon in one of their earlier papers noted that it was suggestive that this effect was not caused by changes in salivary sodium

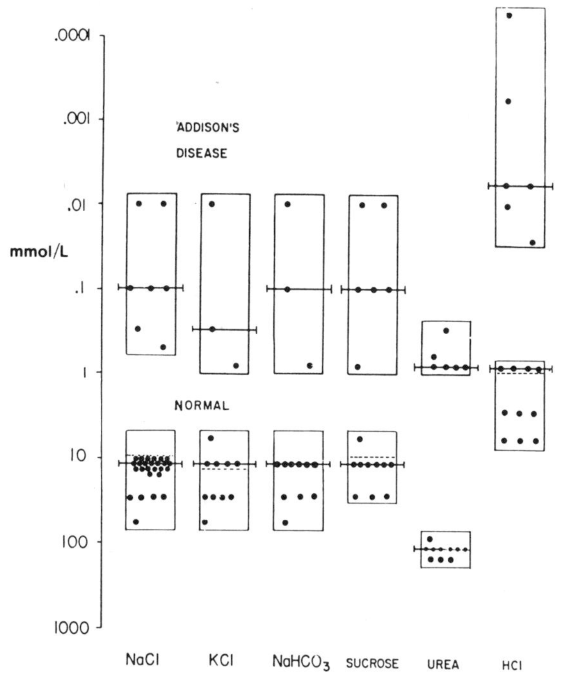

Fig. 10-11. Detection thresholds in normal volunteers compared with those in patients with untreated adrenal cortical insufficiency (Addison's disease) for representatives of each of four taste qualities. Lower filled circles, enclosures and lines represent individual detection thresholds, range of responses, and median detection thresholds, respectively, for the normal volunteers. The dashed lines within the lower enclosures indicate the detection thresholds determined by other investigators. Upper filled circles, enclosures and lines represent individual detection thresholds, range of responses, and median detection thresholds, respectively, in patients with untreated adrenal cortical insufficiency. The detection thresholds of the patients are significantly lower than those of the normal volunteers; there is also no overlap between the data obtained in each group. (From Henkin 1969, with permission.)

concentrations, since sodium deficiency and adrenal insufficiency cause opposite changes in salivary sodium but have the same effect on threshold. When adrenally insufficient patients were treated with 20 mg desoxycorticosterone acetate (DOCA) for 2–7 days, which was considerably greater than the amount needed for normal therapy, there was no return of the detection threshold to normal. Thus although this mineralocorticoid had a significant effect on the serum sodium and potassium concentration and the extracellular volume, it did not return the taste sensitivity to normal. However, treatment of the patients with carbohydrate-active steroids such as prednisolone did return the taste acuity to the normal range. This occurred within 24–48 h according to the route of administration. When steroids were withdrawn the change in taste acuity occurred over 3–4 days.

Coincident with the onset of decreased threshold of taste acuity with adrenal insufficiency there was a significant decrease in acuity of taste recognition. Patients called sodium chloride bitter at concentrations they called salty when on treatment. Recognition acuity was one-fifth to one-tenth normal. Henkin raises the question of whether this decreased acuity of recognition may be determinant of salt craving in adrenal insufficiency. Primary aldosteronism or administration of DOCA do not appear to influence taste acuity in man. Taste recognition acuity is significantly depressed in patients with Cushing's syndrome (hyperadrenocorticism), where inability to obtain normal taste from food is sometimes a presenting symptom.

Henkin (1969) proposes that there are variations in taste acuity in the normal individual, the maximum acuity being in the late evening hours. He finds the variations in normal subjects over 24 h may be between 60 and 0.5 mmol/l sodium chloride. He suggests these changes may be related to the circadian variation in adrenocortical levels. He also notes that these variations in taste are consistently greater for women than men, greater acuity occurring during the follicular phase of the menstrual cycle.

In the introduction to his consideration of the influence of adrenal cortical steroids on olfactory acuity Henkin (1969) states that his interest was initially aroused by seeing a film from our Institute in which a sodium-deficient sheep chose between a series of containers, one of which held sodium chloride solution, by a process of sniffing above them and then vigorously drinking the sodium chloride. 'After viewing the film I wondered whether patients with untreated adrenal cortical insufficiency had increases in olfactory detection acuity as well as in their taste detection acuity.'

In a test to determine the capacity of normal subjects to detect the vapour above a solution of sodium chloride it was found that many were unable to discriminate between water and sodium chloride solutions. At best some could distinguish between water and a solution of sodium chloride as concentrated as 150 mmol/l. Patients with untreated adrenal cortical insufficiency could detect, using olfactory clues alone, differences between water and solutions of sodium chloride one ten-thousandth as dilute as could normal subjects. The median detection threshold with the untreated patients was approximately one-hundredth the concentration they required to detect the difference between water and sodium chloride through the sense of taste. This effect was also seen with other salts dissolved in water, and there was increased olfactory acuity for other substances of an odorous nature. Acuity was restored to normal by the administration of carbohydrate-active steroid.

When 10 ml 10^{-4}M ^{24}NaCl was placed in one 1000-ml graduated cylinder and 10 ml of water placed into each of two other identical cylinders, a double-blind experiment showed that the adrenally insufficient patients could detect the difference easily. However, a Geiger–Müller counter placed over the top of the cylinder containing the ^{24}Na showed no difference over background activity. But when the experiment was repeated with a 10^{-4}M solution of Na^{36}Cl significant increase in radioactivity was recorded within 4 h after the radiolabelled material was placed in the cylinder. Henkin thus considers the vapour is probably composed largely of chlorine gas. Adrenally insufficient patients were also shown to have an increased range of frequency responsiveness to sound, and an increased nerve conduction velocity: that is, the effect is not specific to taste and smell but is some general change in neural excitability. Miller and Erickson (1966) have shown that rats also can detect sodium chloride solutions by smell at approximately the level of the taste detection threshold (3 mmol/l).

It is of interest here to draw attention to the experiments of Bell and Sly (1980) showing that adrenal-intact sodium-deficient steers can detect sodium- and lithium-containing solutions by smell. The steers were allowed access to a line of 12 identical buckets placed 6 m from a starting gate. The buckets had a cut-out lid so that they could be tasted only by pushing the muzzle through the hole. One bucket contained 6 litres of 0.5 M sodium bicarbonate, while the rest contained 6 litres of water. Normal sodium-replete calves never sought out a sodium solution, but the sodium-deficient calves found the sodium bicarbonate solution in 99% of trials with a mean time of 24±4 s. The sodium-depleted calves also located sodium chloride, lithium chloride and sodium acetate. Potassium bicarbonate was detected in 60±23% of tests in significantly longer time. Some animals were able to detect sodium bicarbonate at low concentration: 0.004–0.007 M. Steers made anosmic were unable to detect the solutions though they would work for sodium salts in an operant situation.

Species differences in taste behaviour, including birds

Differences in taste behaviour between species have been alluded to in various parts of the text above, and it is perhaps predictable that differences in response of chemoreceptors may occur. Beidler (1954) points out that ions are bound very weakly to receptors. The binding of various salts should be dependent on overall configurations of the receptor molecules, the available side chains, the proximity of neighbouring molecules, etc. It is clear that animals do not share man's taste world (Kare 1971). This is illustrated by the response to stimulants described by man as sweet, such as 10%–12% solutions of sucrose and the taste equivalent of saccharin. The laboratory rat avidly takes both of these solutions, the calf is much more sensitive to sucrose than is man but generally indifferent to saccharin, dogs like sucrose but many reject saccharin, and the cat is indifferent to both solutions. Beidler's (1961) consideration of the electrophysiological responses to various stimuli in a variety of animals indicates clear species differences, and with not a great deal of overlap in some areas. Kare (1971) notes that a broad body of data indicates distinct as well as indistinct boundaries between the taste worlds of different species, strains and even individuals. He suggests that if all species, or even all members of a single species, had an identical sense of taste, competition would be narrowed to a few food sources and the pressure would be enormous. There are definite ecological advantages to a diversity in taste perception in terms of exploiting the food available in an environment.

In relation to birds, a specific appetite for sodium chloride has been demonstrated with the domestic chicken. Further, chicks will delay drinking for extended periods in order to avoid consuming sodium chloride solution if the concentration exceeds the capacity of their excretory system (Kare and Biely 1948). Duncan (1960a, b) has shown a significant preference for weak sodium chloride solutions (0.043–0.085 M) by feral pigeons. Above isotonic, saline was 'rejected' when consumption was compared with water. Starlings, grackles and herring gulls failed to exhibit a preference for any of the concentrations of sodium chloride solution between 0.005 M and 4.0 M that were tested (in contrast to mammals). Rensch and Neunzig (1925) studied the rejection threshold for sodium chloride in 58 avian species and found that it ranged from 0.35% in the parrot to 3.75% in the siskin. The common tern has a high threshold for salt which has been associated with an intake of brackish water with food.

Beidler, Fishman and Hardiman (1955) studied whole chorda nerve and also some single fibres in various species. They found that the resting discharge was very small in the rat, guinea pig, hamster and dog; large in the rabbit; moderate in the cat. This spontaneous activity was reduced considerably in the rabbit when low concentrations of sodium chloride solution flowed over the tongue—albeit after a rapid burst of firing. The rabbit, dog and cat were not as responsive to salts as were the rat, hamster and guinea pig, either in terms of threshold or the ionic concentration required to give saturation level of response. If the carnivores (cat, dog and racoon) were compared with the rodents (rat, hamster and guinea pig), the lower sodium response of the carnivores is evident; response to potassium is uniformly low in both groups. The authors note that these differences correlate with the red cell Na/K ratio, the carnivores having cells with high sodium content. This taste difference as regards salt is also interesting in relation to a thesis put forward in earlier chapters: the absence of phylogenetic selection pressure on carnivores in relation to sodium balance. However, the rabbit does not fit their grouping on sodium response mentioned above, being a lagomorph not a carnivore.

Zotterman (1949) demonstrated specific fibres in the tongue of the frog that responded to water; similar fibres were found in the cat, dog and pig (Liljestrand and Zotterman 1954) and he thought that 'water taste' may be present in all animals. But he later (1956) showed the absence of such receptors in the rat, and that, in the rabbit, these responses to water were delayed. Bell and Kitchell (1966) found that in the ruminants—goat, sheep and calf—there are gustatory chemoreceptors which respond to the four taste modalities. They also found a response to sodium bicarbonate, which was best in calves, intermediate in the goat and slight in the sheep—species which, as will be recounted, may accept concentrated sodium bicarbonate solutions hedonically when sodium-replete. They found distilled water reduced the background activity in the isolated gustatory nerves, indicating the absence in these ungulates of receptors which respond to stimulation by distilled water. In the sodium bicarbonate preference test by the two-bottle technique in the goat, Bell (1963) found a preference for sodium bicarbonate over water that extended between 0.04% (5 mmol/l) and 1% (120 mmol/l). A 2% solution (240 mmol/l) was relatively aversive, though some intake did occur at highest concentrations (6%–8%). The daily administration of sodium bicarbonate by rumen tube (25 g or 50 g) proportionately reduced voluntary intake of the otherwise preferred 1% sodium bicarbonate to approximately 30% and 10% respectively. In a study of monozygotic twin calves, Bell and Williams (1959) showed a fairly close parallel in the behaviour of the twin calves separated and tested at different locations relative to the mean of the group. The data suggested that genetic factors determined the variation of taste behaviour of these animals. In the ruminants (goat, sheep and calf), the chorda tympani nerve was most responsive to salt and acid while the glossopharyngeal was more responsive to sugars, quinine and acid (Bell and Kitchell 1966).

Fig. 10–12 represents a study done in our

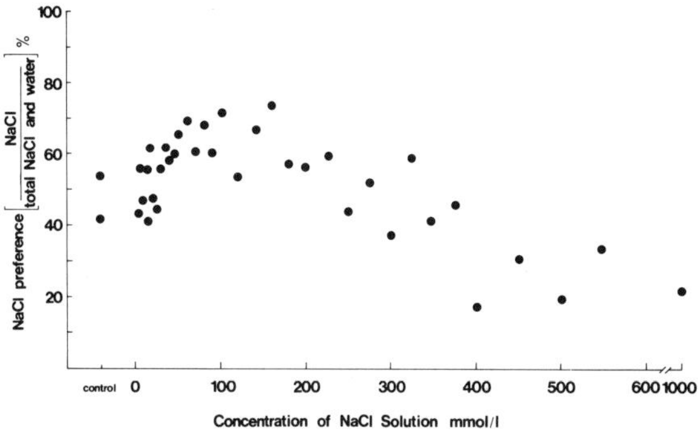

Fig. 10-12. Two-choice sodium chloride preference for two normal sheep offered various concentrations of salt solution (presented in ascending order) and tap water. Each concentration of sodium chloride was offered for two consecutive days and each mean value plotted is the total sodium chloride intake/total fluid intake expressed in per cent for the two days. During the control period water was available from the two bins. (From Abraham et al. 1976, with permission.)

Institute on the preference–aversion behaviour of two sheep to sodium chloride solution. The relative aversion at the higher concentrations (Abraham et al. 1976) and the preference at the lower are seen. Goatcher and Church (1970) carried out preference–aversion studies for a variety of substances on pygmy goats, normal goats, sheep and cattle. With relation to sodium chloride, all species were relatively aversive at 2.5% (425 mmol/l). In these studies pygmy goats showed a conspicuous preference for salt in the range 0.63% up to isotonic. Cattle showed none and, in fact, an aversion to 0.32% (55 mmol/l) and above, whereas their sheep showed no preference up to 1.25% (212 mmol/l). Sheep and cattle showed a pronounced aversion to quinine relative to goats (Goatcher and Church, 1970).

Bell and Sly (1979), in study of calves, showed considerable variation between animals with regard to ingestion of sodium salts. A clear difference was shown in preference for 1% sodium bicarbonate over water between weaned calves allowed access to sodium salts while at the milk feeding stage and weaned calves denied supplementary sodium salts while being milk-fed only. When tested with 1% sodium bicarbonate after weaning, the naive calves showed a strong preference for it and the sodium-exposed calves an indifference or rejection.

In relation to other species, Carpenter (1956) has shown that there is a strong sodium chloride preference in the rabbit, a slight one in the cat and no preference in the hamster. Porcupines in the laboratory are indifferent to various salt concentrations despite the attested avid salt appetite shown by feral porcupines in sodium-deficient areas (Bloom et al. 1973). Mice show a slight preference for sodium chloride, and a significant aversion when concentration reaches 0.33 M (Wolf and Lawrence 1963). C. Pfaffmann (personal communication) has remarked that Carpenter's observation has been followed up by Zucker et al. (1972) and Salter and Zucker (1974) who found that hamsters did not show enhanced salt preference after adrenalectomy. It was suggested that the hamster probably had accessory adrenal tissue but this was not established morphologically. The hamster also does not increase salt intake when treated with DOCA or after administration of thiazide diuretic (Wong and Jones 1978). The hamster requires little water and the question arises of whether these findings could reflect its desert phylogeny.

An interesting ecological aspect of aversion to high salt is represented in the behaviour of *Dipodomys microps*, the kangaroo rat of Western North America (Kenagy 1972), which eats the leaves of the halophyte shrub *Atriplex confertifolia* (saltbush). The animal has evolved lower incisors which are broad and flattened anteriorly and in winter it uses these to shave the epidermis off the leaves before eating them. Analysis has shown that in winter the epidermis is 32.2% water with a sodium content of 4261 mmol/l, compared with 250 mmol/l in spring. At this time of year the leaves are eaten *in toto*. The author draws attention to the fact that in the Sahara, a gerbelline rodent (*Psammomys obesus*) also feeds on saltbush. But it differs from the kangaroo rat in that it apparently consumes the leaves *in toto* and produces a much more concentrated urine (about 5000 mosmol/kg).

By virtue of the chance fact that the chorda tympani nerve runs through the middle ear and is often exposed in surgery, Borg et al. (Zotterman 1970) have had the opportunity to record from it at operation and have shown that it responds to the four modalities of taste. The application of water to the human tongue was followed by a reduction in spontaneous activity of the nerve in exactly the same fashion as previously found in the rat, which does not possess any taste fibres which respond to water. These authors also took the opportunity to compare the quantitative relation between stimulus and electrical response with the relation of stimulus to perception in psychophysical experiments performed on the same patients before operation. That is, the question at issue was whether the rectified, summated and filtered transformation of the activity of the whole chorda tympani is related to the intensities and qualities of taste sensations as we experience them. The electrical record and the psychophysical experiments closely approximated one another. It suggests that the recordings are a meaningful reflection of input.

Salt intake of sodium-replete rats

Some further general consideration of the phenomenon of salt intake by sodium-replete rats should precede analysis of behaviour on sodium deficiency.

In general terms, Young (1948) points out that intake as a criterion of appetite, as used by Richter, reveals differences in palatability as truly as it reveals differences in appetite and bodily need. An animal accepts what it likes as well as what it needs. It is an open question as to how far what it likes agrees with what it needs. Intake depends on a variety of complex interrelated factors (Young 1967). The four groups of determinants known to influence the intake of foodstuffs—constitutional, intraorganic, environmental and acquired disposition—are noted in Chapter 11.

In relation to the characteristics of salt ingestion by rats on a diet with adequate sodium content, Richter (1956) has represented the parameters of the behaviour as shown in Fig. 10-13. A typical set of observations is shown in Fig. 10-8. Similar types of preference behaviour are shown towards other substances, for example sucrose and alcohol. Clearly this intake does not reflect any bodily

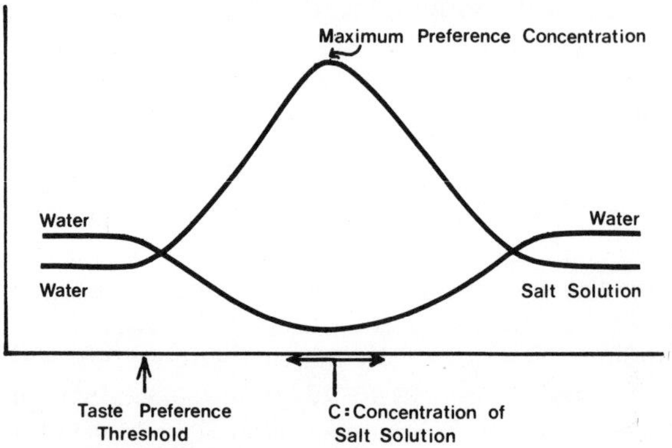

Fig. 10-13. Diagram of taste preference threshold curves. (Drawn from Richter, 1956.)

deficit but a liking for the substance (Young 1948).

Young (1948) has pointed out that in terms of the theory that intake indicates bodily need, one would expect the average daily intake of a substance to be approximately constant under the same set of intraorganic conditions. However, in the normal rat the quantity of fluid ingested varies markedly with concentration. Weiner and Stellar (1951) have also pointed out that the ingestion of some constant amount of salt is not the goal of animals studied in relation to salt preference. Their experimental technique also revealed the important fact that the rats drink the preferred concentrations much more rapidly. This was evident immediately upon access. The preference behaviour, as originally revealed in experiments allowing 24–48 h of access, did not appear to depend crucially on any physiological effect of salt being absorbed from the gut, for it showed too soon after drinking began. Furthermore, salt preference did not depend on previous experience of salt solutions as it appeared too early, naive rats drinking more rapidly from preferred solutions than from tap water within 5 min of their first experience of salt solutions.

These workers agreed with Bare's (1949) formulation that stimulation of taste buds produces salt ingestion. He suggested that as stimulation increases to a given amount, the intake of salt solution increases. Beyond this critical amount of stimulation, produced by concentrations of salt solution which correspond closely to the extracellular fluid salt concentration, further increase in stimulation results in a decrease in ingestion of salt solution. In relation to the biological importance of this behaviour Bare (1949) makes the interesting suggestion that perhaps the normal animal is protected against specific food deprivation by this unlearned stimulus–response connection. Whenever sodium chloride is available in the environment it is ingested, the amount depending on the strength of the gustatory stimulus.

Pfaffmann (1960) in his review of 'The pleasures of sensation' draws particular attention to the similarity of the preference–aversion curve to that of hedonic rating of sodium chloride solutions in man as determined by Engel (1928).

References

Abraham SF, Denton DA, Weisinger RS (1976) The specificity of the dipsogenic effect of angiotensin II. Pharmacol Biochem Behav 4: 363

Adrian ED, Lord (1963) Opening address. In Zotterman Y (ed) Proceedings of the First International Symposium on Olfaction and Taste. Pergamon Press, Oxford

Arms K, Feeny E, Lederhouse RC, (1973) Sodium: Stimulus for puddling behaviour by tiger swallow tail butterflies *Papillo glaucus*. Science 185: 372

Bare JK (1949) The specific hunger for sodium chloride in normal and adrenalectomized white rats. J Comp Physiol Psychol 42: 242

Bartoshuk LM, McBurney EH, Pfaffmann C (1964) Taste for sodium chloride solutions after adaptation to sodium chloride: Implications for the water taste. Science 143: 967

Beidler LM (1954) A theory of taste stimulation. J Gen Physiol 38: 133

Beidler LM (1961) Taste receptor stimulation. In: Progress in physics and biophysical chemistry, vol 12. Pergamon Press, Oxford, p 107

Beidler LM (1969) Innervation of rat fungiform papilla. In: Pfaffman C (ed) Proceedings of the Third International Symposium on Olfaction and Taste. Rockefeller University Press, New York, p 352

Beidler LM (1970) Physiological properties of mammalian taste receptors. In: Wolstenholme GEW, Knight J (eds) Taste and smell in vertebrates. Churchill, London (Ciba Foundation Symposium), p 51

Beidler LM (1971) Taste receptor stimulation with salts and acids. In: Beidler LM (ed) Handbook of sensory physiology, vol IV: Chemical senses, Sect 2: taste. Springer, Berlin Heidelberg New York, p 200

Beidler LM (1978) In: Carterette E, Friedman M (eds) Handbook of perception, vol VIA. Academic Press, New York

Beidler LM, Fishman IY, Hardiman CW (1955) Species differences in taste responses. Am J Physiol 181: 235

Beilharz S, Bott EA, Denton DA, Sabine JR, (1965) The effect of intracarotid infusions of 4M-NaCl on the sodium drinking of sheep with a parotid fistula. J Physiol (Lond) 178: 80

Bekesy G von (1964) Sweetness produced electrically on the tongue and its relation to taste theories. J Appl Physiol 19: 1105

Bekesy G von (1966) Taste theories and the chemical stimulation of single papillae. J Appl Physiol 21: 1

Bell FR (1963) Alkaline taste in goats assessed by the preference test technique. J Comp Physiol Psychol 56: 174

Bell FR, Kitchell RL (1966) Taste reception in the goat, sheep and calf. J Physiol (Lond) 183:145

Bell FR, Sly J (1979) Sodium appetite and sodium metabolism in ruminants assessed by preference tests. J Agric Sci 92: 5

Bell FR, Sly J (1980) Cattle can smell salt. J Physiol (Lond) 305: 68P

Bell FR, Williams HL (1959) Threshold values for taste in monozygotic twin calves. Nature 183: 345

Bell FR, Williams HL (1960) The effect of sodium depletion on the taste threshold of calves. J Physiol (Lond) 151: 42P

Bernard RA (1964) An electrophysiological study of taste reception in peripheral nerves of the calf. Am J Physiol 206: 827

Bloom JC, Rogers JG, Maller O (1973) Taste responses of the North American porcupine (*Erethizon dorsatum*) Physiol Behav 11: 95

Borg G, Diamant H, Zotterman Y (1970) Neural and perceptual responses to taste stimuli. In: Wolstenholme GW, Knight J (eds) Taste and smell in vertebrates. Churchill, London (Ciba Foundation Symposium), p 99

Bradley RM (1973) Electrophysiological investigations of intravascular taste using perfused rat tongue. Am J Physiol 224: 300

Carpenter JA (1956) Species differences in taste preference. J Comp Physiol Psychol 49: 139

Carr WJ (1952) The effect of adrenalectomy upon the NaCl taste threshold in rat. J Comp Physiol Psychol 45: 377

Cohen MJ, Hagiwara S, Zotterman Y (1955) The response spectrum of taste fibres in the cat: A single fibre analysis. Acta Physiol Scand 33: 316
Contreras RJ (1977) Changes in gustatory nerve discharges with sodium deficiency: A single unit analysis. Brain Res 121: 373
Contreras RJ, Frank M (1977) Gustatory nerve discharges in sodium deficient rats. In: Le Magnen J, MacLeod P (eds) Proceedings of the Sixth International Symposium on Olfaction and Taste. Information Retrieval, London, p 278
Contreras RJ, Frank M (1979) Sodium deprivation alters neural responses to gustatory stimuli. J Gen Physiol 73: 569
Contreras RJ, Hatton GI (1975) Gustatory adaptation as an explanation for dietary-induced sodium appetite. Physiol Behav 15: 569
Dethier VG (1977) The taste of salt. Am Sci 65: 744
Dethier VG (1978) A matter of taste. Am Sci 66: 278
Duncan CJ (1960a) The sense taste in birds. Ann Appl Biol 48: 409
Duncan CJ (1960b) Preference tests and the sense taste in the feral pigeon. Ann Behav 8: 54
Engel R (1928) Experimentelle Untersuchungen über die Abhängigkeit der Lust und Unlust von der Reizstärke beim Geschmackssinn. Arch Psychol (Frankf) 64: 1
Erickson RP (1967) Neural coding of taste quality. In: Kare MR, Maller O (eds) The chemical senses and nutrition. Johns Hopkins Press, Baltimore, p 313
Erickson RP (1977) The role of 'primaries' in taste research. In: Le Magnen J, MacLeod P (eds) Proceedings of the Sixth International Symposium on Olfaction and Taste. Information Retrieval, London, p 369
Farbman AI, Hellekant G (1976) Efferent chorda tympani fibres to fungiform papillae of tongue. Anat Rec 184: 400
Fregly MJ, Waters IW (1965) Effect of propylthiouracil on preference threshold of rats for NaCl solutions. Proc Soc Exp Biol Med 120: 637
Garcia J, Hankins WG (1975) The evolution of bitter and the acquisition of toxiphobia. In: Denton DA, Coghlan JP (eds) Proceedings of the Fifth International Symposium on Olfaction and Taste. Academic Press, New York, p 39
Goatcher WD, Church DC (1970) Taste responses in ruminants. III. Reactions of pygmy goats, normal goats, sheep and cattle to sucrose and sodium chloride. J Anim Sci 31: 364
Harper R (1972) Human senses in action. Churchill Livingstone, Edinburgh
Harriman AE, MacLeod RB (1953) Discriminative thresholds of salt for normal and adrenalectomized rats. Am J Psychol 66: 465
Henkin RI (1969) The metabolic regulation of taste acuity. In: Pfaffmann C (ed) Proceedings of the Third International Symposium on Olfaction and Taste. Rockefeller University Press, New York, p 574
Henkin RI (1970) Neuroendocrine control of sensation. In: Bosma JF (ed) Second Symposium on Oral Sensation and Perception. Thomas, Springfield, Ill
Henkin RI (1975) The role of adrenal corticosteroids in sensory processes. Handbook of physiology, Sect 7, endocrinology, vol VI: Adrenal gland. American Physiological Society, Washington, p 209
Henkin RI, Bartter FC (1966) Studies on olfactory thresholds in normal man and in patients with adrenal cortical insufficiency: The role of adrenal cortical steroids and of serum sodium concentration. J Clin Invest 45: 1631
Henkin RI, Gill JR, Jr, Bartter FC (1963) Studies on taste thresholds in normal man and in patients with adrenal cortical insufficiency: The role of adrenal cortical steroids and of serum sodium concentration. J Clin Invest 42: 727
Henkin RI, McGlore RE, Daly RL, Bartter FC (1967) Studies on auditory thresholds in normal man and patients with adrenal cortical insufficiency: The role of adrenal cortical steroids. J Clin Invest 46: 429
Herrick CJ (1903) The organ and sense of taste in fishes. Bull US Fisheries Comm for 1902
Howard E (1963) Effects of corticosterone on the developing brain. Fed Proc 22: 270
Kare M (1971) Comparative study of taste. In: Beidler LM (ed) The handbook of sensory physiology, vol 4, chemical senses 2, taste. Springer, Berlin
Kare M, Biely J (1948) The toxicity of sodium chloride and its relation to water intake in baby chicks. Poultry Sci 27: 751
Kawamura Y, Kasahara Y, Funakoshi M (1970) A possible brain mechanism for rejection behaviour to strong salt solution. Physiol Behav 5: 67
Kenagy GJ (1972) Saltbush leaves: Excision of hypersaline tissue by a kangaroo rat. Science 178: 1094
Kimura K, Beidler LM (1961) Microelectrode study of taste receptors of rat and hamster. J Cell Comp Physiol 58: 131
Kusano T (1965) Stimulating effectiveness of cations and anions on the tarsal chemoreceptors of the cabbage butterfly (*Pieris rapae crucivora*). Trans Tottori Soc Ag Sci 18: 1
Landis EM, Pappenheimer JR (1963) Exchange of substances through the capillary walls. In: Handbook of physiology, Sect 2: Circulation, vol II. American Physiological Society, Washington, p 961
Liljestrand G, Zotterman W (1954) The water taste in mammals. Acta Physiol Scand 32: 291
Liljestrand G, Zotterman W (1956) The alkaline taste. Acta Physiol Scand 35: 380
Le Magnen J (1967) Habits and food intake. In: Code CF (ed) Handbook of physiology, sect 6: alimentary canal, vol 1. American Physiological Society, Washington, p 11
McCance RA (1936) Medical problems in mineral metabolism. Lancet I: 643
McMurray TM, Snoden CT (1977) Sodium preferences and responses to sodium deficiency in rhesus monkeys. J Comp Physiol Psychol 5: 477
Miller SD, Erickson RP (1966) The odour of taste solutions. Physiol Behav 1: 145
Nachman M, Cole LP (1971) Role of taste in specific hungers. In: Beidler LM (ed) Handbook of sensory physiology, vol 4: chemical senses, Part 2: Taste. Springer, Berlin p 337
Nachman M, Pfaffmann C (1963) Gustatory nerve discharge in normal and sodium-deficient rats. J Comp Physiol Psychol 56: 1007
Norgren R (1977) Gustatory neuroanatomy. In: Le Magnen J, MacLeod P (eds) Proceedings of the Sixth International Symposium on Olfaction and Taste. Information Retrieval, London, p 225
Norgren R, Pfaffmann C (1975) The pontine taste area in the rat. Brain Res 91: 99
Nowlis GH (1977) From reflex to representation: Taste elicited tongue movements in the human newborn. In: Weiffenbach JM (ed) Taste and development—the genesis of sweet preference. US Dept Health Education and Welfare, Bethesda (DHEW publication no (NIH) 77–1068)
Nowlis GH, Frank M (1977) Qualities enhanced to taste: Behavioural neural evidence. In: Le Magnen J and MacLeod P (eds) Sixth International Symposium on Olfaction and Taste. Information Retrieval, London, p 241
Ogawa H, Sato M, Yamashita S (1968) Multiple sensitivity of chorda tympani fibres of the rat and hamster to gustatory and thermal stimuli. J Physiol (Lond) 199: 223
Pfaffmann C (1941) Gustatory afferent impulses. J Cell Comp Physiol 17: 243
Pfaffmann C (1952) Taste preference and aversion following lingual denervation. J Comp Physiol Psychol 45: 393
Pfaffmann C (1955) Gustatory nerve impulses in rat, cat and

rabbit. J Neurophysiol 18: 429

Pfaffmann C (1960) The pleasures of sensation. Psychol Rev 67: 253

Pfaffmann C (1963) Taste stimulation and preference behaviour. In: Zotterman Y (ed) Proceedings of the First International Symposium on Olfaction and Taste. Pergamon Press, Oxford, p 257

Pfaffmann C (1970) Physiological and behavioural processes of the sense of taste. In: Wolstenholme GEW, Knight J (eds) Taste and smell in vertebrates. Churchill, London (Ciba Foundation Symposium), p 31

Pfaffmann C (1975) Phylogenetic origins of sweet sensitivity. In: Denton DA, Coghlan JP (eds) Proceedings of the Fifth International Symposium on Olfaction and Taste. Academic Press, New York, p 3

Pfaffmann C, Bare JK (1950) Gustatory nerve discharges in normal and adrenalectomized rats. J Comp Physiol Psychol 43: 320

Pfaffmann C, Fisher GL, Frank ME (1967) The sensory and behavioural factors in taste preferences. In: Hayashi T (ed) Proceedings of the Second International Symposium on Olfaction and Taste. Pergamon, New York, p 361

Pfaffmann C, Frank M, Bartoshuk LM, Snell TC (1976) Coding gustatory information in the squirrel monkey chorda tympani. In: Sprague JM, Epstein AN (eds) Progr Psychobiol Physiol Psychol vol 6. Academic Press, New York, p 1

Pfaffmann C, Norgren R, Grill HJ (1977) Sensory affect and motivation. Ann NY Acad Sci 290: 18

Rensch B, Neunzig R (1925) Experimentelle Untersuchungen über den Geschmackssinn der Vogel 2. J Ornithol 73: 633

Richter CP (1939) Salt taste thresholds of normal and adrenalectomized rats. Endocrinology 24: 367

Richter CP (1954) The effects of domestication and selection on the behaviour of the Norway rat. J Natl Cancer Inst 15: 727

Richter CP (1956) Salt appetite of mammals: Its dependence on instinct and metabolism. In: L'instinct dans le comportement des animaux et de l'homme. Masson et Cie, Paris, p 577

Richter CP, Eckert JF (1938) Mineral metabolism of adrenalectomized rats studied by the appetite method. Endocrinology 22: 214

Richter CP, MacLean A (1939) Salt taste threshold of humans. Am J Physiol 126: 1

Salter P, Zucker I (1974) Absence of salt appetite in adrenalectomized hamsters. Behav Biol 10: 295

Sato M, Yamashita S, Ogawa H (1969) Afferent specificity in taste. In: Pfaffmann C (ed) Proceedings of the Third International Symposium on Olfaction and Taste. Rockefeller University Press, New York, p 470

Scott JR (1977) Information processing in the taste system. In: Le Magnen J, MacLeod P (eds) Proceedings of the Sixth International Symposium on Olfaction and Taste. Information Retrieval, London, p 249

Smith DF, Stricker EM, Morrison GR (1969) NaCl solution acceptability by sodium deficient rats. Physiol Behav 4: 239

Strong FC (1978) A matter of taste. Am Sci 66: 277

Titlebaum LF, Falk JL, Mayer J (1960) Altered acceptance and rejection of NaCl in rats with diabetes insipidus. Am J Physiol 199: 22

Travers JB, Smith DV (1976) Response properties of taste neurons in the nucleus tractus solitarius of the hamster. Neurosci Abstr 2: 165

Vinnikov YA (1965) Structural and cytochemical organization of receptor cells for the sense organs in the light of their functional evolution. Zh Evol Biokhim Fiziol 1: 67

Vinnikov YA (1975) The evolution of olfaction and taste. In: Denton D, Coghlan J (eds) Olfaction and taste V. Academic Press, New York

Wardener HE de, Herxheimer A (1957) The effect of a high water intake on salt consumption, taste thresholds and salivary secretion in man. J Physiol (Lond) 139: 53

Weiner IH, Stellar E (1951) Salt preference of the rat determined by a single-stimulus method. J Comp Physiol Psychol 44: 394

Wolf G, Lawrence GH (1963) Saline preference curve for mice: Lack of relationship to pigmentation. Nature 200: 1025

Wong R, Jones W (1978) Effects of aldactazide and DOCA injections on saline preference in Gerbils (*Meriones unguiculatus*). Behav Biol 23: 460

Yensen R (1959) Some factors affecting taste sensitivity in man. II. Depletion of body salt. Q J Exp Psychol 11: 230

Young PT (1948) Appetite, palatability and feeding habit: A critical review. Psychol Bull 45: 289

Young PT (1967) Palatability: The hedonic response to foodstuffs. In: Code CF (ed) Handbook of physiology, Sect 6: alimentary canal, vol 1. American Physiological Society, Washington, p 353

Young PT, Chaplin JP (1949) Studies of food preference, appetite, and dietary habit. 10. Preferences of adrenalectomized rats for salt solutions of different concentrations. Comp Psychol Monogr 19: 45

Zotterman Y (1949) The response of the frog's taste fibres to the application of pure water. Acta Physiol Scand 18: 181

Zotterman Y (1956) Species differences in water taste. Acta Physiol Scand 37: 60

Zotterman Y (1970) Physiological and behavioural processes of the sense of taste. In: Wolstenholme GEW, Knight J (eds) Taste and smell in vertebrates. Churchill, London, p 46 (Ciba Foundation Symposium)

Zucker I, Wade G, Zeigler R (1972) Sexual and hormonal influences in eating taste preferences and body weight of hamsters. Physiol Behav 8: 101

11 The evidence that salt appetite induced by sodium deficiency is instinctive

Summary

1. Salt appetite is innate and emerges upon an animal's first experience of sodium deficiency. This is strongly suggested by observations in the wild, and shown unequivocally by laboratory studies.

2. The appetite is specific for sodium. There is immediate acceptance of sodium salts by sodium-deficient rats long before any retroactive learned benefit is feasible. There is an innately determined relationship between degree of sodium deficiency and motivation to work to obtain sodium. The behaviour drive was observed even when gustatory satisfaction was not permitted and benefit consequent on absorption was avoided.

3. Experiments on naive sheep under cafeteria conditions where a parotid fistula was made and behaviour was observed on a minute to minute basis during one hour of access per day during 2–4 weeks revealed a delay of 1–6 days in development of appetite and showed the innate character of the drive. The data, showing all solutions were sampled before specific ingestion of sodium either immediately afterwards, or at a later time, emphasized the improbability of any retroactive association of benefit with a single episode of tasting which involved several different solutions. In the initial instance, intake of sodium was sometimes commensurate with body deficit. The data from differing experimental conditions showed rats and sheep have capacity to store information from prior experience, and when the innate system is first activated, associate motivation with that prior experience. It was also evident from the individual variation between animals that learning may play an important role in the development of behaviour of the animal in the face of sodium deficiency. This may be involved in the eventual capacity to rapidly choose sodium solution in a cafeteria, the precise adjustment of intake to deficit with frequent experience, and the choice of sodium bicarbonate rather than sodium chloride in the sheep losing sodium bicarbonate from a parotid fistula.

4. In animals with an established behaviour pattern of drinking salt, in response to continual loss from a fistula, intrarumen administration of sodium or surgical closure of a fistula may terminate or usually substantially reduces sodium appetite. Desoxycorticosterone acetate (DOCA) does this in the adrenalectomized rat.

5. Sodium appetite can be elicited by a number of different hormones in naive sodium-replete animals. There is little basis for the retroactive learned benefit proposal since the animals may sometimes become very ill as a result of sodium intake. Hormones activate the innate neural organization subserving sodium appetite.

6. Salt appetite is discussed in the broader context of constitutional determination of pattern of food ingestion. The question of ontogeny of genetically determined propensities is also noted in this context.

Introduction

I will not atempt any definition of instinct. It would be easy to show that several distinct mental actions are commonly embraced by this term; but everyone understands what is meant, when it is said that instinct impels the cuckoo to migrate and to lay her eggs in other birds' nests. An action, which we ourselves require experience to enable us to perform, when performed by an animal, more especially by a very young one, without experience, and when performed by many individuals in the same way, without their knowing for what purpose it is performed, is usually said to be instinctive. But I could show that none of these characters are universal. A little dose of judgement or reason, as Pierre Huber expresses it, often comes into play, even with animals low in the scale of nature.

(Charles Darwin, *Origin of Species*)

Another special group is formed by complicated organic reflexes (known in literature as 'instincts'). There lies at the basis of these reflexes an inherited biological tendency guaranteeing the life of the individual and species (reflexes of nutrition, reproduction, social reflexes, etc.). Yet one may suppose that the manifestation of these reflexes in many cases (especially in the superior stages of development) takes place under the guidance of the precedent individual experience. In other words, the instinctive actions are due not only to the innate but partly also to the acquired reflexes.

(Alexander L. Schniermann, 'Bekhṭerev's reflexological school'. In Murchison 1930)

For any one species the kinds of goals sought are characteristic and specific and all members of the species seek these goals independently of example and of prior experience of attainment of them though the course of action pursued in the course of striving towards the goal may vary much and may be profoundly modified by experience. We are justified then in inferring that each member of the species inherits the tendencies of the species to seek goals of these several types.

(W. McDougall, 'The hormic psychology'. In Murchison 1930)

There is good evidence that animals without previous experience may give specific reactions to biologically significant objects and that the recognition or discrimination of these objects may be quite precise.

(K. S. Lashley, 'Experimental analysis of instinctive behaviour', 1938).

In this chapter we come to consider the proposition that there is a substantial inbuilt neurophysiological system subserving behaviour which is activated upon the first experience of sodium deficiency. That is, salt appetite emerges spontaneously on the animal's very first experience of sodium deficiency. It is not an instance of the animal learning by trial and error that salt ingestion relieves the sensations induced by deficiency.

Wild animals

On the face of it, the diversity of observations on many species of wild animals makes it likely that the behaviour is innate. Similar predilections of, for example, elephant, moose, kangaroo, elk, wild rat and rabbit, are suggestive of an inbuilt system that is analogous to thirst or hunger. However, it cannot be excluded that learning by trial and error with retroactive sense of benefit and well-being is involved, and that the animals gradually learn the behaviour. It could also be socially transmitted in nature from mother to young and between members of gregarious species. This has been beautifully shown in the instance of washing potatoes in salt water by the Japanese monkeys, as recounted in Chapter 3. But it should be noted that this is hedonic behaviour transmitted in sodium-replete animals, and is not a response to body sodium deficit.

Nonetheless, the eating by wild mountain kangaroos of filter paper pulp blocks impregnated with sodium chloride (after a short delay for 'strange object' investigation) and not of other blocks is strongly suggestive. Myers (1970) has shown that in the rabbit warrens in the Australian alps over 65% of the animals are less than 12 months of age, i.e. could not have learnt the previous season. Here, placing out salt-

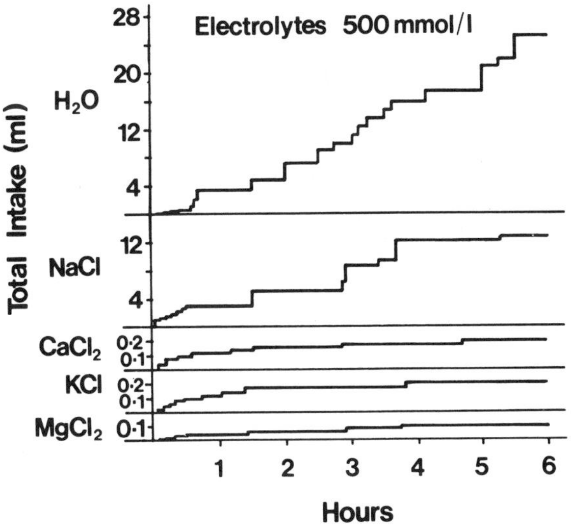

Fig. 11-1. Drinking behaviour study of a wild rat (*Rattus fuscipes*) captured in the sodium-deficient environment of the Snowy Mountains in southern Australia. On initial access to the 'drinkometer' cage with the variety of solutions, intake of sodium chloride occurred in the first few minutes and was predominant over other salts which were also tasted (note ordinate scale for sodium chloride is 20 times that for other solutions). (From Abraham et al. 1975, with permission.)

impregnated pegs may induce fairly immediate specific attack on the ones impregnated with sodium.

A perhaps somewhat more persuasive approach is represented by experiments where native wild rats (*Rattus fuscipes*) in the sodium-deficient Australian alpine regions have been trapped and placed in a 'drinkometer' where intake from a cafeteria of 500 mequiv/l solutions of sodium, potassium, magnesium and calcium chlorides and water could be recorded. The rats usually taste all solutions within a few minutes. But they show a large specific intake of sodium chloride (Fig 11-1: note ordinate scale for sodium chloride is 20 times that for other solutions). By 30 min, sodium equivalent to half the extracellular content was ingested. By 4 h, intake was 1.5 times the sodium content of extracellular fluid. This large specific and immediate intake by wild animals most likely, but not certainly, naive as far as salt ingestion experience is concerned, cannot be attributed to retroactive appreciation of benefit of one of the four substances tasted. Even if there were time for such a reaction before the obvious predominant sodium chloride intake, it would be difficult to see how the animals would recognize which of the four solutions tasted had conferred the sense of well-being.

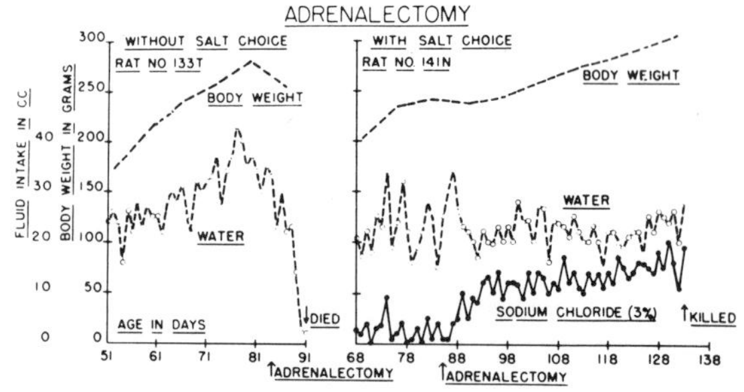

Fig. 11-2. Record of two adrenalectomized rats, one without and one with access to a 3% salt solution. Both rats were kept on low-salt diet. (From Richter 1956, with permission.)

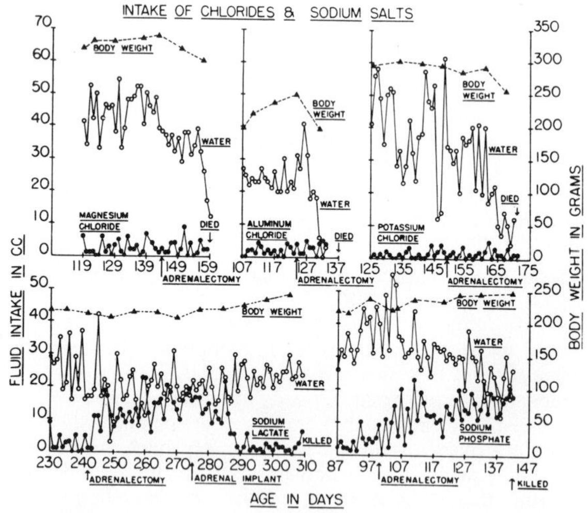

Fig. 11-3. Typical records showing appetite of adrenalectomized rats for three chloride and two sodium solutions. (From Richter 1956, with permission.)

Laboratory rats

Adrenalectomy

As well as suggestive field data instanced above and in Chapter 3, the question is evidently one for stringently controlled laboratory experiment. Richter (1942–43, 1956) in his pioneering work observed that young rats for whom the mother's milk had been the sole source of nourishment, consistently showed an active appetite for salt when confronted with salt for the first time. They took salt voluntarily and promptly, whether it was offered in crystalline form or in solution. Similarly, recently weaned rats which had been exposed to only a small amount of salt in their diet (1%), freely ate or drank salt under experimental conditions in which nothing else was available.

Richter and Eckert (1938) showed that adrenalectomized animals had an increased appetite for sodium salts (Fig. 11-2), including sodium lactate and sodium phosphate as well as sodium chloride (Fig. 11-3). The animals survived best with sodium chloride when the experiments were based on a single choice of the sodium solution and water. The intake of chloride solutions of minerals other than sodium was not changed by adrenalectomy. A small short-lived intake of potassium chloride was sometimes observed. The effect was seen clearly in multiple mineral choice experiments in which rats on a low-mineral food were given a choice of tap water and five different solutions (Fig. 11-4). Adrenalectomized animals on this regime survived the duration of the experiment, whereas the mortality of untreated animals was 100%. After adrenalectomy, sodium chloride intake increased the most, mean intake increasing from 2 mmol/day before adrenalectomy to 6.7 mmol/day in the 30–40 day period after adrenalectomy. The extracellular sodium content of these 200–300 g rats would have been approximately 6–9 mmol of sodium; therefore a substantial portion of the sodium content of the animal was turned over each day. In 38 animals the mean latency of onset of sodium appetite after adrenalectomy was 3.3 days, some animals showing the increase within 24 h or

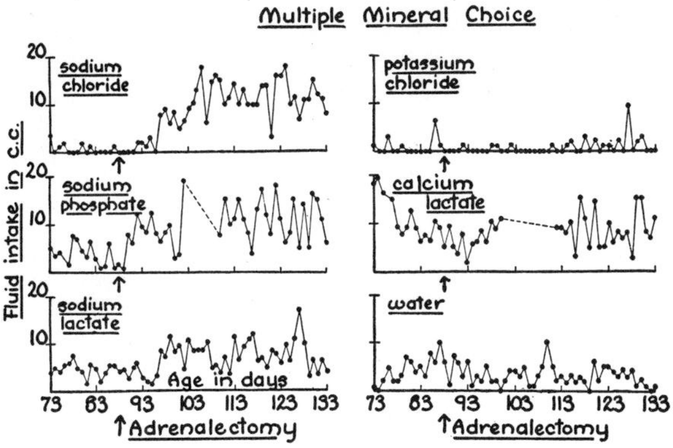

Fig. 11-4. Record from one animal showing effect of adrenalectomy on intake of water and five mineral solutions. (From Richter and Eckert 1938, with permission.)

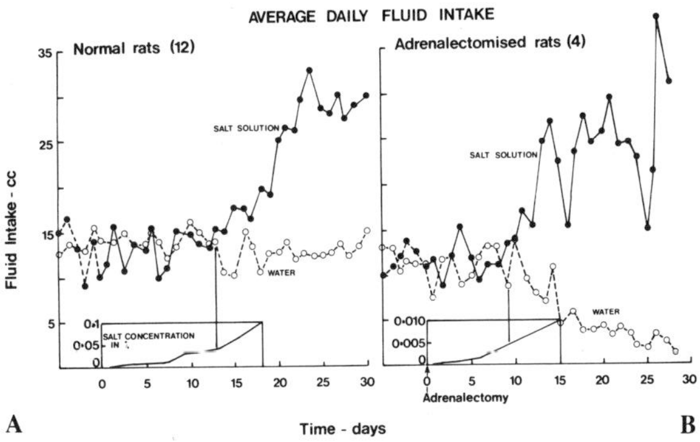

Fig. 11-5. Daily intake of water and salt solutions of increasing concentration for 12 normal rats (A) and for four adrenalectomized rats (B). Threshold concentration is designated by vertical lines. (After Richter, 1956.)

less after operation while others gave no indication until 5–9 days; maximum intake did not occur for 20 days or more. The increased appetite was completely reversed by daily injections of 0.5 mg of DOCA.

Studies of the adrenalectomized animals conducted by the preference method showed that there was a preference for the salt solution at far lower concentrations: 0.003% or 0.5 mmol/l (Fig. 11-5). Richter suggests that the preference for salt preceded the possibility of discovering the benefits of ingesting large amounts. The amount received per day at the lowest concentration (0.0005 g) was very small compared with the amount in food (0.1 g). This indicated that the preference was not based on trial and error resulting in relief of insufficiency symptoms, but on altered appetite.

A landmark investigation was made by Alan Epstein and Eliot Stellar in 1955. They aimed at delineating the influence of learning and taste, as distinct from the salt level of the internal environment, in the causation of increased salt appetite following adrenalectomy. First, they showed that experience of salt solutions was not a necessary condition for the development of the post-operative preference for salt by the adrenalectomized rat. The gradual increase in specific hunger for salt exhibited by the newly operated animal is a gradually increasing response to a developing salt need. When adrenalectomized animals were forced to suffer severe salt deprivation over 10 days before the first post-operative test, the initial intake of salt reached the exaggerated post-operative levels immediately. The animals took 6 ml of 3% sodium chloride offered during the first hour and the average total for the first day was 14 ml (Fig. 11-6). This is typical

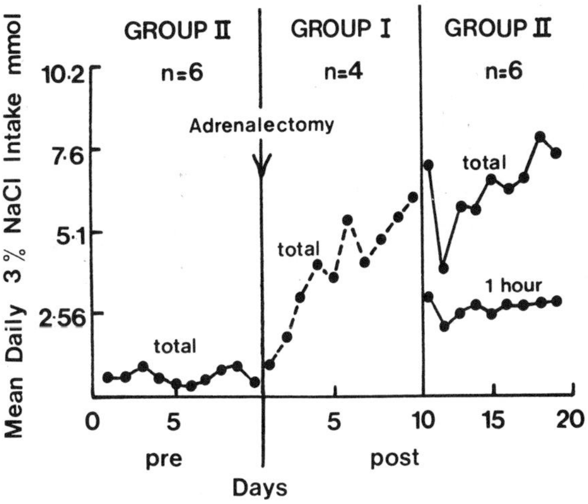

Fig. 11-6. The intake of 3% salt solution by two groups of adrenalectomized rats. Group I was given immediate post-operative access to salt solutions, group II was deprived of salt solutions post-operatively until severe salt need developed. (After Epstein and Stellar 1955.)

of the experienced adrenalectomized animal. Under conditions where they tested preference by the short-term access method used by Weiner and Stellar in 1951, the extent of sodium deficiency was shown to determine the height of the preference–aversion function. It also determined the shape of the curve by moving the maximum preference point towards higher concentration when need was greater. Finally, they examined the effect of pitting the amount of salt in the internal environment against taste stimulation in the adrenalectomized rat. They did this by adding a 10% ion exchange resin to the diet, which caused approximately half the sodium ingested to be exchanged with potassium and ammonium in the duodenum, so that only about half was absorbed. The rats had continuous access to water and 3% sodium solution. One day following addition of the resin, the intake of salt solution approximately doubled (Fig. 11-7). It thus appeared that the amount of salt supplied to the internal environment of the adrenalectomized rat is much more important in overall determination of motivation to drink a 3% sodium chloride solution than the taste stimulation provided by the solution. The authors note further that since the ion exchange did not take place until the duodenum was reached, it is clear that effects of salt solution in the stomach were overshadowed by the conditions of the internal environment. When the resin was given to normal sodium-replete animals, intake of 3% sodium chloride and also of 0.7% and 1.5% solutions was not affected. This was considered to confirm that intake under these conditions depends predominantly on taste stimulation and is unrelated to any body need acting upon the central nervous centres.

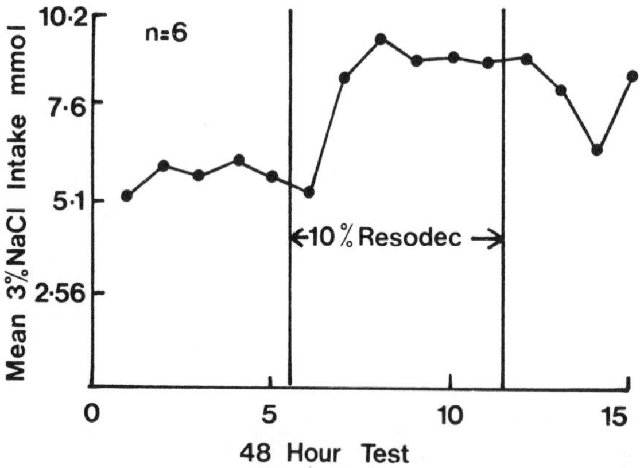

Fig. 11-7. The intake of 3% salt solution by six adrenalectomized rats under treatment with an ion exchange resin (Resodec). (After Epstein and Stellar 1955.)

Nachman (1962) commented on the importance of this paper. He did note, however, that the augmented intake 10 days after adrenalectomy was measured over a period of 1 h. Thus there may have been adequate time for learning to occur, since sodium chloride is relatively rapidly absorbed. Nachman showed that sodium-deplete adrenalectomized rats which had no previous experience of sodium solutions showed an immediate preference for these solutions. They were tested on the fifth day after adrenalectomy and the preference appeared within 1 min. Within 5 min an amount of sodium equivalent to half or more of the normal content of circulating plasma was drunk. The results clearly indicated that the behaviour was not an example of drive-reduction learning. In Nachman's view it was based on a change of taste mechanism, or change of the central reaction to an afferent discharge which was essentially similar to that in the normal sodium-replete animal. He also found that naive animals made salt-deplete by a deficient diet showed essentially the same behaviour. In relation to the animal's capacity to distinguish between various cations, under Nachman's experimental conditions involving short-term access, the only instance where the animals did not make a clear choice was between sodium salts and lithium salts. This could be consistent with their very similar chemical properties. He notes the data of Beidler et al. (1955) recording from the chorda tympani nerve and Fishman (1957) recording from single fibres within the chorda tympani. They found that lithium chloride and sodium chloride applied to the tongue of the rat result in similar neural responses.

At this stage, in the early 1960s, experiments in our laboratory were directed to the specific issue of precise observation of the individual naive animal at the first onset of sodium deficiency as produced by a parotid fistula. These were under cafeteria conditions of multiple choice where each element of the animal's behaviour was carefully recorded over an hour of access. However, we will defer discussion of these experiments to a later section of the chapter, in order to maintain some continuity in the chronology of the analysis in the rat.

In 1963 Pfaffmann summarized the state of knowledge, including his own results from a study of rats rendered sodium deficient by peritoneal dialysis, by emphasizing the following points:

1) Adrenalectomized animals show a preference for weak sodium chloride solutions at a concentration below that taken by normal animals and they take more of stronger sodium chloride solutions.
2) The preference for sodium salts in sodium-deprived animals is highly specific: sodium chloride is taken in preference to potassium, ammonium or calcium chlorides or other taste stimuli.
3) In many situations the preferential response to sodium chloride after deprivation is immediate. It appears to depend on some change in the taste, or response to taste of the stimulus, without prior learning.
4) Efforts to assess the locus or nature of these changes, i.e. whether they are in the peripheral receptor sensitivity or in the central neural mechanisms, point to a central process.

Pfaffmann made a very important point on the relation between taste sensitivity and preference behaviour motivated by salt deficiency. Changes in sensitivity, however induced, might increase the detectability of taste stimuli. However, there is no necessary relation between enhanced sensitivity and preference. It should be remembered that preferences in salt-needing animals are enhanced over the whole range of concentrations tested. Mere change in sensitivity should increase the apparent strength of the higher, but more aversive salt stimuli. It is just these solutions that are taken with increased avidity.

Behaviour study of naive rats upon initial experience of sodium deficiency

In 1965 Handal used the technique of subcutaneous formalin injection for rapid production of sodium deficiency in rats. The aim was to determine the minimum amount of time necessary for differential appetite for sodium to develop upon access to solutions. Following the formalin injection, which sequestrated body sodium and water at the injection site, the animals were washed to remove all salts from the surface of the their bodies. They were put in clean cages with distilled water and salt-free food ad libitum. The rats had no opportunity to taste salt from the time of injection to the time of testing, 24 h later. The test was a 'single stimulus' test, that is each rat was given only one solution which contained either a sodium or a non-sodium salt. Their intake was measured by a drinkometer. Fig. 11-8 shows the results of the experiment. The sodium-deplete rats given sodium salts drank rapidly and continuously, while those given non-sodium salts stopped drinking after a few licks. The difference between the two groups was statistically significant within 5 s after the commencement of drinking. A control group of non-deplete rats (shown in the bottom panel) rejected the strong sodium solution, which was accepted avidly by the deplete rats. The data are consistent with the view that the sodium-deficient rats immediately recognized sodium by taste. They ingested it without previous experience that such behaviour would relieve the deficiency.

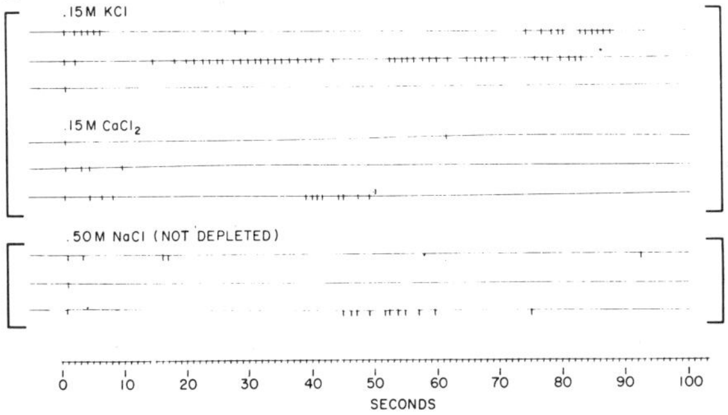

Fig. 11-8. Drinkometer records of individual rats. Every fifth lick is represented by a vertical mark. Records within the upper brackets are from nine sodium-deplete rats drinking solutions of sodium salts. Records within the middle brackets are from six sodium-deplete rats drinking solutions of non-sodium salts. Records within the lower brackets are from three non-deplete rats drinking a sodium salt solution. (From Handal 1965, with permission.)

In these and subsequent experiments to be recounted, there is the assumption that the experimental animals never experience sodium deficiency before the experimental depletion of body sodium. The animals used were adult male laboratory rats which had been fed standard rat chows ad libitum from the time of weaning. There is a surfeit of sodium in these foods in comparison with the minimal amounts necessary for growth, and whilst it cannot be certain that the rats never experienced a transient deficiency of body sodium, it is quite plausible that they never did. As Wolf (1969) points out in discussing this issue in relation to this and later experiments, rigorous verification would require continuous records of body sodium levels from infancy to the time of experimentation. Simply increasing the sodium content of the diet or otherwise administering supplementary sodium would not provide an unequivocal control. It still remains possible that relative decrease in the activity of the hormone system for sodium retention could occasionally result in sodium loss in excess of intake. Notwithstanding this unlikely consideration, it was known in this and the following experiments that the rats had never previously tasted sodium salts in isolation from other taste stimuli. This was because their only source of sodium was their laboratory chow and their excretions, both of which contained many other effective taste stimuli. Under such maintenance conditions, the only way the rat could specifically associate the sodium with a reduction of the need is by some innate mechanism which made it especially sensitive to the taste of sodium with onset of deficiency.

Quartermain et al. (1967) decided to examine whether the naive animal on first experience of sodium deficiency would carry out a learned arbitrary act to obtain access to sodium. This was in terms of studying the behaviour in a fashion contrasting with immediate free-access drinking which might be argued to be of substantially reflexive nature. The rats were pretrained in bar-pressing (by giving water as reinforcement when thirsty). Subsequently they were injected with various amounts of formalin to induce varying degrees of sodium depletion by sequestration. The design was similar to Handal's experiment in that they were prevented from tasting sodium before testing, being given only water and salt-free food. Twenty-four hours later they were put in a Skinner box where bar-presses now yielded a saline solution. It was found that the rate of pressing increased with the degree of sodium depletion (Fig. 11-9). This effect was statistically significant within 10 min after beginning pressing. At that time only a small amount of sodium (0.3 mmol) had been delivered, so that the differences between the groups could hardly have been due to repletion of body sodium. If a group of sodium-deplete rats were given water instead of saline under this regime, they did not continue to press the bar. The results of the experiment suggest that the behaviour of the sodium-deficient rats in ingesting sodium salts is not merely some form of reflexive conjunction of taste stimulus and adequate need. There is an innately determined relationship between the degree of deficiency and the motivation to work to obtain sodium.

Wolf, in commenting on this experiment, notes that whereas it does show the above relation, it does not clearly determine the degree to which increased propensity to work for sodium is due to increased internal drive stimulus or to the increased gratification from the taste of sodium.

Therefore, Krieckhaus and Wolf (1968) carried out an experiment to determine whether onset of sodium deficiency elicits a specific desire for sodium that precedes, and hence is not dependent upon, reinforcement from the taste of sodium. The experiment was designed to show that inexperienced sodium-deficient rats are motivated to obtain sodium before experiencing the effects of ingesting it while in the deficient state. Thus, six groups of rats which, on the basis of the reasoning outlined above, had never experienced sodium deficiency, were deprived of water. They were trained to press a bar to receive as reinforcement one of six aqueous solutions. Three of the solutions contained sodium salts (sodium chloride, sodium phosphate and sodium acetate), while the other three did not (potassium chloride, calcium chloride and pure water). After a few days of bar-press training, when all the rats in the groups were pressing at similar rates, the rats were injected with formalin to induce sodium depletion. As in the previous experiments, they were all washed and given water ad libitum until the next day, when they were put back in the Skinner boxes under extinction conditions (that is, bar-presses did not yield any reinforcement). The groups which had previously learned that bar-presses yield sodium salts, pressed the bar at a rate two or three times faster than the groups in the other conditions, even though they now received no reinforcement (Fig. 11-10). With two additional groups of rats trained with sodium chloride solution or with water reinforcement and then sham-injected to serve as non-depleted controls, the rate of pressing was very much lower than in those that had been trained with sodium and were sodium-deplete. Thus, the experiment showed that the motivated behaviour could not be attributed simply to an increase in the rewarding properties of sodium (for example, increased palatability). Sodium defi-

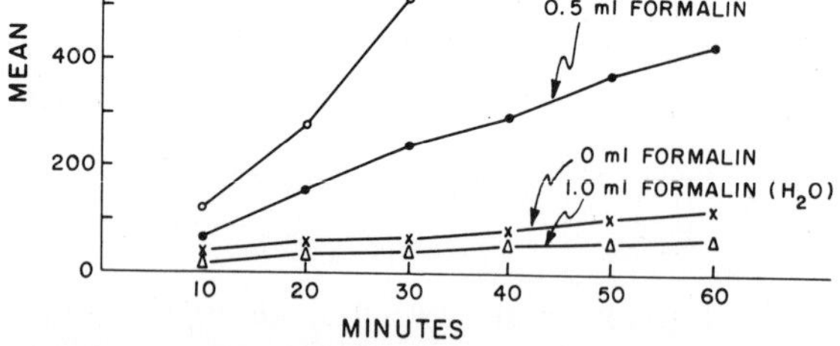

Fig. 11-9. Mean cumulative bar-presses for four groups of rats injected with 0 ml, 0.5 ml or 1.0 ml of 1.5% formalin and given 0.33 M sodium chloride solution as reinforcement, and for one group injected with 1.0 ml of 1.5% formalin and given water as reinforcement. There were four rats in each group. (From Quartermain et al. 1967, with permission.)

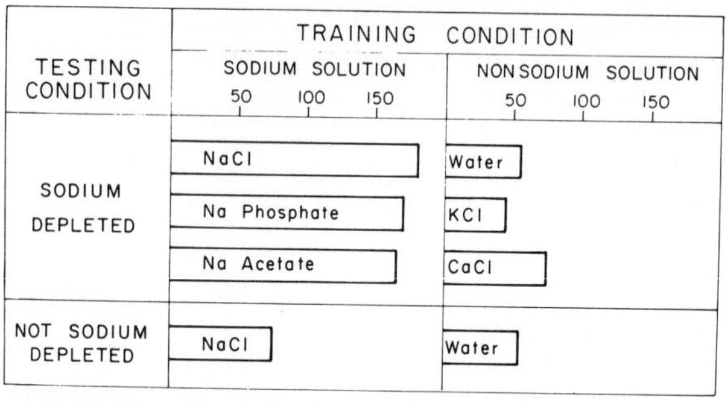

Fig. 11-10. Horizontal bars indicate mean number of bar-presses during a 1-h extinction session for various groups of 12 rats as a function of the solution used as reinforcement during training (denoted inside individual bars) and of sodium balance during testing. The salt solutions are all 0.15 M and sodium and non-sodium solutions are balanced for palatability. (From Wolf 1969, with permission.)

ciency elicited an internal drive state which guided behaviour to acquisition of sodium.

Wolf (1969), in commenting on the data to this point, notes that the fact that the motivational response to sodium deficiency appears to be mediated in large part by innate mechanisms does not necessarily imply the existence of a special neural system whose sole function is the regulation of sodium intake. While he thinks it likely that certain correlates of body sodium deficiency are monitored by a specialized receptor system, all that is necessary to account for the major phenomenon described above is some innate relationship between the coding of neural signals of body sodium deficiency and the coding of neural signals of the taste of sodium. To me, however, Wolf's proposition would imply a system of genetically determined interneuronal connections, with the cell mass specifically excited by sodium deficiency having a consequent irradiating influence on behaviour—though the cellular elements involved need not be solely dedicated to this response. But the existence of a special sense modality for salt taste seems to me consistent with the possibility of dedicated central organization, as appears to be the case with thirst and hunger.

With regard to the Krieckhaus and Wolf experiment, it is interesting to recall experiments of Falk and Herman working at Harvard in 1961. These involved rats running in a shuttle box where they were allowed access to fluid but no significant intake was possible. Depletion of body sodium by intraperitoneal dialysis reversed a previously established preference for water over 3% sodium chloride solution. The preference change occurred rapidly during the post-depletion test and operated in the absence of the post-ingestional consequence of fluid choice. The authors suggested that gustatory afferent signals initiated by sodium chloride solution activate some central mechanism previously sensitized by sodium depletion to produce directly the adjustive intake behaviour.

At the same time as the experiments conducted by George Wolf and colleagues were proceeding, Rodgers (1967a, b) of the University of Pennsylvania was studying the issue of whether or not needed substances have any special unlearned significance to deficient rats other than as novel stimuli. His important contribution to the elucidation of the appetites for vitamins, calcium, potassium and other substances will be discussed in detail in Chapter 25. The main point emerging, in his view, was that a number of hungers that were possibly specific were, in fact, learned, and that initial preferences were based on the novelty of the taste or smell of the particular substance. That is, the animals that had suffered a deficiency on a particular food preferred a new food with a different flavour from the one that had caused their sickness: or, as an alternative hypothesis, they had an aversion to the food which they had eaten during the period of becoming deficient and sick. If the substance they were deficient in was associated with the novel diet, they learned to prefer this diet. They did not make the switch back if the substance was then incorporated in the original diet. Though rats could, as it were, be fooled in this fashion in relation to vitamins, this was not the case when it came to a similar experiment with sodium deficiency. Here the addition of sodium to the originally sodium-free diet caused sodium-deficient rats to turn their attention back to the familiar diet upon which they had become deficient. In other words they had an unlearned preference for the taste of sodium that was much more than a preference for arbitrary novel tastes.

An ingenious experiment on the issue of whether or not learned benefit enters into the sodium appetite of the adrenalectomized rat was reported by Mook (1969). Rats were prepared with an oesophageal fistula, and a gastric cannula which was connected to a fluid reservoir operated by a solenoid valve—an electronic oesophagus. Thus oral and post-ingestional mechanisms could be dissociated and independently varied. Volume delivery to the stomach could be made equivalent to that which passed through the mouth. The principle is that of the Pavlov (1902) 'double fistula' preparation.

In Mook's experiment rats were allowed to drink from two cylinders ad libitum, one of which contained water the other 0.5 M sodium chloride. But whichever one was drunk from, only water was ingested into the stomach. Fig. 11-11 shows that, as is usual, before adrenalectomy the rat showed little

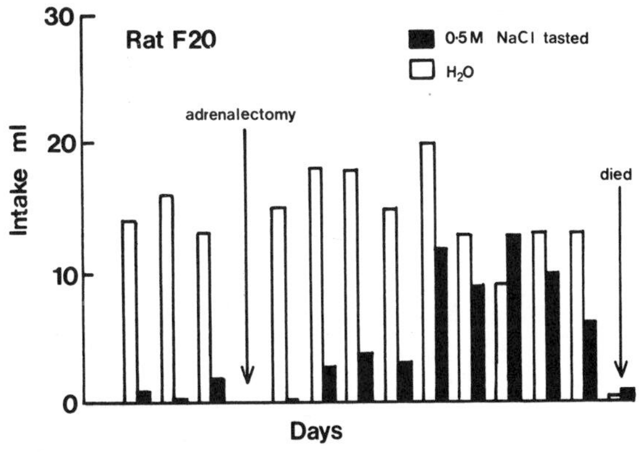

Fig. 11-11. Amount drunk per day of water and 0.5 M sodium chloride before and after adrenalectomy when only water enters the stomach during drinking of either substance. (After Mook 1969.)

drinking of the concentrated saline solution. After adrenalectomy it did more and more drinking from the saline cylinder until large quantities passed through the mouth (ca 6 mmol Na/day, equivalent to 70% of the sodium in the extracellular fluid). Since only water entered the stomach, on the tenth day the rat stopped drinking and died. Mook remarks that while it is well known that a rat will select dilute saline in preference to water in circumstances where that selection can depend only on taste, the data here show that sodium deficiency can extend the selection to a concentration of saline that is normally aversive, again in a situation where selection must depend on taste. The rat uses taste as a means of selection but what is selected can depend on the state of the internal environment.

The data add to the other strong evidence that an adrenalectomized rat does not learn to increase salt intake on the basis of the relief from deficiency symptoms which would follow saline ingestion. Such relief did not occur in these experiments. Thus increased ingestion of hypertonic saline must reflect an influence of the internal environment on the response to saline taste, independent of any post-ingestional consequences of salt drinking. Given that the history of these rats was such that no previous episodes of sodium deficiency with access to salt occurred, this experiment is most illuminating.

The drinking of lithium solutions in sodium deficiency

Cullen and Scarborough (1969) have shown using a bar-press method that rats which exhibited a marked pre-operative preference for sugar over salt, reversed this dramatically following adrenalectomy. If sham-adrenalectomized, or adrenalectomized and given a salt loading into the stomach, the animals retained their sugar preference until they were made sodium deficient. Cullen (1969) has also demonstrated that rats which received sublethal doses of gamma radiation while drinking sodium chloride showed attenuated saline consumption for a week after adrenalectomy. Frumkin (1971) studied the effect of lithium chloride poisoning on subsequent sodium chloride preference in groups of adrenalectomized and sham-adrenalectomized rats. Sham-operated animals exposed to lithium chloride consistently avoided sodium chloride whereas adrenalectomized animals did not. Thus although an aversion to sodium chloride can be conditioned in the adrenalectomized rat, it cannot be maintained in the face of increasing sodium deficiency (see also Smith et al. 1970). Rozin (1976) places considerable emphasis on the lithium data, since in view of lithium's rapid toxic action sodium-deficient rats cannot prefer it for its effects. Because lithium tastes like sodium, the innate mechanism can be triggered into making a dreadful mistake by consuming a poison that tastes like the built-in 'target' substance.

Stricker and Wilson (1970) made a different approach to the study of rats which were without previous experience of sodium deficiency. The rats were first trained to avoid salty-tasting fluids completely by ingestion being associated with aversive after-effects. In this case it was the after-effects of ingestion of lithium salts, which were learned after one trial. However, the rats resumed drinking salty solutions following acute sodium depletion by subcutaneous formalin injection. On the other hand, increased thirst elicited by injected hypertonic saline solution did not disrupt the learned aversion, nor did formalin treatment disrupt a learned aversion to a non-salty type of solution (e.g. saccharin). The results together with the results from other control groups suggested to them that sodium-deficient rats do not select salty fluids serendipitously. Instead, it would appear they have a strong drive to seek and ingest salt, and that this drive precedes the reinforcement associated with taste stimulation or post-ingestional sodium repletion. Under the conditions of their experiments, this motivation must have overridden the rats' memories of the aversive post-ingestional consequences of drinking salty fluids. Specifically, they showed that 22 of 24 rats drank lithium chloride, lithium carbonate or sodium chloride solution following formalin treatment, whereas only two of the rats had even approached the toxic lithium solutions after a few earlier experiences with them. It appeared clear that it was not an accidental intake as a result of the heightened drive state elicited by formalin treatment and thirst. Not one rat made thirsty by hypertonic saline solution treatment drank any of the 0.12 M lithium chloride solution.

The Krieckhaus experiments

An outstanding and definitive paper amongst these enquiries was published by Krieckhaus in 1970. He introduced it with a quotation from Proust which I will repeat here.

> But when from a long distant past nothing subsists, after the people are dead, after the things are broken and scattered, still, alone, more fragile, but with more vitality, more unsubstantial, more persistent, more faithful, the smell and taste of things remain poised a long time, like souls ready to remind us, waiting and hoping for their moment, among the ruins of all the

rest; and bear unfaltering, in the tiny and almost impalpable drop of their essence, the vast structure of recollection.

Krieckhaus remarks that thus did Proust describe most compellingly the unusual properties of gustatory and olfactory experiences in calling up memories from the distant past. He recalled the earlier experiments of Wolf and himself (Krieckhaus and Wolf 1968) where rats that had previously learned before any experience of sodium deficiency that lever-pressing would deliver salt, continued this activity when they received no salt solution as a reward. And he noted that this high level of lever-pressing in the experimental group could have resulted from the animals' need for sodium which, coupled with their not receiving the expected sodium, produced frustration. Thus, the increased rate of bar-pressing may represent an increase in activity produced by frustration rather than an attempt to procure sodium.

In a further series of experiments, thirsty rats with no need for sodium were taught for several weeks to run to one arm of a T-maze for distilled water and to the other arm for a much less acceptable hypertonic sodium chloride solution (0.33 M or greater). This solution would normally be aversive and also would have increased the animals' thirst; thus there was a strong relative aversion to the arm of the maze containing the sodium. The animals were then tested for which arm they preferred when not thirsty, with both solutions removed and under conditions of either sodium depletion or no sodium depletion. It emerged that there was a highly reliable tendency for the animals to run to the arm which had previously contained sodium more often when sodium-deplete than when not sodium-deplete. Animals which were trained in a similar manner with potassium chloride vs water showed no tendency when subsequently sodium-deplete to run more often to the arm which had contained potassium.

There is a crucial assumption in the argument that the phenomenon described here and in the previous study is not an example of latent learning. This is that the animals had never experienced a substantial sodium need that could be associated with a distinctive ingestive experience which reduced the need. If this assumption is correct, the rat appears capable of learning where and what sodium is, without at the time wanting sodium, and later, when it does want sodium and it is available, of knowing what it wants and attempting to procure it. Krieckhaus, in making this interpretation, draws a striking parallel with a description by Thorpe (1961) of an ethological experiment. Chaffinches which were exposed to their species-specific song while fledglings and thus not yet mature enough to sing were able, when mature, to sing a song much more closely resembling that of their species than were birds without this prior experience, even though at maturity they had no opportunity to hear a 'proper' song. Thus, in the language of the present study, birds can also innately recognize, learn, and store information (in this case concerning singing) and later innately recognize the meaning of their motivation (the song) and correctly associate this with their prior experiences. Overall, Krieckhaus sees his data as indicative that rats are capable of innately attending to or recognizing and remembering the meaning of a wide variety of internal and external stimuli.

Investigation of sheep

As mentioned earlier in this chapter, in 1962 we also began an investigation of sodium appetite in the naive sheep. The approach, however, was different. The data on the rat involve a statistical comparison of groups of animals under different regimes, while in the sheep experiments we carried out detailed studies on individual animals. Their behaviour was observed on a minute to minute basis during the time of access to electrolyte solutions. We wanted to follow exactly what the naive animal did on first experience of sodium deficiency in the hope that in certain ways this might provide some insight into the likely behaviour of the wild animal on first experience of deficit (though, of course, the basic element of the experiment, the cafeteria of solutions, is a rather harder immediate test of specificity than conditions likely to hold in nature).

The results of these experiments conducted between 1962 and 1965 with Elspeth Orchard and Sigrid Weller will be described fairly fully as the data have not been presented hitherto, apart from some brief reference with figures in review articles (Denton 1965, 1967), and no study like it exists for any other species.

The sheep were wethers and ewes of Merino and Corriedale breeds, 6 months to 3 years old. They came from a property of a member of our experimental group (E.O.) and were known never to have had access to bore (i.e. subterranean) water or to any form of salt lick. The grass they grazed had the low but adequate sodium content of 7 mmol/kg wet wt (ca 30 mmol Na/kg dry wt); potassium content was 132 mmol/kg wet wt. Nearby river water had <1 mmol/l of sodium. Thus, with the same reservations noted by Wolf

(1969) above, it was presumed the animals had never experienced sodium deficiency until our experimental procedure.

When brought to the laboratory, the animals were placed in individual stainless steel metabolism cages that allowed separate collection of urine, faeces and saliva (Denton 1956). They were fed on a mixed oaten and lucerne chaff diet, 0.8–1 kg/day, which contained 50–100 mmol Na/kg dry wt. They had free access to water.

There were three experimental groups:

Group A. Six animals (1 year old) were placed in metabolism cages 4–7 days before an operation to make a permanent unilateral parotid fistula. On the day following operation and thereafter, at the same time each day, they were allowed access for 1 h to a cafeteria of 300 mequiv/l solutions of sodium bicarbonate, sodium chloride, potassium chloride, calcium chloride and water. These were in galvanized bins, and were distributed around the front and two sides of the cage on the basis of a random number system. As noted above, the water bin, which was similar to the others, was on the cage always, but its position during the hour was determined by the random system. Galvanized sheets of metal acted as guards preventing access while the solutions were placed in position. At zero time all guards were removed and the animal's exact behaviour during the hour in relation to smelling, tasting and actual drinking was recorded. At the end of the access time the volume of each solution drunk was measured. Saliva and urine volumes were measured just before the time of presentation of the solutions. Though the observer was relatively unobtrusive at a distance, the sheep would have been aware of her; however, laboratory personnel normally passed between cages frequently and at all times of the day, so the sheep were habituated to human presence.

Group B. Nine animals, of which four were 6 months old and recently weaned and the other five were 1 year old, were placed in individual cages. The procedure was as for group A except that at the same time each day for 11–14 days before the parotid fistula operation, the animals were offered the cafeteria for an hour. This appeared to familiarize or accustom them to the procedure of guards and multiple bins. This guards regime did, as it turned out, appear to cause some initial apprehension in the animals, as the account of animals in group A will indicate. Though, with two exceptions, the animals drank no solution in the pre-period as determined by no discernible change of bin weight on a Mettler balance, there was quite a lot of tasting initially. In this respect, as it turned out, the circumstances bore some parallel to those of the later experiments of Krieckhaus (1970), in that these animals may have known what the tastes of sodium and other salts were when their first experience of sodium deficiency occurred.

Group C. A third group, included for comparison, consisted of two animals from a different property whose history was known to have included access to bore water pumped to drinking troughs, and two other animals whose history was not known. These four received regime B above. A few studies were made under regime B also, where lithium chloride was offered in the cafeteria in place of calcium chloride.

Group A: No access to electrolyte solutions before parotid fistula operation (Figs. 11-12 to 11-17)

Jasper (Fig. 11-12)

Weight, 23 kg; pre-operative sodium excretion ca 40 mmol/day.

Day 1: The five solutions were offered 24 h after the parotid fistula operation. The sheep had lost 1800 ml saliva containing 268 mmol sodium and 200 ml urine containing 3 mmol sodium, reflecting sodium depletion. On access to the solutions the sheep immediately tasted and drank NaCl for a few seconds. It then sniffed $NaHCO_3$, tasted NaCl, sniffed and tasted water twice and tasted $CaCl_2$. In the following minute it stood quietly, but in the 3rd minute drank water. It then stood quietly and ruminated. Total intake in 60 min: water, 770 ml; NaCl, 10 ml.

Day 2: 240 mmol sodium lost in saliva and urine (total for 2 days=511 mmol). Allowing that sodium intake in food less loss in faeces resulted in net input of ca 50–60 mmol/day, the negative balance at this point was ca 400 mmol. On access to the solutions, the sheep immediately sniffed and tasted $NaHCO_3$ and then NaCl. It drank NaCl over the next 4 min, tasting KCl once at the beginning of the 3rd minute. It then stood quietly for the rest of the hour, only sniffing water and $CaCl_2$ during the 16th minute. Intake of NaCl during the 60 min was 1690 ml (507 mmol). The fact that this corrected the deficit was reflected in the day 3 urinary sodium output.

In terms of the effect of a sudden influx of sodium adequate to correct deficiency in a sodium-deplete animal on the two indices of status—salivary Na/K ratio and urinary sodium excretion—the following facts have to be borne in mind. The collections of urine and saliva are at 24-h intervals, and the salivary loss is continuous. Thus, if at the beginning of the collection period

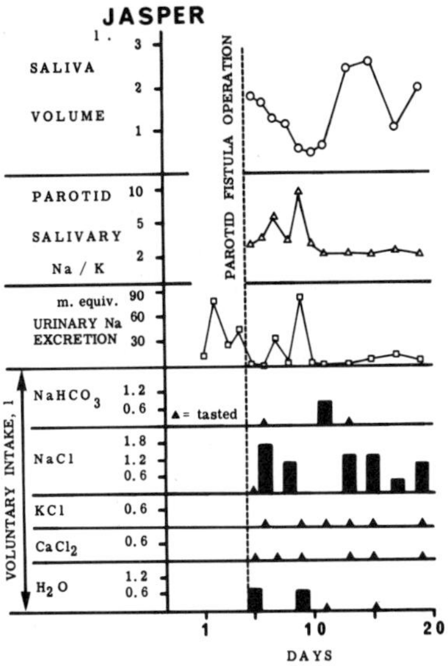

Fig. 11-12. Jasper: naive 1-year-old sheep. The voluntary intake (litres) during 1 h of access to a cafeteria of 300 mequiv/l solutions of sodium bicarbonate, sodium chloride, potassium chloride, calcium chloride and water offered each day following an operation to make a permanent unilateral parotid fistula is shown. Water was continuously available, but its position during the hour test was, as with the solutions, determined by a random numbers system. The saliva volume (litres), the parotid salivary Na/K ratio (normal ca 20) and urinary sodium output are shown also. With this animal, after 7 days the solutions were offered every 48 h, and collections of saliva and urine also made at this interval.

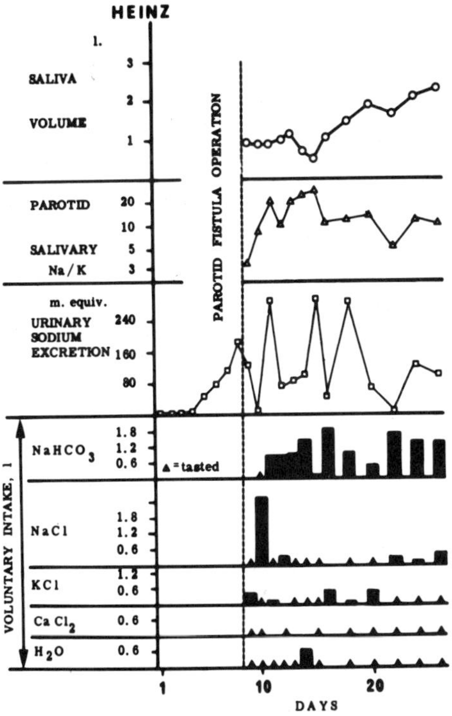

Fig. 11-13. Heinz: naive 1-year-old sheep. Regime as for Jasper. After 8 days tests were performed every 48 h.

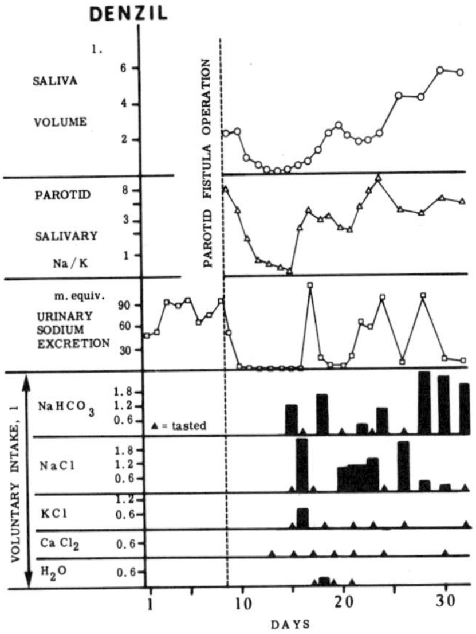

Fig. 11-14. Denzil: naive 1-year-old sheep. Regime as for Jasper. After 16 days tests were performed every 48 h.

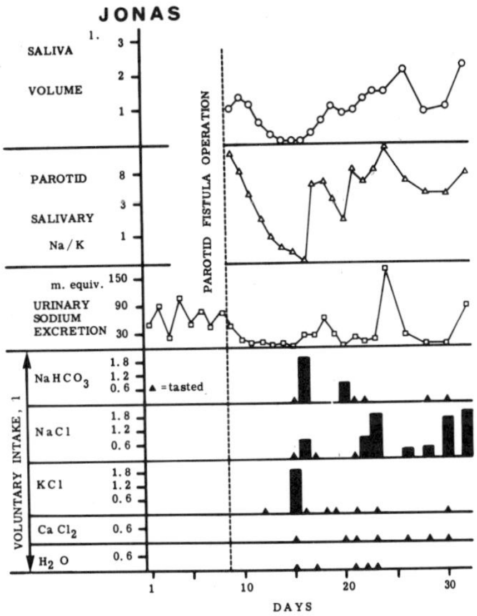

Fig. 11-15. Jonas: naive 1-year-old sheep. Regime as for Jasper. After 16 days tests were performed every 48 h.

the sheep drinks adequate sodium to correct its deficit, there may be a significant urinary excretion, e.g. for 4–8 h, and then a fall in the face of continuous salivary loss. Similarly, with correction of deficiency salivary Na/K ratio may approach or become normal for 4–6 h, but fall progressively again over 12–18 h as a result of the fistula loss. Analysis being only on 24 h collections, the results will reflect a period of positive sodium status in terms of some sodium excretion in urine, but not the fact that the salivary Na/K ratio reached normal for a brief period.

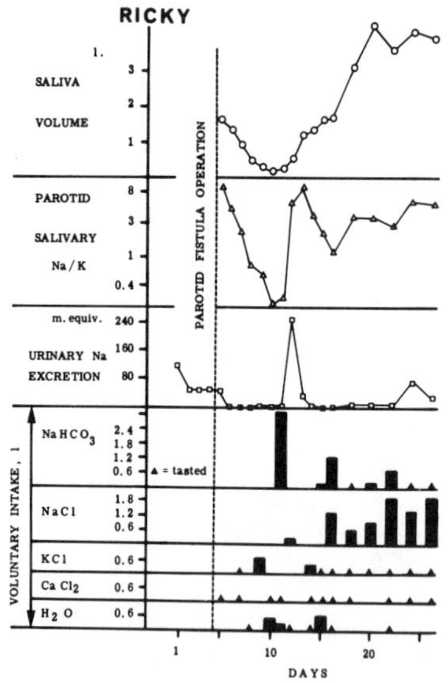

Fig. 11-16. Ricky: naive 1-year-old sheep. Regime as for Jasper. After 12 days tests were performed every 48 h.

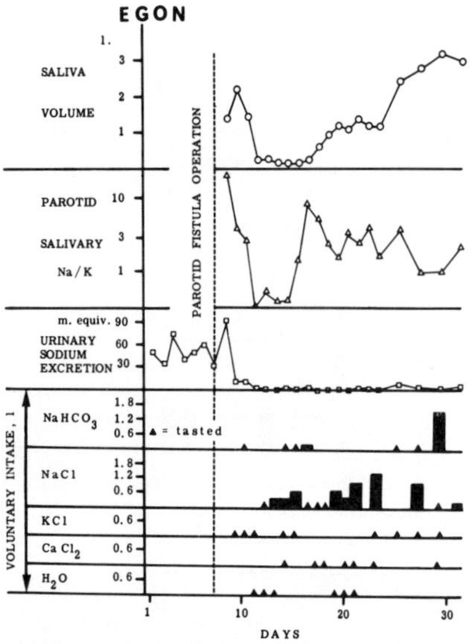

Fig. 11-17. Egon: naive 1-year-old sheep. Regime as for Jasper. After 16 days the tests were performed every 48 h.

Day 3: The sheep took no interest in the solutions except to taste $CaCl_2$ during the 20th minute of access.

Day 4: Salivary and urinary sodium loss since day 2 was ca 426 mmol. The sheep immediately tasted NaCl when the solutions were presented and drank this solution over the next 2 min. It then stood quietly for the rest of the hour and sipped NaCl in the 60th minute. Intake of NaCl during the 60 min was 1060 ml (318 mmol Na). The food intake of the sheep was only 500 g for days 2, 3 and 4, and it fell to hardly any on the fifth day. It picked up after that, but not to the whole 800 g.

Day 5: The sheep immediately sniffed water on access to the solutions, and tasted $CaCl_2$, sniffed NaCl and sniffed and tasted KCl. It then stood and ruminated. It later drank 730 ml water.

Day 6: The sheep took little interest in the solutions; it was looking apathetic, although its food intake had been 500 g overnight.

Day 7: It drank 810 ml $NaHCO_3$ (243 mmol Na) in the first 3 min of access, and from then on it was offered the solutions every 48 h. It drank NaCl only on all these occasions (Fig. 11–12).

Heinz (Fig. 11-13)

Weight, 25 kg; pre-operative sodium excretion ca 100 mmol/day.

Day 1: Sodium lost in the saliva was 150 mmol and in the urine 125 mmol, reflecting no sodium depletion yet. On access, the sheep immediately sniffed $CaCl_2$, and tasted water and KCl. For the following 2 min it drank KCl in short draughts. In the 4th minute it tasted NaCl and continued sipping KCl for the rest of the minute and then again in the 13th and 14th minutes before lying down. It stood up in the 33rd minute and sipped KCl, and in the 37th minute tasted both NaCl and $CaCl_2$. It sipped at KCl twice more before lying down in the 59th minute. Total KCl drunk in 60 min was 460 ml (138 mmol).

Day 2: Sodium lost in the last 24 h was 173 mmol; urinary composition indicated onset of sodium deficiency. On access to the solutions the sheep immediately started drinking NaCl for 140 s, then 15 s. In the 4th minute it tasted KCl, water, $CaCl_2$ and NaCl, and in the 6th minute it tasted $NaHCO_3$ and KCl. It continued drinking NaCl between the 9th and 22nd minutes when it lay down, but stood up again in the 28th minute and sipped NaCl intermittently until the 42nd minute. NaCl drunk in 60 min was 2250 ml (675 mmol); KCl drunk in 60 min was 30 ml.

Day 3: On access to the solutions the sheep immediately tasted NaCl followed by KCl, water and $NaHCO_3$. In the 4th and 6th minutes it drank KCl, and in the 11th minute and later in the hour it drank $NaHCO_3$. $NaHCO_3$ drunk in 60 min was 840 ml (252 mmol); KCl drunk was 80 ml (24 mmol).

Day 4: The sheep immediately tasted $CaCl_2$, and then sipped at NaCl, followed by $NaHCO_3$. In the 2nd minute it drank $NaHCO_3$ and NaCl, and the latter again in the 4th minute. It lay down in the 7th minute and ruminated. $NaHCO_3$ drunk in 60 min was 860 ml (258 mmol); NaCl drunk was 300 ml (90 mmol). The sheep continued to drink $NaHCO_3$ daily (Fig. 11-13), showing a preference for this solution over NaCl. It had always finished its drinking within 20 min of access, and often very much sooner.

Denzil (Fig. 11-14)

Pre-operative sodium excretion ca 90 mmol/day.

Day 1: 433 mmol of sodium had been lost since operation, 383 in the saliva and 50 in the urine. The sheep was very nervous when the solutions were offered, and it stood well back in the cage for the first 11 min. It then moved forward and sniffed KCl. In the 21st minute it sniffed $CaCl_2$ and $NaHCO_3$. It was still very nervous. It started ruminating in the 32nd minute and lay down in the 55th minute.

Day 2: A further 365 mmol sodium had been lost in the saliva and urine. Food intake had decreased to 600 g. The sheep was standing at the back of the cage when the solutions were presented, but came forward in the 1st minute. It took no interest in the solutions until the 11th minute when it sniffed NaCl and KCl. It lay down in the 19th minute.

Over the next 4 days the sheep took little interest in the solutions during the hour they were presented; it tasted $CaCl_2$ on the 5th day. It ate very little food (between 10–50 g daily) and became very apathetic (dull eyes and slow movements). It started grinding its teeth on day 4.

Day 7: During the hour the solutions were offered, the sheep made no attempt to taste any of them. Salivary volume had fallen to 260 ml/day and total sodium loss from saliva and urine since the fistula operation was 996 mmol. The solutions were replaced on the cage for 2 h and 1200 ml $NaHCO_3$ were drunk (360 mmol).

Day 8: The sheep stood quietly for the first minute the solutions were offered and then tasted KCl and drank it in short draughts over the following 7 min. (This solution was in the same position as $NaHCO_3$ on day 7). During the 9th minute it tasted NaCl, and the following minute drank KCl and then tasted NaCl several times. It then drank this solution avidly for 45 s followed by 30 s, 4 s and 68 s draughts. In the 18th minute it tasted $NaHCO_3$ twice, but continued sipping NaCl for most of the remaining part of the hour. In all 2180 ml of NaCl (654 mmol) and 820 ml KCl (246 mmol) were drunk. The urine composition on day 9 reflected that this intake had brought the animal towards normal sodium status.

Day 9: The sheep sipped at NaCl during the first minute of access to the solutions, and during the next 5 min tasted both $CaCl_2$ and water. It stood quietly for the remainder of the hour.

Day 10: The sheep was restless before the solutions were offered and on access immediately tasted KCl. During the next 9 min before lying down, it tasted water and KCl and then drank water for 25 s. It stood up again during the 31st minute and in the 36th minute drank $NaHCO_3$. It continued sipping $NaHCO_3$ and also tasted KCl and water for 3 min. In the 60 min, intake of water was 300 ml and of $NaHCO_3$ 1640 ml (492 mmol). The following day the sheep had regained its appetite completely, but it did not drink any sodium during the hour of access to the solutions. Over the next 5 days the sheep drank predominantly NaCl, but from there on, when presentation was every 48 h, it preferred $NaHCO_3$ (Fig. 11-14). As indicated by urinary sodium excretion and Na/K ratio of 48 h collections of parotid saliva, Heinz and Denzil drank adequate sodium solution under conditions of short-term access every 48 h to have sodium balance oscillate near neutral status.

Jonas (Fig. 11-15)

Pre-operative excretion of sodium ca 60 mmol/day.

Day 1: It had lost 240 mmol sodium, 195 in saliva and 45 in urine. On access it immediately sniffed water, NaCl, $CaCl_2$ and KCl, looking around in between sniffing the solutions. It continued sniffing the solutions and the cage for another 8 min and then lay down in the 22nd minute. It stood up again in thc 36th minute and ruminated.

Day 2: Only 400 g food were eaten, and a further 245 mmol sodium lost. The sheep sniffed all the solutions within the first 5 min of access and in the 6th minute actually dipped its chin into the $NaHCO_3$ solution, shook its head and then sniffed NaCl. It lay down in the 9th minute and ruminated.

Day 3: A further 180 mmol sodium were lost, bringing the total to this point to 665 mmol. The 24-h urine volume was only 80 ml and food eaten 50 g. The sheep looked very apathetic, and vaguely sniffed at the solutions over 15 min before it lay down.

Day 4: The sheep sniffed all the solutions during the first few minutes of access and tasted KCl in the 8th minute.

It took little interest in the solutions on the following 2 days. It ate virtually no food and on the sixth day was grinding its teeth every now and then; when it stood up it hung its head low.

Day 7: The sheep took little interest in the solutions when they were offered, but it remained standing with its head low and eyes closed. The solutions were weighed at 60 min and then replaced on the cage. Two and a half hours later they were re-weighed and replaced as the sheep still had not taken any interest in them. However, 3½ h later, 1930 ml of KCl (579 mmol) had been drunk. It is not known whether the sheep had tasted the other solutions or not.

Day 8: Total sodium loss from saliva and urine since the fistula operation was 943 mmol. The 24-h saliva volume had fallen to 160 ml. During the first 5 min of access to the solutions the sheep sniffed $CaCl_2$, NaCl and KCl and it tasted KCl. It lay down in the 7th minute. In the 35th minute it stood up and in the following minute it tasted $NaHCO_3$ and then drank avidly for 25 s, licked its lips and continued drinking in shorter draughts for 2 min. It tasted NaCl, and drank this solution in short draughts and then drank, back and forth, at NaCl and $NaHCO_3$. It became much less apathetic looking, and from the 47th minute to the 60th minute it stood quietly (total sodium intake=828 mmol).

For the next 3 days the sheep only tasted a few of the solutions on access and then lay down for most of the hour. Its food intake had returned to normal i.e. 800 g, by day 11, and its 24-h salivary volume had increased to 1120 ml. On day 13 the sheep drank 850 ml $NaHCO_3$ and on day 15, 910 ml NaCl. From then on it showed a preference for NaCl over $NaHCO_3$.

Ricky (Fig. 11-16)

Weight, 28 kg; pre-operative sodium excretion ca 40 mmol/day.

Day 1: 1600 ml saliva had been lost containing 285 mmol sodium, and 1100 ml urine containing 45 mmol sodium. On access to the solutions the sheep sniffed them all briefly and tasted $CaCl_2$. In the 5th minute it sniffed water and then $CaCl_2$ and then stood quietly ruminating.

Day 2: A further 226 mmol sodium had been lost as saliva, and urinary sodium reflected sodium deficit. In the first minute of access the sheep sniffed NaCl and KCl, and then NaCl in the following minute. It took very little interest in the solutions for the remainder of the hour.

Day 3: The sheep had eaten very little food. On access it took very little interest, sniffing and tasting $CaCl_2$ and sniffing KCl in the 2nd minute and occasionally grinding its teeth. During the 30th minute it started sipping $CaCl_2$ and in the 48th minute tasted KCl. Amount of $CaCl_2$ drunk during 60 s was 80 ml.

Day 4: Saliva volume for the 24 h was 500 ml and urine volume 390 ml. No food was eaten and the sheep looked very apathetic. It showed no interest in the solutions until it sipped water in the 33rd minute.

Day 5: Approximately 200 g food had been eaten. On access to the solutions the sheep tasted KCl and drank 620 ml over 2 min (186 mmol). It then stood quietly and took no further interest in any of the solutions.

Day 6 : About 600 g food had been eaten. On access to the solutions the sheep immediately drank water for 2 min and then chewed at and licked the cage and the water bin. In the 8th minute it chewed at the KCl and $NaHCO_3$ bins, tasted $CaCl_2$ twice, and sniffed NaCl. It continued chewing at the cage and licking the cage floor for most of the remainder of the hour, tasting water twice.

Day 7: Total sodium lost since operation was 775 mmol. The sheep immediately sniffed $NaHCO_3$ on access to the solutions, and then tasted water. It sniffed water and KCl and stood quietly until the 6th minute when it sniffed and tasted $NaHCO_3$ twice. In the 7th minute it tasted $CaCl_2$ twice and drank water. It then stood quietly and ruminated. In the 56th minute the sheep was licking the cage, when it suddenly tasted $NaHCO_3$. It drank for 75 s when the bin was removed and refilled, and then for 160 s. Total amount of $NaHCO_3$ drunk was 3220 ml (966 mmol), of water 250 ml. This single episode of sodium intake corrected deficit as reflected in urine and saliva.

Day 8: The sheep sniffed and drank a small amount of NaCl immediately the solutions were offered. It sniffed $NaHCO_3$, water and KCl, but ruminated for most of the hour.

After day 7 the sheep ate all its food. On day 9 it did not taste any of the solutions; day 10 it tasted $CaCl_2$ and water, and drank 330 ml of KCl within the first 4 min of access to the solutions; on day 11 it drank 130 ml $NaHCO_3$, 30 ml $CaCl_2$ and 600 ml of water in the first 10 min of the hour.

Day 12: The sheep showed excitement before the solutions were presented and immediately drank $NaHCO_3$ for 47 s. The following 2 min it drank NaCl. Total intakes during the hour for these two solutions were 1250 and 1320 ml respectively (771 mmol).

The solutions were then offered every 48 h for 30 min and the sheep showed a consistent preference for sodium, and for NaCl over $NaHCO_3$.

Egon (Fig. 11-17)

Weight, 21 kg; pre-operative sodium excretion ca 40 mmol/day.

Day 1: The sheep had lost 283 mmol sodium in saliva and 94 mmol in urine. It was obviously upset by the placing of the solutions on the cage and after sniffing KCl and NaCl, it bleated, looked around, sniffed water and $CaCl_2$, bleated, looked from side to side and continued bleating frequently. In the 4th minute, it sniffed NaCl and water and after that it lay down.

Day 2: A further 362 mmol sodium were lost, 348 in saliva and 14 in urine. Food intake was reduced to 550 g. The sheep stood up and bleated as the solutions were placed on the cage, but then immediately lay down again. It stood up again in the 43rd minute and sniffed water and $CaCl_2$ and tasted KCl. It then sniffed $NaHCO_3$ and NaCl, bleated and stood quietly. In the 51st minute it lay down.

Day 3: The sheep was observed licking the front and side bars of the cage for long periods. Total sodium loss was 950 mmol. Food intake over the last 24 h was only 100 g. When the solutions were offered the sheep remained lying, licking the wire floor of the cage where the saliva dripped through, and also the brackets of the solution bins. After 13 min it stood up, sniffed NaCl, licked the KCl bin and then tasted KCl, licked the $NaHCO_3$ bin and then tasted $NaHCO_3$. In the 15th minute it licked the NaCl bin, and continued licking the cage, and then various solution bins. In the 22nd minute the sheep tasted $NaHCO_3$ and then lay down.

Day 4: The sheep remained lying as the solutions were presented, and licked the floor of the cage, and occasionally ground its teeth. In the 16th minute it stood up and during the following minute tasted and then drank water. It continued sipping water on and off, tasting KCl in the 21st minute. It lay down in the 37th minute.

Day 5: The sheep was standing licking the cage floor as the solutions were presented. It sniffed $CaCl_2$, licked the cage and later sniffed KCl and then continued licking the cage until it lay down in the 9th minute. It stood up in the 25th minute and tasted NaCl and then drank this solution noisily over 2 min. It sniffed at this solution as well as licking the cage floor until it lay down in the 39th minute. Only 90 ml NaCl (27 mmol) were drunk.

During the access periods of the sixth and seventh days the sheep was much more interested in the solutions and drank NaCl, 370 ml (111 mmol) and 390 ml (117 mmol) respectively, over long periods. On the seventh day it also tasted all other solutions except water. On the eighth day it drank 282 mmol NaCl and from the twelfth day continued drinking NaCl—and, though in increasing amounts, never adequate to bring it into positive balance except on the eighteenth day. Its salivary secretion gradually increased.

Group B: Access to electrolyte solutions for 1 h per day for 11–14 days before parotid fistula operation (Figs. 11-18 to 11-25)

We will consider the four recently weaned 6-month-old animals first (Figs. 11-18 to 11-21, and Table 11-1).

Simon (Fig. 11-18 and Table 11-1)

On the first day the solutions were offered the sheep tasted both NaCl and $CaCl_2$ and sniffed KCl. After 5 min it lay down and took no further interest. Over the following 13 days when solutions were offered it lay down most of the time and took little interest, apart from tasting KCl on day 4. On 9 of the 14 days the sheep did not even stand up when they were presented. Pre-operative sodium excretion was ca 30 mmol/day.

On day 1 after the parotid fistula operation it had lost 160 mmol sodium in its saliva and urine. It stood up as the solutions were presented and tasted all except $NaHCO_3$. It lay down after 3 min and ruminated. By the following day the sheep had lost 330 mmol sodium. It took little interest in the solutions. On day 3 it smelled all the solutions within 1 min of access. During the 2nd minute it tasted $CaCl_2$ and sipped the solution over 2 min (intake 120 mmol Ca). The sheep licked the cage vigorously for most of the hour, especially the wire mesh floor through which the saliva dripped. This was novel behaviour for the animal, and it was obviously restless. By day 4 the sodium loss was 570 mmol, and it was eating very little food. It was restless and excited when the solutions were presented and it immediately tasted $NaHCO_3$. The sheep then drank this solution for 2.5 min (intake 660 mmol sodium), and then sipped at it for the next 10 min. It took no interest in the other solutions, except that it did taste water. This sodium intake repaired deficit. During the following 2 days the sheep drank predominantly $NaHCO_3$. Thereafter the solutions were offered every 48 h. As shown in Fig. 11-18 the intake of solutions other than sodium was negligible and $NaHCO_3$ predominated from the outset. The urinary sodium excretion and salivary Na/K ratio (normal sodium-replete ratio 20–30) indicated that the drinking behaviour maintained a sodium status such that the animal probably repaired its sodium deficit during each brief period of access to sodium and then passed into deficit over the ensuing 24- or 48-h period. Voluntary sodium intake was never greatly in excess of loss, and therefore the animal did not have a substantial period with high urinary sodium excretion and salivary Na/K ratio sustained in the sodium-replete range.

Table 11-1. Voluntary sodium intake (mequiv) during 1 h of access to a cafeteria containing 300 mequiv/l solutions of sodium chloride, sodium bicarbonate, potassium chloride, calcium chloride and water.

		Before parotid fistula operation							
Sheep	Previous access to salt	Day −7	Day −6	Day −5	Day −4	Day −3	Day −2	Day −1	Mean/ day
Maxine	None	0	0	0	0	0	0	0	0
Marcel	None	0	0	0	0	0	0	0	0
Mark	None	0	0	359	0	0	0	0	51
Mahler	None	0	0	0	0	0	0	0	0
Nicol	None	177	174	78	0	99	0	261	112
Simon	None	0	0	0	0	0	0	0	0
Dominic	None	0	0	0	0	0	0	0	0
Foster	None	0	0	0	0	0	0	0	0
Montagu	None	0	0	0	0	0	0	0	0
Lola	Known[a]	27	318	144	213	NA	477	0	196
Gigi	Known[a]	54	132	48	135	NA	138	6	86
Finochio	Unknown[b]	0	0	0	0	NA	NA	NA	0
Tess	Unknown[b]	0	0	0	0	0	0	0	0

[a]From property with access to bore water (74 mequiv/l of sodium).
[b]Bought at saleyard.
NA, not offered.

Montagu (Fig. 11-19 and Table 11-1)

On the first two control days the solutions were offered, the sheep sniffed all the solutions except NaCl and then lay down, taking no further interest. Over the next 4 days it tasted all the solutions except NaCl and then took little interest in them until the tenth day when it tasted all five solutions. On the day before the operation the sheep tasted NaCl and KCl within the first 2 min of access. Sodium excretion during the pre-operative period was ca 10–20 mmol/day.

On day 1 after the fistula operation (sodium loss 280 mmol) the sheep sniffed all the solutions except $CaCl_2$, and then lay down. On day 2 (sodium loss 460 mmol) the sheep immediately tasted $CaCl_2$ and then drank NaCl for 3 min (intake 1330 ml, or 400 mmol sodium). At the 23rd minute it drank NaCl again and the total intake for the hour was 580 mmol sodium, which approximated the deficit. On most days following this, all solutions were tasted, but except for day 14, when $NaHCO_3$ was drunk (intake 280 mmol sodium), NaCl intake predominated.

Foster (Fig. 11-20 and Table 11-1)

This sheep was offered the solutions for 11 days before the parotid fistula operation. On the first day the solutions were offered the sheep stood up after 4 min and sniffed water and $CaCl_2$ and then tasted both. Thereafter, it showed little interest, and on seven of the days before the operation it did not taste any solutions. On one of the days (eighth) it tasted $NaHCO_3$ five times (intake 10 ml), and then drank 70 ml water. Pre-operative sodium excretion was ca 20 mmol/day.

On day 1 after the parotid fistula operation the sheep immediately sniffed $NaHCO_3$, NaCl and KCl when the solutions were presented, and after 6 min tasted KCl. It tasted KCl twice more and then stood quietly. On day 2 there were many more sampling episodes. It sniffed all solutions within 6 min and tasted water, NaCl and KCl. At the 9th minute it sniffed $NaHCO_3$ and started to drink it. It took several small draughts over the next 6 min (intake 60 mmol sodium). Fistula flow had been small and sodium loss ca 200 mmol. On day 3 it immediately sampled KCl several times. $CaCl_2$ and water were also sampled and at the 15th minute NaCl was tasted and drunk (450 ml or 135 mmol sodium). On day 4 the sheep stood up after 5 min and began drinking $NaHCO_3$ (intake 21 mmol sodium), but then turned to NaCl (intake 96 mmol sodium). The solutions were offered every 48 h after day 4, and on day 6 the sheep drank water (1100 ml) and $NaHCO_3$ (intake 33 mmol sodium). On day 8 the sheep immediately drank 290 mmol NaCl and on day 10 $NaHCO_3$ (intake 234 mmol). It usually showed a preference for $NaHCO_3$ over NaCl thereafter.

Dominic (Fig. 11-21 and Table 11-1)

In the pre-operative period it tasted $CaCl_2$ and water in the 1st minute the solutions were presented. The sheep sniffed KCl and $NaHCO_3$

Table 11-1. (*continued*)

After parotid fistula operation														
Day 1	Day 2	Day 3	Day 4	Day 5	Day 6	Day 7	Day 8	Day 9	Day 10	Day 11	Day 12	Day 13	Day 14	Day 15
0	0	638	215	120	322	477	463	152	306	412	NA	332	NA	300
0	0	0	0	0	0	450	0	443	450	131	461	18	294	344
355	165	414	0	77	0	116	0	403	89	NA	134	NA	1178	NA
0	0	0	0	0	0	0	0	0	0	0	437	713	903	0
0	558	390	0	296	423	282	NA	249	NA	336	NA	561	NA	372
0	0	0	660	27	273	NA	405	NA	639	NA	192	NA	450	NA
0	0	0	381	204	0	6	225	141	18	42	NA	315	NA	360
0	60	135	117	NA	33	NA	288	NA	249	NA	324	NA	417	NA
0	579	3	90	252	9	NA	252	NA	171	NA	294	NA	324	NA
126	NA	375	NA	570	NA	567	NA	282	NA	462	NA	927	NA	579
0	96	NA	231	NA	273	NA	552	NA	374	NA	462	NA	517	NA
0	0	238	NA	123	NA	820	NA	416	NA	414	NA	796	NA	728
108	297	364	466	130	235	395	311	NA	458	NA	477	NA	665	NA

and then stood quietly for the remainder of the period. On the fourth day it tasted $NaHCO_3$ and on the thirteenth day of the pre-operative period it tasted $CaCl_2$ several times (intake 30 ml), and NaCl once. Pre-operative urinary sodium excretion was ca 20 mmol/day.

On day 1 after the parotid fistula operation the sheep showed more interest in the solutions and sniffed them all except NaCl. It tasted $CaCl_2$ in the 5th minute. The salivary sodium and urine loss at this stage was 175 mmol. On day 2 it stood up immediately the solutions were presented, and tasted KCl twice and $CaCl_2$. The sheep lay down for most of the hour and stood up at the 46th minute, sniffed all solutions and tasted NaCl. By day 3 the sheep's salivary and urinary sodium loss was 410 mmol. It was very restless and lay down and stood up often. It sampled KCl frequently. On day 4 (sodium loss 490 mmol) the sheep tasted NaCl and KCl immediately the solutions were placed on the cage. It sniffed at both several times during the next 3 min and at the 5th minute drank KCl (intake 18 mmol). The following minute the sheep began to drink NaCl in long draughts and continued to drink for several minutes. It tasted NaCl frequently over the remainder of the hour (total NaCl intake 1260 ml, or 380 mmol). On day 5 the sheep immediately drank NaCl and tasted $NaHCO_3$. It continued drinking NaCl over the following 5 min (total NaCl intake 204 mmol) and also tasted all the other solutions. On days 6 and 7 the sheep took little interest in the solutions and only tasted them. However, on the eighth day it drank KCl (intake 36 mmol), and after 15 min drank $NaHCO_3$ almost continuously for the next 3 min (intake 225 mmol sodium). From day 11 onwards the solutions were offered every second day and the sheep drank mainly NaCl until the twenty-first day, when intake of $NaHCO_3$ started to increase. On most days it tasted other solutions, but there was no significant intake of them.

The five other sheep in this group were 1-year-old. Only one (Nicol), drank appreciable quantities of sodium before the parotid fistula operation. Examples of behaviour are given below.

Mark (Fig. 11-22 and Table 11-1)

The first pre-operative day the solutions were offered, the sheep took great interest in them, sniffing and tasting all. For the next 7 days the sheep either failed to stand up when the solutions were offered, or stood up for a few minutes and tasted some of them. On the ninth day it tasted water and KCl early in the hour the solutions were available, but then drank 360 mmol NaCl and more water between 36 and 46 min. The sheep took no particular interest in the NaCl solution during the remaining 4 days before parotid fistula operation. The day after the operation it drank 1.18 litres NaCl (354 mmol) in the first 4 min of access to the solutions, and stood quietly taking little interest in the solutions except to taste $CaCl_2$ and water. The sheep continued to drink sodium daily, predominantly NaCl, and kept salivary Na/K normal. There was substantial urinary sodium excretion on two days. Ten days after the

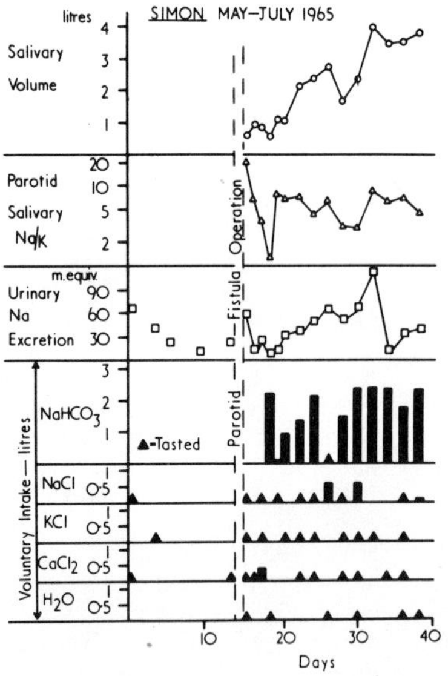

Fig. 11-18. Simon: naive 6-month-old recently weaned sheep. The voluntary intake during 1 h of access to a cafeteria of 300 mequiv/l solutions of sodium chloride, sodium bicarbonate, potassium chloride, calcium chloride and water offered each day for 14 days before operation to make a permanent unilateral parotid fistula is shown. The presentation regime continued after the fistula operation. Water was continuously available but its position during the hour test was, as with the solutions, determined by a random numbers system. The saliva volume (litres), the parotid salivary Na/K ratio (normal ca 20) and urinary sodium output are shown also. With this animal, after the sixth post-operative day the solutions were offered every 48 h and collections of saliva and urine also made at this interval.

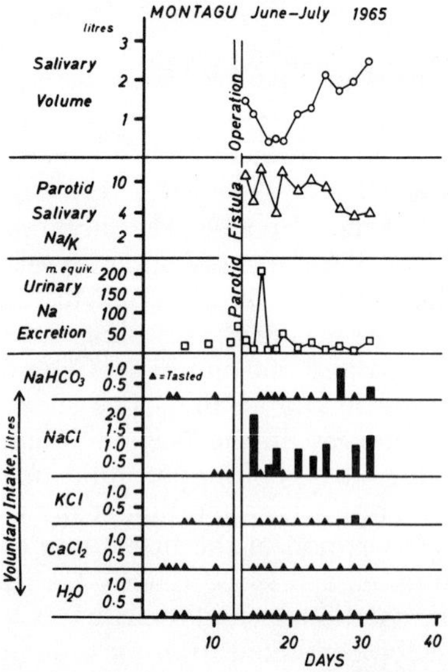

Fig. 11-19. Montagu: naive 6-month-old recently weaned sheep. Regime as for Simon. After 6 days post-operation, the solutions were offered every 48 h.

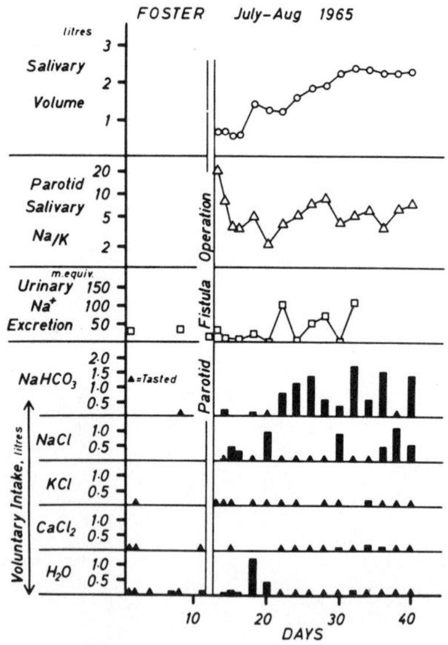

Fig. 11-20. Foster: naive 6-month-old recently weaned sheep. Regime as for Simon. After 4 days post-operation, the solutions were offered every 48 h.

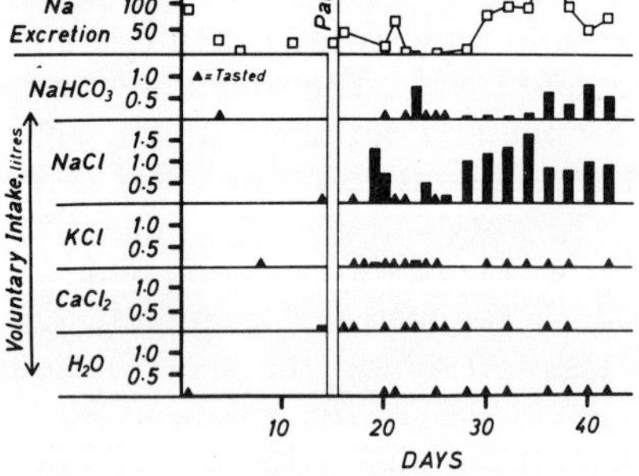

Fig. 11-21. Dominic: naive 6-month-old recently weaned sheep. Regime as for Simon. After 11 days post-operation, the solutions were offered every 48 h.

parotid fistula operation the solutions were offered every second day. The sheep altered its preference to $NaHCO_3$ over the following fortnight.

Maxine (Fig. 11-23 and Table 11-1)

The sheep tasted the solutions frequently before the parotid fistula operation and often drank water

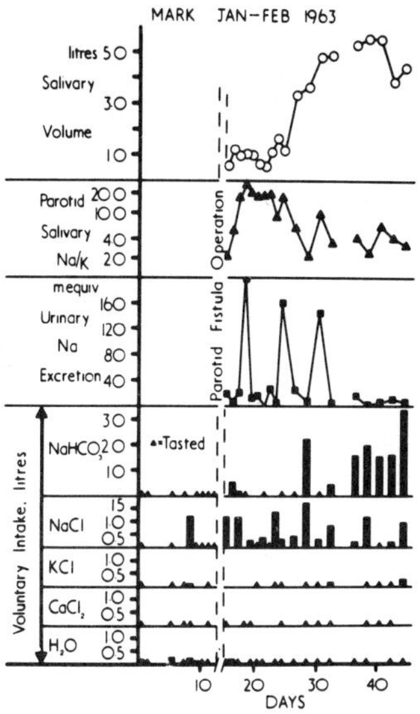

Fig. 11-22. Mark: naive 1-year-old sheep. Regime as for Simon. After 10 days post-operation, the solutions were offered every 48 h.

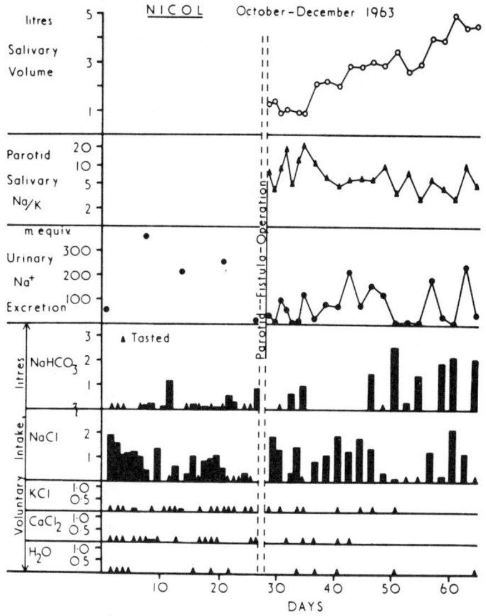

Fig. 11-24. Nicol: naive 1-year-old sheep. Regime as for Simon. after 7 days post-operation, the solutions were offered every 48 h.

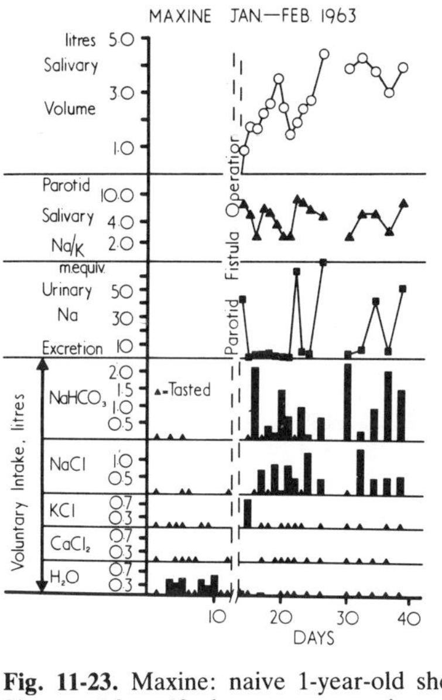

Fig. 11-23. Maxine: naive 1-year-old sheep. Regime as for Simon. After 10 days post-operation, the solutions were offered every 48 h.

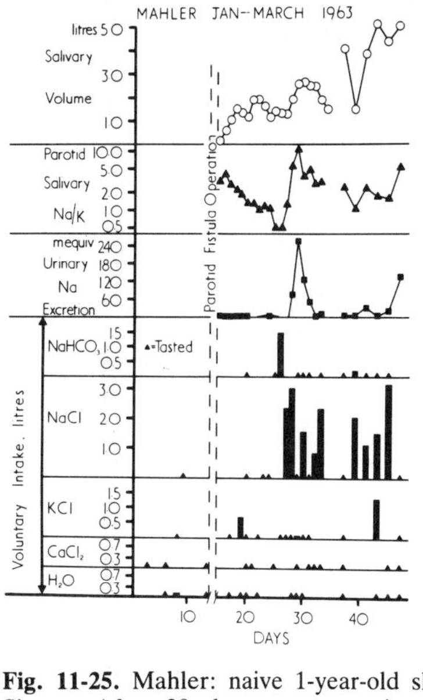

Fig. 11-25. Mahler: naive 1-year-old sheep. Regime as for Simon. After 20 days post-operation, the solutions were offered every 48 h.

during the hour of access to the solutions. On the second post-operative day it drank 240 mmol KCl. The following morning it showed great excitement as the solutions were placed around the cage, salivary sodium loss now being 620 mmol. On access, it immediately drank $NaHCO_3$ (660 mmol) and continued to drink sodium daily, mainly as $NaHCO_3$.

Nicol (Fig. 11-24 and Table 11-1)

The sheep took no interest in the solutions on the first day of observation before parotid fistula operation, but for the next week drank large quantities of NaCl. When the solutions were placed on the cage the sheep would immediately sniff all, taste one or two and then drink NaCl. During the hour the solutions were available it would investigate most of the solutions, either by sniffing or tasting a few times. In most instances there was a definite termination of exploratory behaviour and drinking well before the end of the hour, the activity being mainly in the first 15 min. The day after the operation the sheep took no interest in the solutions but drank 558 mmol NaCl on day 2 (salivary sodium loss for the two days being 410 mmol). After day 7 the sheep was offered the solutions every second day. Saliva volume increased up to 4.5 litres per 48 h. The sheep's preference for NaCl continued until 3 weeks after the parotid fistula operation, when it began to prefer $NaHCO_3$.

Mahler (Fig. 11-25 and Table 11-1)

The sheep took very little interest in the solutions except for sniffing them before the parotid fistula operation. Days 1 to 3 after operation it had diarrhoea, which would have increased sodium loss. It was very lethargic. It sniffed the solutions and tasted water and KCl on one occasion. On day 5 the sheep drank 210 mmol KCl in short sips and on the following day tasted the five solutions but did not drink any. On day 8 it ate very little food but exhibited an appetitive behaviour which became more pronounced over the following 3 days: i.e. the animal was extremely restless, licked and nibbled the cage and solution bins, and when fed in the afternoons, restlessly nibbled food, food bin and water. Salivary Na/K had fallen to <0.5. On the thirteenth day, after 50 min access to the solutions, the sheep started sipping KCl. It then drank $NaHCO_3$ (430 mmol) until the solutions were removed at 60 min. Thereafter it drank adequate sodium solution to bring it to near normal sodium status. The animal showed a preference for NaCl over the rest of the experimental period.

Group C: Long-term access to solutions before parotid fistula operation (Figs. 11-26 to 11-30)

Two sheep, Gigi and Lola (Figs. 11-26 and 11-27), came from a property that pumped bore water (sodium content 74 mmol/l) for stock to drink. They both drank large amounts of sodium solution (Table 11-1) before operation. Intake increased after operation, with preference changing to $NaHCO_3$. There was no interest in the other solutions.

The history of two other sheep, Tess and Finochio (Figs. 11-28 and 11-29), was not known.

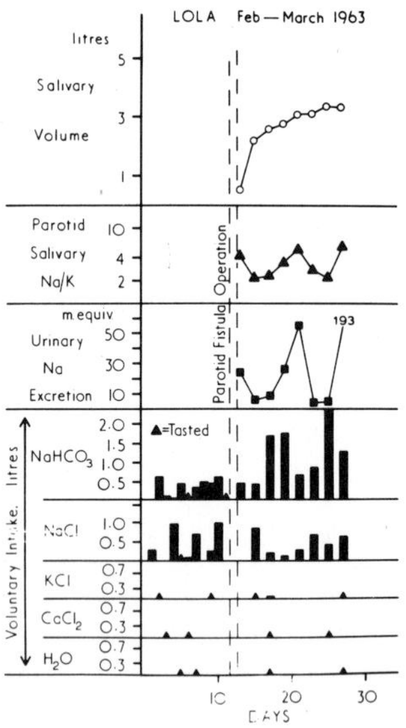

Fig. 11-26. Lola: sheep with a history of access to bore water (74 mmol/l sodium) since weaning. Regime as for Simon. The solutions were offered to the animal every 48 h from the time of parotid fistula operation.

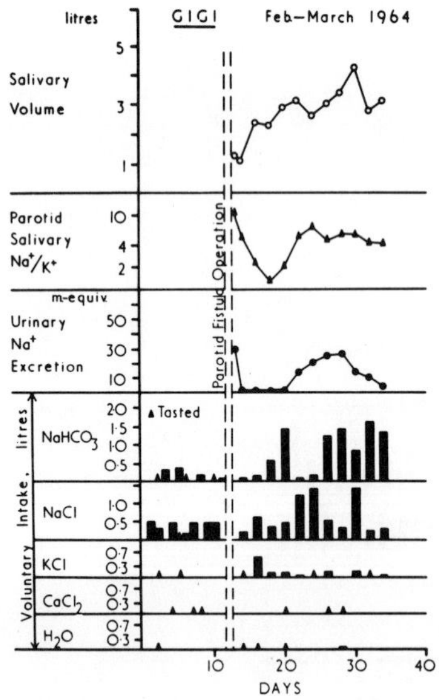

Fig. 11-27. Gigi: sheep with history of access to bore water (74 mmol/l sodium) since weaning. Regime as for Simon. The solutions were offered to the animal every 48 h from the second post-operative day onwards.

Tess showed no interest in the solutions before operation but drank $NaHCO_3$ on the first day afterwards (Table 11-1). On one occasion, over a month before operation, Finochio (Fig. 11-29) drank a large amount of NaCl (470 mmol). On the third day after operation sodium intake began and, on the seventh day, the episode of intake corrected deficit. Thereafter intake was sufficient to result in

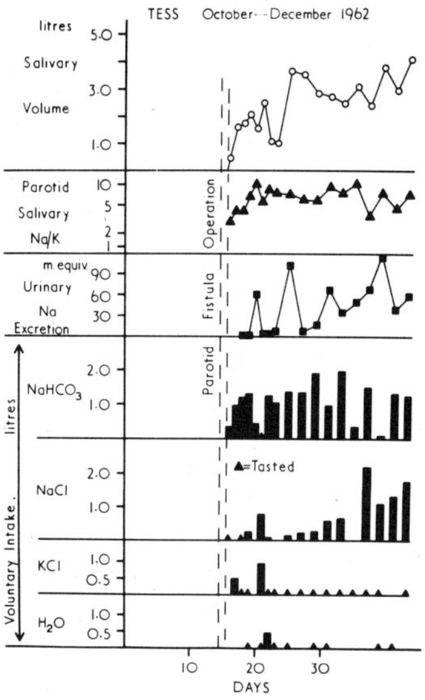

Fig. 11-28. Tess: sheep with unknown history in relation to access to salt sources. Regime as for Simon. The solutions were offered to the animal every 48 h from the eighth post-operative day onwards.

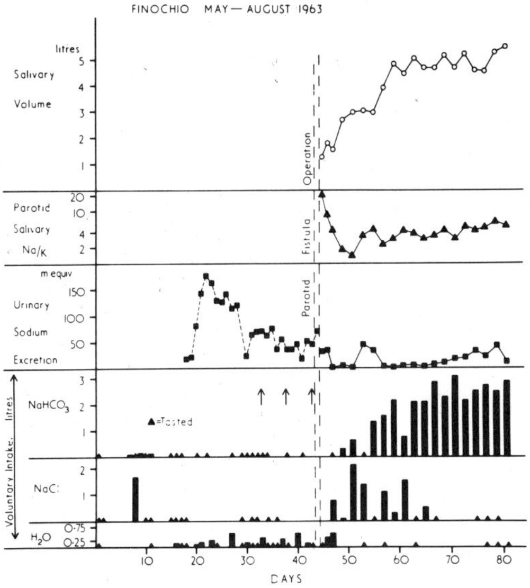

Fig. 11-29. Finochio: sheep with unknown history in relation to access to salt sources. Regime as for Simon. The solutions were offered to the animal every 48 h from the third post-operative day onwards.

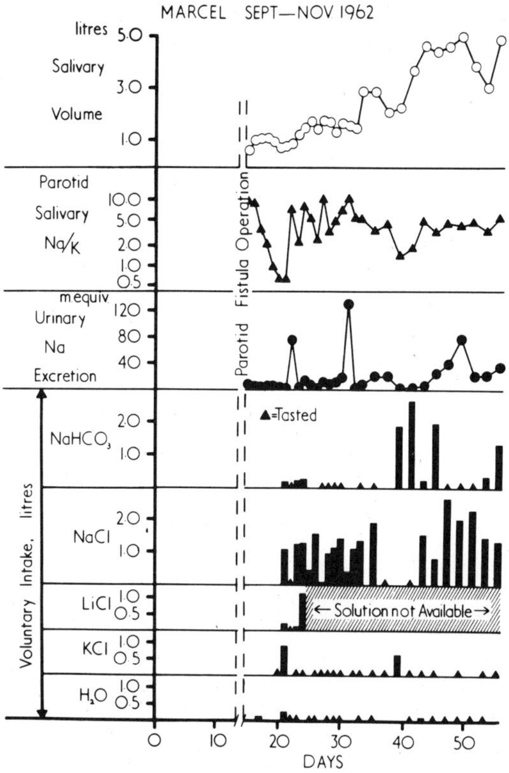

Fig. 11-30. Marcel: naive 1-year-old sheep. Regime as for Simon except that lithium chloride (300 mmol/l) was offered instead of calcium chloride. After the animal drank a substantial amount of it on the tenth post-operative day the solution was withdrawn. The solutions were offered to the animal every 48 h from the nineteenth post-operative day onwards.

a small positive balance after each 48-h episode, as reflected in urinary composition.

Fig. 11-30 records an experiment on Marcel, representative of several animals, which shows that when LiCl was included in the cafeteria in place of $CaCl_2$ the sheep ingested it. Phillip, who drank a large amount on one day, became ill and anorexic for several days afterwards.

Summary of sheep experiments

These experiments had certain clear results.

1. In all animals development of sodium deficiency caused a large voluntary intake of sodium solutions. Onset of the appetite, whether indicated by clear change of behaviour or intake of solution, was usually delayed 2–5 days.

2. Sodium intake was eventually adequate to maintain the animals in good condition physiologically. There was variation, however, in whether the sodium status was maintained on the positive side with a sizable amount of sodium in the urine, or whether near neutral or slightly on the negative side.

3. Specificity for sodium salts was unequivocal, though there were instances when this specificity appeared to emerge more slowly than the attraction to salty taste: i.e. on a few occasions KCl was drunk first. Usually, $NaHCO_3$ was preferred, which was a more appropriate choice in the light of $NaHCO_3$ loss from the fistula. $NaHCO_3$ preference sometimes developed gradually over a period of weeks.

4. This experiment was a stringent test, in that the experimental circumstance of access was for only 1 h per day to a cafeteria choice under conditions which initially may have been strange and disturbing to the animal.

5. There were instances where a small intake of sodium on one day was followed by larger intake the next, or cage licking developed in some animals before sodium drinking. This could be reconciled with some element of learning entering into appetite. However, of signal importance was the observation that sometimes the first occasion of sodium intake was rapid and large enough to correct deficit effectively and was not preceded by any instance of prior intake or licking of the cage. Sodium was the first choice and it was taken in large amounts. In the cases where some tasting of two or three solutions occurred on days before this episode of large sodium intake in response to deficit, the term tasting implies that intake was negligible on weight difference of bins, using a sensitive Mettler balance. In any event, if there had been intake it presumably would have been difficult for the animal to learn which of the several solutions tasted in a short period had conferred benefit.

Discussion and further analysis

Taste

The role of taste in specifically directing behaviour is highlighted by the animals which had no access to the solutions before operation. In the case of those which had access to the solutions before operation, and tasted each or most solutions, there was no state of deficiency existing which would cause any sense of benefit to follow, had significant intake occurred. But as the result of this experience the animals may, at least, have known the taste of sodium solution. This could have facilitated specific association when the central drive state developed as a result of body deficit.

One interpretation is that with onset of sodium deficiency the sheep did develop a preference for salty-tasting substances. This general preference, by learning of benefit, became refined to be specifically for sodium salts. That is, the basic event was an appetite for salty tastes, not a sodium appetite. The several instances of potassium chloride intake in early stages, which was rarely, if ever, repeated, could be consistent with this hypothesis. However, the experimental instances where there was tasting of all solutions, but the eventual large initial intake was of sodium, commensurate with deficit, points to specificity at outset.

Probably there is some individual variation between sheep in the accuracy of manifestation of innate propensities at initial experience. With some the behaviour is clearly specific in 2–4 days. With others, elements of experience add to innate propensity before the characteristic behaviour pattern of the sodium-deficient animal is established. The quotations at the beginning of the chapter are apposite.

Bell et al. (1979) propose that sodium-deficient calves without prior exposure to pure salt detected sodium salts at a distance, and showed agitation when sodium chloride or bicarbonate were brought into view. Since this behaviour occurred on initial exposure to the sodium salt, the response was probably not learned or dependent on visual cues. Olfaction in ruminants, functioning as a teloreceptor, may allow these animals when in a deficient state in natural situations to close on a source of metabolically required material.

Innate and learned behaviour

The general behaviour of the animals as sodium deficiency developed (restlessness and greater exploratory behaviour) is, within the restrictions of the experimental situation, consistent with the appetitive phase of an instinctive behaviour pattern, as characterized by the ethologists (Tinbergen 1951). It is not clear how this state of deficiency, which activates the drive involved, *on the first experience directs awareness and behaviour specifically to actions related to ingestion* (tasting and licking, and sensory aspects related to the realm of ingestion) rather than to many other possible actions. This is to be emphasized in the few instances where cage licking, presumably of dried salty-tasting saliva, developed as a novel pattern. Presumably it must reflect very substantial genetically programmed interconnections within the basal areas of the brain, just as the ultimate specificity of choice of sodium over other salts does. The extent of innately programmed organization is further suggested by the experiments where the very first instance of sodium intake by the naive animal was related to the extent of body deficit. If the extent of body deficit determines the extent of chemical change in the specific neuronal pool which subserves the excitation of salt appetite (and thus drive is related to deficit), the problem presents as to how, if the animal has had no previous experience of drinking salt solution, it integrates fairly accurately in the initial act of intake the taste impulses signalling concentration of salt solution with the pharyngo-oesophageal impulses metering volume swallowed. Whereas it is true that the naive animal may drink on the first occasion an amount approximating deficit, and thereafter closely related to deficit, it is important, however, to stress that considerable variation is seen between animals in this regard.

The relation of intake to deficit

For example, let us consider Lola and Gigi whose histories are part of the record above and who were tested 3–4 years later in this regard. They were offered 300 mmol/l sodium bicarbonate solution for 30 min after either 1, 2 or 3 days of sodium depletion, these episodes occurring without any order. As the graphs of individual observations made over several months show (Figs. 11-31 and

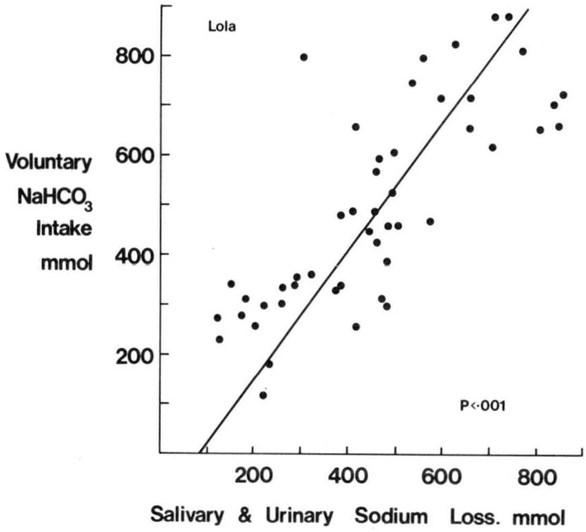

Fig. 11-31. Lola: the highly significant relation between voluntary intake of sodium bicarbonate (300 mmol/l) and body sodium deficit during 15 min of access following 1, 2 or 3 days deprivation of sodium solution. These studies on Lola and Gigi (Fig. 11-32) were carried out over 4 months.

11-32), the commensurate relation was highly significant ($P<0.001$). That is, the animals did not capriciously drink e.g. 4 litres of sodium bicarbonate on one day, none the next, two the next and so on—a behaviour which may well have kept them alive indefinitely though in poor condition in the face of a continuous fistula loss—but on the contrary the state of body balance determined what they drank.

An experiment on an animal (PF33) with very large salivary loss was typical of early observations in our laboratories on sheep with a parotid fistula. The experiment lasted 34 days. During the first week when the sheep was given 830 mmol of

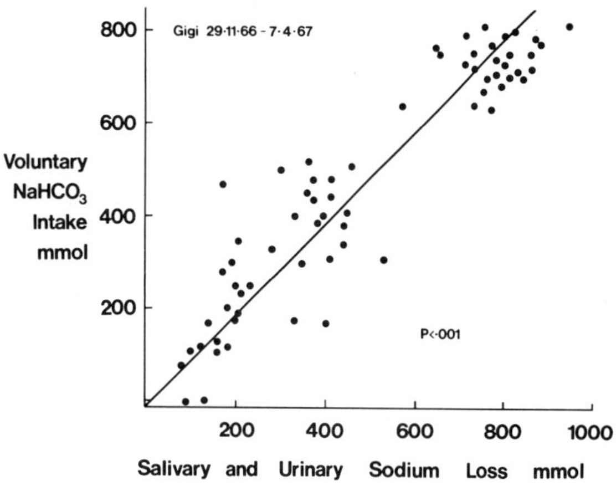

Fig. 11-32. Gigi: the highly significant relation between voluntary intake of sodium bicarbonate (300 mmol/l) and body sodium deficit during 15 min of access following 1, 2 or 3 days deprivation of sodium solution.

sodium bicarbonate in 1.5 litres of water by rumen tube daily to compensate fistula loss, mean sodium intake from a cafeteria of sodium chloride (420 mmol/l), sodium bicarbonate (420 mmol/l), potassium chloride (134 mmol/l) and water presented for 1 h was negligible (<5 mmol/day). For the second week no sodium supplement was given and the cafeteria was available all day. Sodium intake in food was 113 mmol/day and average voluntary sodium intake was 602 mmol/day. Salivary loss averaged 676 mmol, urinary 22 mmol, and faecal loss 3 mmol/day, so that intake approximated deficit. Intake was almost entirely as sodium bicarbonate.

For the third week the solutions were available for only 1 h per day, and average daily sodium intake was 560 mmol, predominantly as sodium bicarbonate. For the next 8 days the solutions were available for 1 h every second day. Average sodium deficit was estimated at 930 mmol and mean intake was 878 mmol, two-thirds of which was as sodium bicarbonate. For the last 6 days the cafeteria was available for 1 h every third day. The average deficit for these two episodes was 1070 mmol and mean intake was 1137 mmol, made up of equal parts of sodium chloride and bicarbonate.

However, despite the naive animal sometimes approximating intake to the extent of its deficit from first instances of sodium deficiency, it is obvious that with many animals a large element of experience and learning antecedes such behaviour. Beilharz and Kay (1963) described sodium bicarbonate intake in sodium-deficient sheep under standard control conditions where saliva loss was relatively constant. They noted that in their series some did not show great variation in amount drunk after 24 h of depletion whereas, with others, the variation was too great to make them suitable for testing the effect of various procedures. By increasing the size of the deficit by 48 rather than 24 h of depletion and by repetition of the procedure at intervals, the animals' ability to match intake to deficit improved (Table 11-2). The record of a single sheep, Satan, illustrates this point.

Table 11-2. The relation of voluntary sodium bicarbonate intake and salivary sodium loss observed in Satan during a series of test periods over 4 years.

Year	Period of loss (h)	No. of tests	Salivary Na loss (mean±s.d.) (mmol)	Mean $NaHCO_3$ solution drunk (mmol)
1960	24	5	324±36	234±332
1960	48	14	769±157	665±473
1960 (later)	48	9	1023±171	807±180
1963	48	45	980±109	959±155

The first studies on this animal (Table 11-2) were made in early 1960 and show more accurate replacement of deficit by the second series of tests with 48-h deficit. In 1963 when Satan was used for studies (Beilharz et al. 1965) on the influence of intracarotid 4 M sodium chloride infusions on sodium appetite it was noted that, under the conditions of control 48-h depletion, mean salivary sodium loss was 980±109 mmol (n=45) and voluntary intake was 959±155 mmol (n=45) (Table 11-2). Reducing the depletion period to 24 h caused commensurate reduction of voluntary intake. Furthermore, Satan was used also for an experimental series in late 1961/early 1962 involving study of the effect of varying the concentration of sodium bicarbonate solution offered during 30 min of access after 48-h depletion upon volume of solution drunk. The sheep showed a very close correlation of amount drunk with deficit over a wide range of concentrations (see Figs. 14-11 and 14-12). The exception was the 100 mmol/l solution. Though, on occasions, 6 litres of this were drunk, apparently rumen distension was a limiting factor. Overall Satan exemplifies the improvement with learning often seen with those animals submitted to protracted experience of variation of deficit, and of the concentration of solution offered. In a sense it is rather curious that they should improve. It can be recognized that 'underdrinking' with residual deficiency may act as a stimulus which provokes them to learn greater intake for a given sensation on later occasions. However, it is not immediately obvious how overshoot would carry any negative reinforcement for learning to adjust intake downwards in the future. Perhaps gross overshoot may act this way. But learning of adjustment would seem more likely from undershoot experiences.

It is interesting to consider this question in the light of the striking findings of Fitzsimons and Le Magnen (1969). They suggest rats may learn to anticipate impending water requirements. Their subjects drank most of their water in close association with meals. But, on transition from a carbohydrate to a protein-rich diet which required more water, they drank more water and much of the additional intake occurred between meals. This may have related to increased post-prandial water loss. Later on, however, the animals' drinking moved back to close association with meals but remained high. The rats appeared to have learned that the diet they were now eating required a higher water intake and were providing for these requirements long before they were actually incurred. In some senses there is a parallel here to food hoarding in animals, though this is no doubt innate rather than legislation for an imagined eventuality.

The strict analogy from the Fitzsimons and Le Magnen (1969) experiment to the sheep would perhaps be an animal which drank twice its deficit in anticipation of the ensuing salivary loss over the next 24 h. There is not any suggestion of animals on *episodic free access* doing this—as evidenced by the fact that most have little or no sodium in the urine. However, when we come to consider sheep maintaining themselves in the face of constant salivary loss by free access to a bar-press which delivers small amounts of sodium bicarbonate a curious spread of behaviour is observed (Chapter 13). Some consistently have little or no sodium in the urine, suggesting that they become deficient and then work and drink to bring themselves to a 'set point' near neutral sodium status. Others, however, appear to work to an equilibrium state well on the positive side of sodium balance and always have a large amount of sodium in the urine (Figs. 13-6 to 13-9). This may simply reflect different hedonic reactions to salt conjoint with the response to deficiency, or it could conceivably reflect anticipatory avoidance of the sensory consequences of sodium deficiency.

A few animals with a parotid fistula on episodic free access appear to remain capricious indefinitely, with intake wildly hunting around deficit as long as the experiment persists. Possibly some of this variation might be eliminated by a strict regime with abolition of disturbing environmental events associated with other laboratory experiments, differences between weekend regime and weekdays, and other extraneous influences. Overall, for practical purposes of experiment, it is simplest to choose animals which show least variation when replete and at various levels of depletion than to accept the higher random element and, thus, the much more extensive experimental series involved in defining a significant influence upon such behaviour.

Hedonic intake of sodium solutions and design of experiments

Sheep vary in their hedonic or 'need-free' intake of sodium solutions under control conditions. Though, as Richter (1956) proposes, there is a universal liking for salt in mammals, and the capacity to taste and be attracted to nutritive substances is important for survival, there is also variation between species in the acceptability of solutions of differing concentration. For example some sheep like 300–900 mmol/l sodium bicarbonate solution even though sodium-replete, and before any laboratory experience of sodium deficiency. On the evidence available this appears to be much more evident for sheep than for goats or cattle. Even in sheep, however, there is a difference in so far as 300 mmol/l sodium bicarbonate is more likely to be drunk in the sodium-replete state than is 900 mmol/l.

It is also evident that in terms of establishing a baseline against which a procedure inducing intake might be evaluated, there are virtues in presenting a cafeteria for limited time (e.g. 1 h per day). Often the sodium-replete animal will display only mild interest, sniffing and perhaps tasting, and then sit down; thus intake over a fortnight of tests may be negligible. Against this protocol, with the same 1 h of access continued, it may then be seen that sodium depletion by fistula, or for that matter specific intracerebroventricular infusions, may cause a radical, easily assessed, change of behaviour and large voluntary intake. At the other extreme, if solutions are available 24 h a day in an environment which is impoverished or restricted relative to field conditions in terms of behavioural repertoire open to the animal, the likelihood of hedonic ingestion appears to increase (though a fair proportion of animals still appear not to like the concentrated solutions in the absence of sodium deficiency). We do not have the data to determine whether this reflects genetic differences in the animals in terms of hedonic preferences: e.g. whether British-bred sheep are different to the Merino. Michell and Bell (1969) showed in Cheviot sheep that the life history of the animals before coming to the laboratory in relation to habituation to bore water or other sources of high sodium content may be determinant.

In animals which exhibit a large hedonic intake before fistula operation, the simple continuity of

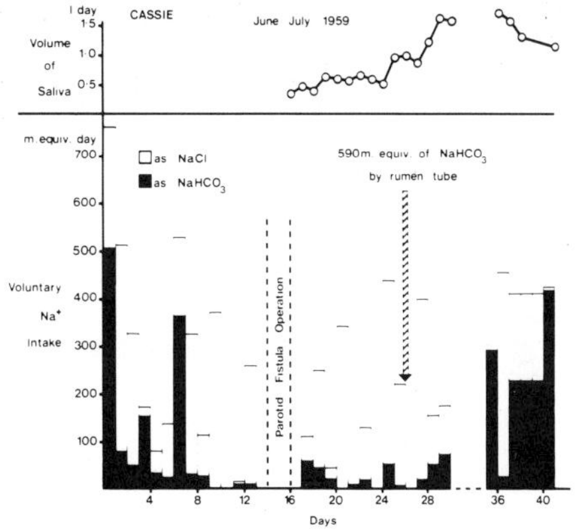

Fig. 11-33. Cassie: voluntary intake of solutions of 420 mmol/l sodium bicarbonate and sodium chloride continuously available before and after operation to prepare a permanent unilateral parotid fistula. On the twenty-sixth day of experimental observation the animal was given 590 mmol of sodium bicarbonate in 1 litre of water by rumen tube, and voluntary intake of sodium bicarbonate ceased for 24 h.

this behaviour will often cover sodium loss post-operatively (Fig. 11-33), or at least it will until fistula flow becomes very large (3–4 l/day). However, notwithstanding this history of development of sodium drinking, it may be seen that administration of sodium to such an animal (e.g. by rumen tube) will greatly reduce intake.

A clear example of how an animal's appetite may adjust to need is shown in Fig. 11-34. In this experiment (Beilharz and Kay 1963), to be discussed in detail in Chapter 14, sodium appetite was observed each 24 h during 30 min access. Mean daily salivary sodium loss was 400±31 mmol. On each day the animal received either 1.5 litres of water by rumen tube 10 min before access, or 450–600 mmol of sodium bicarbonate in 1.5 litres of water either 10 or 120 min before access. The sodium administered had little influence on sodium intake on the same day. But the point relevant here is that this augmented intake on the day of the experiment caused a large decline in, or abolished, voluntary sodium intake on the following day when the animal was sodium-replete (Fig. 11-34).

Another example of reduction of appetite with return to the sodium-replete state arises with surgical closure of a parotid fistula which has been present for some years. Again there is variation between animals, but with some the cessation of sodium loss results in virtual cessation of sodium drinking, despite hundreds of preceding daily episodes of sodium intake (Figs. 11-35 to 11-38). With two sheep, Romy (Fig. 11-35) and Thirza (Fig. 11-36), which were adrenalectomized, closure of the fistula and denervation of the gland was done after 1–2 years (see Fig. 18-6 for the close

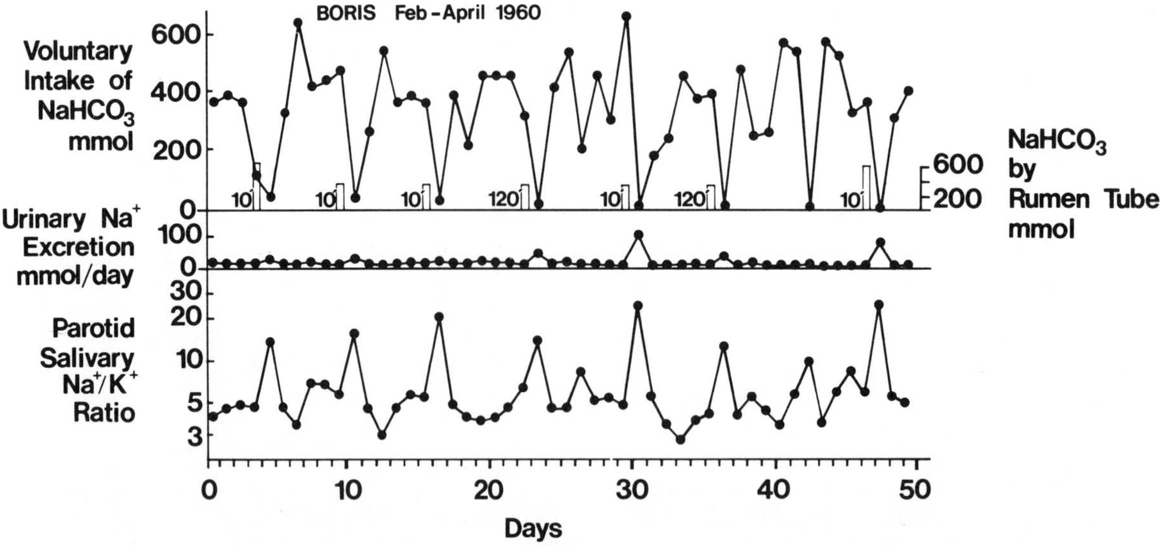

Fig. 11-34. Boris: sheep with a permanent unilateral parotid fistula offered sodium bicarbonate solution (300 mmol/l) every day for 30 min. Each day the animal received 1.5 litres of water by rumen tube 10 min before presentation of sodium bicarbonate solution, or 450–600 mmol of sodium bicarbonate in 1.5 litres of water either 10 or 120 min before access. The voluntary sodium intake was not usually influenced on the day of the experiment but in the face of the large load resulting that day it was usually greatly reduced or abolished the following day. (After Beilharz and Kay 1963.)

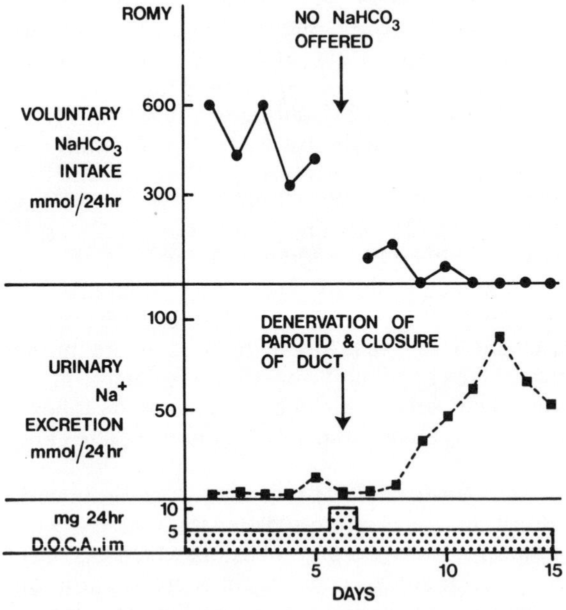

Fig. 11-35. Romy: bilaterally adrenalectomized sheep with a permanent unilateral parotid fistula maintained on basal DOCA dosage. After more than a year the ipsilateral parotid gland was denervated and the parotid duct ligated. The salivary loss of 2–3 l/day stopped. The effect on voluntary intake of 300 mmol/l sodium bicarbonate solution offered for 30 min each day and urinary sodium excretion is shown. (After Denton et al. 1969.)

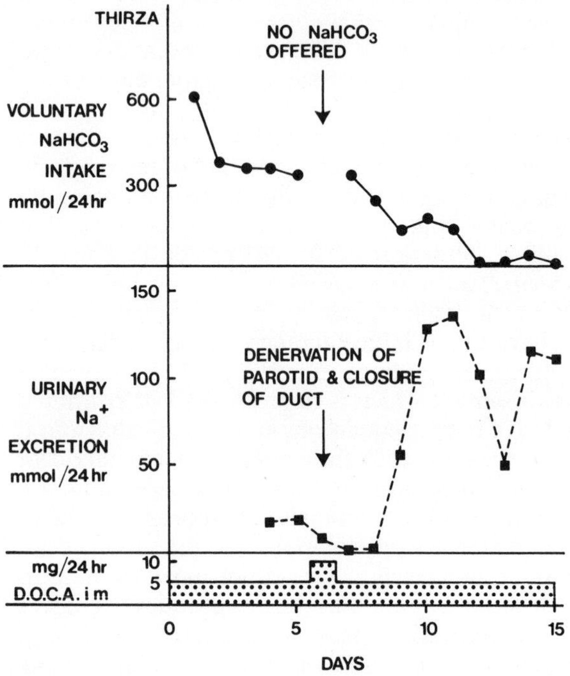

Fig. 11-36. Thirza: regime as for Romy. (After Denton et al. 1969.)

relation between salivary loss and voluntary intake seen in these two animals). Urinary sodium rose from near zero to moderately high levels. Consistent with this, voluntary sodium bicarbonate intake virtually ceased. DOCA administration was constant throughout except for a supportive doubling on the day of operation. The same procedure was carried out on Finochio and Maxine, which had had a parotid fistula for 4 and 7 years respectively. They were also hypertensive as a result of Goldblatt clamps applied to the renal artery (Figs. 11-37 and 11-38). Closure of the fistula in both cases reduced voluntary sodium intake substantially, but with Finochio residual voluntary intake was larger, and correspondingly there was a larger post-operative rise in blood pressure and a larger urinary sodium excretion.

Sly and Bell (1979) studied salt appetite in sodium deficiency in calves by exteriorizing the parotid duct and using the salivary Na/K ratio as an index of sodium status. In a T-maze, normal calves were indifferent to sodium reinforcement whereas sodium-deficient calves readily obtained 0.6 M sodium bicarbonate solution. The development of motivation and learning of the position of the sodium salt were closely related. Sodium deficiency also caused development of licking behaviour. In operant conditioning experiments, calves rapidly learned to push a panel for 0.48 M sodium bicarbonate or 0.51 M sodium chloride rewards. There was a highly significant correlation between the level of operant behaviour and sodium status as reflected in salivary sodium concentration (Fig. 11-39). Bell and Sly (1977) showed the appetite was specific for sodium except that there was an aversion to sodium carbonate, and that lithium chloride was accepted by the animals.

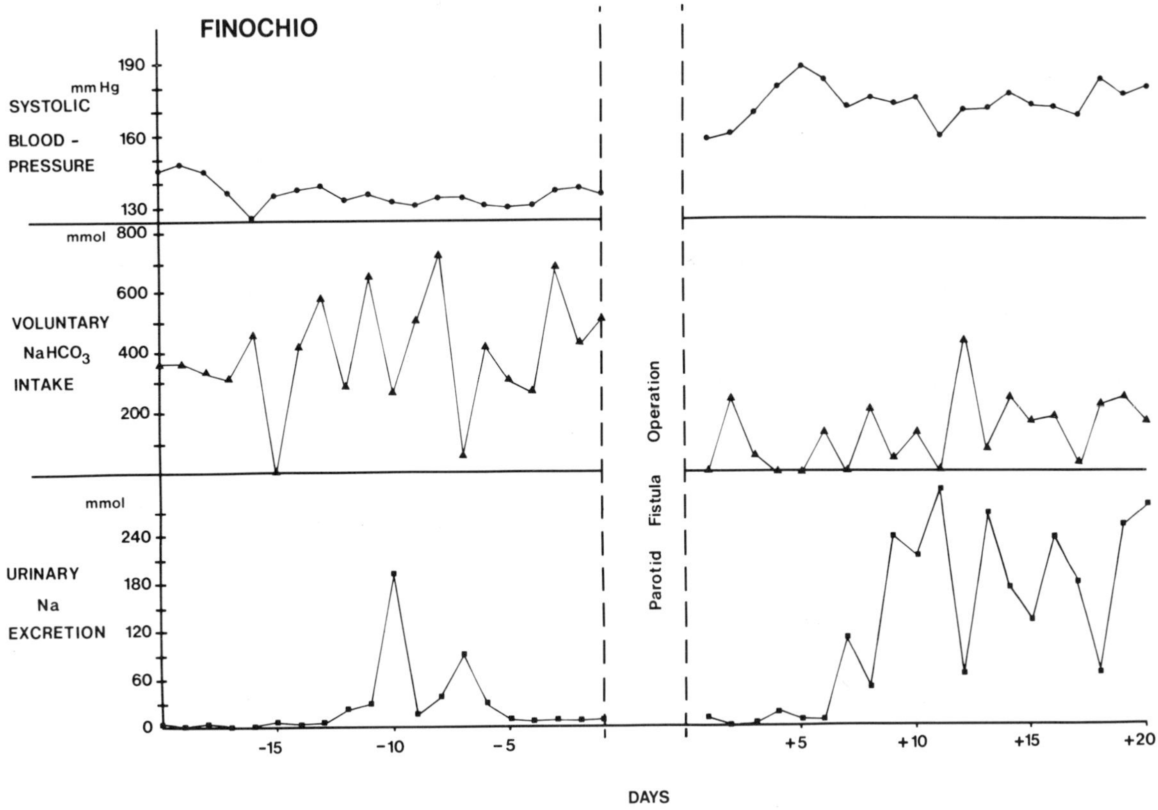

Fig. 11-37. Finochio: sheep with a permanent unilateral parotid fistula for 4 years and hypertension as a result of Goldblatt clamps applied to the renal arteries. Denervation of the ipsilateral parotid gland and ligation of the parotid duct caused salivary loss of 2–3 l/day to stop. The effects on voluntary intake of 300 mmol/l sodium bicarbonate solution presented for 30 min each day, and on systolic blood pressure and urinary sodium excretion are shown.

Hormones and innate appetite

Both DOCA and aldosterone effectively generate salt appetite on injection in rabbits and rats.

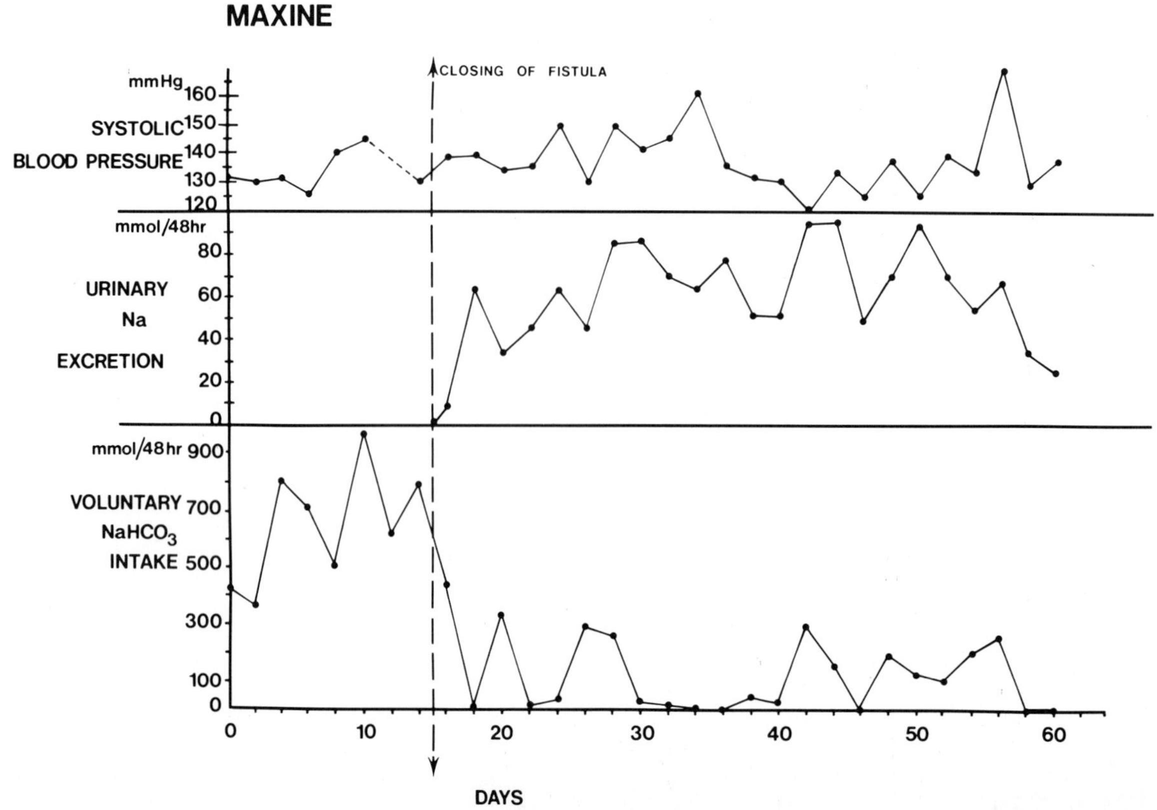

Fig. 11-38. Maxine: sheep with a permanent unilateral parotid fistula for 7 years and hypertension as a result of Goldblatt clamps applied to the renal arteries. Denervation of the ipsilateral parotid gland and ligation of the parotid duct caused salivary loss of 2–4 l/day to stop. The effects on voluntary intake of 300 mmol/l sodium bicarbonate solution offered for 30 min each 48 h, and on systolic blood pressure and urinary sodium excretion are shown. (The Goldblatt clamps used in these experiments were originally made and used by Dr Walter Cannon and were given generously to me by Dr Eugene Landis of Harvard University Department of Physiology.)

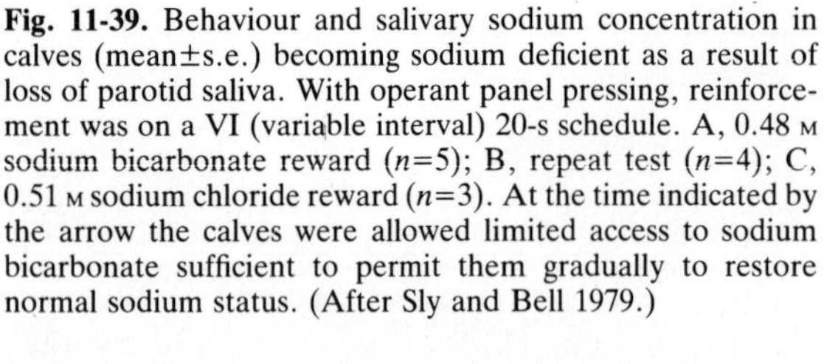

Fig. 11-39. Behaviour and salivary sodium concentration in calves (mean±s.e.) becoming sodium deficient as a result of loss of parotid saliva. With operant panel pressing, reinforcement was on a VI (variable interval) 20-s schedule. A, 0.48 M sodium bicarbonate reward (n=5); B, repeat test (n=4); C, 0.51 M sodium chloride reward (n=3). At the time indicated by the arrow the calves were allowed limited access to sodium bicarbonate sufficient to permit them gradually to restore normal sodium status. (After Sly and Bell 1979.)

Similar effects are seen with physiological amounts of glucocorticoid adrenal hormones in sheep, as recounted in later chapters. The effect has been shown to be specific for sodium salts under cafeteria conditions where salts of other cations were presented. It is difficult to attribute this specific salt appetite to retroactive learned benefit, since the phenomenon is demonstrable in animals which have always been on a diet of adequate sodium content. Further, the increased sodium intake may confer little sense of benefit and well-being, since in combination with DOCA it can evidently be toxic and cause death. It appears most likely that the hormone acts in the same neural areas to produce chemical effects similar to those caused by sodium deficiency, and which entail appetite of a characteristic specificity.

It should, however, be noted that Weisinger and Woods (1971) have reported data indicative of the possibility that aldosterone-induced appetite might be learned. That is, the rise in blood aldosterone with minor fluctuations of sodium balance becomes associated in the nervous system with the sensation of sodium deficiency, and this learned association is the basis of salt appetite inducing effects of aldosterone. However, in relation to this argument, DOCA is under ACTH control and is not increased significantly in sodium deficiency. Thus, such a learned association is unlikely as an explanation of its powerful salt-appetite-inducing effect. This action, and those of other reproductive hormones to be described in detail further on, point to an action of the hormones on the neural systems subserving salt appetite. Furthermore, Shulkin (1978) was unable to reproduce the results of Weisinger and Woods (1971) in that rats maintained on high sodium intake from weaning onwards showed an increased sodium appetite when either DOCA or aldosterone were administered.

Naive adrenalectomized animals, when given their first choice between sodium and non-sodium solutions—such as sodium chloride and either potassium or ammonium chloride; sodium nitrate and calcium nitrate; sodium acetate and potassium acetate; and sodium iodide and water—immediately choose the sodium solutions (Nachman 1962). The fact that the animals are responding to the taste of the solutions, rather than to beneficial consequences, is particularly underlined by the preference shown for such toxic materials as sodium iodide or lithium chloride. As Nachman points out, when long-term tests are used no preference is found for these toxic substances (Richter and Eckert 1938; Fregly 1958), presumably because the animals learn to avoid drinking the solutions. In short-term tests, however, when the animals have not had previous experience, they clearly prefer the taste of these toxic solutions and will ingest lethal amounts (Nachman 1962). In limited studies on the naive sheep under conditions of parotid fistula operation described in the studies above, we also found that sheep would choose lithium chloride, and return to it on subsequent days if it was offered. If permitted, they drank enough to make themselves ill.

Bell and Sly (1979) showed that three to five experiences of intake of lithium salts by sodium-deficient calves which initially accepted them readily, resulted in development of aversion, though intake was not very large. The aversion also involved sodium salts, with the exception of sodium bicarbonate which still caused a significant behavioural response in the sodium-deplete animal.

It is also just possible, as Rozin (1976) remarks, that learning about the positive effects of sodium would be difficult. It might be that the initial effects of sodium ingestion by a sodium-deficient animal are negative, particularly with a hypertonic solution. Rodgers could not induce a preference for a neutral substance in sodium-deficient rats when ingestion was immediately followed by intragastric intubation of sodium chloride. Since intragastric intubation of food or water has been shown to serve as reinforcement, the concentration of sodium chloride used may produce negative effects in the gut. Overall, however, it would not seem these arguments of Rozin get much support from the instance where direct experience of sensation is available—viz. McCance's (1936) classic experiments on man. Very shortly after taking salt there were large changes in sense of taste and flavour and disappearance of weakness and languor, though some gastrointestinal upset did occur in one subject and bounding pulse sensations were experienced.

Salt appetite in the general context of ingestive behaviour

To place salt appetite in a broader context, we can note Young's (1967) categorization of determinants of intake and acceptability.

1. *Constitutional conditions*: dependent on innate bodily mechanisms and differing between species and individuals. In this category would reside, presumably, the basis of carnivores choosing meat and herbivores grass.

2. *Intraorganic conditions*: produced by deprivation and operating through innate mechanisms. This category would include, presumably, the influence of blood osmotic pressure on thirst, blood sugar on hunger and, as Young suggests, the influence on intake of chemical implants or surgical operations on the central nervous system.

3. *Environmental conditions*, including the complex of stimuli arising from a foodstuff and its surroundings. This involves, for example, the simple taste stimulus–response relation which Bare (1949) sees as the basis of the preference–aversion curve of intake of saline solutions according to concentration in the sodium-replete, 'need-free' animal.

4. *Acquired dispositions*: attitudes, motives, expectancies, preferential habits, and similar effects of learning and experience.

We have been asking particular categories of questions about innate programming of salt hunger in the deprived animal. These might also be asked in slightly different experimental context in relation to the 'constitutional' element in acceptability of foodstuff in, for example, the species which has been the principal focus of our study—the sheep.

Thus, a lamb may be reared from birth on nothing other than mother's milk. This might be done with or without the mother's presence and, in the latter instance, feeding could be episodic. So the lamb could be deprived of all experience of the act of the mother eating any variety of food. Maternal rumination or other maternal activities might give olfactory cues, so that bottle feeding could be preferable for the experiment. At an appropriate age, as a first experience of food other than milk, would it be possible to induce the lamb to eat meat in some variety of preparation? Alternatively, on first experience of presentation, would it accept new-mown grass or a stand of some fresh grass and immediately ingest significant amounts? Would, in the case of meat, the sensory inflow, primarily olfactory and taste, on first experience have unpleasant appetitive elements leading to aversion, or would the animal simply fail to register meat as potential food? In the case of grass, would the sensory inflow (sight and smell), activate innate recognition systems involved in the organization of ingestion, so that 'potential food' would register, tasting would occur, and thus release of eating behaviour would follow in a hungry animal? In this case would the onset of such behaviour depend on a critical point of maturation of cerebral systems subserving the behaviour?

Or is it that lambs learn to eat grass by being in the presence of the mother and other members of the flock or acquire some taste or smell association through the mother's milk? At all events, it would be a most interesting matter to examine further experimentally, bearing in mind that some herbivores in a specific circumstance without prior experience can, in fact, be carnivorous (i.e. placentiphagia). Furthermore, species which are entirely herbivorous in the wild natural state will in captivity accept a large meat component of diet (e.g. the gorilla: Schaller 1963). Though, as Rozin (1976) points out, the specialists in relation to food in the animal kingdom are likely to have detection and recognition of foods under tight genetic control, it remains to elucidate the 'ontogeny' of this behaviour in relation to both acceptance and rejection. The settling of the monotreme, the echidna, on ants is a case in point and raises for example the possibility of formic acid as an innate olfactory cue (cf. McAdam and Way 1967).

However Rozin notes that unlike intraspecies recognition, where imprinting seems to be a common mechanism for determining selection, irreversible effects of early contact with a particular food are not common. But Hess (1964) reported evidence for a critical period in chicks for acquisition of preferences for stimuli that, when pecked, led to a food reward. Chicks which received such an experience on days 3 and 4 of life showed a continued preference for this stimulus whereas, if the critical reinforced experience occurred before day 3 or on day 7, little effect was seen. Other experience with chicks is probably highly relevant to the lamb question raised above. With chicks the taste of water and the regulation of water intake (Stricker and Sterritt 1967) are apparently preprogrammed, but neither can come into play until water is visually recognized in the outside world and ingested. Young chicks without experience of water and dehydrated would run through water puddles without recognizing them. But they have an inbuilt propensity to peck at small irregular objects. When this happened to occur at an irregularity in water, the association between visual water and 'prewired' water taste was rapidly made (Morgan 1894) in that the chicks immediately began drinking, and drank an amount approximately equal to water deficit (Stricker and Sterritt 1967). From then on, water was recognized visually. Thus, as Rozin (1976) remarks, everything but the visual recognition of water is prewired here.

In terms of animals choosing what foods make them feel better and avoiding what makes them feel sick, there is a great advantage as Rozin (1976) notes in having one basic system with learning. A whole host of nutritional and poison avoidance problems can be solved in this way. He sees it as unthinkable to have a specific innate mechanism to handle the full range of food selection in rats. For each component there would have to be a unique specific sensory message and a unique central state characterizing the deficiency and sensed by a specific detector. Much of the incredible amount of machinery needed would remain unused, as a given animal does not experience most nutritional deficiencies in its lifetime. And in most deficiency states animals do not show the certainty and rapidity in making up the deficit that is seen in sodium deficiency.

The advantage in evolution of the basic mechanism of learning the effects of food is crystal clear. But it is perhaps not so much the complexity of required neural organization for the innate which has been determinant as the fact that the majority of the many substances which might be relevant do not, in contrast to salt, occur free in natural circumstances. Notwithstanding, it is remarkable what capacities are emerging to be programmed in the brains of animals. For example bone appetite may be evoked and turned off in naive phosphate-deficient British breeds of cattle in a laboratory setting. The behaviour closely parallels that seen, for example, in the African giraffe in the wild (see account in Chapter 25). As the range of species investigated increases from the presently rather restricted repertoire, some interesting surprises about innate capacities may emerge. And further to Richter's demonstration that rats maintain themselves in good health and grow faster when they are given a cafeteria of 17 components than they do on a McCollum diet, it would be vastly interesting to have precise behaviour records of sampling and intake over a limited time of access per day for the first few weeks of study in the same way that we studied naive sheep at the onset of sodium deficiency. As Rozin himself remarks in discussing Clara Davis' classic experiments (1928) on children reared after weaning under cafeteria conditions with a choice of natural food, raw food or simply cooked food without seasoning, the food preferred by all children, at the one meal each day in which it was offered (supper), was prototypical chimpanzee food—raw bananas. Perhaps after supper they then went and built cubbyholes! Davis does not tell us.

References

Abraham SF, Blaine EH, Denton DA, McKinley MJ, Nelson JF, Shulkes A, Weisinger RS, Whipp GT (1975) Phylogenetic emergence of salt taste and appetite. In: Denton DA, Coghlan JP (eds) Proceedings of the Fifth International Symposium on Olfaction and Taste. Academic Press, New York, p 253

Bare JK (1949) The specific hunger for sodium chloride in normal and adrenalectomized white rats. J Comp Physiol Psychol 42: 242

Beidler LM, Fishman IY, Hardiman CW (1955) Species differences in taste responses. Am J Physiol 181: 235

Beilharz S, Kay RNB (1963) The effects of ruminal and plasma sodium concentrations on the sodium appetite of sheep. J Physiol (Lond) 165: 468

Beilharz S, Bott E, Denton DA, Sabine JR (1965) The effect of intracarotid infusions of 4M-NaCl on the sodium drinking of sheep with a parotid fistula. J Physiol (Lond) 178: 80

Bell FR, Sly J (1977) The specificity of sodium appetite in calves. J Physiol (Lond) 272: 60P

Bell FR, Sly J (1979) The effect of lithium intake on the sodium and lithium appetite of calves. J Physiol (Lond) 296: 32P

Bell FR, Dennis B, Sly J (1979) A study of olfaction and gustatory senses in sheep after olfactory bulbectomy. Physiol Behav 23: 919

Cullen JW (1969) Modification of salt seeking behaviour in the adrenalectomized rat via gamma-ray irradiation. J Comp Physiol Psychol 68: 524

Cullen JW, Scarborough BB (1969) Effect of a preoperative sugar preference on bar pressing for salt by the adrenalectomized rat. J Comp Physiol Psychol 67: 415

Davis C (1928) Self-selection of diets by newly weaned infants: An experimental study. Am J Dis Child 36: 651

Denton DA (1956) Effect of sodium depletion on the Na:K ratio of the parotid saliva of the sheep. J Physiol (Lond) 131: 516

Denton DA (1965) Evolutionary aspects of the emergence of aldosterone secretion and salt appetite. Physiol Rev 45: 245

Denton DA (1967) Salt appetite. In: Code CF (ed) Handbook of physiology, Sect 6: alimentary canal, vol 1: food and water intake. American Physiological Society, Washington, p 453

Denton DA, Nelson JF, Orchard E, Weller S (1969) The role of adrenal cortical hormones secretion in salt appetite. In: Pfaffmann C (ed) Proceedings of the Third International Symposium on Olfaction and Taste. Rockefeller University Press, New York

Epstein AN, Stellar E (1955) The control of salt preference in the adrenalectomized rat. J Comp Physiol Psychol 48: 167

Falk JL, Herman TS (1961) Specific appetite for NaCl without post-ingestational repletion. J Comp Physiol Psychol 54: 405

Fishman IY (1957) Single fibre gustatory impulses in rat and hamster. J Cell Comp Physiol 49: 319

Fitzsimons JR, Le Magnen J (1969) Eating as a regulatory control of drinking in the rat. J Comp Physiol Psychol 67: 273

Fregly MJ (1958) Specificity of sodium chloride appetite of adrenalectomized rats: Substitution of lithium chloride for sodium chloride. Am J Physiol 195: 645

Frumkin K (1971) Interaction of LiCl aversion and sodium specific hunger in the adrenalectomized rat. J Comp Physiol Psychol 75: 32

Handal PJ (1965) Immediate acceptance of sodium salts by sodium deficient rats. Psychonomic Sci 3: 315

Hess E (1964) Imprinting in birds. Science 146: 1128

Krieckhaus EE (1970) Innate recognition aids rats in sodium regulation. J Comp Physiol Psychol 73: 117

Krieckhaus EE, Wolf G (1968) Acquisition of sodium by rats: Interaction of innate mechanisms and latent learning. J Comp Physiol Psychol 65: 197

Lashley KS (1938) Experimental analysis of instinctive behaviour. Psychol Rev 45: 452

Le Magnen J (1953) L'intervention des stimulations orales dans la régulation quantitative de la prise d'eau spontanée chez le rat blanc. CR Soc Biol (Paris) 147: 614

Le Magnen J (1955) Le rôle de la réceptivité gustative au chlorure de sodium dans le mécanisme de régulation de la prise d'eau chez le rat blanc. J Physiol (Paris) 47: 405

McAdam DW, Way JS (1967) Olfactory discrimination in the giant anteater. Nature 214: 316

McCance RA (1936) Experimental sodium chloride deficiency in man. Proc R Soc Lond [Biol] 119: 245

Michell AR, Bell FR (1969) Spontaneous sodium appetite and red cell type in sheep. In: Pfaffmann C (ed) Proceedings of the Third International Symposium on Olfaction and Taste. Rockefeller University Press, New York, p 562

Mook DG (1969) Some determinants of preference and aversion in the rat. Ann NY Acad Sci 157: 1158

Morgan CL (1894) An introduction to comparative psychology. Walter Scott, London

Murchison C (ed) (1930) Psychologies of 1930. Clark University Press, Worcester, Mass.

Myers K (1970) The rabbit in Australia. In: Den Boer PJ, Gradwell GR (eds) Dynamics of populations. Proceedings of the advanced study institute on dynamics of numbers in populations. Oosterbuk, The Netherlands, p 478

Nachman M (1962) Taste preferences for sodium salts by adrenalectomized rats. J Comp Physiol Psychol 55: 1124

Pavlov IP (1902) The work of the digestive glands (Translated by WH Thompson). Charles Griffin, London

Pfaffmann C (1963) Taste stimulation and preference behaviour. In: Zotterman Y (ed) Proceedings of the First International Symposium on Olfaction and Taste. Pergamon Press, Oxford, p 257

Proust M (1934) Remembrance of things past. (Translated by CKS Moncrieff). Random House, New York

Quartermain D, Miller NE, Wolf G (1967) Role of experience in relationship between sodium deficiency and rate of bar pressing for salt. J Comp Physiol Psychol 63: 417

Richter CP (1942/43) Total self-regulatory functions in animals and human beings. Harvey Lect 38: 63

Richter CP (1956) Salt appetite of mammals: Its dependence on instinct and metabolism. In: L'instinct dans le comportement des animaux et de l'homme. Masson et Cie, Paris, p 577
Richter CP, Eckert JF (1938) Mineral metabolism of adrenalectomized rats studied by the appetite method. Endocrinology 22: 214
Rodgers WL (1967a) Specificity of specific hungers. J Comp Physiol Psychol 64: 49
Rodgers WL (1967b) Thiamine specific hunger. PhD thesis, University of Pennsylvania, Philadelphia
Rozin P (1976) The selection of foods by rats, humans and other animals. In: Rosenblatt JS et al. (eds) Advances in the study of behaviour, vol 6. Academic Press, New York, p 21
Schaller GB (1963) The mountain gorilla—ecology and behaviour. University of Chicago Press, Chicago
Shulkin J (1978) Mineralocorticoids, dietary conditions and sodium appetite. Behav Biol 23: 197
Sly J, Bell FR (1979) Experimental analysis of the seeking behaviour observed in ruminants when they are sodium deficient. Physiol Behav 22: 499
Smith DF, Balagura S, Lubran M (1970) Some effects of adrenalectomy on LiCl intake and excretion in the rat. Am J Physiol 218: 751
Stricker EM, Sterritt GM (1967) Osmoregulation in the newly hatched domestic chick. Physiol Behav 2: 117
Stricker EM, Wilson NE (1970) Salt seeking behaviour in rats following acute sodium deficiency. J Comp Physiol Psychol 72: 416
Thorpe WH (1961) Birdsong: The biology of vocal communication and expression in birds. Cambridge University Press, Cambridge
Tinbergen N (1951) The study of instinct. Oxford University Press, Oxford
Weiner IH, Stellar E (1951) Salt preference of the rat determined by a single-stimulus method. J Comp Physiol Psychol 44: 394
Weisinger RS, Woods SC (1971) Aldosterone-elicited sodium appetite. Endocrinology 89: 538
Wolf G (1969) Innate mechanisms for regulation of sodium intake. In: Pfaffmann C (ed) Proceedings of the Third International Symposium on Olfaction and Taste. Rockefeller University Press, New York, p 548
Young PT (1967) Palatability: The hedonic response to foodstuffs. In: Code CF (ed) Handbook of physiology, Sect 6: alimentary canal, vol 1: food and water intake. American Physiological Society, Washington, p 353

12 Physiological analysis of salt appetite behaviour

Summary

1. A characteristic feature of salt appetite is the delay in its appearance despite large rapid loss of sodium. With sheep the delay is usually 24 h, with rats 4–8 h. This is in striking contrast to thirst.

2. This delay in onset of sodium appetite with sodium depletion or hypovolaemia caused by extracellular fluid sequestration is also seen with appetite induced by hormones involved in the reproductive process (see Chapter 23) or adrenal hormones. Possibly it takes a considerable time before extracellular chemical changes have influence at threshold level upon the control system. The control system may be slow to access, being inside the blood–brain barrier. Alternatively the delay would reflect the involvement of a complex slow neurochemical process (e.g. transcription) in the induction of appetitive drive. This would be consistent with the known mode of action of hormones on target cells (Chapter 24). Or both factors could be involved in the delay.

3. The appetite is shown to be specific for sodium when examined under cafeteria conditions of free choice.

4. Sodium deficiency produces behaviour changes in sheep. Distance receptor stimuli presaging the imminent presentation of salt solutions evoke a range of visceral conditioned reflexes, both salivary and cardiovascular.

5. Detailed study of sampling behaviour with a cafeteria of solutions showed the effect of onset of sodium deficiency in generating a pre-eminent interest in sodium solutions.

6. A commensurate relation was shown between body sodium deficit and voluntary intake of hypertonic sodium bicarbonate in normal sheep with a parotid fistula and in adrenalectomized sheep with a parotid fistula. The data indicated the absence of any essential role of adrenal hormones in the genesis of salt appetite in response to sodium deficiency in sheep.

7. In rats with sodium deficit as a result of adrenal insufficiency, the characteristic behaviour was to overdrink considerably relative to deficit. Behaviour studies showing episodic intake by adrenalectomized rats with free access to salt solution indicated a behavioural reaction to about 3% loss of body sodium, a finding comparable to that made in sheep studied by bar-press procedures (see Chapter 13).

Time delay of onset of appetite

In the preceding chapter we noted that the mean time of onset of sodium appetite after adrenalectomy in the rat was 3.3 days. The data on naive sheep indicated that whereas there were instances of sodium intake on days 2 and 3 after parotid fistula operation, usually substantial intake began another 1 to 3 days later. In appraising this, account must be taken of the fact that the animal is in positive sodium status at the time of operation, i.e. in an equilibrium such that there is urinary sodium excretion of 30–100 mmol/day. In some instances, initial fistula loss was only 0.75–1.0 l/day. Thus given that there is continued food intake, it may be 2–3 days before fistula loss causes sodium deficit of any consequence to develop.

Experiments were carried out by Dr Richard Weisinger in order to resolve whether the delay might be caused by the time taken for the amount of deficit to reach a threshold. Sodium depletion was caused by parotid duct cannulation, and sodium loss in the first 48 h was 500–1000 mmol. This initial impact was very much greater (4–10-fold) than that with surgical preparation of a parotid fistula. The animals were believed to be

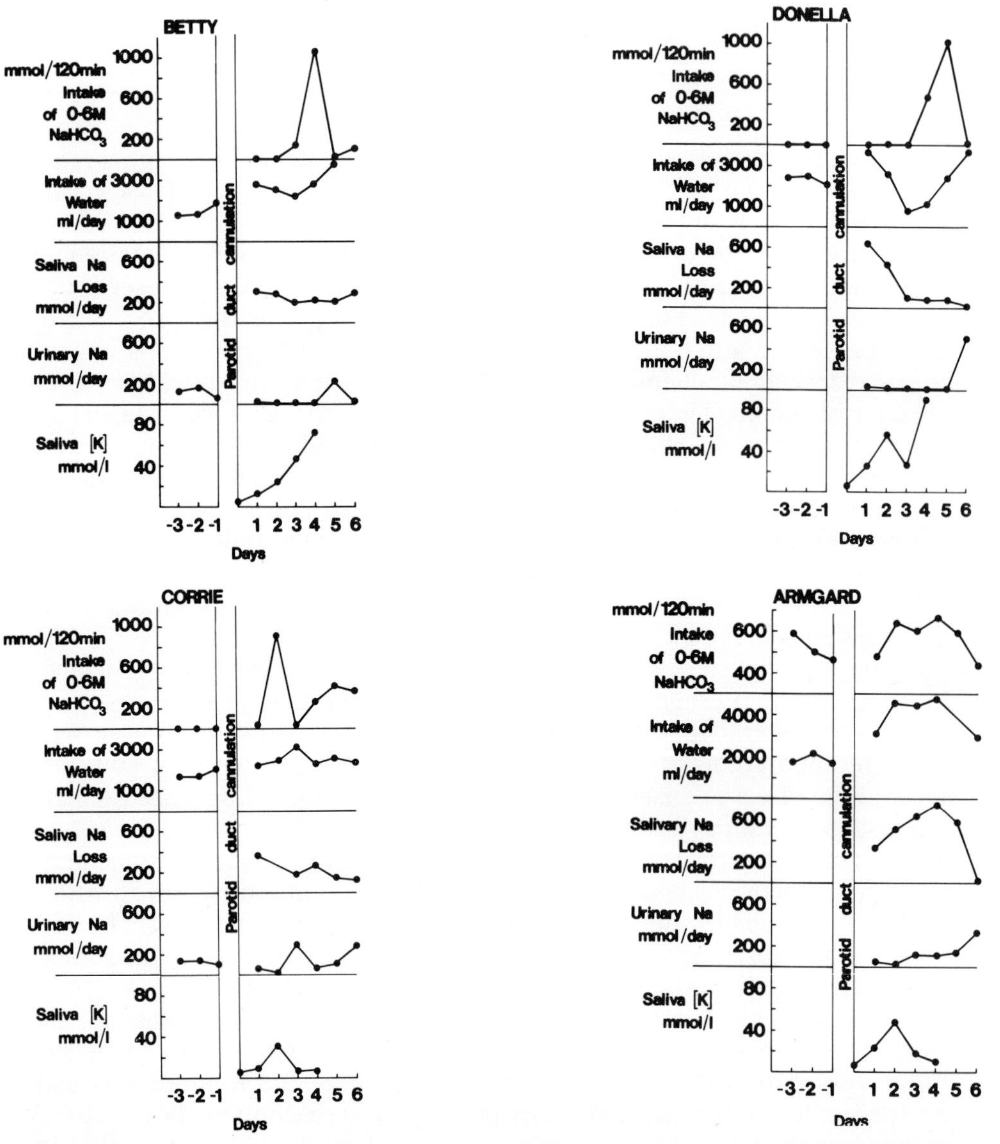

Fig. 12-1. Betty, Donella, Corrie and Armgard: naive normal sheep given access to 0.6 M sodium bicarbonate solution for 120 min/day for 3 days before cannulation of a parotid duct and for 6 days following cannulation of the duct. The voluntary intake of sodium bicarbonate solution in response to rapid onset of sodium deficiency is shown. Daily water intake, loss of sodium in the saliva, urinary sodium output and salivary [K] are shown also.

naive in relation to previous experience of sodium deficit but in the control period they had access to salt. In all but one of ten animals there was little or no voluntary intake. There were two experimental protocols: (i) access by free choice for 2 h of each day to a 600 mmol/l sodium bicarbonate solution; or (ii) access for 2 h each day to a bar-press delivering 15 ml of 600 mmol/l sodium bicarbonate (=9 mmol sodium). Water was continually available, by free choice or by bar-press in (i) and (ii) respectively.

The results of experiment (i) on five animals were clear-cut. As Fig. 12-1 (results on four animals) shows, any significant increase in voluntary drinking of sodium was delayed 2–4 days, by which time there was severe deficit. The intake then was of an order of magnitude sufficient to repair deficit; this was shown by a sharp rise in sodium in the urine. A single animal which before cannulation drank 300–500 mmol of sodium per day increased its intake to 600–700 mmol/day from day 2 onwards. Whereas sodium intake was

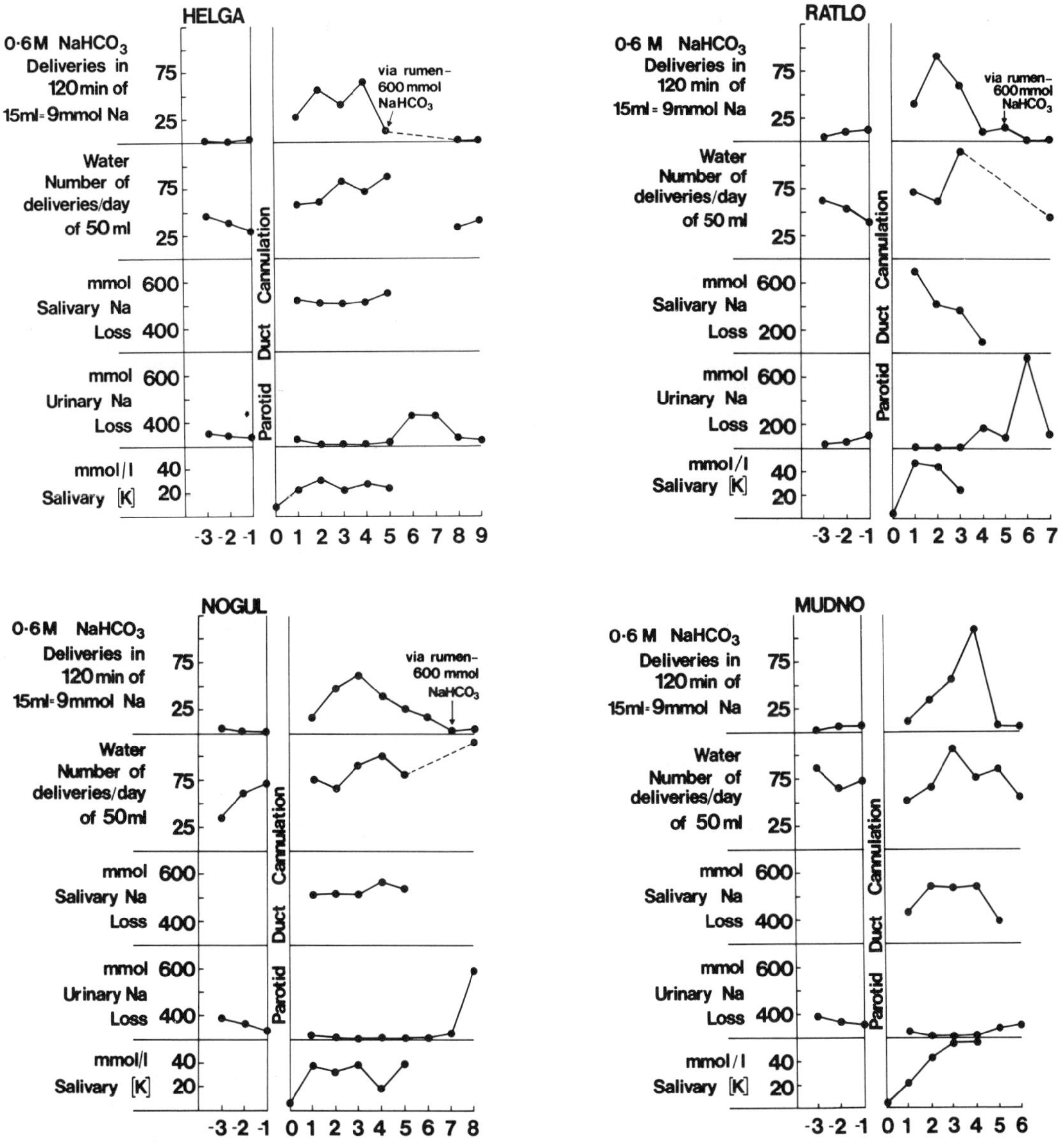

Fig. 12-2. Helga, Ratlo, Nogul and Mudno: naive normal sheep given access for 120 min/day to 0.6 M sodium bicarbonate by bar-press for 3 days before cannulation of a parotid duct and for 6–9 days following. The animals had been trained to bar-press for water, and for 3 days before cannulation one of the two bar-press pedals in the cage delivered 0.6 M sodium bicarbonate for 120 min/day. The number of deliveries (50 ml) of water per day by bar-press is shown also. Salivary sodium loss and urinary sodium output per day and salivary [K] are shown. Following 4–7 days of salivary loss the cannula was removed and 600 mmol sodium bicarbonate given by rumen tube to ensure positive sodium status, a procedure which caused bar-pressing for sodium to cease.

delayed, water intake usually rose on the first or second day (Abraham et al. 1976).

With the bar-press animals (Fig. 12-2) the rise in sodium intake occurred sooner—usually on day 2 and, in instances, on day 1, though this increase was not commensurate with deficit. Possibly bar-pressing, with learned choice between water and sodium bicarbonate levers, gave more precannulation experience of sodium bicarbonate taste, which facilitated goal-motivated behaviour when internal chemical change generated it (Krieckhaus 1970). Overall, however, the data confirm the delayed onset of appetite despite very rapid large loss of salt.

Other studies, with other methods, also indicate this delayed onset. The contrast with thirst is striking. Thirst may be evoked over 30–120 s by intracarotid or intracerebroventricular injection of hypertonic saline or angiotensin II (Fitzsimons 1972; Andersson 1972, 1977; Abraham et al. 1975). Fig. 12-3 shows such experiments in sheep.

Stricker and Wolf (1966) have studied sodium appetite induced by hypovolaemia in rats. They injected subcutaneously 5 ml of a 10%–20% solution of polyethylene glycol (PEG), thereby withdrawing protein-free plasma from the intravascular compartment. Since the sequestrated fluid is isosmotic with plasma, no alteration in intracellular fluid volume is produced. A second technique of subcutaneous injection of 2.5 ml of 1.5% formalin diminished intravascular fluid volume. Extracellular sodium accumulated in the damaged cells, reducing the osmolarity of extracellular fluid and increasing the volume of intracellular fluid (see Chapter 8).

Stricker and Wolf found that PEG injection elicited considerable water intake in 8 h with onset in 1–2 h. It had no observable effect within 8 h upon intake of either 2% saline (Stricker and Wolf 1966) or 0.9% saline (Stricker and Wolf 1969). If the test is made at 24 h, though, a striking generation of sodium appetite is evident. This result is summarized in Fig. 12-4. However, with

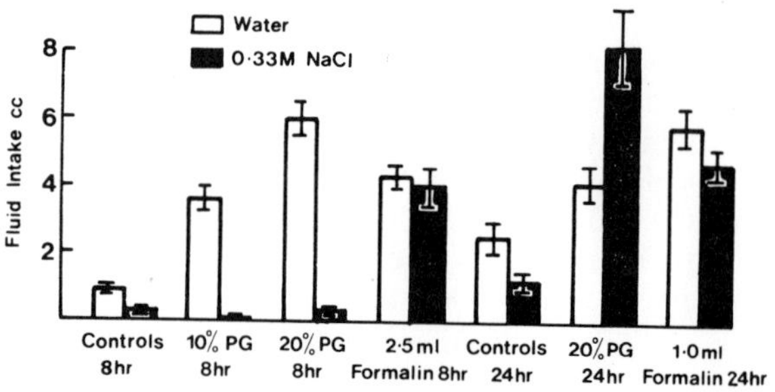

Fig. 12-4. Intake of water and 0.33 M sodium chloride by rats during a 15 min test after various tests and treatments with formalin and polyethylene glycol (PG) (n=8–9 rats in each group). (After Stricker and Wolf 1969.)

formalin treatment, which also causes hyponatraemia, salt appetite is manifest by 6–8 h. Commenting on this result Stricker and Wolf (1969) also draw attention to the slow onset of sodium appetite with mineralocorticoid treatment, citing incidental observations of Braun-Menendez and Brandt (1952) and some preliminary studies of their own. Later studies (Stricker 1980) indicate onset of salt appetite with formalin may be much faster still. This is when the hypovolaemic stress is superimposed after a period of some days of a low-sodium diet.

An important and definitive study of the phenomenon has been made by Jalowiec and Stricker (1970a). First, they followed the changes within the blood produced by the administration of formalin. Analysis of blood samples from formalin-treated rats, deprived of fluid from time of injection, indicated that hyponatraemia and hypovolaemia occurred rapidly after the injection (Fig. 12-5). Haematocrits were well above normal 1 h after injection (P<0.01), and increased still further in subsequent hours. This indicated a pronounced reduction in blood volume. Plasma protein concentration was decreased at 4 h, as would be expected from the loss of protein-rich fluid from the interstitium. Plasma [Na] (sodium

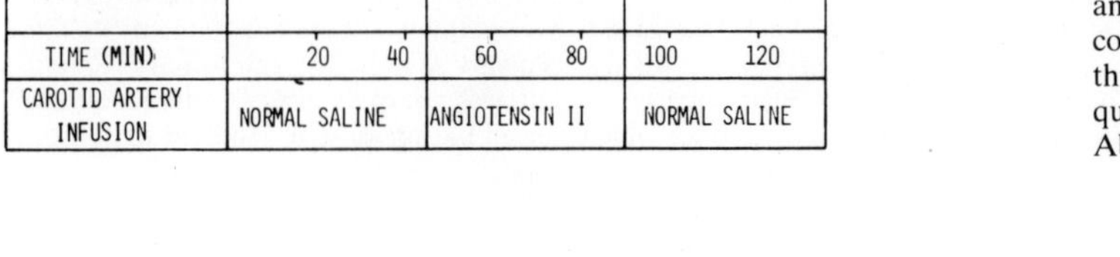

Fig. 12-3. Pattern of water drinking in Priscilla during intracarotid infusion of angiotensin II at 1200, 800, 400 and 100 ng/min and during intracarotid infusion of normal saline for 45 min before and after the angiotensin infusion. The amount of water drunk is shown also. The contralateral carotid artery was occluded throughout the infusions. When ingestion was questionable it is shown with a triangle. (After Abraham et al. 1975.)

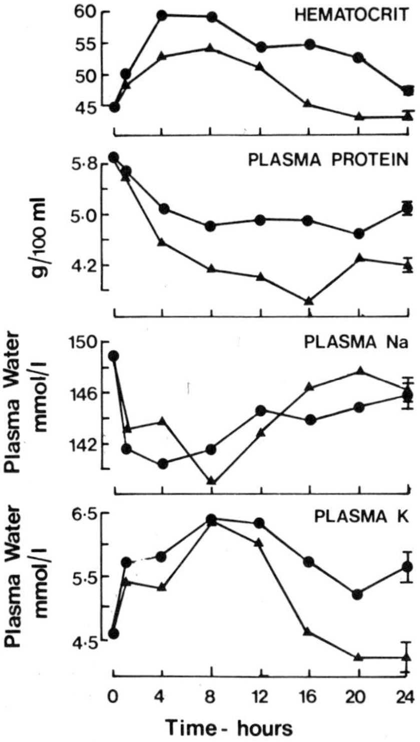

Fig. 12-5. Mean values of haematocrit and of plasma protein, sodium and potassium concentrations of formalin-treated rats deprived of fluids (circles), or permitted access to water and 0.51 M sodium chloride solution ad libitum (triangle). Each point records mean values from 6–8 rats. (After Jalowiec and Stricker 1970a.)

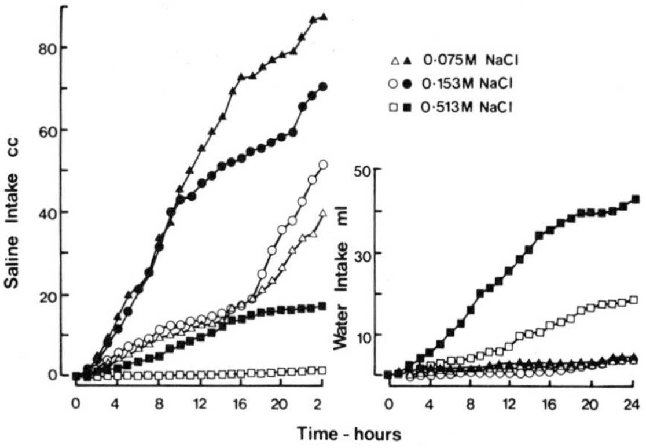

Fig. 12-6. Cumulative mean volumes of saline and water ingested by rats during control test (open symbols) or following subcutaneous formalin treatment (filled symbols). Each point represents mean values from 8–12 rats. (After Jalowiec and Stricker 1970a.)

concentration) decreased while [K] (potassium concentration) was increased, due to the cellular damage by the formalin. Both these effects were large within an hour of the injection. Maximum changes were observed 8 h after injection and thereafter all values tended to return to normal, though the plasma protein and plasma [Na] remained significantly below normal after 24 h. Against this background, behavioural tests were conducted under different conditions. These essentially involved the animals having access to either hypotonic, isotonic or a hypertonic saline solutions. Water was continuously available. The animals were studied for the 24 h following a formalin injection, and the results compared with control experiments in which no formalin was given. In the control experiments it was clear that very little intake of the hypertonic (0.51 M) sodium chloride solution occurred. (Fig. 12-6).

The intake of sodium chloride solution by all rats was significantly increased after treatment with formalin ($P<0.001$ within 4 h in all instances). Rats which drank the hypotonic or the isotonic solutions did so at comparably high rates and consumed little water, whereas those which drank the hypertonic saline did so at a much slower rate and consumed large amounts of water; thus the total water ingested by all rats was similar (Fig. 12-6). Urine sodium decreased to low levels within a few hours of formalin treatment and remained low for 8–20 h, but then abruptly increased to normal levels. This presumably reflected the fact that intake had corrected the deviations in the internal milieu caused by the formalin injection. The magnitude of the appetite produced by formalin treatment was quite large. A positive sodium balance in these animals of 5–6 mmol together with the 30–40 ml of water required to retain isotonicity, represented an expansion in extracellular fluid volume of approximately 50%. Jalowiec and Stricker note that the persistence of the appetite following the correction of body deficit, as reflected in the sodium excretion in the urine, contrasts with the data reported from our laboratories in the study of acute sodium deficiency in sheep. Here there is satiation and cessation of the appetite when voluntary access to sodium solution is permitted. They say also that many features of the behaviour of these formalin-treated rats are similar to those seen in rats receiving subcutaneous treatment with PEG, with the notable difference that sodium appetite appears earlier with formalin treatment. As a possible explanation of this, they consider that the hyponatraemia associated with formalin may potentiate the early appearance of appetite. Alternatively, the increased secretion of aldosterone that occurs in response to the lowering of plasma sodium concentration, as determined by experiments on sheep with adrenal transplants (Blair-West et al. 1963), could have been effective in the situation. A third possibility they considered was the more rapid induction of plasma deficits with formalin.

In relation to both the issue of persistence of appetite and the more rapid effects with formalin than with PEG, experimental data have accumulated in the interim on the powerful appetite-inducing effects of adrenal glucocorticoids in rats, rabbits and sheep; and in one species—the rabbit—ACTH (adrenocorticotrophic hormone) has been shown to have such an effect by direct action on the brain (see Chapter 23). The treatment with formalin could be a powerful stimulus to the release of ACTH and adrenal corticoid activity (Blaine et al. 1975), and ACTH causes a spectacular specific appetite for sodium chloride in the rat (Weisinger et al. 1978). Thus these hormones, in the instance of a traumatic technique like formalin, could synergize with the other mechanisms acting to induce sodium appetite. Jalowiec and Stricker (1970a) also note that the slow onset and the delayed offset of the appetite may reflect the slow depletion and slow replenishment of a reservoir system which is involved in the detection of body deficit and the initiation of the mechanism of appetite.

However, it would seem the continuity of the appetite after repletion could reflect continued action of the stress hormones. Indeed the offset of appetite after ceasing injection of such hormones may be very slow over days (see Chapter 23). This consideration also applies to Jalowiec and Stricker's study (1970b) showing persistence of sodium appetite 24 h after formalin injection when there had been apparent recovery from the treatment. Blood levels of glucocorticoids would be a critical element in interpretation of these data (see Chapter 8). Appetite also persists for days or weeks following its evocation by massive infusion of angiotensin II into the third ventricle (Bryant et al. 1980).

In 1972 McCutcheon and Levy studied rats injected subcutaneously with formalin to determine the onset of operant behaviour to deliver 0.4 M sodium chloride. Water was continuously available. Having determined behaviour on the control day and arranged to follow urinary sodium excretion at hourly intervals, they showed that the onset of bar-pressing began about 7 h after injection. Often the onset was quite sudden and at a high rate. It was preceded by several hours by a fall in urinary sodium excretion. They note that their time of onset of 7 h was 3 h longer than that reported by Jalowiec and Stricker (1970a), which they attribute to the fact that a higher drive level was required to initiate salt-seeking behaviour that involved considerable effort. The time of onset of the appetitive behaviour approximately coincided with the maximum fall in plasma [Na] caused by the formalin injection: viz. 8 h (Jalowiec and Stricker 1970a). The authors also note that the abruptness of onset of behaviour indicated that the restorative effects of salt were not necessary to sustain the initial behaviour—the taste was sufficient. When urinary excretion indicated salt balance was restored, the bar-pressing ceased. This result contrasted with other experiments where, when free intake of saline solutions was allowed, salt appetite persisted. Again the difference seemed attributable to the high motivation required for a regime involving considerable effort. The degree of effort needed in this situation may be increased by pain factors associated with a motor task subsequent to formalin injection (Wolf et al. 1974). Thus onset of strong motivation and satiation of deficit were more clearly delineated.

Stricker (1980) has also examined the question of whether a threshold in reduction of plasma volume is required for induction of appetite and that this causes the delay. Plasma volume is reduced by 25% by 6 h after treatment with 30% PEG. If 20% PEG is given it takes 8–9 h to achieve the same volume reduction, and with 10% PEG the 25% reduction is never reached. But sodium appetite began about 6–10 h after injection in all cases and intake was commensurate with concentration used (Fig. 12-7). Stricker also showed that a tourniquet round one limb caused loss of protein-rich fluid into the interstitial space of the limb and a reduction in plasma volume of 40% within 2 h. The rats drank water as soon as the tourniquet was removed after 5 h but appetite for 0.51 M saline solution did not appear for 8–10 h. He suggests this dissociation of time of profound hypovolaemia and onset of salt appetite offers no support to the idea that baroreceptors provide a critical stimulus for

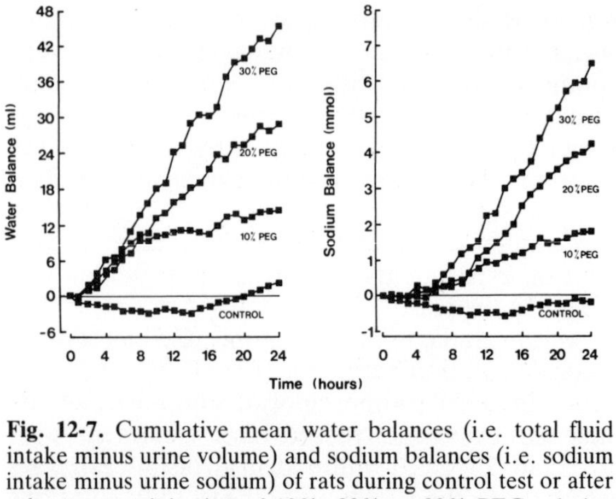

Fig. 12-7. Cumulative mean water balances (i.e. total fluid intake minus urine volume) and sodium balances (i.e. sodium intake minus urine sodium) of rats during control test or after subcutaneous injection of 10%, 20% or 30% PEG solution (n=5 in each group). (From Stricker 1980, with permission.)

sodium appetite. In another experiment Stricker maintained rats for 4 days on a low-sodium diet before giving 30% PEG treatment. This enhanced salt appetite and the animals began drinking salt when it was first offered at 4 h. Blood aldosterone rose to the very high range of 300–500 ng/100 ml as a result of the treatment, but plasma volume deficits and renin activities were only slightly greater than those seen in uncomplicated PEG treatment, and plasma [Na] was unchanged. Thus of all the variables plasma aldosterone concentration was the only one conspicuously altered.

The possibility that the high aldosterone was causative of the early onset of salt appetite was investigated by parallel studies in which high aldosterone levels were produced by ligating the inferior vena cava below the liver but above the kidneys. Blood aldosterone rose to 75–150 ng/100 ml, and when this was done after 4 days on a low-sodium diet the levels reached 350–400 ng/100 ml. But in neither instance did the rats show sodium appetite. Stricker, in summarizing his studies, notes that with PEG injection the effective volume of extracellular fluid is reduced by sequestration, which may be considerable. This reduction in available sodium throughout the body should ultimately affect cellular function and 'such effects in the brain may serve to trigger Na appetite which he says may involve new mechanisms and variables (e.g. Denton, 1966)'.

Ferreyra and Chiaraviglio (1977) depleted rats of sodium by intraperitoneal injection of 10% body weight of isotonic glucose for 30 min. Blood volume decreased 16% immediately after treatment and returned to normal in 2–4 h. There was an immediate decrease in plasma [Na] which returned to normal after 10–12 h. Salt appetite developed 10–12 h after dialysis and was maximal at 16 h, long after blood volume and plasma [Na] had returned to normal.

Zimmerman et al. (1978) have also shown a clear delay in onset of salt appetite in sheep. The animals were rapidly depleted of sodium and water by administration of the diuretic frusemide and urinary loss over 4–5 h was 37–82 mmol. They had access to both water and hypertonic saline. Thirst, as evidenced by water intake, developed in 1–2 h, but no sodium intake was evident at 6–8 h. However, by 24 h there was a clear sodium appetite.

In general, this delay in onset of sodium appetite with sodium depletion or hypovolaemia caused by extracellular fluid sequestration is also seen with appetite induced by the hormones involved in the reproductive process (see Chapter 23), or adrenal hormones. The result may indicate that it takes a considerable time before extracellular chemical changes have influence at threshold level upon the control system. The control system may be slow to access, being inside the blood–brain barrier. Alternatively, the delay could reflect the involvement of a complex slow neurochemical process (e.g. transcription) in the induction of appetitive drive; this would be consistent with the known mode of action of hormones on target cells, as will be discussed in detail in Chapter 24. Or both factors could be involved in the delay.

Specificity of sodium appetite with body sodium deficit

Specificity of sodium appetite with body sodium deficit has been shown in the experiments on the innate nature of appetite in naive sheep and rats described in Chapter 11. A further point may be noted here. The specificity of the appetite for sodium salts has also been demonstrated when the cause of the appetite is isonatraemic hypovolaemia induced by PEG injection, as distinct from other types of sodium deficiency.

Jalowiec et al. (1966) injected rats either with PEG or the Ringer vehicle, and 23 h later tested their choice of water or one of the following solutions: 0.15 M sodium phosphate, 0.15 M sodium acetate, 0.15 M sodium chloride, 0.15 M calcium chloride, 0.15 M potassium chloride and 0.30 M dextrose. No food or water was allowed in the previous 23 h. The PEG treatment resulted in large statistically significant increases in intake of each of the three solutions containing the sodium ion. These effects were apparent early in the test and generally reached statistical significance by the end of the first 5 min of drinking. PEG treatment had no effect on intake of calcium chloride or dextrose, but induced a small but significant increase in the intake of potassium chloride. Water intake was increased in all PEG-treated groups: mean of 1.8 ml in controls, 3.3 ml after PEG ($P<0.01$). The clear influence of PEG on preference for sodium solutions is shown by the ratio of intake of test solution to the total fluid intake. A relative aversion to sodium phosphate and sodium acetate in control animals was reversed to a clear preference. With potassium chloride the aversion was decreased but not to the stage of a preference.

Falk and Herman (1961) showed that sodium depletion produced by intraperitoneal dialysis reversed a previously established preference for water over 3% saline. The preference developed in the post-depletion test and under experimental conditions which prevented any post-ingestional effects of fluid choice. The authors concluded that the gustatory afferent signals produced by sodium chloride in the maze test utilized could have activated a central mechanism previously sensitized by sodium depletion and that this produced the altered behaviour. However, Falk (1965) reported further studies on sodium deficiency with intraperitoneal dialysis (IPD). The dialysing fluid consisted of isotonic glucose, 90 mmol/l ammonium chloride and 5 mmol/l potassium chloride, which presumably would have produced extracellular acidaemia and possibly some effects attributable to the ammonium ion. In these experiments the animals showed no clear-cut preference for sodium chloride over potassium chloride solutions in concentrations which were equivalently acceptable before dialysis. In relation to the intake of potassium chloride, Falk notes the reports of Laragh and Cepeci (1955) that this produces appreciable rises in serum sodium of dogs after IPD, and in the instance of humans suffering from hyponatraemia (Laragh 1954). However, in a behavioural study of this type some account must be taken of possible effects on fine discrimination which may result from disturbance intrinsic to a stressful procedure. The report suggests only 11 of 16 animals survived by the third IPD of the series. Thus, in the light of the likely acidaemia produced by the IPD, it would have been interesting—given the usual preference for sodium bicarbonate over sodium chloride of sheep losing an alkaline fluid from a parotid fistula—to have tested the preference for sodium bicarbonate over potassium bicarbonate.

This choice of sodium bicarbonate over chloride shown by the majority of sheep with parotid fistula was observed very early in our studies. Withdrawal of the daily supplement of cortisone and DOCA from an adrenalectomized sheep causes both sodium and chloride to be lost in the urine in roughly extracellular proportions. In such sheep, adrenal insufficiency caused the choice of sodium chloride to predominate.

The behaviour of sodium-deficient sheep

A study of our colony of parotid-fistulated sheep in relation to routine presentation of sodium bicarbonate solution showed that distance receptor stimuli presaging the imminent presentation of sodium solution to a sodium-deficient animal evoked a wide spectrum of visceral responses which were not observed in the sodium-replete animal (Denton and Sabine 1963).

Before discussing this observation, some general reflections, historical and methodological, on parotid fistulae in ruminant animals are relevant. This is because not only is the fistula the technique for producing severe sodium deficit, but the rate of flow of saliva from it reflects excitatory and inhibitory states within the central nervous system, as initially recognized by Pavlov.

Lorenz (1950) pointed out considerable limitations in the knowledge likely to accrue from the Pavlovian type of experiment on conditioned reflexes. Principally, these stemmed from the fact that the animal was isolated from stimuli other than the one chosen by the experimenter, and this restraint impeded the recognition of innate patterns of behaviour. In emphasizing the importance of observing animals at liberty in their natural environment, Katz (1953) stated that 'the more one restricts the conditions of response, the more specialized will the animal's behaviour become: finally one gets to the point where the animal, so to speak, no longer behaves as a complete organism but only as a bit of itself. In the Pavlovian "conditioned reflex" method the salivary gland of dogs is isolated by a surgical operation. The reactions of the gland to certain situations can thus be followed exactly. But can one still call this a response of the animal? Is it not really the reaction of an isolated psychophysical mechanism to which the animal's body happens to be attached?' While Lorenz's comments on the limitations of the orthodox experimental situation used to study conditioned reflexes are essentially valid (as are Katz's remarks as they bear on this point), the above statement would appear to reflect a misunderstanding of the potentialities inherent in the Pavlovian method of systematically preparing an animal by aseptic surgery so that physiological functions hitherto concealed from easy access may be easily measured in the conscious undisturbed animal. Field observations of an animal's behaviour may reveal a complex motor pattern; but it is quite conceivable that the response of the animal in a specific field situation may also involve, for example, change of motor and secretory activity of viscera. Though concealed, such changes would nevertheless be part of the total response of the animal to a natural situation, no less than that of the skeletal motor system.

As indicated earlier, the development of a new type of parotid fistula by Professor Wright in which the papilla and duct were implanted into a skin flap teat, allowed secretion rate of saliva to be unobtrusively recorded without the attachment of any apparatus to the animal. With the Wright operation, the important advantage accrues that this index of response to distance receptor information can be conveniently used when the animal is in the company of other sheep. Further, parotid secretion in the sheep is continuous. Whereas in the dog a response may be measured by whether or not secretion occurs, in the sheep an environmental event may cause increase, decrease or no change of the continuous secretion. The direction of change may be related to the cortical and affective reaction to information from the particular distance receptor. Thus, the sight or sound of other sheep eating may cause a large increase in salivary secretion rate, whereas the entry of a dog into the room may cause precipitate decline in rate (Fig. 12-8). The latter is due to parasympathetic inhibition, as it occurs also in animals with the superior cervical ganglion extirpated (Fig. 12-8).

While Colin (1898) denied a psychic effect on secretion in ruminants even though the animal was very hungry, Ellenberger and Hofmeister (1887) took a contrary view. On the basis of experiments on sheep with cannulated ducts, Scheunert et al. (1929) agreed with Colin. Kudriavtsev and Anurov (1940) found no psychic secretion of parotid saliva when sheep were teased with food. Blokh (1939) found that parotid secretion in the bull was continuous; but the submandibular secreted only during eating. He demonstrated that the sight of food or the arrival of the usual attendant evoked mandibular secretion, and a large decrease in continuous parotid secretion. This type of reflex effect on parotid secretion was observed in the goat by Tikhomirov and Fomin (1950), and also in cattle by Krinitsin (1940). Fomin (1941) reported that, in elaboration of a conditioned parotid secretion in the calf, there were three stages of effect in which the selected artificial stimulus (metronome sound) respectively decreased, had no effect and then increased parotid secretion.

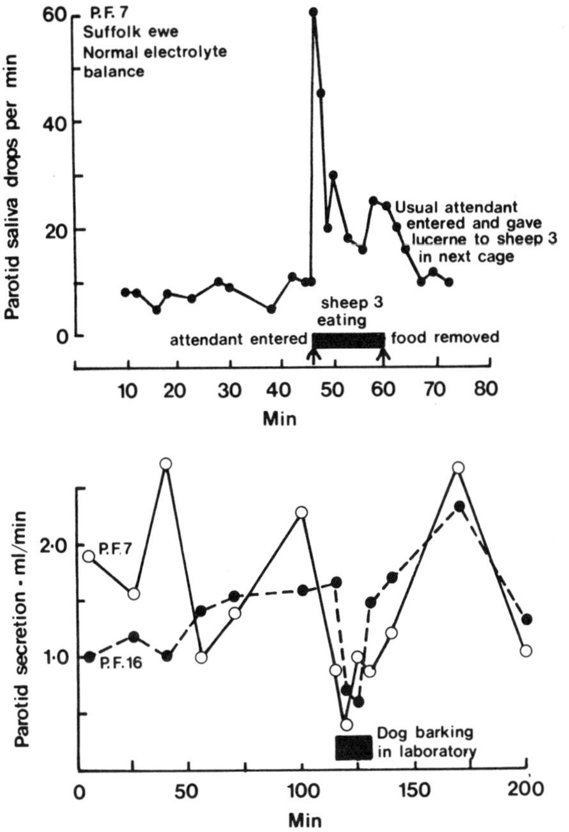

Fig. 12-8. Upper: PF7, the effect on parotid secretion rate of the entry of the usual attendant into the laboratory. The attendant gave fresh lucerne to sheep 3 in an adjacent cage and left the laboratory within 1 min. He re-entered 15 min later and removed the food from sheep 3.

Lower: The effect on salivary secretion rate of bringing a dog which barked into the laboratory for 15 min. PF16 had the ipsilateral superior cervical ganglion extirpated and the cervical sympathetic had also been sectioned. (After Denton 1957.)

The oft-quoted view of Colin, linked with the erroneous belief of Eckhard (1893) that cutting the parasympathetic nerve to the parotid gland of the sheep does not affect the secretion rate, influenced the direction of some important investigations. For example, Liddell (1954), who did outstanding work on the induction of experimental neurosis in sheep, after describing his initial acquaintance with Pavlov's work as a result of a visit by Anrep in 1923, remarks that 'sheep and goats, as cud-chewing animals, salivate continuously; hence we were unable to study salivary conditioned reflexes'. N. F. Popov (1954) also states that, in studying the conditioned reflexes of farm animals, the method by which conditioned reflexes were elaborated on a motor defensive basis was more applicable because of the peculiar aspects of horses' and ruminants' parotid secretion.

Behaviour changes evoked during experiments on voluntary sodium intake

In an individual animal laboratory of the Howard Florey Institute, 6–20 sheep with a permanent unilateral parotid fistula are kept in metabolism cages placed about 1 m apart. There are openings at the front and sides of the cages sufficiently wide for the animals to see easily what is going on in the laboratory. Two to three months after arrival the majority of sheep are quite unperturbed by members of the staff moving close by them or spending all day in close proximity when this is necessitated by an experiment on a sheep in an adjacent cage. A fairly regular routine of feeding, and provision of water and sodium supplement is followed each morning or afternoon. The sheep take a lively interest in the proceedings. Some weeks after the regime of self-selection of sodium supplement is begun, it is evident that preparation of these solutions usually provokes great excitement in a sodium-deplete sheep. When the usual person lifts the sodium bicarbonate container from the shelf, commences to use the scales and add water to the usual tins, the sight and characteristic sounds may cause bleating, stamping in the cage, considerable increase in the rate of salivary secretion, and sometimes an attempt to stand on the back legs and peer over the top of the cage. The specificity of the interest and expectation is indicated by the fact that if a sodium-deplete animal is fed and is provided with water first, its behaviour is essentially the same during subsequent preparation of the sodium solutions. It returns its attention to the food bin and eats only after it has drunk enough of the sodium bicarbonate offered to satisfy itself.

As with the gregarious influence on eating and salivary secretion mentioned above (Denton 1957), the sight and sound of another animal drinking a sodium solution often will excite a large increase in salivary secretion rate in a deplete sheep. The act of drinking sodium solution is different from drinking water in that the sheep usually drinks the sodium solution more slowly, will often 'roll it around' in the mouth, savour and chew it, the end result being a considerably more audible performance than with drinking water. On some occasions it has been observed that the act of drinking a sodium solution by an excited deplete sheep causes an apparently satiated sheep in normal sodium-balance to return its attention to the sodium container and drink some.

In particular experiments a specific daily routine was followed which included the placing of steel guard sheets on each side of the cage before offering the electrolyte solutions at zero time. After several repetitions, the placement of the steel guards acted as a specific stimulus to a conditioned salivary reflex. The steel guards occluded the animal's view while the four fluid-containing bins were placed in randomly determined positions. On removal, the animal had immediate access to all four, and choice procedures could be observed. It was seen that the animal's excitement mounted as the attendant proceeded through the standard sequence of events leading to presentation of solutions.

The degree to which these behaviour changes and conditioned salivary reflexes were manifest was related to the degree of sodium deficiency. This was repeatedly demonstrated over a period of 3 months with experiments on animal PF48. The containers held sodium solutions only on every third day. On the morning following 24 h of access to the sodium solutions the sheep was restless, showed interest in the routine of preparing food and solutions, but, when the food was given, it quietened and was too occupied with eating to sample the fluid in the containers. By 2 days of depletion the sheep was very restless and upon removal of the guards usually sampled the four containers, which all contained water, several times before returning to eat food. By the third day of depletion the sheep jumped up and down in its cage, bleated, pawed the sides, and ignored food until it had drunk a large amount of sodium following the removal of the steel guards.

Some of the sheep studied have shown considerable individuality in their behaviour. PF33, an alert, apprehensive, and rather curious ewe who spent 5 years in our laboratory developed after 3 years a curious stepping activity with the head lowered, a stereotyped feature of which was a lateral swing of the left hoof so that it banged the saliva shield on the side of the cage each time. The rate of 'stepping' was, at times, as rapid as 50 per minute. At first this pattern occurred only when the animal was excited by distance receptor information heralding food or sodium-containing fluid. However, after some months the animal sometimes spent several hours a day 'stepping', though the food bin and sodium bicarbonate container were full. The frequency of occurrence was greatest, however, when the animal was sodium-deficient or hungry.

Visceral conditioned reflexes evoked by distance receptor stimuli

An experiment was made on PF33 when sodium deficient to measure some of the conditioned physiological responses evoked by events in the laboratory. Thirty-five minutes before the record shown in Fig. 12-9 commenced, an 18-gauge needle leading to a pressure transducer was inserted into the left carotid artery. A record of mean systemic arterial pressure was provided. At this point measurement of parotid salivary secretion rate, cardiac and respiratory rate were begun also. At the same time the food bin, which still contained a substantial amount of uneaten food, and the water bin were removed from the cage. Thus the animal was sodium-deplete but had had the opportunity to satisfy desire for food or water. The animal had been calm and the readings relatively constant for the last 10 min of the 35-min period indicated above. Fig. 12-9 shows the observations from there on. In the pre-period there had been one or two episodes of 2 or 3 min duration of 'stepping' and there was a short

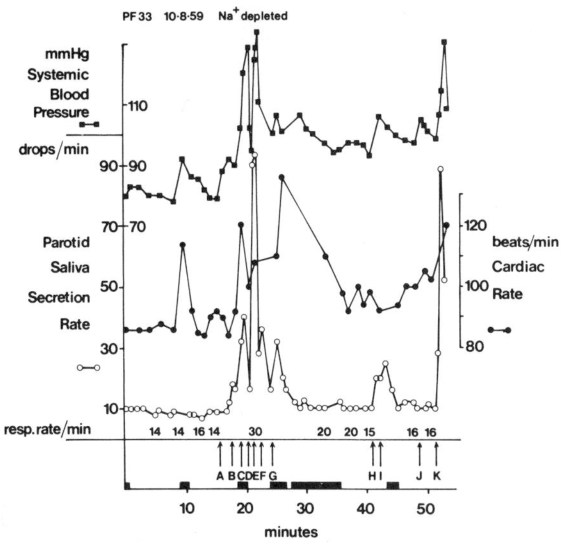

Fig. 12-9. PF33, sodium-deplete Merino cross-bred ewe. The effect of environmental events on a number of physiological functions. The top section records mean systemic blood pressure in mmHg (filled squares). The middle section records cardiac rate in beats/min (filled circles). The bottom section records parotid salivary secretion rate (open circles). The respiratory rate per minute is noted also, and the solid blocks at the bottom of the figure record the duration of episodes of 'stepping' behaviour. The letters A–K designate events occurring in the laboratory (see text). (After Denton and Sabine 1963.)

episode just as the record began (the duration is shown as a solid block at the bottom of the figure). A second episode lasting about 1 min occurred at the tenth minute (Fig. 12-9), and there was an effect on blood pressure and cardiac rate. Otherwise these factors, and the salivary secretion rate and respiratory rate, remained constant.

At the time marked A, one of the observers got up from a chair, walked to the corner of the laboratory where sodium bicarbonate was kept, lifted down the tin, took measuring cylinders and scales out of the cupboard, and at B started to mix a sodium bicarbonate solution. The sheep bleated. These procedures caused a moderate rise in salivary secretion rate and blood pressure. At C the usual container holding sodium bicarbonate solution was picked up and the experimenter walked across the laboratory, but passed PF33's cage and presented the fluid to Transplant 9 in another cage. This action caused a large rise in blood pressure, cardiac rate and salivary secretion rate. There was an episode of 'stepping' at this time, but this stopped when the container was removed from Transplant 9's cage and taken back to the scales for weighing at D. At this point there was a large decrease in salivary secretion rate, blood pressure and cardiac rate. At E the sodium bicarbonate was presented to PF33. As the person carrying the tin approached to within a metre of the cage, the saliva commenced to squirt and the rate record ceased temporarily at 90 drops/min.

Blood pressure reached a level 40 mmHg above basal and respiratory rate doubled. The animal drank approximately 400 mmol of sodium bicarbonate between E and F. At G it was permitted to drink again and the container was withdrawn immediately it ceased drinking. During the following 10 min there were two long episodes of 'stepping' at 40–60 per minute. Salivary secretion rate returned to the control level, but blood pressure remained significantly elevated though there was not enough time for absorption of sufficient sodium bicarbonate solution from the rumen to affect plasma volume. At 40 min the blood pressure was still elevated, and at H food was weighed out on the scales; at I it was presented to Transplant 9. There was a small effect on salivary secretion rate and blood pressure and the 'stepping' recommenced for a brief interval. At J the food was presented to PF33, but it declined to eat and there was less effect on blood pressure and salivary secretion than had followed presenting food to the neighbouring sheep. At K the sodium bicarbonate solution was presented again to PF33, and once again there was a very large effect on blood pressure and salivary secretion rate.

This type of observation, of which the experiment described in detail above is typical, indicates that the state of sodium deficiency invests certain events in the environment with high significance. Distance receptor information activates memory patterns related to previous satiation of salt hunger and the initial state of central excitation and drive engendered by sodium deficiency is thereupon greatly amplified. There is a strong presumption, from direct observation of the animals' response to a wide variety of stimuli which can be presented at this time, that the reticular activating system has attention acutely orientated to stimuli relevant to preparation and presentation of sodium solution. Some stimuli, at other times significant, such as food or water, may now be ignored. But the movement of a tin which usually contains salt elicits bleating and stamping. The direct offer of water, potassium chloride or food fails to satisfy the animal, and it continues to bleat, stamp, and jump up and down in an excited fashion until sodium is given. The state of excitation irradiates generally within the brain. For example, widespread medullary effects occur such as those on salivary secretion, cardiac rate, blood pressure and respiration. In Chapter 21 we have described also effects on aldosterone secretion by the adrenal gland, but whether this is initiated by events in the medulla or in the hypothalamus is not known.

The rapid increase in parotid salivary secretion, reflecting parasympathetic discharge, which has been observed in dozens of animals over the years, is a classic conditioned reflex. Also it is interesting that many of the animals develop Type II conditioned reflexes (Konorski 1950). That is, the occurrence of the procedures which usually indicated imminent presentation of salt solutions provoked characteristic motor responses. Then, eventually, a stage was reached when the animals, when sodium-deplete, would make these movements as soon as any person entered the laboratory, even though the actions of the person (e.g. sweeping the floor) were unrelated to preparation of sodium solutions. It was also noteworthy that the whole spectrum of these central nervous events—vocal, motor, salivary secretory and cardiovascular—were related to the degree of sodium deficiency. This is with the qualification that a grossly deplete animal, concurrent with evidence of circulatory deterioration, occasionally became apathetic. However, it usually drank a large amount of sodium solution when it was offered.

Another interesting aspect of behaviour shown occasionally by sheep, was observed also in PF33 in another experimental context. In an experiment (Denton and Sabine 1961) lasting over 30 days, described in Chapter 14, after a control period in which a sodium bicarbonate solution of 237 mmol/l was available all day to the animal, the sodium solution was available for only 1 h each day. The concentration was changed every fifth day over the next 20 days. The observation of interest here began on the sixteenth day, when the concentration of the solution was reduced from 952 to 119 mmol/l. The large parotid fistula loss caused a negative external sodium balance of 500 mmol/day in this sheep, and during the preceding 5-day period when the concentration of sodium bicarbonate solution offered was 952 mmol/l the deficit was repaired by the animal drinking ca 0.6 litres in the hour. However, with concentration reduced to 119 mmol/l it would have necessitated the animal drinking some 4–5 litres within the hour to correct its sodium deficiency. It was of great interest to observe the restlessness that developed during the hour periods of this 5-day section of the experiment. The sheep sampled the sodium solution throughout the hour, the number of sampling episodes being increased 2- to 4-fold relative to previous observations. The total volume drunk in the hour was often over 3 litres and many small sips were taken in the latter part of the hour. The striking and unique feature was that between these episodes the animal devoted its time to busily nibbling at the stainless steel floor of the cage and its side supports. The whole appearance was that of eating grass. On the first day sodium bicarbonate at a concentration of 476 mmol/l was available, the animal consumed 3.28 litres immediately. This amount greatly exceeded deficiency, or any other episode of intake of the experiment.

The 'grass-eating' behaviour appeared to be a displacement activity. That is, as described by the ethologists, the energy specific for one behaviour pattern is transposed into another behavioural context which is usually incongruous or irrelevant (Armstrong 1950; Lorenz 1950). This usually arises when a drive is thwarted. In this instance it seems that the genesis of this behaviour could have been that alimentary proprioceptor impulses signalling distension from volume drunk, collided centrally with the only partly discharged drive for sodium ingestion which was usually rapidly satiated. Displacement activity of this type in sheep was observed by Liddell (1954) during maze experiments, where animals were observed nibbling the concrete floor. There was evidence also of other effects from some days' experience of the situation of low concentration of sodium bicarbonate. The usual roughly accurate adjustment of intake to state of sodium balance exhibited by this particular animal was overridden when at the end of the period of access to 119 mmol/l sodium a solution of 476 mmol/l was offered. The animal immediately drank 3.28 litres (=1561 mmol of sodium), an amount greatly in excess of need.

Study of sampling behaviour as an index of salt appetite drive

The experiments recorded here on two parotid-fistulated sheep, PF33 and PF48, are representative of observations over many years. The sheep had access to four randomly placed solutions (water, sodium chloride and sodium bicarbonate (420 mmol/l) and potassium chloride (134 mmol/l)) at the same time each day for 1 h only (test period). Water was available continuously. In those control periods during which the animals were kept in normal sodium balance by administration of 595–830 mmol of sodium bicarbonate in 1 litre of water via rumen tube, this dose was given immediately following the test period, i.e. the test period was 23 h after the last dose of sodium. During sodium depletion periods the same procedure was followed except that the intra-rumen dose was 1 litre of water. When the guards were removed at zero time, allowing the sheep access to all four solutions, the exact sequence of actions was carefully recorded. The act of drinking, tasting, dipping the head inside a tin was recorded as 'sampling episode', but clearly a record in this respect is limited by lack of exact knowledge of the olfactory capacities of sheep. Observations recorded on cine film have suggested that the experienced sheep can recognize the contents of containers by smell. In fact, viewing this film with its evidence of sheep recognizing sodium solution by smell was the stimulus to Henkin's study of olfactory acuity of adrenalectomized humans (Henkin 1970). This followed on the studies of Henkin et al. (1963) on taste acuity of adrenalectomized humans.

The results on sheep (e.g. Fig. 12-10, PF33) showed that during sodium deficiency there was a large increase in sodium intake with a preference for sodium bicarbonate, and a large increase in number of sampling episodes occurred. When the animal was deplete the behaviour before removal of the guards at zero time resembled that described and shown in Fig. 12-9. On the first day of depletion drinking episodes continued throughout the hour and there were three times as many as during the control period. On the second day the sheep drank 0.91 litre of sodium fluids in the first 5 min. It tasted potassium chloride four times during the period but did not drink it. Otherwise, its attention oscillated between sodium chloride and bicarbonate, with intake of the former predominating. There were nine times as many sampling episodes with the sodium solutions as during the control period, and this was mainly in the first 10 min. The animal ignored the water until 42 min, when there were a number of sampling episodes and 0.96 litre was drunk. On the other days of depletion all fluids were sampled within 1 min 25 s, and the bulk of the sodium bicarbonate intake was within the first 5 min. On the third and fourth days of sodium deficiency it happened that the sequence of examining the solutions was such that sodium bicarbonate was encountered last, and having tasted and rejected the others, the animal drank this. On the fifth day 0.81 litre of sodium bicarbonate was drunk in 1 min, and from 3 min until 40 min the animal took no interest in any solution, at which point it drank another 140 ml of sodium bicarbonate, going directly to this container without sampling the others at all.

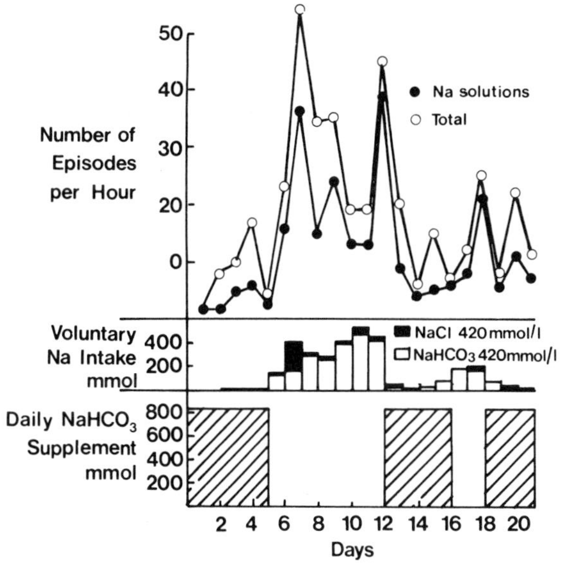

Fig. 12-10. PF33, Merino cross-bred ewe. The upper section shows (i) total number of episodes of sampling four solutions (sodium chloride, sodium bicarbonate, potassium chloride and water) during 1 h of access permitted each day, and (ii) the number of sampling episodes which involved the two sodium solutions. The middle section shows the voluntary intake (mmol of sodium) of sodium chloride and bicarbonate during the hour. The bottom section shows the days on which an intrarumen supplement of 833 mmol sodium bicarbonate in 1 litre of water was given. (After Denton and Sabine 1963.)

However, the choice was not always categorical in that on the seventh day of depletion the animal's attention oscillated between the two sodium solutions over the entire hour, and of the 39 episodes involving these solutions, 19 were with sodium chloride (amount drunk, 0.09 litre) and 20 with bicarbonate (amount drunk: 1.01 litre). The fact that the increase in sampling activity during the period was confined largely to the sodium solutions is shown in Fig. 12-10, and the absence of any significant increase in episodes involving the other containers during this time suggests that once identified they were rejected from further attention. During the first day of the recovery period (i.e. sodium bicarbonate given 23 h beforehand), 9 min passed before all electrolyte solutions were sampled. On the second day 27 min elapsed. During the second episode of sodium deficiency all solutions were sampled within 1 min 3 s but there was little interest in the solutions after 10 min, by which time intake of sodium bicarbonate had occurred. During the second period of recovery from sodium deficiency the animal took little interest in the solutions. On two days it spent most of the test period lying down. On no day during control or deficiency periods did the intake of potassium chloride solution exceed 20 ml.

With PF40 (Fig. 12-11) the results were the same as with PF33 in relation to greatly increased sodium intake during deficiency, preference for bicarbonate and increased sampling of sodium solutions. On some days when it was not in deficit the sheep took no interest in the solutions until 15–45 min had passed. Water was always sampled. But once it was the only container examined, and there was no attempt to smell, taste or drink the other solutions. During depletion, with one

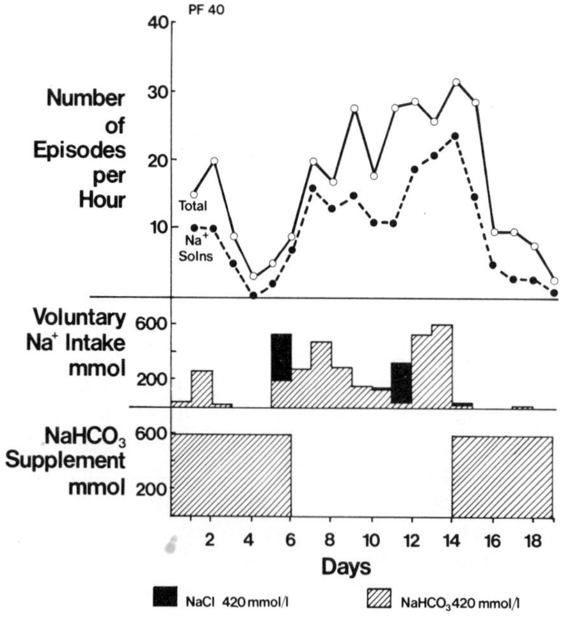

Fig. 12-11. PF40. A similar experiment to the one recorded in Fig. 12-10. In the control period the sheep was given 600 mmol sodium bicarbonate a day to compensate for salivary loss.

exception, all solutions were sampled within 1 min 25 s. On three occasions it encountered the sodium bicarbonate first or third in the series, but it sampled all four solutions and then returned to the bicarbonate and drank a substantial amount immediately. On other occasions it drank the bicarbonate immediately it encountered it. On one occasion, as with PF33, it encountered the other three solutions first, and paid no attention to the fourth bin until 18 min later.

With both PF33 and PF40, the rate of drinking the fluids was recorded. The main finding was that they usually drank water at a rate of 10–20 ml/s while sodium chloride and bicarbonate were drunk much slower, at about 1–5 ml/s. However, when sodium deficient the animal sometimes drank sodium solutions at 10–20 ml/s during the first intake. The rate was reduced to the range cited above during the remainder of the hour.

The relation between voluntary sodium intake and body sodium balance in normal and adrenalectomized animals

Sheep

The relation between voluntary intake of sodium solution (300 mmol/l sodium bicarbonate) and body sodium balance was studied in normal and adrenalectomized sheep.

In both adrenal-intact and adrenalectomized animals, voluntary sodium intake was significantly correlated with sodium balance. The results indicate that salt appetite may be generated commensurately with body deficit independently of the contemporary level of aldosterone in circulating blood.

Eight sheep with a parotid fistula were studied, of which four were adrenalectomized. Sodium deficiency was produced by loss of saliva.

Experimental plan

(1) Daily routine treatment. Water was available continuously. Food, 0.8 kg of chaff, provided about 60–80 mmol sodium per day. Sodium bicarbonate solution (300 mmol/l) was offered for 15 min at about the same time each morning. This was designated as the 'sodium drinking period'. After the sodium drinking period, 5 mg DOCA and 25 mg cortisone were given to the adrenalectomized sheep by intramuscular injection (Goding and Denton 1957). Goding and Denton (1959) showed that the influence of a large (20 mg) intramuscular DOCA injection, as evidenced by the effect on salivary Na/K ratio, had disappeared by 28 h, and we frequently confirmed that the influence of a 5 mg intramuscular DOCA injection had disappeared within 24 h, i.e. by the next drinking period.

(2) Sodium bicarbonate intake after 48 h of depletion. About once a week the routine control procedure set out under (1) above was interrupted and the sodium bicarbonate instead of being offered after a 24-h interval, was offered after 48 h. Hormones were given as usual every 24 h. In this series, therefore, the sheep had the same adrenal insufficiency but a greater sodium deficiency compared with series (1).

(3) Sodium bicarbonate intake when hormones were withheld for a 48-h period. After conclusion of series (2), about once a week routine procedure (1) was interrupted and the hormones, instead of being given every 24 h, were withheld for 48 h. Sodium bicarbonate was offered as usual every 24 h. In this series the sheep had the same order of sodium deficiency but a longer period of deprivation of adrenal hormones compared with series (1).

(4) Sodium bicarbonate intake when both hormones and sodium were withheld for a 48-h period. Subsequent to series (3), the routine procedure (1) was interrupted and, about once a week, sodium bicarbonate and hormones given after a 48-h interval instead of after 24 h. In this

series the sheep had both a greater sodium deficiency and a longer period of deprivation of hormone compared with series (1).

(5) Sodium bicarbonate intake when saliva was returned by rumen tube so that sodium deficiency was small relative to routine procedure (1). In this series the saliva which had dripped from the fistula during the day was given back to the sheep by a rumen tube in the evening, and also in the morning about 3–4 h before the sodium drinking period. The sodium bicarbonate was offered as usual every 24 h and DOCA and cortisone given as usual after the drinking period. The small amount of saliva which collected in the 3–4 h between giving the rumen tube and offering the sodium solution was measured and analysed, as was the 24-h urine collection. In this series, therefore, the sodium deficit was much less than in any other series.

At least six tests were made in each of series (2), (3), (4) and (5). The four sheep which had intact adrenal glands (Gigi, Lola, Ricky and Cecil) were, except for the fact that they did not receive hormone injections, maintained in a similar manner as the adrenalectomized sheep. Various degrees of sodium deficiency were obtained by providing access to the sodium bicarbonate solution for 15 min every 24 h, as in series (1); every 48 h, as in series (2); and every 24 h after the saliva was returned by rumen tube, as in series (5).

For Ricky and Cecil, sodium bicarbonate was also offered after a 72-h period of deficit. For Cecil, the series where the saliva was returned by rumen tube was omitted. In this sheep also, there were two series in the 24-h sodium depletion experiments, one of which was performed at the beginning, the other some months later at the end of the general experiment. By coincidence the mean volume of saliva being secreted daily differed in the two series, so that two separate points could be obtained on the graph. There were 10–25 test episodes in each experimental series on the control non-adrenalectomized sheep.

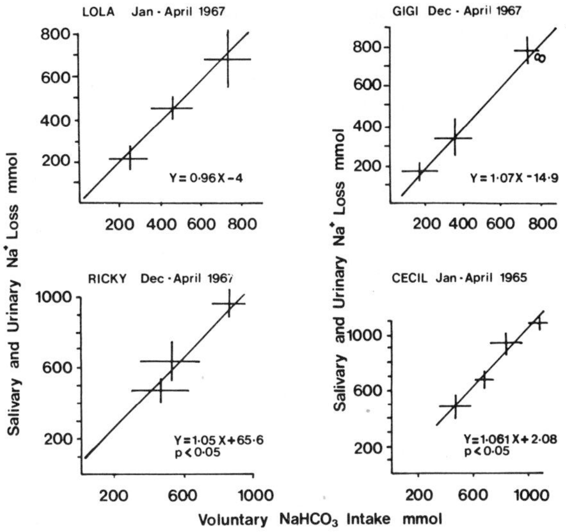

Fig. 12-12. The relation between voluntary intake of 300 mmol/l sodium bicarbonate solution and body sodium loss through saliva and urine in four normal sheep with a permanent unilateral parotid fistula. The sheep had access to the solution for 15 min after being deprived of sodium for 1, 2 or 3 days. Water was continuously available. Observations at near normal balance were contrived by returning saliva for 24 h. Each point of intersection represents the mean ±1 s.d. of 10–25 observations at each level of deficit. (After Denton et al. 1969.)

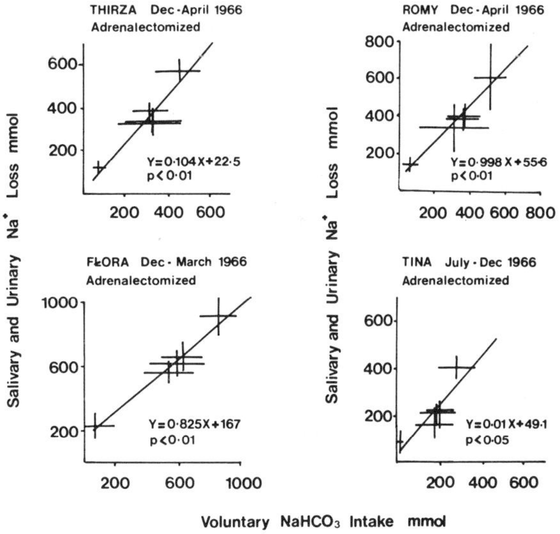

Fig. 12-13. The same study as that shown in Fig. 12-12 but using four adrenalectomized sheep maintained on basal steroid supplement, the steroid maintenance having been given 24 h before the test of appetite. (After Denton et al. 1969.)

Results

The results of this analysis were as follows:

The relation between sodium bicarbonate intake and sodium deficit

Fig. 12-12 shows the close relation between sodium bicarbonate intake in 15 min and mean sodium loss (saliva plus urine) in sheep with adrenals intact. Each point represents the mean of 10–25 observations.

Fig. 12-13 and Table 12-1, using the results from series (1), (2) and (5) above, show the relation between sodium intake and sodium deficit in adrenalectomized sheep. Three separate sets of values for the 24-h depletion period in each animal were obtained by including the control observations from series (3) (Table 12-2) and (4) (Table 12-3) as well as series (1) (Table 12-1). In all instances the same DOCA dosage (5 mg) had been

Table 12-1. The mean voluntary intake of sodium bicarbonate solution (300 mmol/l) related to mean salivary and urinary loss in four adrenalectomized sheep. The number of experimental episodes in each series is recorded also (*n*).

Sheep	Mean salivary Na loss (mmol)	Mean urinary Na loss (mmol)	Mean $NaHCO_3$ intake in 15 min/48 h (mmol)	Mean $NaHCO_3$ intake in 15 min/24 h (mmol)	Mean $NaHCO_3$ intake in 15 min/24 h when saliva was returned by rumen tube (mmol)
Thirza	545±55 (8) 48 h	25±10 (8)	452±104 (8)		
	306±40 (30) 24 h	27±26 (30)		319±145 (30)	
	78±11(6) 24 h	45±28 (6)			70±20 (6)
Romy	602±185 (6) 48 h	7±4 (6)	515±101 (6)		
	324±122 (26) 24 h	11±14 (25)		307±198 (26)	
	79±12 (6) 24 h	61±50 (6)			62±48 (6)
Flora	878±103 (6) 48 h	34±25 (6)	867±114 (6)		
	529±77 (25) 24 h	32±21 (25)		537±160 (25)	
	117±30 (6) 24 h	117±93 (6)			65±113 (6)
Tina	398±45 (6) 48 h	16±7 (5)	276±86 (6)		
	205±47 (25) 24 h	12±15 (25)		200±83 (25)	
	36±5 (6) 24 h	54±39 (6)			5±8 (6)

Table 12-2. The effect of 48 h of hormone withdrawal on mean salivary and urinary sodium loss/24 h and voluntary sodium intake compared with control series where hormone was administered 24 h before test period.

Sheep	Mean salivary Na loss (mmol/24 h)	Mean urinary Na loss (mmol/24 h)	Mean $NaHCO_3$ intake in 15 min 24 h after last hormone supplement and access to Na (mmol)	Mean $NaHCO_3$ intake in 15 min 48 h after last hormone supplement, and 24 h after last access to Na (mmol)
Thirza	326±60 (20)	11±9 (20)	327±132 (20)	
	307±50 (6)	139±55 (5)		467±114 (6)
Romy	391±73 (15)	8±4 (15)	364±97 (15)	
	274±39 (6)	62±22 (6)		354±135 (6)
Flora	581±81 (25)	36±30 (25)	590±180 (25)	
	422±122 (6)	262±106 (6)		606±71 (6)
Tina	213±59 (24)	3±7 (24)	186±98 (24)	
	176±92 (6)	99±59 (6)		194±69 (6)

Table 12-3. The effect of 48 h of withdrawal of hormone and access to sodium solution on mean salivary and urinary sodium loss and voluntary sodium intake compared with control series where hormone was administered 24 h beforehand and previous access to sodium was 24 h beforehand.

Sheep	Mean salivary Na loss (mmol/24 h)	Mean urinary Na loss (mmol/24 h)	Mean $NaHCO_3$ intake in 15 min 24 h after last access to Na and last hormone supplement (mmol)	Mean $NaHCO_3$ intake in 15 min 48 h after last access to Na and last hormone supplement (mmol)
Thirza	377±36 (16) 24 h	9±7 (16)	316±79 (16)	
	688±68 (6) 48 h	46±22 (6)		470±124 (6)
Romy	377±59 (14) 24 h	9±4 (14)	363±100 (14)	
	607±149 (6) 48 h	27±15 (6)		364±106 (6)
Flora	631±92 (17) 24 h	33±30 (17)	622±138 (17)	
	958±149 (6) 48 h	103±34 (6)		783±304 (6)
Tina	165±50 (25) 24 h	6±7 (23)	177±97 (25)	
	322±69 (6) 48 h	119±100 (6)		238±66 (6)

Table 12-4. The effect of the various procedures of hormone withdrawal and sodium depletion on plasma [Na] and [K] and haematocrit.

Treatment	[Na] (mmol/l)	[K] (mmol/l)	Haematocrit (%)
Sheep with adrenals intact	145±3.7 (22)	3.6±0.3 (20)	29±3 (20)
24 h after hormones and last access to $NaHCO_3$ (series 1)	143±6 (72)	5.3±0.6 (72)	31±4 (68)
48 h after last access to $NaHCO_3$. Hormones 24 h before (series 2)	136±5 (17)	6.6±1.3 (17)	35±4 (16)
48 h after hormone supplement. $NaHCO_3$ 24 h before (series 3)	137±5 (19)	6.5±1.1 (19)	35±6 (18)
48 h after last hormone supplement and last access to $NaHCO_3$ (series 4)	133±4 (22)	6.9±0.9 (22)	37±6 (22)
When saliva was returned by rumen tube. Hormones 24 h before (series 5)	148±5 (38)	4.5±0.6 (38)	29±2 (38)

given 24 h before the test period. Although the standard deviations were often greater than for sheep with intact adrenals, there is a clear relationship between sodium intake and sodium deficit. Changes in the plasma are shown in rows 2, 3 and 6 of Table 12-4.

The relation between sodium bicarbonate intake and sodium deficit when hormone supplement was withheld for 48 h

Table 12-3 shows the effect of 24-h sodium depletion when hormone was given 48 h beforehand (series (3) above). Salivary sodium loss was slightly smaller and urinary sodium loss larger than in series (1) where hormone is given after a 24-h interval. The total sodium loss was approximately the same or slightly greater than in series (1). The sodium bicarbonate intake (last column, Table 12-2) bears a good relation to the total sodium deficit. The plasma changes, shown in Table 12-4 (series (3)), were the same as those produced by the experiments of series (2), where hormone was given every 24 h and sodium every 48 h. The salivary [K] of the spot saliva sample taken just before the drinking period was 11.5±2.2 (22) mmol/l compared with 9.2±1.7 (22) mmol/l for series (1). This series shows that there is no significant alteration in the capacity of sheep to regulate sodium intake in accordance with deficit, whether the effect of injected DOCA disappeared approximately 6 or 30 h before the drinking period.

Table 12-3 also shows the effect of sodium bicarbonate intake when both sodium and hormones were given after a 48-h interval. All the sheep failed to balance their deficit by ca 200 mmol sodium. The plasma changes (Table 12-4) show the lowest [Na] and highest [K] and haematocrit of any of the experiments. The salivary [K] of the spot saliva samples was 13.4±3.6 (26) mmol/l. The saliva volume in the second 24-h collection was always lower than in the first 24 h, but otherwise there was no clearly defined evidence for a deterioration in the general condition of the sheep.

This study showed a close relationship between body deficit and rapid drinking of sodium bicarbonate, even though in the instances of severe deficit a 30-kg animal was required to drink 2.5–3.0 litres of the hypertonic solution in the 15 min of access in order to correct the body deficit. The figures, as constructed, show the relation between salivary and urinary loss on the one hand and voluntary intake of sodium on the other. These are the principal factors involved in variation of sodium turnover. The food intake and faecal excretion were not included. Other studies where we have made external electrolyte balances (Denton 1957) have indicated that daily food intake probably exceeded faecal loss by 30–60 mmol. This would be a roughly constant factor over the range of the study. In the 2–3 day depletion episodes with moderate to severe deficit, faecal sodium decreases on the second or third day of deficit, but this effect on balance is usually offset by some decline in food intake. If the small positive balance attributable to this aspect of sodium turnover is accounted for over the study, the intercept of the regression line on the axes will move in some cases to the region consistent with small need-free intake at the point of nil deficit.

The results with adrenalectomized sheep do not

differ substantially from those with adrenals intact. This was seen under conditions of all animals having received the same standard maintenance dose of DOCA 24 h before the test. The action of the DOCA, as evidenced by salivary [K], had disappeared at the time of the test. The salivary [K] values found (9–14 mmol/l) were in the range characteristic of the sodium-deficient adrenally insufficient sheep (Goding and Denton 1957; Blair-West et al. 1964). Thus the commensurate relation between sodium deficit and voluntary drinking was seen independently of variation in the blood concentration of salt-retaining hormone.

Unless one assumes in the adrenalectomized animal that a causal component normally attributable to change in aldosterone concentration was commensurately replaced by other vectors causal of salt appetite, the results could be taken as indicating that blood aldosterone level has little or no role in the causation of salt appetite in sodium-deficient sheep.

The results of the experiments where hormone was withdrawn for 48 h and the effect of 24-h of salivary loss tested, were the same as for 24-h hormone withdrawal. In the instances where both hormone and sodium were withdrawn for 48 h, voluntary sodium intake in the 15-min period of access fell below deficit. The changes in plasma ionic composition were consistent with a greater degree of adrenal insufficiency, and saliva volume was considerably reduced in the second 24-h period. It seems possible that the reduced intake in the limited period of access was due to some deterioration in the animal's condition, but measurements of other physiological parameters which could give information on the question were not made.

The results where sodium solution was withheld for 72 h in the normal animals, and the results where sodium solution was withheld for 48 h and hormone for 24 h in the adrenalectomized animals are relevant in this regard. The capability of the animals to drink the requisite volume of sodium solution (2–3 litres) in the time (15 min) was clearly established. Where both hormones and sodium solution were withheld for 48 h it may have been the case that if access to the solutions had been permitted for a longer period, e.g. 1 h, the animals would have corrected the deficit adequately. That is, the 48-h period of withdrawal of both hormone and sodium solution may have impaired the particular ability tested here of being able to correct salt deficiency rapidly by drinking a large volume over 15 min. But it may not have impaired ability to repair deficit given adequate time to allow for factors such as increased muscular weakness and lethargy.

Plasma [Na] fell according to the extent of deficiency. However, there is other experimental evidence against there being a simple and direct relation between the contemporary concentration of sodium in blood, and salt appetite. This will be discussed in later chapters.

Rats

The outstanding study of this question has been made by Jalowiec and Stricker (1973). Their analysis was preceded, as they note, by several important papers. Epstein and Stellar (1955), as already described in Chapter 11, showed first that the gradually increased intake of sodium chloride by the adrenalectomized rat reflected response to increasing need. If salt was withheld for 10 days, the initial intake of the inexperienced animal over an hour reached the exaggerated post-operative levels immediately.

Falk (1966) studied rats which were depleted of sodium by peritoneal dialysis. Sodium intake (0.51 M saline) was double the deficit when drinking was allowed for 13 h commencing 4 h after dialysis fluid was withdrawn under light anaesthesia. The procedure was carried out again 3 days later. Presumably the rats were in normal sodium balance before the second dialysis, though this was not shown by urinary analysis. Now, intake of sodium was three times deficit. This greater intake after the second dialysis also occurred if the rats were not allowed to drink after the first dialysis, but the deficit was replaced by stomach tube at the time of withdrawal of dialysis fluid. Thus, learning did not appear to determine the effect. The possibility of the long-delayed repair of sodium loss from bone (Bergstrom 1955) influencing the availability of this sodium source to mitigate depletion at the second dialysis was considered. At all events, intake considerably exceeded deficit. However, in the light of data to be cited in later chapters on the large effect of ACTH on salt appetite in rats (Weisinger et al. 1978), dialysis and anaesthesia may have caused sustained ACTH and glucocorticoid release which could have contributed to salt intake independently of body sodium balance change.

Jalowiec and Stricker (1973) see these data as indicative that normal rats, in contrast to normal or adrenalectomized sheep, always ingest considerably more sodium than they require. Noting that normal sheep appear to have little salt appetite response to mineralocorticoids (Denton et al. 1969), they raise the question of whether in rats, since aldosterone clearly stimulates salt appetite, the overdrinking in sodium deficiency is potentiated by aldosterone. Thus in adrenalectomized rats, intake might show a precise relation to body sodium deficit. Accordingly, they studied sodium intake of adrenalectomized rats at varying degrees of sodium deficiency, and further examined sodium intake of intact rats made deficient by dietary deprivation.

First, they found that during ad libitum maintenance, hourly monitoring of fluid intakes of adrenalectomized rats showed they ingested saline (0.15 M) at frequent intervals during a 24-h period. Fewer than 10% of drinking episodes occurred later than 1–3 h after the preceding episode. Mean intakes of 0.51 M saline solution were small (mean, 0.81 ml) and were usually associated with larger intakes of water (mean, 1.75 ml), which reduced ingested saline to isotonic range (0.16 M). If it be assumed that this rate of drinking (0.4 mmol per episode) reflected an equilibrium intake to replace loss, the data are roughly indicative that a laboratory rat's sodium appetite mechanism reacts to a body error of about 3%. (This assumes the exchangeable body sodium of a 300-g rat to be 13.5 mmol, i.e. 45 mmol/kg: Norman 1963.) This might suggest that they are about as sensitive as sheep (which also respond to ca 3% body error: see Chapter 13), but this matter requires comprehensive study with precise activity records of the individual rats, coupled with the urinary excretion data. As with sheep, the pattern may be complicated in individual animals by hedonic and other factors.

The appetitive response to various levels of sodium deficit was studied in the sham-operated and adrenalectomized rats as either an ascending series of deficits or a descending series, the periods of deprivation of saline on the very low sodium diet being 3, 8, 24, 48, 72 and 96 h. Access was allowed for 180 min on conclusion of the deprivation period. As regards the very small deficits incurred over 3–8 h, the intact sham-operated animals replaced them fairly accurately, but with deprivations of 24, 48, 72 and 96 h they drank 2- to 4-fold the actual deficit as estimated by urinary loss. Over these same time intervals the deficits of the adrenalectomized rats, by virtue of the failure of urinary sodium conservation, were much larger, reaching 2–3 mmol in 24–48 h and 3–4 mmol in 72–96 h (i.e. ca 50% of the sodium content of the extracellular fluid). At 3, 8, 24 and 48 h the drinking of saline substantially overcompensated loss (2- to 3-fold) and the large part of the intake was during the first 30 min of access (Fig. 12-14). In the 72-h and 96-h depletion periods intake more closely approximated deficit. The graphs of progressive change of balance, with access, indicate that drinking was much slower. The

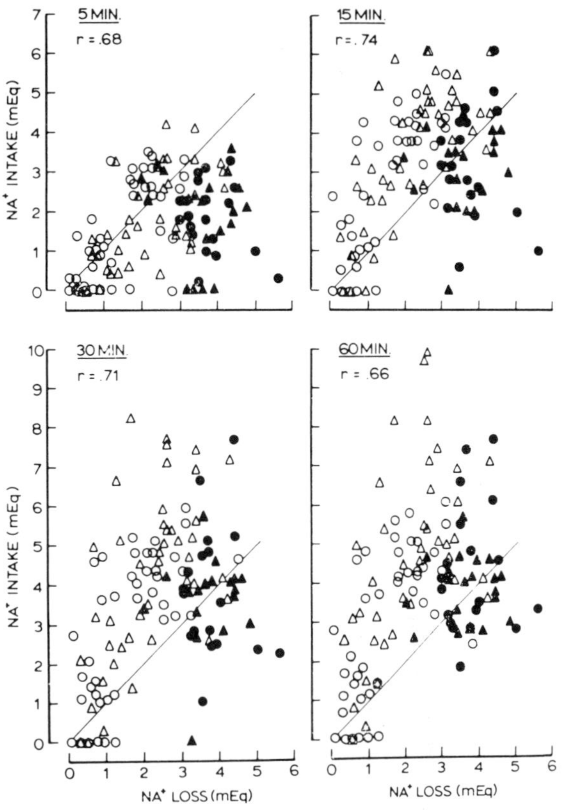

Fig. 12-14. Sodium intake by adrenalectomized rats after 5, 15, 30 and 60 min of access to 0.51 M sodium chloride solution as a function of prior sodium deficit (each point represents one observation, with circles and triangles indicating data from the ascending and descending series of experiments respectively). In ascending experiments the rats were submitted in sequence to progressively larger deficits and in the descending series the alternative procedure held. Correlations are for values obtained after 3, 8, 24 and 48 h of sodium deprivation (open symbols) only and exclude values obtained after 72 and 96 h of sodium deprivation (filled symbols). Diagonal lines indicate perfect precision when sodium intake equals sodium deficit. (From Jalowiec and Stricker 1973, with permission.)

authors draw attention to the weakened state of the animals at this stage, and the haematocrit and plasma [K] and [Na] concentrations recorded substantiate the greater adrenal insufficiency and depletion, though the difference from 48-h depletion is not large. They make the interpretation that the weakened state of the rats, which had difficulty in standing and moving at this stage, led to a fortuitous precise replacement of deficit. In light of this, when summarizing the results from adrenalectomized rats in Fig. 12-14, they have not included the 72-h and 96-h results (filled symbols) in calculating the correlations between deficit and intake. The blood analysis on the intact and adrenalectomized animals showed large fall in [Na] and rise in [K] in plasma. Jalowiec and Stricker (1973) conclude that adrenalectomized rats adjust intake to continuous loss effectively, the threshold for stimulation of intake being similar to that in sodium-deprived intact rats. With more pronounced losses, adrenalectomized rats drink much more than required to replace losses, as do normal rats. This point of overshoot is also strongly emphasized by Pfaffmann (1967). Thus it seems evident that mineralocorticoids need not have any vital role in either the initial salt drinking response of intact rats to minor sodium deficits, or their overcompensation for moderate deficits.

Finally, since it sets out in summary a viewpoint on a number of matters, aspects of which we will examine and discuss in somewhat different light in succeeding chapters, it is appropriate to quote here the last paragraph of Jalowiec and Stricker's discussion:

The present results add to the growing list of important differences between the sodium appetite of rats and sheep, the two species that have been used almost exclusively as experimental subjects in studies of the physiological bases of sodium appetite. First, the capacity of mineralocorticoids to stimulate and/or potentiate sodium appetite in rats represents a striking contrast to their ineffectiveness in sheep (Denton et al. 1969). Second, neither intact nor adrenalectomized rats adjust their sodium intake with precision, in contrast to sodium-deficient sheep (Denton et al. 1969; Denton and Sabine 1961). In this regard, it is interesting to note that sheep also replace water deficits rapidly and with precision (Bott, Denton and Weller 1965), whereas rats rehydrate themselves much more slowly (Adolph 1950), and thus sheep generally may have a greater capacity for an oral 'metering' of their input than do rats, either through special physiological feedback mechanisms or through conditioned responses learned from previous associations of appetite, intake, and post-ingestional satiety. As for a third difference, the apparent relationship between sodium appetite in rats and changes in their plasma volume and electrolyte concentrations that has been emphasized in the present report may be contrasted with the absence of a clear causal relationship between these individual blood parameters and sodium appetite in sheep that Denton and his colleagues have demonstrated in a series of experiments (see review by Bott et al. 1967). However, it was on the basis of Denton's results, as well as similar findings in rats, that a 'sodium reservoir' hypothesis was proposed (Stricker and Wolf 1969) that suggested that sodium appetite was directly affected by the fluctuating content of some body sodium reservoir and was only indirectly affected, through their effects on this reservoir, by acute changes in haemodynamic factors. Thus, sodium appetite in rats and sheep may be mediated by similar control mechanisms and differ simply in the relative roles played by the multiple contributing stimuli. On the other hand, direct and substantial support for this hypothetical reservoir remains elusive, and in the absence of a clear understanding of the physiological bases of sodium appetite in either animal one can only notice the obvious differences between rats and sheep in their diets and feeding habits, in their gastrointestinal apparatus and their control of mineralocorticoid secretion, and in their present environmental circumstances and their evolutionary histories (Denton 1965; Mulrow 1966), and wonder whether there might not also be basic differences in the respective systems controlling their sodium appetite behaviours.

Rabbits

The intriguing data on the capacity of adrenalectomized rabbits to repair body deficit precisely without excess intake are dealt with in Chapter 14 bearing upon satiation behaviour.

References

Abraham SF, Baker RM, Blaine EH, Denton DA, McKinley MJ (1975) Water drinking induced in sheep by angiotensin—a physiological or pharmacological effect? J Comp Physiol Psychol 88: 503

Abraham S, Coghlan JP, Denton DA, McDougall JG, Mouw DR, Scoggins BA (1976) Increased water drinking induced by sodium depletion in sheep. Q J Exp Physiol 62: 185

Adolph EF (1950) Thirst and its inhibition in the stomach. Am J Physiol 161: 374

Andersson B (1972) Receptors subserving hunger and thirst. In: Neal E (ed) Handbook of sensory physiology, vol 3, Enteroreceptors. Springer, Berlin Heidelberg New York, Chap 6

Andersson B (1977) Regulation of body fluids. Annu Rev Physiol 39: 185

Armstrong EA (1950) The nature and function of displacement activities. In: Physiological mechanisms in animal behaviour. Cambridge University Press, London, p 361 (Symposia of the Society for Experimental Biology, no. 4)

Bergstrom WH (1955) The participation of bone in total body sodium metabolism of the rat. J Clin Invest 34: 997

Blaine EH, Covelli MD, Denton DA, Nelson JF, Shulkes AA (1975) The role of ACTH and adrenal glucocorticoids in the salt appetite of wild rabbits (*Oryctolagus cuniculus*(L.)). Endocrinology 97: 793

Blair-West JR, Coghlan JP, Denton DA, Goding JR, Wintour EM, Wright RD (1963) The control of aldosterone secretion. Recent Prog Horm Res 19: 311

Blair-West JR, Coghlan JP, Denton DA, Goding JR, Wright RD (1964) The effect of adrenal cortical steroids on parotid salivary secretion. In: Sreebny LM, Meyer J (eds) Salivary glands and their secretions. Pergamon Press, Oxford, p 253

Blokh EL (1939) The work of the parotid and submaxillary glands in cattle. Fiziol Zh SSSR 27: 200

Bott E, Denton DA, Weller S (1965) Water drinking in sheep with oesophageal fistulae. J Physiol (Lond) 176: 323

Bott E, Denton DA, Weller S (1967) The effects of angiotensin II infusion, renal hypertension and nephrectomy on salt appetite of sodium deficient sheep. Aust J Exp Biol Med Sci 45: 595

Braun-Menendez E, Brandt P (1952) Aumento del apetito especifico para la sal provocado por la desoxicorticosterona. Revista de la Sociedad Argentina de Biologia 28: 15

Bryant RW, Epstein AN, Fitzsimons JT, Fluharty SJ (1980) Arousal of specific and persistent sodium appetite with continuous intracerebroventricular infusion of angiotensin II. J Physiol (Lond) 301: 365

Colin G (1898) Traité de physiologie comparée des animaux, 3e éd. Cited in Schafer's textbook of physiology. Pendland, Edinburgh, p 477

Denton DA (1957) The study of sheep with permanent unilateral parotid fistulae. Q J Exp Physiol 42: 72

Denton DA (1965) Evolutionary aspects of the emergence of aldosterone secretion and salt appetite. Physiol Rev 45: 245

Denton DA (1966) Some theoretical considerations in relation to innate appetite for salt. Conditional Reflex I/3: 144

Denton DA, Sabine JR (1961) The selective appetite for Na

shown by Na deficient sheep. J Physiol (Lond) 157: 97
Denton DA, Sabine JR (1963) The behaviour of Na deficient sheep. Behaviour 20: 364
Denton DA, Nelson JF, Orchard E, Weller S (1969) The role of adrenocortical hormone secretion in salt appetite. In: Pfaffmann C (ed) Proceedings of the Third International Symposium on Olfaction and Taste. Rockefeller University Press, New York, p 535
Eckhard C (1893) Noch Einmal die Parotis des Schafes. Centralbl Physiol 7: 365
Ellenberger W, Hofmeister V (1887) Beitrag zur Lehre von der Speichelsekretion. Arch Anat Physiol Lpz [Suppl] 138
Epstein AN, Stellar E (1955) The control of salt preference in the adrenalectomized rat. J Comp Physiol Psychol 48: 167
Falk JL (1965) Limitations to the specificity of NaCl appetite in sodium depleted rats. J Comp Physiol Psychol 60: 393
Falk JL (1966) Serial sodium depletion and NaCl solution intake. Physiol Behav 1: 75
Falk JL, Herman TS (1961) Specific appetite for NaCl without post-ingestational repletion. J Comp Physiol Psychol 54: 405
Ferreyra MC, Chiaraviglio E (1977) Changes in volemia and natremia and onset of sodium appetite in sodium depleted rats. Physiol Behav 19: 197
Fitzsimons JT (1972) Thirst. Physiol Rev 52: 468
Fomin DA (1941) Secretion of calves' salivary glands. Fiziol Zh SSSR 30: 656
Goding JR, Denton DA (1957) The effects of adrenal insufficiency and overdosage with DOCA on bilaterally adrenalectomized sheep. Aust J Exp Biol Med Sci 35: 301
Goding JR, Denton DA (1959) The response to sodium depletion in adrenalectomized sheep with a parotid fistula. Aust J Exp Biol Med Sci 37: 211
Henkin RI (1970) The neuroendocrine control of sensation. In: Bosma JF (ed) Second Symposium on Oral Sensation and Perception. Thomas, Springfield, Ill, p 493
Henkin RI, Gill JR, Bartter FC (1963) Studies on taste thresholds in normal man and in patients with adrenal cortical insufficiency: The role of adrenal cortical steroids and serum sodium concentration. J Clin Invest 42: 5
Jalowiec JE, Stricker EM (1970a) Restoration of fluid balance following acute sodium deficiency in rats. J Comp Physiol Psychol 70: 94
Jalowiec JE, Stricker EM (1970b) Sodium appetite in rats after apparent recovery from acute sodium deficiency. J Comp Physiol Psychol 73: 238
Jalowiec JE, Stricker EM (1973) Sodium appetite in adrenalectomized rats following dietary sodium deprivation. J Comp Physiol Psychol 83: 66
Jalowiec JE, Crapanzano JE, Stricker EM (1966) Specificity of salt appetite elicited by hypovolaemia. Psychonom Sci 6: 331
Katz D (1953) Animals and men. Penguin, London
Konorski J (1950) Mechanism of learning. In: Physiological mechanisms in animal behaviour. Cambridge University Press, London, p 409 (Symposia of the Society for Experimental Biology, no. 4)
Krieckhaus EE (1970) Innate recognition aids rats in sodium regulation. J Comp Physiol Psychol 73: 117
Krinitsin D Ya (1940) Continuous secretion of ruminants' parotid glands. Fiziol Zh SSSR 29: 384
Kudriavtsev AA, Anurov SN (1940) Fiziol Zh SSSR 28: 372
Laragh JH (1954) The effect of potassium chloride on hyponatraemia. J Clin Invest 33: 807
Laragh JH, Cepeci NE (1955) Effect of administration of potassium chloride on serum sodium and potassium concentration. Am J Physiol 180: 539
Liddell HS (1954) Conditioning and emotions: An account of a long range study in which neuroses are induced in sheep and goats to clarify how irrational emotional behaviour originates and ultimately to indicate how it may be prevented. Sci Am 190: 48
Lorenz KZ (1950) The comparative method of studying innate behaviour patterns. In: Physiological mechanisms in animal behaviour. Cambridge University Press, London, p 221 (Symposia of the Society for Experimental Biology, no. 4)
McCutcheon B, Levy C (1972) Relationship between NaCl rewarded bar pressing and duration of sodium deficiency. Physiol Behav 8: 761
Mulrow PJ (1966) Neural and other mechanisms regulating aldosterone secretion. In: Martini L, Ganong WF (eds) Neuroendocrinology 1. Academic Press, New York
Norman N (1963) The participation of bone in sodium and potassium metabolism in rats, Parts I and II. Acta Physiol Scand 57: 363 (Part I), 373 (Part II)
Pfaffmann C (1967) Critic's comments. In: Kare MR, Maller O (eds) The chemical senses and nutrition. Johns Hopkins Press, Baltimore, p 243
Popov NF (1954) Farm animal physiology. Moscow
Scheunert A, Krzywanek W, Zimmerman K (1929) Zum studium der Speichelsekretion. Pflügers Arch 192: 33
Stricker EM (1980) The physiological basis of sodium appetite: A new look at the 'depletion–repletion' model. In: Kare M, Bernard R, Fregly M (eds) Biological and behavioural aspects of NaCl intake. Academic Press, New York, p 185
Stricker EM, Wolf G (1966) Blood volume and tonicity in relation to sodium appetite. J Comp Physiol Psychol 62: 275
Stricker EM, Wolf G (1969) Behavioural control of intravascular fluid volume: Thirst and sodium appetite. Ann NY Acad Sci 157: 553
Tikhomirov NP, Fomin DA (1950) Cited in Serelryakov PM (1950) IP Pavlov's theories and farm animal physiology. Selkhozgiz, Moscow
Weisinger RS, Denton DA, McKinley MJ, Nelson JF (1978) ACTH-induced sodium appetite in the rat. Pharmacol Biochem Behav 8: 339
Wolf G, McGovern JF, Dicara LV (1974) Sodium appetite: Some conceptual and methodologic aspects of a model drive system. Behav Biol 10: 27
Zimmerman MB, Stricker EM, Blaine EH (1978) Water and NaCl intake after furosemide treatment in sheep (*Ovis aries*). J Comp Physiol Psychol 92: 501

13 The study of salt appetite in sodium deficiency by operant behaviour

Summary

1. Sheep can be easily trained to press levers in their cages which deliver a small volume of either sodium bicarbonate solution or water. Experiments are done either on a basis of free access, or the animals learn that the lever will deliver following the sound of a bell and they are allowed access, for example, for 2 h per day under this regime.

2. In sodium deficiency operant behaviour, as is rapid consummatory drinking when sodium bicarbonate is presented, is commensurately related to body deficit. The sheep will work to get the appropriate amount of salt. There is an innate physiological control system which is reactive to some change of body chemical state which has yet to be clearly defined (Denton 1966; Quartermain et al. 1967; Denton and Nelson 1971; Stricker 1980). This mechanism probably plays an important role in the fine hour-by-hour maintenance in most animals and involves detection of some change or 'error' by the monitoring system. The data from operant studies on animals with free access to sodium bicarbonate suggest reaction to loss from a parotid fistula of 50–100 mmol, which is ca 3% of body sodium content. Data from the study of natriuretic response to injection of hypertonic saline into the third ventricle and also from monitoring aldosterone secretion in response to onset of sodium deficiency confirm physiological reaction in the sheep to a change in sodium status of 50–100 mmol.

3. However, the basic reaction to sodium deficit may be modified by or overlaid with other mechanisms. These include learning of the benefit of salt with possibly anticipatory drinking thereafter, habituation to operant behaviour, and the variable hedonic response to salt taste including the pleasure of taking it while eating. Thus, the motivation behind the ultimate pattern observed with operant behaviour in response to continuous sodium loss from a fistula may be most complex. The overriding component which is the innate drive in response to biochemical error is only clearly revealed by producing significant body sodium deficit.

4. On the other hand, evidence has accrued from experiments examining spontaneous water drinking in dogs given access to water ad libitum which points to a major determinant of intake being the physiological stimulus of an increase in plasma sodium concentration, rather than there being anticipatory drinking.

Introduction

Miller (1967) has shown that in the study of appetitive drive, the use of several different techniques may reveal features of behaviour which would not be apparent with one method of study alone. He notes that the value of using a diversity of rigorous behavioural tests is that it is possible to determine whether unusual interventions such as lesions, electrical stimulation or drugs have all the functional properties of normal increases or reductions in drive.

He illustrates this by citing the effects of bilateral lesions in the region of the ventromedial nucleus which would cause rats to overeat until they became fat. Because the lesion produces such a large increase in food intake, one might assume that it has the more general effect of increased hunger. To test this, rats were first trained to press a bar that caused automatic delivery of food pellets. Then the device was set so that pressing the bar caused delivery only at unpredictable intervals although average frequency of delivery remained constant. Under these conditions it is observed that the actual rate of bar-pressing is relatively constant at a given level of food deprivation, but changes with level of deprivation. Thus rate of working for food represents a relevant measure of hunger. It was found, surprisingly, that the animals with lesions who were eating reliably more, worked reliably less at the bar to secure food. When lids were placed on the food dishes the lesioned animals ate more than controls, but when the lids were weighted with 75-g weights the reverse held. Adulteration of food with quinine also caused lesioned animals to eat less than controls. Similarly the lesioned animals ran more slowly down an alley, pulled less on a restraining harness and were stopped at lower levels of electric shock when attempting to get food. Thus the lesions did not have the same motivational effects as a normal increase in hunger. It appeared that the greater amount consumed by lesioned rats allowed food ad libitum is not determined by how hungry the animals get after deprivation, but by the low levels of hunger which keep the animal nibbling before it is completely satiated.

Bar-press experiments with sheep

Against this background it was of importance and interest to compare the results of satiation of salt appetite in sheep by rapid consummatory drinking of large volumes of sodium solution with satiation by operant behaviour. Also the comparison with the large body of data acquired on the white laboratory rat by this bar-press methodology could be informative.

The instrumental conditioning of the sheep was done as follows. We placed a large lever (bar) either low down on the side wall near the front of the cage or on the floor at the front of the cage (Fig. 13-2). The container to which sodium solution was delivered as a result of the bar-pressing was a standard galvanized iron fluid bin at the side of the cage (Abraham et al. 1973). The sheep learned to press the bar to cause delivery of a small volume of sodium bicarbonate solution to the drinking container. Parotid salivary loss is continuous and amounts to 400–600 mmol Na/day. Thus if each delivery resulting from the bar-press was arranged to provide 15 mmol of sodium, then it follows that 25–50 such deliveries per day are required if the animal is to remain in sodium balance. The precise time characteristics of the way the animal works to maintain sodium balance throw light on the type of error to which the body reacts.

The standard conditions were that each bar-press delivered 50 ml of 300 mmol/l sodium bicarbonate (15 mmol sodium) and that there was a 25-s time-lag before a succeeding bar-press would activate the delivery system. All presses the animal made were registered on a polygraph, but those initiating delivery were registered separately. The animals had free access to water. They were fed at the same time each day and the electric lights were never turned off.

Some sheep learned to press the lever within a week; others never learned or did so only with great difficulty after manoeuvres such as our placing its foot on the bar. An instance of learning is shown here with Brad (Fig. 13-1). As

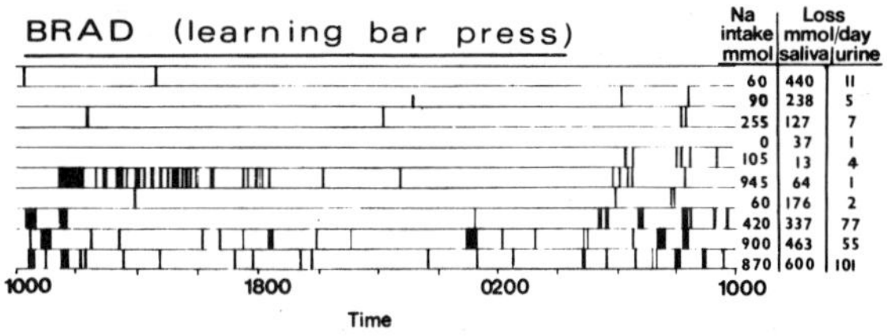

Fig. 13-1. Brad: a sheep with continuous daily loss of saliva from a parotid fistula. The animal had a lever placed in its cage, and pressing the lever caused delivery of 50 ml of 300 mmol/l sodium bicarbonate solution. The figure is constructed from a polygraph record of bar-pressing activity. Each horizontal division represents a 24-h period from 10.00 hours to 10.00 hours and each vertical stroke represents a delivery of 50 ml of sodium bicarbonate. The graphing method has limitations in that when deliveries follow closely upon one another as a result of frequent bar-pressing, individual episodes coalesce to form a block (though they register separately on the original polygraph record). The daily sodium intake by bar-press and the daily salivary and urinary losses are recorded also.

sheep become sodium deficient they often become restless, paw the floor of their cage and jump up and down. A large lever placed in the cage will be hit episodically at random. Here it is seen how, during the first 5 days after withdrawal of free access to sodium bicarbonate and the placing of the lever in the cage, Brad hit it episodically. On one occasion he appeared to work it several times. (Each vertical line on Fig. 13-1 is a delivery by lever, and each horizontal division covers one day, i.e. 10.00 hours to 10.00 hours.) However, the animal became more sodium deficient. Urinary sodium excretion was negligible. Then, on the sixth day, the association between lever-pressing and sodium availability suddenly developed and frenetic pressing and intake followed. By 4 days later a regular pattern had developed. Dr Richard Weisinger has shown subsequently that sheep are trained easily to bar-press for sodium solution if they are first trained to bar-press for water, and have episodes of thirst. Then with sodium deficiency the animals rapidly learn over one episode that bar-pressing delivers salt solution, and do this to achieve satiation. Weisinger has now carried out many experiments where sheep have two levers in the metabolism cage, one delivering water and the other sodium bicarbonate solution. The animals with a fistula learn rapidly which to press to deliver fluid to a cup on the same side of the cage (Fig. 13-2), and in the instance of specific deficiency will concentrate on one lever. They also learn easily that the ringing of a bell signals that the lever which delivers sodium bicarbonate solution has become operative. Thus water may be continuously available from one lever, and sodium bicarbonate solution available for 2 h per day following the sound of the bell.

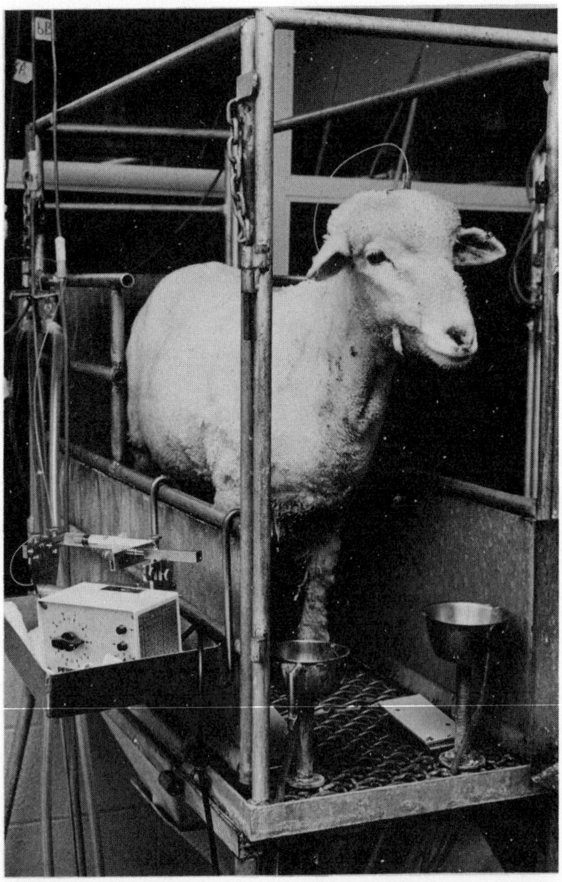

Fig. 13-2. Sheep in a cage designed with operant system. The right lever delivered 50 ml of water to the right cup, and the left 50 ml of a sodium bicarbonate solution to the left cup. The bar-presses were recorded automatically.

Reaction to small deviation of body sodium content

To study the temporal relation between the operant behaviour and the stress of salivary loss, saliva was collected by a device designed by Professor L. Kraintz which emptied automatically when it filled to 10 ml, the event being recorded. Fig. 13-3 gives results from three animals showing the essentially continuous flow of saliva over the 24 h (10.00 hours to 10.00 hours). The record was interrupted for 2–3 h by the feeding of the sheep. By contrast, the operant delivery of sodium bicarbonate solution by bar-press was episodic, and the pattern is suggestive that some quantitative deviation from a norm could be involved in activation of behaviour.

Before dealing with the fine analysis of this question, it is important to mention some basic data on operant behaviour in response to body sodium deficit. Fig. 13-4, covering a 3-week record on Dorothy, shows that if the lever is removed from the cage for a period of 6–48 h, its replacement is followed by intense bar-pressing activity over the next 2–3 h. In fact, the operant behaviour is highly significantly related to body balance, as Fig. 13-5 shows. Variable sodium deficit can be produced by removal of the bar from the cage for 6–72 h. The ordinate scale shows the number of deliveries by bar-press in the first 2 h after the bar is replaced. The zero deficiency points were obtained in experiments where the normal sodium balance was maintained by administration of sodium bicarbonate to the animal.

Baldwin (1968) showed that goats with a parotid fistula which were trained to press an illuminated panel with their muzzle to obtain 5 ml of 500 mmol/l sodium bicarbonate solution would work for sodium solution when only 1 in 60 presses was reinforced by delivery. However, if the animals

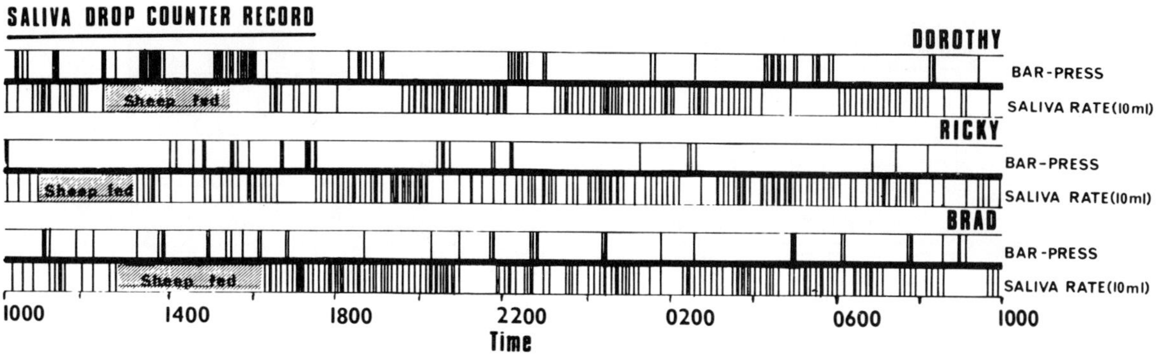

Fig. 13-3. Dorothy, Ricky and Brad: sheep with a permanent unilateral parotid fistula. The daily record of salivary loss (each vertical bar represents loss of 10 ml) was made concurrently with the record of delivery of 50 ml aliquots of sodium bicarbonate solution (300 mmol/l by bar-press. (From Abraham et al. 1973, with permission.)

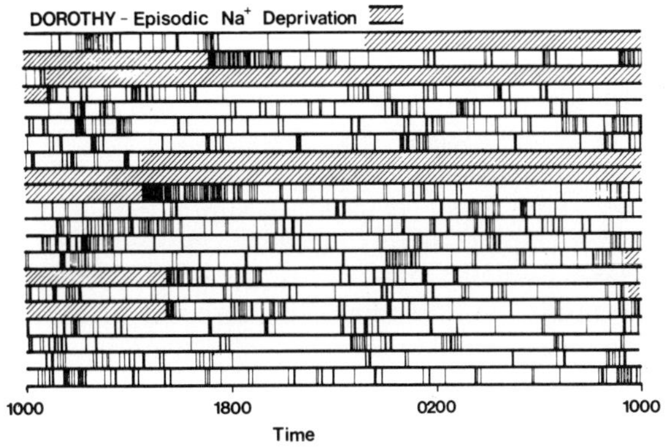

Fig. 13-4. Dorothy: sheep with continuous salivary loss from a parotid fistula. The record of bar-pressing behaviour for 50 ml of 300 mmol/l of sodium bicarbonate over 21 days is shown, during which time there were several episodes where the lever was removed from the cage (hatched areas) to produce sodium deficiency. (From Abraham et al. 1973, with permission.)

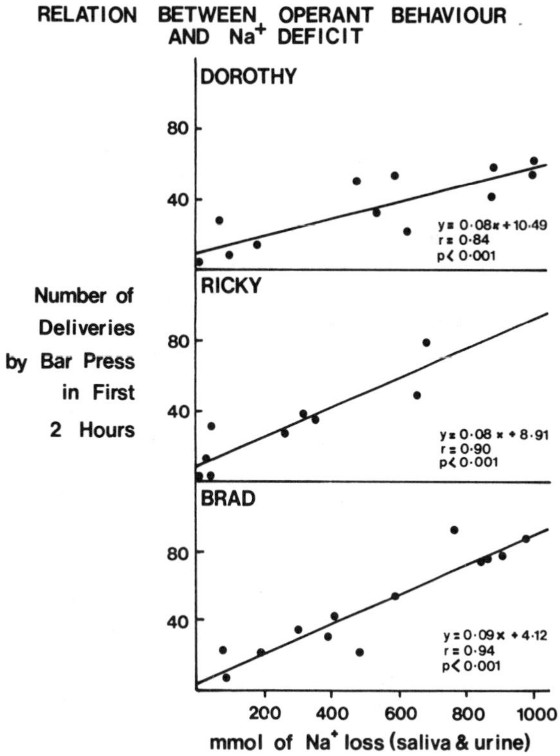

Fig. 13-5. The relation between loss of sodium from the body and operant behaviour for sodium in three parotid-fistulated sheep. The number of deliveries of 50 ml of a 300 mmol/l sodium bicarbonate solution during the 2 h following replacement of the lever in the cage after it had been removed for periods varying between 6 and 72 h is shown. (After Abraham et al. 1973.)

were maintained on 1% sodium bicarbonate ad libitum, with a salt lick in their pen, and salivary Na/K ratio reflected sodium-replete status, the goats no longer tried to obtain sodium bicarbonate by operant behaviour. A subsequent full report of Baldwin's work (1976) is discussed in Chapter 20.

Amongst other conclusions, the data in Fig. 13-5 indicate that the increase in sodium deficit involved a commensurate increase in a motor drive. This argues against the commensurate relation between intake and deficit during free access being simply some taste change with an immediate increase in acceptability of hypertonic sodium during the consummatory act of drinking. On the contrary, the animal is prepared to work to get it.

In terms of the daily maintenance of sodium balance, there were striking and provocative differences between animals. As Fig. 13-6 shows, Ricky always had a low sodium excretion in the urine. Daily collections of saliva had a lowered salivary Na/K ratio showing that during at least part of the day the animal was sodium-deplete. Thus, the metabolic evidence indicated that the episodes of operant behaviour, variable but at 2–4-h intervals, were in response to sodium deficiency: the animal worked to bring itself to

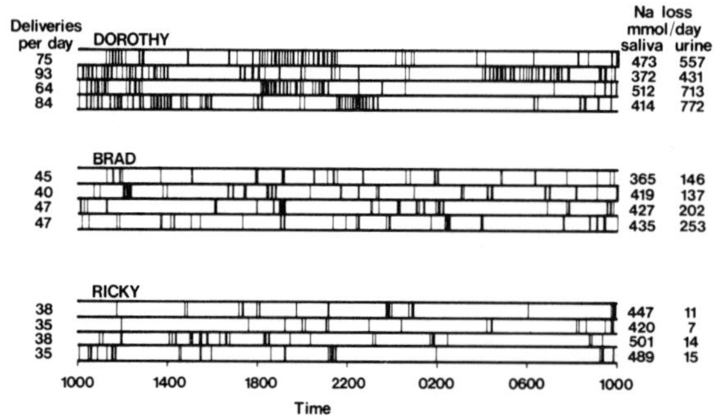

Fig. 13-6. Daily records of operant behaviour for sodium (50 ml of 300 mmol/l sodium bicarbonate) in the three parotid-fistulated sheep Dorothy, Brad and Ricky. Daily loss of sodium in saliva and urine are shown also. (After Abraham et al. 1973.)

near-neutral status. The fact that it maintained itself in good condition for months with low level of sodium excretion in the urine is good evidence that this adjustment was accurate. The salivary loss of about 20 mmol/h and operant episodes every 2–4 h suggest response to a body deficit or error of 40–80 mmol.

By contrast, the sheep Dorothy always had a high urinary sodium output and normal salivary potassium concentration consistent with sodium intake greatly in excess of salivary loss. Brad's behaviour was intermediate between Ricky and Dorothy in terms of balance. In the case of Dorothy, it appeared the animal reacted to a 'set point' well on the positive side of neutral sodium status.

Talilah (Fig. 13-7) appeared to maintain balance

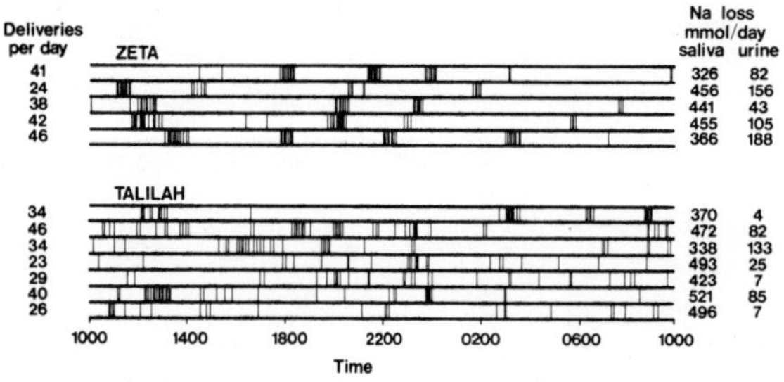

Fig. 13-7. Daily records of operant behaviour for sodium (50 ml of 300 mmol/l sodium bicarbonate) in the two parotid-fistulated sheep Zeta and Talilah. (After Abraham et al. 1973.)

in a fashion similar to Ricky, with response to deficit but with rather more oscillation as evidenced by urinary sodium. Zeta (Fig. 13-7) showed the most clearly defined episodic pattern of sodium replacement, with intense periods of bar-pressing every 4–6 h. During this interval salivary sodium loss approximated 70–100 mmol. Figs. 13-8 to 13-10 show records of daily activity of

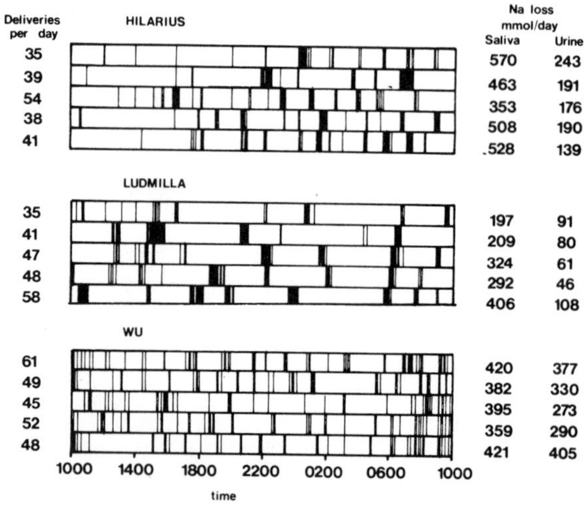

Fig. 13-8. Daily records of operant behaviour for sodium (50 ml of 300 mmol/l sodium bicarbonate) in the three parotid-fistulated sheep Hilarius, Ludmilla and Wu.

other animals—Hilarius, Ludmilla, Wu (Fig. 13-8), Zenni, Isobel (Fig. 13-9), Supra, Esther and Kasimir (Fig. 13-10).

Having shown the linear relation between substantial body sodium deficit and operant behaviour, it would be a reasonable hypothesis to

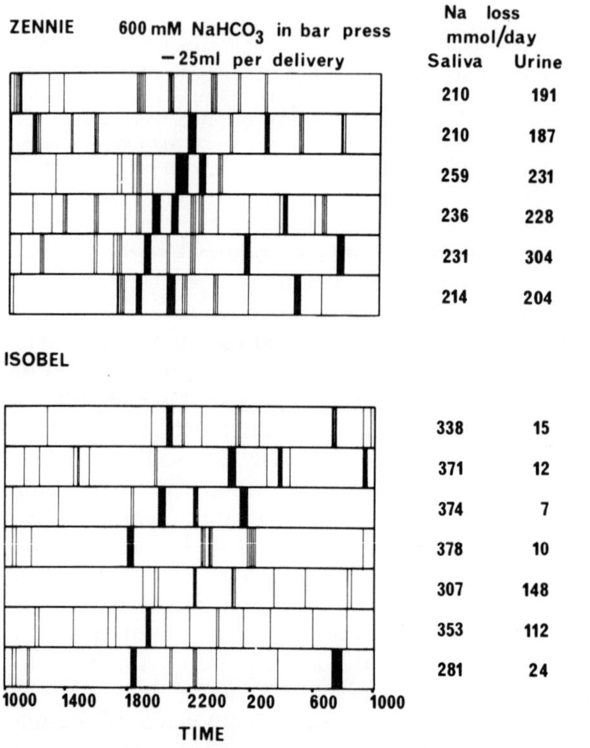

Fig. 13-9. Daily record of operant behaviour for sodium (25 ml of 600 mmol/l sodium bicarbonate) in the two parotid-fistulated sheep Zennie and Isobel.

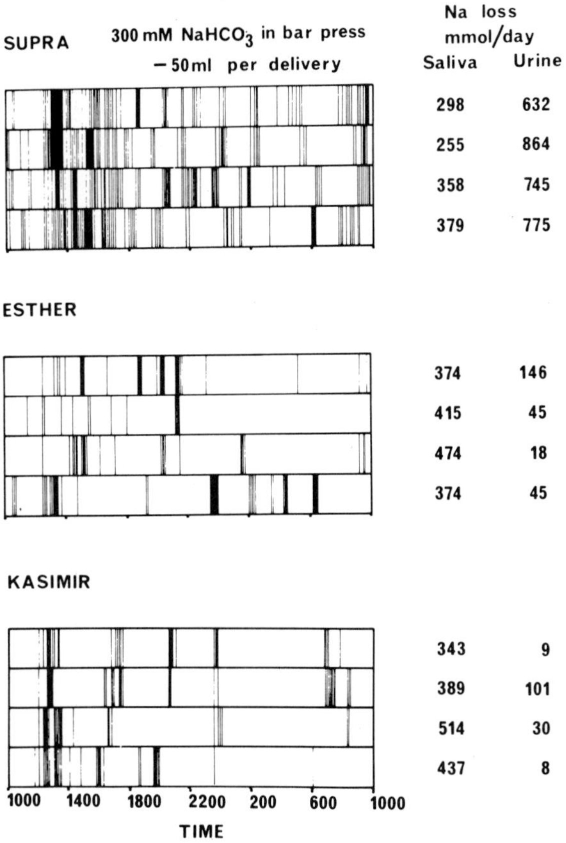

Fig. 13-10. Daily record of operant behaviour for sodium (50 ml of 300 mmol/l sodium bicarbonate) in the three parotid-fistulated sheep Supra, Esther and Kasimir.

regard the 2–4-h interval between each episode of bar-pressing and intake by e.g. Ricky (Fig. 13-6) as a reaction to body deficit of 40–50 mmol. It would remain a puzzling question as to how such a small error would be registered within the context of the dynamics of the large sodium turnover in a sheep with a parotid fistula.

We do, however, have independently determined experimental evidence that a change in body sodium status of this order can be sensed physiologically. Andersson, Jobin and Olsson (1967) have shown that injection of 0.1 ml of 0.85 M sodium chloride into the third ventricle of the goat caused drinking of 1.5–2.0 litres of water within 2 min of injection, ADH (antidiuretic hormone) secretion and increased sodium and chloride excretion. As Andersson (1977) notes, attempts to elucidate the mechanism of the natriuresis have been inconclusive. In the goat the natriuresis was associated with an increase in glomerular filtration rate (Andersson, Dallman and Olsson, 1969) and some rise in arterial blood pressure (Andersson et al. 1972). A consistent relation between vasopressor and natriuretic responses has also been observed in the cat (Chiu and Sawyer 1974) where guanethidine blocks both responses, thus indicating involvement of increased sympathetic nerve activity. However, no change in glomerular filtration rate has been observed during natriuresis induced by intraventricular administration of hypertonic saline in the dog (Dorn et al. 1969). The effect is evidently too fast for it to be related to decreased aldosterone secretion, and, in any event, it is not prevented by prior administration of the hormone (Andersson et al. 1967).

As Figs. 13-11 and 13-12 show, the same natriuresis may be produced by injection of 1 M saline into the third ventricle of a sheep that is sodium-replete on normal laboratory diet. The effect is large and the onset is rapid (10–15 min). By 60–90 min it has reversed. If, however, the animal had a parotid cannula inserted before injection and only 54–75 mmol of sodium were lost in the hour preceding the injection, the natriuresis was clearly blunted in the four animals studied at this level of deficit (Fig. 13-11 shows an individual experiment). Depletion of 150–170 mmol of sodium (involving loss of a litre of saliva) abolished the response in four animals though the antidiuresis still occurred (Fig. 13-12). Furthermore, in such animals the administration of 300 ml of 4.5% dextran in isotonic dextran saline, which only partly repaired sodium deficit (i.e. 200–300 mmol) but expanded plasma volume above that in the sodium-replete state (Blair-West et al. 1967), did not restore the natriuretic response.

Whereas we have not determined the cause of the natriuresis nor therefore the vector of this blunting or abolition of the effect, the data do show that sodium loss of 50–70 mmol can have physiological effect—much as this degree of 'error' of body status would appear to be able to induce salt-drinking behaviour. In terms of physiological reaction to change of sodium status, there is also evidence that the aldosterone control system of the sheep will respond to small changes of this order. Thus Fig. 3-5 records the measurements made sequentially with the onset of sodium deficiency in sheep with an adrenal transplant and a parotid fistula. Upon ceasing replacement of sodium, salivary sodium loss caused virtual cessation of urinary sodium excretion by 6–8 h, and by the time 100 mmol of sodium had been lost at 6 h, the rate of aldosterone secretion by the autotransplanted adrenal gland had trebled. No change in plasma sodium concentration had occurred. Fig. 7-6 of grouped data from a number of animals studied similarly, and which include some measurements of adrenal secretion rate made earlier at the onset of sodium deficiency, indicates also adrenal

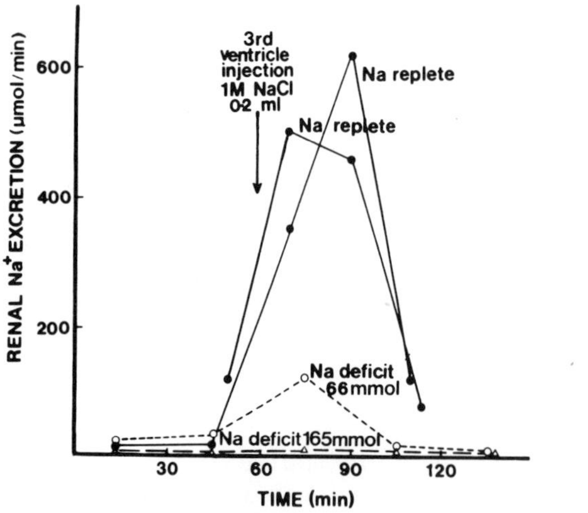

Fig. 13-11. The effect on renal sodium excretion in a conscious sheep of injection into the third ventricle of 0.2 ml of 1 M saline when the animal is sodium-replete (filled circles), and after depletion of 66 mmol (open circles) and 165 mmol (triangles) of sodium following parotid cannulation. (After Abraham et al. 1975.)

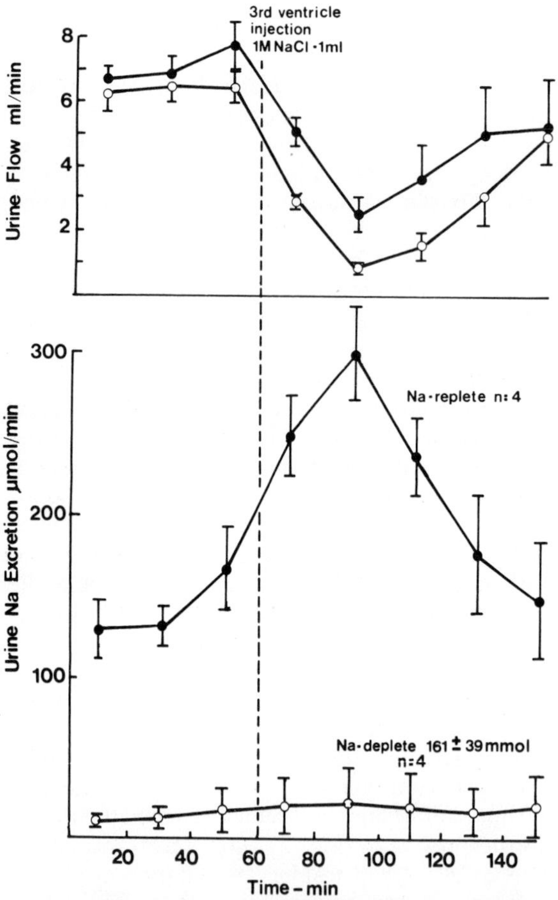

Fig. 13-12. The effect of injection of 0.1 ml of 1 M saline into the third ventricle on urinary sodium excretion and urine flow in four sheep when sodium-replete and when depleted of 161±39 mmol of sodium as a result of short-term parotid cannulation. (M. McKinley, R. S. Weisinger and D. Denton, unpublished).

reaction to deficits of ca 70–100 mmol of sodium (Blair-West et al. 1963).

Overall then, determinations on two other parameters are consistent with the fact that a body deficit of 50–100 mmol of sodium can be sensed and that there is appetitive response to this degree of body error.

Comparison with reaction to water deficit: Body error detected

The threshold for the thirst response has been calculated on an experimental basis in a number of species—man, dog and rat—by infusion of hypertonic saline (Wolf 1958; see review of Fitzsimons 1972). It involves a 1%–3% increase in plasma osmolality. As Fitzsimons points out, a threshold for thirst is useful because it means that small, i.e. subthreshold, losses of fluid do not arouse thirst. The animal is therefore not continually being distracted from its everyday business by the necessity of having to ingest small quantities of water in order to restore minimal changes in cell water. Between the state of water balance, i.e. shortly after ingestion to satiety, and the onset of cellular thirst, i.e. thirst threshold, between 1%–3% of cellular water must be lost. Lesser losses do not cause thirst, though, of course, drinking may occur for quite different reasons.

In sheep (McDougall et al. 1974) the total amount of exchangeable sodium is significantly different according to whether the animal is sodium-replete and non-fistulated or whether it has a long-established fistula and is sodium-replete at the time of measurement (non-fistulated sodium-replete=54±9.8 mmol/kg body wt, n=6; fistulated sodium-replete=64.9±12.5 mmol/kg, n=20) (P<0.01). This probably is due to fistulated animals being thinner and having much less fat than non-fistulated sheep. Both values are higher than for other species (cf. 42.7 and 46 mmol/kg for man and 45 mmol/kg for rat: Miller and Wilson 1953; Norman 1963; Skrabal et al. 1970), reflecting the large sodium circulation in the gut of the ruminant.

The data, then, would indicate that in the 30–40-kg sheep with continuous parotid salivary loss, the salt appetite mechanism reacts to a body error of 2.5%–3% of exchangeable sodium—though the activity records do not allow the precision of interpretation which is feasible in the thirst response with its induction by intravenous infusion of hypertonic saline. As with thirst, there are also other factors inducing sodium drinking, as we will discuss below, but the overall order of the error in body sodium necessary to induce salt-seeking behaviour would appear clear from the data.

However, when we consider the situation of the animal maintaining itself on the positive side of balance, the problem becomes more complex. Does the animal detect an error relative to a physiological 'set point' which is far on the positive side of neutral status? This cannot be excluded. For example, the large mass of clinical experimental work on what is questionably termed sodium deficiency (Chapter 7) involves the study of low-sodium diet—i.e. reduction of intake from 200 mmol/day to 10 mmol/day—this, in fact, being the normal dietetic sodium intake of vegetarian populations as described in Chapter 27. This change in level of intake evokes several physiological reactions, for example rise in aldosterone secretion and renin and angiotensin concentrations and fall in urinary sodium excretion. That is, the set of several physiological functions changes with the establishment of a new equilibrium state where sodium output equals intake but at a new lower level of turnover. This is well illustrated by the fact that the peripheral blood aldosterone level of vegetarian New Guinea Highlanders was a mean of 28 ng/100 ml compared with the 6–8 ng/100 ml of New Guineans living on the coast on a Western diet (Denton et al. 1969). The situation differs from that of an actual body deficit produced by a fistula loss, but the difference is probably quantitative along a continuum of sodium status ranging from positive status with daily turnover of e.g. 500 mmol to severe deficit with negative balance of 700–800 mmol (Chapter 7).

However, allowing that we accept the liking for salt per se as an important component in the behaviour of some animals, it would remain to explain the regularity of episodes of operant activity every few hours. This is more suggestive of detection of a change, even if the animal is in positive sodium status.

The influence of learning in bar-press behaviour

Another consideration is that the animals had experienced sodium deficiency on many occasions during the periods when the lever was withdrawn or bar-pressing did not yield sodium bicarbonate. Thus, they may have learned to bar-press preferentially and frequently to avoid development of effects of sodium deficiency. With Dorothy and Wu, which had highest levels of urinary sodium excretion under the regime where sodium was continuously available from the bar-press (Figs. 13-6 and 13-8), the more than adequate compensation for salivary loss by intravenous saline infusion cut bar-pressing and drinking activity over 3 days to one-sixth or one-seventh of control. However, in sheep with a parotid fistula where access to bar-press was for 2 h every 24 h, administration of 600 mmol of sodium bicarbonate in 1 litre of water by rumen tube 3–4 h before access almost eliminated bar-pressing whereas administration of water alone had no effect (Chapter 22).

When first potassium chloride (300 mmol/l) and then potassium chloride plus quinine were substituted in the bar-press machine delivery system, whilst free access to water and sodium bicarbonate solution (300 mmol/l) were permitted, bar-pressing fell to one-fifteenth to one-twentieth of initial activity. There was considerable variation of response to this manoeuvre. Both Dorothy and Brad drank significant potassium chloride on the first day or two days of access whereas Talilah virtually rejected it from the outset. Zeta ceased to drink it on the second day. The addition of quinine virtually abolished intake in all animals. However, it can be noted that despite liberal access to sodium solution on the side of the cage, the animals habituated to this operant situation still delivered some fluid by bar-press even though they did not drink it.

Also it was found that increasing the volume of solution (300 mmol/l sodium bicarbonate) delivered with each bar-press from 50 ml to 250 ml reduced the rate of bar-pressing, though not in strict proportion (Fig. 13-13). With Ricky, for example, some deliveries were not drunk, and this was more pronounced with the experiments on Brad and Zeta. That is, intake was maintained at roughly the same level as reflected by urinary excretion and bar-pressing was reduced, but some occurred where deliveries were not drunk; this

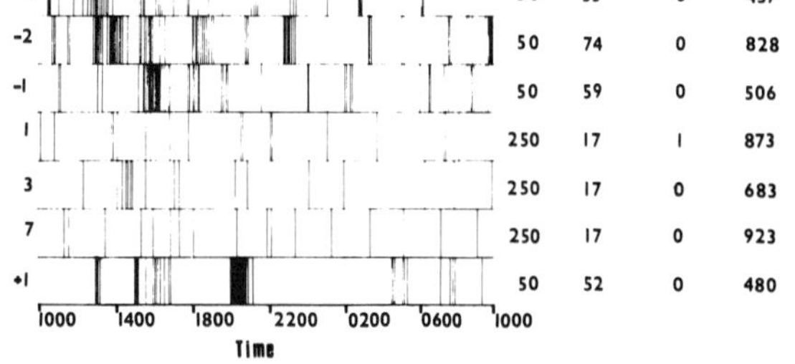

Fig. 13-13. Dorothy: the effect of increasing the volume of 300 mmol/l sodium bicarbonate solution delivered with each bar-press from 50 to 250 ml in a sheep losing saliva continuously from a parotid fistula.

suggested that the learned activity itself carried some reward for the animal.

Overall, these data indicate that the operant activity was not solely an episodic hedonic satisfaction independent of body status, and nor was it conditioned motor activity independent of what the delivery system actually contained. However, the fact of some residual behaviour in these three experimental instances would suggest that these factors (hedonic, learned benefit and conditioned act) could have contributed to the total behaviour observed under conditions of continuous fistula loss and constant access to bar-press. But it is not behaviour in any way independent of the state of body sodium balance. The linear relation between frank deficiency and operant behaviour (Fig. 13-5) testifies to a paramount role of body sodium status.

Bar-pressing during slow fluctuation of sodium status

Further data consistent with the analysis come from experiments of Dr Richard Weisinger and Dr David Mouw. Five sheep (Fig. 13-14) with a parotid fistula were maintained by allowing bar-pressing for sodium bicarbonate (15 ml of a 0.6 M solution=9 mmol sodium) for 2 h each day. On this regime the potassium concentration (50–60 mmol/l) in saliva collected immediately preceding

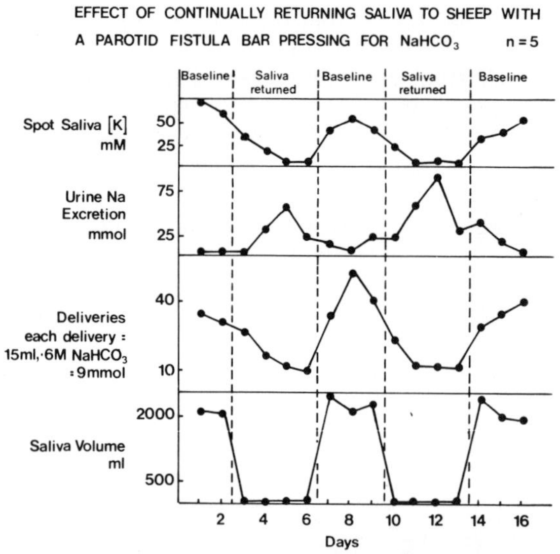

Fig. 13-14. Study of bar-pressing behaviour in five sheep with a parotid fistula when saliva was immediately pumped back into the rumen via an intranasal tube. The bar-press was operative for the same 2 h each day following the ringing of a bell. Mean salivary potassium concentration, daily urinary sodium excretion and salivary volume are shown also.

bar-press access indicated significant sodium depletion since the last access to bar-press. Mean deliveries in the 2 h were 30–35, providing 270–315 mmol of sodium, and mean salivary loss was 2 litres. There was little or no sodium in the urine under these conditions. It was then arranged that for 4 days the saliva lost was immediately pumped back into the animals via an intranasal tube passed to the rumen. As the animals' sodium balance became positive as indicated by an eventual fall of mean salivary potassium concentration to 5 mmol/l and a rise in urinary sodium excretion, bar-pressing decreased to a residual level of ca 10 per 2-h period. Immediately following cessation of return of saliva, bar-pressing rose to 40–50 per 2 h each day, and again this reversed over 2 days to 10 per 2 h when saliva was returned to the animals. These slow changes of balance caused corresponding changes in operant behaviour but a residual operant behaviour was observed when the animals achieved the replete state. This represented 20%–35% of intake occurring in response to uncompensated salivary loss.

Bar-pressing for water on a regime of continuous access to sodium bicarbonate for days

Some further light on behaviour under these conditions comes from experiments on a different regime. The same animals now had free access to food and sodium bicarbonate, and water was placed in the bar-press (50 ml per delivery). Under our experimental conditions it seemed likely that external water loss would be continuous and uniform throughout the day, more uniform even than salivary loss. Fig. 13-15 shows that with Ricky intake was episodic over the 24 h but was concentrated in 10.00–18.00 hours period. Zeta, however, had most operant activity and intake over the 14.00–22.00 hours period, i.e. during and following eating. The behaviour contrasted with Zeta's regularly spaced operant activity for sodium bicarbonate solution (Fig. 13-7). Fig. 13-16 shows Brad's intake to be regular, with 2–4-h episodes over the 24 h, whereas virtually all of Dorothy's was within a 12-h period concentrated on the eating period.

Clearly, the results in two animals are not consistent with continuous external loss causing an error which is detected by thirst osmoreceptors, with resultant drinking at intervals spaced according to the sensitivity of the osmoreceptor. It is known that when sheep are fed once every 24 h and eat rapidly, a large sudden flux of fluid occurs into the gut (Blair-West and Brook 1969). The volume

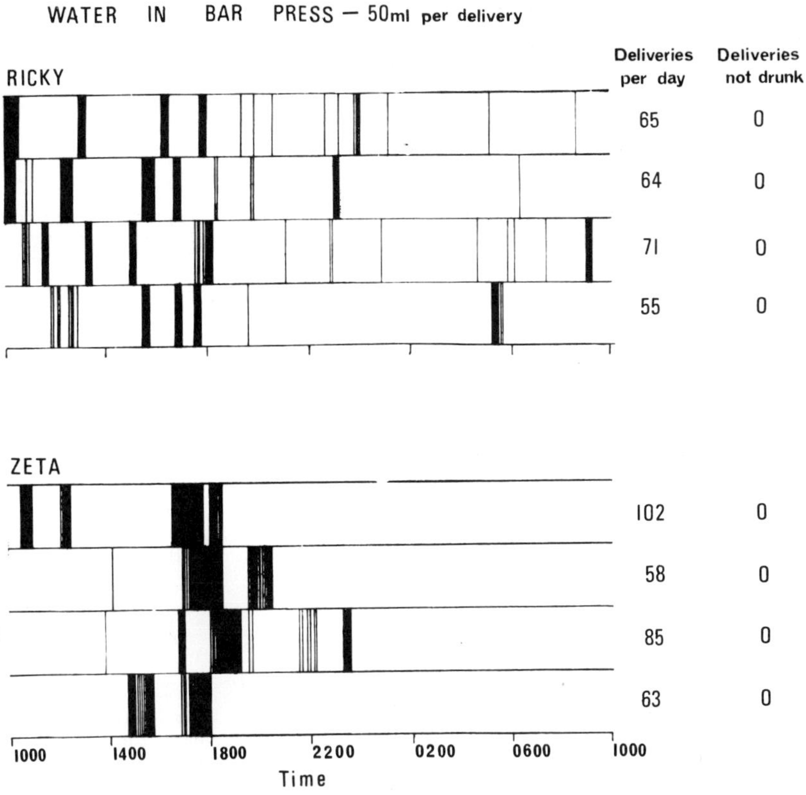

Fig. 13-15. Ricky and Zeta: sheep with a permanent unilateral parotid fistula. The sheep had free access to sodium bicarbonate solution (300 mmol/l). Water was available by bar-press (50 ml per delivery). (From Abraham et al. 1973, with permission.)

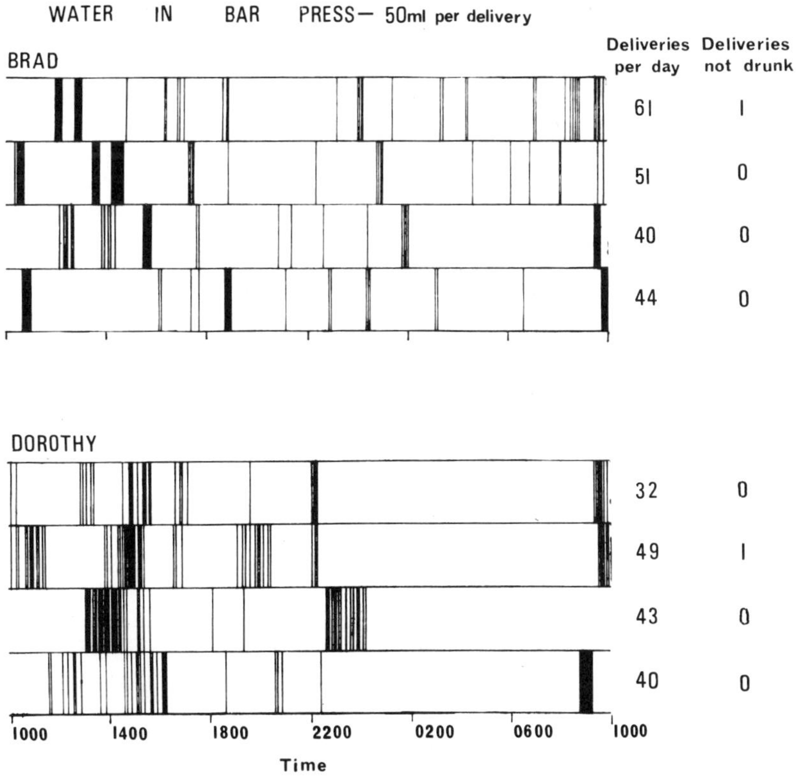

Fig. 13-16. Brad and Dorothy: sheep with a permanent unilateral parotid fistula. The sheep had free access to sodium bicarbonate solution (300 mmol/l). Water was available by bar press (50 ml per delivery).
(From Abraham et al. 1973, with permission.)

effects could generate thirst, or, as Fitzsimons and Le Magnen (1969) have suggested, the learned experience of thirst *following* eating may generate a learned behaviour of drinking with food. This could obviate this sensation. The animal may also drink to moisten food and facilitate swallowing. Oatley (1971) has drawn attention to the circadian rhythm rats show in drinking and the close association of water intake with eating. The major part of water intake is at night. Whilst recognizing factors including those noted above in causation of drinking and its relation to eating, he suggests drinking is not primarily controlled by physiologial needs and deficits but by an independent (presumably neural) programmer with a marked circadian rhythm which is responsive to reversal of the light–dark cycle. The meal programmer has similar properties and in ad libitum conditions both mechanisms are entrained so as to bring about a close association of eating and drinking.

Whereas with substantial water depletion in sheep it is established that the animal will within

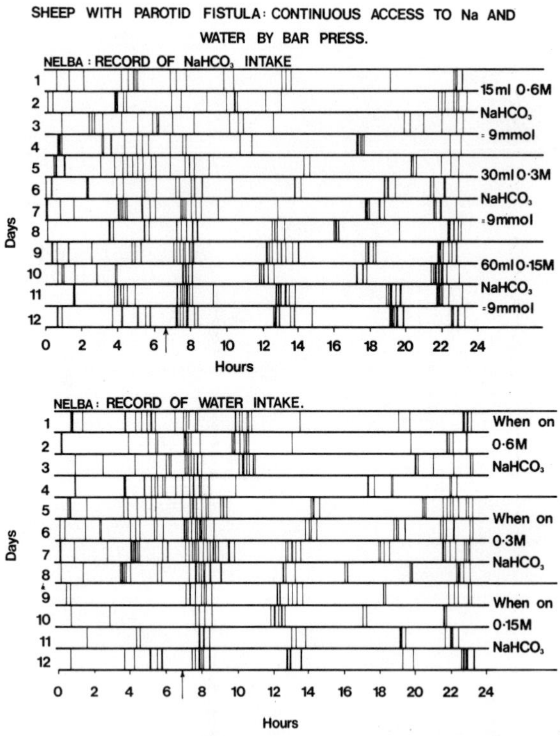

Fig. 13-17. Nelba: sheep losing saliva continuously from a permanent unilateral parotid fistula. The animal had two levers in the cage, one of which delivered 50 ml of water and the other sodium bicarbonate solution. For 4 days each press of the lever for sodium delivered 15 ml of 0.6 M sodium bicarbonate (=9 mmol sodium), for the next 4 days 30 ml of 0.3 M solution (=9 mmol sodium) and for the last 4 days 60 ml of 0.15 M solution (=9 mmol sodium). The upper portion of the figure is a computer record of bar-pressing for sodium bicarbonate over the 12 days, and the lower portion the bar-pressing for water.

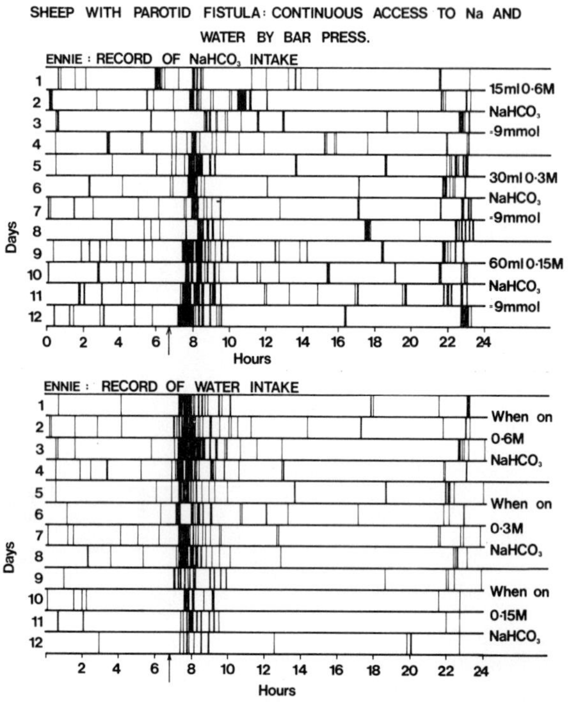

Fig. 13-18. Ennie: sheep losing saliva continuously from a parotid fistula. Experimental record of the same procedure as for Fig. 13-17.

2–3 min of access to water rapidly drink an amount accurately replacing body weight loss (Bott et al. 1965), it is clear from the adjustment during continuous access through bar-press that several factors are operative. That is, control is by a mechanism other than simple immediate reactions to change in the osmotic pressure of the *milieu intérieur*.

Further evidence on this is forthcoming from studies on animals with a permanent unilateral parotid fistula where access to water and sodium bicarbonate was continuous, by way of two levers in the animal's cage. The bar-pressing behaviour for each fluid during some weeks was recorded by computer. During the course of one experiment over 12 days, the concentration of sodium bicarbonate, and thus the volume delivered with each bar-press to contrive a constant amount of 9 mmol reaching the cup, was changed each 4 days. Fig. 13-17 records the pattern of behaviour characteristic of several animals, where overall there were episodes of bar-pressing for sodium bicarbonate and water at regular intervals about 1–3 h apart. Fig. 13-18 records behaviour characteristic of other animals, where there was clearly a concentration of bar-pressing for both water and sodium bicarbonate at the time of feeding and the 2–4-h period following this. Feeding was at 06.45 hours on the abscissca. The mean daily intake of sodium of the eight animals was not changed significantly by the variation in concentration of sodium solution offered. Water intake was reduced when 0.15 M sodium bicarbonate was offered.

Physiological mechanisms in ad libitum drinking in the dog

Rolls, Wood and Rolls (1980) have made an illuminating study of ad libitum water drinking in the dog. Typical patterns of daily water intake are shown in Fig. 13-19. One feature was that little water was drunk before and during feeding, but significant drinking usually started some time between 20 and 60 min after a meal. The food eaten was not dry. Since their observations on urine composition did not give a clear indication of stimuli initiating thirst, dogs were prepared with indwelling venous cannulae and blood was withdrawn every half hour. It was also sampled immediately after every drinking bout, and 90 min after the previous drinking bout to ensure that any plasma overdilution due to drinking had been eliminated by excretion. They found (Table 13-1) plasma sodium was higher at the time of drinking (148.4 mmol/l) than in non-drinking periods (144.2 mmol/l). To determine whether this increase in cellular dehydration (equal to 4.2 mmol/l of sodium) was of sufficient magnitude to initiate drinking they tested the effects of plasma sodium change produced by intravenous hypertonic saline. They found a change of 3.6 mmol/l would initiate drinking and thus concluded that the cellular dehydration present when the dog drinks is of sufficient magnitude to initiate drinking. Thus ad libitum drinking may not be anticipatory, and fluid deficits of sufficient magnitude to produce drinking do occur. It was also shown that the effect of feeding on dogs was to increase plasma sodium by 4.9 mmol/l, which is greater than the threshold to induce drinking. Following this observation, dogs were fed a low-sodium meal to minimize the osmotic effects of feeding. It was found that the latency time for dogs to drink following a meal was increased from 48±5 to 244±44 min (n=10) and total daily water ingested decreased from 42±5 to 21±5 ml/kg. The authors draw attention to the fact that species differences in drinking apparent in the laboratory may reflect adaptation to environmental constraints during evolution, in particular the nature of the diet and water resources.

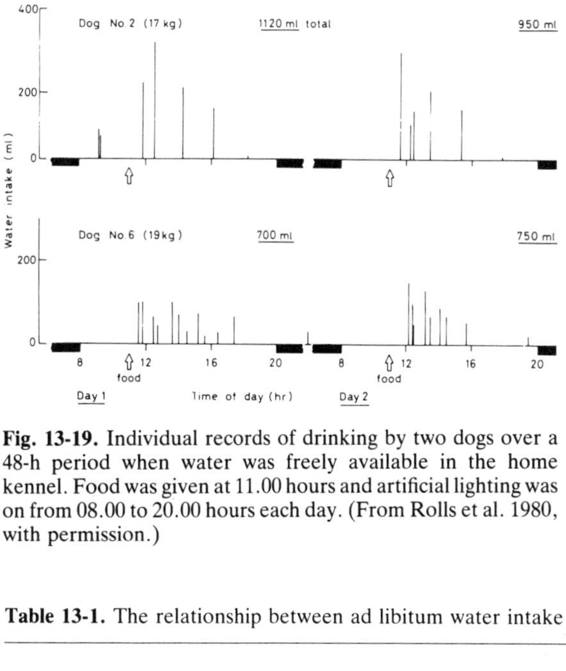

Fig. 13-19. Individual records of drinking by two dogs over a 48-h period when water was freely available in the home kennel. Food was given at 11.00 hours and artificial lighting was on from 08.00 to 20.00 hours each day. (From Rolls et al. 1980, with permission.)

Table 13-1. The relationship between ad libitum water intake and plasma sodium changes in the dog.

	Feeding time		First drink after feeding	All drinking bouts		All non-drinking periods	Drinking to threshold dose of intravenous NaCl
Water intake (ml/kg)			8.0±1.1	7.6±0.8			2.5±0.1
Plasma Na (mmol/l)	144.3±1.4		149.3±1.1	148.4±1.1		144.2±1.3	149.5±0.7
Plasma Na change (mmol/l)		4.9±0.6			4.2±0.4		3.6±0.4

From Rolls et al. (1980), with permission.

Mean (±s.e.m.) are shown for water intake, plasma sodium concentration and plasma sodium changes between conditions, for 11 dogs over 24-h periods with free access to water in the home kennel. The minimum change in plasma sodium concentration necessary to evoke drinking in the same animals (n=10) during intravenous sodium chloride infusions is also shown. Individual data making up the mean figures in the table were derived from at least three complete daily records of drinking, with associated plasma changes measured for each animal. The water intake data refer only to drinking bouts considered to be of a significant volume, excluding any less than 2 ml/kg body wt. In the dog, very little of the total water intake is due to drinking of less than this magnitude.

References

Abraham SF, Baker R, Denton DA, Kraintz F, Kraintz L, Purser L (1973) Components in the regulation of salt balance: Salt appetite studied by operant behaviour. Aust J Exp Biol Med Sci 51: 65

Abraham SF, Blaine EH, Denton DA, McKinley MJ, Nelson JF, Shulkes A, Weisinger RS, Whipp GT (1975) Phylogenetic emergence of salt taste and appetite. In: Denton DA, Coghlan JP (eds) Proceedings of the Fifth International Symposium on Olfaction and Taste. Academic Press, New York, p 253

Andersson B (1977) Central sodium–angiotensin interaction. In: Buckley JP, Ferrario CM (eds) Central actions of angiotensin and related hormones. Pergamon Press, New York, p 463

Andersson B, Jobin M, Olsson K (1967) A study of thirst and other effects of an increased sodium concentration in 3rd brain ventricle. Acta Physiol Scand 69: 29

Andersson B, Dallman M, Olsson K (1969) Observations on central control of drinking and the release of antidiuretic hormone. Life Sci 8: 425

Andersson B, Erikkson L, Fernandez O, Kolmodin CG, Oltner R (1972) Centrally mediated effects of sodium and angiotensin II on arterial blood pressure and fluid balance. Acta Physiol Scand 85: 398

Baldwin BA (1968) The use of operant conditioning in the study of sodium appetite in goats. J Physiol (Lond) 200: 20

Baldwin BA (1976) Effects of intracarotid or intraruminal injections of NaCl or $NaHCO_3$ on sodium appetite in goats. Physiol Behav 16: 59

Blair-West JR, Brook AH (1969) Circulatory changes and renin secretion in sheep in response to feeding. J Physiol (Lond) 204: 15

Blair-West JR, Coghlan JP, Denton DA, Goding JR, Wintour M, Wright RD (1963) The control of aldosterone secretion. Recent Prog Horm Res 19: 311

Blair-West JR, Coghlan JP, Denton DA, Goding JR, Wright RD (1964) The effect of adrenal cortical steroids on parotid salivary secretion. In: Sreebny LM, Meyer J (eds) Salivary glands and their secretions. Pergamon Press, New York, p 253

Blair-West JR, Boyd GW, Coghlan JP, Denton DA, Wintour M, Wright RD (1967) The role of circulating fluid volume and related physiological parameters in the control of aldosterone secretion. Aust J Exp Biol Med Sci 45: 1

Bott E, Denton DA, Weller S (1965) Water drinking in sheep with oesophageal fistulae. J Physiol (Lond) 176: 323

Chiu PJS, Sawyer WH (1974) Third ventricular injection of hypertonic NaCl and natriuresis in cats. Am J Physiol 226: 463

Denton DA (1966) Some theoretical considerations in relation to innate appetite for salt. Conditional Reflex 1: 144

Denton DA, Nelson JF (1971) Effects of pregnancy and lactation on the mineral appetites of wild rabbits (*Oryctolagus cuniculus*(L.)). Endocrinology 88: 31

Denton DA, Nelson JF, Orchard E, Weller S (1969) The role of adrenocortical hormone secretion in salt appetite. In: Pfaffmann C (ed) Proceedings of the Third International Symposium on Olfaction and Taste. Rockefeller University Press, New York, p 535

Dorn JB, Levine N, Kaley G, Rothballer AB (1969) Natriuresis induced by injection of hypertonic saline into the third cerebral ventricle of dogs. Proc Soc Exp Biol Med 131: 240

Fitzsimons JT (1972) Thirst. Physiol Rev 52: 468

Fitzsimons JT, Le Magnen J (1969) Eating as a regulatory control of drinking in the rat. J Comp Physiol Psychol 67: 273

McDougall JG, Coghlan JP, Scoggins BA, Wright RD (1974) The effect of sodium depletion on bone sodium and total exchangeable sodium in sheep. Am J Vet Res 35: 923

Miller NE (1967) Behavioural and physiological techniques: Rationale and experimental designs for combining their use. In: Code CF (ed) Handbook of physiology Sect 6: alimentary canal, vol 1. American Physiological Society, Washington, p 51

Miller H, Wilson GM (1953) The measurement of exchangeable sodium in man using the isotope ^{24}Na. Clin Sci 12: 97

Norman N (1963) The participation of bone in sodium and potassium metabolism in the rat. I. Simultaneous determination of the exchangeable body sodium and potassium and the exchangeable and inexchangeable fractions of these ions in bone in the normal rat. Acta Physiol Scand 57: 363

Oatley K (1971) Dissociation of the circadian drinking pattern from eating. Nature 229: 494

Quartermain D, Miller NE, Wolf G (1967) Role of experience in relationship between sodium deficiency and rate of bar pressing for salt. J Comp Physiol Psychol 63: 417

Rolls BJ, Wood RJ, Rolls ET (1980) Thirst: The initiation, maintenance and termination of drinking. Academic Press, New York, p 263 (Progress in Psychobiology and Physiological Psychology, no. 9)

Skrabal F, Arnot RN, Helus F, Glass HT, Joplin GF (1970) A method for simultaneous electrolyte investigation in man using ^{77}Br, ^{43}K and ^{24} Na. Int J Appl Radiat Isot 21: 183

Stricker EM (1980) The physiological basis of Na appetite: A new look at the 'Depletion–repletion' model. In: Kare MR, Fregly MJ, Bernard RA (eds) Biological and behavioural aspects of salt intake. Academic Press, New York, p 185

Wolf AV (1958) Thirst: Physiology of the urge to drink and problems of water lack. Thomas, Springfield, Ill

14 The consummatory act of satiation of salt appetite in sodium deficiency

Summary

1. Water-deplete sheep precisely correct deficits of 10% of body weight in a rapid consummatory drinking act over 1–3 min. Rats drink much more slowly and drinking behaviour diminishes after 8–10 min as a result of effects of absorbed water.

2. Sodium-deplete sheep, when offered a solution of sodium bicarbonate solution, drink an amount commensurate with deficit rapidly over 1–10 min. Laboratory rats drink more slowly but usually overcorrect deficit substantially except when severely depleted, when intake may approximate loss. Wild rabbits, sodium-deficient as a result of adrenal insufficiency, take 9–12 h to correct deficit whether large or small. That is, the rate of drinking is determined by the amount of deficit. The precise correlation between deficit and intake is striking.

3. With sheep with a parotid fistula, variation every 4–5 days of the concentration of sodium bicarbonate solution offered, over the range of 100–900 mmol/l, on a basis of free access resulted in accurate adjustment of the volume drunk. If solutions were offered for only 15 min every 48 h and the concentrations on different days was varied over this range many animals varied the volume drunk inversely with concentration quite accurately, with the limitation that most animals drank less 100 mmol/l solution than required to correct deficit. Presumably afferent inflow from foregut after 4–6 litres were drunk modified drinking behaviour. The results show that sodium-deficient sheep have the capacity to integrate centrally taste impulses signalling concentration with pharyngo-oesophageal impulses metering volume swallowed in the process of adjusting rapid intake to deficit. Rats have also been shown to adjust volume drunk to concentration of salt solution offered.

4. Tubing either 1.5 litres of water or 300–600 mmol of sodium bicarbonate in 1.5 litres of water into the rumen 10 min or 120 min before offer of sodium bicarbonate solution for voluntary drinking had little or no influence on sodium bicarbonate intake of sodium-deplete sheep; however, as a result of the sodium overload intake was greatly reduced or abolished the following day. Thus this degree of rumen distension did not modify the consummatory act nor was evidence forthcoming of any sodium or osmosensors on the rumen surface. With rats, tubing experiments showed gastric loads could have large effects on drinking of both water and sodium chloride solution according to the animals' hydrational state and the tonicity of the fluid tubed.

5. With water-deplete sheep, tubing 3 litres of water into the rumen 5–30 min before offering water did reduce voluntary intake. If water-deplete sheep were allowed to drink with an open oesophageal fistula, the animals

drank 2.5–9 times deficit in 120 min. The animal continued to drink after the fistula was closed and eventually corrected deficit. Tubing water into the rumen equivalent to deficit (ca 3 litres) immediately before sham-drinking with open oesophageal fistula reduced voluntary intake substantially.

6. Sodium-deficient sheep offered sodium bicarbonate to drink when an oesophageal fistula was open drank nearly double the control intake within 15 min and a mean of 2.7 times control within 120 min. When the concentration of sodium bicarbonate offered was varied over the range 100–900 mmol/l, there was usually an inverse effect on the volume sham-drunk, with some evidence that high concentrations have a greater satiating effect than low. Oesophageal fistula experiments on sodium-deficient animals involve great technical difficulties.

Introduction

In the day-to-day laboratory study of sheep losing sodium through a parotid fistula the two main questions which present are:

1) How does sodium loss from the body generate an appetitive drive which is commensurate with the body loss?
2) When presented with a sodium bicarbonate or sodium chloride solution, how does the animal correct this deficit accurately over a period as short as 2–10 min and 'know' to stop drinking long before the ingested material could enter the bloodstream and change any blood, extracellular or intracellular factor which might be postulated as causal of the appetite? How does the consummatory act of rapid drinking of the appropriate amount cause a precipitate decline in motivation?

Notwithstanding the inevitable variation seen in study of a population drawn from many different backgrounds, as has been the case with our sheep over the course of 20 years, the behaviour of a very large number of animals puts these questions into sharp focus. The satiation behaviour often càn present a parallel to orgasmic phenomena in the mounting excitement and restlessness, the eventual rapid motor act, and then loss of motivation. As shown in Fig. 12-9, changes in a spectrum of visceral functions with satiation of salt appetite follow closely the same time course.

Expressing this in more general terms, Owen Maller (1967) notes that specific appetite encompasses detective as well as correctional functions. Components of detection involve the ability to recognize a change in the internal state associated with depletion of a specific nutrient or nutrients followed by location and selection of the appropriate nutrients in the environment. Correctional aspects involve the consumption of sufficient quantities of the nutrient, the association of relief from the need state with the ingestion of some discriminable part of the intake, and the storage of this information for future use.

The two species in which salt appetite has been most extensively studied, the sheep and the laboratory rat, appear to differ basically in their response to water deprivation. Hatton and Bennett (1970), Feider (1972) and Blass and Hall (1976) have studied drinking in water-deprived rats and precisely correlated drinking behaviour with the time course of changes in plasma sodium concentration, osmolality and haematocrit, and analyses of gut contents for residual water. The effects of prior intubation of water, isotonic or hypertonic saline etc., and of oesophageal fistula were studied also. Consummatory behaviour lasted over 30 min, but clear change in behaviour at 8–10 min (Hatton and Bennett 1970) was related to a change in plasma composition. In contrast, the sheep when deprived of water for 48 h with loss of 9% body weight drank 80%–100% of deficit in the first 2 min of access. Drinking was usually completed in 10 min. Plasma analyses did not indicate any absorption of water from the capacious rumen or abomasum at this stage. Significant absorption was apparent by 30–60 min (Bott et al. 1965). In the sheep, therefore, satiation over 1–10 min does not appear to involve any feedback effect of absorbed fluid whereas this appears to be a significant factor in the rat. As will be discussed below, similar differences are observed in the time characteristics of satiation of salt appetite in the sheep and the rat.

Time characteristics of satiation of salt appetite

Sheep

Fig. 14-1 shows examples of two somewhat different types of behaviour over a week of observation in two sheep with a parotid fistula. The

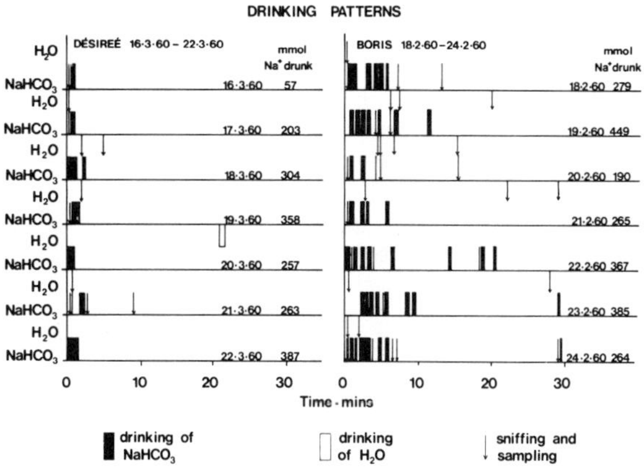

Fig. 14-1. Désirée and Boris: sheep with continuous salivary loss from a unilateral parotid fistula. The pattern of drinking sodium bicarbonate solution (300 mmol/l) when the fluid was presented for 30 min each day. Water was continuously available. (From Denton 1966, with permission.)

animals, which had free access to water, were sodium-deficient for only 24 h. Sodium bicarbonate (300 mmol/l) and also water were presented for 30 min. With Désirée, sodium drinking was completed within 2 min in virtually all instances, with little or no interest in water. With Boris, the intake was usually complete within 10 min, but occasionally interest was sustained longer. Water was not drunk. The same essential pattern is recorded in the instance of much larger deficits with Anselm (Fig. 14-2). Here intake of 2–3 litres of 300 mmol/l sodium bicarbonate was usually complete in 5–10 min. The same pattern with large deficits of either sodium or water is recorded in Fig. 16-1.

The effect of sodium bicarbonate concentration on intake was investigated in an experiment where a sheep was offered different strengths of solution each 24 h over a period of several weeks. Salivary loss was ca 3 l/day. With solutions of 476–952

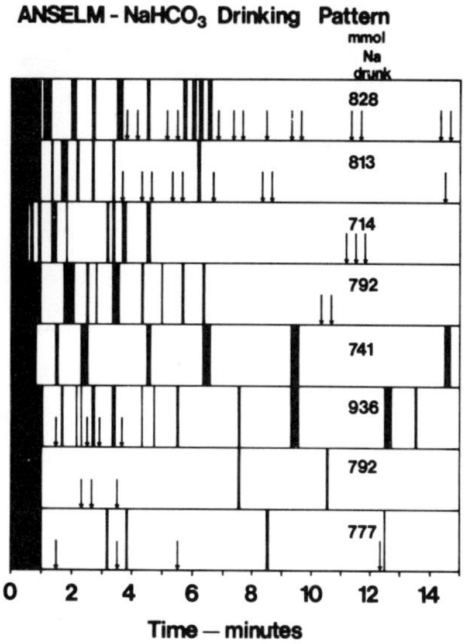

Fig. 14-2. Anselm: sheep with continuous salivary loss from parotid fistula. The pattern of drinking sodium bicarbonate solution (300 mmol/l) when the fluid was presented for 15 min after 48 h of sodium deprivation.

mmol/l, 90% of intake was in the first 5 min, and with 952 mmol/l in most instances there was subsequently some drinking of the water concurrently offered. With lower concentrations the drinking was slower though the bulk of intake was within 15 min. This may result less from any distension, or fatigue from volume drunk, than the fact that the lower concentrations may be less attractive to the individual animal when sodium deficient.

Rats

The comprehensive study of the time course of satiation of salt appetite in adrenalectomized rats deprived of sodium for 3–96 h showed that the drinking usually went on for 30–60 min before asymptote of intake was reached (e.g. Fig. 14-3) (Jalowiec and Stricker 1973). It appeared to be faster (10–20 min) in normal animals deprived of sodium for the same period, but in this case deficit was trivial. The adrenalectomized animals (24–48 h deplete), however, greatly overshot their body deficit as determined on external balance. The actual deficit was repaired by 10 min of drinking. The experiments were done in two series. In the first the rats were tested with an increased degree of deficit on each occasion (ascending) and in the second the maximum deficit was tested first (descending). The adrenalectomized animals seemed to anticipate salt presentation by sleeping in the front of the cage and licking

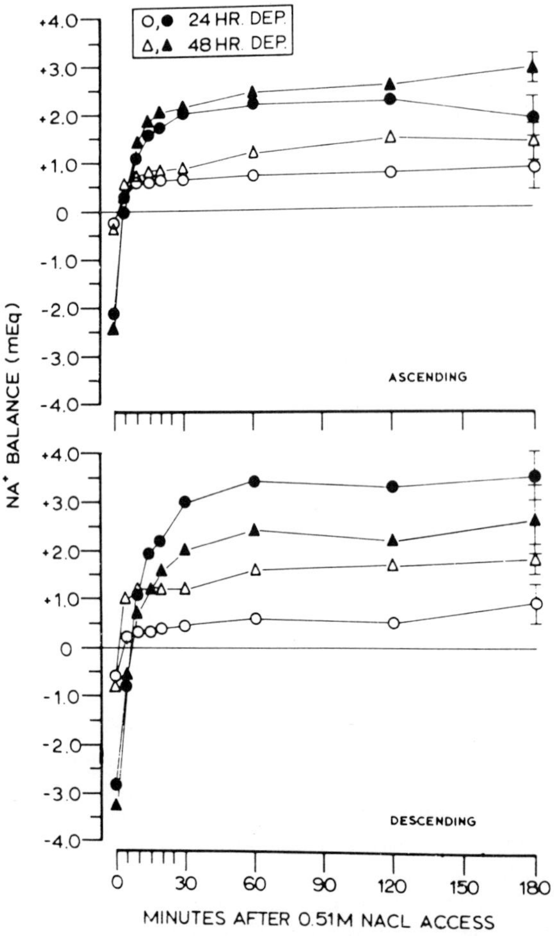

Fig. 14-3. Sodium balance during 3-h drinking tests following 24 h (circles) or 48 h (triangles) of sodium deprivation. Open symbols represent mean values from sham-operated rats; filled symbols represent mean values from adrenalectomized rats; vertical lines represent standard errors of the means. (From Jalowiec and Stricker 1973, with permission.)

the section of the cage where the drinking tube was to be attached. Observation made when 0.51 M sodium chloride was presented 24 h after formalin treatment showed drinking lasting over 6 h without asymptote (Jalowiec and Stricker 1970), but this situation is likely to be complicated by hormones released by stress (Blaine et al. 1975). Adrenalectomized rats are much the better comparison.

When wild rats (*Rattus fuscipes*) trapped in the sodium-deficient Snowy Mountains of Australia, where sodium deficit was probably severe, were allowed access to a drinkometer cafeteria with 500 mequiv/l solutions of sodium chloride, potassium chloride, calcium chloride, magnesium chloride and water the animals drank sodium chloride for 4 h before asymptote was reached (see Fig. 11-1).

Rabbits

The results of studies by Dr J. Nelson on the time course of satiation of sodium deficiency by wild rabbits have been rather remarkable. Five wild rabbits were adrenalectomized and maintained on daily administration of 0.25 mg cortisol acetate and 0.1 mg desoxycorticosterone acetate (DOCA). Precise external sodium balance studies were made. Depletion was produced by withholding maintenance mineralocorticoid for 1, 2 or 3 days before a test. Water and food were continually available but not sodium chloride.

Two hours before presentation of 0.5 M sodium chloride 0.1 mg DOCA was given. This ensured that urinary sodium loss stopped, and that the saline drunk in response to extant deficit was retained. This amount of DOCA has no effect on salt appetite, the dose being one-twentieth of that which gives small stimulating effects in wild rabbits (Denton and Nelson 1970).

Fig. 14-4 shows the time course of correction of

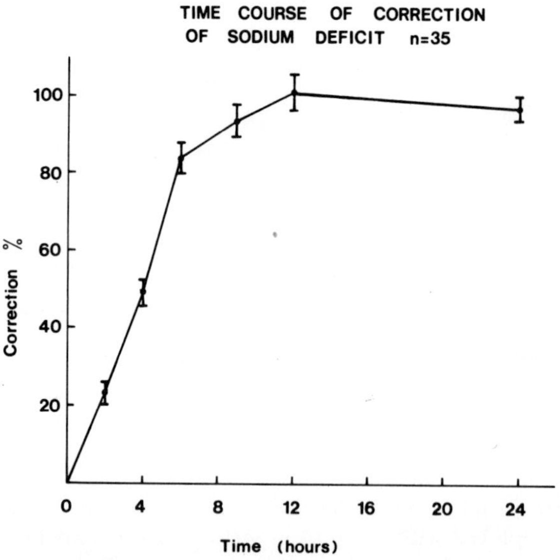

Fig. 14-4. The time course of correction of sodium deficit in adrenalectomized rabbits after presentation of 0.5 M sodium chloride solution (mean±s.e.m., n=35). (From Denton and Nelson 1980, with permission.)

sodium deficit. The deficit was completely replaced by 12 h. The animals were still in exact balance at 24 h, indicating there was no overshoot as with rats. The striking feature was that the same linear correlation between intake and initial deficit was seen at 2, 4, 6 and 9 h (Fig. 14-5) and the correlations at all times were highly significant ($P<0.01$ at all stages). Whereas the animals were slow to repair deficit, the rate of drinking was always related to the degree of the initial deficit,

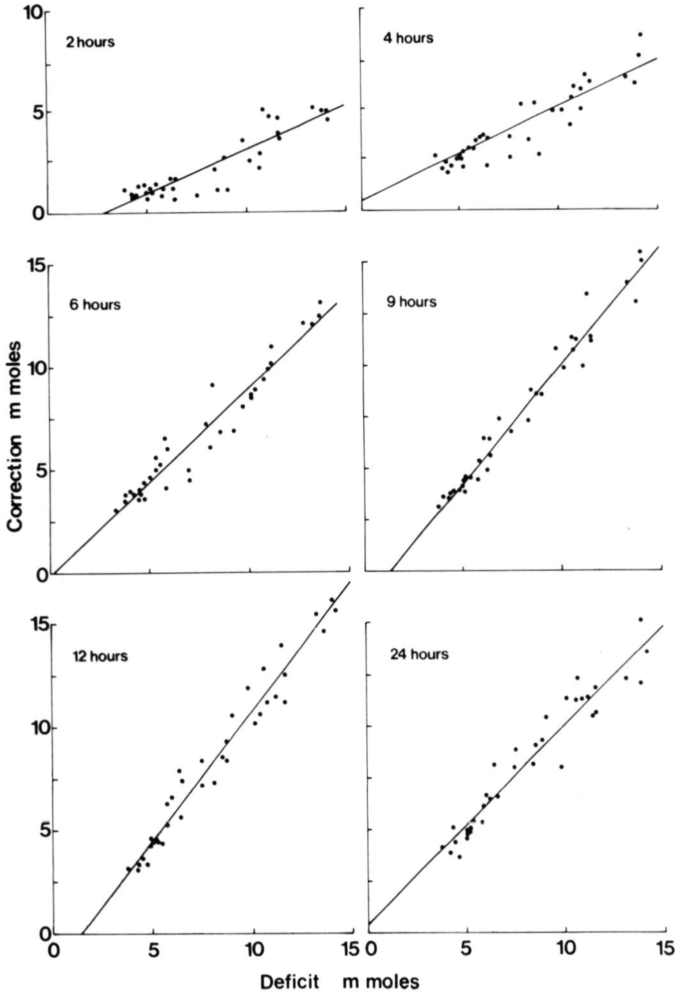

Fig. 14-5. The highly significant correlation between sodium deficit and amount of sodium intake observed at 2, 4, 6, 9, 12 and 24 h after access to 0.5M sodium chloride solution in 35 experiments on nine adrenalectomized rabbits.

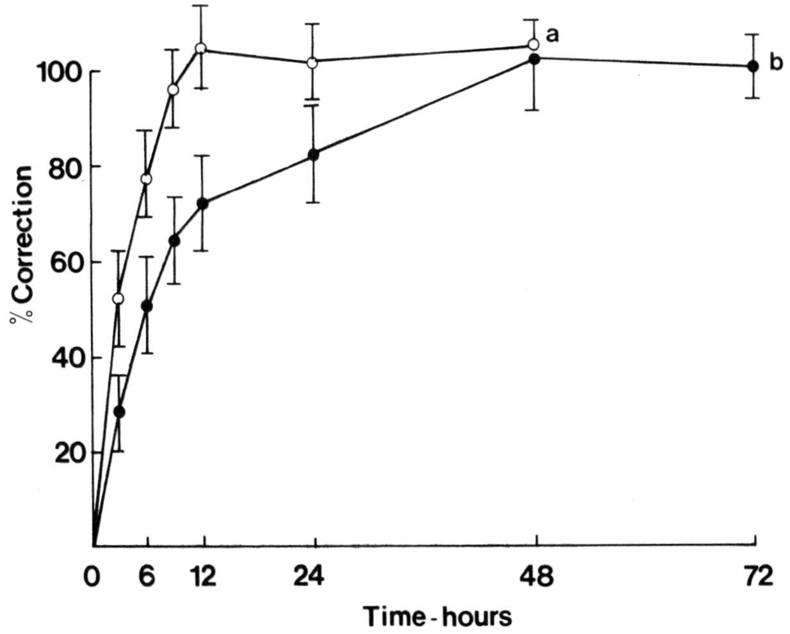

Fig. 14-6. Time course of correction of sodium deficit by frusemide-treated wild rabbits (open circles) and laboratory New Zealand rabbits (filled circles) when 0.5 M sodium chloride was presented 24 h after frusemide administration (mean±s.d.).

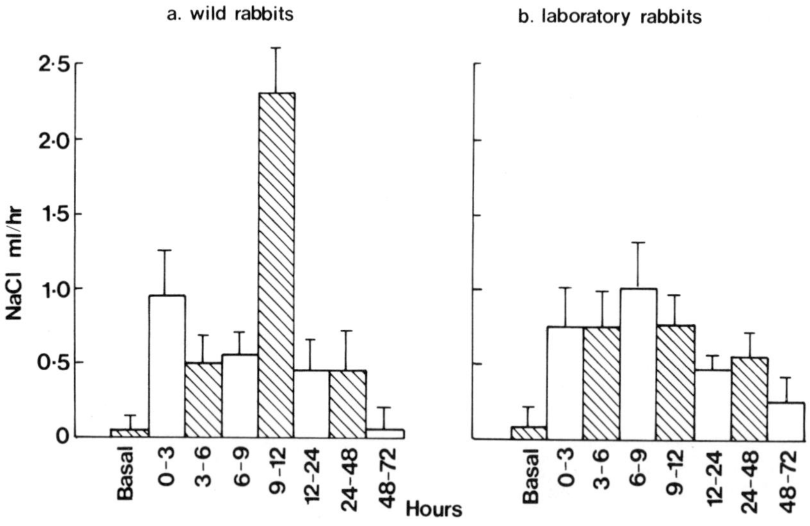

Fig. 4-7. The rate of intake of 0.5 M sodium chloride solution when presented 2 h after administration of frusemide to rabbits. (a) Wild rabbits (n=17, mean sodium loss 8.6 mmol); (b) laboratory rabbits (n=21, mean sodium loss 14.2 mmol). The hatched and open columns are to delineate the time scale and have no other significance.

the more deplete animals drinking faster over the 12 h. If the animal were less deplete it drank slower though the rate of drinking with large deficit showed it could have repaired the small deficit in much less than 9–12 h. Additional to the observation on no overshoot noted above, the rabbits were also in balance at 48 h. The results could be interpreted as indicating that the characteristics of the satiation procedure were overwhelmingly determined by the extent of central drive in response to depletion, and that feedback from different rates of entry of fluid into the gut had little influence.

In another study of satiating behaviour in rabbits, adrenal-intact animals were given 20 mg/kg of frusemide (Lasix; Hoechst), which produced a natriuresis lasting 1½ h. For wild rabbits mean sodium loss over the period was 7.6 mmol and for laboratory rabbits 14.6 mmol. Sodium solution (0.5 M) was offered 24 h later. In the wild rabbits (n=18), deficit was corrected in 9–12 h with initial rate of drinking 50% higher than in adrenalectomized rabbits with similar deficits. There was no overdrinking (Fig. 14-6). In laboratory rabbits (New Zealand type, n=15) drinking was slower than for wild rabbits (Fig. 14-6). It was also observed in other experiments of this series that if the sodium solution was offered 2 h after frusemide, the rate of drinking did not peak until 6–12 h (Fig. 14-7), indicating the characteristic time delay before full activation of appetite; it was noteworthy, though, that there was increase over control by 2–5 h.

Effect of variation of concentration of sodium solution on intake

Sodium-deficient sheep: Voluntary drinking

The animals had a parotid fistula and large daily salivary loss of 3–4.5 litres. Allowing for food, nett negative balance daily was 400–600 mmol of sodium. The sodium bicarbonate was offered continuously, concentration being changed at intervals of 7–10 days.

As shown in Fig. 14-8 with TP4, the adaptation to change of concentration was very often made on the first day. Adding adequate sodium to the food to replace fistula loss caused intake to reduce greatly. With PF33 (Fig. 14-9) the adaptation, though intake was not always commensurate on the first day, was quite precise over the period. With a third animal, TP1, adjustment was in some periods inaccurate with substantial overdrinking at higher concentrations (Denton and Sabine 1961).

If access to sodium solution was reduced to 1 h/day (PF33: Fig. 14-10), the maintenance of balance in the face of change of concentration was much less precise than with access 24 h/day. With 119 mmol/l sodium bicarbonate, intake fell short in that the animal only approached neutral sodium status immediately after drinking over 3 litres of solution. In the previous period, with 952 mmol/l sodium solution, it had oscillated to the positive side of neutral sodium status and at one point had established clear positive balance as reflected by urinary sodium excretion and negligible intake the following day. The first day concentrated sodium

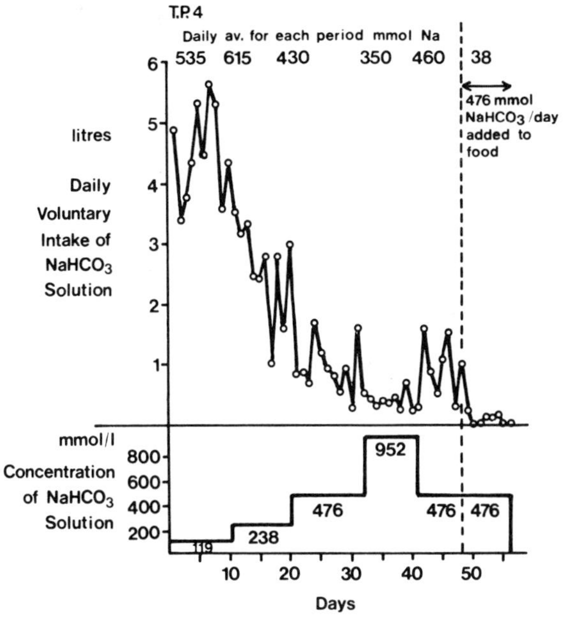

Fig. 14-8. TP4: Merino wether with a parotid fistula. The volume drunk each day of a solution of sodium bicarbonate the concentration of which was varied as shown in the lower section. Water was available continuously. The average intake (mmol Na/day) for each period is shown also at the top of the figure. After 50 days, 476 mmol of sodium bicarbonate per day was added to the food, and voluntary saline fluid intake was greatly reduced. (From Denton and Sabine 1961, with permission.)

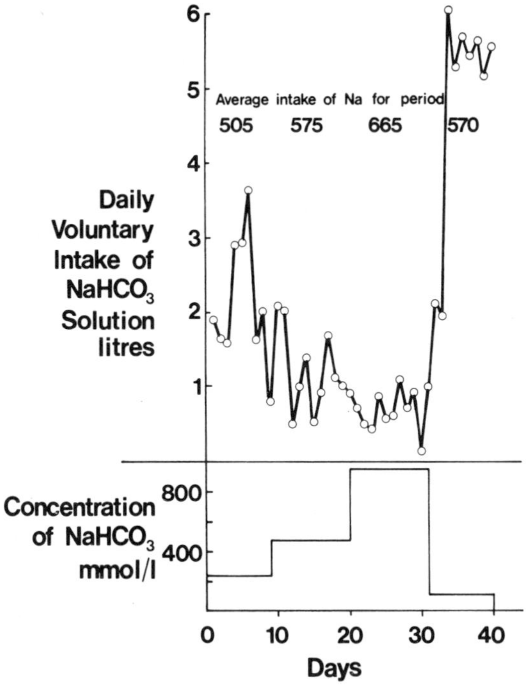

Fig. 14-9. PF33: Merino cross-bred ewe with a parotid fistula. Effect of variation of concentration of sodium bicarbonate offered by free access on the volume drunk each day.

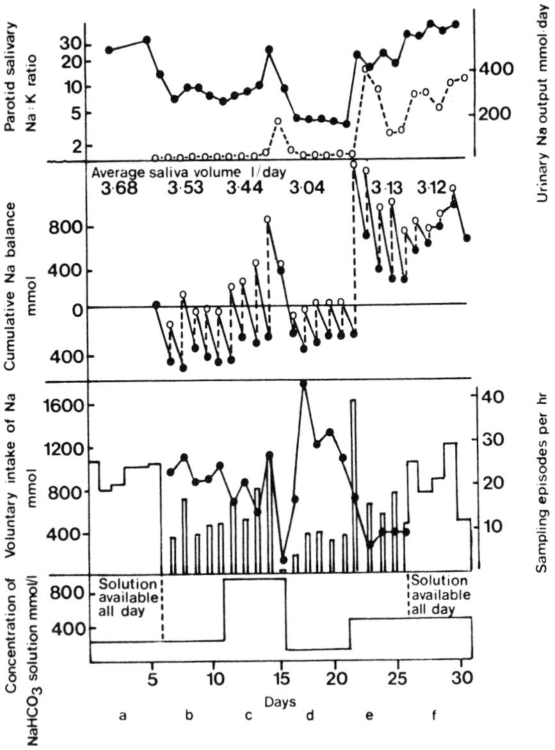

Fig. 14-10. PF33: Merino cross-bred ewe with a parotid fistula. The top section records the parotid salivary Na/K ratio (filled circles) and the urinary sodium excretion per day (open circles). The second section records the cumulative sodium balance. The continuous line shows the change of balance (mmol Na/day) and the interrupted line the change in balance produced by voluntary intake during the hour when access to sodium bicarbonate was permitted. The average salivary volume per day for each period of the experiment is shown also. The third section gives the voluntary intake of sodium when the animal had access all day, and (thin columns) when there was access for 1 h only. The number of sampling episodes during the hour of access (filled circles) is shown also. The bottom section records the concentration of the sodium bicarbonate solution which was offered. Water was available continuously. (After Denton and Sabine 1961.)

bicarbonate (476 mmol/l) was available after the period of dilute solution, the animal grossly overcompensated. A similar study over 67 days with Ned gave the same result: failure to compensate when 119 mmol/l solution was available for 1 h only. However, a third animal, TP12, adjusted volume of intake to variations in concentration of sodium bicarbonate presented between 880 and 220 mmol/l, and this held from the first day of access.

An experiment was made on Satan in which sodium bicarbonate solution at concentrations varying from 100 to 900 mmol/l was offered for 30 min following 48 h of sodium loss. The sheep had a very large salivary flow so loss over this time was in the range of 900–1000 mmol. Fig. 14-11 shows that

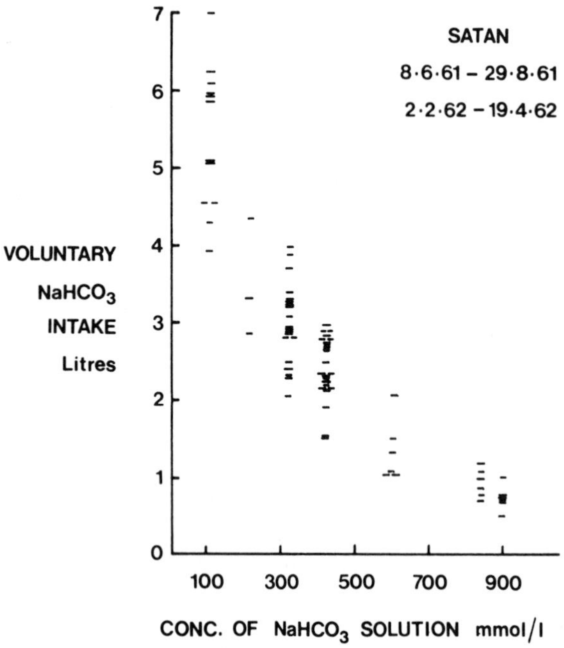

Fig. 14-11. Satan. Effect of varying the concentration of sodium bicarbonate solution offered on the volume of solution drunk in 30 min after 48 h of sodium deprivation. The mean salivary sodium loss over the 48-h period was 1012 mmol. Intake was consistently less than deficit when 100 mmol sodium bicarbonate was offered. (From Denton 1966, with permission.)

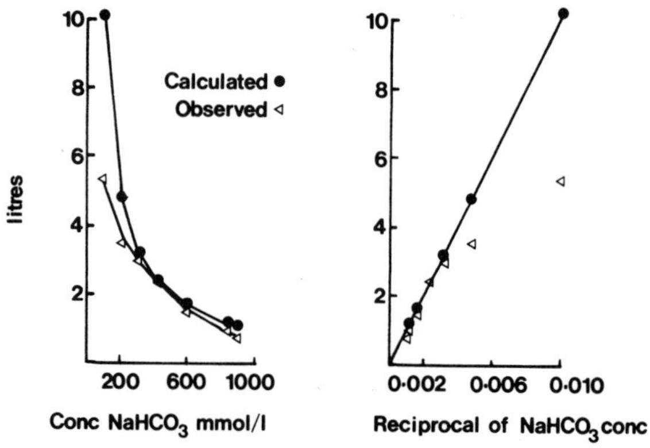

Fig. 14-12. Satan. Variation in volume of sodium bicarbonate drunk with variation in concentration of the solution offered for 30 min after 48 h of sodium depletion. The concentration most commonly offered to the sheep was 420 mmol. Using the animal's intake at this concentration as the starting point (intake being very close to loss at this concentration) a curve showing the volumes theoretically expected to be drunk to equal the deficit at the various concentrations was calculated. This is shown on the left (curve drawn between circles). The results actually observed are shown also (curve drawn between triangles). On the right, both curves are linearized by plotting the independent variable concentration of sodium bicarbonate solutions as the reciprocal form. The graphs show clearly that there is good agreement between observed and expected values at all except the most dilute concentrations. (After Bott et al. 1967.)

the sheep made a remarkable adjustment, matching the volume rapidly drunk in the consummatory act to the concentration of the sodium bicarbonate solution offered. Intake was consistently less than deficit on the days 100 mmol/l solution was offered, though the animal did usually drink 5–6 litres of fluid. The relation between intake at various concentrations and calculated intake necessary in face of deficit is shown in Fig. 14-12. A group of experiments on Tacitus under the same regime are shown in Fig. 14-13, and records of the sequential behaviour with this regime in Figs. 14-14 and 14-15.

There was, thus, a variation between animals, some being precise, others making some adjustment though obviously 'hunting around' neutral sodium status. There was underdrinking with dilute solutions when access was for 30–60 min. This may have been influenced by distension of the rumen as a result of the large volume drunk, or by fatigue of drinking (though ensuing data on studies with oesophageal fistulae do not support this).

Some animals show little capacity to adjust volume drunk to concentration change even after a substantial number of experiences. They over- or underdrink and 'hunt' in relation to a null point of neutral sodium status. But this in no way obviates

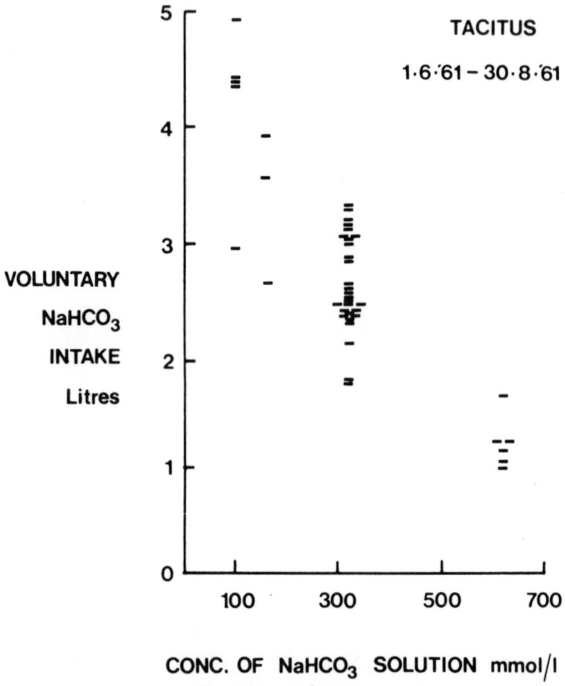

Fig. 14-13. Tacitus. The effect of variation of concentration of sodium bicarbonate offered for 30 min every 48 h on volume of the solution drunk. The animal was sodium deficient because of parotid salivary loss.

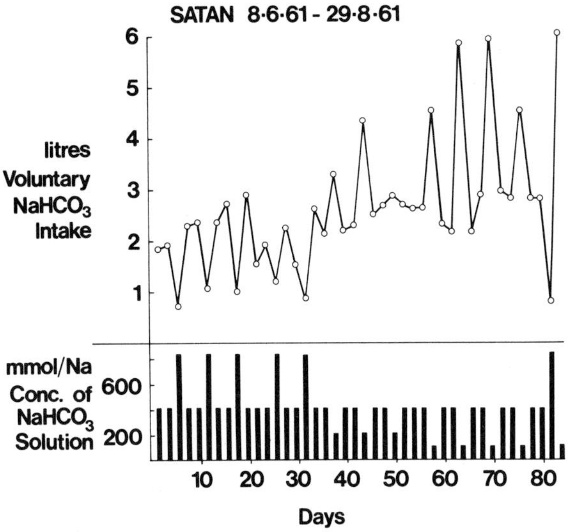

Fig. 14-14. Satan. The influence of variation of concentration of sodium bicarbonate offered for 30 min after 48 h of salivary loss on volume of solution drunk. A sequence of observations over 82 days showing overall adaptation and occasional error.

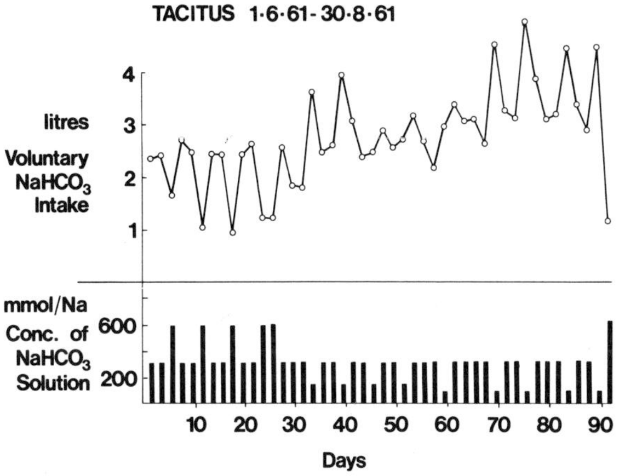

Fig. 14-15. Tacitus. The influence of variation of concentration of sodium bicarbonate offered for 30 min after 48 h of salivary loss on volume of solution drunk. A sequence of observations over 92 days.

the fact of, or the intense interest in analysing physiologically, those other animals which do indeed make the adjustment with considerable accuracy over periods of study (see for example Fig. 14-11). In these, there would appear to be a precise central integration of taste impulses signalling the character of the fluid and its concentration with pharyngo-oesophageal impulses metering the volume swallowed. These animals appear to multiply concentration by volume fairly accurately over a wide range in the course of the rapid consummatory act.

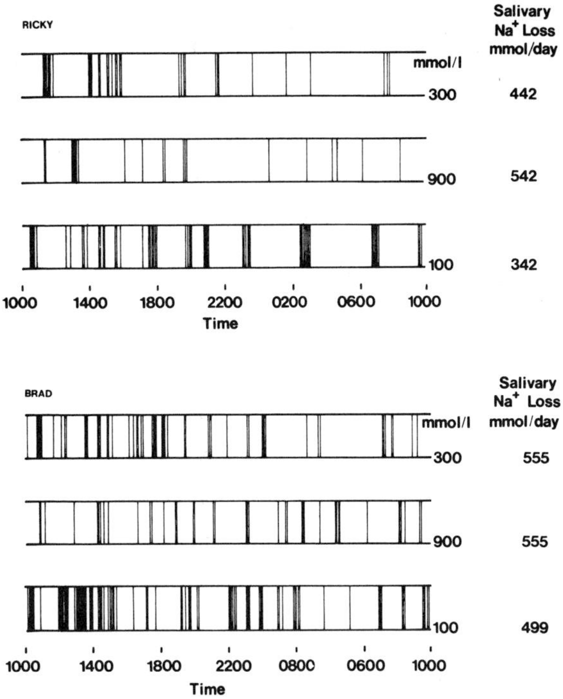

Fig. 14-16. Ricky and Brad. The effect of changing concentration of sodium bicarbonate solution in a bar-press system on the operant behaviour of sheep with a parotid fistula. Each horizontal segment represents a 24-h record, and each vertical stroke a delivery of 50 ml of sodium solution by bar-press. Salivary loss was continuous (2–3 l/day) and water was continuously available.

Sodium-deficient sheep: Sodium bicarbonate access by bar-pressing

Fig. 14-16 records an example of bar-press patterns in two sheep with continuous access to lever when concentration of sodium bicarbonate in the delivery system was changed. The number of deliveries made and drunk increased substantially when concentration was reduced from 300 to 100 mmol/l. There was a reduction when a 900 mmol/l solution was supplied but often the sheep made more deliveries than they drank.

Another experiment was made on four sheep which were depleted for 22 h as a result of salivary loss and then had access to the bar-press for 2 h. In one experimental series each bar-press delivered 50 ml of 150 mmol/l sodium bicarbonate (7.5 mmol). In a second, the delivery was 50 ml of 600 mmol/l solution (30 mmol). There was some difference in mean sodium deficiency between the two series (341 ± 10 mmol and 436 ± 26 mmol respectively, mean $\pm$ s.e.m.).

The cumulative number of deliveries and pattern of behaviour differed with the concentration of sodium bicarbonate delivered (Fig. 14-17).

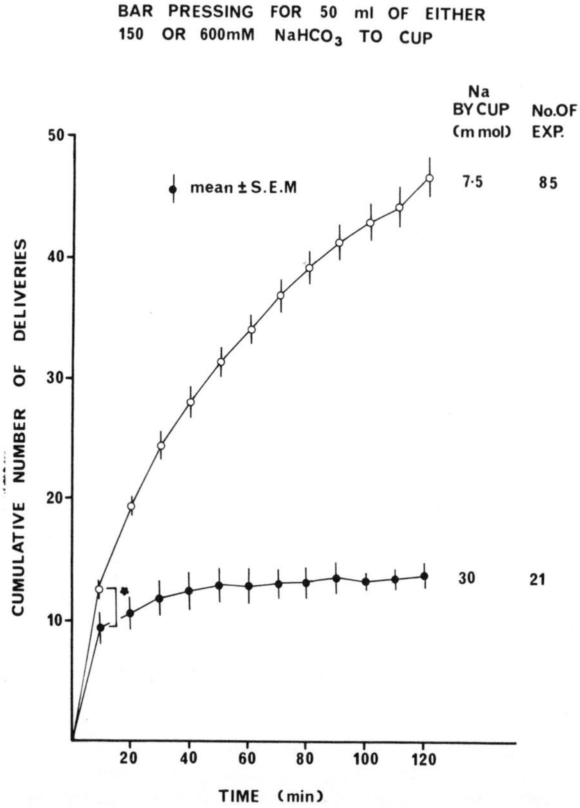

Fig. 14-17. The effect on bar-pressing for sodium bicarbonate solution of variation in the concentration of solution offered. The four sheep were allowed access to the bar-press after 22 h of uncompensated saliva loss. In one group of experiments (n=21) each delivery was 50 ml of 600 mmol/l solution (30 mmol) and in the other (n=85) 50 ml of 150 mmol/lsolution (7.5 mmol). An analysis of variance showed that there was a significant difference in cumulative number of deliveries by 10 min.

When 7.5 mmol of sodium were delivered to the cup the animals made 46±1.5 deliveries, a total of 345 mmol sodium (85 observations on four sheep). With 30 mmol delivered to the cup there were 13.9±1.3 deliveries, a total of 417 mmol (21 observations on four sheep). Thus there was evidence of adjustment to the different mean sodium deficiency in the two series, and this was despite the 4-fold difference in concentration of sodium bicarbonate offered. The difference between the rates of bar-pressing was significant by 10 min. With the lesser concentration bar-pressing was frenetic initially and declined gradually. With the greater concentration, response behaviour reached asymptote at 50–60 min. Fig. 14-18 shows the effect of varying the volume delivered with each bar-press from 10 to 50 ml of a 300 mmol/l solution of sodium bicarbonate. There was a clear difference in behaviour by 20 min, and with the 10 ml volume rapid deliveries continued throughout the 120 min access period.

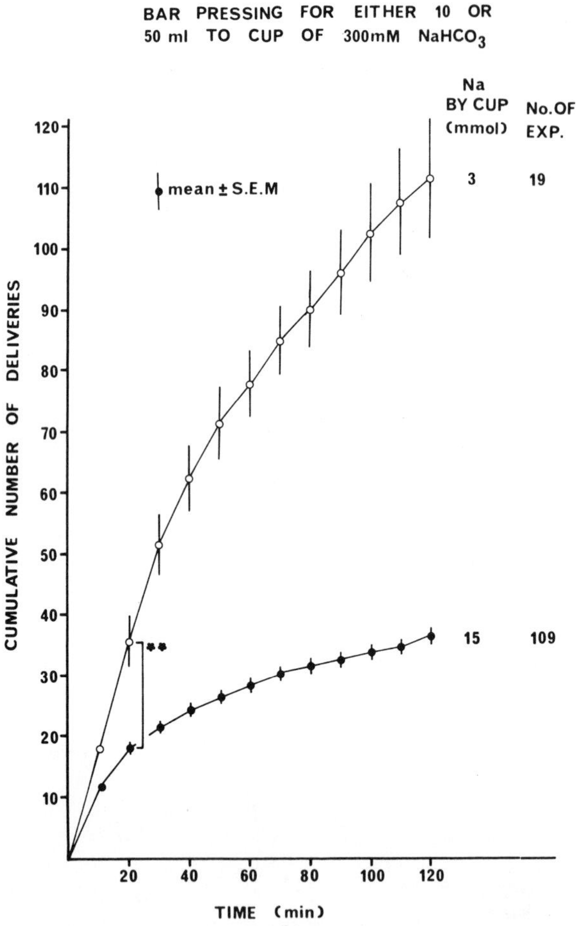

Fig. 14-18. The effect of variation of the volume delivered with each bar-press from 10 ml (open circles) to 50 ml (filled circles) of a 300 mmol/l sodium bicarbonate solution. The four sheep were allowed access for 2 h after 22 h of uncompensated salivary loss.

Sodium-deficient rats: Access to sodium solution by bar-pressing

Handal (1965) and Wolf (1969) present experimental records of naive rats without any previous experience of sodium deficiency. The records of first instance of exposure to sodium deficiency and availability of sodium solution by bar-press show that the rate of bar-pressing was much faster within 10–20 s with sodium salts than it was for non-sodium salts, but also that the rate of pressing and drinking was considerably less when 0.5 M rather than 0.15 M sodium chloride was offered.

Sodium-deficient rats: Voluntary drinking

Fregly (1958) studied the sodium intake of adrenalectomized rats in the face of variation of sodium concentration. With two different regimes it was clear that when sodium chloride was offered as 0.15 M, 0.25 M, 0.3 M, 0.35 M and 0.5 M solutions, daily intake was roughly constant, whereas with 0.025 M, 0.05 M and 0.075 M solutions intake was less.

Jalowiec and Stricker (1970), studying the effect of formalin injection on sodium appetite in rats, offered them either 0.075 M, 0.153 M or 0.513 M saline. Over 12–16 h of drinking, the intakes of these different concentrations, as reflected in mean cumulative sodium balance of the animals, closely approximated one another. However, as with other experiments, the eventual intake achieved with the dilute solution (0.075 M) was the least. It was still being drunk vigorously at the end of 20–24 h.

Influence of prior introduction of sodium solution into stomach or rumen

The aim of these experiments was to determine whether a sudden change in rumen volume or sodium concentration would alter drinking by the sodium-deplete sheep (Beilharz and Kay 1963).

The first series of experiments, on Désirée, Boris and PF33, involved introducing 1.5 litres of either water or sodium bicarbonate solution into the rumen 10 min before the beginning of the sodium drinking period. The solution was at 38 °C and sodium bicarbonate content was either 297 or 595 mmol.

Water caused only a transient reduction in Désirée's voluntary sodium intake. Boris and PF33 showed small but longer-lasting reductions, Boris reducing mean intake from 405 to 328 mmol/day and PF33 from 781 to 675 mmol/day. The differences were not statistically significant. They also drank their sodium solutions more slowly, Boris's rate dropping from 13 to 7 ml/s and PF33's from 20 to 16 ml/s. The sheep showed the usual appetitive behaviour when the bins were presented.

The introduction of 297 or 595 mmol of sodium bicarbonate into the rumen on about every fifth day generally had no greater immediate effect on sodium appetite and drinking behaviour than did the introduction of water alone on the intervening days (Fig. 11-34, Table 14-1). The first trial on Boris was atypical because the introduction of 595 mmol sodium bicarbonate by rumen tube caused a large reduction in sodium appetite. However, usually the animal drank the normal amount of

Table 14-1. The effect of giving 297 or 595 mmol of sodium bicarbonate by rumen tube, either 10 or 120 min before offering sodium bicarbonate solution to drink. Water was given by rumen tube on the intervening control days.

Sheep	Amount of $NaHCO_3$ solution drunk (mmol±s.d.)		
	Control days	'10 min' days	'120 min' days
Boris	390±132 (36)	400±204 (5)	323, 401 (2)
Désirée	648±60 (40)	575±142 (7)	486±171 (6)

From Beilharz and Kay (1963), with permission.
The number of days on which the means are based are shown in brackets. The amount of sodium drunk by Désirée was reduced significantly (P=0.06) on '120 min' days.

sodium solution on days when sodium was given by rumen tube, and became loaded with sodium. This was largely retained throughout the subsequent 24 h. He was not deplete next day and drank little or no sodium solution (Fig. 11-34). Désirée eliminated less than half of the additional sodium in the urine and often drank a reduced amount next day. PF33 was given 595 mmol sodium bicarbonate by rumen tube 10 min before access to sodium solution in four trials. Its sodium appetite was not diminished on three of these days.

In a second series of experiments (Fig. 11-34, Table 14-1) sodium bicarbonate solution was given by rumen tube 2 h before the sodium solution was offered for drinking. The tube was introduced again 10 min before the drinking period but no more fluid was given. This procedure reduced Désirée's sodium appetite by a small amount (P=0.06) only when 595 mmol were given.

The rapid introduction of 1.5 litres of tap water or of sodium bicarbonate solution into the rumen would produce an increase in the volume of rumen fluid, with either a decrease or increase in the sodium concentration. But as these procedures caused only a slight and statistically insignificant decrease in sodium appetite, sodium bicarbonate solution being no more effective than tap water, it is clear that sudden changes in the amount and concentration of sodium in the rumen fluid have no immediate influence on sodium appetite in the great majority of trials. The slight decreases in appetite that did occur may have been due to distension of the reticulo-rumen, although 1.5 litres would be expected to have only a small effect in an organ which normally holds 4–5 litres. The decrease in Désirée's voluntary sodium intake when 2 h elapsed between giving sodium into the rumen and offering sodium to drink is consistent with the view that in the sheep sodium must be absorbed from the alimentary tract before it can affect the appetite for sodium. In this regard Dobson (1959) has estimated that, in normal sheep, sodium is absorbed from the reticulorumen at a rate of 720 mmol/day (60 mmol in 2 h). This rate could be increased after the introduction of additional sodium by rumen tube and by absorption elsewhere in the alimentary tract.

Nachman and Valentino's (1966) studies in sodium-deficient adrenalectomized rats also emphasized the role of gustation in the consummatory process. Twenty-four adrenalectomized rats, sodium-deprived for 9 h, were found to drink 10 ml of 0.4 M sodium chloride when originally offered and 4.2 ml when the solution was offered again 2–2½ h later. They found that if 10 ml of 0.4 M sodium chloride were tubed into the stomach 2–2½ hr before the second test, the reduction in sodium drinking was no greater than if 10 ml of water were tubed, but that both groups of animals drank less than those in a sham-tubing group. That the drop in saline drinking on the second test was due to the taste of salt and not to the consummatory activity involved in drinking was further established in a second series of experiments where at the first test 0.3 M sucrose was either drunk or tubed. Those animals which drank 0.4 M saline in the first test drank significantly less ($P<0.01$) than any other group. However, the group which had had sucrose tubed did drink more than the group where 0.3 M sucrose was drunk ($P<0.05$). A third series of experiments established that 2 h after tubing 0.4 M sodium chloride plasma sodium concentration had risen to the level in normal rats. Nachman and Valentino (1966) note that this makes it unlikely that a receptor system sensitive to plasma sodium concentration is directly responsible for regulation of sodium appetite; this is consistent in their view with our data on sheep. Overall the results suggested to them clearly that the gustatory input produced by salt is monitored and the effects of it stored, so that there is less intake in a test administered 2–2½ h later. They note that similar evidence on the importance of gustation is forthcoming from Le Magnen's (1955) work. He found that drinking 2% saline had a greater effect on drinking of water 2 h later than did tubing the saline. Le Magnen (1953) also showed that oral intake of water was more effective in reducing thirst than was tubed water.

Levy and McCutcheon (1974) tested sodium appetite in rats 24 h after formalin injection. Gastric infusion of 0.15 M sodium chloride at 4.3 ml/h for 9 h commencing at the fourteenth hour reduced appetite more than did 0.15 M glucose or distilled water. The saline-infused rats had reached normal sodium balance before testing as indicated by urinary sodium excretion. Presumably with sodium infusion over a protracted period adequate to restore sodium balance, central appetitive drive is decreased. The result of giving the same infusion intravenously would need to be known before the effect can be attributed to the gastric passage of the sodium chloride over this time interval.

Landmark studies on the post-ingestional effects of drinking various solutions were made by workers at Johns Hopkins University and the University of Pennsylvania (McCleary 1953; Stellar et al. 1954; Mook 1963, 1969; Stellar 1967). McCleary (1953) showed that hypotonic loads into the stomach cause a rapid withdrawal of hypertonic fluid from the blood as such solutions are diluted to isotonicity before being passed on to the small intestine. This fluid withdrawal has a crucial role in the post-ingestion effect. It is non-specific to the type of solute and is rapid in effect, influencing an animal's drinking in 6 min. Similarly, Stellar et al. (1954) found, against the background of the characteristic preference–aversion curve with a peak at isotonicity, that hypertonic gastric loads given immediately before access to fluid increase the drinking of water and hypotonic solutions and depress the drinking of hypertonic solutions; comparable water loads decrease the drinking of water and hypotonic solutions and elevate the drinking of hypertonic solutions. They also noted that animals with an oesophageal fistula showed a characteristic preference–aversion curve for saline but that they drank more of the hypertonic solutions; this points to the inhibitory role of hypertonic solutions in the stomach in the normal animal. Gastric loads of various fluids as small as 0.5 cc had an effect on drinking but loads as large as 10 cc tended to reduce drinking at all concentrations, except when 3% saline was loaded, in which case water drinking was stimulated. Stellar et al. (1954) proposed that drinking of water and salt solutions was under the control of at least three mechanisms: taste and other sensory mechanisms in the mouth, gastric distension, and dehydration produced by osmotic effects.

The ingenious 'electronic oesophagus' used by Mook (1963) has been described elsewhere (Chapter 11). With drinking and loss through an oesophageal fistula, the equivalent amount of fluid is delivered synchronously to the stomach. When the fluid delivered to the stomach is the same as that which is drunk, the normal preference–aversion for glucose or saline is obtained. The same holds if one solution is drunk and the other is delivered to the stomach. The most interesting result is that when either glucose, saline or sucrose is drunk and only water delivered to the stomach, the preference for saline disappears and only the aversion remains. The preference for hypotonic glucose also disappears, but there is an increasing intake of the hypertonic glucose solutions. With sucrose, intake was a positive, negatively accelerated function of the concentration. The findings suggest that once hydrational requirements are met by letting water flow into the stomach, drinking of various concentrations is almost completely regulated by oral factors, particularly taste. In this regard the relatively aversive effect of concentrated saline is confirmed though the effect is not sharp. These tests were made under conditions of some measure of water deprivation, i.e. overnight withdrawal. As Stellar (1967) remarks, the simple preference–aversion curve is complicated indeed.

A series of most interesting experiments on dogs was carried out in Leningrad by Chernigovsky and colleagues (Chernigovsky 1963; Kassil et al. 1956; Kassil 1959). The animals were not adrenalectomized or sodium-deficient. However, these results on normal animals, pointing to stomach receptors feeding back on voluntary selection, highlight the potential interest of any comparable data from sodium-deficient dogs. The stomach of each dog was divided into two parts. The cardiac portion was anastomosed to the intestine and also had a fistula so that, when desired, all fluid flowed to the outside. Between experiments the fistula was closed. The pyloric portion of the stomach formed a closed sac with a fistula.

The dogs were offered milk containing various concentrations of saline (1%–3.5%) when the cardiac fistula was open. Whereas introduction of 200 ml of water into the pyloric pouch did not alter intake of milk solutions within a certain elective range of sodium concentrations, 200 ml of 3% saline resulted in rejection of solutions containing greater than 1% saline. The effect was rapid and preceded any absorption of sodium chloride from the pouch. The aversion produced by high concentrations was abolished by subdiaphragmatic vagotomy. The introduction of hypertonic glucose (27%–60%) did not have a comparable effect, which suggests that the receptor elements were not responding simply to osmotic stimuli. Kassil noted that the introduction of hypertonic salt solutions into the pouch caused a thorough sniffing of the salt solutions, testing of their taste and slowing down of the act of lapping. Further frequent experiences of having concentrated salt introduced into the pouch conditioned the animals to select the milk solutions in ascending order of concentration when they were made available, though no pouch stimulus was given. This behaviour was overridden as soon as some specific stimulus was given in the pouch and order of the bowls containing the solutions was changed. Kassil (1959) remarks in terms of the significance of these data that presumably, in natural surroundings, internal impulses continually change the specialized direction of orientative exploratory reaction, playing an important role in stabilizing the organism in relation to its external surroundings.

Water and sodium drinking in sheep with an oesophageal fistula

Before dealing with the data on satiation of sodium appetite in sheep with an open oesophageal fistula, an account of experiments on satiation of water deficit in sheep and in rats will be given as they give valuable insight into the data on sodium-deficient sheep.

As indicated earlier, sheep will usually satiate a water deficit of 9%–10% of body weight in 10 min, and often in 2 min (Tables 14-2 and 14-3). The animals showed typical appetitive behaviour, jumping up and down before the water was presented and scratching the floor of the cage, and bleating. Often one long draught satisfied the sheep. Other times they tasted the water, threw their heads up and down a few times, seemed to chew the water and then drank. The sheep then lay down, or sometimes took shorter drinks and sips.

Plasma sodium concentration changes as a result of water absorption by 30–60 min (see Table 14-4). Similar findings came from comparable experiments on Norman.

Effect of placing water in the rumen on water intake after 48 h of water deprivation

Water was offered 5, 15 and 30 min after about 3 litres of water had been given by rumen tube. The usual appetitive behaviour was displayed in all instances before the water was offered and the sheep drank rapidly in the first minute, but in most instances the drinking was completed in the first minute by one or two drinking episodes. Only 30%–40% as much was drunk as in the control series (Table 14-5).

Elliot was an exception. He showed very little interest in the water when given access 30 min after 3 litres of water had been given by rumen tube. During the first 3 min the sheep leisurely tasted the water and licked its lips before drinking 450 ml. Nessim drank approximately the same amount of water when it was offered 30 min after return of water deficit by rumen tube as when it was offered after 5 and 15 min. In experiments in which the

Table 14-2. Average water intake (ml) during 120 min following 48 h of water deprivation.

Sheep	No. of expts.	Average water intake (ml) in (min) 2	10	30	120	Range of total intake (ml)	Ave normal daily water intake (ml)
Elliot	14	2500	2710	2960	3090	2440–3930	2610
Lawrence	2	2580	2580	2580	3280	3250–3300	2520
Norman	4	3420	3590	3650	3650	3100–3940	2840
Nessim	7	3280	3290	3290	3330	2650–3630	2360
Rory	1	1700	3920	3920	3920		
Tess	1	2780	3360	3360	3360		

From Bott et al. (1965), with permission.

Table 14-3. Nessim. The volume of water drunk following 2, 10, 30 and 120 min access after 48 h of water deprivation. The weight loss of the animal during water deprivation and the percentage correction of deficiency after 10 min of access to water are shown also.

Expt.	Water drunk after 48 h of deprivation (ml): Voluntary intake in (min) 2	10	30	120	Wt of sheep at start of expt. (kg)	Wt lost over 48 h of water depriv-ation (kg)	% correction of body wt deficit in 10 min
1	3440	3440	3440	3440	36.0	3.3	103
2	3450	3450	3450	3630	36.7	3.2	106

From Bott et al. (1965), with permission.

Table 14-4. Nessim. The effect in two experiments of 48 h of water deprivation on plasma concentration of sodium and potassium and the haematocrit, and the effect on these parameters of free access to water. Water intake in these experiments is shown in Table 14-3.

	Plasma [Na] (mmol/l)		Plasma [K] (mmol/l)		Haematocrit (%)	
Expt.	1	2	1	2	1	2
Day 1						
(water withdrawn)	140	136	4.6	4.0	35	33
Day 2	151	147	4.8	4.5	38	32
Day 3						
−45 min	157	154	4.8	4.5	36	Lost
−30 min	156	155	4.8	4.9	33	32
Water offered						
+15 min	155	156	4.4	3.9	33.5	30
+30 min	153	156	4.2	4.0	32	29.5
+45 min		152	4.3	4.1	31	28.5
+60 min	154	150	4.2	4.4	30	29.5
+120 min	147	147	4.5	4.2	32	26.5
+180 min	145	144	4.8	4.4	32	27

From Bott et al. (1965), with permission.

Table 14-5. Nessim and Elliot. The volume of water drunk following 1, 10, 60, 120 and 240 min access after 48 h of water deprivation. Nessim: the weight loss over the 48 h was given as litres of water by rumen tube 5, 15 and 30 min before water was offered. Elliot: the average volume of water drunk after 48 h of water deprivation (3 litres) was given by rumen tube at the same times as with Nessim.

Expt.	Wt lost over 48 h of water deprivation (kg)	Time water given by rumen tube before access to water (min)	Voluntary water intake (ml) in (min)				
			1	10	60	120	240
Nessim							
1	3.3	5	1030	1030	1030	1030	1030
2	4.0	5	1170	1170	1510	1510	1510
3	3.2	5	680	680	680	680	680
4	3.2	15	860	860	860	860	860
5	2.6	15	910	910	910	910	910
6	3.0	30	1530	1530	1530	1530	1530
7	3.2	30	1140	1140	1140	1140	1140
Elliot							
1		30	Not weighed	450	450	450	450
2		15	1410	1410	1410	1410	1410
3		5	1160	1160	1160	1160	1160
4		5	0	930	1280	1280	1280

From Bott et al. (1965), with permission.

rumen tube was passed without water being given into the rumen, the voluntary intake was the same as in the control experiments.

Water drinking after 48 h of water deprivation when the oesophageal fistula was open

In 1856 Claude Bernard described an experiment on a gastric-fistulated dog which was deprived of water for several days and then allowed to drink: 'In spite of this intake of water through all the upper parts of the intestine, the thirst was not abated. The animal was reduced to a sort of Danaides water wheel and drank until it was fatigued. After it had rested, it recommenced drinking and so on.'

The drinking and behaviour of the water-deplete sheep was observed for 120 min when the oesophageal fistula was open (Fig. 14-19) (Bott et al. 1965). A small amount of water is propelled past the large opening of the fistula and enters the rumen. This can be calculated from the difference between that drunk and that collected from the fistula. Before water was offered the sheep displayed their typical anticipatory behaviour. On access they gulped the water rapidly and continued to drink vigorously for 5–10 min. After this period they reduced intake and rested between drinks. During 120 min the total intake ranged from 2.5 to 9 times that observed in the control experiments (Table 14-6 and Fig. 14-20). The volume entering the rumen was usually much less than that drunk normally in the control experiments with fistula closed. The amount drunk in the 120–160 min after the fistula was closed in these experiments generally increased the total volume entering the rumen to the amount observed in the control experiments. There was no alteration in plasma sodium concentration during the first 60 min of sham drinking. This confirmed that little water entered the rumen past the open fistula. With Tess, the cervical part of the oesophagus was enclosed in a skin tube. No water entered the lower oesophagus and rumen during the sham-drinking. The manner and quantity of drinking were similar to the rest of the sheep with open oesophageal fistulae.

Fig. 14-19. Elliot. Print from a cine film of an animal drinking with an oesophageal fistula open. Water is shown leaving the fistula in a continuous stream. (From Bott et al. 1965, with permission.)

Table 14-6. The volume of water drunk with the oesophageal fistula open following 2, 5, 10, 30 and 120 min access after 48 h of water deprivation. The fistula was closed at 120 min and further drinking noted.

		← Oesophageal fistula open → Cumulative water intake (ml) during (min)					← Oesophageal fistula closed →			Estimated water in rumen (ml) at[a]	
Sheep	Expt.	2	5	10	30	120	140	180	240	120 min	End of expt.
Elliot	2	2000	3470	7270	8880	15190	15330	Water not offered		1500	1650
	3	4580	7030	10700	13370	20180	21260	21580	—	1700	3100
	4[b]	4020	5160	5570	11050	20730	22560	22560	22560	1450	3280
	5	4000	7910	9750	16470	27690	29180	29670	29670	740	2720
	6	2520	4780	5750	11710	17820	17820	19530	20030	Not measurable	2200
	7	3650	7790	12500	19110	22950	24380	24380	26570	Not measurable	3620
Norman	1	2220	2760	4540	10080	11570	Water not offered			3200	3200
	2	4490	8770	11020	13050	13870	13870	—	—	3860	3860
	3	3520	4500	8640	9930	11100	Water not offered			2100	2100
	4[c]	2100	4540	6780	7850	9020	9020	9970	9970	1570	2520
	5	4220	6770	8860	12470	14870	14900	15430	15430	2340	2900
Rory	1	2220	4440	6280	6280	8830	10100	10280	10830	1450	3450
Tess	1	2000	5400	8220	12380	17300	17300	18800	19500	Nil	2300
Lawrence (24-h water depletion)	1	3140	3140	3140	4060	5970	6630	6630	6690	970	1670

From Bott et al. (1965), with permission.

[a]Water entering the rumen was calculated as the difference between the total water intake in 120 min and the water issuing from the fistula during this period plus the water drunk after 120 min when the fistula was closed.

[b]Demonstration experiment: the sheep was disturbed during the first 20 min.

[c]The jugular vein of the sheep was cannulated and blood samples taken throughout the experiment.

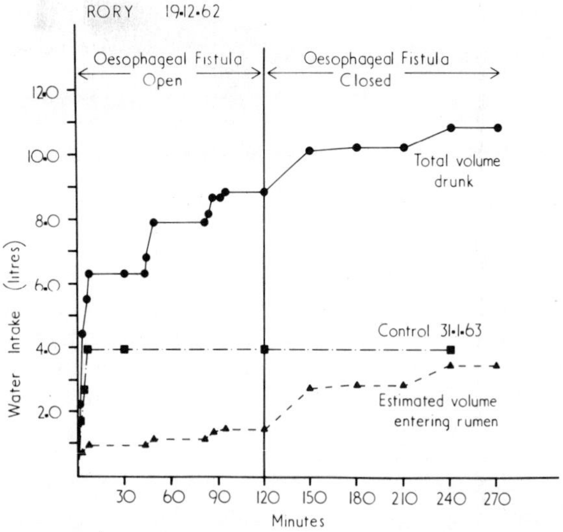

Fig. 14-20. Rory. The voluntary water intake after 48 h of water deprivation. The oesophageal fistula was open during the first 120 min, and was then closed and water intake measured for a further 160 min (continuous line). The interrupted line represents the estimated amount of water entering the rumen, and the line designated 'control' is the water intake after 48 h of water deprivation with the oesophageal fistula closed. (From Bott et al. 1965, with permission.)

Several experiments involving water drinking with the oesophageal fistula open were performed on Elliot (Fig. 14-21). Of the total water sham-drunk, 30%–50% of the intake was in the first 10 min of access, compared with an average of 88% in the control experiments (Table 14-2). There was a time lag of 5–30 min before the sheep drank its residual water deficit once the oesophageal fistula was closed. There was no significant fall in plasma sodium or potassium concentration or in haematocrit during the sham-drinking period.

The effect of placing water in the rumen on water intake after 48 h of water deprivation when the oesophageal fistula was open

The design of this experiment was similar to that for water drinking with oesophageal fistula open, with the additional procedure that 5 min before the bung was removed from the fistula and water offered, the sheep was given into the rumen the

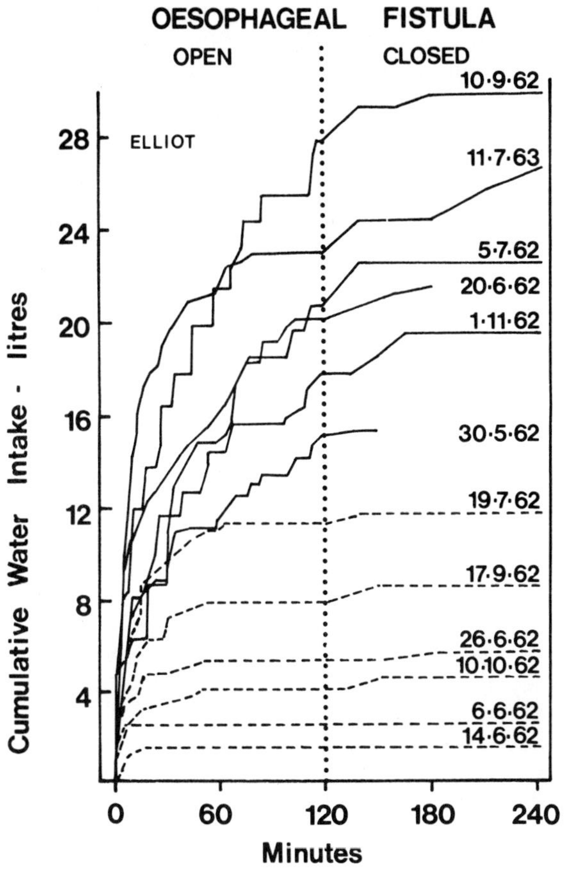

Fig. 14-21. Elliot. Twelve experiments showing the volumes of water drunk by the sheep after 48 h of water deprivation. In each experiment the oesophageal fistula was open for the first 120 min of access, and then closed. In six experiments shown with interrupted lines, the sheep was given 3 litres of water by tube into the rumen 5 min before water was offered. The water drunk 30–120 min after the oesophageal fistula was closed was measured in the experiments. (From Bott et al. 1965, with permission.)

average amount of water drunk after 48 h of water deprivation (see Fig. 14-21 for series on Elliot). Frequent regurgitation by both the sheep used, especially during the second hour of the experiment, made conduct of these experiments difficult.

There was a clear reduction in water drinking in both Elliot and Norman when water was put into the rumen before sham-drinking (Fig. 14-21, interrupted line). In Elliot's case appetitive behaviour was also reduced by this procedure, though the sheep still showed definite interest in the water when it was offered to drink. Norman, on the other hand, displayed appetitive behaviour similar to that when offered water without having had the water given into the rumen. Drinking motivation in both sheep declined rapidly after 2 min, compared with 5–10 min in the experiments without water in the rumen (Table 14-6). When 270 mmol/l sodium chloride was placed in the rumen 5 min before sham-drinking the amount sham-drunk was increased, particularly after 30 min (Table 14-7).

In summary, water-deplete sheep repair their water deficits within the first 10 min of drinking. This satiation of thirst was not attributable to a fall in plasma sodium concentration as this did not begin to decrease until 30–45 min after drinking. Administration, by tube introduced into the rumen immediately before offering water, of water equivalent to the weight deficit did reduce drinking but failed to inhibit the appetitive behaviour. Some drinking occurred even when the water was offered 30 min after the deficit was given into the rumen. Adolph (1950) reported that the water-deplete dog drank its deficit if offered water not later than 15 min after water equivalent to the weight deficit was given by stomach tube. However, after 15 min the intake decreased, owing presumably to water being absorbed from the small gut into the bloodstream (Klisiecki et al. 1933).

Healthy sheep with oesophageal fistulae closed drank like normal animals when water-deplete. If the water-deprived sheep were offered water to drink when the oesophageal fistula was open, they drank 2.5 to 9 times more than with closed fistula and continued to drink over a long period, resting

Table 14-7. Elliot. The volume of water drunk following 2, 5, 10, 30, 60, 90 and 120 min access with the oesophageal fistula open and during the following 180 min when the fistula was closed. The sheep, which had been deprived of water for 48 h, received 3 litres of sodium chloride (270 mmol/l) into the rumen by tube 5 min before access to water.

	← Oesophageal fistula open →							← Oesophageal fistula closed →			
	Cumulative water intake (ml) during (min)										
Expt.	2	5	10	30	60	90	120	140	180	240	300
1[a]	2360	4260	5660	7040	14550	21280	28200	29070	29770	30500	30850
2	3860	4620	5670	9940	16200	20820	27840	28920	29530	30700	30700
3	2010	2240	3570	6330	9380	11310	14630	14630	15680	16000	16350

From Bott et al. (1965), with permission.
[a]The jugular vein of the sheep was cannulated and blood samples taken throughout the experiment.

between episodes. There was no normal termination of drinking. Therefore, imbibing and the act of swallowing with passage of water through the pharynx and the upper oesophagus did not constitute a consummatory act: the animals failed to be satiated. This was not attributable to the failure of water to arrive in the rumen and be absorbed, since the normal sheep stops drinking long before there is any significant absorption from the gut. The large increase over normal in volume sham-drunk was evident in the first 5–10 min. This suggests that the sensory impulses arising from the passage of water through the lower oesophagus and entry into the rumen are important in determining normal satiation. However, it is interesting that when the oesophageal fistulae were closed after 2 h of sham-drinking the animals did not correct their deficits immediately, or with the next drink as they did in the initial stage of the control experiments. Though sham-drinking did not discharge the drive for water ingestion, it did nevertheless modify it.

Towbin (1949) stated that water-deplete dogs become temporarily satiated when sham-drinking after 2.5 times the animal's deficit has passed through the mouth, pharynx and upper oesophagus. Adolph (1950) suggested that the dog may have some pharyngeal metering device which terminates the act of drinking.

Mechanisms of satiation of thirst

Mook (1969) has remarked that of the events that accompany or follow the act of ingestion and that could therefore serve as satiety signals in the regulation of intake, two major classes are distinguishable:

1) Oropharyngeal events such as taste, smell, and proprioceptor feedback from the consummatory response.
2) Post-ingestional events such as distension of the stomach by the material swallowed, and the hydrational and/or metabolic changes that follow its absorption.

Both sets of factors are recognized to influence feeding and drinking.

The point which emerges from experiments with sheep is that whereas distension of the forestomach clearly contributes to satiation, post-absorptive changes can, in the light of the rapidity of consummatory behaviour, play little or no role. Adolph (1939) has shown that water-deprived dogs will satisfy their thirst with a single draught of water in 5 min or less, and in so doing will ingest accurately the amount of their deficit. This confirms the small influence of absorptive factors in satiation. Similar capacity to repair a large deficit in a few minutes has been seen in the young rabbit (Adolph 1950), goat (Andersson et al. 1958) and camel (Schmidt-Nielsen et al. 1956).

Blass and Hall (1976) introduce their substantial experimental dissection of the mechanisms which cause the cessation of drinking in response to water deprivation with some general observations. They suggest that drinking in response to cellular dehydration as a result of loss of water may be initiated by osmoreceptors in the basal forebrain or by periventricular sodium receptors in the third ventricle (Andersson and Olsson 1973). Thirst in response to depletion of extracellular fluid is though, they say, *triggered* by two mechanisms: the renin–angiotensin system and the mechanism residing in the low pressure distensible great veins at the heart. Less well understood, however, are the central and peripheral events that modulate and terminate drinking. In terms of influences that are not attributable to rehydration of the animal, they note Bellows (1939) and later Towbin (1949) reported that sham-drinking in which ingested water is lost through a fistula is proportional to normal drinking. This implicates a role for oropharyngeal metering. Moreover, Adolph et al. (1954) demonstrated that stomach distension by balloon or water preload arrests drinking in a number of species, suggesting that gastric distension also reduces the urge to drink. Rolls et al. (1980) report experiments in collaboration with Maddison involving study of monkeys with gastric and duodenal fistulae. When either cannula was open there was massive overdrinking in the water-deplete animal. Further, if after initial rapid termination of drinking in the water-deplete monkey the gastric cannula was opened allowing drainage of the gastric contents, drinking resumed almost immediately. Thus an important factor in initial rapid termination of drinking in the monkey may be gastric distension.

Blass and Hall have taken as their point of departure for experiment the phenomenon called by its discoverer Adolph (1943), voluntary dehydration. Adolph refers to the fact that certain water-deprived mammals, including rats, do not drink enough water to restore their fluid deficit. Fluid balance studies in water-deprived rats allowed to drink water indicate that the water deficit was incompletely restored for all levels of deprivation at the end of a 5-h test (Fig. 14-22). Drinking, which was vigorous at first, became intermittent and finally stopped after about an hour. The deficit at the end of the test was directly related to that at the start. The authors noted the

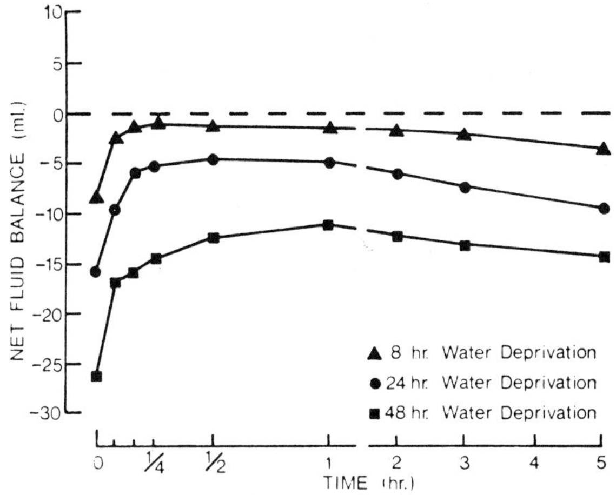

Fig. 14-22. Net fluid balance of rats drinking water after 8, 24 or 48 h of water deprivation. (From Blass and Hall 1976, with permission.)

peculiar behaviour of the rats after they had been drinking for a period of time. There were bizarre elements suggestive of conflict and displacement activity. Concurrent measurement of blood composition indicated that reversal of cellular deficit began within 5 min after initiation of drinking. By 15 min enough water was absorbed to cause cellular overhydration. The authors note Stricker was the first to show that drinking in response to extracellular depletion or to stimuli ostensibly mediated by angiotensin, e.g. ligation of the inferior vena cava, is inhibited by low levels of osmotic dilution (cellular overhydration) (Stricker 1969). In the instance here of water deprivation, Blass and Hall suggest that if drinking were to continue, calamitous results would ensue, the most severe being water intoxication, which becomes profound when serum sodium reaches 125 mmol/l. Even if drinking stopped before this, the temporary alterations in cellular metabolism when plasma sodium reached 133 mmol/l would probably suffice to cause confusion and some degree of motor impairment—conditions inimical to defence in the wild.

Thus the mechanism through which drinking is inhibited may be of considerable survival value. In this author's eyes, it also raises the paradox of the natural selection advantage of termination of drinking when a substantial amount of the body deficit remains unrepaired. For an animal in the wild with safe access to water possibly only episodically, this rapid absorption from the gut and premature termination could sometimes be disadvantageous. It would be interesting to know whether this is true of wild rats as well as the white inbred laboratory variety. In this regard Adolph (1950) studied the water-deplete rat, hamster, guinea pig, dog and rabbit. The first three species drank slowly, and were inhibited by water placed in the stomach, whereas dogs and rabbits drank in one draught to satiation. Dogs were not inhibited by intragastric load. Young rabbits were like dogs but adult animals drank more slowly and were inhibited by intragastric load.

To determine whether the change in cellular state was a necessary and sufficient event to terminate drinking, Blass and Hall preloaded water-deprived rats with their entire deficit to see whether this would prevent drinking. When water was made available 1 h later the rats drank an additional amount (3.5 ml). Thus the production of the body fluid profile, especially cellular overhydration, found at the termination of drinking was not sufficient to prevent drinking. But it substantially reduced it. This indicates that non-hydrational types of controls appeared to contribute to termination of drinking. By studying dehydrated rats with gastric fistulae they showed that the volume and rate of sham-drinking were related to length of deprivation. This supported an oropharyngeal element in the control of drinking in rats, i.e. classical mouth metering. The amount sham-drunk far exceeded that drunk normally. In further experiments on the nature of satiation and why the state of satiation endured, Blass and Hall compared the satiating influence of preloads of water that were tubed versus those drunk, in a manner analogous to that followed by Nachman and Valentino (1966). The drinking of the preload had a greater satiating effect than its tubed counterpart. They remark that a physiological state (here cellular overhydration) becomes especially meaningful within its behavioural context.

Feider (1972) studied water-deprived rats to compare the effect on thirst of stomach distension and concentration of fluid offered. Injection of large volumes of isotonic saline into the stomach (up to 15 ml in 1 min) produced distension but had no effect on drinking. Intragastric injections of water decreased drinking within 1–4 min, whereas injections of hypertonic saline increased it. With bar-pressing for water, similar results were produced except that there was a brief small decrease in operant activity immediately following injections of large volumes. Feider concluded that fluid concentration changes, but not stomach distension of itself, seemed important in the negative feedback control of thirst in the rat. Also, the fact that vagotomy had no influence on the processes counted against stomach osmoreceptors or stretch receptors being relevant.

In conclusion, Blass and Hall suggest it seems that neither behavioural (consider the sham-overdrinking) nor physiological events alone are sufficient to account fully for the endurance of drinking termination. The necessary and sufficient relationship for it to be sustained is that both behavioural and physiological changes must be closely related in time. When this temporal relation does not exist, and physiological changes occur outside of the behavioural context, control is forfeited causing behavioural excess. As will be set out in detail below, we have raised the same issue in the discussion of aspects of the equally complicated problem of satiation of salt appetite, particularly in the light of studies on sheep with oesophageal fistula. The thesis advanced (Denton 1966; Denton et al. 1977) has been that for the consummatory act to be satiating with a resultant precipitate decline of motivation there needs be a 'gestalt' of sensory inflow and this involves an ordered sequence.

Voluntary drinking of sodium bicarbonate solution by sodium-deficient sheep with an open oesophageal fistula

Voluntary drinking of sodium bicarbonate was studied in six sheep with a parotid fistula and an oesophageal fistula, largely as a result of the efforts of Dr Elspeth Orchard (E. Orchard, D. Denton and S. Weller, unpublished).

In experiments with the oesophageal fistula open, the bung was removed 2–5 min before the sodium solution was offered for 120 min. After 120 min the fistula was then closed, and the sodium left available for a further 120 min. With the four sheep with the fistulated oesophagus in the normal position in the neck tissues, very little sodium solution entered the rumen while drinking with the fistula open, but regurgitation of rumen contents occurred (Bott et al. 1965). With the other two sheep, which had the oesophageal fistula in the skin loop in the neck (Eldred and Tess) (Bott and Goding 1965), regurgitation did not occur. Nor did sodium solution enter the rumen when drinking with the oesophageal fistula open. All the animals showed hypersalivation during the episodes of drinking sodium solution, particularly with the high concentrations.

The conduct of the experiments was more difficult than with the earlier study of water drinking. The episodes of sodium deficiency appeared to induce a change in the tone of the muscle of the oesophagus so that the fistula enlarged. This phenomenon became more apparent over succeeding weeks of recurrent episodes of sodium deficiency, and limited the number of experiments which were feasible in the one animal. Considerable practical problems of management of the animals arose, in that if the bung fell out during the night when the animal was unattended, there was a large loss of saliva and of ruminal fluid. This occurred occasionally, and appropriate replacement measures had to be taken. The animal was given an adequate number of days to recover before any further experiments were made. However, notwithstanding these difficulties, useful results which were clear-cut in certain regards were obtained on the main questions under consideration. The change in tone of oesophageal muscle with sodium deficiency was interesting in the light of McCance's (1938) evidence on fatigue of muscle with experimental sodium deficiency in man, and the findings of Streeten (1950) on loss of propulsive power and responsiveness in intestinal muscle with salt deficiency.

The animals were made sodium deficient by loss of 4–7 litres of parotid saliva over a period of 48 h. In control experiments with the oesophageal fistula closed, they were offered a solution of sodium bicarbonate (300 or 317 mmol/l) to drink for 15–120 min at the end of the depletion period (Table 14-8). Also, periodically in the 52 experiments with four of these animals, the concentration of the sodium bicarbonate was altered to 100, 600 and 900 mmol/l. The results were similar to those recorded earlier in this chapter. The amount of sodium solution drunk in 15 min of access was consistently less than the deficit when 100 mmol/l solution was offered. Elliot and Eldred otherwise adjusted volume to concentration, but Norman and Franz overdrank at 900 mmol/l.

The sheep were given sodium bicarbonate solution to drink with the oesophageal fistula open after 48 h of sodium depletion. The results of 16 experiments carried out on the six sheep are shown in Table 14-9. The cumulative amount of sodium drunk over 2, 5, 15, 30 and 120 min with oesophageal fistula open, and then over the following 20 and 120 min with the fistula closed, is shown for each experiment. The total volume of solution sham-drunk during the first 120 min is recorded. In the two columns on the right, the estimated amount of sodium which entered the rumen during the 120 min of sham-drinking, and the total sodium in the rumen after the 120 min of drinking with the fistula closed, are given.

The results show considerable variation in the amount of sodium solution sham-drunk, both between sheep and between experiments on individual sheep. This variation of amount was not related to the degree of sodium deficiency, as this was reasonably constant in the individual animal: e.g. Elliot, in experiments 1 to 5, had a sodium loss of between 590 and 665 mmol but the total sodium

Table 14-8. The mean sodium intake (300 or 317 mmol/l sodium bicarbonate) during 15 or 120 min following 48 h of sodium depletion of six sheep. The mean salivary sodium loss over 48 h and the s.d. for the total sodium intake and the salivary loss for each sheep are recorded also.

		Mean Na intake (mmol)					Mean salivary Na loss during 48 h	
Sheep	No. of expts.	2 min	5 min	15 min	120 min	s.d.	mmol	s.d.
Elliot	30	354	460	536	590	170	636	93
Rory	14		420	460		96	412	87
Eldred	30	410	590	740		180	661	121
Tess	13	440	492	562	608	182	629	89
Franz	33		536	595		217	628	112
Norman[a]	7		555	660		145	702	140

[a]Mean sodium bicarbonate intake after 72 h of sodium depletion.

Table 14-9. The amount of sodium bicarbonate solution (300 mmol/l) sham-drunk in 16 experiments on six sheep during 2, 5, 15 and 120 min after the solution was offered. The fistula was closed after 120 min and voluntary intake observed during the following 120 min. The total volume (litres) sham-drunk is recorded and the two columns on the right record the sodium (mmol) estimated to have entered the rumen after 120 min of sham drinking, and after a further 120 min of voluntary drinking with the fistula closed. The sheep were depleted of sodium for 48 h.

		← Oesophageal fistula open →				← Oesophageal fistula closed →			Estimated Na in rumen (mmol) at	
		Cumulative Na intake (mmol) in (min)						Volume of Na solution sham-drunk in 120 min (ml)		
Sheep	Expt. no.	2	5	15	120	140	240		120 min	240 min
Elliot	1	730	1125	1395	1795	1850	2285	5980	110	600
	2	465	540	620	1035	1475	1590	3450	35	590
	3	245	585	660	960	960	1205	3210	30	275
	4	710	1140	1230	1260	1505	1570	4190	20	330
	5	10	300	550	630	630	935	2100	80	385
Rory	1	180	180	180	180	180	395	540	30	245
	2	55	205	380	520	520	960	1730	65	505
	3	110	300	475	630	765	1015	2100	120	505
	4	3	145	400	485	485	1085	1620	35	635
Eldred	1	240	570	1255	1980	1990	2450	6600	0	470
	2	290	725	1415	2440	2550	2890	8140	0	450
Tess	1	50	180	250	570	610	740	1900	0	170
	2	485	620	970	1430	1525	1710	4770	0	280
Franz	1	1010	1010	1595	2020	2060	2160	6450	145	285
	2	750	900	1260	2250	2300	2540	7100	150	440
Norman[a]	1	900	1000	1385	2580	2760	3140	8230	285	845

[a]Sheep 72 h sodium-deficient.

sham-drunk in 120 min ranged from 630 to 1795 mmol. Fig. 14-23 shows five sham-drinking experiments on Elliot and the 'mean' of the 30 control experiments. The large variation in the amount of sodium sham-drunk is evident in the first 10 min of drinking.

For sheep with the oesophageal fistula open, the mean amount of sodium sham-drunk in the first 15 min (Table 14-9) was 1000 mmol. The mean amount of sodium drunk in the first 15 min in the control experiments was 590 mmol (Table 14-8: $P<0.05$). Thus the results of the experiments on voluntary drinking of sodium bicarbonate solution with the oesophageal fistula open were clear, as in most instances the animals drank more than in control experiments. Drinking usually continued episodically throughout the 120-min period the fistula was open, although the sodium intake was most rapid in the first 5–15 min. On this point, in a number of the experiments it was evident that

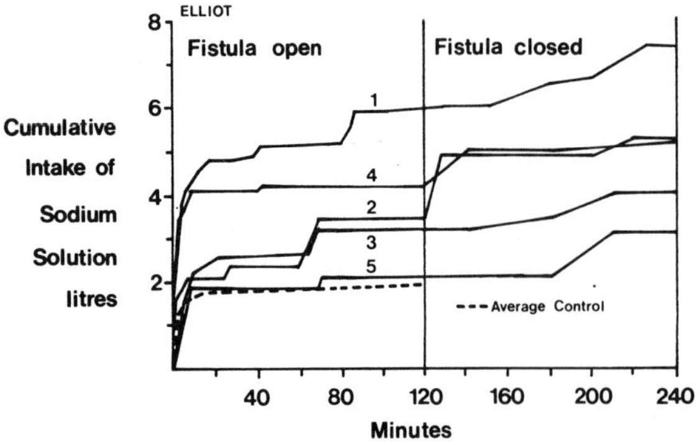

Fig. 14-23. Elliot: sodium-deficient sheep with an oesophageal fistula. The volume of 300 mmol/l sodium bicarbonate solution drunk in five experiments when drinking was permitted with the oesophageal fistula open for 120 min and in the 120 min after the fistula was closed. The mean intake observed in 30 control experiments with the fistula closed is shown also.

intake by 2–5 min was much greater than the control mean, and in some instances greatly exceeded deficit. Thus, the absence of sensory feedback from lower oesophagus and possibly rumen appeared to be immediately effective in influencing drinking behaviour.

An important quantitative difference between sodium drinking (from sodium depletion) and water drinking (from water depletion) arises from the results. After 15 min of access to water, the water-deplete sheep sham-drank 1.6–5.3 (a mean of 3.3) times their deficit, and after 120 min, 2.5–9 times their deficit. On the other hand, the sodium-deficient sheep sham-drank a mean amount of 300 mmol/l sodium bicarbonate equal to 1.7 times their deficit in 15 min, and 2.7 times deficit in 120 min. This suggests that the sham-drinking of sodium solution is more effective in satiation than the sham-drinking of water. The fact that during the 120 min the fistula was closed the sheep did not usually ingest sufficient sodium bicarbonate to correct the sodium deficiency (cf. Tables 14-8 and 14-9), also supports this view. This was not due to the small amounts of sodium solution which passed the fistula and entered the rumen during sham-drinking, since it was seen also in the two sheep with oesophageal loops where all the solution drunk came out of the fistula. With the water-deficient sheep, sufficient water was drunk after the fistula was closed to correct the water deficiency (Bott et al. 1965).

In varying the concentration of sodium bicar-

Table 14-10. The amount of sodium bicarbonate (mmol) drunk with the oesophageal fistula open following 2, 5, 15 and 120 min access to 100, 600 and 900 mmol/l solutions. The fistula was closed after 120 min and drinking over the following 2 h measured. The total volume of sodium bicarbonate solution sham-drunk at each concentration is recorded and the two columns on the right record the sodium estimated to have entered the rumen after 120 min of sham-drinking, and after a further 120 min with the fistula closed. The sheep were depleted of sodium for 48 h.

Sheep	Conc. of Na solution offered (mequiv/l)	⟵ Oesophageal fistula open ⟶ Cumulative Na intake (mmol) in (min) 2	5	15	120	⟵ Oesophageal fistula closed ⟶ 140	240	Volume of $NaHCO_3$ sham-drunk in 120 min (ml)	Estimated Na in rumen (mmol) at 120 min	240 min
Elliot	100	235	400	610	705	705	895	7060	Not measurable	190
	100	240	485	700	790	855	990	7910	10	210
	100	240	340	490	730	785	885	7280	10	165
	600	550	1080	1645	1830	1830	1830	3050	60	60
	600	320	760	1265	1310	1310	1480	2180	Not measurable	170
	900	170	440	540	620	620	620	690	Not measurable	0
	900	260	640	1055	1070	1070	1105	1190	Not measurable	35
Eldred	100	165	420	920	1580	1595	1955	15800	0	375
	600	480	1510	4800	8135	8310	8410	13560	0	275
	900	215	1045	2430	3275	3285	3635	3640	0	360
Franz	100	245	435	900	2195	2205	2430	21940	15	250
	600	600	655	860	2350	2440	2670	3920	180	500
	900	1505	1755	2045	3305	3330	3715	3670	115	525
Norman[a]	100	135	155	185	540	565	665	5380	45	170
	600	810	1030	1220	1620	1790	2140	2700	170	690
	900	1460	1550	1745	4185	4385	4970	4650	335	1120

[a]Sheep 72 h sodium-deficient.

bonate offered to the sheep it was hoped that some insight might be obtained on the extent to which taste may predominate in the consummatory act of satiation of salt appetite. The correction of deficit by adjustment of volume drunk to concentration of solution presented varies between animals, some being relatively accurate over the range of 200–900 mmol/l (see Figs. 14-9 to 14-15). With 100 mmol/l sodium bicarbonate underdrinking occurs (Fig. 14-10). Because of individual variation and also because of the difficulties referred to above as regards experiments on sodium deficiency in sheep with an oesophageal fistula, it was not possible to make a significant series. Notwithstanding, some very interesting findings emerged. Overdrinking, sometimes to a spectacular degree, occurred very often. The sham-drinking of very large volumes was more associated with the 100 mmol/l than with the 600–900 mmol/l solution. There were, however, two instances of extremely high sham-drinking with the 600 and the 900 mmol/l sodium bicarbonate (8135 and 4185 mmol sodium respectively) (Table 14-10).

Fig. 14-24 shows the cumulative volume of 100, 317, 600 and 900 mmol/l sodium bicarbonate solution drunk by a sodium-deficient sheep (Franz) with oesophageal fistula open. The mean control intake of these four sodium solutions is shown on the extreme right. The animal sham-drank approximately the same number of millimoles of sodium during the 120 min for 100, 300 and 600 mmol/l solutions (2020–2350 mmol). However, in control experiments the amount of 900 mmol/l sodium bicarbonate drunk by the sheep tended to exceed its sodium deficit. In fact, it drank approximately equal volumes of 600 and 900 mmol/l solutions in both control and sham-drinking experiments. In another sheep (Eldred), the volume sham-drunk was clearly affected by concentration when the animal was offered 100 and 900 mmol/l sodium bicarbonate in that sequence after the experiments with 300 mmol/l (Fig. 14-25). However, 600 mmol/l solution was offered for sham-drinking 6 days after the experience of sham-drinking the 900 mmol/l concentration and the volume drunk greatly exceeded that observed with 300 mmol/l. A second experiment with 600 mmol/l sodium bicarbonate was not possible with this sheep as the oesophageal fistula became very large, and experimental work could not be continued.

When sodium solution (100, 300, 600 and 900 mmol/l) was sham-drunk over 120 min, and then drunk with the fistula closed for the following 120 min, insufficient solution was drunk after the closure of the fistula to correct the sodium deficit. The sheep often showed very little interest in the sodium bicarbonate solutions of high concentration after the 120 min of sham-drinking (Table 14-10).

The data suggest that the physiological factors operating to modify drinking at the high and low concentrations are different. At low sodium concentrations, the large volume which must be

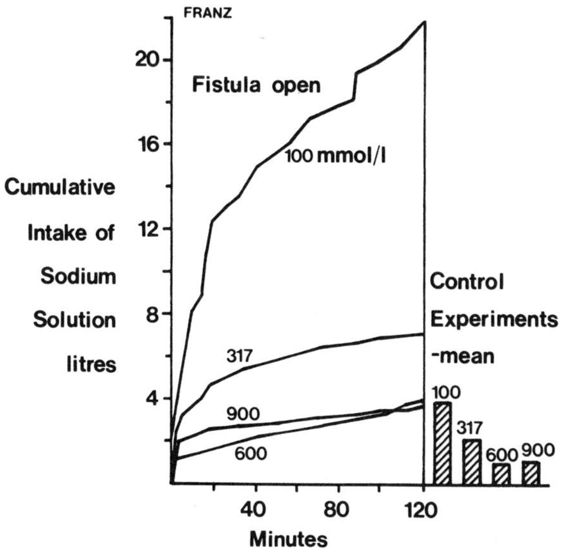

Fig. 14-24. Franz: sodium-deficient sheep with an oesophageal fistula. The volume of sodium bicarbonate solution (100, 317, 600, 900 mmol/l) sham-drunk during 120 min in individual experiments with oesophageal fistula open. The mean volume in control experiments under the same conditions but with oesophageal fistula closed is shown in the blocks on the right.

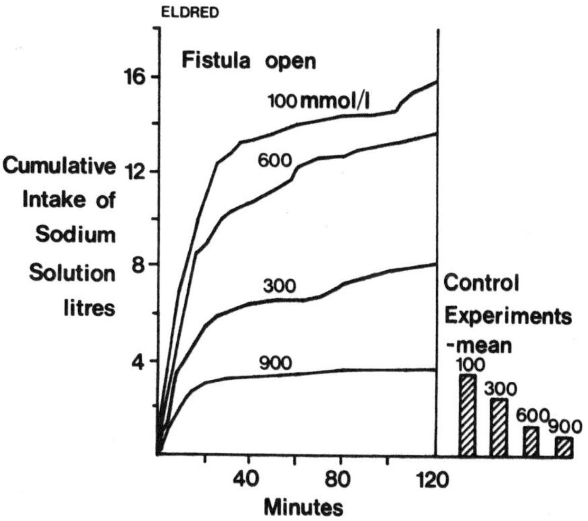

Fig. 14-25. Eldred: sodium-deficient sheep with an oesophageal fistula. The volume of sodium bicarbonate solution (100, 300, 600, 900 mmol/l) sham-drunk during 120 min in individual experiments with the oesophageal fistula open. The mean volume in control experiments under the same conditions but with the oesophageal fistula closed is shown in the blocks on the right.

drunk to replete the sheep may act to inhibit further drinking at the latter stages of drinking. At high sodium concentrations, taste aversion may be an important factor in modifying sodium intake. It has been demonstrated with twin calves (Bell 1963) and with rats (Carr 1952; Fregly 1958; Nachman and Pfaffmann 1963) that the sodium-deficient animal has a higher acceptance threshold for sodium than its normal control. Pfaffmann and Bare (1950), Carr (1952) and Nachman and Pfaffmann (1963) observed that the taste threshold was the same for sodium-deficient and normal rats. Therefore, although the sodium-deficient animal will drink a sodium solution to which it has an aversion under sodium-replete control conditions, individual variation in aversion, as well as the extent of the deficiency, may modify the amount of sodium ingested.

It has been demonstrated in the anaesthetized rat (Zotterman 1956; Fromer 1961; Halpern 1963) that the rate of discharge of impulses along the chorda tympani nerve increased with increasing concentrations of sodium chloride applied to the tongue, and Pfaffmann (1959) observed in the rat commensurate electrical response to sodium chloride over the range of sodium bicarbonate concentrations used in the study on sheep described above. It is likely that the electrical response of the chorda tympani and glossopharyngeal nerve which Baldwin et al. (1959) have demonstrated in the sheep, goat and calf when sodium bicarbonate solution was applied to the tongue, increases commensurately with increasing concentrations of sodium bicarbonate. Presumably when the sodium-deficient sheep sham-drank sodium bicarbonate solution the alteration in the rate of firing of impulses along the chorda tympani (and other gustatory nerves) was effective in reducing the volume of sodium solution drunk as the concentration offered was increased. But it did not cause termination of drinking when an amount equal to the deficit had passed through the pharynx and upper oesophagus. Sham-drinking continued intermittently.

Factors contingent on adaptation to taste stimuli may have modified drinking if there are processes in the sheep analogous to those Hahn (1934) described in man. Adaptation occurs more rapidly at low concentrations than at high, and Pfaffmann (1959) concluded that it is of central origin. Thus, taste adaptation may be a contributing cause of the cessation of the drinking episode when the sheep sham-drink solutions of low sodium concentrations.

Thus it was evident that sensory inflow from tasting and swallowing did not suffice to satiate the animals, though in aggregate the more concentrated solutions slowed intake more than did the dilute. But it was apparent that despite taste inflow, the absence of inflow from proprioceptor elements of the normal sensory pattern during consummation affected the animal virtually at the outset, i.e. in 2–5 min.

Sensory impulses from the lower oesophagus, cardia and rumen are an essential component in the process of satiation. *The results support the hypothesis that during the drinking of sodium solution there is a central integration of taste impulses signalling concentration of sodium solution with pharyngo-oesophageal impulses metering volume swallowed. A 'gestalt' of sensory inflow, involving a specific sequence in time, is the basis of satiation.*

References

Adolph EF (1939) Measurements of drinking water in dogs. Am J Physiol 125: 75

Adolph EF (1943) Physiological regulations. Cattell Press, Lancaster

Adolph EF (1950) Thirst and its inhibition in the stomach. Am J Physiol 161: 374

Adolph EF, Barker JP, Hoy PA (1954) Multiple factors in thirst. Am J Physiol 178: 538

Andersson B, Olsson K (1973) On central control of body fluid homeostasis. Conditional Reflex 8: 148

Andersson B, Jewell PA, Larsson S (1958) An appraisal of the effects of diencephalic stimulation of conscious animals in terms of normal behaviour. In: Wolstenholme GEW, O'Connor CM (eds) Neurological basis of behaviour. Churchill, London, p 76

Baldwin BA, Bell FR, Kitchell RL (1959) Gustatory nerve impulses in ruminant ungulates. J Physiol (Lond) 146: 14P

Beilharz S, Kay RNB (1963) The effects of ruminal and plasma sodium concentrations on the sodium appetite of sheep. J Physiol (Lond) 165: 468

Bell FR (1963) The variation in taste thresholds of ruminants associated with sodium depletion. In: Zotterman Y (ed) Proceedings of the First International Symposium on Olfaction and Taste. Pergamon Press, Oxford, p 299

Bellows RT (1939) Time factors in water drinking in dogs. Am J Physiol 125: 87

Blaine EH, Covelli MD, Denton DA, Nelson JF, Shulkes AA (1975) The role of ACTH and adrenal glucocorticoids in the salt appetite of wild rabbits (*Oryctolagus cuniculus*(L.)). Endocrinology 97: 793

Blass EM, Hall WG (1976) Drinking termination: Interactions among hydrational orogastric and behavioural control in rats. Psychol Rev 83: 356

Bott E, Goding JR (1965) A method of oesophageal fistulation in the sheep. Aust Vet J 41: 326

Bott E, Denton DA, Weller S (1965) Water drinking in sheep with oesophageal fistulae. J Physiol (Lond) 176: 323

Bott E, Denton DA, Weller S (1967) The innate appetite for salt exhibited by sodium deficient sheep. In: T Hayashi (ed) Proceedings of the Second International Symposium on Olfaction and Taste. Pergamon Press, Oxford, New York, p 415

Carr WJ (1952) The effect of adrenalectomy upon the NaCl taste threshold in rat. J Comp Physiol Psychol 45: 377

Chernigovsky V (1963) Neurophysiology of visceral afferent systems and the behaviour of animals towards food (feeding behaviour). Nova Acta Leopoldina NF 28: 317

Denton DA (1966) Some theoretical considerations in relation to innate appetite for salt. Conditional Reflex 1: 144

Denton DA, Nelson JF (1970) Effect of deoxycorticosterone acetate and aldosterone on the salt appetite of wild rabbits (*Oryctolagus cuniculus* (L.)). Endocrinology 87: 970

Denton DA, Nelson JF (1980) The influence of reproductive processes on salt appetite. In: Kare MR, Fregly MJ, Bernard RA (eds) Biological and behavioural aspects of salt intake. Academic Press, New York, p 229

Denton DA, Sabine JR (1961) The selective appetite for Na^+ shown by Na^+-deficient sheep. J Physiol (Lond) 157: 97

Denton DA, McKinley MJ, Nelson JF, Weisinger RS (1977) The hormonal genesis of mineral appetites. XIth Acta Endocrinologica Congress (Lausanne). Acta Endocrinol [Suppl] (Copenh) 85 (212): 3

Dobson A (1959) Active transport through the epithelium of the reticulo-rumen sac. J Physiol (Lond) 146: 235

Feider A (1972) Feedback control of thirst in rats. Physiol Behav 8: 1005

Fregly MJ (1958) NaCl appetite of adrenalectomized rats. Proc Soc Exp Biol NY 97: 144

Fromer GP (1961) Gustatory afferent responses in the thalamus. In: Kare MR, Halpern BP (eds) The physiological and behavioural aspects of taste. Chicago University Press, Chicago, p 50

Hahn HZ (1934) Sinnesphysiol 65: 105. Cited in Pfaffmann C (1959)

Halpern BP (1963) Chemical coding in taste-temporal patterns. In: Zotterman Y (ed) Proceedings of the First International Symposium on Olfaction and Taste. Pergamon Press, Oxford, p 275

Handal PJ (1965) Immediate acceptance of sodium salts by sodium deficient rats. Psychonom Sci 3: 315

Hatton GI, Bennett CT (1970) Satiation of thirst and termination of drinking: Roles of plasma, osmolality, and absorption. Physiol Behav 5: 487

Jalowiec JE, Stricker EM (1970) Restoration of body fluid balance following acute sodium deficiency in rats. J Comp Physiol Psychol 70: 94

Jalowiec JE, Stricker EM (1973) Sodium appetite in adrenalectomized rats following dietary sodium deprivation. J Comp Physiol Psychol 83: 66

Kassil VG (1959) Conditioned influences from stomach receptors on the salt appetite in higher animals. Lect Acad Sci USSR 129(2)

Kassil VG, Ugolev AM, Chernigovsky VN (1956) Gastric receptors and control of food selection in the dog. CR Acad Sci URSS 126: 692

Klisiecki A, Pickford M, Rothschild P, Verney EB (1933) The absorption and excretion of water by the mammal. Proc R Soc Lond [Biol] 112: 496, 521

Le Magnen J (1953) L'intervention des stimulations orales dans la régulation quantitative de la prise d'eau spontanée chez le rat blanc. CR Soc Biol (Paris) 147: 614

Le Magnen J (1955) Le rôle de la réceptivité gustative au chlorure de sodium dans le mécanisme de régulation de la prise d'eau chez le rat blanc. J Physiol (Paris) 47: 405

Levy CJ, McCutcheon B (1974) Importance of post-ingestional factors in the satiation of sodium appetite in rats. Physiol Behav 13: 621

Maller O (1967) Specific appetite. In: Kare MR, Maller O (eds) The chemical senses and nutrition. Johns Hopkins Press, Baltimore, p 201

McCance RA (1938) The effect of salt deficiency in man on the volume of the extracellular fluids, and on the composition of sweat, saliva, gastric juice and cerebrospinal fluid. J Physiol (Lond) 92: 208

McCleary RA (1953) Taste and postingestion factors in specific-hunger behaviour. J Comp Physiol Psychol 46: 411

Mook DG (1963) Oral and postingestional determinants of the intake of various solutions in rats with oesophageal fistulas. J Comp Physiol Psychol 56: 645

Mook DG (1969) Some determinants of preference and aversion in the rat. Ann NY Acad Sci 157: 1158

Nachman M, Pfaffmann C (1963) Gustatory nerve discharge in normal and sodium deficient rats. J Comp Physiol Psychol 56: 1007

Nachman M, Valentino DA (1966) The role of taste and post-ingestional factors in the satiation of sodium appetite in rats. J Comp Physiol Psychol 62: 280

Pfaffmann C (1959) The sense of taste. In: Handbook of physiology, neurophysiology I. American Physiological Society, Washington, p 528

Pfaffmann C, Bare JK (1950) Gustatory nerve discharges in normal and adrenalectomized rats. J Comp Physiol Psychol 43: 320

Rolls BJ, Wood RJ, Rolls ET (1980) Thirst: The initiation, maintenance and termination of drinking (Progress in Psychobiology and Physiological Psychology 9.) Academic Press, New York, p 263

Schmidt-Nielsen B, Schmidt-Nielsen K, Houpt TR, Jarnum SA (1956) Water balance of the camel. Am J Physiol 185: 185

Stellar E (1967) Hunger in man: Comparative and physiological studies. Am Psychol 22: 105

Stellar E, Hyman R, Samet S (1954) Gastric factors controlling water- and salt-solution drinking. J Comp Physiol Psychol 47: 220

Streeten DHP (1950) The effects of sodium and chloride lack on intestine motility and their significance in paralytic ileus. Surg Gynecol Obstet 91: 421

Stricker EM (1969) Osmoregulation and volume regulation in rats: Inhibition of hypovolemic thirst by water. Am J Physiol 217: 98

Towbin EJ (1949) Gastric distension as a factor in the satiation of thirst in esophagostomized dogs. Am J Physiol 159: 533

Wolf G (1969) Innate mechanisms for regulation of sodium intake. In: Pfaffmann C (ed) Proceedings of the Third International Congress on Olfaction and Taste. Rockefeller University Press, New York, p 548

Zotterman Y (1956) Species differences in the water taste. Acta Physiol Scand 37: 60

15 Plasma volume change and the influence on salt appetite

Summary

1. Experimental data implicating decrease in plasma volume in the genesis of thirst are discussed. Afferent impulses from receptors on the low pressure side of the circulation appear to be involved.

2. Experiments where reduced plasma volume of sodium-deficient sheep was rapidly expanded to the normal range or above by intravenous infusion of 500 or 750 ml of 6% dextran did not reduce sodium appetite. Intravenous infusion of dextran for 48 h over the course of development of sodium deficit, a procedure which prevented a fall in plasma volume, did not prevent development of sodium appetite. The data contrast with results of infusion of substantial amount of saline in sodium-deficient sheep, which produced a comparable increase in plasma volume but also reduced sodium appetite.

Plasma volume change and thirst in rats

In the early work on the influence of reduced circulatory volume on ingestive behaviour, controlled haemorrhage was used as a stimulus (see Stricker and MacArthur 1974). However, the data from several laboratories indicated that blood loss did not consistently increase water intake (Holmes and Montgomery 1953; Fitzsimons 1961; Schneiden 1962; Oatley 1964). Factors influential in this lack of consistency included the fact that when losses were small, plasma volume might be restored within minutes by fluid transfers from the interstitial space. And with larger haemorrhages animals might become seriously anaemic or hypotensive, and therefore their behaviour be modified. Accordingly, the alternative technique of sequestration of isosmotic protein-free plasma fluid extruded from local capillaries was applied. This procedure, using hyperoncotic polyethylene glycol (PEG), avoids anaemia as well as compensatory repletion from interstitial fluid reservoirs (see Chapter 8).

Stricker and Wolf (1969) showed a clear commensurate relationship between increasing concentration of PEG injected and water consumption by rats (Fig. 12-7). Increased intake was evident by an hour. On the face of it, these data indicated that thirst could be elicited by hypovolaemia, and that the amount of water consumed was directly related to the degree of reduction in intravascular fluid volume. The supposition that intravascular volume needed to be restored for the associated thirst to be more than transiently satiated was tested by giving intragastric preloads of isotonic (0.9%) saline or water to rats made hypovolaemic by 10% PEG injections (Stricker and Wolf 1967). It was found that isotonic saline preloads corrected the hypovolaemia of PEG-treated rats and reduced their thirst, whereas water preloads had little effect on the intravascular volume or on associated thirst. The data showed, however, that sufficient overhydration and hypo-osmolarity could serve to inhibit drinking though marked hypovolaemia persisted. Since a third to a half of the 15 ml of water preload given in these experiments was excreted within the first 2 h after intubation, pitressin was used to inhibit the renal water loss, or alternatively larger doses of water were given. The inhibition was, however, only transient. Drinking reappeared after renal excretion of the excess water. Thus the data indicated that hypovolaemic thirst is permanently satiated only when sodium is available to permit repletion of intravascular fluid volume. In relation to this hypothesis, as we will discuss further in the consideration of salt appetite and the influence of hypovolaemia, experiments in which plasma volume was expanded to normal, or above normal, by dextran or other agents would be valuable in the logic of analysis.

Fitzsimons (1972), in discussing the vector of extracellular thirst, suggests that the vascular receptors that underly extracellular fluid volume control are stretch receptors, especially those in the walls of the low pressure capacitance vessels near the heart. These veins are 100–200 times more distensible than the arteries and they hold 80% of the blood volume (Gauer and Henry 1963). The moderate day-to-day changes in extracellular fluid volume mainly affect filling of these capacitance vessels and therefore the amount of sensory information from this part of the vasculature. These changes evoke appropriate volume-regulatory responses of either thirst and oliguria or increased urine flow. However, Fitzsimons (1972) states that direct evidence that stretch receptors in the capacitance vessels are concerned in thirst induced by extracellular volume changes is not available. Evidence of tonic inhibitory impulses via the vagus nerves has come from Sobocinska (1969). In weightlessness in man about 2 litres of blood and extracellular fluid are lost from the lower part of the body causing severe engorgement of the circulation and tissues in the upper part of the body (Gauer and Henry 1976). Cardiac output is increased and peripheral resistance and venous tone reduced. There is a fluid loss of 1 litre within a few days of entry to weightlessness which is attributable to a drinking deficit of 300 ml/day and not to increased urinary excretion. The relative adipsia is presumably caused by distension of the capacitance vessels (Fitzsimons 1979).

The issue of a renal component in the regulatory organization of thirst, through the renin–angiotensin system, is discussed elsewhere. Some forms of water drinking are greatly reduced by nephrectomy but as far as the response to subcutaneous injection of hyperoncotic colloid is concerned, the response is much the same in nephrectomized as in normal rats (Stricker and Wolf 1969; Fitzsimons 1972; Stricker et al. 1979).

Plasma volume and salt appetite

Coming to the question of whether change in intravascular volume is a direct vector of genesis of sodium appetite, data on sheep and the rat can be compared. Certain common ground may be noted at the outset. That is, the sheep also responds to a reduction in intravascular fluid volume by increased water intake (this is under conditions where osmotic change can be presumed to be without influence). Briefly, external balance studies have been made on sheep in the 24–48 h following unilateral parotid papilla cannulation. The animals lose from the one parotid 2–4 litres of saliva in the first 24 h. The composition of this saliva (Denton 1957) in the animal which is initially sodium-replete is 160–175 mmol/l sodium and 4–5 mmol/l potassium. Thus the animal is losing sodium in excess of water relative to extracellular proportions. As it becomes sodium deficient, salivary sodium concentration falls and the potassium concentration rises. But monitoring over the course of the 24 h of salivary loss indicates that the salivary sodium concentration only reduces to that of the plasma near the end of the 24-h period. That is, the net loss is of sodium in excess of water, which means that any change in extracellular sodium concentration will be in the direction of a fall not a rise. Indeed, by the second day of salivary loss, plasma sodium concentration is falling. Mean water intake of a group of animals in the first 24 h of loss of saliva from the cannula was increased by approximately a litre (Fig. 20-1) relative to the control period intake. The cause of thirst is clearly not plasma sodium concentration or osmotic pressure and is unknown.

Recently, Zimmerman (1979) has done experiments which throw considerable light on the phenomenon. In conscious sheep he produced deficits of 515–3208 ml of protein-free isosmotic extracellular fluid by haemo-ultrafiltration. This resulted in plasma deficits of 10%–30%. Increased water drinking followed, but apparently a threshold deficit of approximately 2000 ml of extracellular fluid was required to elicit this. Damage to the left atrial appendage by crushing with surgical forceps at a prior operation abolished the increased water intake but did not interfere with either normal daily water intake or increased water intake in response to hyperosmotic stimuli. Zimmerman suggests low pressure volume receptors which are situated in the left atrium of the heart detect changes in quantity of intravascular fluids, and activate appropriate behavioural responses through afferent projections to the central nervous system. Damage to the atrial appendage has also been shown to interfere with renal response to change in left atrial pressure (Kappagoda et al. 1972).

With sodium deficiency as a result of uncompensated parotid saliva loss in the sheep with free access to water, there is reduction in plasma volume and hyponatraemia (Blair-West et al. 1967). Similarly in the rat, when deficiency of available sodium is produced by subcutaneous formalin injection, the decrease in intravascular volume is associated with hyponatraemia, because of entry of sodium to damaged cells. It has been suggested that hyponatraemia potentiates the sodium appetite of sodium-deficient rats (Stricker and Wolf 1966; Jalowiec and Stricker 1970). Hyponatraemia produced in the absence of sodium deficiency by intragastric water loads does not induce sodium appetite in rats (Stricker and Wolf 1966). But in this latter situation water is distributed equally across intracellular and extracellular fluid, and would not induce any concentration differential between intracellular and extracellular concentrations of sodium or any other element (Denton 1957).

With subcutaneous injection of a hyperoncotic solution of PEG (5 ml of 10%–30% solution), intravascular volume decreases without hyponatraemia. This induces sodium appetite, but in contrast to the relatively rapid effect on thirst (1–2 h), the appetite is not manifest until 8–12 h (Fig. 12-7). If the PEG is injected intraperitoneally, thirst occurs but not sodium appetite. This is attributable to the fact that PEG enters the circulation from the peritoneum, approximately 25%–30% entering within 6 h of treatment (Stricker and MacArthur 1974). Thus whereas effects on haematocrit and renin levels with subcutaneous and intraperitoneal routes are comparable at 2–3 h, the situation is quite different at 6–12 h. By 12–24 h, values in intraperitoneally injected animals had returned to normal.

There are no direct data on the role of capacitance vessel receptors in the slow induction of sodium appetite by PEG. These receptors have been implicated in release of antidiuretic hormone and, also, in the stimulation of aldosterone secretion (Gauer 1977). The possibility that aldosterone hypersecretion caused the sodium appetite with PEG has been examined in experiments in adrenalectomized rats. The sodium intake was not reduced relative to that in the intact animals (Wolf and Stricker 1967). Stricker (1980) has reviewed data involving a number of different manipulations of the physiological effects of PEG and resultant salt appetite, and has concluded plasma volume is not a major determinant (see Chapters 12 and 24).

The study of the role of plasma volume change in the genesis of sodium appetite in the sheep has been approached in a different fashion. As baseline data there was the comprehensive study of Blair-West et al. (1967) showing the effect of 24–36 h of parotid salivary loss on the [^{131}I]gamma-globulin space of sheep (Boyd 1967). The mean sodium-deplete volume was 245 ml less than that observed in the same animal in the sodium-replete state. The infusion of 300 or 500 ml of 6% dextran in normal saline increased plasma volume by approximately 400–600 ml (Fig. 15-1) (the solution

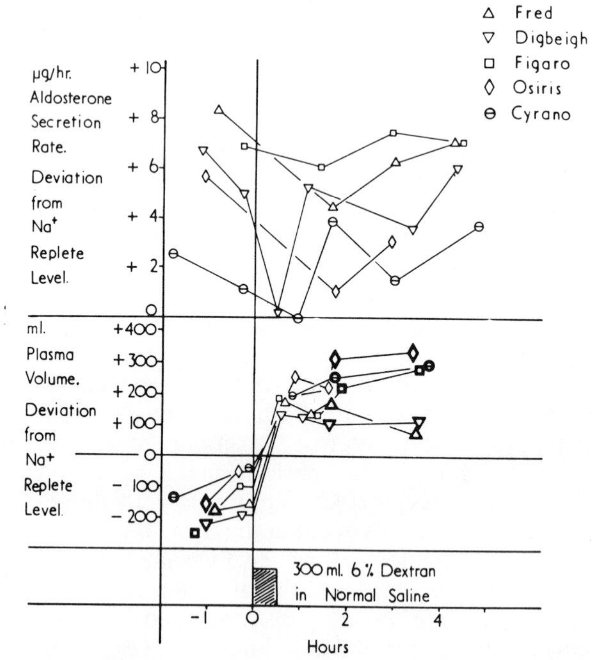

Fig. 15-1. The time course of effect of intravenous infusion of 300 ml of 6% dextran in normal saline in 30–40 min on aldosterone secretion rate and plasma volume of sodium-deficient sheep with an adrenal transplant. The heavy symbols represent direct measurements of plasma volume in terms of [^{131}I]gamma-globulin space and the light symbols indirect measurements (see Blair-West et al. 1967). (From Blair-West et al. 1967, with permission.)

is 1.4 times the oncotic pressure of sodium-deplete sheep plasma), so that in all instances plasma volume was increased above the sodium-replete control level. Systolic, diastolic and central venous pressure were increased over 2–4 h following commencement of infusion and the expansion of plasma volume was sustained over the same period (Fig. 15-2).

In a series of experiments described below sheep were first made sodium deficient by loss of parotid saliva for 48 h. Water was continuously available. At the end of this depletion period voluntary intake was measured during 15 min of access to 300 mmol/l sodium bicarbonate solution (Denton et al. 1971).

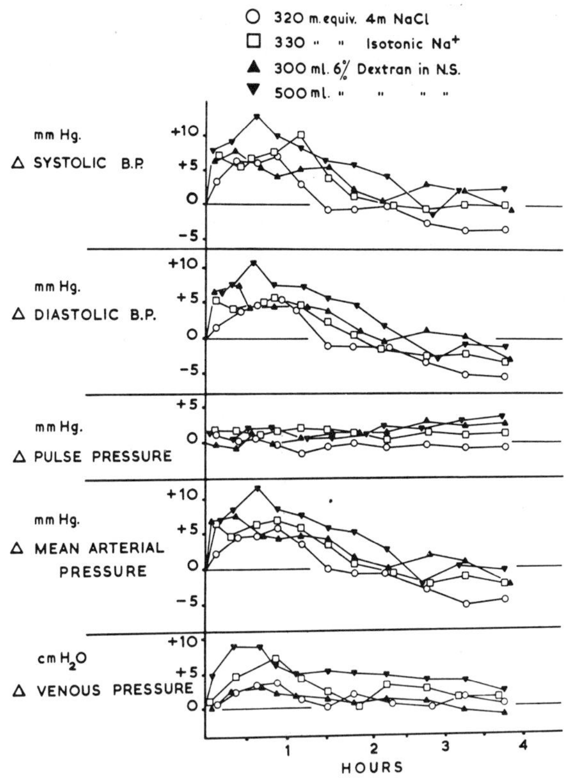

Fig. 15-2. The time course of mean change in haemodynamic parameters following intravenous infusion of 300 and 500 ml of 6% dextran in isotonic saline. The effects of 320 mmol of 4 M sodium chloride and 330 mmol of isotonic saline are shown also. Each point on the figure represents the average of four to nine individual mean animal responses. (From Blair-West et al. 1967, with permission.)

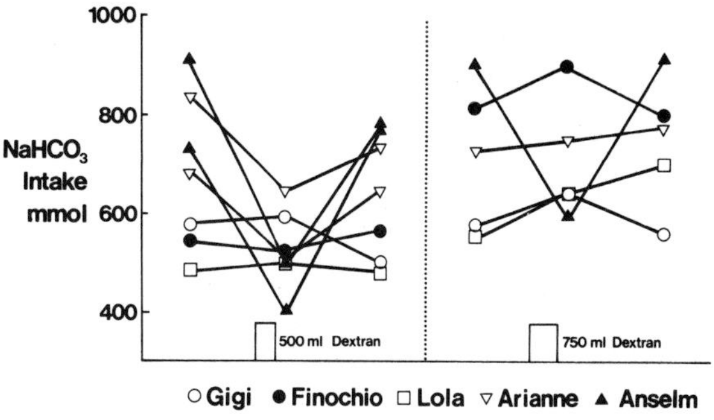

Fig. 15-3. The sodium appetite of sodium-deficient sheep during 15 min of access to 300 mmol/l sodium bicarbonate solution given 10 min after intravenous infusion of either 500 ml or 750 ml of 6% dextran in normal saline over 40 min. The voluntary intakes on the control sodium deficiency day 48 h before infusion and on the day 48 h after the infusion are shown also.

Effect of rapid infusion of 500 or 750 ml of 6% dextran in isotonic saline on sodium appetite of sodium-deficient sheep

On experimental days the 6% dextran in isotonic saline was infused in 40–60 min. The isotonic fluid contained 70 or 100 mmol of sodium which was, in the majority of animals, a small fraction of the deficit occurring over 48 h as indicated by voluntary intake on the control days. The infusion produced no changes in plasma sodium or potassium concentration but a large decrease in haematocrit occurred.

Ten minutes after the end of the infusion, the sodium bicarbonate solution (300 mmol/l) was offered for 15 min. Appetitive behaviour was normal for all sheep except the sheep Anselm which showed less interest. Five of the nine sheep studied received both 500 and 750 ml of dextran and Fig. 15-3 shows that the amount of sodium bicarbonate drunk was not reduced compared with intake on the control day 48 h before infusion. With Arianne, there was some reduction in intake after infusion of 500 ml dextran but not after an infusion of 750 ml. With Anselm, the sodium bicarbonate intake was reduced each time it was tested after dextran infusion. However, the solution was made available to the animal for an additional 15 min in each case and a further 150–250 mmol drunk. This suggests that intake was slowed rather than that there had been a very large reduction in the sodium appetite in this sheep. Of four other sheep (Brad, Dorothy, Cyrano and Ivor) given 500 ml dextran, there was no decrease or increase in intake in three and in the fourth intake was halved.

Another approach to this question involved four experiments in which 1.5 litres of dextran solution were infused over the 2 days during which the animal became sodium-deplete as a result of salivary loss. The saline content of the dextran solution (approximately 210 mmol of sodium) reduced the extent of the resultant sodium deficiency relative to control days. In an experiment where during this procedure there was serial measurement of plasma volume using ^{131}I-labelled sheep globulin, it was shown that the dextran infusion prevented a fall in plasma volume and that after 48 h the volume was higher (1.38 litres) than at the beginning when the animal was sodium-replete (1.32 litres). Making allowance for the saline content of the dextran infusion, the results of the four experiments were indicative that maintenance of normal plasma volume during sodium deficiency did not prevent the development of salt appetite. Immediately after the 48-h depletion period Brad drank 660 mmol of sodium, and on the experimental day 48 h later, 537 mmol; Ricky drank in one experiment 579 mmol on the control day and 561 mmol after dextran, and in another experiment 468 and 678 mmol respectively; and Dorothy drank 972 mmol on the control day and 594 mmol after dextran.

In summary, the infusion of 500 or 750 ml of 6% dextran in isotonic saline, which rapidly expanded plasma volume above the sodium-replete level in sodium-deficient sheep, did not abolish sodium appetite. In six of nine animals no reduction in sodium intake occurred. There was no statistically significant difference between intake on control days and experimental days for the 16 experiments. This was true even when no allowance was made for the small amount of sodium (75 or 100 mmol) infused with the dextran. Furthermore, if the usual plasma volume decrease seen in sodium deficiency was prevented by dextran infusion over 48 h, it did not prevent the genesis of salt appetite commensurate with sodium deficiency. There was no evident explanation of the reduction in intake occurring in some instances except to note that in the animal which showed the effect consistently, the change was more a slowing of intake than a major reduction of appetite. The evidence is against salt appetite being simply and directly determined by decrease in plasma volume or for that matter against plasma volume change being a major immediate causal factor.

This conclusion was supported by experiments which contrasted the effect of plasma volume change per se with change in sodium balance with concomitant increase in plasma volume. That is, under conditions of sodium depletion, 2–2.6 litres of Ringer saline were infused into sheep over 40–60 min, causing a rapid increase in plasma volume to the sodium-replete range or above (Blair-West et al. 1967). The infusion replaced about half the sodium deficit of the animals without causing a significant increase in plasma sodium concentration. It substantially reduced the sodium bicarbonate drinking ($P<0.001$) by the same order in all sheep studied, the mean reduction of 238 mmol representing approximately 70% of the sodium infused (Denton et al. 1971: see Chapter 20). The data were also against any effect of blood pressure change per se on salt appetite. Both dextran and isotonic Ringer infusion caused sustained increase in systolic and diastolic blood pressure—the effect with dextran being somewhat larger (Blair-West et al. 1967)—but only isotonic or 4 M saline significantly influenced salt appetite (Denton et al. 1971). Thus effects on appetite appear attributable to the sodium ion given, and the vector is not haemodynamic change.

The point could be raised that as onset of sodium appetite in the PEG experiments on rats is slow—over many hours—then the mechanism of accessing the control system or the reaction of the system itself is slow. Correspondingly a rapid stimulation of receptors in capacitance vessels and firing of vagal afferents may not necessarily turn appetite off quickly. It could be that if salt appetite were tested 4–5 h after the expansion in the sodium-deficient animal an inhibitory effect would be seen. Whereas the formal experiment has not been done, it is noteworthy that the isotonic infusion with its significant content of sodium was able to access the mechanism in 60 min. As recounted in full in Chapter 20, more recent data show that the control system can be accessed and inhibited in 10–20 min by a large influx of hypertonic saline into the circulation. These results pointing against an effect of plasma volume are reinforced by the finding that development of salt appetite with onset of sodium deficiency was not prevented when plasma volume reduction was stopped by the dextran infusion.

References

Blair-West JR, Boyd GW, Coghlan JP, Denton DA, Wintour EM, Wright RD (1967) The role of circulating fluid volume and related physiological parameters in the control of aldosterone secretion. Aust J Exp Biol Med Sci 45: 1

Boyd GW (1967) The reproducibility and accuracy of plasma volume estimation in the sheep with both ^{131}I gamma globulin and Evan's blue. Aust J Exp Biol Med Sci 45: 51

Denton DA (1957) The effect of variation of the concentration of individual extracellular electrolytes on the response of the sheep's parotid gland to Na depletion. J Physiol (Lond) 140: 129

Denton DA, Orchard E, Weller S (1971) The effect of rapid change of sodium balance and expansion of plasma volume on sodium appetite of sodium-deficient sheep. Commun Behav Biol 6: 245

Fitzsimons JT (1961) Drinking by rats depleted of body fluid without increase in osmotic pressure. J Physiol (Lond) 159: 297

Fitzsimons JT (1972) Thirst. Physiol Rev 52: 2

Fitzsimons JT (1979) The physiology of thirst and sodium appetite. Cambridge University Press, Cambridge

Gauer OH (1977) Volume control. Proceedings of the International Union of Physiological Sciences (Paris) 12: 42 (Abstracts and lectures)

Gauer OH, Henry JP (1963) Circulatory basis of fluid volume control. Physiol Rev 43: 423

Gauer OH, Henry JP (1976) Neurohumoral control of plasma volume. In: Guyton AC, Cowley AW (eds) International review of physiology, vol 9: cardiovascular physiology II. University Park Press, Baltimore, p 145

Holmes JH, Montgomery AV (1953) Thirst as a symptom. Am J Med Sci 225: 281

Jalowiec JE, Stricker EM (1970) Sodium appetite in rats after apparent recovery from acute sodium deficiency. J Comp Physiol Psychol 73: 238

Kappagoda CT, Linden RJ, Snow HM (1972) The effect of distending the atrial appendages on urine flow in the dog. J Physiol (Lond) 227: 233

Oatley K (1964) Changes in blood volume and osmotic pressure in the production of thirst. Nature 202: 1341

Schneiden H (1962) Solution drinking in rats after dehydration and after hemorrhage. Am J Physiol 203: 560
Sobocinska J (1969) Abolition of the effect of hypovolemia on the thirst threshold after cervical vagosympathectomy in dogs. Bulletin de l'Academie Polonaise des Sciences 17: 341
Stricker EM (1980) The physiological basis of sodium appetite: A new look at the 'depletion–repletion' model. In: Kare MR, Fregly MJ, Bernard RA (eds) Biological and behavioural aspects of salt intake. Academic Press, New York, p 185
Stricker EM, MacArthur JP (1974) Physiological bases for different effects of extravascular colloid treatments on water and NaCl solution drinking by rats. Physiol Behav 13: 389
Stricker EM, Wolf G (1966) Blood volume and tonicity in relation to sodium appetite. J Comp Physiol Psychol 62: 275
Stricker EM, Wolf G (1967) Hypovolemic thirst in comparison with thirst produced by hyperosmolarity. Physiol Behav 2: 33
Stricker EM, Wolf G (1969) Behavioural control of intravascular fluid volume: Thirst and sodium appetite. Ann NY Acad Sci 157: 553
Stricker EM, Vagrucci AH, McDonald RH, Leenen FH (1979) Renin and aldosterone secretions during hypovolaemia in rats: Relation to NaCl intake. Am J Physiol 237: R45
Wolf G, Stricker EM (1967) Sodium appetite elicited by hypovolaemia in adrenalectomized rats: Re-evaluation of the 'reservoir' hypothesis. J Comp Physiol Psychol 63: 252
Zimmerman MB (1979) The possible role of left atrial receptors in mediating thirst during hypovolaemia. PhD Thesis, University of Pittsburgh

16 Influence of concurrent water depletion on salt appetite during sodium deficiency

Summary

1. With concurrent depletion of sodium and water in sheep or in adrenalectomized rats, the animals satisfy both deficits when water and hypertonic salt solution are presented.

2. Sodium appetite occurs though plasma sodium concentration is unchanged or increased relative to control.

3. This does not necessarily preclude a theory of a decrease in the intracellular sodium concentration of specific neural elements subserving sodium appetite being the vector of excitation of sodium appetite. Water depletion would affect intracellular and extracellular fluid equivalently and sodium loss from extracellular fluid could still cause a differential concentration change relative to intracellular fluid with resulting intracellular sodium loss.

4. The rapid satiation of both appetites, in close sequence, with satiation of one not eliminating the other drive even though there are a number of motor acts and pharyngo-oesophageal proprioceptor elements in common, highlights the role of taste in the satiation process. Further, the fact that ruminal distension of 2–4 litres in the sheep as a result of drinking one fluid does not preclude the immediate drinking of the other suggests that it was unlikely that ruminal distension was the reason for termination of the initial consummatory act. The metering occurred largely rostral to it, i.e. via taste and pharyngo-oesophageal neural inflow. However, both free access and operant experiments suggest that the act of prior satiation of water deficiency may significantly reduce satiation of sodium appetite in sheep.

Preceding this study of appetite, it has been shown in relation to the control of aldosterone secretion (Denton 1957) that concurrent water depletion, causing a rise in plasma sodium concentration instead of the usual fall, did not significantly impair the normal aldosterone hypersecretion in response to sodium deficiency. Sodium deficiency, as in the experiments here, was caused by salivary loss from a permanent unilateral parotid fistula.

Physiological influences on acceptability of salt solutions

The relative acceptability of sodium chloride solutions as a function of concentration and water need has been studied. Young and Chaplin (1949) maintained rats upon a 24-component free-selection diet including solutions of sodium chloride of 0.1, 0.2, 0.4, 0.7, 1.5, 3.0, 6.0 and

12.0%. The rats ingested more of the 0.7% solution than all other solutions combined (62%–82% of total). Moreover, when 0.7% solution was present the rats ingested 1.5 to 3.5 times as much sodium chloride as when a single 3.0% solution was available. Adrenalectomy did not make them prefer either a more or less salty solution. Many workers have found the optimum concentration near this value. Stellar et al. (1954) found a stomach load of 3% sodium chloride introduced beforehand served to shift the optimum to a lower concentration. Young and Falk (1956) using a brief-exposure preference test showed rats preferred sodium chloride solutions of between 0.75% and 1.5%. When rats are water-deplete the choice of water is enhanced: thirsty rats, more frequently than non-thirsty ones, prefer water to salt solution, and when the choice is between two salt solutions the thirsty animals tend to select the less salty. Thirsty rats completely reject 3.0% and 6.0% saline solutions (513 and 1026 mmol/l respectively), while non-thirsty animals occasionally accept them.

In relation to the data to be recounted below, it has been shown that sodium-replete sheep preferentially drink sodium bicarbonate and sodium chloride solutions of higher concentration than do rats (Abraham et al. 1976). Water-deprived animals may sometimes accept 300 mmol/l sodium chloride solutions, or even higher.

Studies on concurrent water and sodium deficiency in sheep

The effects on fluid ingestion and behaviour of sodium deficiency, water deficiency, and concurrent deficiency of sodium and water were compared in detail on three animals, Satan, Tacitus and Somnos, and also on PF48. The effect of concurrent deficiency was examined in five other sheep.

Sodium deficiency: Experimental conditions

At the same time on every second day the sheep, which all had a parotid fistula, were presented for 30 min with two cans, one containing water the other sodium bicarbonate solution (317 or 420 mmol/l). The relative positions of the cans were varied randomly. The behaviour of the sheep was recorded. Water was supplied ad libitum during the 47.5-h interval between the test periods. During the 2-day period between tests the sheep lost from 4 to 7 litres of saliva. Therefore, at the beginning of each test period the sheep were depleted of about 400–900 mmol of sodium.

Water depletion: Experimental conditions

The experimental procedure was similar to that outlined for sodium deficiency, except that at 12-h intervals the saliva lost from the parotid fistula was returned to the sheep by rumen tube, and the animal had access to sodium bicarbonate solution (317 or 420 mmol/l), but not water, in the 47.5-h between the test periods.

Concurrent depletion of sodium and water: Experimental conditions

The experimental conditions were similar to those described under sodium deficiency, except that the water bin was removed approximately 24 h before the test period.

The detailed experiments on Satan, Tacitus and Somnos will be described first.

Sodium deficiency

As Table 16-1 shows, the sheep drank little or no water but drank sodium bicarbonate eagerly. The table also shows the mean salivary loss during the preceding 48 h and the mean voluntary sodium bicarbonate intake for 25 tests on the three sheep. Allowing for a slight excess of sodium in food over faecal sodium loss and loss of sodium into the saliva bin, the voluntary intake approximated the deficiency.

The mean sodium concentration of arterial blood of sodium-replete sheep is 147.0±3.70 (s.d.) mmol/l, n=15 (Coghlan et al. 1960). In another study under very carefully controlled conditions the findings were 145±0.4 (s.e.) mmol/l, n=22 (Scoggins et al. 1974). Table 16-1 shows that mean arterial sodium concentration was reduced in sodium deficiency (mean=141.9±4.41 (s.d.) mmol/l), and that the salivary Na/K ratio was reduced from a normal of 20–30 to 1.6–1.8. Table 16-2 shows the individual results of the nine experiments on Satan.

Water deficiency

All three sheep drank water in the test period following 48 h of water deficiency. They usually identified the water and drank continuously for

Table 16-1. The voluntary intake of sodium bicarbonate and water after sodium depletion, water depletion or concurrent sodium and water depletion in three sheep (Satan, Tacitus and Somnos). The plasma sodium concentration, plasma protein and the salivary Na/K ratio are recorded also. Each figure is a mean of the number of tests indicated.

Sheep	No. of tests	Salivary Na/K	Volume ingested during test period (ml): $NaHCO_3$ solution	Volume ingested during test period (ml): Water	Salivary Na loss in previous 48 h (mmol)	Voluntary intake of $NaHCO_3$ during test period (mmol)	Water intake in previous 48 h (ml)	Blood sample before appetite test: Plasma [Na] (mmol/l)	Blood sample before appetite test: Plasma protein (g/100 ml)
Sodium depletion									
Satan	9	1.8	1722	11	842	723	8902	143	6.96
Tacitus	10	1.6	2792	14	936	885	10098	139	6.63
Somnos	6	1.8	1728	5	561	564	5585	145	7.37
Water depletion									
Satan	4	26.0	27	4158	0	11	0	160	7.41
Tacitus	5	27.7	1546	4150	0	490	0	161	7.33
Somnos	3	29.0	680	2670	0	216	0	153	7.52
Concurrent sodium and water depletion									
Satan	5	2.7	1050	1495	824	441	7010	149	7.08
Tacitus	5	2.9	1370	1303	778	434	6095	147	6.78
Somnos	3	1.0	950	2100	415	301	3270	168	7.68

After Beilharz et al. (1962)

Table 16-2. Satan. The voluntary intake of sodium bicarbonate (420 mmol/l) and water when offered for 30 min after 47.5 h of sodium depletion, water depletion or concurrent depletion of sodium and water.

Salivary Na/K	Volume ingested during test period (ml): $NaHCO_3$ solution	Volume ingested during test period (ml): Water	Salivary Na loss in previous 48 h (mmol/l)	Voluntary intake of $NaHCO_3$ during test period (mmol/l)	Water intake in previous 48 h (ml)	Blood sample before appetite test: Plasma [Na] (mmol/l)	Blood sample before appetite test: Plasma protein (g/100 ml)	Total urine Na in previous 48 h (mmol)
Sodium depletion								
1.6	2100	20	903	882	8680	142	7.24	3
1.7	2130	20	969	895	9210	144	6.67	6
1.3	1480	10	804	622	8710	146	7.52	9
1.7	1800	0	753	756	7000	143	7.24	7
1.8	1510	0	760	634	9660	138	6.67	7
3.2	770	20	1116	323	9140	142	6.67	25
1.4	1690	10	852	709	9340	144	7.04	3
1.8	1720	10	896	722	9340	148	6.67	5
1.6	2300	10	523	966	9040	142	7.04	3
Water depletion								
32.4	60	4080	0	25	0	161	7.45	—
19.3	20	4380	0	8	0	161	7.37	228
26.7	20	4460	0	8	0	164	7.45	27
25.7	10	3710	0	4	0	153	7.45	37
Concurrent sodium and water depletion								
3.6	30	5300	590	12	0	152	7.45	6
3.2	60	3940	630	25	5070	151	7.12	3
1.8	1070	800	892	449	7600	149	7.24	22
2.1	1610	1220	882	676	7660	149	6.86	5
2.8	1460	20	1126	613	7710	148	7.16	10

After Beilharz et al. (1962).

Each horizontal line in the table refers to one such experiment. The experiments were done in the order listed for each category but no systematic plan was followed in relation to sequence of categories.

about 1 min. Satan (Table 16-2) took no interest in the 420 mmol/l sodium bicarbonate solution during this time. On two occasions Tacitus drank ca 3.6 litres of the 317 mmol/l sodium bicarbonate but in the other three tests drank little or none. On one occasion Somnos drank 1.9 litres of the 317 mmol/l solution. Table 16-1 shows that mean sodium concentration in the plasma (158.8±4.84 s.d. mmol/l) was increased considerably above normal and that the salivary Na/K ratio was normal. The plasma protein concentration was higher than with sodium depletion. (The normal plasma protein of sodium-replete sheep is 6.95 g/100 ml (mean of 21 determinations; range 5.4–7.5).)

Concurrent water and sodium deficiency

In the 13 tests on concurrent deficiency the sheep drank a large amount of both fluids on ten occasions (Table 16-1). With Satan (Table 16-2), in two tests water only was drunk. In one of these water was withdrawn 48 h beforehand, instead of the usual 24 h. The salivary sodium loss was slightly less in concurrent deficiency than in simple sodium deficiency. However, voluntary sodium intake was reduced considerably in concurrent deficiency. Plasma sodium concentration was slightly elevated in this condition (149.8±4.14 s.d. mmol/l) and the salivary Na/K ratio was reduced as in simple sodium deficiency.

Fig. 16-1 records the drinking behaviour of two

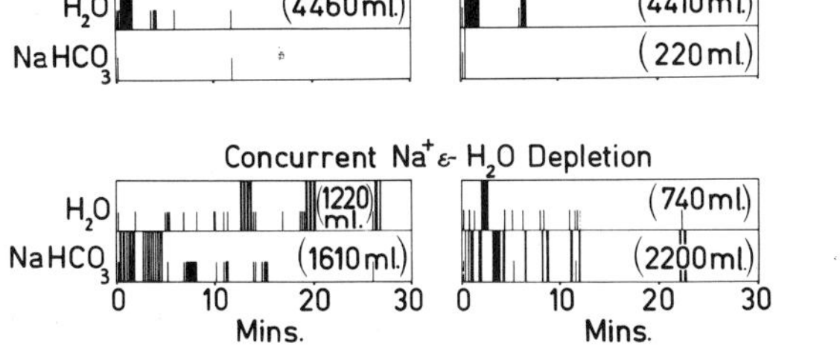

Fig. 16-1. The drinking patterns of Satan and Tacitus during a period of 30 min in which sodium bicarbonate solutions (420 and 317 mmol/l respectively) and water were offered after 47.5 h of depletion of sodium, water, or concurrent depletion of both. The figures in brackets indicate the volume consumed of a particular solution. Full-length vertical lines represent drinking acts for the period of time indicated on the abscissa; half lines indicate acts of sampling the solution, i.e. smelling and tasting but not drinking. (After Beilharz et al. 1962.)

of the animals, Tacitus and Satan, during the test period as an example of each type of depletion. It shows that, with the simple deficiency states, the specific deficiency was paralleled by predominant attention to the container holding the appropriate fluid. There was a considerable reduction of sampling and hence presumably motivation following a large initial intake of the fluid. With concurrent deficiency the attention was directed to both containers.

In an experiment with PF48 in which greater deficiencies were developed over 72 h, similar drinking behaviour resulted (Beilharz et al. 1962). In concurrent water and sodium deficiency plasma sodium concentration rose, and plasma volume decreased from 2.07 to 1.51 litres. In sodium deficiency plasma volume decreased from 2.02 to 1.69 litres.

In four other animals—TP9, Ned, Jack and TP1—with concurrent depletion the behaviour was examined by sequential presentation of solutions. For example, Ned had a total salivary loss of 4.9 litres containing 730 mmol sodium. When offered water he immediately drank 0.52 litre at 26 ml/s. The sheep appeared to be still excited and restless but did not drink any more water. Two minutes later the sodium bicarbonate solution was offered and he drank 1.28 litres at 41 ml/s, and shortly afterwards 0.24 litre at 22 ml/s. The total sodium intake was 456 mmol. The experiment on TP1 gave further evidence of the large volume of fluid a sheep will ingest voluntarily in a short time. At the end of the period of depletion it was offered water and immediately drank 3.32 litres in 72 s without pause, i.e. 47 ml/s. The water was removed as soon as drinking ceased. When it was offered again 1 min later the sheep tasted it but did not drink. Thus it had satisfied its thirst with the one long draught. Sodium bicarbonate solution was then offered and the sheep immediately drank 1.25 litres (529 mmol sodium) in 27 s, i.e. 47 ml/s.

These experiments have shown that a sodium-deficient sheep still exhibited an appetite for sodium solutions when the plasma sodium concentration was increased by concurrent depletion of water. It voluntarily ingested a sodium solution with two to three times greater osmotic pressure than the blood. Thus the simple proposition that a high osmotic pressure causes a water appetite and a low osmotic pressure a salt appetite, would not explain the observation of concurrent appetite for water and sodium solutions. The two mechanisms are separate.

The observation of a sodium appetite when plasma sodium concentration was raised above normal by concurrent water depletion can be interpreted as evidence against plasma sodium concentration being the main immediate stimulus of sodium appetite. A similar conclusion came from data of Beilharz and Kay (1963) and Beilharz et al. (1965), as discussed in Chapter 20.

Voluntary sodium ingestion occurred despite a considerable increase above normal of plasma sodium concentration. However, a different issue enters into interpretation of the findings in concurrent deficiency as has been advanced with the parallel problem of the stimulus to aldosterone secretion in concurrent sodium and water depletion (Denton 1957). It is possible that a significant sodium deficit normally causes a fall in extracellular sodium and a change in the concentration gradient between extracellular sodium and an intracellular component (e.g. sodium content of specific neural elements). Change in the intracellular component results. Water depletion, on the other hand, would cause an equivalent concentration change in both intracellular and extracellular compartments. Thus the superimposed sodium deficit could still distort the gradient in the concurrently water-deplete animal, as in uncomplicated sodium deficiency.

In the three sheep Tacitus, Somnos and Satan, with simple sodium deficiency the voluntary intake of sodium replaced the deficit. But the data on these three animals suggest that with concurrent deficiency voluntary sodium intake was less than the sodium deficit. The results with PF48, Ned and Jack indicate this also, but with PF33 there was no difference, and TP9 drank an amount of sodium greater than the deficit.

Many workers have discussed whether in sodium deficiency some consequence of decreased intravascular volume stimulates aldosterone secretion. The same stimulus can be considered in relation to evocation of sodium appetite. The studies of water depletion and sodium depletion by Elkinton, Danowski and Winkler (1946) showed that a decrease in plasma and extracellular volume occurred with each deficiency. Circulatory deterioration was greater with sodium deficiency. In water deficiency plasma volume declined with extracellular volume, but in sodium deficiency the decline in plasma volume was proportionately greater. The haemodynamic measurements made in our experiments were not comprehensive. In the series on PF48 it was shown that a 15% decrease in plasma volume occurred in sodium deficiency and a 25% decrease with concurrent water and sodium deficiency. No measurement of plasma volume was made in the water deficiency experiment, but the changes in the haematocrit and plasma protein, as in the series with Tacitus, Somnos and Satan, indicate a reduction in plasma volume (Blair-West et al. 1972). Thus this effect is common to all three experimental situations. If decrease in blood volume, causing either stimulation, or reduction of inhibitory tone, of a stretch receptor were proposed as the stimulus of sodium appetite in sodium deficiency, it would be necessary to suppose the effect to be inhibited in water deficiency by the central osmotic effect. However, this argument would not seem plausible as regards concurrent sodium and water deficiency, where sodium ingestion takes place despite the central effect of increased osmotic pressure. These considerations are supplementary to more direct evidence on manipulation of plasma volume where there is absence of direct effect on sodium appetite (Denton et al. 1971; see Chapter 15).

In relation to consummatory behaviour, some quite important general facts emerge. In the case of depletion of either water or sodium alone the usual finding was that the drive was satisfied over a period of 10 min. Attention was predominantly or exclusively directed to the fluid in which the animal was deficient. The critical point emerging here from analysis of drinking pattern, including the instances where the solutions were presented in sequence, was that in concurrent deficiency the act of drinking either fluid, which caused oesophageal proprioception and then ruminal distension, eventually decreased drinking rate and interest in that fluid. But it did not remove the drive for satiation of the other deficiency. The proprioceptor and motor elements were common to the consummatory act in both instances, but the specificity of the two appetites was clear. The difference in the satiating process appeared to reside in recognition through taste. Additional evidence on specificity came from the behaviour of an animal with concurrent depletion when the fluids were presented in sequence. After ingesting a large volume of the one offered first, interest in it was lost. But the animal continued to be restless and to show the characteristic anticipatory behaviour when the possible presentation of a second container was indicated. These concurrent depletion experiments in which the solutions were presented in sequence raise an additional proposition. The instances where the animal drank immediately 3–4 litres of water but refused any more, and then drank 1–1.5 litres of sodium bicarbonate solution as soon as it was offered, suggest that the factor determining the end point of water ingestion was not simple ruminal distension. In any event, in the course of experiments on severe dehydration (Blair-West et al. 1972) or satiation of severe sodium deficiency when dilute sodium bicarbonate (100 mmol/l) was presented (Denton 1966), sheep were observed to drink rapidly as much as 5.5–6.0 litres. But it cannot be concluded that the existence of the concurrent deficiency was without influence on satiation of one or the other deficit. The decrease in sodium intake below usual for deficit in a number but not all of the experiments on concurrent deficiency was noted above.

Bar-press experiments on sheep with concurrent water and sodium deficiency

This question has been further examined by Dr Richard Weisinger of the Howard Florey Institute, and the outcome is consistent with the proposal of Neal Miller (1967) that looking at consummatory behaviour in several ways has evident virtues.

Weisinger (unpublished) studied concurrent water and sodium deficiency by bar-press technique. The methods were similar to those described in Chapters 13 and 20 where operant behaviour of sodium-deficient animals and the influence on it of rapid change in sodium balance are described. The basic experimental condition was that after 22 h of sodium deficiency as a result of parotid salivary loss, the animals were allowed to bar-press for 2 h for 15 ml deliveries of 600 mmol/l sodium bicarbonate (=9 mmol per delivery). They usually delivered and drank ca 40 deliveries (=360 mmol), which replaced salivary loss of 2–3 litres.

The condition on experimental days was that the animals were also water-deprived for 24 or 48 h beforehand. There were three experimental series:

1) Fifteen minutes before bar-pressing for sodium bicarbonate commenced the animals were allowed partly to satiate water deficiency by drinking a fixed amount of 2.0 litres.
2) Under the same conditions they were allowed fully to satiate water deficiency. The mean intake of the six animals was 3.108±0.197 (s.e.m.) litres.
3) Bar-pressing was allowed for the hypertonic sodium bicarbonate (600 mmol/l) in the presence of either 22- or 46-h concurrent water deficiency. In the case of 46-h deficiency, free access to water was permitted immediately after cessation of bar-pressing for 120 min (mean water intake 3.704±0.076 (s.e.m.) litres).

Figs. 16-2 and 16-3 record the results with regimes 1, 2 and 3. Allowing the water-deplete animal to drink water immediately before the operant test reduced intake of sodium over the time permitted by about 60% (Fig. 16-2). There

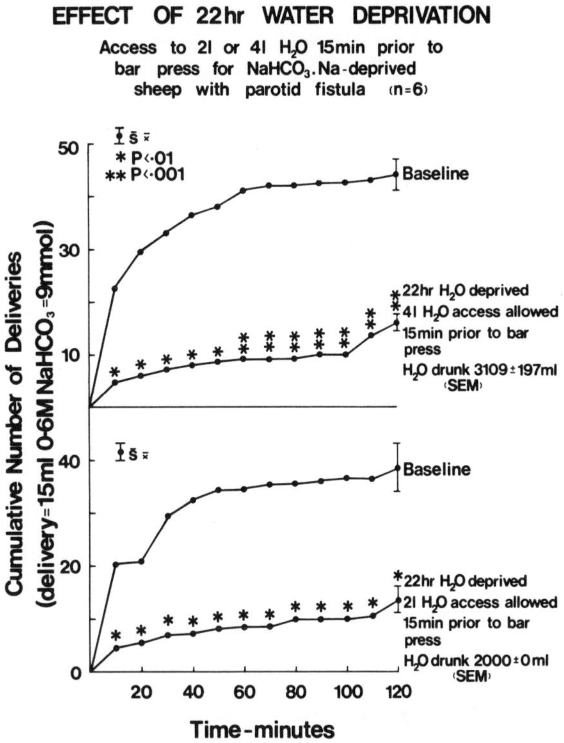

Fig. 16-2. The effect of 22 h of water deprivation on voluntary sodium intake by bar-pressing in sheep depleted of sodium for 22 h by parotid fistula loss. In the upper portion of the graph the animals were permitted access to water 15 min before bar-pressing and mean intake was 3.1 litres. A highly significant reduction in sodium intake occurred. In the lower portion the animals were allowed access to only 2 litres of water 15 min before bar-pressing, and mean intake was 2 litres. A significant reduction in bar-pressing for sodium solution again occurred.

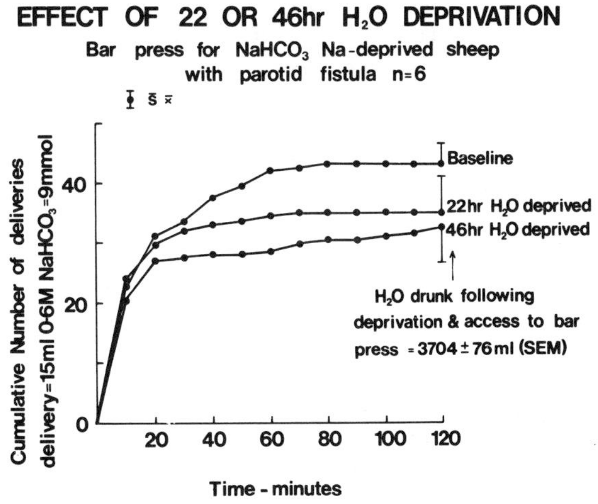

Fig. 16-3. The effect of 22 or 46 h of water deprivation on voluntary sodium intake by bar-pressing in sheep depleted of sodium for 22 h by parotid fistula loss. The change in bar-pressing relative to the water-replete baseline was not statistically significant in either instance. The amount of water drunk following the end of bar-pressing is shown.

was no difference whether water intake was 2.0 litres or full satiation (where the mean was 3.1 litres) (Fig. 16-2). This might point to the act of prior satiation, full or partial, being more influential than the actual degree of distension of the rumen. The concurrent condition of 22 or 46 h of water deprivation influenced bar-pressing for 600 mmol/l sodium bicarbonate much less, reduction of sodium appetite being by about 25% (Fig. 16-3); the effect was not statistically significant. The partial satiation of sodium appetite did not obviously change the satiation of water deficiency when free access was permitted at the end of bar-pressing, when a mean of 3.7 litres was drunk (Fig. 16-3). Though greater numbers of experiments may be required to allow a firm quantitative conclusion, the data indicate that the act of prior satiation of water deficiency has more influence on sodium appetite in the instance of concurrent deficiency than has the contemporaneous existence of the thirst drive itself. The emphasis on the act of satiation rather than volume in rumen is consistent with Beilharz and Kay's (1963) data on the insignificant influence on satiation of sodium appetite of tubing of 1.5 litres of water into the rumen 15 min before allowing access to sodium.

Overall these operant behaviour data are consistent with those from free access experiments in regard to there being some reduction of sodium intake in concurrent water and sodium deficiency, but they emphasize the influence of the act of prior satiation of thirst on operant behaviour for sodium.

Concurrent water and sodium deficiency in the rat

An examination of sodium appetite in concurrent water and sodium deficiency in rats was made by Wolf and Stricker (1967) in the course of examining their 'reservoir' hypothesis of sodium appetite. Sodium appetite is postulated as being controlled by a receptor with the properties of a sodium reservoir. Appetite is stimulated when the sodium content of a receptor is reduced by hypovolaemia or hyponatraemia. It is postulated that in hypovolaemia the vector is increased blood aldosterone concentration which potentiates extrusion of sodium from cells (Crabbé 1961). Hyponatraemia may also increase aldosterone secretion, as well as reducing the sodium of the 'reservoir' receptor by increasing nett passive diffusion of sodium out of it. The hypothesis postulates an important function for aldosterone but allows regulation without the adrenal glands via hyponatraemic influences. Wolf and Stricker (1967) showed that polyethylene glycol (PEG) elicited sodium appetite and thirst without hyponatraemia in intact rats. The first experiment in their 1967 paper showed that PEG elicited sodium appetite in adrenalectomized rats. As the rats had to have maintenance steroid replacement to withstand PEG treatment (2.5 mg of desoxycorticosterone trimethylacetate every 2 weeks and 1.0 mg of cortisone daily) they noted that reduced metabolic clearance rate of steroid and thus increased plasma level could not be excluded in the absence of direct measurement.

They then examined concurrent water and sodium deficiency in adrenalectomized rats, and compared the result with sodium deficiency alone and with control conditions. The rats had no corticoid replacement therapy. Solutions offered were water and 1.0 M sodium chloride, which is normally aversive to the rat. Approximately the same amount of sodium was lost under the two deprivation conditions and there was no significant difference between the amounts of saline ingested. Both were significantly greater than with no deprivation (Table 16-3). Water intake was significantly increased in the concurrent deprivation (Table 16-3). With sodium deprivation and free access to water a significant hyponatraemia developed with a small reduction in blood volume. With concurrent depletion blood volume reduction was apparently greater, but no significant reduction in plasma sodium concentration occurred.

The authors suggest that the results of both experiments weigh against the hypothesis that hypovolaemia elicits sodium appetite solely via aldosterone, and that adrenalectomized rats regulate their sodium intake on the basis of extracellular sodium concentration alone. At this point they raised the question of renal renin as a vector of sodium appetite, a theory not supported by some later evidence (Bott et al. 1967; Fitzsimons and Stricker 1971). The issue is considerered in Chapter 22. However, the data would not seem to preclude the hypothesis (Denton 1957; Beilharz et al. 1962) noted above—that in water-deplete animals concurrently sodium-deplete, a differential gradient resulting in loss of intracellular sodium from brain cells could occur as in uncomplicated sodium deficiency.

A detailed study of concurrent deficiency has been made by Contreras and Hatton (1975), and it clearly reveals interesting facets of drinking behaviour. They note that consummatory behaviour in rats is intermittent, as they drink in bursts,

Table 16-3. Cumulative intake of 1.00 M sodium chloride solution and water following three deprivation conditions in adrenalectomized rats.

	Saline intake (ml)						Water intake (ml)					
	5 min		10 min		15 min		5 min		10 min		15 min	
Deprivation condition	Mean	s.e.	Mean	s.e.	Mean	s.e.	Mean	s.e.	Mean	s.e.	Mean	s.e.
None	0.00	0.00	0.00	0.00	0.00	0.00	0.60	0.48	0.90	0.79	0.90	0.70
Sodium	0.64	0.18	0.81	0.18	0.98	0.21	0.25	0.15	0.50	0.23	1.30	0.40
Sodium+water	0.47	0.15	0.75	0.22	1.01	0.28	2.20	0.71	4.40	1.11	6.25	1.08

From Wolf and Stricker (1967), with permission.

pause and drink again. In their experiment rats on a sodium-deficient diet and those with adequate sodium intake were compared when water-replete and after 23 h of water deprivation. They found that when the animals had 1 h of access to water and 0.4 M sodium chloride, the total time spent drinking salt solution, and the total salt intake, were significantly greater for sodium-deficient animals ($P<0.01$). The water-deprived sodium-deplete animals did not spend less time drinking nor drink less 0.4 M sodium chloride than the sodium-deplete animals with free access to water (Table 16-4). When drinking time for first encounter with salt solution was computed it was found to be significantly longer for sodium-deprived animals whether water-replete or water-deplete. The distribution of time spent drinking salt and water solutions is set out in Fig. 16-4, the analysis being based on whether the animal started drinking salt or water first, and its metabolic status. From a total of 12 water-replete rats, nine started with salt solution. There were nine water-deprived rats out of 16 which did the same. Qualitatively, the time spent with saline gradually declined during the test period. The decline, generally, was more sudden in controls. Contreras and Hatton (1975) remark that sodium-deficient rats drank more often and consecutively for longer stretches of time than controls.

Table 16-4. Characteristics of drinking by water-replete and water-deprived rats.

	Control (sodium-replete)		Experimental (sodium-deficient)	
	Water	NaCl	Water	NaCl
Water-replete				
A	2.008	3.492	2.933	11.158
B	3.000	2.833	4.167	11.833
Water-deprived				
A	9.344	3.969	8.163	10.188
B	15.375	6.000	12.500	14.875

From Contreras and Hatton (1975), with permission.
A, Total time spent drinking (min).
B, Total fluid intake (ml).

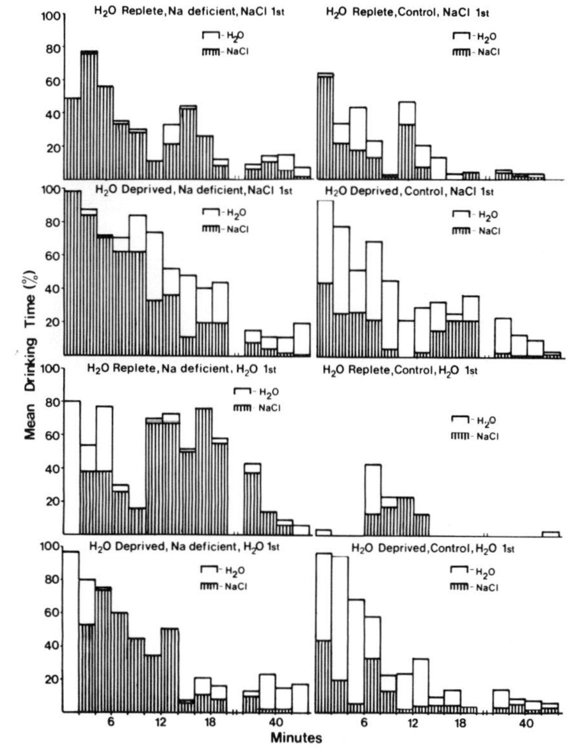

Fig. 16-4. The mean proportion of time spent drinking distilled water and 0.4 M sodium chloride as a function of time, plotted for animals that were either sodium-deficient or normal controls, and were either water-deprived or water-replete. The animals are grouped according to which fluid they drank first. (From Contreras and Hatton 1975, with permission.)

References

Abraham SF, Denton DA, Weisinger RS (1976) The specificity of the dipsogenic effect of angiotensin II. Pharmacol Biochem Behav 4: 363

Beilharz S, Kay RNB (1963) The effects of ruminal and plasma sodium concentrations on the sodium appetite of sheep. J Physiol (Lond) 165: 468

Beilharz S, Denton DA, Sabine JR (1962) The effect of concurrent deficiency of water and sodium on the sodium appetite of sheep. J Physiol (Lond) 163: 378

Beilharz S, Bott E, Denton DA, Sabine JR (1965) The effect of intracarotid infusions of 4M NaCl on the sodium drinking of sheep with a parotid fistula. J Physiol (Lond) 178: 80

Blair-West JR, Brook AH, Simpson PA (1972) Renin responses to water restriction and rehydration. J Physiol (Lond) 226: 1

Bott E, Denton DA, Weller S (1967) The effect of angiotensin II infusion, renal hypertension and nephrectomy on salt appetite of sodium deficient sheep. Aust J Exp Biol Med Sci 45: 595

Coghlan JP, Denton DA, Goding JR, Wright RD (1960) The control of aldosterone secretion. Postgrad Med J 35: 76

Contreras RJ, Hatton GI (1975) Gustatory adaptation as an explanation for dietary-induced sodium appetite. Physiol Behav 15: 569

Crabbé J (1961) Stimulation of active sodium transport across the isolated toad bladder after injection of aldosterone. Endocrinology 69: 673

Denton DA (1957) The effect of variation of the concentration of individual extracellular electrolytes on the response of the sheep's parotid gland to sodium depletion. J Physiol (Lond) 140: 129

Denton DA (1966) Some theoretical considerations in relation to innate appetite for salt. Conditional Reflex 1: 144

Denton DA, Orchard E, Weller S (1971) The effect of rapid change of sodium balance and expansion of plasma volume on sodium appetite of sodium-deficient sheep. Commun Behav Biol 6: 245

Elkinton JR, Danowski TS, Winkler AW (1946) Haemodynamic changes in salt depletion and in dehydration. J Clin Invest 25: 120

Fitzsimons JT, Stricker EM (1971) Sodium appetite and the renin–angiotensin system. Nature New Biol 231: 58

Miller NE (1967) Behavioural and physiological techniques: Rationale and experimental designs for combining their use. In: Code C (ed) Handbook of physiology, Sect 6: alimentary canal, vol 1. American Physiological Society, Washington, p 51

Scoggins B, Coghlan JP, Denton DA, Fan JSK, McDougall JG, Oddie CJ, Shulkes AA (1974) Metabolic effects of ACTH in sheep. Am J Physiol 226: 198

Stellar E, Hyman R, Samet S (1954) Gastric factors controlling water and salt solution drinking. J Comp Physiol Psychol 47: 220

Wolf G, Stricker EM (1967) Sodium appetite elicited by hypovolaemia in adrenalectomized rats: Re-evaluation of the 'reservoir' hypothesis. J Comp Physiol Psychol 63: 252

Young PT, Chaplin JP (1949) Studies of food preference, appetite, and dietary habit. 10. Preferences of adrenalectomized rats for salt solutions of different concentrations J Comp Physiol Psychol 19: 45

Young PT, Falk JL (1956) The relative acceptability of sodium chloride solutions as a function of concentration and water need. J Comp Physiol Psychol 49: 469

17 Hepatic sodium receptors and their possible influence on salt appetite

Summary

1. There is evidence for receptor elements in the hepatic portal circulation which influence sodium excretion, and also the intake of water and isotonic saline by water-deprived animals.

2. To date there are no data bearing upon their influence, if any, on satiation of salt appetite by sodium-deficient animals.

Hepatic influence on sodium excretion

The concept of receptors in the hepatic portal circulation as a first line of defence against sudden rises in sodium intake (Passo et al. 1973) is one facet of a general hypothesis of hepatic influence on sodium regulation. Investigation has embraced the possibility of a humoral factor of hepatic origin influencing sodium excretion. Glaubach and Molitor (1928) observed an increase in sodium excretion following injections of liver extracts. Oettel (1943) noted injection of liver extract caused a diuresis in patients with liver cirrhosis, and Milies (1960) obtained an increase in sodium excretion in dogs infused with concentrated liver perfusate.

Daly et al. (1967) compared the natriuretic effects of infusion of 5% saline into the portal and systemic veins of anaesthetized dogs. On the calculated rate of liver blood flow, the portal infusion would have raised portal plasma sodium concentration by about 4 mmol/l. In 14 of 16 experiments, sodium excretion after portal infusion exceeded that following similar systemic infusion. Moreover in nine of these 14 experiments inulin clearance was significantly reduced. The authors considered these results evidence for a humoral influence and that sodium concentration in the portal vein was the stimulus to its production. Ratification would lend support to the idea that hepatic damage might result in failure of such a mechanism with subsequent sodium retention and oedema.

Another line of evidence suggestive of splanchnic input in sodium regulation comes from the work of Peart, Carey and colleagues. First it was shown with sodium-deplete conscious rabbits that gastric sodium loading produced a greater natriuresis than intravenous administration (Lennane et al. 1975; Carey et al. 1976). Carey (1978) then showed that normal male subjects on a 10 mmol Na/day diet excreted a greater amount of sodium when sodium load was given orally than when the same load was given intravenously. The effect was significant by 2 h, and was not dependent on any difference in aldosterone response. A similar result was seen when the same experiments were made on six patients with primary hyperaldosteronism. However, Gordon and Peart (1979) showed that an intravenous load of sodium was completely excreted by the kidney in most subjects even when oral sodium intake was being reduced. The change in oral sodium intake might be expected to attentuate or extinguish the renal response to intravenous sodium if a sodium intake monitor (Lennane et al. 1975) were relevant to the response of the kidney at the time when dietary sodium is reduced.

Passo et al. (1973) found that in anaesthetized male cats, urine flow, sodium excretion rate and percentage filtered sodium excreted rose significantly 30 min after the initiation of portal vein infusion of 1 M sodium chloride, whereas there was no change with systemic infusion of the same load. Portal vein infusion of 1.68 M sucrose caused some rise in sodium excretion but the effect was not different from that caused by systemic infusion. As will be noted further on, the effect of infusing 1 M sodium chloride into the portal vein was abolished by bilateral vagotomy. Blake and Ulfendahl (1979) obtained some evidence that hepatic portal vein infusion of 1 M glucose increased sodium excretion by the kidney, and the effect was abolished by bilateral vagotomy. These effects of portal infusion were not consistent with earlier work in conscious dogs by Schneider et al. (1970) or Potkay and Gilmore (1970).

Receptors in the portal vein

Alternative or complementary to the notion of a humoral factor, there is evidence for neural receptors located in the liver which are sensitive to increased sodium concentration in the blood of the hepatic portal vein. Electrophysiological studies point to this. Niijima (1969) perfused isolated livers of guinea pigs while monitoring nerve bundles in the hepatic branch of the vagus. Altering the osmolarity of the perfusing fluid by changing the sodium chloride concentration caused a proportionate change in the frequency of nerve impulses.

Andrews and Orbach (1974) perfused rabbit livers through the portal vein and monitored afferent impulses in fine bundles of the hepatic nerves. If the molarity of the solution was varied over a 15 mM range by keeping salts in the same relative proportions and dissolving them in 5% more or less water, the rate of spontaneous discharge was unaltered. With a solution containing 8.5 mM extra sodium chloride the rate of nervous discharge was increased in three of 40 bundles tested, and with a solution deficient in sodium chloride by 11.4 mM the rate diminished in ten of 60 bundles. Addition of mannitol or sucrose did not restore the rate of discharge but sodium sulphate or lithium chloride did. The authors suggest the data support the idea of hepatic sodium receptors activated by imbalance of sodium with other cations and acting through nerves. Schmitt (1973) and Schmitt and Koizumi (1971) reported that injections of hypertonic sodium chloride or glucose into the portal vein of the rat increased or decreased the firing rate of neurones in the lateral hypothalamus. Section of the splanchnic nerves bilaterally or cord transection at the level of T5 abolished the response. Severing both vagi did not abolish but enhanced the response. No response to the injections was found in supraoptic or paraventricular or ventromedial nuclei.

There has been controversy in this area of electrophysiology in addition to that about the effects of portal infusion upon sodium excretion. Adachi et al. (1976) maintain the receptors are not specifically sodium-sensitive but respond to changes in osmotic pressure.

Rogers et al. (1979) report physiological and anatomical evidence of hepatovagal afferent flow through the dorsal part of the gustatory pathway. Numerous cells in or near the ventroposteromedial nucleus of the thalamus (VPM) responded differentially to hepatic portal infusions. The responses were of two general types. The first (Fig. 17-1) type of neuronal unit (Fig. 17-1B) responded only to hepatic influx of sodium or choline chloride. Potassium chloride and non-ionic, hyperosmotic solutions were without effect. The second

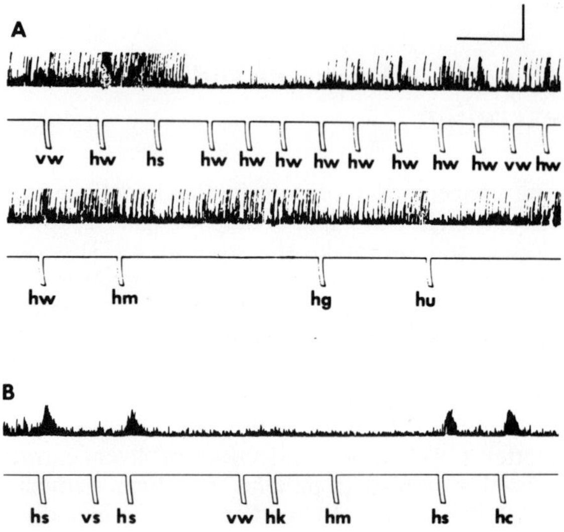

Fig. 17-1. Typical responses of hepatically activated VPM units. A: record of firing rate vs time and infusion condition. VPM neurone activated by hepatic water infusions, inhibited generally by hepatic hyperosmotic solutions. Note inhibition of water activation by preceding infusion of hypertonic sodium chloride. Abbreviations: vw, vena cava water; hw, hepatic water; hs, hepatic 0.3 M sodium chloride; hm, hepatic 0.6 M mannose; hg, hepatic 0.6 M glucose; hu, hepatic 0.6 M sucrose. Infusion volume 0.15 ml in all cases. B: VPM unit activated only by hepatic infusions of sodium chloride and choline chloride. Infusion volume was 0.15 ml in all cases. Abbreviations: hs, hepatic 0.3 M sodium chloride; vs, vena cava 0.3 M sodium chloride; vw, vena cava water; hk, hepatic 0.3 M potassium chloride; hm, hepatic 0.6 M mannose; hc, hepatic 0.3 M choline chloride. Calibration bars: 30 spikes/s, 5 min. (From Rogers et al. 1979, with permission.)

type of thalamic cell (Fig. 17-1A) responded to changes in portal osmolarity. That is, the cell was inhibited by both ionic and non-ionic hyperosmotic solutions infused into the portal vein but was activated by the presence of water in the hepatic portal vein. Both types of cells seemed evenly distributed throughout the VPM. The results suggest that both osmoreceptors and sodium receptors in the liver activate neurones in the ventral sensory thalamus.

Hepatic receptors and thirst

In relation to thirst there are some rather contradictory findings. Kozlowski and Drzewiecki (1973) found in dogs that infusion of water into the hepatic portal vein elevated the threshold for drinking to cellular dehydration, and that this effect was abolished by vagotomy. On the other hand, Wood et al. (1977) measured water intake in dogs given intravenous infusion of hypertonic saline and found that with bilateral concurrent intracarotid infusion of water, which prevented a rise in central osmolarity during the infusion, water drinking did not occur. Thus peripheral hyperosmolality alone at levels which occur after water deprivation is not sufficient for drinking. Rolls et al. (1980) found that in water-deplete dogs, bilateral intracarotid infusion of water at a rate which reduced osmolality to the predeprivation level reduced water intake by 72%. They found in addition that infusion of 0.15 M (isotonic) saline in an amount to restore extracellular volume to predeprivation level reduced drinking by 27%. They suggest that in rat, dog and monkey about 64%–85% of the drinking that follows water deprivation is due to cellular dehydration, hypovolaemia accounting for only 5%–27% of the intake. This is substantiated by the finding by Adolph et al. (1954) that for an equivalent rise in plasma sodium concentration, water deprivation causes more water drinking than does intravenous infusion of hypertonic saline. Furthermore, in other experiments involving nephrectomy or ureteric ligation after water deprivation, Rolls and Wood (1977) found no evidence for participation of the renin–angiotensin system in this category of thirst. Infusions of saralasin into the cerebral ventricles of water-deprived dog (Ramsay and Reid 1975), sheep (Abraham et al. 1976) and rat (Lee et al. 1978) support this view.

These data provide important background to the studies on the influence of hepatic portal infusion upon saline and water intake.

Influence of hepatic portal infusion on saline and water intake

Lin and Blake (1971) made very interesting observations on rats with indwelling catheters implanted in the hepatic portal vein. The intake of water and 0.15 M sodium chloride was studied after 24 h of deprivation of all fluids and infusion of hypertonic (2 M) sodium chloride or equiosmolar (3.8 M) sodium-free solution for 30 min either by portal vein or into the inferior vena cava (IVC). The rate of infusion was calculated to cause an increase in sodium concentration of hepatic portal blood within the physiological range. Drinking was allowed 15 min after beginning the infusion, and access was allowed for 30 min. It was shown that infusion of hypertonic sodium chloride into the hepatic portal vein depressed drinking of 0.15 M sodium chloride solution, but not of water. IVC infusion of the same quantity of hypertonic saline did not have this effect, and neither did the infusion of equiosmolar sodium-free solution into the hepatic portal circulation. Other controls included infusions of isotonic saline and glucose solutions into the hepatic portal vein, and also no infusion at all. There was a general reduction in fluid intake during infusion of hyperosmotic solutions into the portal vein and the authors note that this was difficult to understand. An increase in the fluid intake might have been expected and they raise other considerations such as the stimulation of pain receptors in the liver capsule. They had no behavioural indications of pain with 2 M solutions, but occasionally opisthotonus and fixation of position occurred when a 4 M sodium chloride solution was used.

Following this initial study, Blake and Lin (1978) reported a more extensive investigation of the phenomenon where five groups of rats were submitted to different procedures. The conditions of experiment were essentially similar to the earlier study. They calculate, on the basis of the most recent estimates of hepatic portal plasma flow in the rat of about 6 ml/min (Larsen et al. 1976), that infusion of 2 M sodium chloride solution into the hepatic portal vein would have increased sodium concentration by at least 5.6 mmol/l above the control level. Hepatic portal vein infusion of 2 M and 1 M sodium chloride consistently decreased saline intake and preference, compared either with sham or IVC infusion of the same solution. There were no consistent effects on water intake (Fig. 17-2).

After right cervical vagotomy, hepatic portal infusion of 2 M sodium chloride failed to decrease saline drinking or saline preference (Fig. 17-2).

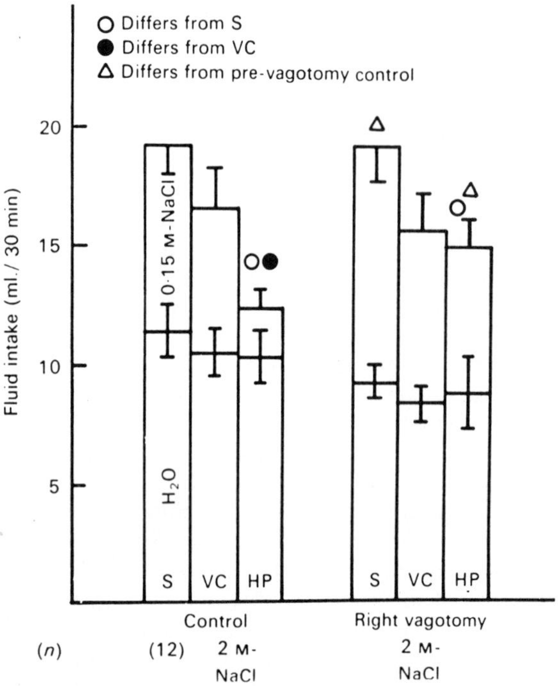

Fig. 17-2. Effect of right cervical vagotomy on saline and water drinking responses to sham (S), vena cava (VC), and hepatic portal (HP) infusion of 2 M sodium chloride solution. Values are the mean±s.e.m. (vertical lines). Differences were considered significant when $P<0.05$ by the method of paired comparisons. (From Blake and Lin 1978, with permission.)

IVC infusion of 2 M or 1 M saline solution decreased saline drinking compared with sham infusion in one group of rats, but not in two others. This effect, then, was inconsistent. Hepatic portal infusion of 2 M glucose consistently increased saline intake and preference, and decreased water intake. When given by IVC, neither 2 M nor 1 M glucose had any consistent effect on saline or water drinking. However when 16 mM or 32 mM potassium chloride was added to the (hypertonic) 1 M sodium chloride solution, hepatic portal infusion had no significant effect on saline drinking or preference compared with sham or IVC infusion. When 32 mM potassium chloride was added instead to isotonic saline, IVC infusion increased saline drinking and preference and decreased water intake compared with sham infusion. That is, potassium chloride blocked the effect of hepatic portal infusion of 1 M sodium chloride in reducing saline intake but, at sufficiently high concentration, increased saline intake when the liver was bypassed.

Hepatic portal infusions of 0.7 mM ouabain, which was calculated on the basis of portal blood flow to give a concentration reaching the tissues of at least 1.9×10^{-6}M, sufficient partially to inhibit neural Na,K-dependent ATPase (Ahmed and Judah 1964), depressed saline intake and preference compared with sham or IVC infusion. So did infusion of 20 mM EDTA (a calcium-chelating agent). Water intake was also depressed by both relative to sham infusion. IVC infusion of either drug also depressed saline intake, but not to the same extent as did hepatic portal infusion, and had no significant effect on saline preference. If it be accepted that these effects on intake were mediated by neural receptors which transmitted information to the central nervous system via the right vagus, the question arises as to the nature of the nerve terminals or receptors involved. As indicated above, there has been considerable controversy as to whether such receptors are sodium-sensitive specifically, or are osmoreceptors. Lack of an effect with 2 M fructose and 2 M sucrose (Blake and Lin 1978) speaks against a hepatic portal osmoreceptor in control of saline drinking.

Blake and Lin (1978) suggest on the basis of the findings of Andrews and Orbach (1973) that these putative receptors are spontaneously discharging neurones. Thus the rate of firing might be altered in a number of ways: e.g. a depolarizing effect due to sodium influx, or change in membrane permeability to sodium, or by inhibition of an electrogenic sodium pump. They remark on the parallel issue of the postulated hypothalamic sodium receptors and note it has been proposed (Denton 1966; Mouw and Vander 1970) that changes in intracellular sodium concentration might alter hypothalamic cell excitability. Thus influx of sodium would inhibit cell activity in the neurones stimulatory to sodium intake. Conversely loss of sodium from the cells, caused by sodium depletion, would excite cell activity. Blake and Lin note that the data on the inhibition by ouabain of sodium bicarbonate drinking in sodium-deplete sheep are consistent with the concept (Denton et al. 1970), but suggest it would be equally plausible to suggest that sodium influx increases cell excitability in neurones inhibitory to sodium intake. This would be consistent with the findings on peripheral receptors and other data on the excitatory action of ouabain on neurones or their terminals (Saum et al. 1976; Sokolove and Cooke 1971).

In relation to the hepatic peripheral receptors, they suggest that the spontaneous discharge rate of the receptors would be determined, at least in part, by the degree of hyperpolarization of the receptor membrane, created by an electrogenic pump. The inhibition of saline drinking by ouabain may be explained as the result of increased receptor discharge secondary to membrane depolarization by partial inhibition of the electrogenic sodium pump. In relation to the action of potassium chloride in blocking the action of sodium chloride, a stimulatory effect of slightly elevated external potassium on the sodium pump could have hyperpolarized the receptor membrane and suppressed spontaneous activity.

Though these experiments break interesting new ground, they do not, as Blake and Lin point out, throw light on the physiological significance of the postulated receptors in the regulation of water and salt balance. The authors note that the variable effect seen on water intake with infusions of glucose, potassium chloride or any of the three drugs used may have been non-specific, i.e. an inconsistent and generalized behavioural response to a 'disturbing' experience—especially when decreased water intake occurred with decreased saline intake to decrease the total fluid intake. Even the clear-cut decreased water intake associated with increased saline intake (as with hepatic portal infusion of 2 M glucose or IVC infusion of 32 mM potassium chloride in 0.15 M sodium chloride) may have been non-specific, in that water intake was neglected if saline drinking was stimulated. There was frequent suppression of water intake without increase in saline intake, so that it would not appear that saline intake was increased by depression of water intake. Tentatively, the receptors might be looked upon as controlling saline drinking rather than water drinking. As the authors emphasize, experiments on the drinking of aversive hypertonic saline solutions under conditions of sodium depletion would be required before any firm conclusions regarding receptor specificity could be made. In this regard, it may be noted that the interesting results have to date been obtained under the special conditions of offer of water and isotonic saline following 24 h of fluid deprivation.

There are other questions that arise from consideration of possible physiological significance of these data as regards the control of salt appetite. There is logical attraction in the idea that an increasing sodium concentration in portal vein blood might inhibit the consummatory act of satiation in a sodium-deficient animal. But there are data on behaviour which are difficult to reconcile immediately with this. For instance, sodium-deficient sheep offered, for example, 300 mmol/l sodium bicarbonate solution, may drink adequate to correct deficit in 2–3 min and a precipitate decline in motivation follows. Control of the satiating act is evidently central with feedback from tongue, pharynx and oesophagus. The time course would be too short for any influence of a change in sodium concentration in portal blood to be operative. Furthermore, administering 600 mmol of sodium bicarbonate in 1.5 litres of water into the rumen 10 min before offer of sodium bicarbonate has little influence on the consummatory act of satiation of appetite commensurate with deficit (Beilharz and Kay 1963). Whether or not by this stage any rise in portal sodium concentration would have occurred can only be determined by sampling from indwelling cannulae in the conscious animal. But, the data tend to substantiate strongly that the control of sodium drinking is governed by afferent flow from the mouth, pharynx and oesophageal receptors. In the case of the rat, Jalowiec and Stricker (1973) have shown drinking in the sodium-deficient animal is rapid. Indeed the rat usually overcompensates considerably compared with deficit, and this is clear-cut after 15 min of access to solutions (Fig. 14-3). The rise in portal sodium concentration could be an element in turning off drinking in this species after adequate time has elapsed. In Chapter 14, which deals with the termination of the consummatory act and the satiation of thirst in water-deplete animals, the evidence of Feider (1972) and Blass and Hall (1976) is shown to point to changes in peripheral blood 5 min after the start of drinking. With rabbits, which correct a sodium deficit slowly and regularly over a period of 8–12 h and then show no signs of overcompensation, it is possible that a mechanism based on a hepatic sodium receptor might be influential. However, the data from this species show that they are more effective in correcting a given sodium deficit if the sodium solution offered is hypertonic than if it is isotonic. In the latter case, deficit is not fully corrected by 24 h. Drinking hypertonic solution would seem more likely to produce an increase in hepatic portal vein sodium concentration than the drinking of an isotonic solution. Indeed, it would seem possible that drinking an isotonic solution would have little or no influence on portal sodium concentration. Therefore any receptors would not be influential in determining the cessation of drinking.

These questions at least focus aspects of these interesting data of Blake and Lin. The necessity of direct sampling from indwelling hepatic or portal vein cannulae in the conscious animal which is rapidly satiating itself is clear. This will determine the time course of change with the drinking of various solutions. Such studies could parallel those determining any effects of direct infusion of hypertonic and hypotonic sodium solutions into the portal vein, mesenteric artery or coeliac axis in the sodium-deficient animal.

References

Abraham SF, Denton DA, Weisinger RS (1976) Effect of an angiotensin antagonist, Sar^1-Ala^8-Angiotensin II on physiological thirst. Pharmac Biochem Behav 4: 243

Adachi A, Niijima A, Jacobs HL (1976) An hepatic osmoreceptor mechanism in the rat: Electrophysiological and behavioural studies. Am J Physiol 231: 1043

Adolph EF, Barker JP, Hoy PA (1954) Multiple factors in thirst. Am J Physiol 178: 538

Ahmed K, Judah JD (1964) Preparation of lipoproteins containing cation-dependent ATPase. Biochim Biophys Acta 93: 603

Andrews WHH, Orbach J (1973) A search for nerves which are sensitive to changes of oncotic pressure and changes in the concentration of sodium and glucose. Digestion 8: 453

Andrews WHH, Orbach J (1974) Sodium receptors activating some nerves of perfused rabbit livers. Am J Physiol 227: 1273

Beilharz S, Kay RNB (1963) The effects of ruminal and plasma sodium concentrations on the sodium appetite of sheep. J Physiol (Lond) 165: 468

Blake WD, Lin KK (1978) Hepatic portal vein infusion of glucose and sodium solutions on the control of saline drinking in the rat. J Physiol (Lond) 274: 129

Blake WD, Ulfendahl HR (1979) Hepatic portal vein infusion of glucose on sodium excretion in the rat. Acta Physiol Scand 105:304

Blass EM, Hall WG (1976) Drinking termination: Interactions among hydrational, orogastric and behavioural controls in rats. Psychol Rev 83: 356

Carey RM (1978) Evidence for a splanchnic sodium input monitor regulating renal sodium excretion in man: Lack of dependence upon aldosterone. Circ Res 43: 19–23

Carey RM, Smith JR, Ortt EM (1976) Gastro-intestinal control of sodium excretion in sodium depleted conscious rabbits. Am J Physiol 230: 1504

Daly JJ, Roe JW, Horrocks P (1967) A comparison of sodium excretion following an infusion of saline into systemic and portal veins in the dog: Evidence for a hepatic role in the control of sodium excretion. Clin Sci 33: 481

Denton DA (1966) Some theoretical considerations in relation to innate appetite for salt. Conditional Reflex 1: 144

Denton DA, Kraintz F, Kraintz L (1970) The inhibition of salt appetite of sodium deficient sheep by intracarotid infusion of ouabain. Commun Behav Biol 5: 183

Feider A (1972) Feedback control of thirst in rats. Physiol Behav 8: 1005

Glaubach S, Molitor H (1928) Untersuchungen über die hormonal diureseregelnde Tätigkeit der Leber. Arch Exp Path Pharmak 131: 31

Gordon D, Peart WS (1979) Sodium excretion in man, and adaptation to a low-sodium diet: Effect of intravenous sodium chloride. Clin Sci 57: 225

Jalowiec J, Stricker EM (1973) Sodium appetite in adrenalectomized rats following dietary sodium deprivation. J Comp Physiol Psychol 83: 66

Kozlowski S, Drzewiecki K (1973) The role of osmoreception in portal circulation in control of water intake in dogs. Acta Physiol Polon 24: 325

Larsen JA, Krarup N, Munck A (1976) Liver haemodynamics and liver function in cats during graded hypoxic hypoxemia. Acta Physiol Scand 98: 257

Lee MC, Thrasher TN, Ramsay DJ (1978) The effect of intracerebroventricular infusion of saralasin on drinking following caval ligation, hypertonic saline and water deprivation in rats. Abstr Soc Neurosci.

Lennane RJ, Peart WS, Carey RM, Shaw J (1975) A comparison of natriuresis after oral and intravenous sodium loading in sodium depleted rabbits: Evidence for a gastrointestinal or portal monitor of sodium intake. Clin Sci Molec Med 49: 433

Lin K, Blake WD (1971) Hepatic sodium receptor in control of saline drinking behaviour. Commun Behav Biol 5: 359

Millies E (1960) A new diuretic factor of hepatic origin. Acta Physiol Lat Am 10: 179

Mouw DR, Vander AJ (1970) Evidence of brain sodium receptors controlling renal sodium excretion and plasma renal activity. Am J Physiol 219: 822

Niijima A (1969) Afferent discharges from osmoreceptors in the liver of the guinea pig. Science 166: 1519

Oettel H (1943) Einwirkung von Leberextrakten auf die Diurese. Z Gesamte Inn Med 111: 613

Passo SS, Thornborough JR, Rothballer AB (1973) Hepatic receptors in control of sodium excretion in anaesthetized cats. Am J Physiol 224: 373

Potkay S, Gilmore JP (1970) Renal responses to vena cava and portal venous infusions of sodium chloride in unanaesthetized dogs. Clin Sci 39: 13

Ramsay DJ, Reid IA (1975) Some central mechanisms of thirst in the dog. J Physiol (Lond) 253: 517

Rogers RC, Novin D, Butcher LL (1979) Hepatic sodium and osmoreceptors activate neurons in the ventrobasal thalamus. Brain Res 168: 398

Rolls BJ, Wood RJ (1977) The role of angiotensin in thirst. Pharmacol Biochem Behav 6: 245

Rolls BJ, Wood RJ, Rolls ET (1980) Thirst: The initiation, maintenance and termination of drinking. Academic Press, New York, p 263 (Progress in Psychobiology and Physiological Psychology, no. 9)

Saum WR, Brown AM, Tuley FH (1976) An electrogenic sodium pump and baroreceptor function in normotensive and spontaneously hypertensive rats. Circ Res 39: 497

Schmitt M (1973) Influences of hepatic portal receptors on hypothalamic feeding and satiety centres. Am J Physiol 225: 1089

Schmitt M, Koizumi K (1971) Possible existence of portal receptors effecting hypothalamic neuron activity. Physiologist 14: 225

Schneider EG, Davis JO, Robb CA, Baumber JS, Johnson JA, Wright FS (1970) Lack of evidence for hepatic osmoreceptor mechanism in conscious dogs. Am J Physiol 218: 42

Sokolove PG, Cooke IM (1971) Inhibition of impulse activity in the sensory neuron by an electrogenic pump. J Gen Physiol 57: 125

Wood RJ, Rolls BJ, Ramsay DJ (1977) Drinking following intracarotid infusions of hypertonic solutions in dogs. Am J Physiol 232: R88

18 The stimulating effect on salt appetite of desoxycorticosterone, aldosterone and other adrenal steroids

Summary

1. Apart from their role in countering sodium loss in the adrenalectomized animal and, in appropriate dosage, reducing sodium appetite, mineralocorticoid hormones in larger dosage specifically stimulate sodium appetite in the sodium-replete animal.

2. Desoxycorticosterone (DOC) has long been recognized to have a powerful stimulating effect on salt appetite in the rat. The effect in rabbit and sheep appears to be much smaller. Aldosterone also has this effect in the rat, but not in the rabbit, sheep or man.

3. It is possible that endogenous aldosterone is a contributory factor to salt appetite seen in sodium deficiency in the rat. But comparison of appetite in response to sodium deficiency of normal and adrenalectomized rats does not suggest it is a major factor. Indeed the voluntary sodium intake of sodium-deficient adrenalectomized rats usually overcompensates deficit.

4. The question of DOC and aldosterone producing their effect by reduction of intracellular sodium concentration is discussed. The case is not clear. An alternative possibility can be considered that they have a direct action by inducing transcription and protein synthesis in specific brain cells subserving appetite.

5. The fact that adrenocorticotrophic hormone (ACTH) has a powerful salt appetite inducing effect in rabbit, rat and sheep is considered. This action of ACTH appears to be glucocorticoid-mediated but in the rabbit there is also a direct effect of the peptide hormone, presumably on the brain. It suggests that the hormonal response to stress which may be a concomitant of several modes of production of body sodium deficit could act to augment the sodium appetite generated by body sodium deficit alone.

The effect of desoxycorticosterone acetate

Rice and Richter (1943) and Braun-Menendez (1952) made the unexpected finding that dosages of desoxycorticosterone acetate (DOCA) above 1 mg/day in normal rats resulted in an increase in spontaneous appetite for sodium chloride. An increased appetite also for potassium chloride was reported by Richter (1956), but the increased intakes of sodium chloride and potassium chloride were not parallel (Fig. 18-1). An increased water intake occurred, but control experiments comparing the effects of DOCA in animals on a low-salt and a high-salt diet showed that this was contingent

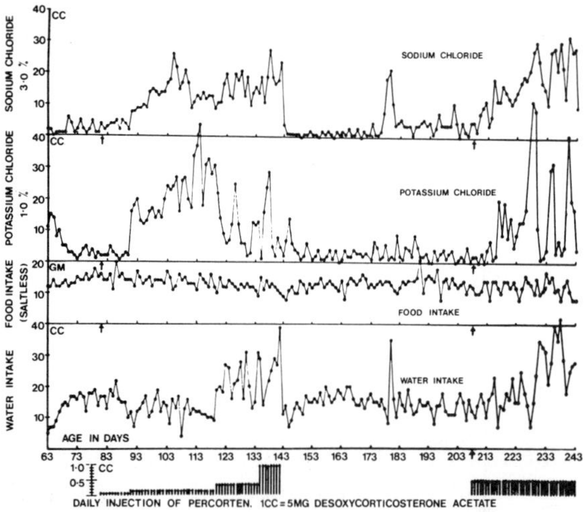

Fig. 18-1. Typical record showing the changes in appetite for sodium chloride and potassium chloride solutions, food and water, produced in a normal rat by daily injections of DOCA (Percorten). (From Richter 1956, with permission.)

on the DOCA being associated with a high salt intake. That is, high water intake was secondary. In Richter's (1956) experiment the sodium chloride was a 3% solution and the potassium chloride a 1% solution. In sheep we have seen often that a thirsty animal will accept 1% potassium chloride solution, i.e. it does not appear aversive when encountered (Beilharz et al. 1962). In this regard, inspection of Fig. 18-1 (Richter 1956), particularly the second phase of DOCA dosage, does suggest a reciprocal relation between intake of water and of potassium chloride, as though 1% potassium chloride may have been an acceptable alternative to the thirsty animal if it happened to encounter it. It is, however, interesting that DOCA would have reduced plasma potassium concentration.

With 5 mg DOCA per day, intake approximated 20 ml of 3% sodium chloride, which is equivalent to the sodium content of extracellular fluid. Braun-Menendez and Brandt (1952) were able to obtain a salt appetite effect with 17-OH corticosterone but not with cortisone, progesterone, 17β-oestradiol, testosterone or stilboestrol. Braun-Menendez (1952) also showed that the DOCA effect was potentiated by ACTH and cortisone, and inhibited by 17β-oestradiol, stilboestrol and testosterone. As indicated in Chapter 23, dealing with salt appetite in the reproductive process, certain of these findings differ from the data derived in the rabbit.

The seemingly paradoxical finding of DOCA, a salt-retaining hormone, inducing a powerful salt appetite drive, remained without much further scrutiny until work by Fregly and colleagues, and Wolf and colleagues, in the mid-1960s. The issue of paradox (see discussion by Richter 1956) gave way to the consideration that this could well be a homeostatic role of the mineralocorticoid: i.e. it generates intake in the face of body deficit. However, the validity of such a hypothesis, as will be recounted, turns on quantitative considerations that require critical appraisal (Denton and Nelson 1970).

Fregly and Waters (1966) and Fregly (1967a, b) found that graded doses of the mineralocorticoid hormones D-aldosterone acetate or DOCA to adrenalectomized rats produced a U-shaped dose–response curve to sodium chloride. With initial increasing dose the animals progressively reduced their sodium chloride intake. (The rats had access to water as well as 0.15 M sodium chloride, and the U-shaped effect on sodium chloride intake was accompanied by a reciprocal effect on water intake.) Basal sodium chloride intake occurred with an aldosterone dosage of 32 μg/day and with a DOCA dosage of 400 μg/day, while increasing dosage of aldosterone to 64 μg/day caused salt intake to increase again and provided the ascending limb of the curve.

In relation to the salt appetite effect of protracted dosage of mineralocorticoid, it is noteworthy that whereas after a period of dosage there is renal escape with natriuresis, no such escape occurs in relation to salt appetite behaviour. There is also no escape from the influence of mineralocorticoids on salivary Na/K ratio.

Fregly and Waters (1966) proposed the effect on sodium chloride intake was specific to mineralocorticoids. Other naturally occurring hormones, including cortisone acetate, oestrone, thyroxine and testosterone propionate were ineffective at a series of different doses. Fregly notes that spironolactone prevents the influence of DOCA on spontaneous sodium chloride intake of adrenalectomized rats. Fregly (1968) reported the effect of administration of metyrapone (SU-4885) on salt appetite of normal rats. Assuming that the effect of this is to inhibit the enzymes of the adrenal cortex responsible for hydroxylation of the steroid molecule at the 11β position, there would be an accumulation of the steroids in the biosynthetic chain prior to this step: e.g., in the human, DOC, 17α-hydroxy DOC (compound S), Δ4-androstenedione and possibly 18-hydroxy DOC. It was found that metyrapone produced an increase in sodium chloride intake in the rat, but compound S and Δ4-androstenedione did not. As the authors point out, the 17-hydroxylated compounds would not have been expected to increase in the rat as the 17-hydroxylase enzyme is lacking in the adrenal cortex.

Fregly (1967a) showed that 9α-fluorocortisol also produced the U-shaped dose–response curve in the adrenalectomized rat, the steroid being approximately twice as potent as aldosterone on a weight basis in relation to this effect.

Anti-thyroid drugs

Before discussing Wolf's approach to the influence of mineralocorticoids on salt appetite, there is another aspect of hormonal influence on this behaviour explored by Fregly which is of interest. He found that propylthiouracil (PTU), an anti-thyroid drug and mild diuretic agent, induced a spontaneous salt appetite in rats given a choice between water and 0.15 M sodium chloride solution (Fregly 1969). The compound also reduced the preference threshold for sodium chloride solution to a level one-sixth that of controls (i.e. from 23 to 4 mmol/l when tested by the two-bottle preference technique). The administration of graded doses of DOCA to PTU-treated animals reduced the spontaneous sodium chloride intake in a graded fashion. Aldosterone secretion rate of rats treated by PTU was reduced to as little as a third that of control animals and this was associated with a reduction in adrenal blood flow (Fregly et al. 1965). Administration of thyroxine restored aldosterone secretion rate to normal. It was shown (Fregly 1969) that administration of DOCA first restored the preference threshold for sodium chloride in PTU-treated rats to the normal range and then, as dosage was increased, reduced it again. Fregly interprets the data as suggesting an interrelationship between sodium chloride intake, blood level of mineralocorticoid and preference threshold for sodium chloride solution.

Investigations by George Wolf on DOCA- and aldosterone-induced salt appetite

George Wolf of Yale University (Wolf 1965a, b) reported that in adrenalectomized rats low doses of DOCA produced a decrease in saline intake (restoration of sodium-retaining ability) whereas high doses produced an increase in saline intake (stimulation of sodium appetite). At high doses, however, intact rats consumed more saline, and manifested a greater preference for it, than did similarly treated adrenalectomized rats. Treatment with corticosterone increased both the absolute saline intake and saline preference of DOCA-treated adrenalectomized rats. The latter finding was interpreted on the basis that adrenalectomized DOCA-treated rats may retain sodium when sodium-deprived as efficiently as do intact rats but, because of inadequate glomerular filtration rate, do not excrete it as rapidly under load. By facilitating the excretion of sodium in the DOCA-treated rat corticosterone could allow sodium to be ingested at a higher rate. Wolf also noted in this early paper that very high doses of mineralocorticoid were required to produce the appetite-stimulating effect, but reasoned that these may be necessary to overcome the antagonistic effects of sodium retention. But this would not necessarily be the case under natural conditions where the high level of mineralocorticoid would be associated with sodium deficiency.

Quartermain and Wolf (1967) compared the strength of the sodium drive of DOC-treated rats with that of adrenalectomized sodium-deficient rats by using bar-pressing and quinine tolerance tests. They found the two groups bar-pressed for sodium chloride reinforcement at approximately equal rates and manifested the same resistance to extinction of bar-pressing response. Saline intake of DOCA-treated rats was not as greatly suppressed by quinine adulteration as was that of adrenalectomized rats. It was proposed that this was not due to a difference in drive state so much as the increased sensitivity and greater intolerance of adrenalectomized rats to aversive-tasting substances. This would be consistent with the data of Henkin et al. (1963).

Wolf (1964) reported that when rats given access to 0.25 M saline in addition to drinking water were injected with either 0.5-mg or 1-mg doses of D-aldosterone acetate, consistent increases in saline intake were observed in two or three rats injected with the lower dose and in all six rats injected with the 1-mg dose. Wolf and Handal (1966) examined the dose–response relation of aldosterone-induced sodium appetite, and also the specificity of the appetite. Bearing in mind the argument cited above concerning the accumulation of sodium within the body acting as an inhibitory factor, the experiments on dose–response relation were conducted with the animals on a low-sodium diet and they were allowed access to the 0.15 M sodium chloride solution for 30 min only. The various doses of aldosterone were given in three divided doses during the 24 h before test, and thus the test was made 8 h after the last injection. Four doses were chosen: 0, 30, 60 or 120 μg of D-aldosterone in 0.3 ml of sesame oil.

It was found (Fig. 18-2) that the 30-μg dose produced no significant increase in saline intake in comparison with the 0-μg dose. The 60-μg dose

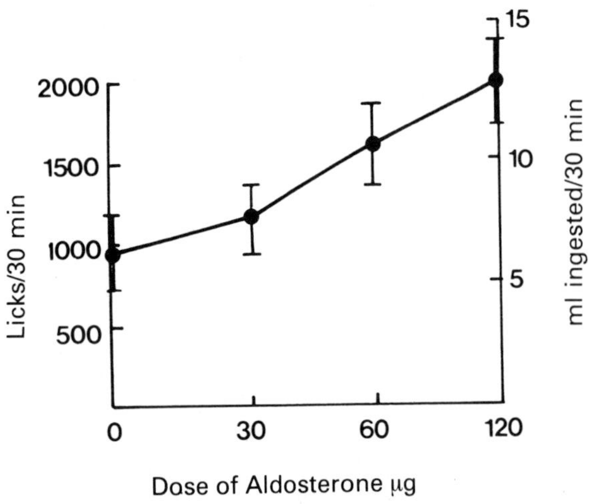

Fig. 18-2. Intake of 0.15 M sodium chloride as a function of dosage of D-aldosterone acetate. Number of licks on left ordinate and millilitres ingested on right ordinate. Vertical lines represent s.e. at each dose. (From Wolf and Handal 1966, with permission.)

elicited an increase in 13 of 16 rats, and the 120-μg dose elicited an increase in 14 of 16 rats. The increases were highly significant. The overall intake did not change significantly over the 4 weeks of the experiment, indicating there was no cumulative effect of the aldosterone injections or, presumably, learning. In a test on specificity under somewhat similar conditions but with a high dose of aldosterone (120 μg) and 0.15 M solutions of sodium bicarbonate, potassium chloride, calcium chloride, magnesium chloride and dextrose it was found that there was a significant increase in the intake of sodium bicarbonate solution but not in those of the other solutions (Fig. 18-3). The authors noted that these results do not rule out the possibility that aldosterone administration may increase potassium intake under other conditions which potentiate its kaliuretic effect. The data

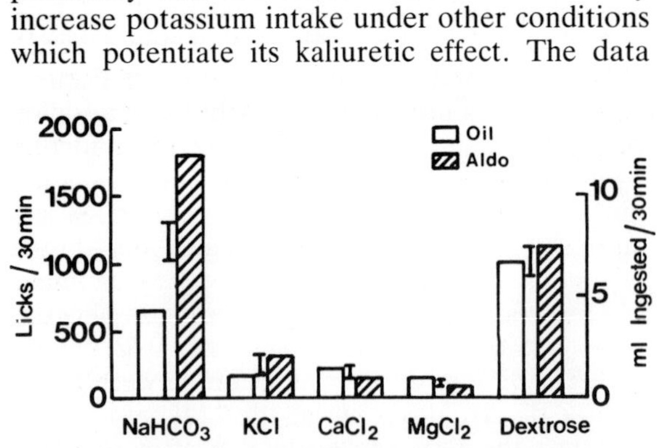

Fig. 18-3. Intake of various 0.15 M solutions after injection of either 0 or 120 μg of D-aldosterone acetate (open and hatched columns respectively). Number of licks on left ordinate and millilitres ingested on right ordinate. Vertical lines between the columns represent s.e. of different scores. (From Wolf and Handal 1966, with permission.)

could be compatible with the proposal made above in relation to Richter's original experiments showing increased potassium intake: i.e. dilute potassium chloride may be acceptable to a thirsty animal.

Physiological levels of desoxycorticosterone secretion

To put the above studies into physiological context, we will give here some quantitative data on corticosteroid secretion. The amount of aldosterone required to maintain adrenalectomized rats is from 1 to 4 μg/day according to strain (J.J. Chart, personal communication). Mean DOC secretion rate in man is 85 μg/24 h (range 34–120, n=10: New et al. 1969); in the anaesthetized rat it is 6 μg/h, n=8 (presumably under ACTH stimulation) (Rapp 1969); in the sheep the basal level is 6 μg/24 h, n=8, and with ACTH infusion 121 μg/24 h (Oddie 1970).

In sheep, blood DOC concentration is 7.0±2.1 ng/100 ml as basal and 9.6±2.5 ng/100 ml after 48 h of sodium depletion (n=6; difference not significant). In man, plasma DOC (9.8±1.8 ng/100 ml) is unchanged by a low-sodium diet (7.0±1.7 ng/100 ml, n=6: Oddie et al. 1972).

Studies on wild rabbits

In the light of the above data on the rat, and certain interpretative issues which emerged (and will be discussed further on), we studied the appetite-inducing effects of mineralocorticoids in wild rabbits brought from the Snowy Mountains region of Australia. To us, particular interest accrued insofar as these creatures showed an avid sodium appetite under sodium-deficient conditions in nature. Also, methodologically, there was the additional advantage emerging from laboratory experiments on them, in that in contrast to the rat, under baseline study with a cafeteria they exhibited only a small hedonic 'need-free' intake of sodium (Denton and Nelson 1970). Analysis of the laboratory diet offered the rabbits showed that it had a liberal content of sodium, potassium, magnesium and calcium (71, 214, 11 and 330 mmol/kg respectively).

The experimental regime was for 150 or 500 mequiv/l solutions of sodium chloride, potassium chloride, magnesium chloride and calcium chloride, and water, to be continuously available to the animals. All fluids were in identical inverted

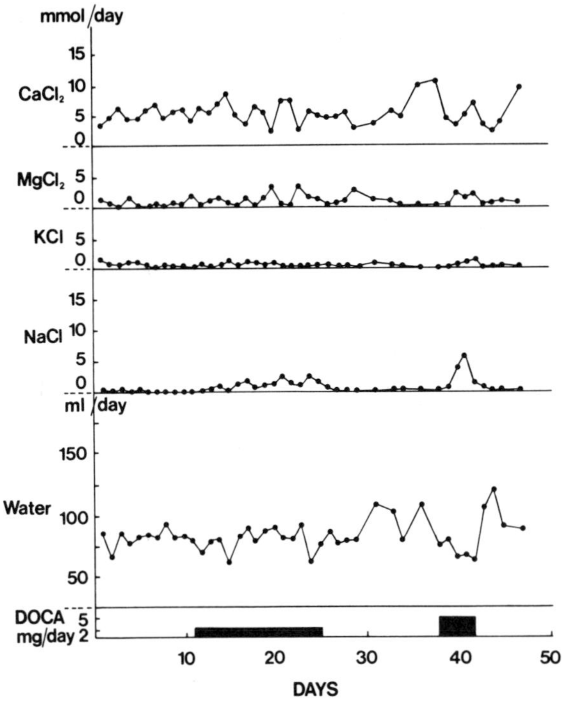

Fig. 18-4. Effect of DOCA on mean daily water intake and intake of 150 mequiv/l electrolyte solutions by wild rabbits on a diet with a liberal sodium content. (From Denton and Nelson 1970, with permission.)

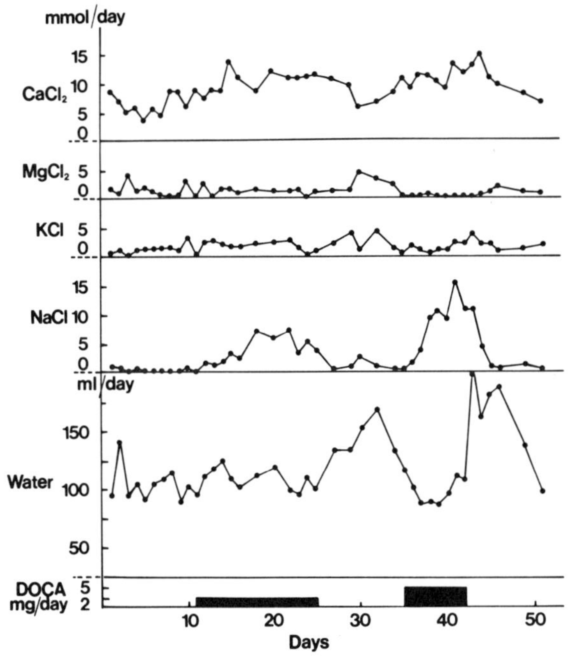

Fig. 18-5. Effect of DOCA on the mean daily water intake and electrolyte intake of 500 mequiv/l solutions by wild rabbits on a diet with a liberal sodium content. (From Denton and Nelson 1970, with permission.)

bottles whose relative positions were determined each day from a table of random numbers. The results for intake of 150 mequiv/l solutions with dosages of 2 and 5 mg of DOCA per day are shown in Fig. 18-4. Significant increase in voluntary intake occurred only with the sodium solution. Fig. 18-5 shows the effect of DOCA at 2 and 5 mg/day on intake of the 500 mequiv/l solutions. Significant increases in sodium and calcium intake occurred at both 2 and 5 mg dosage whereas magnesium intake was significantly decreased at the 5-mg dosage. The DOCA dosage, particularly at the higher levels, caused a decline in the animals' condition with weight loss. Aldosterone administration at 0.25 and 0.5 mg/day for 8 days had no significant effect on the intake of any of the ions when offered in the 500 mequiv/l solutions, and nor did 1 mg/day of DOCA.

These data differ from those from the rat in that the sodium intake with DOCA is greater in the rabbit drinking 500 mequiv/l solutions whilst the reverse holds true for the rat. The appetite-stimulating effect of DOCA was smaller in the rabbit than the rat if intakes are compared relative to extracellular sodium content of the animal. The sodium-retaining effect of DOCA and aldosterone in the rabbit was short-lived, and the overall balances indicated little or no increase in body sodium content.

The question of a physiological role of mineralocorticoids in salt appetite

The hypothesis that aldosterone hypersecretion may be the physiological vector of salt appetite in sodium deficiency has been based on experiments in domesticated laboratory rats (*Rattus norvegicus*). It is not known what effects these steroids have in the wild rat. The data of Wolf and Handal (1966) and Fregly and Waters (1967) indicate an effect on sodium appetite of laboratory rats when aldosterone dosage is of the order of 50–60 μg/day. Bojeson's (1966) experiments on peripheral blood aldosterone and metabolic clearance rate in the rat indicate that daily aldosterone secretion may reach this level in sodium deficiency. Data from adrenal vein cannulations in anaesthetized animals indicate a similar order of output (Wolf and Handal 1966; Cade and Perenich 1965).

The experiments of Denton and Nelson (1970) do not suggest mineralocorticoid as the primary cause of salt appetite in the wild rabbit. There was little effect with DOCA until dosage caused deterioration in condition, and aldosterone dosage equal to the daily secretion of sodium-deficient sheep, an animal 10–20 times larger, did not increase voluntary intake of sodium solutions. This supraphysiological dosage of aldosterone is similar on a body weight basis to that used by Wolf and Handal (1966) in the domesticated rat.

The results on specificity, and the quantitative considerations mentioned above, suggest, however, that a role for aldosterone in salt appetite in the domestic rat is feasible. But Wolf and Handal note it can only have an ancillary role. That is, appetite is manifest in adrenalectomized rats, and after procedures such as induction of sodium appetite by formalin injection, adrenalectomized rats did not ingest significantly less saline than intact rats. This would not indicate that increased aldosterone secretion was contributing greatly to the appetite in intact rats. Moreover, it has been clearly shown using sheep with a parotid fistula that intact and adrenalectomized animals show the same essential relation between body deficit and intake (Fig. 12-13).

Wolf and Handal (1966) found also that although physiological doses of aldosterone elicited intake of palatable saline solutions they did not consistently elicit intake of highly concentrated unpalatable solutions such as occurs in sodium deficiency. Furthermore, the largest doses of aldosterone induced an increased ingestion of only 1–1.5 mmol of sodium, even though a highly preferred saline concentration was used and the rats were tested under near-optimal conditions. This increase was small relative to the effects of sodium deficiency. Further, from a species difference viewpoint, it is noteworthy that in sheep DOCA has little effect on sodium appetite, though large effects on other electrolyte parameters occur (Denton et al. 1969). Increased sodium appetite is not noted in patients with pathologically increased aldosterone secretion (Streeten 1960; Denton 1967).

The question of learning

Experiments of Weisinger and Woods (1971) drew attention to the possibility that salt appetite in response to aldosterone might be a learned association between high aldosterone occurring naturally as a result of sodium deficiency, and the sodium appetite elicited. They found that rats raised on a high-sodium diet from birth did not manifest increased salt appetite after aldosterone treatment as did rats raised on a normal-sodium diet. Their experiments involved administration of doses of 15, 150 and 1500 μg/kg of aldosterone in an acetone vehicle into the ventroabdominal area 24 h before appetite for 0.33 M saline and water was tested for 20 min. This was after deprivation of water, saline and food during the intervening 24 h. Given the 30 min half-life of aldosterone, the possible variations in absorption and metabolic clearance rate of aldosterone under the various experimental conditions, and the absence of any evidence of a dose–response curve under the initial control conditions, the data of the several procedures are difficult to interpret. Schulkin (1978) was unable to reproduce the data.

ACTH and salt appetite

Before considering the mechanism by which mineralocorticoids influence salt appetite, there are other data on the influence of adrenal steroids on salt appetite that should be mentioned. ACTH is one of a group of steroid and peptide hormones which cause the spectacular increase in voluntary salt intake observed in pregnancy and lactation in the wild rabbit. The dissection of the phenomenon in the rabbit showed that part of the action of ACTH was attributable to direct influence of the peptide, presumably on the central nervous system. Another component was due to the increased blood levels of cortisol and corticosterone stimulated by the action of ACTH (Blaine et al. 1975). Following this, it was shown by Weisinger et al. (1978) that ACTH (5 international units/day: Synacthen depot, Ciba) increased daily intake of sodium in rats receiving a diet adequate in sodium. Intake was raised to a level involving turnover of approximately twice the extracellular sodium content of the animals each day. In rabbits (Blaine et al. 1975) ACTH (5 international units/day) increased daily intake to turnover of approximately 20% of the total extracellular fluid sodium.

It was shown in both rats and rabbits that the influence of ACTH was specific for sodium intake. In the rat, in contrast to the rabbit, ACTH had no action in the adrenalectomized animals, i.e. there was no evidence of an extra adrenal action of the peptide hormone.

Weisinger et al. (1980) examined the influence of ACTH on salt appetite in the sheep. ACTH has a most interesting action in the sheep in terms of causing regularly, and reproducibly, a moderate hypertension within 24–48 h of administration. Detailed endocrine and metabolic examinations established the changes which occurred as a result of ACTH administration (Scoggins et al. 1974). It was found that as well as hypertension, ACTH (80 international units/day: Synacthen depot, Ciba) caused increases in water turnover, plasma volume, and plasma sodium, blood cortisol, corticosterone, DOC and desoxycortisol concentrations (Scoggins et al. 1975). Plasma potassium concentration decreased, and aldosterone level in peripheral blood, following an initial rise, did not change or was decreased slightly by the fifth day. By the third day of ACTH administration the mean voluntary sodium intake ($n=8$) was significantly elevated over baseline levels. The final intake resulting represented turnover of approximately half the extracellular fluid sodium content per day.

A detailed study of voluntary intake and urinary excretion of sodium in the sheep (Fig. 18-6)

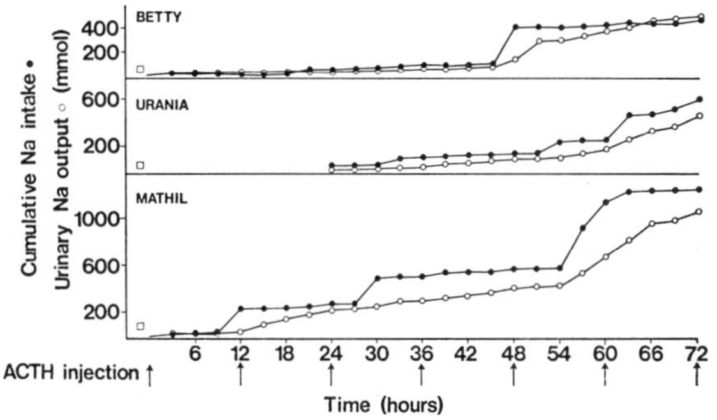

Fig. 18-6. Cumulative sodium intake (filled circles) or urinary sodium excretion (open circles) over time for three sheep. Arrows along the abscissa indicate intramuscular injections of 45 international units ACTH. Baseline intake (i.e. intake during previous 24 h) is shown (open squares). From Weisinger et al. 1980, with permission.)

showed clearly that the increased intake preceded any increase in urinary excretion. In contrast to the rabbit, but as in the rat, adrenalectomy abolished the sodium appetite stimulating influence of ACTH. It was found, however, in the adrenalectomized animals, that the infusion of a combination of steroid hormones (cortisol, 5 mg/h; corticosterone, 0.5 mg/h; 11-desoxycortisol, 1 mg/h; DOC, 20 μg/h; and aldosterone, 5 μg/h day 1, 2 μg/h days 2–4, 1 μg/h day 5) caused salt appetite to increase as in ACTH administration to the normal animal (Fig. 18-7). Adminstration of this steroid cocktail to an adrenal-intact animal also caused increase in salt appetite.

These data in three species show that as well as ACTH having a role in the initiation of lactation and of salt appetite at that time, cortical hormones possibly have some more general influence in stimulating salt appetite. Although the doses of exogenous ACTH administered were high, the levels of adrenal cortical hormones in peripheral blood were, self-evidently, within the functional capacity of the gland to produce these steroids. In

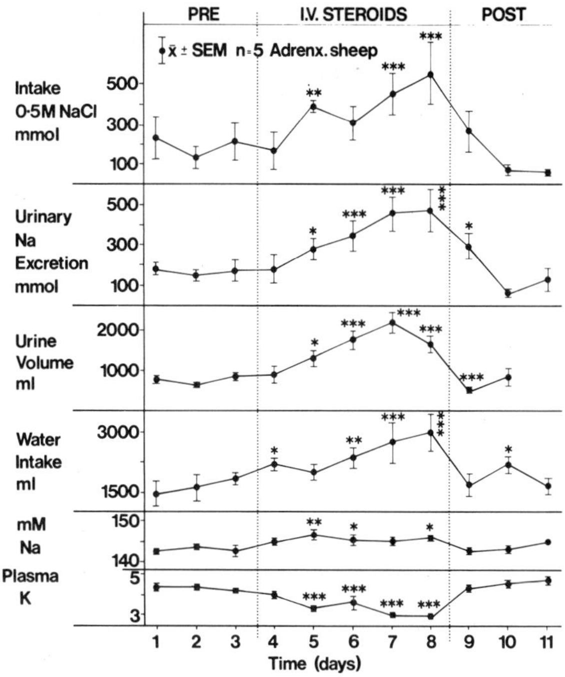

Fig. 18-7. Mean (±s.e.) intake of 0.5 M sodium chloride (mmol), urinary sodium excretion (mmol), urine volume and water intake, and plasma sodium and potassium concentrations, before, during and after infusion of a cocktail of steroid hormones (cortisol, 5 mg/h; corticosterone, 0.5 mg/h; 11-desoxycortisol, 1 mg/h; DOC, 20 μg/h; and aldosterone 5 μg/h day 1, 2 μg/h days 2–4 and 1 μg/h day 5). $n=5$ experiments on four adrenalectomized sheep. $^{*}P<0.05$, $^{**}P<0.01$, $^{***}P<0.001$. (From Weisinger et al. 1980, with permission.)

the case of the rat and the sheep, there is no extra adrenal effect of the exogenous ACTH to account for. In the case of the sheep, it is unlikely on the evidence cited (Denton et al. 1969) that DOC or aldosterone contributed to the salt appetite observed. It could not, however, be excluded that the mineralocorticoids play some role in the appetite effect in the rat. In this regard, it would be interesting to administer aldactone contemporaneously with ACTH in this species.

A clear possibility arising from the data is that stress, associated with a number of conditions which could produce body sodium deficit, might amplify the sodium appetite over and above that contingent upon deficit. This, as has been said in several chapters, could be relevant to the interpretation of data where a stressful procedure was used to induce sodium appetite (e.g. formalin treatment), but it could also reflect augmentation of sodium appetite under natural conditions. Myers (1967) has drawn attention to the production of accessory adrenals in the rabbits under conditions of severe sodium deficiency in spring in the Snowy Mountains of Australia. This phenomenon can be reproduced in these animals by administration of ACTH. The glucocorticoid component of adrenal secretion could amplify sodium appetite under natural conditions in several species of animals, including the grazing ruminants, but additional observations are necessary to establish this fact and put it into some quantitative perspective.

Jalowiec et al. (1970) showed in rats that hypophysectomy attenuated sodium appetite in response to formalin treatment to some extent. Water intake was not influenced.

Mechanism of salt appetite induction by mineralocorticoids

The mechanism by which mineralocorticoids generate salt appetite remains an intriguing issue. But since the initial papers in this field, it has taken on a different complexion. It is part of the general problem of how steroid and peptide hormones are able to generate salt appetite in the sodium-replete animal (Denton et al. 1977). And this will be discussed in Chapter 24. However, there are special aspects to the role of mineralocorticoids in this context, as was brought up originally by Wolf (1965a) in his outstanding PhD thesis for Yale University, and by Wolf and Handal (1966). The arguments against mineralocorticoid-induced changes in salivary Na/K ratio causing appetite were noted. Further, the possibility exists that both sodium deficiency and mineralocorticoids stimulate appetite by the same mechanism: a reduction in the sodium content of special receptor cells. Thus mineralocorticoids could reduce intracellular sodium by stimulating sodium transport, whereas reduction of extracellular sodium by sodium deficiency could reduce passive back-diffusion of sodium into the same cells. However, there is no escape from the appetite-stimulating effect of mineralocorticoids, voluntary sodium intake remains elevated, and the sodium content of most tissues is increased by them associated with hypernatraemia and potassium depletion (Friedman et al. 1948; Green et al. 1955; Knowlton and Loeb 1957).

Herxheimer and Woodbury (1960) proposed that DOC elicitation of sodium appetite was due to lowering of intracellular sodium in the brain. Woodbury, Timiras and Vernadakis (1957) obtained results indicating that DOC in acute or chronic dosage reduced intracellular sodium of brain tissue, and also brain excitability. However, Woodbury (1958) showed glucocorticoids had effects exactly opposite to DOC in relation to brain sodium and electrical excitability—yet glucocorticoids enhance the DOC effect (Wolf 1965a, b) and furthermore have salt appetite inducing effects of their own (Blaine et al. 1975; Weisinger et al. 1978, 1980). Finally Wolf (1965a, b) notes that both mineralocorticoids and sodium deficiency are sufficient conditions for shifting sodium out of certain tissues (Barle et al. 1960; Rovner et al. 1963). Taste bud cells in the tongue, brain cells or other cells may respond similarly because of physico-chemical equilibria or potentiation of active transport (Porter and Edelman 1964). The cells would activate sodium appetite when their sodium content decreased.

References

Barle AB, Nichols G Jr, Karnovsky MJ (1960) Some effects of adrenalectomy and prednisolone administration on extracellular fluid and bone composition in the rat. Endocrinology 66: 508

Beilharz S, Denton DA, Sabine JR (1962) The effect of concurrent deficiency of water and sodium on the sodium appetite of sheep. J Physiol (Lond) 163: 378

Blaine EH, Covelli MD, Denton DA, Nelson JF, Shulkes AA (1975) The role of ACTH and adrenal glucocorticoids in the salt appetite of wild rabbits (*Oryctolagus cuniculus* (L.)). Endocrinology 97: 793

Bojeson E (1966) Concentration of aldosterone and corticosterone in peripheral plasma of rats. The effects of salt depletion, salt repletion, potassium loading and intravenous injections of renin and angiotensin II. Eur J Steroids 1: 145

Braun-Menendez E (1952) Aumento del apetito especifico para la sal provocado por la desoxicorticosterona: sustancias que potencian o inhibiten esta accion. Rev Soc Arg Biol 28: 23

Braun-Menendez E, Brandt P (1952) Aumento del apetito especifico para la sal provocado por la desoxicorticosterona. Rev Soc Arg Biol 28: 15

Cade R, Perenich T (1965) Secretion of aldosterone by rats. Am J Physiol 208: 1026

Denton DA (1967) Salt appetite. In: Code CF (ed) Handbook of physiology, Sect 6: alimentary canal, vol 1. American Physiological Society, Washington, p 433

Denton DA, Nelson JF (1970) Effect of desoxycorticosterone acetate and aldosterone on the salt appetite of wild rabbits (*Oryctolagus cuniculus* (L.)). Endocrinology 87: 970

Denton DA, Nelson JF, Orchard E, Weller S (1969) The role of adrenocortical hormone secretion in salt appetite. In: Pfaffmann C (ed) Proceedings of the Third International Symposium on Olfaction and Taste. Rockefeller University Press, New York, p 535

Denton DA, McKinley MJ, Nelson JF, Weisinger RS (1977) The hormonal genesis of mineral appetites. Proc XI European Congress of Endocrinology, Acta Endocrinol (Copenh) 85: 3

Fregly MJ (1967a) Effect of 9α fluorocortisol on spontaneous sodium chloride intake of adrenalectomized rats. Physiol Behav 2: 127

Fregly MJ (1967b) The role of hormones in the regulation of salt intake. In: Kare MR, Maller O (eds) The chemical senses and nutrition. Johns Hopkins Press, Baltimore, p 115

Fregly MJ (1968) Effect of metyrapone (SU-4885) on spontaneous NaCl intake by rats. Physiol Behav 3: 637

Fregly MJ (1969) Preference threshold and appetite for NaCl solution as affected by propylthiouracil and desoxycorticosterone acetate in rats. In: Pfaffmann C (ed) Proceedings of the Third International Congress on Olfaction and Taste. Rockefeller University Press, New York, p 554
Fregly MJ, Waters IW (1966) Effect of mineralocorticoids on spontaneous sodium chloride appetite of adrenalectomized rats. Physiol Behav 1: 65
Fregly MJ, Waters IW (1967) Hormonal regulation of spontaneous sodium chloride appetite of rats. In: Hayashi T (ed) Proceedings of the Second International Congress on Olfaction and Taste. Pergamon Press, Oxford, p 439
Fregly MJ, Kay JR, Waters IW, Straw JA, Taylor RE (1965) Secretion of aldosterone by adrenal glands of propylthiouracil treated rats. Endocrinology 77: 777
Friedman SM, Polley JR, Friedman CL (1948) The effect of desoxycorticosterone acetate on blood pressure, renal function and electrolyte pattern in the intact rat. J Exp Med 87: 329
Green DM, Reynolds TB, Girerd RJ (1955) Mechanisms of desoxycorticosterone action. X. Effects on tissue sodium concentration. Am J Physiol 181: 105
Henkin RI, Gill JR Jr, Bartter FC (1963) Studies on taste thresholds in normal man and in patients with adrenal cortical insufficiency: The role of adrenal cortical steroids and of serum sodium concentration. J Clin Invest 42: 727
Herxheimer A, Woodbury DM (1960) The effect of desoxycorticosterone on salt and sucrose taste preference thresholds and drinking behaviour in rats. J Physiol (Lond) 151: 253
Jalowiec JE, Stricker EM (1973) Sodium appetite in adrenalectomized rats following dietary sodium deprivation. J Comp Physiol Psychol 83: 66
Jalowiec JE, Stricker EM, Wolf G (1970) Restoration of sodium balance in hypophysectomized rats after acute sodium deficiency. Physiol Behav 5: 1145
Knowlton AI, Loeb EN (1957) Depletion of carcass potassium in rats made hypertensive with desoxycorticosterone acetate (DCA) and with cortisone. J Clin Invest 36: 1295
Myers K (1967) Morphological changes in the adrenal glands of wild rabbits. Nature 213: 147
New MI, Seaman MP, Peterson RE (1969) A method for the simultaneous determination of the secretion rates of cortisol, 11-desoxycortisol, corticosterone, 11-desoxycorticosterone, and aldosterone. J Clin Endocrinol Metab 29: 514
Oddie CJ (1970) Measurement of deoxycorticosterone. MSc thesis, University of Melbourne, p 82
Oddie CJ, Coghlan JP, Scoggins BA (1972) Plasma deoxycorticosterone levels in man with simultaneous measurement of aldosterone, corticosterone, cortisol and 11-deoxycortisol. J Clin Endocrinol Metab 34: 939
Porter GA, Edelman IS (1964) The action of aldosterone and related corticosteroids on sodium transport across the toad bladder. J Clin Invest 43: 611
Quartermain D, Wolf G (1967) Drive properties of mineralocorticoid induced sodium appetite. Physiol Behav 2: 261
Rapp JP (1969) Deoxycorticosterone production in adrenal regeneration hypertension: In vitro vs in vivo comparison. Endocrinology 84: 1409
Rice KK, Richter CP (1943) Increased sodium chloride and water intake of normal rats treated with desoxycorticosterone acetate. Endocrinology 33: 106
Richter CP (1956) Salt appetite of mammals: Its dependence on instinct and metabolism. In: L'instinct dans le comportement des animaux et de l'homme. Masson et Cie, Paris
Rovner DR, Streeten DHP, Louis LH, Stevenson CT, Conn JW (1963) Content and uptake of sodium and potassium in bone. Influence of adrenalectomy, aldosterone, desoxycorticosterone, and spironolactone. J Clin Endocrinol Metab 23: 938
Schulkin J (1978) Mineralocorticoids, dietary conditions and sodium appetite. Behav Biol 23: 197
Scoggins BA, Coghlan JP, Denton DA, Fan JSK, McDougall JG, Oddie CJ, Shulkes AA (1974) The metabolic effects of ACTH in sheep. Am J Physiol 226: 198
Scoggins BA, Coghlan JP, Denton DA, Fan JK, McDougall JG, Oddie CJ, Shulkes AA (1975) The role of adrenocortical hormones in ACTH induced hypertension. J Clin Exp Pharmacol Physiol 2: 119
Streeten DHP (1960) Primary and secondary aldosteronism: Definition and diagnosis. In: Moyer JH, Fuchs M (eds) Edema. Mechanics and management. Saunders, Philadelphia, p 121
Weisinger RS, Woods SC (1971) Aldosterone elicited sodium appetite. Endocrinology 89: 538
Weisinger RS, Denton DA, McKinley MJ, Nelson JF (1978) ACTH induced sodium appetite in the rat. Pharmacol Biochem Behav 8: 339
Weisinger RS, Coghlan JP, Denton DA, Fan JSK, Hatzikostas S, McKinley MJ, Nelson JF, Scoggins BA (1980) ACTH elicited sodium appetite in sheep. Am J Physiol 239: E45
Wolf G (1964) Sodium appetite elicited by aldosterone. Psychonom Sci 1: 211
Wolf G (1965a) Sodium appetite elicited by desoxycorticosterone. PhD thesis, Yale University
Wolf G (1965b) Effect of desoxycorticosterone on sodium appetite of intact and adrenalectomized rats. Am J Physiol 208: 1281
Wolf G, Handal PJ (1966) Aldosterone induced sodium appetite: Dose response and specificity. Endocrinology 78: 1120
Woodbury DM (1958) Relation between the adrenal cortex and the central nervous system. Pharmacol Rev 10: 275
Woodbury DM, Timiras PS, Vernadakis A (1957) Influence of adrenocortical steroids on brain function and metabolism. In: Hoagland H (ed) Hormones, brain function and behaviour. Academic Press, New York, p 27

19 The effect of electrical stimulation and lesions of the central nervous system on salt appetite

Summary

1. Cogent arguments have been advanced by some neurophysiologists against simple interpretation of results of electrical or chemical stimulation of the central nervous system (CNS) in terms of evocation of the natural state of motivation of the animal. On the other hand, clinical observations with brain electrodes in man involving report of subjective feelings do not preclude that deep-seated brain stimulation in animals could alter the stream of consciousness in an analogous fashion.

2. Some general aspects of the issue of conscious awareness in animals are discussed. There appear to be no compelling reasons for departing from a Darwinian approach to consciousness as a brain function. Great survival advantage has accrued to animals with increasing complexity and diversity of this neural function in an environment of complementary intricacy. This is particularly related to development of the capacity to exercise options of action in the light of both contemporary exteroceptor and interoceptor inflow and stored information. The stored information may be phyletic memory or instinct, or derived from impressive experience of the individual. The emergent survival value of plasticity in exercise of options over rigid stereotype is evident. The remarkable data of Richter on 'hopelessness' and sudden death in wild rats, of Gallup on self-awareness in chimpanzees as derived from experiments with mirrors, and of Sperry in split-brain experiments are discussed. Sperry has stated that the lack of unity of consciousness so revealed shows it is not some kind of entity of itself but a functional property of brain in action. The results of Richter and Gallup as well as other data analysed by Thorpe and Griffin are consonant with the existence of consciousness in animals—an evolutionary continuity with progressively emergent levels of self-awareness. This viewpoint is contrary to the views of Eccles, which are discussed in the context of his discourse with Popper (Popper and Eccles 1977). Popper proposes that if the evolutionary story applies to life and consciousness there ought to be degrees of life and degrees of consciousness.

3. Lesion studies provide knowledge on brain regions which organize or transmit information involved in normal regulation, but the nature of the function is not fully elucidated. That is, the data do not show in the specific instance with which we are concerned the degree to which the function is directly related to the neural patterns regulating sodium ingestion or the manner in which sodium ingestion is related to other brain functions.

4. Given these strictures, an impressive body of evidence has accumulated from stimulation or lesion studies to show that neural structures in ventromedial and lateral hypothalamus, the amygdala and septum,

mesencephalic tegmentum, and gustatory thalamus subserve the sodium appetite drive. Large stimulus-bound intake of sodium solution has been produced in the behaviourly satiated sheep by electrodes in the perifornical hypothalamus, or medial forebrain bundle. In rats medial and lateral hypothalamic lesions have resulted in a large decrease in the sodium appetite which normally follows sodium deficiency or administration of desoxycorticosterone acetate (DOCA).

Introductory considerations on electrical stimulation

A monumental body of neuroanatomical knowledge on similar features of cellular aggregations or nuclei, and fibre pathways—indeed the whole spatial organization of the prosencephalon of the mammals—testifies to the genetic determination of the relationships in this structure. Given this determination of morphology, it is difficult to conceive that the vegetative life functions involved will not also be under substantial genetic influence. In Teitelbaum's (1973) view, the evidence is overwhelmingly in support of the notion that fixed inborn specific motivational systems exist in the brain. They are anatomically closely interwoven in the diencephalon. From studies of brain damage it is clear that there is a great deal of inborn specificity of neural circuitry in the brain, as localized lesions in a given area of the brain will lead to a characteristic loss of function: e.g. lateral hypothalamic damage leads to adipsia and aphagia.

However, Teitelbaum states that early studies of localization by electrical stimulation were greatly over-simplified as a result of the use of anaesthesia, which eliminates activity of many cells highly sensitive to it. This became apparent when the brain was explored in the conscious animal using chronically implanted electrodes, as pioneered by Hess (1954). The enormous complexities in correlating structure and function by this method were then clear. For example, systems controlling autonomic functions and instinctive activities in the cat diencephalon are closely interdigitated. Also, a simplistic view of the unanaesthetized brain may result from restricting the behaviours it can manifest (Teitelbaum 1973). As Valenstein et al. (1970) point out, many studies of rat behaviour in response to electrical stimulation merely observe behaviour of an isolated animal in a bare cage (equipped only with food pellets and a water bottle). The patterns elicited are thus often difficult to interpret in terms of an animal in a natural setting.

Stimulation may provide useful data on a given behavioural function, such as the sites which evoke food intake, but there may be too much overlap for such maps to be useful when several systems are considered simultaneously. Mechanisms of funnelling or channelling between mutually inhibitory systems allow an 'indiscriminate' electrical stimulus to evoke only one kind of behaviour. One system may be ascendant by virtue of anatomical proximity, of sensitivity to frequency of stimulation, or of differential motivational tuning through selective activation by, for example, prior water or

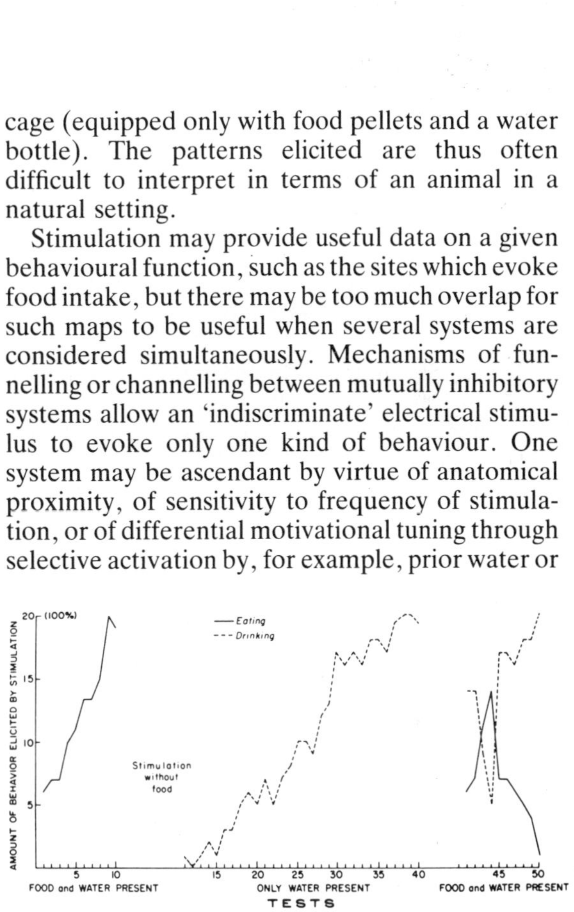

Fig. 19-1. An example of the gradual emergence of drinking elicited by hypothalamic stimulation. Initially, when food and water were both present, the animal displayed eating in response to stimulation with increasing reliability over the course of the ten tests. No drinking at all was observed during these tests. Following a prolonged period of intermittent stimulation with only water present, the animal started to lick the drinking tube occasionally during stimulation periods. At this point the animal was given 40 tests with only water present. It can be seen that during these tests drinking came to be elicited with increasing reliability. Ten additional tests with food and water present revealed that the stimulation now elicited drinking more reliably than eating. (Stimulation intensity was constant for all tests. Each test consisted of twenty, 20-s stimulation periods. It was possible for the stimulation to elicit both eating and drinking during a single period.) (From Valenstein 1973, with permission.)

food deprivation. Physical stimuli in the environment, or the animal's past experience, may also be determinant (see Teitelbaum 1973 for bibliographical review).

Valenstein (1973) notes that examples of stimulation eliciting the gradual emergence of additional behaviours, and the evidence that the execution of the behaviours can become reinforcing, may be extremely important. Fig. 19-1 illustrates the gradual emergence and strengthening of elicited water drinking. This was not seen at all initially during prolonged stimulation bouts with either food or water present—or even with water only present. An association appears to have been established (by means not yet understood) between mechanisms underlying reinforcement and the neural substrate for built-in responses, i.e. a process of 'channelling' of responses. Valenstein summarizes a body of his findings as follows:

1) The hypothalamic sites from which eating and drinking can be obtained by stimulation are much more widespread than previously suggested.
2) The eating and drinking elicited by stimulation differ in many significant ways from the same behaviour consequent on deprivation of food and water.
3) Most hypothalamic electrodes elicit several behaviours if sufficient behavioural opportunities are offered.
4) Many of the behaviours evoked by hypothalamic stimulation can be better understood by an analysis of the pre-potent responses of the species and the individual animal rather than by assuming underlying emotional states such as hunger and thirst.

As Valenstein sees it, 'hypothalamic stimulation at different regions may activate several different states, but these states are not sufficiently designated to preclude response substitution or to justify the application of terms that imply specific drive states. If testing is appropriately arranged, animals can demonstrate that they are able to distinguish between hunger and thirst. It follows that this ability must be reflected in distinctive neural activity. Whether such specific ability can be duplicated by electrical stimulation of discrete hypothalamic sites is another matter.'

Brain stimulation and access to the stream of consciousness. The question of consciousness in animals

Before discussing salt appetite, and bearing in mind some of the strictures on interpretation of data that arise from the questions touched on briefly above, it is interesting to contemplate at a more general level the issue of consciousness or awareness in animals, and the capacity to access the stream of consciousness by electrical or chemical intervention as revealed by such evidence as is available from human study.

Penfield (1966) describes how gentle application of an electrode to a discrete area of the temporal cortex during craniotomy under local anaesthesia may evoke one of two things: an altered interpretation of present experience whereby what the patient sees or hears becomes familiar or frightening, or a sudden flashback when a long-forgotten experience, a stream of former consciousness, moves forward again in great detail. He sees this as the discovery of an anatomical record of previous flow of consciousness, an unaltered record of previous awareness. For example, D.F. heard a song played by an orchestra each time the stimulating electrode was applied to a specific point on the temporal lobe. The point was stimulated over and over again. Attention determines the context of each conscious experience, and this searchlight of attention determines also the content of the record of the flow of consciousness in memory. The rest is lost. Penfield suggests the electrode activates the mechanism in distant grey matter by dromic conduction similar to that which normally activates it. When the stream of consciousness of a previous period of time is caused to flow again, the induced electrical excitation follows a path through a seemingly unending sequence of nerve cells, fibres and synapses that is formed and made permanent by neuronal facilitation.

On this issue, and in the instance of biologically significant experiences in terms of pain or pleasurable reinforcement, Livingston (1967) conceives of what he terms a 'Now print' mechanism. The steps are postulated as (i) reticular recognition of novelty, (ii) limbic discrimination of biological meaning for that individual at that moment, (iii) limbic discharge into the reticular formation, (iv) a diffusely projecting reticular formation discharge distributed throughout both hemispheres, conceived to be a 'Now print' order for memory, and (v) a 'printing' of all recent brain events and all recent conduction activities to facilitate repetition of similar conduction patterns. Only occurrences which are biologically meaningful and receive reinforcement can activate the last three stages of the postulated sequence. He cites as an example of the 'Now print' order being activated by a highly significant event, the fact that probably nearly everyone can remember in exact detail where they were and what they were doing when they heard the news of President Kennedy's assassination.

Reverting to the question of consciousness in animals and the legitimacy of experimental description implying this, we will consider Griffin's (1976) penetrating and provocative analysis. He states that the reductionist behaviourist viewpoint has served our science well for more than 50 years by constraining speculations and directing attention towards phenomena amenable to experimental analysis. Seventy years ago this disciplined restraint was a healthy reaction against ascribing human feelings to a wide variety of animals on the basis of anecdotal evidence. But in the light of the most recent experimental data, with particular emphasis on those of the Gardners (1975) with the chimpanzee Washoe, he sees the communication behaviour of certain animals as complex, versatile, and to a limited degree symbolic. The assumption has hitherto been that animals lack any conscious intent to communicate whilst men know what they are doing.

Investigators of behaviour have attempted to formulate explanatory concepts such as motivation, drives, or action-specific energy. But Griffin asks whether we have been overlooking more directly pertinent concepts lying close at hand or even closer—inside our own heads. When considering Washoe in the act of exchanging information about objects, actions or desires via coded manual gestures he submits it may actually clarify thinking to entertain such thoughts as 'Washoe *hopes* to go out for a romp, and *intends* to *influence* her human companions to that end' (though of course such terms as 'want' or 'like' do not explain the basic causes of the observed behaviour or any mental experiences accompanying it). In fact most people not indoctrinated by the behaviourist tradition take it for granted from their experience of animal behaviour that animals do have sensations, feelings and intentions, sufficiently analogous to some of our own to permit us to empathize. Even Descartes, who was the fountainhead of the view that animals are merely machines, admitted they could feel pain or pleasure and express passions. Aristotle in *Historia animalium* held the view that animals can perceive and form images, taking as proof the fact that they dream and possess memory. In the case of Washoe and other chimpanzees, simple questions about desires for the near future have been asked by the human investigator and answered by the chimpanzee.

In Griffin's view, the customary response of many biologists or psychologists to the question of subjective feelings or conscious intent in animals is to take a cautiously agnostic position, maintaining that subjective or mental states are beyond the reach of scientific enquiry. The behaviourist position stated at its scholarly best (Lashley 1923) is essentially agnostic. But there remains a tendency for what was originally an agnostic position to drift implicitly into a sort of *de facto* denial that mental states or consciousness exist outside our own species. It is a great difference from the viewpoint of Charles Darwin, who took it for granted that animals had mental experiences and emotions, as reflected in the title of one of his major books (*The Expression of the Emotions in Man and Animals*, 1872). Bierens de Haan (1947–1948) in his paper on animal psychology and the science of animal behaviour, which analyses some elements in the emergence of objectivism, states that there is more in an animal's world than stimuli in its surroundings and movements of its limbs. And as we know this, we also wish to know what it is that goes on in its mind.

Griffin (1976) summarizes his viewpoint by saying that the available evidence concerning communication behaviour in animals suggests there may be no qualitative dichotomy, but rather a large quantitative difference in complexity of signals and range of intentions, between animal communication and human language. Strict behaviourists have said it is more parsimonious to explain animal behaviour without postulating that animals have any mental experiences. Yet, he says, the behaviourists hold mental experiences to be identical with neurophysiological processes. He continues:

> Neurophysiologists have so far discovered no fundamental differences between the structure or function of neurons and synapses in men and other animals. Hence, unless one denies the reality of human mental experiences, it is actually parsimonious to assume that mental experiences are similar from species to species as are the neurophysiological processes with which they are held to be identical. This, in turn, implies qualitative evolutionary continuity (though not identity) of mental experiences among multicellular animals.
>
> The possibility that animals have mental experiences is often dismissed as anthropomorphic because it is held to imply that other species have the same mental experiences a man might have under comparable circumstances. But this widespread view itself contains the questionable assumption that human mental experiences are the only kind that can conceivably exist. This belief that mental experiences are a unique attribute of a single species is not only unparsimonious; it is conceited. It seems more likely than not that mental experiences, like many other characters, are widespread, at least among multicellular animals, but differ greatly in nature and complexity.
>
> Awareness probably confers a significant adaptive advantage by enabling animals to react appropriately to physical,

biological, and social events and signals from the surrounding world with which their behaviour interacts.

Huxley (1960) accepted the view that consciousness was adaptive with survival value and would be favoured by natural selection.

Data of seeming great moment have come from studies by Gallup (1979), in relation to a prevailing view in psychology that the issues of self-awareness and consciousness are not amenable to objective analysis in animals. That is, the view that no falsifiable hypotheses have been presented which are susceptible to experimental tests leading to a verdict of true or false. Gallup has used the technique of mirrors to approach the logical problems of self-awareness and render them more objectively analysable. He remarks that in front of a mirror any visually capable organism is an audience to its own behaviour. But most animals react to themselves in a mirror as if they were seeing other animals, engaging in a variety of species-typical social responses directed towards the reflection. While mirrors permit organisms to see themselves as seen by others, many animals seem incapable of recognizing the dualism implicit in such stimulation. Even after prolonged exposure they fail to discover the relationship between their behaviour and the reflection of that behaviour in the mirror. Yet mirrors hold a peculiar fascination for them.

An interesting question would seem to emerge from Gallup's affirmation that most animals react to themselves in a mirror as if they were seeing other animals. Certainly some birds in the bush when encountering a glass reflecting surface will attack their own reflection, seemingly to exhaustion. But why do wild animals coming to a pond or any smooth water surface to drink, when often a clear reflection of themselves must result, not react in this way? Obviously it would be biologically disadvantageous if this provoked the aggressive reaction to a conspecific when their head approached the water. Possibly in herd animals, the behaviour of the group in initiating drinking would counteract any startle and aggressive reaction in the young.

Gallup states that self-recognition in humans is learned. People with congenital visual defects who undergo corrective surgery later in life react initially to themselves in mirrors in the same way as animals. Children younger than 6 months of age also react in this way, and do not show signs of self-recognition until they reach 18–24 months. In pondering an evolutionary discontinuity between animals and man and their ability to recognize themselves, Gallup studied whether chimpanzees were able to decipher mirrored information about themselves. Adolescent chimpanzees exposed to a mirror for the first time invariably responded initially as though in the presence of a number of chimpanzees, engaging in a variety of social gestures while watching the reflection. But after 3 days the tendency to treat the reflection as a companion disappeared. They then began using the mirror to respond to themselves, the reflection being employed to gain visual access to and to experiment with otherwise inaccessible information about themselves. This involved grooming parts of the body they had not seen before, inspecting the insides of their mouths and making faces at the mirror. Following this Gallup made another series of experiments in which he anaesthetized chimpanzees and painted red odourless dye on the uppermost part of one eyebrow ridge and on the top of the opposite ear. These marks were impossible for the animals to see without the use of a mirror. When presented with the mirror the chimpanzees on seeing their reflection attempted to touch the marked areas on themselves while looking into the mirror and showing their renewed interest in the reflection. There were noteworthy attempts to examine the marks visually and to smell the fingers which had been used to touch the marked portions of the face.

These observations, which were begun in 1970, were extended by Lethmate and Dücker (1973) with studies on orang-utans. It has emerged that all primates except man and the great apes fail to exhibit self-recognition even after extended exposure to mirrors. Gallup therefore suggests that in attempting to resolve one apparent evolutionary discontinuity, he appears to have uncovered another. His interpretation of the data taken collectively is that most primates lack a cognitive category which is essential for processing mirrored information about themselves. The nature of this cognitive deficit may be tied to the sense of self, for if you do not know who you are, how could you possibly know it is yourself you are seeing when you look in a mirror? The capacity to infer correctly the identity of the reflection presupposes a sense of identity on the part of the organism making that inference, so perhaps the monkey's inability to recognize itself is because of the absence of a sufficiently well integrated self-concept. Self-recognition may represent an emergent phenomenon which occurs only when a species acquires a certain number of cortical neurones with sufficiently complex interconnections. Alternatively one might propose a threshold model. Different organisms could have different degrees of self-awareness but only with an explicit sense of self does self-recognition become possible. Thus the threshold of self-recognition might be quite high compared with the threshold for other ways of conceiving of oneself.

The existence of consciousness in animals is one central question in the dialogue between Popper and Eccles (1977) in their book *The Self and its Brain*. Popper proposes that if the evolutionary story applies to life and consciousness there ought to be degrees of life and degrees of consciousness. Degree of consciousness is a fact of human experience, as shown by the various states between deep sleep with complete loss of consciousness, and full awakening with self-awareness. This may involve making a link within our brain with memories of previous periods, which is more understandable than the creation of these states of consciousness from nothing. He goes on to say that a new-born child probably has nothing we would call memory but does, of course, have some sort of knowledge or information or expectations and it has to synthesize consciousness. He goes on to propose that consciousness has come to exist by degrees and that anything like conscious awareness—not self-consciousness—can probably only be attributed to animals with a nervous system. Whereas the evidence that other people have minds is infinitely better than the evidence we have that animals have minds, the evolutionary hypothesis more or less forces us to attribute lower degrees of consciousness to animals. Popper disagrees with Eccles' essentially agnostic viewpoint on the issue of animal consciousness, but adds that the only evidence in support of animals having experiences as we do is the evolutionary hypothesis and the degrees of consciousness we find in ourselves. Thus he would describe the problem of consciousness in animals as metaphysical in the sense that any hypothesis about it is not falsifiable—at any rate not at present.

It would seem the experimental work of Gallup (1979) on pongids with mirrors cited above would allow a review of this position.

Popper sees full consciousness as depending on having an abstract theory which is linguistically formulated. Its beginning in animals is an extension of the state of curiosity beyond the sensory stimuli which brings the state of curiosity about—to a lasting curiosity which leads to exploration.

Eccles warns that it is necessary to be very careful to avoid anthropomorphizing the situation when we imagine what is happening to an animal in its waking life. There is the important point that they do not have a proper sense of time, and that they live in the present. Of course they learn from experience, constantly enquire, and give evidence of many purposive performances, but Eccles thinks that at best this is only an indication of consciousness. The complex social life of animals gives bias towards thinking they have some consciousness. They also give evidence of pain when injured, which we think is like the pain we suffer, and of happiness and anticipation (e.g. a dog going for a walk with its master). But in Eccles' view it is uncertain whether they are having experiences as we do. For example, the commonest reaction cited is that to pain, but a decorticate animal can still react to pain, and show rage and fear—in fact the whole range of adversive reactions—so the higher levels of the cerebral cortex are not necessary for reacting to injury. This can all be done when you are unconscious (Popper and Eccles, 1977).

However, in my view it is doubtful whether the classic happiness and anticipation in the dog setting out for a walk with its master and its unfavourable reaction to the path he chose to take to his experimental hot-house—an example cited graphically by Darwin in *The Expression of Emotions in Man and Animals*—could be approached by the above line of reasoning.

Eccles goes on to suggest that the possibility of consciousness in animals should be based on more subtle things such as their relationships with one another and with human beings. Thus, he says, it is nice to see how they get on together as living beings in company with one another and with other species, but how do they care for their sick and their dead? He notes that elephants have high intelligence, and that there is some evidence that they care for their dead (though this may be imitative). There is anecdotal evidence that when an elephant dies the other animals cover the body with leaves and even tend the bones. The dolphins, which have a brain at least as large as man's, apparently do show some feeling for one another and help when a cow gives birth. Eccles says these could be ordinary instinctive animal actions, but human qualities could be read into their actions because they have such large brains and quite clearly have very complex performances. (Parenthetically, it seems to me that there seems no reason why 'ordinary instinctive animal actions' such as courtship and copulation, or satiation of thirst, should not be conscious and involve exercise of options.)

He places particular stress on ceremonial burial customs and agrees with Dobzhansky that these give the first clear indication that primitive man has developed some spirituality, some self-consciousness which he is not only experiencing in himself but also recognizing in his fellow beings. In relation to this very important question, Eccles suggests that if animals have consciousness, they don't have self-consciousness even at a minor level. He elaborates by stating that if one wanted to be difficult about it one could perfectly well be a reductionist identity theorist, saying all the performance of animals of every kind you can imagine is simply the performance of their neural machinery and there is no need to be superimposed upon this something which is a spin-off from the brain action. Thus with animals, we may become parallelists stating that their conscious experiences are a spin-off from the neural actions, but in fact cannot act back and cause any change in the operations of the neural machinery.

The phenomenon of psychogenic sudden death

In relation to this line of reasoning of Eccles, and the evidence generally on consciousness in animals, it is both pertinent and fascinating to consider here a series of experimental investigations by the great psychobiologist Curt Richter.

Richter's enquiry into the phenomenon of sudden death in animals has its parallel in Walter Cannon's interest in voodoo death. As Richter points out in the introduction to the account of his studies on domestic and wild rats (Richter 1957), Cannon (1942) cites many instances of mysterious, sudden, apparently psychogenic death occurring in primitive peoples in many parts of the world. For example, a Brazilian Indian condemned by a so-called medicine man is helpless against his own emotional response to this sentence and dies within hours. In Africa a young black unknowingly eats a wild hen, something which is inviolately banned. On discovery of what he has done, he trembles, is overcome by fear and dies within 24 hours. A classic instance cited is bone-pointing in the Australian aborigines. A man having been 'boned' by an enemy believes that nothing can save him. Basedow (1925) gave a vivid description of this in his book on the Australian aboriginal. A man who has been 'boned' stands aghast with eyes staring at the treacherous pointer. His hands are lifted to ward off the lethal medium which he thinks is pouring into his body. His cheeks go white, his eyes become glassy and his facial expression is distorted. He attempts to shriek but the sound usually chokes in his throat and all that is seen is froth at his mouth. He trembles and twitches. He sways backward and falls to the ground and appears to be in a swoon. He finally composes himself, goes to his hut and there frets to death.

Cannon concluded that the phenomenon is characteristic of aborigines, that is, human beings so primitive, so superstitious as to feel themselves bewildered strangers in a hostile world containing all manner of evil spirits capable of affecting their lives disastrously. Cannon, in asking the question how an ominous and persistent state of fear can end the life of a man, turned for explanation to his experiments on the sympatheticoadrenal system, proposing that death would come in a state of shock produced by continuous outpouring of adrenaline. Individuals would be expected to breathe very rapidly, have a rapid pulse and haemoconcentration. There would be increasing cardiac rate and death would ultimately occur with the heart in systole.

Richter himself studied the effects of extremely severe stress on wild and domestic rats. This was done by measuring endurance in the face of having to swim in specially designed glass cylinders 20 cm in diameter and 75 cm in depth. A jet of water at any desired temperature played on the surface to preclude the animal from floating, while the collar of the cylinder prevented escape. Richter found (Fig. 19-2) that the survival swimming times were directly related to the temperature of the water: thus they were from 10 to 15 min at 17–23 °C, up to 60 h at 35 °C, and 20 min at 40.5 °C. These figures were applicable in general to domestic rats and the major point emerging from Richter's enquiry was that the outcome with fierce aggressive and suspicious wild rats recently trapped was usually quite different.

Richter had observed some years earlier that cutting off the whiskers of a tame domestic rat had been followed, surprisingly, by evidence of disorientation of the animal and death about 8 min later. He found with domestic rats in this test described above that if the whiskers were trimmed before putting the animal in the swimming tank at 35 °C it swam around excitedly on the surface for a

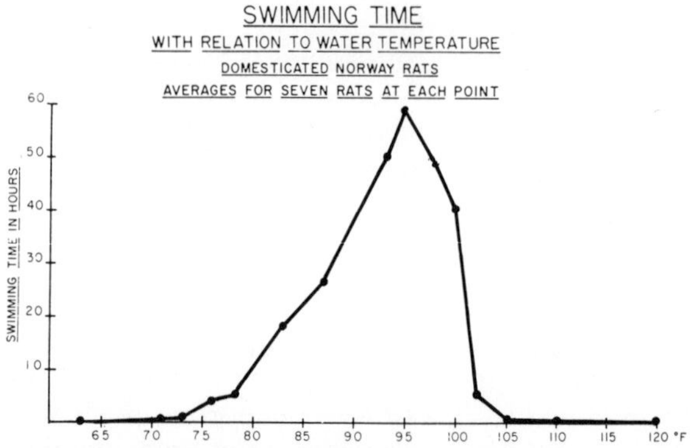

Fig. 19-2. Curve showing average swimming time (end-point is drowning) of unconditioned tame domesticated Norway rats with relation to water temperature. Average for seven rats at each point. (From Richter 1957, with permission.)

short time then dived to the bottom where it began nosing its way around the glass wall and died within 2 min. It seemed with the wild rats that very often they died within a few minutes of being placed in the swimming tank. Of 34 wild rats which had their whiskers and facial hair trimmed in this fashion all died within 1–15 min after immersion. In the course of taking wild rats out of their cages it was also observed that some died simply while being held in the hand. Some died when being put into the water directly from their cages without being held. Being held in the hand and then put in the water was more often lethal. The trimming of the whiskers added to this situation.

In examining the mode of death of these animals, electrocardiogram records (Fig. 19-3) showed that, contrary to Cannon's impression, there was a gradual slowing of the heart and at autopsy a large heart distended with blood was found. Whereas there was a short-lived initial increase in heart rate, the data indicated that the rats had died a central vagus death resulting from overstimulation of the parasympathetic rather than the sympatheticoadrenal system. In some rats the slowing of the heart occurred very promptly, in others only after a few minutes. The influence of parasympathetic over-activity was supported by experiments in wild and domestic rats involving blockage with drugs or accentuation of parasympathetic activity by other pharmacological agents. Richter proposes that the situation of these rats is essentially one of hopelessness: whether they are restrained in the hand or confined in the swimming jar they are in a situation against which they have no defence. This reaction of hopelessness is particularly shown by wild rats and can occur very soon after being grasped in the hand and prevented from moving. The rats seemed to literally 'give up'.

Support for the assumption that the sudden-death phenomenon depends largely on emotional

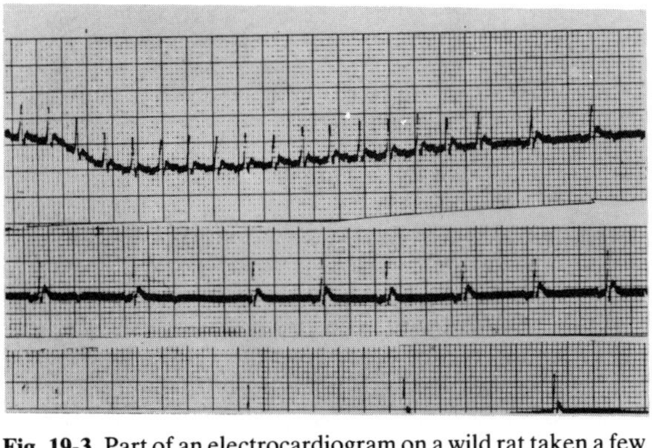

Fig. 19-3. Part of an electrocardiogram on a wild rat taken a few minutes after the rat's immersion in a swimming jar. (From Richter 1957, with permission.)

reactions to restraint or immersion comes from the critical observation that after elimination of the hopelessness the rats do not die. That is, if they are repeatedly held briefly or immersed in water for a few minutes and then retrieved the rats learn quickly that the situation is not actually hopeless. Thereafter they again become aggressive, try to escape and show no signs of giving up. Wild rats after such experience swim just as long as domestic rats or longer. The remarkable speed of recovery of such wild rats is also noteworthy. Once freed from confinement in the glass jar a rat that quite surely would have died in another minute or two becomes normally active and aggressive in only a few minutes. A primitive man, when freed from voodoo, is said to recover almost instantaneously even though he had recently seemed more dead than alive.

Richter suggests that the situation which may result in sudden death may be a one time occurrence both in rats and man, and in any particular circumstance ending either in death or in immunity from this particular kind of death. In human beings as well as in rats we see the possibility that hopelessness and death may result from the effects of a combination of reactions, all of which may operate in the same direction and increase the vagal tone. The 'boned' victim, like the wild rat, is not set for a fight or flight but simply seems resigned to his fate. His situation seems to him hopeless. Richter remarks that the incidence of response varies inversely with the degree of civilization or domestication since it occurs more frequently in wild than in domestic rats and, so far, certainly has been described chiefly in primitive man—that is in creatures living in precarious situations. However he notes that there are many instances in medical practice of sudden death from fright, sight of blood, hypodermic injections or from sudden immersion in water, and during the Korean War unaccountable deaths were reported among US soldiers. At autopsy no pathology was observed and they were apparently in good health. In the same vein Dr R. S. Fisher, coroner of the city of Baltimore, states that a number of individuals die each year after taking small, definitely sublethal doses of poison or after inflicting small non-lethal wounds on themselves, apparently dying as a result of belief in their doom.

In relation to this investigation by Richter, it might be reasoned that sudden death following manual restraint in the wild rat could be a reflex autonomic discharge consequent on extreme fear. But the death after some minutes of swimming round the tank implies the possible operation of some configurational and situational perception. The fact of swimming by the wild creatures for

50–60 h, after having previously experienced rescue from the situation, supports the contention of the animals not merely being awake, but able to make some evaluation of their predicament based upon perception of the situation and past experience (if there were any). It could be argued that this is coming close to a sense of existence of self and expectation of outcome, plausibly to a process involving at least conscious awareness in the animals if not self-awareness. On the face of it this is contrary to the view of Eccles (Popper and Eccles 1977). Certainly the animals' apparent perception of a situation profoundly influences neural events and visceral processes, which again is much at variance with Eccles' view that 'their conscious experiences are a spin-off from neural actions but, in fact, cannot act back and cause any change in the operations of the neural machinery'.

Components of awareness

Thorpe (1966) proposes that the term consciousness involves perhaps three basic components. First an inward awareness of sensibility, an internal perception. Second an awareness of self, of one's own existence. Third the idea of unity: that is, it implies in some rather vague sense the fusion of the totality of impressions, thoughts and feelings which make up a person's conscious being into a single whole. As Lashley (1954) points out, the process of awareness implied in the belief in an internal perceiving agent, an 'I' or 'self' which does the perceiving, leads inevitably to the conclusion that this agent selects and unifies elements into a unique single field of consciousness. Thorpe notes that the fact of attention and the shifting of attention, like the beam of a searchlight, from one aspect of the perceptual world to another is one of the most obvious features of human conscious mental organization. Also there is the question of integration of the present. Fessard (1954) has defined consciousness as the integrated perception of the present and this may be supposed as involving the matching or comparison of incoming current information with previously stored information. That is, we have a congruence between an incoming pattern of impulses and what may be called a filing card of past experience. Thorpe quotes Rioch as suggesting that anticipation of one or other of several possible responses is a necessity for consciousness. Thorpe remarks in relation to consciousness and animals that whereas we cannot give final proof, we can bring evidence to bear which is cumulatively highly impressive and does, he believes, give powerful reasons for concluding that consciousness is a widespread feature of animal life.

From the many instances from several lines of evidence which Thorpe cites, it is of interest to mention one or two examples here. He suggests that striking evidence of anticipation comes from work with captive monkeys in the course of experiments to reveal what can be termed 'reward expectancy'. If a rhesus monkey is allowed to see some food placed under one of two or more containers, for example under inverted cups, and is subsequently allowed access to these cups, it shows a high ability to choose the correct one. If in a subsequent experiment a piece of banana is hidden under one cup while the monkey is watching, and then with a screen placed in front of the monkey the experimenter quickly substitutes for the banana a less desirable food such as a lettuce leaf, the monkey will again go to the cup where he expects to find the preferred food. On finding not banana but lettuce he may not only reject the lettuce (though normally in the absence of banana he would have eaten it) but may even throw a temper tantrum as a result of his disappointment. This and similar experiments on rats and chimpanzees seem to provide overwhelming evidence for a learnt searching image. That is, there is a definite and precise expectancy and a disappointment at failing to find what is expected and sought. In discussing the behaviour of dolphins, Thorpe notes that it is presumed that the female gives birth in isolation. A female giving birth in captivity proved to be very exciting to the males, some of which became aggressive. The other females gathered round the one in labour and helped to ward off the attacking mate. When the newborn infant began its first gradual ascent to the surface to breathe, another female accompanied the mother, swimming just below the infant as if in readiness to support it (McBride and Hebb 1948).

Thorpe cites these analogies to ethical values together with evidence of aesthetic values, anticipation and expectancy, self-awareness, ideation and manipulation of abstract ideas, and attention mechanisms as providing increasingly suggestive data as we go up the evolutionary scale for the existence of consciousness and self-consciousness in animals. In short, the production of consciousness may have been an evolutionary necessity. He remarks on William James' supposition that consciousness is what might be expected in an organ added for the sake of steering a nervous system grown too complex to regulate itself. The evolutionary justification for expectancy is that it cuts down the cost of adjustment. If an event that calls for action is part of a regularly occurring sequence of events, its predecessors can serve as warning signals. Expectations thus open the way for preparatory and avoidance responses, or to prior selection of response by reasoning, if the animal is capable of such reasoning. Expectations may diminish stress in that they make adjustment less abrupt. Some anticipatory mechanisms can be envisaged as functioning without the necessity of anything we would call consciousness, but the anticipatory behaviour of some of the higher animals seems to any intelligent observer to give an overwhelming impression of conscious and deliberate foresight. At lower levels consciousness, if it exists, may be of a very generalized, so to say unstructured, kind. But with the development of purposive behaviour and a powerful faculty of attention, consciousness associated with expectation will become more and more vivid and precise (Thorpe 1966).

A major discovery by R. W. Sperry (1974) has advanced knowledge in this area which, perhaps, is the greatest intellectual challenge in biological science. This is that there are two separate streams of consciousness in the central nervous system. His animal experiments involving brain bisection led to the conclusion that each of the surgically separated hemispheres must sense, perceive, learn and remember quite independently of the other. In humans where disconnection of the two hemispheres has been carried out for treatment of epilepsy by division of corpus callosum, anterior and hippocampal commissures and massa intermedia, the results indicate that each hemisphere is largely oblivious of the cognitive experience of the other. This is illustrated in many ways. For example, when a right-handed subject (thus with dominant left speech hemisphere) is blindfolded and a familiar object such as a pencil, cigarette or coin is placed in the left hand, the right (subordinate, mute or minor) hemisphere connected to the left hand feeling the object perceives and appears to know quite well what the object is. It cannot express this knowledge in speech or in writing, but can manipulate the object correctly and demonstrate how the object is supposed to be used. It can remember the object and go out and retrieve it with the same hand from among an array of other objects either by touch or by sight. The left (dominant, or speech) hemisphere has no conception of what the object is and, if asked, will say so or guess. But if the right hand crosses over and touches the object, or it makes some distinct sound such the rattle of keys, the speech hemisphere gives the correct answer. The evidence indicates the subordinate hemisphere has its own perceptions, reactions and memories. Sperry reports emotional feelings generated by the subordinate hemisphere, such as a smile on completing a task with the left hand, or frowning at an incorrect verbal response or inept action by the right hand when only the subordinate hemisphere knows the correct action.

In other studies on subjects with disconnected hemispheres, two sensory images perceived in right and left hemispheres were arranged to be different and conflicting (Levy et al. 1972). This was done by splitting visual images down the middle and then combining different half-images at the midline to make composite right–left chimaeras. The chimaeric figures were flashed to the subject while the gaze was centred. With rival perceptual processes set up on the left and right sides, it was possible to determine which hemisphere was more proficient in dominating the response (Fig. 19-4). In general, if linguistic processing was involved (i.e. the subject was asked to name the image) the response was dominated by the left hemisphere selecting in favour of the right half of the composite stimulus. On the other hand, in non-linguistic manual readout involving perception of faces and complex patterns and in direct visual–visual matching of shape or pattern the right hemisphere was dominant. The results with chimaeras also demonstrated that the disconnected minor hemisphere is capable of capturing and controlling the motor system under conditions where the two hemispheres are in equal and free competition, that is where the sensory input is equated and the subject is quite free to use either left or right hand.

Sperry's interpretation is that the disconnected minor hemisphere is indeed a conscious system in its own right, perceiving, thinking, remembering, reasoning, willing and emoting, all at a characteristically human level, and that both the left and the right hemisphere may be conscious simultaneously in different, even in mutually conflicting, mental experiences that run along in parallel. Though predominantly mute and generally inferior in all performances involving language or linguistic or mathematical reasoning, the minor hemisphere is nevertheless clearly the superior cerebral member for certain types of tasks. That is, with non-linguistic manual readout either hemisphere may dominate. With perception of faces and of complex or nondescript patterns—or any direct visual–visual matching—the right (mute) hemisphere is dominant. When a conceptual translation or verbal reply is required, the dominance promptly shifts to the left hemisphere.

Sperry believes that the lack of unity in consciousness points to the basic truth that consciousness is not some kind of entity in itself but a functional property of brain in action.

In my view, there would need to be compelling reasons, not immediately apparent objectively, to

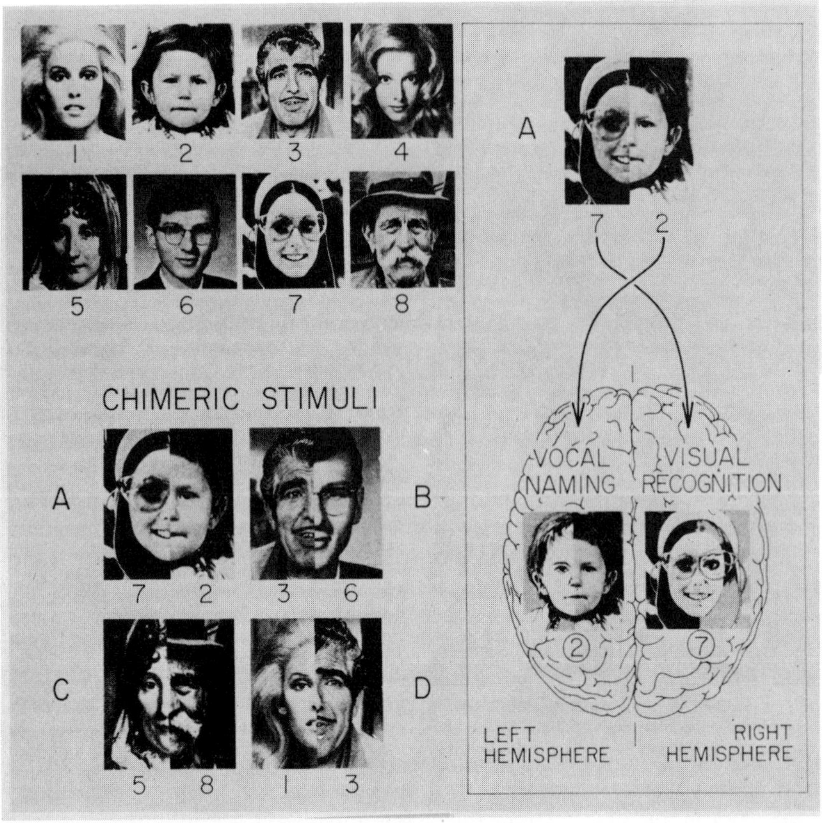

Fig. 19-4. Test for simultaneous perception of faces by the two hemispheres. The diagram on the right shows the projection of a chimaeric stimulus to both halves of the brain to produce two complete dual different percepts. See text for details. (From Levy et al. 1972, with permission.)

depart from a Darwinian approach of consciousness as a brain function. That is, great survival advantage has accrued to animals with the increasing complexity and diversity of this neural function. This is particularly apparent in the development of the capacity to exercise options of action in the light of both contemporary sensory inflow and stored information. The stored information may be phyletic memory or instinct, or derived from the impressive experience of the individual. The emergent survival advantage of plasticity over rigid stereotype as increasingly complex animal forms colonized more diverse ecological situations can be strongly argued, just as it appears obvious when we consider the issue of survival in man and other higher animal forms. Thus the gradual development of consciousness and situational awareness with phylogenetic progression would parallel that of other functions, with spectacular acceleration relative to all others at the stages of evolution of the hominids. Here quantal jumps in development might be proposed: for example the emergence of self-awareness in high pongids as shown by Gallup's (1979) experiments with mirrors. And the accounts of Marais (1939, 1969) of baboon behaviour warrant some consideration on the same grounds. He described the care and defence of the baboon troop by the big males, who showed reckless courage in warding off attacks by leopards and shared food with a female of choice, 'the high water mark of baboon unselfishness' (Marais 1969). There is the emergence of tool-making in the australopithecines, which is the shaping of material for an imagined eventuality and is therefore akin to Aristotle's definition of the artist. And there is the ceremonial burial of the dead that is first seen with Neanderthal man, which speaks of identity with ones fellows in awareness of the ephemeral nature of life and the inevitability of death, the sadness of breaking personal bonds and perhaps the beginning of the making of myths.

Deep-seated brain stimulation in the human and other species

Deep-seated brain stimulation in the human has been done in the course of stereotaxic operations for hyperkinesis or Parkinsonism in awake co-operative patients. Sometimes a smile or definite laughing can be produced and reproduced by electrical stimulation (Hassler and Riechert

1961). The patient cannot suppress this if asked to. Afterwards he says that everything seemed to be funny or that the same amusing situation in his former life came into his mind each time. Hess (1964) remarks in relation to these experiments that one has to conclude that stimulation does not directly affect the facial musculature which has to be co-ordinated for laughing but determines an objective expression of certain contents of consciousness. The effect has been produced from the inner margin of the ventral oral nucleus of the thalamus which projects to the motor cortex. A large number of patients reacted in the manner described, and about 9% of these repeatedly.

Heath (1972) described some remarkable studies involving both electrical and chemical stimulation and also electrical recording from electrodes and cannulae placed deep in the brain of two patients. Recordings from multiple sites were made also. One patient was undergoing treatment for severe mental illness, and the other for intractable epilepsy. In the female epileptic patient, injection of acetylcholine or noradrenaline was made at weekly intervals on 16 occasions into the septal region. On 12, there were characteristic changes independent of which chemical was used. One to two minutes after bilateral injection, the patient's mood would gradually begin to change with elevation to mild euphoria within 10–15 min. The change led to a sexual motive state and within another 5–10 min this culminated in repetitive orgasms. The patient described this when questioned and her sensuous appearance and movements confirmed it. In another patient, a male, with temporal lobe abnormality, similar septal injections of acetylcholine evoked intense feelings of pleasure and sexual motive state.

In the patient with severe mental illness, stainless steel electrodes were placed in right mid-septal region, right hippocampus, left and right amygdalae, right anterior hypothalamus, right posterior ventral lateral thalamus, left caudate nucleus, and at two subcortical sites in the left cerebellum. Cortical leads were placed under the dura at left and right frontal and parietal regions, and right temporal regions. Silver ball electrodes were implanted into left anterior and left posterior septal regions. The patient responded to passive stimulation with pleasure only when electrical stimulation was applied to the septal region. When supplied with a three-button self-stimulating device, based on the technique used by Olds and Milner (1954), it was found he repetitively stimulated only the septal region—1000 to 1500 times in 3 h—and protested when the unit was taken from him. The stimulation caused feelings of pleasure, alertness and warmth, with sexual arousal and compulsion to masturbate. The protracted treatment appeared to change his sexual orientation in that previously he had been exclusively homosexual whereas after 10 days of treatment he had a continually growing interest in women and wish for heterosexual activity, and eventually achieved sexual intercourse with orgastic response with a prostitute. As well as contriving these changes in conscious processes and behaviour, the studies involved comprehensive electrical recordings during sexual arousal and orgasm (Figs. 19-5 and 19-6). Both patients displayed the same

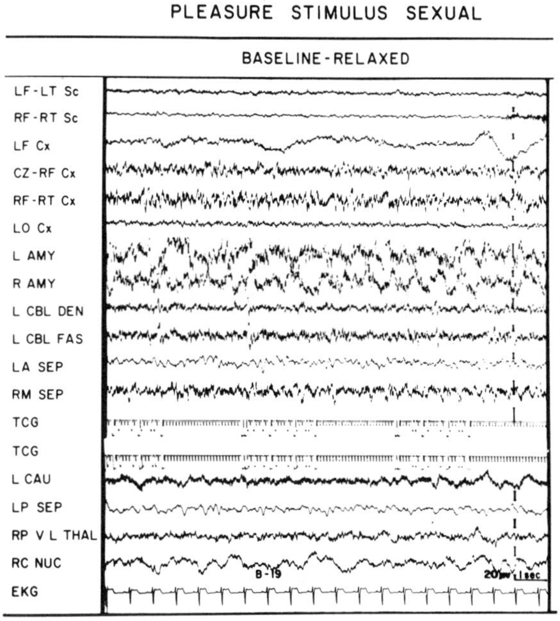

Fig. 19-5. Deep, cortical, and scalp electroencephalograms obtained from patient B-19 during a period of relaxation. LF-LT Sc, left frontal to left temporal scalp; RF-RT Sc, right frontal to right temporal scalp; LF Cx, left frontal cortex; CZ-RF Cx, central zero to right frontal cortex: RF-RT Cx, right front to right temporal cortex; LO Cx, left occipital cortex; L AMY, left amygdala; R AMY, right amygdala; L CBL DEN, left cerebellar dentate; L CBL FAS, left cerebellar fastigius; LA SEP, left anterior septal region; RM SEP, right midseptal region; TCG, time code generator machines 1 and 2; L CAU, left caudate nucleus; LP SEP, left posterior septal region; RP V L THAL, right posterior ventral lateral thalamus; RC NUC, right central nucleus; EKG, electrocardiogram. (From Heath 1972, with permission.)

electroencephalogram pattern with orgasm. The most striking and consistent changes were recorded from the septal region (Fig. 19-6). There was appearance of spike and slow-wave with superimposed fast activity. In one patient, distinct, but less dramatic, changes also occurred in amygdalae, thalamic nuclei, and deep cerebellar nuclei. The occurrence of septal spiking with

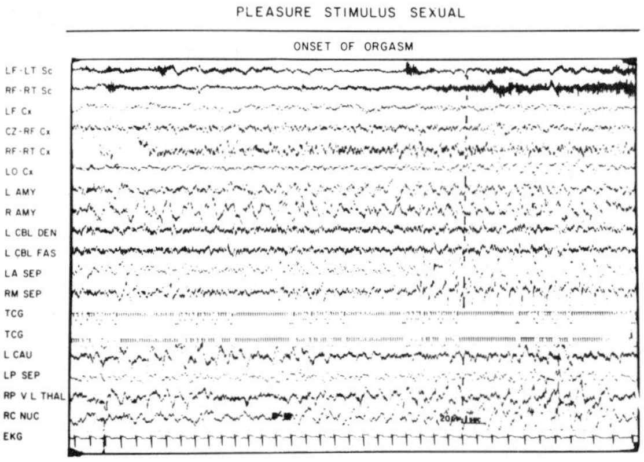

Fig. 19-6. Deep, cortical, and scalp electroencephalograms obtained from patient B-19 with onset of orgasm. See key for leads in legend to Fig. 19-5. (From Heath 1972, with permission.)

psychotic states has been noted by this Tulane group as well as workers at the UCLA Brain Research Institute (Heath 1972).

Thus the efficacy of local electrical stimulation in evoking highly organized neural activity involving change of consciousness in the human is clear: a specific content of consciousness may be correlated with a definite cerebral substrate (see also Sem-Jacobsen and Torkildsen 1960, who showed the evocation of various smell sensations by deep-seated stimulation).

In the instance of the sudden delivery of a chemical stimulus to the human brain via the carotid artery blood as contrived by rapid intravenous injection of concentrated salt solution, the stream of consciousness is altered by emergence of a powerful sensation of thirst concomitant with an urge to drink. This thirst was often preceded by a drying of the mouth dissociated from the sense of thirst. Conversely, with glucose injections in hyperdipsic men, the mouth moistened considerably before the desire to drink abated (Holmes and Gregersen 1947; Wolf 1950). With the same experiment in the sheep, except that the hypertonic salt stimulus is delivered directly into the cerebral arterial blood, drinking of water occurs in 30–120 s. It is very often preceded by mouth movement and lip-smacking and anticipatory behaviour quite similar to that seen immediately before water is presented to a water-deprived animal. However, it needs be recognized that the tongue is also being perfused by the high-salt-content blood, so taste sensations of saltiness could arise. This precludes secure attribution of this behaviour in the animal to thirst analogous to the sensation established in the human experiment. But the specificity of choice of water by such experimental animals in a cafeteria situation (Abraham et al. 1976), in contrast to the choice of concentrated salt solution in similar experimental context when the animal is sodium-deplete, is presumptive evidence of some specificity of conscious sensation being evoked, as occurred in the human experiment. Thus, we are not dealing with a simple stimulus–response reflexive ingestion of fluid from the most proximate source but one of choice between alternatives, where the outcome is both reproducible and apt for the internal chemical state. Immediate drinking of water can also be elicited in sheep by injection of hypertonic saline into lateral or third ventricle.

It would not seem to me too great a hurdle to attribute to the sheep the capacity of conscious recognition of a particular significance amongst diverse sensory input as, for example, the class of situation where it frenetically searches from container to container in a cafeteria type of experiment, or for that matter, in the case where what have been termed type II conditioned reflexes emerge under conditions of laboratory life. Here animals which have become sodium-deplete numerous times over months as a result of parotid fistula loss will come to associate the entry of specific persons into the laboratory, and their actions, with the imminent presentation of salt solution. Eventually they will, when sodium-deplete, stamp, jump, kick, bleat and behave noisily when that person arrives. This is independent of whether he or she shows any activities premonitory of salt presentation or not. When sodium-replete the animal does not show the behaviour. The behaviour, of itself, has had no influence on whether or not sodium solution has been presented during experimental regimes of sodium deprivation, or in normal maintenance of the animals, so that no direct association between behaviour and reward held, though the animal might have made such an anticipatory association.

With the goat, immediate drinking of water can be caused by microinjection of hypertonic saline in the hypothalamus (Andersson 1953). Also electrical stimulation between the fornix and the tractus Vicq d'Azyr within the hypothalamus is capable of initiating a very large water intake in previously water-satiated animals (Andersson and McCann 1955). Presumptive evidence that the need for water—that is, thirst—is the drive evoked is provided by a more complex experiment. By letting the animals become water-deplete, Andersson and Wyrwicka (1957) trained goats to climb a staircase to obtain water to satisfy their desire to drink. However, they were not given enough to satiate themselves and to get further water they had to go down the staircase and then up again once more when they could get the next ration (100 ml each time). The water-deplete animal repeated these movements over 10–15 trials every 1–2 min. If now the goats were stimulated in the hypothalamus in the region of the 'drinking centre' as noted above, even though fully satiated with water they went up the staircase and drank. They exhibited exactly the same motor behaviour and, in one animal, impatient stamping with the hoof on the top stair occurred. As the procedure was repeated and the goat became overloaded with water, the latency increased and amount drunk on each stimulation decreased. But as much as 6.5 litres was drunk in one experiment and increasing the stimulus caused the reaction to grow stronger again. In between stimulation episodes, the goat refused water and stimulation of other diencephalic areas did not produce the response.

Hess (1964) cites this experiment as a sure sign that the diencephalic stimulation evoked a specific feeling corresponding to the one that comes about if an animal has remained without water. He emphasizes that with the stimulation experiments the feeling was so strong that the animals would accept water contaminated with bitter substances, acid or salt. He likens the situation to central polydipsia in man where the patient reports he is tortured by an unquenchable thirst, and tries to get fluid in all possible ways. The relatively complicated behaviour of the goat, which involved a sequential performance, gives support to the idea of it being truly motivated as distinct from some motor automatism. However, following the line of reasoning above on evocation of a central state of thirst, it would be interesting to know the outcome with a cafeteria choice situation at the top of the stairs. Given that a water-deplete goat would choose water in preference to hypertonic saline and other salts, would the drinking with electrical stimulation be indiscriminate, i.e. from the first bin encountered, or would tasting and choice of water occur? If predominantly the latter it should give further support to the electrical stimulation having engendered a specific state. This view would be reinforced if it were shown that under differing conditions, i.e. those of sodium deficiency, the same goat, unstimulated, would choose a very hypertonic salt solution from the same cafeteria of different solutions as sheep do (Abraham et al. 1976). That is, the more complex the situation conditional upon arriving at a specific goal, the stronger the presumption that discrete electrical stimulation which reproducibly contrives it is influencing inbuilt elements which subserve the motivation and produce the characteristic conscious state.

In this regard Stricker and Miller (1968) showed that drinking evoked by microinjection of carbachol in the anterior hypothalamus mimicked closely, in relation to preference–aversion for sodium chloride solutions of differing concentration, the drinking produced by 16 h of water deprivation. They suggest the drinking is due to direct activation of a thirst motivational system. However, in Valenstein's view (1970) it cannot be known that the natural motivational drive is evoked, that is that the natural concomitant conscious sensation is evoked.

Hess (1964) considered this issue in detail in his book *The Biology of Mind*, as indicated above. He remarks that one frequently gets symptoms of a biological nature which enable one to guess what is going on in another person even when one does not rely on language. He instances that a man may get symptoms such as heat, thirst and hunger arising after a climb in the mountains in the sun. He takes off his pack and goes to a nearby brook to get water and then sits down in a shady place to eat. His companion does likewise without comment. There is no reason to doubt the interpretation that similar conditions of consciousness prevailed. One cannot give positive proof but it seems close to certainty. In the same way he would attribute to animals a similar psychic state to that in man in the case of elementary goals such as satisfying the needs of hunger and thirst and regulation of body temperature. Whilst emphasizing the precautions to such transference, such as with intellectual functions, he suggests it would be wrong to deny a feeling of thirst when a dog goes to water in hot weather and drinks. One would also be correct in assuming hunger if, after a long fast, a dog greedily devours its food. Because of different organization of sensory organs in man and animals the information which comes to consciousness may be somewhat different, but the psychic motivation in accord with organic needs plays an analogous role in trying to diminish the tensions based on vegetative functions (viz. feelings of overheating, thirst and hunger).

Hess instances affective defence as one of the instinctual behaviours elicited by electrical stimulation. He treats the point as to whether it is only unconsciousness playing off a neural disposition or whether it is a manifestation of an instinctual content of consciousness. He and his colleagues found that a few seconds after the beginning of diencephalic stimulation in a cat the animal begins to spit and hiss. Hair rises and pupils enlarge, the complex corresponding to a threatening behaviour. The animal keeps its eyes on the person close to it. If this person holds his hand out and the cat is stimulated more strongly it will attack with its paw with its claws extended. If stimulation continues it may get ready to jump at 'the enemy', which Hess suggests can hardly be understood unless a reactive attacking mood is brought about and, according to the environment of the moment, there is a reaction appropriate to the mental state. A deeper insight into the mental forces is given by the fact that the affective defence reaction may lead to a behaviour opposite to attack, the animal seeking to get out of a dangerous situation by fleeing. If there is no way out, it attempts to avoid danger by threat or swiftly executed attack. The same motive, due to the tension of being threatened in its existence, can give one or other behaviour.

These issues discussed may seem very general in relation to the limited area of enquiry involved in endeavouring to generate specific salt-seeking behaviour in animals by cerebral electrical or chemical stimulation. However, they are omnipresent in such experimental procedures in conscious animals (see Chapter 24 also) and should not be ignored.

Salt ingestion responses to diencephalic electrical stimulation in the unrestrained conscious sheep

In 20 conscious sheep, evidence was sought for localized brain mechanisms controlling ingestion of sodium salts, using electrical stimulation of diencephalic and neighbouring regions. The sheep lost saliva continuously through a parotid fistula and were permitted to replace the resultant sodium deficits episodically by drinking sodium bicarbonate solution (300 mmol/l). Focal cathodal stimulation at 40 Hz was delivered through needle electrodes, permanently fixed or roving in depth, when the sheep were behaviourally satiated for available salt, water and food. Response maps were constructed for approximately 1000 brain loci tested (McKenzie and Denton 1974).

Methods

Testing procedures

Before testing the behavioural effects of electrical stimulation, sheep were offered fresh solutions of sodium bicarbonate (300 mmol/l) to drink in order to ensure appetitive satiation for salt.

Principal attention was paid to the induced ingestion of sodium bicarbonate (300 mmol/l) and tap water from bins which were hung on the cage in positions equally accessible to the animal. The relative positions of the two bins were alternated randomly. Feed was available on the opposite side. On occasions, especially if stimulation induced licking at the cage or bins, a piece of salt block was placed in the cage as another available goal-object.

Stimulation

In operative procedures with full aseptic precautions, stimulation electrodes were introduced into the diencephalon in the Horsley–Clarke vertical orientation, using stereotaxic methods and data for Merino ewes developed at Melbourne University (McKenzie and Smith 1973).

At least 2 weeks after surgical implantation, stimulation was applied through each electrode in turn at a range of currents, and at various frequencies in addition to the standard 40 Hz. If no ingestion occurred at 0.1 mA, current intensity was augmented in steps up to 1 mA, or until the animal became disturbed or agitated.

Roving electrode stimulation

In the animals where roving electrode stimulation was used, an assembly of electrode guides was cemented stereotaxically into a cranial defect above the projection zone of the hypothalamus (Robinson 1964, 1967). Each experiment consisted of a single penetration of a sterilized stimulating electrode which was withdrawn at the end of the experiment. Effects of stimulation were tested every 2 mm until the electrode tip was estimated to have reached the hypothalamus, and then every 1 mm.

Results

Fixed stimulation positions

The results are summarized in the composite diagrams of Fig. 19-7, based on 15 sheep. Of 90 positions distributed from the preoptic region to the caudal diencephalon, one point in each of four animals elicited stimulus-bound ingestion (see below for details). From another seven positions, licking of external objects (cage, metal bins, animal's own wool) was elicited. These latter positions were near the base of the mammillothalamic tract (4), in the diagonal band (2) and in periventricular grey (1). In one sheep, stimulation of the lateral mammillary area on either side of the brain produced occasional drinking or licking of water or sodium bicarbonate and frequent chewing of bins and cage, but salt-block licking was not tested.

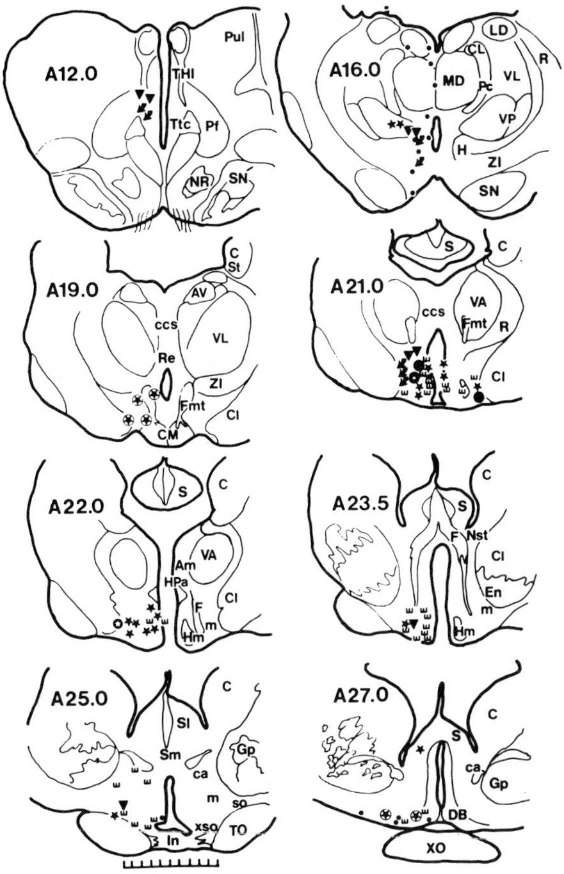

Fig. 19-7. Effects of electrical stimulation through permanent diencephalic electrodes in 15 sheep. Stimulation points are projected onto nearest equivalent loci in standard stereotaxic coronal planes. Explanation of symbols: ● sodium bicarbonate drinking; ✪ salt-block licking; ○ sodium bicarbonate and water drinking; ◉ sodium bicarbonate tasted; ⊛ licking or chewing bins and cage; ★ licking or chewing movements; ш arousal or agitation; ⬋ tonic movement; ▼ crouching; ● no detectable response. Scale in mm. Key to abbreviations: AV, nucleus anterior thalami ventralis; C, nucleus caudatus; ca, commissura anterior; ccs, nuclei centralis thalami; CI, capsula interna; CL, nucleus centralis thalami lateralis; CM, corpus mammillare; DB, lemniscus diagonalis; En, nucleus entopeduncularis; F, fornix (columnae fornicis); Fmt, fasciculus mammillothalamicus; Gp, globus pallidus; H, campus 'H' Foreli; Hm, nucleus ventromedialis hypothalami; Hpa, nucleus paraventricularis hypothalami; In, infundibulum; LD, nucleus lateralis thalami dorsalis; m, fasciculus prosencephali medialis (medial forebrain bundle); MD, nucleus medialis thalami dorsalis; NR, nucleus ruber; Nst, nucleus striae terminalis; Pc, nucleus paracentralis thalami; Pf, nucleus parafascicularis thalami; Pul, nucleus pulvinaris thalami; R, nucleus reticularis thalami; Re, nucleus reuniens thalami; S, area septalis; Sl, nuclei laterales septi; Sm, nuclei mediales septi; SN, substantia nigra; SO, nucleus supraopticus; St, stria terminalis; THI, tractus habenulo-interpeduncularis (fasciculus retroflexus); TO, tractus opticus; Ttc, tractus tegmentalis centralis; VA, nucleus ventralis thalami anterior; VL, nucleus ventralis thalami lateralis; VP, nucleus ventralis thalami posterior; XO, chiasma opticum; xso, commissura supraoptica; ZI, zona incerta. (From McKenzie and Denton 1974, with permission.)

Non-directed licking or chewing movements were elicited from a variety of sites: septal area, supraoptic nucleus, ventromedial thalamus, and hypothalamus in tuberal perifornical, lateral and periventricular regions. Excitation ranging from increased alertness to strong agitation was the dominant response from anterior hypothalamic sites, but also from more caudal sites among those producing licking or ingestion. Tonic postural responses (neck torsion or flexion, extensor hypotonia, integrated crouching movements) were obtained from points in the habenulo-interpeduncular tract (fasciculus retroflexus), medial subthalamus, anterior hypothalamus and supraoptic region.

Ingestion of salt in preference to water or food was elicited from the medial perifornical zone in one sheep and the medial forebrain bundle in two. In a fourth animal, dorsolateral juxtafornical stimulation at first induced sodium bicarbonate drinking, but later the choice between sodium solution and water came to depend on situational factors. Since the induced ingestion responses were studied repeatedly in these four sheep, the results for each will be considered in detail.

Medial perifornical area

In this sheep (Meana), unilateral stimulation of the medial hypothalamus (Fig. 19-8) elicited drinking of sodium bicarbonate solution. Stimulation in the lateral hypothalamic-peduncular border zone of the opposite side elicited rhythmic stepping of the contralateral forelimb, and chewing, but no drinking. The ingestion of sodium solution was stimulus-bound and selective in preference to water. If the positions of sodium solution and water were exchanged between trials, the sheep sometimes approached the water bin at the onset of the next stimulation, then transferred to the sodium bicarbonate bin and drank from it. In several trials, sodium drinking began within 2 s of the onset of stimulation, persisted during continuous stimulation for up to 60 s, and ceased on termination of stimulus current. On other occasions drinking began only after 30 s of arousal and motor restlessness. The amount of sodium bicarbonate ingested during intermittent stimulation trials across several hours exceeded average spontaneous daily intake by up to 1500 mmol. On days following elicited ingestion, urinary sodium output increased several hundred-fold above the higher outputs observed when sodium bicarbonate was continuously available (Fig. 19-8). A natriuresis was observed 1–2 days after stimulation. There were also reduced levels of spontaneous sodium intake.

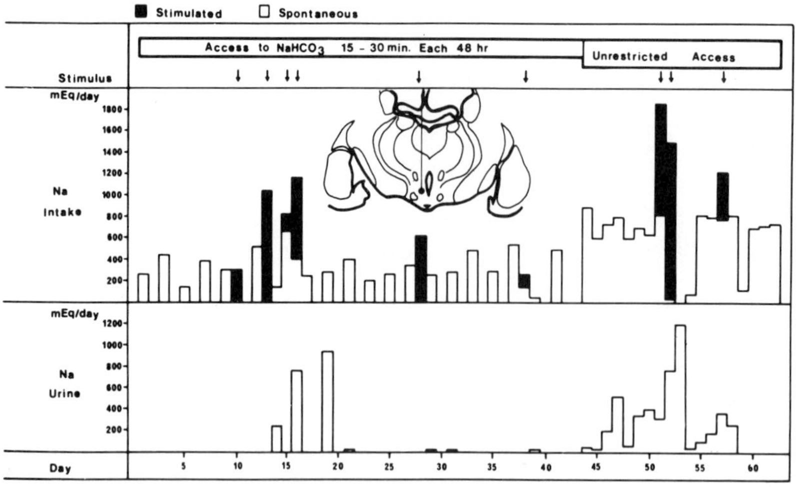

Fig. 19-8. Intake and urinary output of sodium in Meana, with effects of electrical stimulation in medial hypothalamus at locus shown by the filled circle in the inset diagram. Stimulus-bound sodium intake is shown by black columns. Urinary excretion was usually too low to register on the scale during the regime of limited access to sodium bicarbonate, except following stimulation-induced ingestion. During unrestricted access to sodium bicarbonate, urinary sodium excretion varied about higher levels, was further augmented following stimulation, but subsequently displayed a precipitous fall associated with that of sodium intake. For further details see text. (From McKenzie and Denton 1974, with permission.)

Medial forebrain bundle

In another sheep (Raelene), unilateral stimulation at a similar position between the fornix and the third ventricle induced no drinking, but excitation and licking of lips. However, stimulation of the medial forebrain bundle area elicited sodium bicarbonate drinking greatly in excess of spontaneous intake, with natriuresis on following days.

Another sheep responded to stimulation of the medial forebrain bundle by licking a salt block or the cage metal, and to perifornical stimulation with arousal and lip-licking, but never by drinking.

Dorsolateral juxtafornical area

Stimulation immediately dorsal and lateral to the caudal fornix columns elicited lip-licking in several sheep. In one animal (Fimke) the site was strongly dipsogenic, 3 litres of sodium bicarbonate being drunk at the first trial. The response was repeated on succeeding days. But, during stimulation, the ingestion of sodium solution was often interrupted after 5–15 s, when the animal drank instead from the water bin. During intermittent testing over several weeks, the initial drinking preference was lost; on different days the choice between sodium bicarbonate and water was unpredictable. The choice was not modified by variation in stimulus parameters (rate, intensity, pulse duration), nor by stimulating simultaneously through other brain electrodes which alone produced excitation. If no choice of fluid were given, the sheep drank whichever of sodium bicarbonate or water was available during stimulation.

In order to study any positional influences on this behaviour which had emerged, repeated trials each consisting of 5 s of induced drinking, at least 15 s apart, were conducted with adjacent bins of sodium bicarbonate and water interchanged at random between trials. In sessions spaced over one week, response preference varied, but finally stabilized to a reproducible positional preference for the contents of the rear bin, whether sodium solution or water (Fig. 19-9). In succeeding experiments over several weeks, it was sufficient to interchange the two bins for the drinking response to transfer from one to the other liquid. This study of positional preference was further elaborated as follows: inter-trial loading, by ruminal intubation

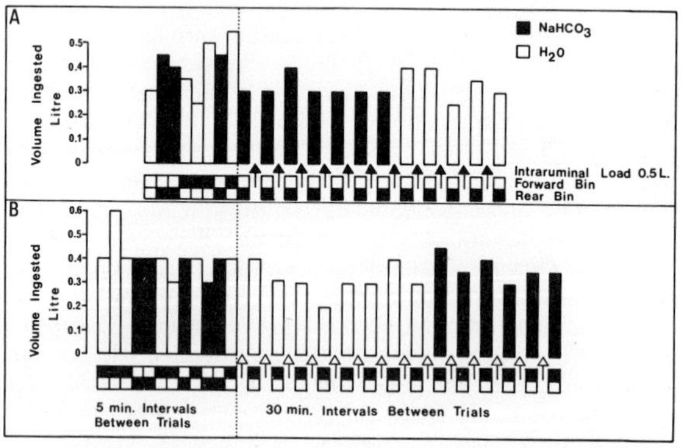

Fig. 19-9. Positional preference for drinking in response to juxtafornical hypothalamic stimulation in Fimke, and influence of forced intake between trials on preference. Left of dotted line: stimulus-bound drinking from rear bin shown by random interchange of sodium bicarbonate and water positions between trials. Right of dotted line: stimulation applied at 30-min intervals with relative positions of sodium bicarbonate and water constant during trials. Between trials, 0.5 litre of liquid in rear bin were administered by ruminal intubation, while both bins were removed. A, sodium bicarbonate in rear bin; B, water in rear bin. After total consumption of several litres of liquid in preferred rear bin position, the direction of stimulus-bound drinking switches to forward bin. Stimulation was applied for 5 s of induced drinking throughout experiments A and B at intervals of 48 h (From McKenzie and Denton 1974, with permission.)

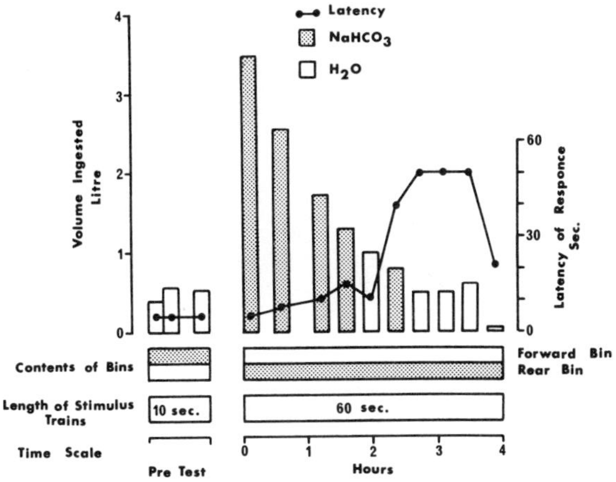

Fig. 19-10. Effect of stimulation in the juxtafornical hypothalamus of Fimke, using trains of long duration. Pre-test sequence of brief (10-s) stimulation trains establishes that drinking response is directed towards preferred rear bin, then bins of sodium bicarbonate and water are interchanged. Stimulation then delivered as 60-s trains at approximately 30-min intervals, inducing drinking of sodium bicarbonate in rear bin. Columns represent volumes ingested during stimulus-bound drinking. Connected filled circles represent latency to onset of drinking after start of stimulation. (From McKenzie and Denton 1974, with permission.)

using the liquid in the rear position, resulted in transfer of the induced response to the contents of the forward bin after a total forced load of 3.5–4.5 litres (Fig. 19-9). Response preference was also reversed by the repetition of 60-s stimulation trials at approximately 30-min intervals (Fig. 19-10). With sodium bicarbonate in the rear bin, the volume drunk diminished on successive trials, and after more than 10 litres of sodium (300 mmol/l) had been ingested the response transferred to water in the forward bin. The decrement in response volume could be attributed in part to increased response latency, associated with increasing restlessness as the trials were repeated.

Character of elicited drinking

At the onset of weak but effective stimulation, an animal characteristically looked about, extended its head over the bin, and began to drink while appearing to remain alert to its surroundings. The drinking was more easily interrupted by extraneous sounds or movements than was the spontaneous drinking of sodium bicarbonate to replace accumulated sodium deficit. With strong effective stimulation, the onset of drinking was delayed only by the time required to reach the liquid. Drinking was not easily distracted, and draughts were swallowed more vigorously than with weak stimulation.

Roving electrodes

In four sheep the effects of stimulation were tested at 933 diencephalic and neighbouring loci. Drinking responses to stimulation were observed at six points in the hypothalamus (Fig. 19-11: planes 19, 22) and two points in the fimbria (Fig. 19-11). Licking of bins or their liquid contents, without drinking, was elicited from three points in the caudate nucleus (Fig. 19-11: planes 23.5, 27), from the subcallosal fasciculus, from the internal capsule (Fig. 19-11: plane 22), and from the hypothalamic ventromedial nucleus (Fig. 19-11: plane 22).

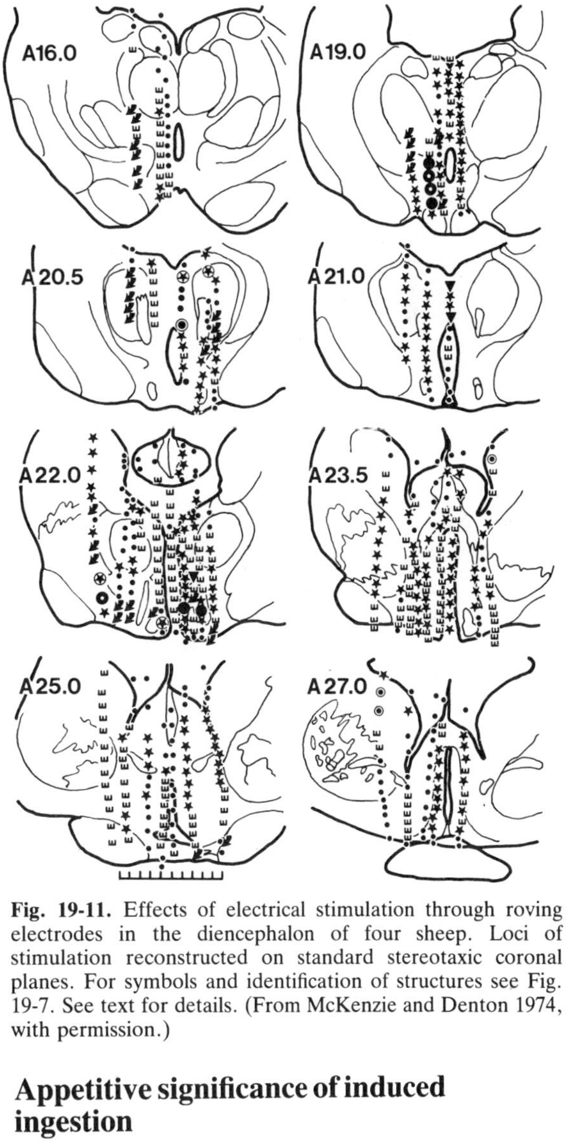

Fig. 19-11. Effects of electrical stimulation through roving electrodes in the diencephalon of four sheep. Loci of stimulation reconstructed on standard stereotaxic coronal planes. For symbols and identification of structures see Fig. 19-7. See text for details. (From McKenzie and Denton 1974, with permission.)

Appetitive significance of induced ingestion

The results showed that, in some instances, electrical stimulation of the hypothalamus elicits drinking of sodium bicarbonate solution by sheep satiated behaviourally for salt. The induced salt ingestion was selective in preference to water upon stimulation in some areas, and in some sheep. In other instances an elicited salt ingestion was not reproduced in consistent preference to water ingestion, from day to day or week to week, or was not reproducible on stimulating closely similar sites from animal to animal. In relation to the choice between salt solution and water, it would be advantageous to amplify such experiments using e.g. a 900 mmol/l solution of sodium bicarbonate. This is relatively aversive to the sheep (this is not the case for many animals with the 300 mmol/l solution: Abraham et al. 1976), but a sodium-deficient animal will drink it avidly whereas a water-deprived animal will reject it.

The induction of salt intake was compatible with an artificial arousal of salt appetite, in that (*a*) it occurred in sheep which had displayed and satiated a spontaneous salt appetite, (*b*) it could drive such animals into a salt-loaded state as judged by subsequent natriuresis, (*c*) its motor pattern closely resembled that of spontaneous ingestion, and could similarly be interrupted by distracting sensory stimuli, (*d*) it was, in some animals, selective relative to concurrent choice of water or food.

However, on repeated stimulation of drinking behaviour in any one animal, the procedure appeared to gain aversive significance, as judged by the animals' attempts to avoid the manipulations necessary before stimulation could begin and by arousal and sometimes threat behaviour (stamping on cage floor) on approach of the experimenter. Thus, stimulation which repeatedly induced drinking, whether of water or sodium solution, never appeared to act as positive reinforcement as has been inferred for ingestion-inducing stimulation in the rat hypothalamus (Mogenson 1969; Mendelson 1970; Wayner 1970) nor did the routine manipulations preceding stimulation become conditioned signals for drinking.

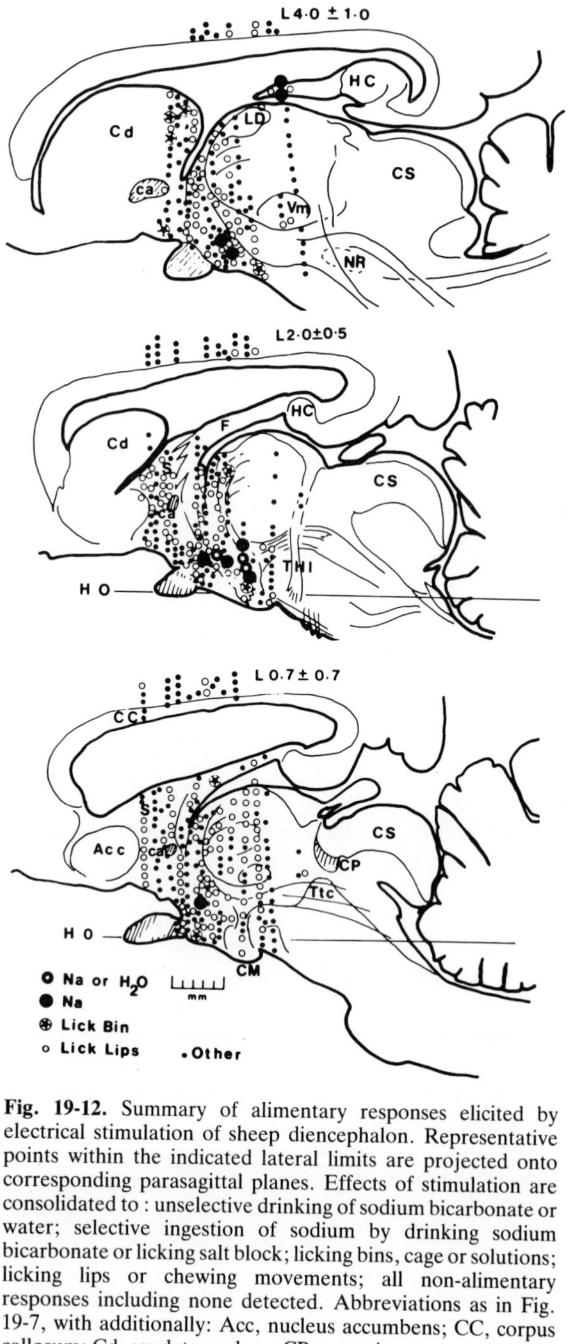

Fig. 19-12. Summary of alimentary responses elicited by electrical stimulation of sheep diencephalon. Representative points within the indicated lateral limits are projected onto corresponding parasagittal planes. Effects of stimulation are consolidated to : unselective drinking of sodium bicarbonate or water; selective ingestion of sodium by drinking sodium bicarbonate or licking salt block; licking bins, cage or solutions; licking lips or chewing movements; all non-alimentary responses including none detected. Abbreviations as in Fig. 19-7, with additionally: Acc, nucleus accumbens; CC, corpus callosum; Cd, caudate nucleus; CP, commissura posterior; CS, colliculus superior; HC, hippocampus; HO, stereotaxic horizontal zero plane; Vm, nucleus ventralis thalami medialis. (From McKenzie and Denton 1974, with permission.)

Anatomical distribution of effective stimulation sites

Salt ingestion was elicited by stimulation of the hypothalamus in periventricular, perifornical and lateral (medial forebrain bundle) positions at tuberal levels, and in posterior positions contiguous with mammillothalamic tract. The positions were in or adjacent to pathways relating the hypothalamus to limbic forebrain structures (Fig. 19-12); in addition, a brief salt-drinking response occurred on stimulation of the fimbria with a roving electrode in one sheep. The distribution suggests that components of the Papez circuit and the medial forebrain bundle, each associating limbic forebrain with hypothalamic and tegmental structures (Nauta 1958), form part of the neural mechanisms subserving salt ingestion behaviour. As only small zones in these pathways were found effective, it may be inferred that the ingestion responses were not due solely to activation of long association pathways.

Points from which ingestion was elicited were closely intermingled with others producing only lip-licking or chewing, or no alimentary behaviour. Licking or chewing were obtained from points distributed widely through the diencephalon, the lateral mammillary area, the lateral hypothalamus and the caudate nucleus (Fig. 19-12).

Brain stimulation maps published for other species have shown that the distributions of points eliciting alimentary or other behaviour patterns embrace several distinct anatomical structures. Among nearly 6000 positions in and around the diencephalon of rhesus monkeys (Robinson and Mishkin 1968), food intake (without swallowing) and water drinking were obtained from stimulation in medial and lateral hypothalamus, medial thalamus, mid-brain tegmentum, and striatum, at a small proportion of points among others from which no alimentary intake was elicited. In the opossum brain, maps of stimulation points eliciting male mating behaviour, biting attack, or eating and grooming (Roberts 1969), indicate, for each behaviour, a statistically distinctive distribution of points in preoptic and hypothalamic regions overlapping with points for investigatory or escape behaviour and with ineffective points. In the lateral hypothalamus of rats, Mogenson and Stevenson (1967) found little overlap between 10 'drinking' points and 40 'non-drinking' points but the maps of Valenstein et al. (1970) for this area show that of 4000 points in various nuclei and pathways the 200 which induced alimentary behaviour were almost entirely interspersed among the remaining ineffective points, frequently with complete superposition when repetitive tests were used.

From electrical stimulation in the hypothalamus of goats, Andersson and McCann (1955) described a 'drinking area' located between the fornix columns and the mammillothalamic tract, with extensions to the anterior hypothalamus and the preoptic area (Andersson 1951; Andersson et al. 1958). Composite maps indicating the variability of response between animals were not presented: in an early publication (Andersson 1951) none of 32 positions, including some in the mid-lateral hypothalamus, elicited drinking on electrical stimulation, but in a later study (Andersson and McCann 1955; Andersson and Wyrwicka 1957) polydipsia occurred on stimulation of 33 points in this zone, in nine goats. Andersson, Jewell and Larsson (1958) reported that electrical stimulation of the lateral border of the drinking area induced vigorous licking of an offered salt block.

Significance of results with electrical stimulation

Together with the present results, these examples suggest that, while differences between species and method may influence the probability of obtaining the responses, electrical stimulation of the same brain structure in a single species does not reliably elicit a particular form of alimentary or other motivational behaviour. The larger the number of stimulation points sampled, the more this suggestion appears to be substantiated. Furthermore, the idea of narrowly circumscribed centres or systems in the hypothalamus, acting as specific substrates of alimentary behaviour, receives equivocal support from the results of detailed stimulation mapping. This is so even in the case of eating behaviour frequently elicited by hypothalamic stimulation in rats (Valenstein et al. 1969; Valenstein 1970).

Recourse to interspecific difference in the anatomical organization of the neural systems subserving alimentary behaviour is unsatisfactory for explaining the above-cited variations in efficacy of hypothalamic stimulation. Such an explanation suggests that the pathways in a phylogenetically ancient brain structure, supporting motivational behaviour of fundamental survival value, are organized differently in brains as similar in morphological characteristics as those of sheep and goats. Further this would appear to imply comparable differences between members of the one species.

Presumably electrical stimulation produces artificial patterns of excitation in neuronal assemblies. The evocation of organized behavioural responses by its action suggests proximity of the electrode to neural elements which can trigger a particular behaviour. This is despite unphysiological patterns of impulse traffic. Failure to produce a behavioural response by stimulation does not exclude the possibility that neurones in the vicinity of the electrode are normally involved in the behaviour. Electrical stimulation within the brain might be expected to activate, indifferently and chaotically, both excitatory and inhibitory neural elements. It is perhaps surprising that organized, goal-specific behaviour ever results from it. In the face of the fact that such responses are indeed observed, it would seem to demand that the neural domains activated by localized stimulation have a particular relationship, of connectivity and synaptic potency, with other neural systems primarily responsible for organizing the behaviour concerned. It is only in this manner that the responses to hypothalamic stimulation imply the concept of 'centres', distributed or compact, with the capability for selective facilitation of other mechanisms determining, for example, the intake of sodium.

It may be argued that such problems associated with electrical stimulation are avoided when different selective behaviours can be elicited by intrahypothalamic application of specific pharmacological agents at one and the same point (Fisher and Coury 1964; Grossman 1964). However the problems of diffusional (Myers et al. 1971) and intravascular (Routtenberg 1972) spread, together with the capacity of parenteral pharmacological antagonists to block behaviours elicited by intrahypothalamic drugs (Grossman 1964), indicate that intracerebral drug effects do not necessarily imply activation of local neural substrates for specific behaviour patterns. With this mode of approach there is also the problem that the microinjections used (0.1–1 μl) usually contrive in the vicinity of a limited number of neurones a grossly pharmacological concentration of the agent employed (Abraham et al. 1975).

These results on electrical stimulation of sheep might allow either of two interpretations:

1) The hypothalamus contains neural structures whose activation produces salt ingestion specifically, and others producing general drinking behaviour. On this basis, a hypothalamic substrate of the drive to ingest salt would appear to include components of both the Papez circuit and the medial forebrain bundle, with a low probability of selective activation by electrical stimuli.
2) There is no specific salt ingestion system in the hypothalamus, but rather structures activating broad response tendencies towards alimentary intake. Again the probability that their stimulation leads to actual intake, rather than fragments of its pattern such as licking or orientation towards alimentary goal-objects, is low with the animals and laboratory conditions used. However, such stimulation leads to organized, sometimes specific ingestive behaviour if activation of the generalized 'alimentary' circuits should interact appropriately with excitation patterns in cortical circuits, under the influence of exteroceptive and interoceptive information.

The large body of data in several species showing salt appetite is innate, and therefore implying there is a substantial neural organization dedicated to it, would seem to favour the first interpretation.

The influence of brain lesions on salt appetite in sodium deficiency, and also saline preference of normal animals

The studies on the influence of brain lesions have been made mainly in the rat and these data will be reviewed first. Some observations of goat and sheep will then be noted.

An early study was made by Chul Kim (1960), which showed that adrenalectomy resulted in increased salt intake in rats with bilateral hippocampal ablation or with neocortex ablation as it did in control animals. Cort (1963) reported that in rats mildly salt-deplete as a result of a constant salt-poor diet, abnormal salt losses were induced as a result of bilateral lesions in the posterior hypothalamus. The animals responded to deficiency by drinking increased amount of saline. In 1966, Novakova and Cort reported that rats with bilateral lesions in the region of the ventromedial nucleus of the hypothalamus failed to increase salt intake after sodium depletion caused by hydrochlorthiazide administration and a salt-free synthetic diet. With other lesions, even in the close vicinity of the ventromedial nucleus, this effect was not observed. They also noted that there was no change in plasma sodium concentration as a result of the experimental procedures. This parameter could not be easily incriminated as an afferent signal generating appetite.

South American studies

An important body of data bearing upon neural organization subserving salt appetite has come from experiments over a period of years conducted by the South American group of Miguel Covian and colleagues. The basic condition of study was that rats were allowed free access to a diet which contained adequate sodium, and to two drinking bottles one of which was filled with tap water and the other with 2% saline. Initially Covian and Antunes-Rodrigues (1963) showed that lesions in the anterior hypothalamus involving supraoptic or paraventricular nuclei, or both, resulted in a conspicuous fall in voluntary saline intake. The decreased consumption remained to the end of the experiments which in some rats lasted 105 days. On the contrary, lesions placed in the central hypothalamus determined a specific increase in saline intake together with an augmented urinary sodium excretion. The increase was up to 3-fold. The authors raised the question whether there could be an organization for salt appetite in the hypothalamus similar to that for feeding behaviour. Following this, in 1967, Negro-Vilar et al. found that bilateral lesions in the septal area resulted on average in a 5-fold increase in voluntary intake of 1.5% saline. They queried whether the increased intake was secondary to a disturbance of endocrine function causing increased sodium loss. Sterc et al. (1975) also found that septal lesions in rats resulted in a higher daily saline intake.

Proceeding further in the analysis, Gentil et al. (1968) examined the effects of the amygdaloid lesions. They found that bilateral lesions in the corticomedial complex, especially the cortical nucleus, resulted in an increase in saline intake. But bilateral lesions placed principally in the lateral nucleus but also reaching medial basolateral and basomedial nuclei resulted in a decrease in saline intake. In these experiments sodium balance studies were done and it was verified that the increase in sodium consumption preceded any changes in urinary output. The authors suggested that the data were pointing to an intricate circuit related to hydromineral metabolism which involves the septal area, the amygdala and the hypothalamus.

The influence of adrenalectomy was examined in this context (Antunes-Rodrigues et al. 1970). It was found that adrenalectomy in rats that had increased saline ingestion due to hypothalamic or amygdaloid lesions resulted in no change, or a decrease, in intake. In rats which had diminished consumptions of sodium chloride due to hypothalamic or amygdaloid lesions, adrenalectomy yielded an increased intake but of lesser degree than in intact rats. If the adrenalectomized animals were given corticoid substitutive therapy to give normal sodium-retaining ability and normal baseline intake, lesions in the hypothalamus or amygdala showed the same changes as in non-adrenalectomized animals. The authors conclude there is a central control of sodium intake independent of fluctuations in blood mineralocorticoid level.

In relation to amplifying this investigation in terms of response during sodium deficiency, Chiaraviglio (1971) reported that unilateral application of acetylcholine to the amygdala of sodium-deprived rats increased saline intake, though it decreased it in thirsty and satiated subjects. Water intake was not affected by the acetylcholine implants. She found also that bilateral transection of the stria terminalis or ventral amygdalofugal tract decreased saline intake in sodium-deprived rats. She concluded that impulses from the amygdala reaching the hypothalamus through the stria terminalis or the ventral amygdalofugal tracts modulate sodium chloride intake. In these experiments 1% saline solution was used, and it would be of considerable interest to see these data amplified with the choice of a sodium chloride solution which is normally aversive for the rat.

The amygdala

The influence of the amygdala on sodium appetite has also been studied by Nachman and Ashe (1974), in the context of a more general study of neophobia and learned taste aversions. These authors found that lesions to the basolateral amygdala produced permanent impairment in the ability of rats to learn a taste aversion. When the lesions were made after rats had learned an aversion there was a loss of this aversion. The animals also had a diminished neophobic response when presented with a novel solution relative to control. Rats with amygdala damage also showed less sodium appetite than normal in response to DOCA injection. Intake of 0.4 M saline was significantly increased relative to control but was significantly less than in the control animals given DOCA. Nachman and Ashe proposed that a normal rat responds to a sodium need state or a DOCA injection by innately recognizing and selecting the sodium taste. It is this perceptual process which they hypothesize as impaired in the rat with the amygdala lesion. In the case of sodium appetite, the perceptual deficit results in an inability to recognize the taste in response to a current need state. But in learned taste aversions and neophobia the perceptual deficit results in an inability to recognize tastes in relation to past experience. They note that the deficits seen in rats with the amygdala lesions have some measure of specificity in this regard in that Rolls and Rolls (1973) have shown that amygdala lesions in rats do not influence water intake in response to thirst-inducing stimuli. Nachman and Ashe (1974) see the rat with an amygdala lesion as having a general perceptual deficiency with inability to recognize the significance of stimuli whether rewarding or punishing and whether learned or unlearned.

The zona incerta

Recently Walsh and Grossman (1977) have examined the effects on sodium appetite of electrolytic lesions and knife cuts in the region of the zona incerta. They note that gustatory information appears to reach the lateral hypothalamus only via the dorsal route. The ascending gustatory pathway from the pontine taste area courses to the ventral thalamus via the mid-brain reticular formation and subthalamus, and also has terminal ramifications in the zona incerta and also lateral hypothalamus. Thus the zona incerta carries neural pathways interconnecting three areas that have been related to regulatory sodium intake: the lateral hypothalamus, ventral thalamus and mid-brain reticular formation. They also draw attention to their data showing that lesions of the zona incerta decrease drinking in response to hypertonic saline or angiotensin but not to hypovolaemia induced by polyethylene glycol.

In their experiments they compared bilateral zona incerta lesions and ventral thalamus lesions in relation to response to subcutaneous formalin injection. Water intake of both groups of animals was reduced, but did increase in response to formalin injections. In contrast to earlier findings of Wolf (1971), the rats with zona incerta lesions did not exhibit normal sodium appetite. The average increase in sodium intake was slightly more than one-quarter the average increase in sodium intake of controls. The lesions in the ventral thalamus damaging the thalamic-gustatory nucleus and adjacent tissue did not prevent the appearance of sodium appetite following the formalin injection, but the response was reliably smaller than that of the sham-operated controls. They also studied a group of animals with bilaterally symmetrical knife cuts in the horizontal plane just above the anterior zona incerta and a second group of animals with similar cuts beneath the zona incerta. Both types of cuts, produced significant reductions in sodium intake following formalin injection. The effect was more pronounced with the ventral cuts, the dorsal cuts having effects which resembled those of the ventral thalamic lesions. In both cases there was some impairment also of water intake following formalin injection. The interpretation of these extremely interesting data, which are examined from many aspects by the authors, is, however, made difficult by the use of formalin as the mechanism for induction of sodium deficiency. If it be that this represents a combination of sodium deficiency and severe stress, then it is not defined whether the impairment is in the response to sodium deficiency or in the response to the appetite-inducing effects of adrenocorticotrophic hormone (ACTH). There is also the consideration that the lesions may have interfered with the ACTH-releasing effect of formalin and this could quantitatively modify the salt appetite producing effects of formalin.

The studies of George Wolf and colleagues

A major experimental study of the influence of hypothalamic and thalamic structures on sodium appetite has also been made by George Wolf and colleagues. Wolf (1964) showed that lesions of the dorsolateral hypothalamus and the zona incerta had no observable effect upon the need-free preference of the rat for weak saline solutions or upon the aversion to strong saline solutions. However, the bilateral lesions, even if not symmetrical, caused a pronounced interference with the elicitation of increased saline intake by DOCA treatment or by acute hyponatraemia caused by peritoneal dialysis. He suggests that whereas this region would not appear to be important in involvement in the mechanisms which govern the appetite for saline in the absence of need, they may be critically involved in the appetite response to internal alteration produced by DOC or the depletion of electrolytes. It should be noted that the animals had recovered from the effects of these lesions on food and water intake at the time that the studies of sodium appetite were carried out.

Wolf and Quartermain (1967) showed that rats with lateral hypothalamic lesions did not ingest increased amounts of saline after adrenalectomy. In rats which were given 0.5 M sodium chloride, 0.3 M potassium chloride solution and distilled water to drink, appetite was tested after formalin injection or after the administration of DOC. In rats which had bilateral lesions of the gustatory nucleus of the thalamus, or the tegmental region immediately caudal to this nucleus, there was a significant impairment of the sodium intake response which normally follows sodium depletion or DOCA administration (Wolf 1968). If the lesions were rostral, dorsal, medial or lateral to the gustatory nucleus there was no effect. The tegmental lesions had some influence on the water intake which characteristically follows formalin injection, a reduction in response occurring.

In discussing these results in relation to his earlier data, Wolf notes that studies such as these serve to delineate brain regions which organize or transmit information important for normal regulation of sodium intake. But they do not clarify the nature of that function—that is, the degree to which it is specifically and directly related to sodium ingestion or the manner in which it is related to other relevant brain functions. Lesions confined to the ventromedial nucleus of the hypothalamus had no effect on the sodium intake provoked by formalin injection, but Quartermain et al. (1969) found that larger lesions which included the major extent of the medial hypothalamic zone did reduce the sodium appetite. The issue Wolf and colleagues had in mind in this study was whether the earlier reported data on influence of medial hypothalamic lesions on sodium appetite were attributable to their extending into the lateral hypothalamic area. This study, in relation to its delineation of the extent of the lesions, was confirmatory of the influence of the medial hypothalamic structures on appetite.

Wolf et al. (1970) studied the sodium appetite in rats after neocortical ablation. Rats with ablations of either the entire neocortex or just the sensory cortex including the taste projection area, responded to sodium depletion or mineralocorticoid treatment by increasing their intake of sodium chloride solution (0.5 M). The response to sodium depletion was somewhat attenuated or delayed in the decorticate rats compared with intact controls but the response to mineralocorticoid treatment was unchanged. The authors conclude that the neural system leading from receptor mechanisms for detecting alterations in body sodium and mineralocorticoid levels to the final pathways for sodium ingestion can function at an entirely subcortical level. They note also that since retrograde degeneration of cells in the thalamic taste nucleus was very extensive if not complete, it seems that the classical taste pathway at and above the level of the thalamus is not essential for basic regulation of sodium intake. Apparently the deficit induced by the ventromedial thalamic lesions reported by Wolf (1968) was not exclusively due to the destruction of the taste nucleus. Possibly the impairment was attributable to fibres of passage or to intralaminar nuclei which are associated with the reticular activating system and which were encroached upon by the lesions (Wolf et al. 1970). Wolf and Di Cara (1974) in a further study confirmed the effect of lesions on the ventral posterior complex of the thalamus in impairing sodium appetite, and also showed that this effect occurred whether the tests were made 2–3 weeks after the lesions, or 2 months later.

Bealer and Johnson (1979) made anteroventral lesions in the third ventricle of rats. Animals which recovered from the adipsia after lesion consumed significantly smaller volumes of 2% saline than sham-lesioned controls. At this stage the lesioned animals were consuming as much water as the sham-lesioned controls. Formalin injection induced sodium appetite though not to the same level as in the sham-operated animals.

Recently, Contreras and Stetson (1981) have shown that lesions of the area postrema which did not cause appreciable damage to the adjacent nucleus of the solitary tract, a sensory relay for gustatory and visceral afferents, caused a supranormal intake of sodium chloride solutions (0.03–0.5 M) but not of glucose or potassium chloride solutions. The change did not appear to be secondary to increased sodium output. Water intake was doubled in the lesioned animals. Assuming that the intake was not secondary to natriuresis the authors postulate that the well circumscribed lesions of the area postrema may result in hypotension—in contrast to the hypertension Reis's group (Nathan and Reis 1977) have shown to follow lesions of the nucleus of the solitary tract. Chronic hypotension, they suggest, may cause a compensatory increase in drinking of saline.

Studies in Stockholm on goats

Andersson, Leksell and Lishajko (1975) reported the results of radiofrequency forebrain lesions in the goat. Medial lesions which involved practically the entire anterior wall of the third cerebral ventricle caused persistent loss of thirst and lack of significant antidiuretic hormone secretion in response to hypernatraemia. Water diuresis occurred immediately following the lesions and this caused pronounced hypernatraemia and hypovolaemia. Forebrain damage, mainly restricted to the septal region, caused hyperdipsia in their animals.

The conditions of these experiments were that the goats had continuous access to chaff, hay and water and in order to maintain them in positive sodium balance they were given 6 g of sodium chloride per day. Nine goats voluntarily drank salt dissolved in 300 ml of warm water. The other four animals in their study received the salt added to the daily ration of food. It was observed that immediately after lesioning and throughout the following observation periods, three of the goats that were rendered adipsic and one hyperdipsic animal showed a remarkably increased appetite for salt. These four animals had voluntarily drunk their salt ration as a sodium chloride solution during the control period but without particular eagerness, but the lesioning induced a craving for salt. The goats started to become very excited when the salt solution was offered to them, drank it very eagerly and licked the bin frenetically when it was empty. This, apparently abnormal, salt appetite also remained when the goats became strongly hypernatraemic during periods of dehydration. The authors note that the four goats which developed increased salt appetite all had fairly extensive forebrain lesions which involved much of the septum. However, septal damage to the same extent, or greater, was seen also in hyperdipsic animals which did not display any obvious increase in salt appetite. They found no evident correlation between the augmented salt appetite and the extent of the forebrain lesions.

To date, limited data only are available on the sheep, from some 11 animals. Studies with Dr J. McKenzie showed lesions involving septum, striatum, anterior hypothalamus and mammillary area including the mammillothalamic tract did not significantly disrupt the capacity to repair sodium deficiency when the animals were offered sodium bicarbonate solution. However, uniform bilateral lesions were not always achieved and the relevance of previous experience of deficiency to interpretation has been emphasized by Ahern et al. (1978) (see below). In one animal with a lesion in the anterior wall of the third ventricle, studied by Dr M. McKinley, there was complete adipsia for 2 weeks and plasma sodium concentration rose to 170–180 mmol/l. Food intake was normal and, although the animal would not drink water, 0.3–0.6 M potassium chloride or potassium bicarbonate or glucose, 0.3–0.6 M sodium chloride or sodium bicarbonate was readily drunk. A comprehensive study in the sheep remains to be done.

Review of lesion work by George Wolf and colleagues

Wolf, McGovern and Di Cara (1974) have endeavoured in a comprehensive review to synthesize the data coming from work on brain lesioning procedures and effects on salt appetite. There is consistency between some elements of their conclusions and the considerations arising from the stimulation studies cited above. They conclude from the lesion data that certain structures of the limbic system possibly involved in motivation, and certain structures of the mesencephalic tegmentum and thalamus possibly involved in gustation, appear critical for normal regulation of sodium intake. In relation to the role of pathways, they note the data cited above on the effect of sectioning the ventral amygdaloid pathway or the stria terminalis. But they record their own failure with lesions of known major hypothalamic pathways to produce observable decrement in sodium appetite (Wolf 1967; Jalowiec et al. 1970; Wolf 1971). They see further research on effective pathways of transmission of hypothalamic regulatory functions to extrahypothalamic sites as an interesting problem for the future.

In relation to the importance of the amygdala they note the data of Gentil et al. (1971) showing that electrical stimulation of the basolateral amygdala and lateral hypothalamus enhance salt intake while electrical stimulation of the corticomedial amygdala diminishes salt intake. This latter observation parallels the fact that lesions of the corticomedial amygdala may cause increased salt intake (Gentil et al. 1968).

As regards the role of the gustatory system, they note that the lesions produce impairments in sodium appetite but that the effects are less consistent than seen after hypothalamic lesions. The finding of Chiaraviglio (1972) that lesions in the caudal mesencephalic reticular formation at the level of decussation of the brachium conjunctivum and ventrolateral to the central grey area decrease sodium appetite could be interpreted on the basis of interruption of the ascending gustatory pathway from the pontine taste area (Norgren 1977).

Saad et al. (1972) summarize their data by concluding that the hypothalamus is the main structure controlling sodium intake and the amygdala and septal area have modulating influences on the hypothalamus. The septal area overcomes the action of the amygdala.

In considering what they see to be the predominant role of the lateral hypothalamus in sodium appetite, Wolf et al. (1974) state they have repeatedly observed that while rats with lateral hypothalamic lesions may recover relatively normal food and water intake within a few days after lesioning, they show little evidence of recovery of sodium appetite in spite of prolonged postoperative survival. They think it possible that the alimentary functions which recover after lateral hypothalamic damage are mediated by other parts of the brain which contain information on or programmes for food and water ingestion as a result of repeated pre-operative feeding and drinking. That is, there is a lifetime of preoperative experience with a satisfaction of the urge to eat and drink. However, the failure of recovery of sodium appetite in comparable circumstances is due to the absence of such learned programmes for sodium intake.

In a further study, Ahern et al. (1978) showed clearly that pre-operative taste experience of saline was critical in protecting a rat from the effect of specifically localized brain lesions which largely abolished salt appetite. Pre-operative experience of salt deficiency as induced by a diuretic but without experience of saline taste did not have this clear-cut effect. *The critical locus of lesions affecting salt appetite in this way was the thalamic taste relay* (ventral posterior medial thalamic nucleus). Lesions in the mesencephalic tegmentum directly caudal to the taste relay and disjoining the ascending thalamopetal taste fibres also produced an all or none disruption of sodium regulation. It would appear that a brain structure necessary for the initial excecution of an innate behavioural process can be bypassed if specific critical information has been fed into the system—in this case prior experience of need-free saline ingestion. Whereas intake of sodium by sodium-deficient rats appears entirely dependent on gustatory information, the lesion might leave intact the memory of the taste of sodium and the association of the taste with an ingestive act at a particular place (Ahern et al. 1978). Thus the memory of sodium taste might be sufficient to cause ingestion though the taste is not currently experienced (this is analogous to findings of Krieckhaus (1970): see Chapter 11). Schwartz and Teitelbaum (1974) found rats with lateral hypothalamic lesions could remember a conditioned taste aversion acquired before operation but could not learn a new taste aversion post-operatively.

In an examination of this complex problem of the influence of pre-operative experience on the behavioural consequences of lesions, Wolf and Schulkin (1980) found that *pre-operative experience of sodium drive reduction by saline ingestion substantially reduced the impairment of salt appetite produced by lateral hypothalamic lesions*, whereas mere experience of tasting saline did not have this effect. The critical pre-operative experience thus appears to differ in the two brain regions. Lateral hypothalamic function appears necessary for elicitation of a sodium appetite drive, and the ventroposterior medial nucleus of the thalamus mediates response to sodium taste.

References

Abraham SF, Baker RM, Blaine EH, Denton DA, McKinley MJ (1975) Water drinking induced in sheep by angiotensin: A physiological or pharmacological effect. J Comp Physiol Psychol 88: 503

Abraham SF, Denton DA, Weisinger RS (1976) The specificity of the dipsogenic effect of angiotensin II. Pharmacol Biochem Behav 4: 363

Ahern GL, Landin ML, Wolf G (1978) Escape from deficits in sodium intake after thalamic lesions as a function of preoperative experience. J Comp Physiol Psychol 92: 544

Andersson B (1951) The effect and localization of electrical stimulation of certain parts of the brainstem in sheep and goat. Acta Physiol Scand 23: 8

Andersson B (1953) The effect of injections of hypertonic NaCl-solutions in different parts of the hypothalamus of goats. Acta Physiol Scand 28: 188

Andersson B, McCann SM (1955) A further study of polydipsia evoked by hypothalamic stimulation in the goat. Acta Physiol Scand 33: 333

Andersson B, Wyrwicka W (1957) The elicitation of a drinking motor conditioned reaction by electrical stimulation of the hypothalamic drinking area in the goat. Acta Physiol Scand 41: 194

Andersson B, Jewell PA, Larsson S (1958) An appraisal of the effects of diencephalic stimulation of conscious animals in terms of normal behaviour. In: Wolstenholme GWW, O'Connor CM (eds) Neurological basis of behaviour. Churchill, London, p 76

Andersson B, Gale CC, Sundsten JW (1964) Preoptic influences on water intake. In: Wayner MJ (ed) Proceedings of the First International Symposium on Thirst. Pergamon Press, Oxford, p 361

Andersson B, Leksell LG, Lishajko F (1975) Perturbations in fluid balance induced by medially placed forebrain lesions. Brain Res 99: 261

Antunes-Rodrigues J, Gentil CG, Negro-Vilar AN, Covian MR (1970) Role of adrenals in the changes of sodium chloride intake following lesions in the central nervous system. Physiol Behav 5: 89

Basedow H (1925) The Australian aboriginal. Preece, Adelaide

Bealer SL, Johnson AK (1979) Sodium consumption following lesions surrounding the anteroventral third ventricle. Brain Res Bull 4: 287

Bierens de Haan JA (1947/48) Animal psychology and the science of animal behaviour. Behaviour 1: 71

Cannon WB (1942) 'Voodoo' death. Am Anthrop 44: 169

Chiaraviglio E (1971) Amygdaloid modulation of sodium chloride and water intake in the rat. J Comp Physiol Psychol 76: 401

Chiaraviglio E (1972) Mesencephalic influence on the intake of sodium chloride and water in the rat. Brain Res 44: 73

Contreras RJ, Stetson PW (1981) Changes in salt intake after lesions of the area postrema and the nucleus of the solitary tract in rats. Brain Res 209: in press

Contreras RJ, Stetson PW, Kosten T (1981) Lesion-produced changes in salt intake: Interdependence between the area postrema and the nucleus of the solitary tract in rats. In: Van der Starre H (ed) Olfaction and Taste VII. IRL Press, London, p 339

Cort JH (1963) Spontaneous salt intake in the rat following lesions in the posterior hypothalamus. Physiol Bohemoslov 12: 502

Covian MR, Antunes-Rodrigues J (1963) Specific alterations in sodium chloride intake after hypothalamic lesions in the rat. Am J Physiol 205: 922

Darwin C (1872) The expression of the emotions in man and animals. Murray, London

Fessard A (1954) Mechanisms of nervous integration and conscious experience. In: Delafresnaye JF (ed) Brain mechanisms and consciousness. Blackwell, Oxford, p 200

Fisher AE, Coury JN (1964) Chemical tracing of neural pathways mediating the thirst drive. In: Wayner MJ (ed) Proceedings of the First International Symposium on Thirst. Pergamon Press, Oxford, p 515

Gallup GG (1979) Self-awareness in primates. Am Sci 67: 417

Gardner BT, Gardner RA (1975) Evidence for sentence constituents in the early utterances of child and chimpanzee. J Exp Psychol 104: 244

Gentil CG, Antunes-Rodrigues J, Negro-Vilar A, Covian M (1968) Role of amygdaloid complex and sodium chloride and water intake in the rat. Physiol Behav 3: 981

Gentil CG, Mogenson GJ, Stevenson JAF (1971) Electrical stimulation of septum, hypothalamus and amygdala and saline preference. Am J Physiol 200: 1172

Griffin DR (1976) The question of animal awareness. Rockefeller University Press, New York

Grossman SP (1964) Some neurochemical aspects of the central regulation of thirst. In: Wayner MJ (ed) Proceedings of the First International Symposium on Thirst. Pergamon Press, Oxford, p 487

Hassler R, Riechert T (1961) Wirkungen der Reizungen und Koagulationen in den Stammganglien bei stereotaktischen Hirnoperationem. Nervenarzt 32: 97
Heath RG (1972) Pleasure and brain activity in man. J Nerv Ment Dis 54: 3
Hess WR (1954) Diencephalon: Autonomic and extrapyramidal functions. Grune and Stratton, New York
Hess WR (1957) The functional organization of the diencephalon. Grune and Stratton, New York, p 180
Hess WR (1964) The biology of mind. University of Chicago Press
Holmes JH, Gregersen MI (1947) Relation of the salivary flow to the thirst produced in man by intravenous injection of hypertonic salt solution. Am J Physiol 51: 252
Huxley J (1960) The evolutionary vision. In: Tacks S, Callender C (eds) Evolution after Darwin. Issues in Evolution, Vol 3. University of Chicago Press, p 249
Jalowiec KE, Stricker EM, Wolf G (1970) Restoration of sodium balance in hypophysectomized rats after acute sodium deficiency. Physiol Behav 5: 1145
Kim C (1960) Nest building, general activity, and salt preference of rats following hippocampal ablation. J Comp Physiol Psychol 53: 11
Krieckhaus EE (1970) 'Innate recognition' aids rats in sodium regulation. J Comp Physiol Psychol 73: 117
Lashley KS (1923) The behaviouristic interpretation of consciousness. Psychol Rev 30: 237, 329
Lashley KS (1954) Dynamic processes in perception. In: Adrian ED, Bremer F, Delafresnaye JF, Jasper H (eds) Brain mechanisms and consciousness. Blackwell Scientific Publ, Oxford, p 427
Lashley KS (1960) Cerebral organization and behaviour. In: Beach FA, Hebb DO, Morgan CT, Nissen HW (eds) The neuropsychology of Lashley. McGraw-Hill, New York, p 529
Lethmate J, Dücker G (1973) Untersuchungen zum Selbsterkennen im Spiegel bei Orang-Utans und einigen anderen Affenarten. Z Tierpsychol 33: 248
Levy J, Trevarthen C, Sperry RW (1972) Perception of bilateral chimeric figures following hemispheric deconnexion. Brain 95: 61
Livingston RB (1967) Reinforcement. In: Quarton G, Melnechuk T, Schmitt F (eds) The neurosciences. Rockefeller University Press, New York, p 568
Marais E (1939) My friends the baboons. Methuen, London
Marais E (1969) The soul of the ape. Anthony Blond, London
McBride AF, Hebb DO (1948) Behaviour of captive bottlenose dolphins. J Comp Physiol Psychol 41: 111
McKenzie JS, Denton DA (1974) Salt ingestion response to diencephalic electrical stimulation in the unrestrained conscious sheep. Brain Res 70: 449
McKenzie JS, Smith MH (1973) Stereotaxic method and variability data for the brain of the merino sheep. J Hirnforsch 14: 355
Mendelson J (1970) Self-induced drinking in rats: The qualitative identity of drive and reward systems in the lateral hypothalamus. Physiol Behav 5: 925
Mogenson GJ (1969) General and specific reinforcement systems for drinking behaviour. Ann NY Acad Sci 157: 779
Mogenson GJ, Stevenson JAF (1967) Drinking induced by electrical stimulation of the lateral hypothalamus. Exp Neurol 17: 119
Myers RD, Tytell M, Kawa A, Rudy T (1971) Microinjection of ^{3}H-acetylcholine, ^{14}C-serotonin and ^{3}H-norepinephrine into the hypothalamus of the rat: Diffusion into tissues and ventricles. Physiol Behav 7: 743
Nachman M, Ashe JH (1974) Effects of basolateral amygdala lesions on neophobia, learned taste aversions and sodium appetites in rats. J Comp Physiol Psychol 87: 622
Nathan MA, Reis DJ (1977) Chronic labile hypertension produced by lesions of the nucleus tractus solitarii in the cat. Circ Res 40: 72
Nauta WJH (1958) Hippocampal projections and related neural pathways to the mid-brain in the cat. Brain 81: 319
Negro-Vilar AN, Gentil CG, Covian MR (1967) Alterations in sodium chloride and water intake after septal lesions in the rat. Physiol Behav 2: 167
Norgren R (1977) Gustatory neuroanatomy. In: Le Magnen JT (ed) Proceedings of the Sixth International Symposium on Olfaction and Taste. Information Retrieval, London, p 225
Novakova A, Cort JH (1966) Hypothalamic regulation of spontaneous salt intake in the rat. Am J Physiol 211: 919
Olds J, Milner P (1954) Positive reinforcement produced by electrical stimulation of septal area and other regions of rat brain. J Comp Physiol Psychol 47: 419
Penfield W (1966) Speech perception and the uncommitted cortex. In: Eccles JC (ed) Brain and conscious experience. Springer-Verlag, Berlin, p 217
Popper KR, Eccles JC (1977) The self and its brain. Springer-Verlag, Berlin
Quartermain D, Wolf G, Keselica J (1969) Relation between medial hypothalamic damage and impairments in regulation of sodium intake. Physiol Behav 4: 101
Richter CP (1957) On the phenomenon of sudden death in man and animals. Psychosom Med 19: 191
Roberts WW (1969) Are hypothalamic motivational mechanisms functionally and anatomically specific? Brain Behav Evol 2: 317
Robinson BW (1964) Forebrain alimentary responses: Some organizational principles. In: Wayner MJ (ed) Proceedings of the First International Symposium on Thirst. Pergamon Press, Oxford, p 411
Robinson BW (1967) Head-mounted electrode platform: Construction and error analysis. Physiol Behav 2: 455
Robinson BW, Mishkin M (1968) Alimentary responses to forebrain stimulation in the monkey. Exp Brain Res 4: 330
Rolls BJ, Rolls ET (1973) Effects of lesions in the basolateral amygdala on fluid intake in the rat. J Comp Physiol Psychol 83: 240
Routtenberg A (1972) Intracranial chemical injection and behaviour: A critical review. Behav Neural Biol 7: 601
Saad WA, Antunes-Rodrigues J, Gentil CG, Covian MR (1972) Interaction between hypothalamus, amygdala and septal area in the control of sodium chloride intake. Physiol Behav 9: 629
Schwartz M, Teitelbaum P (1974) Dissociation between learning and remembering in rats with lesions in the lateral hypothalamus. J Comp Physiol Psychol 87: 384
Sem-Jacobsen CW, Torkildsen A (1960) Depth recording and electrical stimulation in the human brain. In: Ramsey ER, O'Doherty DS (eds) Electrical studies on the unanaesthetized human brain. Paul Hoeber, New York, p 275
Sperry RW (1974) Lateral specialization in the surgically separated hemispheres. In: Schmitt FO, Worden FG (eds) The neurosciences: Third study program. MIT Press, Cambridge, Mass, p 5
Sterc J, Novakova V, Golda V (1975) Effects of dorsal, mediobasal and laterobasal septal lesions in the rat: Water and sodium chloride intake. Exp Neurol 48: 175
Stricker EM, Miller NE (1968) Saline preference and body fluid analyses in rats after intrahypothalamic injections of carbachol. Physiol Behav 3: 471
Teitelbaum P (1973) Discussion on the use of electrical stimulation to study hypothalamic structure and function. In: Epstein AN, Kissileff HR, Stellar E (eds) The neuropsychology of thirst: New findings and advances in concepts. Winston & Sons, Washington, p 143
Thorpe WH (1966) Ethology and consciousness. In: Eccles JC (ed) Brain and conscious experience. Springer-Verlag, Berlin, p 470
Valenstein ES (1970) Stability and plasticity of motivation

systems. In: Schmitt FO (ed) The neurosciences. Second study program. Rockefeller University Press, New York, p 207
Valenstein ES (1973) Invited comment: Electrical stimulation and hypothalamic function: Historical perspective. In: Epstein AN, Kissileff HR, Stellar E (eds) The neuropsychology of thirst: New findings and advances in concept. Winston & Sons, Washington, p 155
Valenstein ES, Cox VC, Kakolewski JW (1969) The hypothalamus and motivated behaviour. In: Tapp J (ed) Reinforcement. Academic Press, New York, p 242
Valenstein ES, Cox VC, Kakolewski JW (1970) Re-examination of the role of the hypothalamus in motivation. Psychol Rev 77: 16
Walsh LL, Grossman SP (1977) Electrolytic lesions and knife cuts in the region of the zona incerta impair sodium appetite. Physiol Behav 18: 587
Wayner MJ (1970) Motor control functions of the lateral hypothalamus and adjunctive behaviour. Physiol Behav 5: 1319
Wolf AV (1950) Osmetric analysis of thirst in man and dog. Am J Physiol 161: 75
Wolf G (1964) Effect of dorsolateral hypothalamic lesions on sodium appetite elicited by desoxycorticosterone and by acute hyponatraemia. J Comp Physiol Psychol 58: 396
Wolf G (1967) Hypothalamic regulation of sodium intake: Relations to preoptic and tegmental function. Am J Physiol 213: 1433
Wolf G (1968) Thalamic and tegmental mechanisms for sodium intake: Anatomical and functional relations to lateral hypothalamus. Physiol Behav 3: 997
Wolf G (1971) Neural mechanisms for sodium appetite: Hypothalamus positive—hypothalamofugal pathways negative. Physiol Behav 6: 381
Wolf G, Di Cara LV (1974) Impairments in sodium appetite after lesions of gustatory thalamus: Replication and extension. Behav Neural Biol 10: 105
Wolf G, Quartermain D (1967) Sodium chloride intake of adrenalectomized rats with lateral hypothalamic lesions. Am J Physiol 212: 113
Wolf G, Schulkin J (1980) Brain lesions and sodium appetite: An approach to the neurological analysis of homeostatic drives. In: Fregly MJ, Kare MR, Bernard RA (eds) Biological and behavioural aspects of salt intake. Academic Press, NY, p 331
Wolf G, Di Cara LV, Braun JJ (1970) Sodium appetite in rats after neocortical ablation. Physiol Behav 5: 1265
Wolf G, McGovern JF, Di Cara LV (1974) Sodium appetite: Some conceptual and methodological aspects of a model drive system. Behav Neural Biol 10: 27

20 The effect of rapid systemic correction of sodium deficiency on salt appetite

Summary

1. In sodium-deficient sheep it is a simple experimental procedure to contrive rapid correction of sodium deficiency by intracarotid or intravenous infusion of hypertonic or isotonic saline begun before or during access to salt solution. The change in voluntary intake of sodium bicarbonate solution so produced provides insight into the ease of access to the system of control of appetite, and also the vectors of physiological effect upon it.

2. Infusion of 4 M sodium chloride into one carotid artery with the contralateral artery occluded, which causes a 7–20 mmol/l rise in the sodium concentration of cerebral arterial blood on both sides of the brain and in lingual blood, does not influence sodium appetite when infusion is begun 7 min before and continued during 8 min of access to sodium bicarbonate solution. There have been similar findings in the sodium-deficient goat.

3. Intravenous infusion of 115 ml 2.7–4.4 M sodium chloride, given 30 min before the drinking period, increased plasma [Na] (sodium concentration) by 15–30 mmol/l, to well above normal range. The sodium appetite was usually reduced but seldom abolished and the effects bore little relation to plasma [Na]. Infusion of 115 ml of 2.7 M glucose reduced plasma [Na] by ca 10 mmol/l, and caused no significant effect on sodium appetite.

4. By 120 min after cessation of intravenous infusion of 4 M sodium chloride, sodium appetite was reduced commensurately with the amount of sodium retained by the animal. Rapid infusion of 2–2.5 litres of isotonic Ringer saline also reduced sodium appetite but not commensurately with the amount retained in the body, whether tested 10 min or 120 min after cessation of rapid infusion.

5. Sodium appetite was also intensively studied in animals trained to bar-press to deliver small volumes of sodium bicarbonate to a drinking cup, access being permitted for 2 h each day. Satiation behaviour after 22 h of sodium deficiency was consistent and commensurate with deficit. Experiments were made where each bar-press automatically caused a short-duration intravenous injection of isotonic or concentrated saline, or hyperosmotic solutions of non-electrolytes. The rapid bar-pressing of the depleted animals resulted in a pulsed influx of large amounts of salt to the circulation. Reduction of appetite relative to a control with no infusion, or pulsed isotonic saline, was seen in 10–20 min when hypertonic saline was infused. This indicated a relatively short latent period of initial influence on the control system compared with the characteristic 6–48-h delay in activation of salt appetite when sodium is rapidly lost from the body. It was found that infusion of 1 M mannitol solution had a comparable effect to equiosmotic saline in this experimental context. The inhibition of sodium

appetite produced by mannitol was much shorter lived than that resulting from systemic influx of hypertonic saline. Self-determined injection of 1 M urea and 1 M glucose also reduced sodium appetite.

6. These results of the effects of salt compared with non-electrolytes are analysed in terms of changes in the [Na] of the cerebrospinal fluid (CSF), the relation between CSF [Na] and plasma [Na], pulsed influx into the circulation versus continuous infusion, plasma volume and circulatory changes, and concurrent effects on osmoregulatory and sodium sensors involved in thirst. The question of the potential influence of thirst and water drinking on the process of satiation of salt appetite is also discussed, the evidence being against significant effect in the experimental conditions here. The data show that the rapid change of sodium balance produced by systemic infusion begins to influence the system controlling sodium appetite by 20 min, and the effect may be commensurate with sodium influx by 2 h.

Introduction

The time characteristics of change of appetite in response to rapid systemic infusion of sodium in the deplete animal could give insight into the characteristics of the appetite control system. Furthermore, since rapid correction of sodium deficiency can be achieved by infusion of either isotonic or hypertonic saline, critical information can be gained on the role of a change in plasma [Na] (sodium concentration) as a signal to the control system.

The sheep is particularly suitable as an experimental animal. Precisely controlled alterations of balance can be made with frequent monitoring of the physiological changes produced. Partial or total correction of large deficits can be made rapidly by intravenous infusion of isotonic or 4 M saline. Alternatively, a larger change in cerebral blood [Na] for a given systemic input of sodium can be achieved by intracarotid infusion of 4 M saline under conditions of occlusion of the contralateral carotid artery loop. The change in blood composition produced in the ipsilateral carotid artery affects both sides of the brain, and also the tongue and other tissues of the head. The same degree of change as such manoeuvres produce in CSF [Na] and other ions, may also be produced locally in the CSF, without systemic change, by lateral or third ventricular infusions of appropriate types of artificial CSF (see Chapter 24). Further elaborations of approach include producing local changes in CSF which counter the direction of change contrived by concurrent systemic infusion.

Other experiments can be designed in the context where a sheep is permitted to correct its sodium deficiency by bar-pressing repetitively for a small quantity of sodium bicarbonate to drink, and it is contrived that each bar-press automatically gives the animal also an intravenous injection of isotonic or hypertonic saline (or another osmotically active agent such as urea, mannitol or glucose).

Overall, the way is opened for systematic dissection of the response characteristics of the control system. Whereas some of these manoeuvres may be technically feasible in the rat, to date they have been little utilized. Analogous methodological approaches have been used on goats (which have the same advantage of size that the sheep does) by Andersson, Olsson and colleagues in Stockholm in their outstanding experiments unravelling the central control of thirst, and by Baldwin at Cambridge in England.

The experimental model: The sodium-deficient sheep

With onset of sodium deficiency as a result of uncompensated salivary loss in the sheep, the plasma [Na] falls. This reflects the fact that the initial [Na] of saliva lost is ca 170 mmol/l, and that therefore there is a loss of sodium in excess of water relative to extracellular proportions. Also, water intake increases as a result of salivary loss.

The changes which resulted from uncomplicated

Table 20-1. Effect of salivary sodium loss produced by parotid cannulation on plasma, urinary and salivary composition and water intake (mean±s.e.m.; *n*=10).

	Days				
	1	2	3	4	5
Urinary Na output (mmol)	50±11	46±7	48±10	16±5	2±0
Salivary Na output (mmol)				485±46	233±32
Plasma [Na] (mmol/l)			146±0.1	143±1.0	141±1.0
Plasma [K] (mmol)			4.5±0.1	4.5±0.1	4.4±0.1
Water intake (ml)	2131±420	1888±261	1588±277	2769±112	
Salivary Na[a] (mmol/l)			170±3	145±3	116±4
Salivary K[a] (mmol/l)			4.7±0.3	24±2.2	50.8±5.1

After Abraham et al. (1975).
[a]Specimens collected at time of parotid cannulation, and at 24 and 48 h.

loss of saliva by parotid cannulation in ten experiments are shown in Table 20-1 and Fig. 20-1 (Abraham et al. 1976). There were three control days (days 1–3) followed by two days of salivary loss. There was a small significant decrease in plasma [Na] after 24 h ($P<0.001$) and 48 h ($P<0.001$) of salivary loss (days 4 and 5) when compared with predepletion levels. The mean [Na] for total saliva collected on day 4 was 154±2.5 mmol/l, representing loss of sodium in excess of water. The mean plasma [Na] of the sodium-replete sheep reported here is in agreement with earlier figures from sodium-replete sheep with a parotid fistula, viz. 147±3.7 mmol/l (Coghlan et al. 1960).

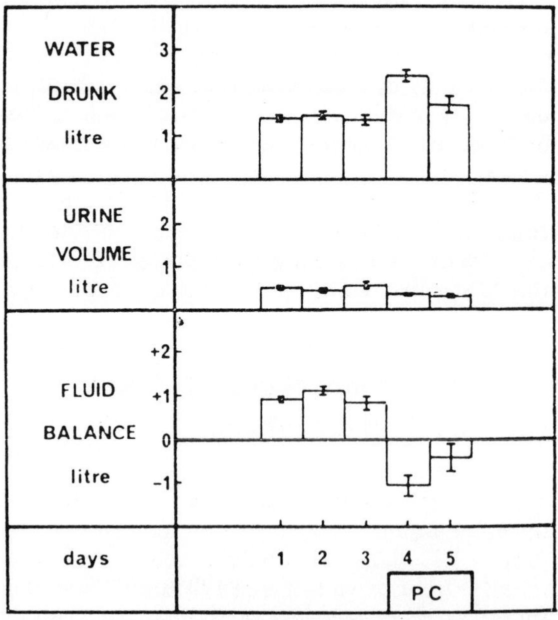

Fig. 20-1. The mean ±s.e.m. water intake, urine volume and fluid balance for three days before (days 1, 2 and 3) and the two days of (days 4 and 5) acute cannulation of one parotid duct (PC) (*n*=10). (From Abraham et al. 1976, with permission.)

It is clear that sodium deficiency due to parotid salivary loss results in a progressive fall in plasma [Na], but that in the first 1–2 days this may not be very large. In very severe deficiency plasma [Na] can fall to between 125 and 135 mmol/l. It is feasible that this change could be the principal direct-acting stimulus in the induction of sodium appetite. Or it could be the initiating vector of other changes, such as loss of sodium from intracellular fluid, which in turn cause appetite. This could be the proximate cause of excitation of appetitive drive. In any event, the data which follow indicate that there is no simple and direct relation between contemporary plasma [Na] and response of the salt appetite control system. In a wide range of experiments it has been shown sodium appetite may be present at times when plasma [Na] is in the sodium-replete range or higher.

Intracarotid infusion in sheep

A direct approach to the influence of plasma [Na] on salt appetite in the sodium-deficient animal was to increase the [Na] of cerebral arterial blood immediately before and during the period the animal had access to sodium bicarbonate solution and water (Beilharz et al. 1965). The aim was to increase the [Na] of cerebral blood up to or above the plasma [Na] of the sodium-replete sheep, but with as little effect on sodium balance as feasible. This could test for a receptor in the cerebral circulation reacting to contemporary change in [Na] of the blood. This was done by infusing hypertonic saline directly into the cerebral circulation. In these and similar experiments to be recounted all cannulations were made 60 min or more before any experimental procedure, so that the animal was undisturbed and without any evident apprehension at the time of the experiment. Infusion was made into one carotid artery with the contralateral carotid loop occluded by a pneumatic cuff at 300 mmHg. In these circumstances the blood on the ipsilateral side perfuses both sides of the brain (Andersson and Jewell 1956; Denton 1957; Baldwin and Bell 1960; Beilharz et al. 1965).

The animals in the experimental series were depleted of sodium for 48 h and had external deficits of 600–900 mmol. Water was continuously available. An infusion of 4 M saline (30 ml) was made over 15–17 min (Fig. 20-2). The infusion was begun 7 min before sodium bicarbonate (300 mmol/l) was presented to the animal, and was continued for the first 8 min of access. Therefore at the time of presentation of sodium solution 60 mmol of sodium had been given (10% or less of deficit). The analysis of blood drawn from ipsilateral and contralateral jugular veins during the infusion, and from the carotid artery at the end of infusion, determined the local rise in cerebral arterial plasma [Na] (7–20 mmol/l) relative to the systemic level (allowing for the effect of recirculation), and the rise in cerebral arterial plasma [Na] relative to the preinfusion concentration (15–29 mmol/l). The differences between concentrations in ipsilateral and contralateral jugular veins were usually not large. The changes reflected elevation of cerebral plasma [Na] to the replete level or above by the time the sodium bicarbonate solution was presented (Beilharz et al. 1965).

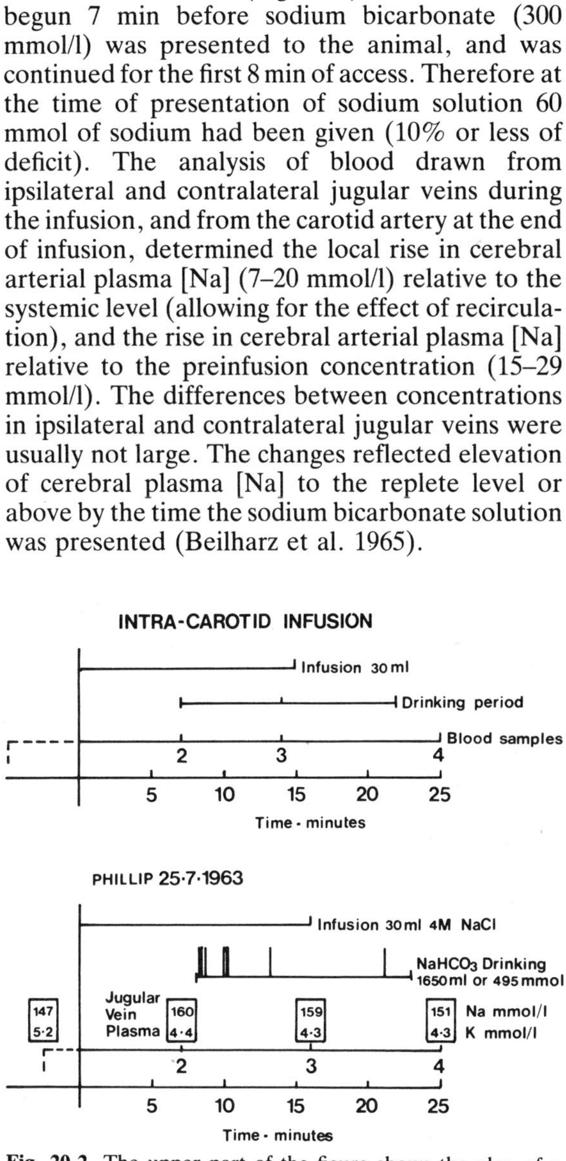

Fig. 20-2. The upper part of the figure shows the plan of a typical infusion experiment, and the lower part the result of infusing 4 M sodium chloride into the carotid artery of sheep Phillip. The composition of jugular blood samples is shown and also the periods in which the sheep drank sodium bicarbonate solutions. The contralateral carotid artery was occluded for 15 min before and during the infusion. (After Beilharz et al. 1965.)

A typical experiment of this series is shown in Fig. 20-2. Despite the large increase (13 mmol/l) in plasma [Na] immediate drinking of sodium bicarbonate occurred. The blood sample taken after cessation of the intracarotid infusion reflects a rise in systemic plasma [Na] as a result of recirculation.

Overall the results (Beilharz et al. 1965) were as follows:

In 135 control experiments on four sheep given 15 min access to sodium bicarbonate, each sheep drank a fairly constant amount of sodium solution and, furthermore, the amount drunk approximated to the loss of sodium from the parotid fistula over 48 h (Table 20-2). Moreover, it was shown formally in the case of Satan that if access to sodium bicarbonate were permitted every 24 h, after only 1 day's salivary loss (average 554 mmol sodium) the voluntary intake in eight experiments averaged 568 mmol. That is, intake was reduced commensurately in response to reduced deficit.

The infusion of 4 M sodium chloride did not result overall in any significant change in the volume of sodium bicarbonate solution drunk (Table 20-2), although changes were seen in early experiments on some sheep. These early experiments are responsible for the overall lowered mean and the large standard deviation (Table 20-2).

The intracarotid 4 M sodium chloride infusion did not diminish the appetitive behaviour normally displayed by sodium-deficient sheep at the time of presentation of the sodium solution. Also there was profuse conditioned salivary flow.

When a sheep which has had free access to water during the period of development of sodium deficiency is offered water at the same time as the sodium bicarbonate solution during the 'sodium drinking period' it will generally sniff at the water but rarely taste or drink it (Beilharz et al. 1962). It was thought unnecessary, therefore, to continue to offer water with sodium bicarbonate with every experiment on sodium deficiency here. However, water was offered with sodium bicarbonate to Satan on five of the days when 4 M sodium chloride was infused, on each of the six days with 4 M sodium chloride infusions to Tacitus, and on eight of the 4 M sodium chloride infusion days with Anselm. Satan and Tacitus never took any interest in the water and Anselm drank a small amount of (270 ml) on one occasion only.

Intracarotid infusion of isotonic saline did not affect either appetitive behaviour or the voluntary sodium bicarbonate intake during the 'sodium drinking period' (Table 20-2).

When 4 M sodium chloride was infused into the carotid artery of sodium-replete sheep and sodium

Table 20-2. Phillip, Anselm, Satan and Tacitus. The effect of intracarotid infusion of 4 M sodium chloride on the voluntary intake of sodium bicarbonate solution. The number of experiments is shown in brackets. The mean loss of sodium in the saliva during each 2-day period of depletion, and mean voluntary intake of sodium bicarbonate solution on control days when no intracarotid infusion was given, and on days when 0.15 M sodium chloride was infused into the carotid artery is shown also. During the intracarotid infusion, the opposite carotid artery was occluded.

		Mean $NaHCO_3$ intake (mmol Na±s.d.)		
Sheep	Mean salivary Na loss (mmol) ±s.d.	Control days	4 M NaCl infusion	0.15 M NaCl infusion
Phillip	644±98 (40)	676±118 (40)	631±99 (7)	708, 660 (2)
Anselm	872±105 (30)	782±140 (30)	924±101 (9)	795, 816 (2)
Satan	980±109 (45)	959±155 (45)	753±300 (13)	969±238 (6)
Tacitus	907±66 (20)	882±153 (20)	743±258 (6)	729, 777 (2)

After Beilharz et al. (1965).

bicarbonate and water were offered in the manner described, the animals did not drink the sodium solution in any experiment. They differed from each other with regard to water intake, in that Satan did not drink water, Phillip drank 270 ml water on one occasion, and Anselm and PF33 drank considerable quantities of water at each trial. In two infusions of 0.15 M sodium chloride Anselm, when sodium-replete, drank neither sodium bicarbonate nor water.

Discussion of intracarotid infusion data

The results show that it is unlikely that sodium appetite is stimulated by an intracranial receptor which responds rapidly to the contemporary sodium concentration of arterial blood in a manner analogous to the osmoreceptor or sodium receptor controlling antidiuretic hormone (ADH) secretion and thirst. With intracarotid infusion of hypertonic saline in the dog, reduction of an established water diuresis begins almost as rapidly as with an intravenous injection of ADH, indicating that the response of the receptor elements is very rapid (Verney 1947). We have confirmed this in the sheep, and the immediate effects on thirst have been discussed elsewhere. In the light of this overall result on intake, and the fact that appetitive behaviour was not reduced, a possible explanation of the decreased voluntary intake in the initial experiments in some animals only was that the increased [Na] of blood in the tongue and oral tissues produced sensations sufficiently unfamiliar to modify intake. However, after several experiences the animal became habituated to this effect which was essentially non-specific.

A consideration of the blood supply to the sheep's brain is important in these experiments. The blood analyses from simultaneous jugular bleeds have indicated that material infused into one carotid while the other one was completely occluded distributed fairly evenly to both sides of the head (Beilharz et al. 1965). Notwithstanding some variation, a clear-cut rise in concentration was produced on both sides. Baldwin and Bell (1960) have shown cinephotographically that dye injected into one carotid with the other one clamped results in bilateral distribution of the dye, whereas injection without clamping the other carotid causes coloration of the ipsilateral cerebral hemisphere only. This was confirmed in relation to plasma [Na] by jugular vein sampling in an experiment on Anselm in which the contralateral carotid was not clamped during 4 M sodium chloride infusion.

Baldwin and Bell state that in the sheep the brain is supplied almost entirely by blood from the carotid arteries. The arterial input to the intracranial carotid rete through which blood passes before reaching the circle of Willis is via branches of the internal maxillary artery arising from the external carotid. The internal carotid has no functional significance. They have also shown that blood from the vertebral arteries plays no part in the supply to the cerebral hemispheres. However, when both carotids are clamped, vertebral blood can maintain cerebral function in the sheep by entering the external carotid arteries via the occipital arteries (Denton 1957; Baldwin and Bell 1960). Daniel et al. (1953) have shown that blood flows caudally in the basilar artery of the sheep. Andersson and Jewell (1956) demonstrated an essentially similar blood distribution in the brain of the goat. They also found in the goat that blood from the vertebral arteries may extend forward in the cervical cord to reach the posterior medulla and that the boundary between carotid and vertebral fields varied between individuals.

The procedure of intracarotid infusion may modify the [Na] and [Cl] of the blood perfusing the tongue, taste buds and salivary glands and also the composition of the CSF (McKinley et al. 1978), as well as that of the blood passing through the brain. Because of the reduced pressure in the section of contralateral carotid artery superior to the occlusion (Denton 1957), it is probable that both sides of the tongue were perfused with blood with increased [Na].

One possibility in considering the genesis of salt appetite in sodium deficiency is that the receptor elements are associated with the taste buds in the tongue and that a change in afferent impulses acts centrally to provoke appetite by eliciting a sensation localized to the mouth. This would be analogous to the dry mouth component in the spectrum of thirst sensation. McCance (1936) described significant alterations in sensations of taste associated with salt deficiency in man. Richter (1956) stated that it always appeared to him that the appetite changes of deficiency occurred at the taste bud membrane. Possibly some changed taste sensation could result from the high [Na] in the blood in the lingual artery in these experiments, and this might be reinforced by the rise in [Na] of parotid saliva of 10–15 mmol/l because of the rise in [Na] in the carotid blood (Denton and McDonald 1957). However, since these immediate changes in [Na] in lingual blood did not abolish or significantly modify appetite, the results do not support a hypothesis of the receptor elements generating sodium appetite being situated in the tongue.

Furthermore, it might be anticipated that intracarotid infusion of 4 M sodium chloride in sodium-deplete sheep would cause them to drink water, yet in a total of 19 instances in which sodium bicarbonate and water were offered simultaneously during infusion of hypertonic saline there was only one instance where a sheep drank a small amount of water. Possibly the thirst mechanism is modified by the sodium-deficient state, or, during the short period of access, attention is directed exclusively to the sodium bicarbonate solution.

Intravenous infusions in sheep

Parallel to these experiments, observations were made in our laboratories on the influence on sodium appetite of much larger intravenous infusions over a longer time period (Beilharz and Kay 1963). About 115 ml of either 0.16 M sodium chloride, 2.7 M glucose or 2.6–4.4 M sodium chloride were infused over 33 min. Ten minutes after the infusion had ended the sheep were offered sodium bicarbonate (317 or 420 mmol/l)

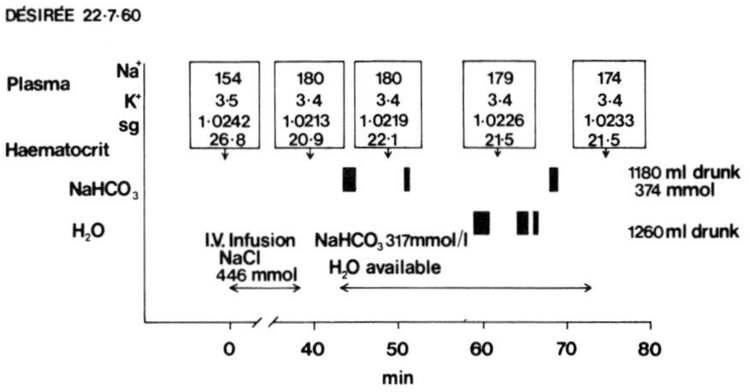

Fig. 20-3. Désirée: sheep depleted of sodium by 24 h of parotid salivary loss. A typical experiment in which 4.4 M sodium chloride was infused into the jugular vein is shown. Ten minutes after cessation of the infusion the sheep was offered sodium bicarbonate (317 mmol/l) and water. The periods during which the sheep drank water and sodium bicarbonate solution are recorded in the middle and the composition of jugular blood samples in the upper panels. Water was not available during the infusion.

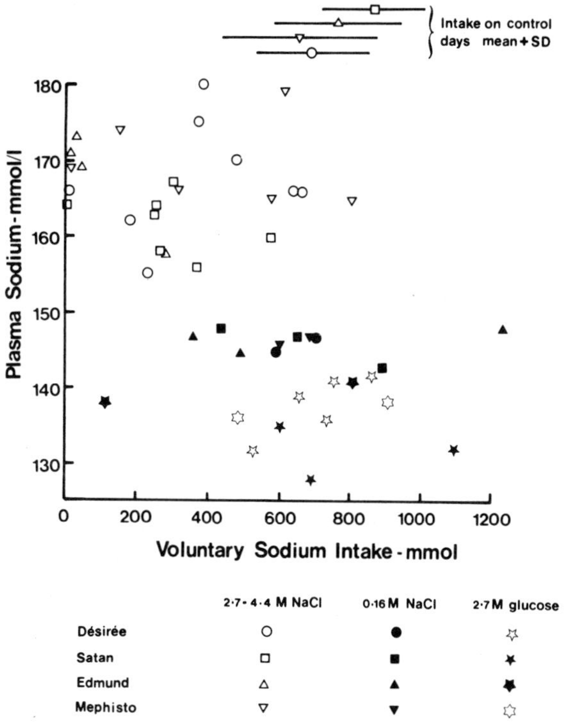

Fig. 20-4. Sodium appetite and plasma [Na] after intravenous infusions. The ordinate shows the [Na] in plasma 5 min after completion of various intravenous infusions (i.e. 5 min before the beginning of the 30-min 'sodium drinking period'. The abscissa shows the amount of sodium drunk during the 'sodium drinking period'. The mean and s.d. of the amount of sodium drunk on control days are shown at the top. (After Beilharz and Kay 1963.)

and water to drink. Water was normally continuously available but was removed during the infusion. Four sheep (Désirée, Satan, Mephisto and Edmund) were used in a series of experiments.

Fig. 20-3 shows the course of a typical

Table 20-3. The effects of intravenous infusions of sodium chloride and glucose solutions on voluntary sodium intake and blood composition. The 'normal values' shown represent the mean sodium intake on the control day preceding the infusion and the mean composition of the blood drawn a minute before the infusion was started. The values shown against the infusions are the means of the differences between normal and post-infusion values. Further explanation is given in the text.

	Désirée	Satan	Mephisto	Edmund	Mean value
No. of infusions					
Control days before infusions	15	13	10	9	—
0.16 M NaCl	2	3	2	3	—
2.7 M glucose	5	3	2	2	—
2.6–4.4 M NaCl	8	7	6	4	—
Voluntary Na intake (mmol $NaHCO_3$ drunk)					
Normal values	714	888	679	788	—
0.16 M NaCl	−73	−236	−43	−101	−113
2.7 M glucose	−11	− 97	12	−330	−106
2.6–4.4 M NaCl	−350	−603	−272	−697	−480
Plasma [Na] (mmol/l)					
Normal values	146	141	147	147	—
0.16 M NaCl	1	1	1	−1	0.5
2.7 M glucose	−8	−6	−12	−9	−8.8
2.6–4.4 M NaCl	22	20	23	23	22.0
Plasma specific gravity					
Normal values	1.0252	1.0261	1.0256	1.0259	—
0.16 M NaCl	−0.0003	0	−0.0002	−0.0004	−0.0002
2.7 M glucose	−0.0010	−0.001	−0.0009	−0.0009	−0.0010
2.6–4.4 M NaCl	−0.0028	−0.0035	−0.0032	−0.0039	−0.0034
Haematocrit					
Normal values	28.3	29.4	28.6	33.9	
0.16 M NaCl	2.4	0.4	−1.7	−2.9	−0.4
2.7 M glucose	−6.2	−4.1	−11.1	−6.8	−7.0
2.6–4.4 M NaCl	−7.0	−7.5	−7.3	−10.0	−8.0

From Beilharz and Kay (1963), with permission.

experiment, and Fig. 20-4 illustrates the full series of experiments. The effects of the infusions on voluntary sodium intake and on blood composition are compared in Table 20-3.

The infusions of 0.16 M sodium chloride, which provided only 18 mmol of sodium, caused no significant changes in blood composition. Although sodium intake was usually reduced slightly, the mean reduction being 113 mmol, it was not statistically significant (Beilharz and Kay 1963). The 2.7 M glucose infusions did not affect the sodium balance but caused significant reductions in the plasma [Na], specific gravity of the plasma and haematocrit value. However, they did not increase the voluntary sodium intake; rather there was a slight (106 mmol) but statistically insignificant fall.

The infusions of 2.6–4.4 M sodium chloride supplied between 290 and 511 mmol of sodium and caused significant reductions in the mean sodium intake. Generally the sheep drank rapidly from the solution for shorter periods, rather than at a slower rate. The plasma [Na] rose by about 22 mmol/l (Table 20-3). The specific gravity of the plasma fell sharply; the haematocrit value also fell, but no more than after glucose infusions.

The hypertonic saline infusions reduced the mean voluntary sodium intake of the four sheep to rather different extents. This was despite relatively uniform changes in blood composition. Désirée, into whom an average of 375 mmol of sodium was infused, drank 350 mmol less sodium than on control days. Thus her total sodium intake, that infused plus that drunk, approximated that on control days. But the other sheep did not achieve this balance. Mephisto drank about 100 mmol too much sodium, Satan drank about 200 mmol too little, and Edmund almost 300 mmol too little. The variability of intake at individual trials under these conditions was not related to plasma [Na] (Fig. 20-4). Intake and plasma [Na] could vary independently.

The infusions of hypertonic saline caused water intake. Satan, Mephisto and Edmund eagerly drank up to 3 litres of water during the ensuing 'sodium drinking period'. An example is seen in Fig. 20-3. Désirée also drank after three out of the eight infusions. Little or no water was drunk after the infusions of 0.16 M sodium chloride or 2.7 M glucose.

Plasma volume was expanded by both hypertonic saline and 2.7 M glucose infusions as evidenced by the decrease in specific gravity and haematocrit. Beilharz and Kay (1963) determined that the haematocrit fall was much greater than accounted for by the red cell shrinkage as a result of the increased osmotic pressure.

A rapid fall in the plasma [Na] caused by glucose infusions does not stimulate the appetite; a rapid rise caused by sodium chloride infusions does not inhibit the appetite. Therefore, Beilharz and Kay concluded that it is unlikely that the plasma [Na], by itself, is an important stimulus to sodium appetite.

This conclusion is compatible with the observation that during the first day or two of sodium depletion a pronounced appetite for sodium develops although there may often be little or no fall in plasma [Na]. For example, in three of the four sheep featured in Table 20-3 as showing a large sodium appetite, the mean plasma [Na] was 146 or 147 mmol/l before infusion. This is still within the range found for sheep in normal sodium balance. Also, in support of this hypothesis, Beilharz and Kay draw attention to the experiments on concurrent water and sodium depletion reported by Denton et al. (1959), where it was noted there was little evidence of a relation between readiness to drink sodium solutions and the plasma [Na]. Similarly there is little evidence from the data that plasma volume influences appetite to a great extent, since although both glucose and sodium chloride infusions caused a considerable increase in plasma volume they had variable effects on sodium appetite.

Intracarotid infusion in goats

Baldwin (1976) has studied sodium appetite in the goat. He confirms the data on intracarotid infusion in sheep, and adds much to the description of the basic behaviour in sodium deficiency. Goats were prepared with a parotid fistula made by dividing the duct in the cheek and explanting the cut end. These fistulae remained patent for months and in some cases for over a year. The animals were allowed access to a mineral block which contained salt and were given 1% (0.12 M) sodium bicarbonate solution instead of water. They all remained in normal sodium balance as evidenced by the salivary Na/K ratio. Sodium deficiency was produced by withdrawing the sodium supplement. When the goats were in normal sodium balance they showed no interest in drinking concentrated (4%) (0.5 M) solutions of sodium bicarbonate, but when depleted of sodium they learned to press panels with their muzzles in order to obtain small volumes (5 ml) of the solution as reinforcement. During the sessions to study operant behaviour the animals stood on a wooden stand and faced a black Perspex panel with two transilluminated panel switches. One switch provided 5 ml of the 0.5 M sodium bicarbonate solution whereas the other delivered 5 ml of water to another drinking bowl. When the animals were fully trained to operate the panels, reinforcements were available on a fixed-ratio 10 schedule (i.e. every tenth press caused delivery).

On withdrawal of sodium supplement, saliva [Na] fell while [K] rose, and the goats were highly motivated to work for the concentrated sodium bicarbonate solution during trials lasting 20 min each day (Fig. 20-5). This occurred with little

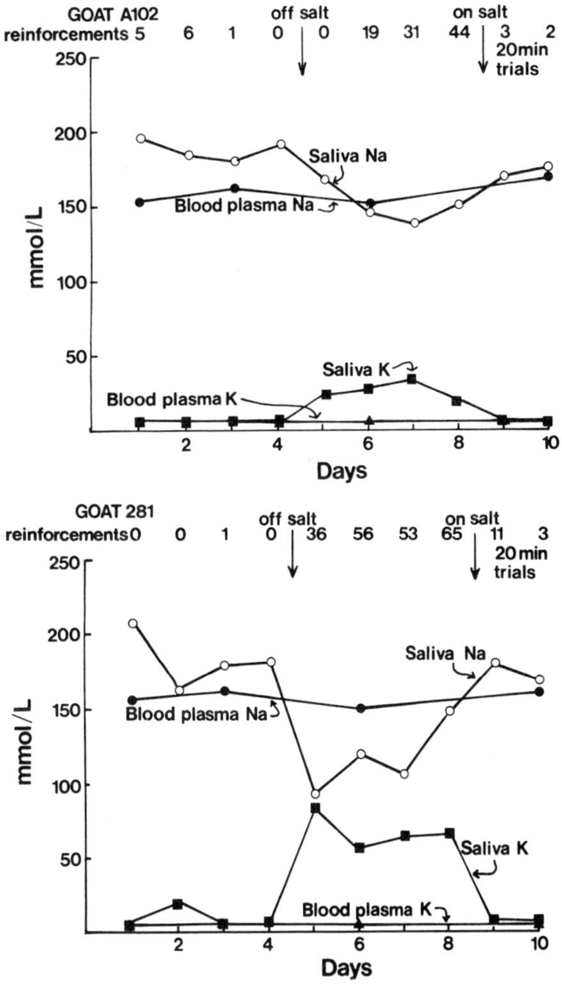

Fig. 20-5. Two goats losing parotid saliva from a unilateral fistula. The effect of withdrawal of salt supplement for 4 days on number of bar-presses during 20 min each day to obtain 5 ml of 4% (0.5 M) sodium bicarbonate. The plasma and salivary [Na] and [K] are shown also. (After Baldwin 1976.)

change in plasma [Na], as in the sheep data. Baldwin notes that sodium depletion appears to result in the taste of this normally aversive concentrated solution becoming very attractive, because it is consumed avidly on first exposure before the goats have benefited from its physiological effects.

In a series of 17 experiments using three goats, the effects of continuous intracarotid infusions of concentrated salt solutions (3.9 M) on operant behaviour were studied. The goats were responding to obtain 5 ml of 0.5 M sodium bicarbonate solution when intracarotid infusions (3.8 ml/min) were made for periods of 5–10 min (Fig. 20-6). In five of the experiments blood samples were taken from the jugular vein before, during and after intracarotid injection to determine the changes in

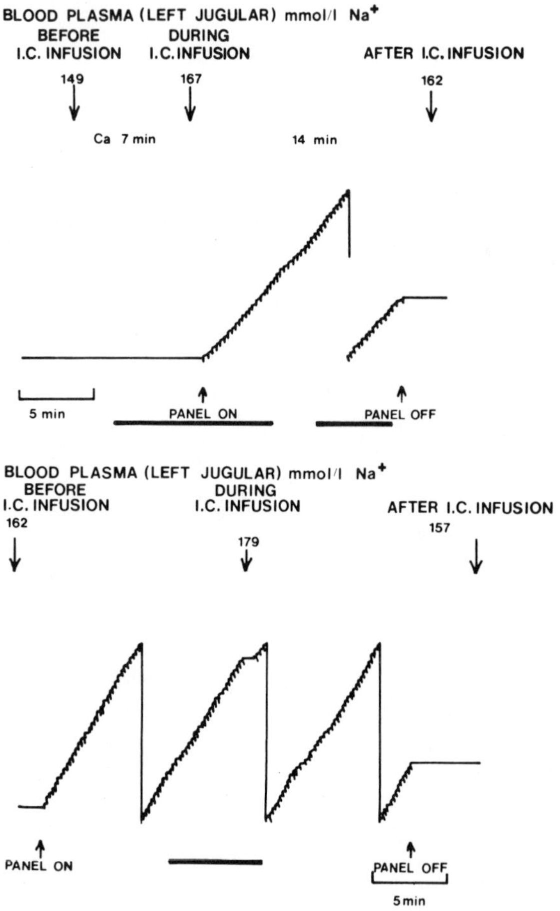

Fig. 20-6. Cumulative records from a goat responding on a fixed-interval 10 schedule to obtain 5 ml of 0.5 M sodium bicarbonate. Each downward deflection of the pen indicates delivery of a reinforcement. During the time indicated by the black bars, intracarotid infusions of 3.9 M sodium chloride were made at 3.8 ml/min. In the upper record the injection began before the response panel was made operative. The changes in jugular plasma [Na] are shown also. The records are consecutive. (After Baldwin 1976.)

sodium level in the blood perfusing the brain. As in the sheep experiments the infusions were made with the contralateral carotid loop clamped in order to give bilateral distribution. The intracarotid infusions had no evident effect on the operant responses of the goats for sodium bicarbonate (Fig. 20-6).

In the first ten experiments only the hypertonic sodium bicarbonate was available. Since it was considered possible that the goats might be continuing to respond during the infusion because the hypertonic saline made them thirsty, a further seven experiments were carried out in which water was continuously available as well as the sodium bicarbonate. It was found that although the goats occasionally paused briefly to drink small volumes of water both before, during and after the intracarotid infusions, there was no consistent cessation of response for the bicarbonate in order to drink water during the intracarotid infusions. It was clear that they were not simply made thirsty by the hypertonic infusion so that any liquid would be reinforcing. Baldwin suggests that the data confirm the results of Beilharz et al. (1965) on lack of effect of intracarotid infusion in the sheep.

Further analysis of effect of intravenous infusion in sheep

In comparing the data on intravenous infusion in sheep, which reduced sodium intake, with those on intracarotid infusion in both sheep and goats, where no clear-cut effect on sodium appetite occurred, the difference in time interval between commencing infusion and appetite test (i.e. carotid 7 min vs intravenous 40 min) may be influential. In the Beilharz and Kay data there is evidence of a significant reduction in sodium appetite by the intravenous infusion of hypertonic saline. Given the greater time interval and the greater amount of sodium infused, it seemed clear that the control system was being accessed, though the variability of result suggested that this effect was only just emerging. Further experiments were therefore initiated.

In this group of experiments (Denton et al. 1971) the translation of effect of intravenous infusion was studied over a 1–3-h period. In order to evaluate the role of an increase in plasma [Na] in any decrease of appetite observed, comparison was made between the effect of intravenous infusion of 2–2.6 litres of Ringer saline with that of 80 ml of 4 M sodium chloride.

The experimental plan involved, in the first instance, examining the influence of rapid infusion of 2–2.6 litres of Ringer solution over 40–60 min on appetite tested 10 min after cessation of infusion. In a second series the effect on appetite was studied 120 min after cessation of infusion, and compared with the effect of 80 ml of 4 M sodium chloride, given over 50–55 min, studied 120 min after the end of the infusion. The influence on appetite of rapid infusion of a significant amount of isotonic saline had never previously been looked at. As quantitative comparisons were the aim of this study, considerations in the assessment of sodium balance are set out below.

The amounts of sodium lost in the saliva by each sheep in the three series of experiments during each 48-h depletion period are shown in Tables 20-4, 20-5 and 20-6. Loss in the saliva and intake as sodium bicarbonate solution are the major factors in sodium turnover in these animals. Other factors such as food, urine, faeces and unmeasured salivary loss outside the cage or on the sides of bins play a smaller role. The food provided 50–100 mmol Na/day. The sodium gain in food would be to a large extent offset by faecal loss and unmeasured salivary loss (Denton 1957); we have verified this approximation during total electrolyte balance studies including distilled water washings of the cage. The residual difference is small relative to the other factors in electrolyte turnover. For purposes of analysis, therefore, we have compared voluntary sodium bicarbonate intake with total sodium loss in saliva and urine. In the control experiments the mean intake of sodium bicarbonate during 15 min of access approximated the sodium lost in the preceding 48 h. The extent of variation in intake of sodium bicarbonate solution on control days may be seen from Tables 20-4, 20-5 and 20-6. The continued survival of the animals in good health with maintained weight in the face of salivary sodium loss which, if uncompensated, would cause rapid circulatory deterioration and death indicates that, although some day-to-day variation occurred, voluntary sodium bicarbonate intake compensated sodium loss. As shown in Figs. 20-7 and 20-8, there was no significant difference (NS, paired t-test) between sodium loss and voluntary drinking of sodium solution during a control 48-h period.

Rapid intravenous infusion of Ringer saline: Appetite tested 10 min later

Twenty experiments of this type were performed on five sheep. The sheep Arianne and Gigi were smaller (28 and 30 kg) than the others and received 2 litres of Ringer saline in 40–60 min. The other sheep (38–40 kg) received 2.6 litres of Ringer saline (375 mmol Na) in most experiments (Table 20-4).

Mean plasma [Na] of sodium-replete sheep is 147±3.7 (s.d.) mmol/l. Depletion caused plasma [Na] to fall to 141±4 mmol/l and the rise produced by Ringer saline infusion was small—ca 2 mmol/l. The infusion caused the plasma [K] to change from 3.9±0.7 mmol/l to 3.1±0.3 mmol/l. Haematocrit and plasma specific gravity fell 30%–33%. Blood pressure usually increased 10–20 mmHg.

Ten minutes after the end of the infusion, the sodium bicarbonate solution (300 mmol/l) was offered for 15 min. The appetitive behaviour shown by the sheep at the time of offering the sodium bicarbonate was not distinguishable from that shown on control days. However (Fig. 20-7),

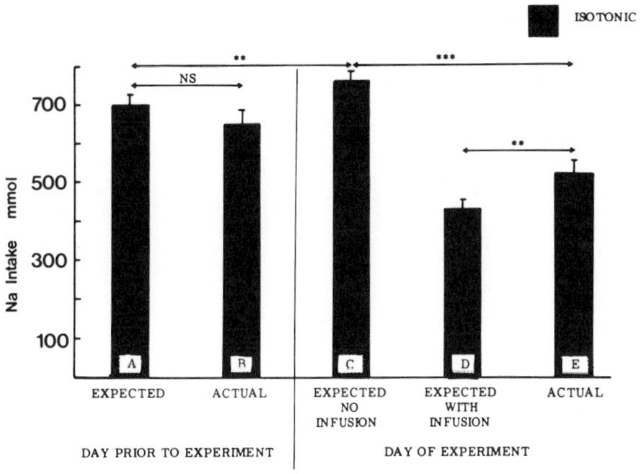

Fig. 20-7. The left side shows the relation between mean expected voluntary sodium intake and mean voluntary intake found with five sheep (mean±s.e.). The right side shows the mean expected intake on the day of the Ringer saline experiment, the mean expected intake allowing for the sodium infused, and the mean voluntary sodium intake actually found in the five sheep. The statistical relationships are shown (***$P<0.001$; **$P<0.01$; NS, not significant), and analysed in the text. (After Denton et al. 1971.)

the isotonic Ringer saline infusion caused a highly significant reduction in voluntary sodium intake (E<C, $P<0.001$; paired t-test). The mean reduction in intake was 238 mmol, which was approximately 70% of the sodium infused. The reduction was of the same order in all sheep studied. Actual intake during the test was significantly greater than expected intake taking into account the amount of sodium infused (E>D, $P<0.01$). On the control day there was no significant difference between expected (A) and actual intake (B). Fortuitously, the sodium loss over 48 h preceding the experimental day, i.e. the expected intake, was significantly greater than on the control day (C>A, $P<0.01$).

In the light of this effect on appetite within 50–70

Table 20-4. The effect of rapid intravenous infusion of 2–2.6 litres of isotonic Ringer saline on the voluntary sodium bicarbonate intake of five sodium-deficient sheep when tested 10 min after stopping the infusion. The total sodium loss in saliva and urine over the preceding 48 h is shown also. The sodium loss in saliva and urine over the immediately preceding 48-h control period and the voluntary sodium intake at the end of this period are shown on the left-hand side of the table.

	Control day (mmol Na)				Experimental day (mmol Na)				
Sheep	Saliva	Urine	Total loss in 48 h	Voluntary intake of $NaHCO_3$ during 15 min	Saliva	Urine	Total loss in 48 h	Voluntary intake of $NaHCO_3$ during 15 min	Amount of isotonic NaCl infused
Arianne	720	57	777	780	756	96	852	588	300
	647	177	824	756	781	86	867	510	300
	634	107	741	849	715	163	878	561	300
	705	36	741	693	690	54	744	864	300
	648	74	722	867	672	132	804	609	300
	671	56	727	867	652	154	806	639	300
Finochio	842	50	892	537	769	5	774	663	375
Gigi	622	2	624	462	602	2	604	342	300
	585	12	597	396	612	2	614	234	300
	727	16	743	588	703	17	720	480	300
	684	65	749	450	612	3	615	519	300
Anselm	791	4	795	738	946	3	949	693	375
	870	2	872	882	942	15	957	690	375
Lola	531	2	533	783	789	15	804	435	300
	546	2	548	513	755	2	757	501	300
	624	5	629	486	632	9	641	450	375
	676	3	679	570	733	2	735	315	375
	713	4	717	600	750	2	752	576	375
	493	4	497	543	630	3	633	324	375
	607	5	612	609	737	4	741	480	375

After Denton et al. (1971).

min of beginning an infusion which had a very small effect only on plasma [Na] (+2 mmol/l), the influence on plasma volume, as indicated by the fall in haematocrit from 29±3.5 to 20±2 (n=31), needs comment.

Comprehensive studies by Boyd (1967) and Blair-West et al. (1967) in which ^{131}I-labelled sheep globulin was used to measure plasma volume of sheep have shown that a sodium deficit of 300–500 mmol was associated with a mean reduction of plasma volume of 250 ml. The infusion of 2–2.5 litres of isotonic saline in sodium-deplete sheep expanded plasma volume to a level above the control sodium-replete range. This effect declined to a small extent during the next 100–200 min. Sodium deficit was greater in the experiments recorded here than in the volume experiments cited above. However, the data on the effect of intravenous infusion (Blair-West et al. 1967) indicate that the Ringer saline infusions described above (2–2.5 litres) would probably elevate plasma volume to the sodium-replete level.

Against this background it was shown (Denton et al. 1971) in the same animals used in the series here, that expansion of plasma volume to an extent verified to be above sodium-replete range by infusion of 6% dextran in normal saline (500 or 750 ml) did not abolish or significantly reduce sodium appetite. Also, preventing the usual fall in plasma volume by slow dextran infusion over the 48-h course of onset of sodium deficiency did not reduce the sodium appetite tested at the end of 48 h (Denton et al. 1971). Thus the comparison tended to emphasize that some consequence of the sodium infused other than the change in plasma volume was influential on appetite in this time and quantitative context (see Chapter 15).

Rapid infusion of Ringer saline: Appetite tested 120 min later

There was no change in plasma [Na] 5 min after the end of infusion or after 120 min. There was a 0.7 mmol/l fall in the [K] in the plasma. The haematocrit and plasma protein levels had begun to increase again at 120 min but had not returned to the preinfusion level.

The relation between the intake of sodium bicarbonate and the loss of sodium in saliva and urine over the 48 h preceding the experimental test is shown on the right-hand side of Table 20-5 for six sheep studied, together with the amount of normal saline infused 2 h before testing voluntary intake. The amount of sodium lost in the urine during the interval between the end of the infusion and the appetite test 2 h later is shown also. The left-hand side of Table 20-5 records the relation between sodium loss and voluntary intake over the 48-h control period which immediately preceded the experimental period.

The result was generally the same as when appetite was tested 10 min after the end of Ringer saline infusion except that the reduction of voluntary intake was not so great. As shown in Fig. 20-8, which records the statistical comparison of the effect on appetite tested 2 h after the end of

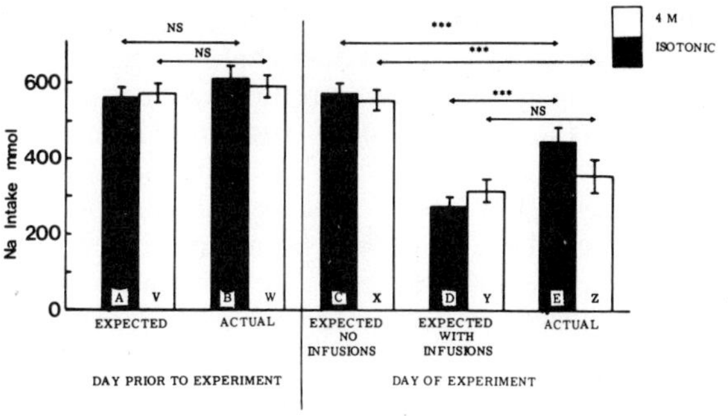

Fig. 20-8. The effect of isotonic Ringer saline infusion (filled columns) and 4 M sodium chloride infusion (open columns) compared when voluntary intake of sodium bicarbonate was examined 120 min after the end of the infusion. The mean expected intake after infusion in the six sheep was calculated allowing for both the amount infused and urinary sodium loss over the 120 min between the end of infusion and testing of appetite. See legend to Fig. 20-7 for other details. (After Denton et al. 1971.)

Table 20-5. The effect of rapid intravenous infusion of 2–2.4 litres of isotonic Ringer saline on the voluntary sodium bicarbonate intake of six sodium-deficient sheep when tested 2 h after stopping the infusion. The total sodium loss in saliva and urine over the preceding 48 h is shown also. The sodium loss in saliva and urine over the immediately preceding 48-h control period and the voluntary sodium intake at the end of this period are shown on the left-hand side of the table.

	Control day (mmol Na)				Experimental day (mmol Na)					
Sheep	Saliva	Urine	Total loss in 48 h	Voluntary intake of $NaHCO_3$ during 15 min	Saliva	Urine	Total loss in 48 h	Voluntary intake of $NaHCO_3$ during 15 min	Amount of isotonic NaCl infused	Urinary Na loss over 2 h between infusion and test
Ricky	685	6	691	693	678	12	690	601	300	9
	387	5	392	822	584	4	588	465	300	34
	459	7	466	765	579	8	587	375	300	22
	531	5	536	840	689	123	812	372	300	21
	521	5	526	606	508	5	513	555	300	12
Lola	624	1	625	687	620	6	626	627	350	12
	658	25	683	588	482	8	490	369	350	8
	613	1	614	639	610	19	629	522	350	10
	395	3	398	420	475	7	482	213	350	10
Tina	301	8	309	519	554	5	559	180	300	3
	251	24	275	456	282	14	296	327	300	1
	413	9	422	453	244	5	249	24	300	5
	447	3	450	246	330	2	332	267	300	2
	417	3	420	291	359	2	361	33	300	1
Dorothy	705	9	714	684	550	1	551	633	300	0
	653	32	685	525	551	5	556	477	300	0
	636	3	639	630	611	6	617	240	300	0
	574	2	576	696	585	2	587	501	300	0
	612	3	615	813	692	45	737	321	300	30
Arianne	577	74	651	441	614	117	731	615	300	15
	474	78	552	801	585	86	671	582	300	a
	642	121	763	768	664	71	735	483	300	a
	490	69	559	657	480	95	575	570	300	a
Gigi	710	1	711	669	684	3	687	720	300	9
	823	5	828	606	651	8	659	657	300	a
	589	14	603	723	588	16	604	624	300	a

After Denton et al. (1971).
[a]Not collected separately; included in 48-h urine loss.

Ringer saline or 4 M sodium chloride infusion, and which will be described in detail below, the Ringer infusion highly significantly reduced actual intake (E) after the infusion, by a mean of 130 mmol, relative to expected intake (C) if no infusions were given (E<C, $P<0.001$). The sodium lost in the urine over the 2-h period between the end of Ringer saline infusion and the presentation of sodium bicarbonate solution was small (mean loss 10 mmol, range 0–38 mmol). This excretion of infused sodium has been accounted for in each experiment of both the Ringer and the 4 M sodium chloride series in the course of calculating the expected intake allowing for the infusion (i.e. D and Y of Fig. 20-8). Thus, actual intake (E) in the case of the Ringer infusion was reduced by about 40% of the amount of infused sodium retained in the body (D>C, $P<0.001$).

Infusion of 4 M sodium chloride: Appetite tested 120 min later

An infusion of 80 ml of 3.8–4.2 M sodium chloride (300–340 mmol sodium) was given in 50–55 min. Twenty-four experiments were performed on the same six sheep used for the Ringer saline infusion.

Two hours after the end of the infusion the mean increase in plasma [Na] was 16 mmol/l. Plasma [K] and the haematocrit value were changed to the same extent as after the Ringer saline infusion. Plasma protein had almost returned to the preinfusion level.

There was a reduction in sodium bicarbonate intake in most experiments except those with Lola (Table 20-6). As shown in Fig. 20-8, the mean total loss of sodium in saliva and urine (X) was no different from the loss in the corresponding

Table 20-6. The effect of rapid intravenous infusion of 80 ml of 4 M sodium chloride on the voluntary sodium bicarbonate intake of six sodium-deficient sheep when tested 2 h after stopping the infusion. The total sodium loss in saliva and urine over the preceding 48 h is shown also. The sodium loss in saliva and urine over the immediately preceding 48-h control period and the voluntary sodium intake at the end of this period are shown on the left-hand side of the table.

	Control day (mmol Na)				Experimental day (mmol Na)					
Sheep	Saliva	Urine	Total loss in 48 h	Voluntary intake of $NaHCO_3$ during 15 min	Saliva	Urine	Total loss in 48 h	Voluntary intake of $NaCOH_3$ during 15 min	Amount of 4 M NaCl infused	Urinary Na loss over 2 h between infusion and test
Ricky	578	7	585	614	682	8	690	189	310	70
	761	25	786	744	734	21	755	228	310	58
	562	4	566	591	626	12	638	294	310	75
	507	7	514	540	511	7	518	547	311	77
	618	6	624	651	561	5	566	230	310	74
Lola	504	4	508	591	528	4	532	828	310	64
	589	9	598	432	274	2	276	510	300	80
	559	12	571	456	483	4	487	660	310	62
Tina	339	4	343	345	401	8	409	12	315	85
	415	8	423	528	481	18	499	114	320	83
	365	5	370	330	361	2	363	456	320	50
	361	5	366	516	372	23	395	9	320	120
	483	5	488	393	342	6	348	51	320	49
Dorothy	712	3	715	711	593	2	595	504	310	1
	689	4	693	750	635	7	642	414	310	20
	458	5	463	567	453	1	454	315	310	62
	600	2	602	786	572	2	574	450	310	20
	656	6	662	816	693	11	704	252	310	20
Arianne	547	77	624	507	522	97	619	513	300	85
	504	79	503	780	423	74	497	507	307	93
	548	195	743	681	477	83	560	417	307	100
Gigi	663	19	682	702	804	3	807	474	315	86
	607	3	610	672	626	16	642	420	310	58
	508	21	529	573	518	19	537	66	315	95

After Denton et al. (1971).

isotonic Ringer saline experiments (C). The mean actual intake of sodium bicarbonate solution (Z) was reduced from that expected without infusion ($Z<X$, $P<0.001$). The mean sodium lost in the urine over the 2-h interval between end of the 4 M sodium chloride infusion and presentation of sodium bicarbonate was 66 mmol (range 1–120 mmol). Taking this loss of infused sodium into account, there was no significant difference between the actual intake of sodium bicarbonate (Z) and the intake expected after infusion (Y). The mean reduction of voluntary sodium intake of 200 mmol represents a reduction equivalent to approximately 84% of the amount of infused sodium retained in the body.

Thus, though there was considerable variation in both groups, the effect of 4 M sodium chloride on appetite as tested after this 2-h time interval was greater than that of isotonic saline, where the difference between actual intake and intake expected after infusion was highly significant (i.e. $D<E$, $P<0.001$ whereas $Y<Z$, NS). However, the difference between E and Z was not statistically significant. This, in part at least, results from the different amounts of infused sodium retained in the body.

Two other sheep, Foster and Benowski, were each tested also on four occasions for the effect of 4 M sodium chloride on voluntary intake when presentation was delayed 2 h. They were not included in the statistical analysis because corresponding isotonic infusion experiments were not made. These experiments confirmed the trend of the other 4 M sodium chloride experiments. In seven of eight experiments, voluntary intake was reduced by more than the amount of sodium retained in the body.

Summary

The results of the experiments show that there is, over comparable time intervals, a larger effect on appetite when sodium is administered as a hypertonic solution than as an isotonic solution. This may be attributable to the increase in plasma [Na] produced by the former procedure, which may accelerate the movement of sodium into areas which constitute the receptor elements controlling salt appetite in deficiency. If these regions are in the brain, the delay in translation of a change in sodium balance to a change in appetite may be that involved in penetration across the blood–brain barrier (Davson and Pollay 1963) and into cerebral cells. If a process of this type is involved, there would be a complex group of rate determinants operating. This may explain in part the variability of result over a standard time interval of a single procedure such as intravenous saline infusion. Notwithstanding, the greater effect of 4 M sodium chloride than of isotonic Ringer saline was statistically clear-cut, though less administered saline was retained in the body in the case of the 4 M sodium chloride infusion. The results set out in Chapter 24 strongly suggest that effect on CSF [Na] may have been a major factor.

It is interesting that the isotonic saline infusion caused no greater decrease of appetite when tested 2 h after infusion than when tested 10 min after the infusion finished. Little water or sodium was excreted over this period, and presumably the sodium infused, which represented approximately 50% of the deficit, was largely retained in the extracellular compartment. With infusion of 500–750 ml dextran, the effect of which in haemodynamic terms and on expanding the plasma volume was probably greater, the change in extracellular volume would be negligible compared with that produced by the isotonic infusion. Whether this expansion of volume of the tissue fluids independently of sodium content could facilitate sodium transfer and appetite change is not known. The fact that infusion of 2.7 M glucose, which would expand extracellular volume at the expense of intracellular fluid, did not influence salt appetite (Beilharz and Kay 1963) can be noted, but the conditions are distinctly different in several respects.

The sheep did not drink the water that was concurrently available in the test periods which followed 120 min after Ringer saline infusion. However, the majority drank 1–2 litres after 4 M sodium chloride infusion, presumably as a result of thirst evoked by the sharp rise in plasma [Na]. The possibility has to be considered that the water drinking may have contributed to lower sodium intake with 4 M sodium chloride infusion. However, during concurrent water and sodium deficiency (Beilharz et al. 1962) the animals satiated themselves with one fluid and then proceeded to drink the other. Furthermore, the fact of which fluid was encountered first did not appear to determine the amount drunk. The sheep's rumen is capacious, and distension over a substantial range does not appear to inhibit drinking in the way it does in other species (Bott et al. 1965). In the experiments of Beilharz and Kay (1963), where 1.5 litres of water were introduced into the rumen 10 min before a sodium drinking test, similar to the experiments reported here, there was no statistically significant influence on sodium intake. Furthermore, the same reduction in sodium intake was seen in experiments here with two animals that consistently failed to drink water.

Also, the reduction in sodium intake was seen in many experiments in which the sodium solution was sampled and drunk first, before the animal paid attention to the water container and drank from it. The evidence suggests that concurrent water intake did not cause the greater reduction in sodium intake in the 4 M sodium chloride experiments.

Bar-press experiments on sodium-deficient sheep: Effect of concurrent automatic intravenous infusion of sodium chloride on sodium bicarbonate intake

Following the clues and some measure of insight given by the procedures described above, it was decided to take another approach to explore the system controlling sodium intake in sheep. In the experiments recounted so far, the essential condition was that sodium balance was changed by rapid systemic infusion. The appetite was tested by free access to hypertonic sodium bicarbonate solution. Most often the consummatory act of satiation lasted 2–10 min. Drinking stopped, or at least the bulk of intake ceased, long before any systemic change due to absorption of sodium could influence any central mechanisms or peripheral sensors. Other evidence (Beilharz and Kay 1963) indicates it is unlikely that the rumen wall contains any sodium sensors influential on appetite.

In the two experiments now to be recounted (Weisinger et al. 1978) the animals were trained to bar-press for 120 min daily to repair a sodium deficit of 300–500 mmol caused by salivary and urinary loss in the preceding 22 h (Fig. 20-9).

The two experiments were done in two parts. In the first part, the amount of sodium bicarbonate solution delivered to a drinking cup was varied. A small volume was given with each bar-press on some days and a larger volume on others. The amounts were selected to produce widely divergent patterns of behaviour. Thus a significant difference in number of bar-press deliveries between the two treatments was obtained early in the 2-h period. The time at which the difference occurred gave a reference point with which the results of the second part of the experiments could be compared.

In the second part of the experiments the small volume was delivered to the cup, but also a short-duration intravenous infusion was given automatically contemporaneously with every delivery to the cup. The infusion was hypertonic saline on some days and isotonic saline (control)

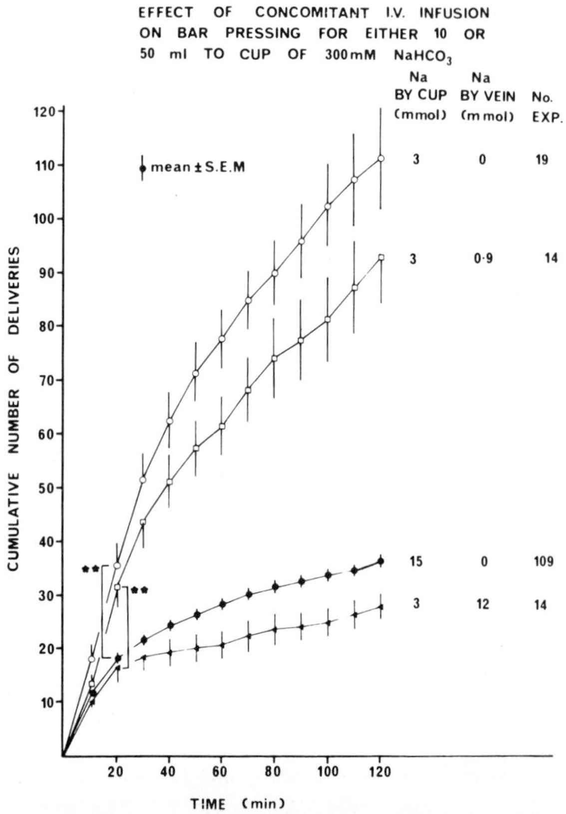

Fig. 20-9. Effect of concomitant intravenous (i.v.) infusion on bar-pressing for sodium bicarbonate solution. Cumulative mean ±s.e.m. of number of deliveries over the 120-min period for the various treatments of Experiment 1: filled circles, 15 mmol (cup), n=109; open circles, 3 mmol (cup), n=19; open squares, 3 mmol (cup)+0.9 mmol (i.v.), n=14; filled triangles, 3 mmol (cup)+12 mmol (i.v.), n=14. (From Weisinger et al. 1978, with permission.)

on others. The concentration and volume of the hypertonic saline solution were such that the total amount of sodium per delivery, i.e. cup plus intravenous infusion, was equal to that of the *large* amount of sodium bicarbonate delivered to the drinking cup in the first part of the experiment. The time at which a significant difference in number of deliveries between treatments occurred would be indicative of the time necessary for recognition of, and response to, the systemic sodium chloride.

The regime for all experiments was that the urine and saliva volume and water intake for the preceding 24 h were measured at 10.30 hours. At 11.00 hours the animals were fed and offered 4 litres of water. Water and food were thereafter available continuously from separate bins on the outside of the cage. The sodium bicarbonate solution was available from 12.00 to 14.00 hours via bar-pressing, and delivery was to a drinking cup on the inside of the cage. Bar-pressing was on a fixed-interval schedule of 10 s (FI 10 s).

Detail of the experimental design

Experiment 1

Four sheep were used to determine response patterns. Table 20-7 shows the total number of observations per animal per treatment. The sodium deficit before each of the treatments was similar (400–500 mmol).

Table 20-7. Number of observations of each of the four animals used in the determination of the response patterns with and without involuntary intravenous injection of saline with each bar-press.

	Experiment 1			
	Part 1		Part 2	
Animal	15 mmol (cup)	3 mmol (cup)	15 mmol (cup+i.v.)	3.9 mmol (cup+i.v.)
Brad	35	4	4	4
Kasimir	32	5	3	3
Hilarius	24	5	3	3
Supra	18	5	4	4
Total	109	19	14	14

	Experiment 2			
	Part 1		Part 2	
Animal	30 mmol (cup)	7.5 mmol (cup)	30 mmol (cup+i.v.)	8.85 mmol (cup+i.v.)
Supra	6	27	8	12
Zelda	7	20	8	16
Kasimir	3	22	5	7
Zigmas	5	16	6	14
Total	21	85	27	49

From Weisinger et al. (1978), with permission.

In the first part of the experiment the amount of sodium bicarbonate delivered to the drinking cup was varied. Each delivery was 15 mmol on some days, or 3 mmol on others (50 or 10 ml of 300 mmol/l sodium bicarbonate). (See 15 mmol (cup) or 3 mmol (cup) in Table 20-7.)

In the second part of the experiment, contemporaneously with each 3 mmol delivery to the drinking cup the animal was automatically given a 10-s intravenous infusion of sodium chloride solution. The infusion was 12 mmol on some days and 0.9 mmol on others (6 ml of either 2 M or 0.15 M sodium chloride). (See 15 mmol (cup+i.v.) or 3.9 mmol (cup+i.v.) in Table 20-7.)

Experiment 2

Four sheep were used to determine response patterns. Table 20-7 shows the total number of observations per animal per treatment. Five additional animals were included in some of the other measurements made.

In the first part of the experiment each delivery to the drinking cup was 30 mmol on some days, or 7.5 mmol on others (50 ml of either 600 or 150 mmol/l sodium bicarbonate). (See 30 mmol (cup) or 7.5 mmol (cup) in Table 20-7.) Sodium deficit was 330–440 mmol before each of the treatments.

In the second part of the experiment, concurrently with each 7.5 mmol of delivery to the drinking cup the animal was automatically given a 10-s intravenous infusion of sodium chloride solution. The infusion was 22.5 mmol on some days and 1.35 mmol on others (9 ml of either 2.5 M or 0.15 M sodium chloride). (See 30 mmol (cup+i.v.) or 8.85 mmol (cup+i.v.) in Table 20-7.)

In order to determine the time at which there was a significant difference in cumulative number of deliveries between treatments an analysis of variance (repeated measures design) was used. The data were first made orthogonal, i.e. for any treatment every animal contributed the same number of observations. The maximum number of observations were used. Other data were analysed by *t*-test for independent observations, unless otherwise stated. Results are expressed as mean ± standard error of the mean.

Results

Experiment 1

In the first part of the experiment 15 mmol or 3 mmol of sodium were delivered to the drinking cup. With 15 mmol of sodium per delivery (Fig. 20-9) the animals obtained and drank 36.4±1.2 deliveries, equivalent to 548 mmol, in the 120-min period. Reduction to 3 mmol in the amount of sodium bicarbonate per delivery (Fig. 20-9) caused a large increase to 111.7±9.2 deliveries, equivalent to 335 mmol. The increase, however, was not great enough to equal the total sodium intake seen with 15 mmol per delivery, i.e. total sodium intake was 213 mmol less. The shape of the curve of cumulative deliveries suggests that if another hour or so had been available it is likely that the difference may have disappeared. However, the 2-h period provided a wholly satisfactory basis for comparison. By 10 min there was a significant difference in cumulative number of deliveries between the 3 mmol and the 15 mmol conditions: 17.5±3.2=52 mmol vs 12.7±0.8 deliveries=190 mmol, respectively. Substantial water intake was not seen during the 0–60 min period of 15 mmol (cup) treatment (Table 20-8) even though it was noted that plasma [Na] was clearly elevated by 60 min (Table 20-9).

Table 20-8. Mean±s.e.m. water intake (ml) and urine volume (ml) over various intervals before and during the different conditions.

	Experiment 1			Experiment 2		
	15 mmol (cup) n=15	3.9 mmol (cup+i.v) n=14	15 mmol (cup+i.v.) n=14	30 mmol (cup) n=17	8.85 mmol (cup+i.v.) n=49	30 mmol (cup+i.v.) n=27
Water intake (ml)						
−60–0 min	441±124	316±99	539±149	326±104	235±49	368±92
0–20 min	3±2	77±38	1102±175***	277±92	15±10	668±116***
20–40 min	0±0	157±72	494±141	214±57	11±5	212±56**
40–60 min	17±10	39±23	191±77	160±66	7±5	450±89***
60–120 min	540±157	336±137	1091±175**	907±142	20±13	734±118***
Total: 0–120 min	560±157	608±193	2878±186***	1560±128	54±17	2064±149***
Urine volume (ml)						
0–20 min	1.7±1.6	4.2±1.2	30.8±10.8*	4±3	4±4	11±4
Total: 0–120 min	33.3±9.8	41.7±15.0	276.2±47.4**	13±7	27±9	253±37***

From Weisinger et al. (1978), with permission.
*=P<0.05, **=P<0.01, ***=P<0.001; t-test for correlated (Experiment 1) or independent means (Experiment 2). Statistical comparisons were made between the two 'cup+i.v.' conditions.

Table 20-9. Mean±s.e.m. plasma [Na] and [K] (mmol/l) before and at various times during the different conditions.

			Time (min)				
Condition		Pre	5	10	20	60	120
30 mmol (cup+i.v.) n=3	Plasma [Na]	142.3±1.2	158.0±0.6**	158.3±4.1	158.3±4.9	—	154.3±3.8
	[K]	3.6±0.1	3.4±0.1	3.2±0.1	3.1±0.2**	—	3.4±0.1
8.85 mmol (cup+i.v.) n=3	Plasma [Na]	143.0±0.5	143.7±0.3	143.3±0.7	144.0±0.6	—	148.3±0.9
	[K]	4.0±0.2	4.1±0.2	4.1±0.1	4.0±0.1	—	4.0±0.2
30 mmol (cup) n=5	Plasma [Na]	145.8±1.2	—	—	148.8±1.2*	151.0±2.5	155.6±1.6**
	[K]	4.7±0.2	—	—	4.3±0.1	4.6±0.2	4.4±0.2
15 mmol (cup) n=6	Plasma [Na]	141.7±0.7	—	—	143.8±1.1*	146.4±1.2**	149.5±1.1***
	[K]	4.4±0.1	—	—	4.4±0.1	4.2±0.1	4.0±0.2
15 mmol (cup+i.v.) n=1	Plasma [Na]	142	—	—	170	157	157
	[K]	4.5	—	—	3.8	3.7	3.7

From Weisinger et al. (1978), with permission.
*= P<0.05, **= P<0.01, ***= P<0.001; t-test for correlated means. Statistical comparisons made between the values obtained at each time interval and the value obtained before treatment.

In the second part of the experiment 3 mmol of sodium were delivered to the drinking cup with either 12 or 0.9 mmol of sodium intravenously. The intravenous sodium chloride infusion of 12 mmol contemporaneously with each 3 mmol delivery of sodium bicarbonate resulted in a pattern of behaviour similar to that observed when 15 mmol of sodium was drunk from the cup. In 14 observations, the animals obtained 28.0±2.2 deliveries (range=16–46)=420 mmol of sodium. In 14 control observations with intravenous sodium chloride infusion of 0.9 mmol paired with 3 mmol drunk, a high rate of response lasted throughout the 120 min, the animals obtaining 93.6±8.9 deliveries (range=50–155)=364 mmol (Fig. 20-9).

The time for systemic influx of hypertonic sodium to affect sodium appetite was 20 min or less. Even though animals within both conditions tasted and drank the same amount of sodium per delivery (3 mmol), by 20 min there was a significant difference in the cumulative number of deliveries.

With the isotonic saline infusion there were 29.6±3.9 deliveries=115 mmol, compared with 17.8±2.5 deliveries=267 mmol with the hypertonic infusion.

Water intake ranged from 0 to 1500 ml coincident with food intake during the 60 min which preceded the bar-press period. By 20 min after the start of bar-pressing the animals receiving 12 mmol intravenously had ingested 1102±175 ml of water (Table 20-8). The latency to first drink was 2–5 min. Little water was ingested by animals when they received 0.9 mmol of sodium intravenously.

Experiment 2

The whole experimental design was repeated under conditions where the volume delivered to the cup was constant but the concentration of sodium bicarbonate was varied (50 ml of either 600 mmol/l or 150 mmol/l sodium bicarbonate). The intake from the cup was higher (7.5 or 30 mmol to the cup, and 22.5 or 1.35 mmol i.v.: Table 20-7). The results were essentially the same (Fig. 20-10). The time delay for the systemic influx to affect the appetite was less, being 10 min. Fig. 20-11 shows

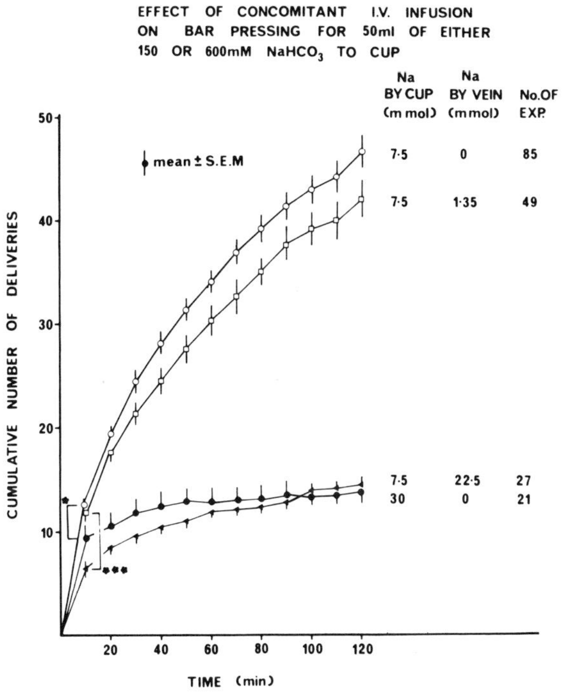

Fig. 20-10. Effect of concomitant intravenous (i.v.) infusion on bar-pressing for sodium bicarbonate solution. Cumulative mean ±s.e.m. of number of deliveries over the 120-min period for the various treatments of Experiment 2: filled circles, 30 mmol (cup), *n*=21; open circles, 7.5 mmol (cup), *n*=85; open squares, 7.5 mmol (cup)+1.35 mmol (i.v.), *n*=49; filled triangles, 7.5 mmol (cup)+22.5 mmol (i.v.), *n*=27. (From Weisinger et al. 1978, with permission.)

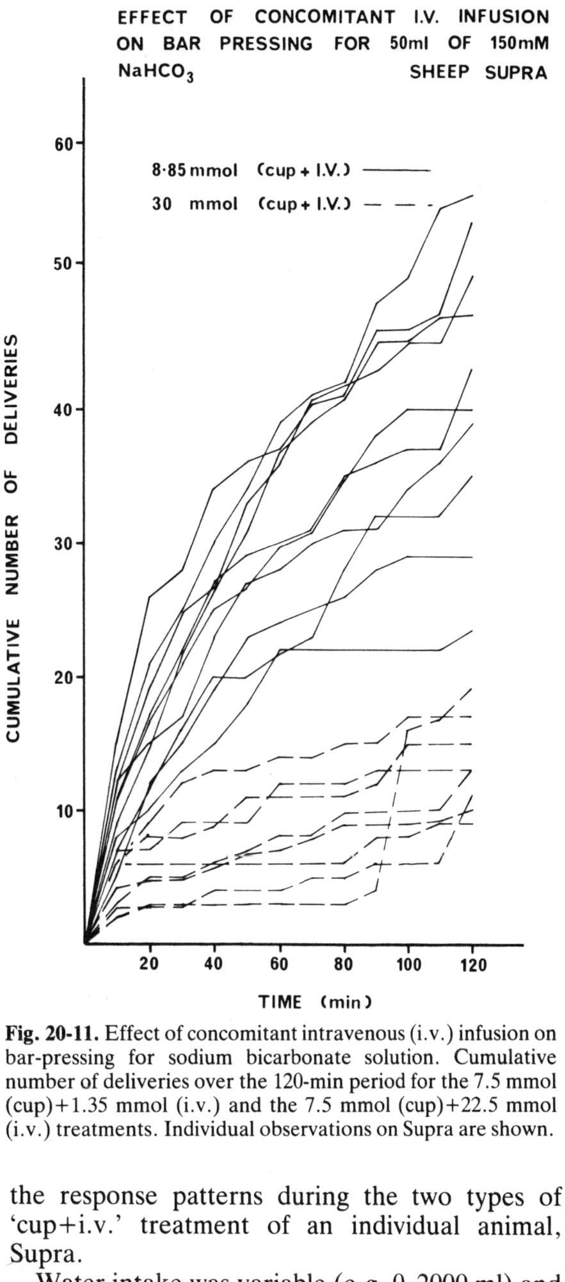

Fig. 20-11. Effect of concomitant intravenous (i.v.) infusion on bar-pressing for sodium bicarbonate solution. Cumulative number of deliveries over the 120-min period for the 7.5 mmol (cup)+1.35 mmol (i.v.) and the 7.5 mmol (cup)+22.5 mmol (i.v.) treatments. Individual observations on Supra are shown.

the response patterns during the two types of 'cup+i.v.' treatment of an individual animal, Supra.

Water intake was variable (e.g. 0–2000 ml) and coincident with the food intake during the 60 min which preceded the bar-press period. By 20 min after the start of bar-pressing the animals which received 22.5 mmol sodium intravenously ingested 668±112 ml of water (Table 20-8). The median latency to first drink was 5 min. Little water was ingested by animals when they received 1.35 mmol intravenously as isotonic fluid. The intravenous infusion of 22.5 mmol sodium chloride caused an increase in plasma [Na], evident at 5 min, and a decrease in plasma [K], evident at 20 min (Table 20-9). The infusion of 1.35 mmol had little effect on any of these parameters.

A notable feature with the cup+22.5 mmol intravenous experiments was that bar-pressing continued progressively over 120 min though, from 5 min onwards, plasma [Na] was increased by 16 mmol/l to well above sodium-replete concentration. This is consistent with other data on lack of immediate effect of high plasma [Na] on salt-seeking behaviour.

Discussion

In these experiments the rate of sodium acquisition (i.e. number of deliveries per time) varied inversely with the amount of sodium delivered to the cup. Thus, by varying the amount of sodium per delivery, widely divergent curves of behaviour were produced. This facilitated pinpointing the effect of systemic influx of sodium.

The systemically produced change in sodium balance affected sodium appetite within 20 min. That is, the difference in cumulative number of deliveries between the two 'cup+i.v.' treatments (control isotonic vs hypertonic influx) was significant by 20 min (Experiment 1) or 10 min (Experiment 2). At the time this happened, the amount of sodium infused into animals in both experiments was greater than 140 mmol. This fact could suggest that amount of sodium infused has an important role in determining the time for a change in the response, since 140 mmol corresponds roughly to the extent of body error to which animals appear to react under conditions of free access to bar-press and constant loss of saliva from a parotid fistula (Abraham et al. 1973). The animals with free access to bar-press lose about 20–30 mmol/h and manifest episodes of bar-pressing activity every 3–5 h. It is also important to note that here the change in plasma [Na] and sodium balance was produced by a series of large 'pulses' rather than the uniform rise in plasma [Na] produced by constant intravenous infusion of hypertonic sodium chloride. This may have an effect on a sensor system such as to facilitate earlier reaction to change.

These systemic 'pulses' were contemporaneous with repetitive acts of drinking sodium bicarbonate; whether this had any influence on facilitating their recognition by a central control system remains open. However, Baldwin (1976) found with sodium-deplete goats that if he combined intracarotid injection of 0.5 M sodium chloride with oral reinforcement of operant response, and then after 10 min withdrew the reinforcement, the operant response extinguished over 5–7 min. The intracarotid injections appeared to have no reinforcing value though the goats showed licking and chewing movements in response to intracarotid injection. A similar extinction occurred when each bar-press was paired with intrarumen injection of concentrated sodium bicarbonate and then after 10 min the oral reinforcement was withdrawn. Nachman and Valentino (1966) have shown that sodium-deficient adrenalectomized rats drink significantly less sodium chloride 120 min after being allowed to drink 10 ml of 0.4 M sodium chloride than after having the same amount of sodium chloride tubed into the stomach. In fact, stomach tubing 10 ml of 0.4 M sodium chloride was no more effective than stomach tubing the same volume of water even though the stomach loading with sodium chloride resulted in a normal plasma [Na]. The importance of stored information 120 min beforehand on gustatory inflow was emphasized. However, in the experiments reported here the behaviour curves with drinking plus intravenous injection were similar to those where intake was by drinking alone. Systemic administration was as effective in satiating the animal as the actual tasting and drinking of the additional amount of sodium involved. Indeed, the overall statistically significant lower intake with intravenous infusion first noted at 50 min with the 15 mmol experiment may be attributable to the faster systemic effects with intravenous infusion than with oral intake.

Some other matters require note in interpretation. First, it is possible that the results were related to the infusion of a volume of fluid itself. This seems unlikely since the same volume of isotonic fluid at the same temperature did not inhibit appetite. And when this isotonic volume was given, the curve of response was essentially similar to the behaviour when the small amount alone was delivered to the cup. When the concentrated sodium chloride was given intravenously the behaviour was essentially similar to that when the large amount was delivered to the cup.

As with other experiments of this type, it has to be taken into account whether the increased water intake observed during these infusions decreased sodium intake. Either the time spent drinking or rumen distension caused by the increased water intake might have suppressed sodium intake. In regard to reduction of sodium appetite by rumen distension, it has been shown that prior rumen water loads of 1500 ml or prior ingestion of larger amounts of water do not have any significant effect on sodium appetite when tested 10 min or 120 min later (Beilharz et al. 1962; Beilharz and Kay 1963). The fluid capacity of the sheep rumen is over 5–6 litres (Bell 1972), and it is known that sheep can consume 4–5.6 litres of water within 4 min (Blair-West et al. 1972). At the 20-min point, none of the intakes was much over 1.0 litre (Table 20-8). The animals were not seen to spend any great amount of time drinking water. In addition, by 20 min the animals in the 8.85 mmol (cup+i.v.) condition had consumed over 890 ml of fluid, yet they continued rapidly to bar-press for, and consume, sodium bicarbonate solution. It might be argued that with operant satiation of drive, rumen distension with a moderate amount of water might be more influential by its weight than in the situation of free access to sodium solution and rapid satiation. However, the matter seemed resolved by a further series of experiments. In these water was not available. Exactly the same result with respect to bar-pressing behaviour was seen at 10–20 min when the effects of isotonic and hypertonic sodium chloride infusions were compared.

It seems unlikely that the signal which altered behaviour with the systemic influx of concentrated sodium chloride was change in blood pressure or central venous pressure since there were negligible differences in either between the 8.85 mmol (cup+i.v.) and 30 mmol (cup+i.v.) conditions over the duration of the experiment. Furthermore, increased blood pressure caused by rapid intravenous dextran infusion does not significantly alter the sodium intake of sodium-deficient sheep (Denton et al. 1971).

These data indicated to us that the system of central control of salt appetite, or sensors directly influencing it, was more rapidly accessible than we had earlier thought. Questions present on the influence of a rapid systemic influx of other osmotically active materials, and the effect of changes in cerebrospinal fluid composition. And it can be noted here, parenthetically, that this whole logic of systematic analysis bearing upon the nature of the satiation process and the influence of systemic sodium influx and ventricular infusion, may also be applied to scrutiny of the salt appetite generated by the hormones of the reproductive process. This conjoint action of several steroid and peptide hormones provokes a powerful salt appetite drive in the sodium-replete animal (see Chapter 23), but the quantitative mechanisms in satiation have, as yet, not been investigated.

An interesting recent study of the influence of rapid correction of sodium deficiency on salt appetite in cattle has been made by Bell et al. (1981). They used operant behaviour for delivery of 10 ml of 0.3 M sodium bicarbonate. They found that infusion of 4 litres of hypertonic saline at 50 ml/min in sodium-deplete calves caused a 90% decrease in rate of operant behaviour by 2 h, and the rate remained decreased when tested over the following 4 h. Intravenous infusion of 4 litres of Ringer saline over the same interval had no influence on sodium appetite. Allowing the animals to voluntarily drink 2 litres of hypertonic saline also reduced operant behaviour for sodium bicarbonate, by 80% over the next 2 h, whereas drinking 4 litres of isotonic Ringer saline had no effect on appetite. If the animals were allowed to voluntarily ingest up to 16 litres of isotonic Ringer saline there was a variable but large reduction in operant behaviour for the 0.3 M sodium bicarbonate. Concurrent measurements of blood pH, [Na], aldosterone and plasma renin activity (PRA) gave no evidence of these factors having any determinant effect on appetite. Unfortunately in this study, data on the rate of loss of saliva and changes in electrolyte balances are not available for correlation with appetite.

Influence of self-determined intravenous infusion of osmotically active substances on operant behaviour of sodium-deficient sheep

The conditions of these experiments (R. Weisinger, D. Denton and M. McKinley unpublished) were similar to those described in the previous section except that the experiment was performed two ways: once with water concurrently available and once without. Also each delivery to the cup was 15 ml of 0.6 M sodium bicarbonate (9 mmol sodium).

The first experiment was designed to determine the influence of intravenous infusion of various osmotically active substances on sodium appetite. With each bar-press by the sodium-deplete animal an intravenous infusion (10 ml in 10 s) was given concurrently. The different solutions used were 0.15 M saline, 0.65 M saline, 1 M mannitol, 1 M urea, and 1 M glucose. The last three were made up in 0.15 M saline. Only one of the solutions was used on a given day and the solution used on a given day was randomly determined.

No water available during experimental period

Fig. 20-12 shows the cumulative number of deliveries over the 120-min period for the various conditions when no water was available during the experimental period. There was very little variation in the mean sodium loss between the various

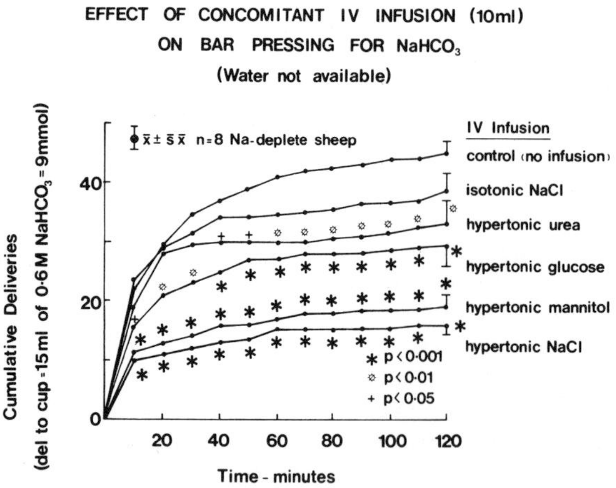

Fig. 20-12. Sheep with a parotid fistula depleted of sodium for 22 h. The influence on bar-pressing for 15 ml of 0.6 M sodium bicarbonate (9 mmol sodium) of concurrent automatic infusion with each bar-press of 10 ml of 0.65 M sodium chloride, 1 M mannitol in 0.15 M saline, 1 M glucose in 0.15 M saline, 1 M urea in 0.15 M saline and 0.15 M saline. The significance of the change in the rate of bar-pressing relative to the control conditions with no infusion is designated.

conditions (342–385 mmol). Under baseline conditions (i.e. 22 h of sodium deprivation, 2 h of sodium access, no infusion) the sheep bar-pressed for and consumed 45.1±2.1 deliveries=406 mmol (n=40 observations on eight sheep). Concurrent intravenous infusion (10 ml) of 0.65 M saline with each delivery of sodium bicarbonate to the drinking cup (i.e. cup+i.v.=9 mmol+6.5 mmol =15.5 mmol sodium) caused a marked reduction in bar-pressing to 16.2±1.8 deliveries=251 mmol sodium ($P<0.001$, n=8). Concurrent intravenous infusion of 1 M mannitol in 0.15 M saline (i.e. cup+i.v.=9 mmol+1.5 mmol=10.5 mmol) caused a similar reduction in bar-pressing for sodium to 18.9±2.4 deliveries ($P<0.001$, n=8).

Concurrent intravenous infusion of either 1 M urea in 0.15 M saline, or 1 M glucose in 0.15 M saline caused smaller, but significant ($P<0.01$ and 0.001, respectively) reduction in bar-pressing to 32.8±3.6 and 29.0±3.3 deliveries, respectively. Concurrent intravenous infusion of 0.15 M saline did not significantly alter bar-pressing for sodium solution.

The difference in cumulative number of deliveries between the baseline and the different concurrent intravenous infusion conditions was significant by 10 (glucose, mannitol, 0.65 M saline) or 40 min (urea) after the start of bar-pressing.

A feature of the experiment above was that, with both 0.65 M saline and 1 M mannitol intravenous infusions, the total intake by 120 min fell short of the body sodium deficit. It was decided to examine, therefore, the time for reactivation of appetite (i.e. end of satiety) under these two conditions. The criterion arbitrarily chosen as indicating end of satiety was the obtaining of an additional 100 mmol of sodium (i.e. 11 deliveries). All eight of the sheep tested reached this criterion faster after concurrent intravenous infusion of mannitol than after concurrent intravenous infusion of 0.65 M saline ($P<0.01$). The time required to reach the criterion was 3.1±0.4 h and 11.6±2.6 h respectively for the two conditions.

Water available during experimental period

The results (Fig. 20-13) when water was available during the experimental period showed the presence and ingestion of water did not significantly alter the influence of the concurrent intravenous infusions on sodium appetite. Under baseline conditions the sheep bar-pressed for and consumed 47.6±2.3 deliveries=428 mmol sodium (n=7). Concurrent intravenous infusion of 0.65 M saline greatly decreased sodium intake to 18.4±3.1 deliveries=284 mmol ($P<0.001$, n=7). Concurrent intravenous infusion of 1 M mannitol in 0.15 M saline caused a similar reduction in bar-pressing for sodium to 25.7±4.0 deliveries=270 mmol ($P<0.001$, n=7). Concurrent intravenous infusion of either 1 M urea in 0.15 M saline or 1 M glucose in 0.15 M saline caused a smaller, but significant ($P<0.05$ and 0.001, respectively) reduction in bar-pressing to 37.1±4.4 and 32.4±6.1 deliveries, respectively. Concurrent intravenous infusion of 0.15 M saline did not significantly alter bar-pressing for sodium.

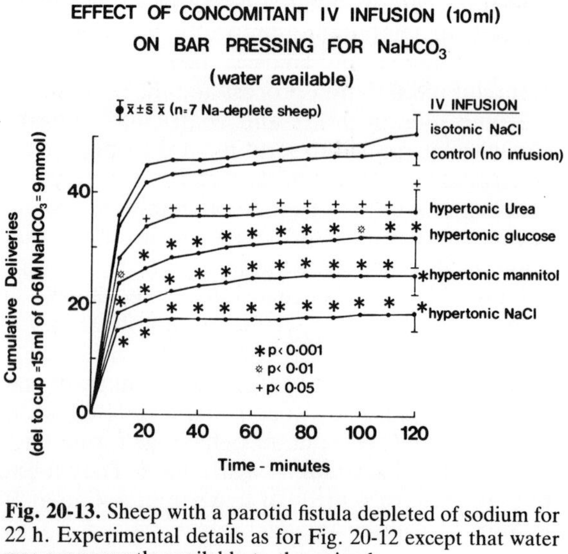

Fig. 20-13. Sheep with a parotid fistula depleted of sodium for 22 h. Experimental details as for Fig. 20-12 except that water was concurrently available to the animals.

The difference in cumulative number of deliveries between the baseline and the different concurrent intravenous infusion conditions was significant by 10 (glucose, mannitol, 0.65 M saline) or 20 min (urea) after the start of bar-pressing.

Under baseline conditions, water intake during the 2 h period of bar-pressing for sodium was 6.0±1.0 deliveries=300 ml. This intake was significantly increased during concurrent intravenous infusion of hypertonic mannitol (21.1±4.4 deliveries=1055 ml; $P<0.001$, $n=7$) or 0.65 M saline (17.0±4.7 deliveries=850 ml: $P<0.001$, $n=7$). There were no other significant differences. At the end of the 2-h period plasma [Na] was decreased by concurrent intravenous infusion of either hypertonic mannitol or glucose ($P<0.001$ and $P<0.05$, respectively). Plasma osmolality was increased by concurrent intravenous infusion of hypertonic urea ($P<0.001$). There were no other significant changes measured in plasma.

The data indicated that concurrent intravenous infusion of solutions which would have increased the osmolality of the extracellular fluid (ECF) (i.e. within the first 20 min) without changing sodium balance caused decreased bar-pressing for sodium bicarbonate. In fact, concurrent intravenous infusion of hypertonic mannitol was as effective as concurrent infusion of hypertonic saline. The decreased bar-pressing observed during hypertonic glucose and hypertonic urea infusion was not as great. Since glucose and urea would, at least in part, distribute into the intracellular fluid (ICF) as well as the ECF, it is likely that the increase in the osmolality of the ECF would not be as great as that produced by hypertonic saline or mannitol. The concurrent intravenous infusion of isotonic saline did not significantly affect total sodium ingested, indicative that the intravenous infusion itself was not responsible for the changes in behaviour observed when hypertonic solutions were infused. It can be calculated that the increase in ECF volume with isotonic saline probably approximated the increase in ECF volume caused by the hypertonic solutions as a result in some instances of their drawing water from ICF plus the 0.15 M saline component of the infusion.

Fitzsimons and Wirth (1976) obtained similar results in sodium-deplete adrenalectomized rats injected subcutaneously with equal osmotic loads of various solutes 15 min before a 3-h drinking test. Injection of 2 M sodium chloride and 2 M sucrose caused significant reduction in 2.7% saline intake whereas 2 M glucose and 2 M urea caused small but insignificant change. By comparison, when sodium-deplete adrenalectomized rats were injected subcutaneously with equal amounts of sodium as hypertonic (4 M or 0.462 M), isotonic (0.154 M) or hypotonic (0.104 M) saline 60 min before offer of 2.7% saline and water for 3 h, the hypertonic solutions approximately halved voluntary intake of saline whereas the other fluids did not have any significant effect. This latter result differs from the data on sheep with regard to isotonic saline. One factor in this may reside in the intravenous infusion used in sheep experiments as against the indeterminant systemic influence of subcutaneous injection of isotonic fluid 30 min before test in the rat experiments.

The data focus consideration on the blood–brain barrier and the influence of systemic changes in ions and osmolarity on composition of the cerebrospinal fluid (CSF).

As Rapoport (1975) points out, the blood–brain barrier at cerebral blood vessels is due to a continuous layer of endothelial cells that are connected by tight junctions. It resembles a lipoid plasma membrane. Rates of transfer of non-electrolytes across cerebral capillaries are related to lipid solubility if they are not facilitated by special mechanisms as in the case of glucose. The barrier closes to sucrose and inulin after 50 days of gestation in the sheep (Bradbury 1975). Rapoport (1975) suggests these junctions deform and increase their permeability when subjected to tensile stresses. He adds that hypertonic solutions injected into the carotid artery may reversibly open the blood–brain barrier by shrinking endothelial cells and widening tight junctions, but most of the experimental demonstration of this involves gross change of osmotic pressure (see Chapter 24).

In sheep the influence of intracarotid infusion of a number of substances has been tested in relation to the effect on CSF [Na] and also plasma [Na] (Fig. 20-14 and Table 20-10) (McKinley et al. 1978). As well as the substances recorded on the figure it was shown that 2 M glucose and 2 M galactose had no effect on CSF [Na] and both caused a highly significant fall in plasma [Na]. These results are in agreement with data from several species that a far more effective blood–CSF barrier exists for sucrose, sodium chloride and urea than for glucose and galactose (Coppen 1960; Davson 1967; Coxon 1968). Since there is adequate evidence that water passes rapidly across the choroid plexus (Sweet et al. 1950; Bering 1952) the increased CSF [Na] was probably mainly caused by osmotic withdrawal of water from CSF to blood. Katzman and Pappius (1973) state that the choroid plexuses are the only sites which can be considered as a CSF–blood interface at which the two fluids are in contact with each other across a capillary endothelium and the choroid plexus epithelium without the intervention of other cells

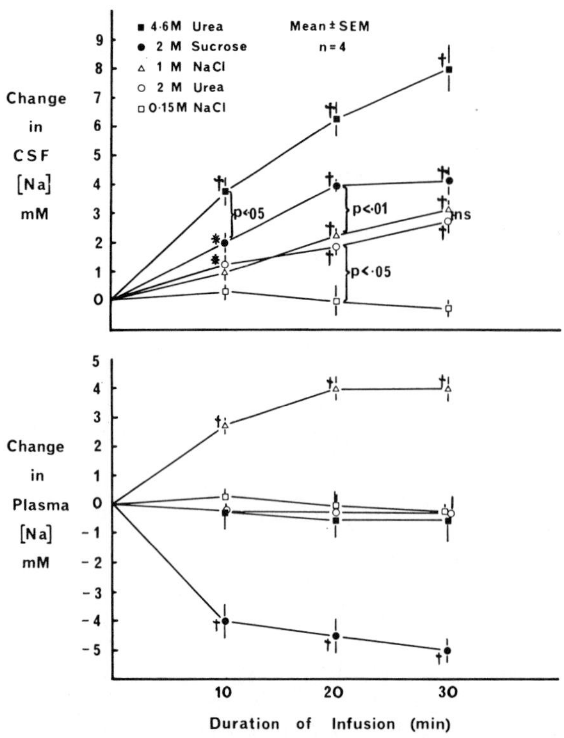

Fig. 20-14. Effect of intracarotid infusion of 0.15 M saline, 1 M saline, 2 M sucrose, 2 M urea and 4.6 M urea at 1.6 ml/min for 30 min on CSF and jugular vein plasma [Na]. The contralateral carotid artery was occluded to allow bilateral distribution of infusate. A significant change from final preinfusion value is denoted by *$P<0.05$, †$P<0.01$ (paired *t*-test). Other *P* levels were obtained by Student's *t*-test. Results with 2 M glucose and 2 M galactose (not included on the graph) showed no change in CSF [Na] but both caused highly significant reduction of plasma [Na]. (From McKinley et al. 1978, with permission.)

of the cerebral tissues. They note capillaries lying close to the ependymal surfaces may also play a role in exchanges between blood and CSF.

Clinical and experimental observations show that intravascular infusion of hypertonic solutions of sucrose, sodium chloride, urea or mannitol reduces intracranial pressure (Weed and McKibben 1919; Javid and Settlage 1956; Javid and Anderson 1959; Wise and Chater 1962; Wise et al. 1964). This is because of osmotic withdrawal of brain water across a blood–brain barrier which exists for these molecules but not for glucose (Crone 1965; Yudilevich and de Rose 1971) On the data derived from direct sampling (McKinley et al. 1978), it is likely that intravenous infusion of hypertonic urea as in the experiments recounted above increases—as do mannitol, sucrose and sodium chloride—the CSF [Na] and osmotic pressure of brain interstitial fluid in the areas where there is a blood–brain barrier. In the experiments of McKinley et al. (1978) it was found that only intracarotid hypertonic sucrose and sodium chloride cause rapid antidiuresis and thirst. Hypertonic urea caused slow antidiuresis and inconsistent water intake. It was proposed that the osmoreceptors which at least partly cause the responses may be in regions of the central nervous system lacking a blood–brain barrier, but there are other elements inside the barrier which determine the slow responses—possibly sodium sensors in this instance—as Andersson and colleagues originally postulated. Intraventricular infusion of hypertonic urea in artificial CSF had no thirst or antidiuretic effects. These data and associated findings raised the issue of whether there are osmoreceptors (Verney 1947) or sodium receptors (Andersson and Olsson 1973) determining thirst and ADH release, or both, as McKinley et al. (1978) suggested and discussed fully. Strong support for this latter theory comes from recent experiments by Leksell et al. (1981) where, in sheep dehydrated for 44 h, the intraventricular infusion of 0.7 M mannitol in artificial CSF at 1 ml/h caused a clear-cut significant rise in free water clearance. Direct measurement of CSF composition showed a large rise in osmolality but a fall in [Na] (Fig. 20-15).

Table 20-10. Effect of intracarotid infusion (1.6 ml/min) of various hypertonic solutions on CSF and plasma [Na] and [K].

Solution infused	*n*	CSF [Na] Control	CSF [Na] 25 min infusion	Plasma [Na] Control	Plasma [Na] 25 min infusion	CSF [K] Control	CSF [K] 25 min infusion	Plasma [K] Control	Plasma [K] 25 min infusion
0.15 M NaCl	7	149.9 (0.9)	150.3 (0.7)*	142.0 (1.1)	142.5 (1.1)*	2.73 (0.03)	2.73 (0.02)*	4.12 (0.20)	4.05 (0.20)*
1 M NaCl	5	151.6 (0.7)	154.6 (0.7)***	145.0 (0.7)	146.2 (0.8***	2.78 (0.08)	2.82 (0.07)*	4.52 (0.9)	3.98 (0.06)***
2 M Sucrose	6	151.0 (1.2)	156.2 (1.4)***	146.0 (1.2)	140.8 (1.1)***	2.80 (0.06)	2.88 (0.4)*	4.37 (0.14)	3.88 (0.11)***
2 M Urea	8	151.0 (0.8)	153.1 (0.5)**	143.6 (1.1)	143.1 (1.0)*	2.74 (0.03)	2.84 (0.04)**	4.28 (0.13)	4.01 (0.08)*
2 M Glucose	4	149.8 (1.7)	150.7 (1.7)*	143.0 (1.7)	137.5 (1.3)***	2.88 (0.08)	2.88 (0.10)*	4.55 (0.13)	3.98 (0.12)***
2 M Galactose	5	153.4 (0.9)	154.0 (1.2)*	148.8 (0.7)	139.8 (1.0)***	2.88 (0.05)	2.82 (0.06)*	4.55 (0.12)	3.96 (0.10)***

From McKinley et al. (1978), with permission.

Values are expressed as mean (s.e.m.) in mmol/l. Statistical analysis compared value at 25 min after commencement of infusion with preinfusion control value using Student's *t*-test (paired) analysis. *=Not significant; **=$P<0.05$; ***=$P<0.01$.

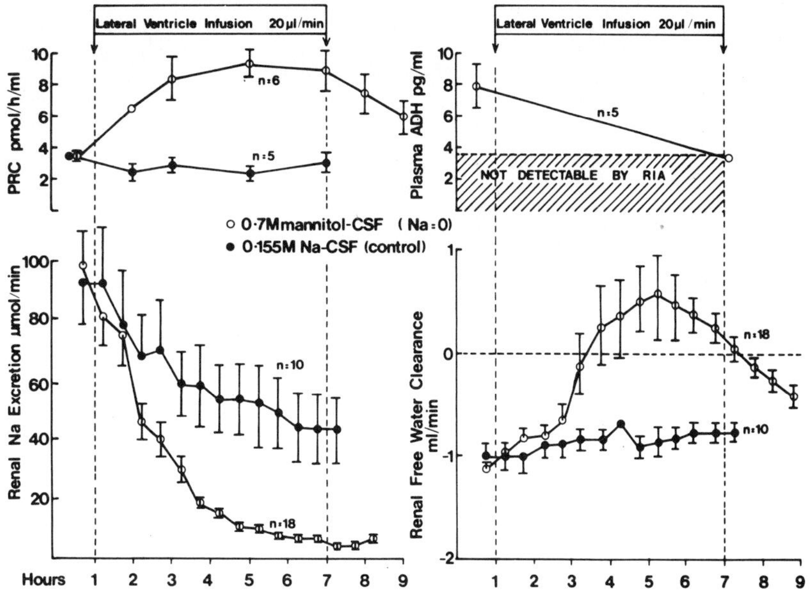

Fig. 20-15. Effect of infusion into the lateral cerebral ventricle at 20 µl/min of hypteronic low-sodium CSF (0.7 M mannitol CSF, [NaCl] = 0 mM) (open circles) or control CSF ([NaCl] = 155 mM) (filled circles) on plasma renin concentration (PRC), plasma ADH, renal sodium excretion, and renal free water clearance in sheep deprived of water for 44 h. The CSF [Na] fell from 157±1 to 126±4 mM, and CSF osmolality increased from 311±3 to 340±5 mosmol/kg over the course of the infusion of 0.7 M mannitol CSF. Mean and s.e.m. are shown. (From Leksell et al. 1981, with permission.)

In the concurrent automatic intravenous infusion experiments only the infusion of hypertonic saline or mannitol stimulated increased intake of water. Neither hypertonic glucose nor urea increased water intake. These results are consistent with our previous results (McKinley et al. 1978) and suggest that the stimulus to increased water intake could be either increased [Na] or osmolality, provided that the sensors involved in thirst are in an area of the brain lacking the blood–brain barrier (e.g. organum vasculosum of the lamina terminalis, subfornical organ). The fact, however, that all of the hyperosmotic infusions decreased sodium intake would suggest, in contrast, that the sensors involved in sodium appetite are located in an area of the brain with a blood–brain barrier. The sodium appetite data with glucose remain somewhat paradoxical in this regard though, since, as noted above, intracarotid infusion of glucose did not cause significant rise in CSF [Na]. On the other hand it did cause plasma [Na] to fall as a result of withdrawal of water from ICF so that it still caused a differential between CSF [Na] and plasma [Na] with CSF [Na] relatively higher (Table 20-10), and this differential may be the relevant factor even if there is no absolute rise of CSF [Na].

Whereas these latter data provide further evidence on systemic changes influencing salt appetite in the short term, they give only limited insight into mechanisms. Evidently, influences of pulsed versus continuous change of sodium balance and concentration, rate of change of CSF [Na] *and this relative to concurrent alteration of plasma [Na]*, concurrent effects on osmoreceptors and evocation of thirst, all arise. The problem remains as to how large volumes of isotonic saline, without concomitant plasma [Na] change, eventually significantly reduce sodium appetite. Much more direct and intriguing data in this regard come from studies in which the composition of CSF has been altered by lateral ventricular infusion of solutes before and during bar-pressing episodes by sodium-deficient animals and in sodium-replete sheep. These will be described in Chapter 24, which deals with theories as to how sodium appetite is generated.

References

Abraham SF, Baker R, Denton DA, Kraintz F, Kraintz L, Purser L (1973) Components in the regulation of salt balance: Salt appetite studied by operant behaviour. Aust J Exp Biol Med Sci 51: 65

Abraham SF, Coghlan JP, Denton DA, McDougall JG, Mouw DR, Scoggins BA (1976) Increased water drinking induced by sodium depletion in sheep. Q J Exp Physiol 61: 185

Andersson B, Jewell PA (1956) The distribution of carotid and vertebral blood in the brain and spinal cord of the goat. Q J Exp Physiol 41: 462

Andersson B, Olsson K (1973) On central control of body fluid homeostasis. Conditioned Reflex 8: 147

Baldwin BA (1976) Effects of intracarotid or intraruminal injections of NaCl or $NaHCO_3$ on sodium appetite in goats. Physiol Behav 16: 59

Baldwin BA, Bell FR (1960) The contribution of the carotid and vertebral arteries to the blood supply of the cerebral cortex of sheep and calves. J Physiol (Lond) 151: 9P

Beilharz S, Kay RNB (1963) The effect of ruminal and plasma

sodium concentration on the sodium appetite of sheep. J Physiol (Lond) 165: 468
Beilharz S, Denton DA, Sabine JR (1962) The effect of concurrent deficiency of water and sodium on sodium appetite of sheep. J Physiol (Lond) 163: 378
Beilharz S, Bott EA, Denton DA, Sabine JR (1965) The effect of intracarotid infusions of 4 M NaCl on the sodium drinking of sheep with a parotid fistula. J Physiol (Lond) 178: 80
Bell FR (1972) The relative importance of the sodium ion in homeostatic mechanisms in ruminant animals. Proc R Soc Lond [Biol] 65: 631
Bell FR, Drury PL, Sly J (1981) The effect on salt appetite and the renin-aldosterone system on replacing the depleted ions to sodium deficient cattle. In press
Bering EA (1952) Water exchange of central nervous system and cerebrospinal fluid. J Neurosurg 9: 275
Blair-West JR, Boyd GW, Coghlan JP, Denton DA, Wintour M, Wright RD (1967) The role of circulating fluid volume and related physiological parameters in the control of aldosterone secretion. Aust J Exp Biol Med Sci 45: 1
Blair-West JR, Brook AH, Simpson PA (1972) Renin responses to water restriction and rehydration. J Physiol (Lond) 226: 1
Bott EA, Denton DA, Weller S (1965) Water drinking in sheep with oesophageal fistulae. J Physiol (Lond) 176: 323
Boyd GW (1967) The reproducibility and accuracy of plasma volume estimation in the sheep with both ^{131}I gamma globulin and Evans blue. Aust J Exp Biol Med Sci 45: 51
Bradbury MWB (1975) Ontogeny of mammalian brain-barrier systems. In: Cserr HF et al. (eds) Fluid environment of the brain. Academic Press, New York, p 81
Coghlan JP, Denton DA, Goding JR, Wright RD (1960) The control of aldosterone secretion. Postgrad Med J 35: 76
Coppen AJ (1960) Abnormality of the blood cerebrospinal fluid barrier of patients suffering from depressive illness. J Neurol 23: 156
Coxon RV (1968) Cerebrospinal fluid transport. In: Lajtha A, Ford DH (eds) Progress in brain research, Vol 29. Brain barrier systems. Elsevier, Amsterdam, p 134
Crone C (1965) Facilitated transfer of glucose from blood into brain tissue. J Physiol (Lond) 181: 103
Daniel PM, Dawes JDK, Pritchard MML (1953) Studies of the carotid rete and its associated arteries. Philos Trans R Soc Lond [Biol] 237: 173
Davson H (1967) Physiology of the cerebrospinal fluid. Churchill, London
Davson H, Pollay M (1963) The turnover of ^{24}Na in the cerebrospinal fluid and its bearing on the blood–brain barrier. J Physiol (Lond) 167: 247
Denton DA (1957) The effect of variations in blood supply on the secretion rate and composition of parotid saliva in sodium depleted sheep. J Physiol (Lond) 135: 227
Denton DA (1966) Some theoretical considerations in relation to innate appetite for salt. Conditional Reflex 1: 144
Denton DA, McDonald IR (1957) The effect of rapid change in sodium balance on the salivary Na:K concentration ratio of sodium depleted sheep. J Physiol (Lond) 138: 44
Denton DA, Goding JR, Wright RD (1959) Control of adrenal secretion of electrolyte active steroids. Br Med J ii: 447, 522
Denton DA, Orchard E, Weller S (1971) The effect of rapid change of sodium balance and expansion of plasma volume on sodium appetite of sodium deficient sheep. Commun Behav Biol 6: 245
Fencl V, Miller TB, Pappenheimer JR (1966) Studies on the respiratory response to disturbances of acid–base balance with deductions concerning the ionic composition of cerebral interstitial fluid. Am J Physiol 210: 459
Fitzsimons JT, Wirth JB (1976) The neuroendocrinology of thirst and sodium appetite. In: Kaufmann W, Krause DK (eds) Central nervous control of Na balance: Relations to the renin–angiotensin system. George Thieme, Stuttgart, p 80
Javid M, Anderson J (1959) The effect of urea on cerebrospinal fluid pressure in monkeys before and after bilateral nephrectomy. J Lab Clin Med 532: 484
Javid M, Settlage P (1956) Effect of urea on cerebrospinal fluid pressure in human subjects. JAMA 160: 943
Katzman R, Pappius HM (1973) Brain electrolytes and fluid metabolism. Williams and Wilkins, Baltimore
Katzman R, Schimmel H, Wilson CE (1968) Diffusion of inulin as a measure of extracellular fluid space in brain. Proc Virchow Pirquet Med Soc 26: 254
Leksell LG, Congiu M, Denton DA, Fei DTW, McKinley MJ, Tarjan E, Weisinger RS (1981) Influence of mannitol-induced reduction of CSF Na on nervous and endocrine mechanisms involved in the control of fluid and sodium balance. Acta Physiol Scand 112: 33
McCance RA (1936) Experimental sodium chloride deficiency in man. Proc R Soc Lond [Biol] 119: 245
McKinley MJ, Denton DA, Weisinger RS (1978) Sensors for antidiuresis and thirst—osmoreceptors or CSF sodium detectors? Brain Res 141: 89
Nachman M, Valentino DA (1966) The role of taste and post-ingestional factors in the satiation of sodium appetite in rats. J Comp Physiol Psychol 62: 280
Rall DP, Oppelt WW, Patlak CS (1962) Extracellular space of brain as determined by diffusion of inulin from the ventricular system. Life Sci 2: 43
Rapoport SI (1975) Experimental modification of blood brain barrier permeability by hypertonic solutions, convulsions, hypercapnia and acute hypertension. In: Cserr HF et al. (eds) Fluid environment of the brain. Academic Press, New York, p 61
Richter CP (1956) Salt appetite of mammals. In: L'Instinct dans le comportement des animaux et de l'homme. Masson et Cie, Paris, p 577
Sweet WH, Selverstone B, Solloway S, Stetten D (1950) Studies of formation, flow and absorption of cerebrospinal fluid. II. Studies with heavy water in the normal man. Surg Forum 1: 376
Verney EB (1947) The antidiuretic hormone and the factors which determine its release. Proc R Soc Lond [Biol] 135: 25
Weed LH, McKibben PS (1919) Pressure changes in the cerebrospinal fluid following intravenous injections of solutions of various concentrations. Am J Physiol 48: 512
Weisinger RS, Denton DA, McKinley MJ (1978) Effect of self-determined intravenous infusion of hypertonic NaCl on Na appetite of sheep. J Comp Physiol Psychol 92: 522
Wise BA, Chater N (1962) The value of hypertonic mannitol solution in decreasing brain mass and lowering cerebrospinal fluid pressure. J Neurosurg 19: 1038
Wise BA, Perkins RK, Stevenson E, Scott KG (1964) Penetration of ^{14}C-labelled mannitol from serum into cerebrospinal fluid and brain. Exp Neurol 10: 264
Yudilevich DL, de Rose N (1971) Blood–brain transfer of glucose and other molecules measured by rapid indicator dilution. Am J Physiol 220: 841

21 The endocrine effects of rapid satiation of salt appetite in sodium deficiency

Summary

1. The phylogenetic emergence of salt appetite and the adrenal secretion of aldosterone have occurred in parallel under the same selection pressure of environmental sodium deficiency. This has resulted in some functional interlinking of the two systems, but no very substantial interaction.

2. Rapid satiation of salt appetite can have large but evanescent influence on aldosterone secretion.

3. The phenomenon has particular interest, since the vector of effect is not one of the factors presently known to be involved in control of aldosterone: namely, plasma concentration of sodium, potassium, angiotensin II or III, or adrenocorticotrophic hormone (ACTH).

4. Other experimental data described, including cerebral ventricular perfusion studies, are consistent with a central nervous effect on aldosterone secretion.

5. The phenomenon is a striking example of the 'gustatory metabolic reflexes', which are anticipatory in the sense that they precede and are in the same direction as the changes which are produced by the substances ingested as a result of post-absorptive systemic influences. Other examples of this interesting physiological mechanism, such as rapid effects on antidiuretic hormone secretion following drinking, are discussed.

Introduction

The brain systems which react to depletion of the body salt content, and the adrenal capacity to secrete the salt-retaining hormone aldosterone, have evolved in response to the same selection pressure: environmental sodium deficiency. A question of major interest in relation to the overall organization of the physiological control of sodium homeostasis is thus whether these two systems, behavioural and hormonal, are in any way functionally integrated and directly influence one another, or whether they are largely disjoined and independent though they subserve the same homeostatic end.

Corollary questions include whether the same type of change within the *milieu intérieur*, e.g. alterations in plasma [Na] and [K] (sodium and potassium concentrations), ACTH (adrenocorticotrophic hormone), and angiotensin II and III concentrations, which control aldosterone secretion, are physiological stimuli of salt appetite. In the case of water metabolism, for example, there is evidence in several mammalian species that the same stimulus, a rise in osmotic pressure of the arterial plasma, and/or a rise in [Na] of the CSF, acts to cause thirst and secretion of antidiuretic hormone (ADH) (Gilman 1937; Verney 1947; Holmes and Gregerson 1950; Wolf 1958; Fitzsimons 1972; Andersson and Olsson 1973; McKinley et al. 1978; Andersson 1978). From the data in the preceding chapters and also those in Chapter 24, it is clear that direct effects of sodium play the dominant role in the reactions of the sodium appetite system, in contrast to the control of aldosterone secretion where plasma [K] and angiotensin II and III are major influences.

The phylogenetic emergence of the capacity of a herbivorous species to correct body deficit of water or salt relatively accurately in a rapid consummatory act has conferred considerable survival advantage. This is in the face of the need of animals to go to water holes and saline spring licks, where they may be exposed to carnivorous predators. The anatomical structures associated with the ruminant mode of digestion, being capacious and less likely to initiate inhibitory feedback until a large volume is taken, would be conducive to this central nervous organization of rapid satiation behaviour.

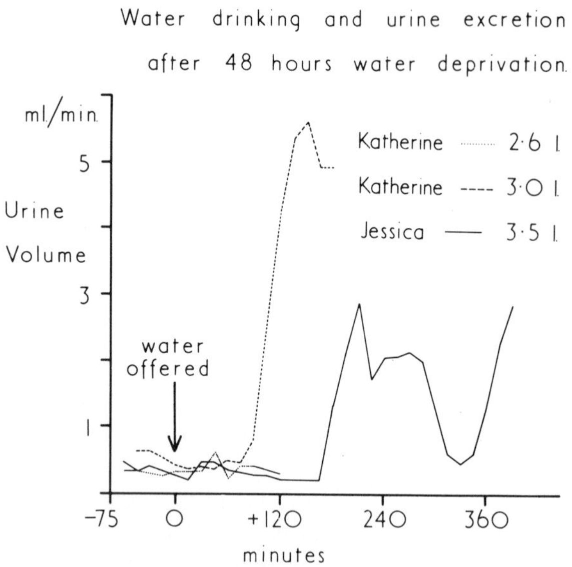

Fig. 21-1. The effect of rapid drinking of 2.6–3.5 litres of water in two sheep depleted of water for 48 h. Urine flow rose significantly after 90 min.

Hormonal changes following fluid ingestion

Adolph's (1939) observations on urine excretion in water-deficient dogs that were drinking large volumes which were lost through an oesophageal fistula, did not suggest that the drinking act increased urine flow by inhibiting ADH secretion. But it might be argued that in these conditions the actions were not satiating. With water-deficient sheep which satiate a 10% body deficit in one drinking act over 2–4 min, urine flow observations have not indicated that there is a sustained and large inhibition of ADH secretion until 90–140 min (Fig. 21-1). By this time a significant amount of water has been absorbed from the gut. However, this is not to say that with frequent repetition of the same experimental situation, an immediate effect of rapid satiation upon hormone secretion might not eventually occur, analogous to effects described below for salt appetite.

Rougon-Rapuzzi et al. (1978) have studied arginine-vasopressin (AVP) content of a number of hypothalamic nuclei and the posterior lobe of the hypophysis (PLH) of rats during dehydration and rehydration. The AVP content of the hypothalamus and of the PLH move in parallel and deplete markedly with 2–4 days of dehydration. But with rehydration two dramatic increases in AVP occur in these structures 15 min and 3 h after the onset of drinking. A brief period of rehydration results in a significant increase in AVP in the PLH which reaches a maximum after 15 min of free drinking. This, along with comparable changes in hypothalamic nuclei, may correspond to a decrease in or even to transient interruption of AVP release, most probably associated with slowing of AVP axonal flow. These events synchronous with onset of water intake are unlikely to arise from nervous reflexes coming from central or peripheral osmoceptors or volume receptors since at this early stage osmolality is still elevated and probably blood volume is reduced. But it is possible that a neuronal reflex arc with sensory buccopharyngeal afferents is operative and modifies instantly, but briefly, the discharge pattern of neurosecretory cell bodies. These could decrease AVP release and axonal transport.

Indeed, during recent years as the result of studies of several workers, a body of evidence has built up which indicates that the alimentary tract is not a passive canal in metabolic regulation. Stimulation of gustatory as well as alimentary afferents may, through neural systems, have direct consequences on metabolic processes. These 'gustatory metabolic reflexes' are anticipatory in the sense that they precede and are in the same direction as the changes which are produced by the substances as a result of post-absorptive systemic influences (Nachman and Cole 1971; Nicolaidis 1969).

It has been shown that stimulation of stomach receptors directly affects neural activity in hypothalamic feeding centres (Sharma et al. 1961). Glucose placed directly in the stomach of man produces a greater and more sustained level of plasma insulin than glucose administered intravenously (Elrich et al. 1964).

Andersen et al. (1978) note that gastric inhibitory polypeptide (GIP) has been postulated as a major enteric hormonal mediator of insulin release. The release of immunoreactive GIP (IR-GIP) after oral glucose and its role in insulin release was studied in normal men by the glucose clamp technique. In 24 subjects studied with the hyperglycaemic clamp, blood glucose was maintained at 125 mg/dl above basal for 2 h via a primed-continuous intravenous glucose infusion coupled to a servo-controlled negative feedback system. Forty grams of glucose per square metre surface area were ingested at 60 min, and the blood glucose was maintained at the steady-state hyperglycaemic level. Plasma IR-GIP and immunoreactive insulin (IRI) levels were measured throughout the 2-h period. IR-GIP levels changed little when intravenous glucose alone was given; the mean basal value was 305 ± 34 (s.e.m.) pg/ml. After oral glucose, plasma IR-GIP began to rise within 10 min and reached a peak within 40 min of 752 ± 105 pg/ml. Plasma IRI responded initially to the square wave of hyperglycaemia in the typical biphasic pattern. After oral glucose, plasma IRI levels rose strikingly above the elevated levels produced by hyperglycaemia alone, reaching a peak of 170 ± 15 μU/ml within 45 min. The time courses of the rise in IR-GIP and IRI were nearly identical.

The class of effect produced by a stimulus may depend directly on the pre-existing metabolic state of the animal, i.e. whether food-deprived, dehydrated, salt-deplete and so on. Anokhin and Sudakov (1966) have found that when milk is injected into the mouth of the cat, stimulation of the oral receptors produces a rapid decrease in neural excitation in the frontal areas. During hunger, they had found that the activity of hypothalamic centres produces an ascending excitatory influence on the cortex, particularly frontal areas. The duration of effect of oral milk is prolonged when milk enters the stomach after stimulating the oral receptors.

Nicolaidis (1969) has studied anticipatory metabolic reflexes resulting from stimulation of the alimentary canal with fluids and food. His data include (i) a rapid change in respiratory quotient in response to eating carbohydrates, (ii) hyperglycaemic responses to oral stimulation with saccharin or sucrose in hungry rats, (iii) the demonstration of almost immediate sweating in dehydrated human subjects in a hot room when they drank fluids.

The background of this last experiment, as described by Nicolaidis (1969), is as follows. It is well known that the threshold for sweating response, which is modulated by central temperature, rises as a function of water and electrolyte depletion. Experiments were performed on eight men and two women, 17 to 32 years old. Dehydration was produced by water deprivation for 42 h and the rehydration was accomplished by having the subjects ingest 1.5 to 2.0 litres of water. Twenty-four recordings were obtained. After a 30-min stay in a warm chamber at 33°C when baseline measurements of sweating rate were taken, a stimulation with 150 ml of water at 12°C was given. In three experiments the water was given at 37°C and in three, isotonic saline was given. The experimental procedure was identical when the subjects were dehydrated and when they were rehydrated.

In dehydrated subjects there was a pronounced sweating (up to 500 ml m^{-2} min^{-1}) over a period of 5 min, which was rapidly effective in restitution of temperature regulation (which becomes defective in dehydrated subjects). In the same subjects when rehydrated, the drink did not provoke sweating or had a small effect only. The relatively constant and very short latency times of onset strongly suggest the existence of a ‘poto-hidrotic’ reflex with an orogastric origin. The results were similar for the two temperatures of water and saline. Before ingested liquids have reached the cellular level, the sweating rate of the dehydrated subject reaches, and even passes, that of the rehydrated or normal subject. The importance of the existing metabolic state in allowing the reflex to function is clear. Another facet of this is the data of Kenney and Miller (1949) which show that the diuretic response to drinking a litre of water by resting men is greatly modified by change in ambient temperature, diuresis being suppressed by a rise in temperature.

In rats stomach-loaded with water to give a constant diuresis, stimulation of the mouth with water increased diuresis, whereas stimulation with 5% saline inhibited diuresis. Parallel to this latter finding, Nicolaidis found buccal stimulation with water decreased integrated neural activity in the supraoptic nucleus, whereas buccal stimulation with 1 M saline increased it.

Vincent et al. (1972a, b) made studies of great interest on the osmosensitive cells in the hypothalamus of the monkey, which responded to osmotic stimulation by an intracarotid pulse of hypertonic sodium chloride. Three types of osmosensitive cells were identified. All the osmosensitive cells studied established a significant variation of their firing rate during drinking which differed accord-

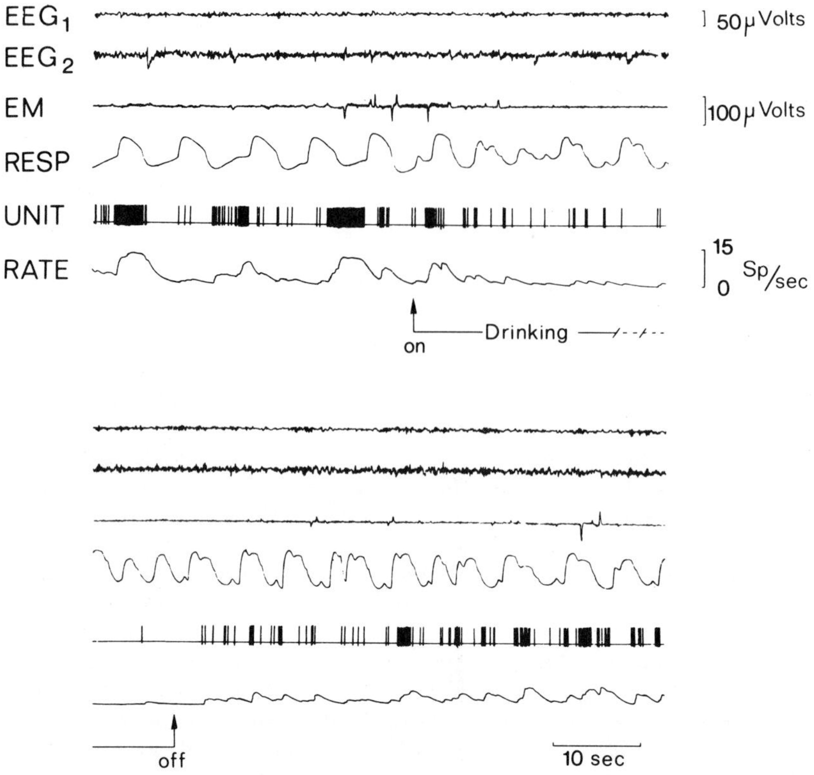

Fig. 21-2. Firing pattern of a specific osmosensitive neurone type I in the nucleus supraopticus. A slow oscillatory excitatory–inhibitory pattern is observed before the initiation of drinking. When water is supplied, the cell shows a progressive and complete inhibition and does not return to the initial level of firing after cessation of drinking. EEG_1, biparietal electrocorticogram; EEG_2, occipital electrocorticogram; EM, eye movements; RESP, nasal respiration; UNIT, pulse output from pulse height discriminator triggered by action potentials of the cell; RATE, analogue output proportional to the rate of unit discharge. (From Vincent et al. 1972a, with permission.)

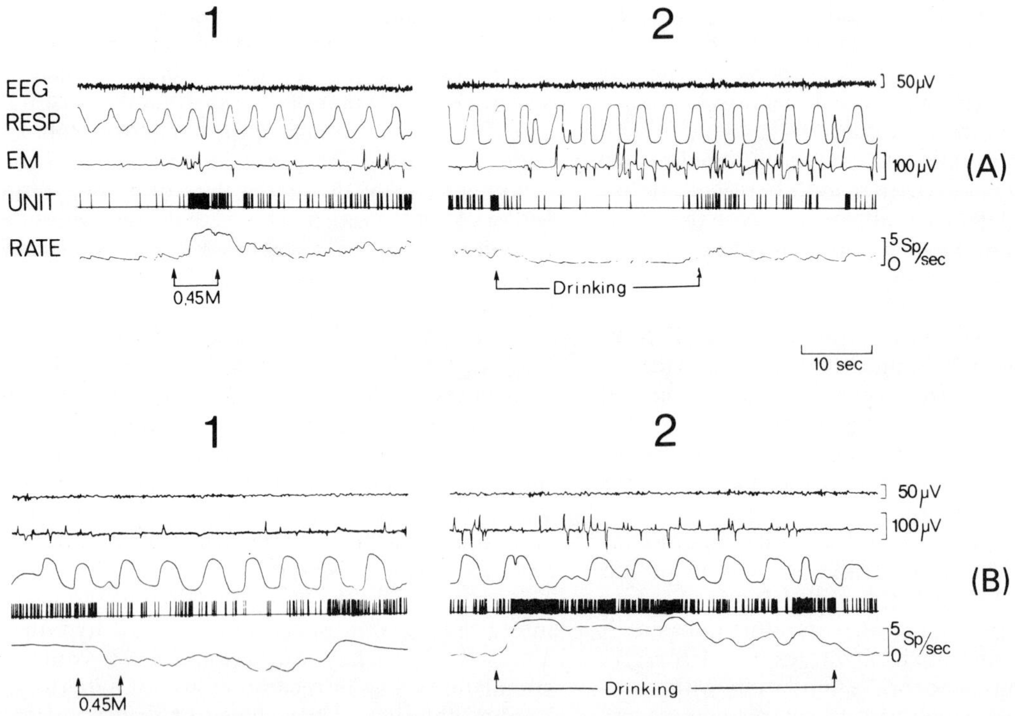

Fig. 21-3. Electrical responses of two specific osmosensitive neurones A and B located in the perinuclear zone of the nucleus supraopticus to (1) hypertonic stimulation (intracarotid injection of 1 ml of 0.45 M sodium chloride) and (2) drinking of tap water (20 ml), in unanaesthetized rhesus monkey. EEG, biparietal electrocorticogram; RESP, respiration; EM, eye movements. UNIT and RATE as in Fig. 21-2 above. (From Vincent et al. 1972b, with permission.)

ing to the type of cell (Figs. 21-2 and 21-3). The neurosecretory cells in the nucleus supraopticus, identified by an antidromic stimulation from the posterior pituitary, were inhibited during drinking (Fig. 21-2). A dramatic decrease a few seconds after the beginning of drinking was seen. The authors suggest that this may give a neurophysiological explanation for the rapid increase of diuresis after drinking described in, for example, the rat. The receptors in mouth, throat and stomach may inform the osmoreceptor-neurosecretory system of the volume of water drunk before correction of the altered plasma osmotic pressure caused by absorption of water. They found it was only actual drinking of water which was able to produce such an inhibition independently of motor behaviour, and did not observe inhibition after sham-drinking induced by intracarotid injections of saline in the absence of water. The neuronal observations provide some support for the existence of a mechanism anticipating satiation by drinking, and explain the cessation of specific behaviour before correction of the internal imbalance which has triggered the response.

J. D. Vincent (personal communication) has subsequently shown that with water drinking in monkeys (Fig. 21-4) depleted of water for 5 days, there is a dramatic fall in AVP over 5–15 min towards basal level followed by a rebound. The fall precedes a significant fall in plasma osmolality. Ramsay et al. (1980) and Thrasher et al. (1981) also report that in water-deplete dogs plasma vasopressin falls 50% by 3 min after voluntary water drinking, whereas plasma osmolality takes 12 min to fall significantly. This reduction still occurs when the ingested water is immediately removed via a gastric fistula. When the same volume of water is placed directly into the stomach via the fistula, there is no rapid phase in the reduction of plasma AVP, its return to normal following redilution of the plasma. Thus Ramsay et al. propose that the inhibition of AVP secretion after water drinking depends on oropharyngeal factors.

The functional anticipatory nature of these various physiological effects cited is evident in comparison with the relatively long delay in feedback from post-absorptive influences.

Against this background of the general biological question posed at the outset, and the experimental data cited above on other physiological systems, I wish to record some unique observations made since 1962 on our colony of sheep with adrenal autotransplants and parotid fistulae. Specifically we have observed that the rapid satiation of salt hunger by ingesting sodium bicarbonate solution may cause a rapid, large evanescent inhibition of aldosterone hypersecretion by the adrenal gland (Denton et al. 1977). The phenomenon has great intrinsic interest in relation to the pathway of effect on the zona glomerulosa of the transplanted adrenal. This is because the inhibition may occur without an appropriate change in any of the factors currently known to contribute to secretion of aldosterone in response to sodium deficiency.

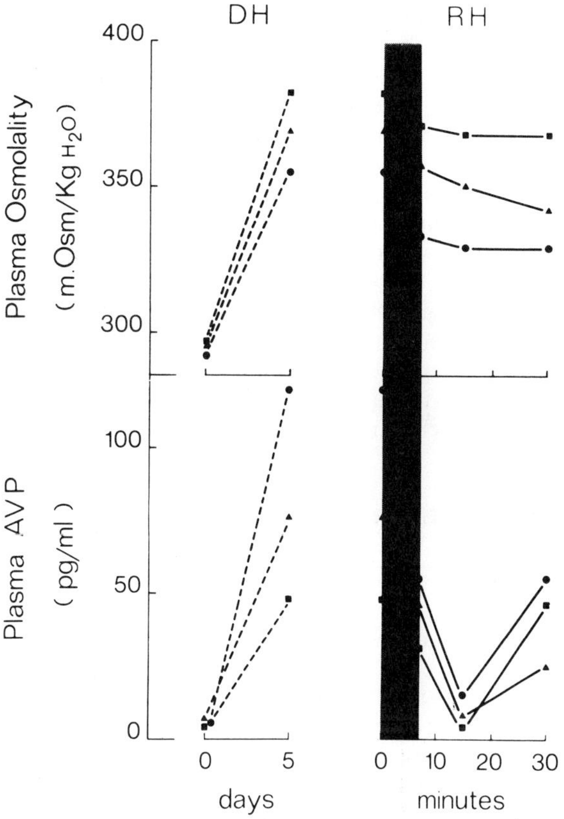

Fig. 21-4. The effect of 5 days of water deprivation on the plasma osmolality and plasma AVP of three monkeys. On the fifth day 50 ml of water was drunk over 7 min. A dramatic decrease in AVP followed over the period, without a preceding comparable fall in plasma osmolality. DH, dehydration; RH, rehydration. (J. D. Vincent and colleagues, unpublished, with permission.)

Rapid inhibition of aldosterone secretion produced by satiation of salt appetite in sodium-deplete sheep with adrenal autotransplants

The parotid-fistulated sheep were depleted of sodium by withdrawal of sodium bicarbonate supplement for 24–72 h, so that deficits of 300–900 mmol of sodium were produced. One or two control collections of adrenal venous blood were taken for corticosteroid and electrolyte analyses, and in later experiments, where these factors were measured, control collections of venous blood for renin, and arterial blood for angiotensin II, were made.

Behaviour

Following control blood collections, the solutions of 300 mmol/l sodium bicarbonate were prepared in the laboratory within sight and hearing of the sheep. The animals, hitherto standing quietly in their cages, became restless, very excited and stamped and bleated. Under these conditions increased salivary secretion, cardiac rate and blood pressure were observed: see Chapter 12 (Denton and Sabine 1963). On presentation of 2 litres of the solution, in some instances the animals finished the lot in 3–4 min. In these cases a further 2 litres were usually presented. The animals usually drank an amount related to the extent of body deficit (Denton et al. 1969). Drinking was usually completed within 5–15 min and thereafter no further interest in the solution was shown.

Initial individual experiments

It is a technically exacting procedure to manage, without disturbing the animal, sequentially to sample the venous effluent of an autotransplanted endocrine gland while the creature is satiating an ingestive drive in its usual way. Therefore, before considering grouped data, it is illustrative to describe some individual experiments. These first brought the rapid inhibition of aldosterone secretion to our notice.

Fig. 21-5 records an initial observation with TP9 when sodium-deplete. The aim of the first part of the experiment was to study the influence of increased adrenal arterial [Na] (sodium concentration) on aldosterone control, and provides a striking contrast to the result which was subsequently obtained. After control collection of adrenal venous blood, the pneumatic cuff on the carotid jugular loop cranial to the adrenal autotransplant was inflated to 300 mmHg. An infusion of sodium chloride (600 mmol/l) was begun into the carotid artery caudal to the transplant with a graded increase of rate up to 25 mmol/h (Fig. 21-5). The arterial occlusion cranial to the transplant contrived that the carotid artery became, in effect, the adrenal artery with initial measured blood flow of 6.8 ml/min. This infusion, although having little effect in correcting the

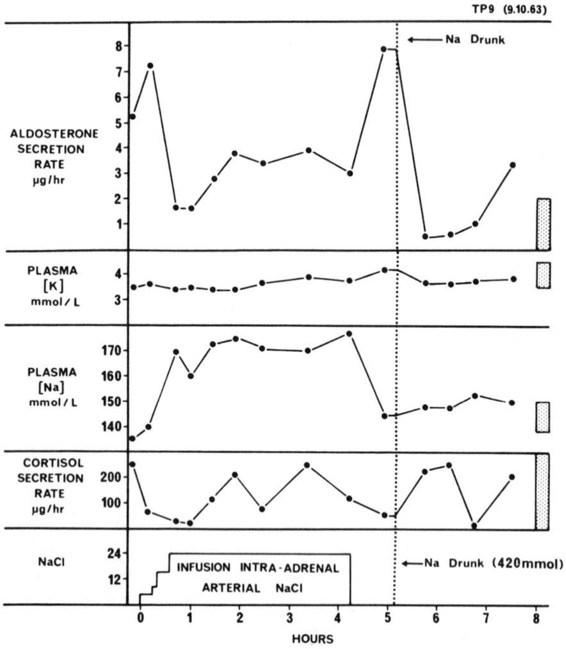

Fig. 21-5. TP9, sodium-deficient sheep with an adrenal transplant. The effect on aldosterone and cortisol secretion rate, adrenal plasma [Na] and plasma [K] of adrenal arterial infusion of a 600 mmol/l sodium chloride solution with stepwise increase in rate. Following cessation of infusion the animal was offered a solution of 300 mmol/l sodium bicarbonate to drink and consumed 420 mmol in 10 min. The effects on the same parameters are recorded. The blocks on the right record the normal range of the parameters in the sodium-replete animal.

animal's sodium deficiency, produced a local rise in [Na] of adrenal arterial blood of 5–31 mmol/l in rapid steps. Plasma [K] rose 0.2 mmol/l. The rise in [Na] of 5 mmol/l by 10 min did not reduce aldosterone secretion, but the rise of 31 mmol/l by 30 min reduced it by 70% (Fig. 21-5). This rapid increase in plasma [Na], which was sustained over 4 h, was to a concentration considerably above the normal sodium-replete range. Aldosterone secretion rate recovered to 50%–60% of control though the infusion continued. When the infusion was stopped the rate returned to control level.

Following this the animal was offered sodium bicarbonate (300 mmol/l) to drink, and it rapidly drank 420 mmol over 10 min. Within 25 min of beginning to drink, aldosterone secretion was reduced by 94%. This dramatic effect was sustained in this very low sodium-replete range for an hour. The rise in plasma [Na], and thus adrenal arterial [Na], was 3 mmol/l. This contrasted with the 31 mmol/l rise contrived by adrenal arterial infusion which had much less effect on aldosterone. Plasma [K] fell by 0.5 mmol/l. The profound fall in aldosterone was associated with a four-fold rise in cortisol secretion and no change in corticosterone secretion rate. At the time this

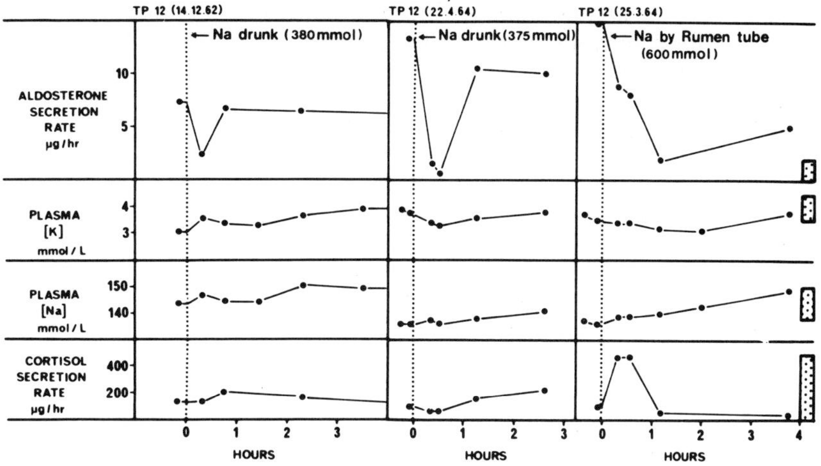

Fig. 21-6. TP12, sodium-deficient sheep with an adrenal transplant. Left and centre panels: The effect on aldosterone and cortisol secretion rate and plasma [Na] and [K] of permitting rapid drinking of hypertonic sodium solutions (400 mmol/l sodium bicarbonate and sodium chloride on left, and 300 mmol/l sodium bicarbonate in centre). The right panel shows the effect on the same parameters when a larger amount of hypertonic sodium solution (300 mmol/l) sodium bicarbonate) was administered by tube into the rumen. The blocks on the right indicate the normal range of the parameters in the sodium-replete animal.

observation was made, the animal TP9 had had a parotid fistula and experience of episodic satiation of salt appetite for 6 years.

An instance of major evanescent inhibition of aldosterone secretion following drinking of a 400 mmol/l solution of equal parts of sodium bicarbonate and sodium chloride was observed also with TP12 (Fig. 21-6, left panel). Aldosterone secretion rate fell 66% in 16 min. In this instance a rise in plasma [K] of 0.4 mmol/l occurred and there was no change in cortisol secretion. Two further experiments strengthened the impression that rapid satiation of salt appetite caused some novel effect on aldosterone secretion. Rapid drinking of 1.25 litres of 300 mmol/l sodium bicarbonate solution caused aldosterone secretion to fall by 88% of control in 20 min and by 95% by 30 min, which was, in fact, to the sodium-replete range (1.0±0.6 µg/h, n=107). By 70 min a large rebound occurred (Fig. 21-6, middle panel). Fig. 21-6 (right panel) records that the administration of a larger amount of sodium (600 mmol in 1.5 litres of water) over 2–3 min by intrarumen tube caused a much smaller fall in aldosterone secretion rate, by 43% of control by 15 min.

In the light of the suggestive individual experiments, systematic examinaion of adrenal secretion and other physiological parameters associated with the act of rapid satiation of salt appetite was made over succeeding years in our colony of animals with adrenal transplant and parotid fistula.

Fig. 21-7 records 22 experiments made on nine sodium-deplete animals with an adrenal transplant: TP9, TP12, Digbeigh, Cyrano, Ivor, Figaro, Benedict, Fred C and Malvina. The results show in this series that in the period 18–36 min (period 2) following drinking of sodium bicarbonate solution there was a highly significant fall in aldosterone secretion rate (P<0.001, period 1 vs 2). In several instances the fall was to the sodium-replete range (1.0±0.6 µg/h, n=107). This was followed by a highly significant rebound towards control level over the period 36–120 min (period 3) (P<0.001, period 2 vs 3). By 240 min and after (period 4), aldosterone secretion rate had fallen substantially as a result of absorption of the sodium ingested.

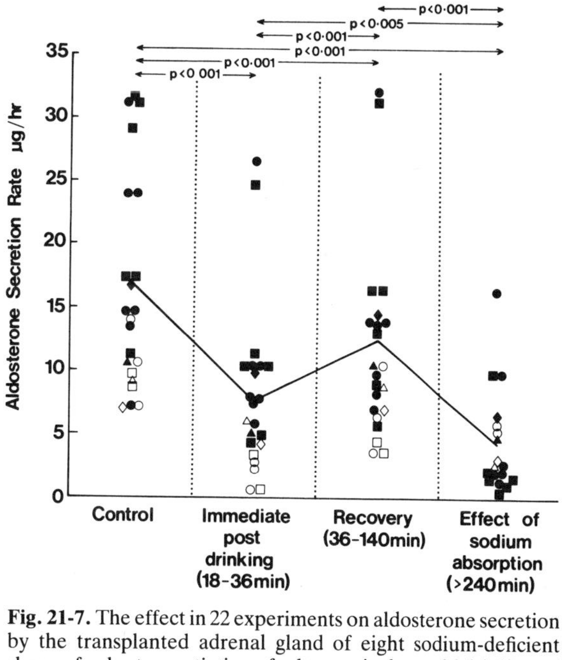

Fig. 21-7. The effect in 22 experiments on aldosterone secretion by the transplanted adrenal gland of eight sodium-deficient sheep of voluntary satiation of salt appetite by rapid drinking of 300–900 mmol of sodium bicarbonate solution (300 mmol/l). The data show a highly significant (P<0.001, paired t-test) decrease in mean aldosterone secretion by 50% within 18–36 min of beginning drinking, a highly significant partial recovery (P<0.001) towards the control pre-drinking level by 36–140 min, and a highly significant (P<0.001) later fall, relative to the 36–140 min observations, due to absorption from the gut of a substantial amount of the sodium bicarbonate solution drunk. Mean values are shown by the uninterrupted line. (From Denton et al. 1977, with permission.)

Many of the final determinations in this period of observation fell in the sodium-replete range.

In recording this phenomenon, it can be noted firstly that this effect of inhibition with satiation, *was reproducible over years in the one animal.* For example, *it was first observed in Cyrano in 1963 and still occurred consistently in 1970–71.* Secondly, there were some animals in our colony which did not show this biphasic response to rapid ingestion of sodium bicarbonate when tested. The results on them are not included in the Tables 21-1 and 21-2 which are concerned with examining concurrent effects on other variables occurring with the satiation inhibition. These non-responding animals were not followed systematically over a period of years in order to determine whether the phenomenon did eventually develop.

Table 21-1 records the changes in aldosterone secretion compared with other variables in the 22 experiments. The mean decrease in aldosterone secretion rate by more than 50% occurred without any significant change in plasma [K] or in the rate of adrenal blood flow. There was a highly significant but small rise in plasma [Na] in each period. However, the recovery rise of aldosterone secretion rate back towards pre-drinking level between periods 2 and 3, occurred *despite a further rise in plasma [Na] similar to that which occurred between periods 1 and 2.* Thus the small change in plasma [Na] did not seem determinant of the inhibition of aldosterone secretion between periods 1 and 2. Cortisol and corticosterone secretion rate were both in the normal range for the undisturbed animal in the sodium-replete and sodium-deficient states. Cortisol secretion decreased significantly in the immediate post-drinking period and was still reduced in period 3 though aldosterone secretion rate had rebounded significantly back towards pre-drinking level. Corticosterone secretion rate showed a highly significant decrease immediately following drinking. The pattern of secretion rate generally followed that of aldosterone, but this correspondence was not obligatory and in several individual experiments the immediate post-satiation inhibition of aldosterone occurred without a corresponding fall of cortisol and corticosterone, or in face of a small rise (e.g. Figs. 21-5 and 21-11).

Fig. 21-8 and Table 21-2 record effects on the same physiological parameters when sodium deficiency was corrected by 450–600 mmol sodium bicarbonate solution (1.5–2.0 litres) administered over 1–3 min by oral tube passed into the rumen. This was done with as little disturbance to the animal as feasible. There were nine experiments, eight of which were on five of the animals used in the series recorded in Table 21-1 (TP12, Cyrano, Ivor, Figaro and TP9), and one on Warner. Fig. 21-8 shows that the starting point involved a similar level of sodium deficiency. No large rapid

Table 21-1. The effect of voluntary satiation of salt appetite by sodium-deplete sheep on the rate of secretion (mean±s.d.) by the autotransplanted adrenal gland of aldosterone, corticosterone and cortisol, the rate of adrenal blood flow and the plasma [Na] and [K].

		Period 1 (control)	Period 2 (18–36 min)	Period 3 (36–140 min)	Period 4 (>240 min)
Aldosterone secretion (μg/h)		16.5±6.4	7.5±6.1	12.0±7.4	4.4±3.6
	P	1 vs 2 <0.001	2 vs 3 <0.001	3 vs 4 <0.001	
		1 vs 3 <0.001	2 vs 4 <0.005		
Corticosterone secretion (μg/h)		23.1±13.1	10.8±20.4	13.6±17.6	4.9±4.8
	P	1 vs 2 <0.001	2 vs 3 NS	3 vs 4 <0.05	
		1 vs 3 <0.01	2 vs 4 NS		
Cortisol secretion (μg/h)		283±193	143±190	179±193	134±134
	P	1 vs 2 <0.05	2 vs 3 NS	3 vs 4 NS	
		1 vs 3 <0.05	2 vs 4 NS		
Plasma [K] (mmol/l)		4.2±0.5	4.1±0.6	4.2±0.6	4.0±0.9
	P	No significant differences			
Plasma [Na] (mmol/l)		137±4	141±4	143±4	147±4
	P	1 vs 2 <0.001	2 vs 3 <0.001	3 vs 4 <0.001	
		1 vs 3 <0.001	2 vs 4 <0.001		
Adrenal blood flow (ml/min)		11.03±4.6	11.15±4.8	11.36±5.11	12.02±5.44
	P	1 vs 2 NS	1 vs 4 NS	2 vs 3 NS	
		1 vs 3 NS	3 vs 4 NS	2 vs 4 NS	

From Denton et al. (1977), with permission.

Statistical analyses comparing effects at different times were made by paired *t*-test ($n=22$ on 9 animals). Observations were made immediately before drinking, in a period 18–36 min after drinking, 36–140 min after drinking and 240 min or longer after drinking.

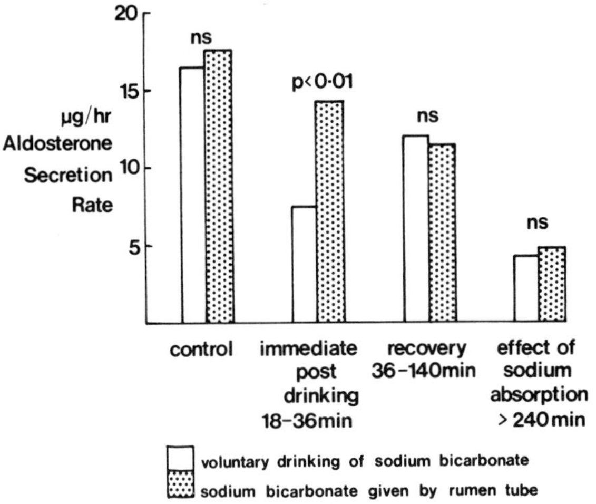

Fig. 21-8. Comparison of the effect on mean aldosterone secretion rate of correction of sodium deficiency by voluntary drinking of sodium bicarbonate solution (see Fig. 21-7) with the effect of administering approximately the same amount of sodium bicarbonate into the rumen by tube. The aldosterone secretion rate at 18–36 min following voluntary satiation of salt appetite was significantly less ($P<0.01$) than with rumen tube.

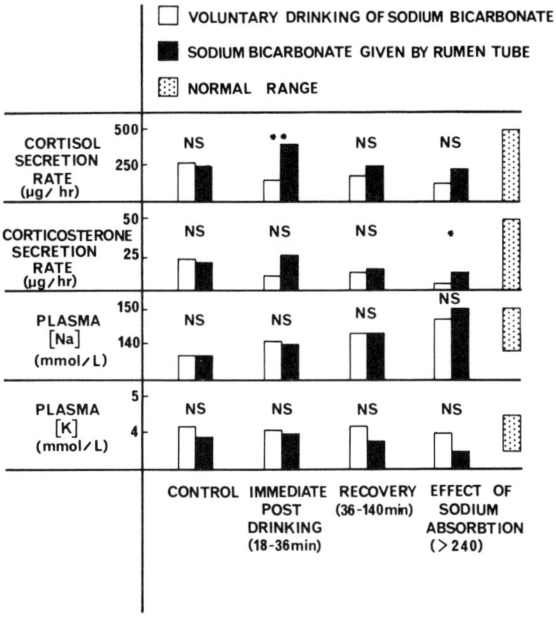

Fig. 21-9. Comparison of the effect on cortisol and corticosterone secretion rate and plasma [Na] and [K] of rapid correction of sodium deficiency by voluntary drinking of sodium bicarbonate (see Fig. 21-7) with the effect of administering approximately the same amount of sodium bicarbonate into the rumen by tube over 2–3 min. The cortisol secretion rate at 18–36 min was significantly less than with the rumen tube. The normal range of the parameters for the conscious sodium-replete sheep is shown by the blocks on the right.

reversible fall of aldosterone secretion occurred with rumen tube administration. There was a significant difference ($P<0.01$) in the fall of aldosterone secretion rate at the 18–36 min period (period 1) when compared with voluntary drinking

Table 21-2. The effect of administration by rumen tube of 450–600 mmol of sodium bicarbonate on the rate of secretion (mean±s.d.) by the autotransplanted adrenal gland of aldosterone, corticosterone and cortisol, the rate of adrenal blood flow and the plasma [Na] and [K].

		Period 1 (control)	Period 2 (18–36 min)	Period 3 (36–140 min)	Period 4 (>240 min)
Aldosterone secretion (µg/h)		17.6±8.3	14.2±7.1	11.5±7.0	4.8±3.4
	P	1 vs 2 <0.05	2 vs 3 <0.01	3 vs 4 <0.05	
		1 vs 3 <0.01	2 vs 4 <0.01		
Corticosterone secretion (µg/h)		20.3±14.5	24.3±20.5	16.2±16.2	12.9±11.6
	P	No significant differences			
Cortisol secretion (µg/h)		260±188	398±253	246±260	220±175
	P	No significant differences			
Plasma [K] (mmol/l)		3.8±0.4	3.9±0.4	3.7±0.4	3.6±0.3
	P	1 vs 2 NS	2 vs 3 <0.05	3 vs 4 NS	
		1 vs 3 NS	2 vs 4 <0.05		
Plasma [Na] (mmol/l)		137±5	140±5	143±5	149±6
	P	1 vs 2 <0.001	2 vs 3 <0.01	3 vs 4 <0.001	
		1 vs 3 <0.001	2 vs 4 <0.001		
Adrenal blood flow (ml/min)		9.05±4.45	9.30±4.31	9.23±2.59	9.99±3.51
	P	No significant differences			

From Denton et al. (1977), with permission.
Statistical analyses comparing effects at different times were made by paired *t*-test (n=9 on 6 animals). Observations were made immediately before administration by rumen tube, in a period 18–36 min after administration, 36–140 min after administration and 240 min or longer after administration.

but no difference at the other periods. The decrease with rumen tube administration was progressive, reaching near the sodium-replete range after 240–480 min. Table 21-2 shows that this progressive fall in aldosterone was not associated with any significant change in cortisol or corticosterone secretion rate, or in the rate of adrenal blood flow. There was a small decrease in plasma [K] which involved a significant change between periods 2 and 3, and 2 and 4. Plasma [Na] increased significantly in a progressive fashion essentially similar to that seen when sodium bicarbonate was voluntarily ingested.

Fig. 21-9 records statistical comparison by *t*-test of the changes with time of variables other than aldosterone. There were no relevant differences in the two procedures of correction of sodium deficiency except with the secretion rate of cortisol ($P<0.01$), where satiation by voluntary drinking produced a lower level in the 18–36 min period.

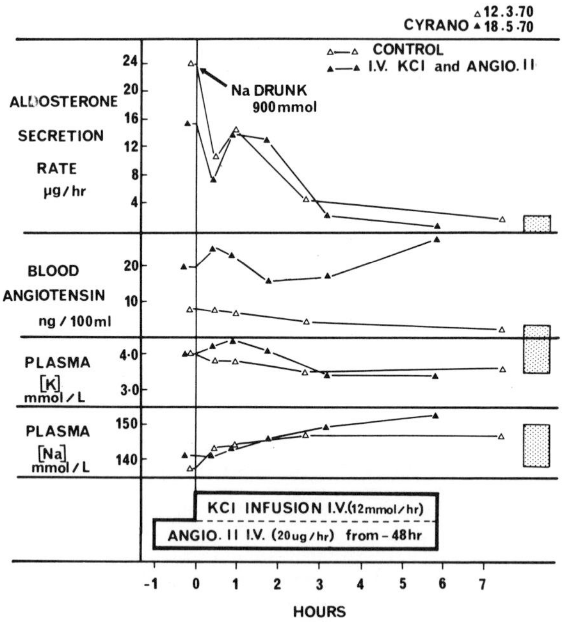

Fig. 21.10. Cyrano, sodium-deficient sheep with an adrenal transplant. The effect on aldosterone secretion rate, plasma [K], plasma [Na] and blood angiotensin II concentration of rapid voluntary drinking of 900 mmol of sodium bicarbonate solution. A control experiment (open triangles) is compared with an experiment (filled triangles) where an intravenous angiotensin II infusion at 20 μg/h, begun at −48 h when sodium deficiency commenced, was continued at constant rate after drinking, and an intravenous potassium chloride infusion (12 mmol/h) was begun also as drinking commenced. (From Denton et al. 1977, with permission.)

Experiments to examine the mechanism of rapid evanescent inhibition of aldosterone secretion

Angiotensin II concentration in arterial blood

The development of the radioimmunoassay method for assay for angiotensin II in peripheral blood in 1968 allowed study to determine whether the satiation inhibition of aldosterone secretion was caused by a precipitate fall in the peripheral blood level of angiotensin II.

Fig. 21-10 records two experiments on Cyrano when severely sodium-deplete. In both instances the animal was allowed to drink 3 litres of 300 mmol/l sodium bicarbonate solution over 5–15 min. In the control experiment under usual conditions of observation, aldosterone secretion rate fell by over 50% from 24 μg/h to 10 μg/h by 25 min and then rebounded to 14 μg/h by 60 min. Systemic arterial blood angiotensin II concentration was unchanged. Plasma [K] fell 0.2 mmol/l in 25 min and remained at this level at the time of rebound (60 min). There was a rise of 5 mmol/l in plasma [Na] by 25 min and by 60 min it was 6 mmol/l higher than the pre-drinking concentration. Cortisol secretion fell from the low level of 153 μg/h to 39 μg/h at 25 min and corticosterone fell from 28 μg/h to 5 μg/h. Aldosterone fell to the sodium-replete range in the 140–400 min period, associated with a small decline in angiotensin II to the sodium-replete range, a further 0.3 mmol/l fall in plasma [K] and a 2–3 mmol/l rise in plasma [Na].

In the second experiment the animal had received a continuous intravenous infusion of angiotensin II at 20 μg/h for the 48 h during which it had become sodium-deplete as a result of parotid salivary loss. This infusion was continued at the same rate during the post-drinking period. Furthermore, 2–3 min before sodium bicarbonate solution was presented to the animal, an intravenous infusion of potassium chloride was begun at 12 mmol/h and continued at this rate throughout the post-drinking period. In this experiment the sheep drank the same amount of sodium bicarbonate in the same time as in the first experiment. Aldosterone secretion rate fell 50% in 22 min and rebounded to a larger extent to just below pre-drinking level at 60 min. The secretion rate was still close to this level at 100 min but then fell rapidly to be at the sodium-replete range by 180–330 min. Peripheral blood angiotensin II concentration, *initially at a high concentration near 20 ng/100 ml increased over the period of the satiation inhibition as did the plasma [K].*

Angiotensin did not fall any substantial amount below control during the infusion, and in fact the level was highest in the last measurement though aldosterone secretion rate was then at the sodium-replete level. Plasma [K] was increased or unchanged in the first 100 min and then fell despite continuation of the intravenous potassium chloride infusion. Plasma [Na] did not change during the satiation inhibition of aldosterone secretion at 22 min, but rose slightly, coincident with the rebound rise of aldosterone secretion rate at 60 min. Cortisol and corticosterone secretion rates were low and fell a small amount with the satiation inhibition.

A similar experiment was made on Ivor (Fig. 21-11), except that the rate of intravenous potassium chloride infusion was 6 mmol/h. The post-satiation inhibition of aldosterone secretion was observed at 24 min (Fig. 21-11), with a smaller rebound at 60 min than other experiments with

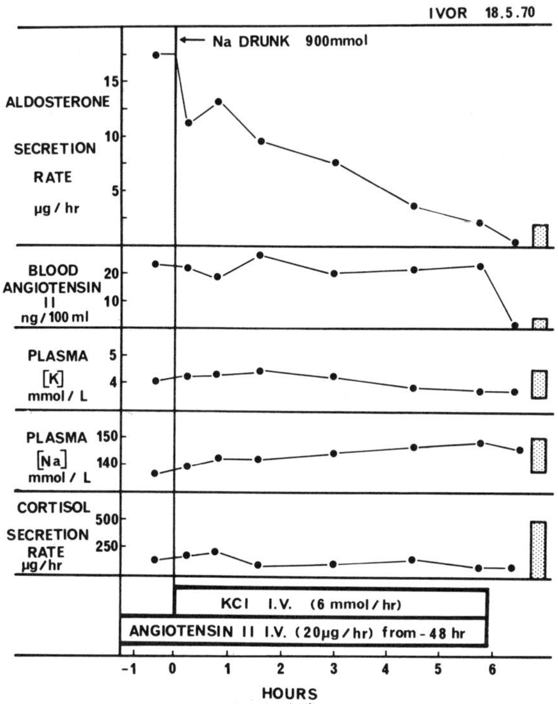

Fig. 21-11. Ivor, sodium-deficient sheep with an adrenal transplant. The effect of rapidly drinking 900 mmol of sodium bicarbonate solution (300 mmol/l) on the rate of secretion of aldosterone and cortisol. The peripheral blood angiotensin II concentration and the systemic peripheral plasma [Na] and [K] are shown also. The normal sodium-replete range of these parameters is shown by the blocks on the right. The experiment was conducted under conditions where an intravenous infusion of angiotensin II amide at 20 μg/h was begun at -48 h and continued for 350 min in the post-drinking period. An intravenous infusion of potassium chloride was begun immediately before offer of sodium bicarbonate solution and continued for 350 min also. (From Denton et al. 1977, with permission.)

Ivor. Thereafter, a progressive fall of aldosterone secretion to the sodium-replete range followed over 330 min. Angiotensin II concentration in peripheral blood was unchanged during the satiation inhibition, and fell a very small amount during the rebound rise of aldosterone secretion. Blood angiotensin II level then was essentially unchanged until the infusion was stopped at 350 min, even though aldosterone fell to the sodium-replete level. Angiotensin level measured at 360 min, 10 min after infusion stopped, was in the lower part of the sodium-replete range. Aldosterone secretion rate was very low also. Plasma [K] level was sustained at a slightly increased level relative to pre-drinking concentration for 130 min and then fell a small amount. Plasma [Na] rose 2 mmol/l associated with the satiation inhibition, and continued to rise progressively by a small amount over 330 min. Cortisol secretion rose slightly during the satiation inhibition but remained in the low normal range throughout the experiment. Corticosterone secretion rate fell from 18 μg/h to 5 μg/h with drinking and rebounded to 14 μg/h by 60 min.

Two other experiments on Digbeigh and Ivor with the same protocol of intravenous angiotensin II and potassium chloride infusions produced essentially similar results. With Digbeigh, satiation produced a 50% fall in aldosterone measured at 30 min. Rebound to control level occurred at 60 min. Arterial blood angiotensin II and plasma [K] were unchanged and a 1 mmol/l rise in plasma [Na] was observed at 30 min and a 2 mmol/l rise by 60 min. With Ivor, a 64% fall in aldosterone secretion rate at 20 min with rebound to 80% of pre-drinking control at 60 min, was associated with no change in angiotensin II, plasma [K] or renin concentration at 20 min, or at 60 min. Plasma [Na] rose 2 mmol/l by 20 min and 4 mmol/l by 60 min.

Plasma sodium concentration and neural effects on the adrenal gland

Further experiments were made to exclude a change in plasma [Na] as a cause of the phenomenon (Fig. 21-12). Commencing 100 min before presentation of sodium bicarbonate solution, an adrenal arterial infusion of hypertonic sodium chloride was begun to TP9, adequate to raise adrenal plasma [Na] by 3 mmol/l. This resulted in aldosterone secretion falling from 13 to 11 μg/h by 20 min, but it returned to 15 μg/h immediately before presenting sodium bicarbonate. Following presentation and drinking of the sodium solution, the characteristic fall in aldosterone secretion by 55% occurred by 30 min. By 60

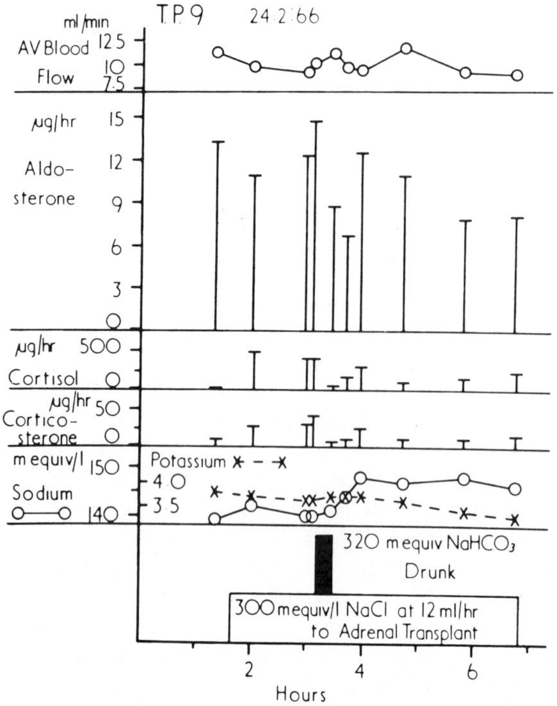

Fig. 21-12. TP9, sodium-deficient sheep with an adrenal transplant. The effect of rapid drinking of sodium bicarbonate solution on the rate of aldosterone, cortisol and corticosterone secretion when, before presentation of the solution, a slow adrenal arterial infusion of hypertonic sodium chloride was commenced to increase locally the plasma [Na] of adrenal arterial blood. The 'psychic' inhibition of aldosterone occurred as in control experiments. (From Blair-West et al. 1972, with permission.)

min the rate had returned to 85% of pre-drinking level. The preceding and continuing adrenal arterial infusion of sodium chloride had no discernible effect on the phenomenon. Similar experiments on Warner and TP12 gave the same result.

A histological and histochemical study of adrenal transplants made up to 10 years after the operation has shown that reinnervation of the transplant with sympathetic catecholaminergic nerves occurs. Clear-cut evidence of cholinergic reinnervation has not been forthcoming (Wright et al. 1972). The medulla survives and has a histologically normal appearance. Against this background, an experiment was made on TP9, an animal with a normal adrenal medulla, involving extensive infiltration of the tissues around the adrenal transplant and the vascular anastomosis area with 2 ml of 5% procaine hydrochloride 30 min before sodium bicarbonate solution was offered. This procedure caused an increase in cardiac rate and secretion of ACTH as evidenced by a rise in cortisol and corticosterone over this 30-min period. Nevertheless, rapid drinking of 450 mmol of sodium bicarbonate solution reduced aldosterone secretion from 18 μg/h to the near sodium-replete level of 3.0 μg/h by 18 min. There were also large falls in cortisol and corticosterone secretion rate from 611 to 35 μg/h and 59 to 3 μg/h respectively. Aldosterone secretion rebounded to 8 μg/h by 60 min, though cortisol and corticosterone secretion rates remained low at 57 and 5 μg/h respectively. Ionic changes during the inhibition were minimal: 0.1–0.2 mmol/l fall in plasma [K]. A second experiment on TP9 had a similar result.

The effect of dexamethasone suppression of ACTH secretion on aldosterone secretion in the sodium-deplete animal

An initial experiment was made on Digbeigh, an animal which exhibited the characteristic inhibition of aldosterone secretion with drinking. Following intravenous administration of a large dose (8 mg) of dexamethasone (Decadron) (Fig. 21-13), corticosterone and cortisol fell in 20 min, and by 60 min were secreted at a very low basal rate. This indicated a large suppression of ACTH, which persisted over 7 h. This suppression had no effect on the hypersecretion of aldosterone caused by sodium deficiency, and nor did a further 4 mg injection of Decadron given 5 h later have any effect.

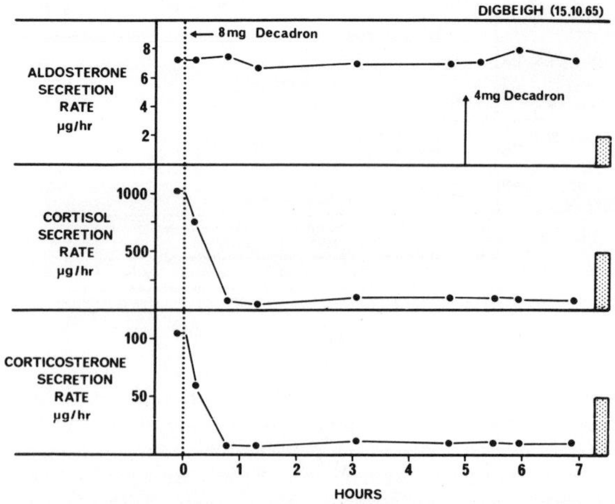

Fig. 21-13. Digbeigh, sodium-deficient sheep with an adrenal transplant. The effect on aldosterone, cortisol and corticosterone secretion rate of the intravenous injection of 8 mg of Decadron (dexamethasone phosphate, MSD) followed 5 h later by injection of a further 4 mg. The blocks on the right show the normal range of these parameters in the conscious sodium-replete animal. (From Denton et al. 1977, with permission.)

Following this, five other experiments were made in which an intravenous infusion of Decadron was given at a stepwise rate of 1, 4 and 12 mg/h. The effect on corticosteroid secretion was observed at 2, 4 and 6 h, which followed 2 h of infusion at the given step of rate. Infusion at 1 mg/h reduced cortisol to the same extent as voluntary drinking and had no effect on aldosterone. The higher infusion rates, as in Fig. 21-13 above, reduced mean cortisol to approximately 30% of control, and halved mean corticosterone secretion rate. After Decadron infusion rate of 12 mg/h for 2 h, mean aldosterone secretion rate was reduced by 24%.

Implications of gustatory-alimentary inhibition of aldosterone hypersecretion in relation to current ideas of control of aldosterone secretion

The large inhibition of aldosterone secretion occurring shortly after rapid drinking of sodium bicarbonate solution, indicates that oral afferents and central nervous events associated with rapid satiation initiated the phenomenon. The effect on aldosterone secretion, which may be very large, is in the same direction as the post-absorptive effect of the sodium bicarbonate ingested. In the light of the known metabolic clearance rate of aldosterone (Tait et al. 1962; Scoggins 1968), the duration of the fall in aldosterone secretion (15–45 min) would have produced a significant fall also in peripheral blood aldosterone concentration at target organs. However, taking into consideration the substantial rebound effect occurring by 60 min, the influence on the overall progress of the fall in aldosterone concentration in peripheral blood was probably not large.

Thus the physiological importance of the mechanism is not comparable to that of, for example, the poto-hidrotic reflex following oral intake of water in dehydrated subjects (Nicolaidis 1969). Also, comparable to the variability of the rapid response reported in blood glucose levels following oral stimulation with sweet solutions, it should be noted that not all sheep investigated showed the response. It occurred in about 50%–60% of animals studied. Generally, these were animals with the moderate to high levels of daily salivary sodium loss, which repaired deficit rapidly and had been doing this repetitively over several years. In individual animals, the phenomenon was reproduced in a consistent fashion over as long as 3 to 7 years of study. The striking association with the behavioural stereotype of the excited animal engaged in the rapid consummatory act over 2–5 min (Denton and Sabine 1963) is suggestive that the aldosterone inhibition may be a conditioned neurohumoral reaction. It may take a long period to develop. Our first observations of the phenomenon involved animals which had already had a parotid fistula and adrenal transplant for 5 years and had been through the behavioural sequence many hundreds of times.

Given these considerations on the nature of the phenomenon and its rather limited homeostatic role it can, however, be noted that it is the first instance of behavioural-induced inhibition of aldosterone secretion reported in the conscious animal. There have been a number of reports suggestive of an increase in aldosterone secretion generated by central nervous mechanisms, usually apprehension (Lamson et al. 1956; Venning et al. 1957; Elmadjian 1962; Blair-West et al. 1970a).

There is an aspect of great interest in this study. This is the question of the pathway of effect on aldosterone secretion. The experimental evidence would appear to rule out an evanescent precipitate fall in either plasma [K] or arterial blood angiotensin II as the causal vector. No significant change in plasma [K] occurred. Furthermore, there was satiation inhibition despite a contemporaneous small increase in plasma [K] contrived by intravenous potassium chloride infusion. The same considerations apply to angiotensin II. This is generally agreed to be essential for aldosterone hypersecretion in sodium deficiency and some investigators (Davis 1975; Laragh 1975) assign it the direct commensurate causal role. Again, intravenous infusion also contrived a small increase in concentration contemporaneous with the satiation inhibition of aldosterone.

There is ancillary evidence that interference with the action of arterial blood angiotensin II had little or nothing to do with the satiation inhibition. It has been shown that infusion of [Sar^1, Ala^8]angiotensin II (P113) or other antagonists (Blair-West et al. 1976, 1979) into the arterial supply of the transplanted adrenal of the sodium-replete sheep blocks the aldosterone response to infused angiotensin II when the ratio of antagonist to agonist is 100-200:1. However, infusion of P113 or other antagonists in a ratio of up to 5000–10 000:1 of the contemporary angiotensin II level in the sodium-deplete animal does not cause any significant reduction of aldosterone secretion (Blair-West et al. 1976, 1979). The agonist activity of these antagonists is such that in the sodium-replete animal the relevant doses increase aldosterone secretion by about 10% of the effect caused by sodium deficiency. Thus if the angiotensin-like agonist activity of the antagonist is what sustains high aldosterone secretion it follows that all that is required is a permissive level of influence of angiotensin or little more.

Plasma [Na] usually, but not invariably, rose 2–3 mmol/l at the time of the satiation inhibition. The change was highly significant. However, a further similar rise occurred during the course of the rebound rise of aldosterone. The direct effect of arterial plasma [Na] on aldosterone secretion has been studied extensively by adrenal arterial hypertonic sodium bicarbonate and sodium chloride infusions (Blair-West et al. 1966, 1971). The small 2–3 mmol/l rise would have very small or no effect in contrast to the 30%–80% fall in aldosterone seen with the satiation inhibition. That is, a 7 mmol/l rise in adrenal arterial [Na] caused a 9% fall in aldosterone secretion rate (Blair-West et al. 1971). Also, no satiation inhibition occurred in the instance of intraruminal administration of sodium bicarbonate where the same significant rise in plasma [Na] occurred. The commencement of an adrenal arterial infusion of hypertonic sodium chloride which contrived a similar rise in adrenal arterial plasma [Na] before and during drinking, did not influence the occurrence of the satiation inhibition.

Having eliminated the effects of angiotensin, and of plasma [K] or [Na], then a rapid fall in plasma ACTH, the fourth factor recognized to be involved in control of aldosterone secretion, remains to be considered. The fall in aldosterone was associated with a significant fall in cortisol, and a highly significant fall in corticosterone. These changes persisted, however, at a significant level of reduction during the period of rebound rise of aldosterone secretion at 60 min. The observation could be indicative of a large fall in ACTH secretion or, alternatively, that the humoral inhibitory agent which reduces aldosterone secretion impinges on the biosynthetic pathway in a manner which reduces production of these two steroids as well as aldosterone—that is, at an early site (Muller 1971).

ACTH does play an important role in aldosterone secretion in experiments conducted under anaesthesia and with surgical trauma and blood loss (Davis et al. 1960; Davis 1961; Blair-West et al. 1964). However, it is established by a great deal of data (Forsham et al. 1961; Blair-West et al. 1963; Ganong et al. 1966) that ACTH stimulation of aldosterone secretion commences when the systemic level is adequate to have stimulated cortisol and corticosterone secretion to moderately high levels: 35%–50% of adrenal functional capacity to produce these steroids in the case of the sheep. Thus in these experiments on trained sodium-deficient animals, the control rates of cortisol and corticosterone secretion were low, suggestive that ACTH was playing little or no part in the stimulation of aldosterone hypersecretion unless an exquisite hypersensitivity to ACTH accompanied the sodium-deficient state (Muller et al. 1956; Thorn et al. 1957; Blair-West et al. 1963; Ganong et al. 1966).

Whereas a virtual cessation of ACTH secretion as a result of the central nervous events of satiation may be possible, there was considerable evidence here against such an event. However, even if it occurred, there was evidence against it having caused the satiation inhibition. Thus, a fall in the glucocorticoids was not an invariable concurrent event. In several instances, cortisol secretion rose during the satiation inhibition whereas corticosterone secretion almost always did fall. Secondly, hypophysectomy has a small effect only on the aldosterone secretory response to sodium deficiency in the sheep (Blair-West et al. 1968). Thirdly, when dexamethasone was administered to reduce ACTH, there was no reduction in aldosterone secretion at the stage cortisol was reduced to the same degree as seen to occur concurrently during the satiation inhibition of aldosterone secretion. Further, a very high dose of dexamethasone which reduced cortisol and corticosterone to very low levels had a very small effect only on aldosterone, which was not comparable to the satiation inhibition.

Overall, the data indicate that the central nervous events associated with rapid satiation of salt appetite can cause a large reduction in aldosterone secretion by an unknown mechanism. The failure of procaine infiltration around the transplanted adrenal to alter the effect is consistent with evidence against neural effects on the corticosteroid secretion of the adrenal (Fleming and Farrell 1956; Denton et al. 1959; Davis 1961). Thus, events in the brain could influence zona glomerulosa secretion in sodium deficiency either by release of an adrenal inhibitor from the brain or by the decrease of secretion of an adrenal stimulating agent. Alternatively, neural outflow from the brain could influence such secretory change in another organ with subsequent humoral effect on the adrenal.

The data support a number of lines of evidence accumulated since 1966, indicative of a fifth factor in the aldosterone control system (Blair-West et al. 1966, 1970b, 1972; Chin et al. 1970; Muller 1971; McCaa et al. 1973; Fichman et al. 1974; Palkovits et al. 1974; Boulter et al. 1974).

Two other lines of experimental enquiry are appropriate to mention here within this general theme of brain mechanisms and effects on adrenal secretion of aldosterone. In the first, a colony of sheep were prepared with permanent indwelling cannulae in the lateral ventricles and cisterna magna, as described by Pappenheimer et al. (1962). Following depletion of 400–500 mmol of sodium, the ventricles of the conscious animals were perfused with artificial cerebrospinal fluid (CSF) with high [Na] and the results compared with control experiments. Perfusion was made for 9 h and in the three groups of experiments the initial sodium deficit was similar. Group I (n=5) received perfusion with artificial CSF containing 170 mmol/l of sodium, Group II (n=7) received perfusion with artificial CSF containing 145 mmol/l of sodium, and Group III (n=7) received no perfusion.

The results (Abraham et al. 1976) were as follows. In Group I, the blood aldosterone fell from 26.4±7.4 to 8.6±2.3 ng/100 ml by 9 h after perfusion (Fig. 21-14 and Table 21-3). There were

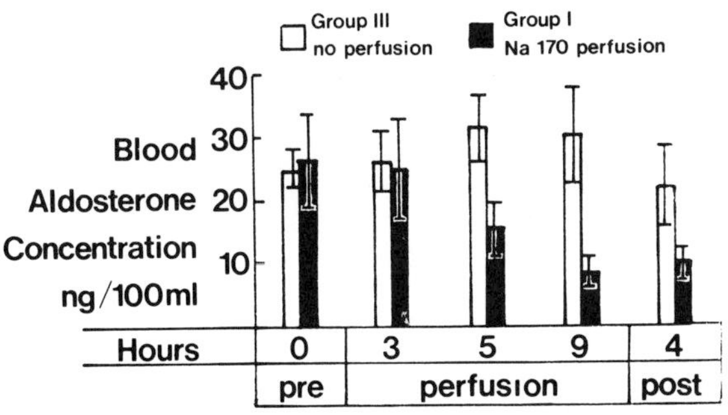

Fig. 21-14. The effect of ventriculo-cisternal perfusion of sodium-deficient sheep for 9 h at 1.2 ml/min on blood aldosterone concentration. Group I was perfused with an artificial cerebrospinal fluid containing 170 mmol/l sodium (n=5). Group III received no perfusion (n=7). (From Abraham et al. 1976, with permission.)

no significant changes in arterial plasma [Na] or [K], blood angiotensin II or plasma renin concentration (Table 21-4). Blood cortisol and corticosterone rose, indicative that the influence of ACTH, if any, would have been to increase aldosterone secretion (Table 21-3). There was also a fall in blood aldosterone in Group II. However, this was not significant until 4 hours post-perfusion and was correlated with a statistically significant fall in plasma [K]. Group III showed no significant change in aldosterone concentration (Fig. 21-14). Multivariate statistical analysis showed that the fall in aldosterone levels during perfusion with 170 mmol/l sodium could not be accounted for by changes, either alone or together, of ACTH as evidenced by alteration of blood cortisol or corticosterone, or by plasma [Na], [K] or renin concentrations.

Thus, the data derived from an entirely different approach from that involved in study of the process of satiation of sodium appetite, suggest also an unidentified central nervous influence in aldosterone control. They are also consistent, in the light of the results with the satiation study, with the possibility of a central nervous mechanism subserving appetite also influencing aldosterone—though, taken by itself, no such suggestion would be warranted.

Secondly, apart from data already cited, suggestive that states of apprehension may influence aldosterone secretion, we have made a few observations indicative that customary laboratory events premonitory of the impending presentation of sodium solution can influence aldosterone secretion rate in sodium-deficient sheep. Fig. 21-15 is representative of these observations.

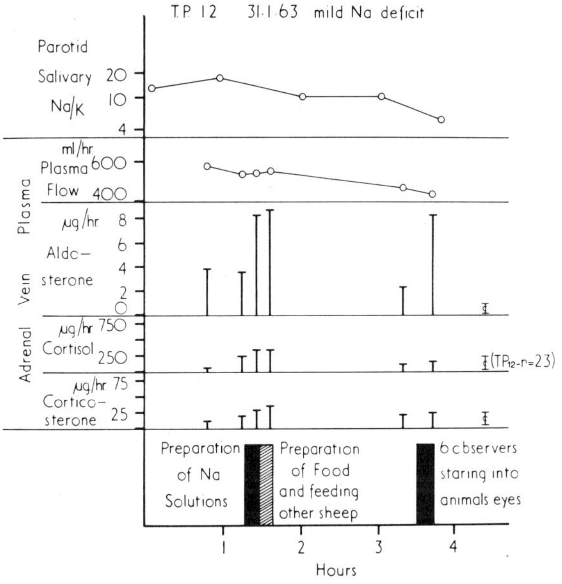

Fig. 21-15. TP12, mildly sodium-deficient sheep with an adrenal transplant. The effect of environmental events on adrenal secretion of aldosterone, cortisol and corticosterone and adrenal plasma flow rate. First sodium solutions were prepared in the laboratory within sight and hearing of the animal, and then six observers closely gathered around the cage staring into the animal's eyes.

Here, preparation of sodium solutions rapidly doubled the aldosterone secretion rate of a mildly deficient animal; there was a small effect only on cortisol and corticosterone secretion rate, which remained in the normal range. Two hours following this procedure, to which we had added the additional exciting factor of feeding other sheep without causing further increase of aldosterone, the aldosterone secretion rate had returned to control level. Now it was arranged that six observers sat around the front of the sheep's cage

Table 21-3. Effect of ventriculo-cisternal perfusion on blood aldosterone, cortisol and corticosterone levels. Results expressed as mean±s.e. Group I, perfusion with 170 mmol/l sodium (n=5); Group II, perfusion with 145 mmol/l sodium (n=7); Group III, no perfusion (n=7).

	Pre	3 h	5 h	9 h	Post
Aldosterone (ng/100 ml)					
Group I	26.4±7.4	24.8±8.1	15.3±4.3	8.6±2.3	10.1±2.6
II	23.3±3.6	26.5±3.7	20.0±2.7	16.7±2.8	11.5±3.7
III	25.1±3.1	26.0±4.7	31.4±5.2	30.4±7.6	22.2±6.3
Cortisol (μg/100 ml)					
Group I	0.91±0.29	1.99±0.48	2.53±0.55	2.35±0.38	2.74±0.62
II	1.19±0.27	1.78±0.57	4.19±0.96	4.17±1.00	2.19±0.37
III	1.05±0.25	0.35±0.12	0.42±0.10	0.79±0.35	0.26±0.05
Corticosterone (μg/100 ml)					
Group I	0.06±0.01	0.10±0.03	0.12±0.04	0.09±0.03	0.05±0.03
II	0.08±0.02	0.11±0.03	0.25±0.07	0.20±0.05	0.08±0.04
III	0.09±0.02	0.06±0.01	0.06±0.01	0.06±0.02	0.03±0.01

From Abraham et al. (1976), with permission.

Table 21-4. Effect of ventriculo-cisternal perfusion on plasma [K]. [Na], renin concentration and blood angiotension II. Results expressed as mean±s.e. Group I, perfusion with 170 mmol/l sodium (n=5); Group II, perfusion with 145 mmol/l sodium (n=7); Group III, no perfusion (n=7).

	Pre	3 h	5 h	9 h	Post
Plasma [K] (mmol/l)					
Group I	4.2±0.2	4.1±0.2	3.9±0.2	3.8±0.2	3.8±0.1
II	4.3±0.2	4.2±0.1	3.9±0.2	3.5±0.1	3.5±0.2
III	4.5±0.2	4.5±0.2	4.4±0.2	4.1±0.1	3.9±0.1
Plasma [Na] (mmol/l)					
Group I	142±2	142±2	141±2	142±2	143±2
II	142±1	142±2	142±1	143±2	144±2
III	143±1	143±1	143±1	144±1	144±1
Plasma renin (ng h^{-1} ml^{-1})					
Group I	2.7±0.5	2.5±0.5	3.0±0.5	3.4±0.7	3.6±0.6
II	3.2±0.6	3.5±0.5	3.2±0.6	3.6±0.7	3.5±0.5
III	3.6±0.5	3.9±0.4	3.8±0.5	3.9±0.7	4.4±0.9
Blood angiotensin II (ng/100 ml)					
Group I	8.3±1.7[a]	—	7.8±1.5	10.1±1.6	15.5±1.1[a]
II	11.4±3.1[b]	—	13.5±3.8	12.0±2.8	15.3±3.5
III	13.2±2.6[c]	—	12.4±2.3	12.1±2.5	13.4±1.4

From Abraham et al. (1976), with permission.
[a]n=4; [b]n=5; [c]n=6.

and silently stared directly into its eyes—a procedure which the sheep appears to find disturbing, as evidenced by episodes of yawning and shifting its feet. Schaller (1963) has also observed in watching gorillas that they do not like a direct stare. Surprisingly, no effect on cortisol and corticosterone secretion by the transplanted adrenal occurred, but aldosterone secretion rate almost quadrupled. The effect had reversed approximately 1 h after ceasing the staring procedure. Again, the data are congruent with other material (Blair-West et al. 1970b, 1972) suggestive of a non-ACTH-mediated central nervous influence on aldosterone secretion.

It should be noted here that Williams and colleagues (Williams et al. 1972; Tuck et al. 1974) have made carefully controlled experiments on the correlation of renin, angiotensin II and plasma aldosterone in humans when aldosterone hypersecretion as a result of a low sodium diet (10 mmol/day) was corrected by intravenous isotonic saline infusion, or by drinking of 3 litres of isotonic bouillon over 3 h. The results were compared with the effects of infusion of dextran in 5% glucose, and 5% glucose infusion. The major point emerging from the study was that with both routes of sodium intake, renin and angiotensin decrease was highly significantly correlated with aldosterone decrease and large decrease occurred within an hour. Also there was no disjunction with dextran, which caused expansion of plasma volume at the expense of interstitial fluid volume. No decline in renin or aldosterone occurred within 60 min, but a decline occurred later. In the sodium-deplete sheep it has been observed that rapid expansion of plasma volume with dextran does not cause significant reduction of aldosterone secretion.

In their discussion, Williams et al. (1972) do take into account that there may be substantial difference in the study of a human in positive sodium status on a low-sodium diet, and a sheep in negative sodium status having been substantially depleted of body sodium by a parotid fistula. Furthermore, the circumstances of a patient being persuaded to drink 3 litres of isotonic bouillon over 3 h does not involve the same central nervous events as an excited severely sodium-deplete sheep drinking 2–3 litres of 300 mmol/l sodium bicarbonate in 3–5 min. As indicated above, no change in plasma [K] occurred in the sheep over the time course of the 'psychic inhibition' of aldosterone secretion, and, in any event, the fall occurred even if intravenous potassium chloride infusion were given concurrently. Williams and colleagues (1972) suggest that in the circumstances of simple 'correction' of sodium status from that of a low-sodium diet, there is no dissociation of the relation between renin and aldosterone, and that plasma aldosterone is dominantly under renin control in these circumstances. However, as with sheep, where we would propose the experiments have been conducted over a quite different quantitative range, there are also human observations where plasma renin and aldosterone secretion are dissociated and other factors appear operative (Collins et al. 1970; Chin et al. 1970; Luetscher et al. 1969).

Overall, the data from these experiments show that two of the major systems involved in sodium homeostasis, though not intimately linked, are functionally interrelated to some extent. In the light of substantial evidence from other investigations pointing to the possibility of sodium sensors in or near the walls of the ventricular system having direct influence on sodium excretion (Abraham et al. 1975: see Fig. 13-11; Leksell et al. 1981: see Fig. 20-15) and that a natriuretic hormone may be involved, one speculative indulgence is that the neural events of satiation of salt appetite release such a substance (anticipatory of metabolic change analogous to other instances described above), and that this natriuretic substance also acts as an inhibitor of aldosterone secretion.

The implications of the data reported here in relation to control of aldosterone secretion are important. Regrettably there is no comparable experimental material to date from any other species and there are clearly formidable technical problems in trying to derive such information in any of the other species so far studied in relation to control of aldosterone secretion. Parallel to the fact that salt appetite mechanisms can influence aldosterone secretion, it is discussed elsewhere in this book (Chapter 18) that aldosterone secretion may have a measure of influence on salt appetite in some species.

References

Abraham SF, Blaine EH, Denton DA, McKinley MJ, Nelson JF, Shulkes AA, Weisinger RS, Whipp GT (1975) Phylogenetic emergence of salt taste and appetite. In: Denton DA, Coghlan JP (eds) Olfaction and taste V. Academic Press, New York, p 253

Abraham SF, Blair-West JR, Coghlan JP, Denton DA, Mouw DR, Scoggins BA (1976) Aldosterone secretion during high sodium cerebrospinal fluid perfusion of the brain ventricles. Acta Endocrinol 81: 120

Adolph EF (1939) Measurement of water drinking in dogs. Am J Physiol 125: 75

Andersen DK, Dariush E, Brown JC, Tobin JD, Andres R (1978) Oral glucose augmentation of insulin secretion. J Clin Invest 62: 152

Andersson B (1978) Regulation of water intake. Physiol Rev 58: 3

Andersson B, Olsson K (1973) On central control of body fluid homeostasis. Conditional Reflex 8: 147

Anokhin PK, Sudakov KV (1966) Sensory mechanisms of satiety. In: Proceedings of the Seventh International Congress on Nutrition, Vol 2. Regulation of hunger and satiety. Hamburg

Blair-West JR, Coghlan JP, Denton DA, Goding JR, Wintour M, Wright RD (1963) The control of aldosterone secretion. Recent Prog Horm Res 19: 311

Blair-West JR, Coghlan JP, Denton DA, Munro JA, Wintour M, Wright RD (1964) The effect of bilateral nephrectomy and midcollicular decerebration with pinealectomy and hypophysectomy on the corticosteroid secretion of sodium-deficient sheep. J Clin Invest 43: 1576

Blair-West JR, Coghlan JP, Denton DA, Goding JR, Wintour M, Wright RD (1966) The direct effect of increased sodium concentration in adrenal arterial blood on corticosteroid secretion in sodium-deficient sheep. Aust J Exp Biol Med Sci 44: 455

Blair-West JR, Coghlan JP, Denton DA, Goding JR, Wintour M, Wright RD (1968) The effect of nephrectomy on aldosterone secretion in the conscious sodium-depleted hypophysectomised sheep. Aust J Exp Biol Med Sci 46: 295

Blair-West JR, Coghlan JP, Denton DA, Funder JW, Nelson JF, Scoggins BA, Wright RD (1970a) Sodium homeostasis, salt appetite and hypertension. Circ Res [Suppl] 26/27: 11–251

Blair-West JR, Coghlan JP, Denton DA, Scoggins BA, Wintour M, Wright RD (1970b) Effect of change of sodium balance on the corticosteroid response to angiotensin II. Aust J Exp Biol Med Sci 48: 253

Blair-West JR, Coghlan JP, Denton DA, Funder JW, Scoggins BA, Wright RD (1971) The effect of adrenal arterial infusion of hypertonic $NaHCO_3$ solution on aldosterone secretion in sodium deficient sheep. Acta Endocrinol (Copenh) 66: 448
Blair-West JR, Coghlan JP, Denton DA, Funder JW, Scoggins BA (1972) The role of the renin–angiotensin system in control of aldosterone. Adv Exp Med 17: 167
Blair-West JR, Coghlan JP, Denton DA, Niall HD, Scoggins BA, Tregear GW (1976) Effect of angiotensin II analogues on aldosterone secretion in sodium-deficient sheep. In: Sambhi MP (ed) Systemic effects of antihypertensive agents. Stratton Intercontinental Medical Book Corporation, New York, p 441
Blair-West JR, Coghlan JP, Denton DA, Hardy KJ, Scoggins BA, Wright RD (1979) Effect of adrenal arterial infusion of P113 (Sar^1Ala^8 angiotensin II) on aldosterone secretion in sodium deficient sheep. Am J Physiol 236: F333–F341
Boulter PR, Spark RF, Arky RA (1974) Dissociation of the renin–aldosterone system and refractoriness to the sodium-retaining action of mineralocorticoid during starvation in man. J Clin Endocrinol 38: 248
Chin RH, Brown JJ, Fraser R, Heron SM, Lever AF, Murchison L, Robertson JIS (1970) The natriuresis of fasting: Relationship to changes in plasma renin and plasma aldosterone concentrations. Clin Sci 39: 437
Collins RD, Weinberger MH, Dowdy AJ, Nokes GW, Gonzales CM, Luetscher JA (1970) Abnormally sustained aldosterone secretion during salt loading in patients with various forms of benign hypertension: Relation to plasma renin activity. J Clin Invest 49: 1415
Davis JO (1961) Mechanisms regulating the secretion and metabolism of aldosterone in experimental secondary hyperaldosteronism. Recent Prog Horm Res 17: 293
Davis JO (1975) Regulation of aldosterone secretion. In: Greep RO, Astwood EB (eds) Handbook of physiology, Section 7: Endocrinology, Vol VI. American Physiological Society, Washington DC, p 77
Davis JO, Yankopoulos NA, Lieberman F, Holman J, Bahn RC (1960) The role of the anterior pituitary in the control of aldosterone secretion in experimental secondary hyperaldosteronism. J Clin Invest 39: 765
Denton DA, Sabine JR (1963) The behaviour of Na deficient sheep. Behaviour 20: 364
Denton DA, Goding JR, Wright RD (1959) Control of adrenal secretion of electrolyte-active steroids. Br Med J ii: 447–456, 522–530
Denton DA, Orchard E, Weller S (1969) The relation between voluntary sodium intake and body sodium balance in normal and adrenalectomised sheep. Commun Behav Biol 3: 213
Denton DA, Blair-West JR, Coghlan JP, Scoggins BA, Wright RD (1977) Gustatory–alimentary reflex inhibition of aldosterone hypersecretion by the autotransplanted adrenal gland of sodium deficient sheep. Acta Endocrinol (Copenh) 84: 119
Elmadjian F (1962) Aldosterone excretion in behavioural disorders. In: Korey SR, Pope A, Robins E (ed) Ultrastructure and metabolism of the nervous system. Williams and Wilkins, Baltimore, p 414
Elrich H, Stimmler L, Hlad CI, Azai Y (1964) Plasma insulin response to oral and intravenous glucose administration. J Clin Endocrinol 24: 1076
Fichman MP, Michelakis AM, Horton R (1974) Regulation of aldosterone in the syndrome of inappropriate antidiuretic hormone secretion (SIADH). J Clin Endocrinol 39: 136
Fitzsimons JT (1972) Thirst. Physiol Rev 52: 468
Fleming R, Farrell G (1956) Aldosterone and hydrocortisone secretion by the denervated adrenal. Endocrinology 56: 360
Forsham PH, Di Raimondo VC, Biglieri EG, Li CH (1961) Adrenocorticotrophic activity of a synthetic nonadecapeptide (NDP) in man. Metabolism 10: 335
Ganong WF, Biglieri EK, Mulrow PJ (1966) Mechanisms regulating adrenocortical secretion of aldosterone and glucocorticoids. Recent Prog Horm Res 22: 381
Gilman A (1937) The relation between blood osmotic pressure, fluid distribution and voluntary water intake. Am J Physiol 120: 323
Holmes JH, Gregerson MI (1950) Observations on drinking induced by hypertonic solutions. Am J Physiol 162: 326
Kenney RA, Miller DM (1949) Effects of environmental temperature on water output and pattern of chloride excretion. Acta Med Scand 135: 87
Lamson ET, Elmadjian F, Hope JM, Pincus G, Jorjorian D (1956) Aldosterone excretion of normal, schizophrenic and psycho-neurotic subjects. J Clin Endocrinol 16: 954
Laragh J (1975) Report on round table on renin as a risk factor in hypertension. Clin Sci Mol Med 48: 123
Leksell LG, Congiu M, Denton DA, Fei DTW, McKinley MJ, Tarjan E, Weisinger RS (1981) Influence of mannitol-induced reduction of CSF Na on nervous and endocrine mechanisms involved in the control of fluid and sodium balance. Acta Physiol Scand 112: 33
Luetscher JA, Weinberger MH, Dowdy AJ, Nokes GW, Balikian H, Brodie A, Willoughby S (1969) Effects of sodium loading, sodium depletion and posture on plasma aldosterone concentration and renin activity in hypertensive patients. J Clin Endocrinol Metab 29: 1310
McCaa RE, Young DB, Guyton AC, McCaa CS (1973) Evidence for a role of an unidentified pituitary factor in regulating aldosterone secretion during altered sodium balance. Int Res Commun Syst 15: 7–10
McKinley MJ, Denton DA, Weisinger RS (1978) Sensors for antidiuresis and thirst: Osmoreceptors or CSF sodium detectors? Brain Res 141: 89
Muller AF, Riondel AM, Manning EL (1956) Mécanismes régulateurs de l'aldosterone chez l'homme. Helv Chir Acta 23: 610
Muller J (1971) Monographs on endocrinology. In: Gross F, Labhart A, Mann T, Samuels LT, Zander J (eds) Regulation of aldosterone biosynthesis, Vol 5. Springer-Verlag, Berlin
Nachman M, Cole LP (1971) Role of taste in specific hungers. In: Beidler LM (ed) Handbook of sensory physiology, Vol IV. Springer-Verlag, Berlin, p 2
Nicolaidis S (1969) Early systemic responses to orogastric stimulation in the regulation of food and water balance: Functional and electrophysiological data. Ann NY Acad Sci 157: 1176
Palkovits M, de Jong W, de Wied D (1974) Hypothalamic control of aldosterone production in sodium-deficient rats. Neuroendocrinology 14: 297
Pappenheimer JR, Heisey SR, Jordan EF, Downer JC (1962) Perfusion of the cerebral ventricular system in unanaesthetised goats. Am J Physiol 203: 763
Ramsay DJ, Thrasher TN, Keil LC (1980) Mechanisms controlling the stimulation and inhibition of drinking and vasopressin secretion in dogs. In: Proceedings of the Sixth International Congress on Endocrinology, Melbourne. Australian Academy of Science, Canberra, p 442
Rougon-Rapuzzi G, Cau P, Boudier JA, Cupo A (1978) Evolution of vasopressin levels in the hypothalamo-posthypophyseal system of the rat during rehydration following water deprivation. Neuroendocrinology 27: 46
Schaller GB (1963) The mountain gorilla. Ecology and behaviour. University of Chicago Press, Chicago, p 149
Scoggins BA (1968) The application of radioisotopic procedures to the physiology of steroid hormones in blood with particular reference to aldosterone. PhD Thesis, University of Melbourne
Sharma KN, Anand BK, Dua S, Singh B (1961) Role of

stomach in regulation of activities of hypothalamic feeding centres. Am J Physiol 201: 593
Tait JF, Little B, Tait SAS, Flood CJ (1962) The metabolic clearance rate of aldosterone in pregnant and non-pregnant subjects estimated by both single-injection and constant-infusion methods. J Clin Invest 41: 2093
Thorn GW, Ross EJ, Crabbe J, Van't Hoff W (1957) Studies on aldosterone secretion in man. Br Med J ii: 955
Thrasher TN, Nistal-Herrera JF, Keil LC, Ramsay DJ (1981) Satiety and inhibition of vasopressin secretion after drinking in dehydrated dogs. Am J Physiol 240 (Endocrinol Metab 3): E394
Tuck ML, Dluhy RG, Williams GH (1974) A specific role for saline or the sodium ion in the regulation of renin and aldosterone secretion. J Clin Invest 53: 988
Venning EH, Dyrenfurth I, Beck JC (1957) Effect of anxiety upon aldosterone excretion in man. J Clin Endocrinol Metab 17: 1005
Verney EB (1947) The antidiuretic hormone and the factors which determine its release. Proc R Soc (Lond) [Biol] 135: 25
Vincent JD, Arnauld E, Bioulac B (1972a) Activity of osmosensitive single cells in the hypothalamus of the behaving monkey during drinking. Brain Res 44: 371
Vincent JD, Arnauld E, Bioulac B, Faure JMA (1972b) Unitary neuronal activities of the hypothalamus and hydromineral homeostasis in non-anaesthetized monkeys (*Macaca mulatta*). Med Primatol 2: 332
Williams GH, Tuck ML, Rose LI, Dluhy RG, Underwood RH (1972) Studies of the control of plasma aldosterone concentration in normal man. III. Response to sodium chloride infusion. J Clin Invest 51: 2645
Wolf AV (1958) Thirst: Physiology of the urge to drink and problems of water lack. Charles Thomas, Springfield Ill
Wright RD, Blair-West JR, Coghlan JP, Denton DA, Goding JR, Nelson JF, Scoggins BA (1972) The structure and function of adrenal transplants. Aust J Exp Biol Med Sci 50: 873

22 The influence of the renin–angiotensin system and experimental hypertension on salt appetite

Summary

1. A great diversity of actions have been ascribed to the ubiquitous potent octapeptide angiotensin II, a product of the action of the enzyme renin upon its glycoprotein substrate. Of some 50 effects reported in the literature, some indubitably involve fine regulation of physiological processes. Others probably reflect pharmacological consequences of very high ambient concentrations contrived by the experimental procedures used. Pharmacological effects of high dosage of hormones have been extensively discussed by Parsons (1976). There are stringent criteria which need to be met before ascription of a particular physiological role to a peptide such as angiotensin.

2. Angiotensin given intravenously or intracranially causes water drinking in a wide variety of species of animals. The role of angiotensin II in the physiological genesis of thirst has been controversial. Whereas initial experiments suggestive of the hypothesis involved administration of unphysiological amounts, more recent data with intravenous infusion at lower rates support the proposal that increased production of the hormone plays a contributory causal role in thirst, and possibly, under some conditions, can be the primary cause. The dog appears more sensitive in this regard than other species such as the rat or sheep. Excessive water drinking in dogs with circulatory impairment is reduced by administration of saralasin, a competitive antagonist of angiotensin.

3. Comparative physiological studies show that a variety of avian and mammalian species which live in arid areas and do not have free access to water, or depend on food for most of their ingested water, are relatively insensitive to angiotensin, as are carnivorous birds. These findings are consonant with the hypothesis of a physiological role of angiotensin in thirst in sensitive species which freely imbibe water.

4. Using immunohistochemical methods, angiotensin II and III immunoreactivity has been shown to be widespread in the brain and spinal cord, and it has been proposed the peptide may act as a transmitter in several regions. The subfornical organ (SFO) and the organum vasculosum of the lamina terminalis (OVLT) appear to be more sensitive to angiotensin than other areas of brain and it has been demonstrated within the SFO itself that there are neurones specifically sensitive to angiotensin II. There is immediate and specific selection of water in a cafeteria choice situation when angiotensin II is injected intracranially.

5. Measurements of angiotensin in cerebrospinal fluid show the concentrations to be lower than in contemporaneously sampled arterial blood including when the latter concentration is raised, for example by

sodium depletion. With intravenous infusion of angiotensin to produce high physiological blood concentration in the sodium-replete animal, there is no significant change in cerebrospinal fluid angiotensin concentration after 6 h of infusion. The concentrations of angiotensin II which have been produced by intracerebral ventricular infusions or injections into brain substance to evoke water drinking involve sudden changes of angiotensin II concentration in the environment of neurones to the very high physiological or pharmacological range. The data from this method of approach are difficult to reconcile quantitatively with the hypothesis of renin of renal origin as a vector of thirst, but are more persuasive when considered in conjunction with the evidence outlined in 2, 3 and 4 above.

6. Evidence from studies in sheep and rats involving systemic infusion of angiotensin and nephrectomy is against a role of renal renin in salt appetite.

7. Intraventricular injection of renin or infusion of angiotensin II causes an immediate increase in water intake and, after a time delay, a very large appetite for salt. Occasionally the onset of salt appetite is rapid: it occurs within minutes in rats. With sheep the levels of angiotensin concentration in cerebrospinal fluid which produce the effect are grossly pharmacological relative to concentrations of angiotensin measured under a variety of conditions. Lower rates of infusion of angiotensin, albeit still pharmacological, have no effect on salt appetite. In rats, the rates used so far to cause the effect are also grossly pharmacological. It could be argued that this reflects the necessity to produce sufficient concentration for penetration at a distance from the ventricle. That is, angiotensin II may be produced physiologically within the brain tissue at high concentration from a local cerebral renin–angiotensin system responding to sodium deficiency and the hormone may act as a transmitter (see also Chapter 7). However, intraventricular infusion of converting enzyme inhibitor or a competitive antagonist of angiotensin, which prevent the salt-appetite-inducing effect of intraventricular renin, do not reduce the salt appetite which occurs physiologically in response to sodium deficiency. Furthermore, changes of sodium concentration of cerebrospinal fluid within the physiological range produced by infusion of specific saccharides (see Chapter 24) have a large effect on salt appetite indicative of relative ease of accessing the neuronal elements subserving the behaviour. It is possible that infusions causing pharmacological concentrations of angiotensin in the immediate milieu of the relevant neurones could, in the light of known effects of angiotensin on sodium, potassium and calcium fluxes in other tissues, contrive ionic changes similar to those occurring physiologically as a consequence of sodium deficiency, and accordingly evoke salt appetite. At the time of writing, the physiological implications of the striking effect of angiotensin on salt appetite are *sub judice*. Resolution may require methods quite different from intraventricular injection and infusion of renin and angiotensin which have been used to date (see Chapter 24).

8. The effect of hypertension on salt appetite is variable and, when present, would appear to depend on the mechanism of production of hypertension rather than direct influence of high blood pressure per se. With kidney wrap hypertension, impaired ability to excrete salt may underlie the relative aversion to salt observed. With renal hypertension involving constriction of one artery with the other kidney untouched, increased salt appetite may follow pressure natriuresis from the untouched kidney. Rats genetically susceptible to hypertension as a result of increased salt intake show aversion

to salt relative to strains resistant to effect of increased salt intake. Rats of a strain which develops hypertension spontaneously show increased appetite for salt.

At the outset, it needs to be said that the grouping of these considerations in one chapter is a matter of convenience in relation to some aspects of subject matter. It carries, of course, no implication that the renin–angiotensin system is activated, for example, in chronic renal hypertension.

The renin–angiotensin system

In 1898, Tigerstedt and Bergman showed that crude saline extracts of kidney were pressor when injected into anaesthetized rabbits. They thought the substance was directly vasoconstrictor and named it renin. The discovery by Goldblatt et al. (1934) that experimental renal ischaemia caused hypertension, greatly intensified work on renal pressor substances. Page and Helmer (1939, 1940) and Braun-Menendez et al. (1940) identified the active agent produced by the action of the enzyme renin on plasma substrate, and it was named angiotensin. The composition and amino acid sequence of angiotensin I (AI) and mechanism of conversion to angiotensin II (AII) by converting enzyme were described subsequently (Elliott and Peart 1956; Lentz et al. 1956; Skeggs et al. 1956; see review by Page and Bumpus 1961, and Page and Bumpus 1974).

Actions of angiotensin

In considering the formal possibility of a physiological role of the renin–angiotensin system in the genesis of salt appetite, it is, perhaps, salutary at the outset to note that some 21 actions, at least, have been described for this potent ubiquitous tissue-active substance. Consideration of physiological function involves not only the issue of angiotensin arising in the circulation from renin of renal origin, but also of the putative function of angiotensin arising from renin, possibly of renal origin, in tissue (Coghlan et al. 1979; Fei et al. 1980; Aguilera et al. 1981) and from isorenins such as for example, those in the brain (Ganten et al. 1971).

Included amongst the actions of AII and the heptapeptide angiotensin III (AIII) which have been described, with some representative review references, are:

1) Constriction of resistance vessels (Page and Bumpus 1961) including circulatory adaptation to sodium deficiency (Johnson and Davis 1973).
2) Constriction of veins (Blair-West, McKenzie and McKinley 1971c; Sutter 1965).
3) Essential for, or commensurately regulating, aldosterone biosynthesis in sodium deficiency (Gross 1958; Davis 1959, 1961; Genest et al. 1960; Laragh 1960; Blair-West et al. 1963; Baniukiewicz et al. 1968).
4) Stimulation of antidiuretic hormone (ADH) secretion (Bonjour and Malvin 1970).
5) Genesis of thirst (Fitzsimons 1972; Andersson 1977).
6) Potentiation of action of sodium concentration in genesis of thirst and ADH release (Andersson and Olsson 1977).
7) Anti-natriuretic (low doses) and natriuretic (high doses) (Navar and Langford 1973).
8) Natriuresis mediated by brain (Andersson, Dallman and Olsson, 1969).
9) Pressor response mediated by either area postrema of brain, midbrain or hypothalamus (Joy and Lowe 1970; Severs and Daniels-Severs 1973).
10) Reduction of parotid salivary secretion in sodium deficiency (McKinley et al. 1979).
11) Stimulation of adrenocorticotrophic hormone (ACTH) secretion (Daniels-Severs et al. 1971; Ramsay et al. 1978).
12) Altered transport of fluid and sodium in jejunum (Bolton et al. 1975).
13) Release of acetylcholine from post-ganglionic nerve endings and contraction of intestinal smooth muscle (Khairallah and Page 1961).
14) Inhibition of uptake of catecholamines (Khairallah 1972) and stimulation of sympathetic ganglia (Lewis and Reit 1966).
15) Potentiation of adrenergic vasoconstriction (Sakurai and Hashimoto 1965).
16) Increased permeability of vascular wall due to

platelet aggregation (Robertson and Khairallah 1972).
17) Increased myocardial contractility (Koch-Weser 1965).
18) Stimulation of the adrenal medulla (Feldberg and Lewis 1964; Peach 1971).
19) Inhibition of renin secretion (Vander and Geelhoed 1965; Blair-West et al. 1971b).
20) Regulation of glomerular filtration rate by tubular feedback mechanisms (Thurau et al. 1967).
21) Decrease of body temperature following intracranial injection (Sharpe et al. 1979; Lin 1980).

In relation to these many actions, the question arises as to which effects of this potent hormone are elements of fine physiological regulation. Indubitably, some are well validated as physiological regulation. Pharmacologic effects could be a result of sufficiently high concentration contriving changes in sodium movement between extracellular and intracellular fluid. This effect has been demonstrated on frog skin and jejunal cells (McAfee and Locke 1967; Crocker and Munday 1967, 1970), and toad bladder (Coviello and Crabbé 1965). Such sodium movement clearly could alter cellular activity.

Other actions in this category include stimulation of sodium pump in kidney cortex slices, and stimulation of fluid transport in the rat colon (Munday et al. 1971, 1972; Davies et al. 1969, 1970), stimulation of sodium uptake in arterial wall (Villamil et al. 1970), inhibition of potassium uptake in adrenal slices (Szalay 1969), influence on calcium distribution (Daniels et al. 1967), release of calcium from membranes (Baudouin et al. 1972) and influx of extracellular calcium in papillary muscle (Peach 1977). AII causes a substantial efflux of calcium from zona glomerulosa cells whereas ACTH and change in [K] had no effect though all stimulated aldosterone secretion (Williams et al. 1980). A most extensive tabulation of some 50 categories of effect in different tissues (Vecsei et al. 1978) includes a variety of influences on DNA and RNA mechanisms and protein synthesis, as well as upon cyclic AMP and other elements of intracellular metabolic systems.

Angiotensin in the brain

Fuxe et al. (1976), using indirect immunohistochemical methods, found it possible to demonstrate AII or AII-like immunoreactivity resistant to bilateral nephrectomy in probable axons and nerve terminals of the brain and spinal cord of rats. Nerve terminals containing specific AII immunofluorescence were found all over the brain and spinal cord. The specific fluorescence disappeared after absorbing the specific antibodies with AII, whereas somatostatin was without effect. The nerve terminals had a distinctly varicose appearance. Axons with specific fluorescence were only observed in the retrochiasmatic area. No cell bodies with specific immunofluorescence were demonstrated. High density was found in four regions: substantia gelatinosa of the spinal cord, substantia gelatinosa of the spinal nucleus of the trigeminal nerve, the sympathetic lateral column of the spinal cord, and the medial external layer at all levels of the median eminence. Moderate density was found in the dorsomedial hypothalamic nucleus, in the ventral hypothalamus, in the locus coeruleus, and the nucleus amygdaloideus centralis. Single nerve terminals were found in all parts of the brain. The authors propose that AII is probably synthesized in the neurones, as nephrectomy did not reduce the stores.

In considering the implications of these findings the authors go on to propose a variety of actions for angiotensin. The presence of high density of terminals in the sympathetic lateral columns suggests angiotensin may influence blood pressure by effect on preganglionic sympathetic neurones and on the hypothalamus. The concentration in the median eminence suggested to them AII may act possibly as a hypothalamic hormone upon the pituitary or influence release of other hypothalamic hormones. The noradrenergic cell bodies of the locus coeruleus give rise to the coerulocortical noradrenergic pathway subserving tonic arousal and positive reinforcement in self-stimulation studies, as well as learning functions, and their angiotensin innervation suggests that these functions may be influenced by AII. Also some noradrenergic pathways are considered to be vasopressor, and again angiotensin II neurones could increase blood pressure partly by direct activation of central noradrenergic vasopressor pathways. In the light of the dense innervation of the spinal cord and medulla oblongata it must be seriously considered that AII or an AII-like peptide may also be involved in the control of sensory information (Fuxe et al. 1976). Thus the authors see AII or an AII-like peptide as a possible transmitter in several regions of the brain and spinal cord.

Changaris et al. (1978) studied localization of AII by immunocytochemical methods. Neurones within the deep cerebellar nuclei and brain stem were studded with dense synaptic AII immunoprecipitate. AII reaction product was demonstrated in pericapillary pinealocytes, posterior pituicytes, third ventricular tanycytes and some cells within the spinal trigeminal tract. The antiserum showed 70% cross-reactivity with AIII, and therefore distribution could also correlate with the heptapeptide. There was significant cross-reactivity with AI. The authors draw attention to the fact that the fluorescin studies of the Swedish workers did not emphasize the dense reaction product especially in the deep cerebellum shown by the unlabelled antibody method used by Changaris et al. (1978). The cerebellum has not been focused as being a cerebral site of angiotensin action, though receptors are present there. In calf brain high-affinity AII binding was almost entirely restricted to the cerebellum, especially the cortex (Bennett and Snyder 1976).

There is clearly a teleological attraction in the idea of the hormone which is formed as a result of extracellular and plasma volume depletion regulating the homeostatic response by a spectrum of actions ranging from support of blood pressure, a critical role in aldosterone biosynthesis and thus sodium conservation, release of ADH, and also stimulation of intake of water and salt to repair the deficits (Fitzsimons 1978; Epstein 1978). However useful this idea may be as a stimulus to experiment, the facts are decided on other grounds. Pivotal in this exercise will be quantitative considerations—the physiologically feasible—since the catalogue of actions angiotensin II can have, especially on excitable tissues, should engender considerable caution in interpretation.

In relation to such a teleological concept, there is currently no clear evidence that ADH evoked by water deprivation and other stimuli stimulates thirst. That is, ADH administration in the absence of water intake suppresses thirst in diabetes insipidus (Pasqualini and Codevilla 1959). When given systemically in the rat over a wide range of infusion rates, it has no effect on drinking or is antidipsogenic (Adolph et al. 1954; Radford 1959; Rolls 1971). The data on the dog are contradictory. Bellows (1939) found it antidipsogenic in that it delayed onset of sham-drinking induced by hypertonic saline. Adolph et al. (1954) found it increased drinking in dogs deprived of water—when, presumably, the endogenous level was already high. Szczepanska-Sadowska et al. (1974) report that the intravenous infusions of subpressor amounts of ADH lowered the threshold of drinking in response to intravenous hypertonic saline although the hormone never caused drinking on its own. Pressor amounts of ADH inhibited drinking. Epstein et al. (1970) found that very large amounts of ADH (10–20) mU) injected directly into the septal region, the preoptic region, the anterior hypothalamic region, the lateral hypothalamic region, the ventromedial nucleus or the median eminence of the tuber cinereum through a chronically implanted cannula never stimulated drinking. More recently Sadowski et al. (1980) have reported that in dogs with cannulae in the third ventricle injection of 100 and 300 μU of ADH increased spontaneous water intake about ten-fold over a 1-h period. Injection of 30, 50 and 3000 μU did not have a significant effect nor did injection of 300–30 000 μU into the lateral ventricle. The concentrations produced in the third ventricle (volume 0.1–0.2 ml) would have been very large relative to normal plasma concentration in the dog (1–2 μU/ml) (Fitzsimons 1979). At present, the position would seem appropriately summarized by Fitzsimons' statement (1972) that the evidence that ADH has a direct action on thirst mechanisms remains slender.

Angiotensin II in the genesis of thirst and water intake

The quantitative issue is perhaps best illustrated by consideration of the putative role of angiotensin II in physiological control of water intake. The initial observations of Linazasoro et al. (1954) pointed to a renal factor in control of water drinking, though their subsequent enquiries did not suggest it to be renin. The concept of a renal control of water drinking, specifically via the renin–angiotensin system, has been elaborated systematically with important, indeed landmark, contributions to physiological knowledge of brain mechanisms subserving drinking behaviour by Fitzsimons and colleagues (Fitzsimons 1972, 1979).

Fregly (1978) in introducing a symposium on the matter gives an even-handed précis of present controversy. It is established, beyond cavil, that AII administered intravenously, intracarotid, or intracranially, is a potent dipsogen. The molecular requirements for dipsogenic activity in the rat are similar to pressor effects (Fitzsimons 1980). The (1–7) heptapeptide is completely without effect on drinking. The (2–8) heptapeptide (AIII) retains dipsogenic activity in rat and pigeon, but has little in the dog, whereas AIII is equipotent with AII on steroid secretion in the sheep (Blair-West et al. 1971b, 1980).

Epstein (1978) notes that the phenomenon of angiotensin-induced drinking is biologically ubiquitous. As Fitzsimons (1978) has pointed out, vertebrates from reptiles to monkeys have been found to drink in response to angiotensin, and Epstein remarks that new species are added to the list from time to time as investigators are stimulated to try such experiments on their favourite experimental subjects. As a matter of fact, one of the best teleological lines of evidence that angiotensin may be physiologically involved in water intake is that its effect is *not* ubiquitous. Comparative studies by Japanese workers have shown by intraperitoneal administration of angiotensin in a wide range of avian and mammalian species that animals which live in arid areas and do not have free access to water, and those which depend on food for most of their ingested water in nature, are relatively insensitive to angiotensin. Carnivorous birds and amphibians do not respond either. Angiotensin stimulates drinking in sea-water eels which naturally drink water to compensate for osmotic loss in much smaller dose than in fresh-water eels (Kobayashi et al. 1977, 1979; Nishimura 1978; Hirano et al. 1978). In all these comparative studies, however, angiotensin doses have been very large.

One of the many lines of evidence suggestive of a physiological role is that in the subfornical organ (SFO), which has been postulated as the receptor site for the control of AII-mediated drinking (Simpson and Routtenberg 1973), there are neurones which are responsive specifically to AII (Fig. 22-1), whereas others are responsive to acetylcholine or both (Felix and Akert 1974; Felix 1976).

Systemic infusion of angiotensin

On the other hand, Abraham et al. (1975) reported that the levels of AII necessary to induce drinking in sheep by intravenous, intracarotid, or cerebral ventricular infusion or injection exceeded the levels which are directly measured in arterial blood or cerebrospinal fluid (CSF) under a variety of physiological conditions where water drinking is stimulated. They analysed the data published for the rat. It was concluded that the dosages employed up to that time to mimic naturally occurring states where water drinking was induced were pharmacological. Subsequently, Stricker, Bradshaw and McDonald (1976) did directly measure plasma renin activity (PRA) and reported that in rats renin levels required to induce drinking greatly exceed those caused by dipsogenic procedures such as constriction of the inferior vena cava.

Stricker (1977, 1978) derived a regression equation for the relation between PRA and water intake. The equation showed that PRA must exceed 18 ng/ml per 90 min before water intake was stimulated by AII in a 1-h drinking test, and increases of PRA of 91 ng/ml per 90 min were required for each millilitre of water consumption. To put these figures into physiological perspective he notes that when rats were maintained on a sodium-deficient diet for 2 weeks, or when plasma volume was reduced by 13%, PRA was elevated by 15–25 and 50–60 ng/ml per 90 min respectively. Thus PRA would rarely exceed these levels, and water intakes of 2–4 ml would require values in the range of 200–382 ng/ml per 90 min for stimulation. He proposes that the PRA levels which result from caval ligation, polyethylene glycol or isoproterenol treatment indicate that AII contributes less than 30% of the stimulus for actual water intake observed in each case. Other stimuli such as from baroreceptors must be responsible. He cites reports that either caval constriction or isoproterenol will still elicit thirst while rats are infused with angiotensin receptor blockers (Tang and Falk 1974; Rolls and Wood 1977).

On this latter issue, Ramsay (1978) has shown that this capacity, in the instance of intraventricular saralasin (P113), is dependent on the dose of isoproterenol given. At lower dosage saralasin is effective, but at higher, though plasma renin does not rise to a much greater level, it is not. Ramsay thinks this is due to the much greater cardiovascular effects at the higher doses which may influence drinking. Stricker (1978) also suggests that the findings on abolition of water intake after

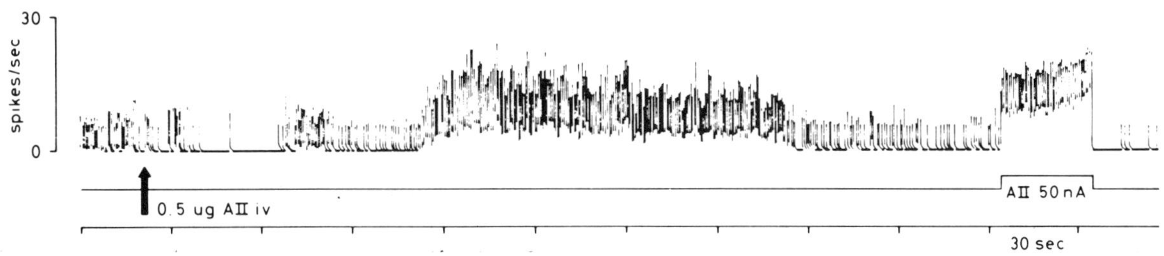

Fig. 22-1. Response of a SFO neurone to intravenous administration of 0.5 μg AII (Hypertensin, CIBA). The beginning of injection (indicated by arrow) is accompanied by a short-lasting decrease followed by an increased firing rate after a delay of approximately 1.5 min. After recovery from the intravenous injection additional microiontophoretic injection of AII has still an excitatory effect (illustrated at the end of the record). Time scale 30 s. (From Felix 1976, with permission.)

nephrectomy in caval-constricted or isoproterenol-treated rats may be attributable to the fact that blood pressure is so low as to preclude drinking behaviour. Further, though drinking may be stimulated in such rats by various treatments, in each case a pressor response also is observed and this may have restored the ability to drink.

Since our initial criticism (Abraham et al. 1975) new data have appeared, proposing dipsogenic action of angiotensin II at physiologically feasible infusion rates. For example, Fitzsimons et al. (1977), reported intracarotid infusion experiments in the dog which, on the basis of carotid blood flow of 100–200 ml/min, could be calculated to give cerebral blood angiotensin II concentrations of 20–40 ng/100 ml—within the range seen, for example, in moderate to severe sodium deficiency. The dogs drank modest amounts of water and intravenous injections which would have contrived the same order of change were also effective.

Epstein and Hsiao (1975) obtained lower values for drinking threshold in response to intravenous AII infusion than those of Fitzsimons and Simons (1969). The latter had been analysed by Abraham et al. (1975) on the basis of clearance data and found to be substantially supraphysiological. Taking the Epstein and Hsiao (1975) data with drinking in 100% of animals at 80 ng $kg^{-1}min^{-1}$ and with assumed clearance of 60% of the cardiac output of approx 50 ml/min, the peripheral blood level would have been in the range of 100 ng/100 ml. The normal range reported in the rat is 3–13 ng/100 ml. In Hsiao et al.'s (1977) later data, the threshold was found to be 25 ng $kg^{-1}min^{-1}$, and if the rats had a minor degree of concurrent cellular dehydration the threshold producing a small effect of about 1 ml intake was 10 ng $kg^{-1}min^{-1}$. These infusions can be calculated to give systemic concentrations of 30 and 12 ng/100 ml respectively, which are within the physiological range. In contrast to these findings Avrith et al. (1980a) report that 4-day infusion of AII in rats at 25 or 50 ng $kg^{-1}min^{-1}$ caused no increased intake of water or 2.7% saline. Infusion of AII at 75 or 100 ng $kg^{-1}min^{-1}$ produced a marked increase in water and 2.7% saline intake but secondary to the induction of a large natriuresis and diuresis.

In the goat, Andersson and Westbye (1970) did not induce drinking with infusions of 5 ng $kg^{-1}min^{-1}$ AII which would produce blood levels in the low physiological range. Kozlowski et al. (1972) reported that intravenous infusion of AII lowered the threshold of drinking response in dogs to cellular dehydration produced by hypertonic saline. Trippodo, McCaa and Guyton (1976) gave intravenous AII infusion over 10 days in dogs at rates confirmed to give blood AII concentration in the range of 20–30 ng/100 ml which is the order of rise seen in sodium deficiency. This doubled voluntary water intake. Blood pressure was elevated 30–35 mmHg. In this study it was reported that urine output also increased by a mean of about 100 ml/day (voluntary water intake increased 151±45 ml/day) so, though there was no initial natriuresis, it would have been desirable to know whether the increased water drinking preceded any diuresis. Overall it would appear, as Fitzsimons suggests in a detailed analysis (Fitzsimons 1979), that the dog is more sensitive than the rat to dipsogenic influence of AII, and it may also be that it is more sensitive than the ruminants (goat and sheep). Stricker (1978) notes that the water intake induced in dogs was very small and would dilute body fluids by less than 1%–2% even if all the water were retained. He suggests both dog and rat data provide no evidence that either renin or AII in physiological amounts elicits substantial drinking in any animal. He suggests that, by way of contrast, the 'threshold doses' of infused AII that elicited drinking in rats (Hsiao et al. 1977) were 15 times greater than those found to elevate aldosterone secretion (Campbell et al. 1974).

Fitzsimons (1979) correctly points out that in some of the experiments of Abraham et al. (1975) the sheep did drink at the very high physiological concentrations, and that it is quite feasible that with concurrent operation of other dipsogenic conditions such as raised osmotic pressure, angiotensin II could be a contributory cause of physiologically induced water intake. However, on this theme, Abraham et al. (1976a) showed infusion of P113 into the third ventricle (50–100 μg/ml at 1.1 ml/h) effectively abolished the large water intake induced 1–2 min after starting intracarotid infusion of AII at 800 ng/min (which gives a supraphysiological AII concentration in cerebral arterial blood). This same intraventricular infusion begun 20 min before and during water presentation to sheep deprived of water for 48 h did not influence intake and it did not influence water intake with rapid food intake or intracarotid infusion of 4 M sodium chloride.

On the other hand, Malvin, Mouw and Vander (1977) found that in water-deplete rats, lateral ventricular infusion of P113, begun 30 min before presentation of water, increased water intake significantly; but, if infusion was begun 75 min before water was available, P113 attenuated drinking in all but two of their animals. These two drank twice the average of controls whereas the others drank only after a delay of 10–25 min. They suggested that the longer delay period before presentation may have avoided the agonist action of P113 and allowed the antagonist action to be evident. The infusion 75 min before water did not influence food intake after deprivation. The data are interesting but leave questions unanswered in relation to the transition of effect; observations in the interval between 30 and 75 min would have been informative. There is also the question of the relative agonist effects in the water-replete animal at 30 and 75 min, and whether the same total dose received by 75 min would have an inhibiting effect if it had been received over 30 min. It should incidentally be remarked that in introducing their study the authors attribute to Abraham et al. (1975) the statement that drinking induced by angiotensin is a pharmacological effect rather than a physiological one. This is not what was said. The argument was that all experiments to that point in time involved pharmacological concentration of AII and that the question of physiological effect was, therefore, *sub judice*.

Intracranial injection of angiotensin II

Fitzsimons (1979) remarks that there are a number of possible ways the angiotensin system could play a role in drinking. For example, it might reduce the discharge in the receptors in the low pressure capacitance vessels which are inhibitory to thirst. The evidence, on a time basis, is obviously against any effect via the adrenal cortex. The results point to a central action.

Booth (1968) was the first to show that rostral hypothalamic injections of angiotensin II caused drinking, and the phenomenon was comprehensively reported in 1970 (Epstein et al. 1969, 1970; Severs et al. 1970). The neuroanatomical structures from which angiotensin-induced drinking can be obtained are fairly widespread, and include SFO, organum vasculosum of the lamina terminalis (OVLT) and the preoptic area. Phillips (1978) proposes that both SFO and OVLT contain angiotensin receptors, with the SFO responding to AII in the blood and not the CSF and with the OVLT responding to AII in both blood and CSF. The data accumulated have suggested that the areas are separate from those giving neurogenic pressor responses. This latter consideration is relevant to the issue of whether angiotensin acts directly on a neural thirst system or whether drinking is secondary to neurologically mediated changes in the peripheral circulation. Severs et al. (1970) and Hoffman and Phillips (1976) found that intraventricular AII was pressor as well as dipsogenic in the rat whereas Kucharczyk and Mogenson (1976) found otherwise.

Fitzsimons (1979) suggests the following arguments to justify the conclusion that intracranial AII plays a physiological role in thirst. First, as drinking proceeds after intracranial injection of AII in the animal which is in water balance, it becomes overhydrated which would tend to bring drinking to a halt. Secondly, the signals from the low pressure capacitance vessels do not indicate a need to take water, and thus AII is causing drinking unaided which is not what would occur normally. Thirdly, he suggests that a small amount of active material is being introduced by single injection into a small portion of a larger and bilaterally sensitive region that would normally be perfused by lower endogenous concentrations of hormone for longer periods of time. He goes on to suggest that given the difference between a single injection of exogenous material intracranially and the more physiological circumstances in which the elevated level of circulating angiotensin is maintained by increased endogenous renin secretion in an animal with a concurrent stimulus for thirst arising from the vasculature, the high intracranial dose needed to initiate drinking was not a serious objection. To me, as indicated below, a contrary line of reasoning is feasible in that the very sudden rise in concentration (rate of change of stimulus) with intracranial injection could provoke reaction at much lower concentration than might be the case with slow rise of stimulus under physiological conditions. This issue is critical to the evaluation of physiological relevance of intracranial injections.

In 1975 Abraham et al. pointed out that intracranial injections of 100 pg of AII (the smallest amount used up to that time) may seem a small amount but, in fact, this amount dissolved in 1 μl of injection fluid represents a solution of 10 000 ng/100 ml flooding the environment of a few hundred neurones: for example about 1000 times the highest level measured in CSF of severely sodium-deficient sheep (Chapter 7). The question of the physiologically feasible has now been given closer attention, as in the systemic infusion experiments of Fitzsimons et al. (1977) recorded above. There is the report of Simpson et al. (1975) that the threshold at which 50% of rats drank after injection of AII into the SFO was 0.1–1 pg (10^{-16}–10^{-15} mol) (see Fig. 22-2). This represents injection of a concentration of 10–100 ng/100 ml, which is within the high physiological to supraphysiological range as measured in CSF (Abraham et al. 1980). The difference of a positive result with 1 pg in the SFO vs the 100 times higher threshold in the third ventricle is strongly suggestive of specific action. The drinking data at 0.1 pg (10^{-16} mol) are marginal in allowing of any interpretation. In the dog (Fitzsimons and Kucharczyk 1978) it required 10 ng to produce a significant effect, and similarly, in the monkey, with injections in the septum,

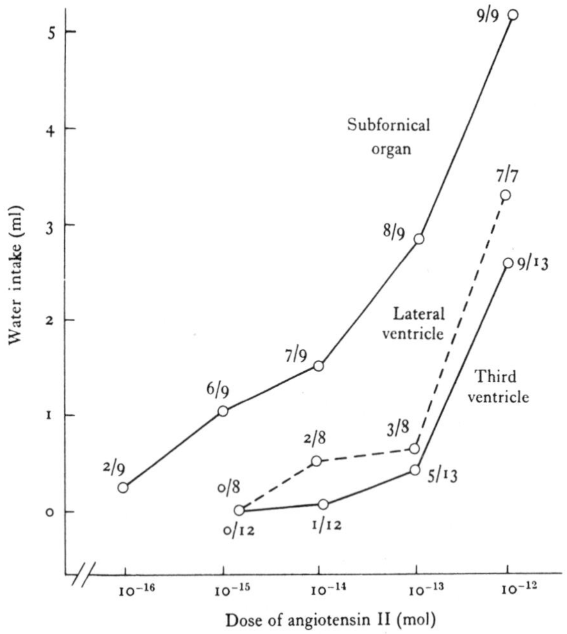

Fig. 22-2. The increase in water intake above the baseline for rats injected with angiotensin into the SFO, lateral ventricle and third ventricle. The numerator in each fraction is the number of rats responding and the denominator is the number of animals injected for the particular dose and cannula position. (After Fitzsimons 1979, from unpublished results of Simpson with permission.)

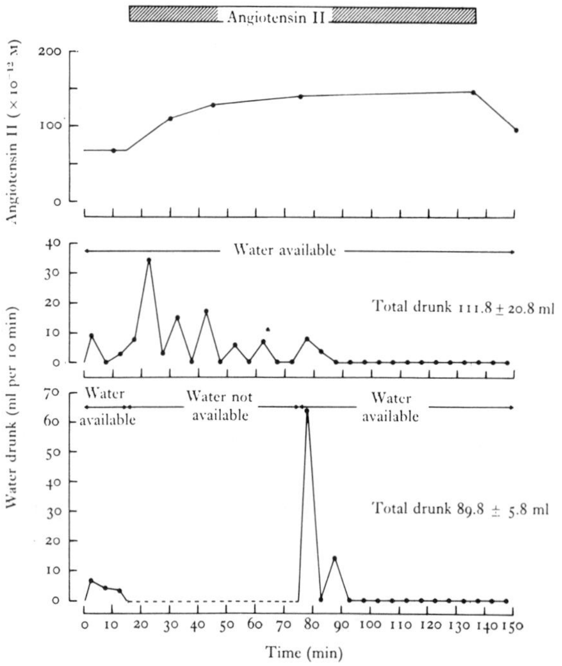

Fig. 22-3. Rate of water intake in 10-min periods of five dogs infused intravenously with angiotensin II at 250×10^{-12} mol/min for 2 h with immediate access to water (middle panel) and when access to water was not allowed until 1 h after the start of the AII infusion (bottom panel). Before and after the infusion of angiotensin, 0.9% saline was infused at 2.5 ml/min for 15 min. The mean total amounts of water drunk did not differ significantly. Note that in each case the rate of drinking was maximal when water was first made available and then fell off markedly even though the infusion continued at the same rate. AII levels produced by a similar infusion schedule in other experiments are shown in the upper panel. (From Fitzsimons 1979, with permission.)

anterior hypothalamus and preoptic area, 10 ng were required on bilateral injection to give any notable effect on water intake—though 1 ng did appear to have an influence relative to unilateral injection of angiotensin (Sharpe and Swanson 1974).

However, at the time of writing, the problem remains with the intracarotid, intravenous and intracranial injections that whereas under physiological conditions when AII rises and water drinking occurs the rise is usually gradual over many hours to a plateau level, with these injection–infusion techniques the contemporary level of angiotensin rises to a plateau over seconds to a minute. The rate of change of stimulus with time may have a significant bearing on physiological response (Parsons 1976), and thus the procedures do not mimic the physiological circumstances if the aim is to see AII as the prime mover, or even as a contributory cause, in drinking.

Fitzsimons et al. (1978), in intravenous AII infusion experiments on dogs, have withheld water until 60 min after beginning the infusion, when a mean of 89±5.8 ml was taken almost immediately (Fig. 22-3); this procedure might be said in some ways to obviate the criticism cited above. In sheep, infusion of AII in stepwise fashion, with a gradual increase of blood concentration within the physiological range, has been made. The rates were 2, 4, 6, 10, 15 and 30 μg/h, with the step change made each hour (n=6). One animal drank at 15 μg/h, and four at 30 μg/h (110–700 ml). This latter rate would give a blood AII concentration of ca 30 ng/100 ml, which is a high physiological level. It caused a pressor response of 20–30 mmHg. M. J. McKinley (unpublished) has also made experiments in the goat using rates of 6, 12, 30 and 60 μg/h. Drinking occurred (28–250 ml) at 30 μg/h with latency of 3–12 min. Some animals had a small intake at 12 μg/h. The drinking was no greater at 60 μg/h. Blood pressure at 30 μg/h rose by 30–64 mmHg. When the AII infusions at 6 and 30 μg/h were combined with intravenous infusion of 3 M sodium chloride at 90 or 180 mmol/h, water drinking was not increased above that caused by infusion of 3 M sodium chloride alone. Results with other species must be awaited with interest.

The report of Ramsay et al. (1975) showing a doubling of water drinking in dogs with congestive cardiac failure produced by thoracic caval constriction, and the reduction of drinking by P113 infusion suggests AII to be playing a role in the exaggerated water intake. Similarly Fitzsimons (1980) reports a commensurate relation between graded constriction of the inferior vena cava, with reduction of venous return to the heart and a fall in systemic arterial blood pressure, and volume of water drunk in 1 h by dogs. The water drinking was reduced by half if large amounts of saralasin were infused intravenously.

Mode of action of angiotensin II in the brain

Turning to the mechanism by which the action of angiotensin on the brain causes drinking, Fitzsimons (1979) considers four possibilities. First, that it causes drinking by increasing ATPase activity which allows sodium to gain access to some appropriate receptor. Drinking is reduced when CSF sodium is reduced or ATPase activity inhibited. He remarks that it is difficult to conceive of such a universal attribute of the cell as its ATPase activity as playing a specific role in thirst. In relation to the synergistic action of AII and CSF [Na] in causing drinking, and the absence of effects of AII on drinking in the goat when it was infused into the ventricular system dissolved in isotonic glucose, Andersson, Olsson and colleagues (Andersson and Westbye 1970; Andersson and Ericsson 1971; Olsson and Kolmodin 1974) suggested also that it may facilitate transependymal movement of sodium from CSF into brain, or sensitize receptors to the existing [Na] in the extracellular fluid of the brain.

Second is the possibility that angiotensin may modulate the activity of the neurotransmitters known to be involved in drinking behaviour. Angiotensin facilitates the release of acetylcholine which is dipsogenic in the rat. Angiotensin acts presynaptically, and enhances synthesis, facilitates release, and depresses re-uptake of catecholamines. The distribution of angiotensin and noradrenaline in the brain are correlated closely.

Third is the possibility that AII is a neurotransmitter in its own right released by neurones concerned in thirst and drinking behaviour. Thus injected AII would be mimicking the action of endogenously released AII. In this regard, Fitzsimons considers that the proposed cerebral isorenin–angiotensin system could generate AII for release. It would seem to me, however, that account must also be taken of the possibility that peptidergic neurones might synthesize an angiotensin precursor directly as a transmitter.

Fourth is the fact that the structures most sensitive to dipsogenic action of AII—the SFO and OVLT—are highly vascularized, and Nicolaïdis and Fitzsimons (1975) have suggested that drinking is mediated by vasoconstrictor effect on the blood vessels in these structures. Thus they would act normally in this regard as sensitive extracellular volume receptors by virtue of stretch receptors in their walls. To some extent confirmatory of such a notion is the fact that acute removal of CSF induces some drinking in the rat and causes thirst in man after a short interval (Nicolaïdis 1974). Fitzsimons sees the evidence in favour of such an hypothesis as follows:

1) The vascular circumventricular organs are the structures most sensitive to AII.
2) The peptide specificity of dipsogenic response is similar to myotrophic receptors, but differs from receptors for steroidogenesis (where AII and AIII are equivalent: Blair-West et al. 1971b, 1980) and sodium transport. On this point, there are data difficult to reconcile with the Fitzsimons hypothesis since AIII is much less potent as a dipsogen than AII. Bennett and Snyder (1976), in finding AII binding an order of magnitude more potent in brain than in adrenal cortex, found AIII had about three times the binding affinity of AII. Schlegel and Felix (1976) in their studies of action of microelectrophoretically applied analogues on SFO unit discharges discovered AIII showed a slightly shorter latency and significantly higher stimulation of firing rate compared with AII. Both AII and AIII action was blocked by [Sar^1, Ala^8]AII.
3) The effects of antagonists of the renin–angiotensin cascade support the dipsogenic receptor resembling the vascular receptor.
4) Many substances with little in common except an action on blood vessels cause some drinking on injection into the brain.
5) Vasoplegic drugs reduce angiotensin-induced drinking.
6) The drinking responses to combinations of vasoactive drugs suggest that individual drugs activate the same mechanisms.

Though angiotensin is pre-eminent, a variety of drugs which affect smooth muscles of arterioles elsewhere in the body sometimes cause drinking when injected into the brain: e.g. adrenaline, noradrenaline, isoprenaline, dopamine, serotonin, histamine, prostaglandin E_1, eledoisin and substance P. Different drinking responses in different species are attributed to species differences of effects on blood vessels. In these data there would presumably be wide variation in proximity of injection site to the putative vascular stretch receptors in the circumventricular organs.

In the case of prostaglandin E_1 it is reported in the goat that it amplifies the drinking response caused by AII (Leksell 1976). In the rat it sometimes causes an immediate burst of drinking with latency comparable to that of angiotensin, but it reduced the drinking caused by angiotensin. Kenney and Epstein (1975) found that 1 μg of prostaglandin E_1 injected into the lateral ventricle inhibited drinking in response to intracranial angiotensin and other dipsogenic influences. Nicolaïdis and Fitzsimons (1975), and Fitzsimons (1979) see these data as consistent with their vasomotor theory of action of angiotensin.

In relation to such a hypothesis, and the notion of these localized receptors in very vascular tissue reacting by stretch response to the volume of extracellular fluid, presumably the pressure and flow of blood in them would be directly influential in causing this reaction. Pertinent to this, in sheep we have completely occluded one carotid artery in its skin loop and applied graded constriction to near or above diastolic pressure in the other. This is shown by measurement to cause a fall in blood pressure in the cranial circulation distal to the constriction (Denton 1957). There is no basilar artery in the sheep, and the occipito-vertebral anastomoses carry only limited inflow. This is clearly shown by the fact that in sodium-replete sheep sudden complete occlusion of both carotids may cause no disturbance of the animal though blood pressure falls in the head circulation, whereas the sodium-deplete animal falls unconscious after this manoeuvre. These procedures of graded constriction in the sodium-replete animal have been singularly ineffective in causing spontaneous water drinking in a large number of sheep. If a dipsogenic response were caused by a sensitive system of reaction to small pressure changes in the cranial circulation in specific regions, it would seem some effect would be likely with such experimental procedures. This is notwithstanding that the prime reaction of the SFO and OVLT vessels may be to the contemporary concentration of vasoactive material in cerebral blood, and the fact that some measure of autoregulation of flow may occur.

Summary of evidence on role of angiotensin II in thirst

Overall, it would seem the most suggestive evidence for a physiological role of angiotensin in thirst is that intravenous AII infusion, particularly in stepwise fashion but in any event, causes specific drinking of water. The experiments on the dog suggest this is the most sensitive species studied to date. The blood concentrations required appear to be in the high physiological range, which is compatible with a contributory causal role such as in a system which comes into action when stress on volume regulation is moderately severe. The drinking evoked, for example, by caval constriction in the dog, is reduced by intravenous infusion of competitive blockers of action of AII. Any role of AII in thirst arising from water deprivation is probably small. The argument for a physiological role is strongly supported by comparative data where evocation of the complex behavioural pattern involved in drinking by angiotensin is correlated with whether or not the animals have a propensity to drink water in natural circumstances. The synergistic actions of angiotensin and [Na] are consistent with the proposal of a contributory causal role. The sensitivity of the SFO and OVLT, together with the demonstration of SFO neurones specifically reactive to AII, are highly suggestive of a physiological role. Against the background of these facts, the effects of intracranial injections which very suddenly elevate AII concentrations to unphysiological levels in the environment of neurone populations are rather more persuasive than would, perhaps, otherwise be the case.

In the study of the specificity of AII-induced drinking in sheep under conditions of cafeteria choice, it was shown that intracarotid infusion of AII elicited an immediate preference for water, as did 4 M sodium chloride infusion or for that matter, water deprivation, whereas with sodium depletion the sheep chose hypertonic sodium bicarbonate solution (900 mmol/l) (Abraham et al. 1976b). The evidence on specificity of choice of water induced by angiotensin is relevant to our main question of whether it has any direct causal role in salt appetite. With sodium deficiency there is a rise in renin and angiotensin in peripheral blood commensurate with negative balance (Fig. 9-16). There is also a rise in the AII concentration in CSF (Fig. 7-7).

The discussion of thirst above represents a brief consideration only of some of the issues which will be relevant to salt appetite. A contemporary in-depth analysis of the role of AII in thirst, with comprehensive discussion of physiological mechanisms involved, is, as indicated above, to be found in Fitzsimons (1979).

Angiotensin in the genesis of salt appetite

Effect of intravenous angiotensin II infusion on the sodium appetite of sodium-deficient sheep

In six sheep, uncompensated sodium loss was permitted for 48 h. The animals were allowed access to sodium solution (300 mmol/l sodium bicarbonate) for 15 min at the end of this time. Against a background of 24–60 control experiments on each animal under these conditions, experiments were made in which AII was infused at 60 or 120 μg/h for 15 min before and during the 15 min of access. The infusions raised systolic blood pressure 30–40 mmHg, and blood pressure rose a further 5–15 mmHg during the excitement of imminent presentation of the solutions. Control experiments were also conducted with normal saline infusions, instead of angiotensin. The results (Bott et al. 1967a) were that in three animals AII reduced mean sodium intake by at least 100 mmol ($P<0.005$, $n=25$) relative to control experiments where intake ranged in different animals between means of 322 and 692 mmol of sodium.

It should be noted, however, that any conclusion from this experiment is limited. Though the AII infusion was short-term, the rate of infusion—120 or 60 μg/h—has been shown by clearance rate studies and direct measurement of AII in peripheral arterial blood to raise AII concentration by 120–60 ng/100 ml of blood. Thus, the initial AII concentration—in the range of moderate deficiency—was elevated to the very high physiological to pharmacological range by the experimental infusion. That is, the result is qualified in terms of acuteness of rise and supraphysiological level of angiotensin, in the same manner as in the water-drinking experiments discussed above.

This same experimental study did include similar infusions to three sheep with a parotid fistula, which had chronic hypertension due to renal artery constriction. The animals' control mean systolic blood pressures ranged from 148 to 158 mmHg (the normal range in sheep is 70 to 100 mmHg). The angiotensin infusion raised systolic pressure to 174–206 mmHg and the imminent presentation of sodium solution increased it a further 8–26 mmHg. Thus, severe hypertension existed at the time of the test. In 12 experiments, voluntary intake in response to 48 h of sodium deficiency was not changed relative to control intake established in 27–60 experiments on the individual animals (Bott et al. 1967a).

As far as they go, the data do not suggest angiotensin has a direct stimulating effect on salt appetite. It does not increase it from the 300–700 mmol seen after 2 days to the 500–900 mmol levels seen after 3 days of depletion (Denton et al. 1969). As noted earlier, Avrith et al. (1980a) found in rats that AII infused for 4 days at 25–50 ng $kg^{-1}min^{-1}$ had no influence on intake of water or 2.7% saline but at 75–100 ng $kg^{-1}min^{-1}$, which caused diuresis and a natriuresis, there was, after 24 h, an increased intake of both. Rats which received a 100 ng $kg^{-1}min^{-1}$ infusion of AII for 24 h also showed a large increased intake when the infusion was terminated. Fregly (1980) has shown that the addition of angiotensin converting enzyme blocker captopril to the food of rats caused, after a week, the development of salt appetite. This was accompanied by a 50-fold reduction in the threshold of detection of salt relative to control animals (from 0.030 mmol/l to 0.0006 mmol/l). The acceptance of glucose and saccharin was also changed by captopril. The interpretation of these interesting data is difficult in the absence of total external balance studies to determine whether sodium loss occurred in the captopril-treated animals as it did in the experiments of Avrith et al. (1980a).

The effect of bilateral nephrectomy on the sodium appetite of sodium-deficient sheep

Six sheep were used to study the effect of bilateral nephrectomy on salt appetite in sodium deficiency. They had the right kidney removed some months before the experimental period began. Table 22-1 shows the effect of removal of the second kidney on mean voluntary sodium bicarbonate intake during 15 min.

The studies on nephrectomy provided evidence on the question whether the renal renin–angiotensin system has any role in determining salt appetite. Interpretation of such experiments encounters the difficulty that failure to exhibit salt appetite after nephrectomy might reflect general non-specific depressive effects of the surgery, coupled with the possibility of similar effects due to progressive accumulation of waste products. Therefore, a finding on the retention of salt appetite and normal behaviour after nephrectomy would appear to carry more weight than the converse.

In sheep, as in other species, the pressor effect of AII infusion disappears within 2–4 min of stopping the infusion (Blair-West et al. 1962). In sodium-deficient nephrectomized sheep the pressor effect

Table 22-1. The effect of total nephrectomy on the voluntary intake of sodium bicarbonate solution by sodium-deficient sheep when tested 3–6 and 18–30 h after nephrectomy. The control observations (mean±s.d.) when only one kidney (right nephrectomy) had been removed are shown on the left. The upper group of four sheep was studied after 48 h of sodium deficiency; the lower group of two after 24 h. The plasma sodium and potassium concentrations at the time of the final episode of test drinking are shown also.

		Right nephrectomy		Left nephrectomy					
								Plasma electrolytes following final drinking episode (mmol/l)	
Sheep	Wt. (kg)	Mean sodium loss/48 h (saliva and urine) (mmol)	$NaHCO_3$ intake/48 h (mmol)	Sodium loss over 48 h (saliva and urine) (mmol)	$NaHCO_3$ intake 3–6 h after nephrectomy (mmol)	Sodium loss over following 21–27 h (saliva) (mmol)	$NaHCO_3$ intake 24–30 h after nephrectomy (mmol)	Na	K
Jessica	46	789±112 $n=22$	726±169 $n=22$	942	Nil	115	Nil	143	5.0
Napoleon	43	771±166 $n=15$	774±159 $n=15$	701	Nil	113	114	141	5.3
Peggy	33	628±70 $n=22$	662±124 $n=22$	564	270	241	420	143	7.9
Wyatt	50	685±97 $n=14$	602±128 $n=14$	643	714	300	285	155	6.2
		Mean sodium loss/24 h (saliva and urine) (mmol)	$NaHCO_3$ intake/24 h (mmol)	Sodium loss over 24 h (saliva and urine) (mmol)	$NaHCO_3$ intake 18 h after nephrectomy (mmol)				
Cecil	42	667±66 $n=24$	660±76 $n=24$	450	396			144	5.9
Nessim	35	456±67 $n=25$	498±139 $n=25$	204	Nil			152	5.5

of an intravenous infusion of sheep renin disappears within approximately 1–2 h of stopping the infusions (Blair-West et al. 1965). It seems likely in the experiments here that renin would have greatly decreased in the circulation 4–6 h after nephrectomy, and would have almost completely disappeared by 18–24 h. In dogs the half-life of renin has been estimated at 45–79 min (Assaykeen et al. 1968), in man 5–6 min (Oparil et al. 1970), 35–45 min (Devaux et al. 1968) and 42–120 min (Hannon et al. 1969), in rats at 30 min (Schaechtelin et al. 1964), and in sheep 40 min for a first component after nephrectomy, and ca 7 h for a second component (Blair-West 1976).

Two of the four sheep exhibited an appetite for sodium bicarbonate when tested 3–6 h after nephrectomy (Table 22-1) and in three of six instances the expected appetite in proportion to loss was observed 18–24 h after nephrectomy. Of the other three sheep, appetite was abolished in two, and a third did not drink in the first test but ingested some sodium solution when tested 24 h post-nephrectomy. These three sheep which did not drink were apathetic after operation, and in contrast to the other three they all showed little appetite for food. The question should be considered whether a process of learning by the sheep to drink a certain volume of sodium solution might, despite the hypothetical absence of the primary stimulus—the renin–angiotensin system—have determined the results in these experiments. Against this is the fact, from experiments on salt appetite in sheep, that appetite is related to deficit: the voluntary intake after 1 day's uncompensated salivary loss is significantly different from that after 2 or 3 days' loss (Bott et al. 1967). Furthermore, the correction of sodium deficit by intravenous infusion of sodium or by giving the animal its own saliva back by rumen tube substantially reduces or abolishes sodium appetite according to the amount administered and the time elapsed between sodium administration and the testing of appetite. These observations are against a hypothesis that a sheep drinks some habitual volume, reflecting learned behaviour independent of the state of sodium balance, and there is no evidence suggesting that such a process operated here. Overall, the results may be interpreted as indicating that the level of renin and angiotensin in circulating blood does not have any immediate, direct influence in stimulating salt appetite in sodium deficiency.

In contrast to sodium appetite, with the thirst evoked by constriction of the inferior vena cava in the rat, nephrectomy causes a substantial reduction in water ingestion. It does not reduce water ingestion in response to increased extracellular osmotic pressure produced by hypertonic saline injection (Fitzsimons 1966). It is also relevant to note here in relation to these experiments in conscious sheep that, by 12–24 h after nephrectomy, aldosterone secretion, and thus peripheral plasma concentration, would have been at a low level (Blair-West et al. 1965). Thus, the data also provide evidence which indicates salt appetite is not simply determined by the level of aldosterone in circulating blood (see Chapter 18).

Studies on the renin–angiotensin system and salt appetite in the rat

Nephrectomy

Fitzsimons and Stricker (1971) approached this problem with a series of operative procedures in the rat. The animals were anaesthetized and bilaterally nephrectomized. Thirty minutes after the operation acute redistribution of body sodium was produced by subcutaneous injection of 2.5 ml of buffered 1.5% formalin into the loose tissue at the back of the shoulders. The animals were then given access to water and either 0.9% or 3% saline solutions. Results of the experiments and their controls are set out in Table 22-2. Perhaps, the data with the unpalatable 3% saline are the more pertinent. One of the several facets of the data which are noteworthy is that the formalin-injected sham-operated controls drank twice as much water as their nephrectomized counterparts. The 3% saline intake of the nephrectomized animals was very much reduced relative to the sham-operated controls injected with formalin. However, the formalin-injected nephrectomized animals did increase sodium chloride intake relative to the nephrectomized controls—approximately fourfold—though this intake was no greater than the normal sham-operated animals drank over the 24-h period. In the light of this clear reduction of sodium appetite of the formalin-treated nephrectomized animals, they then injected formalin subcutaneously 30 min after bilateral nephrectomy, and either 2.5–5 units of hog renin or an aqueous homogenate from two rat kidneys was injected intraperitoneally 0–3 h later. Only 1 out of the 13 rats showed substantial fluid intake, in this instance 12 ml of 3% saline during the 24 h. The other 12 consumed an average of 28.5 ml water and 3.1 ml of saline solution. Thus the administered renin was a potent stimulus of water intake, but not of 3% saline intake. Specificity of AII on water intake under cafeteria conditions involving concurrent presentation with hypertonic salt has been demonstrated in the sheep (Abraham et al. 1976b).

Fitzsimons and Stricker (1971) also considered the possibility that nephrectomy influenced sodium appetite by a reduction of aldosterone levels. They note, however, that sodium appetite is observed in formalin-treated rats, even when aldosterone secretion is abolished by adrenalectomy. Under their experimental conditions here, they were unable to restitute appetite by the administration intramuscularly of 3–40 mg of desoxycorticosterone acetate (DOCA) 4 h after formalin treatment. In order to determine whether the failure of the nephrectomized animal to drink saline was attributable to anuria, they also studied a group of animals with ureters ligated near the kidneys and another group where the fundus of the urinary bladder was amputated and urine allowed to drain into the peritoneal cavity, but not be lost from the body. In these conditions, formalin again increased water intake, but sodium intake did not increase. Thus, eliminating normal kidney function, but leaving the kidneys in situ, prevented the appearance of sodium appetite in sodium-deficient rats—though presumably the animals could still

Table 22-2. The amounts of water and saline drunk (mean±s.e.) by normal or by nephrectomized rats offered either water and 0.9% saline, or water and 3.0% saline, after injection with 2.5 ml of 1.5% formalin subcutaneously. Controls received 2.5 ml of 0.9% saline subcutaneously. Amounts in ml.

	6 h		24 h		6 h		24 h	
	Water	0.9% NaCl	Water	0.9% NaCl	Water	3.0% NaCl	Water	3.0%NaCl
Sham-operated								
Control (n=15)	0.8±0.3	6.0±1.8	4.4±1.2	41.3±5.1	2.9±0.9	0.4±0.2	19.4±3.2	2.6±0.9
Formalin-injected (n=15)	3.6±0.8	14.6±2.9	9.0±2.2	65.3±5.6	11.8±1.0	4.1±0.6	41.0±2.6	18.4±1.6
Nephrectomized								
Control (n=7)	1.3±0.8	0.3±0.1	19.0±2.0	1.9±1.1	4.5±1.0	0.2±0.1	21.9±2.6	0.6±0.3
Formalin-injected (n=14)	3.3±0.8	1.5±0.5	14.3±1.7	7.3±2.5	9.6±1.7	1.9±0.7	23.9±3.2	2.5±0.7

From Fitzsimons and Stricker (1971), with permission.

release renin. Such procedure does not abolish the normal dipsogenic response to cellular dehydration and hypovolaemia.

They also studied the influence of isoprenaline administered subcutaneously on the drinking behaviour of rats. They found that rats so treated showed a marked preference for water over 0.9% saline. The lack of interest in saline shown by isoprenaline-treated rats contrasted with the increased saline preference of formalin-treated rats. They suggested the data were consistent with reports that elevation of plasma renin activity by infrahepatic ligature of the inferior vena cava or by intraperitoneal administration of renin or renal extracts does not stimulate sodium appetite in rats (Fitzsimons 1969).

Although nephrectomy greatly reduced the salt appetite with formalin injection, the authors did not see in their data evidence for a direct and obvious role of the renin–angiotensin system in sodium appetite, the result being generally consistent with the data on nephrectomy in sheep published earlier. Why nephrectomy significantly reduces sodium appetite in rats has never become quite clear (Stricker 1978), though blood pressure decrease and deterioration of the animals in experiments along the lines described above may be important.

In experimental conditions where choice was between water and highly preferred 1% saline, Chiaraviglio (1976) showed that 60–90 min after nephrectomy the intake of saline was unaffected, but that 3 h later it was substantially reduced. This did not occur after sham nephrectomy or unilateral nephrectomy. Urethral ligature reduced intake by half only. In contrast to the results of Fitzsimons and Stricker, it was found that renin injected intraperitoneally re-established the sodium appetite abolished by nephrectomy. Comparison between these experiments, however, is difficult in the light of use of the highly preferred isotonic saline along with water. Injection of spectacularly high amounts of AII (5–50 ng) into the third ventricle restored sodium appetite in the sodium-deplete rat in which it had been abolished by nephrectomy.

Intracranial injection of angiotensin II and salt appetite

Fisher and Buggy (1975) were the first to show, using cannulae implanted in the preoptic and septal region, that intracranial angiotensin could provoke increased sodium intake. With very high doses (10, 100, 1000 ng of AII) they found an immediate increase in preference for isotonic saline and also 1.8% saline over water following injection. With 1.8% saline the dose used was 500 ng AII. With animals on a low-sodium diet, angiotensin also caused a substantial immediate increase in sodium intake. With 8 h of infusion of 500 ng/μl per 8 min into the ventricle, 'drinkometer' records revealed that intake of saline often began as soon as water intake. Sodium intake was greater than loss in the urine, and the animals ended the period in positive balance. Increased saline intake occurred with both 1.8 and 2.7% saline. Calculations of ventricular concentrations of AII likely to have been achieved with doses used suggest levels four or five magnitudes above the highest measured in CSF.

Fitzsimons and Wirth (1978) found that pharmacological activation of the renin–angiotensin system with isoprenaline or phentolamine caused increased water intake but did not stimulate sodium appetite in sodium-replete adrenalectomized rats, and decreased it in sodium-deplete adrenalectomized animals. Intravenous infusion of AII at 58.5 ng $kg^{-1}min^{-1}$ for as long as 24 h (which would give a systemic AII concentration of ca 60 ng/100 ml, a high physiological level: Abraham et al. 1975) into intact rats caused a significant increase in water intake but had no effect on intake of 2.7% saline. Neither the normal group nor the sodium-replete adrenalectomized group showed any increase of intake of 2.7% saline. The authors state that as with pharmacological activation of the renin–angiotensin system, exogenous renin and AII did not stimulate sodium intake even in the adrenalectomized rat which is highly experienced at taking aversive concentrations of saline. Fitzsimons and Wirth (1978) did not appear to confirm the immediate preference for hypertonic saline reported by Fisher and Buggy (1975) following preoptic intracranial injection of AII in intact rats. However, with sodium-replete adrenalectomized rats the components of the renin–angiotensin system given intracranially increased intake of 2.7% saline as well as water. With 100 ng AII there was significant intake of saline as early as 5 min. In sodium-deplete adrenalectomized rats the components of the renin–angiotensin system given intracranially never affected sodium intake. The authors suggest that even though the adrenalectomized rat is highly experienced in taking aversive concentrations, the effect on saline intake was unremarkable compared with that on water intake.

The papers of Avrith and Fitzsimons (1980), Bryant et al. (1980) and Coghlan et al. (1981) greatly change emphasis in regard to a possible role of angiotensin in salt appetite, though the physiological interpretation of the effects produced experimentally remains an open question.

Avrith and Fitzsimons (1980) found single pulse injections of AII into the preoptic area caused an initial increase in water intake and delayed sodium appetite. However, significant increases in sodium appetite (for 2.7% saline) were seen within an hour with two of the three dose levels used (10, 100 and 1000 pmol dissolved in 1 μl). The lowest dose (10 pmol) represents ca 10 ng in 1 μl=1 mg/100 ml, which is about 100 000 times the largest concentration measured in CSF. Renin (1–10 mU) also had the same powerful effect, increasing intake of 2.7% saline 10–30-fold over 18 h relative to control conditions. The influence of continuous infusion of AII over some days was then examined and the results were clear-cut. Infusion of 1–10 pmol/h into the third ventricle produced large intake of 2.7% saline as well as water, which in one animal reached over 100 ml of 2.7% saline per day. Carbachol did not have this effect on sodium intake. The drinking returned to baseline the day after the infusion was terminated (Fig. 22-4).

The authors state that these results of the long-term infusion experiments vividly demonstrate that prolonged exposure to very low amounts of AII can produce a sodium appetite that gradually builds up to levels observed after adrenalectomy or any other technique used for generating sodium appetite. The results are proposed as evidence that AII may play a physiological role in the development of salt appetite under natural conditions where AII levels are elevated.

On this issue of very low amount, it would be open to disagree. The volume of the third ventricle of the rat has been estimated at 1–2 μl by Prof. R.D. Wright using the stereotaxic atlas of the rat brain of Konig and Klipper (1963). Concentrations of CSF AII in the rat as high as 35.9 ng/100 ml have been reported (Simpson, Saad and Epstein 1976) though problems of accurate measurement in volumes of CSF feasible in the rat are formidable. The lowest rate used here (1 pmol/h=16.6 fmol/min) produced an increase in intake of 2.7% saline in two of three animals. Assuming the volume of the third ventricle to be 0.001–0.002 ml, 16 pg entering this in the first minute would contrive a concentration of ca 800–1600 ng/100 ml. By 5 min the concentration would have been extremely high. Furthermore, the half-life of AII in CSF may be considerably prolonged relative to that found in peripheral blood (D. Ramsay personal communication). Indeed, Avrith and Fitzsimons (1980) consider ascribing the greater effect of long-term infusion than pulse injection to a build-up of AII in the brain. Whereas it is entirely feasible that the local concentration of AII may be very high if produced endogenously at specific sites

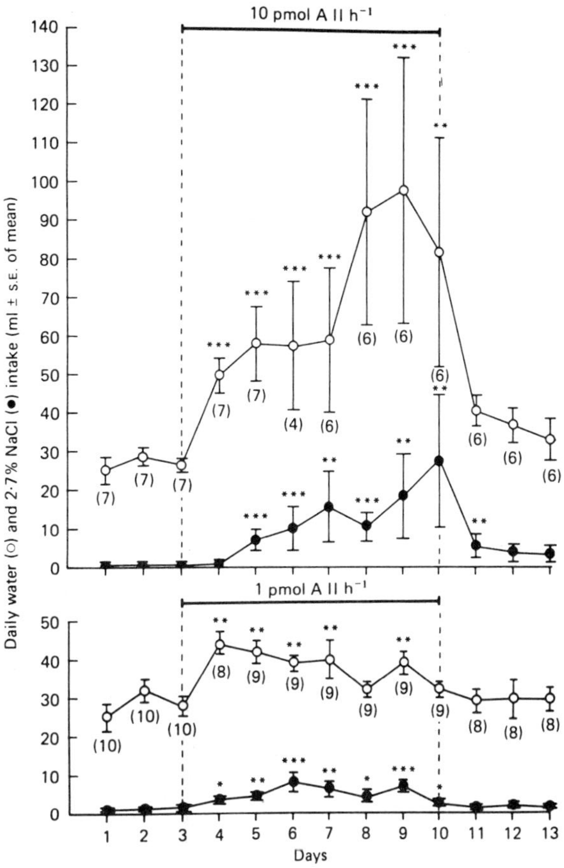

Fig. 22-4. Mean daily water (open circles) and 2.7% saline (filled circles) intakes (±s.e. of mean, number of rats in parentheses) of rats with access to both solutions in response to infusion of AII (1 or 10 pmol/h infused at 2 or 1 μl/h respectively) into the third cerebral ventricle over a period of 7 days. **P*<0.05, ***P*<0.01, ****P*<0.001 compared with the three preinfusion baseline days. (From Avrith and Fitzsimons 1980, with permission.)

in the brain, and that high ventricular concentration may be required to contrive such concentration at the site (as will be discussed further on), it should not be overlooked at the outset that these concentrations produced by the very lowest rate of infusion are pharmacological. This is if levels presently measured in body fluids in the instance of proven physiological effects are any criterion.

Avrith and Fitzsimons (1980) showed that the sodium appetite was specific and not secondary to sodium loss in the urine. Independent of the quantitative considerations raised above, the data are evidently very interesting in terms of accessing and activating the systems subserving salt appetite. The authors remark on the delayed onset of appetite: this is discussed in full in Chapter 24 in relation to possible mechanisms (Denton 1966). They suggest here the delay may be accounted for by delays in activation of other endocrine systems, or induction of delayed biochemical changes which are the immediate cause of sodium appetite. For example, AII increases noradrenaline activity in the nervous system by blocking reuptake, and it has been shown that noradrenaline stimulates sodium appetite in the rat (Chiaraviglio 1976). In our experience, intraventricular infusion of noradrenaline (E. Tarjan, unpublished) at 50 μg/h for 1 h before and during the first 30 min of sodium presentation to mildly sodium-deficient sheep does not influence sodium appetite. Finally, Avrith and Fitzsimons (1980) consider that prolonged changes in neural activity brought about by an AII-induced activation of brain Na,K-ATPase activity might also contribute to the stimulation of sodium appetite. In considering the matter of stimulation of thirst, however, Fitzsimons (1979) wrote that it is difficult to conceive of such a universal attribute of the cell as its Na,K-ATPase activity as playing a specific role in thirst.

Bryant et al. (1980) described similar experiments with higher doses of AII which revealed other facets of its effect. Pulse injections of 10 ng AII into the brain (it made no difference whether into the preoptic area or the lateral ventricle) resulted, upon repetition 2–3 days apart, in an increase in 2.7% saline drunk in the hour after injection. By the fifth injection saline intake had increased six-fold whereas carbachol had no effect of this type. Furthermore, both isoprenaline subcutaneously (100 μg/kg) and intraperitoneal injection of 10 Goldblatt units of renin when repeated in the same fashion had the same effect on saline intake. Repeated episodes of cellular dehydration produced by intraperitoneal 2 M sodium chloride or 2 M sucrose did not have this effect. In a second series of experiments infusions of AII were made for 4 days into the antero-ventral third ventricle at rates of 6, 60, 600 and 6000 ng/h in 1 μl/h. I have commented above on the implications of this order of concentration change in CSF. Under control conditions there was little or no intake of 3.0% saline (Fig. 22-5). With 6 ng/h of AII a small (mean=7.8 ml) but statistically insignificant amount of saline was consumed. At rates of 60, 600 and 6000 ng/h marked ingestion of 3% saline occurred (means of 58.1, 39.0 and 100.9 ml/day respectively). The rate of 6 ng/h was proposed to be very close to threshold. However, three out of five rats on at least 2 days during AII infusion did not drink any saline although excessive water intake occurred. There was no difference between 60 and 600 ng/h.

The right-hand panel of Fig. 22-5 shows another characteristic phenomenon in the experiments: the persistence of dose-dependent intake of 3% saline solution after cessation of infusion. Six rats which

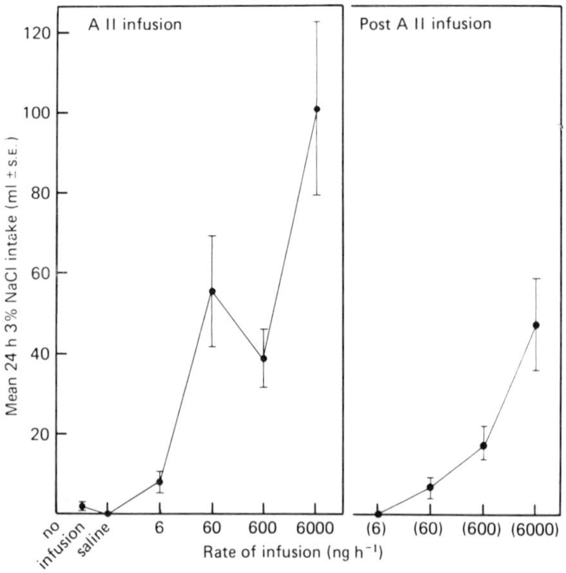

Fig. 22-5. Dose–response curves showing mean 24-h 3.0% saline intake during no infusion, 0.9% saline infusion, and the last 2 days of a 4-day infusion of AII at 6, 60, 600 or 6000 ng/h (left-hand panel, AII infusion). The right-hand panel (post AII infusion) shows mean 3.0% saline intake for the 4 days following termination of AII infusion in the same animals arranged according to previous dose of AII (doses in parentheses). (From Bryant et al. 1980, with permission.)

were infused with the 6 μg/h dose into the third ventricle were observed for periods of 21–65 days after infusion, during which time intakes of 3% saline were never less than 5 to 66 times their preinfusion baseline intakes. Fig. 22–6 shows the

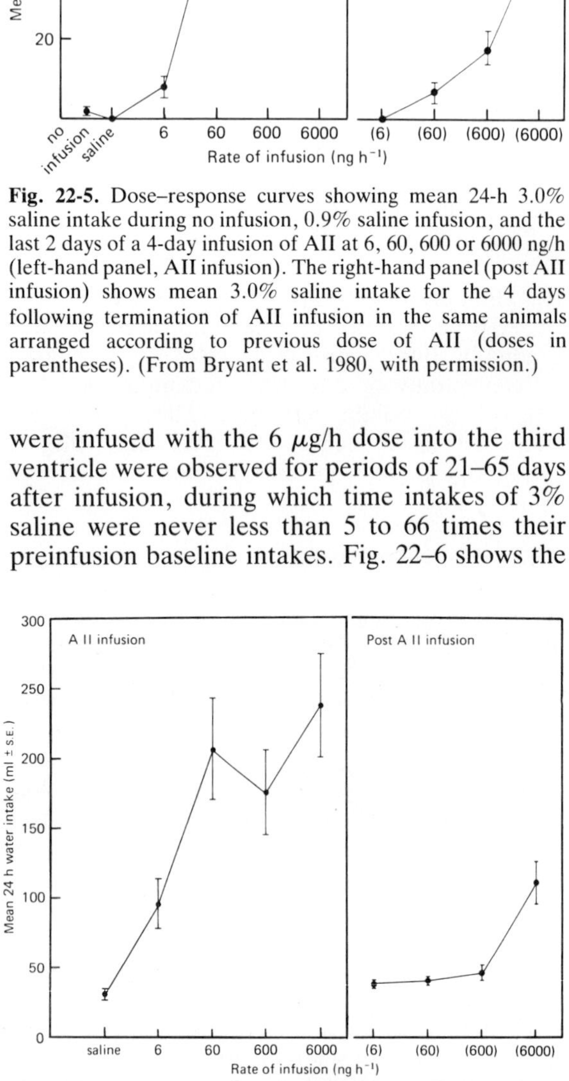

Fig. 22-6. Dose–response curves showing mean 24-h water intake during the last 2 days of 0.9% saline infusion and AII infusion at 6, 60, 600, 6000 ng/h (left-hand panel, AII infusion). The right-hand panel (post AII infusion) shows mean water intake for the 4 days following termination of AII infusion in the same animals arranged according to previous dose of AII (doses in parentheses). (From Bryant et al. 1980, with permission.)

water intake of the animals under these conditions. Bryant et al. propose that the dose–response curve is similar to the salt intake curve (Fig. 22-5). The lowest rate of infusion (6 ng/h) evoked a proportionately larger increase in intake of water than of 3% saline, although the increase was not statistically significant. There were no significant differences between the intakes of water with infusions of 6000 ng/h and of 60 and 600 ng/h. All these levels were, however, significantly different from the 0.9% saline control ($P<0.01$). The rats infused with AII usually ate and gained weight normally through the experiment, and according to the authors showed no signs of upset. The persistence of the water appetite after infusion was presumably secondary to the persistent intake of saline. The authors showed that the appetite was specific for sodium chloride and also not secondary to consumption of large amounts of water. With drinkometer experiments it was evident that there was wide variation in the onset of saline drinking with intraventricular infusion of angiotensin. Some started drinking saline immediately whereas in others the onset was delayed 6 h. Most of the drinking occurred in the dark, notwithstanding continuous infusion of AII.

Interpretation of dosages used

The data show that the brain mechanisms determining sodium appetite can be driven, as Bryant et al. (1980) remark, to extremes of excess by the sustained exposure of the brain to AII. At rates of 60 ng/h to 6 μg/h, 46% of animals drank at least 100 ml/day (ca 50 mmol sodium), which represents turnover of about five times the sodium content of the extracellular fluid per day. The appetite had characteristics of the natural phenomenon though the delay in onset was clearly not consistent. The capacity of repeated systemic elevation of AII concentration to generate an effect along these lines is noteworthy, though there is no information on the levels of peripheral AII generated by the very large dosages used to contrive these effects (e.g. 10 Goldblatt units in a rat). Nor were effects on blood pressure recorded. Sudden marked increase of systemic blood pressure with excessive doses of AII may allow the peptide through the cerebral capillaries (Wislocki and King 1936). Evidence for temporary breakdown of the blood–CSF barrier and blood–brain barrier came from work of Deshmukh and Phillips (1977), who combined horseradish peroxidase and a 2 μg/kg intravenous infusion of AII. After 5 min horseradish peroxidase could be detected in periventricular tissue and other brain sites whereas at normal blood pressure there was crossing of horseradish peroxidase only at the SFO and OVLT.

The data do, however, bring to focus the fact that infusion of AII at 6 ng/h (100 pg/min initially into the volume of the third ventricle) was near threshold. The striking way the appetite is sustained after cessation of infusion suggested to Bryant et al. (1980) that this might occur in the absence of endocrinological determinants by a conditioning process during which the taste of strong saline is associated with the benefits of sodium ingestion. Whether the ingestion of, and, as balances show, loading with very large amounts of hypertonic saline in these animals would give any retroactive sense of benefit is problematical. Though laboratory rats are more inclined to overdrink with sodium deficiency, and appetite may not cut out with satiation as with other species, persistence of this type would seem quite uncommon, and contrary to much experimental data cited by Richter (1956): as, for example, when DOCA was given to adrenalectomized rats or when eye implants of cortical cells were made in adrenalectomized animals. It raises the issue of whether some element of the hedonic reaction to salt is specifically affected by the procedures as distinct from the mechanisms responsive to sodium deficiency. Against this interpretation, however, is the fact that with the powerful salt appetite drive induced by *physiological* amounts of the hormones involved in normal pregnancy and lactation an 'after discharge' effect persisting for days to weeks is frequently seen (Chapter 23). In the light of known mechanisms of transcription and translation involved in hormone action on target cells this could well reflect the fact that the neuronal chemistry has been profoundly modified and the effect persists for some time. Similar processes may be involved here with AII and, for that matter, it is reported (Davies et al. 1969; Khairallah et al. 1972; Ganten et al. 1976) that AII can influence DNA, RNA and protein synthesis.

Bryant et al. (1980) remark that specification of an effective dose of AII for induction of appetite is impossible since the hormone has been infused at an unknown distance from unknown receptors, and the route of CSF is probably an unphysiological route of access to the brain for AII. They say, however, that the rates are not larger than those used to elicit thirst by intraventricular injection of the hormone, and that lowest effective rate (100 pg/min) (it was nearer 20 pg/min in Avrith and Fitzsimons' (1980) experiments) was not an excessive dose for production of dipsogenic response—a reflection which allows more than one interpretation as the discussion on quantitative aspects of intracerebral angiotensin injection and water drinking set out earlier suggests.

Nerve growth factor

Before the appearance of these two important papers discussed above, Lewis, Avrith and Fitzsimons (1979) reported short-latency drinking and increased sodium appetite in rats following intracerebral microinjections of nerve growth factor (NGF). This polypeptide is trophic for peripheral sympathetic and sensory neurones, and both NGF and NGF receptors have been identified in brain (Johnson et al. 1971; Szutowicz et al. 1976). Following the report of Perkins et al. (1979) of crystalline NGF in the diencephalon causing intense polydipsia, Lewis et al. (1979) made preoptic injection of NGF in 1, 10 and 100 pmol doses. A very large dose-related water ingestion was produced in the first hour and, with later onset, a very large intake of 2.7% saline occurred over 18 h. This salt appetite effect was specific. With subsequent injections intake of both water and saline became greater. The data with experiments when only one fluid was available showed that water intake was not necessary to generate a sodium appetite, nor was overnight water intake secondary to the sodium appetite. The increased intakes with repeated injections of NGF were attributed to a powerful learning or sensitization process. At the time, the authors drew attention to the marked similarity between the effects of intracranial injections of NGF and of renin, both with respect to immediacy of influence on water intake and delay of effect on sodium appetite. They raised the issue of whether effect of NGF is mediated by the cerebral isorenin–angiotensin system or by an unidentified neural system involved in behavioural regulation of fluid and electrolyte balance.

Levi-Montalcini (1980) pointed out that isorenin in 2.5 S NGF preparations is separated only after laborious additional purification steps (Cozzari et al. 1973). She proposed that NGF prepared with the same method as that used in the studies on polydipsia and sodium appetite still showed considerable renin activity, and it should be clearly established that preparations are entirely free from contaminants, particularly preparations from mouse salivary glands, where NGF is tightly bound to other proteins with enzymatic or other biological activities. In response, Lewis et al. (1980) concurred, and reiterated their having drawn attention in their initial report to the marked similarity between effects of intracranial renin and those of 2.5 S NGF, and that the effects of 2.5 S NGF might be mediated by the cerebral isorenin–angiotensin system. Theoretically this might mean that NGF in some way activated this system, or alternatively that there was an isorenin contaminant. They showed pressor activity with 2.5 S and 7 S NGF in rats which was abolished by converting enzyme inhibitor and saralasin, and also NGF antiserum. Somewhat paradoxically they reported earlier (Fitzsimons et al. 1979) that the response to NGF but not that to renin was attenuated by infusing NGF antiserum intracranially for 1 h before injecting NGF. In the 18 h after intracranial injection of 25 pmol NGF rats drank 33.52±6.08 ml of water and 1.47± 1.25 ml of 2.7% saline (n=8) when pretreated with NGF antiserum, compared with 146±33.15 ml water and 13.03±6.64 ml 2.7% saline when pretreated with control serum. Responses to 100 mU renin were little affected by NGF antiserum. But intracranial injection of converting enzyme inhibitor abolished both responses. It is not clear in the report whether the antiserum was prepared with purified NGF, and secondly whether the antiserum would block the pressor response to renin as well as that to 2.5 S and 7 S NGF. Chapman et al. (1979) found that immunogenically pure β NGF subunit from mouse submandibular gland was devoid of pressor activity and had a much smaller and more variable effect on thirst and sodium appetite than other preparations of NGF.

Lewis et al. (1980) conclude in relation to other effects of NGF reported that unless renin-free preparations of NGF are employed, the use of AII antagonists such as SQ 14225 and saralasin should be considered obligatory, particularly when a biological effect attributed to NGF is known also to be produced by renin or AII.

In 1979, in the light of the first report on the action of NGF, our group decided in collaboration with Dr Ralph Bradshaw to investigate its effect on salt appetite in sheep. We also examined the effect of renin and angiotensin and whether there was any clear evidence of a cerebral renin–angiotensin system being involved in genesis of the salt appetite of sodium deficiency (Coghlan et al. 1981).

Infusions at 0.38 ml/h or injection of 0.25 ml of test substance dissolved in an artificial CSF vehicle were made into a lateral ventricle. We confirmed in the sheep that intraventricular (i.v.t.) administration of 0.025 Goldblatt units (U) of partially pure sheep renin (1 U/mg) elicits rapid water drinking within 10 min and, after a longer latency (6–12 h), increased intake of a hypertonic sodium solution (Fig. 22-7). Lower dosage of renin had no effect on sodium appetite or thirst (Table 22-3). The group of animals had a moderate baseline

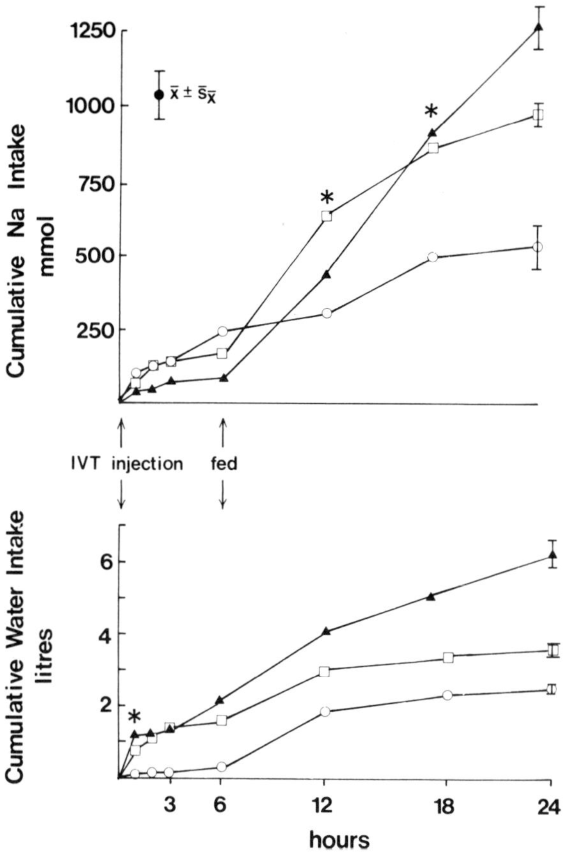

Fig. 22-7. Time course of effect over 24 h on intake of 600 mM sodium bicarbonate and water after i.v.t. injection of 50 μg 2.5 S NGF (filled triangles, n=5) or 25 μg sheep renin (assayed at 1.0 U/mg) (open squares, n=7). Normal intake (control) on a day where no injection was made is also represented (open circles, n=7). Mean intakes and the mean standard error of the mean are also shown $\bar{s}_{\bar{x}}$. The asterisks indicate the point at which intake of sodium or water became significantly greater ($P<0.05$, Student's t-test) than the control intake.

hedonic intake of saline. The effects of i.v.t. renin were blocked by central administration of the converting enzyme inhibitor teprotide (SQ 20881, Squibb) (Table 22-3). Both effects were attenuated or blocked also by i.v.t. infusion of saralasin at 10 or 30 μg/h. Intraventricular AII, at 1.9 μg/h for 24 h, more than doubled water intake but there was no effect on sodium appetite despite the fact that within only 5 min the concentration of AII entering the third ventricle probably exceeded by a hundred-fold the highest physiological level measured in sheep CSF. This was calculated as follows. Measurements of casts of the lateral ventricle in the sheep give a volume of 2–3 ml. CSF production rate is probably about 10 ml/h (cf. Heisey et al. 1962) and there is little or no angiotensinase activity in CSF (Reid 1977). Thus AII infusion at 1.9 μg/h would give a concentration of at least 1200 ng/100 ml after 5 min. The highest AII concentration in CSF in severe sodium deficiency recorded in sheep was 12 ng/100 ml. However infusion of twice as much AII (3.8 μg/h for 24 h) more than doubled daily sodium bicarbonate and water intake. This effect on sodium appetite was not reduced by i.v.t. teprotide

Table 22-3. Effect on mean intake of 0.6 M sodium bicarbonate or water of intraventricular (i.v.t.) injection of renin or nerve growth factor preparation or i.v.t. infusion of angiotensin II. These agents have been administered with or without i.v.t. treatment with converting enzyme inhibitor injections (teprotide, 100 μg each 6 h for 24 h), i.v.t. infusions of saralasin (10 or 30 μg/h for 24 h). Baseline values represent the mean intake on the pre-treatment days (mean±s.e.m.).

Treatment	n	Mean±s.e.m. intake 0.6 M $NaHCO_3$ (mmol/24 h)	Mean±s.e.m. intake water (ml/24 h)
(*a*) Baseline	10	331±122	1780±127
0.0025 U renin	5	426±193^{ns}	1610±152^{ns}
(*b*) Baseline	14	407±65	1802±99
0.025 U renin	7	979±112***	3611±564***
(*c*) Baseline	8	551±91	1681±77
0.1 U renin	4	1432±262***	6013±469***
(*d*) Baseline	10	395±72	1715±116
0.025 U renin+			
100 μg/6 h teprotide	5	210±66*	1350±127^{ns}
(*e*) Baseline	10	258±51	1770±310
0.025 U renin+			
10 μg/h saralasin	5	559±153*	2130±536^{ns}
(*f*) Baseline	10	183±63	2462±252
50 μg NGF	5	1266±103***	6310±1026***
(*g*) Baseline	12	460±88	1814±176
50 μg NGF+			
100 μg/6 h teprotide	6	287±119^{ns}	1366±83^{ns}
(*h*) Baseline	4	135±53	1450±215
1.9 μg/h AII	4	180±122^{ns}	4150±618**
(*i*) Baseline	18	344±67	1822±179
3.8 μg/h AII	9	780±242**	4205±570***
(*j*) Baseline	10	210±74	1812±110
3.8 μg/h AII+			
100 μg/6 h teprotide	5	817±264***	4455±863***
(*k*) Baseline	8	179±60	1638±137
3.8 μg/h AII+			
10 μg/h saralasin	4	585±220*	2388±613^{ns}
(*l*) Baseline	10	234±68	1300±59
3.8 μg/h AII+			
30 μg/h saralasin	5	426±137^{ns}	1980±313**

After Coghlan et al. 1981, with permission.
*** $P<0.001$, ** $P<0.01$, * $P<0.05$; ns, not significant.
Data were analysed by two-way analysis of variance and subsequent t-test.

(100 μg every 6 h for 24 h) but was eliminated by i.v.t. saralasin (30 μg/h for 24 h) (Table 22-3). The increased water intake with this dose of saralasin but not with 10 μg/h may represent agonist activity on water intake.

Sheep were also injected intraventricularly with 50 μg of 2.5 S NGF or 0.025 U sheep renin, and 75 min later CSF was withdrawn and assayed for AII. Levels were greatly elevated over the normal AII level in CSF of 15.1±8.9 fmol/ml (n=18), to 639±119 fmol/ml with i.v.t. NGF and to 816±161 fmol/ml with i.v.t. renin. On another day, i.v.t. injection of 50 μg 2.5 S NGF was found to cause rapid water drinking within 5–20 min and large sodium appetite after 6–18 h (Table 22-3 and Fig. 22-7). Both effects were abolished by 6-hourly i.v.t. administration of teprotide (Table 22-3). The 2.5 S NGF used in these experiments and those of Lewis et al. (1979) was prepared by the method of Bocchini and Angeletti (1969) and is contaminated by mouse submandibular gland isorenin (Cozzari et al. 1973). The dipsogenic and salt-appetite-inducing actions of centrally administered crude NGF preparations in sheep, rats (Avrith et al. 1980) and dogs (Ramsay et al. 1980) therefore appear to be due to generation of AII subsequent to the action of an isorenin contaminant of the NGF preparation.

Significance of results with intracranial infusion of renin and angiotensin

What then is the significance of this striking salt appetite generated by intracerebrally administered renin or AII? The evidence against circulating renal renin and angiotensin as a determinant of salt appetite in sodium deficiency has been summarized earlier in this chapter.

The possibility clearly exists that AII measured in CSF is locally generated in the brain. Persistence of AII in CSF after nephrectomy supports this (Ganten et al. 1976; see also Fig. 7-10). Accordingly, the very high concentration required in CSF for appetite effect may reflect that required to cause appropriate concentration at a distant site in the brain, such as is generated by local production under some stimulus associated with sodium deficiency.

An alternative explanation is that a precursor of AII is directly synthesized in neurones and AII acts as the neurotransmitter or a modulator of peptidergic neurones (Fuxe et al. 1976) which subserve salt appetite. The massive i.v.t. doses required may reflect penetration at sufficient concentration to a distant synaptic site in the brain substance whereas the rapid action of AII on thirst may be because of the juxtaventricular situation of the reactive elements. However, the hypothesis of a massive dose being needed in order to reach a distant brain site is not supported by the fact that a 10 mmol/l rise or fall in [Na] of CSF can have a large effect on salt appetite in 15 and 30 min respectively. That is, for example, an infusion of isotonic mannitol in artificial CSF which reduces CSF [Na] by 10–20 mmol/l appears to contrive a fluid 'front' which moves relatively rapidly through ependyma and the tortuous intercellular channels to the neurones controlling salt appetite. It does not require that the change in [Na] of CSF be outside the physiological range to give a clear effect. In fact, a change of 4–6 mmol/l in CSF [Na] is effective (see Chapter 24).

In terms of probing the relevance of angiotensin-induced sodium appetite to physiological salt appetite caused by sodium deficiency, the blocking by saralasin infusion of thirst and salt appetite effects caused by i.v.t. renin in sheep, and also rats (Lewis et al. 1979), shows that the competitive antagonist can effectively access sites where AII acts.

The effect of the i.v.t. administration of drugs which block the renin–angiotensin system was investigated in sheep with a parotid fistula made sodium deficient by 22 h of salivary loss (deficits 350–550 mmol sodium). They were then allowed to bar-press for sodium bicarbonate for 2 h (each delivery to drinking cup was 15 ml of 600 mmol/l sodium bicarbonate=9 mmol of sodium: see Chapter 13). It was found that i.v.t. infusion of saralasin at 10 or 30 μg/h for 24 h had no effect on sodium or water intake during sodium deficiency (Fig. 22-8). As well as giving saralasin at the dosage which over 24 or 48 h blocked the effect of i.v.t. AII, teprotide was given i.v.t. for 24 h with the same regime which blocked completely the effect of i.v.t. renin or 2.5 S NGF. Similarly i.v.t. teprotide or captopril (SQ 14225, Squibb) had no effect on sodium or water intake in sodium deficiency (Fig. 22-8).

Additional to these findings, it was shown that the same sodium-deplete animals responded predictably to physiological manipulation of sodium status. If 1.0 litre of 600 mM sodium bicarbonate was given into the rumen by tube 3–4 h before access to bar-press, mean intake was reduced to very low level (Fig. 22-8). Giving 1.0 litre of water in the same fashion on three preceding days had no effect. Allowing the animals to become sodium-deplete for 46 h instead of 22 h almost doubled the number of bar-press deliveries (Fig. 22-8). For comparison the effect of i.v.t. infusion of isotonic or 0.7 M mannitol in artificial CSF begun 60 min beforehand (Weisinger et al.

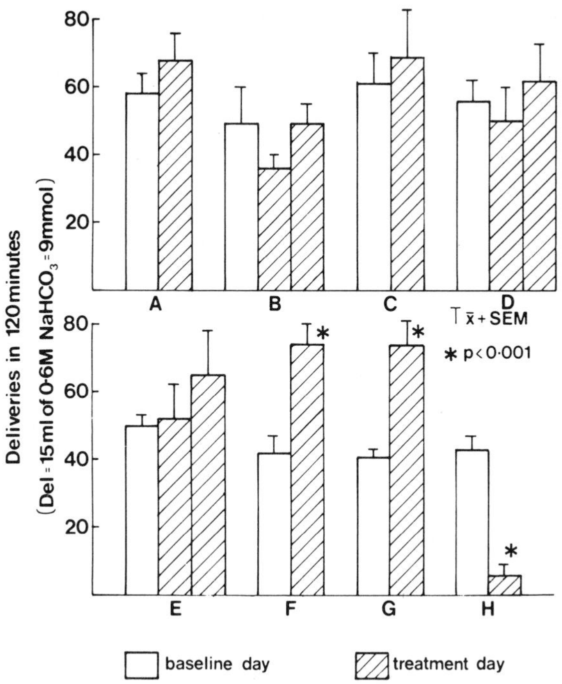

Fig. 22-8. The effect of various treatments on the bar-pressing behaviour of sodium-deficient sheep. Animals were made sodium-deficient by loss of saliva for 22 h, then were allowed to bar-press for 0.6 M sodium bicarbonate (15 ml per delivery=9 mmol). The baseline shown by the open columns is the number of deliveries on the preceding day. Treatments (shown by the hatched bars) were: A, injection of 100 μg teprotide into a lateral ventricle every 6 h for 24 h, n=4; B, injection of captopril (20 μg/6 h) for 48 h into a lateral ventricle, n=6; C, infusion of saralasin (10 μg/h for 24 h) into a lateral ventricle, n=4; D, infusion of saralasin (30 μg/h for 48 h) into a lateral ventricle, n=5; E, infusion of angiotensin II (3.8 μg/h for 48 h) into a lateral ventricle, n=5; F, sodium deprivation for 48 h, n=7; G, infusion of 0.7 M mannitol-CSF at 1 ml/h for 3 h into a lateral ventricle commencing 1 h before bar-pressing, n=6; H, administration of a sodium load of 600 mmol sodium bicarbonate into the rumen 3 h before bar-pressing. Mean and s.e.m. are shown. Statistical analyses were made by two-way analysis of variance (repeated measures design) and/or t-test. (From Coghlan et al. 1981, with permission. Copyright 1981 the American Association for the Advancement of Science.)

1979), and which lowered CSF [Na] by 10–20 mmol/l, is also shown on Fig. 22-8. It doubled sodium intake whereas an increase in CSF [Na] by 10–20 mmol/l produced by i.v.t. infusion at 1 ml/h of 500 mM sodium chloride in artificial CSF more than halved sodium intake (see Chapter 24 for detailed results). However i.v.t. AII (3.8 μg/h for 48 h) in sodium-deplete animals had no influence on salt appetite in 22 h, and caused an increase which was not statistically significant in the 22–46-h period (Fig. 22-8). There was significant effect on water intake. The i.v.t. AII infusion given to the same fistulated animals when maintained sodium-replete by immediate pumping of saliva back into the rumen similarly caused a large increase in water intake and an increase in sodium intake, the increase being significant during the access period after 46 h of infusion.

Thus, whereas physiological change of sodium status produced large change of appetite as expected, the i.v.t. infusions of blockers of the renin–angiotensin system at doses attested to eliminate effects of i.v.t. renin and NGF, had no significant quantitative effect on salt appetite in response to sodium deficiency. In addition, AII added to CSF of sodium-deficient sheep did not cause a further increase in sodium intake despite the fact that more severe sodium depletion did. By contrast, lowering of CSF [Na] with i.v.t. mannitol consistently and rapidly doubled sodium intake. The explanation of lack of effect of AII in moderate sodium deficiency is not clear.

These data bring into question a role for cerebral renin–angiotensin II in the sodium intake resulting from sodium deficiency. While not being conclusive, the results are suggestive of the possibility that the sodium appetite caused by i.v.t. renin or AII may not be a physiological regulatory action. It could result from direct pharmacological actions of AII, or be secondary to effects AII may have on systemic sodium balance and ionic concentrations.

Thus, the data with i.v.t. renin and AII may reflect the attested in vitro effect of AII to alter a variety of ionic fluxes between cells and tissue fluid (Vecsei et al. 1978) including sodium fluxes between extracellular and intracellular fluid as described for jejunum and skin (Crocker and Munday 1970; McAfee and Locke 1967). In grossly supraphysiological concentration the induction of such sodium movements in the neuronal systems subserving sodium appetite may mimic events occurring physiologically with sodium deficiency, and possibly approximated by ionic movements within physiological range contrived by i.v.t. infusion of mannitol, a proven stimulus for sodium appetite in sheep.

With regard to this latter point, we have measured [Na] of CSF and plasma in sodium-replete sheep treated with i.v.t. AII (3.8 μg/h), and having access to water and 0.6 M sodium bicarbonate. CSF and plasma [Na] fell 11±1 and 13.5±2.6 mmol/l (n=4) respectively over 24 h. This occurs also in control experiments in which AII-treated animals were subjected to water restriction (0.6 litres in the 24 h) and not allowed access to 0.6 M sodium bicarbonate. Therefore the fall in [Na] is not solely due to increased water intake (Fig. 22-9). Because i.v.t. AII causes a rapid onset of natriuresis in which the [Na] of the urine is usually greater than that of plasma, this phenomenon will contribute to the lowering of CSF and

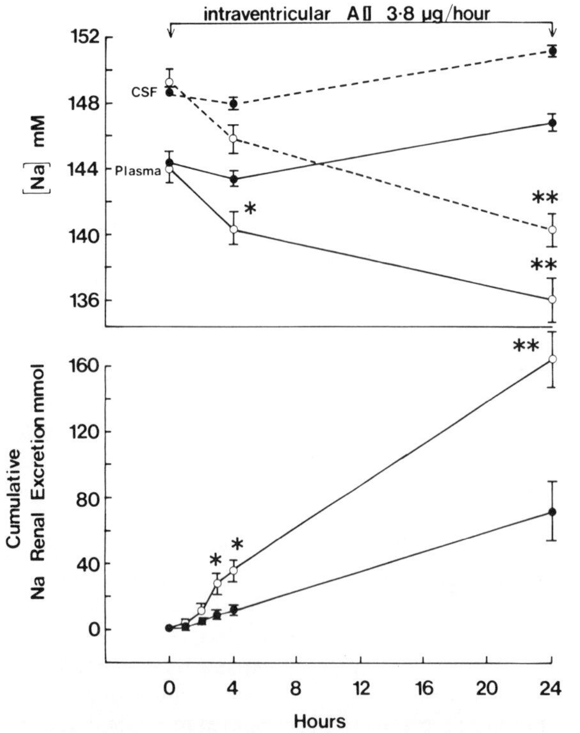

Fig. 22-9. Effect of infusion of AII (3.8 μg/h) for 24 h (open circles, n=7) or control artificial CSF (filled circles, n=5) at 0.38 ml/h into a lateral ventricle on the CSF [Na] (interrupted lines) and on plasma [Na] (continuous lines) and renal sodium excretion of water-restricted sheep. Animals were allowed access to food and 0.5 litre of water 4 h after the start of the i.v.t. infusions. Normal daily water intake of these sheep was 1–2 litres. Mean and s.e.m. are shown. Comparison of the effects of i.v.t. AII with control i.v.t. artificial CSF by Student's t-test is denoted by $^*P<0.05$, $^{**}P<0.01$. (From Coghlan et al. 1981, with permission. Copyright 1981 the American Association for the Advancement of Science.)

plasma [Na] caused by i.v.t. AII (Fig. 22-9). The increased water drinking and ADH release attributable to AII will augment this effect. In the light of the powerful sodium-appetite-inducing effect of a lowering of CSF [Na], this may be an important consideration in sodium appetite induced by i.v.t. renin or AII in sheep. However, Avrith and Fitzsimons (1980) produce data, including the action of i.v.t. renin in nephrectomized rats, arguing against the salt appetite in their experiments being secondary to renal loss. Fischer and Buggy (1975) in their original report also showed appetite preceded urinary loss in the rat. As did Bryant et al. (1980) they noted that sometimes onset of salt appetite followed, in the way that water intake did, very rapidly after i.v.t. injection, the usual delay not being evident. In a number of these experimental series it was shown that carbachol increased water intake but not salt intake.

Avrith et al. (1980b) studied this phenomenon in adrenalectomized and hypophysectomized rats. There was no attenuation of the salt appetite effect of preoptic injection of renin and AII in the adrenalectomized animals. With hypophysectomized rats, the intake of 2.7% saline in the 18 h after injection was 6.8±1.6 mol compared with 1.2±0.6 ml for control ($P<0.01$). On the other hand, from their Fig. 1 it would appear that renin caused ca 15 ml intake of 2.7% saline in sham-adrenalectomized and adrenalectomized rats so that a significant reduction of intake by hypophysectomized animals seems to have occurred. That this is attributable to absence of ACTH effect on salt appetite seems unlikely. Though angiotensin and renin release ACTH (Maran and Yates 1977), and ACTH causes salt appetite, it does not do so in the adrenalectomized rat (Weisinger et al. 1978), which indicates there is no direct action on the brain as there appears to be in the rabbit (Blaine et al. 1975).

At this time, it would seem the question of the role of angiotensin in salt appetite is an open one. Possible modes of neuronal action are noted in Chapter 24, and the proposition is advanced that the problem is unlikely to be solved by experiments involving introduction of pharmacological amounts into the brain. Solution may feasibly turn on newer techniques involving demonstration of messenger RNA species associated with production of AII in the brain, and commensurate activation of the transcription processes by sodium deficiency (see Chapter 24). But a negative finding would still not be conclusive. Thus J.P. Coghlan (personal communication) points out as a result of issues arising from studies showing the peripheral production of a proportion of AII and AIII in blood from tissue renin stores, that the brain renin may represent the slow passage of renal renin across the blood–brain barrier and its accumulation in specific sites. It could be activated by changes associated with sodium deficiency.

The effect of hypertension on the salt appetite of sodium-replete and sodium-deficient animals

Salt appetite in hypertensive rats

In 1949, Abrams, De Friez, Tosteson and Landis considered that in the light of Richter's demonstration of self-selection capacities in a wide variety of metabolic disorders, it would be interesting to look at whether hypertensive rats, when given systematically a choice of a wide variety of salts, might alter their intake in a fashion that indicated metabolic abnormalities associated with the hypertensive state. When rats were offered a cafeteria of water and 0.17 M solutions of various salts, they found that the mild polydipsia of hypertensive rats was confirmed. The rats elected to take only one-third to one-half as much sodium chloride and sodium bicarbonate as did normal rats tested concurrently. This effect was more marked in the moderately hypertensive rats than a rat with severe hypertension. However, the abstinence was not adequate to restore their blood pressure to normal. Intake of potassium chloride, calcium chloride and magnesium chloride was small and not altered. They showed no special appetite for substances described as producing temporary alleviation of hypertension, viz. sodium bromide, nitrite, iodide and ammonium chloride. Sodium bromide was taken in increased amounts by a few moderately hypertensive rats. Opportunity to select salts of certain trace elements—boron, manganese, cobalt, lithium, strontium and cadmium—did not reveal any significant change with hypertension. In a subsequent analysis the same group found that hypertensive rats continued to drink less sodium chloride than did control rats following adrenalectomy and during the administration of DOCA. Increasing the fluid intake of normal rats did not reduce but indeed increased sodium chloride intake. Thus, the increased fluid intake of the hypertensive rats did not appear to be a determinant of their reduction in sodium intake. The possibility of an alteration in renal function causing this effect was raised by the authors. Subsequently, Fregly, Yates and Landis (1955) showed that the small (2 mmol/l increase in plasma sodium in hypertensive animals could not be related consistently to the relative aversion to salt. Fregly (1955) found that encapsulating the kidneys of both young and old rats caused a significant rise in blood pressure 2–4 weeks after operation in both groups. A relative aversion to sodium chloride developed in the young hypertensive rats 8 weeks after operation, the time at which blood pressure had attained its highest level. The effects with the older hypertensive rats were not as clear-cut. Growth rate and food intake of the rats of both age groups were similar to those of corresponding controls, but the life span of the younger rats was appreciably shortened by the procedure.

Fregly (1955, 1959), who has devoted a great deal of effort to analysis of this phenomenon, examined (1959) the specificity of the aversion to potassium chloride, sodium sulphate, lithium chloride and sodium saccharin, which was similar in character to the sodium chloride aversion. Hence, he concluded that the sodium chloride aversion manifest in hypertension in rats is not specific, but part of a general salt aversion. Before this finding, Fregly (1956) had examined the preference threshold of normal and hypertensive rats and found that they were the same, lying between 0.008 and 0.016 M sodium chloride. Normotensive rats preferred sodium chloride to water at all concentrations above threshold that he studied, whilst hypertensive rats showed the same preference, only in concentrations ranging from 0.016–0.090 M; above this, an aversion developed. The data indicated that the sodium chloride aversion in hypertensive rats could not be attributed to a difference in acuity in differentiating between sodium chloride solutions and water.

Fregly (1962), having shown rats rendered hypothyroid by the goitrogenic agents thiouracil, propylthiouracil and methimazole had a spontaneous sodium chloride appetite when offered a choice between water and 0.15 M saline to drink, tested the effect of these agents in hypertension. The rats so treated at the time of kidney encapsulation increased their spontaneous sodium chloride appetite and their blood pressure did not rise in the usual fashion. He found the sodium chloride appetite was similarly altered by thyroidectomy (Fregly, 1962).

In further experiments, Fregly (1967) was able to show that propylthiouracil reduced the renal capacity to conserve sodium with a sodium-deficient diet, and by direct measurement showed that the drug reduced the secretion of aldosterone by the adrenal gland, an effect which was reversed by the administration of thyroxine.

The effect of experimental renal hypertension on the sodium appetite of sodium-deficient sheep

The partial occlusion of the right renal artery caused the systolic blood pressure to begin to rise within 24–48 h, as observed originally in dogs by Goldblatt et al. (1934); it reached a peak within 2 weeks, oscillated around the peak for variable lengths of time and then began to fall slowly over 2–3 months.

A striking feature of experiments in sheep (Bott et al. 1967a) was the occurrence of a diuresis and natriuresis with constriction of one renal artery with the other kidney in situ. Usually blood pressure rose in two phases—the second phase beginning on the third or fourth post-operative day—and the natriuresis began concurrent with, or just before the second phase and continued for 18–36 days. There was a parallel fall in salivary volume. The overall data for five sheep with a parotid fistula before and after renal hypertension produced in two stages in two animals are shown in Fig. 22-10. Voluntary sodium intake was related to deficit notwithstanding the large changes in blood pressure.

In sheep made hypertensive in this fashion, the reduction in salivary volume following renal artery constriction is of interest. This is in the light of the demonstration by McKinley et al. (1979) that the infusion of saralasin into the carotid artery ipsilateral to a parotid fistula in the sodium-deficient animal increases the rate of salivary secretion; this finding is consistent with a physiological peripheral vascular action of AII. Vasoconstriction of parotid vasculature will reduce salivary secretion (Coats et al. 1956).

A similar diuresis and natriuresis to that noted above was seen in other experiments on renal

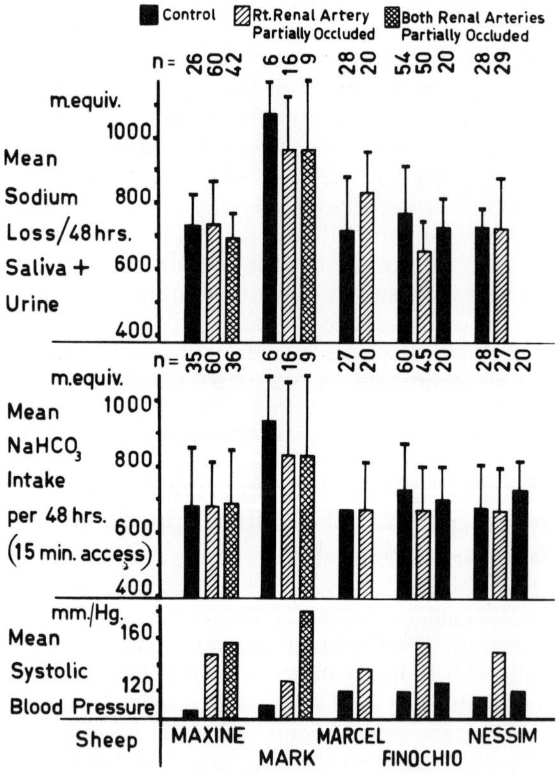

Fig. 22-10. Effect of production of renal hypertension by chronic partial occlusion of one or both renal arteries on the voluntary intake of sodium bicarbonate solution (300 mmol/l) during 15 min. The solution was offered after 48 h of sodium deprivation. Mean blood pressure is recorded on the lowest block. In Finochio and Nessim, the clamped kidney was eventually removed and a subsequent series of control observations were made. (From Bott et al. 1967b, with permission.)

hypertension in sodium-replete sheep without a parotid fistula which were not being presented with sodium solutions for voluntary intake (Blair-West et al. 1968). The production of severe sodium deficiency as a result of natriuresis, occurring with acute onset of hypertension due to renal artery stenosis, has been described in man (Barraclough 1966). Mohring et al. (1976a) showed in rats with arterial supply constricted to one of two kidneys that there were instances of malignant hypertension with extremely high blood pressure, marked activity of the renin–angiotensin system, sodium and water loss, polydipsia and evidence of vascular complications. The animals developed malignant nephrosclerosis in the unclipped kidney and blood urea rose. When offered 0.9% saline to drink, they consumed large amounts and blood pressure regressed. The high salt intake appeared to have at least temporary benefit on the malignant course of hypertension and suggested the natriuresis may have helped to trigger onset of the malignant hypertension. Mohring et al. (1976b) also reported studies on onset of hypertension with the 'two kidney–one clipped' model and noted increased salt appetite during this phase despite sodium and water retention.

In summary, it would appear that renal hypertension does not impair salt appetite in response to sodium deficiency, though it may reduce the need-free intake of salt exhibited by rats. As Tosteson et al. (1951) suggested, this may be attributable to the renal encapsulation method reducing renal capacity to excrete ingested saline.

The influence of genetic susceptibility to hypertension on salt appetite

Wolf, Dahl and Miller (1965) in their investigation of the influence of genetic susceptibility to hypertension on salt appetite noted that it is not known to what extent innate differences in salt appetite, if such exist, are related to the development of experimental or human hypertension. By selective inbreeding, it has been possible to develop a strain of rats with a strong genetic disposition to hypertension, and a strain genetically resistant to developing hypertension. Whereas this selection was made initially on the basis of respective capacities to develop hypertension with sodium chloride ingestion, it has emerged subsequently that the susceptible group show a similar disposition to develop hypertension after DOCA plus sodium chloride, as well as after unilateral renal artery compression. The other strain proved much more resistant using these techniques (Dahl et al. 1962, 1963).

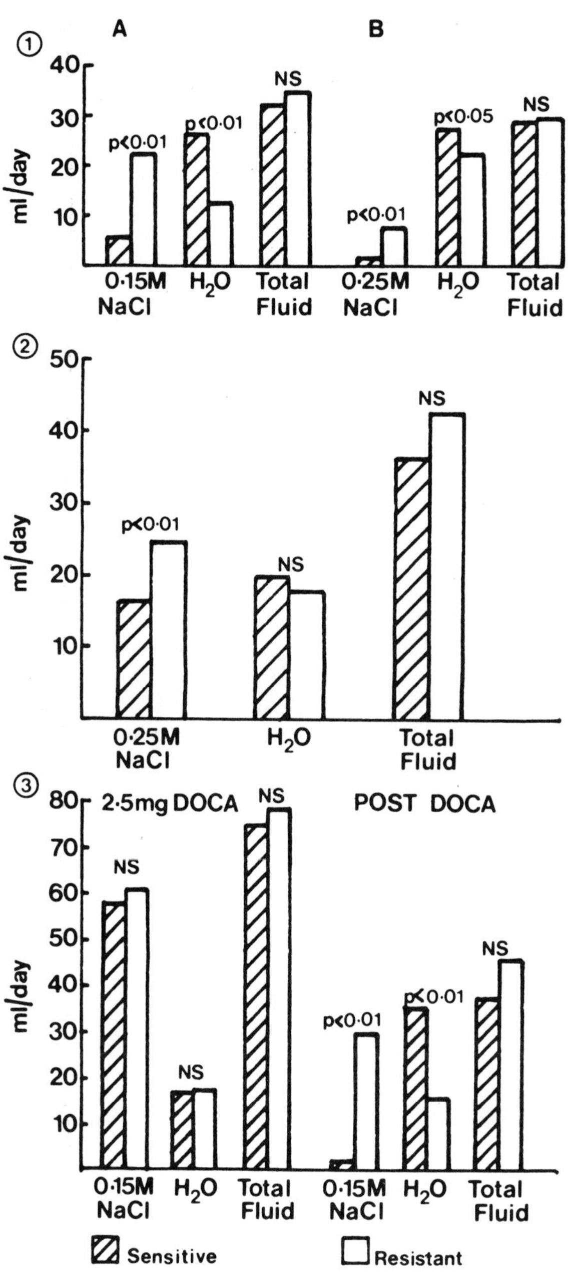

Fig. 22-11. (1) Mean daily saline, water, and total fluid intake of intact, untreated rats during last 4 days of 6-day period. Data from six pairs of Sensitive and six pairs of Resistant rats given 0.15 M saline are shown in A. Data from six Sensitive and six Resistant rats given 0.25 M saline are shown in B. (2) Mean daily 0.25 M saline, water and total fluid intake of ten Sensitive and ten Resistant adrenalectomized rats during last 10 days of 12-day period. (3) Mean daily 0.15 M saline, water and total fluid intake of six Sensitive and six Resistant rats during last 4 days of 6-day injection period (2.5 mg DOCA/day) and during last 4 days of immediately following 6-day period. (After Wolf et al. 1965.)

The study of Wolf et al. (1965) showed that given a free choice of either saline solution or water to drink, rats from a strain with a strong genetic predisposition to hypertension ingested less saline solution than those from a strain genetically resistant to developing hypertension (Fig. 22-11). Adrenalectomy did not abolish this difference, but the administration of DOCA did (Fig. 22-11). The authors note that the blood pressure of the animals with the genetic susceptibility was probably 10–20 mmHg higher than that of the resistant strain. Thus their data might be consistent with the reports of early hypertension causing an aversion to saline. The mechanism of disparity between the two behaviours appeared difficult to interpret. It seemed that differences in adrenocortical function would not be responsible because the difference in appetite persisted after adrenalectomy. They also claimed that the two strains of rats showed no differences in capacity to excrete sodium after a sodium chloride load, or in total carcass sodium or total exchangeable sodium. By contrast, Tobian et al. (1978) found clear differences in the capacity of isolated perfused kidneys of resistant and hypertensive strains to excrete a salt load (see Chapter 27). Wolf et al. (1965) proposed that the differences in both vascular reactivity and sodium appetite in the two strains are due to genetically determined differences in hypothalamic function. They note taste sensitivities and preferences can be transmitted on a genetic basis, and there is a large body of evidence on the role of hypothalamus and taste preferences in regulation of nutrient intake.

Salt appetite in the spontaneously hypertensive rat

Catalanotto, Schechter and Henkin (1972) and McConnell and Henkin (1973) have shown that spontaneously hypertensive rats exhibited an increased preference for sodium chloride (0.15 M and 0.3 M) when compared with either Holtzman or Wistar normotensive controls. Similarly, total fluid intake (sodium chloride solution and water) was increased in the spontaneously hypertensive rats compared with the controls. The spontaneously hypertensive rats were derived from the original Japanese strain of Wistar rats; their systolic blood pressure was 195±8 mmHg (mean±1 s.e.m.). The systolic blood pressure of the controls ranged between 116 and 120 mmHg. The results of preference–aversion studies in relation to not only sodium salts but also potassium, quinine and hydrochloric acid show that the spontaneously hypertensive rats prefer not only solutions of sodium salts more than controls, but also potassium salt. However, the absence of any difference in preference for quinine sulphate or hydrochloric acid between the hypertensive animals and the controls suggested to the authors that altered taste acuity in hypertensive animals cannot account for increased preference for salts. Indeed, the data indicated increased preference for salts without change in preferences for sour or bitter substances in any rat.

These observations take on a different and intriguing aspect as a result of the studies of Phillips, Ganten and colleagues (Phillips et al. 1978, 1979a, b; Mann et al. 1978). Stroke-prone spontaneously hypertensive rats (SP-SHR) were tested for response to AII given intravenously and intracerebroventricularly, and also to the antagonist saralasin. The outcome of a series of experiments on SP-SHR rats was that:

i) plasma AII levels were not elevated;
ii) AII given intraventricularly raised blood pressure more than in Wistar Kyoto control rats;
iii) saralasin intravenously did not lower blood pressure;
iv) nephrectomy did not lower blood pressure;
v) intraventricular saralasin lowered blood pressure before and after nephrectomy.

This latter effect even in the nephrectomized SP-SHR rats suggested that the AII, presumably inducing hypertension, is derived from the brain isorenin–angiotensin system. The authors (Phillips et al. 1978) suggest that large increases in brain AII are not necessary because of increased central receptor sensitivity to AII. Ganten et al. (1975) reported that the levels of AII in CSF of SHR rats were 263 pg/ml compared with 169 pg/ml in control animals: both these values are considerably higher than those we obtain in sheep. Reid (1977) notes that central administration of AII in normal rats at rates 1000 to 1 million times the claimed brain AII secretion rate failed to raise blood pressure to levels seen in SHR rats. Phillips et al. (1979a) respond that the answer to this paradox can be given in the fact that receptor binding studies have shown that in SHR rats the OVLT region has a 100% higher binding level than in normotensive controls, and that in SHR rats not only the endogenous AII level but also the AII binding is increased. The CSF level of AII may only reflect the overspill of AII formed in brain tissue.

However, it would not seem consistent with this view that Gordon et al. (1979) found that lesions of the anterior third ventricle, which were stated to destroy the OVLT along with other structures in the region, did not affect the development of hypertension in the SHR rats. This result differs from what is alleged to occur with such lesions in the case of hypertension caused by DOC, salt or renal hypertension. The lesions were ratified as causing attenuation of drinking response to cellular dehydration as well as intraventricular AII. In evaluating such lesion studies and specifically the effects on blood pressure, the confirmation of satisfactory general condition of the animals is of great importance. Such lesions do not influence the hypertension induced by ACTH in sheep (McKinley et al. 1980). Reid (1977) pointed out that there have been conflicting reports on the effect of AII antagonists in reducing the blood pressure of SHR rats.

The potential importance of these reports in relation to the matter under consideration here lies in the question of whether intraventricular saralasin would reduce the increased salt intake reported to occur in SHR rats. This increased intake may be in response to increased urinary sodium loss in the face of hypertension but it would be of great interest if, on the other hand, saralasin blocked it. However, no increased water intake was reported (McConnell and Henkin 1973) and the preference was not specific. The action of intraventricular renin and angiotensin in specifically increasing salt appetite in rats and sheep has been described above, and this brings the question into focus.

Endocrine pharmacology

In an outstanding and salutary review of endocrine pharmacology Parsons (1976) of Mill Hill, London, has pointed out that because hormones are natural constituents of the body it is often assumed that they are freer from side effects than synthetic medicines and have specific actions. That is, these actions can be described without the pharmacologist's usual insistence on including the dose in practically every statement. But, he says, there is not a single peptide hormone of which this is true. Multiple effects of ADH, parathyroid hormone and insulin provide striking examples, but even apparently more specific hormones such as the glycoprotein group have effects remote from their recognized physiological function. Parsons continues his analysis along the following lines of reasoning. It seems safe to conclude that all hormonal peptides have multiple actions and that this can be added to the list of properties which make it clear that hormones are drugs. Like drugs, the pattern of response very often changes qualitatively with dose. With hormones each action in a complex pattern is characteristically evoked by a separate set of receptors with its own dose–response curve, When such curves lie far apart, different actions will dominate the in vivo response at different levels of the circulating concentration.

Parsons cites the dose dependence in the character of the in vivo response to ADH. The normal circulating level of this hormone is about 4 pg/ml rising to 40 pg/ml during water deprivation (Lauson 1967). In this concentration range ADH seems to act exclusively on the kidney, yet it is, at present, most often known as vasopressin, a name which implies a vasoconstrictor and pressor agent. It does indeed produce pressor effects but they are first seen after injecting doses as high as 10–40 μg in man to produce blood levels 20–100 times the normal circulating concentration (Fig. 22-12). Only in severe shock do such levels ever result from endogenous secretion. Fig. 22-12 gives the

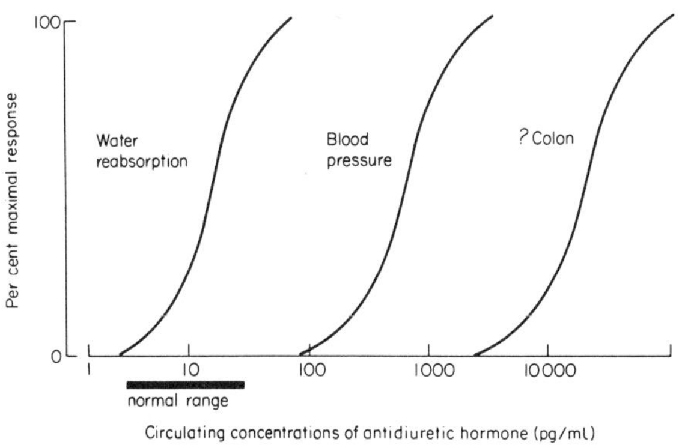

Fig. 22-12. Estimated log dose–response curves of three effects of ADH, shown in relation to the normal range of circulating concentrations. For details see text. The position of the curve representing constriction of the colon has been estimated from the initial plasma concentration corresponding to the doses of ADH which used to be used in cholecystography (20 U, i.e. 40 μg) and from unpublished in vitro experiments by T. Moriya and A. Gunn (personal communication to J. Parsons). The latter found that muscle from the transverse (but not the descending) human colon responded to 200 ng/ml of ADH, about four times the concentration required for half-maximal contraction by angiotensin. (From Parsons 1976, with permission.)

dose–response curves for three different actions of ADH. One of the effects is indeed 'pharmacological', i.e. the contraction of intestinal smooth muscle. This effect was at one time used by radiologists to constrict the colon and eliminate gas shadows during cholecystography. Clearly the doses required to constrict the colon as indicated on the Fig. 22-12 will produce severe vasoconstriction and indeed deaths did occur during the time that ADH was used for this purpose in radiography as a result of decrease in coronary blood flow and myocardial infarction.

Parsons places great emphasis on the significance of the rate of entry of a hormone to the circulation. Any hormone with multiple actions entering the circulation abnormally fast may cause an unphysiological pattern of response. Even though, following redistribution, excretion, or metabolism, the circulating level may rapidly return to the physiological range, the effects of a brief elevation may be important if it triggers any lasting response. While some side effects, as for example a rise in blood pressure, may be transient others (such as lipolysis, osteolysis or other metabolic action) may long outlast the concentration spike.

The disappearance of a hormone from the circulation can be considered as the sum of exponential processes. Each hormone has a characteristic half-life. The slowest process of disappearance of biological activity for any hormone presumably represents the physiologically significant rate of excretion or metabolic destruction. For interpretation of this to be safe, it has to be done by bioassay of a series of plasma samples after interruption of endogenous secretion in the physiologically normal subject. This has been done in only a few cases. Measurement by immunoassay and studies of disappearance of exogenous hormone have to be treated with reserve for several reasons. Immunoactive fragments may circulate longer than the bioactive molecular species. Exogenous hormone may not circulate in a form identical with the endogenous. The use of iodinated hormone has in Parsons' view given misleading results too often to be worth serious consideration. The only method of administration not liable to introduce mixing artefacts is termination of a long infusion. The use of high doses to facilitate measurement may induce physiological effects (such as redistribution of intestinal or renal blood flow) which change the rate of disappearance or may saturate disappearance processes which are important at physiological levels. These considerations may be set against the background of the fact that peptide hormones differ very greatly in their rates of metabolic destruction. Some, the most labile, disappear with a single circulation, whereas for others (chorionic gonadotrophin) the half-time is 24 h. Certain implications of these considerations are set out in Fig. 22-13.

Parsons goes on to illustrate some of these general points by specific consideration of the multiple actions of parathyroid hormone (PTH). Most of the multiple actions of PTH play a part in what is apparently its principal function: raising the calcium concentration in plasma and extracellular fluid (Parsons and Potts 1972). It

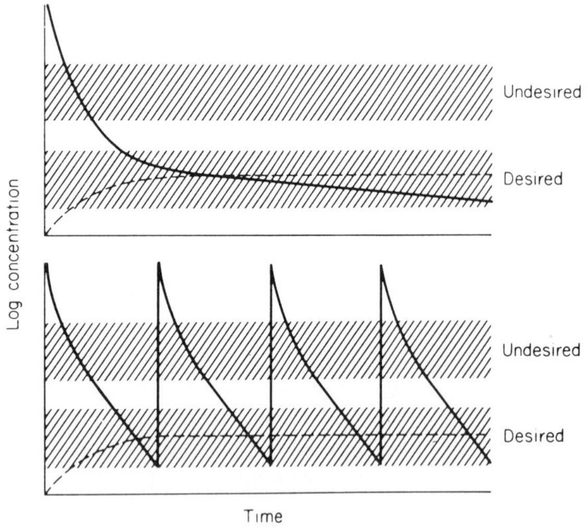

Fig. 22-13. Diagrams comparing the dose schedules needed to maintain therapeutically effective blood levels of a persistent drug (upper panel) and a labile one (lower panel). In both panels bands of oblique hatching indicate the blood levels causing a desired and undesired effect. It is assumed that both drugs have a good therapeutic index, that is, the levels causing the undesired effects are above and well separated from the therapeutic.

As shown by the continuous lines in each panel, rapidly absorbed injections lead to high initial blood levels, reaching and even surpassing the undesired range. After each injection, blood levels at first fall rapidly owing to redistribution and then exponentially owing to excretion and metabolic destruction, as indicated by straight-line portions of these semilogarithmic plots.

The frequent injections required to maintain the effects of the labile drug entail exposure to the undesired blood level for a high proportion of the time. By limiting the rate of entry of either drug to the circulation (whether by infusion or by injection of delayed-release preparations) it is possible to maintain the blood levels illustrated by broken lines, entirely within the therapeutic range. The advantage gained by controlling entry rate is greatest in the case of the labile drug. (From Parsons 1976, with permission.)

accelerates bone breakdown and increases renal calcium retention and intestinal calcium absorption. This is the order of relative importance in which these actions contribute to the hypercalcaemia elicited by injecting a large dose. It was not surprising that it was also the historical order of their discovery. Following the observations of Nordin and colleagues on clinical ground (Nordin and Peacock 1969), Parsons and Potts examined whether PTH-induced bone breakdown could play the dominant role in normal calcium homeostasis that had formerly been ascribed to it. Applying a new bioassay to examine the effect of PTH which was more sensitive than previous in vivo bioassays, simple calculations showed that the doses required for the osteoclastic response to PTH corresponded to initial blood levels of more than 100 ng/ml. It would appear that the initial concentration was the most relevant, because although maximum hypercalcaemia was not reached for several hours (depending on the size of the animal) the calcium mobilization process is evidently triggered within minutes of an intravenous PTH injection. On the other hand the application of sequence-specific radioimmunoassays showed that circulating forms of PTH containing the N-terminal sequence believed to be essential for biological activity were unmeasurable by current methods and their level probably did not exceed 10 pg/ml (Parsons et al. 1975).

In the light of this 10,000-fold discrepancy, the effect of giving PTH in a much more physiological way was studied with the aim of imitating the normal secretory pattern of the parathyroid glands. A slow continuous infusion was made into a vein. When PTH was infused into dogs at 100 ng $kg^{-1}h^{-1}$ (a rate calculated to provide an equilibrium concentration of about 50 pg/ml in the circulation), PTH caused marked hypercalcaemia persisting throughout infusions lasting up to 4 weeks (Parsons and Reit 1974). Confirming the suspected insensitivity of the osteoclastic response, the hypercalcemia was not accompanied by any significant increase in bone breakdown. This was so whether judged by examination of bone biopsies, or by measurements of alcohol-soluble hydroxyproline in plasma, or by studies involving skeletal labelling with radiocalcium and the use of ethane-1-hydroxy-1,1-diphosphonate (EHDP), an agent which prevents mobilization of calcium from bone. Hypercalcaemia resulting from the infusion with very small dosage must therefore have depended on more sensitive responses to PTH. This was found for both increased intestinal calcium absorption and decreased urinary calcium excretion.

The evidence cited (of which the account here is only a brief summary) indicates that the actions of PTH on the principal calcium fluxes can be represented by multiple dose–response curves, the approximate positions of which are illustrated in Fig. 22-14. Varying levels of PTH signal effects on various sources of calcium. The rapid response to a low PTH level is hypercalcuria and that to a higher (but still normal) level is renal calcium retention. Intestinal calcium absorption probably varies from about 15% to over 85% in a similar range of hormone levels. Parsons remarks that extension and refinement of the hypothesis must await methods to measure the circulating levels of bioactive PTH in various physiological states. Much work will be required to assign precise shapes and positions for the various curves shown. It will probably be necessary to insert one or more additional curves for other hypercalcaemic ac-

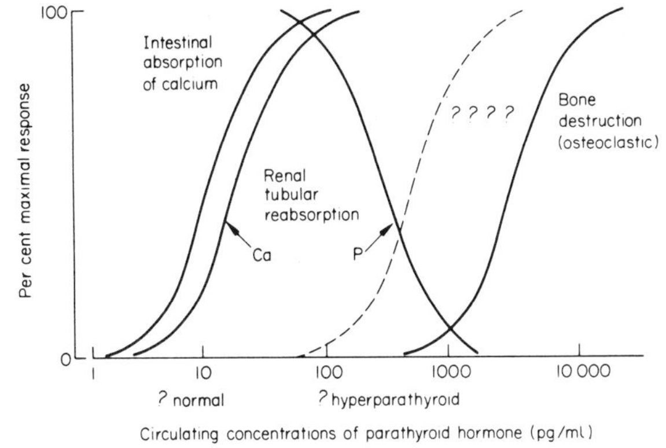

Fig. 22-14. Hypothetical log dose–response curves of four effects of PTH, shown in relation to estimated values of the circulating level of the bioactive hormone in health and disease. For details see text. (From Parsons 1976, with permission.)

tions. The broken line with question marks, for example, might represent non-osteoclastic bone reabsorption or inhibition of bone mineralization. Both may be early hypercalcaemic responses to PTH.

The results of low-dose PTH infusion made it clear that renal and intestinal effects predominate at slightly supraphysiological blood levels. The combination would of course lead to a positive calcium balance so that the results of administration in a physiological manner actually reverse the formerly accepted picture of PTH as an agent of bone destruction and demineralization. Its normal role may be to promote calcium retention. The data focus attention on earlier studies in which PTH given chronically to rats in low doses was found to have an anabolic effect on bone (Selye 1932; Kalu et al. 1970; Walker 1971). The demineralizing condition currently regarded as typical may result only when there is gross oversecretion of PTH by the parathyroid gland.

Parsons concluded by remarking that the pharmacological re-evaluation on lines discussed for PTH (and also for insulin in his account) is needed in the case of other peptides including angiotensin and ADH. The assumption that hormones do not cause side effects to the same extent as other drugs is unfounded, and the character of response as seen in normal physiology depends on a precisely regulated circulating concentration as much as on the selectivity of the receptors. Because many peptides disappear from the blood with half-lives of a few minutes, the overall patterns of response in vivo vary qualitatively not only with the dose but also with the rate and site at which the hormone enters the circulation. These kinetic considerations are fundamental to understanding the biological significance of a hormone. It is clear that Parsons' analysis is highly pertinent to the current explosion of data on biological effects produced by the administration of renin and angiotensin—often in extremely high doses.

References

Abraham SF, Baker RM, Blaine EH, Denton DA, McKinley MJ (1975) Water drinking induced in sheep by angiotensin: A physiological or pharmacological effect? J Comp Physiol Psychol 88: 503

Abraham SF, Denton DA, McKinley MJ, Weisinger RS (1976a) Effect of an angiotensin antagonist, sar^1-ala^8-angiotensin II on physiological thirst. Pharmacol Biochem Behav 4: 243

Abraham SF, Denton DA, Weisinger RS (1976b) The specificity of the dipsogenic effect of angiotensin II. Pharmacol Biochem Behav 4: 363

Abraham SF, Coghlan JP, Denton DA, Fei DTW, McKinley MJ, Scoggins BA (1980) Correlation of cerebrospinal fluid and blood angiotensin II in sheep. In: Proceedings of the Sixth International Congress on Endocrinology, Melbourne. Australian Academy of Science, Canberra, p 572

Abrams M, De Friez AIC, Tosteson DC, Landis EM (1949) Self-selection of salt solutions and water by normal and hypertensive rats. Am J Physiol 156: 233

Adolph EF, Barker JP, Hoy PA (1954) Multiple factors in thirst. Am J Physiol 178: 538

Aguilera G, Schirar A, Baukal A, Catt KJ (1981) Circulating angiotensin II and adrenal receptors after nephrectomy. Nature 289: 507

Andersson B (1977) Regulation of body fluids. Annu Rev Physiol 39: 185

Andersson B, Ericsson L (1971) Conjoint action of sodium and angiotensin on brain mechanisms controlling water and salt balances. Acta Physiol Scand 81: 18

Andersson B, Olsson K (1977) Evidence for periventricular sodium sensitive receptors of importance in the regulation of ADH secretion. In: Proceedings of the Second International Conference on the Neurohypophysis. S Karger, Basel

Andersson B, Westbye O (1970) Synergistic action of sodium and angiotensin on brain mechanisms controlling water and salt balance. Nature 228: 75

Andersson B, Dallman MF, Olsson K (1969) Evidence for a hypothalamic control of renal sodium excretion. Acta Physiol Scand 75: 496

Assaykeen TA, Otsuka K, Ganong WF (1968) Rate of disappearance of endogenous dog renin from the plasma of nephrectomized dogs. Proc Soc Exp Biol Med 127: 306

Avrith DB, Fitzsimons JT (1980) Increased sodium appetite in the rat induced by intracranial administration of components of the renin–angiotensin system. J Physiol (Lond) 301: 349

Avrith DB, Fitzsimons JT, Nicolaïdis S (1980a) The effects of long term intravenous infusions of angiotensin II on thirst, sodium appetite, and water and sodium balance in the rat. In: Proceedings of the Seventh International Conference on the Physiology of Food and Fluid Intake. IUPS, Warsaw

Avrith DB, Wiselka MJ, Fitzsimons JT (1980b) Increased sodium appetite in adrenalectomized or hypophysectomized rats after intracranial injections of renin or angiotensin II. J Endocrinol 87: 109

Baniukiewicz S, Brodie A, Flood C, Motta M, Okamoto M, Gut M, Tait JF, Tait SAS, Blair-West JR, Coghlan JP, Denton DA, Goding JR, Scoggins BA, Wintour EM, Wright RD (1968) Adrenal biosynthesis of steroids in vitro and in

vivo using continuous superfusion and infusion procedures. In: McKerns L (ed) Functions of the adrenal cortex, Vol 1. Appleton-Century-Crofts, New York, p 153
Barraclough MA (1966) Sodium and water depletion with acute malignant hypertension. Am J Med 40: 265
Baudouin M, Meyer P, Fermandjian S, Morgan JL (1972) Calcium release induced by interaction of angiotensin with its receptors in smooth muscle cell microsomes. Nature 235: 336
Bellows RT (1939) Time factors in water drinking in dogs. Am J Physiol 125: 87
Bennett JT, Snyder SH (1976) Angiotensin II binding to mammalian brain membranes. J Biol Chem 251: 7423
Blaine EH, Covelli MD, Denton DA, Nelson JF, Shulkes AA (1975) The role of ACTH and adrenal glucorticoids in salt appetite of wild rabbits (*Oryctolagus cuniculus* (L.)). Endocrinol 97: 793
Blair-West JR (1976) Renin–angiotensin system and sodium metabolism. In: Thurau K (ed) International review of physiology, Vol 11. Kidney and urinary tract physiology II. University Park Press, Baltimore, p 95
Blair-West JR, Coghlan JP, Denton DA, Goding JR, Munro J, Peterson RE, Wintour EM (1962) Humoral stimulation of adrenal cortical secretion. J Clin Invest 41: 1606
Blair-West JR, Coghlan JP, Denton DA, Goding JR, Wintour M, Wright RD (1963) The control of aldosterone secretion. Recent Prog Horm Res 19: 311
Blair-West JR, Boyd GW, Coghlan JP, Denton DA, Goding JR, Wintour M, Wright RD (1965) Aldosterone secretion in relation to homeostasis of body fluids. Excerpta Medica Int Congr Series 87, p 207
Blair-West JR, Coghlan JP, Denton DA, Orchard E, Scoggins BA, Wright RD (1968) The renin–angiotensin aldosterone system and sodium balance in experimental renal hypertension. Endocrinology 83: 1199
Blair-West JR, Coghlan JP, Denton DA, Funder JW, Scoggins BA, Wright RD (1971a) Inhibition of renin secretion by systemic and intra-renal angiotensin infusion during onset of sodium depletion. Am J Physiol 220: 1309
Blair-West JR, Coghlan JP, Denton DA, Funder JW, Scoggins BA, Wright RD (1971b) Effect of the heptapeptide 2–8 and the hexapeptide 3–8 fragments of angiotensin II on aldosterone secretion. J Clin Endocrinol Metab 40: 540
Blair-West JR, McKenzie JS, McKinley MJ (1971c) The actions of angiotensin II on the isolated portal vein of the rat. Eur J Pharmacol 15: 221
Blair-West JR, Coghlan JP, Denton DA, Fei DTW, Hardy KJ, Scoggins BA, Wright RD (1980) A dose response comparison of the actions of angiotensin II and angiotensin III in sheep. J Endocrinol 87: 409
Bocchini V, Angeletti PU (1969) The Nerve Growth Factor: Purification as a 30000 molecular weight protein. Proc Natl Acad Sci USA 64: 787
Bolton JE, Munday KA, Parsons BJ, York BG (1975) Effects of angiotensin II on fluid transport, transmural potential difference and blood flow by rat jejunum in vivo. J Physiol (Lond) 253: 411
Bonjour JP, Malvin RL (1970) Stimulation of ADH release by the renin angiotensin system. Am J Physiol 218: 1555
Booth DA (1968) Mechanism of action of norepinephrine in eliciting an eating response on injection into the rat hypothalamus. J Pharmacol Exp Ther 160: 336
Bott E, Denton DA, Weller S (1967a) The effect of angiotensin II infusion, renal hypertension and nephrectomy on salt appetite of sodium deficient sheep. Aust J Exp Biol Med Sci 45: 595
Bott E, Denton DA, Weller S (1967b) The innate appetite for salt exhibited by sodium-deficient sheep. In: Proceedings of the Second International Symposium on Olfaction and Taste. Pergamon Press, Oxford, p 415
Braun-Menendez E, Fasciolo JE, Leloir LF, Munoz JM (1940) The substance causing renal hypertension. J Physiol (Lond) 98: 283
Bryant RW, Epstein AN, Fitzsimons JT, Fluharty SJ (1980) Arousal of a specific and persistent salt appetite with continuous intracerebroventricular infusion of angiotensin II. J Physiol (Lond) 301: 365
Campbell WB, Baux SN, Pettinger WA (1974) Angiotensin II and angiotensin III induced aldosterone release in vivo in the rat. Science 184: 994
Catalanotto F, Schechter PJ, Henkin RI (1972) Preference for NaCl in the spontaneously hypertensive rat. Life Sci 11: 557
Changaris DG, Severs WB, Keil LC (1978) Localization of angiotensin in rat brain. J Histochem Cytochem 26: 593
Chapman CA, Banks BEC, Carstairs JR, Pearce FL, Vernon CA (1979) The preparation of nerve growth factor from the prostate of the guinea-pig and isolation of immunogenically pure material from mouse submandibular gland. FEBS Letts 105: 341
Chiaraviglio E (1976) Effect of renin angiotensin system on sodium intake. J Physiol (Lond) 255: 57
Coats DA, Denton DA, Goding JR, Wright RD (1956) Secretion by the parotid gland of the sheep. J Physiol (Lond) 131: 13
Coghlan JP, Blair-West JR, Denton DA, Fei DT, Fernley RT, Hardy KJ, McDougall JG, Puy R, Robinson PM, Scoggins BA, Wright RD (1979) Control of aldosterone secretion. J Endocrinol 81: 55P
Coghlan JP, Considine PJ, Denton DA, Fei DTW, Leksell LG, McKinley MJ, Muller AF, Tarjan E, Weisinger RS, Bradshaw RA (1981) The effect of nerve growth factor with isorenin contamination on thirst and sodium appetite. Science (in press)
Coviello A, Crabbé J (1965) Effect of angiotensin II on active transport of sodium by toad bladder and skin. Biochem Pharmacol 14: 1739
Cozzari C, Angeletti PU, Lazar J, Orth H, Gross F (1973) Separation of isorenin activity from nerve growth factor (NGF) activity in mouse submaxillary gland extracts. Biochem Pharmacol 22: 1321
Crocker AD, Munday KA (1967) Aldosterone and angiotensin action on water absorption in rat jejunum. J Physiol (Lond) 192: 36
Crocker AD, Munday KA (1970) The effect of the renin angiotensin system on mucosal water and sodium transfer in everted sacs of rat jejunum. J Physiol (Lond) 206: 323
Dahl LK, Heine M, Tassinari L (1962) Role of genetic factors in susceptibility to experimental hypertension due to chronic excess salt ingestion. Nature 194: 480
Dahl LK, Heine M, Tassinari L (1963) Effects of chronic excess salt ingestion. Role of genetic factors in both DOCA and salt and renal hypertension. J Exp Med 118: 605
Daniels AE, Severs WF, Buckley JP (1967) The effects of angiotensin II on the ^{45}calcium distribution of the rat. Life Sci 6: 545
Daniels-Severs A, Ogden E, Vernikos-Danellis J (1971) Centrally mediated effects of angiotensin II in the unanaesthetized rat. Physiol Behav 7: 785
Davies NT, Munday KA, Parsons BJ (1969) Dose dependent biphasic effects of angiotensin on fluid transfer by rat colon. J Physiol (Lond) 204: 142P
Davies NT, Munday KA, Parsons BJ (1970) The effect of angiotensin on the rat intestinal fluid transfer. J Endocrinol 48: 39
Davis JO (1959) Mechanisms regulating the secretion and metabolism of aldosterone in experimental secondary hyperaldosteronism. Recent Prog Horm Res 15: 298
Davis JO (1961) Mechanisms regulating the secretion and metabolism of aldosterone in experimental secondary

hyperaldosteronism. Recent Prog Horm Res 17: 293
Denton DA (1957) The study of sheep with permanent unilateral parotid fistulae. Q J Exp Physiol 135: 227
Denton DA (1966) Some theoretical considerations in relation to innate appetite for salt. Conditional Reflex I: 144
Denton DA, Orchard E, Weller S (1969) The relation between voluntary sodium intake and body sodium balance in normal and adrenalectomized sheep. Behav Neural Biol 3: 213
Deshmukh P, Phillips MI (1977) The effect of acute arterial hypertension on the transport of exogenous horseradish peroxidase in rat brain. Anat Rec 187: 566
Devaux C, Menard J, Alexandre J, Idatte J, Meyer P, Milliez P (1968) Variations in renin and its substrate after nephrectomy. Lancet I: 300
Elliott DF, Peart WS (1956) Amino-acid sequence in a hypertensin. Nature 177: 527
Epstein AN (1978) Consensus, controversies and curiosities. Fed Proc 37: 2711
Epstein AN, Hsiao S (1975) Angiotensin as a dipsogen. In: Peters G, Fitzsimons JT, Peters-Haefeli L (eds) Control mechanisms of drinking. Springer-Verlag, Berlin, p 108
Epstein AN, Fitzsimons JT, Simons BJ (1969) Drinking caused by the intracranial injection of angiotensin into the rat. J Physiol (Lond) 200: 98P
Epstein AN, Fitzsimons JT, Rolls BJ (1970) Drinking induced by injection of angiotensin into the brain of the rat. J Physiol (Lond) 210: 457
Fei DTW, Coghlan JP, Fernley RT, Scoggins BA, Tregear GW (1980) Peripheral production of angiotensin II and III in sheep. Circ Res 46/1: I-135
Feldberg W, Lewis GP (1964) The action of peptides on the adrenal medulla. Release of adrenaline by bradykinin and angiotensin. J Physiol (Lond) 171: 98
Felix D (1976) Peptide and acetyl choline action on neurones of the cat's subfornical organ. Arch Pharm (Weinhein) 292: 15
Felix D, Akert K (1974) The effect of angiotensin II on neurones of the cat's subfornical organ. Brain Res 76: 350
Fisher AE, Buggy J (1975) Central mediation of water and sodium intake: A dual role for angiotensin. In: Peters G, Fitzsimons JT, Peters-Haefeli L (eds) Control mechanisms of drinking. Springer-Verlag, Berlin, p 139
Fitzsimons JT (1966) Hypovolaemic drinking and renin. J Physiol (Lond) 186: 130P
Fitzsimons JT (1969) The role of a renal thirst factor in drinking induced by extracellular stimuli. J Physiol (Lond) 201: 349
Fitzsimons JT (1972) Thirst. Physiol Rev 52: 468
Fitzsimons JT (1978) Angiotensin, thirst, and sodium appetite: Retrospect and prospect. Fed Proc 37: 2669
Fitzsimons JT (1979) The physiology of thirst and sodium appetite. Cambridge University Press, Cambridge
Fitzsimons JT (1980) Angiotensin and other peptides in control of water and sodium intake. Proc R Soc Lond [Biol] 210: 165
Fitzsimons JT, Kucharczyk J (1978) Drinking and haemodynamic changes induced in the dog by intercranial injection of components of the renin–angiotensin system. J Physiol (Lond) 276: 419
Fitzsimons JT, Simons BJ (1969) The effect on drinking in the rat of intravenous angiotensin, given alone or in combination with other stimuli of thirst. J Physiol (Lond) 203: 45
Fitzsimons JT, Stricker EM (1971) Sodium appetite and the renin angiotensin system. Nature 231: 58
Fitzsimons JT, Wirth JB (1978) The renin angiotensin system and sodium appetite. J Physiol (Lond) 274: 63
Fitzsimons JT, Richards G, Kucharczyk J (1977) Systemic angiotensin-induced drinking in the dog: A physiologically relevant phenomenon. In: Proceedings of the Sixth International Symposium on Food and Fluid Intake. Jouy-en-Josas, Paris
Fitzsimons JT, Avrith DB, Lewis ME (1979) Interrelation between nerve growth factor (NGF) and renin and thirst and sodium appetite. Neurosci Lett [Suppl] 3: S327
Fitzsimons JT, Kucharczyk J, Richards G (1978) Systemic angiotensin-induced drinking in the dog: a physiological phenomenon. J Physiol (Lond) 276: 435
Fregly MJ (1955) Hypertension, NaCl aversion and polydipsia in rats: Time course in relation to age. Am J Physiol 182: 139
Fregly MJ (1956) Effect of renal hypertension on the preference threshold of rats for sodium chloride. Am J Physiol 187: 288
Fregly MJ (1959) Specificity of sodium chloride aversion of hypertensive rats. Am J Physiol 196: 1326
Fregly MJ (1962) Effect of hypothyroidism on sodium chloride aversion of renal hypertensive rats. Endocrinology 71: 683
Fregly MJ (1967) The role of hormones in the regulation of salt intake in rats. In: Kare MR, Maller O (eds) The chemical senses and nutrition. John Hopkins Press, Baltimore, p 115
Fregly MJ (1978) Angiotensin-induced thirst: Peripheral and central mechanisms. Fed Proc 37: 2667
Fregly MJ (1980) Effects of angiotensin converting enzyme inhibitor, captopril, on NaCl appetite in rats. J Pharmacol Exp Ther 215: 407
Fregly MJ, Yates FE, Landis EM (1955) Serum sodium concentration of hypertensive rats: Relation to NaCl intake, blood pressure and age. Proc Soc Exp Biol Med 90: 695
Fuxe K, Ganten D, Hokfelt T, Bolme P (1976) Immunohistochemical evidence for the existence of angiotensin II containing nerve terminals in the brain and spinal cord of the rat. Neurosci Lett 2: 229
Ganten D, Minnich JL, Grainger P, Hayduk K, Brecht HM, Barbeau A, Boucher R, Genest J (1971) Angiotensin forming enzyme in the brain tissue. Science 173: 64
Ganten D, Hutchinson J, Schelling P (1975) The intrinsic brain iso-renin angiotensin system: Its possible role in central mechanisms of blood pressure regulation. Clin Sci 48: 265
Ganten D, Hutchinson JS, Schelling P, Ganten U, Fischer H (1976) The iso-renin-angiotensin systems in extrarenal tissue. Clin Exp Pharmacol Physiol 3: 103
Genest J, Nowaczynski W, Koiw E, Sandor T, Biron P (1960) Adrenocortical function in essential hypertension. In: Bock KD, Cottier TP (eds) Essential hypertension: An international symposium. Springer-Verlag, Berlin, p 126
Goldblatt H, Lynch J, Hanzal RF, Summerville WW (1934) Studies on experimental hypertension. I. Production of persistent elevation of systolic blood pressure by means of renal ischemia. J Exp Med 59: 347
Gordon FJ, Haywood JR, Brodie MJ, Mann JFE, Ganten D, Johnson AK (1979) The effect of anteroventral third ventrical (AV3V) lesions on the development of hypertension in spontaneously hypertensive rats. Jpn Heart J 20/1: 116
Gross F (1958) Renin und Hypertensin, physiologische oder pathologische Wirkstoffe? Klin Wochenschr 36: 193
Hannon RC, Derryck RP, Joosens JV, Amery AK (1969) Disappearance rate of endogenous renin from the plasma after bilateral nephrectomy in humans. J Clin Endocrinol Metab 29: 1420
Heisey SR, Held D, Pappenheimer JR (1962) Bulk flow and diffusion in the cerebrospinal fluid system of the goat. Am J Physiol 203: 775
Hirano T, Takei Y, Kobayashi H (1978) Angiotensin and drinking in the eel and the frog. In: Jørgenson CB, Skadhauge E (eds) Osmotic and volume regulation, Alfred Benzon Symposium XI. Academic Press, New York, p 123
Hoffman WE, Phillips MI (1976) Regional study of the cerebral ventricle sensitive sites to angiotensin II. Brain Res 110: 313
Hsiao S, Epstein AN, Camardo JS (1977) The dipsogenic potency of peripheral angiotensin II. Horm Behav 8: 129
Johnson DG, Gorden P, Kopin IJ (1971) A sensitive radioimmunoassay for 7S nerve growth factor antigens in

serum and tissues. J Neurochem 18: 2355
Johnson JA, Davis JO (1973) Effects of a specific antagonist of angiotensin II on arterial pressure and adrenal steroid secretion in dogs. Circ Res Suppl I 32/33: 159
Joy MD, Lowe RD (1970) Evidence that the area postrema mediates the central cardiovascular response to angiotensin II. Nature 228: 1303
Kalu DN, Pennock J, Doyle FH, Foster GV (1970) Parathyroid hormone and experimental osteosclerosis. Lancet I: 1363
Kenney NJ, Epstein AN (1975) The antidipsogenic action of prostaglandin E (PGE). Neurosci Abstr 1: 469
Khairallah PA (1972) Action of angiotensin on adrenergic nerve endings. Inhibition of norepinephrine uptake. Fed Proc 31: 1351
Khairallah PA, Page IH (1961) Concerning the mechanism of action of angiotensin and bradykinin on smooth muscle. Am J Physiol 200: 51
Khairallah PA, Robertson AL Jr, Davilla D (1972) Effects of angiotensin II on DNA, RNA and protein synthesis. In: Genest J, Koiw L (eds) Hypertension 1972. Springer, Berlin Heidelberg New York, p 212
Kobayashi H, Uemura H, Wada M, Takei Y (1977) Angiotensin and drinking behaviour: A comparative survey. In: Japan–US co-operative science programme. Seminar: Comparative studies of the renin angiotensin system. Tochigi, Japan
Kobayashi H, Uemura H, Wada M, Takei Y (1979) Ecological adaption of angiotensin induced thirst mechanism in tetropods. Gen Comp Endocrinol 38: 93
Koch-Weser J (1965) Nature of the inotropic action of angiotensin on ventricular myocardium. Circ Res 16: 230
Konig JFR, Klipper RN (1963) The rat-brain stereotaxic atlas. Williams and Wilkins, Baltimore
Kozlowski S, Drzewiecki K, Zurawaski W (1972) Relationship between osmotic reactivity of the thirst mechanism and the angiotensin and aldosterone level in the blood of dogs. Acta Physiol Pol 23: 369
Kucharczyk J, Mogenson GJ (1976) Specific deficits in regulatory drinking following electrolytic lesions of the lateral hypothalamus. Exp Neurol 53: 371
Laragh JH (1960) The role of aldosterone in man. Evidence for regulation of electrolyte balance and arterial pressure by a renal–adrenal system which may be involved in malignant hypertension. JAMA 174: 293
Lauson HD (1967) Metabolism of antidiuretic hormones. Am J Med 42: 713
Leksell LG (1976) Influence of prostaglandin E_1 on cerebral mechanisms involved in the control of fluid balance. Acta Physiol Scand 96: 1
Lentz KE, Skeggs LT Jr, Woods KR, Kahn JR, Shumway NP (1956) The amino acid composition of hypertensin II and its biochemical relationship to hypertensin I. J Exp Med 104: 183
Levi-Montalcini R (1980) Polydipsia after intracranial injections: A property of NGF or a contaminant? Nature 284: 577
Lewis GP, Reit E (1966) Further studies on the actions of peptides on the superior cervical ganglion and suprarenal medulla. Br J Pharmacol 26: 444
Lewis ME, Avrith DB, Fitzsimons JT (1979) Short-latency drinking and increased sodium appetite after intracerebral microinjections of NGF in rats. Nature 279: 440
Lewis ME, Avrith DB, Fitzsimons JT (1980) Reply to R Levi-Montalcini. Polydipsia after intracranial injections: A property of NGF or a contaminant? Nature 284: 577
Lin MT (1980) Effects of angiotensin II on metabolic, respiratory and vasomotor activities as well as body temperature in the rabbit. J Neural Transm 49:197
Linazasoro JM, Diaz Jimenez C, Castro Mendoza HJ (1954) The kidney and thirst regulation. Bull Inst Med Res (Madrid) 7: 53
Malvin RI, Mouw D, Vander AJ (1977) Angiotensin: Physiological role in water-deprivation induced thirst of rats. Science 197: 171
Mann JFE, Phillips MI, Dietz R, Haebara H, Ganten D (1978) Effects of central and peripheral angiotensin blockade in hypertensive rats. Am J Physiol 234: 629
Maran JW, Yates FE (1977) Cortisol secretion during intrapituitary infusion of angiotensin II in conscious dogs. Am J Physiol 233: 273
McAfee RD, Locke W (1967) Effect of angiotensin amide on sodium isotope flux and short circuit current of isolated frog skin. Endocrinology 81: 1301
McConnell SD, Henkin RI (1973) Increased preference for sodium and potassium salts in spontaneously hypertensive (SH) rats. Proc Soc Exp Biol Med 143: 185
McKinley MJ, Denton DA, Hatzikostas S, Weisinger RS (1979) Effect of angiotensin II on parotid saliva secretion in conscious sheep. Am J Physiol 237: E56
McKinley MJ, Denton DA, Graham WF, Leksell LG, Mouw DR, Scoggins BA, Smith MH, Weisinger RS, Wright RD (1980) Lesions of the organum vasculosum of the lamina terminalis inhibit water drinking to hypertonicity in sheep. In: Proceedings of the Seventh International Conference on the Physiology of Food and Fluid Intake. Warsaw
Mohring J, Petri M, Szokol M, Haach D, Mohring B (1976a) Effects of saline drinking on malignant course of renal hypertension in rats. Am J Physiol 230: 849
Mohring J, Mohring B, Petri M, Haack D, Hackenthal E (1976b) Sodium and water balance, ECFV and BV and hormones in the pathogenesis of renal hypertension in rats. In: Proceedings of the Sixth International Congress on Nephrology. Karger, Basel, p 255
Munday KA, Parsons BJ, Poat JA (1971) The effect of angiotensin on cation transport by rat kidney cortex slices. J Physiol (Lond) 215: 269
Munday KA, Parsons BJ, Poat JA (1972) Studies on the mechanism of action of angiotensin on ion transport by kidney cortex slices. J Physiol (Lond) 224: 195
Navar LG, Langford HG (1973) Effects of angiotensin on the renal circulation. In: Page IH, Bumpus FM (eds) Handbook of experimental pharmacology, Vol 37. Springer-Verlag, Berlin, p 455
Nicolaïdis S (1974) Y-a-t-il une regulation et des recepteurs distincts d'un éspace hydrique extracellulaire non vasculaire (ECNV)? J Physiol (Paris) 69: 166A
Nicolaïdis S, Fitzsimons JT (1975), La dépendance de la prise d'eau enduite par l'angiotensine II envers la fonction vasomotrice cerebrale locale chez le rat. CR Acad Sci [D] (Paris) 281: 1417
Nishimura H (1978) Physiological evolution of the renin angiotensin system. Jpn Heart J 19: 806
Nordin BEC, Peacock M (1969) Role of kidney in regulation of plasma calcium. Lancet II: 1280
Olsson K, Kolmodin R (1974) Accentuation by angiotensin II of the antidiuretic and dipsogenic responses to intracarotid infusions of NaCl and fructose. Acta Endocrinol (Copenh) 75: 333
Oparil S, Vassaux C, Sanders A, Haber E (1970) Role of renin in acute postural homeostasis. Circulation 41: 89
Page IH, Bumpus FM (1961) Angiotensin. Physiol Rev 41: 331
Page IH, Bumpus FM (eds) (1974) Angiotensin. Springer-Verlag, Berlin
Page IH, Helmer OM (1939) A crystalline pressor substance, angiotonin, resulting from the reaction between renin and renin activator. Proc Central Soc Clin Invest 12: 17
Page IH, Helmer OM (1940) A crystalline pressor substance (angiotonin) resulting from the reaction between renin and renin activator. J Exp Med 71: 29
Parsons JA (1976) Endocrine pharmacology. In: Parsons JA (ed) Peptide hormones. Macmillan Press, London, p 67

Parsons JA, Potts JT Jr (1972) Physiology and chemistry of parathyroid hormone. J Clin Endocrinol Metab 1: 33
Parsons JA, Reit B (1974) Chronic response of dogs to parathyroid hormone infusion. Nature 250: 254
Parsons JA, Rafferty B, Gray D, Reit B, Zanelli JM, Keutmann HT, Tregear GW, Callahan EN, Potts JT Jr (1975) Pharmacology of parathyroid hormone and some of its fragments and analogues. In: Proceedings of the Fifth Parathyroid Conference. Excerpta Medica, Amsterdam, p 33
Pasqualini RQ, Codevilla A (1959) Thirst-suppressing (antidipsetic) effect of pitressin in diabetes insipidus. Acta Endocrinol 30: 37
Peach MJ (1971) Adrenal medulla stimulation induced by angiotensin I, angiotensin II and analogues. Circ Res (Suppl) 28/II: 107
Peach MJ (1977) Renin angiotensin system biochemistry and mechanism of action. Physiol Rev 57: 313
Perkins MS, Margueles DL, Ward IL (1979) Nerve growth factor: Intense polydipsia of long duration produced by central diencephalic application. Brain Res 161: 351
Phillips MI (1978) Angiotensin in the brain. Neuroendocrinology 25: 354
Phillips MI, Quinlan J, Keyser C, Phipps J (1978) Organum vasculosum of the lamina terminalis (OVLT) as a receptor site for ADH release, drinking and blood pressure responses to angiotensin II (AII). Fed Proc 37: 1204
Phillips MI, Mann JH, Hoffmann WE, Haebara A, Schmid PG, Ganten D (1979a) Responses of stroke prone spontaneously hypertensive rats to central and peripheral angiotensin II and saralasin. Jpn Heart J (Suppl) 20/1: 123
Phillips MI, Weyhenmeyer J, Felix B, Ganten D, Hoffman WE (1979b) Evidence for an endogenous brain renin–angiotensin system. Fed Proc 38: 2260
Radford EP (1959) Factors modifying water metabolism in rats fed dry diets. Am J Physiol 196: 1098
Ramsay DJ (1978) Beta-adrenergic thirst and its relation to the renin–angiotensin system. Fed Proc 37: 2689
Ramsay DJ, Rolls BJ, Woods RJ (1975) The relationship between elevated water intake and oedema associated with congestive cardiac failure in the dog. J Physiol (Lond) 244: 303
Ramsay DJ, Keil LC, Sharpe MC, Shinsako J (1978) Angiotensin II infusion increases vasopressin, ACTH, and 11-hydroxycorticosteroid secretion. Am J Physiol 234: R66
Ramsay DJ, Thrasher TN, Keel LC (1980) Nerve growth factor stimulates angiotensin-dependent drinking and vasopressin secretion in dogs. In: Proceedings of the XXVIII International Congress on Physiological Sciences, Budapest, p 655
Reid IA (1977) Is there a brain renin–angiotensin system? Circ Res 41: 147
Richter CP (1956) Salt appetite of mammals: Its dependence on instinct and metabolism. In: L'Instinct dans le comportement des animaux et de l'homme. Masson et Cie, Paris, p 577
Robertson AL, Khairallah PA (1972) Effects of angiotensin II and some analogues on vascular permeability in the rabbit. Circ Res 31: 923
Rolls BJ (1971) The effect of intravenous infusion of antidiuretic hormone on water intake in the rat. J Physiol (Lond) 219: 331
Rolls BJ, Wood RJ (1977) Role of angiotensin in thirst. Pharmacol Biochem Behav 6: 245
Sadowski B, Szczepanska-Sadowska E, Sobocinska J (1980) Spontaneous water intake elicited by vasopressin injected to the third cerebral ventricle of the dog. In: Proceedings of the Seventh International Conference on Physiology of Food and Fluid Intake, Warsaw
Sakurai T, Hashimoto Y (1965) The vasoconstrictor action of angiotensin in relation to catecholamines. Jpn J Pharmacol 15: 223
Schaechtelin C, Regoli D, Gross F (1964) Quantitative assay and disappearance rate of circulating renin. Am J Physiol 206: 1361
Schlegel W, Felix D (1976) Structure–activity relations for angiotensin II action in subfornical organ. Experientia 32: 761
Selye H (1932) On the stimulation of new bone-formation with parathyroid extract and irradiated ergosterol. Endocrinology 16: 547
Severs WB, Daniels-Severs AE (1973) Effects of angiotensin on the central nervous system. Pharmacol Rev 25: 415
Severs WB, Summy-Long J, Taylor JS, Connor JD (1970) A central effect of angiotensin release of pituitary pressor material. J Pharmacol Exp Ther 174: 27
Sharpe LG, Swanson LW (1974) Drinking induced by injections of angiotensin into forebrain and midbrain sites of the monkey. J Physiol (Lond) 239: 595
Sharpe LG, Garnett JE, Olsen NS (1979) Thermoregulatory changes to cholinomimetics and angiotensin II, but not to the monoamines microinjected into the brain stem of the rabbit. Neuropharmacology 18: 117
Simpson JB, Routtenberg A (1973) Subfornical organ: Site of drinking elicitation by angiotensin II. Science 181: 1172
Simpson JB, Epstein AN, Camardo JS (1975) Dose–response analysis of angiotensin-induced drinking in subfornical organ (SFO) and third ventricle. Physiologist 18: 391
Simpson JB, Saad WA, Epstein AN (1976) In: Onesti G, Fernandes M, Kim KE (eds) Regulation of blood pressure by the central nervous system. Grune and Stratton, New York, p 191
Skeggs LT Jr, Lentz KE, Shumway NP, Woods KR (1956) The amino acid sequence of hypertensin II. J Exp Med 104: 193
Stricker EM (1977) The renin–angiotensin system and thirst: A reevaluation II. Drinking elicited in rats by caval ligation or isoproterenol. J Comp Physiol Psychol 91: 1220
Stricker EM (1978) The renin angiotensin system and thirst: Some unanswered questions. Fed Proc 37: 2704
Stricker EM, Bradshaw WG, McDonald RH (1976) The renin–angiotensin system and thirst: A re-evaluation. Science 194: 1169
Sutter MC (1965) The pharmacology of isolated veins. Br J Pharmacol 24: 742
Szalay KS (1969) Inhibiting effect of angiotensin on potassium accumulation of adrenal cortex. Biochem Pharmacol 18: 962
Szczepanska-Sadowska E, Kozlowski S, Sobocinska J (1974) Blood antidiuretic hormone level and osmotic reactivity of thirst mechanism in dogs. Am J Physiol 227: 776
Szutowicz A, Frazier WA, Bradshaw RA (1976) Subcellular localization of nerve growth factor receptors. J Biol Chem 251: 1524
Tang M, Falk J (1974) Sar^1-Ala^8 angiotensin II blocks renin–angiotensin but not beta-adrenergic dipsogenesis. Pharmacol Biochem Behav 2: 401
Thurau K, Schnermann J, Nagel W, Horster M, Wahl M (1967) Composition of tubular fluid in the macula densa segment as a factor regulating the function of the juxtaglomerular apparatus. Circ Res (Suppl II): 20/21 79
Tigerstedt R, Bergman PG (1898) Niere und Kreislauf. Skand Arch Physiol 8: 223
Tobian L, Lange J, Azar S, Iwai J, Koop D, Coffee K, Johnson MA (1978) Reduction of natriuretic capacity and renin release in isolated, blood perfused kidneys of Dahl hypertension prone rats. Circ Res 43/I: 92
Tosteson DC, De Friez AIC, Abrams M, Gottschalk CW, Landis EM (1951) Effects of adrenalectomy, desoxycorticosterone acetate and increased fluid intake on intake of sodium chloride and bicarbonate by hypertensive and normal rats. Am J Physiol 164: 369
Trippodo NC, McCaa RE, Guyton AC (1976) The effect of

prolonged angiotensin II infusion on thirst. Am J Physiol 230: 1063
Vander AJ, Geelhoed GW (1965) Inhibition of renin secretion by angiotensin II. Proc Soc Exp Biol Med 120: 399
Vecsei P, Hackenthal E, Ganten D (1978) The renin–angiotensin–aldosterone system: Past, present and future. Klin Wochenschr (Suppl) 56/1: 5
Villamil MF, Nachev P, Kleeman CR (1970) Effect of prolonged infusion of angiotensin II on ionic composition of the arterial wall. Am J Physiol 5: 1281
Walker DG (1971) The induction of osteopetrotic changes in hypophysectomized, thyroparathyroidectomized, and intact rats of various ages. Endocrinology 89: 1389
Weisinger RS, Denton DA, McKinley MJ, Nelson JF (1978) ACTH induced sodium appetite in the rat. Pharmacol Biochem Behav 8: 339
Weisinger RS, Considine P, Denton DA, McKinley MJ, Mouw D (1979) Rapid effect of change of cerebrospinal fluid sodium concentration on salt appetite. Nature 280: 490
Williams BC, McDougall JG, Tait JF, Tait SAS, Zananiri FAF (1980) Calcium efflux from superfused isolated rat adrenal glomerulosa cells. Proc Sixth Int Cong Endocrinol, Melbourne, p 537
Wislocki GB, King LS (1936) The permeability of the hypophysis and hypothalamus to vital dyes. Am J Anat 58: 421
Wolf G, Dahl LK, Miller NE (1965) Voluntary sodium chloride intake of two strains of rats with opposite genetic susceptibility to experimental hypertension. Proc Soc Exp Biol Med 120: 301

23 Salt appetite during reproduction, including discussion of learned appetites and aversions, and pica

Summary

1. Studies of salt appetite in pregnancy and lactation have opened up a new dimension in our understanding of the response of the neural organization underlying this instinctive behaviour.

2. In pregnancy, as in the non-pregnant animal, salt appetite reflects behaviour selected on the basis of high survival value in the face of the low salt content of the vegetation of large portions of the planet. Sodium for the tissues of the developing foetus and adnexa and for the milk, is crucial to the efficiency of the reproductive process. Induction of a salt hunger by the hormonal elements of reproduction and independently of any preceding sodium deficit provides the maximum possibility of avoiding deficit. Further, if in natural conditions exploratory behaviour is amplified resulting in location of salt sources, or food richer in sodium (Arnold 1964), this learning may be of great value during lactation when the sodium needs are much greater, and severe stress may result.

3. The data may also point to a basis for some seemingly aberrant appetitive behaviour in human pregnancy, which may represent a phylogenetic legacy of hormonal action on brain mechanisms. In the face of diverse cultural overlay, its expression may be very variable in direction and intensity, and sometimes bizarre. Some possible bases of pica are also discussed in terms of reactions to body deficit, and these are considered in the light of knowledge of learned appetites. There are remarkable capacities of animals to associate benefit or gastrointestinal toxicity with the taste or smell of food even if eaten 8 h previously.

4. During pregnancy and lactation a remarkable salt hunger develops in animals. Richter and Barelare (1938) first described it in rats. The analysis of this phenomenon at the Howard Florey Institute by means of external sodium balance studies has shown that onset and persistence of appetite are not determined by any sequestration of sodium or loss from the body. It has brought to light new facts about the action in the brain of a number of steroid and peptide hormones. The effects involve their action in conjoint fashion. To reproduce the natural phenomenon by hormone administration, at least four, and possibly six, hormones are jointly sufficient and severally necessary. Dosage at physiological levels of oestrogen, adrenocorticotrophic hormone (ACTH), prolactin and oxytocin are required in an appropriate time sequence. Progesterone may also have a role. The action of ACTH, at least in the rabbit, involves a direct central effect of the peptide as well as the increased blood level of cortisol which its administration causes.

5. The similarities to induction of maternal behaviour, which is dependent on a number of hormones, are discussed.

6. The hormonal changes also generate an appetite for calcium, which points to the possibility that this appetite is innately programmed rather than learned.

Pica

Because salt appetite is such a striking phenomenon in pregnant animals, the question naturally arises as to the nature of the effects, if any, in the pregnant human. The sometimes bizarre appetites and desires of the pregnant woman are legendary. From earliest times, the literature on pica emphasizes its major incidence in pregnant women.

Thus before proceeding to analysis of specific appetite for sodium and other minerals in pregnancy and lactation, it is interesting to consider the problem of pica generally. There is presumptive evidence that hormones may play a role in its genesis, as indicated by frequent association with pregnancy. Metabolic changes such as iron-deficiency anaemia are also frequently correlated, but, evidently, other factors are also involved. It is a complex problem.

History

Pica, or perversion of appetite with ingestion of strange or unsuitable substances, has intrigued doctors and those interested in nutrition for centuries. It is worldwide in distribution and occurs at all ages. One of the earliest accounts was that of Mergiletus in 1701. The title page of this, shown below, may suggest a tactful approach for contemporary doctoral candidates.

In Christ's Name
Inaugural Medical Dissertation
on the subject of
PICA
which
with the aid of the inspiration of the divine grace
AUGUSTUS FRIDERICUS MERGILETUS
submits
by the authority and degree of
THE BELOVED FACULTY OF MEDICINE
in
THE MOST ILLUSTRIOUS AND FAMOUS
ACADEMIA PATRIA
for
THE HIGHEST HONOURS IN THE ART OF MEDICINE
and the duly following
DOCTORAL PRIVILEGES
to the
SOLEMN AND CALM CONSIDERATION OF THE LEARNED GENTLEMEN
31st day, Month of March, 1701
at the accustomed time and place.

Mergiletus begins:

> Since the correct method for investigating those things which come under the consideration of medical practice demands first of all a definition of terms, as Galen advises . . . we too shall pay attention to the matter . . . This malady is called 'κιττα' in Greek. Its derivation from the common bird is believed by many who think this condition is called Pica because the bird ['pica' is the Latin name for 'magpie'] itself suffers from the same malady . . . because it flits from tree to tree or because it enjoys all sorts of food—thus under the constraint of this malady craving various things. Others try to derive it from 'κιττος' [ivy] because this winds itself around plants, trees and walls just as it comes across them—and troubled by this disease people are wont to crave and eat many different things.

Mergiletus considers other possible derivations from the Greek including 'hunger' and 'pluck', because often the patient tries to snatch what has been forbidden. He goes on to say the term has been variously applied to refer to an insatiable desire for food and drink; or to the desire for things which are ridiculous, difficult to digest and hardly convertible into the substance of the body. This is seen in children and adults of both sexes. Secondly in a stricter usage it is applied to the tendency of pregnant women to desire things which err in their quantity rather than their quality. He largely considers the second usage.

Mergiletus gives his ideas on how pica comes about.

> The appetite is formed by the digestive juices or fermentation of the stomach, dripping from the numerous annulated glands, which tickle or gently excite with their salty pungency the interior lining of the stomach. This movement, created by fermentation and communicated afterwards by the mediating 'animal spirit' through the fibrillae to the brain, generates the animal appetite. If, however, the fermentation wells up with particles which are more than normally acrid, bitter, sharp, salty or what you will, it arouses a correspondingly more violent irritation in the nervous lining . . . causing the mind to think more avidly about food.

He reasons then that there is no ability to discriminate between those things which are good and should be desired and those things which are bad and should be avoided, and that therefore the subject resides in no other part of the body than the brain. In the normal state it is not the stomach, nor its admirable fermentation, which determines our choice in desiring things but the 'imagination' (Latin: *phantasia*). When the body is in an abnormal state, he thinks the same arrangement applies. In fact, from time to time, women, especially when pregnant when their stomach is quite obviously in a healthy and unpoisoned state, are attracted towards all sorts of things simply by intuition or because it crossed their minds. This happens every day. Mergiletus proposes that when menstruation is suppressed, whether in pregnant women or women who are not pregnant, the blood acquires a quite singular and altered nature around the uterine cavity and is eventually carried back to the heart. This blood, greatly altered and contaminated with foreign and diseased bodies is mixed with the rest of the blood, gushes from the left ventricle, and thus becomes inextricably mixed with the 'animal spirits' and the fermentation of the stomach.

He says that pica need not be for food or things eaten in place of food. Examples are cited of a young woman who for days on end breathed in the odour of musty books, which at other times is offensive to the nostrils, with great enjoyment. Another breathed the odour of old shoes, which was repulsive due to the shoes having been worn for ages, and preferred this odour to the fragrance and sweetness of other things. A catalogue of bizarre instances is included. With the male sex, there was one who devoured leather, wood, nestlings and live mice without discomfort; another devoured woollen garments, leather, a live cat, and some mice; and another cats' tails and rotten carcasses riddled with worms. With women the bizarre in the account included the desire for human flesh—with a particularly horrifying story of a woman who lured away children with various enticements, killed them and ate them pickled in brine as her daily food. The homicides were discovered in an extraordinary fashion: the woman's cat furtively stole a child's hand and carried it into the neighbourhood. But the food desires were usually for substances such as mud and mortar scraped off walls; there were also girls who ate their own hair, the threads with which their garments were sewn, cotton, uncooked grain and lizards.

Animals are not immune from the condition: household cats eat charcoal and ashes and a case of a pregnant hunting hound with pica is cited. Parenthetically one might note here Rozin's (1976) reflection that the specialists in the animal kingdom, like the great cats, who eat a narrowly circumscribed group of foods are likely to have detection and recognition of foods under tight genetic control—a restricted repertoire. The generalists or omnivores require a much more plastic system. Extending this reasoning, I would expect that the omnivores would be more susceptible to pica behaviour.

Mergiletus notes that sufferers from pica may not have any ill effects from these otherwise harmful substances.

> They may become harmless because many people have become habituated over a long period of time to detrimental things so that, thenceforth, they never suffer injury . . . comparing Hippocrates 'Things which in the course of time have become habitual even though they are detrimental are less harmful than

things to which one is unaccustomed'. Such was King Mithridates' opinion too. He used to drink poison every day to protect himself against assassination.

Mergiletus concluded the condition is not lethal per se, provided that no extremely harmful or unusual substances are eaten. His treatment consisted of provision of a good healthy diet, and helpful and solicitous behaviour of relatives and friends. While allowing that vein opening may be practised, his approach was conservative, it being safe for the younger and well-fleshed whose strength was unimpaired. The amounts of blood removed from pregnant women should be small. He says

the amounts venerable antiquity used to draw in cases of acute and continual fever (up to several pounds) went so far as to separate soul and body. It will be obvious to anyone that this could scarcely be done without injury to the child concealed in the mother's womb. In the present day we are taught to conserve the precious fluid. In cases of extreme necessity not more than one pound is drawn off. Purgation, vomitives and other measures can be prescribed—but gently not harshly and preferably not for the fourth to last month since the danger of abortion is always present—as Hippocrates 'When things are to be guided, they should be guided where they incline, and via what paths are fitting.'

This thesis of Mergiletus is interesting in nutritional history. Though a multitude of bizarre behaviours feature in it which must have been very puzzling to a man of his era, and some might still be in ours, some relationships are clearly perceived and medical attitudes are expressed which are valid today.

Cooper (1957) in her classic work on pica cites several ancient and medieval authors who emphasized the occurrence of pica in pregnant women. Aetios noted pregnant women to crave various and odd foods, some salty and some acrid; some crave for sand, oyster shells and ashes. He recommends a diet including fruits, green vegetables, pigs' feet, freshly caught fish and old tawny fragrant wine. Boezo (1638) remarks that pica is most often seen in women, particularly the pregnant. He saw the cause of the condition as physiological. Boezo is the first to mention iron preparations as medication and suggests one or one and a half scruples of iron dross taken for many days as wonderfully beneficial for men and women. He quotes from the literature a case of a virgin who was accustomed to devour salt in great quantity, from which chronic behaviour she developed diarrhoea and wasting. Ledelius stressed the importance of determining the basic cause of pica and suggested that when the cause is known we have knowledge of the disease. He states that those of any age or sex may be affected by pica but that the evil chiefly attacks pregnant women, necessitating a close study of the uterus which he refers to as 'this penitentiary of the slaves of love'. Cooper suggests Ledelius foreshadows the theory of homeostasis, as he hypothesises that in the natural appetite when the ferment lacks sufficient salt we desire salty things, when it is too salty we desire substances to mitigate and quench this salty strength, if it is too liquid we abhor liquids and when it is too dry we choose liquids. Why should it not be the same in the unnatural state? He cites prescriptions for treating the condition but warns that all medicaments will be fruitless unless the correct diet follows.

Regarding domestic animals, Cooper cites Le Conte (1846) as suggesting that animals ingesting earth do so because of want of inorganic elements. He reported bone-chewing by cattle in Georgia and he recognized that the animals were probably deficient in phosphates. Foster, who worked in Africa, reported pica in animals and humans and also suggested that the underlying factor in human pica is in deficiency in inorganic constituents of the diet. An interesting observation which has not been followed up is McCollum's report that cows in the United States have been found to eat nails and wire and will lick rusty iron objects. There appears to be little evidence of iron deficiency occurring in cattle; possibly some benefit might accrue from associated trace elements.

Current literature

In current medical literature the substances most frequently reported as being eaten have included dirt, coal, clay, chalk, starch, pebbles, wood, plaster, paint, chimney soot, hair and cloth. Pagophagia or ice-eating is also common. The practice is more frequent in low socioeconomic groups and sometimes has a family history, but David Livingstone observed in Africa that both slaves and rich men were affected (Waller 1874).

Mental deficiency has been cited as a cause (Kanner 1948) but this has been doubted by other workers including Cooper (1957). Keith et al. (1970) note that psychotic persons rarely incorporate pica into the framework of aberration, though it could account for the behaviour of some subjects, including some cited by Mergiletus. Psychiatric aberration could be superimposed on other underlying disorders which are determinant of pica behaviour, with bizarre results.

Infestation with worms has been cited by many authors as a cause; ankylostomiasis, ascariasis, oxyuriasis and others have been implicated. Such parasitism is, of course, frequently associated with anaemia and malnutrition. In Africa, pica is often associated with magical and superstitious beliefs (Gelfand 1945; Lanzkowsky 1959).

Anaemia as cause

There have been frequent suggestions that pica may be associated with mineral deficiency, particularly iron. Cooper (1957) drew attention to the fact that simultaneous existence of pica and anaemia was recognized in medieval times. Hancock (1831) attributed pica to 'paucity of good blood and lack of proper nutrition'. Waller (1874) reports that the explorer Livingstone commented on clay- and earth-eating, a frequent form of pica. He stated it was the cause of bloodlessness amongst the African tribes in Zanzibar. Orr and Gilks (1931) and de Castro (1952) appreciated that edible earths might be rich in sodium, iron and calcium. Dickens and Ford (1942) suggested that black children they studied might eat earth to make up for an iron deficiency in their diet. Lanzkowsky (1959) has reviewed evidence incriminating iron and also cited other possible causes including meagre diet and poor nutrition. Gelfand (1945) noted that pica was common in Kenya amongst African tribes (Kikuyu) living mainly on a vegetarian diet, but absent in the eaters of diets high in animal protein (Maasai).

Earth-eating may involve a presumed personal and spiritual bond between the land and the tribe living on it. The fertile powers of the earth are recognized in the beliefs that eating earth will increase fertility, and continuing to eat it will increase lactation. Children are brought up in these beliefs.

However, whereas custom, education and economic conditions greatly influence the intake of pregnant women, there is, from several sources, evidence that anaemia is associated with the clay-eating. Substantial accounts by Ferguson and Keaton (1950) and Edwards et al. (1959, 1964) of the eating of clay and cornstarch by women in southern United States have described how women obtained clay from a river bank and ate 6–130 g per day. Some baked it in the oven before eating it. The women stated the clay tasted sour, and they were upset if they could not get it. For some, it was a desire or craving. Practically all the clay-eaters had a taste for the substance only during pregnancy, rejecting it after the baby was born. The diets of the pregnant clay- and cornstarch-eaters were substantially low in calories, calcium, iron, thiamine and niacin. The incidence of poor diet was much higher than in control pregnant individuals not showing the behaviour. The analysis of clay eaten showed 20.8 mg/100 g of iron, but experiments on suspension in 0.1 N hydrochloric acid, comparable to gastric juice, suggested only 0.03 mg/100 g were available; calcium availability was also low. However, it is not clear whether this procedure would be equivalent to chewing in small amounts and passage through the digestive tract.

Richter (1955) notes that in 1854 Ehrenberg published analyses of many hundred specimens of edible clays. It was a most extensive study, although it did not arrive at an explanation of why people ate them. He cites also Kirkham and Birch's analysis of 20 samples of clay that was eaten with avidity by all kinds of animals in Africa. They gave the percentages as follows: silica 47.76; aluminium oxide 22.03; sodium oxide 4.87; potassium oxide 2.88; iron oxide 4.98; manganese oxide 0.32; calcium oxide 2.97; magnesium oxide 1.08; phosphorus pentoxide 0.13; carbon dioxide, trace; and water. Richter noted that in general clay contained sodium, calcium, potassium, magnesium and iron in large quantities and that it may be eaten in some places for its salt content. He adds that iron seems to be a fairly constant ingredient of more recent analyses of clays and suggests that it is of interest that anaemic individuals often express a great liking for iron-containing clays. He did not know of these preferred clays being analysed at all for their trace metal content. Women, especially when pregnant, have been known to travel many miles to get their favourite clay. With relation to the iron content of the clays and the taste that it may impart it would evidently be of great interest to know what absorption of iron actually occurs as a result of chewing them. Presumably it cannot be spectacular as anaemia has often been associated with this preference and behaviour. On the other hand it has been shown with anaemic pigs that if, instead of being kept in their sty with access to a particular diet, they are permitted to go outside and fossick for themselves, they are able to cure their anaemia (Hart et al. 1930).

The incidence of pica in children is not easy to ascertain as mothers may withhold information about their children. Estimates include that of Dickens and Ford (1942) that about 25% of children of either sex ate earth. Cooper (1957) gave a 21% incidence in children referred to the Mothers' Advisory Service in Baltimore.

Lanzkowsky (1959) found in 12 children suffering from pica that their haemoglobin ranged from 3 g% to 10.9 g% with a mean of 7.89 g%±2.64 s.d. The institution of iron therapy (intramuscular iron-dextran) resulted in a cure of pica in all children within one or two weeks. The majority had worm infestation but no treatment for this was given over the course of the iron therapy. In considering the iron deficiency anaemia associated with pica, the question evidently arises whether the anaemia is the primary disorder or whether it is a secondary dietetic consequence of the prolonged pica. However, the rapid reversal with iron would favour the former explanation.

Lanzkowsky's assertion on iron cure has been challenged by Gutelius and colleagues (1962) who found a good response to saline injections, and maintained that additional attention and care by parents are the effective factors. However, McDonald and Marshall (1964) supported Lanzkowsky as the result of a double-blind experiment on 25 children who were mainly eating sand. One group received iron injections and the other saline. After 3–4 months, 11 of the 13 given iron had lost their pica compared with only 3 of 12 control patients. A corresponding rise in haemoglobin level, also statistically significant, occurred in the iron-treated group. Some patients suffered a relapse after treatment ceased, this being associated with a fall in haemoglobin.

Reynolds and colleagues (1968) found in 38 patients with anaemia from blood loss that pagophagia (ice-eating) was the most common form in a high incidence of pica. Mean haemoglobin was 8.1 g/100 ml. Fasting serum iron levels were also very low (0–52 μg/100 ml). Twenty-three patients admitted ingestion of two or more glasses of ice per day. In 22 of 25 patients this symptom disappeared with correction of iron deficiency, when serum iron levels rose above 70 μg/100 ml. Woods and Weisinger (1970) reproduced this pagophagia experimentally in rats by withdrawal of blood. The increased preference for ice was eliminated by recovery from the anaemia.

Carlander (1959) studied over 100 patients, mostly women but including a few men with gastric resection. All had sideropenia, and he thinks this to be the major cause of pica. Most also had anaemia, but pica was observed in a few patients with normal haemoglobin levels. All responded to iron treatment.

Without having detailed all the arguments on causation, it would seem legitimate to conclude that a sizeable incidence of pica is caused by anaemia or sideropenia. The question then arises as to how the behaviour is generated. At this stage of our knowledge, perhaps the most apposite considerations are those analysed in Rozin's (1967) review of thiamine-specific hunger and further discussed in Rozin (1976). Though treating this here breaks our account of pregnancy, the knowledge is basic to an evaluation of pica because it focuses the issue of 'neophilia' caused by sickness or deficiency and, thus, the data on propensity to learn appetites and aversions.

Vitamin-specific hunger and learned appetites

Harris et al. (1933) were the first to demonstrate a vitamin-specific hunger. Rats deficient in the B complex were offered a choice of two diets, one lacking the B vitamins and the other supplemented with them. Deficient rats showed a strong preference for the vitamin-rich source (Fig. 23-1),

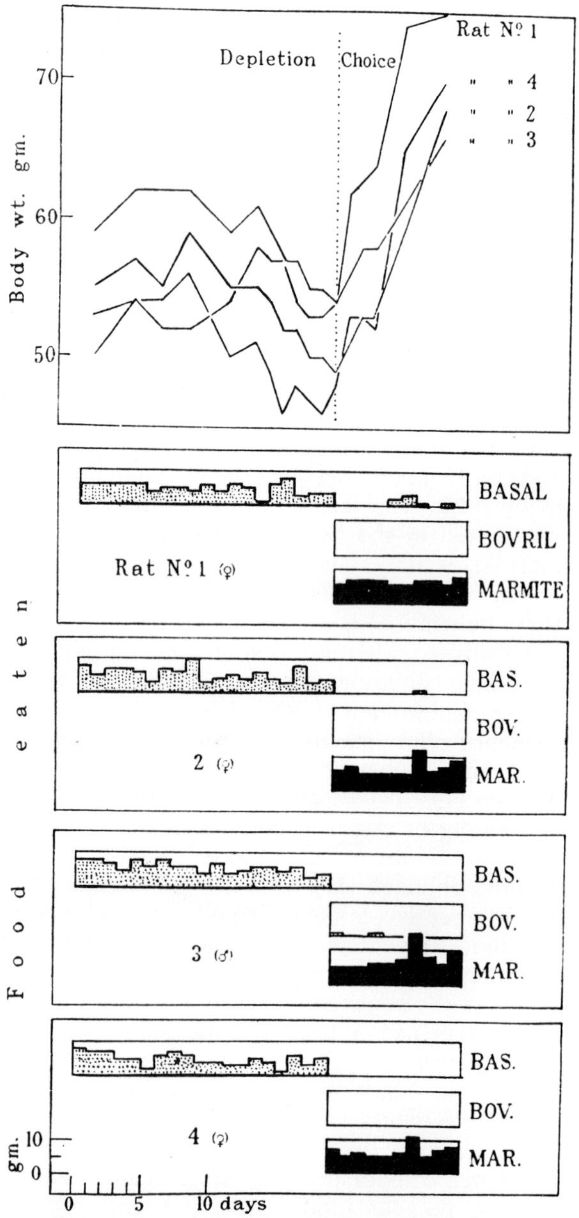

Fig. 23-1. Preference of rats depleted of vitamin B for a diet containing the vitamin. Boxes indicate that the diet was offered during the period, and histograms the amount of the diet eaten day by day. Filled histograms, diets containing adequacy of vitamin; stippled histograms, diets devoid of vitamin. (From Harris et al. 1933, with permission.)

whereas non-deficient rats showed no clear preference. Richter et al. (1937) offered thiamine-deficient rats a full cafeteria with 13 choices, including one tube containing a high concentration of thiamine in water. The rats showed a clear response to this thiamine source when it was presented, as follows:

> One vitamin deficient rat drank 11 cc or 5500 IU in less than half an hour; another drank 29 cc or 14500 IU in 24 hours. The odour of the vitamin as well as its taste arouses great interest. This is shown by the fact that the rats found the bottles at once even when as many as 12 other containers filled with different foods or solutions were presented in the cage at the same time. It was difficult to stop the animals from drinking the substance once they had tasted it. Efforts to remove the bottle were met with fierce resistance. The bottle was held tightly with both paws and even with the teeth. By reaching far up into the bottles the rats made an effort to obtain every remaining drop of the vitamin.

In Harris's experiments the choice of the vitamin-enriched diet was made within the first 24 h. Scott and colleagues (1946, 1947, 1949) and Rozin et al. (1964) confirmed these results in terms of preference of the deficient animal for a thiamine-enriched diet. In the course of their studies, Harris et al. (1933) found that normal rats when offered a choice of foods of which only one contained the vitamin B complex, often developed vitamin B complex deficiency before showing a preference for the enriched diet. Scott and Verney (1949) made essentially the same finding.

Richter and Rice (1945) studied coprophagia in this context. There is bacterial synthesis of many vitamins in the hind gut. They are not absorbed from this site and coprophagia by rats serves as a source of endogenously manufactured vitamins. They found that rats would, in a cafeteria situation, immediately ingest faeces in substantial amounts. Intake was adequate to prevent vitamin B deficiency in circumstances where the faeces were the only source of the B complex. Barnes and his colleagues (Barnes 1962; Barnes et al. 1963) extended these studies and showed that the normal rat consumes 35%–50% of its faeces, apparently by intake of faeces upon their extrusion from the anus. Faecal ingestion rises as high as 100% in the case of vitamin deficiency, suggesting a specific adaptive change in the vitamin-deficient rat. Other data from these studies included the finding that rats deficient in the vitamin B complex or thiamine became hyperactive.

In the light of Tinbergen's (1951) categorization of the elements of appetitive behaviour including increased activity and ranging, this increased restlessness or hyperactivity may be taken to indicate motivation analogous to that in the food-deprived rats. As Rozin (1967) points out, hyperactivity, if related to exploration, could be of a highly adaptive nature.

A further finding in the pioneering experiments of Harris and colleagues was that if the B complex deficient rats were offered six or ten different choices of diet after a period on a single deficient diet, they were not able to select the one food containing adequate amounts of the vitamins (Fig. 23-2). However, it was possible to teach the rats the correct choice. If, after a period of failing to make the correct choice, e.g. some 5–6 days, the animal was offered only the vitamin-rich diet, recovery ensued. If then the remaining five to nine choices were added there was a clear-cut preference for the vitamin-rich source. The subsequent selection of the single diet after the education period appeared only if this diet contained the vitamin B complex. It was concluded that the rats, having eaten the vitamin-rich diet for some period of time, developed a preference for it based upon its beneficial effect. In a further series of free-choice experiments the essential vitamin content was transferred from one diet to another. The diets were marked with particular tastes such

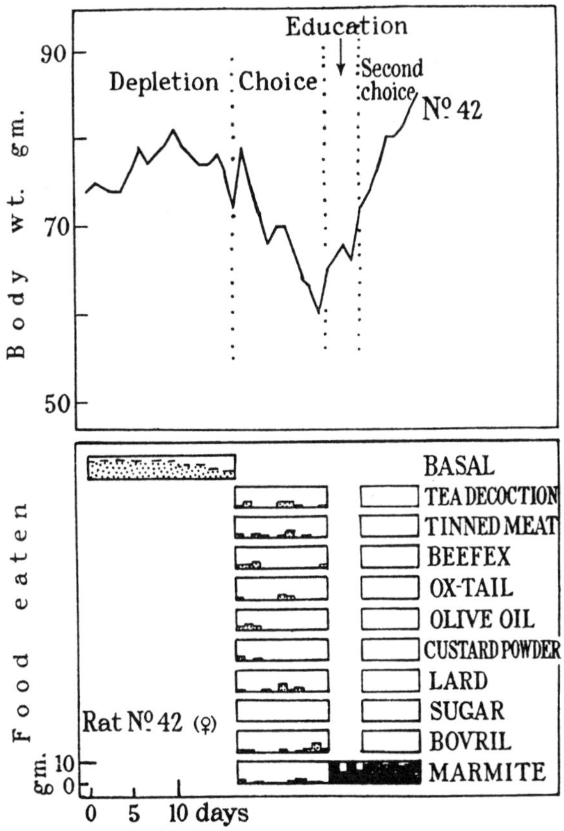

Fig. 23-2. In a vitamin-deplete rat given a choice of ten diets, selection of the vitamin-containing diet occurred only after 'education' or experience of its beneficial action. (From Harris et al. 1933, with permission.)

as cocoa and Bovril. The data indicated that the attraction of the preferred diet was not the vitamin content per se but the distinctive taste which became associated with the benefit and recovery. Scott and colleagues also concluded that thiamine hunger is learned. They noted that once the habit of eating the diet was established, presumably as a result of the feeling of well-being arising from correction of deficiency, it persisted long after the deficiency was corrected.

In the light of these data the observations of Richter et al. (1937) remain unexplained. Possibly some component of the odour of these early preparations acted specifically as a releaser of behaviour in the deprived animals. If some form of preference for food materials with a high content of vitamin B group were genetic and thus caused by natural selection, it is plausible preference could be for the odour of the complex with high vitamin B content, or for the odour of a substance frequently associated with vitamin B, rather than for the pure vitamin itself. Unlike sodium chloride, for which an innate recognition and appetite exists, components of the B complex do not occur in pure form free in nature. In this vein, it might be asked what the stimulus to coprophagia is in the rat. Is it innate or is it learned? At what stage in the life cycle does it develop? Richter and Rice (1945) state that the intake of faeces from normal rats by vitamin B deficient rats in the cafeteria situation was immediate and substantial. The animals ate the daily faecal output of two to four adult rats each day. Tribe and Gordon (1955) showed that rats deficient in vitamins of the B complex that had their olfactory bulbs sectioned showed the usual specific hunger for the vitamin-rich diets. Olfactory cues therefore do not appear to be essential for the development of the specific hunger: i.e. the selection and preference behaviour can be generated by taste alone. But this would not preclude that under normal conditions olfactory cues also are determinant. It would be interesting to know whether the converse is true. That is, whether animal with chorda tympani and glossopharyngeal nerves divided, but with olfactory bulbs intact, can make the necessary choice.

Biological analysis of vitamin-specific hunger has advanced considerably as a result of the work of Rozin (1968). He found that thiamine-deficient rats when presented with a new food, with or without thiamine, show a vigorous initial eating response. This led him to examine the role of novelty in the genesis of thiamine-specific hunger. Rodgers and Rozin (1966) placed rats on a thiamine-deficient diet for 3 weeks. After this time the animals were offered a choice between the diet they were raised on and a new diet differing considerably in taste. Thiamine was included in the familiar diet for half the rats, and in the new diet for the other half. The response of deficient rats to this choice was striking. The rats uniformly readily chose the novel diet whether it contained thiamine or not. If it contained thiamine, the rats of course commenced to recover and a maintained preference over 10 days was established. If the familiar diet contained thiamine, rats ate the deficient novel diet for a few days and then dramatically reversed their preference and ate almost exclusively from the thiamine-rich familiar diet. Control rats showed no marked preference for either diet.

Rogers and Rozin suggest that this 'non-specific neophilia' might reconcile the learning explanations of specific hungers with the learning abilities of the rat by assuming special patterns of eating in deficient rats. Rzoska (1953) showed rats that have been poisoned either developed a reluctance to eat new food or ate in a pattern of a few nibbles followed by waiting for a period of time before eating more. Rodgers and Rozin felt that their thiamine-deficient animals might have a reverse response in that they would readily take a novel diet. If the novel diet contained the needed substance the beneficial effects of ingesting this diet would reinforce the novel preference. The avid response to new foods characteristic of thiamine-deficient rats may be an exaggeration of normal inquisitiveness and exploratory tendencies.

These data may provide an explanation of an aspect of the pioneering work of Harris and colleagues. When rats which had become deficient on a particular diet were offered a choice of ten diets, one of which contained the B complex, if the rats avidly sampled all the diets in the light of their novel character, it could have been exceedingly difficult for them retroactively to associate any sense of benefit with one of the ten substances tested and eaten. But exposure to each of the components singly allowed association of the sense of benefit with a particular cue which then determined subsequent behaviour. By contrast, when sodium-deficient wild mountain rats were provided with a cafeteria containing solutions of sodium chloride, potassium chloride, magnesium chloride, calcium chloride and water, the recording devices showed that they sampled all these substances immediately. But the overwhelming preference for sodium was immediate and large intake of it occurred long before any retroactive benefit might have been established (Chapter 11).

Rozin and Rodgers (1967) have further suggested that the response to new foods by deficient rats represents an aversion to the familiar deficient diet so that perhaps what has been described as a 'neophilia' might be termed a 'palaeophobia'. The anorexia seen in thiamine deficiency is then viewed as stimulus-bound and reflecting a specific aversion to the deficient diet. The new diet has not acquired aversive properties and the rat naturally eats it, thus generating a novelty response. They add that it is possible that both palaeophobia produced by learned aversion and neophilia may be involved, but it is not necessary to assume both to account for the novelty response. They note that it may be difficult to reconcile the learning of beneficial effects or the learned aversion to deficient diet with what might appear to be limited learning capacity in rats, but rats may be particularly good at learning things about their food. In general, it may be recognized that animals have particular abilities to learn particular things. They draw attention to Lorenz's (1957) emphasis on 'learning–instinct interlocking', which suggests that capacities may be innately primed to establish connections between certain stimuli and certain responses. Tolman (1932) has stated 'some characters will organize themselves into "gestalt" wholes more readily than will others ... certain types of to be conditioned stimuli, perhaps odours, should come to evoke salivary secretions, i.e. become assigned for coming food more readily than will others'. In fact, the work of Garcia and colleagues (1966) and Garcia and Koelling (1966), followed by more recent data (Revusky and Garcia 1970; Garcia et al. 1972) shows the remarkable capacities of the rat to associate chemoreceptor stimuli with systemic after-effect notwithstanding delays of some hours in the systemic effect. Garcia has also, with Gustanson, shown this with other animals, e.g. coyotes and cougars (Garcia and Hankins 1975).

The presumption from personal experience of many humans that this ability is present in our own species has recently been confirmed by experiments on children. Some of the drugs used for the treatment of cancer in humans are the same as those used to produce conditioned taste aversions in laboratory animals. Bernstein (1977) examined whether aversion to unusually flavoured food would occur as a result of administration of drugs which cause nausea and emesis in children being treated for neoplastic diseases. The study involved three groups. The experimental group received a paired association between a novel-flavoured food and drug-induced gastrointestinal (GI) toxicity; one control group received drugs which are GI-toxic but were not exposed to the food; and a second control group was exposed to the food but received no GI treatment. The results indicated that patients in the experimental group were significantly less likely to choose that food again than patients in the control groups. The data demonstrated that aversion could be acquired by humans to a uniquely flavoured food when it is consumed before a treatment which induces GI toxicity. Such aversions might contribute to the anorexia seen in patients with cancer. Evidently it could be important to have the consideration in mind in relation to regime of treatment to avoid unnecessary side effects with nutritious foods. It is also of interest here to recall Davis' (1928, 1933) experiments on self-selection of diet by newly weaned infants, and the ability in cafeteria conditions of the child with rickets to choose cod liver oil: the preference largely disappeared when the condition ameliorated.

Long-delay learning

Garcia and colleagues (1966) offered flavoured water for short periods to rats on a restricted-water regime. At varying times after completion of the drink apomorphine was injected to induce unpleasant effects. The clear-cut finding was that rats developed an aversion to the flavour even when the aversive effect was delayed by 1½ h.

The long-delay learning was of great significance in relation to ideas both about learning and the basis of food selection. The results of intensive study of the phenomenon over the following decade by Garcia, Revusky, Rozin and their colleagues has been summarized by Rozin (1976) as follows.

1) Long-delay learning has been demonstrated for intervals as great as 8–12 h. There is evidence that when anaesthesia is administered for the period between the conditioned stimulus (CS) (e.g. flavour) and the unconditioned stimulus (US) (e.g. nausea-producing radiation or apomorphine), the interval could be extended even further.
2) Long-delay learning is limited to taste and smell CS and an unknown class of visceral US that include GI stimuli. In this regard, Rozin notes that the taste visceral field limits have not been defined, and the question of whether, for example, pain in the chest would be a US has not been determined. The general principle appears to be that animals associate food-related stimuli with the type of consequences food produces. Thus taste–GI (or smell–GI) linkage is not invariable: quail, which identify food visually, have preferential association of visual characteristics of food with GI consequences (Wilcoxon et al. 1971).
3) Long-delay learning occurs rapidly, in most cases in one trial.

4) Long-delay learning lasts a long time but can be extinguished rather easily. This might be seen as adaptive in the face of problems of poison avoidance and food selection.
5) Long-delay learning cannot be explained as a peripheral phenomenon as, for example, an after-taste concurrent with sickness. The interval is too long and in any case quail show the phenomenon in terms of an association between food and visual cues.

Parallel to this in an evolutionary context, Garcia and Hankins (1975) at the Fifth International Symposium on Olfaction and Taste proposed that bitter receptors first appeared in coelenterates during the Cambrian age (Hodgson 1967), and that rejection of bitterness represents a phylogenetically ancient response tendency. They pointed out that nearly 10% of natural plant species contain toxic alkaloids, and toxic glycosides are even more prevalent. Synthesis of such bitter-tasting substances involved a natural selection advantage for many plant species. Garcia and Hankins presented a wealth of comparative data on solutions which man would describe as bitter and which were rejected by worms, arthropods, echinoderms, through all classes of chordates up to primate and human neonates. The behaviour ranged from withdrawal, avoidance, cessation of feeding, spitting out, and disgust, to grimacing and conditioned aversions on further encounter with the food.

The innate propensities for rapid learning of metabolic effects of ingested substances represent a major element in the armamentarium of survival. Garcia and Ervin (1968) in their pioneering work showed that stimuli from the oral cavity and olfactory mucosa may be linked more easily to post-ingestional effects than exteroreceptor stimuli of vision and hearing signalling where ingestion occurred. They see the phenomenon as a neurologically programmed association, centring on the nucleus of the fasciculus solitarius in the mammal.

The explanation of long-delay learning is not clear (Rozin 1976). The possibilities considered include that taste memory traces decay more slowly than others, that with taste there are less interfering stimuli of a similar nature during the CS–US interval (limits on length of CS–US interval being set by retroactively interfering stimuli), and that what rats really learn is what is safe.

Returning to the appetite in deficiency, the increased activity and multiple sampling pattern which has been established by Rozin for thiamine-deficient rats could operate under natural conditions as a valuable selective mechanism favouring survival. But there may be a limitation on this process. If the animal samples too many things, it may not have the capacity to associate retroactively a benefit with a particular choice. In the experiment of Harris and colleagues where the ten choices were given (see Fig. 23-2) the animal sampled nine of the ten substances offered on day 1 and was unable to select the specific food with vitamins.

Overall, the evidence would suggest that the learned responses to beneficial substances by animals on a deficient diet, and the aversion to poisonous substances by animals once affected by them, are part of the same general mechanism. Neurophysiological organization has developed which facilitates learned association between chemoreceptor inflow via seventh, ninth and first cranial nerves with alimentary inflow via the tenth nerve and with conscious sensation such as nausea or well-being. There has been pressure favouring the survival of creatures with these abilities programmed in their central nervous organization.

This most interesting body of data is largely derived from rats. Its extrapolation to man was mentioned above. Its application to the phenomenon of pica in man may not be justified, and the common instances of pica may have many explanations. It does, though, provoke a number of questions. For example, is there some element of novelty preference analogous to mechanisms operating in lower animals and perhaps reflecting the phylogenetic legacy? As Richter (1955) pointed out, clay- and earth-eating in pregnant women may sometimes be caused by the attraction to its salt content. He emphasizes that pregnant women may have special appetites for substances such as salt, a matter we will analyse below. This is unlikely to have any relevance to the high incidence of starch-eating and the various satisfactions associated with the filling of the stomach that have been attributed to it. With anaemic individuals, and during pregnancy, the question arises of whether the state of anaemia could induce some form of general neophilia analogous to that seen in experimental animals. This would hardly explain why, with the multiple choices that could be available, the common objects of pica, including clay, should be settled upon. Taste may play a role: Richter (1955) and Fremont-Smith (1955) point out that the reddish clays have a high iron content and further that iron has a quite definite flavour which can be detected in very small concentrations. The question also arises as to whether, if there is occurrence of iron-deficiency anaemia in pregnancy, the various hormones of pregnancy and lactation in some way modify and accentuate the process, analogous to the profound effect on salt appetite.

Though our consideration of these questions here resolves nothing, it does bring to attention the potential interest of considering the whole subject of neophilia, and for that matter poisoning, in the pregnant animal to determine the extent to which this physiological state modulates the process. In particular, there is the issue of whether the hormones of pregnancy will accentuate and generalize neophilic behaviour in the same fashion as they spectacularly increase the appetite for sodium and calcium.

Human pregnancy and salt appetite

Then out spoke Blessed Mary so soft and so mild,
Pick me a cherry, Joseph, for I am with child
(Country songs of Vermont, Helen Hartness Flanders)

Richter (1955) emphasizes the extent to which cultural practices play an important part in preferences. Most of us tend to disregard our tastes in the face of current beliefs about what is good for us or just the opposite. The whole picture is complex in man, but it may be simpler in children. He instances that some children show a ravenous appetite for pure salt, referring to the case of the adrenally insufficient child who survived by taking large amounts of it (Wilkins and Richter 1940). Adults who are adrenally insufficient may not eat pure salt but often freely eat large amounts of salty food. There is a report published in 1691 by Christiani of Frankfurt of a woman who by actual count ate over 1400 salted herrings during her pregnancy. Richter wonders whether if she had been offered just plain salt she could have brought herself to eat it.

In pregnancy, increased appetite and cravings for a wide variety of normal food substances have been reported (Edwards et al. 1954; Taggart 1961). About two-thirds of the 153 pregnant women that Taggart studied developed some cravings. The most common craving was for fruit; there were also frequent cravings for strongly flavoured or savoury foods such as pickles, black puddings (blood sausages), liquorice, potato crisps, cheese and kippers. Aversions to certain foods occurred as often as cravings. In early pregnancy, common drinks like tea and coffee were often found to have a peculiar taste or to be insipid. Trethowan and Dickens (1972) note in their review also that the most popular craving is for fruit. A craving for sweet things (ice cream and jam) and for vegetables, nuts and pickles falls into second place. Marcus (1965), in another survey, studied 225 pregnant women and found 96 had cravings, 34 had aversions and 1 showed pica. Some of the women showed craving for one food, some for a few. Again fruit was found to be the most popular followed by vegetables and sweetmeats. A number of salty and savoury foods were also desired. Most of the craving behaviour was in the first trimester as reported by Taggart. Taggart notes that increased thirst was even more common than increased appetite. Several of the women stated that they craved fruit partly because its juiciness helped to satisfy their thirst. One explanation possible for the increased thirst could be an increase in salt intake.

There are no studies where 24-h urinary excretion of sodium has been determined in a population of women and subsequently during the first trimester of pregnancy. This could show whether some increased intake of sodium occurs. A report in the early literature raises an aspect of this problem: Schmidt (1925) reported a patient who had an almost complete loss of the sense of taste and smell in pregnancy. The faculties recovered after delivery. In the light of this, Schmidt made a study of taste acuity in pregnancy. The threshold detection for the four basic modalities of taste in young healthy non-pregnant women was similar to that reported by other investigators. But in 28 pregnant women aged between 18 and 33 years he found that the threshold of taste for salt was 114% higher, for acid 89%, for sweet 35% and for bitter 60% higher. He raised the question of whether the so-called pregnancy desires—specific preference for salts, spices and sweet things—could be due to reduced sensitivity for the relevant taste qualities.

With relation to the effects of extra intake of salt in pregnancy, Robinson (1958) has shown that salt cured the cramps of pregnancy in a large percentage of cases with little risk of producing toxaemia. She later found that 915 pregnant women with cramps who were given salt had no greater incidence of toxaemia than 2690 untreated women attending ante-natal clinics during the same period. Most of the patients, however, complained that they did not like the salt. In an investigation of the influence of high salt intake in the early part of pregnancy, 2077 pregnant women were investigated. Of these, 1039 were told to take more salt and 1038 to take less salt. The high salt intake was on the basis of advice to use more salt in cooking, to add extra salt to the food at the table and to eat salt bacon, salt fish, and salted nuts, butter etc. The second group were told to take less salt in their diet by using as little salt as possible in cooking, refraining from adding salt to their food at table and eating smoked bacon, unsalted butter and unsalted fish etc. Otherwise diets and other advice were the same in both groups. Patients were considered to have toxaemia if their blood pressure rose above 140/90 mmHg with or without other symptoms. Robinson reports the provocative finding that there was a lower incidence of toxaemia, oedema, perinatal death, ante-partum haemorrhage and bleeding during pregnancy in those told to take more salt. It would have been interesting in these two groups to have some comparison of the 24-h urinary sodium output in order to be able to evaluate what difference in intake occurred as a result of this advice. In the light of these findings, 20 women with early toxaemia (only early cases were available for treatment) were then treated with extra salt. All of them improved. The larger the dose of salt taken the quicker and more complete was the recovery. The extra salt had to be taken up to the time of delivery otherwise the symptoms of toxaemia recurred.

In relation to Robinson's data it is interesting that my own enquiries at Sendai Hospital in relation to the Akita and Aomori provinces of northern Honshu in Japan—where salt intake in food (25–30 g NaCl/day) is as high as anywhere in the world—failed to elicit any story of increased toxaemia. On the other hand there are some reports that toxaemia is a common condition in South Korea, where the addition of pickles and similar substances to a basic rice diet makes salt intake similar to that in northern Honshu.

Pseudocyesis

Pseudocyesis is a condition that has been described since the time of Hippocrates. Accounts by Paddock (1928) and Schopbach, Fried and Rakoff (1952) review its history and some aspects of clinical presentation. No account of any significant change of appetites is included. The data of Schopbach and colleagues suggested some diminution of gonadotrophin secretion. The effect of pseudopregnancy on appetites of experimental animals will be described further on.

Infantiphagia and placentiphagia

Some analysis has been made of these 'appetites' observed in the reproductive period. Richter (1955) notes that in laboratory conditions infantiphagia is much more prevalent in wild than in domestic rats. Even the least disturbance can cause a wild female to eat her young. He does not think it has anything to do with hunger. The larger the cage—particularly if a retreat is provided in its structure—the less likely it is that the animal will eat its young. Enders (1955) reported that infantiphagia caused enormous losses in mink farms and occurred as a result of aeroplanes buzzing the areas. However, if an extra nest-box was placed at the other end of the cage, the mother, instead of killing the young, would transfer them from one nest-box to the other. In nature each female mink has three or four holes, a safe territory into which she will retreat on occasions.

Lehrman (1956) and Bindra (1959) have suggested that increased genital licking seen in the parturient rat could be related to increased preference for sodium chloride, brought about by a salt deficit in the body. Secretions in the genital region are salty in taste, and Steinberg and Bindra (1962) reasoned that genital licking should increase not only at parturition but during a major part of pregnancy; and also that genital licking would occur to a lesser degree in females with increased salt intake than in animals on a normal intake. They did not include a salt-deprived group in their study. Otherwise, the results showed that pregnancy increased the time spent in genital licking and further that increased salt availability and consumption decreased the time of licking in the female rat. Birch (1956) has shown the 'cleaning' activities of the normal parturient female rat to be dependent upon prior experience of genital licking. During the normal birth process the female rat licks the vaginal orifice and the foetuses, eats the foetal membranes, biting off the umbilical cord in the process, and devours the placenta (Wiesner and Sheard 1933). Birch showed that females prevented from licking before parturition tend to eat rather than lick the foetuses. It is, of course, in no way established that prevention of genital licking causes the infantiphagia because of salt deficiency.

Richter (1955) reports that placentiphagia is usual in rats. He raises the question of the significance of the phenomenon in cattle, which are ordinarily herbivorous: some do it, others do not. It may reduce some deficiency or supply a special substance for nursing the young—or cover up the traces of the birth as Page (1955) has suggested. Most ungulates eat the placenta. Hartman has recorded placentiphagia in monkeys on cine film. There is folklore about the human placenta and some primitive people make concoctions of it to give to the parturient mother. Evidently there is a high hormone content in the organ. At present there is little information as to why placentiphagia occurs in many species and whether the behaviour varies according to dietetic status.

Salt appetite in pregnancy and during lactation in experimental animals

In 1937 and 1938 Richter and Barelare decided to apply the method that had been used successfully for the examination of the self-regulatory capacities following adrenalectomy and parathyroidectomy in rats to a study of self-selection behaviour during pregnancy and lactation (Richter and Barelare 1938; Barelare and Richter 1938). The rats were kept in divided cages, one part of which contained the nest and served as a living compartment while the other contained three food cups for solids (casein, sucrose and dried baker's yeast) and eight graduated inverted bottles containing sodium chloride, dibasic sodium phosphate, calcium lactate, potassium chloride, olive oil, cod liver oil, wheat germ oil and water. A partition was placed in the living cage that allowed access to the feeding compartment to adult rats but not the young until they were at least 25 days old and ready to be weaned. Thus the records of intake were not complicated by feeding by the young. The animals were accustomed to the multiple-choice diet for at least a month, then mated. They gave birth to normal young and nursed them through the usual lactation period. The average daily intake of fat (olive oil) increased from 1.2 ml before mating to 2.1 ml at the end of pregnancy and to 6.1 ml at the end of lactation. After weaning it dropped to 2.4 ml. Casein intake also increased. In contrast, the sucrose appetite showed almost no change during pregnancy with only a slight increase during lactation.

The intake of 3% sodium chloride solution showed an increase by 3–4 days after mating. At the end of gestation it had increased from a level of 1.4 to 4 ml/day, and by the end of lactation it had reached a level of 6 ml/day. Weaning produced an immediate decrease to normal levels (Fig. 23-3). The average daily intake of calcium lactate of 4.4 ml increased by about 50% during the second and third 5-day periods of gestation and then reduced somewhat during the last 5-day period of gestation. The appetite increased very sharply during lactation reaching a peak of 14.4 ml in the last 5-day period. In one rat, intake increased forty-fold relative to control. After weaning, the decrease was gradual compared with the abrupt post-lactational drop in sodium chloride intake, normal levels not being reached for several weeks. Appetite for dibasic sodium phosphate (Fig. 23-3) increased during pregnancy and even more markedly during lactation. After weaning, it dropped almost at once to the pre-gestational

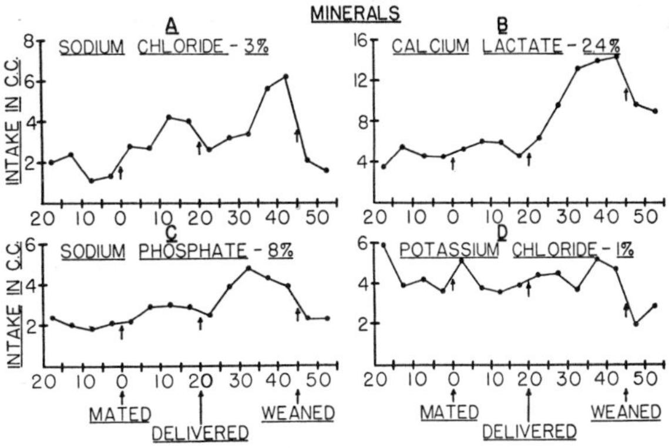

Fig. 23-3. The average daily intake of mineral solutions by rats, in 5-day periods, for 20 days before mating, for 20 days of pregnancy, for 25 days of lactation and for 10 days after weaning. (From Richter and Barelare 1938, with permission.)

level. The appetite for potassium chloride showed irregular fluctuations; after weaning, it dropped abruptly below the pre-gestational level. There was a small increase in yeast intake (vitamin B complex) though this was slight in relation to other components. There was little change in vitamin A and D intake (cod liver oil), but vitamin E intake (wheat germ oil) did increase during the lactational period. There was very little individual difference in the response to pregnancy and lactation.

Richter (1955) confirmed the findings of Wang (1923) that the onset of pregnancy was associated with a marked decrease in spontaneous running activity within hours of mating. A similar effect was caused by the induction of pseudopregnancy by cervical stimulation with a glass rod. There was a great increase in nest-building activity by the rats if the opportunity were provided. Richter and Barelare (1938) remarked that the increased demand for protein and fat during pregnancy and lactation is in keeping with our knowledge of the needs of the growing foetuses and young.

Coming to consideration of the minerals, Richter suggested the greatly increased intake of sodium chloride indicated a need that had not received much attention in metabolism of pregnancy and lactation. The increase appeared within the first few days after mating when the embryos were still minute, indicating that the increased appetite must have a maternal rather than a foetal origin. The knowledge that a similar increase in salt appetite appeared after adrenalectomy suggested to him that the adrenal glands may play an important part even in the first few days of pregnancy. He (Richter 1955) noted the great interest there would be in data on sodium balance and also considered the possibility of ovarian influence, possibly via progesterone. Increased calcium appetite is appropriate when the building of bone and teeth in the growing foetuses and young is considered. The experiments indicated that the calcium needs are much greater during lactation than during pregnancy, and it appeared to Richter and Barelare (1938) that the need was so great at this time that the mothers were unable to supply the necessary calcium from intakes alone, making it necessary for them to draw on endogenous stores. This explained why the appetite did not return to normal until several weeks after weaning. A similar interpretation was applied to the increased phosphate appetite. The absence of an increase in potassium appetite was considered surprising, though the sharp drop after weaning suggested a considerable amount may become available after weaning as a result of involution processes.

Richter and Barelare state that the physiology underlying these changes in appetite is not understood. The amazing consistency of the choices made by the animals indicates that the changes do not depend on a trial and error process. It would seem far more likely that they are due to an altered chemistry of the taste mechanism. This reflects the altered chemistry of the body during pregnancy and lactation.

A significant accumulation of sodium occurs in the foetus and other products of conception. However, the amount would be small relative to the total increase in intake during pregnancy (i.e. approximately 50 mmol or about five times the extracellular sodium content of the mother). Lichton (1961) showed in rats that in the first week of pregnancy an average of 0.8 mmol of sodium accumulated, in the second week 1.61 mmol, and in the third week 4.48 mmol. Thus approximately 60% of the sodium accumulation occurred in the last week. Direct analysis of gravid uteri and observation of post-partum excretion indicated that there was no significant maternal accumulation over and above that accounted for by the gravid uterus and contents. This does, however, involve an increment in maternal blood volume and extracellular volume. Accumulation of sodium has been demonstrated in pregnant women by metabolic balance studies and estimates of total exchangeable sodium. The total foetal and maternal sodium increment of normal pregnancy has been estimated by Hytten and Leitch (1971) to be 947 mmol, or about 23 mmol/week. Cattle are estimated to require 1.3–2.2 g/day of sodium from the seventh to ninth month of pregnancy (see Aitken 1976 for review).

Striking effects of a sodium-deficient diet concurrent with pregnancy have been shown by Pike et al. (1962). The comparison with pregnancy alone or low-salt diet alone revealed marked adrenal hypertrophy and significant reduction in the sodium content of maternal adrenal glands, heart, muscle, brain, plasma, amniotic fluid, placenta and total foetus (Kirksey et al. 1962). The authors suggest the data indicate an increased need for sodium in pregnancy.

Though increased salt intake in pregnancy in the rat was confirmed by Braun-Menendez (1952), rather surprisingly approximately 30 years elapsed before there was intense investigation of the mineral appetite aspects of these pioneering and provocative observations of Richter and Barelare. As indicated in Chapter 3, our group was involved with Myers and colleagues of the Wildlife Division of the CSIRO in Canberra in a collaborative investigation of the biochemistry and behaviour of wild native and introduced species of animals in the sodium-deficient regions of the alps of southern Australia. During studies of self-selection of various mineral salts by wild rabbits and kangaroos, it became evident that it was the females, and at the time when they are pregnant and lactating, that predominated in this behaviour. It was therefore decided to bring wild rabbits from this region and nearby areas to the Howard Florey Institute. Having habituated them to the experimental conditions we desired, the influence of pregnancy and lactation on self-selection of minerals was studied. Central to the thinking was that any hypothesis on what was happening depended on knowledge of the sodium balance which was monitored. It turned out that these data opened up a whole new basis for the genesis of salt appetite. The new knowledge would seem relevant also to pica in pregnancy. Thus methods and results will be described in some detail here.

Study of wild rabbits

During the experiments the animals were fed a diet which provided liberal sodium and other minerals. The control mean urinary sodium excretion of the rabbits represented a turnover of about 15% per day of the extracellular sodium content. The animals were housed individually in metabolism cages 15″×15″×15″ (38 cm×38 cm×38 cm) and partially tamed by handling for a period of many months before the start of experiments. Stainless steel mesh (¼″; 0.6 cm) was placed in the collecting funnels below the cages to separate urine and faeces. Water and electrolyte solutions were available continuously in inverted bottles which held 120 ml. These were fixed outside the cage and the placement was altered each day on the basis of random tables. Urine volume was measured daily and any urine trapped in the faeces or adhering to the funnel was removed by washing with a measured amount of distilled water.

Food in pelleted form was provided from hoppers and was always available. Consumption was measured by weight loss. The pellets contained 18.2% wheat, 17.2% maize meal, 21.2% pollard, 16.2 meat meal, 11.1% coconut meal, 10.1% lucerne meal, 5% powdered milk, 1% calcium (as calcium phosphate) and added vitamin A (9000 IU/kg) and vitamin D (2200 IU/kg). This diet contained carbohydrate, digestible protein and crude fat of 50, 18.7 and 8.8% in that order and about 10% ash and 7% fibre. Laboratory analysis showed the sodium, potassium, magnesium and calcium contents of food were 80, 145, 10 and 310 mequiv/kg respectively. The electrolyte solutions sodium and potassium chlorides (500 mM) and magnesium and calcium chlorides (250 mM) were offered for two months before the beginning of the study and throughout the tests. The water provided contained less than 0.1 mmol of sodium and potassium per litre. All animals, including litters, were weighed at weekly intervals during each phase of the test. Following an initial control period of 15 days the animals were mated. The commencement of pregnancy was determined by observed individual matings. On or after the 25th day of pregnancy, each rabbit was allowed free access to a rectangular plastic nesting box, attached to the back of the cage in such a fashion that any excreta in the box could be collected in the metabolism cage. Strips of well-washed hessian were placed in the box to provide supplementary nesting material. This was removed 21 days post-partum, washed thoroughly with a measured volume of water and the washings analysed for sodium and potassium. The outcome of this was that at no stage were nests obviously fouled with urine and faeces, nor did the hessian contain significant amounts of sodium and potassium.

Pregnancy

Pregnancy caused an increase in sodium chloride intake but in contrast to the rat this only began 10 days after mating (Fig. 23–4). In the second half of pregnancy mean daily intake was substantially increased and reached a peak value ten times greater than control just before parturition. Again in contrast to Richter's findings in the rat, the calcium chloride intake was also increased significantly during pregnancy and the pattern of intake of calcium chloride was similar to that of potassium chloride, increasing within 2–3 days of mating. Food intake decreased at the end of pregnancy.

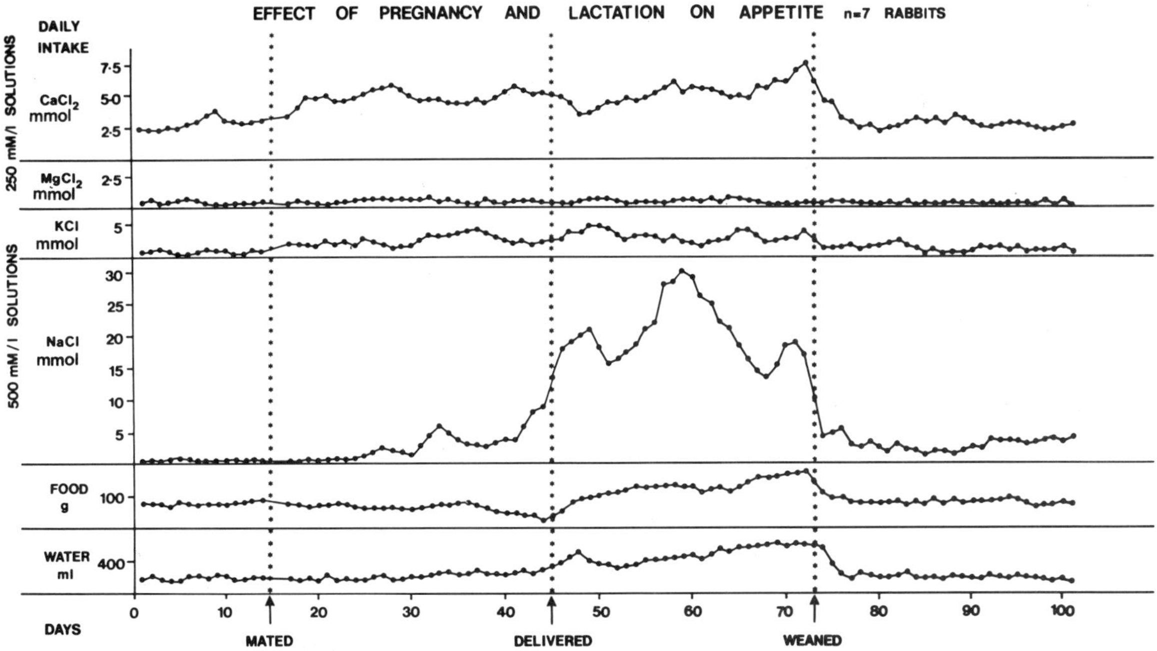

Fig. 23-4. Mean daily intake of food, water, 500 mM solutions of sodium and potassium chlorides, and 250 mM solutions of magnesium and calcium chlorides by seven female rabbits during a control period, during pregnancy, during lactation and subsequent to weaning. (From Denton and Nelson 1971, with permission.)

Lactation

Following parturition, the rise in the amount of sodium chloride drunk was spectacular. The intake reached a maximum on the 14th day of lactation when the mean daily intake was 31 mmol of sodium, over half extracellular content. There was a significant relation between litter size and sodium chloride intake (Fig. 23-5). The mean daily consumption of both potassium and calcium chlorides was approximately the same as that observed during pregnancy. Food intake rose rapidly in lactation. On the 22nd day of lactation small amounts of faeces from the baby rabbits were found in the metabolism area of the several cages. It is possible they contributed to the tertiary peak in sodium chloride intake during lactation, though it would have been difficult for them to reach the levels of the drinking nozzles. After weaning, intake of sodium, potassium and calcium chlorides and of food and water dropped precipitously and became stable about 7 days later. These changes almost always occurred on the first day after removal of the young, although in two animals the intake remained high and was still high when they were tested again after an interval of 6 weeks. Apparently, in these two animals a relatively permanent alteration of the pattern of electrolyte intake had occurred as a result of pregnancy and lactation.

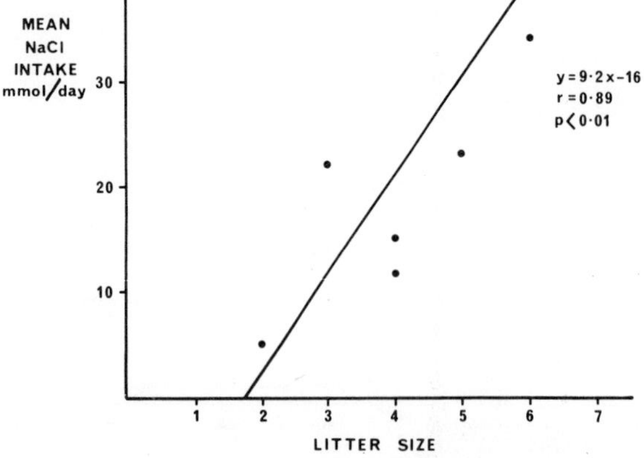

Fig. 23-5. Relationship between number of rabbits in individual litters and the mean daily intake of sodium chloride by the respective mothers during lactation. (From Denton and Nelson 1971, with permission.)

The mean body weight of mothers at approximately weekly intervals and the mean weight of the young during the lactation period is shown in Table 23-1. The body weights of the does rose steadily during pregnancy.

Sodium balance

Fig. 23-6 shows the sodium balance over the course of the experiment. The lower panel records mean total sodium intake (food plus electrolyte solution) and the upper panel the cumulative amount of sodium unaccounted for on the basis of total intake minus urinary output. During the control period the average total sodium intake was 112 mmol per rabbit of which 110 mmol (or 98%) was recovered in the urine (Fig. 23-6). The amount unaccounted for probably represents faecal excretion. During pregnancy the mean total sodium intake per rabbit was 250 mmol with 235 mmol being excreted in the urine. As the rabbits had gained about 250 g in weight over this period, approximately 13 mmol of sodium (Huth and Elkinton 1959) was sequestrated so that only about 2 mmol or approximately 1% was not accounted for and was probably excreted in faeces. Analysis of the balance showed that most of the differences between intake and urinary output occurred over the days just before parturition when the rate of weight gain was greatest.

In the lactation period the mean total sodium intake per rabbit was 860 mmol, and 60 mmol was not excreted in the urine. The mean gain in litter weight during this period (almost 1100 g) would result in the sequestration of approximately 55 mmol, so that less than 1% of total intake was not accounted for. After weaning a similar pattern was still present, the mean total intake and excretion being 280 and 270 mmol respectively over a 28-day period. Fig. 23-7 shows the relation of voluntary intake of sodium chloride solution and urinary sodium excretion in an individual rabbit which had a litter of three. *This representative figure shows there was a substantial urinary sodium excretion throughout the experiment. There was no evidence that the rise in voluntary sodium chloride intake was preceded by a large fall in urinary sodium excretion indicative of depletion of body sodium.* The fluctuations in urinary output followed the variations of voluntary intake.

In relation to the increase in calcium appetite, it may be noted that there was a sizeable calcium content of the food, as was the case with sodium. The relation of sodium intake to litter size may have been determined by the volume of milk secreted but many other possibilities presented,

Table 23-1. Mean body weights of wild rabbit mothers (kg) and the mean total weight of their litters measured at weekly intervals throughout the study.

		Control					*Pregnancy*		
Weeks	1	2	3			1	2	3	4
Mothers	1.710	1.709	1.726			1.757	1.810	1.892	1.975
Litters	—	—	—			—	—	—	—
			Lactation				*Post Weaning*		
Weeks	0	1	2	3	4	1	2	3	4
Mothers	1.735	1.771	1.741	1.750	1.756	1.731	1.728	1.737	1.722
Litters	0.205	0.310	0.552	0.811	1.290	—	—	—	—

including the influence of sensory inflow with suckling and the intensity of any pheromone effects. Calcium intake was not related to litter size. One important difference from the conditions in nature in these laboratory experiments should be mentioned. Under natural conditions the female isolates her litter by blocking a cul-de-sac in the burrow to form the nest. Once daily she removes the earth obstruction, enters and allows the babies to suckle for about 5 min (K. Myers, personal communication). In the laboratory, at least during the hours when observations were made, the doe spent most of her time in the nest and the young were possibly suckled more frequently. The mean daily weight gain of the baby rabbits was 11 g, which is 50% higher than that noted for wild rabbits (Myers and Poole 1962). This implies that the milk output of the experimental animals was greater than in the natural environment.

The central fact that sodium intake during lactation was very much greater than required to replace sodium lost in milk strongly suggested a hormonic or sensory influence on appetite. Cross (1955) reported that maximum milk production observed in six lactating rabbits weighing approximately 4 kg (double the weight of wild rabbits) was 240 ml/day. Even if this rate of production

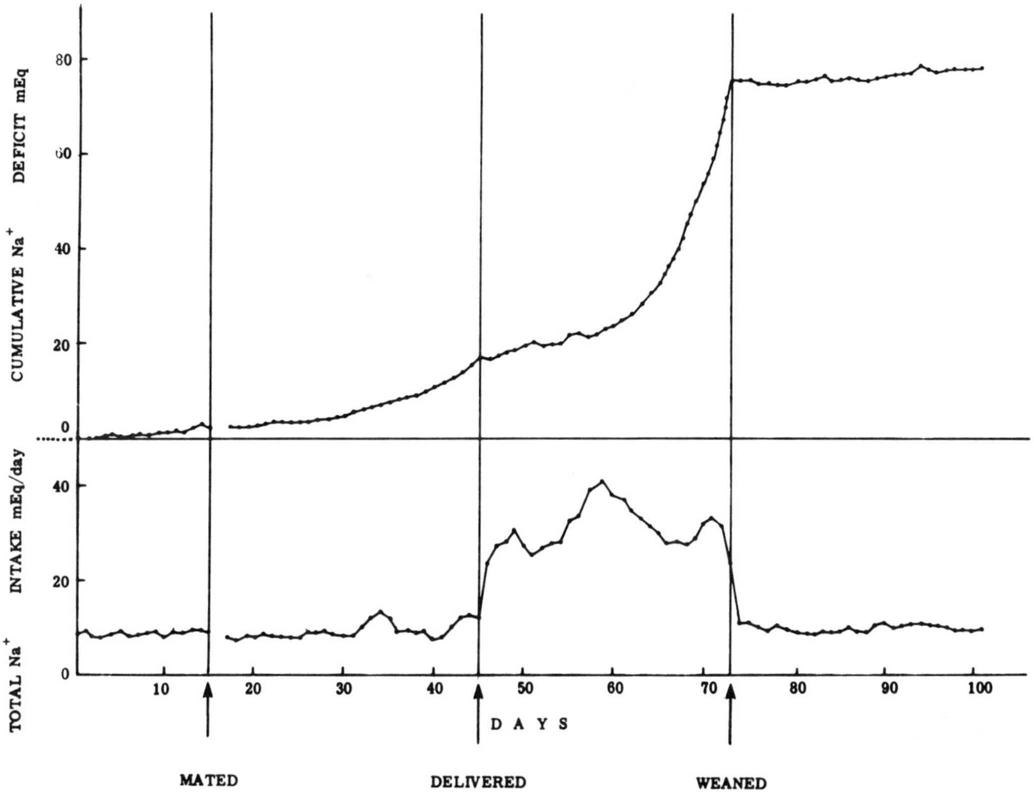

Fig. 23-6. Sodium balance data on seven rabbits during pregnancy and lactation. The mean daily total sodium intake (food and voluntary intake of sodium chloride solution) is shown in the lower panel. The upper panel records the cumulative amount of sodium unaccounted for by the total intake minus the urinary sodium excretion. (From Denton and Nelson 1971, with permission.)

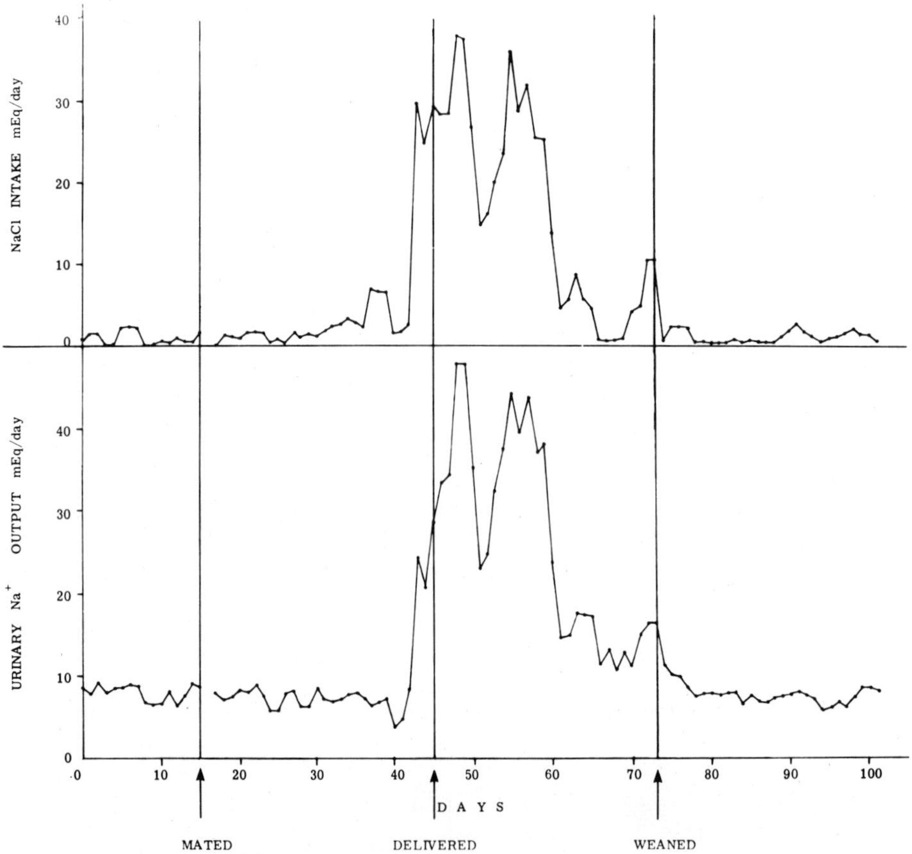

Fig. 23-7. Relationship in an individual rabbit of urinary sodium excretion and daily voluntary intake of sodium chloride solution during a control period, pregnancy, lactation and subsequent to weaning (litter of three). (From Denton and Nelson 1971, with permission.)

occurred throughout lactation in the wild rabbits, direct observations of sodium content of the milk showed that it was 30–40 mmol/l and this would only account for about 10 mmol/day of sodium, or less than half the mean increase in intake. Furthermore, as indicated above, the individual patterns of urinary sodium excretion showed no evidence that the increased voluntary intake of sodium chloride solution was preceded by a drop in sodium excretion that would have been indicative of body deficit in the face of sodium sequestration in the young or sodium loss in lactation. Voluntary daily sodium intake on many occasions exceeded the total extracellular content of sodium.

It was decided, therefore, to investigate comprehensively the salt-appetite-generating effects of the spectrum of steroid and peptide hormones involved in pregnancy and the initiation and sustaining of lactation.

Analysis of hormonal factors influencing salt appetite in pregnancy and lactation

Pseudopregnancy

Covelli et al. (1973) induced pseudopregnancy in five female wild rabbits (*Oryctolagus cuniculus* (L.)) by a single intravenous injection of 200 IU of human chorionic gonadotrophin (Pregnyl, Organon; hCG). The conditions of study, including external sodium balance, were as above.

An increase in appetite for sodium chloride was found in every animal following hCG injection (Fig. 23-8). The increase reached a maximum 17–25 days after treatment (4–5 times control) and a rapid return to normal was noted during the next 17 days. The sodium chloride intake was significantly elevated between days 11 and 25 after commencement of pseudopregnancy ($P<0.05$). Food intake did not alter. Sodium balance

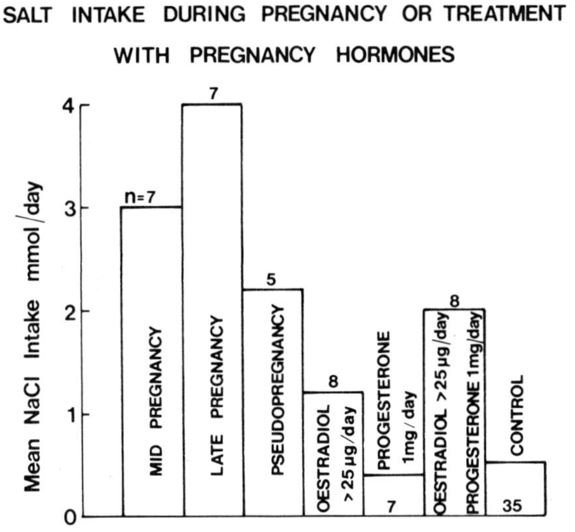

Fig. 23-8. Intake of 500 mM sodium chloride solution during pregnancy, pseudopregnancy or during treatment with oestradiol (25 μg/day) and/or progesterone (1 mg/day). The numbers with each column are the numbers of rabbits in each study.

remained unchanged except for the first week of pseudopregnancy when retained sodium was 2–3 mmol. Body weight increased by 50 g in this week. A small but significant rise in calcium chloride intake was observed between days 6 and 14 of the pseudopregnancy. No significant change in intakes of potassium and magnesium chloride or of food or water was seen.

17β-Oestradiol

Injection of 3 μg of 17β-oestradiol (E_2) per day (Ovocyclin, Ciba) did not alter the intake of electrolytes, but 12.5 or 25 μg E_2/day had a highly significant but small influence on sodium intake, which rose a little over two-fold (Fig. 23-8). The rise occurred within 2 days of beginning injections. On cessation of injections intake returned to normal on the second day with the group on 12.5 μg/day but this took 5–10 days in the group on 25 μg/day. Intake of both potassium chloride and calcium chloride was increased a small amount by the oestrogen injections. The injections produced no effect on food intake at 12.5 μg/day, but at 25 μg/day there was a reduction. However, the increased intake of sodium chloride on some occasions preceded this reduction in food intake.

Progesterone

Daily intramuscular injections of 1 mg progesterone for 14 days had no effect on any of the parameters measured (Fig. 23-8).

17β-Oestradiol plus progesterone

When the rabbits received 14 daily injections of 25 μg 17β-oestradiol (E_2) and 1 mg progesterone, sodium appetite increased on the first day, the mean increase over 14 days being 5 times control (Fig. 23-8). The effect was 50% greater than observed with the same dose of E_2 alone. Intake of sodium chloride solution remained elevated for 3–5 days after injections ceased.

With this group of data, it was noteworthy that the greatest increase in salt appetite following hCG occurred 16–25 days after injection, just before and immediately after pseudopregnancy terminated. Hammond and Marshall (1914) found that milk could be expressed from the nipples of rabbits on the 19th–24th day after induction of pseudopregnancy (the time when mammary hypertrophy appeared to be maximal). In our studies the highest salt intake was noted just before lactation commenced. This strengthens the suggestion that one or more of the hormones involved in lactation (ACTH, glucocorticoids, prolactin, somatotrophin and oxytocin) may be responsible for the phenomenon.

As changes in salt appetite and oestrogen and progesterone levels are similar in pregnancy and pseudopregnancy (Eaton and Hilliard 1971), the findings would appear to confirm the view that the stimulus to salt appetite is of maternal, not of foetal origin. It may be hormonal or neural. As indicated, the increase in salt appetite was not necessarily correlated with any decrease in food intake, a relevant consideration as Cizek (1961) has shown that rabbits offered a choice of saline solution and water, drink more saline during starvation. The experiments using oestrogen and progesterone show that hormonal changes alone are adequate to produce a salt appetite response. There is evidence that the doses used to produce these effects were in the physiological range. Whereas 2–4 μg E_2/day will maintain pregnancy in hypophysectomized rabbits (Robson 1939; Saldarini et al. 1970), at parturition the mammary gland weights of these animals were much less than those of normal pregnancy. Incomplete mammary development has also been observed in rabbits receiving 15 μg E_2 and 1 mg progesterone for 30 days (Lyons and McGinty 1941; Yanmamoto and Turner 1956). Thus the doses of 12.5 or 25 μg E_2/day which were effective in stimulating salt appetite are in the lower level of the range required for full development of the mammary tissue of intact rabbits. We found the salt appetite effect of 25 μg E_2/day was maximal, a tenfold increase in the amount resulting in no further enhancement.

The role of ACTH and adrenal glucocorticoids in salt appetite

The roles of adrenocorticotrophic hormone (ACTH) and adrenal corticosteroids in the initiation and regulation of milk secretion have interested physiologists for many years. In addition to prolactin, ACTH and glucocorticoids are also intimately concerned in lactogenesis (Nelson et al. 1943; Lyons 1958; Chadwick and Folley 1962; Meites et al. 1963a, b; Chadwick 1971; Turkington 1972). For example, studies on explants indicate that insulin, hydrocortisone and prolactin are required to stimulate the formation of transfer RNA in mammary epithelial cells in vitro (Turkington 1972). Nicoll et al. (1960) found that severe stress can induce lactogenesis in oestrogen-primed rats. ACTH alone induces lactogenesis in intact pseudopregnant rabbits (Chadwick and Folley 1962). The synthetic partial ACTH peptide (Synacthen) also has this lactogenic action (Chadwick 1971). In 1961, Talwalker, Nicoll and Meites showed that a dose of 2 mg hydrocortisone actetate/day injected during the 5th–9th or the 9th–13th days of gestation in rats initiated mammary secretion. The same dose also initiated lactation if given at the same phase during pseudopregnancy. A dose of 15 mg hydrocortisone acetate/day initiated copious milk secretion in rabbits on the 20th–21st day of pregnancy when administered during the 16th–19th days of pregnancy. Our analysis of blood taken from wild rabbits shot without initial disturbance in the field indicated that cortisol and corticosterone levels were elevated during lactation (Scoggins et al. 1970).

Given this background, it was evidently important to determine the influence of ACTH and the glucocorticoid hormones on salt appetite (Blaine et al. 1975). The conditions of experiment and the electrolyte solutions offered were the same as in the earlier studies described above.

ACTH

Ten rabbits received 4 IU/day of long-acting ACTH (Corticotrophin Zn, Organon). Voluntary sodium chloride intake rose within 24 h and the tenfold rise in mean daily intake was very highly significant (Fig. 23-9). Water intake increased, but intakes of potassium, magnesium and calcium chlorides remained unaltered.

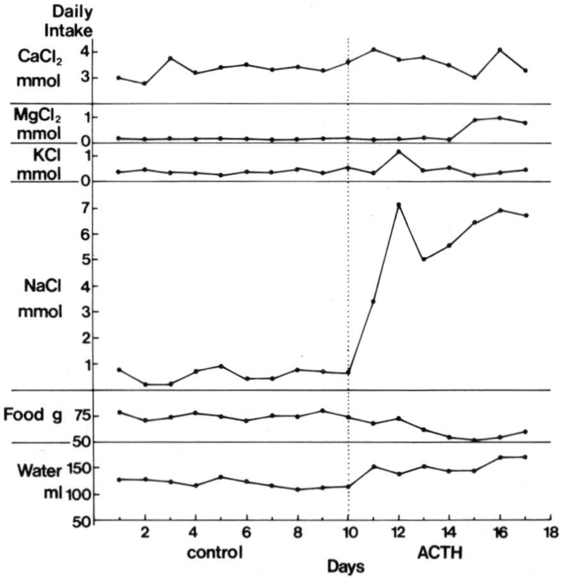

Fig. 23-9. Mean daily intake of food, water, 500 mM solutions of sodium and potassium chlorides, and 250 mM solutions of magnesium and calcium chlorides by ten rabbits during a control period and during treatment with long-acting ACTH (4 IU/day).

Cortisol

The effect of cortisol acetate (1.0 or 2.5 mg/day) was a two- or fourfold increase in voluntary sodium chloride intake respectively. In this instance the increased voluntary sodium chloride intake was associated with an increase in food intake. The intakes of potassium and calcium chlorides both doubled at the higher dose. The increased intake of calcium chloride persisted into the post-administration period, as it did for 10 days with the sodium chloride.

Corticosterone

Corticosterone (2.5 mg/day) also elicited a twofold increase in sodium appetite, the rise beginning within the first 2 days of injections and persisting for 3 days after the injections ceased. Corticosterone also increased the intake of food but had no effect on the intake of the other electrolytes.

Cortisol plus corticosterone

When the two hormones were combined (corticosterone 2.5 mg/day and cortisol 1 mg/day), the sodium chloride intake was 5 times higher than control levels. Again intake in the six animals

Table 23-2. The effect of various hormonal treatments on the peripheral blood concentrations of cortisol and corticosterone (mean±s.d.).

Treatment	Cortisol (μg/100 ml)	Corticosterone (μg/100 ml)
Control (saline-injected) (8)[a]	0.84±0.31	1.45±0.23
Cortisol acetate 1 mg (5)[a]	2.42±0.48	0.25±0.13
Cortisol acetate 2.5 mg (4)[a]	3.18±0.89	0.16±0.03
Corticosterone 2.5 mg (5)[a]	0.38±0.12	1.91±0.58
ACTH 4 IU (5)[a]	3.60±1.14	2.54±0.26
Lactation[b]	3.0	2.0

Animals were injected daily for 4 days. Blood was taken 6 h after the last injection.
[a]Number of animals in treatment groups.
[b]Mean of samples taken in the field from lactating rabbits in high-salt areas.

tested remained elevated for a number of days after injection, but decline had begun by 10 days post-injection.

The effects of steroid and ACTH administration on the circulating levels of cortisol and corticosterone are presented in Table 23-2. The estimations were made by the double-isotope dilution derivative method (Coghlan et al. 1966). The rabbit normally secretes mainly corticosterone (Blair-West et al. 1969), but following ACTH injection the pattern of production is altered and the two steroids are secreted in similar amounts. Doses of cortisol (2.5 mg/day) and corticosterone (2.5 mg/day) were chosen to produce blood levels of the steroids which approximate the range of values found during ACTH injection. These concentrations were similar to those previously reported for sodium-replete pregnant and lactating animals in the field.

Bilateral adrenalectomy

Bilateral adrenalectomy was performed in two stages on five rabbits. Peripheral blood levels of cortisol, corticosterone and aldosterone were determined in a number of these animals following withdrawal of maintenance steroids and treatment with ACTH in order to assess the completeness of adrenalectomy. No significant amounts of any of these hormones were detected. Following removal of the second adrenal, daily injections of 0.1 mg desoxycorticosterone acetate (DOCA, Ciba) and 0.75 mg cortisone acetate were administered. These had been found to be the minimum doses at which food, fluid, electrolyte intakes and sodium balance were not significantly changed from pre-operative levels. In most subsequent experiments the rabbits were treated with ACTH, and cortisol and corticosterone, without alteration of this basal DOCA and cortisone administration.

Treatment of adrenalectomized animals with ACTH resulted in a rapid fourfold increase in sodium chloride intake (Fig. 23-10). This was much less than that obtained in normal rabbits. The increased intake persisted for 4–5 days after cessation of treatment. No changes in the other electrolytes were found. The withdrawal of glucocorticoid supplement from the adrenalectomized animals while maintaining basal DOCA therapy would result in close to maximal release of endogenous ACTH. Under these conditions sodium chloride intake increased to almost the same degree as with exogenous ACTH (Fig. 23-10).

The administration of cortisol and corticosterone to these adrenalectomized rabbits produced an increase in sodium chloride intake similar to

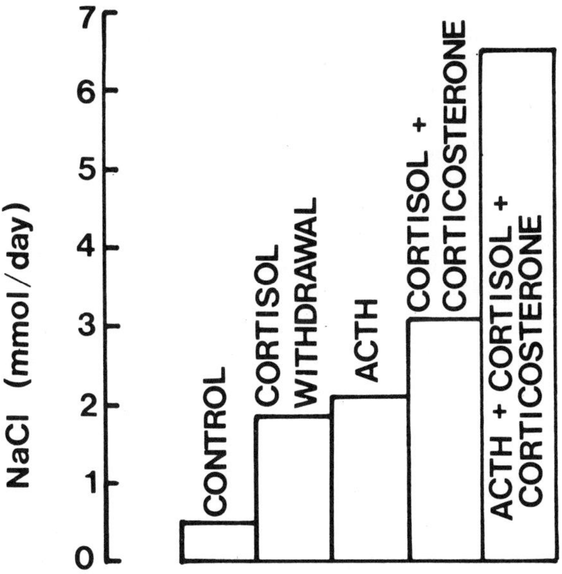

Fig. 23-10. Effect of (*a*) cortisol withdrawal (n=4), (*b*) ACTH (n=5), (*c*) cortisol acetate plus corticosterone (n=4) and (*d*) cortisol acetate plus corticosterone plus ACTH (n=4) on the mean daily intake of 500 mM solution of sodium chloride by adrenalectomized rabbits. Daily hormone doses were cortisol acetate 1 mg, corticosterone 2.5 mg and ACTH 4 IU.

that seen in normal rabbits. If then ACTH was added to this regime of the two glucocorticoids, there was a further large elevation in sodium chloride intake which then reached a level as high as that obtained in normal rabbits following ACTH injections (Fig. 23-10).

In summary, these results with intact rabbits showed clearly that cortisol and corticosterone administered singly in doses which gave circulating blood levels similar to those found in sodium-replete pregnant or lactating rabbits in the field produced a significant stimulation of sodium appetite. In each experiment the enhancement of sodium appetite occurred without any significant change in sodium balance. When cortisol and corticosterone were administered together the effects were added. The combined effect is much less than that obtained with doses of ACTH which produce a comparable alteration of blood glucocorticoid levels. The data thus indicate that ACTH has an effect on salt appetite which is not solely due to increased cortisol and corticosterone levels. The fourfold rise in sodium chloride intake produced by ACTH administered to adrenalectomized rabbits maintained on minimal steroid replacement establishes the extra-adrenal behavioural effect of ACTH. The fact that the increase in sodium chloride intake is larger when circulating glucocorticoid levels are also high indicates that the elevated steroid concentrations are necessary for the mechanism to operate at maximum response. However ACTH can also stimulate sodium appetite in the complete absence of circulating glucocorticoids. In adrenalectomized rabbits when the endogenous secretion is elevated as a result of cortisone withdrawal, sodium appetite is enhanced to the same extent as is seen with ACTH injections (Fig. 23-10). Enhanced mineralocorticoid secretion resulting from ACTH therapy can be excluded as a contributory factor in its effect on sodium chloride appetite. In normal rabbits, aldosterone even in supraphysiological doses of 0.5 mg/day is not an effective stimulus to appetite (Denton and Nelson 1970). Furthermore, the results obtained with adrenalectomized rabbits show that the full response to ACTH can be elicited without change in mineralocorticoid administration.

For two reasons some doubt can be cast on the interpretation of the findings in adrenalectomized rabbits. First, in both studies food intake fell quite dramatically and it could be argued that the response in salt appetite which occurred at that time was due to altered body metabolism associated with this decreased food appetite. Secondly, as reported later in this chapter, the results obtained with ACTH could not be replicated using adrenalectomized sheep or rats. These latter were treated with synthetic (1–24) ACTH, while ACTH prepared from porcine pituitaries was used for the rabbits. This apparent contradiction suggested to us that the response in rabbits may have been due to an impurity in the porcine extract.

Two further studies have now been carried out with adrenalectomized rabbits. In the first, they were treated with synthetic (1–24) ACTH. Here again there was a rapid fall in food intake commencing simultaneously with the rise in salt intake, so that this experiment gave no assistance in resolving the problem. In the second, endogenous ACTH secretion was stimulated by reduction of the daily cortisol dosage to 50% of the maintenance requirement. This less drastic procedure had no effect on food or water intake or on the amounts consumed of the other concurrently available electrolytes. The increase in salt appetite was as great as that found following total glucocorticoid withdrawal even though, under these conditions, the stimulus to ACTH production may not have been as great (Fig. 23-11). The results of this experiment remove any reservations which may have reasonably been held regarding the interpretation of the earlier finding.

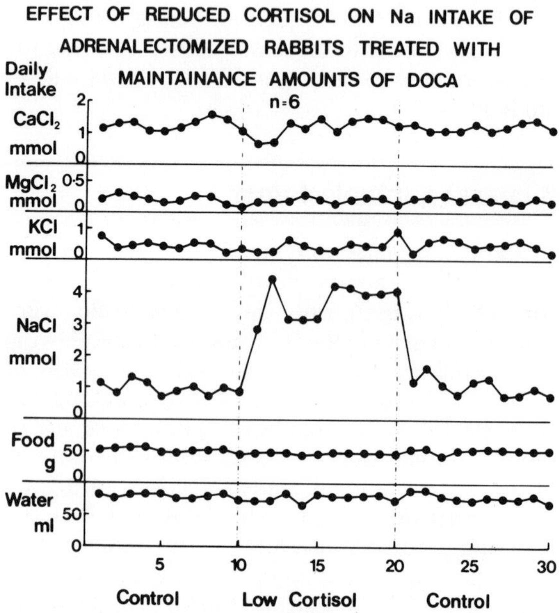

Fig. 23-11. Effect on adrenalectomized wild rabbits of reduction of cortisol administration from the maintenance dose (0.75 mg/day) to 0.375 mg/day on the mean daily intakes of food, water, 500 mM sodium and potassium chlorides, and 250 mM magnesium and calcium chlorides (n=6). Maintenance treatment with DOCA (0.1 mg/day) was continued throughout.

Among the electrolytes tested, the effects of ACTH appear to be specific for sodium chloride. In no case were intakes of the other solutions increased following ACTH injection. However, calcium chloride intake was increased in normal rabbits but not in adrenalectomized rabbits by injection of cortisol with or without corticosterone. Cortisol administration is known to increase urinary calcium excretion (Clark and Roth 1961) and to inhibit its intestinal absorption with consequent hypocalcaemia (Stoerk and Arison 1961). Thus, should a specific calcium appetite exist in the rabbit, it might be expected that increased calcium intake would follow the administration of cortisol. It is not clear why ACTH, which raises the circulating levels of cortisol, does not have the same effect—if indeed the above processes did have any bearing upon it.

Experiments were also made on the appetite-inducing effects of ACTH in other species (Weisinger et al. 1977, 1978, 1980). In sodium-replete rats, subcutaneous injections of ACTH (Synacthen depot, 5 IU/day) caused a large specific increase in sodium chloride intake, and by the fifth day of injection the animals were ingesting and excreting each day an amount of sodium approximating their total body sodium (18 mmol/day) (Fig. 23-12). ACTH was ineffective in adrenalectomized rats, suggesting that the appetite is determined by adrenal hormones. The adrenalectomized rats were maintained with basal DOCA and corticosterone or cortisol supplement so that the exogenous ACTH was not superimposed on an already maximal endogenous level of ACTH. In the early 1950s Braun-Menendez reported that ACTH had no effect on the salt appetite of rats. The effect of DOCA, however, was potentiated by ACTH. He and Brandt (Braun-Menendez and Brandt 1952; Braun-Menendez 1952) showed sodium appetite effects of cortisol, but not of cortisone, progesterone, 17β-oestradiol or testosterone.

In experiments on normal sheep (Weisinger et al. 1980), intramuscular injections of ACTH (90 IU/day in two doses) produced a threefold increase in the intake of hypertonic saline solution by sheep receiving adequate sodium in their diet (50–150 mmol/day) (Fig. 23-13). Observations made every 3 h indicated that increased sodium chloride intake preceded increased urinary output (see Fig. 18-6). ACTH did not increase sodium chloride intake in adrenalectomized sheep maintained on an infusion of basal amounts of cortisol and aldosterone. However, increased sodium chloride intake was obtained in adrenalectomized sheep by intravenous infusion of a combination of steroid

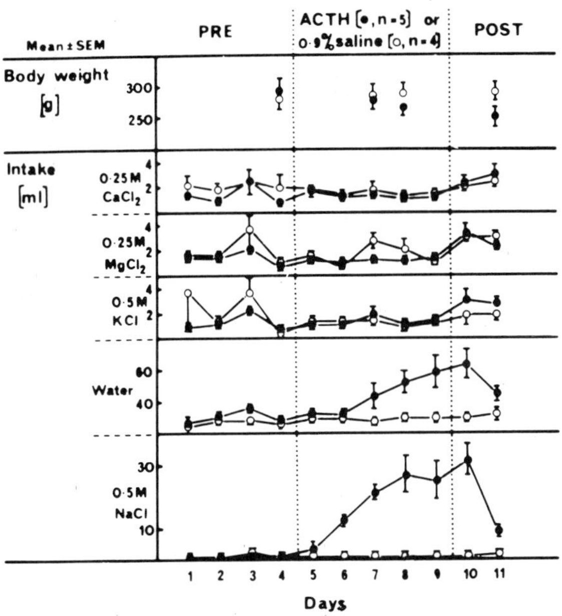

Fig. 23-12. Mean (±s.e.m.) body weight (g), intake of water (ml), of 500 mM solutions of sodium and potassium chlorides (ml) and of 250 mM solutions of magnesium and calcium chlorides (ml) for ACTH-treated (filled circles) and control (open circles) groups of rats. (From Weisinger et al. 1978, with permission.)

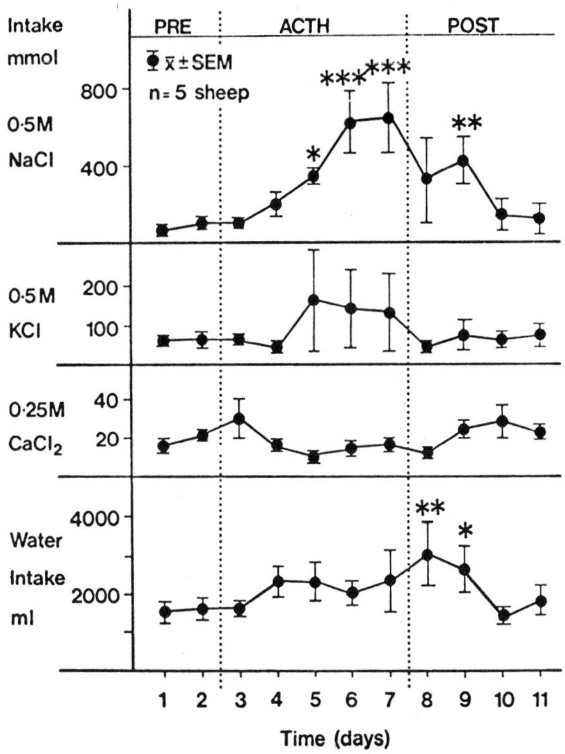

Fig. 23-13. Mean (±s.e.m.) intake of water, 500 mM solutions of sodium and potassium chlorides, and 250 mM solution of calcium chloride before, during and after ACTH treatment (n=5 sheep). $^{*}P<0.05$; $^{**}P<0.01$; $^{***}P<0.001$. (From Weisinger et al. 1980, with permission.)

hormones (desoxycorticosterone, aldosterone, cortisol, corticosterone, 11-desoxycortisol) which achieved peripheral blood concentrations similar to those observed in ACTH-treated intact animals (Scoggins et al. 1974). These data on sheep also indicate that physiological levels of adrenal hormones are effective in induction of appetite. Action of ACTH, presumably directly upon the central nervous system, in genesis of salt appetite was seen only in the wild rabbit. In terms of extra-adrenal action of ACTH, Stark et al. (1967) have shown that ACTH, natural and synthetic, increases ovarian blood flow in the adrenalectomized and normal dog.

The direct central effects of ACTH on behaviour in rabbits have particular interest in the light of the fascinating studies during the last decade indicating direct action on the brain of pituitary and hypothalamic hormones (see de Wied 1977 for review). It has been shown that there is impaired behaviour when there is lack of pituitary components. This may be produced by hypophysectomy or, in the instance of hereditary diabetes insipidus, by inability to synthesize antidiuretic hormone (ADH). De Wied and colleagues have shown there is impaired ability to learn and maintain conditioned avoidance behaviour, such as pole-jumping to avoid electric shock to the feet (de Wied 1977).

This impaired behaviour can be corrected readily by treatment with ACTH and ADH. But also, normal behaviour can be restored by fragments of these polypeptides which are themselves devoid of the classical endocrine effects of the parent hormones. The concept emerges of the hypothalamic–pituitary complex as a source of 'neuropeptides' generated from hypothalamic and pituitary hormones which may thus be regarded as precursor molecules for these entities (de Wied 1977). Hormonally inert fragments of ACTH, melanocyte stimulating hormone (MSH), lipotropin (β-LPH) and ADH have been shown to be active. For example, (4–10) ACTH, ADH and desglycinamide^{9}-lysine^{8}-vasopressin (DG-LVP) have action but the effects of ACTH and ADH appear to differ. Upon cessation of treatment with ACTH learned performance, which was at a high level, deteriorates over a week despite continuing the electric shocks. In contrast, the learned performance persists after ADH withdrawal. It is possible that the (4–10) ACTH sequence represents the actual neurotrophic effector of ACTH, and also MSH and β-LPH which also contain it, although other sequences may be necessary (Reith et al. 1977).

Many studies describe psychoactive effects of corticotrophin and related peptides in intact rats in a number of behavioural situations (Lissak et al. 1957; Dempsey et al. 1972; de Wied 1974). These peptides enhance the retention of learned responses whether the motivation is fear, hunger or sex (de Wied 1974). Learning in many situations is associated with stress; the concomitant release of corticotrophin and other pituitary hormones may provide the biochemical conditions for enhancing the response of the nervous system in the formation of new behaviour. The peptides (prohormones or fragments) might enter the brain by the blood circulation, through transport along the pituitary stalk (Bergland and Page 1977) or by release directly into the cerebrospinal fluid from tanycytes following transport in the pituitary stalk. Synthesis in the brain might also be possible. ACTH, ADH and vasotocin and prolactin are present in cerebrospinal fluid. Intraventricular administration of an antiserum against ADH prevents the formation of passive avoidance behaviour (van Wimersma Greidanus et al. 1975).

There is evidence that the enhancement of motivation produced in rats by peptides related to corticotrophin is also seen in man (Kastin et al. 1975).

The mode of action of ACTH, together with other stimuli, in the genesis of salt appetite will be considered in Chapter 24. At this stage we do not know whether smaller ACTH fragments will produce the effect on salt appetite, though the (1–24) sequence appears as effective as the whole molecule.

Hormones influencing salt appetite in lactation: Prolactin, oxytocin and growth hormone

The effects of oxytocin, prolactin and growth hormone were studied in a series of experiments on wild rabbits of both sexes (Shulkes et al. 1972).

Daily administration of 50 IU ovine prolactin for 10 days resulted in a sixfold increase in sodium chloride intake which began on the first day (Fig. 23-14). There was also a significant increase in the intake of calcium chloride. Intake of both electrolytes remained elevated in the 10-day period following the injections. The intravenous administration of 100 mU of synthetic oxytocin (Pitocin) twice daily for 8 days caused a 2½-fold increase in sodium intake ($P<0.01$) (Fig. 23-14) but no significant changes in the intake of the other electrolytes. ADH (40 ng/day) (Pitressin) had no effect on salt appetite, indicating that the effects of prolactin were not attributable to ADH con-

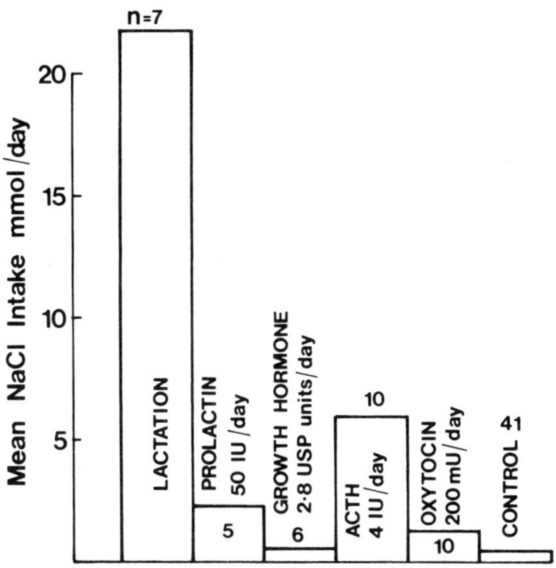

Fig. 23-14. Intake of 500 mM sodium chloride during lactation or treatment with individual lactogenic hormones. The daily hormone doses were ACTH 4 IU, prolactin 50 IU, oxytocin 200 mU, and growth hormone 2.8 USP units. The numbers with each column are the numbers of rabbits receiving each treatment.

tamination of the preparations (Sigma and NIH) used.

Ovine growth hormone (2.8 USP units/day for 10 days) did not cause any effects (Fig. 23-14).

Prolactin and oxytocin cause significant rises in sodium appetite. The data are consistent with the observation that in both pseudopregnancy and pregnancy the greatest increase in salt appetite occurs once milk secretion commences (Denton and Nelson 1971). The absence of any effect of growth hormone is not surprising. Although, in the rat and goat, both prolactin and growth hormone are necessary to produce adequate mammary development before parturition and also for continued efficient lactation, this does not seem to be the case in the rabbit (Cowie 1969).

A critical question in relation to these experiments is whether an effect on salt appetite is seen at levels of dosage related to the physiological output in the pregnant lactating rabbit. The dose of ovine prolactin (50 IU/day) is less than half the amount required to evoke a moderate degree of milk secretion in pregnant rabbits (Meites et al. 1963a) and is the minimum needed for maintenance of lactation in the rabbit hypophysectomized during lactation (Cowie 1969). There is considerable evidence for release of oxytocin during parturition and milk ejection in the rabbit. The output at parturition (Fuchs 1964) and during suckling (Fuchs and Wagner 1963) is 100–600 mU. Thus the daily dose of 200 mU of oxytocin employed here is within the physiological range just before and during lactation. As with ACTH and cortisol and corticosterone, it is clear, however, that neither oxytocin nor prolactin induced appetitive changes comparable with those seen during lactation in the wild rabbits.

Initiation of suckling involves the central nervous release of prolactin, oxytocin and ACTH in amounts which depend on litter size (Mena and Grovener 1968); it is possible that the direct positive relationship between litter size and salt appetite (Denton and Nelson 1971) may be related to this finding.

Synergism of hormones

In our experiments we found that the mean voluntary intake of sodium chloride of lactating rabbits was 22 mmol/day, with a peak at the 14th day of lactation of 31 mmol/day. Litter size is a factor in the observed mean. However, 20–30 mmol/day appeared to be the order of magnitude of intake for the animal with three to six young (Denton and Nelson 1971). This is much higher than the intake produced by any hormone alone, and thus we investigated the role of synergism of hormones involved in lactation. Treatment with a combination of the pituitary hormones—ACTH, oxytocin and prolactin—at physiological doses was given.

First (Fig. 23-15), a combination of ACTH and oxytocin caused sodium chloride intake to rise to about 9 mmol/day, which is a little less than 50% of that seen with lactation. A rise in calcium chloride intake also occurred, and there was a small change in potassium chloride intake. On stopping injections after 13 days, sodium chloride intake fell rapidly over 24 h, and reached control level 4–5 days later. There was also a small decrease in intake of calcium and potassium chlorides.

Secondly (Fig. 23-16), the triplet ACTH, prolactin and oxytocin produced an increase in sodium chloride intake to a mean of 14 mmol/day, which is almost 70% of the effect seen in lactation. A rise in intake of calcium chloride and potassium chloride was also significant but no change in magnesium chloride intake occurred. After 14 days the treatment was continued with ACTH and oxytocin alone for 7 days. On this regime all intakes were maintained at their elevated levels.

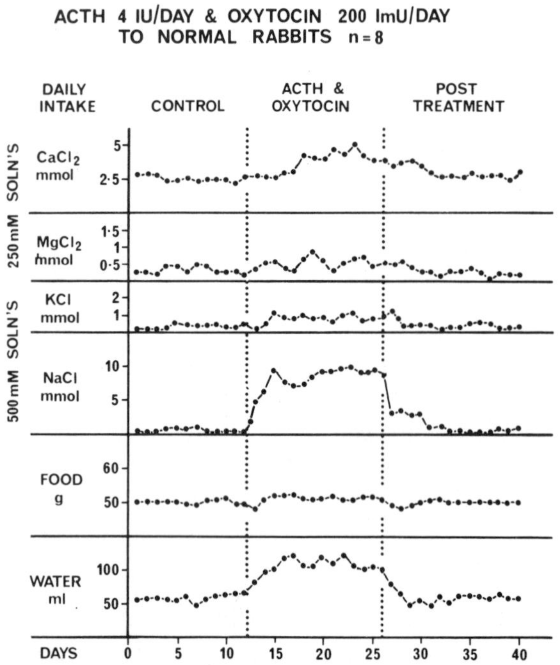

Fig. 23-15. Mean daily intakes of food, water, 500 mM solutions of sodium and potassium chloride and 250 mM solutions of magnesium and calcium chlorides during a control period; during treatment with 4 IU ACTH and 200 mU oxytocin/day; and after injections ceased (eight female rabbits). (From Denton and Nelson 1978, with permission.)

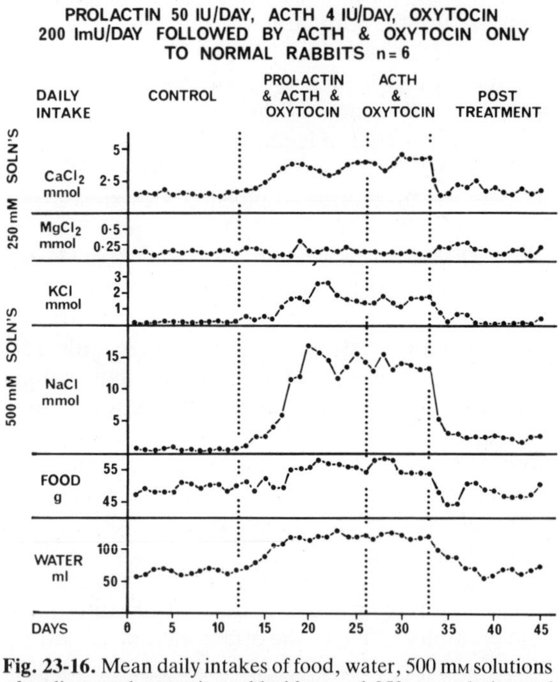

Fig. 23-16. Mean daily intakes of food, water, 500 mM solutions of sodium and potassium chlorides, and 250 mM solutions of magnesium and calcium chlorides during a control period; during treatment with 50 IU prolactin, 4 IU ACTH and 200 mU oxytocin/day; after withdrawal of prolactin from the hormone treatment; and after injections ceased (six female rabbits). (From Denton and Nelson 1978, with permission.)

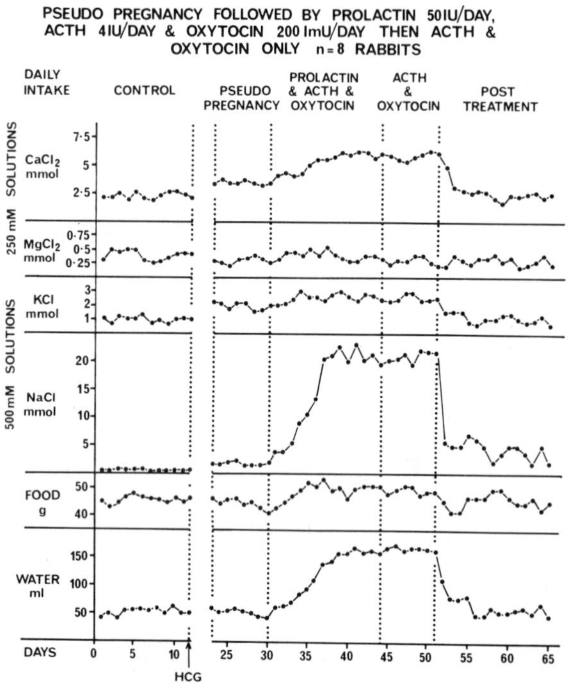

Fig. 23-17. Mean daily intakes of food, water, 500 mM solutions of sodium and potassium chlorides, and 250 mM solutions of magnesium and calcium chlorides during a control period; in late pseudopregnancy; during treatment with 50 IU prolactin, 4 IU ACTH and 200 mU oxytocin/day; after withdrawal of prolactin from the hormone treatment; and after injections ceased (eight female rabbits). (From Denton and Nelson 1978, with permission.)

Upon stopping treatment, calcium chloride and potassium chloride intakes returned rapidly to basal levels. Whereas intake of sodium chloride fell sharply over 24–48 h, it was still at an elevated level 14 days later.

These data suggest that prolactin has an essential priming effect on the appetite mechanism.

Thirdly, we tested the priming role of the hormones of pregnancy upon the mechanism. Since this could also be done with pseudopregnancy (i.e. without products of conception), the influence of parturition on the process could also be examined.

Pseudopregnancy was induced and small increases in the intakes of sodium, potasium and calcium chlorides were seen (Fig. 23-17). On the 18th day administration of the hormone triplet ACTH, oxytocin and prolactin commenced, and produced further large rises (Fig. 23-17). Sodium chloride intake rose in a fashion very similar to that seen at the end of pregnancy. From the 7th day of treatment onwards it exceeded 20 mmol/day, and the mean of 21.1 mmol/day was equivalent to that seen in lactation. Intake of potassium chloride averaged 2.6 mmol/day, calcium chloride 11.8

mmol/day, and water 160 ml/day. These findings were also similar to those in lactation. In this experiment, as in non-pregnant animals, the increases in appetite were maintained on withdrawal of prolactin, which again suggested that prolactin had a priming effect. Upon cessation of injections all intakes fell precipitously, but whereas those of calcium and potassium chlorides reached near control levels, that of sodium chloride was still evidently above control level 10–14 days later.

As the intake of sodium chloride with this treatment was clearly about 50% greater than in the absence of pseudopregnancy, it is clear that pseudopregnancy was determinant of a greater response. Further, since suckling did not occur, it was confirmed that sequestration of sodium in the young was not a necessary condition for the magnitude of effect on salt appetite during lactation.

Fourthly (Fig. 23-18), in order to determine whether the stimulatory effect of pseudopregnancy (and presumptively also that in normal pregnancy) was attributable solely to changes in oestrogen and progesterone, oestradiol and progesterone were given to rabbits for 28 days. A clear rise in intakes of sodium and potassium chlorides was seen. The administration of the hormone triplet ACTH, oxytocin and prolactin at the end of this period (the administration of oestradiol and progesterone ceasing at this point) resulted in the characteristic striking upswing in voluntary sodium chloride intake. Again, this high level, which was of the order seen in lactation in normal rabbits, persisted when both prolactin and oxytocin were withdrawn after 15 days. When hormone treatment stopped (Fig. 23-18), calcium chloride intake fell precipitously to the control pre-ovarian hormone level within a day, and potassium chloride intake, which had been raised, reached control in 3 or 4 days. Intake of sodium chloride fell precipitously over 2 days, but remained elevated above the control pre-ovarian hormone level during the following 12 days of observation.

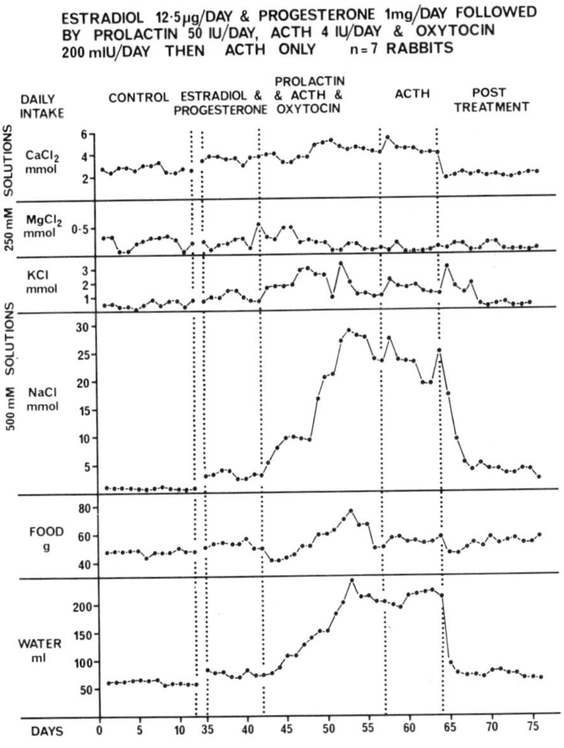

Fig. 23-18. Mean daily intakes of food, water, 500 mM solutions of sodium and potassium chlorides, and 250 mM solutions of magnesium and calcium chlorides during a control period; during the last 8 days of a 28-day treatment with 12.5 µg oestradiol and 1 mg progesterone/day; then 50 IU prolactin, 4 IU ACTH and 20 mU oxytocin/day; followed by 4 IU ACTH/day; and after injections ceased (seven female rabbits). (From Denton and Nelson 1978, with permission.)

In a final experiment of this series the conditions were the same as in the preceding experiment except that five male rabbits were used instead of females. The appetites induced by oestrogen and progesterone were the same. With combined hormone treatment (Fig. 23-19) the rise in sodium chloride intake was much slower but it did eventually reach the same level as with lactation and with the hormone triplet ACTH, oxytocin and prolactin in the female. However, it was then shown that following withdrawal of ACTH from the triplet, prolactin and oxytocin would not sustain a high sodium appetite (Fig. 23-19) as ACTH did (Fig. 23-18).

Summarizing, this behavioural phenomenon of the normal pregnant and lactating female would appear to be the outcome of a number of hormonal stimuli. These are oestrogen, ACTH and probably both cortisol and corticosterone, prolactin and oxytocin. It is possible that progesterone withdrawal at the appropriate time may itself be important if there be any close parallel to maternal behaviour as discussed below. We have not tested this. The evidence as far as we have explored it, as set out here, would suggest that a definite chronological sequence of the hormones in their jointly sufficient and severally necessary capacity is required to reconstruct the behaviour seen in the normal animal during reproduction.

There is an evident and striking feature of the behaviour described which reflects, in these sodium-replete animals, a large difference from the rapid satiation behaviour of sodium-deplete sheep. Here, in our hormone-treated animals, there is a marked 'after discharge' effect persisting for days to weeks after hormone dosage has stopped. The question arises as to whether the

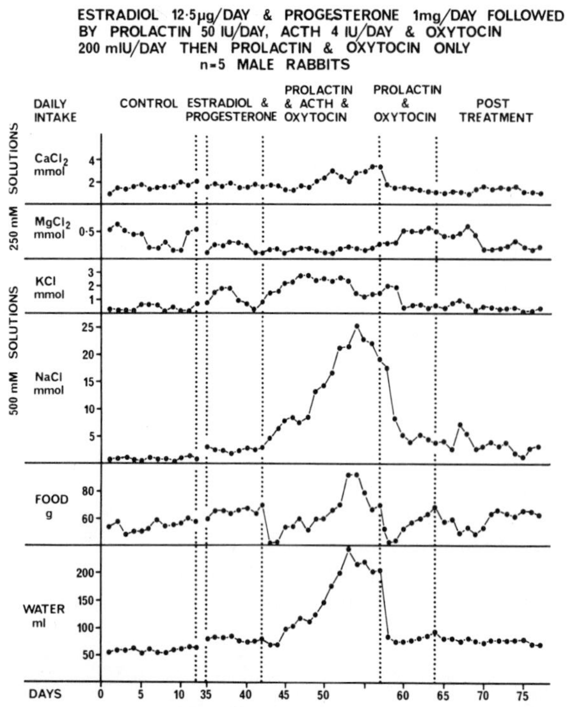

Fig. 23-19. Mean daily intakes of food, water, 500 mM solutions of sodium and potassium chlorides, and 250 mM solutions of magnesium and calcium chlorides during a control period; during the last 8 days of a 28-day treatment period with 12.5 μg oestradiol and 1 mg progesterone/day; then 50 IU prolactin, 4 IU ACTH and 200 mU oxytocin/day; after removal of ACTH from the hormone treatment; and after injections ceased (five male rabbits). (From Denton and Nelson 1978, with permission.)

effector stimuli have induced some persistent change in the neurones subserving salt appetite behaviour. The results also contrast with those on sodium-deplete wild rabbits where depletion was produced by adrenalectomy. In the adrenalectomized animals correction of deficiency was slow (9–12 h) but quite precise, and there was no overdrinking once deficit was corrected. This is described in detail in Chapter 14, and sets the magnitude of effect of hormones described in this chapter in high relief.

The striking fact of 'after discharge' or persistence of effect is brought out in unequivocal fashion in a further experiment in which the young of the pregnant rabbits were removed immediately at birth. No placentiphagia was permitted to occur. One issue under consideration in the experiment was the earlier point of the statistically significant relation of sodium chloride intake to number of young in the litter. The possible role of a pheromone or suckling influence on salt ingestive behaviour was questioned.

Following a control period, seven wild rabbits were mated. All conditions were similar to those in the experiments described earlier. The sodium content of pelleted food was 40 mmol/kg. When litters were born the young were either removed immediately or they, in their first movements, fell through the mesh in the cage to a tray 5 cm below and were then removed. The young were not permitted to suckle at any time.

The average litter size (4.7/litter) was similar to that in the initial study of pregnancy and lactation (4.4/litter) and the ranges (2–7) and gestation period (30 days) were identical. Thus comparison between the two experiments was not biased by these factors.

Pregnancy had the usual effects on electrolyte intake (Fig. 23–20). It turned out that the rise in sodium chloride intake after birth occurred whether the litters were suckled or not. The amount reached a maximum at the 3rd–4th day post-partum, or earlier if the litters were small. During the first 5 days post-partum the sodium chloride intake was not significantly different whether litters were removed or not. From the 5th day on, voluntary intake of sodium fell gradually, but did not reach control level until the 21st day post-partum. The pattern contrasted strikingly to that seen in normal suckling where a very large fall came 24 h after weaning—though it did remain above the pregnancy control level for some weeks.

In the experiments on litter removal the external

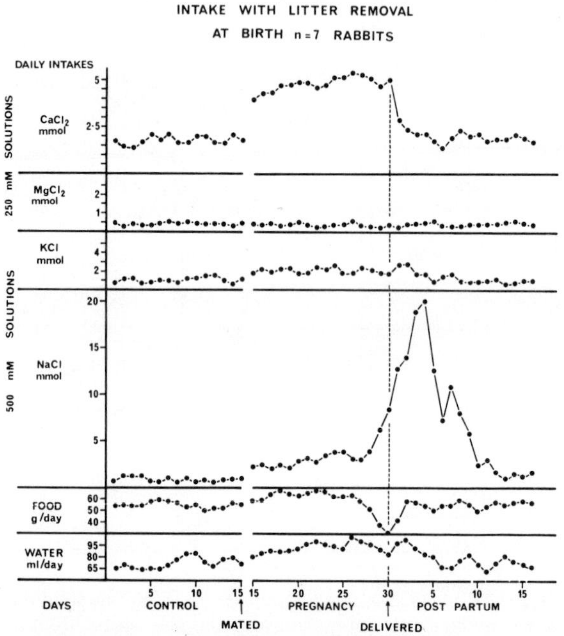

Fig. 23-20. Mean daily intakes of food, water, 500 mM solutions of sodium and potassium chlorides and 250 mM solutions of magnesium and calcium chlorides during a control period; the second half of pregnancy; and post-partum with litter removal at birth (seven rabbits). (From Denton and Nelson 1978, with permission.)

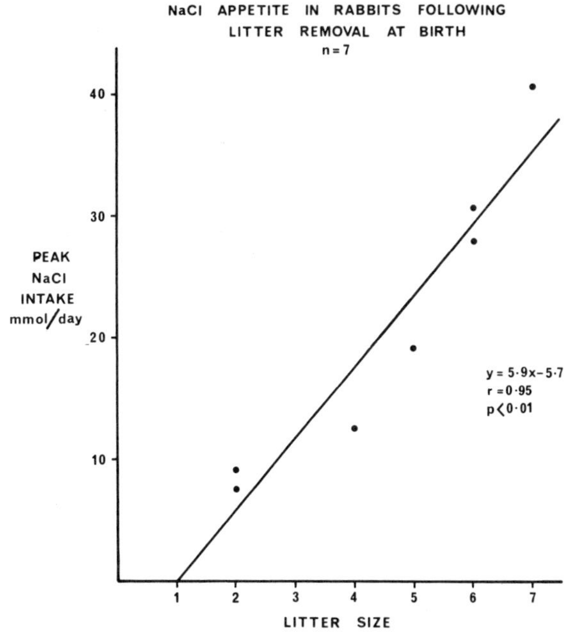

Fig. 23-21. Relationship between number of young in a litter and maximum post-partum daily sodium chloride intake of the respective mothers after litter removal at birth (seven rabbits). (From Denton and Nelson 1978, with permission.)

sodium balance showed no change. That is, approximately 0.2 mmol of ingested sodium was not excreted in the urine each day in control and post-partum periods and probably represents faecal loss. The increased intakes of calcium and potassium chlorides at the end of pregnancy were not maintained after litter removal. They returned to control levels within 3 days, as occurred after normal weaning.

Surprisingly, when the litter was removed there was still a significant correlation between number of young born and peak sodium chloride intake (Fig. 23-21). For the two smallest litters this was 8–9 mmol/day and for the largest 40.8 mmol/day. No such correlation was seen with intakes of calcium chloride, potassium chloride, water or food.

These data indicate that the sensory stimuli associated with the presence of the litter do not have any critical role in the rise in sodium intake following birth. The rise appears to be related to continued secretion of lactogenic hormones. In this regard it has been reported that if rat pups are removed at birth, prolactin does not fall to basal level until day 5. It has also been observed that if the litter is removed, involution of the uterus is delayed. Roth and Rosenblatt (1967) examined the role of self-licking throughout pregnancy as a stimulus to mammary development. They note in their discussion that cervical stimulation also induces lactation (see Meites et al. 1963b for review). In the face of the rather perplexing precise relation of salt appetite to litter size even when the litter is removed, the question emerges whether there is an afferent flow from the cervix and vagina which, commensurate with number of foetuses passing, is reflected in the extent of rise of hormones, or in some other process which influences the neural system subserving appetite. When the level of appetite characteristic of lactation was simulated by the quintet of hormones, there was not a great variation in sodium intake between animals.

Is the sodium appetite behaviour dependent on the number of young carried? There is evidence in mice that the intensity of nest-building from the 4th day of pregnancy onwards is determined by the number of foetuses (Gandelman 1975). Progesterone is apparently responsible for eliciting the building of brood nests. Plasma progesterone levels in mice normally are high throughout most of the gestation period and decline just before parturition (Murr et al. 1974). Maternal nest-building follows a similar course. Progesterone causes similar nest-building activity in intact non-pregnant mice. However, Gandelman (1975) points out that while the amount of progesterone produced in mice may be a function of the amount of placental tissue, this does not appear on the evidence to be the case in the rabbit.

Is the increase in appetite influenced by number of young born, i.e. is the parturition influential?

These questions are being approached by my colleague Dr J. Nelson by experiments involving study of sodium appetite after: (*a*) removal of one horn of the uterus and the associated conceptus 1–2 weeks before parturition, allowing the animal to come to term and studying appetite during the subsequent lactation; and (*b*) delivering the young by Caesarian section at the end of the normal gestation period and following post-partum appetite when either the offspring are returned to the mother or are permanently removed. Against these possibilities is the fact that analysis of the data in Figs. 23-4 and 23-20 shows that the sodium chloride intake on each of the last four days of pregnancy in the individual rabbits was highly significantly related to the number of young eventually born, suggesting a commensurate effect of number of young *in utero* on amount of hormone secreted and thus salt appetite.

Sheep

In sheep, mean sodium chloride intake increased early in pregnancy from a control level of 123 mmol/day to 208 mmol/day, and then remained

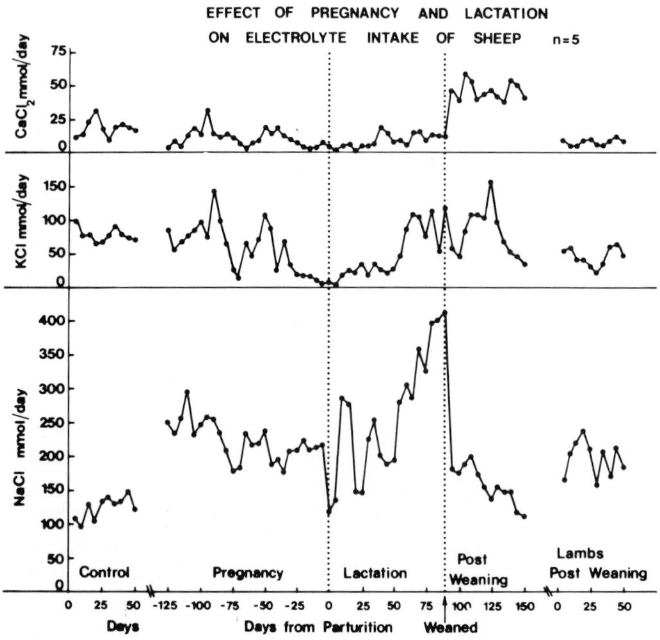

Fig. 23-22. Mean daily intake of 500 mM solutions of sodium and potassium chlorides, and 250 mM solution of calcium chloride by five ewes during a control period, pregnancy, lactation and after weaning. The mean daily intake of these electrolytes by their five lambs after weaning is also shown. Each point represents the average of the previous 5 days of observation.

fairly constant (Fig. 23-22). There was little change in the intakes of potassium or calcium chlorides until the last month of pregnancy, when both fell to levels below those observed during the control period.

The mean intake of sodium chloride during the first 50 days of lactation was maintained at the high level recorded in pregnancy, but wide variations, both from day to day and between animals, make this value not significant. During this phase of lactation the intakes of potassium and calcium chloride remained low. After this time, increases in intakes of sodium and potassium chlorides occurred. At this stage the lambs were eating some solid food, drinking some water and were occasionally seen taking small amounts of electrolyte solutions.

After the lambs were weaned the intake of sodium chloride by the mothers fell to control levels in about 30 days, and calcium chloride intake rose. The electrolyte intake of the lambs was also measured after weaning. The results show that most, if not all, of the increases in sodium and potassium intakes during the later stages of lactation could be attributed to the developing appetites of the lambs.

This experiment confirms the findings of increased sodium intake previously made in rats and rabbits during pregnancy. It also indicates that when suckling a single offspring, the sheep behaves differently from the other two species suckling larger numbers of young in that no dramatic rise in sodium chloride intake occurs.

In this general context of hormonal influences on voluntary intake of electrolytes in the sodium-replete animal, Michell and Bell (1969) and Michell (1978) have shown variations of appetite coincident with the oestrus cycle in sheep. At the time of oestrus both urinary sodium excretion and voluntary intake of 40 mmol/l sodium bicarbonate decreased, as did food intake. Cyclical variations in voluntary intake of 2% saline have also been reported in rats, with a sharp fall in salt and water intake in oestrus (Antunes-Rodrigues and Covian 1963). Oestrogen has a stimulatory effect on sodium appetite in the guinea pig (Middleton and Williamson 1974) but in the rat it is reported to inhibit intake (Thornborough and Passo 1975).

It will be of great interest to determine the role, if any, of the ovarian peptide hormone relaxin in the genesis of salt appetite in the later stages of pregnancy. The full structure of porcine relaxin has recently been determined in our Institute (James et al. 1977) as a result of collaborative endeavours with the group at the Department of Anatomy and Reproductive Biology, University of Hawaii, and Schwabe and colleagues (1976, 1977) have reported on the A and B chain structure. It is a member of the family of insulin-like growth factors, and probably related to insulin through a common ancestral polypeptide resembling proinsulin. It has a molecular weight of 6000 and consists of two non-identical chains joined by disulphide bridges. It is secreted in late pregnancy in pigs, guinea pigs, rabbits, mice, rats and humans. It acts on the cervix, uterus and pubic symphysis, and, as with other peptide hormones, the possibility of direct central neural action exists.

Maternal behaviour

The multifactorial nature of the hormonal induction of salt appetite has a parallel in the induction of maternal behaviour in rats. Maternal behaviour in non-pregnant females is seen in several mammalian species: mouse, hamster, wolf, monkey and human beings. In the mouse it has been shown that hypophysectomy does not prevent it, so it is clearly not based on pituitary hormones. In the rat maternal behaviour is basically non-hormonal (Rosenblatt 1967). Rats were tested for induction of maternal behaviour by exposing them to young pups continuously for 10 to 15 days. Non-pregnant intact, ovariectomized and hypophysectomized females were included as well as intact and castrated males. Retrieving the young, crouching over them, licking and nest-building were used as criteria of maternal performance. It required an average of 6 days to elicit maternal behaviour in these groups; this is not dependent on hormones or sex for its arousal. This latent period, however, contrasts with the immediate response to the young of the post-parturient female rat.

Obviously, the reduction of latency in the parturient female is essential to neonatal survival. Rosenblatt concluded that the hormonal changes of pregnancy and parturition cause an increase in maternal responsiveness since if pregnant females were delivered by Caesarian section at progressively later stages of pregnancy, the latencies of onset of maternal behaviour relative to nulliparous females shortened as pregnancy advanced.

Moltz et al. (1970) found that this reduced latency depended on the action of a triad of hormones near term. These were oestrogen, progesterone and prolactin, and all three were necessary, the absence of one markedly affecting the result. The three together given to the nulliparous female in a specific regime reduced latency to the range of the puerperal female though not entirely so: i.e. to 35–40 h. The data from sequence of administration, including a timed cessation of progesterone, suggested that the progesterone 'rebound' decreased the thresholds for action of oestrogen and prolactin, and such a fall paralleling the physiological situation was conditional for the synergistic action of prolactin and oestrogen.

The problem has been further investigated by Terkel and Rosenblatt (1972) by means of cross-transfusion experiments between freely moving rats. Blood was cross-transfused for 6 h between mothers and virgins beginning at the following times: 24 h before parturition, 30 min after parturition and 24 h after parturition. In addition cross-transfusion was performed between pairs of virgins. Pups remained with the virgins from the beginning of the transfusion. Only blood from newly parturient mothers induced maternal behaviour in a significant proportion of virgins. Cross-transfusion resulted in the shortening of mean latency for the appearance of maternal behaviour (i.e. retrieving etc.) to 14.5 h. Rosenblatt (1975) emphasizes the role of oestrogen in the pre-partum onset of maternal behaviour.

Returning to mineral appetite, a final point is that hormonal and pregnancy influences on taste preference in relation to saccharin and aversive substances such as quinine have been described (Wade and Zucker 1969, 1970). The aversive effect of quinine was blunted by pregnancy. Possibly such effects may contribute to voluntary intake patterns for electrolyte solutions. The issue was reviewed by Hoshishima (1967) at the Conference on the Chemical Senses and Nutrition. Overall the consensus from discussion (Pfaffmann 1967) was that the sensitivity of taste receptors was very hard to budge with sex hormones though structural changes brought about by some other deficiencies and disease states may be influential.

References

Aetios of Amida (1542) The gynaecology and obstetrics of the VIth century AD. (Translated from the Latin edition of Coronarious, 1542, by JV Ricci)

Aitken FC (1976) Sodium and potassium in nutrition of mammals. Commonwealth Agricultural Bureau, Farnham Royal

Antunes-Rodrigues J, Covian MR (1963) Hypothalamic control of sodium chloride and water intake. Acta Physiol Lat Am 13: 94

Arnold GW (1964) Some principles in the investigation of selective grazing. Proc Aust Soc Anim Product 5: 258

Barelare B, Richter CP (1938) Increased sodium chloride appetite in pregnant rats. Am J Physiol 121: 185

Barnes RH (1962) Nutritional implications of coprophagy. Nutr Rev 20: 289

Barnes RH, Fiala G, Kwong E (1963) Decreased growth rate resulting from prevention of coprophagy. Fed Proc 22: 25

Bergland R, Page RB (1977) Can the pituitary secrete directly to the brain? (Affirmative anatomical evidence.) Endocrinology 102: 1325

Bernstein IL (1977) Learned food aversions in children receiving chemotherapy. In: Le Magnen J (ed) Proceedings of the Sixth International Conference on Food and Fluid Intake. Information Retrieval, London

Bindra D (1959) Motivation. A systematic reinterpretation. Ronald Press, New York

Birch HG (1956) Sources of order in the maternal behaviour of animals. Am J Orthopsychiatry 26: 279

Blaine EH, Covelli MD, Denton DA, Nelson JF, Shulkes AA (1975) The role of ACTH and adrenal glucocorticoids in the salt appetite of wild rabbits (*Oryctolagus cuniculus* (L.)). Endocrinology 97: 793

Blair-West JR, Coghlan JP, Denton DA, Scoggins BA, Wintour EM, Wright RD (1969) The onset of effect of ACTH, angiotensin II and raised plasma potassium concentration on the adrenal cortex. Steroids 15: 433

Boezo MH (1638) De Pica. Sm, Leipsig

Braun-Menendez E (1952) Aumento del apetido especifico para la sal provocado por la desoxycorticosterona: Substancias que potenciano inhibiten esta accion. Rev Soc Argent Biol 28: 23

Braun-Menendez E, Brandt P (1952) Aumento del apetido especifico para la sal provocado por la desoxycorticosterona: Caracteristicas. Rev Soc Argent Biol 28: 15

Carlander O (1959) Aetiology of pica. Lancet 277: 569

Castin AJ, Sandman CA, Stratton LD, Schally AB, Miller LH (1974) In: Gispen WH, van Wimersma Greidanus TJ, Bohus B, de Wied D (eds) Hormones homeostasis and the brain. Progress in brain research vol 42, p 143. Elsevier, Amsterdam

Castro J de (1952) Geography of hunger 80. Gollancz, London

Chadwick A (1971) Lactogenesis in pseudopregnant rabbits treated with adrenocorticotrophin and adrenal corticosteroids. J Endocrinol 49: 1

Chadwick A, Folley SJ (1962) Lactogenesis in pseudopregnant rabbits treated with ACTH. J Endocrinol 24: xi

Cizek LJ (1961) Relationship between food and water ingestion in the rabbit. Am J Physiol 201: 557

Clark I, Roth ML (1961) In: Mills LC, Moyer JH (eds) Inflammation and disease of connective tissue. Saunders, Philadelphia, p 404

Coghlan JP, Wintour EM, Scoggins BA (1966) The measurement of corticosteroids in adrenal vein blood of sheep. Aust J Exp Biol Med Sci 44: 639

Cooper M (1957) Pica. A survey of the historical literature as well as reports from the fields of veterinary medicine and anthropology, the present study of pica in young children, and a discussion of its paediatric and physiological implications. Charles Thomas, Springfield, Ill

Covelli MD, Denton DA, Nelson JF, Shulkes AA (1973) Hormonal factors influencing salt appetite in pregnancy. Endocrinology 93: 423

Cowie AT (1969) General hormonal factors involved in lactogenesis. In: Reynolds M, Foley SJ (eds) Lactogenesis: The initiation of milk secretion and parturition. University of Pennsylvania Press, Philadelphia, p 157

Cross BA (1955) Neurohormonal mechanisms in emotional inhibition of milk ejection. J Endocrinol 12: 29

Davis CM (1928) Self-selection of diet by newly weaned infants. Am J Dis Child 36: 651

Davis CM (1933) A practical application of some lessons of the self-selection of diet study to the feeding of children in hospitals. Am J Dis Child 46: 743

Dempsey GL, Kastin AJ, Schally AB (1972) The effects of MSH on a restricted passive avoidance response. Horm Behav 3: 333

de Wied D (1974) Pituitary adrenal system, hormones and behaviour. In: Schmitt FO, Worden FJ (eds) The neurosciences: Third study programme. MIT Press, Cambridge, Mass, p 653

de Wied D (1977) Peptides and behaviour. Life Sci 20: 195

Denton DA, Nelson JF (1970) Effect of desoxycorticosterone acetate and aldosterone on the salt appetite of wild rabbits (*Oryctolagus cuniculus* (L.)). Endocrinology 87: 970

Denton DA, Nelson JF (1971) Effect of pregnancy and lactation on the mineral appetites of wild rabbits (*Oryctolagus cuniculus* (L.)) Endocrinology 88: 31

Denton DA, Nelson JF (1978) The control of salt appetite in wild rabbits during lactation. Endocrinology 103: 1880

Dickens D, Ford RN (1942) Geophagia (dirt-eating) among Mississippi negro school children. Am Sociol Rev 7: 59

Eaton LW, Hilliard J (1971) Estradiol-17β, progesterone and 20α-hydroxypregn-4-en-3-one in rabbit ovarian venous plasma. I. Steroid secretion from paired ovaries with and without corpora lutea: Effect of LH. Endocrinology 89: 105

Edwards CH, Swain H, Hare S (1954) Odd dietary practices of women. J Am Diet Assoc 30: 976

Edwards CH, McDonald S, Mitchell J, Jones L, Mason K, Kemp A, Laing D, Trigg L (1959) Clay and cornstarch eating women. J Am Diet Assoc 35: 810

Edwards CH, McDonald S, Mitchell J, Jones J, Mason L, Trigg L (1964) Effect of clay and cornstarch intake on women and their infants. J Am Diet Assoc 44: 109

Enders (1955) In discussion of CP Richter 'Self-regulatory functions during gestation and lactation'. In: Villee CA (ed) Second Conference on Gestation. Josiah Macy Jr Foundation, New York, p 79

Ferguson JH, Keaton AG (1950) Studies of diets of pregnant women in Mississippi: Ingestion of clay and laundry starch. New Orleans Med Surg J 102: 460

Fremont-Smith F (1955) In discussion of CP Richter 'Self-regulatory functions during gestation and lactation'. In Villee CA (ed) Second Conference on Gestation. Josiah Macy Jr Foundation, New York

Fuchs AR (1964) Oxytocin and the onset of labour in rabbits. J Endocrinol 30: 217

Fuchs AR, Wagner G (1963) Quantitative aspects of release of oxytocin by suckling in unanaesthetized rabbits. Acta Endocrinol (Copenh) 44: 581

Gandelman R (1975) Maternal nest-building performance and fetal number in Rockland-Swiss albino mice. J Reprod Fertil 44: 551

Garcia J, Ervin FR (1968) Gustatory–visceral and telereceptor–cutaneous conditioning: Adaptation in internal and external milieus. Commun Behav Biol Part A 1: 389

Garcia J, Hankins WG (1975) The evolution of bitter and the acquisition of toxiphobia. In: Denton DA, Coghlan JP (eds) Olfaction and taste V. Academic Press, New York, p 39

Garcia J, Koelling RA (1966) Relation of cue to consequence in avoidance learning. Psychonomic Sci 4: 123

Garcia J, Ervin FR, Koelling RA (1966) Learning with prolonged delay of reinforcement. Psychonomic Sci 5: 121

Garcia J, McGowan BK, Green KF (1972) Biological constraints on conditioning. In: Black A, Prokasy WF (eds) Classical conditioning II: Current theory and research Appleton, New York, p 3

Gelfand M (1945) Geophagy and its relation to hookworm disease. East Afr Med J 22: 98

Gutelius MF, Millican FK, Layman EM, Cohen GJ, Dublin CC (1962) Nutritional studies of children with pica. II. Treatment of pica with iron given intramuscularly. Paediatrics 29: 1018

Hammond J, Marshall FHA (1914) The functional correlation between the ovaries, uterus and mammary glands in the rabbit, with observations on the oestrus cycle. Proc R Soc Lond [Biol] 87: 422

Hancock J (1831) Remarks on the common cachexia, or leucophlegmasia, called Mal d'Estomac in the colonies, and its kindred affections, as Dropsy etc. Edinburgh Med Surg J 35: 67

Harris LJ, Clay J, Hargreaves F, Ward A (1933) Appetite and choice of diet. The ability of the vitamin B deficient rat to discriminate between diets containing and lacking the vitamin. Proc R Soc Lond [Biol] 113: 161

Hart EB, Elvehjem CA, Steenbock H (1930) Study of anaemia of young pigs and its prevention. J Nutr 2: 277

Henderson W (1931) Relationship of pica in cattle to trypanosomiasis. Vet J 87: 518

Hilliard J, Spies GH, Sawyer CH (1968) Cholesterol storage and progestin secretion during pregnancy and pseudopreg-

nancy in the rabbit. Endocrinology 82: 157
Hodgson ES (1967) Chemical senses in the invertebrates. In: Kare M, Miller O (eds) The chemical senses and nutrition. Johns Hopkins Press, Baltimore, p 7
Hoshishima K (1967) Endocrines and taste. In: Kare MR, Miller O (eds) The chemical senses and nutrition. Johns Hopkins Press, Baltimore
Huth EJ, Elkinton JR (1959) Effect of acute fasting in the rat on water and electrolyte content of serum and muscle and on total body composition. Am J Physiol 196: 299
Hytten FE, Leitch I (1971) The physiology of human pregnancy, 2nd edn. Blackwell Scientific Publications, Oxford
James R, Niall H, Kwok S, Bryant-Greenwood G (1977) Primary structure of porcine relaxin: Homology with insulin and related growth factors. Nature 267: 544
Kanner L (1948) Child psychiatry. 2nd edn. Charles Thomas, Springfield, Ill, p 475
Kastin AJ, Nissen C, Nikolics K, Medzikradsky DH, Coy I, Schally AV (1975) Behavioural and electrographic changes in rat and man after MSH. In: Gispen WH (ed) Hormones, homeostasis and the brain. Progress in brain research 42. Elsevier, Amsterdam, p 143
Keith L, Brown E, Rosenberg C (1970) Pica: The unfinished story. Background: Correlations with anaemia and pregnancy. Perspectives in Biol Med 13: 626
Kirksey AR, Pike RL, Callahan JA (1962) Some effects of high and low sodium intakes during pregnancy in the rat. II. Electrolyte concentrations in maternal plasma, muscle, bone and brain, and of placenta, amniotic fluid, fetal plasma and total fetus in normal pregnancy. J Nutr 77: 43
Lanzkowsky P (1959) Investigation into the aetiology and treatment of pica. Arch Dis Child 34: 140
Le Conte J (1846) Observations on geophagia. South Med Surg J 1: 417
Ledelius J (1957) Cited in Cooper M. Pica. Charles Thomas, Springfield, Ill
Lehrman DS (1956) On the organization of maternal behaviour and the problem of instinct. In: L'Instinct dans le comportement des animaux et de l'homme. Masson et Cie, Paris, p 475
Lichton IJ (1961) Salt saving in pregnant rats. Am J Physiol 201: 765
Lissak K, Endroczi E, Meggyesy P (1957) Somatisches Verhalten und Nebennierenrindentatigkeit. (Somatic behaviour and adrenal cortex activity.) Pflugers Arch 265: 117
Lorenz K (1957) The nature of instinct. Translation of 1937 paper. In: Schiller CH (ed) Instinctive behaviour. International Universities Press, New York. Reprinted by Marcus, 1965
Lyons WR (1958) Hormonal synergism in mammary growth. Proc R Soc Lond [Biol] 149: 303
Lyons WR, McGinty DA (1941) Effects of estrone and progesterone on male rabbit mammary glands: Varying doses of progesterone. Proc Soc Exp Biol Med 48: 83
Marcus RL (1965) Cravings for food in pregnancy. Manchester Med Gazette 44: 16
McDonald R, Marshall SR (1964) Value of iron therapy in pica. Paediatrics 84: 558
Meites J, Hopkins DF, Talwalker PK (1963a) Induction of lactation in pregnant rabbits with prolactin, cortisol acetate or both. Endocrinology 73: 261
Meites J, Nicoll CS, Talwalker PK (1963b) The central nervous system and the secretion and release of prolactin. In: Nalbandov AV (ed) Advances in neuroendocrinology. University of Illinois Press, Ch. 8
Mena F, Grovener CE (1968) Effect of number of pups on suckling- induced fall in pituitary prolactin concentration and milk ejection in the rat. Endocrinology 82: 623
Mergiletus AF (1701) Pica. Doctoral thesis in medicine submitted to Academia Patria
Michell AR (1978) Sodium need and sodium appetite during the oestrus cycle of sheep. Physiol Behav 21: 519
Michell AR, Bell FR (1969) Spontaneous sodium appetite and red-cell type in sheep. In: Pfaffmann C (ed) Olfaction and taste III. Rockefeller University Press, New York, p 562
Middleton E, Williamson PC (1974) Electrolyte and fluid appetite and behaviour in male guinea pigs: Effects of stilboestrol treatment and renal function tests. J Endocrinol 61: 381
Moltz H, Lubin M, Leon M, Numan M (1970) Hormonal induction of maternal behaviour in the ovariectomized nulliparous rat. Physiol Behav 5: 1373
Murr SM, Stabenfeldt GH, Bradford GE, Geschwind II (1974) Plasma progesterone during pregnancy in the mouse. Endocrinology 94:1209
Myers K, Poole WE (1962) A study of the biology of the wild rabbit in confined populations. III. Reproduction. Aust J Zool 10: 225
Nelson WO, Gaunt R, Schweizer M (1943) Effects of adrenal cortical compounds on lactation. Endocrinology 33: 325
Nicoll CS, Talwalker PK, Meites J (1960) Initiation of lactation in rats by non-specific stresses. Am J Physiol 198: 1103
Orr JB, Gilks JL (1931) Special report. SA Med Research Council (London) p 155
Paddock R (1928) Spurious pregnancy. Am J Obstet Gynecol 16: 845
Page I (1955) In discussion of CP Richter 'Self-regulatory functions during gestation and lactation'. In: Villee CA (ed) Second Conference on Gestation. Josiah Macy Jr Foundation, New York
Pfaffmann C (1967) In: Discussion of K Hoshishima, 'Endocrines and taste'. In: Kare MR, Maller O (eds) The chemical senses and nutrition. Johns Hopkins Press, Baltimore, p 148
Pike RL, Nelson J, Lehmkuhl MJ (1962) Some effects of high and low Na intakes during pregnancy in the rat. III. Sodium, potassium and water in maternal adrenals and hearts. J Nutr 78: 325
Reith MEA, Schotman P, Gispen WH, de Wied D (1977) Pituitary peptides as modulators of neural functioning. Trends Biochem Sci 2: 856
Revusky SH, Garcia J (1970) Learned associations over long delays. In: Bower GH (ed) The psychology of learning and motivation: Advances in research and theory. Academic Press, New York, Vol 4, p 1
Reynolds RD, Binder HJ, Miller MB, Chang WWY, Horan S (1968) Pagophagia and iron deficiency anaemia. Ann Intern Med 69: 435
Richter CP (1955) Self-regulatory functions during gestation and lactation. In: Villee CA (ed) Second Conference on Gestation. Josiah Macy Jr Foundation, New York, p 11
Richter CP, Barelare B (1938) Nutritional requirements of pregnant and lactating rats studied by the self-selection method. Endocrinology 23: 15
Richter CP, Rice KK (1945) Self-selection studies on coprophagy as a source of vitamin B complex. Am J Physiol 143: 344
Richter CP, Holt LE Jr, Barelare B (1937) Vitamin B1 craving in rats. Science 86: 354
Robinson M (1958) Salt in pregnancy. Lancet I: 178
Robson JM (1939) Maintenance of pregnancy in hypophysectomized rabbit by administration of oestrin. J Physiol (Lond) 95: 83
Rodgers W, Rozin P (1966) Novel food preferences in thiamine-deficient rats. J Comp Physiol Psychol 61: 1
Rosenblatt JS (1967) Non-hormonal basis of maternal behaviour in the rat. Science 156: 1512

Rosenblatt JS (1975) Prepartum and postpartum regulation of maternal behaviour in the rat. In: Wolstenholme G (ed) Ciba foundation symposium 33: Parent infant interaction. Associated Scientific Publishers, Amsterdam, p 17
Roth LL, Rosenblatt JS (1967) Self-licking throughout pregnancy as a stimulus for mammary development in the rat. J Endocrinol 37: xlii
Rozin P (1967) Thiamine specific hunger. Handbook of physiology, Vol 1: Alimentary canal. American Physiological Society, Washington, p 411
Rozin P (1968) Specific aversions and neophobia resulting from vitamin deficiency or poisoning in half-wild and domestic rats. J Comp Physiol Psychol 66: 82
Rozin P (1976) The selection of foods by rats, humans and other animals. In: Rosenblatt J (ed) Advances in the study of behaviour, Vol 6. Academic Press, New York, p 21
Rozin P, Rodgers W (1967) Novel diet preferences in vitamin deficient rats and rats recovered from vitamin deficiencies. J Comp Physiol Psychol 63: 421
Rozin P, Wells C, Mayer J (1964) Thiamine specific hunger: Vitamin in water versus vitamin in food. J Comp Physiol Psychol 57: 78
Rzoska J (1953) Bait shyness, a study in rat behaviour. Anim Behav 1: 128
Saldarini RJ, Hilliard J, Abraham GE, Sawyer CH (1970) Relative potencies of 17α- and 17β-estradiol in the rabbit. Biol Reprod 3: 105
Schmidt H (1925) Transient loss of the sense of smell and taste during pregnancy. Clinischwassenschrift p 4
Schopbach RR, Fried PH, Rakoff AE (1952) Pseudocyesis. A psychosomatic disorder. Psychosom Med 14/2: 129
Schwabe C, McDonald JK, Steinetz BG (1976) Primary structure of the A chain of porcine relaxin. Biochem Biophys Res Commun 70: 397
Schwabe C, McDonald JK, Steinetz BG (1977) Primary structure of the B chain of porcine relaxin. Biochem Biophys Res Commun 75: 503
Scoggins BA, Blair-West JR, Coghlan JP, Denton DA, Myers K, Nelson JF, Orchard E, Wright RD (1970) The physiological and morphological response of mammals to changes in their sodium status. Memoirs of the Society of Endocrinology, Vol 18. Hormones and the Environment, Cambridge University Press, p 577
Scoggins BA, Coghlan JP, Denton DA, Fan J, McDougall JG, Oddie Catherine J, Shulkes AA (1974) Metabolic effects of ACTH in sheep. Am J Physiol 226: 198
Scott EM, Quint E (1946) Self-selection of diet. III. Appetites for B vitamins. J Nutr 32: 285
Scott EM, Verney EL (1947) Self-selection of diet. VI. The nature of appetites for B vitamins. J Nutr 34: 471
Scott EM, Verney EL (1949) Self-selection of diet. IX. The appetite for thiamine. J Nutr 37: 81
Shulkes AA, Covelli M, Denton DA, Nelson JF (1972) Hormonal factors influencing salt appetite and lactation. Aust J Exp Biol Med Sci 50: 819
Stark E, Varga B, Acs Zs (1967) An extraadrenal effect of corticotrophin. J Endocrinol 37: 245
Steinberg J, Bindra D (1962) Effects of pregnancy and salt intake on genital licking. J Comp Physiol Psychol 55: 103
Stoerk HC, Arison RN (1961) In: Mills LC, Moyer JH (eds) Inflammation and diseases of connective tissue. Saunders, Philadelphia, p 399
Taggart N (1961) Food habits in pregnancy. Proc Nutr Soc 20: 35
Talwalker PK, Nicoll CS, Meites J (1961) Induction of mammary secretion in pregnant rats and rabbits by hydrocortisone acetate. Endocrinology 69: 802
Terkel J, Rosenblatt JS (1972) Humoral factors underlying maternal behaviour at parturition: Cross-transfusion between freely moving rats. J Comp Physiol Psychol 80: 365
Thornborough JR, Passo SS (1975) The effects of oestrogens on sodium and potassium metabolism in rats. Endocrinology 97: 1528
Tinbergen N (1951) The study of instinct. Oxford University Press, Oxford
Tolman EC (1932) Purposive behaviour in animals and men. Appleton-Century, New York
Trethowan WH, Dickens G (1972) Cravings, aversions, and pica of pregnancy. In: Howells JG (ed) Modern perspectives in psycho-obstetrics. Oliver and Boyd, Edinburgh, p 251
Tribe DE, Gordon JG (1955) Choice of diet by rats. III. The importance of the sense of smell in the choice of diets deficient in the vitamin B complex. Br J Nutr 9: 1
Turkington RW (1972) Molecular biological aspects of prolactin. In: Wolstenholme G, Knight J (eds) Lactogenic hormones. Churchill Livingstone Press, London, p 111
van Wimersma Greidanus Tj B, Dogterom J, de Wied D (1975) Intraventricular administration of anti-vasopressin serum inhibits memory consolidation in rats. Life Sci 16: 637
Wade GN, Zucker I (1969) Hormonal and developmental influences on rat saccharine preferences. J Comp Physiol Psychol 69: 291
Wade GN, Zucker I (1970) Hormonal modulation of responsiveness to an aversive taste stimulus in rats. Physiol Behav 5: 269
Waller H (1874) The last journals of David Livingstone in Central Africa from 1865 to his death. John Murray, London, Vol 2, p 83
Wang GH (1923) The relation between 'spontaneous' activity and oestrous cycle in the white rat. Comp Psychol Monogr 2: 1
Weisinger RS, Coghlan JP, Denton DA, Fan J, Hatzikostas S, McKinley MJ, Nelson JF, Scoggins BA (1977) ACTH-induced sodium appetite. Proc Int Union Physiol Sci Congr, Paris, 13: abstract 2399
Weisinger RS, Denton DA, McKinley MJ, Nelson JF (1978) ACTH induced sodium appetite in the rat. Pharmacol Biochem Behav 8: 339
Weisinger RS, Coghlan JP, Denton DA, Fan J, Hatzikostas S, McKinley MJ, Nelson JF, Scoggins BA (1980) ACTH induced Na appetite in sheep. Am J Physiol 239: 645
Wiesner BP, Sheard NM (1933) Maternal behaviour in the rat. Oliver and Boyd, Edinburgh
Wilcoxon HC, Dragoin WB, Kral PA (1971) Illness induced aversions in rats and quail: Relative salience of visual and gustatory cues. Science 171: 826
Wilkins L, Richter CP (1940) A great craving for salt by a child with cortico-adrenal insufficiency. JAMA 114: 866
Woods SE, Weisinger RS (1970) Pagophagia in the albino rat. Science 169: 1334
Yanmamoto R, Turner CW (1956) Experimental mammary gland growth in rabbits by estrogen and progesterone. Proc Soc Exp Biol Med 92: 30

24 Theories on genesis and satiation of salt appetite

Summary

1. Salt appetite embraces four facets of behaviour: there is hedonic liking for salt unrelated to need, the hunger which follows body sodium deficiency, the hunger engendered by the hormones of the reproductive process, and the appetite evoked by the hormonal response to stress. It is not known whether different neuronal groups subserve the different components of behaviour or the one neural system has several reactive capacities.

2. Parallel to ethological studies, which have revealed a wide diversity of instinctive behaviour with motor patterns transmitted in the genome like physical traits, more biomedically orientated studies show comparable diversity of innate behaviour patterns subserving the vegetative functions of hunger, thirst, temperature control, appetites for specific minerals, and sexual and maternal behaviour. Knowledge emerging from both schools of investigation has underscored the major role of chemical and hormonal change in the *milieu intérieur* in activating innate patterns, and also the manner in which distance receptor information may initiate hormonal changes.

3. The question of how a specific motivation or drive state is activated is discussed. Sensory inflow or chemical change initiates hypothalamic activity which affects the reticular system and state of arousal, and biases attention to sensory inflow relevant to translation of body need into satiation behaviour. Cortical influences, including memory, may also have major influence on the processes. The early psychological and ethological 'reservoir' and 'action-specific energy' models of motivation may be seen as eventually having come to have some physicochemical basis, in the accumulation of specific substances including the neuropeptides in specific cell populations subserving innate behaviour.

4. The characteristic time delay of many hours in the onset of salt appetite in response to rapid production of body sodium deficit may reflect slow access of effect of systemic chemical change to a receptor system inside the blood–brain barrier and/or the fact that genesis of the salt appetite drive involves a complex neurochemical mechanism.

5. Operant behaviour studies in sheep under conditions of continuous loss of sodium from the body from a parotid fistula indicate reaction of the salt appetite system to loss of 2%–3% of body content of sodium.

6. The several theories of genesis of salt appetite are discussed in detail. These include a change in the salt taste receptor, a hypothalamic receptor reacting directly to a change in [Na] (sodium concentration) in cerebral blood, a response to a change in aldosterone concentration in blood, and an alteration in [Na] and [K] in saliva altering taste characteristics and thus

intake. Whereas such changes may influence appetite, evidence against their having a primary causal role is advanced. The theory is proposed that a decrease in intracellular [Na] in the specific neuronal population subserving salt appetite behaviour is the major stimulus of appetite, and that it acts to initiate transcription and translation processes. The protein synthesized alters the ionic or membrane characteristics of the specific neurocytons, and hence excitability, or increases the capacity for neurotransmitter generation at synapses. Excitation and transmission in the neuronal pool are facilitated. The body of data on the genesis of salt appetite by steroid and peptide hormones involved in reproduction or stress would, in the light of the known mode of action of hormones, give additional weight to the theory of activation by genomic processes.

7. Many lines of investigation point to a major influence of changes in intracellular [Na] and [K] on transcription and translation processes, and DNA synthesis associated with mitogenesis. The physiological processes studied include the rapid 'puffing' of polytene chromosomes produced by ecdysone, mitogenesis in cultured neuronal cells, mitogenesis in fibroblast cultures, and the ratio of the four classes of δ-crystallins synthesized in the ocular lens of vertebrates, where the ionic changes appear to determine synthesis at the post-transcriptional or post-translational level.

8. Experiments in sodium-deficient sheep aimed at increasing the intracellular [Na] of cerebral cells caused reduction or abolition of sodium appetite. Intracarotid ouabain infusion, calculated to cause a concentration of 10^{-7} to 10^{-8} M in cerebral blood, virtually abolished salt appetite whereas the same infusion given intravenously had little effect. The intracarotid infusion had no apparent effect on hunger and caused a small reduction only in thirst.

9. Aspects of the physiology of formation and movement of cerebrospinal fluid (CSF) that bear upon effects of cerebral ventricular infusion of solutes are discussed.

10. Slow cerebral ventricular infusion of hypertonic saline in artificial CSF to increase CSF [Na] by 10–20 mmol/l greatly reduced salt appetite in sodium-deficient sheep. Infusion of 0.7 M mannitol in artificial CSF to produce an equivalent rise in osmotic pressure greatly increased salt appetite. The infusion decreased CSF [Na] by 10–20 mmol/l. Infusion of isotonic mannitol-CSF, which also reduced CSF [Na] by 10–20 mmol/l, similarly approximately doubled sodium appetite. This effect of mannitol-CSF also occurred consistently in the sodium-replete animal. The latent period of effect of a decrease in CSF [Na] in the sodium-deplete animal was 25–30 min, the result representing the most rapid consistent induction of salt appetite yet reported. Comparable effects on appetite to those with 0.7 M mannitol-CSF were produced by 0.7 M sucrose-CSF, whereas 0.7 M urea-CSF had a lesser though significant effect. Infusion of 0.7 M glucose-CSF did not increase sodium appetite nor did infusion of an artificial CSF lacking sodium chloride or other osmotically active component (aqua dist.-CSF). The latter two infusions also reduced CSF [Na] by 10–20 mmol/l. The factors determining the influence of a change in CSF [Na] on brain extracellular fluid (ECF) [Na] are discussed in detail. It is proposed that the data from various infusions suggest that the sodium-sensitive receptor elements are located some distance from the ventricular wall in the hypothalamic neuropil, and are reached through tortuous channels of the intercellular space. With isotonic or hypertonic mannitol or sucrose infusion a fluid of low [Na] would move from

the ventricle through the intercellular channels little modified. With glucose infusion uptake in cells would occur and there would be a 'sink' effect of transport into blood capillaries, as would occur with aqua dist.-CSF infusion. There would be little or minimal effect on brain ECF [Na] at a distance from the ventricle. With urea infusion, entry into cells and capillaries would be less, and therefore the effect on brain ECF [Na] intermediate between those of mannitol and glucose. The data on inhibition with hypertonic saline and excitation with saccharide-CSF infusion are compatible with the hypothesis of a reduction in the intracellular [Na] of neurones being causal of sodium appetite.

11. The effects of steroid hormones on salt appetite, particularly in the context of the reproductive process, are discussed in relation to their genomic actions, and also in relation to data on non-genomic actions of steroids on membranes of neurones. As with peptide hormones, iontophoretic application of steroids to neurones may give short-duration electrical responses. The effects of blockers of genetic transcription and translation on the capacity of steroid hormones to induce sexual behaviour are noted. There is the possibility that one influence of oestradiol in the salt appetite of reproduction resides in the induction of prolactin receptors in the brain.

12. The action of peptide hormones involved in the reproductive process, including the extra-adrenal effect of adrenocorticotrophic hormone (ACTH), which is presumably directly on the brain, is discussed in terms of the present limited knowledge of the mode of action of peptides on cells. The mode of action of prolactin on mammary epithelium, involving, as it does, conjoint action of several necessary factors is a speculative parallel to the action of the peptides on the salt appetite neurones. In this regard, it is noteworthy that some evidence on prolactin action suggests a reduction in intracellular [Na] to be a messenger leading to its effects—including protein synthesis. The generation of salt appetite in the adrenalectomized rabbit by reduction of cortisol but not basal desoxycorticosterone (DOCA) maintenance support, and thus without changing the sodium balance, raises the question of how physiological amounts of endogenous ACTH act on the brain. Questions arise on transport of ACTH or active fragments of ACTH across the blood–brain barrier, or retrograde transport up the pituitary stalk, or synthesis within the brain itself giving rise to ACTH precursor and ACTH. Possible scenarios for the mode of action of neuropeptides are discussed, and the actions of angiotensin II are particularly noted in this context. The immediate effect of intracerebral angiotensin II on thirst compared with the time delay before its effect on salt appetite is a provocative finding. The difficulty of resolving the controversy about the actual physiological relevance of effects produced by high dosage of angiotensin II into the CSF is noted, and attention is drawn to the new technique of hybridization histochemistry involving cDNA probes which may resolve whether renin and/or angiotensin II are actually synthesized within the brain in sodium deficiency.

Introduction

In Chapter 19 the survival advantage of the phylogenetic emergence of consciousness was emphasized in terms of increasing the exercise of options by the organism in any situation. Livingston (1967) suggests that somewhere early in the process of evolution there must have emerged some sense of satisfaction or dissatisfaction in response to behaviour successful or unsuccessful. He quotes Coghill and Herrick as visualizing a combination of a sense of effort, as a most primitive sensibility, with that of effect associated with elements of satisfaction and dissatisfaction as providing a basis of consciousness. In the same vein, Aristotle defined appetite as a tendency towards something we do not have and which we need.

The different behavioural elements constituting salt appetite

In the repertoire of behaviour designated as salt appetite, three major facets are evident, and, as such, they underscore that a complex neurophysiological system is involved. These are: (i) the hedonic element; (ii) the hunger for salt which follows frank body deficit of sodium; (iii) the hunger which is engendered by the hormones involved in the reproductive process. A fourth facet or component might be considered to be the appetite evoked by the hormonal response to stress.

As regards the hedonic component it has been noted that its biological value resides in the normal animal being protected against specific food deprivation by the unlearned stimulus–response connection (Bare 1949). Natural selection in this instance determines that the organism likes what it needs, and the genetic system has advantages over trial and error in survival. Though little sodium storage is feasible, the animal may remember the situation of the source and return to it in the circumstances of frank body deficit. We do not understand the neurophysiological nature of the liking or pleasurable process, though anatomical correlates of some aspects have been identified by Olds and colleagues (Olds 1962). It would seem the system might be part of the substrate upon which the salt appetite drive acts when it is evoked by body sodium deficit. In this action the drive clearly alters the hedonic character of salt solution in that high concentrations hitherto aversive are drunk with avidity.

The hormones of the reproductive process evoke a powerful appetitive drive and this may involve their action within the brain upon the same neurones which specifically react to body sodium deficit. Thus the neurones of the system subserving reaction to body sodium deficit may be target cells which also have receptor systems for response to a variety of hormonal agents; in this latter way they would be analogous to, for example, mammary epithelium. Alternatively, there may be an anatomically separate system of neurones which respond to the sequence of steroid and peptide hormone action involved in the appetite of pregnancy and lactation. Such a grouping may also subserve elements of activation of maternal behaviour. This population, when so activated by hormones, may then impinge by predetermined pathways on the cell population subserving reaction to body salt deficit, and correspondingly activate it. The resolution of the question of whether a diversity of reactive capacities reside in one cell type subserving salt appetite or whether there are different types of cell—possibly spatially separate—awaits studies with appropriate existing techniques. In the discussion further on, bearing upon the neurochemical events underlying activation of salt appetite drive, the former will be assumed for simplicity. It also focuses the possibility of some interactions.

Innate behaviour

From a wealth of fascinating observation in the field, the ethologists Craig, Lorenz, Tinbergen, von Frisch, Hinde, Thorpe and many other distinguished workers, mostly with a primarily zoological interest, have shown the manner in which exteroceptor stimuli may evoke complex fixed-action patterns. These motor patterns are transmitted in the genome like physical traits and, as such, may serve as a basis for classification and dissection of phylogeny of behaviour. Concurrently over the last half-century or so, more biomedically orientated studies, generally under laboratory conditions, have determined the way in which chemical and hormonal changes within the organism activate genetically determined behaviour patterns which subserve the major vegetative functions. These maintain the stability of the *milieu intérieur* and optimal functioning of physiological systems, as well as generating the propensities for those behaviours of paramount importance for continuity of the species—sexual and maternal. This hierarchy of systems embraces at, perhaps, the most primitive, the imperious hunger for air which is rapidly generated by derangement or impediment of gas exchange, through the behaviours of temperature control, thirst, hunger and specific mineral appetites, to the complex patterns of sexual and maternal behaviour.

Whereas these bodies of knowledge, ethological and biomedical, grew up initially around different formal scholastic disciplines, they are, in fact, inextricably interwoven and any distinction is arbitrary, as evident from their synthesis over the last two decades. The relevance of distance receptor information in releasing innate behavioural patterns may depend on the hormonal state of the body and thus the reactivity of elements in the central nervous system. Correspondingly, exteroceptor stimuli may initiate activation of hormonal systems which in turn change the neural basis of arousal and detection such that other exteroceptor stimuli, hitherto of indifference, take on an immediate dominant significance. Before discussing how the hunger for salt might be generated in the brain, some brief general consideration of drive and motivation is appropriate.

Motivation and drive

If one watches a sleeping animal wake and begin to go about its affairs it may go to drink, or search for food, groom itself, or search for the opposite sex. Why one and not another, and why sometimes one more intensively than others? What motivates or drives it?

An outstanding review of this vast subject is Eliot Stellar's (1960) chapter in the *Handbook of Physiology*, in the course of which he draws extensively on the analyses of Beach (1942, 1947, 1955) and Lashley (1938); Beach traces the earliest scientific thinking about motivation arising from the instinct doctrine. Stellar proposes that present evidence indicates that the major focus of the neural system or integrating mechanism responsible for arousal, execution and satiation of motivated behaviour lies in the diencephalon, probably the hypothalamus. Thus stimulation and ablation of restricted loci in this region of the brain can result in increase or decrease of motivated behaviour. Analysis of the data indicates, in some instances at least, two kinds of functional areas, which may be termed excitatory and inhibitory. The implication is that arousal of motivated behaviour is determined by heightened activity of the excitatory mechanism whereas satiation is determined by output of the inhibitory mechanism. In relation to the question of what controls the activity of the excitatory system, and thus arousal, execution and satiation of the motivated behaviour, the evidence, as suggested by Beach and Lashley, indicates that three groups of factors are operative:

i) Sensory influences may operate through afferent pathways which are physiologically specific or non-specific, and these activate the diencephalic mechanisms directly. Various sensory influences may be additive in their effects so the sum total of input determines the amount of arousal. These sensory influences must contribute highly specific information because motivated behaviour may be highly discriminative and selective. The influences may be either learned or unlearned, it being established that previously ineffective stimuli can become arousing as a result of past experience.
ii) Chemical and physical changes in the internal environment acting through the circulatory system or cerebrospinal fluid can directly influence the activity of excitatory and inhibitory mechanisms and therefore arousal. Whereas some humoral influences may have general arousing and depressing effects, others may be highly specific to a particular type of motivated behaviour, implying the existence of highly selective central receptors sensitive to changes in the internal environment.
iii) Other neural influences arising, for example, in the cortex and rhinencephalon may contribute excitatory or inhibitory effects to the control of the diencephalic mechanisms. They influence the sequential organization patterning of motivated behaviour as well as its arousal and satiation.

The different levels of organization which may be involved in the arousal of motivation and satiation in an instinctive behavioural pattern are clearly illustrated in experimental studies involving transection of the nervous system at various levels. There is a parallelism to the concept of hierarchical organization of instinctive behaviour proposed by the ethologists (described below).

Philip Bard (1940) studied the sexual behaviour pattern in ovariectomized female cats when the nervous system was transected at various levels. The female cat becomes sexually receptive twice a year. If the genital areas or back are stroked the cat in heat will crouch by bending its forelimbs and tread in place while alternately flexing one hind limb and extending the other. If a rod is inserted into the vagina it will move its tail to one side, tread very vigorously and give a low throaty oestral call. If sufficiently stimulated, the cat will reach a violent climax indicated by a loud cry and followed by an after-reaction of rolling and squirming on its back while purring. When not on heat or if its ovaries have been removed such stimulation of a cat's genitals causes it to move away or display anger. Bard found that in the spinal cat, even if the gonads had been removed, stimulation of genitals elicited the normal pattern of crouching, tail deflection and hind limb treading. Injection of oestrogen had no effect on the threshold of sexual response to genital stimulation. This was also the case with the decerebrate cat where transection was made at level of the midbrain. Only in the instance where transection was above the hypothalamus, allowing it to remain connected to the lower part of the nervous system, did the behaviour pattern show a sensitivity to hormone influence. As Teitelbaum (1967) remarks, the inclusion of the hypothalamic tissue restores hormonal control of sexual behaviour and such tissue, acting as a centre of integration without any brain above, is sufficient to exert control over spinal reflexes.

In the course of early studies of motivation, largely the result of the work of Cannon, there was great emphasis on local events such as gastric contractions, dry mouth and local irritation or pressure from the genitals in giving rise to the motivation of hunger, thirst and sexual behaviour respectively. Similarly, local changes in taste receptor sensitivity were implicated in the genesis of salt hunger. However, the experimental examination of these ideas, which involved denervation or surgical removal of the stomach, study of patients with congenital absence of salivary glands, or excision of the genitalia in animals, provided clear evidence against such local mechanisms being essential for the arousal of these specific motivated behaviours. It is also clear, though, that such local factors may make a contribution to motivation, and provide an important element of the sensations accompanying motivated behaviour. Lashley (1938) drew attention to the complexity of motivated behaviour in a classical definition of instinct as the faculty which animals have instead of intellect which makes their behaviour seem intelligent. He emphasized that instincts and motivated behaviour are not simply complex chains of reflexes represented by stereotyped acts. The detailed responses involved in mating, nesting, retrieving, etc., vary from individual to individual on occurrence to occurrence. It is not feasible to specify a particular motor sequence that characterizes the behaviour, for the same result may be achieved by different behavioural means on different occasions.

Action-specific energy and the reticular activating system

As Cofer and Appley (1967) remarked, the ethological theory of instinct is a motivational theory in every sense of the term. Essential elements in the formulations of Craig (1918), Lorenz (1950) and Tinbergen (1951) are (*a*) accumulation of action-specific energy giving rise to appetitive behaviour, (*b*) appetitive behaviour striving for and attaining a stimulus situation activating the innate releasing mechanism, (*c*) the setting off of the releasing mechanism and discharge of the endogenous energy in a consummatory action. Tinbergen was of the view that motivational energy is generalized to all activities of the major instinct. The energy of each centre is released in sequence by the appropriate innate releasing mechanism until the final consummatory act takes place with discharge of action-specific energy.

Discussing some of the early reservoir and valve analogies of action-specific energy in the writings of McDougall (1923) and Lorenz (1950), Haldane (1956a) suggested that whereas the ideas were metaphorical, the idea of accumulation of a substance in the process of central activation of behaviour may not be a metaphor. The readiness of a centre or small volume of nervous system to produce nervous impulses may depend on the concentration in it of some substance or substances: for example, precursors of acetylcholine or the various neurone stimulating compounds which are found in different parts of the brain stem. In this view, Haldane had the work of Vogt (1954) particularly in mind in relation to the accumulation of acetylcholine, adrenaline, noradrenaline, histamine, serotonin, substance P and the posterior pituitary hormones. Vogt had remarked that from the point of view of manufacture of pharmacologically active compounds, the hypothalamus is the most versatile part of the central nervous system so far recognized. The explosive growth of knowledge about the neuropeptides during the past decade, including that of their individual accumulation in specific areas of the brain, may provide an eventual physiological basis for theoretical constructs of this type. Haldane drew the parallel in the case of human respiration with the effects of carbonic acid or perhaps hydrogen ions in the respiratory centre of the medulla.

Tinbergen (1951) has stressed that instinctive behaviour consists of two successive parts of very different kinds. The animal in which an instinctive urge or drive is activated starts random exploratory or seeking behaviour that when observed seems typically purposive. This is continued until the animal comes into a situation which provides the sign stimuli necessary to release the motor response of one of the nervous centres of the lowest level. He sees the sequence as controlled by a hierarchy of nervous centres. He cites, as a familiar example, the peregrine falcon, which when the hunting drive becomes active searches its hunting territory, possibly purposely visiting special locations where it has met with success before. The sight of the prey releases the motor responses of catching, killing and eating which are a chain of simple, relatively rigid responses.

Thorpe (1956) suggests there are some main characteristics of instinct about which there has been more or less general agreement. First, instincts are innate in the sense of being inherited, and so largely specific. Secondly, they most often involve complex and highly rigid patterns of behaviour in which numerous muscles, muscle groups and whole organs and organ systems are beautifully coordinated. Thirdly, instinctive actions are characteristically evoked by complex environmental situations as, for example, elaborate visual and/or auditory patterns, to which the senses appear to be inherently tuned. Thus the animal tends to pay attention to particular objects, or objects in a particular setting, and often appears to be seeking such objects with great perseverance and sometimes intelligence.

In relation to this latter contention, Dell (1958) emphasized that variations in metabolites and hormones of the *milieu intérieur* produce particular activation and consequently arousal effects, enhanced alertness and motor facilitation—as has been recognized for sensory stimulation. These reticular events have a basic role in building up the non-specific component common to the class of instincts which have a biochemical core, and are concerned with the translation of bodily needs into behaviour. He thinks there can be no doubt that this reticular activation by humoral and hormonal factors constitutes the central core of instinctive behaviour, providing the active source of continued excitation referred to by psychologists as motivation or drive. This non-specific component of behaviour, and the state of central activity which it provides, has two aspects of physiological significance. First, the chance of encountering an adequate external situation for reduction of drive is considerably enhanced by the prolonged exploration by an alert animal over a wide range. Secondly, a non-specific activation is a necessary precondition for the more specific perceptual and motor patterns. Thus the state of vigilance is created, and it depends mainly on the activity of the brain stem reticular formation. The classical consummatory act creates stimuli and causes internal changes which have a depressing effect on reticular activity and on vigilance.

Finally, in this brief reference to some only of the ideas in the history of concepts of motivation and instinctive behaviour, it is essential to emphasize that the rigid dichotomy between innate and learned has been largely abandoned. As Barnett (1967) remarks, some fixed-action patterns have been shown to require little or no special experience for their development. But others, like bird song, though remarkably uniform within a species, depend for normal development, at least in some species, on early opportunity to imitate adults. The ability to learn is confined to an early sensitive period. Thus there are adaptable components in virtually all complex behaviour, however stereotyped it may appear. At a higher level, Harlow (1969) remarks that in a very real sense there is no such thing as a pure human instinct since in animals as intelligent as monkeys and men, new behavioural capabilities are rapidly modified by learning once they emerge. Probably it is equally true that there is no such thing as a purely learned trait since constitutional variables influenced by growth underlie all learned behaviours.

In addition to some of the works mentioned briefly, major contributions to analysis in this field were made also by McDougall (1923, 1930) who, as Haldane (1956b) points out, proposed before Lorenz (1950) an elaborate reservoir-type 'energy' model of instinct to reflect controls and interactions. Examination of motivation, and, in some cases, of energy models including the issue of their limitations has been made by Lehrman (1953), Hinde (1960), Miller (1957), Baerends (1950) and Zeigler (1964). The similarity of the models to Freud's concepts on energy underlying instinct has been discussed by Carthy (1951), Kennedy (1954) and Hinde (1960). In this same general context it is most interesting to recall views of Charles Darwin (1872) in *The Expression of the Emotions in Man and Animals* where he says:

> Experience shows that nerve force is generated and set free whenever the cerebro-spinal system is excited. The direction this nerve force follows is necessarily determined by lines of connection between nerve cells, with each other and with various parts of the body. But the direction is likewise much influenced by habit; inasmuch as nerve force passes readily along accustomed channels ... That the chief expressive actions exhibited by man and by lower animals are now innate or inherited—that is, have not been learnt by the individual—is admitted by everyone. So little has learning or imitation to do with several of them that they are from the earliest days and throughout life quite beyond our control; for instance, the relaxation of the arteries of the skin in blushing, and the increased action of the heart in anger. We may see children,

only two or three years old, and even those born blind, blushing from shame . . . The facts alone suffice to show that many of our most important expressions have not been learnt; but it is remarkable that some, which are certainly innate, require practice in the individual before they are performed in a full and perfect manner; for instance weeping and laughing. The inheritance of most of our expressive actions explains the fact that those born blind display them . . . We can thus also understand the fact that the young and old of widely different races, both with man and animals express the same state of mind by the same movements.

The genesis of salt appetite drive in the brain: The time delay of onset

The signal or outstanding feature of salt appetite, apart from its being innate, is the time delay in its onset. It follows 4–72 h after establishment of body deficit in the naive animal, be that rat, sheep or rabbit. This delay is, perhaps, not so surprising in the case of appetite generated by hormones, or combination of hormones in appropriate sequence, allowing what is known of the cytoplasmic and nuclear mechanisms involved with receptor systems, transcription and translation. With onset of sodium deficiency in sheep, however, even if salivary loss from a fistula is very rapid and 300–700 mmol sodium are lost in the first 24 h, the delay time is evident (Chapter 11).

The character of the behaviour contrasts with thirst. Though the two ingestive behaviours have common elements in the operation of taste factors and oropharyngeal metering of inflow associated with satiation of appetite, the excitation of them is quite different. With thirst a sudden rise in the [Na] (sodium concentration) of cerebral arterial blood contrived by intracarotid infusion, or a rise in angiotensin concentration produced similarly, results in avid drinking within 30–120 s. Similarly with isosmotic volume depletion contrived by sequestration with formalin, polyethylene glycol or haemodialysis, the onset of thirst is rapid. The reaction of the osmoreceptors, sodium receptors, and possibly angiotensin II receptors as well as those transmitting from the left atrium via the vagus is rapid, the redirection of the stream of consciousness as a result of reticular arousal and cortical reaction is rapid, and the seeking and drinking of water is more or less immediate according to circumstances.

As another comparison, it is also apparent that the response of the adrenal gland in terms of aldosterone secretion is rapid. As well as the fact that with infusion into the adrenal artery and an increase in concentration of potassium, angiotensin and adrenocorticotrophic hormone (ACTH), aldosterone secretion rises in 2–3 min with plateau of effect at 7–15 min (Fig. 24-1) (Blair-West et al. 1969), the reaction to postural and circulatory change including sudden reduction of renal arterial pressure is rapid over 15–30 min. Similarly with the consummatory act of satiation of salt appetite in the sodium-deficient sheep, adrenal aldosterone secretion is inhibited in striking manner in 15 min by mechanisms currently unknown (Chapter 21).

The slow reaction to body deficit by emergence of salt hunger could reflect the fact that the receptor of the system is within the blood–brain barrier and the access of effects of systemic change is slow. Or, as well as this factor, the system itself generating the salt appetite drive could involve complex neurochemical mechanisms, and the latent period reflects this fact.

In considering this latent period, however, account must be taken of the influence of experience. Thus an experienced animal may reflect more accurately the threshold of reaction of the system. This is because learning may circumvent the time required for build-up of a strong drive and exploratory behaviour. There is a hint of this in Figs. 12-1 and 12-2 where in one group of animals after parotid cannulation, free access was allowed to sodium solution for 2 h each 22 h whereas in the second group access was allowed by bar-press for 2 h each 22 h. In the latter group the animals had tasted sodium bicarbonate while in the sodium-replete state in the course of learning the operation of the bar-press. There were signs of changed behaviour towards sodium bicarbonate after the first 22 h in some bar-press experiments, in contrast to the 2–4 days with the group of animals with free access. But most important in this regard are the data on voluntary behaviour of sheep with a parotid fistula and constant availability of sodium bicarbonate by bar-press. As shown in Chapter 13 (e.g. Zeta, Fig. 13-7) there were episodes of bar-press behaviour about every 3–4 h. Parotid saliva loss was continuous so that in this time 60–100 mmol of sodium would be lost. The data on several animals could be construed as a threshold of reaction to a body deficit (error) of about 2%–3% occurring every 3–4 h. This order of loss will also change natriuretic response to intraventricular injection of hypertonic saline or will evoke aldosterone hypersecretion. That it is not simply episodic exercise of a learned behaviour is suggested by extinction of the behaviour by giving sodium solution into the rumen, or its augmentation in commensurate fashion when sodium deficit is increased by withdrawal of the lever in the cage for 6, 12 or 24 h. Whereas other factors may come in, the weight of the data does argue for response to loss which implies a reaction of the system (albeit already primed) within 3–4 h.

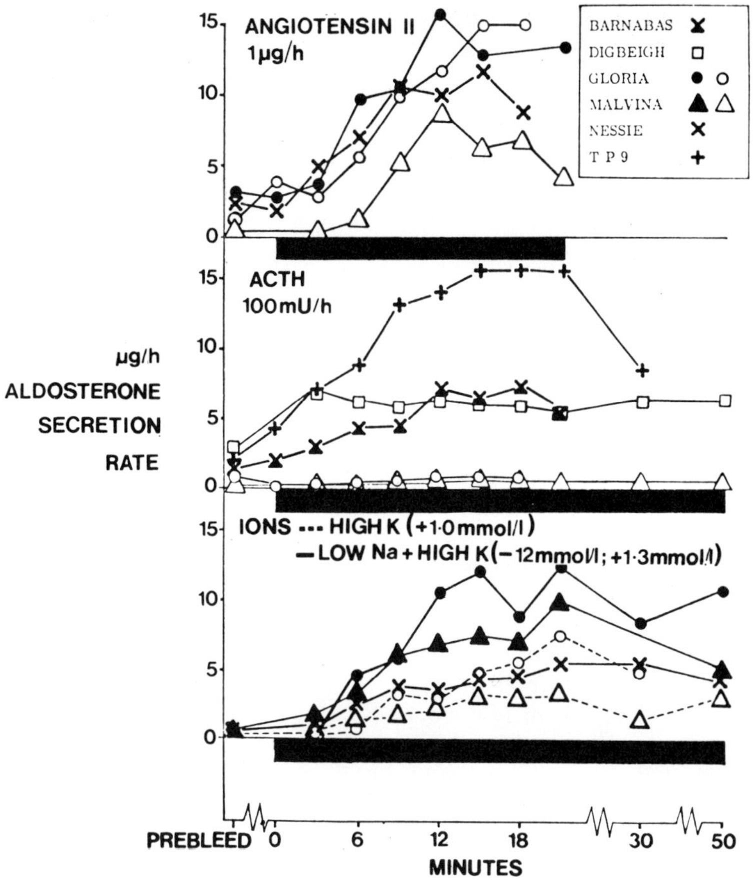

Fig. 24-1. The onset of effect on aldosterone secretion following the start of various infusions into the adrenal artery of six sodium-replete sheep with adrenal transplants. Infusions were made of angiotensin II and ACTH, or adrenal arterial plasma [K] was increased, or [K] increased and [Na] cuncurrently lowered by appropriate infusions of potassium chloride or 5% glucose plus potassium chloride. Maximum effect occurred usually by 12 min. (From Blair-West et al. 1969, with permission.)

Some experiments with formalin injection in rats also suggest that salt appetite response may be more rapid than the usual 6 h (see Stricker 1980). In discussing the basis of delay in his experiments where rats which had been injected with formalin or polyethylene glycol had concurrent access to water and 0.5 M saline and drank water avidly first, Stricker (1980) raised three possibilities: (i) time for plasma volume deficits to reach threshold levels, (ii) time for levels of angiotensin or aldosterone to reach threshold levels, or (iii) time for osmotic dilution such as that resulting from renal retention of water to reach threshold. He did, however, produce evidence against each of these categories. His studies on salt appetite caused by formalin or polyethylene glycol in rats on low-sodium diet or adrenalectomized rats showed much more rapid onset of voluntary intake of hypertonic saline (Stricker 1981). He proposes that rather than sensing interstitial fluid reserves or total body sodium, receptors monitor the availability of sodium in the critical tissue—brain (Denton 1966). Hyponatraemia and/or hypovolaemia could limit sodium delivery to brain, and animals on a low-sodium diet or after adrenalectomy may already have lowered [Na] in the cerebrospinal fluid (CSF). Stricker suggests the interstitial fluid represents a large and labile store of sodium that protects the brain from sodium deficiency. It acts like a buffer, which when finally depleted permits the impact of sodium loss to affect the brain and in so doing precipitates sodium appetite. That is, sodium appetite is associated with sodium deficiency in the brain not in the periphery.

It has become clear from experiments involving rapid influx of hypertonic sodium solution into the systemic circulation of sodium-deficient sheep that the system of control of salt appetite may be accessed and inhibited in 10–20 min (see Chapter 20; Weisinger et al. 1978). But 'turn off' of the system may not necessarily simply reflect reversal of the process which was responsible for 'turn on'.

Different theories of the genesis of salt appetite

With regard to the genesis of salt appetite in sodium deficiency six main theories have been considered.

Change in the taste receptor

Richter (1956) states that it has always seemed likely that appetite changes occur at the taste bud membrane. Whereas his original ideas on the appetite being innate have been fully confirmed, as Rozin (1976) remarks, the idea that salt hunger is mediated by peripheral inflow has had a more uncertain fate. This notion, with its similarity to Cannon's theories, has not been supported by the evidence of electrophysiological studies. The absolute threshold is the same in adrenalectomized and normal animals: it is preference that is lowered. The physiological need must reflect itself elsewhere in the nervous system. Similarly, increased sensitivity will not explain the acceptance of highly concentrated, normally aversive solutions. In fact, the work of Contreras and Frank (1979) points to a diminished response of salt fibres in sodium deficiency, and increased preference for sodium chloride in sodium deficiency may in part be related to reduced response to sodium at the periphery (Contreras 1978). It is an interesting question to ask in relation to Richter's original ideas as to whether this decreased response reported by Contreras and Frank is reversed immediately following the infusion into the lingual artery of hypertonic saline adequate to increase plasma [Na] to a physiological degree. This procedure does not turn off salt appetite.

Altered taste sensitivity, as with the 'dry mouth' theory of thirst, may contribute some elements to the excitation of salt appetite. In this regard, McCance's account of altered taste sensations can be recalled (see Chapter 26). But the experiments of Richter (1956) showed that salt appetite remained after bilateral section of either the seventh or ninth cranial nerve or both, and it was only when the lingual or the pharyngeal branch of the tenth nerve was cut bilaterally as well as the ninth, that the animals' increased intake of sodium in response to adrenalectomy was consistently stopped. Though both Richter (1956) and Pfaffmann (1952) in the course of reporting such studies emphasize, on the basis of post-operative inspection of papillae, the great difficulty of total destruction of the taste buds, some clear quantitative influence of the section of the principal inflow through seventh and ninth nerves would be expected if drive were generated from peripheral inflow. It would seem the primary result of such denervation is to eliminate the system for *detection* of salt, and it is for this reason, in the extreme case, that the animals die. This interpretation is supported by studies on sheep with a parotid fistula in which the ninth, the lingual and buccal nerves were divided bilaterally. This should have denervated tongue, floor of mouth and cheeks and part of pharynx, and largely abolished the sense of taste (Beilharz and Kay 1963). The operation had little effect on the behaviour of the sodium-deficient sheep or their salt intake when sodium bicarbonate was offered each day. They were excited as usual and appeared to be able to distinguish the sodium bicarbonate from water by smell. There was small reduction only in salt intake. At post-mortem the tongues were scarred in many places indicating effective sensory denervation. There was no evidence that the extensive section modified salt appetite drive.

Hypothalamic receptor reacting to changed sodium concentration in cerebral blood

The idea of a hypothalamic receptor responding simply and directly to change in the contemporary [Na] of arterial blood, analogous to the reaction characteristics of an osmoreceptor as conceived by Verney (1947), appeared to be eliminated by experiments where sodium appetite of sheep was not changed by intracarotid infusion of 4 M sodium chloride which increased [Na] of cerebral blood bilaterally by 10–25 mmol/l for 7 min before and during 8 min of offer of sodium bicarbonate solution (Beilharz et al. 1965; see Chapter 20). Similar conclusions emerged from Beilharz and Kay's (1963) experiments where rapid intravenous infusion of 115 ml of 2.7–4.4 M sodium chloride over 30 min did not abolish and sometimes had little or no effect on salt appetite tested 10 min later, though plasma [Na] was increased to 160–180 mmol/l. Baldwin (1976) made similar findings with intracarotid infusions in sodium-deplete goats.

Richter's (1956) concept of organization of salt appetite is shown on the lower portion of Fig. 24-2. The hypothalamic receptors reacting to low blood [Na] are proposed to evoke increased general activity, and, with trial and error, the encounter with salt stimulates the taste buds (less salt on the blood side of the membrane). The sensory cortex responds to this, salt appetite is generated, and the animal eats or drinks more salt.

Increase in blood aldosterone concentration

The logic that the hormone hypersecretion evoked to conserve salt in sodium deficiency should also

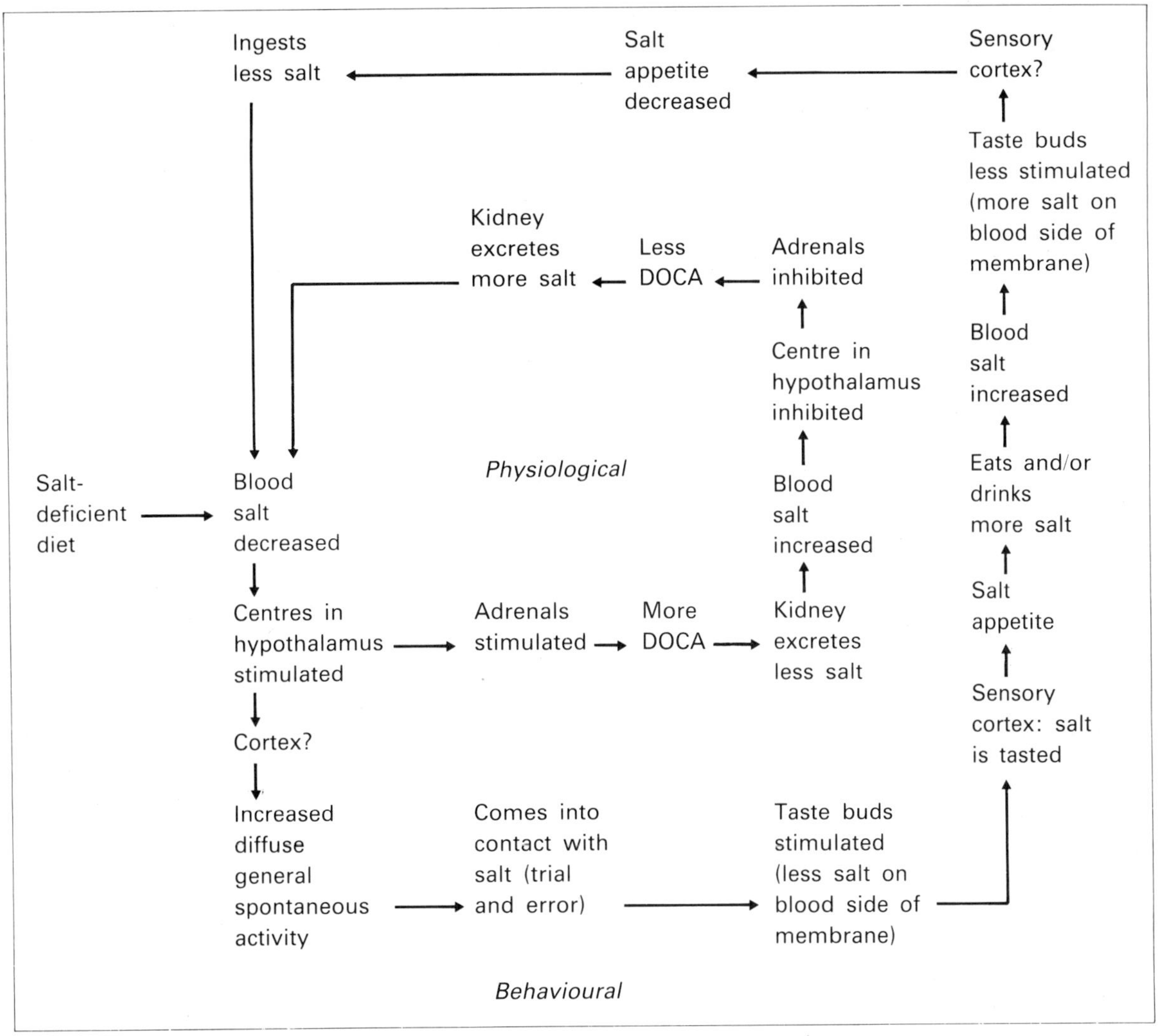

Fig. 24-2. Various possible steps in the operation of physiological and behavioural feedback circuits for the regulation of sodium balance in an animal on a sodium-deficient diet. (From Richter 1956, with permission.)

stimulate salt appetite is sometimes put forward, though, for example, there is no clear evidence that physiological amounts of antidiuretic hormone (ADH) play such a role in thirst. This question is discussed extensively on quantitative grounds in Chapter 18. It is possible that in the rat with severe sodium deficiency, aldosterone is a contributory cause of appetite. It is also theoretically feasible that, in the same manner in which sodium deficiency sensitizes the parotid gland response to aldosterone, and the adrenal to angiotensin, it might sensitize the brain mechanisms of appetite to the action of aldosterone. Aldosterone receptors have been reported to have been identified in the hypothalamus.

However, the main problem in assigning any major role to aldosterone is the fact, well attested, that adrenalectomy does not attenuate sodium appetite in response to sodium deficiency. Studies on adrenalectomized sheep with a parotid fistula show there is the same quantitative response to deficit as in the adrenal-intact animals (Chapter 12). This is true also in adrenalectomized rats. With withdrawal of hormones for a protracted period, and with severe deficit, intake may be reduced somewhat in sheep and rats, but this appears to be related to physical deterioration. In fact, rats then balance deficit rather than overdrinking, as is usual. With adrenalectomized sodium-deficient wild rabbits, appetite is highly significantly related to body deficit. Thus the evidence is against any primary role of aldosterone in the genesis of sodium appetite. The mode of action with unphysiological high doses could be in terms of reducing the sodium content of the brain cells which react to sodium deficiency (Chapter 18).

Change in the concentration of sodium and potassium or Na/K ratio of the saliva

A general background of this idea lies in the fact that in the human sodium chloride solution applied to the tongue, following adaptation to the stimulus, is tasteless. Lower and higher concentrations have a taste—the lower being bitter-sour and the higher salty—and intensity increases with deviation from the adapting concentration. Thus adrenalectomized rats might prefer weak saline solution to water because of the bitter-sour taste of plain water in the face of their elevated salivary [Na] (Contreras 1978). But adrenalectomy would increase the recognition threshold of sodium chloride, not reduce it.

Whereas there are some interesting unresolved problems in relation to detection abilities when the capacities of the adrenalectomized and the adrenal-intact sodium-deficient animal are considered (Chapter 10), it would appear that there are some overriding considerations against assigning any major role to this phenomenon in the actual genesis of salt appetite drive. First and foremost, and best illustrated by consideration of salt appetite in a ruminant such as the sheep, is the very large deviation from normal in the salivary composition of the sodium-deficient adrenal-intact animal. Parotid saliva is the bulk of the secretion, and in the sodium-replete state parotid salivary Na/K is 170/5, whereas in very severe deficiency the ratio may be 10/160. In the adrenalectomized sodium-deficient animal saliva composition remains near the sodium-replete state: 160/10. Yet in both circumstances, as with formalin-treated adrenalectomized and adrenal-intact rats, an equivalent powerful salt appetite drive exists (Chapter 12). Whereas there might be recognition problems with presentation of sodium solutions at concentrations below mixed saliva [Na], there is clearly a powerful drive in both groups of animals. As said above in relation to the sheep, accurate satiation of appetite occurs in both circumstances when a cafeteria of hypertonic solutions including sodium salts is presented.

Further, with intracarotid infusion of 4 M sodium chloride in the sodium-deplete sheep, the immediate rise in arterial blood [Na] includes the lingual and parotid arteries, and a rise of 20–30 mmol/l in salivary [Na] occurs *pari passu* with the arterial blood change, but without influence on appetite. This is also relevant to the idea of [Na] on the blood side of the taste bud being influential in the genesis of appetite. Some preliminary experiments in our laboratory have shown the sodium-deficient adrenalectomized sheep can select 50 mmol/l sodium bicarbonate in a cafeteria of bicarbonate solutions: the [Na] of this is probably less than that of mixed saliva. But the study was on experienced animals and experiments on naive creatures will be relevant to any proper appraisal. The arguments on nerve section also apply to this issue of feedback from the mouth based on altered salivary Na/K ratio. The evidence seems to be against a dominant role of Na/K ratio in the genesis of drive which, in any event as will be recounted, can be generated by changes in CSF where no change in salivary Na/K ratio occurs.

The sodium reservoir hypothesis

Wolf and Stricker (1967) have postulated that sodium appetite is directly responsive to the sodium content of a body sodium reservoir, and, thus, only indirectly responsive to changes in intravascular fluid parameters. They note that bone has been identified as a major reservoir for body sodium, and though they have no reason to believe that this particular reservoir contains the receptor for sodium appetite, they found it useful to use bone as a model for the receptor system (Stricker and Wolf 1969). Hypovolaemia could diminish reservoir sodium by stimulating aldosterone secretion which extrudes sodium from cells (Crabbé 1961). Hyponatraemia could do the same thing by action on aldosterone secretion and directly through passive equilibrium processes. Thus the reservoir hypothesis accounts for the salt-appetite-inducing effects of hypovolaemia, hyponatraemia and mineralocorticoids in a parsimonious manner. Thus, if like bone, the reservoir released sodium quite slowly, it would account for the long latency of onset of sodium appetite. The authors note that with desoxycorticosterone (DOC) the delay may be up to 24 h. Further it would account for the generally poor relation between sodium appetite and intravascular volume or [Na], as these parameters can return to normal values in the presence of a depletion of reservoir sodium. Thus sodium appetite can be seen to persist in formalin-treated rats even after intravascular fluid [Na] has returned to normal (Stricker and Wolf 1966). Such a reservoir would be the ideal place for a receptor since it would allow maintenance of a signal for sodium ingestion while temporary homeostatic processes redistributed sodium and water to restore normal values in the intravascular compartment.

Stricker and Wolf note in relation to this construct that experiments show that a large sodium appetite can be elicited by isosmotic reduction of intravascular volume by 20% polyethylene glycol (PEG) in the adrenalectomized rat. Since they have no evidence how hypovolaemia could deplete the reservoir by an extra-adrenal route they envisage more than one receptor. They have also shown the genesis of salt appetite in the adrenalectomized rat concurrently depleted of water and salt. Though they suggest this is against the hypothesis, we (Beilharz et al. 1962) have reasoned that water depletion affects both compartments—extracellular and intracellular—equivalently and a differential could still exist. Stricker (1980) mentions additional evidence on the role of hyponatraemia in sodium appetite. If water and salt are withheld for 8 h after formalin and PEG treatment, the formalin-treated animals with low plasma [Na] drink salt immediately whereas the PEG-treated animals do not (Stricker and Wolf 1966). If formalin-treated rats are given only water for 24 h, as a result of which they become more hyponatraemic than usual, a greater saline appetite is seen when tested. But PEG-treated rats denied fluid for 24 h drank both fluids, and it is also shown that acute osmotic dilution with water load does not provoke sodium appetite (Stricker and Wolf 1966). On this latter point, the argument of Beilharz et al. (1962) is again relevant. But, in any event, in the light of the data showing that chronic hyponatraemia with water intoxication reduces brain intracellular [Na] and [K], and thus avoids cell swelling (Chapter 7), it would be interesting to test salt appetite after water intoxication lasting several days.

Stricker (1980) concludes that the data do not support a model of salt appetite based on volume of body fluid compartments and levels of related hormones in blood, and that 'availability of sodium to the brain is the stimulus to sodium appetite (cf Denton 1966)'. He suggests new rationales need to be developed as mentioned earlier in this chapter (Stricker 1981).

Discussing this issue of sodium in brain, Wolf et al. (1974) draw attention to the theory that intracellular sodium depletion of specific cells of the central nervous system subserving sodium appetite mediates salt appetite directly (Denton et al. 1969), and suggest that in accordance with this idea of a cerebral site and the reservoir hypothesis, the glial cells might act as sodium reservoirs for neurones, and modify relevant neuronal activity when their sodium content is diminished. There are many more glial cells in the mammalian central nervous system than neurones but the total volumes of the two types are approximately equal (Aidley 1971). The astrocytes have extensive cytoplasmic processes, and the capillaries of the brain are largely surrounded by processes of astrocytes that are swollen to form 'end-feet'. These astrocytes are also in contact with neurones, which do not themselves form any direct link between capillaries. Oligodendrocytes are restricted to the white matter where they are responsible for myelin formation. Aidley (1971) remarks that it is widely believed glial cells act as nutritive reservoirs for neurones, but Kuffler and Nicholls (1966) state that in their opinion there has been no convincing demonstration yet that substances are in fact exchanged between neurones and glia across the narrow intercellular clefts. Orkand, Nicholls and Kuffler (1966) have shown that the resting potential of glial cells appears to be larger than that of neurones, that resting potential may alter following nervous activity, and that the membrane potentials are sensitive to external [K].

Reduction of sodium concentration in the neurones subserving sodium appetite

The transcription–protein synthesis hypothesis

The transcription–protein synthesis hypothesis arose primarily from consideration of the time delay in onset of salt appetite, and also the delay in turning off appetite by systemic infusion of sodium. The possibility of a neurochemical process of some complexity, as distinct from a receptor reacting directly and immediately to some contemporary change in systemic blood, led to the speculation that a change in intracellular [Na] in the neurones subserving sodium appetite (Denton and Sabine 1961) might act by initiation of transcription and protein synthesis (Denton 1966). This was following in the train of the landmark ideas of Jacob and Monod (1961) who, though stating that their conclusions applied strictly to the bacterial systems from which they were derived, suggested they offered a valuable model for the interpretation of biochemical coordination within tissues, and between organs of higher organisms. The discovery of regulator and operator genes, and of repressive regulation of the activity of structural genes, and the fraction of RNA which Jacob and Monod termed 'messenger RNA', revealed that the genome contained not only a series of blueprints, but a coordinated programme of protein synthesis and a means of controlling its execution. In eukaryotes, the chromosomes incorporate both DNA and a variety of special proteins complexed as chromatin. The chromosomal proteins are of two main classes: the alkaline histones, which are similar from tissue to tissue and organism to organism, and the non-histonal chromosomal proteins which are acidic. This difference from bacterial chromosomes has been shown to have major influence on the gene regulatory mechanism, as exemplified beautifully by work of O'Malley and colleagues (O'Malley and Schrader 1976).

This protein synthesis theory of the genesis of salt appetite, as set out below, also embodied some affinity with the ethologists' ideas of 'action-specific energy' to account for mounting appetitive behaviour, and also J. B. S. Haldane's (1956a) reflections on this notion in terms of it having a real neurochemical basis. Of course, as the situation now presents, any theory of salt appetite must needs account not only for the behaviour in response to sodium deficiency, but also for that induced by the sequential action of four to five jointly sufficient and severally necessary steroid and peptide hormones, as occurs with the striking sodium appetite produced in the sodium-replete animal by pregnancy and lactation. This more recent body of data on the action of hormones involved in reproduction (Denton and Nelson 1971, 1978) has, however, if anything, given additional weight to the initial theory in the light of the known modes of action of steroid and peptide hormones on target tissues—where transcription and/or translation may follow receptor binding (e.g. Monroy et al. 1965; Turkington 1972; O'Malley et al. 1975; Dunn and Gispen 1977; Falconer et al. 1978).

In relation to induction of sodium appetite in deficiency, it was proposed (Denton 1966) that within a special population of brain cells which are not necessarily strictly localized, the development of sodium deficiency induces, via a decrease in intracellular [Na], genetic transcription with protein synthesis which sets in train the behavioural mechanisms of appetite. That is, when the intracellular ionic composition of the specific cells changes, the nuclear DNA initiates via messenger RNA the synthesis by ribosomes of a specific protein which may either alter the ionic or membrane characteristics of the neurocyton and hence its excitability, or may increase the capacity of its transmitter generation at synapses. By either means, excitation and transmission in a neuronal pool specifically subserving salt appetite would be facilitated (Fig. 24-3). In terms of the basic Jacob and Monod theory of protein synthesis it was proposed that the extent of ionic change within the

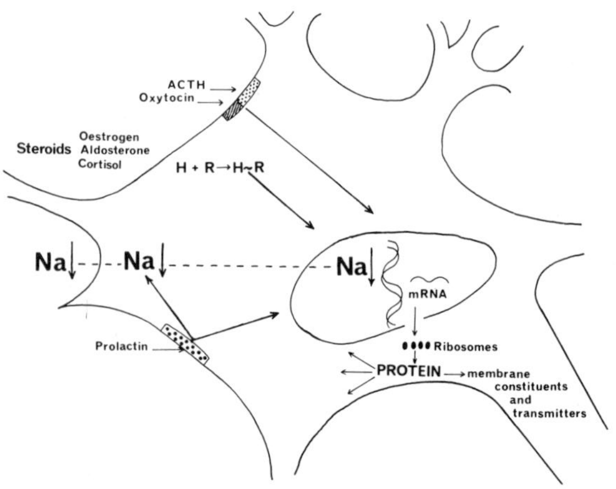

Fig. 24-3. Diagrammatic representation of the mode of excitation of neurocytons subserving salt appetite drive. (From Denton and Nelson 1980, with permission.)

specific cells, commensurate with body sodium deficit, was determinant of the extent of release of the structural gene from the repressor action of its regulator gene and thus of the extent of messenger RNA synthesis. Appetite drive would therefore bear a relation to body deficit. The possibility that the neuronal organization subserving salt appetite might be spatially complex is suggested by data from electrical stimulation on areas with influence on salt appetite, as set out in Chapter 19. There is also the analogy with thirst, where mapping by microinjection (Fisher 1964) indicated relationships extending over the limbic system including Papez circuits.

The state of excitation in the specific cells may induce arousal and the characteristic restlessness of appetitive behaviour by effects on the activity of the midbrain reticular activating system. This could modulate sensory inflow (Hernandez-Peon et al. 1956; Adey et al. 1957; Livingston 1959) with bias of attention towards afferent chemoreceptor and other impulses relevant to the act of ingestion of salt. The changes induced in the neurones subserving appetite might initiate firing behaviour with attendant spreading activation in the manner described by Adrian (1950), who, in analysing spontaneous electrical activity in the brain suggested that a few cells beating at a high rate may excite their neighbours and those in turn would excite others until the rhythm spread throughout the mass. He points out that there are instances of single isolated units, nerve cells, or sensory endings discharging rhythmically for long periods, and a rhythmically active cell may influence its neighbours by means of impulses it sends to them, or conceivably by the electrical or chemical fields engendered by its activity.

With such a hypothetical system, it is correspondingly necessary to postulate how its activity is turned off. It is proposed that the protein is constantly being degraded in the cells concerned. When sodium is administered systemically with repair of intracellular ionic deficit, commensurate with amount and concentration of sodium given, the regulator system depresses messenger RNA release and protein synthesis is reduced or ceases. Thus, in the face of constant protein breakdown, appetite disappears. The consummatory act of satiation by rapid drinking with ensuing precipitate decline in motivation as, for example, in the sheep must involve other processes. In this case a characteristic sequence of afferent impulses generated by the chemosensory and proprioceptor elements of the act is proposed to cause a very rapid to explosive disintegration of the protein. That is, a physicochemical process of this type is involved in the rapid satiation. The degradation products may influence the regulator system to suppress synthesis of activator over a sufficient time interval for sodium drunk to be absorbed from the gut and eventually repair the intracellular sodium deficit initiating the appetite. RNA synthesis can, for example, be inhibited if newly made RNA is not released from RNA polymerase–DNA (Lezzi 1970). Alternatively the central organization may be of an excitation–inhibition type, and the inhibitory elements stimulated by drinking may suppress activity of the excitatory mechanism over a sufficient time interval for salt drunk to be absorbed, alter brain extracellular fluid composition and eventually repair the intracellular deficit.

The main physicochemical mechanisms postulated as subserving the activation of this pattern of instinctive behaviour have been considered earlier in the context of a possible protein molecular basis of memory (Katz and Halstead 1950; Hyden 1960). There has been comprehensive discussion including the question of relation to electrophysiological theories of memory. Considerations raised have included alterations in RNA and protein synthesis in neurones associated with learning (Hyden and Egyhazi 1962) and postulation of very rapid or explosive breakdown of intracellular proteins under specific conditions (Geiger et al. 1956; Vrba 1956; Richter 1965; Hyden 1960). The subject has been reviewed by Mark (1974).

A speculative concept along these lines at least focuses the complexities to be accounted for in, for example, delayed inhibition of appetite with rapid systemic infusion versus the precipitate disappearance with rapid drinking over 2–3 min, and the failure to inhibit appetite with rapid drinking when a component of sensory inflow is absent (as with drinking salt solution with an open oesophageal fistula). As a working hypothesis it has been useful in suggesting experiments. Some other aspects have been discussed in more detail in Denton (1966). There is, in fact, substantial evidence for ionic effects on transcription processes, and this will be briefly discussed before dealing with experimental data on salt appetite relevant to the hypothesis.

Ionic effects on transcription–translation processes

The principle of differential gene activation states that differences between cells having the same genetic content are the result of different sets of genes being turned on and off. A valuable example for study of the effect of steroid hormone on the genome was the induction of puffs in the giant chromosomes of Diptera by ecdysone. The puffs occur at specific bands in these polytene chromosomes and involve uncoiling of chromatin strands with accumulation of acidic proteins and de novo RNA synthesis indicative of increased gene activity (Berendes 1972). Ecdysone induces formation of puffs with latency of 15 min or less to 6 h, and the pattern recapitulates the natural events of the moulting sequence (Ashburner 1972; McEwen et al. 1974).

Lezzi (1970) remarked that histone may be considered *the* repressor of gene activity in chromosomes of higher organisms. If there are other kinds of factors repressing, stimulating or modulating gene activity in metazoan chromosomes they probably cannot function until the repressive effect of histone is reduced, unless the task of these factors is such a reduction. This viewpoint developed from comparisons of the histone content of differentially active chromatins, as well as from studies on the effect of experimental histone removal or addition on the template activity of chromatin or DNA.

Lezzi and Robert (1972) propose the probable sequence of events leading from the inactive to the active state is that ions come into play before RNA polymerase or any other acidic protein. The ions are required for dissociation of ionic bonds between the cationic amino groups of the histone and the anionic phosphate groups of the DNA, and the dissociation of the these bonds is a prerequisite for binding of RNA polymerase to DNA and the initiation of RNA synthesis. Ions may also play a role between growing RNA molecules and ribosomes. Lezzi and Robert state that the fundamental problem of differential ionic sensitivity of chromosome regions has not been solved but a working hypothesis is that chromosomes are preprogrammed so that their genes react differently and specifically to varying ionic conditions in the nucleus. Ionic sensitivity is differential in two respects: for different ionic strengths and for different ionic species. The specificity of sodium and potassium depends on the presence of divalent cations.

Lezzi and Robert (1972) have put forward a model based on experimental data from isolated chromosomes treated with salt solutions which show there exist sodium-sensitive chromosome regions and also potassium-sensitive regions. The selective activation of these regions can be achieved by salt solutions containing sodium and potassium together, the total cation content of such solutions is within the physiological range, i.e. between 120 and 220 mmol/l, and it requires a minimal change in Na/K ratio to switch from exclusive activity in sodium-sensitive regions to exclusive activity in potassium-sensitive regions. In relation to physiological concentrations of ions in the nuclear sap of *Chironomus* salivary glands, they refer to Kroeger's measurements of sodium plus potassium concentration as being 170±50 mmol/l. Langendorf et al. (1961) determined the monovalent cation concentration of wet liver nuclei to be about 200 mmol/l, about twice as high as the corresponding value for wet cytoplasm.

The fact that ionic conditions most favourable for transcription require monovalent cation concentrations of 150–200 mmol/l and magnesium concentrations of 12 mmol/l indicates that the Langendorf et al. (1961) values depict a physiological milieu (Lezzi 1970). Direct determinations of sodium and potassium content of liver of normal and ethionine-treated rats also show a more or less parallel shift of nuclear and cytoplasmic Na/K ratios despite a difference in the absolute amounts of sodium and potassium in the two compartments (Okazaki et al. 1968). Measurements of electropotential of salivary glands of the insects *Chironomus thummi* and *Galleria mellonella* indicate that ecdysone when it acts in vivo or in vitro causes nuclear Na/K ratio to fall whereas juvenile hormone causes it to rise (Kroeger 1966; Baumann 1968).

In relation to the general viewpoint that control of gene activity in higher organisms can take place on more than one level, the level of histone is proposed as primary in gene activation. The interaction between it and DNA is ionic and thus it is primarily influenced by the ionic environment of the chromosome. This hypothesis and experimental data, termed the Kroeger hypothesis by Ashburner and Cherbas (1976), has been criticized by them. At the outset they agree that nucleoprotein complexes are exquisitely sensitive to changes in salt concentration, and that ionic selectivity is a well-known property of proteins. Thus it might not be shocking if cells used this specificity in some general control over chromosome structure. They focus the question by stating that a decrease in intranuclear Na/K, according to Kroeger, acts as a second messenger mediating the effects of ecdysone—much as cyclic AMP apparently mediates the effects of many polypeptide hormones. They doubt the hypothesis on the grounds that ionic concentrations required for specific puffing in isolated nuclei are not reasonable intranuclear concentrations, notwithstanding the ion binding (Palmer and Civan 1975) which could complicate the experiments. Whereas some regions are differentially sensitive to cations under some conditions, and while this says something about underlying differences in structure in these regions, they were, on their analysis of the data, sceptical of Kroeger's interpretation.

Kroeger (1977) has responded to this criticism. He proposes that allowing what Ashburner and Cherbas (1976) appear to agree in relation to differential sensitivity, the issue hinges on whether natural variation of cellular electrolytes occurs within the range in which chromosomal sites respond. On the issue of whether the concentrations used are physiologically reasonable, he suggests Ashburner and Cherbas (1976) using the data of Lezzi (1966) have made a computation error. The fact was that the ionic strength of material added by Lezzi contributed only 25% to the final volume, and thus concentrations did fall within determined ion content of nuclei. Thus it was a relevant experiment.

These data and discussion, though not resolving any question in relation to ionic changes in neurones subserving salt appetite, serve to focus attention on the fact that there is perhaps nothing inherently unlikely in the idea that a change in intracellular [Na], or Na/K ratio in these cells, could influence or determine nuclear transcription, as well as possibly having other vectors of influence on translation. Douzou and Maurel (1977) have discussed this issue also in the light of their experimental work. They propose proteins involved in genetic translation systems act under the influence of nucleic acids, and therefore of strongly negative polyelectrolytes. Some of these proteins interacting with these polyanions are very basic in character and might themselves behave as polycations, suffering and promoting modifications of the local electrostatic potentials which are in turn influenced by any ionic strength and/or pH fluctuation. Variation in pH is shown to influence ribonuclease activity, and, at constant pH, the sequential processes of genetic translation are influenced by ionic strength. They summarize by stating that proteins of a polyelectrolyte nature are specifically influenced by changes in ionic strength. This may well have physiological importance since experimentally induced changes are not usually very far from the possible fluctuations that may arise in vivo. Under these conditions, the physiological changes in proton and cation levels within cell compartments could be understood in terms of their ability to regulate metabolic processes by modulating the electrostatic potentials under which most macromolecular components of cell machinery operate. This 'physiological' control would take place together with the controls already established in systems involved in genetic translation, and in cells of plants and higher animals (Douzou and Maurel 1977).

In this same vein of the profound influence ionic changes may have, a recent conference on 'Growth Regulation by Ionic Fluxes', summarized by Koch and Leffert (1979) in *Nature*, dealt with many findings pointing to increased sodium influx being an early, perhaps initial, mitogenic signal and to subsequent calcium ion movement coupled with cyclic AMP formation being necessary for DNA replication. Whilst this is far from proven, they suggest that an impressive body of data derived from diverse sources focuses attention on such a unifying concept. Experiments by Epel (1980) with sodium influx inhibitors and ^{22}Na showed that burst-like influxes of sodium are required to initiate DNA synthesis in sea urchin eggs after fertilization. The ideas put forward for a mode of action of sodium in stimulating proliferation in cells included activation of membrane Na,K-ATPase. Other consequences of increased sodium influx, suggested by Piatigorsky et al. (1980), included sodium and potassium selective translational and/or post-translational effects. Koch and Leffert state that the existence of two regulatory ionic 'landmarks' has interesting implications for understanding growth regulation in general in that sets of growth factors may act synergistically. These are that certain substances activate sodium influx while others act predominantly upon calcium–cyclic AMP coupling. Interference with the sodium flux system perturbs transitions from resting to growing states differently from interference with calcium–cyclic AMP coupling.

Cone (1980) at this same conference reported studies on 20-day cultures of neurones from the central nervous system aimed at determining whether imposition of an elevated intracellular sodium concentration, $[Na]_i$, would induce mitogenesis. The 20-day cultures were treated with ouabain to inhibit the Na,K-ATPase of the membrane sodium pump and thus cause a leakage influx of sodium and a rapid rise in $[Na]_i$. For example, Thomas (1972) has shown with continuous recording from sodium-specific microelectrodes that $[Na]_i$ in snail neurones doubles within 8 min during exposure to 10^{-4} M ouabain. Cone (1980) found that 30% of neurones initiated DNA synthesis when exposed to ouabain at concentrations of 10^{-4} to 10^{-6} M. Two other depolarizing agents capable of producing sustained elevation of $[Na]_i$, veratridine and gramicidin, had the same effect. Thus a surprisingly rapid mitogenic response was produced in activated neurones which were destined never to divide again under normal culture conditions. As an overview of the data from this and other studies he cites, Cone (1980) reasons as follows (Fig. 24-4).

The cell surface and its ionic partitioning capabilities assume a position of central importance by virtue of their direct control of the overall intracellular ionic hierarchy. This, in turn, is considered to play the key role in determining the specific temporal expression and coordination of cellular metabolism, including that of mitogenesis. The factors of primary importance in ionic partitioning by the cell surface are the membrane permeability to each of the ionic species in the cellular environment, and also the active transport ability to move ion species against concentration gradients. These ionic partitioning capabilities of the membrane which are specific for each cellular phenotype act in the light of the conditions in the external environment to jointly determine the particular overall intracellular ionic hierarchy at any given time. Environmental events that affect the pattern and rate of ionic partitioning include ionic species and their relative and absolute concentrations, hormones, metabolites and various other influences including surface contact with other cells.

Under mitogenically quiescent environmental conditions the cell remains in the G_0 metabolic state. That is, it may be very active metabolically in relation to its basic phenotypic functions but all mitogenic metabolic pathways are inactive because of the absence or inactivation of the required mitogenic enzymes. In these G_0 circumstances the $[Na]_i$ is relatively low, which acts to maintain the repression of the mitogenesis-specific genes. As long as the low $[Na]_i$ level is maintained by the membrane sodium pump, mitogenic metabolism remains turned off. However, when environmental conditions change so that the ability of the membrane to keep down the $[Na]_i$ to a required low level is altered for a sustained period, the $[Na]_i$ rises and the 'mitotic switch' shifts to the G_1

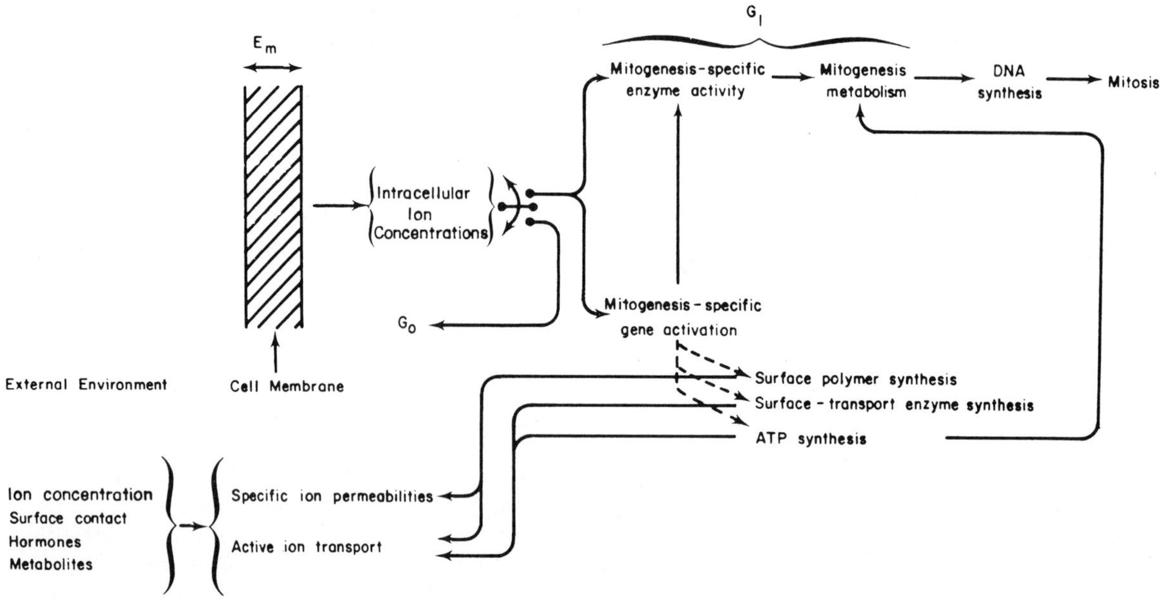

Fig. 24-4. Schematic representation of the primary system of biochemical and biophysical events and interactions underlying control of mitogenesis in somatic cells. The cytogenetically endowed ionic partitioning specificity of the cell membrane acts, in association with influencing factors in the cellular environment, to determine the intracellular hierarchy of ionic species and concentrations. This ionic hierarchy in turn controls the intracellular metabolism, including that of mitogenesis, through regulation of gene-expression patterns and enzyme reaction rates. Mitogenic environmental conditions inhibit the sodium pump of the membrane and thus allow the $[Na]_i$ level to increase, with the consequence that synthesis of the complex of mitogenesis-specific enzymes is elicited and the cell shifts from the G_0 to the G_1 metabolic state. (From Cone 1980, with permission.)

metabolic mode. The increased $[Na]_i$ level causes derepression of the mitogenesis-specific complex genes, and transcription and translation of the mitogenesis-specific messenger RNA generates the full complement of mitogenesis-specific enzymes. This includes enzymes like thymidylate synthetase and ribonucleotide reductase. Genome replication proceeds and ultimately mitosis takes place. Environmental factors which directly lower the functional capacity of the sodium pump, or increase membrane permeability to sodium, or cause inhibition of the Na,K-ATPase will lead to this change if they act for a sufficient period of time to raise the $[Na]_i$ above the mitogenic threshold. Cone proposes that the absolute intracellular concentration of sodium is the key factor involved in the control of mitogenesis activation in somatic cells accompanied, perhaps, by small concomitant changes in the concentration of ancillary ions (calcium, magnesium, hydrogen).

Rozengurt and Mendoza (1980) have studied the manner in which the addition of serum to quiescent cultures of fibroblasts enhances the rates of protein and RNA synthesis and dramatically stimulates DNA synthesis and cell division. They note that the addition of serum or the polypeptide MSA (multiplication-stimulating activity) to quiescent cultures of 3T3 fibroblasts stimulates ^{86}Rb uptake (this isotope serves as a potassium tracer). It was also reported that serum rapidly increased lithium movement across the membranes of cultured fibroblasts. A series of experiments involving use of ouabain, valinomycin (a highly specific ionophore for potassium) and amiloride pointed to alterations in the rate of cation pumping reducing the progression of the proliferative response to serum. Rozengurt and Mendoza conclude that activation of sodium influx is one of the earliest membrane changes in the action of growth factors and suggests a resemblance of fibroblasts to other excitable cells such as muscle in which an inward movement of sodium ions is a primary excitatory process.

Piatigorsky et al. (1980) have produced evidence that changes in concentration of electrolytes affect the ratio of synthesis of the different classes of δ-crystallin polypeptides. Crystallins are structural proteins that constitute 80%–90% of the soluble protein of the ocular lens in vertebrates. There are four classes and they are easily detectable markers for gene expression during differentiation. Chicken δ-crystallin has a native molecular weight of approximately 200 000 daltons and can be resolved into two bands with molecular weights near 48 000 and 50 000 daltons by polyacrylamide gel electrophoresis. Because ouabain-treated lenses are permeable to sodium and potassium, this results in changes in the intracellular concentration of these cations, and it is found that the ratio of synthesis of the two δ-crystallins varies with these changes in [Na] and [K]. In general, intracellular K/Na ratios higher than 1 are correlated with the normal ratio of [^{35}S]methionine incorporation into the bands of δ-crystallin, while K/Na ratios less than 1 are associated with the differential reduction of isotope incorporation into the smaller polypeptide. The only structural differences found between the large and smaller δ-crystallin polypeptides are two acidic methionine-containing tryptic peptides which are absent from the lower molecular weight classes. On the basis of data that the ratio of polypeptides synthesized can be changed in 3 h by alteration of the cation ratios, and that less than 1% of δ-crystallin messenger RNA known to be present in the lens cell would be synthesized over this time interval, the authors propose that even total inhibition of messenger RNA synthesis would not significantly influence protein synthesis, and thus that the alteration in synthesis is regulated at the post-transcriptional or post-translational level. This is currently under intense investigation.

It is most interesting to note parenthetically here that Piatigorsky and colleagues (1980) remark that numerous cataracts are associated with appreciable ion concentration changes, particularly an increase in sodium and decrease in potassium. Piatigorsky (1980) states that in a natural cataract the sodium pump gets 'turned down' and this is followed by hydration as water comes in with the sodium. Vacuoles then form in the lens and cause the opacity of cataract. Changes in protein synthesis follow this and contribute to the irreversibility of the condition. Studies on the Nakano mouse mutant indicate that the cataract results from a specific inhibitor of Na,K-ATPase (Kinoshita 1974; Fukui et al. 1978). It is a polypeptide, and it is not known whether it arises by cleavage or by de novo synthesis, but it turns off the sodium pump. An interesting sidelight to this discussion is the question of why fish do not develop cataract when migrating between salt and fresh water. The explanation apparently resides in the fact that they have unusually thick corneas which are impermeable to ions and thus protect the lens (Edelhauser et al. 1965).

The large body of data from the conference on 'Growth Regulation by Ionic Fluxes', as with the other considerations on transcription above, emphasizes the feasibility of an increase or decrease in $[Na]_i$ having decisive effects on fundamental cellular functions, including expression of DNA and effects at the post-transcriptional level. At present, of course, decisive information is not available on many points which are speculative and, self-evidently, the direct approach in the study of sodium appetite will require the location of the neuronal population which subserves the behaviour by one of several approaches aimed to show changed cellular composition and function in sodium deficiency. In this regard, demonstration of transcriptional changes in a particular population by radiolabelled markers such as [^{3}H]thymidine or [^{3}H]uridine, or blockage of behaviour, with evidence of specificity, by use of puromycin, cycloheximide and actinomycin D represents an important experimental goal, as would proton or electron probe X-ray microanalysis measure of cation composition of the specific neurone population in sodium deficiency.

Experimental procedures aimed at altering sodium concentration in neurones in sodium-deficient and sodium-replete sheep

Whittam (1967) remarks that every cell that has a coupled transport of sodium and potassium has an ATPase that is activated by these two ions. It is widely accepted that enzymatic splitting of ATP is intimately connected with the working of the sodium pump. There is a particularly high activity of transport ATP in membrane preparations from excitable tissues such as electric organ, brain cortex and peripheral nerve as well as in secretory tissues such as kidney and salivary glands. Active transport, as a physiological process, is greater in nervous tissue than in other tissues, as neurones and axons are required to maintain a high conduction velocity. Sodium efflux in cat motor neurones is three times higher than in rat diaphragm and over 100 times greater than in human erythrocytes. Skou (1957) described enzymatic breakdown of ATP that was stimulated synergistically by sodium and potassium. Magnesium was needed for the hydrolysis. Strophanthin inhibits that part of the activity which needs both sodium and potassium. Transport rates can be varied in three ways without directly affecting the supply of ATP from metabolism: (*a*) by altering external [K], (*b*) by altering internal [Na], (*c*) by partial inhibition with ouabain (Whittam 1967).

Experiments with brain cortex have shown the Na,K-activated enzyme is found in cell membranes and not in cytoplasm or intracellular particulate matter. Cummins and Hyden (1962) showed with morphological techniques that it is located in the surface membrane of neurones. Glial cells are relatively low in the enzyme and nerve endings or synaptic vesicles are particularly rich in it. Swanson and McIlwain (1965) demonstrated that G-strophanthin (ouabain) specifically inhibits Na,K-ATPase in brain tissue, causing a rise in intracellular sodium and a decrease in intracellular potassium (Yoshida et al. 1962; Swanson and Ullis 1966). Concentrations required are in the range of 10^{-4} to 10^{-8} M (Glynn 1964). Ouabain in low concentrations (10^{-8} to 10^{-11} M), on the other hand, has been shown to stimulate the enzyme in brain tissue (Repke 1963; Palmer et al. 1966).

Intracarotid ouabain infusion in sodium-deplete sheep

Experiments were done to test the direct effect of the infusion of ouabain into the blood perfusing the head on the salt appetite of sodium-deficient sheep (Denton, Kraintz and Kraintz, 1969). Additional experiments were made on the same animals at the same dosage to determine whether hunger or thirst were affected and thus whether any influence on sodium appetite appeared specific at the dosage used. The infusions were made into one carotid artery, with the other artery occluded, for 4¼ h preceding the presentation of sodium bicarbonate (300 mmol/l) to sheep depleted of sodium by parotid fistula for a period of 48 h. Water was continuously available. The infusion of ouabain at doses from 0.11 mg/h to 0.16 mg/h greatly diminished or abolished the voluntary intake of the sodium bicarbonate solution during the 30-min test period. Fig. 24-5 shows the individual results for four sheep: Bettina, Zeta, Snowdrop and Amy. The mean of 25 control experiments where intake was measured after 48 h of depletion and without ouabain infusion is shown also. Initial studies on Bettina indicated that a total dose as low as 0.46 mg was effective in abolishing the sodium appetite during the 30-min test period. A similar dose in Amy did not completely abolish the sodium appetite and therefore higher doses were employed in the rest of these experiments. Table 24-1 gives the total dose of ouabain used in each experiment and the amount of sodium drunk during the 30-min test period. It is apparent that the effectiveness of ouabain in blocking sodium appetite varied slightly between the individual sheep. It did not appear related to body weight. The influence of the infusion on appetitive behaviour was variable. In some instances, interest in the preparation and presentation of the solution was abolished, whereas in others there was considerable interest, with initial tasting and licking of the lips several times, but little or no intake. When the sodium solution was left on the cage after the test period, the sheep did not begin to drink until 3–6 h later.

Intracarotid infusion of isotonic saline over the same period did not affect either appetitive behaviour or the voluntary sodium bicarbonate intake during the 30-min test period in 14 experiments on the four sheep (Fig. 24-5).

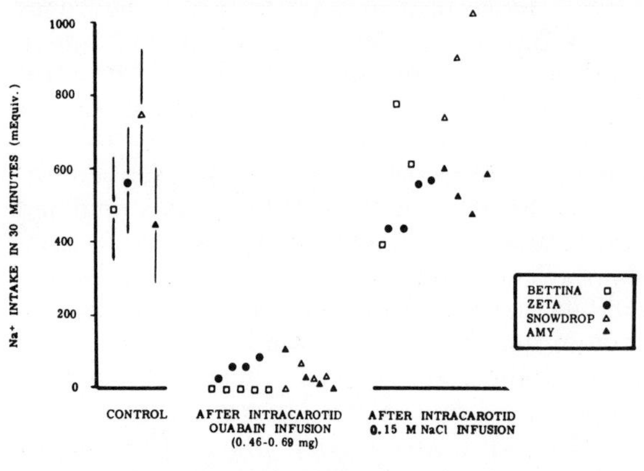

Fig. 24-5. The effect of intracarotid ouabain infusion (0.46–0.69 mg over 4¼ h) on the voluntary intake of 300 mmol/l sodium bicarbonate solution during 30 min of access subsequent to 48 h of sodium depletion caused by parotid salivary loss. The results with four sheep are shown and compared with two control observations (mean±s.d.) on the same animals after 48 h of sodium depletion. The effect of intracarotid infusion of normal saline (0.15 M) over the same period is shown also. (From Denton et al. 1969, with permission.)

Table 24-1. The effect of intracarotid ouabain infusion at various rates on the voluntary sodium intake during 30 min of access in sodium-deficient sheep.

Sheep	Weight (kg)	Amount of ouabain (mg) infused during 255 min	Na drunk in 30-min test period (mmol)	Na loss in 48 h (saliva+ urine) (mmol)
Bettina	34.0	0.46	0	444
		0.48	0	676
		0.51	0	717
		0.64	0	612
		0.64	0	654
Amy	26.7	0.51	105	569
		0.64	0	544
		0.67	30	557
		0.67	15	514
Zeta	24.8	0.64	30	422
		0.64	60	572
		0.64	90	634
		0.69	60	529
Snowdrop	35.1	0.62	75	799
		0.64	37	655
		0.65	0	681
		0.69	30	625

After Denton et al. (1969).

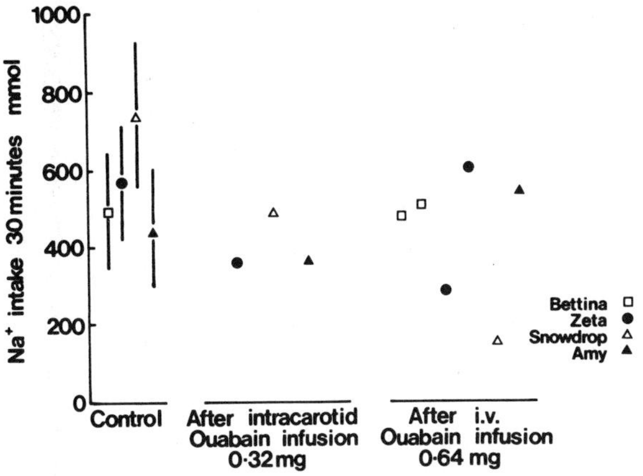

Fig. 24-6. The effect of intracarotid infusion of ouabain (0.32 mg over 4¼ h) and intravenous infusion of ouabain (0.64 mg over 4¼ h) on the voluntary intake of 300 mmol/l sodium bicarbonate solution by four sheep during 30 min of access subsequent to 48 h of sodium depletion. The results are compared with 25 control observations (mean±s.d.) on the same animals after 48 h of sodium depletion. (From Denton et al. 1969, with permission.)

Fig. 24-6 compares the results of three intracarotid ouabain infusions at a dose of 0.32 mg, and seven intravenous ouabain infusions at a dose of 0.64 mg. Neither procedure abolished the sodium drinking during the test period, although there was decreased intake in four instances.

The effect of similar intracarotid ouabain infusion on water or food intake in the same sheep when water-deprived or starved

Water was withheld for 48 h, and saliva was returned to the animals by rumen tube each 12 h with a small sodium supplement (5–10 g sodium bicarbonate) to ensure no sodium depletion. Fig. 24-7 compares the control volume of water drunk during a 30-min test period following water deprivation with the volume drunk during the test period following intracarotid ouabain infusions of 0.64 mg over 4¼ h. In no instance did the ouabain infusion abolish water drinking. However, the intake of water was decreased to some extent in most experiments. In another series of experiments (Weisinger et al. 1977), where the thirst drive was much less, ouabain had a somewhat larger effect on water drinking.

In control experiments, three sheep (Zeta, Snowdrop and Amy) after 48 h of food deprivation showed only normal interest when the food was presented. They ate in the usual fashion without any obvious increase in rate, consuming amounts

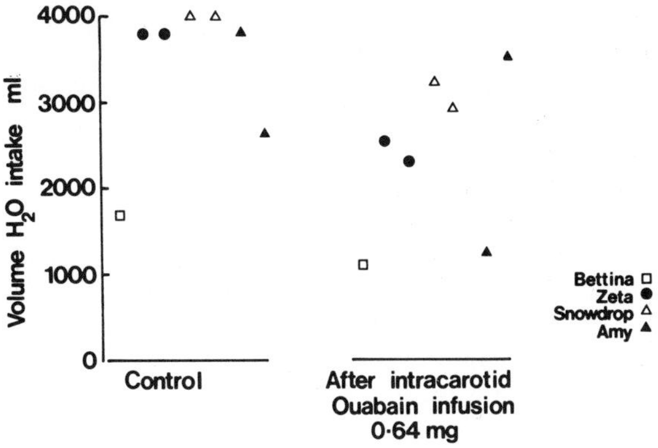

Fig. 24-7. The effect of intracarotid ouabain infusion (0.64 mg over 4¼ h) on water drinking during 30 min in four sheep. The sheep were without access to water for 48 h beforehand. Control experiments without ouabain infusion are shown also. (From Denton et al. 1969, with permission.)

varying from 100 to 300 g during the first 2 h. On three occasions the intracarotid ouabain infusion (0.64 mg over 4¼ h) did not alter this pattern of behaviour nor the amount eaten.

The effect of ouabain on instrumental conditioning related to salt appetite

Fig. 24-8 shows the bar-press activity of Ricky, a sheep which had learned to supply himself at will with metered volumes of sodium bicarbonate solution. The upper record gives the replacement behaviour of this fistulated sheep under control

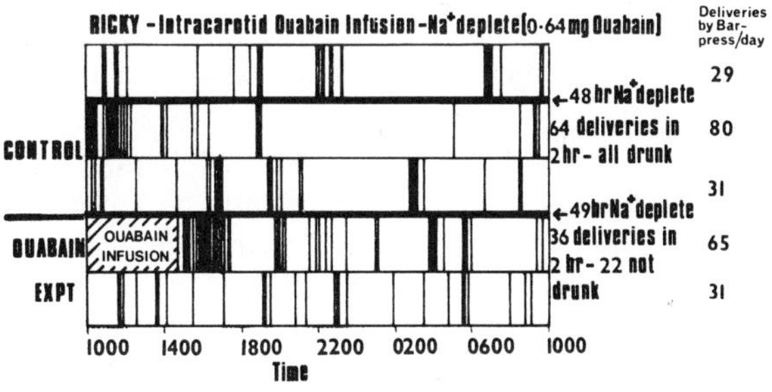

Fig. 24-8. Ricky: sheep with continuous daily loss of saliva from a unilateral parotid fistula. The animal was trained to press a bar in its cage in order to deliver 50 ml of 300 mmol/l sodium bicarbonate solution to a drinking vessel. The top panel records episodes of bar-pressing over a control 24-h period (10.00–10.00 hours). The bar-press behaviour subsequent to replacement of the bar after its absence for 48 h is shown in the panel below. Body deficit was repaired over 2 h. The bar-press behaviour the following day is also recorded. Ouabain (0.64 mg) was infused into the carotid during the last 4¼ h of a 48-h period of absence of the bar. The record of bar-pressing on the day of the experiment and over the following 24 h is shown. In the 2 h after the bar was replaced following intracarotid ouabain infusion, 36 deliveries were made but 22 were not drunk. (From Denton et al. 1969, with permission.)

conditions. He delivered and drank twenty-nine 50-ml volumes of 300 mmol/l sodium bicarbonate (435 mmol) during the 24-h period. The salivary sodium loss was 440 mmol. Following removal of the bar-press lever and 48 h of sodium depletion (thick black line), the lever was replaced. The animal showed concentrated activity, and in a 2-h period, 64 deliveries were made and drunk. The following day Ricky drank 31 volumes of the sodium solution. Following a similar period of sodium depletion, and after intracarotid ouabain infusion with a total dose of 0.64 mg over 4¼ h, Ricky showed considerable bar-pressing activity immediately when sodium bicarbonate was available but sometimes missed hitting the bar with his front leg. This animal made 36 deliveries during the first 2 h, but he drank only 14 of these volumes. Five hours later he had consumed all of the sodium solution that had accumulated in his bin. As before, by the following day he had returned to his normal pattern of behaviour, and delivered and drank 31 volumes of sodium solution.

Fig. 24-9 shows a similar bar-press record for

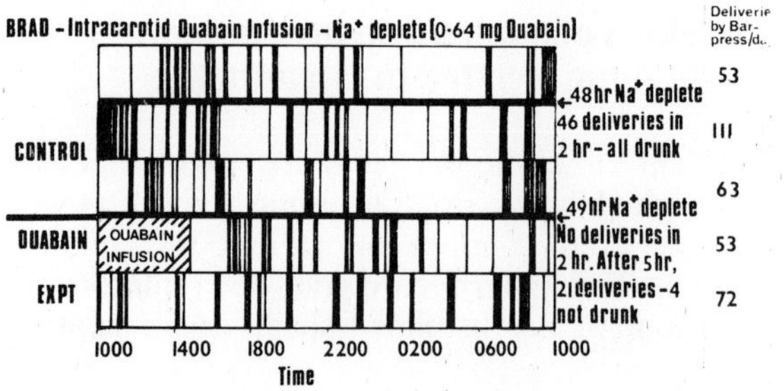

Fig. 24-9. Brad. Sequence and conditions as described in Fig. 24-8 for Ricky. (From Denton et al. 1969, with permission.)

another fistulated sheep, Brad. Brad delivered and drank 53 metered volumes of sodium bicarbonate solution on the control day. During the first 2 h after the lever and sodium solution were again available to the sheep after 48 h of depletion, 46 deliveries were made and drunk. The following day activity returned to normal. After intracarotid ouabain infusion using a dose of 0.64 mg, following a similar period of sodium depletion, there was no bar-pressing in the 2 h subsequent to the cessation of the infusion. At approximately 3 h post-infusion the sheep initiated bar-pressing at intervals shown on the figure, and it drank the solution delivered. By 5 h, 21 deliveries were made and all but four volumes drunk. The deliveries for the following day were more frequent than the activity during a typical 24-h period, presumably because of the delay in the return of sodium appetite after the ouabain infusion.

Discussion of the effect of ouabain on sodium appetite

The effect of ouabain on sodium appetite did not appear to be due to a systemic effect of the drug, but rather a result of perfusion of the head with optimal concentrations of the drug. That some critical concentration in cerebral blood may have been important to the effect on salt appetite is suggested by the fact that there was much less effect with intravenous infusion of the same dose over the same time interval or with intracarotid infusion of half the dose (Fig. 24-6). During intracarotid infusion the concentration of ouabain in the blood perfusing the salivary glands, taste buds and tongue, as well as the brain, would be higher than that achieved by jugular vein administration. From experiments involving bilateral concurrent sampling of jugular vein blood during slow intracarotid infusion of 4 M sodium chloride with the contralateral carotid loop occluded, the blood flow through the head of sodium-deficient sheep of differing weights and degree of sodium deficiency was found to be in the range 300–800 ml/min (Beilharz et al. 1965). Thus, in the experiments here, the ouabain concentration in the cerebral blood, as an immediate result of starting the infusion, could have ranged from 3 to 10 μg/l. With recirculation the concentration would have gradually risen, but a differential of this order in the concentration of cerebral blood relative to systemic blood would have been maintained. Cerebral artery concentration may have ranged from 10^{-8} to 10^{-7} M. The effect of ouabain in inhibiting salt appetite was relatively short-lived, and usually the sheep regained interest and drank the sodium solution 3–6 h later. Okita et al. (1955) determined the half-life of digitoxin in blood to be of the order of 15–30 min.

The central question of this study is whether the inhibition of salt appetite by ouabain is a specific action or merely a generalized non-specific effect also involving other aspects of appetitive behaviour such as drinking and eating. This was tested to a limited degree in sheep that had been deprived of food or water for a period of 48 h. Sheep deprived of water for 48 h lose 3%–4% body weight and correct the deficit within 5 min when presented with water (Bott et al. 1967). The results in the control experiments here on water deprivation were similar to these previous experiments, and ouabain infusion did not substantially reduce water intake. But some decrease occurred on most occasions. The possibility exists that ouabain could specifically affect thirst through action on ionic transfers in the osmoreceptors (Bergmann et al. 1967). It also reduces the water intake induced by intracarotid infusion of angiotensin II or 4 M sodium chloride in dose-related fashion (Weisinger et al. 1977). Food intake subsequent to 48 h of deprivation was not altered in any substantial manner by ouabain. The results were suggestive, though perhaps not conclusive, that the effect on salt appetite was relatively specific and not a consequence of some generalized depression of all forms of appetite and satiation concurrent with a degree of lethargy which was often observed with the infusions. When taken with the positive effect of infusion into the cerebral circulation in contrast to the negligible effect of intravenous infusion of the same amount, the result is consistent with the mechanism of control of salt appetite being situated in the head rather than primarily determined by a peripheral receptor system.

As indicated, the concentration of ouabain produced in these experiments was greater than 10^{-8} M, and perhaps more than that derived from calculation based on blood flow if distribution into blood cells was not complete in the seconds between the intracarotid needle and cerebral tissue. Blaine et al. (1974) have shown in the instance of adrenal arterial infusion of ouabain that substantial release of potassium from adrenal cells occurs at 10^{-5} M. Working with outer cortical slices of calf adrenal tissue incubated in vitro with various concentrations of ouabain, Wellen and Benraad (1969) found a 75% decrease in [K] and a doubling of [Na] intracellularly at 10^{-5} M. Ugolev and Roschina (1965) reported K-strophanthin (0.75–0.375 mg/kg) given intraperitoneally in rats inhibits sodium appetite and increases potassium and water appetite but, as the rats were not deficient, any influence is on the hedonic component of sodium intake.

The data from the experiments here on sheep are consistent with the theory that reduced intraneuronal [Na] causes sodium appetite. Thus an agent which is demonstrated to increase intracellular [Na] when given above a concentration of 10^{-8} M reversibly reduced sodium appetite in sodium-deficient sheep. There are evident limitations to interpretation of experiments with an agent like ouabain, but the results with concentrations over a narrow range in the region of 10^{-7} M are suggestive of a relatively specific effect.

Changes within the physiological range in the sodium concentration of cerebrospinal fluid

A brief consideration of the physiology of cerebrospinal fluid and the blood–brain barrier

Discussion of the physiological effects of intraventricular infusions, as well as those of intracarotid infusions, is facilitated by first recounting some general features of the physiology of CSF and the blood–brain barrier as they emerge from the writings of Davson (1967), Rapoport (1975, 1976), Pappius (1975), Bradbury (1975), Fenstermacher and Patlak (1975), Cserr (1975) and Knigge et al. (1975). Some speculations on unresolved questions are also mentioned. The barrier between the CSF and brain on the one hand, and the blood on the other, occurs at the choroid plexus, the blood vessels of the brain and subarachnoid space, and the arachnoid membrane which overlies the subarachnoid space. All the barrier sites are characterized by cells connected by rings of tight junctions that restrict intercellular diffusion. The effect is like a continuous cell layer since solute exchange can take place mainly by the transcellular route.

Stretching of epithelial cells connected by tight junctions deforms junctional fibrillar patterns and increases junctional permeability. This can occur as a result of hydrostatic pressure but tight junctions can also be deformed by osmotic stress since the cells shrink and pull on the junctional regions. Sufficiently hypertonic solutions whether applied to the surface of the brain or perfused through the carotid artery may reversibly open the blood–brain barrier. These techniques, with demonstration of unilateral marking of the brain with Evans blue have, however, involved occlusion of the external carotid and of the common carotid artery caudal to the injection site (Rapoport 1975). This allows very much higher concentrations to be achieved by perfusion than the alterations within feasible physiological range caused by slow infusion of hypertonic solutions into the blood flowing up the common carotid artery. In, for example, sheep experiments where cerebral blood [Na] is increased by 8–20 mmol/l (Chapter 20), intracarotid infusion of 1–4 M sodium chloride is made at 1–2 ml/min into a carotid flow of 300–800 ml/min. There is, however, clinical evidence that the barrier may open to some extent in conditions of hyperosmolarity (Rapoport 1975).

The blood–brain barrier is a regulatory series of interfaces between blood and nervous system, and complicated and multiple mechanisms exist at the choroid plexus for active transport and facilitated diffusion. Rapoport (1976) remarks that a 'blood–CSF' barrier is distinguished at times from a 'blood–brain barrier' in order to explain why intravascular substances enter CSF and brain extracellular fluid (ECF) at different rates (Davson 1967). However, these kinetic differences can be interpreted by taking into account gross anatomical relations between CSF, blood and brain. The ependymal surface of the ventricles and the pial–glial surface of the brain do not impede exchange between CSF and brain ECF and do not represent a sub-barrier. Thus the distinction is not very useful.

Approximately 70% of CSF is produced by secretion at the choroid plexus, which is located on the roofs of the third and fourth ventricles and in the lateral ventricles. Secreted fluid mixes with ventricular CSF and then circulates. The remaining 30% of CSF is derived from the capillary bed of the brain or by metabolic water production. Normally, the hydrostatic pressure returns CSF to blood through one-way valves or arachnoid villi which protrude from the subarachnoid space into lumens of large dural sinuses and from the subarachnoid space of spinal cord into venous blood (Fig. 24–10). Flow of CSF begins when pressure exceeds venous pressure by 3–6 cm of water, which overcomes interfacial tension between collapsed tubules and overlapping surface endothelium.

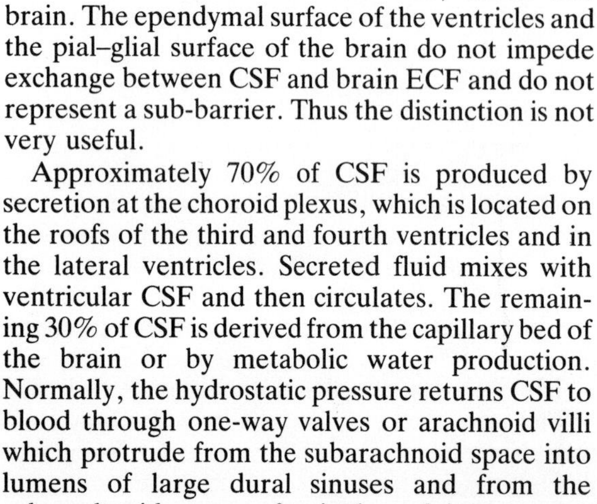

Fig. 24-10. Cross-section of the dorsomedial region of the cerebral hemisphere at the sagittal fissure showing meninges and subarachnoid space. Two arachnoid villi are shown protruding into venous blood of the superior sagittal sinus. The left one is closed and has an overlapping surface layer of endothelial cells. The right one is open to CSF flow from the subarachnoid space. Endothelial cells are partly separated at the surface, and tubular passageways for bulk flow are evident. (From Rapoport 1976, with permission. Adapted from Weed 1923.)

Metabolic products, drugs and other substances which diffuse into the brain from blood are excreted by the 'sink' action of CSF (Davson et al. 1962). The stream of CSF sweeping over ventricles and the pial surface of the brain drains away solutes through arachnoid villi into venous blood. Furthermore some substances, like penicillin, or transmitters such as serotonin and noradrenaline, are removed from CSF rapidly by the choroid plexus (Bárány 1972). If a substance enters the brain slowly it may never reach a plasma/CSF ratio of unity because of the 'sink' action of flowing CSF. Fenstermacher and Patlak (1975) also emphasize the importance of movement of some molecules between brain ECF and blood. If this occurs at an appreciable rate, the tissue profile will be strongly affected by the 'sink' action of the blood. If the rate of exchange across a capillary complex is relatively high, a particular test substance which is being perfused through a part of the CSF system and enters the adjacent central nervous tissue may only penetrate into the tissue for a very short distance. As will be recounted further on, where reduction of CSF [Na] is shown to have striking behavioural effects, its reduction by mannitol or sucrose is found to have quite different effects from those produced by infusion of distilled water or glucose, where uptake across capillaries or cells may greatly reduce any effects at small distance from the ventricular wall.

The 30% of CSF which comes from extrachoroidal sources may contribute to bulk volume movement within the brain parenchyma under normal conditions. However, as shown by Fenstermacher and colleagues (1970) in cerebral grey matter, concentration profiles of tracers placed in ventricular fluid do not appear to be influenced by bulk flow. The concentration profile of extracellular solute like inulin and sucrose, as it diffuses into the brain from CSF, is a complementary error function of distance and does not indicate any hindrance to diffusion at the ependymal level.

The maximum distance for diffusion between CSF and any cerebral site in adult man is 15 mm, and less in smaller animals. Diffusion can take place through extracellular space, which equals about 18% of brain wet weight (Fenstermacher and Rall 1973).

Bulk flow through the brain may be augmented in pathological and experimental manoeuvres. A pressure gradient from CSF to brain capillaries which occurs in experimental hydrocephalus, and when brain is dehydrated by intravenous injection of hypertonic solution, accelerates solute move-

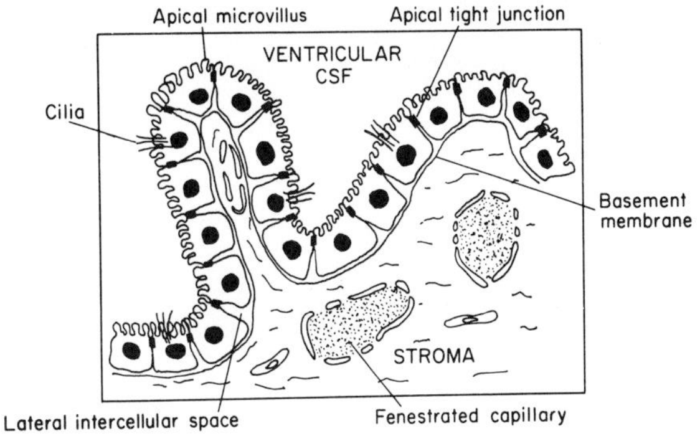

Fig. 24-11. Features of a choroid plexus villus. The villus is covered by a single layer of cuboidal epithelium, with apical microvilli facing on ventricular CSF and the basal cell surface placed on a basement membrane. Tight junctions connect the apical regions of the cell membranes. The lateral intercellular spaces swell and shrink, probably as a function of choroidal secretion rate. The villus separates choroidal stroma from CSF. Capillaries within stroma have a fenestrated endothelium. (From Rapoport 1976, with permission. Adapted from Miller and Woollam 1962, and Wright 1972.)

ment within the brain (Cushing 1914; Wislocki and Putnam 1921; Reulen and Kreysh 1973). Morphological conduits for bulk flow through brain may be the 'Virchow–Robin' spaces that surround intracerebral arterioles and venules. These are continuous with CSF and are bounded on one side by vascular basement membrane and on the other by basement membrane of brain glia (Brightman 1965; Rapoport 1975). Tracers introduced into CSF penetrate these spaces and eventually surround brain capillaries.

Passive solute exchange at choroid plexus between blood and CSF is restricted by tight junctions (Fig. 24-11). The choroid plexus is a villous structure with frond-like processes extending into the CSF. Proteins and small hydrophilic molecules pass easily through the fenestrated endothelia of the capillaries, and the barrier is at the tight junctions between the cuboidal cells. The formation of CSF results primarily from a sodium/potassium exchange pump catalysed by Na,K-ATPase which transports sodium to the ventricular surface of the plexus and potassium in the opposite direction (Fig. 24-12). Water flow is secondary to osmotic gradients set up. The anions chloride and bicarbonate move passively into CSF. The choroid plexus is rich in the enzyme carbonic anhydrase which accelerates the hydration of carbon dioxide to hydrogen and bicarbonate ions (or carbonic acid) within choroid cells and this may facilitate provision of hydrogen or bicarbonate ions to support sodium transport. Inhibition of the enzyme reduces sodium transport, and CSF formation is reduced also by ouabain. There is some controversy on the role of the hydrostatic

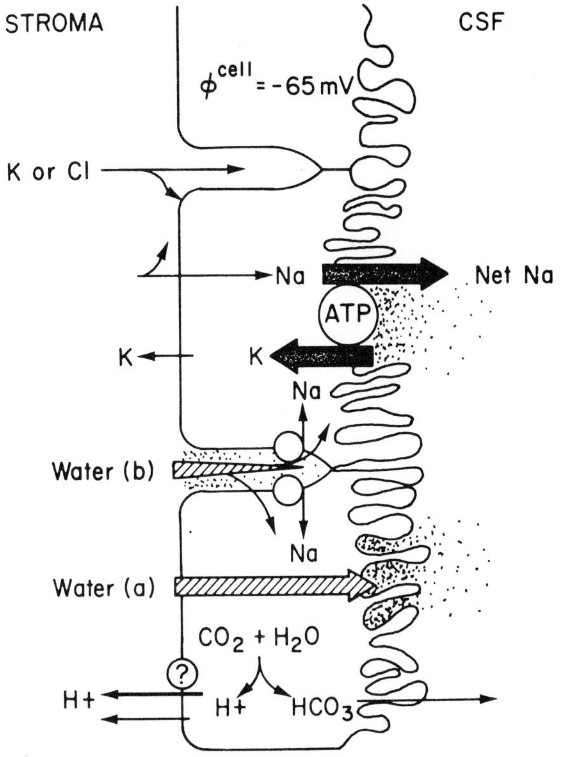

Fig. 24-12. A model for ionic and water movement at choroid plexus. The sodium/potassium pump is placed at the apical villus surface of choroidal epithelium and drives sodium into the CSF and potassium into the cell. It may operate also at the lateral intercellular spaces. Sodium transport into the CSF is optimized if it is accompanied by movement of bicarbonate and chloride ions in the same direction and of hydrogen and possibly potassium in the opposite direction. Bicarbonate and hydrogen ions are formed within the cell by hydration of carbon dioxide (bottom), as catalysed by carbonic anhydrase. Ions may move either through or between the cells. Water secretion arises from standing osmotic gradients set up by the sodium pump either at apical microvilli or within the lateral intercellular spaces. Local differences in dot density represent differences in osmolalities. The electrical potential between CSF and stroma is negligible, and the inside potential of the cell is −65 mV. (From Rapoport 1976, with permission. Modified from Wright 1972, and Woodbury 1971.)

pressure difference between capillary blood and CSF in determining the water component of CSF. Changes in choroid blood flow have been shown to alter CSF formation. Choroidal transport maintains [K] relatively invariant in CSF in the face of large variation in plasma [K].

There are regions in the brain where capillaries with continuous endothelia are replaced by capillaries with fenestrated endothelia and thus proteins and solutes irrespective of lipid solubility can transfer across these specific sites. There are regions outside the brain—at terminals of peripheral nerves, sensory ganglia and olfactory epithelium, and the optic nerve where it penetrates the sclera—which can provide access to brain for proteins and infective agents. In the special regions of the brain where cells have direct contact with blood, and may secrete hormones or contain hormonal or chemoreceptors, the fenestrated capillaries allow exchange of protein and other solutes. The solutes do not remain localized in the special regions such as pineal, median eminence, area postrema, wall of optic recess, etc., but diffuse into surrounding brain parenchyma (Hoffman and Olszewski 1961). The path of entry is, however, tortuous, because the ependymal cells which overlie the region are joined by tight junctions that restrict intercellular diffusion (Reese and Brightman 1968; Knigge and Silverman 1972).

Knigge et al. (1975) remark that ependyma lining the ventricular cavities exhibit unusual regional modifications and in specific circumventricular sites are organized to provide a neurovascular link with the CSF. The fluid milieu of the ventricular cavities must communicate here with the peripheral environment in significantly different ways. The ependyma (tanycytes) of the median eminence represent the anatomical compartment which unites CSF of the third ventricle with the pituitary portal vasculature. From the apical surface bathed in CSF, through the median eminence to their basal end-feet on the neural basement membrane of the portal perivascular space they cover 400–700 μm. The ventricular surface of them, active with microvilli and protrusions, is indicative of CSF–cell exchange (Fig. 24-13). High concentrations of ribosomes and of endoplasmic and Golgi material suggest active protein synthesis. The basal end-feet contain semi-dense pleomorphic granules, and neurofilaments and tubules are packed densely in linear array in the cells. The end-feet occupy 40%–50% of the area of the portal contact surface. The evidence indicates substances move from CSF to pituitary portal blood. Since tight junctions at the ependyma prevent interstitial migration, it can be inferred that movement from CSF to blood occurs through ependyma. For example, when radiolabelled thyroid releasing hormone (TRH) is infused into the third ventricle, equilibrium in CSF is reached in 15–20 min. Uptake into the median eminence follows, with the ratio of TRH between CSF and intracellular space of median eminence reaching 0.35–0.60, and uptake is demonstrable in the pituitary soon after beginning the infusion. The process of uptake is greatly reduced if thyroxine is infused into the third ventricle concurrently with TRH. ^{3}H-labelled dopamine infused into the third ventricle also enters the ependyma of the median eminence.

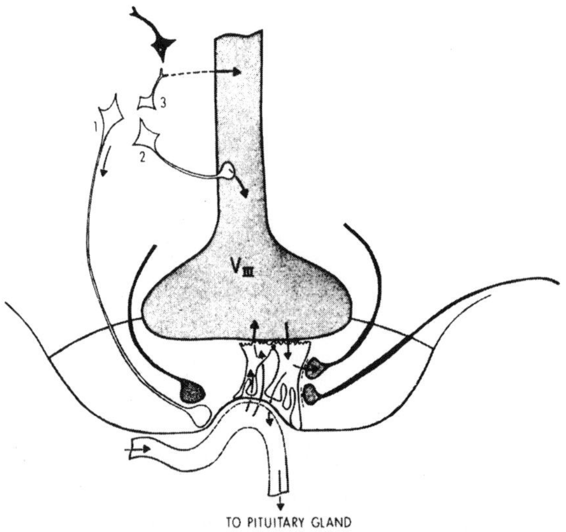

Fig. 24-13. Schema of potential arrangements of organization of releasing-factor-producing neurones in the CNS and delivery of hormones to the pituitary gland. 1, Neurones of hypothalamus synthesize releasing hormones and deliver them by axoplasmic flow to nerve terminals in the median eminence. 2, Releasing-hormone-synthesizing neurones project axons to the third ventricle (V_{III}) wall and release hormone into it. Hormone is transported through ependyma of the ventricular recess to median eminence. 3, Releasing-hormone-synthesizing neurones project upon other neurones with the hormone acting as a regulator neurotransmitter. (From Knigge et al. 1975, with permission.)

These anatomical considerations probably provide the basis of an explanation for a seeming paradox mentioned in Chapter 7 in relation to the fact that angiotensin II does not cross the blood–brain barrier. No rise in CSF concentration is seen with systemic angiotensin II infusion producing high physiological blood level. Intracarotid angiotensin II infusion in sheep causes rapid drinking, an effect blocked by an immediately preceding intraventricular infusion of the competitive antagonist saralasin. Conversely intraventricular infusion of angiotensin II in the sheep causes rapid drinking, an effect blocked by commencing immediately beforehand an intracarotid infusion of saralasin (Fig. 24-14). The explanation could reside in conjoint access to the circumventricular organ of the two substances, with the competitive antagonist blocking agonist action within the organ or close nearby. It is not necessary to invoke the explanation that pressor effects open the blood–brain barrier (Ganten and Speck 1978).

Except via the special regions mentioned, protein entry to CSF is severely restricted, and proteins in CSF are at only 0.4% of their plasma levels. Immunoglobulins IgG and IgA, with molecular weights of approximately 180 000, are

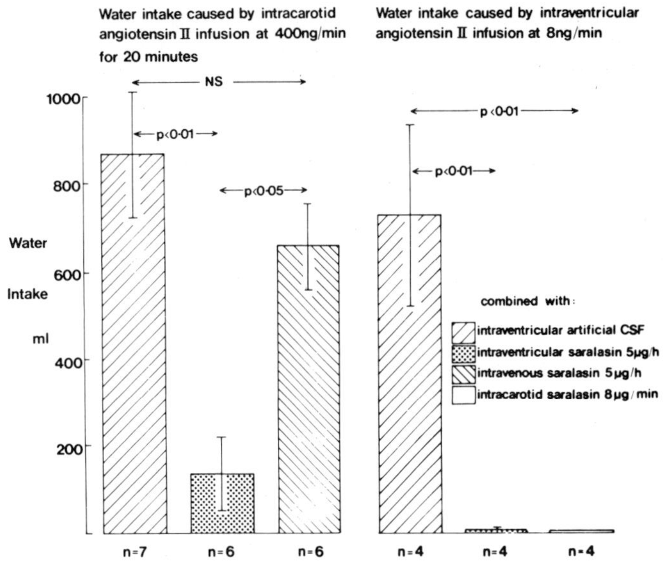

Fig. 24-14. Left panel shows the effective blockade of water drinking in response to intracarotid angiotensin II infusion (400 ng/min for 20 min) by infusion of saralasin acetate (5 μg/h) into a lateral ventricle starting 45 min before the intracarotid infusion. Intravenous infusion of saralasin at a similar rate had no effect. The right panel shows the effective blockade, by infusion of saralasin at 5 μg/h into a lateral ventricle or into a carotid artery at 8 μg/min, of water drinking in response to infusion of angiotensin II into a lateral ventricle at 8 ng/min for 20 min. The infusions of saralasin commenced 15 min before the angiotensin II infusion.

present at only 0.2%–0.4% of their plasma level. Restriction of entry is protective in that elevation of interstitial protein concentration reduces ability to avoid oedema. Rapid release of protein antigens from the central nervous system could provoke peripheral autoimmune mechanisms. Maturation of the peripheral immune system precedes appearance of some brain antigens (Caspary 1968). The equilibration time of immunoglobulins between CSF and blood is 4 days in man (Barlow 1972).

Bradbury (1975) in considering the ontogenesis of astrocytes has put forward a hypothesis in relation to function which is relevant to the consideration here of control of ionic composition of CSF. He remarks that the orientation of the astrocyte with respect to its capillary is the same as that of the choroid epithelium to its capillary plexus. Similarly the perineuronal interstitial fluid relates to the astrocyte in the way that the ventricular CSF does to the choroid epithelium. Bradbury proposes that if the astrocyte were indeed a bipolar epithelial cell it is logical to assign the function of ion secretion to it as Pappenheimer has done on the basis of other evidence (Pappenheimer 1969, 1970). It could be like the epithelial cells of frog's skin, with apical and basal functions. The apical processes investing and even isolating neurones, groups of neurones and synaptic regions (Peters and Palay 1965) might be secreting interstitial fluid on the basis of sodium pumping. Whereas there is little net secretion of cerebral interstitial fluid as evidenced by the small brain contribution to CSF, it is possible that fluid formed in the vicinity of the apical processes might flow under the influence of a hydrostatic pressure gradient through interstitial spaces to the peri-

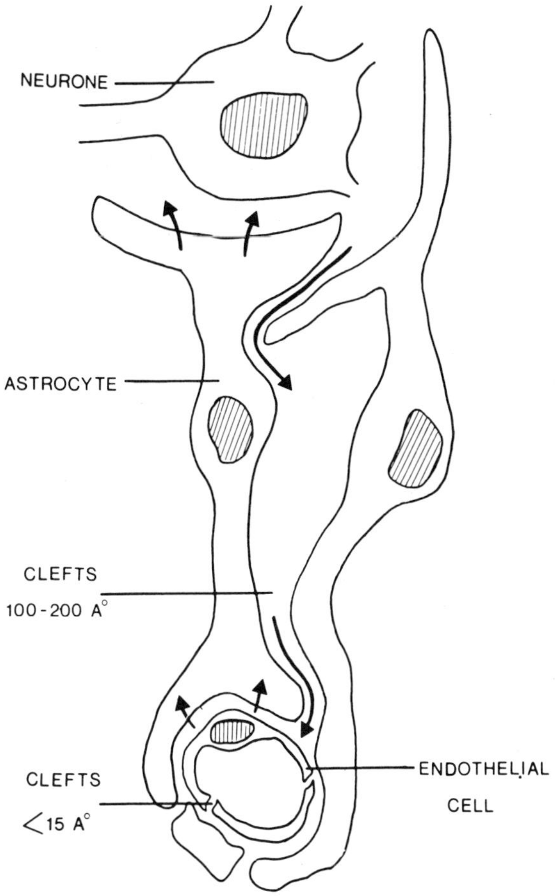

Fig. 24-15. Diagrammatic representation of two astrocytes, their basal processes ensheathing a capillary and their apical processes making contacts with a neurone. The arrows indicate the hypothesized direction of net fluid movement as mediated by secretory processes in the astrocytes. (From Bradbury 1975, with permission.)

vascular region for reabsorption into the basal processes (Fig. 24-15). This idea allows a function for the astrocytic end-feet, as distinct from all barrier and active properties being sited at the capillary wall. Further, the concentration of several ions which are remarkably stable in CSF (potassium, calcium, magnesium and hydrogen) is, on good evidence, probably paralleled by comparable homeostasis of ionic composition of cerebral interstitial fluid (Pappenheimer 1967; Cohen et al. 1968). Such a control would be much simpler to effect in a flowing system than a static one. If entry of a solute from astrocyte to cerebral interstitial fluid were constant irrespective of its plasma concentration or were linked in a constant way to sodium influx, a constant concentration in the interstitial fluid would be aided. The fluid would be continually drained and renewed by fresh passage through the astrocytes. Any solute which did not pass back into the astrocyte through its basal membrane would tend to accumulate outside the capillary wall, and build up a concentration gradient favouring its diffusion back into blood. Bradbury concludes by suggesting the model explains the extreme liability of astrocytes to swell in the face of cold, hypoxia and other stresses impairing the sodium pump. The ready swelling would be explained by a rapid gain of sodium and chloride in intracellular fluid when pumping is reduced.

Intraventricular infusion experiments on sodium-deficient sheep

These experiments (Weisinger et al. 1979 and in press) were aimed at accessing the neuronal systems subserving salt appetite by a direct route without disturbance in the confident, trained animal. The results were clear-cut in showing for the first time that ionic changes within the physiological range could influence sodium appetite in predictable fashion far more rapidly than would have been expected from the aggregate of evidence accumulated hitherto.

Forty-two crossbred Merino and Corriedale ewes, 30–40 kg body weight, were used. Before experimentation all of the animals were surgically prepared with a guide tube (17-g stainless steel needle, 42 mm long) implanted 4–15 mm above each lateral brain ventricle. Sixteen of the sheep had a unilateral parotid fistula. On the day of the infusion a probe of the appropriate length was inserted through the guide tube into the lateral brain ventricle (see Fig. 13-2). The infusion rate was 1 ml/h and, unless specified otherwise, was begun 60 min before and continued until the end of the 120-min sodium access period. CSF (1–2 ml) from the ipsilateral and also occasionally the contralateral ventricle was obtained, and analysed for sodium and potassium concentrations and osmolality. The sodium-deficient sheep were studied by the bar-press method. The bar-press lever delivered 15 ml of 600 mmol/l sodium bicarbonate (=9 mmol sodium). Access to water was either from a water bin on the side of the cage, or via a second lever in the cage that delivered 50 ml with each bar-press. Sodium deficiency was produced by parotid saliva loss for 22 h without access to sodium bicarbonate.

Hypertonic saline-CSF, hypertonic mannitol-CSF and isotonic saline-CSF at 1 ml/h

The preparation of artificial sheep CSF was similar to that used by Mouw, Abraham and McKenzie (1974a). Table 24-2 shows the components of the different artificial CSF solutions infused. The cumulative number of bar-press deliveries over

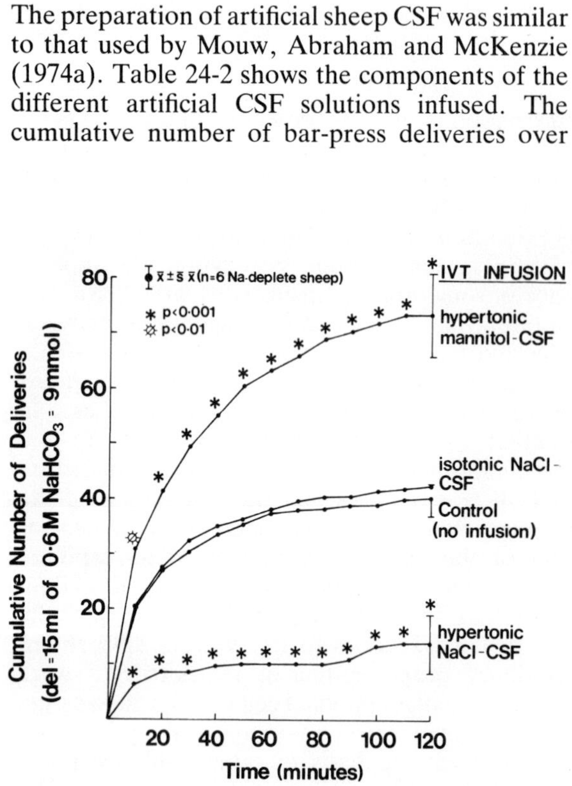

Fig. 24-16. Effect of i.v.t. (lateral ventricular) infusion (1 ml/h) of isotonic saline-CSF, hypertonic 0.5 M saline-CSF or 0.7 M mannitol-CSF on bar-pressing for 600 mM sodium bicarbonate (15 ml per delivery=9 mmol of sodium). The sheep (n=6) were sodium-deficient as a result of 22 h of parotid salivary loss. The 3-h infusion started 1 h before 2 h of access to sodium via bar-pressing. Cumulative mean (and mean s.e.m.) deliveries over the 2-h sodium access period for the baseline (i.e. the no-infusion day preceding each experimental day; n=18 observations on six sheep) and the different i.v.t. infusion conditions are shown. A two-way analysis of variance (repeated measures design) and subsequent t-tests were used to determine the significance of the difference between baseline and experimental values for each 10-min interval.

Table 24-2. Composition of the artificial CSF solutions infused into the lateral ventricle (mmol/l).

	Na	K	Cl	HCO_3	HPO_4	Ca	Mg	Mannitol	Sucrose	Glucose	Urea
Natural sheep CSF[a]	150.7	2.8	135	23	×	1.22	0.89	×	×	×	×
Isotonic saline-CSF	151	2.8	157.5	0	0.5	1.2	0.9	0	0	0	0
0.7 M mannitol-CSF	151	2.8	157.5	0	0.5	1.2	0.9	700	0	0	0
0.7 M glucose-CSF	151	2.8	157.5	0	0.5	1.2	0.9	0	0	700	0
0.7 M urea-CSF	151	2.8	157.5	0	0.5	1.2	0.9	0	0	0	700
0.7 M sucrose-CSF	151	2.8	157.5	0	0.5	1.2	0.9	0	700	0	0
Hypertonic saline-CSF	501	2.8	507.5	0	0.5	1.2	0.9	0	0	0	0
0.3 M mannitol-CSF	1.0	2.8	7.5	0	0.5	1.2	0.9	300	0	0	0
0.27 M mannitol-CSF	1.0	2.8	7.5	0	0.5	1.2	0.9	270	0	0	0
0.14 M mannitol-CSF	76	2.8	82.5	0	0.5	1.2	0.9	140	0	0	0
0.1 M mannitol-CSF	90	2.8	97.5	0	0.5	1.2	0.9	100	0	0	0

[a]From Mouw et al. (1974).
×, not analysed.

120 min for the various infusion conditions is shown in Fig. 24-16. Mean salivary sodium loss over the 22 h in the experimental series detailed in the succeeding sections ranged from 360 to 420 mmol. Under baseline conditions (i.e. the day before each infusion), the sodium-deficient sheep bar-pressed for and consumed 40.7±2.3 deliveries, or 336 mmol of sodium, during the 120-min sodium access period (n=18 observations on six sheep). Compared with this baseline, intraventricular (i.v.t.) infusion of hypertonic (500 mM) saline-CSF decreased sodium intake to 13.8±8.1 deliveries, or 124 mmol of sodium (P<0.001, n=6); i.v.t. infusion of hypertonic 0.7 M mannitol-CSF increased sodium intake to 73.5±7.0 deliveries, or 662 mmol of sodium (P<0.001, n=6); and infusion of isotonic saline-CSF did not significantly alter intake. The difference in cumulative number of deliveries between the baseline and either the hypertonic saline-CSF or the hypertonic mannitol-CSF infusion condition was significant within 10 min of the start of the sodium access period (P<0.001 and P<0.01, respectively).

It can be said that having obtained the result of a reduction in appetite with 500 mM saline-CSF, the 0.7 M mannitol-CSF infusion was made to produce comparable elevation of the osmotic pressure of CSF, and the result of almost a doubling of sodium intake associated with a measured reduction in CSF [Na] was unexpected.

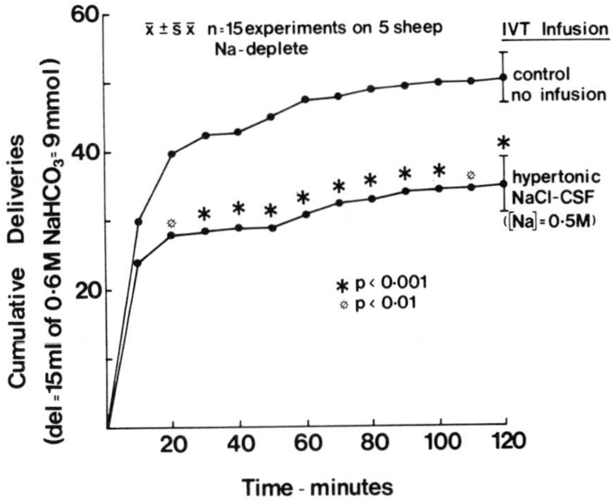

Fig. 24-17. Effect of i.v.t. infusion of hypertonic 0.5 M saline-CSF (n=15 observations on five sodium-deficient sheep) on bar-pressing for sodium bicarbonate solution (as in Fig. 24-16 except that water was available during infusion). Three hours of infusion were begun 1 h before the sodium access period. Cumulative mean (and mean s.e.m.) deliveries over the 2-h sodium access period for baseline (n=15 observations on five sheep) and i.v.t. infusion condition are shown. Statistical analysis similar to that used in Fig. 24-16.

Hypertonic saline-CSF; water available during infusion period

This repeated the previous experiment with 500 mM saline-CSF except that water was available during the infusion (Fig. 24-17). Under baseline conditions the sheep bar-pressed for and consumed 50.4±3.8 deliveries, or 454 mmol sodium, during the 120-min sodium access period (n=15 observations on five sheep). I.v.t. infusion of hypertonic saline-CSF significantly decreased sodium intake to 25.0±4.6 deliveries, or 315 mmol of sodium (P<0.001, n=15), the difference being significant within 20 min of the start of the access period (P<0.01). Thus sodium intake was significantly reduced by i.v.t. infusion of 1–2 mmol of sodium. By contrast, in our experiments where animals gave themselves involuntarily an intravenous injection of hypertonic saline (Chapter 20), more than 140 mmol of sodium were infused before there was a significant reduction in appetite (Weisinger et al. 1978).

Under control conditions, water intake during the 60 min before sodium access was 1.9±0.6 deliveries, or 95 ml of water (n=15 observations on five sheep). Compared with this baseline, infusion of hypertonic saline-CSF caused a marked increase in water intake during the first hour of infusion (P<0.001, n=15): there were 26.8±3.0 deliveries, or 1340 ml of water.

Thus sodium intake was highly significantly reduced whether or not water was concurrently available. But the thirst presumably produced when no access to water was allowed resulted in a greater reduction of intake of 600 mM sodium bicarbonate, which probably was relatively aversive under those conditions. A large amount of data cited earlier shows clearly that the intake of 1–2 litres of water when it was available would have, of itself, negligible effect on salt appetite in an animal with a capacious rumen like the sheep.

Hypertonic 0.7 M mannitol-CSF; infusion begun 60, 15 or 0 min before sodium access period

The cumulative number of deliveries over 120 min for the various conditions is shown in Fig. 24-18. Under baseline conditions the sheep bar-pressed for and consumed 45.3±3.0 deliveries, or 408 mmol of sodium, during the 120-min sodium access period (n=15 observations on five sheep). Infusion of hypertonic mannitol-CSF increased (P values <0.001, n=5) sodium intake to 82.8±12.2 deliveries (745 mmol of sodium), 73.6±6.9 deliveries (662 mmol) and 79.2±9.7 deliveries (713 mmol) for infusions begun 60, 15 and 0 min before the sodium access period, respectively. The difference in cumulative number of deliveries between the baseline and the hypertonic mannitol-CSF infusion conditions was significant at 20, 10 and 40 min from the start of the access period for infusions begun 60, 15 and 0 min before this period respectively. Water intake during infusion of hypertonic mannitol-CSF (n=15 experiments on five sheep) was greater than 100 ml on four occasions only (in two sheep). In the previous group of experiments water was not available during the infusion (see p. 479), and the result with hypertonic mannitol-CSF was no different from these experiments. Water was available during infusion in all subsequent experiments.

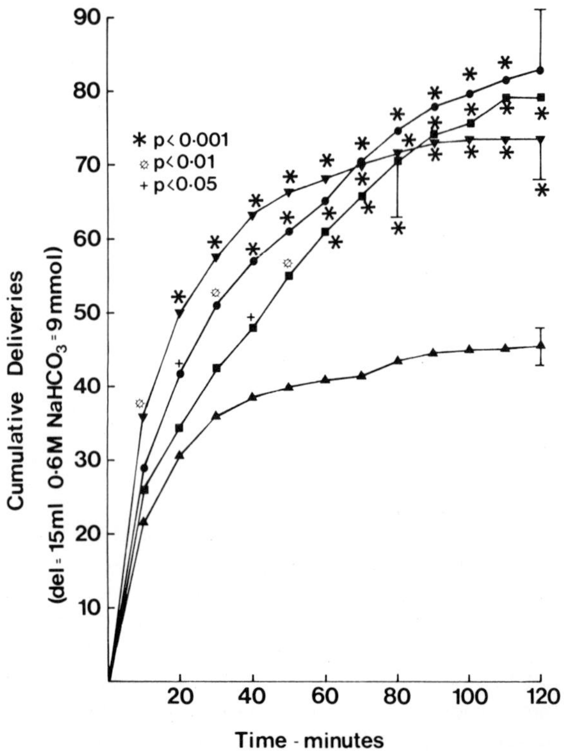

Fig. 24-18. Effect of i.v.t. infusion of 0.7 M mannitol-CSF on bar-pressing for sodium bicarbonate solution in sodium-deficient sheep (n=5). Infusion begun 60 (circles), 15 (inverted triangles) or 0 (squares) min before the start of the sodium access period. Cumulative mean (and mean s.e.m.) deliveries over the 2-h access period for the baseline (upright triangles, n=15 observations on five sheep) and the different i.v.t. infusion conditions are shown. Statistical analysis similar to that used in Fig. 24-16.

Hypertonic 0.7 M mannitol-CSF, hypertonic 0.7 M glucose-CSF, hypertonic 0.7 M urea-CSF, hypertonic 0.7 M sucrose-CSF or isotonic saline-CSF

As Fig. 24-19 shows, under baseline conditions the sheep bar-pressed for and consumed 40.9±3.0 deliveries, or 368 mmol of sodium, during the 120-min sodium access period (n=20 observations on five sheep). Hypertonic mannitol-CSF significantly increased sodium intake to 75.6±17.3 deliveries, or 680 mmol of sodium (P<0.001, n=5), and hypertonic urea-CSF significantly increased it to 58.2±11.5 deliveries or 523 mmol of sodium (P<0.05, n=5). Infusion of either isotonic saline-CSF (n=5) or hypertonic glucose-CSF (n=5) did not significantly alter sodium intake. The effect of either hypertonic mannitol-CSF or hypertonic urea-CSF infusion was significant within 10 min of the start of the sodium access

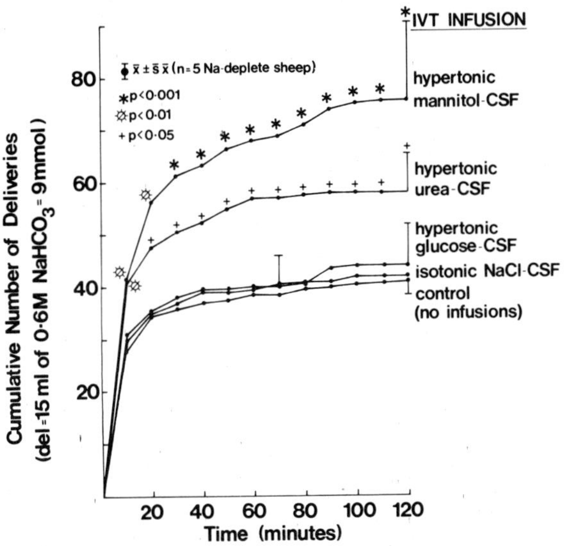

Fig. 24-19. Effect of i.v.t. infusion of isotonic saline-CSF, 0.7 M mannitol-CSF, 0.7 M urea-CSF or 0.7 M glucose-CSF on bar-pressing for sodium bicarbonate solution in sodium-deficient sheep (n=5). Three hours of infusion were begun 1 h before the start of the sodium access period. Cumulative mean (and mean s.e.m.) deliveries over the 2-h access perod for the baseline (n=20 observations on five sheep) and the different i.v.t. infusion conditions are shown. Statistical analysis similar to that used in Fig. 24-16.

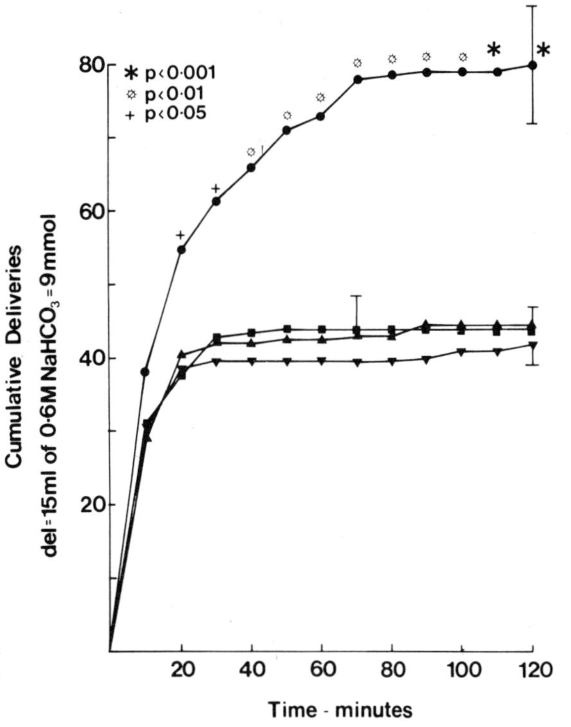

Fig. 24-20. Effect of i.v.t. infusion of isotonic saline-CSF (upright triangles, n=12 observations on four sheep) or 0.7 M sucrose-CSF (circles, n=8 observations on five sheep) on bar-pressing for sodium bicarbonate solution in sodium-deficient sheep. Three hours of infusion were begun 1 h before the sodium access period. Cumulative mean (and mean s.e.m.) deliveries over the 2-h access period for the baselines (inverted triangles, squares) and the different infusion conditions are shown. Statistical analysis similar to that used in Fig. 24-16.

period. Also, in another series, infusion of hypertonic sucrose-CSF (Fig. 24-20) significantly increased bar-pressing for sodium (i.e. baseline=43.9±4.6 deliveries or 395 mmol of sodium vs infusion=79.9±8.2 deliveries or 719 mmol of sodium: P<0.001, n=8 observations on five sheep). Recently, Dr Richard Weisinger and Professor Alex Muller of Geneva have shown that 0.7 M L-glucose-CSF also doubled sodium intake whereas 0.7 M mannose-CSF or 0.7 M fructose-CSF did not influence intake.

Under baseline conditions water intake during the 60 min before sodium access was 1.2±0.3 deliveries or 60 ml of water (n=20 observations on five sheep). Compared with this baseline, water intake during the first 60 min of infusion was significantly increased by either hypertonic mannitol-CSF (P<0.001, n=5) or hypertonic urea-CSF (P<0.05, n=5), intakes being 10.4±3.9 deliveries (520 ml of water) and 5.4±1.7 deliveries (270 ml of water), respectively. Water intake was not significantly altered by infusion of either isotonic saline-CSF or hypertonic glucose-CSF. I.v.t. infusion of hypertonic sucrose-CSF increased water intake from 6.2±2.5 deliveries, or 310 ml, to 14.9±3.2 deliveries or 745 ml (P<0.01, n=8 observations on five sheep).

Composition of cerebrospinal fluid

Table 24-3 shows the [Na] and [K] and osmolality of CSF obtained under baseline conditions (i.e. no infusion) or after the different i.v.t. infusions. Under baseline conditions CSF [Na] was 149.5±0.6 mmol/l. Compared with baseline, CSF [Na] was markedly increased by infusion of hypertonic saline-CSF to 174±4.1 mmol/l, and significantly decreased by infusion of hypertonic mannitol-CSF, hypertonic glucose-CSF, hypertonic urea-CSF and hypertonic sucrose-CSF. Hypertonic glucose caused a smaller change at 60 min (Table 24-3).

Compared with baseline, CSF [K] was decreased by infusion of hypertonic saline-CSF (Table 24-3), hypertonic mannitol-CSF, hypertonic glucose-CSF and hypertonic sucrose-CSF. CSF osmolality was significantly increased, compared to baseline, by infusion of hypertonic saline-CSF, hypertonic mannitol-CSF, hypertonic glucose-CSF, hypertonic urea-CSF and hypertonic sucrose-CSF. Thus a comparable decrease in CSF [K]

Table 24-3. Effect of intraventricular infusions on CSF composition of sodium-deficient sheep.
A. Samples obtained from ipsilateral lateral ventricle at end of 60-min infusion (i.e. before sodium intake) or on day of no infusion (Mean±s.e.m., n=4)

i.v.t. infusion	CSF composition [Na] (mM)	[K] (mM)	Osmolality (mosmol/kg)
None	149.5±0.6	3.0±0.04	294±1
Hypertonic saline-CSF (water available)	162.2±1.6***	2.8±0.05**	318±5**
Hypertonic saline-CSF (water not available)	161.0±3.0**	2.9±0.06**	326±7***
0.7 M mannitol-CSF	133.0±1.2***	2.7±0.05***	342±6***
0.7 M urea-CSF	138.5±3.5**	2.9±0.06***	358±7***
0.7 M glucose-CSF	142.0±1.8*	2.8±0.03**	338±8***

B. Samples obtained from ipsilateral lateral ventricle at end of 180-min infusion (i.e. post-ingestion) (Mean±s.e.m. (n))

i.v.t. infusion	CSF composition [Na] (mM)	[K] (mM)	Osmolality (mosmol/kg)
None[a]	149.5±0.6 (4)	3.0±0.04 (4)	294±1 (4)
Isotonic saline-CSF	151.4±0.2 (8)	2.8±0.03 (8)*	296±3 (4)
Hypertonic saline-CSF	174.4±4.1 (14)***	2.6±0.04 (14)***	340±7 (8)*
0.7 M mannitol-CSF	134.2±2.4 (9)**	2.5±0.04 (9)***	354±4 (5)*
0.7 M urea-CSF	136.4±2.1 (5)*	3.0±0.04 (5)	353±6 (5)*
0.7 M glucose-CSF	139.4±2.4 (5)	2.6±0.07 (5)***	354±10 (5)*
0.7 M sucrose-CSF	135.9±3.0 (7)*	2.5±0.07 (7)***	378±26 (7)***

***$P<0.001$, **$P<0.01$, *$P<0.05$.
[a]Values from part *A*.
In *A* statistical analysis was by two-way analysis of variance (repeated measures design) and subsequent *t*-tests; in *B* data were analysed by *t*-test.

and an increase in CSF osmolality were common to conditions where sodium appetite was nearly doubled or was reduced by one-third to two-thirds.

Control experiments with change of sodium status

Because of the importance of these results it was formally ratified on the same animals that changes in sodium status would produce quantitatively appropriate changes in sodium appetite. Thus it was shown (Fig. 24-21) that giving 1 litre of water by rumen tube 3–4 h before access to sodium solutions on three consecutive days did not influence sodium appetite whereas giving 600 mmol of sodium bicarbonate in 1 litre on the fourth day caused a large reduction (43.4±4 deliveries or 391 mmol of sodium (n=7) vs 5.9±2.9 deliveries or 53 mmol of sodium (n=7): $P<0.001$). After 46 h of sodium deprivation (Fig. 24-22) instead of 22 h the animals bar-pressed for 77.6±7.0 deliveries, or 698 mmol of sodium ($P<0.001$, n=7) compared with 40.8±5.1 deliveries or 367 mmol of sodium. With 46 h of depletion the difference from baseline in cumulative bar-presses was significant at 10 min, as it was with sodium loading.

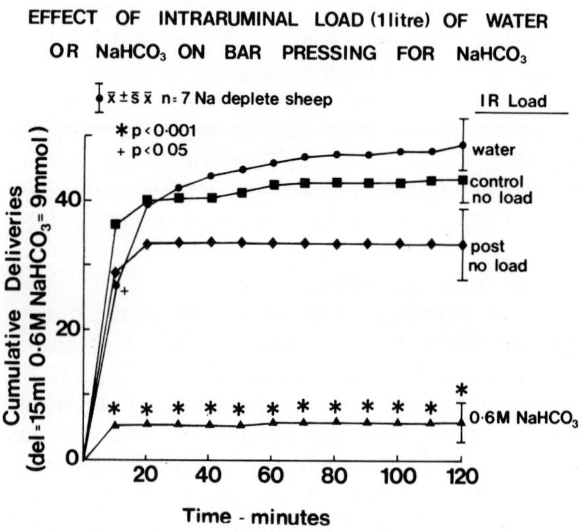

Fig. 24-21. Effect of intraruminal (IR) sodium load on bar-pressing for sodium bicarbonate solution by sheep (n=7) with a parotid fistula deprived of sodium for 22 h. IR load (1 litre) was given 3–4 h before the sodium access period. Cumulative mean (and mean s.e.m.) deliveries over the 2-h access for baseline (squares), IR water (circles, n=21 observations), IR 600 mM sodium bicarbonate (triangles) and post sodium load day (diamonds) are shown. Statistical analysis similar to that used in Fig. 24-16.

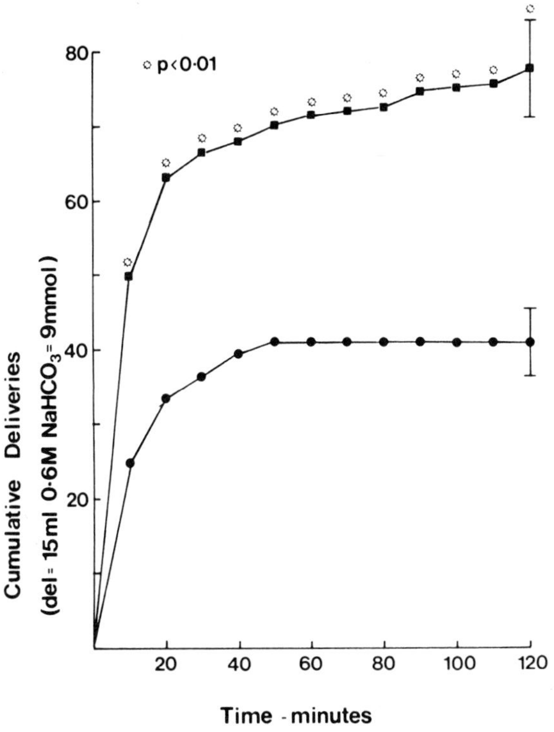

Fig. 24-22. Cumulative mean (and mean s.e.m.) deliveries of sodium bicarbonate over the 2-h sodium access period by sheep (n=7) with a parotid fistula deprived of sodium for 22 h (circles) or 46 h (squares). Statistical analysis by paired t-test.

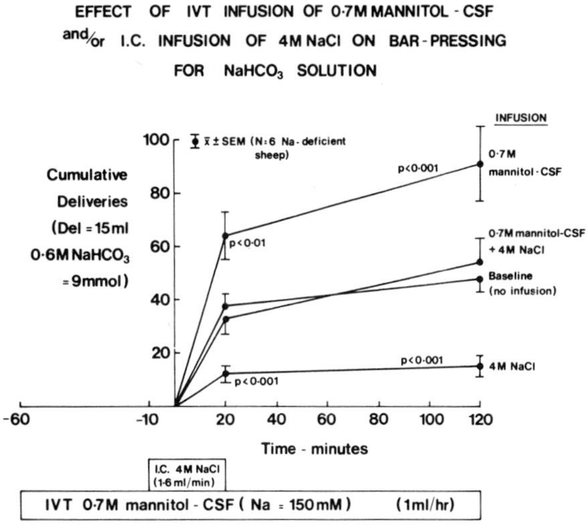

Fig. 24-23. Effect of intracarotid 4 M saline infusion at 1.6 ml/min (contralateral carotid loop occluded) begun 10 min before access to bar-press and continued for 30 min, on the voluntary intake of sodium bicarbonate solution by sodium-deplete sheep. Baseline conditions with no infusion are also shown. For comparison the effect of intraventricular 0.7 M mannitol-CSF infusion at 1 ml/h (−60 to +120 min) on voluntary sodium bicarbonate intake is shown, as is the effect of making this infusion in the experimental situation where intracarotid 4 M saline infusion was made. (Unpublished data of Professor A. Muller and Dr R. Weisinger.)

Depression of salt appetite by 4 M saline infusion counteracted by hypertonic 0.7 M mannitol-CSF infusion

A series of experiments by Dr R. Weisinger and Professor A. Muller have shown that intracarotid infusion of 4 M saline at 1.6 ml/min (contralateral carotid loop occluded) begun 10 min before access to bar-press increased plasma [Na] by 15–20 mmol/l and CSF [Na] by 10–15 mmol/l when measured 10 and 30 min after beginning the infusion. The infusion decreased bar-pressing for salt over 120 min by about 75% (Fig. 24-23). In the same animals intraventricular infusion of 0.7 M mannitol-CSF for 3 h, begun 1 h before access to sodium, nearly doubled salt appetite (Fig. 24-23). If this same infusion were given before and during the intracarotid 4 M saline infusion it prevented an increase in CSF [Na] and abolished the decrease in sodium appetite (Fig. 24-23) consistently caused by systemic 4 M saline infusions.

Intraventricular infusion experiments on sodium-replete sheep

Hypertonic mannitol-CSF

Fig. 24-24 shows the results of this experiment. In sodium-replete sheep which to our knowledge had never experienced a sodium deficit, i.v.t. infusion of hypertonic mannitol-CSF caused a marked increase in voluntary sodium intake. Under baseline conditions, i.e. the 2-day period before each of the infusions, hedonic intake of 600 mM sodium bicarbonate was 251±22 mmol of sodium (n=60 observations on five sheep). On the three consecutive occasions (n=5), spaced a week apart, infusion of hypertonic 0.7 M mannitol-CSF increased sodium intake to 675±97 mmol (P<0.001), 779±50 mmol (P<0.001) and 416±65 mmol (P<0.01), respectively. Intake of sodium returned to baseline level (trial 1) or below (trials 2, 3) on the post-treatment day (Fig. 24-24). Intraventricular infusion of isotonic saline-CSF did not significantly alter sodium intake either before or after the i.v.t. infusions of hypertonic mannitol-CSF (Fig. 24-24).

Water intake was increased a small amount, relative to baseline, during the 60 min before sodium access, in the second (P<0.05) and third

Table 24-4. Effect of intraventricular infusions on CSF composition of sodium-replete sheep. Samples obtained at end of 90 or 180 min of infusion (i.e. post-ingestion) or on day of no infusion (mean±s.e.m., (*n*)).

	CSF composition		
i.v.t. infusion	[Na] (mM)	[K] (mM)	Osmolality (mosmol/kg)
None	149.2±0.6 (15)	2.9±0.04 (15)	294±1 (15)
Isotonic saline-CSF	150.7±0.4 (21)	2.8±0.03 (21)	297±1 (16)
0.7 M mannitol-CSF	128.8±1.4 (24)***	2.5±0.03 (24)***	370±8 (24)***
0.3 M mannitol-CSF	125.1±2.9 (9)***	2.7±0.06 (9)**	316±5 (8)*
0.27 M mannitol-CSF	125.1±2.8 (10)***	2.7±0.07 (10)**	298±1 (10)
0.14 M mannitol-CSF	136.6±2.6 (5)***	2.7±0.02 (5)*	297±3 (5)
0.1 M mannitol-CSF	143.9±1.3 (9)*	2.7±0.03 (9)**	298±1 (9)

***$P<0.001$, **$P<0.01$, *$P<0.05$.
Statistical analysis by *t*-tests.

Table 24-5. Effect of intraventricular infusions on CSF composition. Samples obtained from the infused (I) and contralateral (C) lateral ventricles (mean±s.e.m.).

			CSF composition		
i.v.t. infusion	Length of infusion (min)	Sample from infused or contralateral ventricle	[Na] (mM)	[K] (mM)	Osmolality (mosmol/kg)
0.27 M mannitol-CSF	60	I	134.6±1.9	2.9±0.02	295±2
(*n*=8)		C	138.0±1.4	2.8±0.02	291±1
0.1 M mannitol-CSF	60	I	142.8±2.3	2.7±0.04	299±0.2
(*n*=4)		C	147.0±0.9	2.8±0.04	298±2
0.7 M mannitol-CSF	180	I	127.3±3.8	2.6±1.00	367±20
(*n*=3)		C	130.7±2.3	2.6±0.06	326±15

trial ($P<0.001$) of i.v.t. infusion of hypertonic mannitol-CSF.

Intraventricular infusion of hypertonic mannitol-CSF decreased CSF [Na] and [K] ($P<0.001$) compared with baseline and increased CSF osmolality ($P<0.001$). Intraventricular infusion of isotonic saline-CSF slightly increased CSF [Na] ($P<0.05$) (Table 24-4).

Table 24-5 shows the changes in [Na] and [K] and osmolality of CSF sampled from the contralateral lateral ventricle during and after i.v.t. infusion of hypertonic mannitol-CSF, and of CSF sampled from the infused lateral ventricle at the end of the infusion. [Na] of CSF obtained at the end of the infusion was decreased from 149±1.2 (baseline) to 130.7±2.3 mmol/l and 127.3±3.8 mmol/l respectively for samples obtained from the contralateral and infused ventricles.

It seems reasonable to assume that the infused material was adequately dispersed throughout the ventricular system. That is, the pulsations of the choroid plexus, the movement of fluid into the ventricle from the choroid plexus and brain tissue, the action of cilia present on the ventricular side of the choroid plexus, and postural changes would all contribute to the mixing and dispersion of the infusate. The similarity of the CSF values obtained from the contralateral and infused lateral ventricles would appear to confirm that the CSF values obtained were reasonable approximations of the situation throughout the ventricular system, including the third ventricle.

In experiments where measurements were made of mean arterial pressure, cardiac rate and respiratory rate, these usually did not change during infusion. Little or no change in CSF pressure occurred, the animals' demeanour was quite normal and they would eat if food were allowed. On the face of it, and consistent with the data cited above, it is unlikely infusion at 1 ml/h into a system with ventricular volume of 10 ml, CSF formation rate of 6 ml/h and with free drainage through the arachnoid villi when CSF pressure rises above 3 cm of water into a blood flow

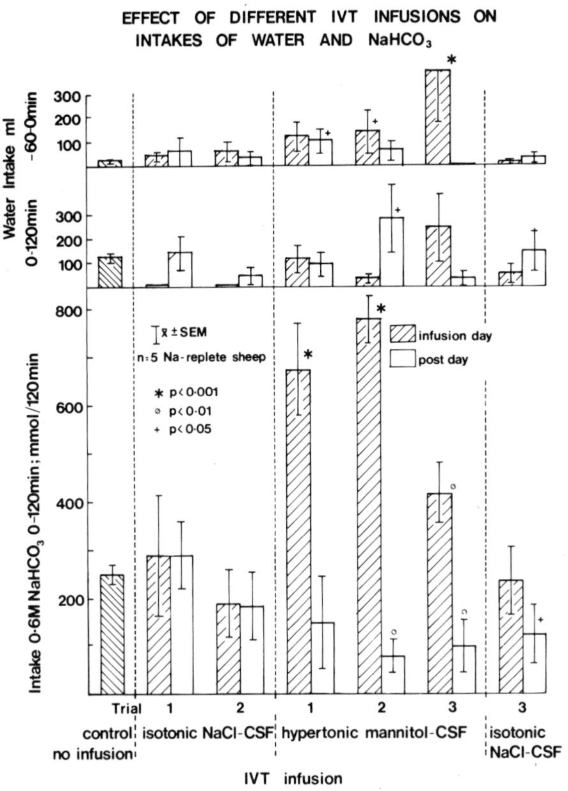

Fig. 24-24. Effect of i.v.t. infusion of 0.7 M mannitol-CSF on intake of 600 mM sodium bicarbonate by sodium-replete sheep (n=5). Three hours of i.v.t. infusion of isotonic saline-CSF (three occasions, two before and one after mannitol treatment) or 0.7 M mannitol-CSF (three consecutive occasions) were begun 1 h before sodium access. Mean (±s.e.m.) intake of water during the 1-h before sodium access (−60 to 0 min), or of water (ml) or sodium bicarbonate (mmol) during the 2-h sodium access period (0–120 min), for the baseline (i.e. two non-infusion days before each experiment; n=60 observations on five sheep), infusion and post-treatment days are shown. A two-way analysis of variance (repeated measures design) and subsequent t-tests were used to determine the significance of the difference between the baseline and experimental or post-treatment values.

of 30–60 l/h, would cause any disturbance to the animal. Furthermore the phenomenon under consideration is a consistent increase in appetite and not an issue of some non-specific reduction. Control experiments with infusion of isotonic saline-CSF had no effects on appetite.

Mannitol-CSF close to or equiosmotic with natural sheep CSF: Threshold of effect

Compared with baseline (i.e. intake on the 2 days before each infusion), sodium intake (Fig. 24-25) was increased by i.v.t. infusion of 0.3 M and 0.27 M mannitol-CSF (Table 24-2) (P values <0.001) and not significantly changed by i.v.t. infusion of isotonic saline-CSF. Intake was at baseline level or below on the post-treatment day following each infusion. Water intake was not altered by any of the infusions.

Examining the threshold of effect (Fig. 24-26), in another series of experiments baseline sodium intake was 166±24 mmol/30 min (n=40 observations on five sheep). Sodium intake was increased by infusion (Table 24-2) of 0.7 M mannitol-CSF (492±58 mmol, P<0.001), 0.27 M mannitol-CSF (324±73 mmol, P<0.01), 0.14 M mannitol-CSF (300±87 mmol, P<0.05) or 0.1 M mannitol-CSF (291±88 mmol, P<0.05). Intake returned to baseline or below following each of the infusions.

Water intake was not significantly altered by any of the infusions. However, two of the five animals drank water (200 and 400 ml) during infusion of hypertonic mannitol-CSF.

Composition of cerebrospinal fluid

Table 24-4 shows that CSF [Na] was decreased by about 24, 12 and 5 mmol/l by i.v.t. infusion of isotonic mannitol-CSF: 0.27, 0.14 and 0.1 M respectively. CSF osmolality was increased by infusion of either 0.7 M or 0.3 M mannitol-CSF, confirming the latter was slightly hypertonic. The CSF values obtained after 60 min of infusion, i.e. before sodium access, are shown in Table 24-5. Changes in CSF [Na] were similar for the infused and contralateral ventricles. There was evidence of correlation between CSF [Na] at the end of the mannitol infusion and an increase in sodium intake (Fig. 24-27).

Specificity of appetite induced by hypertonic or isotonic mannitol-CSF

These infusions in sodium-replete sheep under cafeteria conditions caused an increase in intake predominantly of sodium solutions.

Sheep (n=7) offered 600 mM solutions of sodium, potassium and ammonium bicarbonates and water during i.v.t. infusion of 0.7 M mannitol-CSF increased sodium bicarbonate intake from 92±24 mmol to 510±109 mmol (P<0.001). Intake of potassium bicarbonate was increased from 43±22 mmol to 197±69 mmol (P<0.01). Intakes of ammonium bicarbonate and water were not significantly altered (Fig. 24-28, upper section). Sheep (n=6) offered 600 mM sodium bicarbonate, potassium bicarbonate and sodium chloride, 300 mM magnesium chloride and water, increased their intake of sodium bicarbonate from 65±124 mmol to 307±124 mmol (P<0.01). Intakes of the other solutions were not significantly altered (Fig. 24-28, lower section).

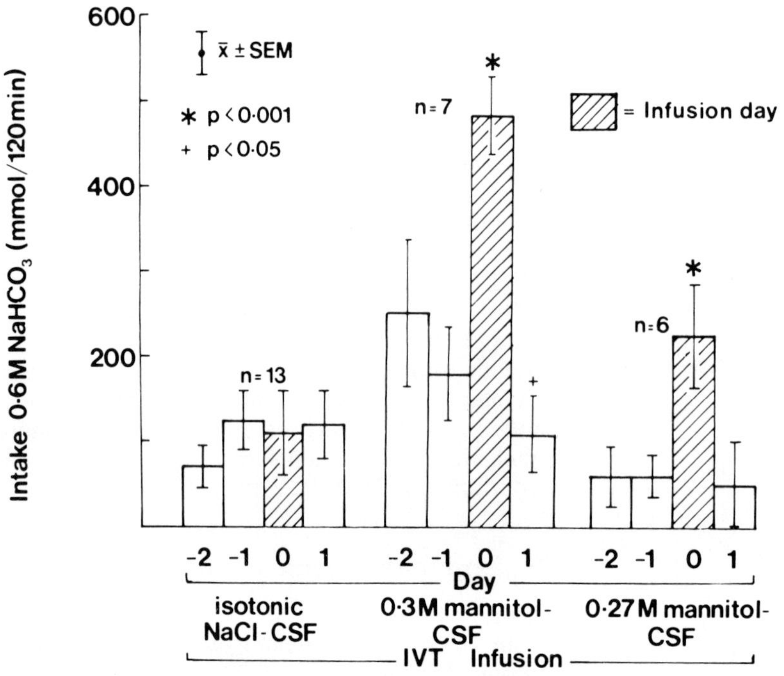

Fig. 24-25. Effect of i.v.t. infusion of isotonic saline-CSF ($n=13$), 0.3 M mannitol-CSF ($n=7$) or 0.27 M mannitol-CSF ($n=6$) on intake of 600 mM sodium bicarbonate (mmol/120 min) in sodium-replete sheep. Three hours of infusion were begun 1 h before the sodium access period on day 0. No infusion was given on days −2, −1 (baseline) or +1 (post-treatment). Statistical analysis similar to that used in Fig. 24-24.

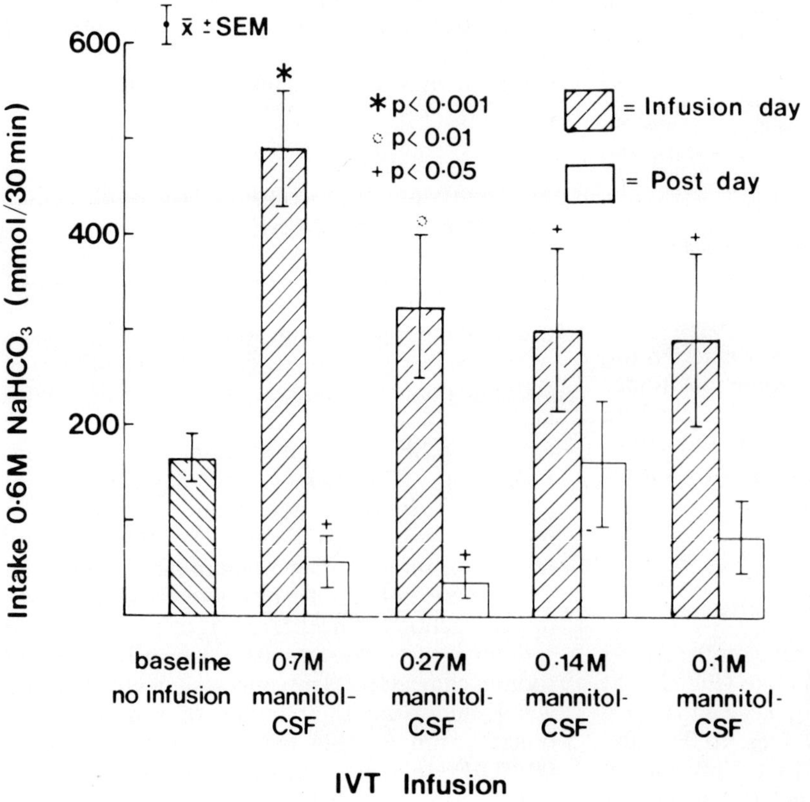

Fig. 24-26. Effect of i.v.t. infusion of 0.7 M, 0.27 M, 0.14 M or 0.1 M mannitol-CSF on intake of sodium bicarbonate solution by sodium-replete sheep ($n=5$). A ninety-minute infusion was begun 1 h before sodium access. Mean (±s.e.m.) intakes of 600 mM sodium bicarbonate (mmol/30 min) for the baseline ($n=40$ observations on five sheep), infusion and post-treatment days are shown. Statistical analysis similar to that used in Fig. 24-24.

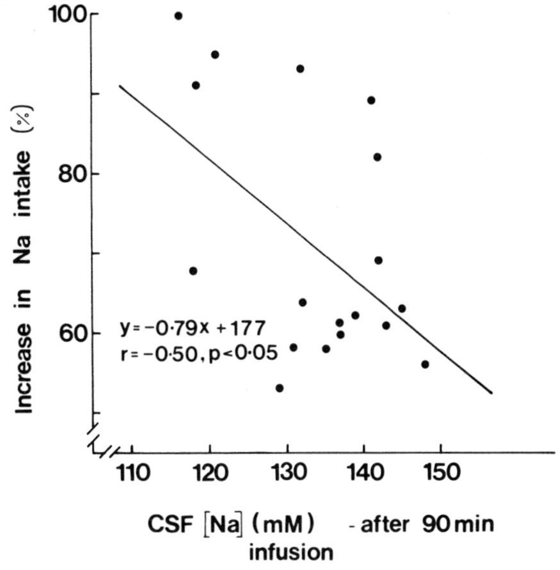

Fig. 24-27. Correlation between CSF [Na] after 90-min mannitol-CSF infusion and increase in sodium intake, (intake on infusion day)/(intake on infusion day + mean baseline intake), expressed as a percentage ($n = 18$ observations in five sodium-replete sheep). Data from Fig. 24-26 replotted.

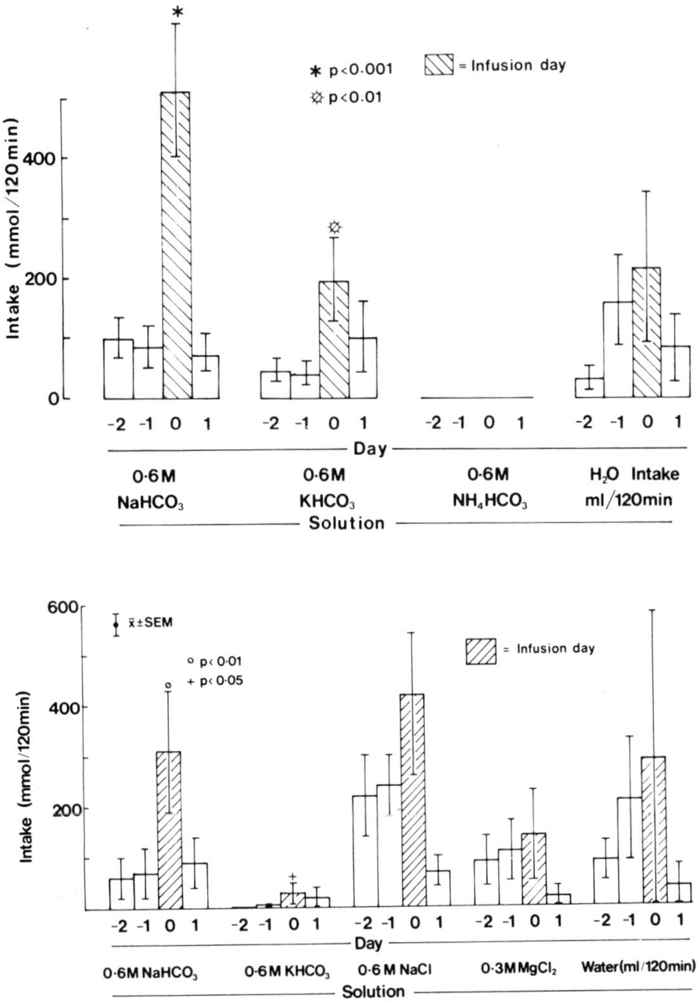

Fig. 24-28. Specificity of appetite induced by i.v.t. infusion of 0.7 M mannitol-CSF in sodium-replete sheep. Three hours of i.v.t. infusion were begun 1 h before access to different salt solutions and water on day 0. There was no infusion on days −2, −1 (baseline) or +1 (post-treatment day). Statistical analysis similar to that used in Fig. 24-24. Upper section: mean (±s.e.m.) intakes (mmol/120 min) of 600 mM solutions of sodium, potassium and ammonium bicarbonates and water (ml/120 min) are shown ($n = 7$). Lower section: mean (±s.e.m.) intakes (mmol/120 min) of 600 mM solutions of sodium bicarbonate, potassium bicarbonate and sodium chloride, and 300 mM magnesium chloride and water (ml/120 min) ($n = 6$).

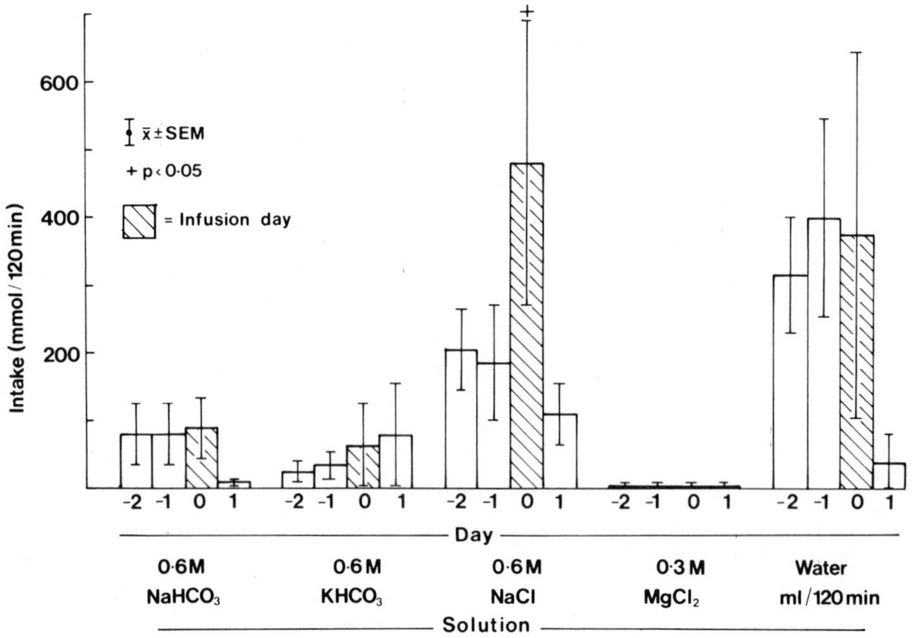

Fig. 24-29. Specificity of appetite induced by i.v.t. infusion of 0.3 M mannitol-CSF in sodium-replete sheep (n=5). Three hours of infusion were begun 1 h before access to the different salt solutions and water on day 0. There was no infusion on days −2, −1 (baseline) or +1 (post-treatment). Mean (±s.e.m.) intakes (mmol/120 min) of 600 mM solutions of sodium bicarbonate, potassium bicarbonate and sodium chloride, and 300 mM magnesium chloride and water (ml/120 min) are shown. Statistical analysis similar to that used in Fig. 24-24.

In sheep (n=5) offered 600 mM sodium bicarbonate, potassium bicarbonate and sodium chloride, 300 mM magnesium chloride and water during i.v.t. infusion of 0.3 M mannitol-CSF (Fig. 24-29), intake of sodium chloride increased from 193±50 mmol to 481±209 mmol (P<0.02) but intakes of the other fluids were unchanged by the infusion.

Summary and discussion of physiological implications of these findings on altered sodium concentration of cerebrospinal fluid

The outcome of these experiments appears clear-cut and of basic importance even if the explanation of some aspects is not immediately apparent. *Physiological increase in CSF [Na] decreased sodium appetite in the sodium-deficient sheep. The increased osmolality produced by infusion of hypertonic saline-CSF did not appear to be the cause since a comparable osmolality increase produced by hypertonic mannitol-CSF did not decrease appetite but did, indeed, elevate it. Thus the relevant factor in turning off the sodium appetite was the increase in [Na].*

The hypertonic mannitol-CSF which reduced [Na] by 15–20 mmol/l did cause a 60–80 mosmol/kg increase in CSF osmolality, but the same increase in sodium intake occurred with isotonic mannitol which caused a comparable CSF [Na] decrease without osmotic change or water transfer across choroid plexus or brain capillaries. Isotonic saline-CSF had no influence. The evidence indicated that a decrease in CSF [Na] was the relevant factor in stimulating sodium appetite. Furthermore the basic nature of the phenomenon was ratified by its consistent evocation in the sodium-replete animal which had not experienced sodium deficiency hitherto: this is compatible with the activation of a 'prewired' system of receptors and neural effectors. It was shown the appetite was specific for sodium salts. Examination of the threshold of effect in the sodium-replete animal showed that the appetite could be generated by a decrease in CSF [Na] within the measured physiological range found in mild to moderate sodium deficiency (Mouw et al. 1974b): i.e. a 4–6 mmol/l decrease in [Na] was effective in one series and a 10 mmol/l decrease was more effective.

The effects produced by 0.7 M mannitol-CSF were reproduced by osmotically equivalent hypertonic urea-CSF and hypertonic sucrose-CSF. However, whereas the sucrose produced an effect similar to mannitol, the effect with urea was less, though significant, and there was no effect with hypertonic glucose-CSF. This was the case even though CSF (Na], as measured at the 120-min period of sodium access, decreased almost as much with glucose as with urea and sucrose (Table 24-3). This latter fact is not consistent with the hypothesis of simple relation between sodium appetite and CSF [Na]. However, an important fact to be taken into account, and a possible basis for an explanation of the lack of effect of glucose and smaller effect of urea is that urea and glucose penetrate cells; glucose is metabolized and there is also the 'sink' effect of transcapillary transport to the blood. Also, with hypertonic glucose-CSF it was shown that though CSF [Na] was reduced at 180 min, the rate of change of [Na] by the first 60 min when bar-press access began was different from the other procedures (Table 24-3). Close study of this and a series of other procedures bearing upon metabolic effects of glucose and penetration of cells is needed. As noted above, 0.7 M L-glucose had the same effect as 0.7 M mannitol-CSF. It may well be that glucose infusion has little effect on brain ECF composition a short distance from the ventricular wall. Considerations bearing upon movement of solutes through the brain are discussed, for example, by Pappenheimer et al. (1965) and Fenstermacher et al. (1970).

Pappenheimer et al. (1965) were concerned with interpretation of results where it was legitimate to assume that changes in composition of CSF exert their effects on pulmonary ventilation by altering hydrogen ion concentration in the tissue fluid bathing respiratory neurones. They note that there are three known anatomical barriers to passage of ions between ventricular fluid and cerebral capillary blood: ventricular ependyma, intracellular channels which penetrate everywhere between glial and neuronal components of the neuropil, and walls of cerebral capillaries with their associated glial investments. These barriers are shown schematically in Fig. 24-30. Ependymal cells are separated by clefts which may be in functional continuity with the channels penetrating between neurones and glial cells deep within the nervous system. Zonulae occludentes are shown in the figure between some ependymal cells to indicate the possibility that at least some of the intracellular channels may be obliterated at this boundary (Brightman and Palay 1963).

Weindl and Joynt (1972), using scanning electron microscopy, have shown that the major part of the ventricular walls is covered by a dense layer of cilia. The surface of the choroid plexus and of the circumventricular organs are free of cilia. In these areas, as for example with the organum

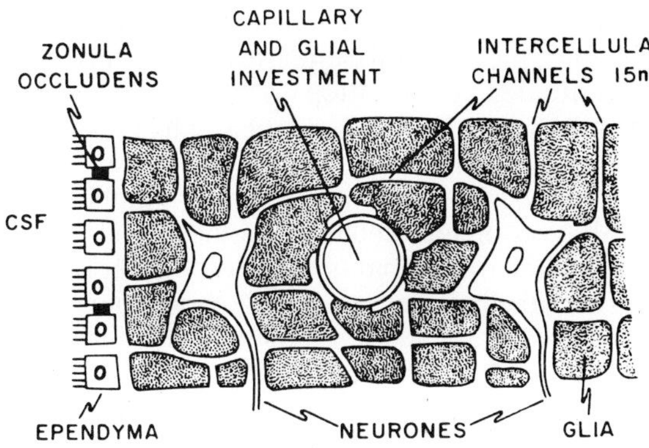

Fig. 24-30. Diagram of cellular boundaries and diffusion pathways interposed between CSF and blood. Ependymal cells are separated by clefts which may be in functional continuity with channels penetrating between neurones and glial cells deep within the nervous system. The width of intercellular channels is of the order of 15–20 nm and comprises about 5% of total cross-sectional area. Channels are wide enough to allow free diffusion of small molecules. Sodium, potassium, sucrose, and even molecules as large as inulin have been shown to penetrate brain substance at speeds consistent with free diffusion. Zonulae occludentes are shown between some ependymal cells to indicate the possibility that at least some of the intercellular channels may be obliterated at this boundary. Molecules which do not enter neurones or glial cells may enter a capillary as an alternative to continued penetration of brain substance through intercellular channels. Relative amounts lost to blood will depend on distribution and permeability of capillary–glial boundaries in relation to number and distribution of intercellular channels. Concentration profile during net flux will therefore be complex. Neurones will be exposed to extracellular fluid at various points along the concentration profile. (From Pappenheimer et al. 1965, with permission.)

vasculosum of the lamina terminalis, there are tight junctions between the non-ciliated ependymal cells. Data on brain ultrastructure show the neuropil is penetrated by a network of intercellular channels or clefts, 15–20 nm in width and occupying some 5%–10% of brain volume or cross-sectional area (Palay 1956; Lasansky and Wald 1962). These channels are too wide to influence diffusion velocities of small ions, and it can be expected that particles less than 1 nm in diameter would diffuse freely through them to reach neuronal surfaces. Nicholls and Kuffler (1964) have shown that both potassium and sodium, and also sucrose, penetrate through intercellular channels (not glia) in the avascular central nervous system of the leech at rates which are consistent with free diffusion. Neurones located 50 μm from the surface responded to changes in extracellular [K] or [Na] within 4 s. Free linear diffusion of an ion the size of bicarbonate occurs at a rate such that 50% of a step change in concentration penetrates 100 μm in 6 s and 1 mm in about 10 min.

Pappenheimer et al. (1965) go on to say that in the vertebrate nervous system, molecules which reach the brain parenchyma from the ependyma may enter the cerebral capillaries as an alternative to continued penetration of brain substance through the intercellular channels. The number of capillaries and the permeability of the capillary–glial boundaries, in relation to the number and distribution of intercellular channels, will determine the relative amounts lost to blood. Molecules such as labelled water, which readily penetrate cerebral capillaries, are removed from ventricular fluid by cerebral blood before they can diffuse any appreciable distance into the brain (Heisey, Held and Pappenheimer, 1962). On the other hand, molecules such as inulin which cannot penetrate capillary glial membranes have been shown to diffuse from the lateral ventricle to the lateral surface of the thalamus, a distance of 1.5–2 cm in the dog, at speeds consistent with free diffusion (Rall et al. 1962).

Pappenheimer et al. (1965) remark that injection or perfusion of the ventricles with abnormal concentrations of potassium, calcium or magnesium causes profound alterations in blood pressure, reflex excitability, somatic muscle tone, and degree of wakefulness. It is possible that all these effects are mediated through superficial chemoreceptors which are especially sensitive to each ionic species and which relay synaptically to nuclei in the reticular formation of the hypothalamus or the medulla. It is simpler, however, to suppose that these ions produce their effects by direct penetration into brain tissue.

Fenstermacher et al. (1970) state that it can be shown mathematically that molecules that distribute only in the extracellular space, do not enter cells, or bind, or cross capillary walls (Fig. 24-31, top row of arrows) appear to move relatively more rapidly through the brain parenchyma than molecules which do a combination of the above possibilities. The distribution space of such 'extracellular' molecules would be expected to be from 10% to 20%. Results in agreement with these two points have been found for inulin and sucrose. Compounds that bind and/or enter cells but do not cross capillary walls to any significant extent (Fig. 24-31, middle row of arrows) give larger tissue distribution spaces and slower relative rates of movement through the parenchyma. This result was found with mannitol, *p*-aminohippuric acid (PAH), fructose and creatinine. Conversely materials which cross the capillary wall more rapidly than those above (Fig. 24-31, bottom row of arrows) give very small distribution spaces and exceedingly low apparent rates of movement into deeper tissue. This was seen experimentally with urea and tritium-labelled water, ^{3}HOH.

Fenstermacher et al. examined tissue concentration profiles, i.e. tissue concentration as a function of distance into the brain parenchyma, following bilateral ventriculocisternal perfusion of dogs with ^{14}C- and ^{3}H-labelled compounds in artificial CSF. They note at the outset that it is possible there is a significant bulk flow of fluid through the extracellular space of the brain. Such a flow would be expected to influence solute transport within the system (Rall 1967). They found that the apparent tissue diffusion coefficient for sucrose and inulin was about 40% of that in free water or agar D for both molecules. Such a lowering of tissue diffusibility in the system can be shown to be caused by factors such as binding, cellular uptake and transcapillary loss. Since both sucrose and inulin appear not to cross brain capillary walls or enter cells, such a lowering of the tissue diffusibility can also be explained by diffusion through tight relatively narrow channels or by diffusion via tortuous channels. The tortuous channel explanation (i.e. the molecules follow long devious paths [the extracellular space] in their movements through the tissue) is argued for by the similarity in the diffusion coefficient ratios for two molecules of such disparate size as inulin and sucrose.

Mannitol and PAH had larger tissue distribution spaces and lower tissue diffusion coefficient ratios relative to those in agar than did inulin and sucrose. This suggests some cellular uptake and/or

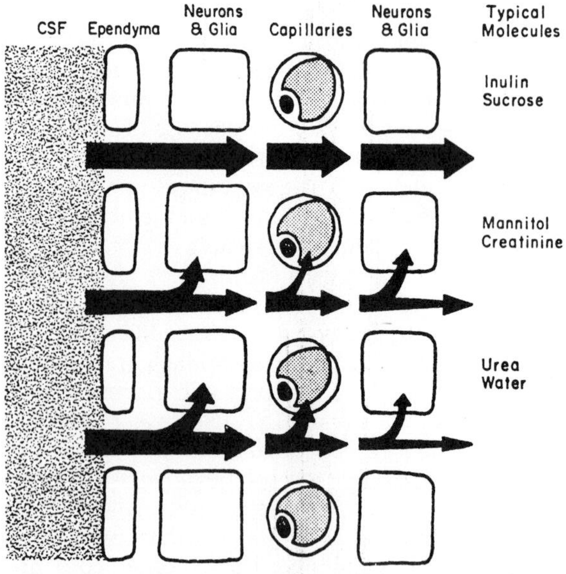

Fig. 24-31. Diagrammatic representation of several of the possible pathways a molecule can follow once it enters the brain tissue from the ventricular CSF. The arrows indicate the pathways; the thickness of the arrow represents the relative importance of that pathway. Experimental molecules which appear to follow a particular diagrammed course are listed on the right. (From Fenstermacher et al. 1970, with permission.)

binding is occurring for these molecules. It was also found that the mannitol space appeared to increase with time, suggesting a slow but significant cellular uptake. Fructose was found to have an even larger tissue distribution space and lower diffusion coefficient ratio than mannitol. This suggested that it entered brain cells, and that it does so more readily than mannitol. The size of the space implies very low capillary permeability to fructose and, also, lack of a special transport mechanism for this monosaccharide at the blood–brain barrier. Results with [^{14}C]creatinine showed a distribution space of almost 70%, which was quite close to that of brain water. This suggests relatively rapid cellular uptake of creatinine and very little loss of the compound from brain into blood. However, urea achieved a distribution space significantly less than creatinine. Fenstermacher et al. (1970) suggest this indicates that urea enters the cells of the brain but that a sizeable amount of urea crosses the capillary walls and leaves the brain via the blood (i.e. the situation illustrated by the bottom row of arrows in Fig. 24-31). Following ventriculocisternal perfusions with ^{3}HOH no significant radioactivity was observed beyond the first millimetre of tissue. Whereas ^{3}HOH enters the brain from the perfusion medium and remains to some extent in the first millimetre or less of tissue, the bulk of what enters is removed by transcapillary movement into the circulating blood.

Heisey et al. (1962) estimate the flow of CSF in the goat at 0.16 ml/min (ca 9.6 ml/h), and it would seem possible that a similar order would hold in the sheep. Ventriculocisternal volume is estimated at about 10 ml. They propose on several grounds that the major fraction of the bulk formation of CSF originates as a secretion from the choroid plexus, but when formation rates are abnormally high (e.g. double) as a result of perfusion of the ventricles with hypertonic fluid, osmotic flow coming through the ependymal walls may be a contributing source of fluid. With the hypertonic 0.7 M mannitol artificial CSF used here, osmotically determined inflow of water with resultant decrease in CSF [Na] may derive from the choroid plexus and also from the capillaries of the brain adjacent to the lateral and third ventricular walls. However, this osmotic process is not essential to the sodium appetite phenomenon since it also occurs with isosmotic mannitol, when decrease in CSF [Na] results from dilution with a low-[Na] fluid.

One explanation of the results with sodium appetite is that the effects are due to stimulation or inhibition of sodium-sensitive receptors very close to or on the ventricular walls. The facts that the appetite could be inhibited rapidly (10 min), was enhanced within 30 min, and further was enhanced by small decreases in sodium (4–10 mmol/l) with minimal changes in other ions and independently of a change in osmotic pressure, might favour this. But the finding that hypertonic mannitol, sucrose, and also urea and glucose have different effects on appetite even though they produce similar changes in CSF [Na], would qualify such an idea.

Another explanation of the data could be that the sodium-sensitive cells—reacting to low [Na] in brain ECF or reacting because intracellular [Na] decreases as a result of reduced [Na] in brain ECF—are located a considerable distance from the ventricular wall in the hypothalamic substance and are reached by tortuous channels. In these circumstances the several processes and factors considered by Pappenheimer et al. (1965) and Fenstermacher et al. (1970) (Figs. 24-30 and 24-31) becomes operative, and different effects on brain ECF [Na] according to distance from ventricle become likely with different hyperosmotic artificial CSFs. Sucrose and mannitol would presumably penetrate furthest into the brain—e.g. as inulin has been shown to penetrate 1.5–2.0 cm in the dog at speeds consistent with free diffusion (Rall et al. 1962). Sodium would probably diffuse out freely and penetrate deeply (Nicholls and Kuffler 1964) in the course of its role in inhibiting salt appetite in the sodium-deplete animal.

The fact that Dr R. Weisinger has found no salt appetite increase in six experiments with i.v.t. infusion of distilled water which lowered [Na] in the CSF by 10–20 mmol/l suggests the water had little or no effect in brain ECF due to its rapid extraction osmotically by brain capillaries at the ventricular wall. This is consistent with the receptor–effector system being in the brain substance away from the ventricular walls. With isotonic or hypertonic mannitol a 'front' of fluid of low [Na] would move through the tortuous channels of intercellular space of the hypothalamus virtually unmodified, whereas no comparable effect would occur with glucose because of cell uptake and the 'sink' effect of transport to blood. The effects of urea would be intermediate between mannitol and glucose. Fructose and mannose are metabolized in brain (Chain et al. 1969) whereas L-glucose is not, and the effect on appetite reported earlier with these infusions is generally consistent with the considerations above.

Although not obvious from the results recorded in Tables 24-3 and 24-4 with different groups of sodium-replete and sodium-deficient animals, it has been shown in a more rigorous series of experiments that CSF [Na] is decreased by 2–3 mmol/l in mildly sodium-deficient sheep (Mouw et al. 1974b). The samples of CSF were large (5–7 ml), collected under paraffin oil, and taken from the same animals (n=17) under sodium-replete and sodium-deplete conditions. A 3 mmol/l decrease in plasma [Na] has also been shown to occur in mildly sodium-deficient sheep (Abraham et al. 1976). Also, in eleven experiments of Dr R. Weisinger, CSF and blood samples were obtained from sheep with a parotid fistula when presumably sodium-replete (i.e. 2–3 h after the start of the sodium bar-press period) and when sodium-deplete (i.e. after 22 h of sodium deprivation). Relative to the former condition, both CSF and plasma [Na] were significantly (P values <0.001) decreased by sodium-depletion (CSF [Na]: 152.2 ± 0.5 vs 149.5 ± 0.8 mmol/l; plasma [Na]: 148.9 ± 0.6 vs 144.4 ± 0.6 mmol/l). These observations are consistent with the fact that the salivary sodium loss by sheep with a parotid fistula (or with a cannula in a parotid duct) represents a loss of sodium in excess of water relative to extracellular proportions (Abraham et al. 1976). In short, a physiological decrease in CSF [Na] may be sufficient to stimulate sodium appetite in sheep.

It can also be noted that mannitol is a valuable experimental material in these studies because of its lack of toxicity (Wise et al. 1964; Guinane 1977). This includes clinical experience of its use (Wise and Chater 1962). However, the effects here were in no way exclusive to its use, and it is also clear-cut that infusion of hypertonic mannitol into the carotid arterial blood of the sodium-deficient animal, as described in Chapter 20, temporarily causes a large reduction in sodium appetite. Thus the effect on sodium appetite of hyperosmolality produced by mannitol was strikingly dependent on which side of the blood–brain barrier the change was produced, and accordingly what effects were produced on ionic parameters—highly probably the [Na] of brain ECF.

These data represent the most consistent rapid induction of salt appetite yet reported. In the sodium-replete animal the effect was evident in 60 min and to date it has not been determined whether it can be faster. In the animal with an existing moderate deficit, augmentation can be produced in 25–30 min. This time delay, though perhaps somewhat shorter than the initiation of some effects of transcription and translation with, for example, steroid hormones in mammalian cells, is compatible with others (Schulster et al. 1976), and also with the data on speed of effects of ions on chromosomes cited in an earlier part of this chapter. For that matter, genomic actions of cortisol in thymus lymphocytes result in onset of suppression of glucose transport within 15 min (Mosher et al. 1971).

It remains, however, somewhat paradoxical that there is normally a very much longer time delay in the induction of salt appetite: often 6–24 h. Also, renin and angiotensin produce their effect on thirst in a few minutes whereas the maximum effect on salt appetite is delayed some hours. Yet the ionic changes produced by i.v.t. mannitol or other procedures reducing CSF [Na] are effective on salt appetite in 30–60 min. This could be because the procedure circumvents some steps normally required for systemic change to influence CSF, brain ECF and intraneuronal [Na]; it may also involve direct effect on the translation mechanism. Further, the characteristic delay in the induction of salt appetite by renin and angiotensin would argue against the effect of angiotensin being simply attributable to an unphysiological concentration producing movement of sodium out of cells and mimicking events which occur physiologically in sodium deficiency. Mannitol may, however, act by mimicking ionic changes which occur in sodium deficiency. The data suggest that angiotensin may be having a much more complex neurochemical effect.

The effect of steroid and peptide hormones on salt appetite

The physiological effect of a jointly sufficient and severally necessary sequence of steroid and peptide hormones in producing the salt appetite behaviour seen with pregnancy and lactation, and certain similarities with the induction of maternal behaviour by hormones, were described in Chapter 23. These findings, in the light of the known mode of cellular action of certain of these hormones, are generally consistent with the hypothesis of induction of protein synthesis underlying the central excitation of appetitive drive. The interactions described, such as the prior necessity of pseudopregnancy or oestrogen and progesterone treatment in order for the triplet of peptide hormones (ACTH, oxytocin and prolactin) to have full physiological effect, could be consistent with the role of oestrogen in induction of prolactin receptors, and with this effect being operative in the brain (Waters, Freisen and Bohnet, 1978). The more recently discovered role of angiotensin II in the genesis of sodium appetite following a time delay raises many questions including a possible mode of action through mitochondrial release of calcium ions (Exton 1980).

Steroid hormones: Mode of action

Steroid hormones circulating in the blood gain rapid and relatively unrestricted access to all parts of the nervous system (McEwen et al. 1979), where the hormone may be metabolized or interact with receptor sites to produce an effect. It has been shown that oestradiol (E_2) and testosterone, when infused intravenously, cross the blood–CSF barrier in the rhesus monkey. After 6 h of infusion the E_2 level was about 3.5% of the concurrent plasma level (Marynick et al. 1976). In sheep made sodium deficient by parotid salivary loss, the ratio of CSF to blood aldosterone concentration for 11 paired samples was 0.38±0.03. There was a significant linear relation between CSF and blood aldosterone (r=0.94, P<0.001: Fig. 7-11). Measurements also showed significant levels of cortisol in CSF, though less than in plasma. It seems likely that such levels would also hold in brain interstitial fluid (Rall et al. 1962). Pfaff et al. (1974) state that steroid sex hormones can directly trigger mating in the female rat by their action on the brain rather than through the pituitary, since oestrogen plus progesterone successfully stimulates lordosis mating behaviour in the ovariectomized, hypophysectomized female.

Following earlier work by Michael (1965), it has been shown that radioactive oestradiol is highly concentrated by cells in specific parts of the limbic system, preoptic area and hypothalamus. Autoradiographs made 2 h after injection show high densities in medial anterior hypothalamus, the ventromedial nucleus, the arcuate nucleus and the ventral premammillary nucleus (Pfaff 1972; Pfaff et al. 1974). In scintillation-counting experiments chromatography showed over 80% of material taken up by the brain to be in the chemical form of oestradiol (Zigmond and McEwen 1970). In most of the experiments an optimum time for detecting the oestradiol was 2–4 h after systemic injection. This is before the earliest time (about 16 h) that lordosis has been detected after systemic injections (Green et al. 1970). Pfaff et al. (1974) remark that it may be hypothesized that oestrogen entering an oestrogen-concentrating cell in the hypothalamus or preoptic area sets into motion a chain of events that takes several hours and which results in the priming of neural circuits having to do with lordosis. An anatomical basis for coordination of action of the oestrogen-sensitive system exists (McEwen and Pfaff 1973) in that well-established pathways link all the structures with the highest concentrations of oestrogen. In relation to effects of the oestrogen on neurones, preoptic and anterior hypothalamic nerve cells studied electrophysiologically showed higher spontaneous activity in pre-oestrus compared with anoestrus female rats (Cross and Dyer 1971; Moss and Law 1971). Also, the preoptic units in oestrogen-treated rats gave more inhibitory reactions than control rats to cervical stimulation but not to other control stimuli (Lincoln and Cross 1967). Kawakami and Sawyer (1959) proposed that gonadal steroids affect the electrical activity of the hypothalamus and reticular activating system.

Over the past two decades the role of intracellular receptors for hormones has been emphasized. The hormone–receptor complex is translated to the nucleus where expression of genetic information is triggered, and ribonucleic acid synthesis and synthesis of specific protein follow. There is evidence for the existence of such mechanisms in the brain for all five major subclasses of steroid hormones in that these can influence neural events underlying behaviour and brain function in adult vertebrates. Intracellular receptors for oestrogens, androgens and glucocorticoids have been identified in the brain. The evidence as regards progestins and mineralocorticoids is less extensive (McEwen et al. 1978). At the same time as the emergence of these data, evidence for the direct action of steroids on neural activity and synaptic function has been found, as will be discussed further on in this section.

Considering the role of receptors in the mechanism of steroid hormone action in the brain, Muldoon (1980) suggests the following sequence in relation to the oestrogen receptor system. He proposes that, for the most part, the considerations are similar for all steroid hormones though there are some significant differences, as, for example, with the transformation of testosterone to 5α-dihydrotestosterone or the aromatization of testosterone to oestradiol which can occur in some cells. The oestradiol molecule appears to encounter little or no resistance as it traverses the plasma membrane of the cell. A limited number of receptor molecules are available in the cytoplasmic compartment for binding of the steroid to form an oestrogen–receptor complex, which effectively eliminates the possibility of the steroid diffusing back out of the cell. The steroid–receptor complex is rapidly transformed to a new form which accumulates in a short period of time within the cell nucleus. Muldoon suggests it is a plausible conceptualization of the intracellular mechanism of steroid hormone action that the sole function of the steroid molecule resides in its ability to modify the cytoplasmic receptor to a form which is capable of transformation into the nucleus (see Gorski and Gannon 1976). Having gained access to the nucleus the oestrogen–receptor complex is strongly attracted to the genomic apparatus, where it affects genetic expression by alteration of the nature and extent of transcription. Differential synthesis of RNA species engenders different patterns of protein synthesis and these translation products could direct the ultimate display of hormonal activity.

With relation to intracellular receptor dynamics, Muldoon suggests that the regulation of the level of cytoplasmic receptor is most powerfully affected by oestradiol itself (Cidlowski and Muldoon 1974) and that there are three mechanisms of control: (i) cells of oestrogen-responsive tissues have the constitutional property of synthesizing oestrogen receptors; (ii) the process is enhanced by oestradiol and it follows the genetic expression route; (iii) as circulating levels of oestradiol increase, a proportion of the nuclear oestrogen–receptor complex leaves the nucleus and becomes associated with the microsomes for a period of time. After this, the intact complex is liberated into the cytoplasm and can immediately re-enter the nucleus thus describing a re-utilization phenomenon. A second nuclear uptake phase of this type has recently been confirmed in the mouse uterus (Korach and Ford 1978). Also, when endogenous titre of oestradiol is high, nuclear oestrogen–receptor complexes are dissociated allowing regeneration of receptor into the cytoplasm where it is capable of being reactivated by another molecule of oestradiol.

Muldoon (1980) stresses that steroid hormones bind with high affinity to proteins other than receptors. These proteins may be intracellular or of plasma origin. They are different for different steroids, and they often vary in amount as a function of the stage of development of the organism. The presence of these binding proteins limits the accessibility of the hormone to its receptor. It also imposes severe methodological restrictions upon accurate analysis of receptor levels and properties. Another problem is the difference in vivo in the rates of penetration of various steroid hormones from the blood into discrete areas such as the brain. These and many other factors dictate the need for caution in attempting to correlate data from studies in vivo with those done in vitro.

Saturable, oestrogen-dependent steroid-specific cytoplasmic and nuclear receptors for progesterone have been found in brain and pituitary (Evans et al. 1978; Kato et al. 1978; Blaustein and Wade 1978) with a distribution similar to that of oestrogen receptors. Oestrogen may also act on cells in ways other than alteration of genetic transcription with primary action on nuclear sites (Ohno and Lyon 1970; Liang and Liao 1975).

In contrast with the distribution of oestradiol and progesterone receptors, corticosterone is preferentially accumulated by nuclei of the rat hippocampus, with lower levels in the amygdala and septum. There is very little binding in pituitary, hypothalamus, preoptic area or cortex and this is consistent with autoradiographic analysis (Warembourg, 1975; McEwen 1977). Muldoon (1980) remarks that mineralocorticoid interactions with brain receptors have not been accorded much attention. Anderson and Fanestil (1976) showed that the antagonist SC-9420, which competes for renal aldosterone receptors, blocked whole-brain cytosol binding of [^{3}H]aldosterone or [^{3}H]dexamethasone when used in large molar excess. Equimolar concentrations of SC-9420, however, competed for [^{3}H]aldosterone binding but not for that of [^{3}H]dexamethasone, which suggested some degree of selectivity for presumably different sites. Autoradiographic analysis of aldosterone retention favours preferential binding in hippocampal nuclei with lesser amounts in anterior pituitary glandular cells, septum, amygdala and pyriform cortex (Ermish and Ruhle 1978). Muldoon is of the view that further analysis is required before these latter studies can be accepted as evidence for mineralocorticoid-specific receptors, since competition by unlabelled steroids was not investigated. He suggests the function of putative aldosterone receptors in the brain remains obscure.

With relation to the intranuclear events of hormone action, O'Malley and Schrader (1976) and O'Malley et al. (1975) have given a lucid account of the outstanding work of their group at Baylor Medical School. With Spelsberg, they were able to show that hormone–receptor complexes were able to bind to chromatin isolated from the nucleus. The receptor–progesterone complex bound preferentially to chromatin from oviduct nuclei compared with chromatin from other tissues (Spelsberg et al. 1971), and receptor protein without hormone did not so bind. The histones of the nucleus seemed unlikely as specific binding sites because of their uniformity between tissues, and within the 500 or more species of non-histone proteins within a single nucleus, Spelsberg determined a selective affinity of the AP_3 fraction of the non-histone proteins for the hormone–receptor complex. The progesterone receptor molecule of the oviduct was found to be a dimer consisting of two subunits, A and B, which are not identical and have a molecular weight somewhat greater than 100 000. Each subunit has a binding site for progesterone. The activated complex enters the nucleus and binds to the chromatin at an acceptor site defined by the position of some specific protein in the AP fraction of the non-histone protein. The binding takes place specifically through the B subunit, as the A subunit has been shown to be incapable of binding to chromatin. The A subunit is presumably liberated to bind with a nearby specific nucleotide sequence of DNA. The gene is activated by the binding of RNA polymerase to the DNA and synthesis of specific messenger RNA follows.

McEwen et al. (1974) state that the effects of oestradiol on protein and nucleic acid formation in the brain implicate activation of the genome by oestradiol in the control of lordosis behaviour and gonadotrophin secretion. Studies with inhibitors of RNA and protein formation tend to support this conclusion by showing prevention by actinomycin D of oestrogen suppression of gonadotrophic hormone secretion from the pituitary (Schally et al. 1969). Also it has been demonstrated that actinomycin D and protein synthesis inhibitors block the oestrogen induction of lordosis behaviour (Quadagno et al. 1971; McEwen et al. 1978). Demonstrated effects of oestradiol on dopamine and noradrenaline turnover may be related to feedback effects of oestrogens on gonadotrophin secretion, since both amines have been implicated in the control of releasing hormone action (Ahren et al. 1971; Anton-Tay and Wurtman 1971; McCann et al. 1972).

Oestrogen induction of prolactin receptors and relevance to salt appetite

Moving from the more general issues of the mode of action of steroid hormones and the direct action of such hormones on the brain, there are data suggestive of a possible way in which oestrogens may synergize with prolactin in contributing to the spectacular salt appetite seen during pregnancy and lactation.

It has been shown that a 9-fold increase in the lactogenic receptor capacity of liver occurs between 20 and 40 days of age in female rats, and a further increase in binding occurs during late pregnancy (Kelly et al. 1974). Day 40 corresponds to the onset of puberty in the rat. Posner et al. (1974) demonstrated that oestradiol injected into normal male rats resulted in a 10–30-fold increase in the number of lactogenic receptor sites after 8–12 days of injection, with significant effect being seen after 4 days. In normal female rats this regime increased prolactin binding to the level seen during pregnancy. Progesterone (500 μg/day) did not result in such an increase, and hypophysectomy decreased prolactin binding to low levels in the mature female rat liver. Further studies (Kelly et al. 1975) showed that a single injection of long-acting oestradiol valerate (2 mg) led to a maximum (35-fold) induction of lactogenic receptors 9–12 days later. Ovariectomy significantly reduced the extent of lactogen binding in females and anti-oestrogens prevented oestradiol induction of the lactogenic receptors in male rats.

Whereas the appearance of receptors in response to oestradiol valerate in vivo is slow, a single injection of cycloheximide reduces binding by more than 90% within 3 h, which provides evidence for a rapid turnover of prolactin receptors (Kelly et al. 1975). While cycloheximide prevented the induction of receptors by oestradiol, actinomycin D did not, even at high doses. This implies a translational control of induction from stable messenger RNA rather than the expected transcriptional control through binding of oestradiol to its receptors in the nucleus. Further evidence for rapid turnover was obtained by determining the disappearance of lactogenic receptors after hypophysectomy in female rats: the half-time of disappearance was about 12 h.

An increase in circulating prolactin is known to result from oestrogen administration in the normal rat (Meites et al. 1972). The hypothesis has been advanced that prolactin induces its own receptor, in contrast to the repressive effect of insulin upon its receptor (Gavin et al. 1974). Costlow et al. (1975) showed prolactin was able to induce its own receptor by injection of 0.5 mg of ovine prolactin into hypophysectomized female rats. An effect of other pituitary hormones—ACTH and growth hormone—on production of prolactin receptors has also been shown (Waters et al. 1978). The numerous paths of effect which have been demonstrated suggest that one way in which the several hormones involved in the high salt intake in pregnancy and lactation may have influence is via the induction of receptors.

There has been little or no work done on the manner in which either sex hormones or adrenal cortical steroids may act within the brain to cause salt appetite. This is in contrast to the large number of experiments on steroid effects on sexual behaviour, including, for example, experimental study of the influence of antibiotics which block steps in transcription and translation. Nor have there been incorporation studies to determine active sites. However, in terms of the possible avenues of effect, this brief account of some aspects of steroid action on tissues serves to draw attention to some other considerations.

Rapid non-genomic effects of steroid hormones in the central nervous system

First, in contrast to the delay of several days in the activation of sexual behaviour by hormone implants as shown by Harris and Michael (1964), there are data showing very rapid induction of such behaviour by microinjection of large amounts of soluble sex hormone in the brain (Fisher 1964; Miller 1965), the speed of which may not be easily reconciled with the protein synthesis hypothesis. The injections (1–5 μg) would have produced local concentrations extremely high relative to physiological, but the specificity of effect was striking.

Fig. 24-32 depicts the two modes of action of steroids on neurones—genomic and non-genomic—and Fig. 24-33 identifies the time characteristics of these two different modes. The rapidity (<5 min) and short duration of electrical responses of neurones to iontophoretically or systemically applied steroids such as oestradiol hemisuccinate, dexamethasone phosphate or cortisol are examples of the non-genomic action (Ruf and Steiner 1967; Feldman and Sarne 1970; Phillips and Dafny 1971; Kelly et al. 1976). Jones and Hillhouse (1976) presented evidence that corticosterone feedback on corticotrophin releasing factor release was via an immediate membrane effect involving a change in calcium flux. The inhibition of pulsatile luteinizing hormone release by oestradiol occurs within minutes of intravenous steroid infusion and is mimicked by the α-adrenergic blocking agent, phenoxybenzamine (Knobil 1974; Blake 1974; Blake et al. 1974). Other effects of steroids in vitro point also to non-genomic actions: for example the fast negative feedback effects of glucocorticoids on the ability of hypothalamic synaptosomes to release corticotrophin releasing factor (Edwardson and Bennett 1974). Glucocorticoid stimulates rapid uptake of tryptophan by synaptosomes in vitro (Neckers and Sze 1975). The authors note that the concentrations used were appropriate to action in vivo, and that the effect was specific to glucocorticoid, androgen and progesterone being ineffective. The data were the first demonstration of non-genomic regulation of neurotransmitter metabolism. On the other hand, McEwen et al. (1978) draw attention to the fast (15 min) genomic actions of cortisol on thymus lymphocytes and note it is conceivable that the action of glucocorticoids on single unit activity in the hippocampus (McGowan-Sass and Timiras 1975) and evoked potential size (Pfaff et al. 1971) may be mediated by a genomic mechanism.

It can be noted parenthetically here that there is

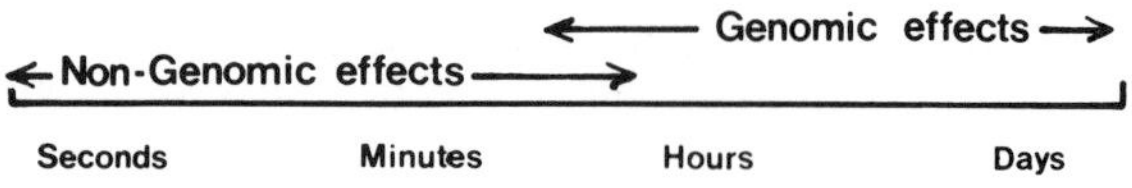

Genomic effects

SLOW IN ONSET - Minutes, Hours

LONG IN DURATION – persist after steroid disappears from tissue

Non-Genomic effects

RAPID IN ONSET – Seconds, Minutes

SHORT IN DURATION – following disappearance of steroid from tissue

Fig. 24-33. Time characteristics of genomic and non-genomic effects. (From McEwen et al. 1978, with permission.)

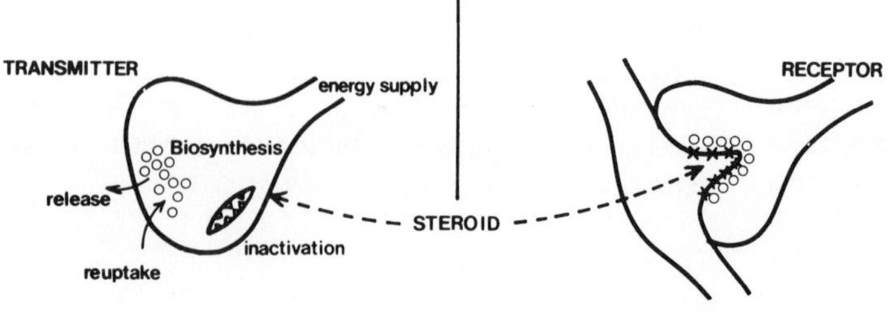

Fig. 24-32. Genomic and non-genomic effects of steroid hormones on pre- and post-synaptic events. Non-genomic effects (dashed line) may involve the action of the hormone on the pre- or post-synaptic membrane to alter permeability to neurotransmitters or their precursors and/or functioning of neurotransmitter receptors. Genomic action of the steroid (unbroken line) leads to altered synthesis of proteins, which, after axonal or dendritic transport, may participate in pre- or post-synaptic events. (From McEwen et al. 1978, with permission.)

a wealth of evidence on an immediate direct effect of peptides on neurones when, for example, they are applied iontophoretically or injected. The reactions are quite selective and only certain neurones will respond to certain peptides (Renaud 1979, 1980). Some act like classical neurotransmitters whereas others have a different role as neuromodulators. Barker (1976) and Renaud (1979, 1980) record data on a direct influence of somatostatin, thyrotrophin releasing hormone, vasoactive intestinal peptide, opiate peptides, substance P, oxytocin and antidiuretic hormone in exciting and inhibiting various neurones. Dufy et al. (1979) have shown that both thyrotrophin releasing hormone and oestrogen applied to the membrane of prolactin-secreting pituitary cells caused action potentials in 1 min as recorded by intracellular microelectrodes. In the case of thyrotrophin releasing hormone, which has a short-term effect on release of prolactin and a long-term effect on its synthesis, the induced electrical activity may be associated with the stimulation of prolactin production. Data on direct action of angiotensin on subfornical neurones were noted in Chapter 22.

There is a body of data which argues for post-transcriptional control of protein synthesis in various systems (Hogan 1974). Evidence for such a process with glucocorticoids comes from experiments showing a paradoxical or 'superinductive' effect of actinomycin D in some circumstances, pointing to synthesis of a labile repressor molecule (Tomkins et al. 1972). In the case of the induction of tyrosine aminotransferase by glucocorticoid, the concept has been challenged by Kenney et al. (1973).

Action of steroid hormones by pathways other than classic transcription is beautifully shown by the action of progesterone in causing meiotic maturation of amphibian oocytes in vitro (Smith and Ecker 1971; Masui and Markert 1971; Masui et al. 1977; Baulieu et al. 1978). Progesterone and other steroids can act at surface membranes to trigger meiosis possibly by calcium movements, or translocation from membrane stores or both. The formation of a maturation promoting factor leads to germinal vesicle breakdown. El-Etr et al. (1979) report that this effect can also be produced by insulin though it is not as rapid as with progesterone.

The group of Baulieu (1981) has recently described the presence of dehydroepiandrosterone sulphate in the brain, and the fact that it persists 15 days after adrenalectomy plus orchidectomy. Stress conditions, produced 2 days after adrenalectomy plus orchidectomy or after sham adrenalectomy, resulted in a rise in brain content of the hormone. The preliminary data, if confirmed, raise some questions parallel to those on the peptides present in the CNS.

The action of peptide hormones

In Chapter 23 dealing with reproduction, the important physiological role of peptide hormones in the induction of a large salt appetite in the sodium-replete animal was described. The role of oestrogen in induction of receptors for prolactin was noted in the preceding section of this chapter as probably being part of the physiological events involved in salt appetite during pregnancy and lactation. There is also to be considered the spectacular sodium-appetite-inducing effects of the octapeptide angiotensin II, and its possible physiological significance in the salt appetite of sodium deficiency.

Reproduction

In pregnancy and lactation the principal peptide actions involved are those of ACTH, prolactin and oxytocin. Following the priming action of oestrogen and progesterone, which appears essential for the full quantitative effect of the peptides, this hormone triplet acts conjointly to give the full salt appetite response seen with lactation. Prolactin appears to have an initiating role since ACTH and oxytocin have only 50% effect in its absence. However, ACTH will, itself, sustain salt appetite once appetite is maximal. By contrast, if ACTH is withdrawn when the triplet is having maximal action, prolactin and oxytocin will not sustain the appetite, and there is a large rapid decrease in salt intake (Denton and Nelson 1978). The action of ACTH is predominantly via the secretion of adrenal steroids, but central direct action of the peptide may also be involved in the rabbit. The consideration of data on the mode of action of prolactin on the mammary epithelium, involving as it does substantial interaction between hormones, may give some insight into mechanisms operating in the brain to activate the neural systems subserving salt appetite. Interaction of steroid and peptide hormones also appears to be involved in the initiation of maternal behaviour.

The mode of action of peptide hormones on differing target cells is a very large and very rapidly growing field of knowledge (Guillemin 1977; Roth et al. 1979; Catt et al. 1979: Schulster and Levitski 1980; Roth 1980; Falconer 1980), as is also that of their behavioural actions (de Wied 1977a, b; Guillemin 1977; Dunn and Gispen 1977; Denton and Nelson 1978; van Wimersma Greidanus et al.

1980). A consideration of some very limited and selected aspects here will serve to indicate the complexity of neurochemical events which may be involved in the conjoint action of these hormones, and the magnitude of the questions to be resolved by experiment when considering how specific neural systems subserving salt appetite are activated in the naive animal. The equal efficacy of the steroids and the peptide triplet in the naive male rabbit suggests that the response capacity of these neural systems is not modified by any sex difference of hormonal environment during a critical period of ontogenesis.

Prolactin

In mammary tissue, prolactin both participates in the regulation of alveolar cell differentiation and induces the secretory milk proteins after parturition. Its role in the mammalian secretory mechanism is set against a phylogenetic background of major osmoregulatory function in sodium conservation in fish and eels moving from salt to fresh water. It also controls sodium loss from the salt gland of marine birds, thus aiding osmoregulation. Falconer (1980) remarks that milk composition more closely resembles intracellular than extracellular fluid. Prolactin rapidly activated reciprocal transport of sodium and potassium in slices of lactating mammary tissue, which resulted in elevated tissue potassium and reduced tissue sodium concentration.

Turkington (1972) studied the basic mechanism of prolactin action in organ cultures of mouse mammary gland using physiological concentrations. Cortisol and insulin were shown to play important roles in contriving the conditions under which prolactin acts. Turkington showed that prolactin covalently linked to Sepharose beads stimulated RNA synthesis in this preparation, data which pointed to the fact prolactin action was not dependent on entry of the hormone into cells. Subsequent work located prolactin by autoradiography on the surface of the cell, and the ^{125}I-labelled hormone has been shown to be associated with the membrane (Falconer 1976). Shiu and Friesen (1974, 1976) isolated and purified the receptors, showed the characteristics of prolactin binding to them and produced an antibody to the receptors. Between 2 days pre-partum and day 6 of lactation in the rabbit, the receptor numbers rise tenfold, due predominantly to increased numbers of receptors per cell (Djiane et al. 1977). In the liver prolactin does enter the cells and is found associated with the Golgi vesicles. Turkington (1972) analysed the action of prolactin on mammary explants pretreated for more than 72 h with insulin and cortisol. An early effect was increased rate of synthesis of rapidly labelled nuclear RNA. The activity of DNA-dependent RNA polymerase was also stimulated, as was the rate of phosphorylation of histones and of certain non-histone nuclear proteins. Following this period of increased biosynthetic activity at the chromosomal level, the milk proteins were synthesized. The synthesis of specific casein phosphoproteins was undetectable until prolactin was added to the medium. The increased formation of ^{32}P-labelled casein or ^{14}C-labelled casein and increased activity of lactose synthetase were prevented by actinomycin D or cycloheximide, indicating that these effects of prolactin stimulation required concomitant synthesis of RNA and protein.

Turkington and colleagues also showed that increases in the intracellular concentrations of cyclic-AMP-activated protein kinase and the cyclic AMP binding protein occur rapidly and in coordination in response to addition of prolactin. Increases were prevented by actinomycin D and cycloheximide, the result being consistent with the conclusion that prolactin induces the formation of new protein molecules and does not merely activate the pre-existing proteins. However, prolactin, unlike several other pituitary polypeptide hormones, does not stimulate adenyl cyclase activity in the cell membranes of a target cell. Furthermore cyclic AMP in the culture medium was not able to replace the action of prolactin. The observations suggest that prolactin acts through a novel regulatory mechanism. It appeared the proteins with which cyclic AMP interacts were rate-limiting since they are rapidly induced by the hormone. Turkington (1972) suggested that the mechanism by which perturbation of the state of the plasma membrane influences the nucleus was the weakest link in our knowledge of prolactin's action.

Falconer and Rowe (1975) concluded from their experiments on mammary gland slices utilizing ouabain that the observed decrease in tissue sodium when prolactin is present in the incubation indicates that prolactin directly stimulates the transport of sodium out of the cell. This is presumably by increasing the activity of the Na,K-ATPase. Further, there is evidence that the prolactin receptors and the Na,K-ATPase are on the same region of the mammary alveolar cell membrane. They propose that it is likely that an early response of alveolar cells to prolactin is the activation of the sodium pump leading to a decreased intracellular sodium concentration and a marked change in the intracellular Na/K ratio. These events may act as an intracellular messenger in the biosynthetic events resulting from prolactin stimulation, as alterations in monovalent cation concentration have been shown to stimulate nuclear activity in, for example, insect salivary glands. Changes in intracellular calcium have also been implicated in the hormone stimulation of cells (Hales et al. 1977). Cyclic GMP has been shown by Rillema (1975) to stimulate RNA synthesis in organ cultures and mammary tissue which were cultured in the presence of insulin and cortisol. Cyclic GMP alone has no effect on casein synthesis but when added in conjunction with spermidine, casein synthesis was stimulated (Rillema et al. 1977).

Falconer (1980) drew attention to the fact that the multiplicity of biochemical and ultrastructural changes brought about by prolactin require gene activation but probably also involve post-transcriptional and other kinds of control in addition. With relation to the possible second messenger function for intracellular sodium and potassium, Falconer points out that the influences of ouabain and valinomycin, which suppress prolactin-induced lactogenesis, could be at a post-transcriptional level through interference with intermediary metabolism or enzyme synthesis. But the data on suppression of fatty acid synthesis by ouabain pointed to interference with transcription or translation. At present it is not clear how prolactin works in relation to gene transcription but the hypothesis of influence through change in intracellular sodium concentration is of interest in relation to the possibility that this change, of itself, in sodium deficiency may be the vector of other chemical events including transcription and translation leading to altered electrical activity in the neurones subserving salt appetite (Fig. 24-3).

ACTH

A direct effect of ACTH in stimulating salt appetite in the adrenalectomized rabbit was shown by Blaine et al. (1975).

The profound effects which ACTH and other pituitary peptides may have on a variety of cerebral processes including learning, memory, motivation and attention have been studied by de Wied and his colleagues (see de Wied 1977a, b, 1979; Bohus 1979; van Wimersma Greidanus and Versteeg 1980 for reviews). As de Wied points out, the influence of ACTH, α-MSH and neurohypophyseal hormones on adaptive behaviour was first suggested by the classical endocrinological method of the removal of the pituitary, the finding that the deficiency behaviour such as shuttle-box avoidance exists, and then the demonstration that this deficiency could be corrected by treatment with hormones produced by the extirpated gland. Furthermore, it was found that the behavioural impairment could be corrected with ACTH but also with α-MSH and fragments of ACTH/MSH which were devoid of corticotrophic effects. The data suggested that the behavioural effects of pituitary hormones were not mediated through an influence on their endocrine target tissues. Pituitary hormones therefore may act as precursor molecules for neuropeptides involved in the acquisition and maintenance of adaptive behaviour. Van Wimersma Greidanus et al. (1980) propose that the effects of ACTH and its congeners on behaviour, which are usually of a short-term nature, may be due to enhancement of the motivational strength of environmental stimuli, such as facilitation of selective arousal state in limbic structures. This would result in a higher probability of generating a stimulus-specific behavioural response.

The amino acid sequence (4–7) ACTH appears to have full intrinsic activity in relation to extinction of pole-jumping avoidance behaviour, but (7–9) ACTH also seems to contain information in this respect although expression of its full activity is apparent only after modifying the structure. These modifications include replacement of amino acids and elongation of the peptide chain yielding products which are more stable and protected against metabolic degradation. Van Wimersma Greidanus et al. have found that the substitution of Arg^8 by Lys is accompanied by loss of steroidogenic activity of (1–24) ACTH and of MSH activity. But this alteration in the molecule does not affect the behavioural activity. If the Trp^9 is replaced by Phe a marked decrease in steroidogenic potency is found but in the presence of D-Lys^8 the behavioural activity rises 100-fold. Substitution of Met^4 by methionine sulphoxide in such combinations (Org 2766) produces a molecule with behavioural activity 1000 times greater than that of (4–10) ACTH.

The action of ACTH on extinction of pole-jumping avoidance behaviour is mimicked also by peptides related to the C terminus of β-lipotropin (β-LPH): the endorphins. In particular, α-endorphin (β-LPH 61–76) appears to be even more potent than (4–10) ACTH in this respect. Influence of endorphins and of ACTH-like peptides on avoidance behaviour is not prevented by a specific opiate antagonist which blocks the anti-pain effects of β-endorphin. Unlike the analgesic effect of β-endorphin, which decreases when the peptide chain is shortened, the influence on avoidance behaviour increases. Thus, it is possible that the activity of β-endorphin on avoidance behaviour depends on the fragmentation of this peptide.

Though the effects of ACTH and endorphins on avoidance behaviour may be elicited after systemic administration, intracerebroventricular administration is more effective. Observations together with the demonstrated presence of ACTH and other pituitary hormones in the CSF and in the brain (Kleerekoper et al. 1972; Allan et al. 1974; Rudman et al. 1974; Krieger et al. 1977a, b; Oliver et al. 1977) raise the issue of production of ACTH or ACTH-like peptides in the brain or the transport of these hormones from the pituitary to the brain. Several authors have observed that blood in some of the vessels in the anterior pituitary flows to the hypothalamus (Torok 1964; Szentagothai et al. 1968). Page and Bergland (1977) found in the dorsum of the rabbit pituitary stalk, vessels which connect the posterior lobe of the pituitary to the hypothalamus, and these data have been elaborated by Bergland and colleagues. Bergland and Page (1978, 1979) suggested that part of the venous output of the anterior pituitary was by way of the vasculature of the posterior pituitary. There could be retrograde transport of pituitary principals to the hypothalamus and other brain areas. Intrapituitary injection of radio-labelled ACTH has been shown to cause uptake of hormone in the hypothalamus, but there was also uptake in a number of other areas of the brain. de Wied (1979) remarks that transport to the hypothalamus presumably is partly vascular via the stalk while transport to the other brain areas may occur via the CSF, though the neural route cannot be excluded.

In relation to the mode of action of ACTH, recent experiments and comments by Bristow et al. (1980) are important in relation to ACTH action on the adrenal gland. From their data using hormone fractions and antagonists, they propose that (1–39) ACTH physiological hormone can act via either of two receptors. Binding to one receptor, it elicits steroidogenesis through the involvement of cyclic AMP production. The binding to the other receptor elicits steroidogenesis through some other mechanism. The details of the other mechanism are as yet unknown, but they speculate that it might involve calcium and/or cyclic GMP, and perhaps explain the current controversy regarding the involvement or non-involvement of cyclic GMP in this response. From their studies they believe that (1–39) ACTH would work via cyclic AMP while (5–24) ACTH would tend to work via the other mechanism.

In relation to this phenomenon Schwyzer (1980) remarks that this is a further example of the pleiotropic action of the opiocortin gene through different mechanisms. Different portions of a peptide sequence can trigger widely different receptors, an ability which can be characteristic of linearly organized and flexible molecules. The fact that various hormones and neurotransmitters can act via different receptor mechanisms is well established and, in this respect, parallels might be drawn with the known mechanism of action of catecholamines on α- and β-adrenergic receptors (see Exton 1980).

Dunn and Gispen (1977) point out that receptors for ACTH have not been identified in brain as they have in the adrenal cortex. There are formidable problems of technique in their eyes as the concentration of any receptors is undoubtedly very low. ACTH has shown considerable structure-specific binding to opiate receptors in rat brain. Plomp and van Ree (1978) have shown the (7–10) fragment of ACTH will occupy the opiate receptor and has agonist activity. Dunn and Gispen (1977) suggested the Trp^9 and Gly^{10} are important in this interaction. They remark that it is to be expected that any profound or long-term effects of ACTH would be mediated by changes in gene expression. This would necessitate changes in RNA metabolism. However, there is little evidence for large effects of ACTH on brain RNA and there are difficulties in interpretation with the data available. There is, however, ample evidence that protein metabolism is affected in the brain by ACTH and this could be correlated with a prolonged effect on the brain. The important question is whether these effects on protein metabolism operate at the genome (transcriptional control) or at the ribosome (translational control). Dunn and Gispen suggest that the rate of protein synthesis in brain can be altered by ACTH or (4–10) ACTH. Since only (1–24) ACTH has been shown to alter RNA metabolism, it can be speculated that the (4–10) ACTH sequence is sufficient to produce translational effects whereas the full (1–24) ACTH sequence is necessary for transcriptional ones.

The blood–brain barrier is rather impermeable to ACTH (Allan et al. 1974), but fragments of the peptide, such as α-MSH or a (4–9) ACTH analogue, are taken up by the brain. A tritium-labelled (4–9) ACTH analogue (Org 2766) is taken up preferentially in the septal area following intraventricular injection. Hypophysectomy resulted in a significantly enhanced uptake in the septum and competitive studies suggested the possibility of receptor sites for the ACTH fragments. Experiments exploring the influence of ACTH fragments on salt appetite will be of great interest.

In relation to the influences of ACTH on salt

appetite, it could well be that breakdown of the molecule occurs before entry to the brain. A similar process may occur when salt appetite is caused by endogenous ACTH release: that is, when cortisol is withdrawn and basal DOC continued in the adrenalectomized animal, and external balance studies show unequivocally that no sodium loss occurs and yet there is a large salt appetite (Chapter 23). This fascinating situation raises three possibilities. First, that endogenous ACTH, or fragments of it, released into the systemic circulation crosses the blood–brain barrier to stimulate appetite. Secondly, that ACTH is transported via the pituitary stalk to the brain. Thirdly, that ACTH is synthesized within the brain. The particular importance of these data on salt appetite in this general field is that *the behaviour is a response to endogenous physiological production of the peptide* as distinct from injected material (with the attendant problems of the validity of concentration used and route).

Imura (1980) reports that he and his colleagues found that patients with Nelson's syndrome or Addison's disease with high plasma ACTH levels had no increase in CSF ACTH. He proposes that the discordance of ACTH, β-endorphin and α-MSH between plasma and CSF suggests that plasma hormones are not transported into the CSF, and that ACTH and related peptides in CSF probably originate from the brain. Rees and colleagues (Clement-Jones et al. 1979) in studies of heroin addicts showed that successful electroacupuncture for withdrawal symptoms was associated with a rise in CSF metencephalin levels in all patients, although concentration in blood did not alter.

As Brownstein (1980) remarks, about 20 years ago adrenocorticotrophic activity was found in extracts of hog and dog hypothalami (Guillemin et al. 1962; Schally et al. 1962). Guillemin et al. argued ACTH might have a hypothalamic origin. Data from bioassays and immunoassays showed ACTH is present in brains of rats, rabbits, cats, monkeys and man (Krieger et al. 1977a; Moldow and Yalow 1978; Orwoll et al. 1979). The amounts are lower than in the pituitary. ACTH precursor has a molecular weight of 31 000, and this glycoprotein undergoes proteolytic cleavage to yield various active products including α-MSH, β-MSH, β-lipotropin, β-endorphin and ACTH. Several immunoactive ACTH-like species have been found in the CNS and regional differences identified (Orwoll et al. 1979). Brownstein (1980) remarks that the 31 000 molecular mass precursor may be processed differently in different neurones, or peptide products may be metabolized at different rates in different regions.

ACTH-like peptides are more concentrated in the hypothalamus than elsewhere in the brain (Krieger et al. 1977a; Orwoll et al. 1979). Preoptic area, septum and amygdala are also moderately rich in ACTH, in contrast to cortex and cerebellum. Immunocytochemical studies have revealed ACTH in axon and nerve terminals throughout the brain (Larsson 1977; Pelletier and Leclerc 1979). The only ACTH-containing cell bodies detected are in and lateral to the arcuate nucleus (Watson et al. 1977; Dube et al. 1978; Matsukuma et al. 1978). ACTH does not disappear from the brain after hypophysectomy (Krieger et al. 1977b; Orwoll et al. 1979). It has been proposed that most (but not all) of ACTH in the central nervous system is in the processes of the arcuate neurones. Destruction of arcuate neurones results in a marked fall in brain ACTH without substantially altering pituitary levels (Krieger et al. 1979). The identification of big forms of ACTH in the brain suggests local synthesis and processing. Arcuate neurones in vitro incorporate ^{3}H-labelled amino acids into a protein of 31 000 molecular weight which reacts with affinity-purified anti-ACTH and anti-endorphin antibodies (Liotta et al. 1979).

In relation to this field, Brownstein (1980) makes some salutary observations regarding the identification of peptides in the brain. He emphasizes the eventual necessity to characterize the peptide products, as immunological methods alone have proven unequal to the task. Antibodies react typically with five or fewer amino acids in a peptide molecule, and can therefore react with peptides other than the one against which an antibody was raised, to give a positive radioimmunoassay reaction or immunochemical staining. Thus biologically distinct structurally related peptides may react with the same antibody (eg α-MSH and ACTH, cholecystokinin and gastrin, bombesin and substance P, thyrotrophin releasing hormone and urinary anorexogenic peptide). Precursors of a peptide can sometimes react with antibodies raised against the peptide itself. Also, post-translationally modified forms of a peptide may be bound by the same antibody that binds unmodified forms of the peptide, but have altogether different biological properties (e.g. sulphated and unsulphated cholecystokinin or gastrin). It is not realistic to ignore how peptides can be modified enzymatically: esterification, amidation, phosphorylation, glycosylation, ADP-ribosylation, acetylation and sulphation. *Thus Brownstein (1980) says it is obvious that until one knows which forms of peptide are present in the brain, and which forms are detected by the available antibodies, it will be impossible to do definitive*

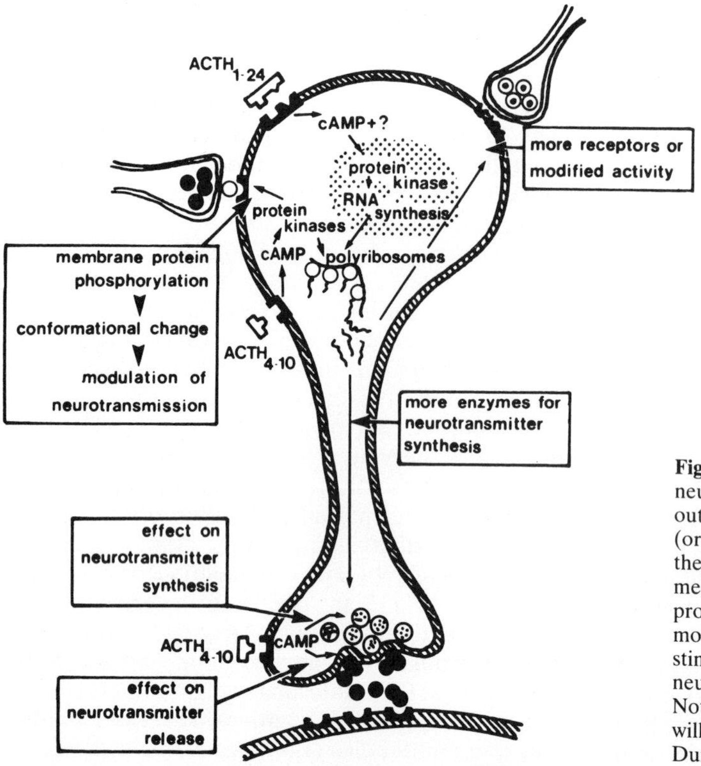

Fig. 24-34. Proposed mechanisms of ACTH action on neurones. ACTH interacts with specific receptors on the outside of the cell membrane, thus activating adenylate (or perhaps guanylate) cyclase. The cyclic nucleotides then activate protein kinases which may activate messenger RNA synthesis in the nucleus, activate protein synthesis by polyribosomes in the perikaryon, modify the properties of post-synaptic receptors, stimulate neurotransmitter synthesis or modulate neurotransmitter release, or any combination of these. Not all cells respond to ACTH and even responsive cells will only show one or a few of these responses. (From Dunn and Gispen 1977, with permission.)

mapping studies or to explore the peptide at defined synapses. It is possible that neurones release more than one active molecule and that they do not always release the same molecular forms. Neurones may conceivably secrete different forms of peptides at different synapses.

Dunn and Gispen (1977) remark that there is no doubt ACTH causes neurochemical actions. Their schematic illustration of possible actions of ACTH on brain cells is given in Fig. 24-34, but they remark that it is important to realize these different actions may not occur in the same cell. It is necessary to account for changes in protein and RNA metabolism in the perikaryon and for neurotransmitter metabolism, most probably at the synapse. By analogy with receptors at the adrenal cortex, they postulate cell surface receptors of which there are at least two types to account for effects of (4–10) ACTH and (1–24) ACTH. Receptor interaction stimulates synthesis of cyclic AMP and/or cyclic GMP and perhaps prostaglandins. The ionic environment may also be affected. The cyclic nucleotides act via protein kinase. Synthesis of RNA could be affected as in *E. coli* (Pastan and Pearlman 1970). Neurotransmitter synthesis may be stimulated, as occurs following protein kinase action on tyrosine hydroxylase. Permeability changes of membrane may occur, as have been reported with bladder and implicated in synaptic transmission. Further, the known interrelationships between cyclic AMP and calcium ion could permit cyclic AMP to alter the rate of release of neurotransmitters in which calcium is intimately involved. Dunn and Gispen (1977) cite substantial literature for the steps in the postulated mechanisms.

Clearly, in relation to the salt appetite of reproduction, where little is known of the role of ACTH, sustained intraventricular infusions of ACTH and ACTH fragments in the presence or absence of other hormones could provide data of great interest. Evidently, in the context of considerations such as those in the immediately preceding discussion of Dunn and Gispen's postulate, there are the issues of genesis of receptors, and the influence of hormone on their loss, or synthesis. These have been analysed *in extenso* by Roth et al. (1979) and Catt et al. (1979).

Oxytocin

That antidiuretic hormone (ADH) may be involved physiologically in memory processes was suggested by the fact that picogram amounts were active upon cerebral administration (de Wied 1976; see review of van Wimersma Greidanus and Versteeg 1980). As little as 25 pg of ADH injected into the lateral ventricle significantly inhibited extinction of the pole-jumping avoidance response, the dosage relating physiologically to

ADH levels in CSF. Systemic administration of oxytocin has a weak effect in delaying extinction of this response compared with ADH, but when given into the cerebral ventricles it was shown that oxytocin could impair retention of performance. This effect was ratified by showing that neutralization of bioavailable oxytocin in the brain by administration of specific oxytocin antiserum induced longer passive avoidance latencies than found in controls.

At this stage little has been done to analyse the role of oxytocin in salt appetite during reproduction. The quantitative data suggest that it has a smaller role than prolactin and ACTH, but definitive differentiating experiments have not been made. In the context of the problem of access to the brain, as was discussed with ACTH, studies indicate neurosecretory axons run to the ependyma of the ventricles (Vorherr et al. 1968; Vigh-Teichmann and Vigh 1974), and it has been suggested that neurohumours are released directly into the CSF from these terminals (Heller and Zaidi 1974). ADH and oxytocin have been measured in the CSF of several species (Dogterom et al. 1977). Possibly there is transport via CSF to sites of action.

Renin–angiotensin

Though, as discussed in Chapter 22, there is, as yet, no clear evidence of a physiological role of angiotensin II (AII) in salt appetite, the delay of 6–12 h in the production of its main effect on salt appetite remains a provocative fact in the light of what is known about the delay with the natural phenomenon of appetite.

Ganten and Speck (1978) propose that increased plasma AII can lead to suppression of the brain renin–angiotensin system. Conversely elevation of brain AII, e.g. by injection of renin into the brain ventricles, can lead to suppression of plasma renin (Reid and Day 1977). The choroid plexus appears to be of importance in this context. Renin and angiotensin I converting enzyme concentrations are very high in the choroid plexus, and preliminary evidence is claimed to show that circulating plasma AII may influence electrolyte handling by the choroid plexus (Ganten et al. 1976). High AII levels are accompanied by increased renin content in the choroid plexus. Further, acute unilateral carotid artery stenosis decreased and chronic unilateral occlusion increased choroid plexus renin—a parallelism to the kidney. Ganten et al. speculate that a feedback of circulating AII on the brain renin–angiotensin system may occur via changes in electrolyte concentrations across membranes such as the choroid plexus. Also, direct contact of the peripheral and central renin–angiotensin system would be possible via tanycytes or via brain sites where the blood–brain barrier was deficient. The subcellular localization of brain renin in the mitochondrial fraction and in purified synaptosomes as well as the intraneuronal occurrence of angiotensin suggests the formation of AII can occur intracellularly (Ganten and Speck 1978). It is possible that the enzyme–prohormone reaction occurs in intracellular compartments, e.g. the renin-containing granules.

Ganten and co-workers suggest that another possibility of angiotensin synthesis in brain tissue is indicated by the report of Kreutszberg et al. (1975) that large molecules such as the enzyme acetylcholinesterase, after being released into intracellular compartments, may be bound to the outer surface of the dendritic membrane to be later taken up by the dendrite. Thus a logical hypothesis would be that renin is released at dendritic points, bound to the membrane, and cleaves angiotensin from angiotensinogen which may be available in extracellular compartments or in the axoplasm. The angiotensin formed would then be directly active locally or could be taken up by the dendrite and transported to a more distant site of action. They note a parallel concept has been proposed for the biosynthesis of opioid peptides (Guillemin 1978). In elaborating, they propose that this model would be in harmony with the concept of intragranular hormone synthesis such as occurs with insulin, where several trypsin-like enzymes are bound to the inner surface of the granule membrane to produce insulin from proinsulin through multiple enzyme steps. Here what would be conceived is a stepwise catalysis of angiotensinogen to angiotensin I and conversion to AII and AIII, and possibly the activation of an inactive renin precursor (prorenin). They draw the parallel of a dual function of the true hormone circulating in blood and a local tissue role not distant from the site of synthesis as probably illustrated by several peptide hormones in addition to angiotensin: for example ACTH, ADH, prolactin, opioid peptides, and also catecholamines.

Within the brain the classification of the peptides as a hormone, neurohormone or neurotransmitter is difficult. Barker and Smith (1977) in drawing attention to the difference between neurohormonal communication and synaptic transmission note that the effective neurohormone concentrations are lower than those of transmitters, and that neurohormone receptors may be localized extrasynaptically and remote from the site of peptide synthesis. In relation to the action of neurohormones the mechanism may be different from classical synaptic transmission: in the latter only rapid kinetics are described while both rapid and slow kinetics appear to occur in neurohormonal communication. The possible direct chemical interaction of peptides with membranes is important (Mayer 1972; Bleich et al. 1976), as discussed above. Ganten and Speck (1978) suggest that this could result in a change of membrane structure and could influence the excitability of neurones in a manner different from that of amine synaptic transmission.

A possible role for dendrites in the synthesis of peptides has been indicated. Guillemin (1977) has discussed their role in low-voltage 'processing of information'. He states that there is already evidence that dendritic traffic of chemicals works both ways, with both release and uptake. Thus, in a reciprocally functioning system, if the endorphins and encephalins were enzymically cleaved extracellularly from β-lipotrophin as the circulating precursor in a fashion analogous to the biogenesis of angiotensin, they could then be picked up by multiple dendritic endings and be carried by retrograde axoplasmic flow whatever distance were necessary for their physiological function. Dendrites may be both pre-synaptic and post-synaptic to each other, as in reciprocal synapses (Schmitt et al. 1976). There is also electron microscopic evidence of gap junctions between dendrites. Electronic couplings may result in several neurones responding synchronously, with extremely low voltages being required. Examples of oscillatory behaviour have been observed when the population of neurones in some invertebrates (Gettings and Willows 1974) was electrically coupled. Schmitt et al. (1976) report that such electronic junctions are frequently observed in immediate proximity to chemical synapses. These phenomena do not require high-voltage spikes and none of the structures produce them. Information transfer by such mechanisms is relatively slow—in seconds or longer, not milliseconds. A moderately wide spread of a peptide so released could result in substantial influence on a large aggregate of nerve cells. This, of course, will depend on some neural aggregates being dedicated to particular functions and the cells concerned having a rich endowment of receptors for the specific peptide or peptides.

The hypothesis outlined above would seem important to take into account, along with others on multiple ionic and cellular metabolic effects of angiotensin set out in Chapter 22, when considering central action of angiotensin. It is noteworthy that concomitant iontophoresis of sodium and angiotensin excited hypothalamic units far more than application of either ion or peptide alone, leading to the suggestion that AII might influence specifically transport mechanims in sodium-sensitive hypothalamic cells (Wayner et al. 1973). Synergistic action of angiotensin and sodium on brain mechanisms controlling water and salt balance has been shown in the goat by Andersson and Westbye (1970) and Andersson and Ericksson (1971). However, at present it is not clear why angiotensin can induce the behaviour of water drinking within 1 min while there is a delay of 6–12 h in maximal induction of salt appetite. As considered in Chapter 22, a factor in this may be the time for effects of intraventricular infusion of angiotensin involving natriuresis and water drinking to produce a fall in CSF [Na], but this could be coincident with other actions as well. The extent of ACTH release (Severs and Daniels-Severs 1973; Phillips 1978) and corticosteroid secretion are other factors to be defined.

The further exploration of the action of hormones in the induction of salt appetite in the sodium-replete animal, as well as the appetite of sodium deficiency, will entail experiments using many of the rapidly developing techniques of molecular biology. As suggested above, the use of antibiotics which block various steps of the transcription–translation process, as has been done with central nervous actions of oestrogen, may give insight, though there will be the inherent difficulty in interpretation of the results of intracerebral administration of such substances. The failure of specific behaviour to develop in the presence of the inhibitor could be due to a general detrimental or toxic effect rather than the blockage of a precise step in a chemical chain of events generating activity in the neural tissues subserving salt appetite.

In the case of angiotensin the issue will be to show its generation within the brain in response to deficiency. This could arise as a result of production of renin in the brain in this situation followed by the classical cascade to derive angiotensin. Alternatively, it is formally possible that there are peptidergic neurones in which angiotensin is directly involved in the transmission process, and that a precursor or prohormone is synthesized in the neurone which is subsequently cleaved. In either instance, a new methodology developed at the Howard Florey Institute gives promise of resolving the matter. Hybridization histochemistry, a technique conceived by Dr John Coghlan (Hudson et al. 1981), allows determination of whether a hormone present in a group of cerebral cells or any other tissue is there by virtue of being stored or whether it is actually manufactured there. The principle involves reacting a cDNA probe radiolabelled with ^{32}P with a freshly

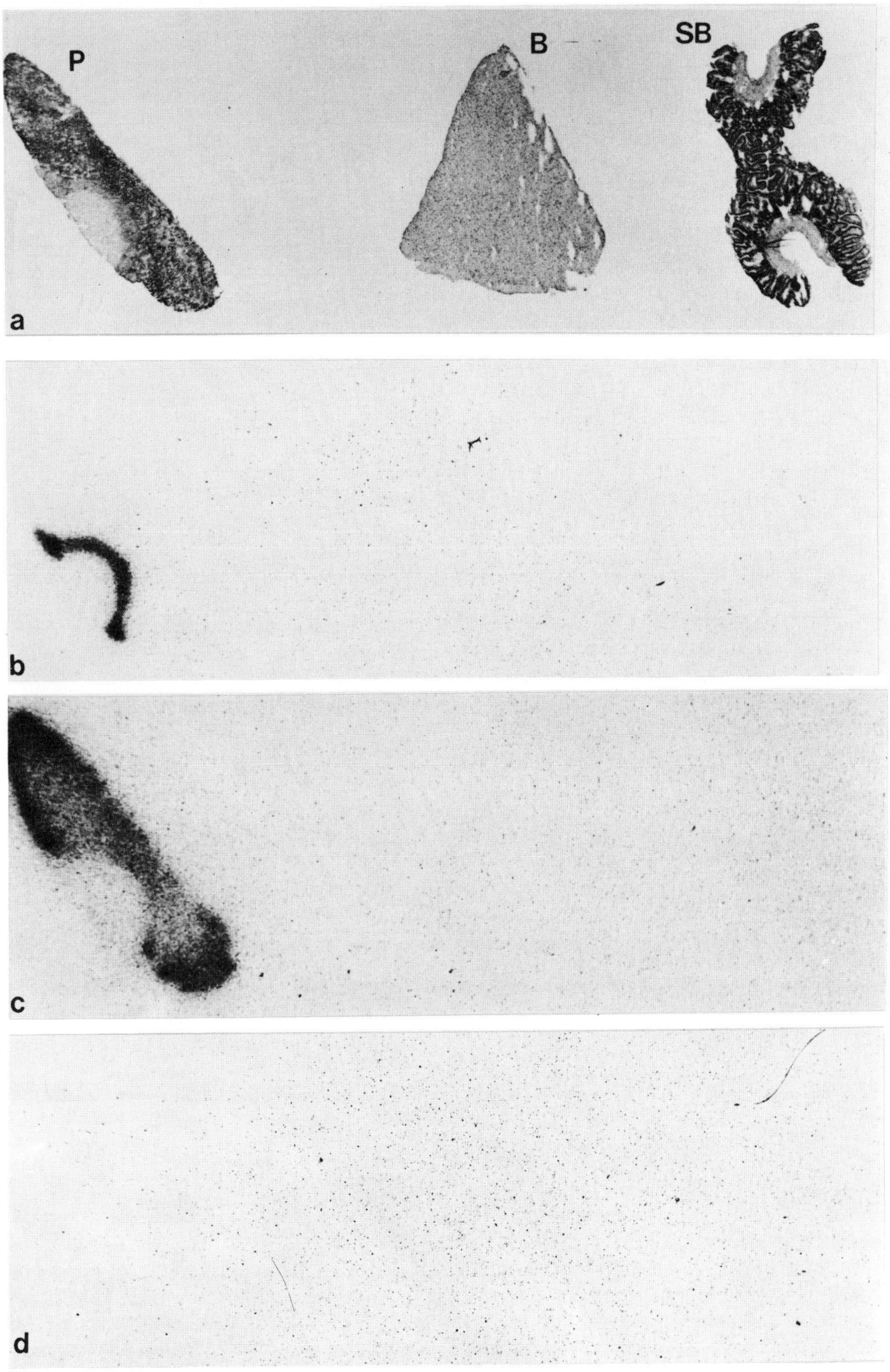

Fig. 24-35. (*a*) Sections (6 μm) of rat pituitary (P), brain (B) and small bowel (SB) stained with haematoxylin and eosin. Subsequent sections cut on the cryomicrotome were hybridized with ^{32}P-labelled cDNA probe corresponding to (*b*) a 150-base-pair fragment complementary to the endorphin-coding region of mouse corticotrophin/β-lipotropin messenger RNA, (*c*) an 800-base-pair fragment corresponding to the structural gene for rat growth hormone, and (*d*) a 3000-base-pair fragment of bacterial DNA (*Rhizobium trifolii*). (From Hudson et al. 1981, with permission.)

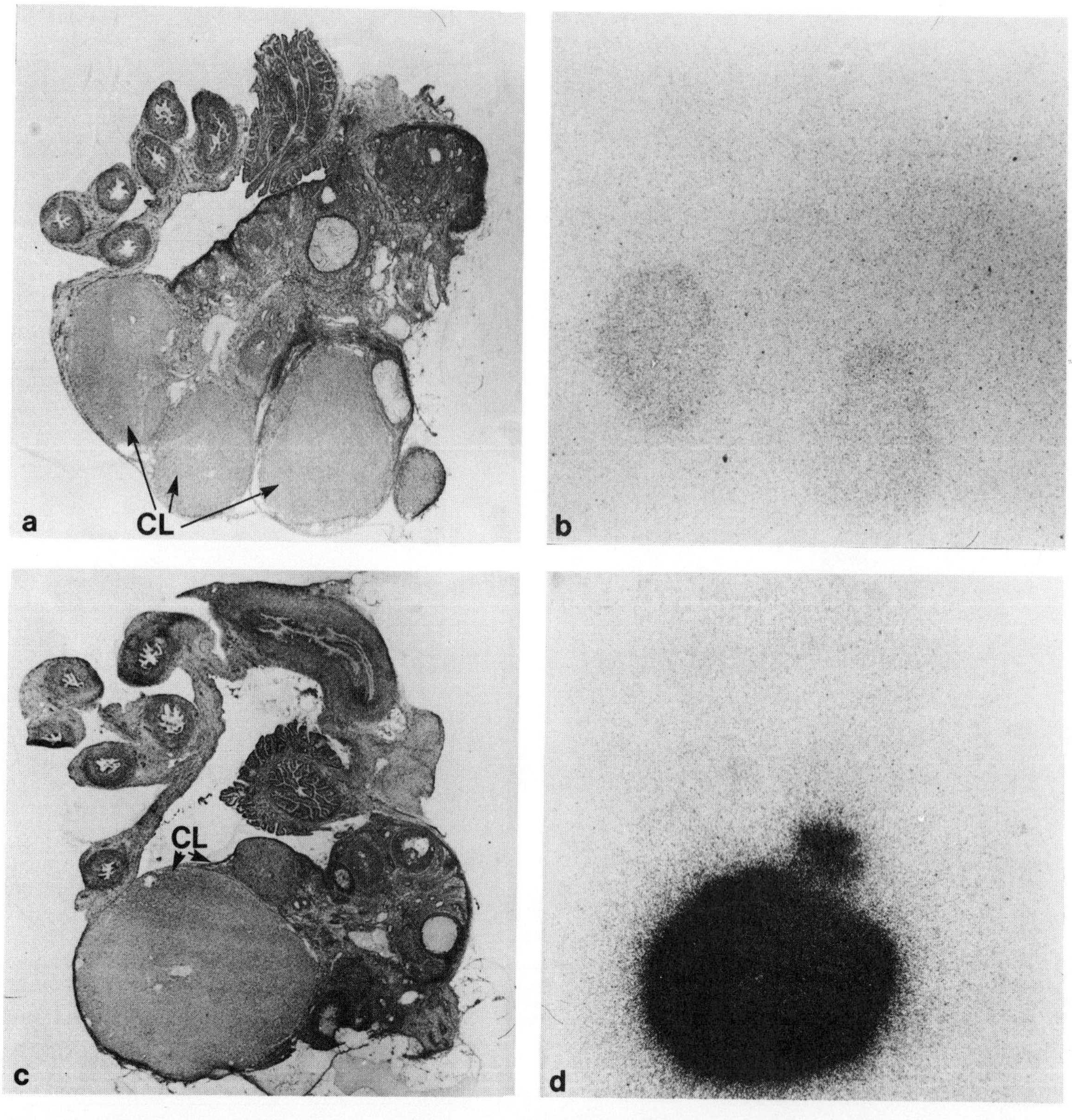

Fig. 24-36. Hybridization histochemistry used to detect the site of hormone synthesis at the messenger RNA level in rat ovaries. A relaxin-specific radiolabelled cDNA probe was synthesized, representing the entire coding region of rat preprorelaxin with only a small portion of the untranslated regions. (*a*) A non-pregnant rat ovary with corpora lutea marked CL (arrowed) and adjacent stroma, follicles and uterine tube. (*b*) Autoradiograph of (*a*) showing background labelling. (*c*) A late-pregnant rat ovary (18th day) with corpora lutea, adjacent follicles and uterine tube. (*d*) Autoradiograph of (*c*) showing heavy labelling of corpora lutea as the site of highest messenger RNA concentration in the ovary. ×10. (From Hudson et al. 1981, with permission.)

cut section of tissue to determine whether specific messenger RNA populations are present. After appropriate washing, the tissue is dried and the specific cell populations or regions binding the probe are identified by autoradiography. Thus hybridization histochemistry is similar in principle to immunohistochemical procedures.

In the first such study three recombinant DNA probes were used: a 150-base-pair fragment complementary to the endorphin-coding region of mouse corticotrophin/β-lipotropin (Roberts et al. 1979), an 800-base-pair fragment corresponding to the structural gene for rat growth hormone (Seeburg et al. 1977), and a 3000-base-pair fragment of bacterial DNA (*Rhizobium trifolii*) (Taylor et al. 1976). Fig. 24-35 shows sections of rat pituitary, brain and small bowel stained with haematoxylin and eosin, and subsequent sections hybridized with the ^{32}P-labelled cDNA probes. Other tissues not shown in the block labelled as did rat brain. The specificity of labelling is clearly demonstrated. Random absorption and entrapment of cDNA within the fixed time is not likely to be a great problem. The labelling of the anterior pituitary by growth hormone but with no labelling of the pars intermedia is consistent with previous immunohistochemical data (Nakane 1975). Adrenalectomy of the rats resulted in some labelling of the anterior pituitary by the endorphin probe. The method with use of a cDNA probe for rat relaxin has provided striking evidence on the changes in messenger RNA for rat relaxin over the course of pregnancy and post-partum in this species (Fig. 24-36). As suggested above, as a histochemical means of distinguishing biosynthesis from storage, the method has great promise, including elucidation of the role of polypeptides which appear in multiple sites (e.g. brain and gut). The authors observe that in a heterogeneous tissue it should be possible to identify the predominant messenger RNA species derived from a particular cell type. This could be achieved by screening clones or groups of clones derived from total messenger RNA for their ability to provide cDNA probes homing onto the cells of interest. Individual messenger RNAs coding for either known or unknown products could be identified through manipulation of the physiological state of the tissue. Further development envisaged is identification of virus infection within cells where virus culture is difficult, and determination of the hormone-producing behaviour of malignant tissues.

The specificity of hybridization histochemistry is inherent, the results are highly reproducible, and the results are not marred by high blanks. The minimum replication frequency which can be detected—i.e. the sensitivity—has yet to be fully explored. It seems possible that with development of appropriate cDNA probes there is good possibility of solution of some of the problems focused in the discussion above.

References

Abraham SF, Coghlan JP, Denton DA, McDougall JG, Mouw DR, Scoggins BA (1976) Increased water drinking induced by sodium depletion in sheep. Q J Exp Physiol 61: 185

Adey WR, Segundo JP, Livingstone RB (1957) Corticifugal influences on intrinsic brainstem conduction in cat and monkey. J Neurophysiol 20: 1

Adrian ED (1950) The control of nerve-cell activity. Symp Soc Exp Biol 4: 85

Ahren K, Fuxe K, Hamburger L, Hokfelt T (1971) Turnover changes in the tubulo-infundibular dopamine neurones during the ovarian cycle of the rat. Endocrinology 88: 1415

Aidley DJ (1971) The electrical properties of glial cells. In: The physiology of excitable cells. Cambridge University Press, London, p 185

Allan JP, Kendall JW, McGilvra, R, Vancura C (1974) Immunoreactive ACTH in cerebrospinal fluid. J Clin Endocrinol Metab 38: 586

Anderson NS, Fanestil DD (1976) Corticoid receptors in rat brain. Evidence for an aldosterone receptor. Endocrinology 98: 676

Andersson B, Ericksson L (1971) Conjoint action of sodium and angiotensin on brain mechanisms controlling water and salt balances. Acta Physiol Scand 81: 18

Andersson B, Westbye O (1970) Synergistic action of sodium and angiotensin on brain mechanisms controlling water and salt balance. Nature 228: 75

Anton-Tay F, Wurtman RJ (1971) Brain monoamines and endocrine functions. In: Martini L, Ganong WF (eds) Frontiers of endocrinology and neuroendocrinology. Oxford University Press, New York, p 45

Ashburner M (1972) Puffing patterns in *Drosophila melanogaster* and related species. Developmental studies on giant chromosomes. Results Probl Cell Differ 4: 101

Ashburner M, Cherbas P (1976) The control of puffing by ions—the Kroeger hypothesis: A critical review. Mol Cell Endocrinol 5: 89

Baerends GP (1950) Specializations in organs and movements with a releasing function. Symp Soc Exp Biol 4: 337

Baldwin BA (1976) Effects of intracarotid or intraruminal injections of NaCl or $NaHCO_3$ on sodium appetite in goats. Physiol Behav 16: 59

Bárány EH (1972) Inhibition by hippurate and probenecid of in vitro uptake of iodipamide and *o*-iodohippurate: A composite uptake system for iodipamide in choroid plexus, kidney cortex and anterior uvea of several species. Acta Physiol Scand 86: 12

Bard P (1940) The hypothalamus and sexual behaviour. Res Publ Assoc Res Nerv Ment Dis 20: 551

Bare JK (1949) A specific hunger for sodium chloride in normal and adrenalectomized white rats. J Comp Physiol Psychol 42: 242

Barker JL (1976) Peptides: Roles in neuronal excitability. Physiol Rev 56: 435

Barker JL, Smith TG Jr (1977) In: Cowan WM, Ferrendelli JA (eds) Approaches to the cell biology of neurones. Society for Neuroscience, Bethesda (Symposia, vol 2, p 340)

Barlow CF (1972) Physiology and pathophysiology of protein permeability in the central nervous system. In: Reulen HJ, Schumann K (eds) Steroids and brain edema. Springer, Berlin, p 139

Barnett SA (1967) Instinct and intelligence. Macgibbon & Kee, London

Baulieu EE (1981) Steroid hormones in the brain: Several mechanisms. In: Fuxe H, Gustafsson JA, Wellerberg L (eds) Steroid hormone regulation of the brain. Pergamon Press, London

Baulieu EE, Godeau F, Schorderet M, Schorderet-Slatkine S (1978) Steroid-induced meiotic division in *Xenopus laevis* oocytes: Surface and calcium. Nature 275: 593

Baumann G (1968) Zur Workung das Juvenilhormons: Elektrophysiologische Messungen an der Zellmembran der Speicheldruse von *Galleria mellonella*. J Insect Physiol 14: 1459

Beach FA (1942) Analysis of factors involved in the arousal, maintenance and manifestation of sexual excitement in male animals. Psychol Med 4: 173

Beach FA (1947) A review of physiological and psychological studies of sexual behaviour in mammals. Physiol Rev 27: 240

Beach FA (1955) The descent of instinct. Psychol Rev 62: 401

Beilharz S, Kay RNB (1963) The effects of ruminal and plasma

sodium concentrations on the sodium appetite of sheep. J Physiol (Lond) 165: 468
Beilharz S, Denton DA, Sabine JR (1962) The effect of concurrent deficiency of water and sodium on the sodium appetite of sheep. J Physiol (Lond) 163: 378
Beilharz S, Bott E, Denton DA, Sabine JR (1965) The effect of intracarotid infusions of 4 M NaCl on the sodium drinking of sheep with a parotid fistula. J Physiol (Lond) 178: 80
Berendes HD (1972) The control of puffing in *Drosophila hydei*. Developmental studies on giant chromosomes. Results Probl Cell Differ 4: 181
Bergland RM, Page RB (1978) Can the pituitary secrete directly to the brain? (Affirmative anatomical evidence.) Endocrinology 102: 1325
Bergland RM, Page RB (1979) Pituitary–brain vascular relations: A new paradigm. Science 204: 18
Bergmann F, Chaimovitz M, Costin A, Gutman Y, Ginath Y (1967) Water intake of rats after implantation of ouabain into the hypothalamus. Am J Physiol 213: 328
Blaine EH, Coghlan JP, Denton DA, Scoggins BA (1974) In vivo effects of ouabain on aldosterone, corticosterone and cortisol secretion in conscious sheep. Endocrinology 94: 1304
Blaine EH, Covelli MD, Denton DA, Nelson JF, Shulkes AA (1975) The role of ACTH and adrenal glucocorticoids in the salt appetite of wild rabbits (*Oryctolagus cuniculus* (L.)). Endocrinology 971: 793
Blair-West JR, Coghlan JP, Denton DA, Scoggins BA, Wintour EM, Wright RD (1969) The onset of effect of ACTH, angiotensin II and raised plasma potassium concentration on the adrenal cortex. Steroids 15: 433
Blake CA (1974) Localization of the inhibitory actions of ovulation blocking drugs on the release of luteinizing hormone in ovariectomized rats. Endocrinology 95: 999
Blake CA, Norman RL, Sawyer CH (1974) Localization of the inhibitory action of estrogen and nicotine on release of luteinizing hormone in rats. Neuroendocrinology 16: 22
Blaustein JD, Wade GN (1978) Progestin binding by brain and pituitary cell nuclei and female rat sexual behaviour. Brain Res 140: 360
Bleich HE, Cutnell JD, Day AR, Freer RJ, Glasel JA, McKelvy JF (1976) NMR observation of the interaction of small oligopeptides with phospholipid vesicles. Biochem Biophys Res Commun 71: 168
Bohus B (1979) Neuropeptide influences on sexual and reproductive behaviour. In: Zichella L, Pancheri P (eds) Psychoneurendocrinology in reproduction. Elsevier North-Holland, Amsterdam
Bott E, Denton DA, Weller S (1967) Innate appetite for salt exhibited by sodium deficient sheep. In: Olfaction and taste, vol II. Pergamon Press, London, p 415 (Proc 2nd Int Symp, Tokyo)
Bradbury MWB (1975) Ontogeny of mammalian blood–brain systems. In: Cserr HF, Fenstermacher JD, Fencl V (eds) Fluid environment of the brain. Academic Press, New York, p 81
Brightman MW (1965) The distribution within the brain of ferritin injected into cerebrospinal fluid compartments. I. Ependymal distribution. J Cell Biol 26: 99
Brightman MW, Palay SL (1963) The fine structure of ependyma in the brain of the rat. J Cell Biol 19: 415
Bristow AF, Geed C, Fauchere JL, Schwyzer R, Schulster D (1980) Effects of ACTH (corticotrophin) analogues on steroidogenesis and cyclic AMP in rat adrenocortical cells. Biochem J 186: 599
Brownstein MJ (1980) Peptidergic pathways in the central nervous system In: Neuroactive peptides. The Royal Society, London, p 79
Carthy J (1951) Instinct. New Biol 10: 95
Caspary RE (1968) Demyelinating diseases and allergic encephalomyelitis: A comparative review with special reference to multiple sclerosis. In: Cumings JN, Kremer M (eds) Biochemical aspects of neurological disorders, 3rd series. Blackwell, Oxford, p 44
Catt KJ, Harwood JP, Aguilera G, Dufau ML (1979) Hormonal regulation of peptide receptors and target cell responses. Nature 280: 109
Chain EB, Rose SPR, Masi I, Pocchiari F (1969) Metabolism of hexoses in rat cerebral cortex slices. J Neurochem 16: 93
Changaris DG, Severs WB, Keil LC (1978) Localization of angiotensin in rat brain. J Histochem Cytochem 26: 593
Cidlowski JA, Muldoon TG (1974) Estrogenic regulation of cytoplasmic receptor populations in estrogen-responsive tissues of the rat. Endocrinology 95: 1621
Clement-Jones V, McLoughlin L, Lowry PJ, Besser GM, Rees LH (1979) Acupuncture in heroin addicts: Changes in met-encephalin and β-endorphin in blood and cerebrospinal fluid. Lancet II: 380
Cofer CN, Appley MH (1967) Motivation: Theory and research. Wiley, New York
Cohen MW, Gerschenfeld HM, Kuffler SW (1968) Ionic environment of neurones and glial cells in the brain of an amphibian. J Physiol (Lond) 197: 363
Cone CD Jr (1980) Ionically mediated induction of mitogenesis in CNS neurons. In: Leffert HL (ed) Growth regulation by ion fluxes. Ann NY Acad Sci 339: 115
Contreras RJ (1978) Salt taste and disease. Am J Clin Nutr 31: 1088
Contreras RJ, Frank M (1979) Sodium deprivation alters neural responses to gustatory stimuli. J Gen Physiol 73: 569
Costlow ME, Buschow RA, McGuire WL (1975) Prolactin stimulation of prolactin receptors in rat liver. Life Sci 17: 1457
Crabbé J (1961) Stimulation of active sodium transport across the isolated toad bladder after injection of aldosterone to the animal. Endocrinology 69: 673
Craig W (1918) Appetites and aversions as constituents of instincts. Biol Bull 2: 91
Cross D, Dyer R (1971) Cyclic changes in neurones of the anterior hypothalamus during the rat oestrus cycle and the effects of anaesthesia. In: Sawyer CH, Gorski RA (eds) Steroid hormones and brain function. University of California Press, Berkeley
Cserr HF (1975) Bulk flow of cerebral extracellular fluid as a possible mechanism of cerebrospinal fluid–brain exchange. In: Cserr HF, Fenstermacher JD, Fencl V (eds) Fluid environment of the brain. Academic Press, New York, p 215
Cummins J, Hyden H (1962) Adenosine triphosphate levels and adenosine triphosphatases in neurones, glia and neuronal membranes of the vestibular nucleus. Biochim Biophys Acta 60: 271
Cushing H (1914) Studies on the cerebrospinal fluid. I. Introduction. J Med Res 31: 1
Darwin C (1872) The expression of the emotions in man and animals. Murray, London
Davson H (1967) Physiology of the cerebrospinal fluid. Churchill, London
Davson H, Kleeman CR, Levin E (1962) Quantitative studies of the passage of different substances out of the cerebrospinal fluid. J Physiol (Lond) 161: 126
Dell PC (1958) Some basic mechanisms of the translation of bodily needs into behaviour. In: Wolstenholme G (ed) The neurological basis of behaviour. Churchill, London (Ciba Foundation Symposium)
Denton DA (1966) Some theoretical considerations in relation to innate appetite for salt. Conditional Reflex 1: 144
Denton DA, Nelson JF (1971) Effect of pregnancy and lactation on the mineral appetites of wild rabbits (*Oryctolagus cuniculus* (L.)). Endocrinology 88: 31

Denton DA, Nelson JF (1978) Control of salt appetite in wild rabbits during lactation. Endocrinology 103: 1880
Denton DA, Nelson JF (1980) The influence of reproductive processes on salt appetite. In: Kare MR, Fregly MJ, Bernard RA (eds) Biological and behavioural aspects of salt intake. Academic Press, New York, p 229
Denton DA, Sabine JR (1961) The selective appetite for Na shown by Na-deficient sheep. J Physiol (Lond) 157: 97
Denton DA, Kraintz FW, Kraintz L (1969) Inhibition of salt appetite of sodium deficient sheep by intracarotid infusion of ouabain. Commun Behav Biol [A] 4: 183
Djiane J, Durand D, Kelly PA (1977) Evolution of prolactin receptors in rabbit mammary gland during pregnancy and lactation. Endocrinology 100: 1348
Dogterom J, van Wimersma Greidanus TjB, Swaab DF (1977) Evidence for the release of vasopressin and oxytocin into cerebrospinal fluid: Measurements in plasma and CSF of intact and hypophysectomized rats. Neuroendocrinology 24: 108
Douzou P, Maurel P (1977) Ionic control of biochemical reactions. Trans Biochem Sci 2: 14
Dube D, Lissitzky JC, Leclerc R, Pelletier G (1978) Localization of α-melanocyte-stimulating hormone in rat brain and pituitary. Endocrinology 102: 1283
Dufy B, Vincent J, Fleury J, Du Pasquier P, Gourdfi D, Tixier-Vidal A (1979) Membrane effects of thyrotropin-releasing hormone and estrogen shown by intracellular recording from pituitary cells. Science 204: 509
Dunn AJ, Gispen WH (1977) How ACTH acts on the brain. Biobehav Rev 1: 15
Edelhauser HF, Hoffert JR, Fromm PO (1965) In vitro ion and water movement in corneas of rainbow trout. Invest Ophthalmol Vis Sci 4: 290
Edwardson JA, Bennett GW (1974) Modulation of corticotrophin-releasing factor release from hypothalamic synaptosomes. Nature 251: 425
El-Etr M, Schorderet-Slatkine S, Baulieu E-E (1979) Meiotic maturation in *Xenopus laevis* oocytes initiated by insulin. Science 205: 1397
Epel D (1980) Ionic triggers in the fertilization of sea urchin eggs. In: Leffert HL (ed) Growth regulation by ion fluxes. Ann NY Acad Sci 339: 74
Ermish A, Ruhle HJ (1978) Autoradiographic demonstration of aldosterone-concentrating neuron populations in rat brain. Brain Res 147: 154
Evans RW, Sholiton LJ, Leavitt WW (1978) Progesterone receptor in the rat anterior pituitary: Effect of oestrogen priming and adrenalectomy. Steroids 31: 69
Exton JH (1980) Hormonal regulation of glycogen metabolism. In: Proceedings of the Sixth International Congress on Endocrinology. Australian Academy of Science, Canberra, p 323
Falconer IR (1976) Prolactin binding to plasma membranes, and its effect on monovalent cation transport in mammary alveolar tissue: A possible mechanism of action. In: Pecile A, Muller EE (eds) Growth hormone and related peptides. Excerpta Medica, Amsterdam, p 433
Falconer IR (1980) Biochemical actions of prolactin on the mammary gland. In: Proceedings of the Sixth International Congress on Endocrinology. Australian Academy of Science, Canberra, p 327
Falconer IR, Rowe, JM (1975) Possible mechanism for action of prolactin on mammary cell sodium transport. Nature 256: 327
Falconer IR, Forsyth IA, Wilson BM, Dils R (1978) Inhibition by low concentrations of ouabain of prolactin-induced lactogenesis in rabbit mammary-gland explants. Biochem J 172: 509
Feldman S, Sarne Y (1970) Effect of cortisol on single cell activity in hypothalamic islands. Brain Res 23: 67
Fenstermacher JD, Patlak CS (1975) The exchange of material between cerebrospinal fluid and brain. In: Cserr HF, Fenstermacher JD, Fencl V (eds) Fluid environment of the brain. Academic Press, New York, p 201
Fenstermacher JD, Rall DP (1973) Physiology and pharmacology of cerebrospinal fluid. In: Pharmacology of the cerebral circulation, vol I. Pergamon Press, Oxford, p 35
Fenstermacher JD, Rall DP, Patlak CS, Levin VA (1970) Ventriculocisternal perfusion as a technique for analysis of brain capillary permeability and extracellular transport. In: Crone C, Lassen N (eds) Capillary permeability. The transfer of molecules and ions between capillary blood and tissue. Scandinavian University Books, Copenhagen
Fisher AE (1964) Chemical stimulation of the brain. Sci Am 210: 60
Fukui HN, Merola LO, Kinoshita JH (1978) A possible cataractogenic factor in the Nakano mouse lens. Exp Eye Res 26: 477
Ganten D, Speck G (1978) The brain-angiotensin system: A model for the synthesis of peptides in brain. Biochem Pharmacol 27: 2379
Ganten D, Hutchinson JS, Schelling P, Ganten U, Fischer H (1976) The iso-renin angiotensin systems in extrarenal tissue. Clin Exp Pharmacol Physiol 3: 103
Ganten D, Fuxe K, Phillips MI, Mann JFE, Ganten U (1978) In: Ganong WF, Martini L (eds) Frontiers in neuroendocrinology, vol 5. Raven Press, New York, p 61
Gavin JR, Roth J, Neville DM, DeMeyts P, Buell DN (1974) Insulin-dependent regulation of insulin receptor concentration: A direct demonstration in cell culture. Proc Natl Acad Sci USA 71: 84
Geiger A, Yamasaki A, Lyons R (1956) Changes in nitrogenous components of brain produced by stimulation of short duration. Am J Physiol 184: 239
Gettings PA, Willows AOD (1974) Modification of neuron properties by electrotonic synapses. II. Burst formation by electrotonic synapses. J Neurophysiol 37: 858
Glynn IM (1964) The action of cardiac glycosides on ion movements. Pharmacol Rev 16: 381
Gorski RA, Gannon F (1976) Current models of steroid hormone action: A critique. Annu Rev Physiol 38: 425
Green R, Lutt GW, Whalen R (1970) Induction of receptivity in ovariectomized female rats by single intravenous injection of oestradiol-17β. Physiol Behav 5: 137
Guillemin R (1977) The expanding significance of hypothalamic peptides, or is endocrinology a branch of neuroendocrinology? Recent Prog Horm Res 33: 1
Guillemin R (1978) Biochemical and physiological correlates of hypothalamic peptides. The new endocrinology of the neuron. In: Reichlin S, Baldessarini RJ, Martin JB (eds) The hypothalamus. Raven Press, New York (Association for Research in Nervous and Mental Disease, vol 56, p 155)
Guillemin R, Schally AV, Lipscomb HS, Anderson RN, Long JM (1962) On the presence in hog hypothalamus of β-corticotropin releasing factor, α- and β-melanocyte stimulating hormones, adrenocorticotropin, lysine-vasopressin, and oxytocin. Endocrinology 70: 471
Guinane JE (1977) Cerebrospinal fluid pressure in mannitol treated rabbits. J Neurol Sci 34: 191
Haldane JBS (1956a) Les aspects physicochemiques des instincts. In: L'Instinct dans le comportement des animaux et de l'homme. Masson et Cie, Paris
Haldane JBS (1956b) The sources of some ethological notions. Br J Anim Behav 4: 162
Hales CN, Campbell AK, Luzio JP, Siddle K (1977) Calcium as a mediator of hormone action. Biochem Soc Trans 5: 866
Harlow HF (1969) William James and instinct theory. In: McLeod RB (ed) William James: Unfinished business.

American Psychological Association, Washington, p 21
Harris GW, Michael RP (1964) The activation of sexual behaviour by hypothalamic implants of oestrogen. J Physiol (Lond) 171: 275
Heisey SR, Held D, Pappenheimer JR (1962) Bulk flow and diffusion in the cerebrospinal fluid system of the goat. Am J Physiol 203: 775
Heller H, Zaidi SMA (1974) The problem of neurohypophysial secretion into the cerebrospinal fluid: Antidiuretic activity in the liquor and choroid plexus. In: Mitro A (ed) Ependyma and neurohormonal regulation. Vida, Bratislava, p 229
Hernandez-Peon R, Scherrer H, Jouvel M (1956) Modification of electric activity in cochlear nucleus during 'attention' in unanesthetized cats. Science 123: 331
Hinde RA (1960) Energy models of motivation. Symp Soc Exp Biol 14: 199
Hoffman HJ, Olsewski J (1961) Spread of sodium fluorescein in normal brain tissue: A study of the mechanism of the blood–brain barrier. Neurology (Minneap) 11: 1081
Hogan B (1974) Post-transcriptional control of protein synthesis. In: Weber R (ed) The biochemistry of animal development, vol III. Academic Press, New York
Hudson P, Penschow J, Shine J, Ryan G, Niall H, Coghlan JP (1981) Hybridization histochemistry: Use of recombinant DNA as a 'homing probe' for tissue localization of specific mRNA populations. Endocrinology 108: 353
Hyden H (1960) The neuron. In: Brachet J, Mirsky A (eds) The cell. Academic Press, New York, p 306
Hyden H, Egyhazi E (1962) Nuclear RNA changes of nerve cells during a learning experiment in rats. Proc Natl Acad Sci USA 48: 1366
Imura H (1980) ACTH, beta endorphin and related peptides. In: Proceedings of the Sixth International Congress on Endocrinology. Australian Academy of Science, Canberra, p 58
Jacob F, Monod J (1961) Genetic regulatory mechanisms in the synthesis of proteins. J Mol Biol 3: 318
Jones MT, Hillhouse EW (1976) Structure activity relationship and the mode of action of corticosteroid feedback on the secretion of corticotrophin-releasing factor (corticoliberin). J Steroid Biochem 7: 1189
Kato J, Onouchi T, Okinaga S (1978) Hypothalamic and hypophyseal progesterone receptors: Estrogen-priming effect, differential localization, 5α-dihydroprogesterone binding, and nuclear receptors. J Steroid Biochem 9: 419
Katz JJ, Halstead WC (1950) Protein organization and mental function. Comp Psychol Monogr 20: 1
Kawakami M, Sawyer CH (1959) Induction of behaviour on electroenkephalographic changes in the rabbit by hormone administration and brain stimulation. Endocrinology 65: 631
Kelly PA, Posner BI, Tsushima T, Friesen HG (1974) Studies of insulin, growth hormone and prolactin binding: Ontogenesis, effects of sex and pregnancy. Endocrinology 95: 532
Kelly PA, Posner BI, Friesen HG (1975) Effects of hypophysectomy, ovariectomy and cycloheximide on specific binding sites for lactogenic hormones in rat liver. Endocrinology 97: 1408
Kelly MJ, Moss RL, Dudley CA (1976) Steroid specific changes in preoptic-septal (POA-S) unit activity in normal cyclic female rats. Endocrine Society Program and Abstracts, 58th Annual Meeting, p 360
Kennedy JS (1954) Is modern ethology objective? Br J Anim Behav 2: 12
Kenney FP, Lee KL, Styles CD, Fritz JE (1973) Further evidence against post-transcriptional control of inducible tyrosine amino transferase synthesis in cultured hepatoma cells. Nature New Biol 246: 208
Kinoshita JH (1974) Mechanisms initiating cataract formation. Invest Ophthalmol Vis Sci 13: 713
Kleerekoper M, Donald RA, Posen S (1972) Corticotrophin in cerebrospinal fluid of patients with Nelson's syndrome. Lancet I: 74
Knigge KM, Silverman AJ (1972) Transport capacity of the median eminence. In: Knigge KM, Scott DE, Weindl A (eds) Brain endocrine interaction. Median eminence: Structure and function. Karger, Basel, p 350
Knigge KM, Morris M, Scott DE, Joseph SA, Notter M, Schock D, Krobisch-Dudley G (1975) Distribution of hormones by cerebrospinal fluid. In: Cserr HF, Fenstermacher JD, Fencl V (eds) Fluid environment of the brain. Academic Press, New York, p 237
Knobil E (1974) On the control of gonadotropin secretion in the rhesus monkey. Recent Prog Horm Res 30: 1
Koch KS, Leffert HL (1979) Ionic landmarks along the mitogenic route. Nature 279: 104
Korach KS, Ford EB (1978) Estrogen action in the mouse uterus: An additional nuclear event. Biochem Biophys Res Commun 83: 327
Kreutszberg GW, Schubert P, Lux HD (1975) Neuroplasmic transport in axons and dendrites. In: Santini MN (ed) Golgi Centennial Symposium. Raven Press, New York, p 161
Krieger DT, Liotta A, Brownstein MH (1977a) Presence of corticotropin in brain of normal and hypophysectomized rats. Proc Natl Acad Sci USA 74: 648
Krieger DT, Liotta A, Brownstein MJ (1977b) Presence of corticotropin in limbic system of normal and hypophysectomized rats. Brain Res 128: 575
Krieger DT, Liotta A, Nicholson G, Kizer J (1979) Brain ACTH and endorphin reduced in rats with monosodium glutamate-induced arcuate nuclear lesions. Nature 278: 562
Kroeger H (1966) Potentialdifferenz und puff-Muster. Elektrophysiologische und cytologische Untersuchungen an den Speicheldrusen von *Chironomus thummi*. Exp Cell Res 41: 64
Kroeger H (1977) The control of puffing by ions: A reply. Mol Cell Endocrinol 7: 105
Kuffler SW, Nicholls JG (1966) The physiology of neuroglial cells. Ergebn Biol 57: 1
Langendorf H, Siebert G, Lorenz I, Hannover R, Beyer R (1961) Kationenverteilung in Zellkern und Cytoplasma der Rattenleber. Biochem Z 335: 273
Larsson L-I (1977) Corticotropin-like peptides in central nerves and in endocrine cells of gut and pancreas. Lancet III: 1321
Lasansky A, Wald F (1962) Extracellular space in the toad retina as defined by the distribution of cerocyanine. A light and electron microscope study. J Cell Biol 15: 463
Lashley KS (1938) Experimental analysis of instinctive behaviour. Psychol Rev 45: 445
Lehrman DS (1953) A critique of Konrad Lorenz's theory of instinctive behaviour. Q Rev Biol 28: 337
Lezzi M (1966) Induktion eines Ecdyson-aktivierbaren Puff in isolierten Zeilkernen von *Chironomus* durch KCl. Exp Cell Res 43: 571
Lezzi M (1970) Differential gene activation in isolated chromosomes. Int Rev Cytol 29: 127
Lezzi M, Robert M (1972) Chromosomes isolated from unfixed salivary glands of *Chironomus*. Results Probl Cell Differ 4: 35
Liang T, Liao S (1975) A very rapid effect of androgen on initiation of protein synthesis in prostate. Proc Natl Acad Sci USA 72: 706
Lincoln DW, Cross B (1967) Effect of oestrogen on the responsiveness of neurones in the hypothalamus, septum and preoptic area of rats with light induced persistent oestrogen. J Endocrinol 37: 191
Liotta AS, Gildersleeve D, Brownstein MJ, Krieger DT (1979) In vitro biosynthesis of immunoreactive 31K ACTH β-endorphin-like activity of bovine hypothalamus. Proc Natl Acad Sci USA 76: 1448

Livingston RB (1959) Central control of receptors and sensory transmission systems. In: Field J, Magoun HW, Hall VE (eds) Handbook of physiology, vol I. American Physiological Society, Washington

Livingston RB (1967) Reinforcement. In: Quarton T, Melnechuk F, Schmitt FO (eds) The neurosciences. Rockefeller University Press, New York, p 568

Lorenz KZ (1950) The comparative method in studying innate behaviour patterns. Symp Soc Exp Biol 4: 221

Mark R (1974) Memory and nerve cell connections: Criticisms and contributions from developmental neurophysiology. Clarendon Press, Oxford

Marynick SP, Havens WW, Ebert MH, Loriaux DL (1976) Study on the transfer of steroid hormones across the blood cerebrospinal fluid barrier in the rhesus monkey. Endocrinology 99: 400

Masui Y, Markert CL (1971) Cytoplasmic control of nuclear behaviour during meiotic maturation of frog oocytes. J Exp Zool 177: 129

Masui Y, Meyerhof PG, Miller MA, Wasserman WJ (1977) Roles of divalent cations in maturation and activation of vertebrate oocytes. Differentiation 9: 49

Matsukuma S, Yoshimi H, Sucoka S, Kataoka K, Ono T, Ohgush N (1978) The regional distribution of immunoreactive β-endorphin in the monkey brain. Brain Res 159: 228

Mayer MM (1972) Mechanism of cytolysis by complement. Proc Natl Acad Sci USA 69: 2954

McCann SM, Kalra PS, Donoso AO, Bishop W, Schneider HPG, Fawcett CP, Krulich L (1972) The role of monoamines in the control of gonadotrophin and prolactin secretion. In: Knigge KM, Scott DE, Weindl A (eds) Brain–endocrine interaction. Median eminence: Structure and function. Karger, Basel, p 224

McDougall W (1923) An outline of psychology. Methuen, London

McDougall W (1930) The hormic psychology. In: Murchison C (ed) The international university series in psychology, II. Psychologies of 1930. Clark University Press, Worcester

McEwen BS (1977) Glucocorticoid receptors in neuroendocrine tissues. In: James VHT (ed) Endocrinology, vol 1. Excerpta Medica, Amsterdam, p 23

McEwen BS, Pfaff DW (1973) Chemical and physiological approaches to neuroendocrine mechanisms: Attempts at integration. In: Ganong WF, Martini L (eds) Frontiers in neuroendocrinology. Oxford University Press, New York

McEwen BS, Denef CJ, Gerlach JL, Plapinger L (1974) Chemical studies of the brain as a steroid hormone target tissue. In: Schmitt FO, Worden FG (eds) The neurosciences—Third study program. MIT Press, Cambridge, Mass.

McEwen BS, Krey LC, Luine VN (1978) Steroid hormone action in the neuroendocrine system: When is the genome involved? In: Reichlin S, Baldessarini RJ, Martin JB (eds) The hypothalamus. Raven Press, New York

McEwen BS, Davis PG, Parsons B, Pfaff DW (1979) The brain as a target for steroid hormone action. Annu Rev Neurosci 2: 65

McGowan-Sass BK, Timiras PS (1975) The hippocampus and hormonal cyclicity. In: Isaacson RL, Pribram KH (eds) The hippocampus, vol 1. Plenum Press, New York, p 355

Meites J, Lu KH, Wuttke W, Welsch CW, Nagasawa H, Quadri SK (1972) Recent studies on functions and control of prolactin secretion in rats. Recent Prog Horm Res 28: 471

Michael RP (1965) Oestrogens in the central nervous system. Br Med Bull 21: 87

Miller JW, Woollam DHM (1962) The anatomy of the cerebrospinal fluid. Oxford University Press, London

Miller NE (1957) Experiments on motivation. Science 126: 1271

Miller NE (1965) Chemical coding of behaviour in the brain. Science 148: 328

Moldow R, Yalow RS (1978) Extrahypophysial distribution of corticotropin as a function of brain size. Proc Natl Acad Sci USA 75: 994

Monroy A, Maggio R, Rinaldi AM (1965) Experimentally induced activation of the ribosomes of the unfertilized sea urchin egg. Proc Natl Acad Sci USA 54: 107

Mosher KM, Young DA, Munck A (1971) Evidence for irreversible Actinomycin-D sensitive, and temperature sensitive steps following binding of cortisol to glucocorticoid receptors and preceding effects on glucose metabolism in rat thymus cells. J Biol Chem 246: 64

Moss RL, Law OT (1971) The oestrus cycle: Its influence on single unit activity in the forebrain. Brain Res 30: 435

Mouw DR, Abraham SF, McKenzie JS (1974a) The use of ventriculo-cisternal perfusion in conscious sheep. Lab Anim Sci 24: 505

Mouw DR, Abraham SF, Blair-West JR, Coghlan JP, Denton DA, McKenzie JS, McKinley MJ, Scoggins BA (1974b) Brain receptors, renin secretion, and renal sodium retention in conscious sheep. Am J Physiol 226: 52

Muldoon TG (1980) The role of receptors in the mechanism of steroid hormone action in the brain. In: Motta M (ed) Neuroendocrine functions of the brain. Raven Press, New York

Nakane PK (1975) Identification of anterior pituitary cells by immunoelectron microscopy. Ultrastructure in Biological Systems 7: 45

Neckers L, Sze PY (1975) Regulation of 5-hydroxytryptamine metabolism in mouse brain by adrenal glucocorticoids. Brain Res 93: 123

Nicholls JG, Kuffler FW (1964) Extracellular space as a pathway for exchange between blood and neurones in the central nervous system of the leech: Ionic composition of glial cells and neurones. J Neurophysiol 27: 645

Ohno S, Lyon MF (1970) X-linked testicular feminization in the mouse as a non-inducible regulatory mutation of the Jacob–Monod type. Clin Genet 1: 121

Okazaki K, Shull KH, Farber E (1968) Effects of ethionine on adenosine triphosphate levels and ionic composition of liver cell nuclei. J Biol Chem 243: 4661

Okita GT, Talso PJ, Curry JH Jr, Smith FB, Geiling EMK (1955) Blood level studies of C-14-digoxin in human subjects of cardiac failure. J Pharmacol Exp Ther 113: 376

Olds J (1962) Hypothalamic substrates of reward. Physiol Rev 42: 554

Oliver C, Mical RS, Porter JC (1977) Hypothalamic–pituitary vasculature: Evidence for retrograde blood flow in the pituitary stalk. Endocrinology 101: 598

O'Malley BW, Schrader WT (1976) The receptors of steroid hormones. Sci Am 234: 32

O'Malley BW, Woo SLC, Harris SD, Rosen JR, Mean AR (1975) Steroid hormone regulation of specific messenger RNA and protein synthesis in eucaryotic cells. J Cell Physiol 85: 343

Orkand RK, Nicholls JG, Kuffler SW (1966) The effect of nerve impulses on the membrane potential of glial cells in the central nervous system of amphibia. J Neurophysiol 29: 788

Orwoll E, Kendall JW, Lamorena L, McGilvra R (1979) Adrenocorticotropin and melanocyte-stimulating hormone in the brain. Endocrinology 104: 1845

Page RB, Bergland RM (1977) The neurohypophyseal capillary bed. I. Anatomy and arterial supply. Am J Anat 148: 345

Palay SL (1956) Synapses in the central nervous system. J Biophys Biochem Cytol 2: 193

Palmer L, Civan MM (1975) Intracellular distribution of free potassium in *Chironomus* salivary glands. Science 188: 1321

Palmer RF, Lasseter KC, Melvin SL (1966) Stimulation of Na and K dependent adenosine triphosphatase by ouabain. Arch Biochem 113: 629

Pappenheimer JR (1967) The ionic composition of cerebral extracellular fluid and its relation to control of breathing. Harvey Lect 61: 71

Pappenheimer JR (1969) Transport of HCO_3^- between brain and blood. In: Crone C, Lassen NA (eds) Capillary permeability. Munksgaard, Copenhagen, p 454

Pappenheimer JR (1970) On the location of the blood–brain barrier. In: Coxon RV (ed) Proceedings of the Wales Symposium on the blood–brain barrier. Truex, Oxford, p 66

Pappenheimer JR, Fencl V, Heisey SR, Held D (1965) Role of cerebral fluid in control of respiration as studied in unanaesthetized goats. Am J Physiol 208: 436

Pappius HM (1975) Normal and pathological distribution of water in brain. In: Cserr HF, Fenstermacher JD, Fencl V (eds) Fluid environment of the brain. Academic Press, New York, p 183

Pastan I, Pearlman RL (1970) Cyclic adenosine monophosphate in bacteria. Science 169: 334

Pelletier G, Leclerc R (1979) Immunohistochemical localization of adrenocorticotropin in the rat brain. Endocrinology 104: 1426

Peters A, Palay SL (1965) An electron microscope study of the distribution and patterns of astroglial processes in the central nervous system. J Anat 99: 419

Pfaff DW (1972) Interactions of steroid sex hormones in brain tissue: Studies of uptake and physiological effects. In: Segal S (ed) The regulation of mammalian reproduction. Charles Thomas, Illinois, p 5

Pfaff DW, Silva MTA, Weiss JM (1971) Telemetered recording of hormone effects on hippocampal neurons. Science 172: 394

Pfaff DW, Diakow C, Zigmond RE, Kow L (1974) Neural and hormonal determinants of female mating behaviour in rats. In: Schmitt FO, Worden FG (eds) The Neurosciences—Third study program. MIT Press, Cambridge, Mass, p 61

Pfaffmann C (1952) Taste preference and aversion following lingual denervation. J Comp Physiol Psychol 45: 393

Phillips ME, Dafny N (1971) Effects of cortisol on unit activity in freely moving rats. Brain Res 25: 651

Phillips MI (1978) Angiotensin in the brain. Neuroendocrinology 25: 354

Piatigorsky J (1980) Discussion of control of cellular differentiation. In: Leffert H (ed) Growth regulation by ion fluxes. Ann NY Acad Sci 339: 308

Piatigorsky J, Shinohara T, Bhat SP, Reszelbach R, Jones RE, Sullivan MA (1980) Correlated changes in δ-crystallin synthesis and ion concentrations in the embryonic chick lens: Summary, current experiments and speculations. In: Leffert HL (ed) Growth regulation by ion fluxes. Ann NY Acad Sci 339: 265

Plomp GJJ, van Ree JM (1978) Adrenocorticotrophic hormone fragments mimic the effect of morphine in vitro. Br J Pharmacol 64: 223

Posner BI, Kelly PA, Freisen HG (1974) Induction of a lactogenic receptor in rat liver: Influence of estrogen and the pituitary. Proc Natl Acad Sci USA 71: 2407

Quadagno DM, Shryne J, Gorski RA (1971) The inhibition of steroid-induced sexual behaviour by intrahypothalamic actinomycin D. Horm Behav 2: 1

Rall DP (1967) Transport through the ependymal linings. In: Laytha A, Ford DM (eds) Progress in brain research, vol 29. Elsevier, Amsterdam, p 159

Rall DP, Oppelt WW, Patlak CS (1962) Extracellular space of brain as determined by diffusion of inulin from the ventricular system. Life Sci 1: 43

Rapoport SI (1975) Experimental modification of blood–brain barrier permeability by hypertonic solutions, convulsions, hypercapnia and acute hypertension. In: Cserr HF, Fenstermacher JD, Fencl V (eds). Fluid environment of the brain. Academic Press, New York, p 61

Rapoport SI (1976) Blood–brain barrier in physiology and medicine. Raven Press, New York

Reese IS, Brightman MW (1968) Similarity in structure and permeability to peroxidase of epithelia overlying fenestrated cerebral capillaries. Anat Rec 160: 414

Reid IA, Day RP (1977) Interactions and properties of some components of the renin–angiotensin system in brain. In: Bukney JP, Ferrario CM (eds) Central actions of angiotensin and related hormones. Pergamon Press, New York, p 267

Reith MEA, Schotman P, Gispen WH, de Wied D (1977) Pituitary peptides as modulators of neural functioning. Trends Biochem Sci 2: 856

Renaud LP (1979) Electrophysiology of brain peptides. In: Gotto AM Jr, Pech EJ Jr, Boyd AE (eds) Brain peptides: A new endocrinology. Elsevier/North-Holland, Amsterdam

Renaud LP (1980) Electrophysiologic effects of peptide hormones. In: Proceedings of the Sixth International Congress on Endocrinology. Australian Academy of Science, Canberra, p 479

Repke K (1963) Metabolism of cardiac glycosides. In: Proceedings of the International Pharmacological Meeting on Thirst vol 3, p 37

Reulen HJ, Kreysh HG (1973) Measurement of brain tissue pressure in cold induced cerebral oedema. Acta Neurochir (Wien) 29: 29

Richter CP (1956) Salt appetite of mammals, its dependence on instinct and metabolism. In: L'Instinct dans le comportement des animaux et de l'homme. Masson et Cie, Paris

Richter D (1965) Factors influencing the protein metabolism of the brain. Br Med Bull 21: 76

Rillema JA (1975) Cyclic nucleotides and the effect of prolactin on uridine incorporation into RNA in mammary gland explants of mice. Horm Metab Res 7: 45

Rillema JA, Linebaugh BE, Mulder JA (1977) Regulation of casein synthesis by polyamines in mammary gland explants of mice. Endocrinology 100: 529

Roberts JL, Seeburg PH, Shine J, Herbert E, Baxter JD, Goodman HM (1979) Corticotrophin and β-endorphin: Construction and analysis of recombinant DNA complementary to mRNA for the common precursor. Proc Natl Acad Sci USA 76: 2153

Roth J (1980) Receptors for peptide hormones. In: Proceedings of the Sixth International Congress on Endocrinology. Australian Academy of Science, Canberra, p 19

Roth J, Lesniak MA, Bar RS, Muggeo M, Megyesi K, Harrison L, Flier JS, Wachslicht-Rodbard H, Gorden P (1979) An introduction to receptors and receptor disorders. Proc Soc Exp Biol Med 162: 3

Rozengurt E, Mendoza S (1980) Monovalent ion fluxes and the control of cell proliferation in cultured fibroblasts. In: Leffert HL (ed) Growth regulation by ion fluxes. Ann NY Acad Sci 339: 175

Rozin P (1976) Selection of foods by rats, humans and other animals. In: Rosenblatt JS (ed) Advances in the study of behaviour, vol 6. Academic Press, New York

Rudman D, Scott JW, Del Rio AE, Houser DH, Sheen S (1974) Effect of melanotropic peptides on protein synthesis in mouse brain. Am J Physiol 226: 687

Ruf K, Steiner FA (1967) Steroid-sensitive single neurons in hypothalamus and midbrain: Identification by microelectrophoresis. Science 16: 667

Schally AV, Lipscomb HS, Long JH, Dear WE, Guillemin R (1962) Chromatography and hormonal activities of dog hypothalamus. Endocrinology 70: 478

Schally AV, Bowers BY, Carter WH, Arimura A, Redding TW, Saito M (1969) Effect of actinomycin D on the inhibitory response of oestrogen on LH release. Endocrinology 85: 290

Schmitt FO, Dev P, Smith BH (1976) Electrotonic processing of information by brain cells. Science 193: 114

Schulster D, Bristow AF (1980) Biochemical and genetic analysis of ACTH mechanism of action. In: Proceedings of the Sixth International Congress on Endocrinology. Australian Academy of Science, Canberra, p 315

Schulster D, Levitski A (1980) Cellular receptors for hormones and neurotransmitters. Wiley, New York

Schulster D, Burstein S, Cooke BA (1976) Molecular endocrinology of the steroid hormones. Wiley, New York

Schwyzer R (1980) Structure and function in neuropeptides. In: Neuroactive peptides. The Royal Society, London, p 5

Seeburg PH, Shine J, Martial SA, Baxter JD, Goodman HM (1977) Nucleotide sequence and amplification in bacteria of structural gene for rat growth hormone. Nature 270: 486

Severs WB, Daniels-Severs AE (1973) Effects of angiotensin on the central nervous system. Pharmacol Rev 25: 415

Shiu RPC, Friesen HG (1974) Solubilization and purification of a prolactin receptor from the rabbit mammary gland. J Biol Chem 249: 7902

Shiu RPC, Friesen HG (1976) Blockade of prolactin action by an antiserum to its receptors. Science 192: 259

Skou JC (1957) The influence of some cations on an adenosine triphosphatase from peripheral nerves. Biochim Biophys Acta 23: 394

Smith L, Ecker R (1971) The interaction of steroids with *Rana pipiens* oocytes in the induction of maturation. Dev Biol 25: 232

Spelsberg DC, Steggles AW, O'Malley BW (1971) Progesterone binding components of chick oviduct. III. Chromatin acceptor sites. J Biol Chem 264: 4188

Stellar E (1960) Drive and motivation. In: Field J, Mazour HW, Hall VE (eds) Handbook of physiology: Neurophysiology, vol 3. American Physiological Society, Washington

Stricker EM (1980) Physiological basis of sodium appetite: A new look at the 'depletion–repletion' model. In: Fregly MJ, Bernard RA, Kare MR (eds) Biological and behavioural aspects of salt intake. Academic Press, New York

Stricker EM (1981) Thirst and sodium appetite after colloid treatment in rats. J Comp Physiol Psychol 95: 1

Stricker EM, Wolf G (1966) Blood volume and tonicity in relation to sodium appetite. J Comp Physiol Psychol 62: 275

Stricker EM, Wolf G (1969) Behavioural control of intravascular fluid volume: Thirst and sodium appetite. Ann NY Acad Sci 157: 553

Swanson PD, McIlwain H (1965) Inhibition of the sodium ion-stimulated adenosine triphosphatase after treatment of isolated guinea pig cerebral cortex with ouabain and other agents. J Neurochem 12: 877

Swanson PD, Ullis K (1966) Ouabain induced changes in sodium and potassium content and respiration of cerebral cortex slices: Dependence on medium calcium concentration and effects of protamine. J Pharmacol Exp Ther 153: 321

Szentagothai J, Flerko B, Mess B, Halasz B (1968) Hypothalamic control of the anterior pituitary. Academiai Kiado, Budapest, 81: 62

Taylor JM, Illmersee R, Summers J (1976) Efficient transcription of RNA into DNA by avian sarcoma virus polymerase. Biochim Biophys Acta 422: 324

Teitelbaum P (1967) The biology of drive. In: Quarton G, Melnechuk T, Schmitt FO (eds) The neurosciences: A study programme. Rockefeller University Press, New York, p 557

Thomas RC (1972) Intracellular sodium activity and the sodium pump in snail neurones. J Physiol (Lond) 220: 55

Thorpe WH (1956) Learning and instinct in animals. Methuen, London

Tinbergen N (1951) The study of instinct. Oxford University Press, London

Tomkins GM, Levinson BB, Baxter JD, Dethlefsen L (1972) Further evidence for post-transcriptional control of inducible tyrosine aminotransferase synthesis in cultured hepatoma cells. Nature New Biol 239: 9

Torok B (1964) Structure of the vascular connections of the hypothalamo-hypophysial region. Acta Anat (Basel) 59: 84

Turkington RW (1972) Molecular biological aspects of prolactin. In: Wolstenholme G, Knight J (eds) Lactogenic hormones. Churchill Livingston, London

Ugolev AM, Roschina GM (1965) Changes in the appetite for water, glucose, sodium and potassium in rats when the sodium pump is depressed by the administration of strophanthin-K. Docl Akad Nauk SSSR 165: 1211

Verney EB (1947) The antidiuretic hormone and the factors which determine its release. Proc R Soc Lond [Biol] 135: 25

Vigh-Teichmann I, Vigh V (1974) Correlation of CSF contacting neuronal elements to neurosecretory and ependymosecretory structures. In: Mitro A (ed) Ependyma and neurohormonal regulation. Vida, Bratislava, p 281

Vogt M (1954) The concentration of sympathin in different parts of the central nervous system under normal conditions and after the administration of drugs. J Physiol (Lond) 123: 451

Vorherr H, Bradbury MWB, Houghough IM, Kleeman CR (1968) Antidiuretic hormone and cerebrospinal fluid during endogenous and exogenous changes in its blood level. Endocrinology 83: 246

Vrba R (1956) On the participation of glutamic acid–glutamine system in metabolic processes in the rat brain during physical exercise. J Neurochem 1: 12

Warembourg M (1975) Radioautographic study of the rat brain and pituitary after injection of ^{3}H-dexamethasone. Cell Tissue Res 161: 183

Waters MJ, Friesen HG, Bohnet HG (1978) Regulation of prolactin receptors by steroid hormones and use of radioligand assays in endocrine research. In: Birnbaumer L, O'Malley B (eds) Receptors and hormone action, vol 3. Academic Press, New York, p 457

Watson SJ, Barchas JD, Li CH (1977) β-Lipotropin: Localization of cells and axons in rat brain by immunocytochemistry. Proc Natl Acad Sci USA 74: 5155

Wayner MJ, Ono T, Nolley D (1973) Effects of angiotensin II on central neurones. Pharmacol Biochem Behav 1: 679

Weed LH (1923) The absorption of cerebrospinal fluid into the venous system. Am J Anat 31: 191

Weindl A, Joynt RJ (1972) Ultrastructure of the ventricular walls. Arch Neurol 26: 420

Weisinger RS, Denton DA, McKinley MJ (1977) Inhibition of water intake by ouabain administration in sheep. Pharmacol Biochem Behav 7: 121

Weisinger RS, Denton DA, McKinley MJ (1978) Effect of self-determined intravenous infusion of hypertonic NaCl on sodium appetite in sheep. J Comp Physiol Psychol 92: 522

Weisinger RS, Considine P, Denton DA, McKinley MJ, Mouw D (1979) Rapid effect of change of cerebrospinal fluid sodium concentration on salt appetite. Nature 280: 490

Weisinger RS, Considine P, Denton DA, Leksell L, McKinley MJ, Mouw D, Muller A, Tarjan E (to be published) Effect of change of sodium concentration and osmolality of cerebrospinal fluid on the salt appetite of sheep. Am J Physiol

Wellen JJ, Benraad ThJ (1969) Effect of ouabain on corticosterone biosynthesis and on potassium and sodium concentration in calf adrenal tissue in vitro. Biochem Biophys Acta 183: 110

Whittam R (1967) The molecular mechanism of active transport. In: Quarton G, Melnechuk T, Schmitt F (eds) The neurosciences. Rockefeller University Press, New York, p 313

Wied D de (1976) Behavioural effects of intraventricularly administered vasopressin and vasopressin fragments. Life Sci 19: 685

Wied D de (1977a) Peptides in behaviour. Life Sci 20: 195

Wied D de (1977b) Behavioural effects of neuropeptides

related to ACTH, MSH and beta LPH. Ann NY Acad Sci 297: 263
Wied D de (1979) Pituitary neuropeptides in behaviour. In: Fuxe K, Hokfelt T, Luft R (eds) Central regulation of the endocrine system. Plenum, New York
Wimersma Greidanus TjB van, Versteeg DHG (1980) Neurohypophysial hormones: Their role in endocrine function and behavioural homeostasis. In: Nemeroff CH, Dunn AJ (eds) Behavioural neuroendocrinology. Spectrum, New York
Wimersma Greidanus TjB van, Bohus B, van Ree JM, de Wied D (1980) Behavioural effects of neuropeptides as related to ACTH, endorphins and neurohypophyseal hormones. In: Proceedings of the Sixth International Congress on Endocrinology. Australian Academy of Science, Canberra
Wise BL, Chater M (1962) The value of hypertonic mannitol solution in decreasing brain mass and lowering cerebrospinal fluid pressure. J Neurosurg 19: 1038
Wise BL, Perkins RK, Stevenson E, Scott KG (1964) Penetration of ^{14}C-labelled mannitol from serum into cerebrospinal fluid and brain. Exp Neurol 10: 264
Wislocki B, Putnam TJ (1921) Absorption from the ventricles in experimentally produced internal hydrocephalus. Am J Anat 23: 313
Wolf G, Stricker EM (1967) Sodium appetite elicited by hypovolaemia in adrenalectomized rats: Re-evaluation of the reservoir hypothesis. J Comp Physiol Psychol 63: 252
Wolf G, McGovern JF, Dicara LV (1974) Sodium appetite: Some conceptual and methodologic aspects of a model drive system. Behav Neurol Biol 10: 27
Woodbury JW (1971) An epilogue. A hypothetical model for CSF formation and blood brain barrier function. In: Siesjö BK, Sorensen SC (eds) Ion homeostasis in the brain. Munksgaard, Copenhagen, p 465
Wright EM (1972) Mechanisms of ion transport across the choroid plexus. J Physiol (Lond) 226: 545
Yoshida H, Kaniike K, Fujisawa H (1962) Studies in the change in ionic permeability of brain slices. Jpn J Pharmacol 12: 146
Zeigler HP (1964) Displacement activity and motivational theory: A case study in the history of ethology. Psychol Bull 61: 362
Zigmond RE, McEwen BS (1970) Selective retention of estradiol by cell nuclei in specified brain regions of the ovariectomized rat. J Neurochem 17: 889

25 The appetite for phosphate, calcium, magnesium and potassium, and the question of learning

Summary

1. Individual built-in systems for intake of all components needed by the body is an unlikely phylogenetic development. It is clear, however, that some such systems of high survival value have emerged. One major determinant of whether this happens is the likelihood of the substance occurring free in nature in easily detectable form. An alternative general-purpose innate regulation resides in the neophilia developed by animals in the face of deterioration of condition on a deficient diet. Neural mechanisms involving chemoreception and the nucleus of the tractus solitarius facilitate greatly the learning of association between ingestion of food with particular taste qualities and subsequent effects, whether they be a sense of well-being, or nausea and toxicity.

2. The craving for bones, virtually the only accidentally available source of phosphate in nature, which is exhibited by cattle and wild game on phosphate-deficient pastures, has been reported in many parts of the world over the past two centuries. It is established that the behaviour is associated with a low blood phosphate concentration. Deer and other wild animals chew long bones in a fashion to create beautifully forked patterns which have resulted in them being mistaken for hominid artefacts. In southern Texas on the phosphate-deficient pastures of the King Ranch, it has been shown cattle will voluntarily ingest adequate disodium hydrogen phosphate from troughs placed in the prairie, and maintain good condition.

3. Parathyroidectomy in rats induces a large calcium appetite and an aversion for phosphate solutions.

4. Experiments on animals with a calcium-deficient diet have led to the conclusion that calcium appetite is learned on the basis of retroactive learned benefit of the calcium. But contrary to this viewpoint are the data on parathyroidectomy, and also those on the calcium appetite induced in calcium-replete animals by pregnancy or administration of the steroid and peptide hormones involved in reproduction. It would seem elements of calcium appetite are preprogrammed.

5. There is little clear-cut evidence of a specific potassium or magnesium appetite.

6. It has been possible in the laboratory to reproduce the appetite for bones manifest by cattle in the field. Cattle were fed a low-phosphate diet. Also a parotid fistula was made and sodium loss from the body replaced; thus the fistula drained phosphate from the body and large fall in blood phosphate occurred. Experiments under cafeteria conditions showed the cattle had a specific appetite for new ground bone or for bird faeces, which might be regarded as another naturally occurring source of phosphate. The centres of

control of phosphate appetite can be accessed, in so far as rapid intravenous infusion of phosphate bringing about an increase in plasma phosphate to normal range causes appetite to disappear after a variable interval—a minimum of ½–1 h and, in the case of lactating animals, usually only after 1–3 days of infusion. The mechanism of induction of appetite, the basis of recognition of bone, the centres of control, and the role of hormones are presently unknown. The data on twin-cattle experiments, where there is immediate recognition of bone by the deplete animal, strongly suggest the appetite is innate.

7. The choice of bone is discussed in the general context of the basis of recognition of needed substances. The Richter experiment where rats choose an adequate diet from a cafeteria of 17 components focuses the issue of how retroactive benefit could be learned if several components are sampled concurrently. Olfactory stimuli may be involved in innate acceptance and rejection behaviour. Given that differing deficiencies may produce quite different changes in blood chemistry (e.g. calcium vs vitamin deficiency), the genesis of the neophilia seen in experimental animals may involve activation of an innate system by a general affective state. This could be caused by disordered metabolism coupled with sensory inflow from gastrointestinal tract via the tenth cranial nerve.

8. In an assessment of the organization of self-regulatory capacities in animals, the extent to which opinion is governed predominantly by experiments on the domesticated inbred albino *Rattus norvegicus* needs to be recognized. Domestication has been shown to have caused profound changes relative to the wild rat. The question thus arises of whether the white rat is the ideal species in which to study naturally selected capacities of high survival value directed to the recognition and ingestion of nutritious substances by the feral animal. As an example, there are 36 genera and 124 species of Old and New World rats in Australia and New Guinea alone, and little is known of their behaviour, endocrinology and physiology.

It is established that metabolic need of minerals other than sodium may result in specific increased intakes of them. Study of how this comes about can throw more light on the organization of sodium appetite, apart from the intrinsic interest of the problems themselves. The situation with sodium, accordingly, is set in the broader biological context of the overall regulation of minerals in the organism. Appetite for vitamins again enters the discussion. Analysis of vitamin appetite experimentally involves many of the same issues, because they are the issues of regulation in the animal living wild in nature.

How does an animal, when it is metabolically deficient, choose the specific substance it needs and thus benefit itself? Two possibilities are:

1) The choice is genetically programmed. When an animal is deficient in a substance the decreased level in the body acts on the nervous system to make some stimulus, most likely taste or smell, much more acceptable, and generative of intake. There is an innate genesis of an avidity, though such mechanisms may be modified by experience to more elaborate behaviour with greater survival value.
2) The increased preference is a consequence of initial experience of need-reduction, and perhaps sense of well-being, which follows ingestion and absorption: a retroactive learning of a sense of benefit.

The innate avidity theory involves the animal's

immediate acceptance of the substance on first encounter when in the deficient state. As a corollary, it has to be recognized that there needs to be a reasonable chance of the animal encountering the substance relatively pure in nature, or in significant to high concentration marked by some regularly associated cue. This is likely to be a necessary condition for natural selection to favour the emergence of such a genetically programmed system. We have presented evidence that this is so with sodium. The same line of reasoning, if we allow that lithium does not occur to even trivial degree free in nature, would suggest that there has been no selection pressure favouring any capacity to distinguish the taste of sodium from lithium. The ingestion of lithium will poison the organism.

Overall, built-in systems for intake of all components needed by the body is too much to expect, as Rozin (1976) notes. But it is clear that some which have high survival importance have emerged. In the case of learned appetites, the neophilia behaviour described by Rozin and Rodgers (1967) would represent a general-purpose type of innate behaviour of very high survival value. It would set the behavioural stage. It would favour learning of benefit from repair of deficiency in any new environment, or a changed and potentially deleterious feeding situation. Storage in memory could allow immediate utilization in the same circumstances in the future. Such learned behaviour may be socially transmissible in animals and certainly is so in humans. Presumably the smorgasbord habits in Scandinavia coupled with pickling of materials, some at the time of the height of their vitamin C content (customs which arose before the scientific information which would ratify them), reflect this process. As already noted, Garcia and Ervin (1968) have shown elegantly that learning of the association of sensation localized to the alimentary canal with taste and smell stimuli presented several hours earlier is greatly facilitated through neural mechanisms involving the nucleus of the tractus solitarius. The efficiency of learning such associations is very much greater than with visual or auditory cues.

Against this background I will consider the appetite for phosphate. Following an account of the history and some early experimental data I will deal with laboratory experiments on calcium, magnesium and vitamin deficiency since they highlight the implications of some recent work on phosphate appetite in cattle which will then be described.

Phosphate appetite in pastoral and wild animals

Depraved appetite in cattle and wild animals in the phosphorus-deficient soils of southern Africa has been recognized for centuries. In 1796 Le Vaillant described the eager search by cattle in the region for bones left by his dogs. When these were lacking the animals gnawed wood or even each other's horns. Theiler et al. (1924) analysed the osteophagia (bone-chewing) and its relation to phosphate deficiency. They noted that phosphorus was unquestionably a limiting factor in the growth rate of cattle and a dominating factor in the maintenance of live weight under ordinary conditions of veld grazing. They state phosphorus deficiency is a general characteristic of South African soils, though it may vary in different districts. Osteophagia was the most obvious outward sign or clinical symptom of phosphorus deficiency in cattle. Though not infallible, it was a very valuable indicator of such deficiency in the majority of veld-bred bovines. It was not shown to the same extent by other animals, and was only feebly manifest in sheep.

In extreme cases the osteophagia may pass over into allotriophagia in which specific discrimination is lost and the animal will chew any sort of rubbish at all. In its mild form the abnormal appetite is quite finely adjusted and only the best sun-bleached bones are eaten. In its more aggravated forms any putrefying bones may be eaten, and it is not an uncommon sight to see extreme cravers crowding round the rotten carcass of a dead animal. Given the nice discrimination displayed in typical osteophagia, it was possible to sort for experimental purpose the marked cravers in a herd from the slight cravers and non-cravers. This was done by offering first sterilized rotten bones, i.e. bones still possessing a distinctly putrid odour. The animals which chewed such bones were classified as exhibiting marked osteophagia. Those animals that did not show this behaviour were then allowed access to sweet bones, i.e. bones that had been well bleached in the sun and had no putrid odour. Any animals that would chew such bones were regarded as having mild osteophagia, and the remaining animals that showed no predilection as having no osteophagia. Provision of only 85 g of bone-meal per animal per day made, over a period of some months, a difference of 48 kg per head in the weight of cattle grazing the veld.

Theiler et al. (1924) noted that while the osteophagia nearly always (with the exception of a few habitual chronic cravers) indicated nutritional need for phosphorus, quite a few animals (especially young stock) failed to display osteophagia even when a nutritional need could be demonstrated by a dietetic experiment. Further, osteophagia only appeared (in general) at a level of deficiency distinctly below the optimum requirements from a nutritional point of view. There was quite a sizeable weight-gain difference of 100 kg in 15 months in favour of animals given excess bone-meal over those given only enough to stop osteophagia. In this context the relation between physiological state (i.e. pregnancy and lactation) and phosphorus requirement was clearly noted. Theiler et al. say that whereas phosphorus deficiency is manifest the whole year round, it may reduce or temporarily disappear in some animals in the spring when the youngest new growth becomes available. Other compounds such as chalk, iron sulphate, sulphur and so on had no influence on the osteophagia. These interesting animal husbandry data indicate that there is a metabolic threshold before the osteophagia appears.

It is an interesting and provocative fact that physiological deficiency of phosphorus should be reflected in the craving for bones, which are almost the only accidentally available source of phosphorus capable of relieving the condition. The behaviour is widespread in a deficient herd (though a small percentage of animals may not show it) which, as Green (1925) suggests, may mean that the animals with some indiscriminate induced craving randomly happen on and chew objects including bones and eventually associate the sense of benefit with the taste. Perhaps some cattle imitate ‘a trick of the herd’ and then learn that it satisfies a need. Green states that young cattle brought up in a deficient area carefully cleaned of all bone debris have been found to manifest osteophagia the first time bones were made available to them. If these observations are valid they may indicate some taste preference for bones under the metabolic condition of phosphorus deficiency.

Bone-chewing has been described in many other species including reindeer, red deer, camels, giraffes and wildebeest. An extensive scholarly examination of the problem has been made by Sutcliffe (1973). He was initially interested in the similarity of bones and antlers gnawed by deer to some proposed human artefacts. He noted that osteophagia in natural conditions has a distinct geographical distribution which depends principally on the phosphorus content of the parent rock on which food plants are growing but also on other factors such as excessive calcium, aluminium or iron which can reduce the availability of phosphorus to plants. Fraser-Darling (cited by Sutcliffe 1973) described how even before modern advances in the understanding of animal nutrition, the inhabitants of the Atlantic townships of the Hebrides ground the backbones of fish to make a meal consisting largely of calcium phosphate which was fed to their cows during the winter. In this way the lime-rich and phosphate-poor soils of the area were able to support cattle. This industry has now disappeared but in the 1940s cattle were often seen beachcombing for bones along the shore.

Sutcliffe describes accounts by various observers he lists of cows eating sheep skulls on the Isle of Skye, of crossbred Highland cattle chewing bones, and of antlers chewed by Scottish red deer. Red deer have been seen chewing rabbit carcasses and chewing their neighbours’ antlers. Kelsall (cited by Sutcliffe 1973) has given a detailed description of antler-gnawing by caribou; in the Bathurst Inlet area of Canada it was unusual to find any caribou antlers, even large ones, which had not been chewed. Cowles (cited by Sutcliffe 1973) described two camels squabbling over the skeleton of a goat already clear of flesh and skin in the mountains of Oman. One camel was standing with its head in upright position with the bones hanging out of its mouth. It was crunching the bones and swallowing the fragments. With giraffe, Nesbit Evans (1970) observed Rothschild’s giraffe eating the stomach contents of an eland carcass and picking up the lower jaw which it did not chew. Western (1971) observed the Maasai giraffe chewing bones in the Maasai Amboseli game reserve. Three male giraffe were noted with their heads bent down over a carcass of an adult male Grant’s gazelle which had been killed a week earlier and picked clean. One giraffe lifted its head and had in its mouth the pelvis and entire rear leg of the gazelle’s skeleton. In a period of 5 min, four of seven giraffe present were noted to pick up bones and were apparently chewing. Splintered fragments of parts of the bones were observed at this site. It was noted that the soils of the area were highly saline and alkaline but the status in relation to phosphorus was not determined. Wyatt (1971) has observed Maasai giraffe chewing long bones over a period of some minutes.

After recording additional instances of bone-chewing in red deer where the bone was held lengthwise in the mouth with more than half projecting forwards and a little to one side, ‘like a cigar’, Sutcliffe (1973) describes the modelling effects which occur from this type of chewing in a ruminant. Most bones and antlers are oval rather than circular in cross-section. In consequence, when a deer holds one with its longest dimension pointing forwards and applies a sideways chewing movement, the bone becomes orientated with its

Fig. 25-1. Reindeer metatarsals chewed by wild Norwegian reindeer to create 'fork' with zig-zag margins. (From Sutcliffe 1973, with permission.)

widest diameter across the width of the mouth. Continued chewing will plane off the top and bottom until the marrow cavity or antler core is reached leaving only the sides intact. The fork- and prong-like remnants which result are a characteristic product of chewing by deer (see Fig. 25-1). Instances of animals damaging themselves by this type of chewing are recorded, in addition to the well-known incidence of botulism amongst the cattle due to the anaerobic bacterium *Clostridium botulinum* type D found with the bones. More recently Sutcliffe (1977) has extended the account with detailed records and beautiful specimens from Canada, South Georgia, Kashmir, New Zealand and Scotland.

Sutcliffe (1973, 1977) observed that since phosphorus deficiency in herbivores is unlikely to be a new phenomenon, evidence for it is to be expected in the palaeontological records. Antlers and bone remains of the pigmy deer (*Cervus cretensis*) of the Pleistocene age, showing apparent chewing have been collected by Sondaar from a cave west of Rethymnon, Crete. Similar Pleistocene remains have been described from other Cretan localities by Kuss (1969), who attributed fragments of antler, radius, ulna and metatarsal all showing forks with prongs at the sides of their planes of flattening to an 'osteokeratic culture of Palaeolithic man'. Tokunaga (1936) described leg bones with forks at both ends and a lower jaw with a fork at the symphyseal end as being bone artefacts used by ancient man in the Riukiu Islands. Sutcliffe (1973, 1977) believes both collections are attributable to chewing by animals and that the interpretation of the authors is incorrect. Both groups of specimens came from limestone caves so that calcium deficiency is unlikely to be a cause in either instance. Brothwell (1976) has recorded bones apparently chewed by sheep on the Island of Ronaldsay off the north of Scotland. The damage is similar to that observed in deer-chewed specimens with quite well developed forks.

Experiments on phosphate-deficient cattle in the field

Further investigation of the feeding behaviour of phosphorus-deficient cattle and sheep was carried out by Gordon, Tribe and Graham (1954). They studied 90 cattle and 500 sheep grazing an area of 5500 acres (2225 ha) on the Isle of Skye in Scotland that although providing adequate food was seriously lacking in phosphate. The blood levels of a number of cattle which were bled ranged from 1.8 to 2.5 mg of inorganic phosphorus/100 ml (0.58–0.81 mmol/l), in comparison with the normal animal which has levels of the order of 5 mg/100 ml (1.61 mmol/l). The animals on the area showed obvious allotriophagia and thus were suitable subjects for investigation. Eight groups of three troughs were put out. In each group, two troughs contained only calcium carbonate (limestone) but the third contained equal parts of calcium carbonate and dibasic calcium phosphate. It was noted that the high humidity of the locality caused the powdered substances to become moist and set firm in a few days and therefore losses from the material being blown away were small. While the average amount of material removed from each trough over the 1-year period of study was 50% greater in those containing calcium carbonate plus dibasic calcium phosphate, the difference was not statistically significant. Calculated as the amount eaten per beast per day it was less than a fifth of the amount necessary to correct aphosphorosis.

Since this result is contrary to some data cited below it is perhaps important to note that a mixture of calcium carbonate and dibasic calcium phosphate when wet would result in the following slow reaction:

$$CaCO_3+2CaHPO_4 \rightarrow Ca_3(PO_4)_2+H_2O+CO_2$$

There is also the consideration that dibasic calcium phosphate is almost insoluble at the pH of saliva and therefore like calcium carbonate would be virtually tasteless to the animals: thus differentiation between troughs would hardly be feasible. In Theiler et al.'s experiments in 1924 when bone-meal, sodium phosphate, phosphoric acid and wheat bran were shown to reverse osteophagia, they noted that precipitated calcium phosphate has a very similar effect to bone-meal but is more expensive, tasteless and more troublesome to administer especially since the cattle do not 'recognize' it and do not take it of their own accord.

On the King Ranch on the south-east Texas sea-coast there was a long history of severe phosphate deficiency and low blood phosphate, associated with gross deleterious effects on cattle. However, as a result of experiments by Robert J. Kleberg Jr and colleagues, disodium hydrogen phosphate is now offered in covered troughs on the prairie. All animals appear to elect to visit these troughs (Fig. 25-2) and ingest adequately

Fig. 25-2. Santa Gertrudis cattle at trough used on the King Ranch, Texas, to provide disodium hydrogen phosphate. (Photographs by courtesy of Mr John Cypher.)

(Reynolds et al. 1953), and the previous widespread physiological disabilities ('the creeps') are not seen. How often the cattle come to the troughs, the nature of the stimulus initiating behaviour and the mode of regulation of adequate intake are unknown. However, the problem in terms of serious bone disorders and diminished fertility has largely disappeared under this self-selection regime. In this high salt area with its proximity to the sea, sodium chloride is not licked by cattle. The introduced nilgai and other buck on this area of the King Ranch, which browse on leaves, are not seen to visit the phosphate troughs. Generally minerals, including phosphate, are more abundant in native plant material such as leaves, stems and pith than elsewhere, with leaves being a particularly good source of calcium.

In a preliminary study of phosphate deficiency which we carried out in Queensland in conjunction with the State Department of Agriculture scientists, cattle which had been fed for some months on a phosphate-deficient diet and consequently had markedly reduced blood phosphate levels were presented with bins containing 300 mmol/l solutions of potassium chloride, potassium bicarbonate, potassium dihydrogen phosphate and water. Most of the six naive animals tasted all bins and three of them did in fact ingest several litres of phosphate solution in preference to other salts within 20–60 min of access. As we will describe further below we were not able to reproduce this behaviour when we produced phosphate deficiency of a severe type in another way in cattle.

Krupski (1946), Krupski and de Quervain (1948) and von Duerst (1946–47) have noted the occurrence of severe phosphate deficiency in cattle in the alpine areas of Switzerland and the Jura Mountains. Their accounts include the association of serious osteoporosis with the behavioural manifestation of allotriophagia, and cure by salts of phosphoric acid. The calf of a cow concerned also had deficient formation of primary spongy bone. There was medullary atrophy with replacement of haemopoietic marrow by fatty marrow.

Klein (1970) studied black-tailed deer and feral reindeer during summer months in southern Alaska and St Matthew Island respectively. He concluded they feed in such a way as to ingest only the highest quality forage available in a given area. Swift (1948) showed that white-tailed deer fed consistently on sections of wheat and clover fields which were 8% higher in protein, 12% higher in fat, 38% higher in calcium and 34% higher in phosphorus than adjacent unfertilized areas, despite the areas being on open elevated ground which increased the risk of predators.

In their pioneering investigations Theiler et al.

(1924) considered the question of whether the appetite for bones was instinctive or whether it was learned: further, in the latter category, whether the learning of benefit was related to herd experience. French (1945), who also worked in Africa, asserted in discussing their work that in picas associated with cobalt and copper deficiencies, miscellaneous objects which where unlikely to satisfy the demand for these elements were chewed indiscriminately. He wondered whether animals would learn by experience to select materials which supplied the required elements if such materials were available. He added that perhaps the only evidence on this point known to him is that of Rigg and Askew (1934) who made available to sheep on 'bush sickness pastures' soil which could cure the disease. After a number of months some animals were still suffering from the disease and it was recorded that certain sheep were not eating the earth. When sheep were 'drenched' with the earth they all progressed well. Perhaps sheep are less capable of profiting by 'flock experience' than cattle. Certainly imitation plays an important part in the feeding habits of cattle because when a completely new food, such as banana leaves or cottonseed, is made available to a herd they usually do not take it straight away. It is only consumed in any quantity when the herd has watched one or two of its bolder members eating it with relish.

Laboratory experiments on mineral appetite in rats

Phosphate appetite

At the laboratory experimental level, data were obtained by Richter and Eckert (1937) and Richter and Helfrick (1943) indicative that phosphate aversion could develop after parathyroidectomy. Eight 74-day-old male rats were put on a diet low in phosphorus. They had access to two drinking fountains, one with tap water the other with 1% dibasic sodium phosphate. After parathyroidectomy the animals reduced phosphate intake by about a half and increased water intake accordingly. The administration daily of parathyroid extract increased phosphate appetite to the normal range again.

In other experiments Richter showed that parathyroidectomized rats survived on a low phosphorus diet but when placed on a high phosphorus diet there was 100% mortality. Earlier Schelling (1932) had shown that parathyroidectomized rats refused diets rich in phosphorus but ate diets with a high calcium content. As with the calcium appetite experiments to be described below, the examination of the time scale of Richter's experiments shows that whereas the changes were sometimes immediate (i.e. within 24 h), the measurements being daily, the data would not allow for any differentiation between learning during this interval or immediate recognition of the substance concerned.

Calcium appetite

McCance (1936) in his Goulstonian lectures pointed out that of the six metals commonly present in living matter calcium was the one with the greatest tendency to form insoluble salts. The carbonate and the phosphate have been very widely used as hardening agents during evolution—the former by the invertebrates for their protective exoskeleton, and the latter by the vertebrate phylum for their bones and teeth. He stated that because of this tendency of calcium there seemed no doubt that the development of a satisfactory method of excreting it must have been a *sine qua non* of evolutionary survival. He notes also that the insolubility of many calcium salts may be a source of danger for another reason, in that it may prevent enough of the calcium in food being absorbed to supply the needs of the animal. Excessive inorganic phosphates in the diet may cause this, expecially if the stomach juices are not acid. In the higher mammals, serum calcium is normally maintained within narrow limits by homeostatic mechanisms which regulate the two roles of calcium: its metabolic function in bone, and metabolic actions of ionized calcium on cell membrane permeability, neuromuscular activity, blood coagulation and enzymic and secretory processes (Catt 1970). Plasma calcium fluctuates daily within $\pm 3\%$ regardless of calcium intake and excretion. The regulation involves integrated actions of parathyroid hormone, calcitonin and vitamin D in controlling calcium absorption of bone, and turnover and excretion of calcium and phosphorus. Parathyroid hormone at high blood concentration affects osteocytes to cause calcium release from mature bone crystals into blood. It also acts directly at lower concentration on the kidney to reduce calcium clearance and enhance phosphate excretion and on the gut to increase calcium absorption. Parathyroidectomy results in a lowering of plasma calcium and an increase in phosphate.

Richter and Eckert (1937) studied young rats which were given a diet deficient in calcium and

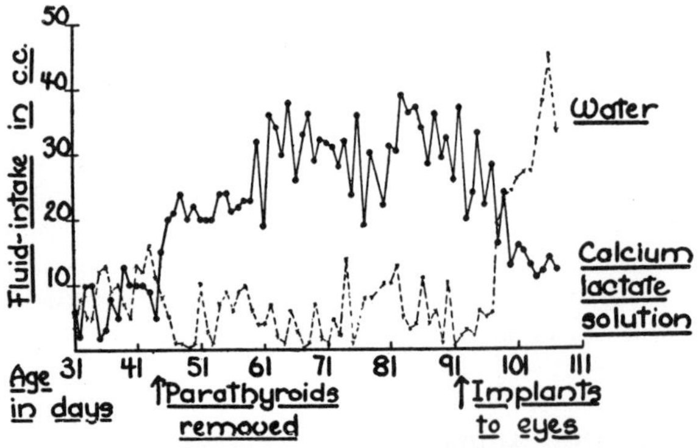

Fig. 25-3. The effect produced on calcium lactate intake and water intake by parathyroidectomy and subsequent implantation of parathyroid glands. (From Richter and Eckert 1937, with permission.)

free choice of water and a 2.4% calcium lactate solution. Following parathyroidectomy calcium intake of the rats increased by a mean of 3.9 times. In one animal it went as high as 12.9 times the control level. As Fig. 25.3 shows, the increase in calcium appetite was immediate following operation. The increase was reversed by making a successful parathyroid implant in the anterior chamber of the eye. Richter and Helfrick (1943) showed also that rats on a phosphate-deficient diet but with access to dibasic sodium phosphate solution took large amounts of the phosphate (mean 13.5 cc/day). Following parathyroidectomy, intake of phosphate solution halved (mean 7.9 cc/day). With daily injection of parathyroid extract (0.3 cc), intake rose to 15.2 cc/day. Parathyroidectomized rats with AT10 (dihydrotachysterol) added to food also increased their phosphate intake. Food intake was not influenced by parathyroidectomy.

Scott et al. (1950) showed also that calcium-deficient rats showed a preference for a high-calcium diet. Lewis (1964) showed that parathyroidectomized rats had a high voluntary intake of calcium lactate solution before operation. Calcium deprivation motivated the parathyroidectomized rats to learn lever-pressing for calcium and the rate of pressing increased with deprivation of calcium solution up to 15 h. She concluded that a drive for calcium could be produced that had similar properties to other appetitive drives.

Rodgers (1967) has made an outstanding contribution to the study of the specificity of hungers by experiments orientated to the issue of whether needed substances have any special unlearned significance, other than as novel stimuli, to deficient rats or rats that have recovered from a deficiency. As a first step Rodgers confirmed that rats on a calcium-deficient diet chose a diet, when offered a choice, which included calcium chloride at levels of 1 or 10 mg/g. This was also true of animals that had recovered from a calcium-deficient diet. He drew attention to Richter and Eckert's (1937) initial suggestion that calcium hunger was not learned but based on changes occurring in the taste receptors of the calcium-deficient animal. He also noted that Scott et al. in their experiments mentioned above had concluded that calcium appetite was learned, on the basis that if a calcium-supplemented diet was associated with a particular flavour, and then the flavour transferred to another diet, the animal's choice was switched with respect to the flavour rather than the calcium. By various combinations of the novel-diet technique Rodgers confirmed this: novel-diet preference was greater than any other specific unlearned preference for calcium that might have been present in the calcium-deficient animals.

It would seem important, in analysing these experiments, to note that the changes that are produced metabolically by parathyroidectomy may be different from those of a low-calcium diet. The blood levels of calcium and phosphate and also of parathormone and calcitonin may be influential on behaviour. The clear experimental demonstration of the capacity of several hormones involved in the reproductive process to induce a large increase in calcium appetite in rabbits within the first 24 h after injection (as described in Chapter 23) requires analysis before the learning theory is accepted as explaining all aspects of calcium appetite: the data on hormonal induction could suggest a contrary conclusion.

Snowdon and Sanderson (1974) have shown that calcium deficiency causes weanling rats to ingest lead acetate solutions even at high concentrations which are extremely aversive to normal animals. They speculate as to whether calcium deficiency may be a component of lead pica in young children. In relation to field data on calcium appetite, Snapper (1955) records the observation that calcium-deficient chickens in areas where access to the substance is very difficult (in places such as the Amazon Basin) have the habit of eating the shells of eggs they have themselves produced. Another interesting example of apparent calcium deficiency in birds, a consequence of the elimination by man of the large carnivores from the ranching areas of Zimbabwe, has been described by Mundy and Ledger (1976): the shortage of suitable-sized bone fragments which are normally collected by parent vultures for their young resulted in rickets among the chicks. Addition of shell grit to the diet of caged chickens not only assists digestion of food but is commonly claimed to lessen egg-shell eating. Orr (1929) reports that horses in Victoria, Australia, circumvent the natural consequences of grazing on inferior pasturage by eating the bark of grey box trees, which is exceedingly rich in lime compared with native grasses.

Schaller (1967) reports on the behaviour of chital and other deer in the Kanya Park in central India and records that they scraped at soil licks with their incisors, sometimes leaving holes over 30 cm deep. Analysis of the licks showed them to be very rich in calcium and phosphate relative to specimens from non-lick areas. Drummond (1934) observed pigs deficient in calcium and vitamin A. Relative to the quiet behaviour of controls, calcium-deficient animals spent a large amount of time trying to lick whitewash off walls or root out a fragment of mortar from the brickwork, whilst those deficient in vitamin A searched for the smallest blade of grass or chance weed in the cracks in the floor. In relation to man, in East Africa the search for edible earths rich in calcium resulted in tribal wars. Ashes from water plants of high calcium content enjoyed the greatest popularity among pregnant women (Orr and Gilks 1931). The Chinese diet being low in milk products may be deficient in calcium and pregnant women have a predilection for chewing bones from sweet and sour dishes. The acidity could make the calcium in bones more readily available (de Castro 1952).

Widmark (1944) discusses the results of Hellwall, who fed calcium-deprived hens either macaroni or macaroni closed at the ends and containing pulverized egg-shell, in which instance the hens received 3–4 g of calcium in the meal. Four hours later both groups were given access to calcium and the untreated group took 4–9 times as much calcium. Thus it might be concluded that the chemical correction in the organism had an almost immediate effect, as the hens swallowed the macaroni whole and presumably had no way of recognizing the calcium with the senses. Widmark also draws attention to the great need for calcium of reindeers, who shed horns with a calcium content of around 600 g annually. Lactating cows, however, may need an additional 36 g/day, or approximately 1 kg/month. He also showed that while growing rats given a choice between a calcium-poor diet and calcium-enriched diet similar in all other respects showed some relative aversion to the rich diet under control conditions of choice, after a period of 12 days of calcium deprivation the choice switched to predominant intake of the calcium-rich diet.

Hughes and Wood-Gush (1971) have shown calcium-deficient chickens show a preference for calcium carbonate, a natural salt, and also for the synthetic salt calcium lactate. They find the preference to be learned and either gustatory or visual cues alone allow selection to occur. Calcium-deficient rats exhibit a preference for strontium salts (Richter and Eckert 1939), but Hughes and Wood-Gush (1971) found the appetite to be more specific in the chicken: there was no generalization to strontium carbonate, which was strongly rejected by control and deprived groups. They reasoned that if calcium appetite were similar to sodium appetite, where the choice and acceptance of lithium occurs until the toxic manifestations become apparant (Nachman 1963), then strontium carbonate might be selected as freely as calcium carbonate in the short term as its gustatory properties are similar to calcium carbonate. The fact that the birds do not take strontium emphasizes the importance of post-ingestional factors. In relation to learning of effects of calcium, the authors note that whereas learning associated with effects of ingestion may be effective when the delay between intake and effect is as long as 7 h, experiments with radiocalcium indicate it is entering the bloodstream of a fowl within 15 min of ingestion. Emmers and Nocenti (1963) showed that thalamic lesions involving the taste area reduced the increased calcium intake of parathyroidectomized rats offered a cafeteria, whereas destruction of the thalamic tactile area for tongue or snout did not. They suggest taste has an important role in self-selection of dietary calcium.

Magnesium appetite

Scott et al. (1950) found no evidence whatever of an appetite for magnesium-containing diets in magnesium-deficient animals. In fact such diets were usually avoided to the extent that the animals died with magnesium deficiency convulsions despite magnesium being present in food in the cage. Rodgers (1967) showed that magnesium-deficient animals did show a novel-diet preference but that if magnesium intake were associated with the diet by its concurrent addition to the drinking water they reverted to the original diet. He conceded Scott et al.'s (1950) initial suggestion that recovery from magnesium deficiency may be a very unpleasant process, so much so that it exceeds the unpleasant features of magnesium deficiency itself which, at least, are of an order sufficient to induce novel-diet preference.

Potassium appetite

On the face of it, potassium deficiency in feral primates or other animals, herbivore or carnivore, is unlikely. The qualification to this statement is, perhaps, disease conditions involving loss of intestinal secretions: for example, infectious diarrhoea in gregarious species will involve direct loss of potassium. Also the disturbance of the acid–base chemistry of the *milieu intérieur* which results from disproportionate electrolyte loss in diarrhoea, and the deterioration of circulation, will result in loss of potassium from intracellular fluid. This loss is essentially secondary to the sodium-depleting process.

Scott et al. (1950) were unable to show any evidence of potassium appetite in rats fed a potassium-deficient diet. Orent-Keiles and McCollum (1940) of Johns Hopkins University showed that potassium-deficient rats had a striking alertness and a peculiar pica. They appeared to be constantly searching and licking different parts of the cage and equipment. They continuously licked each other, as well as their own genitals after urinating. There was a roughness and thinning of the hair. Life span was normal, but ovulation was irregular and spermatozoa were defective and mating did not occur. Magnesium storage was increased. In muscle, the concentration of potassium was lowered and that of sodium increased. Possibly the genital licking reflected impetus to potassium intake. Milner and Zucker (1965) have studied weanlings reared on a potassium deficient diet. When the deficient animals had ceased to grow they tested for a preference for potassium chloride solution relative to controls on the same diet but with potassium chloride added. The control solution they compared for preference relative to potassium chloride was magnesium sulphate, which they stated rats neither avoided nor preferred; the rationale was to exclude the possibility that the subjects were merely attracted to anything which tasted different from their usual regime. Milner and Zucker claim the potassium-deprived subjects showed an initial choice of potassium chloride significantly more frequently than controls, and suggested they were responding to olfactory cues, possibly innate. No comparison tests were made against sodium chloride or any other substances.

Zucker (1965) studied young rats maintained on a potassium-free diet. When offered a choice between a potassium chloride solution and one of five sodium chloride solutions, these animals showed a greater preference than controls for four of the five sodium chloride solutions. However, as Zucker records in his methods section that the potassium-deficient rats usually stopped growing and showed other signs of potassium deficiency such as diarrhoea—which would give sodium as well as potassium loss—the data cannot be interpreted in any simple way. Blake and Jurf (1968) also showed that potassium deprivation stimulated increased sodium chloride intake. Data suggestive of some measure of specificity of potassium appetite came from the work of Adam and Dawborn (1972). They also noted some diarrhoea in their potassium-deplete animals. They found potassium depletion caused increased ingestion of solutions of sodium chloride, potassium chloride, calcium chloride and quinine sulphate in concentrations which were aversive to normal rats. The amount of potassium ingested was related to the degree of potassium depletion, and the animals repleted themselves within 24 h when potassium was offered. But the potassium-deplete animals also drank large quantities of aversive concentrations of sodium chloride which was preferred to potassium chloride. However, this appetitive state was reversed by prior intragastric repletion with potassium but not with sodium salts.

Experimental analysis of phosphate appetite in cattle

The provocative fact that phosphorus-deficient cattle in the field choose the only naturally available source of phosphorus—bones—caused Dr J. Nelson and others of us to analyse this phenomenon further in laboratory experiments. Green (1925) reported that osteophagia could be produced in stalled cattle on artificial rations selected so as to reproduce the phosphorus intake of natural veld conditions, but the process is lengthy and uncertain unless animals are selected which have already shown osteophagia at some time in their past history. We decided to try a somewhat different method.

A herd of young identical twin cattle of various breeds were collected with the assistance of the Victorian State Department of Agriculture. They included Angus, Herefords and Murray Greys. Dietetic history suggested no likelihood of any previous experience of phosphate deficiency.

In the first instance a group of four pairs of twins were placed on a phosphate-deficient diet at our Institute farm. The composition was as follows: barley straw 49%, rice hulls 28%, molasses 10%, *Sorghum* grain 8%, tallow 4%, urea 0.75%, together with added limestone ($CaCO_3$), salt and trace elements zinc, manganese, cobalt, copper and iodine. Animals ate approximately 8 kg/day. The diet was similar to that used previously in phosphate deficiency experiments by Gartner and Murphy of the Queensland Department of Agriculture, but one of each pair of twins had

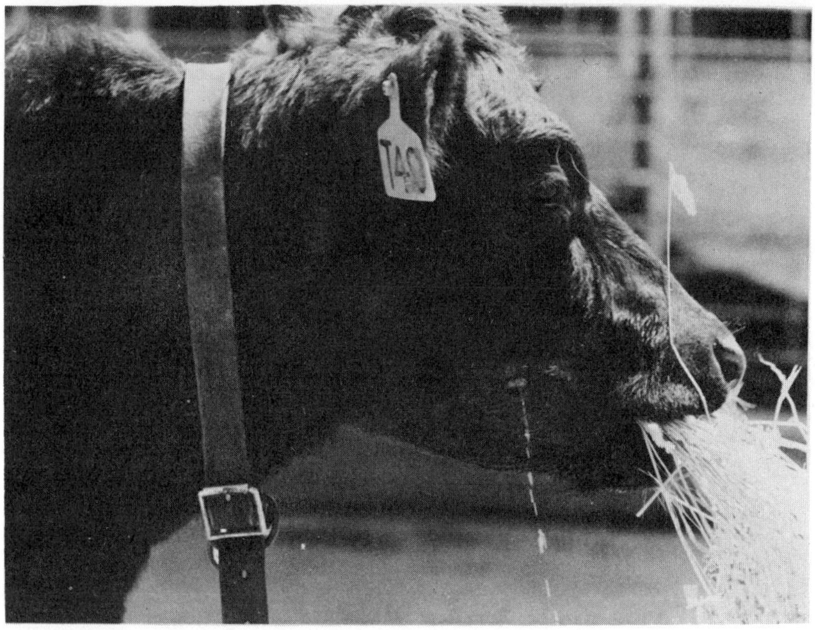

Fig. 25-4. Cow with saliva flowing from a permanent unilateral parotid fistula.

phosphate added to the diet by mixing 55 g/day of monoammonium phosphate with the food. The other twin had a permanent unilateral parotid fistula made (Fig. 25-4), from which a cow loses 20-40 litres of parotid saliva each day. The composition of the saliva is essentially the same as in sheep (phosphate, 15–40 mmol/l) (see Fig. 3-2). Whereas these cattle were in stalls, and not metabolic cages which would allow measurement of saliva loss, the difference in water intake of the fistulated group relative to the control group was consistent with a saliva loss of 20–25 l/day. This group with a fistula also had sodium bicarbonate (200 g/day) and sodium chloride (150 g/day) added to the food, which replaced the sodium component of the saliva lost each day. Thus the fistula, in essence, became equivalent to a phosphate tap draining the body of its supplies. At regular intervals the compositions of saliva, faeces and blood were determined with regard to phosphate. The results from our initial experimental series are shown in Fig. 25-5. Severe phosphate deficiency was produced over a period of several weeks.

The first experiment which was carried out on these animals was to present them for a limited period of time, individually, under free-choice conditions in their yard, with 300 mmol/l solutions of potassium phosphate (K_2HPO_4/KH_2PO_4,pH 7), potassium bicarbonate, potassium chloride and also water. Whereas in some preliminary experiments we made along these lines in Queensland a considerable intake of potassium phosphate was shown by some animals (3–10 litres drunk by four animals overnight for 6 days), in this case no intake was seen. These operated animals had been depleted for 4 weeks by this stage while the Queensland animals had been on a low-phosphate diet for 6 months when the test was carried out. In the light of this lack of response the various potassium solutions were left overnight in the animals' stalls together with the usual water bin. Again no intake occurred.

The animals were allowed to lose phosphate for a further 7 weeks. At this point it appeared that plasma phosphate level was constant at a low level: progressive fall was not occurring (Fig. 25-5). The

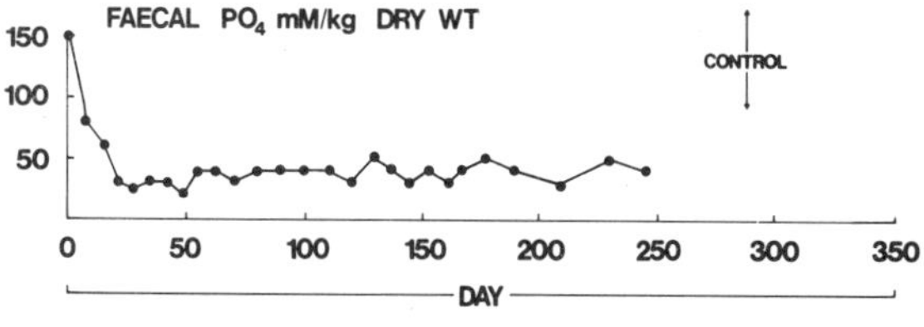

Fig. 25-5. Effects of low-phosphorus diet and loss of parotid saliva from a fistula on the plasma, saliva and faecal phosphate levels of cattle (mean values).

Fig. 25-6. Phosphate-deficient cow in the process of eating a rib bone.

animals were now permitted for the first time in the concrete cattle yards to have access to sun-dried bones. Two of the animals showed an immediate keen interest in the bones, picking them up and chewing them (Fig. 25-6). A third animal also did this with lesser avidity, and a fourth did not do so though it nosed them with some curiosity. The control animals, with phosphate added to the artificial diet, were conspicuously less interested: some nosing of the bones did occur briefly but no picking up in the mouth and chewing. The most avid interest was shown by animals with the largest chemical changes.

This experiment was repeated on several occasions with the same result over the following 1–2 months. At no time were the animals allowed to chew bones to achieve any significant intake. The two animals which earlier had shown less interest in the bones gradually developed behavioural patterns almost identical with the other two, and on the last test both picked up and chewed bone (6 months after fistula).

A further experiment was then made. It was repeated several times and the results were clear-cut. Instead of bones, the animals, one at a time, were presented with a cafeteria of trays. One contained new bone which had been smashed and worked to make a moderately fine powder of chips. A second was made up of old sun-dried bone treated the same way. Another identical container had dibasic calcium phosphate in it, another gypsum (calcium sulphate), another limestone (calcium carbonate) and another disodium hydrogen phosphate (Na_2HPO_4). The identical twins with phosphate added to the diet did not pay any significant attention to this cafeteria of trays, but usually the deficient animals made a rapid choice, preferring the new bone (Fig. 25-7). If this were then removed from the cafeteria they rapidly changed attention to the old bone. Both substances were licked up and chewed with great avidity. The animals made the choice very rapidly, apparently on olfactory cues. Often, in the first stages after licking up the first pieces of bone they protruded the tongue and licked inside each nostril. This behaviour was often repeated. It was also seen at this stage that the animals had an extremely avid appetite for dried bird faeces which were added to the cafeteria assembly. As this is another naturally occurring form of available phosphate its consumption raises interesting questions on innate mechanisms.

In preliminary investigation of this bone appetite, we have tested the influence of rapid intravenous infusion of buffered sodium phosphate (NaH_2PO_4/Na_2HPO_4,pH 7.4) (8.4 ml/min of 750 mmol/l) for 32 min giving a total injection of 202 mmol phosphate. This succeeded in raising plasma phosphate from 0.4 to 2.0 mmol/l by the end of the infusion and saliva phosphate from 4 to 15 mmol/l. Two animals which showed avid appetite for the bone in the cafeteria situation immediately before the intravenous infusion had largely lost interest when tested 35 min after beginning the infusion. When the phosphate infusion was given at twice this rate for 16 min, i.e. same total amount infused, and the animals tested immediately, they showed a transient interest in

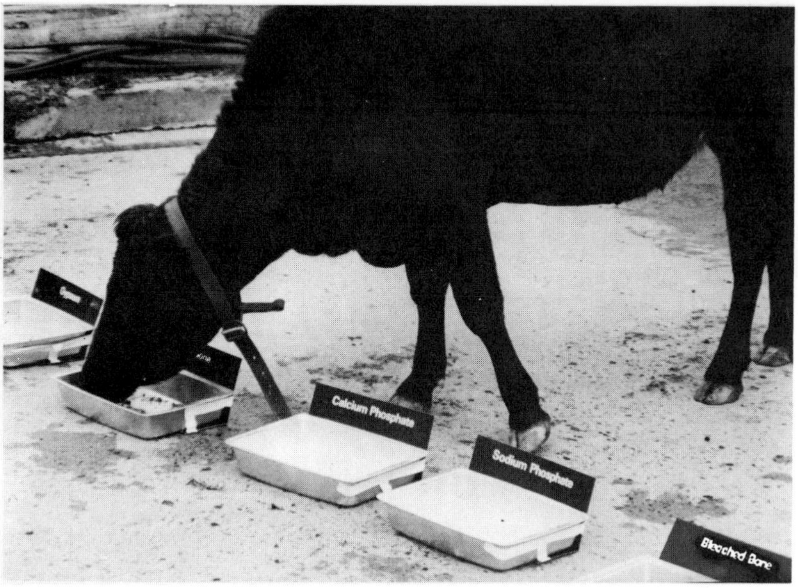

Fig. 25-7. Phosphate-deficient cow selecting ground-up fresh bone from a cafeteria of trays which also included ground-up old bleached bone, disodium hydrogen phosphate, calcium sulphate (gypsum), calcium carbonate (limestone) and dibasic calcium phosphate.

the bones lasting only for 2–3 min. Thus despite the fact that plasma phosphate was much higher at 4 mmol/l and salivary phosphate 30–40 mmol/l the appetite was not suppressed as markedly. Saline infusion did not have this effect. Another experiment showed that 200 mmol of phosphate given over 1 h stopped bone-eating for the next 2 h. Plasma phosphate had risen to 1.6–1.8 mmol/l. By 48 h, when phosphate had fallen to 0.6 mmol/l, all the animals ate bone again. Insofar as such preliminary data may be taken as indicative, they suggest that the mechanisms involved in producing the avid appetite for bone can be accessed in some way as a result of a rapid rise in blood phosphate. But there is time delay involved. We are currently investigating this proposition with further twin-cattle studies where pregnancy and lactation are included as metabolic stresses to aggravate further the phosphate deficiency produced by diet and fistula.

The effect of pregnancy and lactation

The study was started with four pairs of twins but one deficient animal died soon after calving. A parotid fistula was prepared in one of each of the pairs of twins when it was 4–5 months pregnant. All animals were then fed the same diet as had been used previously but the phosphate intake was approximately 50% greater. The non-fistulated twins again had 55 g/day of monoammonium phosphate added to the diet. In these pregnant fistulated cattle the decline in plasma, salivary and faecal phosphate occurred much more slowly than in the previous group (Fig. 25-8). The plasma phosphate level fell to 1.2 mmol/l by the tenth week after surgery, and then remained fairly constant until after calving. During this period the cattle demonstrated no interest in bones. Thereafter a more rapid fall was observed, and within 3–6 weeks all animals actively chewed bones when these were made available.

On presentation of the cafeteria during this early stage of lactation all of the fistulated animals selected only crushed bone, again apparently by olfactory cues. Within a few weeks these cattle

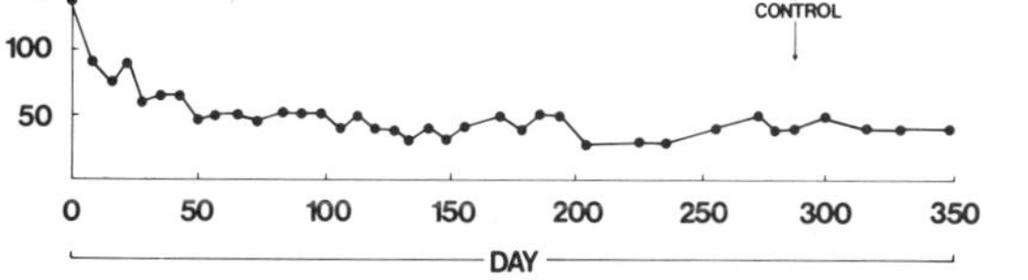

Fig. 25-8. Effect of low-phosphorus diet and parotid fistulation on plasma, saliva and faecal phosphate levels of pregnant cattle (mean values).

became less discriminatory in their selection, and gradually all extended their eating to bird faeces and sodium phosphate. Since then one cow (No. 87) has not changed its pattern of selection, a second (No. 69), after 13–14 months of phosphate depletion, has eaten calcium phosphate, and at about the same time the third (No. 25), in addition to eating calcium phosphate, also ate limestone and gypsum when none of the phosphate-containing components was available. The non-fistulated cattle have never chewed bone throughout this study.

During the period when the cattle would select bone, bird faeces and sodium phosphate, several experiments on the effects of intravenous phosphate infusion were carried out. In the first, 200 mmol of sodium phosphate were infused in 1 h. This had little or no effect on selection from the cafeteria during the next hour despite the fact that at the end of the cafeteria test the plasma phosphate concentration was still in the normal range. Infusion of 350–600 mmol phosphate in 3–4 h, with similar effects on plasma phosphate levels, also had little or no effect on the appetites of these cows for components of the cafeteria.

A third experiment in which increased plasma phosphate was maintained for a more extended period was performed 15 months after fistulation. Plasma and salivary phosphate levels during this treatment are shown in Fig. 25-8. After 24 h during which 1050 mmol phosphate were infused two cows had restricted their interest to bone only, but the third still ate all the substances in the cafeteria. Little change occurred after 2 or 3 days of infusion with 700 mmol and 600 mmol respectively. On the fourth day, after the administration of a further 800 mmol phosphate the infusions were stopped. Testing with the cafeteria then showed that the first cow now had no appetite for any of the cafeteria material, the second still ate bone but with greatly diminished drive, and the third, although eating only bone and sodium phosphate, did so with its original pronounced avidity.

Plasma phosphate levels were maintained for a further 3 days in all animals by the addition of 400 mmol/day of disodium phosphate to the diet. This did not produce any further suppression of the drive of the second or third cows to eat from the cafeteria. The animal which had not eaten bone for the last 4 days was then returned to the low-phosphate diet. It commenced to eat bone 7 days later when its plasma phosphate had fallen to 0.6 mmol/l. After another 4 weeks it also ate sodium phosphate and bird faeces as it had before the infusion.

The provision of the high-phosphate diet to the other two cows was continued for a total of 5 weeks but had no further effect on cafeteria selection. These two cows were then also returned to the low-phosphate diet and when tested 2 weeks later exhibited their initial selection patterns from the cafeteria, now taking components additional to bone and sodium phosphate.

In tests carried out subsequent to this experiment, potassium phosphate has been included in the cafeteria and attempts have been made to disguise the odour of crushed bone by the addition of either butanol or eau de cologne. In four tests none of the cows, after a first tentative lick, has taken any potassium phosphate. The addition of butanol had no effect on the identification of crushed bone as the cattle moved to it immediately on entering the cafeteria area and ate with undiminished intensity. Eau de cologne appeared to be more effective. The animals hesitated for up to a minute in identifying the tray and were more diffident about picking up and eating this bone. When tested again 1 week later they would not eat either bone scented with eau de cologne or crushed dried bone, but would readily consume the other components which they usually ingested. After a further week the refusal to eat crushed bone was no longer apparent but no approach was made to the tray with scented bone.

In relation to the ages of the cows in these groups of experiments, the first was carried out using animals which were approximately 2 years old while in the second the animals were approaching 4 years of age and pregnant, and consequently much heavier. A third study has recently been started on 5-year-old cows of approximately the same weight as the pregnant animals. In this last study the fall in plasma phosphate following fistulation was also slower than in the first, but it was more rapid than in the pregnant animals. About 4 months elapsed between surgery and the establishment of stable low plasma phosphate levels. Again no bone-eating occurred until the low level had been maintained for several weeks.

In this last group of cattle the effect of sodium phosphate infusion on the reversal of bone-eating behaviour was, like the rate of fall of plasma phosphate, also intermediate between its influence on the young cows and on the pregnant animals. In one animal only was bone-eating stopped by a 1-h infusion; in a second 6–8 h of infusion was necessary; the third animal did not stop bone-eating until the infusion had continued for almost 3 days; and the fourth was still showing slight interest in bone 5 days after the infusion started.

These results suggest that when phosphate depletion occurs more slowly, or perhaps it is a characteristic of older cows, the drive to eat bone is stronger. Consequently correction of plasma phosphate for a more extended period is necessary to suppress it. It would appear that total obliteration of the behaviour is very difficult in some animals, and Green (1925) remarked that a small percentage of his animals (chronic cravers) failed to lose the appetite even when dosed with bone-meal in large excess of physiological requirement.

Conclusions from experiments on cattle

These experiments on first access to bones by naive phosphate-deficient cattle have succeeded for the first time in reproducing under laboratory conditions with controlled onset of deficit the phenomenon reported in the field over two centuries. The clear difference in behaviour from the control identical twin, the immediacy of the response on first access, the evident use of olfactory cues in this regard (which we extensively recorded on cine film during the initial experiments), persuade us that the recognition and behaviour are very likely innate. This will be scrutinized further in studies on identical twins in the course of dissecting the physiological organization of the behaviour.

The basis of recognition of a needed substance

Obviously many questions arise as to the basis of recognition of bone by phosphate-deficient cattle. Is there a particular element of the olfactory stimulus of bone which initiates the behaviour? Or is there some gestalt of smell and sight and then taste which initiates the chewing in the animal? Is low blood phosphate the actual stimulus to some neural receptor system outside the blood–brain barrier which initiates the behaviour, or is it the CSF or intraneuronal content of phosphate which initiates it? Given the observations confirmed over years in Texas that animals will take disodium hydrogen phosphate in solid form from troughs and remain in good health, how accurate is the satiation process? And is this behaviour of intake from troughs learned or is it facilitated by cues related to those involved in initiating bone-chewing?

The question of how the animal makes this choice can take us back to consideration of some early and basic data in this field. Richter et al. (1938) gave rats complete freedom of choice among 11 pure food elements: three solid foods and eight liquids each presented in a separate container. The foods were casein, sucrose and dried baker's yeast, and the liquids 3% solution of sodium chloride, 8% solution of dibasic sodium phosphate, 2.4% solution of calcium lactate, 1% solution of potassium chloride, water, olive oil, cod liver oil, and wheat-germ oil. The daily intakes were measured. Under normal free-choice conditions the proportions of food elements ingested from day to day were found to be remarkably constant. The growth curves of rats maintained on a self-selected diet paralleled those on McCollum's standard diet (actually the self-selection group grew a little more rapidly). Davis (1928) also found newly weaned rats grew very well when fed on the free-choice system. She elaborated the study to infants and young children fed on the free-choice system but cautioned against exaggerated claims about the trustworthiness of appetite as a guide (Davis 1933). Young (1941) has reviewed the literature on this behaviour and Richter (1956) elaborated with studies on free choice of 17 components in a cafeteria. He found that some rats grew normally. The issue evidently is: How do they do it?

It seems legitimate to ask whether in some instances the choice is innate. Thus when the animal does not take the substance for some time, a specific drive is engendered by an eventual state of deficiency. Or in the replete state is there some simple innate stimulus–response relation between taste and ingestion for some of the components of the diet—because the mechanism has survival advantage? In other words, the animal takes what it likes and this, in fact, is what it needs because natural selection has established hedonic elements in the taste of nutritious substances. Certainly aversive sensation is coupled with the taste of certain poisonous substances—alkaloids and glycosides—as has been analysed by Garcia and Hankins (1975). Or is intake of some essential substances in this situation learned? If so, when a neophilia is induced because failure to ingest has produced deleterious consequences, how does the animal associate the sense of benefit with intake of the *particular* substance as distinct from all the others it is ingesting? The implication could be that such learning would need to be superimposed on an already fairly standard pattern of intake of the other substances so that the new choice was the only significant variant and the effect would be recognized. Clearly the progressive daily records of the intake of all components from the starting day are important in giving some indication of what occurs in the Richter experiment.

Rodgers (1967) in his experiments on precision of specific hungers included a study in which rats were either thiamine deficient or pyridoxine deficient. With the various diet combinations employed in analysing the behaviour he found the important fact that neither the thiamine-deficient nor the pyridoxine-deficient subjects showed significant preference for the diet containing the needed vitamin over the diet containing the un-needed but equally novel vitamin. These data were against any unlearned cue determining the eventual preference for the diet with the needed vitamin. There was an initial preference for novelty of taste or smell, and the benefit from the chemoreceptor cue was learned.

The question remains as to how Richter's rats achieve this with 11 or 17 components. It would be interesting to have urine analyses on such animals to determine whether in some degree intake of some components showed some relation to metabolic requirement or whether there was evident excess in all instances. Allowing what Richter's rats were able to do, it remains puzzling that in the experiments of Harris et al. (1933) on thiamine-deficient rats the association between Marmite, which exclusively contained thiamine and thus potential benefit, and the nine other components of the diet, which did not, could not be learned without an education period. Rozin (1967) stressed that there is the possibility that the rat, like a rational human being, may ingest one new food at a time, so that the opportunity for association of intake and benefit is maximal. However, another fact which has to be taken into account in the Harris et al. (1933) experiments is their choice of Marmite as the vehicle for thiamine. This 'spread' contains about 10% to 12% sodium chloride and thus, in the first instance, might have been aversive to the rats.

Other general matters emerge for discussion when we consider bone-chewing in cattle. As Young (1967) noted there are constitutional issues, differing from species to species, determining food intake. As I discussed earlier (Chapter 11), lambs which for a period of time feed on ewes' milk eventually take to eating grass. Do they do so because they try some and feel better and learn accordingly? Do they imitate their companions and learn a habit of the herd? Or is it innate programmed behaviour in that at a certain stage of cerebral and physiological maturation the sensory stimulus of grass (sight and taste) releases the ingestion mechanism? Presumably it is the last possibility. An innate organization of selection of at least some elements of food would have much greater survival value, on the face of it, than trial and error learning.

Le Magnen (1971) has reviewed this issue of innate responses to olfactory stimuli in relation to nutrition in human and animal species. His conclusion is that the matter has really been little studied. The giant ant-eater seems to be able to find its specific food by smell (McAdam and Way 1967). After the removal of olfactory bulbs, kittens become unable to find the maternal nipple, and they apparently lose the sucking response (Kovach and Kling 1967). Wild rodents in their natural environment are dependent on smell in their food-seeking for nuts and seeds (Svericenko 1954). Food preferences by wild and domesticated rats have been studied (Barnett and Spencer 1953; Barnett 1956) in relation to innate preferences based on odour of unfamiliar foods. Barnett and Spencer found whole wheat preferred to grains and to white bread; horse liver preferred to grains, etc. Le Magnen also noted that food odours have the capacity to reinforce the sense of hunger when the individual is deprived yet become unpleasant, even nausea-inducing, when the subject is satiated. Other odours, e.g. faecaloid, induce nausea in hungry or satiated subjects. The neural mechanism of these sickening odours is unknown; some slight effects on gastric contraction have been reported (Ginsberg et al. 1948). Le Magnen stated in relation to the olfactory properties of natural nutritious foods that these properties are not necessarily associated with the nutritive components of the food. Such associated odour fractions are very numerous. Olfactory responses are therefore differentiated responses to various associated signals of nutritive components. The neuroanatomy of the olfactory system, with close relations between olfactory projections and both rhinencephalic and diencephalic areas involved in the regulation of food intake, represents the structural basis of the interactions between olfactory afferents and feeding responses.

Rodgers (1967) recognized that his experiments on the learned basis of vitamin appetite did not rule out true specific hungers, because he was using thiamine hydrochloride, a synthetic product rather than a natural source of the vitamin. Perhaps the natural substance has a different taste or smell or perhaps the presence of the vitamin in natural foods is highly correlated with some other tagging substance—much as we might presume with the phosphate of bone. Control experiments done with yeast (Rodgers and Rozin 1966) as the source of vitamin B did not support this: again yeast was chosen only on a novelty basis. Perhaps the experiment could be repeated using as the source of vitamin B, substances a feral rat might be likely to encounter.

As indicated earlier, the data as they stand contrast with the experiments on sodium deprivation where immediate selection of the sodium-supplemented familiar diet occurs. While considering that the reason for this ability may be a stronger selection pressure for reliable sodium intake mechanisms than vitamin intake mechanisms, Rozin (1967) notes Rodgers' suggestion that true sodium-specific hunger may have evolved because it would be particularly difficult for sodium-deficient rats to learn to prefer sodium-rich foods—because sodium does not have rapid reinforcing effects. However, sodium is rapidly absorbed from the gut after ingestion and substantial changes in extracellular chemistry do follow (Denton et al. 1977). There is no doubt that sodium-deficient animals can choose foods with higher sodium content. Arnold (1964) has shown this beautifully in studies on grazing sheep prepared with an oesophageal fistula so that their food selection under field conditions could be accurately determined. Control animals were compared with those which also had a parotid fistula. Following withdrawal of sodium supplement for 10 days the latter animals selected the more sodium-rich species of plants in the plots in which they were allowed to graze. This involved switching their relative preference to two species of the five available.

Neophilia

Finally, it may be noted we know little of the actual mechanism which initiates neophilia in the case of the learned appetites. It seems unlikely that any element of change of blood chemistry is common to vitamin deficiencies, calcium deprivation, potassium depletion, etc., if these are indeed all learned appetites. Thus no common change is likely to act on any receptors in the central nervous system. Possibly some generalized central affective consequence of such states of disordered metabolism occurs and this also involves gastrointestinal function and sensory inflow via the tenth cranial nerve. This, then, in some undetermined fashion could be associated in consciousness with ingestive rather than some other behavioural activity.

In relation to the central organization of such capacity, Nachman and Ashe (1974) have shown bilateral amygdala lesions cause a deficiency in the learning of aversions. This deficit, on the basis of their data, appears to be more a perceptual deficit involving the lack of an appropriate alerting response to novelty and thus ability to learn: the rats lack a neophobia.

The laboratory rat: The species most studied, and the influence of domestication

A discussion of the issues instanced above is not complete without ventilation of one other matter. That is, the extent to which our opinions on these matters may tend to be governed predominantly by experiments on the domestic inbred albino *Rattus norvegicus*, the animal most extensively used in all medical investigation.

Again it is Richter (1954) who has examined the influence of domestication on rats. He cites the classic tract of H. H. Donaldson (1924). The Norway rat, which originated north of India, did not reach Europe until 1730 and America until near the end of that century, but then rapidly replaced the Alexandrine rat. These rats produce albinos in the course of breeding and it was selection of these which gave rise to the domesticated strain. Defects in vision owing to lack of pigmentation in the eyes made them easier to handle and less apt to escape. The first standard strain was developed at the Wistar Institute in the first decade of this century. The process of domestication has caused large anatomical, physiological and behavioural changes from the wild *Rattus norvegicus*. For example, the adrenals of the domestic rat are only one-third to one-fifth as large, as a result of a reduction in cortical tissue; liver, spleen, brain, heart and kidneys are also smaller; the hypophysis and thymus are larger; the gonads develop earlier; and the number of fungiform papillae in the tongue has been reduced. Domesticated rats also mate more freely, have greater susceptibility to audiogenic fits, are much less active, and are much less aggressive in terms of killing other rats and mice and in fighting. The wild rats can survive longer on a low-sodium diet but do not survive well on salt supplement after adrenalectomy. They need large amounts of cortical hormone but even then survival is tenuous.

Overall the wild rat, as Richter points out, lives in an environment where it must be constantly on the alert and ready to fight for its existence against all types of enemies. It is a fierce aggressive suspicious animal. The domestic creature is tame, gentle and trusting and lives placidly, its only contributions to its own survival being its feeding, drinking, grooming and mating activities. Like other domestic animals it has shown numerous mutations; there are 23 strains that breed true. They have been selected for the laboratory environment where the fittest are the most gentle and fertile.

The intense study of one species may be highly

fruitful in many respects, as witness the knowledge on genetics derived from *Drosphila*. However, in a provocative article entitled 'The Snark was a Boojum' Frank Beach cited Lewis Carroll's *The Hunting of the Snark* to question whether the Comparative Physiologist has largely disappeared, in the predominance of behavioural study of one animal—the albino rat. He bluntly asked whether behavioural science, as reflected in the predominance in the journals of publications on this one species, was building a science of rat learning rather than a general science of behaviour. He noted that there are approximately 3500 extant species of mammals, which together represent only less than half of one per cent of all animal species, and yet in the principal journal on comparative psychology 50%–70% of articles for 15 years were on the albino rat. Thorpe (1948) of Cambridge made essentially the same point, noting that Tolman's book *Purpose Behaviour in Animals and Men* is dedicated to the white rat, and there is hardly another animal mentioned in it. Without having done the same quantitative comparison as Beach (1950) the situation today cannot be assessed, but casual perusal of journals and attendance at symposia on experimental studies of ingestion behaviour could give the impression that the situation has not changed much.

The question clearly emerges as to whether the domesticated white rat is the ideal species to study if the search is for naturally selected capacities of high survival value directed to the recognition and ingestion of nutritious substances by the feral animal. The process of domestication may have modified such neural mechanisms no less than it has profoundly modified other mechanisms and physical characters. Of course, there is no ideal species and without any doubt in many ways the domesticated rat is an excellent one to study and a wealth of quite basic knowledge has accrued. Its similarity to the human, except in the rat's capacity to synthesize vitamin C and its inability to vomit, has been eloquently argued by Richter (1968). His contributions have been monumental, and it is noteworthy that he states it has been his practice where feasible to use equal numbers of wild rats in his studies, as well, of course, as many other species of wild animals.

This point can be put into perspective also in another way—even if we for the moment consider 'the rat'. We have drawn attention to the fact (Blair-West et al. 1970) that the Archbold Expedition of the American Museum of Natural History identified 36 genera, 124 species and 276 sub-species of Muridae, i.e. Old and New World rats, on the continent of Australia and New Guinea (Tate 1951). Little is known of the physiology, endocrinology and appetitive behaviour of these creatures, which range from large fruit-eating tree rats in the tropical north to various types of water rats, alpine dwelling rats, etc. Thus when a statement is made, as it is generally and commonly made, that a particular finding is true in the rat, it should be kept in mind, at least if the issue has any comparative overtone, that this usually is a very restricted instance of an inbred albino domesticated—and correspondingly modified—*Rattus norvegicus*.

Owen Maller (1967) in the book *The Chemical Senses and Nutrition* has reviewed a number of the general issues raised here. He stresses the extent to which domestication may have altered the taste and regulatory mechanisms of the rat. He also considers the fact that single components of nutrients, for example vitamins used in laboratory studies, are rarely found in nature. He proposes that an innate value system for the acceptance of items classified as food would have greater survival potential than the risk involved in a trial and error process of learning.

References

Adam WR, Dawborn JK (1972) Effect of potassium depletion on mineral appetite in the rat. J Comp Physiol Psychol 78: 51

Arnold GW (1964) Some principle in the investigation of selective grazing. Proc Aust Soc Anim Product 5: 258

Barnett SA (1956) Behaviour components in the feeding of wild and laboratory rats. Behaviour (Netherlands) 9: 24

Barnett SA, Spencer MM (1953) Experiments on the food preferences of wild rat (*Rattus norvegicus* Berkehhout). J Hyg (Lond) 51: 16

Beach FA (1950) The snark was a boojum. Am Psychol 5: 115

Blair-West JR, Coghlan JP, Denton DA, Funder JW, Nelson JF, Scoggins BA, Wright RD (1970) Sodium homeostasis, salt appetite and hypertension. Circ Res XXVI [Suppl II] 251

Blake WD, Jurf AN (1968) Increased voluntary sodium intake in K deprived rats. Commun Behav Biol [A] 1: 1

Brothwell D (1976) Further evidence of bone chewing by ungulates. The sheep of North Ronaldsay, Orkney. J Arch Sci 3: 179

Catt KJ (1970) Hormonal control of calcium homeostasis. Lancet II: 255

Davis CM (1928) Self-selection of diet by newly weaned infants. Am J Dis Child 36: 651

Davis CM (1933) A practical application of some lessons of the self-selection of diet study to the feeding of children in hospitals. Am J Dis Child 46: 743

de Castro J (1952) The geography of hunger. Little Brown, Boston, Mass

Denton DA, Blair-West JR, Coghlan JP, Scoggins BA, Wright RD (1977) Gustatory–alimentary reflex inhibition of aldosterone hypersecretion by the autotransplanted adrenal gland of sodium deficient sheep. Acta Endocrinol (Copenh) 84: 119

Donaldson HH (1924) The rat. Memoirs of the Wistar Institute No 6, 2nd edition

Drummond JC (1934) Lane medical lectures. Biochemical studies of nutritional problems. Stanford Univ Press
Duerst Von J Ulrich (1946/47) Verschiedenheiten der Ursachen und Beziehungen der Mangelkrankheiten zu den endokrinen Drusen. Bull Schweiz Akad Med Wiss 2: 193
Emmers R, Nocenti MR (1963) Role of thalamic nucleus for taste in modifying calcium intake in rats maintained on a self-selection diet. Physiologist 6: 176
French MH (1945) Geophagia in animals. East Afr Med J 22: 103
Garcia J, Ervin FR (1968) Gustatory–visceral and telereceptor–cutaneous conditioning-adaptation in internal and external milieus. Commun Behav Biol [Part A] 1: 389
Garcia J, Hankins WG (1975) The evolution of bitter and the acquisition of toxiphobia. In: Denton DA, Coghlan JP (eds) Olfaction and taste V. Academic Press, New York, p 39
Ginsberg RS, Feldman M, Necheles H (1948) Effect of odours on appetite. Gastroenterology 10: 281
Gordon JG, Tribe DE, Graham TC (1954) The feeding behaviour of phosphorus deficient cattle and sheep. Anim Behav 2: 72
Green HH (1925) Perverted appetites. Physiol Rev 5: 336
Harris LJ, Clay J, Hargreaves F, Ward A (1933) Appetite and choice of diet. The ability of the vitamin B deficient rat to discriminate between diets containing and lacking the vitamin. Proc R Soc Lond [Biol] 113: 161
Hughes BO, Wood-Gush DGM (1971) A specific appetite for calcium in domestic chickens. Anim Behav 19: 490
Klein DR (1970) Food selection by North American deer and their response to overutilization of preferred plant species. In: Watson A (ed) Animal populations in relation to their food resources. Blackwell, Oxford
Kovach JK, Kling A (1967) Mechanisms of neonate suckling behaviour in the kitten. Anim Behav 15: 91
Krupski A (1946) On the question of the causes of deficiency symptoms of cattle of our high pastures. Bull Schweiz Akad Med Wiss 1: 115
Krupski A, de Quervain F (1948) Über die Haufigkeit des Vorkommens bestimmter Mangelstorungen beim Rind in der Schweiz, deren Ursachen und Bekampfung. Bull Schweiz Akad Med Wiss 4: 228
Kuss SE (1969) Die palaolithische osteokeraticsche Kultur der Insel Kreta (Griechenland). Ber Naturf Ges Freiburg i Br 59: 137
Le Magnen J (1971) Olfaction and nutrition. In: Beidler LM (ed) Handbook of sensory physiology, Vol 4. Chemical senses I, Olfaction. Springer-Verlag, Berlin, p 465
Le Vaillant (1796) Travels into the interior parts of Africa in the years 1781–1785. English Translation, GG and J Robinson, 2nd edn. London
Lewis M (1964) Behaviour resulting from calcium deprivation in parathyroidectomized rats. J Comp Physiol Psychol 57: 348
Maller O (1967) Specific appetite. In: Kare MR, Maller O (eds) The chemical senses and nutrition. Johns Hopkins Press, Baltimore, p 201
McAdam DW, Way JS (1967) Olfactory discrimination in the giant anteater. Nature 214: 316
McCance RA (1936) Medical problems in mineral metabolism. Lancet I: 643
Milner P, Zucker I (1965) Specific hunger for potassium in the rat. Psychonom Sci 2: 17
Mundy PJ, Ledger JA (1976) Griffin vultures, carnivores and bones. S Afr J Sci 72: 106
Nachman M (1963) Taste preferences for lithium chloride by adrenalectomized rats. Am J Physiol 205: 219
Nachman M, Ashe JH (1974) Effects of basolateral amygdala lesions on neophobia, learned taste aversions, and sodium appetite in rats. J Comp Physiol Psychol 87: 622
Nesbit Evans EM (1970) The reaction of a group of Rothschild's giraffe to a new environment. East Afr Wildl J 8: 53
Orent-Keiles E, McCollum EV (1940) Potassium in animal nutrition. J Biol Chem 140: 337
Orr JB (1929) Minerals in pastures. Lewis, London
Orr JB, Gilks JL (1931) Studies on nutrition. The physique and health of two African tribes. Med Res Council Rept, No 155. HMSO, London
Reynolds EB, Jones JM, Jones JH, Fudge JF, Kleberg RJ Jr (1953) Methods of supplying phosphorus to range cattle in south Texas. Texas Agric Exp Stat Bull 773: 3
Richter CP (1954) The effects of domestication and selection on the behaviour of the Norway rat. J Natn Cancer Inst 15· 727
Richter CP (1956) Salt appetite of mammals. Its dependence on instinct and metabolism. In: L'Instinct dans le comportment des animaux et de l'homme. Masson et Cie, Paris, p 577
Richter CP (1968) Experiences of a reluctant rat catcher. The common Norway rat: friend or enemy? Proc Am Philos Soc 112: 403
Richter CP, Eckert JF (1937) Increased calcium appetite of parathyroidectomized rats. Endocrinology 21: 50
Richter CP, Eckert JF (1939) Mineral appetite of parathyroidectomized rats. Am J Med Sci 198: 9
Richter CP, Helfrick S (1943) Decreased phosphorus appetite of parathyroidectomized rats. Endocrinology 33: 349
Richter CP, Holt LE, Barelare B (1938) Nutritional requirements for normal growth and reproduction in rats studied by the self-selection method. Am J Physiol 122: 734
Rigg T, Askew HO (1934) Soil and mineral supplements in the treatment of bush sickness. Emp J Exp Agric 2: 1
Rodgers WL (1967) Specificity of specific hungers. J Comp Physiol Psychol 64: 49
Rodgers W, Rozin P (1966) Novel food preferences in thiamine-deficient rats. J Comp Physiol Psychol 61: 1
Rozin P (1967) Thiamine specific hunger. In: Handbook of physiology, alimentary canal Vol 1. Waverly Press, Baltimore, p 411
Rozin P (1976) The selection of foods by rats, humans and other animals. In: Rosenblatt J et al. (eds) Advances in the study of behaviour, Vol 6. Academic Press, New York, p 21
Rozin P, Rodgers W (1967) Novel-diet preferences in vitamin-deficient rats and rats recovered from vitamin deficiency. J Comp Physiol Psychol 63: 421
Schaller GB (1967) The deer and tiger. University of Chicago Press, Chicago
Schelling DH (1932) Calcium and phosphorus studies. The effect of Ca and P of diet on tetany, serum Ca, and food intake of parathyroidectomized rats. J Biol Chem 96: 195
Scott EM, Verney EL, Morissey PD (1950) Self-selection of diet. II. Appetites for calcium, magnesium and potassium. J Nutr 41: 187
Snapper I (1955) Food preferences in man. Special cravings and aversions. Ann NY Acad Sci 63: 92
Snowdon CT, Sanderson BA (1974) Lead pica produced in rats. Science 183: 92
Sutcliffe AJ (1973) Similarity of bones and antlers gnawed by deer to human artifacts. Nature 246: 428
Sutcliffe AJ (1977) Further notes on bones and antlers chewed by deer and other ungulates. Deer 4: 73
Svericenko PA (1954) The searching by rodents for food in fields, and their conditioned reflexes to non-food odours. Zool Zh 33: 876
Swift RW (1948) Deer select most nutritious forages. J Wildlife Management 12: 109
Tate GHH (1951) Results of the Archbold expeditions No 65. The rodents of Australia and New Guinea. Bull Am Mus Nat Hist 97: 183
Theiler A, Green HH, du Toit PJ (1924) Phosphorus in the

livestock industry. S Afr Dept Agric J 8: 460
Thorpe WH (1948) The modern concept of instinctive behaviour. Bull Anim Behav 7: 1
Tokunaga S (1936) Bone artifacts used by ancient man in the Riukui Islands. Proc Imp Acad Tokyo 12: 352
Uehlinger Von E, Krupski A, Almasy F, Ulrich H (1946) Die pathologische Anatomie der Skelettveranderungen bei der Lecksucht des Rindes. Bull Schweiz Akad Med Wiss 2: 47
Western D (1971) Giraffe chewing a Grant's gazelle carcass. East Afr Wildl J 9: 156
Widmark EMP (1944) The selection of food. III. Calcium. Acta Physiol Scand 7: 322
Wyatt JR (1971) Osteophagia in Masai giraffe. East Afr Wildl J 9: 157
Young PT (1941) The experimental analysis of appetite. Psychol Bull 38: 129
Young PT (1967) Palatability: The hedonic response to foodstuffs. Handbook of physiology, section 6: Alimentary canal. Vol 1. American Physiological Society, Washington, p 353
Zucker I (1965) Short-term salt preference of potassium-deprived rats. Am J Physiol 208: 1071

26 Clinical studies of salt appetite

Summary

1. Instances of the development of avid salt appetite in very young children with salt deficiency have been reported. The precise circumstances of observation in the classic Wilkins and Richter case of deficiency of electrolyte-active steroid at 1–2 years of age suggest that a greatly increased desire for salt taste preceded any likely learned benefit from ingestion.

2. Some patients with Addison's disease present with salt craving or predilection for salty foods, but they represent only about a fifth of patients with the disease.

3. In McCance's study of experimental salt deficiency in man, alteration of sense of taste was a consistent finding though not necessarily associated with any increased salt craving.

4. As well as attested instances of beneficial choice of food substances by humans with metabolic disorder, there are conversely many accounts of behaviour devoid of any relation to metabolic need.

5. At a theoretical level the possibility can be considered, in the face of comparison with herbivores, that a change to meat-eating by the hominids, with the resultant obligatory sodium intake, resulted in some regression over the Pleistocene period of the mechanisms of activation of salt appetite during salt deficiency. The data on tasters and non-tasters of phenylthiocarbamide (PTC) are discussed in terms of variation in populations of different ethnic origin and resultant susceptibility to goitrogens. There is no evidence of the salt taste threshold in man being different from that in animals, but this does not exclude the possibility of some change in central organization as alluded to above.

Against the background of the salt hunger manifest in mammals in particular ecological situations and the imperious desire for salt recorded in human history in various circumstances, the instances of patients presenting clinically with salt craving are rather sparse in the literature. In the author's experience of discussion with clinicians, it is not uncommon to find that they are able to cite cases from their own experience which have not been published. However, the impression emerging overall is that the development of salt hunger when there is attested body deficit in modern man is rather capricious. It appears in some individuals and not in others and, further on, we will discuss possible bases for this fact.

The patient of Wilkins and Richter

A classic instance of great salt craving in adrenal insufficiency was described by Wilkins and Richter in 1940. The case has special importance since it covers behaviour in a very young child. It seems clear from the record that the behaviour emerged before a possible influence of social dietetic factors which, evidently, can influence events with an adult. The child, aged 3½ years, was admitted to hospital with marked development of secondary sex organs. The penis and testes were as large as those of a 12-year-old boy and the prostate was fairly well developed. There was rather abundant dark hair over the pubis, his voice was deep and the laryngeal cartilage was as prominent as that of a full-grown man. His appetite was poor in hospital and he ate little food. In hospital he did not seem especially ill. He behaved like a defective child, growling and snarling in an incoherent manner when attempts were made to examine him. When feedings were forced, he vomited on several occasions. Blood sodium was low, and 7 days after admission he suddenly died. Post-mortem showed that both adrenals were large, with hyperplasia of the androgenic zone and marked diminution of normal cortical cells. The conclusion was that death from adrenal insufficiency resulted from destruction of the electrolyte-controlling elements of the adrenal cortex. After his death it was learned that from the age of 12 months the child had eaten salt in large quantities. During the 2½ years before admission he had kept himself alive in this fashion. In hospital he had had a regular ward diet and had not been given access to salt, and, as a result, he died suddenly.

The parents gave a detailed description of a very difficult feeding history. Their initial observation was that when the boy was around a year old he started licking all the salt off crackers and always asked for more. He had a special sound for signalling his desire of them. He also showed the same interest in bacon, chewing it to get the salt but not swallowing the pieces. At 18 months they gave him a few grains of salt from a salt shaker, thinking he would not like it; but he ate it and asked for more. This was the beginning of his showing that he really craved salt, for this one time was all it took him to learn what was in the shaker. Shortly after this, when being fed, he would keep crying for something that was not on the table and always pointed to the cupboard. The mother did not think of salt so held him up in front of the cupboard to see what he wanted. He picked up the salt at once and in order to see what he would do with it he was allowed to have it. He poured some out and ate it by dipping his finger in it. After this he would not eat any food without having the salt too. The foods he would not otherwise ordinarily eat would be taken after he had salted them. At 18 months he was beginning to say a few words and salt was one of the first. Practically everything he really liked was salty, such as crackers, pretzels, potato crisps, olives, pickles, fresh fish, salt mackerel, crisp bacon and most foods or vegetables if more salt were added. He ate about a teaspoon of table salt a day in addition to the fact that his foods were much saltier than those eaten by the parents. If the mother tried to make him think she had added salt to the food but had not done so he always knew the difference with the first taste. In contrast to the great craving for salt, he had a clear antipathy to sweet substances. As well as salt he also showed a great liking for water from a very early age and frequently chose it in preference to milk.

Wilkins and Richter in discussing the case draw the parallel in relation to taste behaviour with the adrenalectomized rat, which they point out can distinguish distilled water from salt solution in a 1:33 000 concentration, while normal rats do not make the distinction below 1:2000. Such minimal amounts of salt as are received in drinking a 1:33 000 solution could not possibly have any physiological effect. They note this boy also showed a positive reaction to salt when he first tasted it in pure form in various quantities on the top of the crackers. They did not feel it was readily conceivable that such a small amount of salt could have any detectable effect on his deficiency condition. They propose that this emphasized the fact that in the salt-deficient state the boy was very much attracted to salt rather than having experienced that the ingestion of salt made him feel better. The taste of it had changed.

Observations of salt appetite in patients with Addison's disease and other conditions

Reports by Thompson and McQuarrie (1934) and McQuarrie (1935) dealt with four cases of juvenile diabetes mellitus in which there was abnormal craving for and ingestion of large quantities of sodium chloride (up to 80 g daily). These patients presented with hypertension and a high fluid intake and output. This reversed on removing the excess salt from the diet. In 1936 Darley and Doane recorded a patient with pulmonary arteriosclerosis, marked shortness of breath and an almost constant craving for salt, all of which had been present since early childhood. The patient's parents stated that when she became large enough to climb onto a chair and reach objects on the table or sideboard an abnormal hunger for salt became manifest, and the knowledge of certain punishment would not deter her from appropriating the contents of the salt cellar. The patient died at the age of 20 after an exploratory operation of the chest. Unfortunately no post-mortem data were available on the abdominal viscera.

Richter (1941) recorded a 34-year-old patient with Addison's disease who ate salt in enormous amounts, even on oranges, lemons and grapefruit, but had a strong aversion to sugar and candy. Richter (1942–43) reports patients with Addison's disease who manifested a marked craving for food with a high salt content—ham, sauerkraut, etc.—without associating the craving with the high salt content. All they knew was that the food had a particularly pleasant taste for them. Thorn, Dorrance and Day (1942) reported on 158 patients with Addison's disease and the evaluation of desoxycorticosterone acetate (DOCA) therapy in this condition. Of the 64 patients in the group who were seen in the Johns Hopkins Hospital, 10 (16%) presented with salt craving. Because of the supplementary sodium chloride given daily with the DOCA therapy, it was not established whether the salt craving disappeared with the institution of hormone treatment. Henkin and Soloman (1962) and Henkin, Gill and Bartter (1963) note also that only 15%–20% of patients with Addison's disease report with salt craving.

Spiro, Prineas and Moore (1970) report a case of congenital myopathy with mitochondrial and lipid disorders in a 13-year-old boy. The muscle weakness had been present since birth. During the first three years of life the patient had intermittent diarrhoea and the parents were told he had a coeliac-like disease. A feature of the history was that all his life the boy had had a marked craving for salt. Sandwiches of bread with large quantities of salt or salted bananas were consumed ravenously. The mother had to forbid him seasoning his food with salt or eating it directly from the packet. There was, however, no difference in muscle strength between these times of enforced restriction of salt and the times when his cravings were satisfied. The authors note the report of Poskanzer and Kerr (1961) where excessive intake and therapeutic response to salt had been reported in a family with normokalaemic periodic paralysis. In this case there was no response to salt and the explanation of the appetite is obscure unless, perhaps, it was learned behaviour of benefit over the first few years where the diarrhoea reported in the clinical history may have aggravated the weakness of the myopathy.

Grossman et al. (1977) have reported that increased salt appetite is commonly encountered in children with sickle cell haemoglobinopathies. Their preliminary results show 16 of 43 children with the condition had increased or exceedingly increased salt appetite, whereas 20 black children who were controls from other families showed only 1 child which would be so classified on their criteria. The appetite was manifest as early as 2½ years of age and one such young child salted hard candy and licked the salt off and discarded the candy. Other children salted apples, oranges and peaches; members of the family without sickle cell disease did not show this behaviour. The authors propose that the appetite arises as a result of excess sodium loss as distinct from non-osmotic stimulation of antidiuretic hormone (ADH) secretion. Hyponatraemia occurs during sickle cell crises (Radel et al. 1976; DeFronzo et al. 1976), and DeFronzo et al. (1976) suggested ADH release is responsible.

Experimental study of salt appetite in man

The classical experimental study of salt deficiency in man was made by McCance upon himself and three volunteer medical students (McCance 1936a, b). By the use of low-sodium diet and episodes of sweating, severe sodium deficit could be produced in about 7 days and the deprivation periods of the experiment lasted over 11 days. The recovery periods varied from 1 to 7 days according to the plan. The net negative balance of sodium produced in the case of McCance himself was 763 mmol and with one of the students, Robert Niven, 804 mmol. In relation to symptoms and signs, a major feature was the lethargy, excessive fatigue and general sense of exhaustion of the subjects. They all suffered from cramps except one subject in whom the deficiency was relatively slight. These cramps were not of the severe localized type that had been described in miners and stokers; they were mild and easily controllable, but any muscle suddenly brought into action was likely to spasm. Blood pressure did not fall in the subjects. Anorexia and nausea were prominent symptoms.

The sense of flavour and taste was greatly affected. Miss Edwards, the one who was least depleted, interpreted this aberration or lack of sensation as thirst. She complained of it constantly and drank freely but without obtaining any relief. McCance recognized the feeling as distinct from thirst. Food was tasteless—even highly flavoured food—and this was more noticeable because such foods were eagerly sought to make the meals more appetizing. Even cigarettes did not have taste. Robert Niven was less troubled with this symptom. Niven experienced thirst and on occasion drank a lot of water, but it made little difference. Niven longed for salt and often went to sleep thinking about it. McCance felt no specific craving for salt and had difficulty in convincing himself that taking salt would at once make him feel all right again. Recovery from sodium deficiency was quite dramatic. Half an hour after taking 15 g of salt with bread and butter Miss Edwards' sense of flavour and taste had returned, even though no fluid had been taken (possibly because she had drunk water beforehand). She spoke of this as quenching her thirst. Genuine and almost unbearable thirst supervened later and was only satisfied by copious draughts of water. McCance found his sense of flavour returned before he finished his first salt meal. With Niven the return of flavour and taste was not a matter of comment.

De Wardener and Herxheimer (1957) were able to produce negative sodium balance by forced water drinking. They found it lowered the taste theshold and this was not associated with the lowering of the salivary Na/K ratio. An intense craving for salt developed in the subjects.

Another facet of clinical incidence of salt appetite comes from observations of Langford et al. (1977). They suggest that hypertensive patients treated with diuretics increase their sodium intake so partially negating the effect of therapy, and that this change of behaviour is associated with and perhaps due to a change of salt taste threshold. A group on therapy had a mean sodium excretion of 202 mmol/day compared with 126 mmol/day for those not on therapy. Langford et al. state that a decreased taste threshold for salt was demonstrable when tested 2 weeks after commencing diuretic therapy. They note the study of Parijis et al. (1973) also shows patients on thiazide therapy excrete significantly more sodium than controls given a placebo. They suggest that an increased salt appetite engendered by sodium loss could negate the effect of therapy, and with low-dosage therapy such negation of effect has been demonstrated by adding 50 mmol of sodium to the diet.

Other evidence of dietetic wisdom in man or lack of it

As Pangborn (1967) points out in discussing chemoreception in man, the editors of *Nutrition Reviews* in 1944 advanced six generalizations on the self-selection of diets.

1) Nutrition based upon appetites is not universally successful.
2) Appetites are often fickle and unpredictable.
3) Appetites may be trivial in origin.
4) Individual animals vary in their ability to make choices that will improve their nutritional status.
5) Factors affecting human appetites may be expected to be more numerous and more complex than those affecting the appetites of animals.
6) From the evidence, self-selection of diets appears to be inferior to scientific evaluation of diets for the maintenance of good nutrition.

Snapper (1967), in considering this issue, notes that when Minot discovered that pernicious anaemia could be cured by the ingestion of liver, only one or two of the patients described were able to keep themselves alive by eating large amounts of liver. There were observations in the Netherlands during the Second World War of the effects of liquorice, and that this material could be used successfully to treat adrenal cortical insufficiency. Despite this finding, no patient with Addison's disease is recorded as having treated himself with liquorice. However, he says that perhaps instincts for health and nutrients are found in the jungle population. In the eastern part of the Indonesian archipelago and New Guinea, Snapper notes sago flour is the staple food of the population. This is prepared from the marrow of the sago palm and contains practically nothing but carbohydrates with mere traces of protein and vitamins. The same is true of cassava flour. The native population supplements the sago and cassava diets with animals caught in the forest, ranging from lizards, spiders, caterpillars, worms and larvae to small game. These animals, high in vitamin B and protein, form the necessary addition to the sago flour diet. In New Guinea the Papuans put maggots into the marrow of the living sago palm and enclose with mud the opening through which the maggots are introduced. Eventually the large maggots are eaten as a delicacy—presumably providing some protein addition to the diet. Snapper notes that as long as tribes remain in virgin forests and follow their centuries-old dietetic habits their health is relatively satisfactory. But as soon as civilization teaches these natives to despise the animals of the forest, deficiency diseases appear. Jungle inhabitants often have eating habits that conserve their health and these should not be changed.

In the same volume on the *Chemical Senses in*

Nutrition, Henkin (1967a) notes that children with phenylketonuria and galactosaemia will eat diets rich in phenylalanine or galactose respectively and become increasingly ill. The doctor must decrease the amount of these substances in their diets in order to get the children at least metabolically on the road to recovery. Only with a great deal of effort can he do this. Similarly with the alcoholic who is deficient in many nutrients, it is extremely difficult to persuade the patient to take the essential nutrients. On the other hand, the studies of Davis on young children discussed in Chapter 11 point to a notable ability to self-select a nutritious diet when given cafeteria selection of a wide variety of foods.

Set against this brief consideration of the relation of food choice and need, there are parallels with the data on salt appetite in clinical medicine. It seems possible from the evidence on Addison's disease that there could be individual variation, genetically determined, in the generation of salt hunger by body deficit. Further, the outcome of behaviour may be strongly influenced by previous dietetic history. However, as far as animals are concerned, or at least as regards the sheep, whether or not the animal has habitually had a high sodium intake due to environmental circumstances or a low one, it does not seem to obviate the onset of an increased intake when the creatures are made deficient by a parotid fistula. But as well as the instances of increased salt appetite in very young children developing in association with metabolic deficiency, there are observations attesting to absence of appetite under these conditions. Recently, for example, Dr Maria New of Cornell Medical School Pediatric Department (personal communication) has had under her care a baby 1 year of age who appears to have a genetically determined absence of mineralocorticoid receptors. Peripheral blood aldosterone was extremely high and hyperkalaemia was a prominent abnormality. The infant has been sustained by daily salt administration but it has been observed that the addition of salt to milk he is fed results in his rejection of the material. This has been tried at several concentrations. Similar absence of salt liking and preference in children with salt-losing adrenogenital syndromes has been observed in the same clinic.

Possible regression of salt appetite in contemporary man

If the hypothesis that salt appetite was highly developed in the primates and hominoids during the vegetarian stage of phylogeny were valid, it could, perhaps, follow from the advent of substantial carnivorous behaviour that the selection pressure favouring salt appetite may have dwindled to a low level because of adequate sodium intake in meat. Notwithstanding what Clark (1970) has observed in relation to the predominantly vegetarian diet of hunter-gatherer societies, this situation of reduced survival advantage of salt appetite could have held over a million years or more, that is, 50 000 generations. This is adequate time for a characteristic to be greatly modified. If, for example, a characteristic were sexually selected it could, of course, alter quite rapidly (Dobzhansky 1962). But there is no suggestion that the organization subserving salt appetite might be selected against in this fashion in the carnivore or omnivore. It seems possible that in a complex neural organization at the basis of an innate behaviour pattern such as salt appetite, critical elements might be dependent on one or two genes and mutation in such loci might attenuate the capacity of chemical and hormonal changes to activate this drive. This could then vary in individuals according to heredity. Certainly, for example, ethnic differences are known in relation to the capacity to taste certain substances such as phenylthiourea (phenylthiocarbamide, PTC).

Phenylthiocarbamide is a special case since taste thresholds for most compounds follow a Gaussian or monomodal distribution in a population whereas tasting of PTC follows a bimodal distribution in a Caucasian population, differentiating it into 'tasters' and 'non-tasters' (Fischer 1967). The 'non-tasting' dimorphism of PTC-type compounds apparently antedates the separation of the hominids and pongids, since tasters and non-tasters of PTC exist in orang-utan, chimpanzee and gorilla in proportions similar to those in Caucasians (Fisher et al. 1939).

It is found that 100% of the population of Amerindians and some African tribes are able to taste this bitter substance (Kalmus 1971). Eskimos have 40% of non-tasters. Fischer (1967) suggests there may be changes occurring in that whereas most black tribes in Africa have practically no non-tasters (Allison and Blumberg 1959) the difference is diminishing in black communities in the United States. Kalmus (1971) remarks that N—C=S substances occur in wild plants and vegetables and their bitter taste may lead tasters of PTC to avoid eating these plants and thus escape noxious effects. But nobody knows whether these plants are an important food. Bitter taste is not always a deterrent, and the natural concentration of the substances in the leaves or fruits is low, and consequently their taste is not very strong. The food-rejection theory of the PTC polymorphism is, however, if somewhat indirectly, supported by the distribution of non-tasters among human populations. As these are absent in most primitive tribal people and rather frequent in those people who have a long history of agriculture or even urban life, one might argue that the disappearance from the diet of some plant food which was formerly collected in the wild has relaxed the selection against the non-taster gene, which thus could spread from the always-present rare mutants in the population. The data indicate that non-tasters are recessive homozygotes having the genetic constitution *tt*, while the tasters represent two genotypes: the heterozygotes *Tt* and the homozygotes *TT*. Against the hypothesis must be held the fact that tasters and non-tasters occur either separately or polymorphically in many primate species.

The hypothesis is that tasters can successfully avoid bitter-tasting goitrogens, which has adaptive value in geographic areas where goitrogens are potentially significant in the diet and iodine is in short supply (Rozin 1976). Goitrin isolated from *Brassica* seeds as well as from crushed rutabaga contains the same N—C=S grouping as PTC compounds. Foods such as cabbage, turnips, peas, beans and strawberries display antithyroid activity, and turnips and rutabaga actually contain goitrogen in amounts as high as 200 parts per million (Jukes and Shaffer 1960). In apparent support of the advantage of tasting, a four-fold prevalence of non-tasters relative to tasters was found in 962 goitrous Israeli subjects, and in the endemic goitre area of Brazil, nodular goitre proved to be significantly elevated among non-tasters of PTC, implying that goitre in non-tasters is more likely to evolve in nodular form (Fischer 1967).

But there is little evidence on comparable variation in the threshold for tasting salt. Indeed the same threshold level is seen in experiments over a variety of species in the animal kingdom, and from the experiments on man. There is evidence from Henkin's studies summarized in Chapter 10 that the detection threshold may alter substantially in adrenal insufficiency, and in cystic fibrosis similar changes have been observed mainly in males, though, in contrast to Addison's disease, the effect is not reversed by carbohydrate steroids (Henkin 1967b). But overall the evidence is strongly suggestive that salt taste is a basic feature of chemoreception in all higher mammals, and has not been much modified by any recent evolutionary events. This, however, may not necessarily be true with regard to central neural mechanisms associated with genesis of drive in deficiency.

As far as the phenomenon of experience in the modification of taste sensibility is concerned, it is appropriate to consider the issue in the context of ontogenesis. Mistretta and Bradley (1977) and Bradley and Mistretta (1973) note that in both humans and sheep, taste buds on the tongue appear and mature structurally *in utero*. In both species taste buds are morphologically mature by about 14 weeks. In the case of the sheep, by 100 days of gestation individual nerve fibres stain distinctively and branch extensively in the taste buds. In contrast to the sheep and the human, the taste buds of the rat reach maturity about 12 days post-natally. Mistretta and Bradley made single and multi-fibre recordings from sheep foetuses aged between 96 and 147 days, and found that foetal taste receptors over the last third of gestation responded similarly to those of the newborn and the adult.

In order to determine whether foetuses demonstrate taste preferences and aversions before birth they have studied foetal swallowing activity and tried experiments to alter it. Using oesophageal flow transducers they found sheep swallowed amniotic fluid at rates varying from 50 to 785 ml/day, the data being comparable to those reported for human foetuses, which ranged between 200 and 760 ml/day (Pritchard 1965). The great variation in foetal swallowing behaviour made it difficult to establish a baseline which allowed assessment of the influence on this behaviour of injection of substances into the amniotic fluid. They could not obtain consistent alteration of swallowing with either sugar or a bitter-tasting substance.

Mistretta and Bradley note that the human foetus swallows from about the twelfth week of gestation, so that it has 28 weeks of gustatory experience of chemicals in the amniotic fluid. In the sheep, for example, sodium in amniotic fluid decreases from 116 mmol/l at 100 days of gestation to 86 mmol/l at term. There is a rise in urea from 6 to 12 mmol/l over this time. Further, the sheep foetus urinates spontaneously throughout the day and the amniotic fluid composition would be altered. The question of whether these small changes in amniotic fluid composition, or such changes in it which might result from changes in maternal nutritional status, could have any influence on the foetus remains open. Reduction of the sodium concentration of amniotic fluid by 10–15 mmol/l occurs in very severe maternal sodium deficiency in sheep (Phillips and Sundaram 1966; Wintour et al. 1977).

It is, however, clear that in the neonatal state the human baby's taste experience with regard, for example, to salt will alter substantially in that human milk has a sodium content one-tenth or less that of the amniotic fluid. On the other hand, in the instance of an infant reared on cow's milk and commercial baby foods (before the modification of their contents by exclusion of added salt) the experience would be quite different and not far removed from that of swallowing amniotic fluid with a sodium content of approximately 100 mmol/l.

Whereas taste responses may be present *in utero*, the question is unresolved as to at what stage of ontogeny the appetitive drive in response to sodium deficiency emerges in those species which clearly show this behaviour. If ontogenetic development of the neural mechanisms subserving this behaviour occupies a sizeable period of time *in the neonate*, it could be that a high salt intake over this period, quite different from that which would normally hold under conditions of breast feeding, might modify the character of this mechanism in the adult. This might be relevant to clinical experience over the last half century or so in relation to salt appetite and response to metabolic sodium deficiency. Whereas such a possibility may seem remote, it cannot be entirely ruled out.

References

Allison AC, Blumberg BS (1959) Ability to taste phenylthiocarbamide among Alaskan Eskimos and other populations. Hum Biol 31: 352

Bradley RM, Mistretta CM (1973) The gustatory sense in foetal sheep during the last third of gestation. J Physiol (Lond) 231: 271

Clark PJ (1970) The pre-history of Africa. Praeger, New York

Darley W, Doane CA (1936) Primary pulmonary arteriosclerosis with polycythemia: associated with the chronic ingestion of abnormally large quantities of sodium chloride (halophagia). Am J Med Sci 191: 633

DeFronzo RA, Siegel NJ, Pearson HA (1976) More on the pathogenesis and management of hyponatraemia in the crisis of sickle cell disease. J Pediatr 89: 1038

Dobzhansky T (1962) Mankind evolving. Yale University Press, New Haven

Fischer R (1967) Genetics and gustatory chemoreception in man and other primates. In: Kare MR, Maller O (eds) The chemical senses and nutrition. Johns Hopkins Press, Baltimore, p 61

Fisher RA, Ford EB, Huxley J (1939) Taste-testing the anthropoid apes. Nature 144: 750

Grossman H, Kennedy E, McCamman S, Rice K, Hellerstein S (1977) Salt appetite in children with sickle cell disease. J Pediatr 90: 671

Henkin RI (1967a) In discussion of Maller O. In: Kare MR, Maller O (eds) Chemical senses and nutrition. Johns Hopkins Press, Baltimore, p 208

Henkin RI (1967b) In discussion of Fischer R. In: Kare MR, Maller O (eds) Chemical senses and nutrition. Johns Hopkins Press, Baltimore, pp 76, 77, 78

Henkin RI, Soloman DH (1962) Salt taste threshold in adrenal insufficiency in man. J Clin Endocrinol Metab 22: 856

Henkin RI, Gill JR, Bartter FC (1963) Studies on taste thresholds in normal man and in patients with adrenal cortical insufficiency, the role of adrenal cortical steroids and of serum sodium concentration. J Clin Invest 42: 727

Jukes TH, Shaffer CB (1960) Antithyroid effects of Aminotriazole. Science 132: 296

Kalmus H (1971) The genetics of taste. In: Beidler LM (ed) Taste. Springer-Verlag, Berlin, p 165

Langford HG, Watson RL, Thomas JG (1977) Salt intake and treatment of hypertension. Am Heart J 93: 531

McCance RA (1936a) Medical problems in mineral metabolism. Lancet I: 644, 704, 823

McCance RA (1936b) Experimental sodium chloride deficiency in man. Proc R Soc Lond [Biol] 119: 245

McQuarrie I (1935) The effects of excessive salt ingestion on carbohydrate metabolism and arterial pressure in diabetic children. Mayo Clin Proc 10: 239

Mistretta CM, Bradley RM (1977) Taste in utero: Theoretical considerations in taste and development—genesis of sweet preference. US Dept Health Education and Welfare, DHEW Publication No. (NIH) 77-1068

Pangborn RM (1967) Some aspects of chemoreception in human nutrition. In: Kare M, Maller O (eds) Chemical senses and nutrition. Johns Hopkins Press, Baltimore, p 45

Parijis J, Joosens JV, Van der Linden L, Verstreken G, Amery AKPC (1973) Moderate sodium restriction and diuretics in the treatment of hypertension. Am Heart J 85: 22

Phillips GD, Sundaram SK (1966) Sodium depletion of pregnant ewes and its effects on foetuses and foetal fluids. J Physiol (Lond) 184: 889

Poskanzer DC, Kerr DNS (1961) A third type of periodic paralysis, with normakalaemia and favourable response to sodium chloride. Am J Med 31: 328

Pritchard JA (1965) Deglutition by normal and anencephalic fetuses. Obstet Gynecol 25: 289

Radel EG, Kochen JA, Finberg L (1976) Hyponatraemia in sickle cell disease. J Pediatr 88: 800

Richter CP (1941) Behaviour and endocrine regulators of the internal environment. Endocrinology 28: 193

Richter CP (1942/43) Total self-regulatory functions in animals and human beings. Harvey Lectures Series, Vol 38

Rozin P (1976) The selection of foods by rats, humans and other animals. In: Rosenblatt J et al. (eds) Advances in study of behaviour Vol 6. Academic Press, New York, p 21

Snapper I (1967) The etiology of different forms of taste behaviour. In: Kare MR, Maller O (eds) The chemical senses and nutrition. Johns Hopkins Press, Baltimore, p 337

Spiro AJ, Prineas JW, Moore CC (1970) A new mitochondrial myopathy in a patient with salt craving. Arch Neurol 22: 259

Thompson WH, McQuarrie I (1934) Effects of various salts on carbohydrate metabolism and blood pressure in diabetic children. Proc Soc Exp Biol Med 31: 239

Thorn GW, Dorrance SS, Day E (1942) Addison's disease: Evaluation of synthetic desoxycorticosterone acetate therapy in 158 patients. Ann Intern Med 16: 1053

de Wardener HE, Herxheimer A (1957) The effect of a high water intake on salt consumption, taste thesholds and salivary secretion in man. J Physiol (Lond) 139: 53

Wilkins L, Richter CP (1940) A great craving for salt by a child with cortico adrenal insufficiency. JAMA 114: 866

Wintour EM, Brown EH, Denton DA, Hardy KJ, McDougall JG, Robinson PM, Rowe EJ, Whipp GT (1977) In: Conti C (ed) Research on Steroids VII. Elsevier, Amsterdam, p 475

27 Salt intake and high blood pressure in man

Summary

1. The question at issue in this chapter is whether there are gradual damaging effects of long-term exposure to excess of a substance which is a normal, indeed essential, component of diet. Elements in the analysis include the individual variation in hedonic, or nutritionally unnecessary, intake within the population, and variation in the genetic susceptibility to effects of high salt intake.

2. There is large variation in hedonic intake between species just as there is seen to be variation in satiating behaviour in response to body sodium deficit. There is also large individual variation in the 'set' of hedonic or palatability mechanisms within species.

3. Studies in the USA have shown children prefer greater sweetness and saltiness than adults. Furthermore, black children have a greater preference for high salt concentration than white children or black or white adults, an interesting finding in the light of the predominance of hypertension in the black community.

4. Some biological bases of hedonically determined ingestive behaviour are discussed in relation to creatures liking what they need and rejecting harmful substances. Hedonic intake of salt is a definitive component in the overall physiological organization of salt appetite in man and reflects behaviour favourable to survival under the conditions in which the species evolved. The author differs with the viewpoint of Dahl (1960), reiterated by Freis (1976), who proposed salt appetite in man is induced rather than innate. Showing the human has no need of the excess salt, which is clearly true, does not vitiate the proposal that hedonic liking may have a biologically significant basis, though the mechanisms may be modified culturally. The fact that salt is taken liberally in the absence of biological need is not more or less surprising than that sexual activity occurs in humans in the absence of intent to reproduce the species. Behaviour components of both appetitive and motor phases in the latter instance reflect innate neural mechanisms. Konrad Lorenz has noted Craig (1918) exploded the myth of the 'infallibility' of instinct in showing that appetitive behaviour aims at discharge of an instinctive action—consummatory action—not its survival value.

5. There are identified physiological mechanisms which 'turn off' hedonic intake as it proceeds. But the general liking for salt, seen in most mammals, together with its cheapness operates to cause intake greatly in excess of need in man.

6. High blood pressure is a major factor causing atherosclerosis and its complications—mainly coronary and cerebrovascular disease. In Western

society means and variances of systolic and diastolic pressure rise with age, and to greater degree in females. Hypertension is present in 10%–20% of the adult population. There is considerable acceptance of the view that the causation is multifactorial. Inherited characteristics are multifactorial and operative, as are environmental factors. The roles of some psychological and sociocultural influences are discussed in relation to this latter category in the aetiology.

7. There is an impressive list of primitive communities, many within the Pacific basin, where there is no blood pressure rise with age. An outstanding study is that of Sinnett and Whyte (1973a, b) on a total highland population in New Guinea, involving 1489 subjects. In the male, mean blood pressure fell after the age of 30. With females, systolic pressure increased a small amount with age whereas diastolic fell so that pulse pressure increased. There was little evidence of hypertension (only 3% of males over 40 years), or complications of arteriosclerosis. The population was physically fit and, on a specific criterion (Harvard Pack test), better than Royal Australian Air Force personnel and comparable to British Commonwealth divisions in Korea. Mean 24-h urinary sodium excretion was 14 mmol for females and 6 mmol for males. There was no evidence that disease, infection or parasitism prevented some normal blood pressure rise with age or that the data did not represent normality in terms of blood pressure.

Studies by Dutch anthropologists in highland New Guinea show the predominantly vegetarian diet of sweet potato, taro, yam, bananas and cassava results in a urinary K/Na ratio of 200–400:1, with daily sodium excretion often as low as 1–2 mmol. Several studies confirmatory of Sinnett and Whyte's data have been made in New Guinea, and, also, it has been shown in transitional groups and in those living in urban areas that blood pressure increase does occur. This is coincident with access to or eventual change to Western diet and life style. Similar evidence has come from comparison of unacculturated and transitional communities in Melanesia, Australia and Polynesia, with data confirming that failure of blood pressure to rise with age is not determined by disease or low protein intake.

8. Aspects of the salt trade in New Guinea are described, including the zealous use of salt springs in the mountains. The extensive use of plant ash, ratified as potassium salt, attests the general desire for salt taste since there could be no retroactive sense of benefit in the face of the very large surfeit of potassium in the diet.

9. Studies of hunter-gatherer societies in Africa, similarly have shown no rise in blood pressure with age in a healthy population. Sodium excretion measured in the !Kung bushmen was low (less than 40 mmol/day) and urinary K/Na ratio was 3–4:1. Transitional societies show blood pressure rise though there are no definitive data attributing this to diet. Usually an increase in obesity is seen.

In southern Iran, pastoral nomads who are only slightly acculturated had systolic and diastolic pressure increased above 140 and 90 mmHg respectively in over 20% of both males and females over 30 years of age. Ponderal index was negatively correlated with blood pressure in males and not correlated in females. They are lean active people. Sodium excretion was 140–180 mmol/day. This society could represent a strong upward trend of blood pressure in an unacculturated society unrelated to weight change and possibly related to high sodium intake.

10. Of several primitive communities in Central and South America where no rise in blood pressure occurs with age, the study of the Yanomamö of the Amazon Basin has been outstanding. They are highly active, aggressive people. Diet is vegetarian with significant supplementation from game and fish to give a protein intake of approximately 50 g/day. Urinary sodium excretion was extremely low (1 mmol/day) and potassium excretion was in the range of 150 mmol/day. Aldosterone secretion and plasma renin were high, and aldosterone spectacularly high in the pregnant Yanomamö, but during lactation no higher than in males or non-lactating females.

The Yanomamö have particular interest because they are a society which has been studied extensively anthropologically as well as metabolically. This is in the light of Pickering's (1980) view of the unimportance of salt intake in the difference between primitive and Western communities with regard to blood pressure. His opinion was that the great contrast was between certainty of behaviour in a primitive society ruled by ritual and taboo, and the uncertainty in Western societies in which life was a series of individual choices and decisions. Chagnon's (1977) study *Yanomamö: The Fierce People* covered 15 years and records in detail the life of people engaged in chronic warfare and whose existence is charged with violence and tension. Yanomamö goad each other within their own villages to the brink of explosion, and conduct warfare and raids on other villages. Chagnon describes it as politics of brinkmanship. They frequently exercise options involving outcomes which vary from humiliation, through loss of property and women, to severe bodily injury and death. From the evidence, the idea that there is emancipation from mental stress by virtue of rituals which bestow certainty on behaviour is hardly credible.

11. In North America and the Caribbean there is a greater incidence of hypertension in black males and females than in whites, and this is reflected in mortality statistics.

12. Notwithstanding some debate about accuracy of certification, death from apoplexy and incidence of cerebrovascular accident or stroke are very high in Japan, with striking variation between prefectures. Data suggest peak levels may be 3 to 8 times those in Western Europe and USA. Highest incidence is in Akita prefecture in northern Honshu where the incidence of hypertension was reported as 39% in a large group of average age 45, compared with 21% in Hiroshima in southern Japan. Regional differences in incidence of apoplexy parallel differences in blood pressure. Few individuals live over 70 years in villages in northern Honshu. The importance of the results and the comparisons between regions resides in the very high salt intake in northern Honshu (20–30 g/day), coincident with the intake of high-quality rice with abundant miso soup, pickles and condiments. Fishing villages which had a diet without such very high salt components had a much lower incidence of apoplexy and hypertension. Other factors which have been invoked as feasible explanations of the data from Akita include a high sulphate/carbonate ratio in the water due to the volcanic nature of the ground, cadmium accumulation in fish eaten, and calcium deficiency.

13. Attempts to relate incidence of hypertension to variation of salt intake within populations have produced contradictory results. Estimates of salting may be unreliable because of subject unreliability and the fact that the main addition occurs before food reaches the table. Twenty-four hour urine specimens are more reliable but can give 2–4-fold variation in sodium content from day to day in the individual. In any event if sodium intake were

influential, the major effect may have been during infancy to adolescence—years before the short-term adult measurements. In relation to such analysis, if it were true for instance that only 10%–20% of the population were genetically susceptible to development of hypertension with high salt intake and the threshold of this effect is in the 50–100 mmol/day range, the fact that the large majority of the population has intake in this range or above from childhood will make recognition of any relation between salt intake and hypertension within the population difficult. This problem would be greater if variation in salt intake above threshold had little influence on blood pressure. If this were true, however, it would not seem compatible with the reductions in blood pressure claimed to follow relatively small reductions in salt intake: e.g. from 200 to 150–100 mmol/day. It also has to be taken into account that development of hypertension may itself modify salt appetite.

14. Hedonic appetite for salt is reduced in rats made hypertensive by perinephric wrapping, and in rats genetically susceptible to induction of hypertension by salt, and the appetite for salt induced by pregnancy is reduced in rats by renal hypertension. On the other hand, spontaneously hypertensive rats developed an increased preference for hypertonic salt solutions following weaning. In hypertensive patients, there are some reports suggestive of decreased ability to recognize salt taste. The data on whether patients with essential hypertension show increased preference for salt are conflicting, and have been criticized on methodological grounds. Some data suggest hypertensives may perceive suprathreshold salt concentrations as less intense than do controls.

15. Very severe restriction of salt intake (e.g. the Kempner rice diet) reduced blood pressure in the hypertensive patient. This is independent of weight reduction, though weight reduction of itself also reduces blood pressure. The long-term effect of a moderate reduction of salt intake in patients on a characteristic Western diet remains controversial, and the issue is yet to be resolved. Recent clinical studies have suggested that moderate salt restriction combined with high potassium intake may help prevent hypertension, that salt-sensitive subjects exist, and that these subjects may benefit most. Criticisms of the salt hypothesis of causation of hypertension are set out.

The interesting and comprehensive epidemiological studies of Joossens and colleagues relating the overall death rate from stroke and that from stomach cancer in many different countries, and over the course of several decades in many individual countries, are discussed. Highly significant relations are found, and Joossens' contention is that the linking factor is salt. The gradation in effects corresponds closely to the extremely high use of salt in Japan and Korea, intermediate usage in Eastern Europe and Portugal, and lower level in Western Europe and USA. The progressive reduction is related to the introduction of refrigeration of foods, and the reduction in intake of salted meat, fish, vegetables and cereals, and of smoked meats. The linking factor is proposed to be influential by virtue of an influence on blood pressure and stroke, and by the damaging influence of hypertonic salt on gastric mucosa, its effect in delaying gastric emptying time, and in causing atrophic gastritis which favours production of carcinogenic nitrosamines from interaction of nitrate and food amines.

16. Experiments on rats show that hypertension can be induced by high salt intake, that life span is directly affected, and that the influence is greater if begun with young animals. There is variation in individual susceptibility, and

the effect is decisively ameliorated by a high potassium intake. Early experimental data from rats on the influence of high sodium intake or low potassium intake during pregnancy on subsequent blood pressure of the young are recalled. High salt intake does not cause hypertension in rabbits and sheep, but preliminary data suggest an effect in baboons.

17. Human breast milk (ca 6 mmol/l of sodium) provides adequate sodium and nutrition during the first 6–9 months of life. Studies showed that with use of cow's milk and processed infants foods, salt intake of infants and children very often greatly exceeded (5–10-fold) any feasible sodium requirements. Dahl et al. (1970) showed feeding commercial baby foods to salt-sensitive rats caused hypertension and eventually death in about 50% of animals. Salt content of baby food was drastically reduced in Australia in early 1970s and other countries also followed the same policy. Studies on schoolchildren in the USA have suggested a discernible influence on blood pressure of relatively small differences in sodium content of the community water supply. No comparable effect was seen in an Australian study. Consideration, however, of the large variation in sodium content of municipal water as recorded in both USA and Australia highlights problems of contriving salt restriction in patients with cardiac, renal and hypertensive disease in many communities. Whereas refrigeration has greatly reduced salt intake over past decades, analyses suggest that in the same regions the trend may now be significantly counteracted by the practices of food processing firms and the fast-food outlets. There are significant technical impediments to greatly reducing salt levels in or removing salt from some types of foods.

18. Experimental analysis of the mechanism by which high salt intake increases blood pressure delineates the role of increased extracellular volume, increased cardiac output, and an eventual rise in total peripheral resistance due to autoregulatory processes. The direct effect of salt on resistance vessels is also identified. The primary role of the kidney, and individual variation in functional capacity to excrete an excess load of salt has been emphasized; for example, renal transplantation experiments in rats showed that the phenotype response of blood pressure was more influenced by the genotype of the renal graft than the genotype of the recipient. Renal perfusion studies support these data.

A hypothesis accepted by many workers is that in the hypertensive human there is an inherited defect in the handling of sodium such that the kidney requires a higher than normal perfusion pressure to maintain extracellular homeostasis in the face of high sodium intake. Quantitative aspects of this are recounted, and the arguments for significant reduction of salt intake are noted, as well as the opposing case, proposed before the US Congress by the Salt Institute, for not reducing salt intake as a public health measure.

19. In the last sections of the chapter there is an overview of hypertension as a malady of civilization. This is set against considerations of changing selection pressures during evolution in relation to the premium on hedonic salt ingestion and salt conservation mechanisms. For example there are striking differences in the incidence of hypertension between the US and Caribbean blacks and US whites. The ancestors of citizens of the USA reached its shores in many different ways. The passage from hominids of the Pleistocene era in tropical/subtropical Africa to white settlement of America has been circuitous: perhaps twenty to forty thousand generations of existence in Europe and northern Europe occurred. Major changes postulated over the period include regression of skin pigmentation with

attendant effect on vitamin D production, bone growth and facility of reproduction. The cold climate with virtual absence of sweating and with greater meat eating may have ameliorated stress on sodium homeostasis and some regression of the major ingestion and conservation mechanisms ensued.

On the other hand, the ancestors of the black citizens reached the USA and the Caribbean two to four hundred years ago as a result of the slave trade. Their progenitors correspondingly represent twenty to forty thousand generations of existence in tropical/subtropical conditions with considerable stress on sodium homeostasis. Greater hedonic liking of salt and greater facility of conservation with less capacity to excrete any excess could be a legacy of this different history. Comparative studies have given evidence of a lesser capacity of blacks to excrete a sodium load, of a greater rise in blood pressure with it, and of differences in potassium metabolism under these conditions.

Analogous to the situation in tropical/subtropical Australia, where blonde or red-headed people of Celtic descent are advised rigidly to restrict skin exposure to sunlight in an effort to decrease the high incidence of skin cancer in people with this ethnic background, there may be a particular case for restriction of salt intake in blacks.

One fact emergent from the anthropological studies is the physical vigour and health of some low-salt communities, and whereas it may be argued that they may withstand, for example, gastrointestinal infection with consequent fluid loss less well than high-salt communities, this would be marginal since little salt storage occurs in the body.

20. In relation to the general issue of possible intra-uterine influences on preference behaviour arising as a result of altered metabolic state of the mother, it seems unlikely that significant effects could occur, notwithstanding data on foetal swallowing, validated intra-uterine activity of taste receptors, and neonate rejection and acceptance behaviour. This is because, as far as sodium is concerned, little change occurs in amniotic fluid with changed maternal sodium status.

21. At any scientific meeting centred on analysis of the pathophysiology of human and experimental hypertension, the role of sodium is paramount in the discussion, because of its diverse influence on the relevant parameters. This involves sodium balance, sodium distribution in the organism and thus volume of compartments, modulation of response of excitable tissues, and effects on hormone secretion and action including the renin–angiotensin system, antidiuretic hormone, catecholamines and the steroid hormones. Nevertheless, it is formally possible that the role of high sodium intake is permissive of expression of other factors which might contrive the hypertensive state, including influence of obesity and psychogenic factors determined by life-style.

Many human data support a genetic susceptibility to hypertension. The rat experiments on genetic susceptibility to high salt intake, including the data on cross-breeding, kidney perfusion and parabiosis, point to a basic renal defect, and possibly a humoral factor, a natriuretic hormone, which also has the capacity to produce hypertension. A large body of evidence has accrued on disorder of intracellular electrolyte composition, and of several membrane sodium transport systems in essential hypertension. Intracellular sodium of red and white blood cells is increased, and new evidence implicates both ouabain-sensitive and other sodium and potassium transport systems. Some workers propose a genetic difference in non-oubain-sensitive sodium transport which is manifest also in normotensive children of hypertensive parents in contrast to results from controls. However, other workers indict depression of the ouabain-sensitive sodium transport system as a result of high circulating level of a natriuretic hormone which inhibits it. Increase in sodium in arterial smooth muscle cells, also caused by the same agent, is proposed to increase intracellular calcium and thus contractility.

22. Systematic studies of the dynamics of excretion of a sodium load point to greater change in blood pressure in the face of a sodium load in borderline hypertensives, and also in the normotensive relatives of hypertensive parents, than in controls. There was an inverse correlation between age and sodium excretion on the sodium-loading day in the relatives, but not in the control subjects. Normotensive sons of hypertensive fathers show accelerated natriuresis with a sodium load.

23. There are good grounds for proposing excess salt intake, probably associated with reduced potassium intake, as the prime factor, or a major factor, in the aetiology of essential hypertension. This affects the genetically susceptible, the functional defect being in the kidney. However, the proper assessment at this point would appear to be the Scottish court verdict: not proven. If not the direct cause, high salt intake throughout life may be an important permissive factor in the expression of other factors—psychogenic, nutritional, sympathetic nervous and hormonal—which contrive the hypertensive state. It might be wondered, notwithstanding the possibility of extensive controlled experiments on higher primates especially pongids, whether it ever can be proven decisively and whether all measures at a public health level, particularly those aimed at the young, should be withheld until unequivocal proof is available. Given the effect of hypertension on 15%–20% of the population in terms of morbidity and mortality and the enormous costs to the health care system, many would feel the probabilities would make a preventive approach a good national investment. As noted in a recent *Lancet* editorial (1980: II, 459), there will be ample evidence for some, and there will never be enough for others, to recommend action towards dietary change at a public level of food policy and education. At least it is time to consider whether the amount of salt we eat is doing us more harm than good, and whether reducing it would do more good than harm.

In my view, it would still be possible to conduct the decisive controlled experiment on unacculturated societies but it would be unethical to do so in practice, as well as in principle, because in all probability blood pressure would go up in the salt-exposed primitive and hitherto low blood pressure societies.

Introduction

Hence if too much salt is used in food, the pulse hardens, tears make their appearance and the complexion changes.
(Huang Ti Nei Ching Su Wen (1000 B.C.), from the translation by Wang Ping (A.D. 762))

In human communities in the past, salt was often a very rare commodity and highly prized to the point of warfare. With the advent of modern technology and commerce, salt has become common and relatively cheap. In the main the surfeit has arisen over the last 100 years or so, though in some countries such as Sweden which had particular eating habits dependent on, for example, salted fish, intake has probably fallen. Alwall (1958) estimates that the average salt intake of the Swedes may have been about 100 g/day in the middle of the sixteenth century, whereas Josephson (1971) suggests the contemporary level to be about 5.5 g/day. Joossens et al. (1979) report that daily output of sodium in urine determined in a large survey in Belgium declined from 15.3 g in 1966 to 8 g in 1978.

In 1948 the world production of salt was 42 million long tons (Parsons 1951); in 1925 it was 22 million. Overall this increase is probably reflected in dietetic intake in Western communities. In 1978, salt sales in the USA alone were 24 million tonnes, of which 4% was food grade. US usage per capita in the food industry, allowing use in brines and processing where discarding occurs, was estimated at 18 g/day in 1970 by the Committee on Food Protection of the National Academy of Sciences. Actual daily intake per capita has been estimated at 7–8 g. The total sales of salt for discretionary intake in the USA at this time were about 275 000 tons (250 000 000 kg).

The question in relation to salt intake and hypertension is whether there are subtle aggregate damaging effects of long-term exposure to excess of a substance which is a normal, indeed essential, component of diet. There are primitive peoples—unacculturated societies—where hypertension is rare and there is no blood pressure rise with age as is seen in the acculturated Western societies. Low salt intake is the common circumstance of these primitive societies.

When we come to analyse salt intake in modern man a central issue is the individual variation in 'need-free' or nutritionally unnecessary salt intake in the population, and also variation in the genetic susceptibility to effects of high salt intake. Hence I will precede discussion of human hypertension by mention of some aspects of hedonic salt intake in animals and man. This will stress individual variation in behaviour which is much more evident than, for example, with thirst.

Preamble on hedonic behaviour and individual variation

From earlier chapters it is apparent that there is substantial genetic variation between species in the neural organization which subserves salt appetite, and there is also apparently considerable individual variation within species. Thus there are differences between species in the form of the preference–aversion curves for saline solutions. For example, the sheep ingests preferentially much higher concentrations than the rat or calf. As well as this variation in behaviour in the 'need-free' sodium-replete state, large differences between species are seen in ingestion behaviour in sodium deficiency. Wild rabbits repair body deficit slowly and accurately and do not overdrink, whereas intake in great excess of body deficit is characteristic of the inbred white laboratory rat (Chapter 14).

Within a particular species, such as the sheep, there may be large individual variation in the 'set' of the hedonic or palatability mechanisms determining 'need-free' intake of salt. Given the same environment and dietetic conditions from birth, and these, for example, involving low but adequate salt intake in food and water, some animals when given free access to sodium solutions behave as salt gluttons, whereas others ingest little or no salt. Again, when made sodium deficient for the first time by parotid fistula, animals vary in that some soon show evident restlessness, curiosity and exploratory behaviour. Rapid sampling of a cafeteria of solutions occurs when it is made available, with specific ingestion of sodium solution. Others show little behavioural response or ingestion, and may become severely sodium deficient and debilitated before any intake (see Chapter 11). There is also considerable individual variation in learning propensity. This is seen, for example, in the instance of adjusting intake to deficit when the concentration of sodium bicarbonate solution presented to animals is varied.

Other animal and human studies indicate variation in taste sensitivity. Hoshishima et al. (1962) showed, in six different strains of mice, variation in taste threshold to sodium chloride, saccharin, acetic acid and phenylthiocarbamide (PTC). A white strain showed the lowest thresholds to all substances studied whereas a black strain showed the highest. The 'taste-blindness' of certain groups of people to PTC is well documented (Kalmus 1971), as is its transmission as a Mendelian recessive character. Darwin in the *Origin of Species* stated that pigs with black hair are known to have more sensitive gustatory and olfactory sensations than those with white hair.

McConnell and Henkin (1973) showed that salt preference behaviour over the 16 weeks after weaning in spontaneously hypertensive rats differed strikingly from that seen in the normotensive Wistar Kyoto strain. The spontaneously hypertensive rats showed a greater preference for 0.3 M sodium chloride, and this was also manifest if the behaviour was tested for the first time 14 weeks after weaning. This result contrasted with that found when rats genetically predisposed to hypertension in the face of high salt intake were fed a diet high in sodium chloride and were concurrently tested for sodium chloride solution preference. Here an aversion was manifest (Wolf et al. 1965).

Desor et al. (1975) tested the preferences for sweet and salty in 9- to 15-year-old black and white children and adults. They remark that individual preferences persisted over a relatively long time, suggesting that they are characteristic of the person rather than of his immediate metabolic state (Greene et al. 1975). They found the younger subjects preferred greater sweetness and saltiness than did adults. The main finding was that younger blacks differed from other groups in their preference for salt. Only a small percentage of white adults or black adults and white children (10%, 12% and 9% respectively) demonstrated a preference for the highest concentration of salt, whereas 30% of the younger blacks preferred the saltiest sample. The authors reflect on the implications in relation to salt intake and hypertension. They suggest that the data are consistent with the view that human populations are not homogenous in their preferences for sweetness and saltiness.

This is a major question. Given the evidence of substantial variation within animal species in relation to salt ingestion in several circumstances, it might be expected that there would be considerable individual variation within the human species, both in relation to hedonic intake of salt and also in any hunger manifest with metabolic sodium deficiency.

Before proceeding to the analysis of these data, some further aspects of hedonic salt intake are pertinent. There would appear to be, as Bare (1949) first suggested, some selection advantage in the simple stimulus–response connection of the animal ingesting some salt or dilute salt solution when it is encountered in nature. Though sodium cannot be stored, this 'need-free' intake may to some extent protect the animal against deprivation, and, perhaps more important, the animal may remember the situation of the hedonic intake and return to it, particularly if later it becomes salt deficient.

In a broader context of food intake in general, Young's (1944, 1946) studies led him to observe that rats may develop food preferences which do not correspond to bodily needs, and form dietary habits which tend to stabilize selection of food. Young (see Janowitz and Grossman 1949) suggests that animals learn to select and to seek foods which they like rather than foods which they need (require nutritionally). To a considerable extent, however, animals like what they need. In discussing 'The Pleasures of Sensation' Pfaffmann (1960) proposes that sensory stimulation per se, together with its ensuing central nervous events, be considered as the prime determinant in the chain of events culminating in acceptance behaviour, and hedonic effect.

In reflecting on hedonic preferences and the development of the sweet taste, Young (1977) suggests that the reaction to the sweet taste is just one specific instance of the general biological fact of instinctive reactions elicited by specific patterns of stimulation. The sensitivity to sweet, and subsequent ingestion, has evolved through millions of years and is biologically adaptive in terms of recognition and selection of some carbohydrates that are valuable as a source of energy. Ripe fruits are sweet and preferred to green sour fruits which are unpleasant. Similarly, bitter tastes are unpleasant and may indicate the presence of toxic alkaloids. Salty tastes usually indicate the presence of sodium, which is required for homeostasis. Thus the biological value of these reactions is obvious. Chemical, thermal and tactile receptors at the oral end of the alimentary canal stand guard over the acceptance or rejection of foodstuffs. Hedonic processes are both activating and reinforcing.

I wish to stress here the hedonic determination of salt intake and its biological basis as an element of the physiological organization of sodium appetite. This is because in discussion of excessive sodium intake and its role in hypertension, Dahl (1960) has suggested that salt appetite in the human is induced rather than innate. He places much emphasis on the fact that, in humans, intake bears no relation to deficit. However, showing that the human has no need of the excess salt, which is clearly true, does not vitiate the suggestion that hedonic liking may have a biologically significant basis. The existence of salt-specific taste fibres, and the propensity to ingest salt at least over a certain concentration range because the creature likes it, reflect natural selection of behaviour favourable to survival under the conditions in which the species has evolved.

But a neural mechanism which is innate and carries survival advantage may be conditioned, and also modified by complex social circumstances. Such immediate causes of ingestive behaviour may be evident: e.g. ritual and custom. The innate propensity for the behaviour is there—even if there is no contemporaneous biological need. The fact that salt intake may occur in the absence of metabolic need is no more or less surprising than that sexual activity occurs in humans in the absence of intent to increase the numbers of the species. The latter observation does not gainsay the fact that behaviour components of both appetitive and motor phases reflect innate neural mechanisms. Similar considerations apply in relation to socially conditioned drinking for pleasure as against the innate physiological mechanisms of thirst in water deprivation.

The hedonic propensity to take salt, in my view, augments the likelihood of excessive intake in the circumstance of ready and cheap supply. The arguments of Dahl (1957; 1958a, b; 1960), Ball and Meneely (1957), Blair-West et al. (1970), Denton (1973) and Freis (1976) on the possible relation of high blood pressure to excessive salt intake do not depend on the notion of salt appetite being purely acquired behaviour as is the viewpoint of Dahl (1960), reiterated by Freis (1976). As Konrad Lorenz (1950) noted, it was Wallace Craig (1918) who once and for all exploded the myth of the 'infallibility' of instinct. With a wealth of evidence he showed that appetitive behaviour aims at the discharge of an instinctive action—consummatory action—not its survival value. Lorenz remarks that introspective self-observation makes this superlatively obvious in relation to appetitive behaviour. People cannot stop eating though it is their purpose to lose weight. He suggested that however obvious and even commonplace Craig's discovery may seem, it nevertheless undoubtedly was one of the most decisive steps towards a real understanding of behaviour.

Without pre-empting the outcome of the debate on excess salt intake and high blood pressure outlined hereafter, it might be remarked now that the above viewpoint needs adequate recognition. This is so if there emerges any public health or clinical aim to greatly reduce salt intake in the population at large and which, like the Lilac Fairy in *The Sleeping Beauty*, could put things right.

As indicated in earlier chapters, the mechanisms which arrest hedonically determined salt intake are not clearly defined. Certainly Stellar et al. (1954) have shown in rat experiments that the salinity of gastric contents does affect salt preference. On the other hand, this is not the only 'stop' factor since these workers have also found that animals with an oesophageal fistula showed the preference for hypotonic solutions and aversion for hypertonic ones even though the solutions never reached the stomach. Pfaffmann (1960) takes the view that 'stop' of intake may simply reflect a change in 'sign' of afferent input as a consequence of its increasing intensity. A central neural switching occurs when intensity of afferent discharge reaches a critical value. When recording from the chorda tympani nerve there is a steady increase in discharge rate over the whole of the preference–aversion curve. There is no change in the afferent signal at the point of the behaviour change. The rather limited effects on voluntary ingestion of saline solution sometimes seen in animal experiments where body load has been increased by intravenous infusion or high-salt food have pointed to the strong influence of salt taste and its hedonic overtone in determining ingestion. Taste and stomach factors appear to arrest ingestion—with also some systemic influences. It is, however, interesting that in sheep exhibiting 'need-free' intake, or in those with moderate sodium depletion, intraventricular infusion for 1 h of 500 mmol/l sodium chloride at 1 ml/h, which raised the sodium concentration of cerebrospinal fluid by about 20 mmol/l, significantly decreased voluntary sodium intake in both instances.

Overall, the matters touched upon in this preamble are generally consonant with Richter's (1956) view that there is a universal liking for salt in mammals, though I think perhaps we should regard the issue as *sub judice* as far as carnivores are concerned, and there are other exceptions (see Chapter 10). Furthermore, as noted above there is individual variation in ingestive behaviour within species. This obviously holds in humans and there is also the question of individual variation of susceptibility to effects of high salt intake.

Epidemiology of hypertension

Incidence

High blood pressure is a major factor causing atherosclerosis and its complications—mainly coronary and cerebrovascular disease. Normal standards of blood pressure for North America and for Europe are generally recognized to be the same (Shattuck 1937). The tabulations of Symons made in 1922, based on the work of the examiners of 162 000 subjects for the Mutual Life Assurance Company of New York between 1907 and 1919 show, in both men and women, a progressive rise in systolic blood pressure with age.

Lovell et al. (1960) note that casual arterial pressure has been measured in Britain (Hamilton et al. 1954; Miall and Oldham 1955, 1958), Norway (Boe et al. 1957), and in the USA (Gover 1948; Comstock 1957) in populations in which bias due to exclusion of people with high blood pressure was avoided. The mean levels varied between populations but certain other characteristics of blood pressure were common to all. Thus: the means and variances of systolic and diastolic pressures rise with age; and the rise in mean systolic and diastolic pressures with age is greater in females than in males. The frequency distributions of pressures tend to show positive skewness, with more pressures and a larger range of pressures above modal values than below them, and as Gover (1948) and Pickering (1955) have stated, the distributions are continuous. This suggests the absence within these populations of two distinctly separate groups of people, one with high pressures and one with normal pressures.

Lovell et al. (1960) found similar features of blood pressure in their study of Fijians and Indians living in Fiji. The mean and standard deviation of systolic pressure rose with age in both groups, the diastolic pressures rose at least to the sixth decade, and the pressures rose more in females than males. Pressures rose more with age in the Indians. Differences between races in the younger groups were reduced when differences in arm circumference were taken into account. Correction for arm circumference increases the relatively higher pressure of elderly Indians. Comparison with Londoners shows lower relative pressures in Fijians and Indians after the sixth decade.

Figures from the USA indicate that hypertension is very common. In the Framingham study, for example, 10% of 1882 middle-aged men had systolic blood pressures exceeding 160 mmHg and 33% exceeded a diastolic pressure of 86 mmHg (Kagan et al. 1958). Among 2294 middle-aged women, the percentages exceeding the above levels were 13 and 27 respectively. In the total population of a town in Michigan, Epstein (1963) found about 30% of men and 34% of women had pressures of 160/95 or greater. Epstein suggests that 20% of middle-aged men in the USA have a systolic blood pressure of 160 mmHg or over and/or a diastolic pressure of 96 mmHg, or over. In examining the effectiveness of treatment of mild hypertension, the World Health Organization and International Society of Hypertension cite an incidence of 10%–15% of hypertension in the adult population.

The following definition of essential hypertension was put forward in 1960 by the American Heart Association Committee on Criteria for Diagnosis of Disease and Clinical Evaluation (adapted from Goldring and Chasis 1944):

> The term hypertensive disease is synonymous with essential hypertension and should properly be restricted to designate the as yet unidentified physiologic disturbance (or disturbances) characteristic of this disease and which leads ultimately to elevation of diastolic and systolic blood pressure, anatomical changes in the vascular tree, and functional impairment of the involved tissue. Hypertension is the earliest clinically recognizable disturbance and results from constriction of the peripheral arterioles, this constriction leading to an increase in the total effective peripheral resistance and hence to elevation of diastolic and systolic blood pressures. Hypertensive disease is considered to be a clinical entity in which an unknown pressor mechanism initiates arteriolar vasoconstriction, elevated blood pressure and vascular sequelae. Hypertension, as such, like arteriolar changes, is conceived to be a sequel appearing during the progressive development of the disease.

In relation to the issue of whether essential hypertension is an 'entity', one viewpoint has been summarized by Pickering (1961b) as follows:

> 1. blood pressure is inherited as a graded character over the whole range, including that which we commonly call essential hypertension.
> 2. environmental factors are largely responsible for the rate of rise of pressure with age.
> 3. the difference between normal pressure and hypertension is not one of kind, it is one of degree.
> 4. essential hypertension as a disease represents the effects of a quantitative deviation from the norm.
> 5. the aetiology of essential hypertension is almost certainly multifactorial, inherited characteristics are multifactorial and they play a part, and certainly environmental factors also play a part.
> 6. the attempt to divide sharply between normal blood pressure and essential hypertension is not only futile . . . it is wrong.

This position has been opposed by Morris (1961) who put his view as follows:

> To sum up, then, there seem to be two 'populations'—hypertensives and normotensives. These two populations seem to be qualitatively different and the evidence does not support the notion that high blood pressure without evident cause in middle age is merely the tail end of the normal distribution in the population.

At this same conference on hypertension in Prague, Perera (1961) stated:

> . . . many things about essential (or what I choose to call primary) hypertension, seem to me to label it a distinct disease. It shortens life. Those who have an established elevation of diastolic blood pressure develop, at varying rates, characteristic pathologic changes. There is a sex incidence favoring women and some constitutional differences exist. There are various disturbances in hyper-reactivity in response to drugs, in measurements of metabolism, in the responses to salt loads and sodium withdrawal, and in response to desoxycorticosterone and thiocyanates, which are different from those seen in normotensives.

Platt's (1963) theory that hypertension is genetically determined, with the outcome governed by whether the individual receives the hypertensive gene from neither parent (normotensive) or one (mild) or both parents (severe hypertension), has been discussed by Thomas (1973). Twin studies were cited, and also those on Johns Hopkins medical students, in relation to history of parents, uncles, aunts and grandparents. Hypertensive parents had more than twice as much hypertension among their siblings as normotensive parents. Thomas concludes that whereas the case for inheritance is strong the mode of inheritance is uncertain. Genes governing blood pressure may interact with genes governing other factors relevant in cardiovascular disease. Possibly there are several major causes of hypertension rather than one. Pickering's idea of polygenic inheritance is favoured by many, and, as Page (1976) suggests, the genetic influences may be partly if not wholly permissive of action of environmental factors.

In the epidemiology of hypertension, one of the major preoccupations has been with geographical differences, and the mode of their genesis. The literature has been critically reviewed from time to time, in particular, for example, by Bays and Scrimshaw in 1953. They drew attention to the shortcomings of many studies, including the failure to account for variation in arm size. They felt, however, some cautious conclusions could be drawn in terms of racial differences in blood pressure which could not be accounted for by the technical factors. They noted that these may be due to cultural traits or environmental stresses rather than genetic factors. The authors accept Ling's (1936) conclusions from observations on 15 114 Chinese males and 493 females indicating a lower blood pressure in China, with relative underweight being the most important factor. In considering this against other studies from the Orient, they emphasize the absence of age–height–weight data in many reports, but suggest, overall, that all the differences cannot be accounted for on the basis of body build and that mean blood pressure of Mongoloids is lower. They also regard the data available at that time as showing a high incidence of hypertension in blacks in the New World, with pressures consistently higher than in whites.

Epstein (1963), with additional data and with the criticisms of Bays and Scrimshaw (1953) in mind, proposed a discernible pattern in the course of plotting systolic blood pressures with age in 16 different populations: one group where blood pressure was essentially constant with age and the other where a rise occurred. This analysis was updated with much additional data by Epstein and Eckoff (1967) using 34 studies from different parts of the world (Fig. 27-1). Five different grids were constructed, each having three channels with systolic blood pressure intervals of 105–115 mmHg, 115–125 mmHg and 125–135 mmHg at commencing age a little below 10 years. These channels were given five slopes (0–4) representing increasing degrees of rise in blood pressure with age. The age-trend channels for 34 different studies are shown for males and females in Table 27-1. Most of the source references indicated are in the bibliography. Epstein and Eckoff note that the lines of division are purely empirical and each group does not necessarily fit neatly into its own box. However, the same model was obtained when diastolic pressures were used. Large differences between populations are evident.

Some postulated aetiological factors

In embarking on a detailed examination of the data on unacculturated societies where hypertension is rare, a major issue is the diet of such populations, and its comparison with that in societies where blood pressure rise with age is characteristic. But first, it is important to designate the various epidemiological factors which have been indicated or discussed as possible causes in the aetiology of

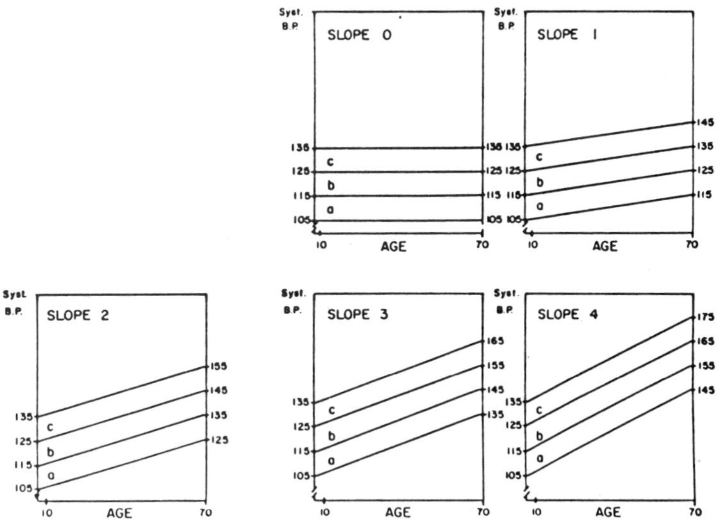

Fig. 27-1. A system for classifying systolic blood pressure: age-trends and levels. (From Epstein and Eckoff 1967, with permission.)

Table 27-1. Systolic blood pressure classification by geographic area.

Slope	Level a	Level b	Level c
		Males	
0	Caracas Indians (Loewenstein) Thailand (ICNND) Ethiopia (ICNND) Africa: Bushmen (Kaminer) Uganda: nomads (Shaper)	Vietnam (ICNND) India: rural (Padmavati) India: urban workers (Padmavati) Africa (Donnison) Uganda: semi-nomads (Shaper)	New Guinea: coast and Highlands (Whyte) Gilbertese (Maddocks) Africans (Williams)
1	Sri Lankans (Bibile)	Colombia (ICNND) Fiji (Maddocks) India: high socio-economic group (Padmavati) W. Africa (Abrahams)	Rural Zulu (Scotch)
2	Mundurucus Indians (Loewenstein) Javanese (Bailey) Formosa: Mainlanders (Liu) Liberia (Moser)	Bahamas: white (Johnson) Fiji (Lovell)	St Kitts: black (Schneckloth)
3	Atiu: Mitiaro (Hunter) Formosa: Taiwanese (Liu) India (Wilson)	Michigan: Tecumseh (Johnson) Jamaica: rural (Miall) Jamaica: urban (Miall) Rarotongans (Hunter) Japan: Nagasaki (ABCC) Japan: Hiroshima (Switzer)	
4	Georgia: white (Comstock) Chile (ICNND) Fiji: Indians (Lovell) Atlas Jews (Dreyfuss) Urban Zulu (Scotch) Uganda: villagers (Shaper) London (Pickering) Wales (Miall) Bergen 2 (Boe)	Georgia: black (Comstock) Bahamas: black (Johnson) Bergen 1(Boe)	Georgia: black (McDonough)
		Females	
0	Mundurucus and Caracas Indians (Loewenstein) Thailand (ICNND) Ethiopia (ICNND) Africa: Bushmen (Kaminer)	Gilbertese (Maddocks) Vietnam (ICNND)	New Guinea: coast (Whyte)
1	India: rural (Padmavati) Sri Lankans (Bibile)	Colombia (ICNND)	New Guinea: Highlands (Whyte)
2	Formosa: Mainlanders (Liu) Rural Zulu (Scotch)	Fiji (Maddocks) Javanese (Bailey)	
3	Formosa: Taiwanese (Liu) India (Wilson)	Bahamas: white (Johnson) Fiji (Lovell) Atiu: Mitiaro (Hunter) Japan: Nagasaki (ABCC)	Bahamas: black (Johnson)
4	Georgia: white (Comstock) Chile (ICNND) Japan: Hiroshima (Switzer) Urban Zulu (Scotch) Uganda: villagers (Shaper) W. Africa (Abrahams) London (Pickering) Wales (Miall) Bergen 2 (Boe)	Georgia: black (Comstock) Michigan: Tecumseh (Johnson) Jamaica: urban (Miall) Jamaica: rural (Miall) St Kitts: black (Schneckloth) Fiji: Indians (Lovell) Rarotongans (Hunter) Atlas Jews (Dreyfuss)	Georgia: black (McDonough) Bergen 1 (Boe)

From Epstein and Eckoff (1967), with permission.

hypertension or as modifying influences on the population. Reference to specific data on these factors will be made in relevant cases as particular societies are considered.

Paul and Ostfeld (1965) remark, in commenting on Pickering's views, that it seems to make sense that a physiological variable such as blood pressure may be altered and controlled by several influences, inherited and acquired. But this does not preclude the possibility that some genetic predisposition—such as one arising in the kidneys—is needed for an environmental influence to give rise to a clinical disease. The mortality rate from hypertensive disease is more than twice as high among non-whites as in whites in all sections of the USA. Higher pressures among US blacks exist in both men and women, and in all but the youngest age group. Epstein (1963) writing on this point emphasizes that both Comstock (1957) and Rose (1962) have shown that the difference is not evident in the young. Thus if there is a hereditary element in the black–white differences the phenotype manifests later in life. The data of Desor et al. (1975) showing the accentuated preference of 9–15-year-old black children relative to white for concentrated salty taste has been mentioned earlier.

Considering the difference in distribution of hypertension between black and white, Lennard and Glock (1957) emphasize the necessity of making comparison in geographic areas other than the USA. They note that research on the African populations appears to contradict the US results. Williams (1944) and Donnison (1929), for example, report lower rates of hypertension among particular African tribes than among comparable white groups. Donnison's data were derived from Kenyan natives living near Lake Victoria in primitive conditions which probably had not changed for centuries. Blood pressure in males fell after the age of 40. Some aspects of Donnison's data have been criticized by Bays and Scrimshaw (1953). Becker (1946) found a high incidence of hypertension among blacks in Johannesburg. However, the comparisons in the West Indies and Central America made by Saunders and Bancroft (1942), Marvin and Smith (1942), Kean (1951) and Schneckloth et al. (1962) are consistent with US data in that the blood pressure of blacks was higher than that of native Indian populations or whites in the same area. However, as Lennard and Glock (1957) point out there is nothing to eliminate the possibility that environmental factors are responsible for the difference. The blacks in the West Indies and Central America were in a social situation comparable to that in the USA and possibly this rather than heredity was the determining influence.

The issue of psychologic and sociocultural influences in the aetiology of hypertension has been the subject of comprehensive and scholarly reviews by Scotch and Geiger (1963) and Geiger and Scotch (1963). They note that observations have ranged from consideration of stress factors to attempts to define a hypertensive personality. They suggest there is much difference of opinion as to how and, indeed, whether psychological phenomena relate to hypertension. But there is a certain amount of agreement that psychological conflict is not a necessary precondition of hypertension and, further, that hypertension does not necessarily develop whenever such conflict is present.

Family size has received attention as a determinant in hypertension. Boe et al. (1957) reported in Norway that the level of parental blood pressure decreased as the number of children increased. Miall and Oldham (1958) made the same finding in Wales. Further, they noted that since the husbands as well as the wives in the larger families had lower pressures than their counterparts in smaller families, a physiological factor associated with pregnancy could be ruled out. They raise the possibilities of selective factors accounting for the finding or of large families counteracting possible stresses to which small families are vulnerable. Scotch (1963) found that a larger number of children is associated with normal blood pressure among rural Zulu women and that the reverse is true of urban Zulu women. Scotch and Geiger (1963) in discussing this, note that while a Zulu woman's status in the rural or traditional value system rests on her ability to produce children, urban women often assume the role of wage-earner, and in this context a larger number of children presents serious obstacles. Thus having more children is stressful in the city but non-stressful in the rural area.

Scotch, an anthropologist, produced further data from ethnographic surveys of two Zulu populations. One was in a rural native reserve and the other an urban location or housing development. Attention was given in the observations and interviews to identification of major problems of life and possible points of stress, the social or individual conflict, and frustration and role incongruities. The data were drawn from 548 adult members of the rural and 505 adult members of the urban population. Compared with an unacculturated population the mean blood pressures of both rural and urban Zulu were relatively high. The rural Zulu in general had mean blood pressures somewhat below those of US white and black groups and the urban Zulu had higher pressures than the American whites but not quite as high as the US blacks. Scotch puts considerable emphasis on the contrast between rural and urban Zulu blood pressures. Both mean blood pressure and frequency of hypertension were greater in the urban than the rural population for all age groups in both sexes. The findings were interpreted as implicating social stress as a causative variable. Scotch states:

> It is undoubtedly true that social stress is found in both the rural and urban areas in which the Zulu live, but it is hypothesized that stress is not only more frequent in the city but is much more severe as well. When informants were questioned in the rural areas as to what were the major problems of life, most all answered by saying that poverty and migratory labour were the big problems. In the city, however, respondents mentioned not only poverty but degradation and humiliation in the treatment of Africans by Europeans, frequent arrest, high rates of illegitimacy, divorce and separation, alcoholism and open competition for jobs. From the foregoing it is quite evident that the Africans themselves recognize that there are greater and more kinds of stress in city life than in rural.

It would have been of great interest to have had data on the salt intake of the two populations also available for comparison.

In a study of individuals in high-stress situations, Cobb and Rose (1973) in the US examined the incidence of hypertension and peptic ulcer in 4325 male air-traffic controllers working in control towers and centres as compared with 8435 second-class airmen. Although differences in age and licensing procedures may have influenced the analysis, the incidence of hypertension was four times higher in the air-traffic controllers. Further, there was some evidence of higher incidence in control towers where traffic density was high, and this was also true of peptic ulcer incidence.

In relation to the issue of which life situations represent stress, it is interesting to recall the study of Lee and Schneider (1958) which showed in the comparison of 1171 male executives with 1203 non-executives (including 563 females) that the incidence of hypertension and cardiovascular disease was higher in the non-executive males than executive males of comparable age. As well as the question of healthier individuals having more career success, and greater financial capacity to care for themselves medically, the issue of insight and organization of life-style to provide avenues of expression outside the work situation was discussed.

The question of different levels of physical exertion in different populations has been raised as an aetiological factor. Miall and Oldham (1958) in a study of a population in South Wales found that those with light occupations with less energy expenditure, and with higher incomes and higher standards of living, had higher blood pressures.

With regard to cigarette smoking, as Paul and Ostfeld (1965) point out, the data of Edwards et al. (1959), Miall (1961) and Karvonen et al. (1959) indicate that smokers have lower blood pressures than non-smokers, but they also draw attention to their own data showing that cigarette smoking appreciably increases the risk of coronary artery disease. Miall found that pressures were highest amongst male ex-smokers, which he attributed to the increase in weight that usually follows cessation of smoking.

Primitive peoples, unacculturated societies: With some comparisons

Shaper (1972) introduces his discussion of cardiovascular disease in the tropics by stating there are communities in which blood pressure does not rise with age and in which the problem of essential hypertension and its complications appears to be virtually non-existent. But he emphasizes that the vast majority of tropical communities show blood pressure patterns similar to those seen in the economically advanced countries of the world: the ones under consideration are in the minority. The list of low-blood-pressure communities which have been described is fairly long and includes New Guinea Highlanders, Kalahari Bushmen, Congo pygmies, Pacific Islanders, South American Indians, Australian Aborigines and nomadic tribesmen in East Africa.

The point emerging from these studies is that blood pressure need not necessarily rise with age. The question is whether all of these findings are artefacts of sampling or whether they reflect the presence of factors such as chronic infection, parasitism or malnutrition which somehow prevent the normal rise of blood pressure with age. Another possibility is that these communities represent normality in terms of blood pressure in the biological sense of conducive to good health and absence of disease. A crucial issue is whether these isolated, often tropical communities are incapable of developing high blood pressure or whether, under changed environmental circumstances, they will show a rise in blood pressure level.

New Guinea

The outstanding epidemiological study made by Sinnett and Whyte (1973a, b) in a total highland population at Tukisenta, New Guinea, is a landmark. They open their account by drawing attention to the fact that the data available from hospital studies and also limited pathological studies show that cardiovascular disease in general, and coronary heart disease in particular, is quite uncommon in Papua New Guinea. Cardiovascular disease has been reported to account for only 0.9% of 2000 admissions to the medical ward of the Port Moresby General Hospital.

The population selected for their study consisted of 1489 people, the entire membership of Murapin, a tribal community. The territory is the hamlet of Tukisenta in the western highlands of New Guinea, 150 km north-west of Mount Hagen. The people live in scattered homesteads along the valleys of three rivers at an altitude of 1800–2600 m. They are primitive horticulturists and pig herders with a subsistence economy based almost entirely on the sweet potato. As wildlife is scarce and tribal fighting a thing of the past their activities centre around garden preparation, harvesting sweet potato crops, collecting firewood and occasional house building. European influences at the time had not significantly disrupted their social structure, subsistence economy or traditional dietary pattern. A total of 1413 subjects were examined, representing a response rate of 95%. The survey carried out on each of 777 persons over the age of 15 included comprehensive physical examination and history, blood chemical and metabolic assessment and cardiovascular tests including chest X-rays.

Blood pressure was measured in the manner recommended by Rose and Blackburn (1969) and their criteria of systolic blood pressure of 160 mmHg or more and/or a diastolic pressure equal to or greater than 95 mmHg were used as designating hypertension. Hypertension was found in 26 males and 21 females. The hypertension was systolic only in 10 subjects, diastolic only in 31 and combined in 6. The age distribution is shown in Table 27-2. Thus only 3% of males over the age of 40 years were hypertensive, in contrast to 20% of middle-aged American men (Epstein and Eckoff 1967). Further, hypertension was more common in males in the younger age groups. Thus 11 of 103 men aged between 30 and 39 had diastolic pressure equal to or greater than 95 mmHg. There is little evidence in the New Guinean population of the clinical features associated with elevated blood pressure levels. Fundoscopic abnormalities were found in 6% of the population over the age of 39 years as compared with 40% in an American group. Evidence of any renal disease similarly was much less than in American groups (McDonough et al. 1967). There was little evidence of the three major clinical complications of arteriosclerosis: ischaemic heart disease, peripheral vascular disease and cerebrovascular accident.

Mean values for systolic and diastolic pressure at different ages in the two sexes are shown in Fig. 27-2. The blood pressure levels were at maximum in males in the third decade, after which there was a progressive decrease in systolic, diastolic and pulse pressure. In females the systolic pressure increases with age, exceeding the male values after the age of 35 (Fig. 27-2). Diastolic pressure of females falls so that pulse pressure increases with advancing age. The transverse diameter of the aorta, as assessed radiographically, showed the same rate of increase with age in males as in females. According to American standards 8% of males and 14% of females in the survey had aortic enlargement, but none of these had hypertension or clinical evidence of arteriosclerotic disease.

The authors suggest that the average salt intake was probably less than 1 g/day. Twenty-four hour urinary sodium excretion measured from one sample in each of 273 subjects gave mean values of

Table 27-2. The age distribution of hypertension in the inhabitants of Tukisenta, New Guinea.

	Males				Females			
		Hypertensives[a]				Hypertensives[a]		
Age (yr)	Total	S	D	C	Total	S	D	C
15–19	47	0	3	1	50	0	3	1
20–29	74	0	5	0	92	0	1	0
30–39	103	1	11	0	107	0	3	1
40–49	100	1	0	1	91	3	4	1
50–59	38	1	0	1	30	2	0	0
60+	29	0	1	0	16	2	0	0
Total	391	3	20	3	386	7	11	3

From Sinnett and Whyte (1973a), with permission.
[a]S, systolic hypertension (160 mmHg); D, diastolic hypertension (95 mmHg); C, combined systolic and diastolic hypertension.

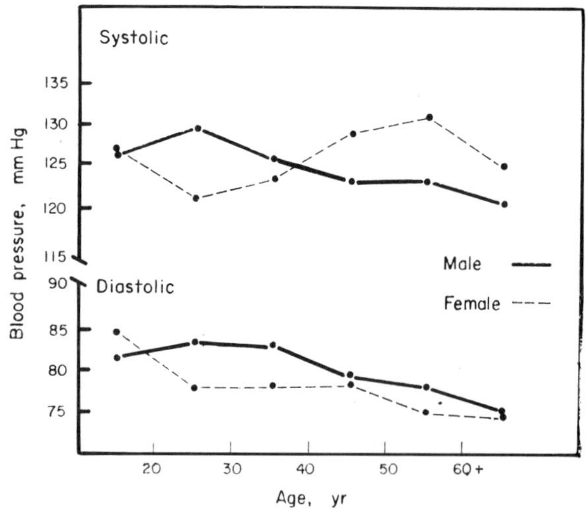

Fig. 27-2. Mean systolic and diastolic blood pressures at different ages in both sexes of a total highland population, Tukisenta, New Guinea. (From Sinnett and Whyte 1973a, with permission.)

13.7 (s.d. ±18.1) mmol for 135 females and 6.4 (s.d. ±12.3) mmol for 138 males. This agreed with earlier reports of Whyte (1958) and Hipsley and Kirk (1965). Sweet potato has a very low sodium content.

Sinnett and Whyte (1973a) remark that, as in the Framingham study, there was no correlation between blood pressure and sodium output in either male or females in their study. If, however, the sodium intake of all was at a very low level, it seems possible to me that virtually all were below a threshold level at which sodium intake has any influence on blood pressure. At least it is very unlikely that any effect would be seen over the small range of these low levels, and with one 24-h urine sample as an index of long-term sodium ingestive behaviour of a given individual.

Serum cholesterol was low by European and American standards and showed no rise with increasing age. The mean value for males was 148 and for females 157 mg/100 ml. The mean value for triglycerides was 135 mg/100 ml in both males and females. Glucose tolerance was high and there was no increase with age. No diabetes or gout were found. Overall the subjects were shorter and lighter than Europeans, muscular and mostly very lean. With increasing age there was a decline in relative weight, which is in contrast to the increase found in European populations. In relation to physical fitness, good or superior scores with the Harvard Pack test were gained by 76% of the men and 36% of the women. The score for males did not fall with age; it was comparable to British Commonwealth Divisions serving in Korea and was superior to Royal Australian Air Force personnel. Daily intake of energy was assessed at 2300 kcal for males and 1770 kcal for females. Carbohydrate, almost exclusively from sweet potato, provided more than 90% of the calories and fat only 3%. Protein intake was estimated to average 25 g/day. The water was soft. The community was judged to have strong social stability and emotional support for individuals in the clan.

Overall the data of Sinnett and Whyte suggest strongly that this is a normal healthy community in which no increase in blood pressure occurs with age.

This matter of salt consumption has been considered comprehensively by the Dutch anthropologists in New Guinea (e.g. Oomen et al. 1961; Oomen 1961a, 1967). Oomen remarks that the amount of salt eaten in the interior is very low. There are situations in which a craving for salt is evident. As it is considered a delicacy and has prestige value, the distribution among consumers is anything but even.

The main food sources are sago, the many varieties of sweet potato (55 varieties in the Wissel Lakes area alone), and taro, yams, cassava, bananas and some native vegetables and nuts. The nature of the food and lack of any significant storage mean that the impact of an interruption of supply may be serious. Oomen (1961b) in his discussion of the Papuan child as a survivor also notes that the labour of collecting and transporting bulky foods from the far-off cultivated plots is part of the physical exertion of women whether they are pregnant, lactating or not. A cause of limited caloric intake is the energy required to get it. Twins often cause insurmountable difficulties. The primitive cooking procedures involve difficulties in preparing food for the very young, and breast-feeding may continue well into the second and third year.

It is also relevant to note that in these marginal biological conditions, the practice of suckling until 2–3 years old will in one way add an additional stress to the sodium and protein balance of the mother. On the other hand, the continued suckling results in high blood prolactin level and a state of lactational amenorrhoea with suppression of ovulation for 2–3 years. Thus the consequent spacing of births by up to 3–4 years may avoid much larger stress. Parenthetically, from an anthropological viewpoint, the move from hunter-gatherer nomadic society to agriculture with domestication of animals and provision of supplementary milk, and later, with industrialization, the distribution of dried milk, has been a major element in the population explosion by reducing drastically the birth interval (Short 1976, 1979).

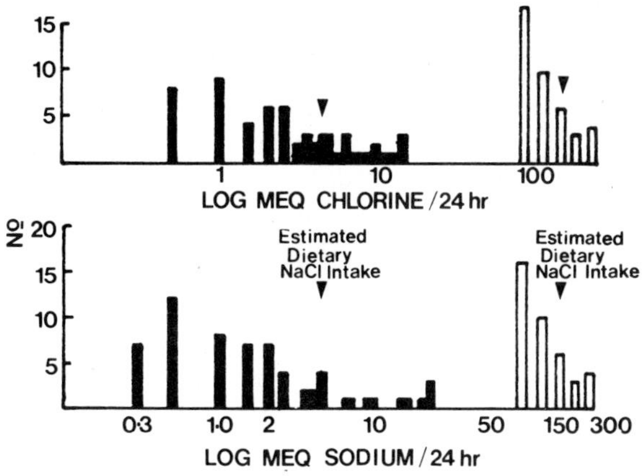

Fig. 27-3. The excretion of sodium and chloride in milliequivalents per day in New Guinean sweet-potato eaters (solid columns) and in European controls (open columns) observed in 63 and 40 24-h urine collections respectively. The New Guineans were from the Western Highlands and the control population from Holland. (From Oomen 1967, with permission.)

Per capita food intake in lowland New Guinea is ca 1500 kcal daily and is somewhat higher in the cooler mountain regions. Protein intake may be very low (12–15 g). In the highlands, sodium intake was found by Oomen to be strikingly low. Food and urine analysis indicated 1–2 mmol/day, whereas potassium intake was 3–4 times the usual European level. Thus the K/Na ratio of urine was 200–300:1 and even as high as 1000:1. Oomen (1967) remarks that the ash percentage of sweet potato tubers in Wissel Lakes region of Indonesian West New Guinea is 3%–4% dry weight of which 0.1%–0.2% is sodium; in the Chimbu region the oven-dry product is as low as 0.001% sodium. Fig. 27-3 shows the urinary sodium excretion of sweet-potato-eaters in the Wissel Lakes region compared with a sample of Europeans. The corresponding potassium output per day ranged from 105 to 180 mmol. The mineral content of diet of these skilful stone age farmers of New Guinea—some 600 000 to 700 000 people throughout the mountains of this huge island over 3000 km long—is indicated in Table 27-3. The plasma Na/K found was normal but sweat K/Na ranged from 1.1 to 2.8 with a mean of 1.8, whereas that of a local Papuan eating imported food was 0.2.

Table 27-3. Electrolyte excretion in sweet-potato-eaters and in European controls (averages per day in milliequivalents).

Subjects	Sodium	Potassium	Chlorine
Highland groups			
10 adolescent males	1.7	105	4.7
7 adult males[a]	6.8	180	4.0
6 adolescent females	1.3	123	3.7
7 adult females	2.1	138	4.1
Control groups			
10 adult males	177	49	174
10 adult females	154	55	149

From Oomen (1967), derived from data collected in the Chimbu region.

[a]Two had some imported salt.

Professor Victor Macfarlane in collaboration with Drs B. Scoggins and J. Coghlan of the Howard Florey Institute and Dr S. L. Skinner (Macfarlane et al. 1970) studied two New Guinea tribal groups. One was still in the primitive feral condition, the other had closer Western contact including diet.

In the primitive highland community the mean aldosterone concentration in peripheral blood was 28 ng/100 ml, which is about 4 times that of the mean in the Australian community or in New Guineans acculturated to Western society. This high aldosterone concentration relative to normal European values was not associated with increased plasma renin activity as determined by Dr Skinner. Within the accepted vectors of stimulation of aldosterone control, the high potassium intake may be important. The composition of sweat and urine in these primitive highlanders is consistent with their low sodium intake and high blood concentration of aldosterone. Macfarlane asserts that the highlander Chimbu drink little water but that intake is ca 3 l/day in food and also that the total body water and extracellular water were 30% higher in Chimbu than in Europeans. Before trade stores entered villages such as Pari and Koinambe, blood pressures were below 100/60, as Maddocks (1967) has shown. Urine pH of adults was usually between 7.5 and 9.0 because of potassium bicarbonate and carbonate excretion. The highlanders produced only half as much arm-bag sweat as Europeans marching with them. K/Na ratio was up to 10 times higher in the highlanders.

Macfarlane states that as coffee and cash crops developed and highlanders moved to salaried jobs, changes in eating habits occurred. Money was used to buy salt as well as sugar, tea, flour and canned meats. He followed the acculturation sequence from labourers in Madang, through cash crop areas in the Chimbu country to untouched Tsenda regions in the upper Jimi. The socioeconomic changes to European customs were associated with a fall in urine pH and K/Na ratio, and a decline in blood aldosterone.

The most important statement from the viewpoint of our preoccupation here is that resting blood pressures, originally around 100/60, rose to 117/70 in trade store contact areas and to 127/75 amongst adult Melanesians under 40 years of age with salaried posts. Unfortunately the data on number of subjects are not included in this account (Macfarlane et al. 1970).

Blood pressure patterns with age similar to those described by Sinnett and Whyte have been found in other surveys in New Guinea. Whyte also made a survey in 1958 on the Chimbu communities in the highlands living at between 1500 and 2800 m. They had had little association with civilization since contact was first made in 1933. They have good physique and powerful muscular development. Haemoglobin levels were found to exceed the average European figures and serum proteins were normal. The daily food intake of adult males provided about 2800 kcal, 30 g or less of protein, and little or no fat and salt. The highlanders were compared with coastal natives who lived on the southern coast of Papua in a sparsely populated delta region. The coastal natives had had contact with Europeans for longer and were affected to a greater extent by malaria, tuberculosis and other diseases. They were found to have low haemoglobin levels, normal protein concentration and a diet that provided about 1600 kcal daily, 40 g of protein and extremely little fat. It was stated that they do not lack salt, but there were no measurements made to give an indication of the order of intake. In making the observations, measurements were made of height, weight, folds of skin and subcutaneous fat, and account was taken of the circumference of the arm in the assessment of blood pressure.

It was found that the blood pressure of the young highland adults was the same as for Europeans (125/80 mmHg at 25 years of age). With advancing age there was a decrease in diastolic pressure with an increase in pulse pressure in both sexes. Systolic pressure increased in females but was constant in males. It was noted above that the more extensive survey of Sinnett and Whyte (1973a, b) did not show the increased pulse pressure of males. At 65 years of age the average blood pressure for males was 130/70 mmHg and for females 140/75 mmHg. These data were derived from 278 men and 169 women.

An additional finding of interest was made on 49 men between 20 and 35 years of age who had returned from service on coastal plantations. For 12–24 months they had been labouring and receiving extra rations such as meat, margarine, sugar, salt, wholemeal bread and rice. The daily energy intake was about 3500 kcal. Compared with 101 village-fed men of the same age, they were of similar height and 4 kg heavier, with a greater arm circumference and with more subcutaneous fat. The systolic blood pressure was significantly higher than for the villagers ($P<0.001$) but the difference in diastolic pressure was not significant. However, using the correction for arm circumference of Pickering et al. (1954), corrected pressures were 131/77 mmHg for ordinary villagers and 136/77 for labourers. The difference in systolic pressure was then only just significant at the 5% level. The resident coastal natives did not have blood pressures significantly different from the highlanders, and again blood pressure did not rise with age.

Whyte agrees with Pickering and colleagues that essential hypertension is merely a term applied to subjects whose blood pressures fall at the right-hand end of a relatively smooth frequency distribution and that there is no clear-cut dividing line between normality and hypertension. However, he is inclined to believe that there are normal limits for blood pressure, that these correspond to the distribution found among healthy young adults, and that increased latitude should not be conceded to the elderly. In fact, as with atherosclerosis, so with hypertension it could be claimed that the disorder is so rife in civilized communities, particularly in the upper age groups, that it becomes difficult to find normal controls. In relation to the similarity of the highland and the coastal natives who are comparable in body bulk, he notes that the highlanders are salt-hungry and the average daily output of sodium in the urine of 20 subjects studied was found to be 30 mmol. As noted by Wills (1958) the salt obtained by highlanders is often a potassium salt. Whyte states that the coastal natives of his study were not desperate for salt and could be assumed to have a greater intake. However, it was not determined what this was. Nor was it clear as to how long they had been in contact with Europeans and whether increased intake could be assumed.

Stanhope (1968) reports an examination of inhabitants in scattered hamlets at an altitude of 1100–1200 m round the headwaters of the Aiyau tributary of the Asai River. The diet of the people consisted of taro, sweet potato and bananas with greens and occasional meat added. Their nearest government mission stations were more than a day's walk away. Salt and other imported items of food were very scarce and Stanhope was besieged by people anxious to buy or give exchanges for salt. Five hundred and fifty-six out of the registered population of 917 were examined and it was noted that the number of people of 50 years or over who appeared for examination was less than those of other age groups and therefore the study was not without the possibility of bias. He found that in both sexes there was little change in blood pressure with age, the data resembling those of the other reports.

Ivinskis et al. (1956) in a medical and anthropological study of the Chimbu people of the central highlands confirmed that though pigs are an important aspect of life with the Chimbu, they contribute little to the nutrition of the natives. The animals are usually killed at periodic *sing-sings*, and these are the only occasions when the Chimbu natives consume pig flesh in any quantity. Sometimes a pig is killed for food or if a pig dies it is eaten, but these events are so infrequent that they can be disregarded as a source of protein. The *sing-sings* are opportunities for discharging obligations to neighbouring tribes. To an observer the impression is given that the hosts hand out lavish gifts to the guests, but careful enquiry reveals that they are merely returning what has been given to them in the past. At the *sing-sing* all debts are settled. Possibly a period as long as 5–10 years may elapse before a tribe has to act the part of host again. In the meantime the tribe is accumulating and fattening more pigs for the occasion but will be guest at other *sing-sings*. The medical survey showed that protein malnutrition was the main cause of death amongst patients admitted to hospitals. Pregnant women, nursing mothers and babies were the groups most affected, but the condition is sometimes seen in male adults.

Maddocks (1967) records a most important survey of blood pressure on a representative sample of the population in New Guinea. The five regions—Mainland region, Highland region including the Chimbu, the Delta and Hanuabada—are shown in Fig. 27-4. He found that individuals

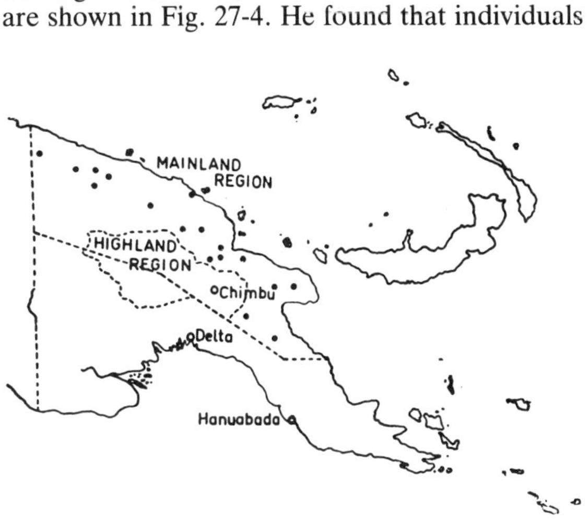

Fig. 27-4. East New Guinea showing location of the five populations studied by Maddocks (1967). (From Maddocks 1967, with permission.)

with high blood pressure are uncommon among New Guineans of any age. The frequency-distribution curves for systolic pressures, unlike those of Europeans, show neither an increased skewness nor a shift to the right with increasing age. There is some broadening of the curve which reflects an increased standard deviation. This is shown in the curves for Chimbu females aged

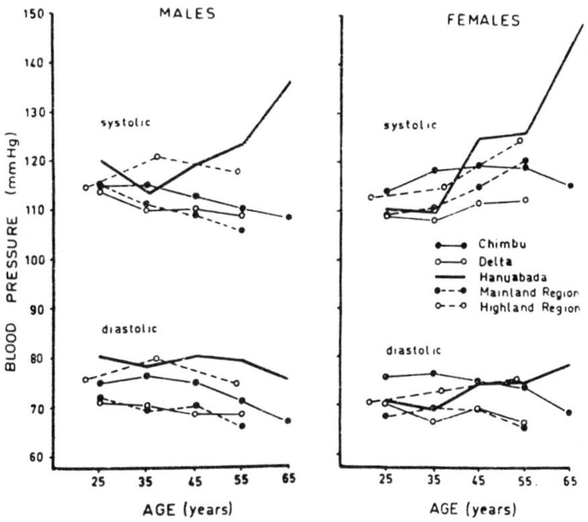

Fig. 27-5. Systolic and diastolic blood pressures by age in five New Guinea populations. (From Maddocks 1967, with permission.)

between 20 and 29 years and over 50 years (Fig. 27-5).

On the other hand, the same curves for Hanuabada females, also illustrated, show all the characteristic changes described in the Europeans. Hanuabadans of all peoples in New Guinea have been for the longest time the most closely associated with Europeans, living within the town of Port Moresby. Nearly all the men and many of the girls are employed as clerks, typists or artisans; some own trucks and cars and many drink alcohol. Like Westerners they spend more time in light activity and do not often undertake strenuous exertion. They are taller and heavier than other New Guineans and stay fat longer, but their body weight difference will not, of itself, account for the greater number of Hanuabadans with high blood pressures. Hanuabadans live over the sea and eat a lot of tinned food. The average K/Na ratio in the urine of 10 Hanuabadans measured was one-tenth of that of the Chimbu measured. In this study Maddocks also notes the influence of splenomegaly associated with a reduction in both systolic and diastolic blood pressure.

The reports of Barnes (1965), Maddocks and Rovin (1965), Maddocks (1967) and Stanhope (1968) are consistent with the Sinnett and Whyte report in terms of the absence of a rise in blood pressure with age. Maddocks and Rovin (1965) remark that there are about 150 000 people in the Chimbu district. In their sample of a single population involving 268 males and 161 females they found blood pressure fell with age, as did Barnes (1965). In discussing this they exclude selective mortality on the basis that the common causes of death (acute respiratory infections, diarrhoea and other infections) are not associated with high blood pressure. They found a steady fall of mean values for all measures concerned with body build (weight, height, skinfold, and arm circumference) for all sexes after the third decade. Since body build has a definitive effect on blood pressure, this might explain the fall in the Chimbu. However, analysis of the data corrected for arm circumference showed that, even after allowing for the reduction in body build measurements, blood pressure still falls with age. The issue of very low protein intake was also considered but it was noted that Indonesian populations with an even lower dietary intake of protein showed a blood pressure rise with age. The authors consider the possibility that the pattern observed is that of Stone Age or Neolithic man and that in these primitive communities we are seeing survival of the 'natural' or 'original' blood pressure pattern.

Sinnett and Whyte (1973a, b) reflect on the viewpoint that with antibiotics and medical care eliminating the infectious diseases some might regard the increased dietary intake of fat, refined carbohydrates and salt, coupled with smoking, as possible factors inducing an increase in arterio-sclerotic disease, hypertensive heart disease and bronchogenic carcinoma. Prediction of effects of sociocultural change is even more hazardous, but disruption of tightly knit kinship systems, loss of social values, and the replacement of physical violence by restrained and inwardly directed aggressive patterns seen in most Western communities, and the introduction of alcohol, could adversely affect psychological health. Delinquency, psychosomatic disease and cardiovascular disease could become more prevalent.

Salt trade in New Guinea

At this point, some remarks on the salt trade and European contact in New Guinea would seem appropriate. In reviewing and studying some aspects of traditional trade in the New Guinea central highlands, I. Hughes of the Australian National University (1969) describes a study on an area extending the 270 km from Astrolabe Bay to the Gulf of Papua. The area includes a great range of altitudes, terrain types, soils, rainfall, and local climates, which have resulted in regions of contrasting ecological resources, including distinctive flora and fauna. Human adaption to these environments has been just as varied. Detailed interviews were conducted with tribes from the sub-coastal hills of the Madang district to the Lake Tebera–middle Purari area of the Gulf district, and from the Upper Asaro to the Upper Wahgi. Groups studied by the anthropologists included the Chimbu, Daribi, Garfuku, Garia, Gende, Karam, Maring, Medlpa, Siane and Wiru.

Hughes notes that whereas there was some penetration of the area in the late nineteenth century, by 1910 German influence extended no more than 15 km from the coast. Ten years later, sub-coastal hills were still uncontrolled, but were occasionally visited by labour recruiters. The main highland valleys were first visited by the Taylor–Leahy expedition in 1933, a year which saw the climax of exploration and some consolidation of European influence in the more populated parts. However, in many parts there was no European influence until the 1950s and 1960s.

Hughes notes that traditional valuables are still traded in ceremonial exchange in many areas, but that the context of exchange and manner of transfer have altered. Relative values have changed. The first 10 years after the Second World War saw the destruction of traditional scales of value for many wealth objects, especially minor shell valuables and stone axes. These years also saw the end of salt manufacture by the people. However, some parts of the area were still uninfluenced by Europeans, and a few pockets had very little European influence even in the later 1960s.

The use of the ash of several different plants as a salt substitute was widely reported. The technique of passing fresh water through plant ash and evaporating it to reduce the soluble solids to a dry precipitate has been noted in many parts of the highlands. Most of the people in the main highland valleys made the substance and probably all knew the techniques. Some of the products were more highly valued than others, depending on the plant type used and the location of the group concerned. All consumed the substance at home, gave it as a luxury to visitors, and carried it as a ceremonial gift to more distant kin. Although it was transferred also by trade, this practice was not important in these parts.

Hughes suggested that it was possible to distinguish six ways in which people satisfied their need for salt. Two were used where they lacked access to salt water, and four were based on salt springs. First, there was the simple burning of plants and collection of the ash, which occurred everywhere except where people had access to salt water. The second required a solution process and an evaporation process: fresh water was passed through the plant ash, the water was then evaporated from the solution and residue collected. This process was used by most of the highlanders in the study area.

Where salt springs existed, the first method was

simply to drink the water and to cook with it. The second was to evaporate the water and collect the salty residue. The third method involved soaking plant material in salt water. It was then dried, burnt and the ash collected. This was not actually seen in the area under study, but was reported from the Western Highlands and the Huon coast. The fourth and most elaborate method was practised by the Gumine-speakers living on either side of the Wahgi River. Here, plant material was soaked in salt water, dried, burnt and the ash collected. Fresh water was then passed through the ash, the resulting solution evaporated and the residue collected. Analysis of salt-water samples used in these processes have indicated considerable salt content of 10 g/l. The combination of techniques used by the Gumine people resulted in their producing a superior product which was the best known in the central highlands and widely traded. As late as the 1940s a parcel of it weighing 2 kg could be exchanged locally for a medium-sized pig. In earlier times, by the time this salt reached the Upper Chimbu, amounts a quarter that size commanded a large pig or skins of the bird of paradise from the Ramu valley.

Wills (1958) made analyses of native salts collected from various areas in New Guinea. Specimens collected from Bougainville Island, Marobe, Sepik and the Eastern Highlands showed a potassium content of 23–49 g/100 g with negligible sodium (less than 1 g/100 g), whereas some speciments collected from the Eastern Highlands, Madang, the Sepik and Dutch New Guinea had sodium content varying from 7 to 32 g/100 g, and small potassium content. Wills describes how in the Madang district Gwiarak natives pour sea water onto a large fire which they build on the beach. After evaporation, the salt and the ashes are carried back to their village—a day's walk inland. Coastal natives use sea water for cooking and some inland natives even carry sea water to their villages. She also describes how in the Eastern Highlands the salt which is made by evaporation is eaten with pig, taro and sweet potato. A mouthful of food is eaten, then a portion of salt. In the Milne Bay area, natives used to prepare their own salt by burning driftwood, but now are able to purchase it. Wills notes that the addition of potassium salt to a vegetarian diet already containing large amounts of potassium can have no physiological basis. The early account of Moszkowski (1911) that the natives refused salted food with distaste is also noted.

Meggitt (1958) also investigated the technique in the highlands of soaking wood or leaves in salt springs, burning the plant material and collecting the ash. He noted that the salt springs are not uniformly distributed geographically so that a considerable trade in salt exists. He describes how one big salt pool in the highlands is owned by a clan, but each plot within it is individually owned. The family jealously guards its rights in the pool's salt production. The salt, which is prepared by burning wood which has been steeped for a long time in these springs, is wrapped in large leaves and bound into packages weighing between 1 and $5\frac{1}{2}$ kg. Meggitt describes how in some areas people walk for 2 days across mountains to buy salt from particular sources, paying for it with pigs and yams. He notes that southern Aruni and Kandep Enga people also procure most of their salt from the Yandapu Springs, carrying net bags, shells, pigs and gourds full of tree oil (for decorative use at *sing-sings*) for 4 or 5 days over steep mountains to pay for it. He also describes trading parties involving as many as 60 people travelling to procure salt.

The Ingha tribe particularly seemed to devote a disproportionate amount of time and energy to the distribution of salt. When they were asked about this, the reply was that salt manufacture and exchange were important for several reasons. First, salt is desired because it improves the flavour of food, especially meat. Secondly, their ancestors did this and they should continue to follow their example. Thirdly, salt can be traded or distributed ceremoniously and thus establish friendly relations with other tribes. Through these relationships, men can enhance their social reputations and find opportunities to acquire more wives and pigs, possession of many wives and pigs being a mark of a really important man in Ingha society.

The apparent good health of populations on a low sodium intake which use potassium rather than sodium salts as flavouring could suggest that the taste is culturally rather than physiologically determined. But the prodigious physical efforts and varied techniques used to obtain salt, and the use of potassium ash salt for taste in circumstances where sodium is not available, could point to physiologically activated behaviour—given the ratified marginal sodium status and its likely aggravation by reproduction. The failure to accept sodium salts immediately by peoples habituated to the bitter-salt taste of the potassium material does not necessarily gainsay a physiologically determined behaviour so much as illustrate the erroneousness of the idea of the 'infallibility of instinct', as set out earlier in this chapter.

Other Pacific areas

Maddocks (1961) noted that on some Pacific Islands, relatively isolated populations can be studied. He examined the casual arterial blood pressures of two such complete Pacific island populations: those of Gau in the Fiji Islands, and of Abaiang in the Gilbert Islands. The Fijians are predominantly Melanesians, the Gilbertese are Micronesians. Gau is a mountainous island with a population of 1546 males and 1536 females. The blood pressure of 99% of the adult population was measured. Abaiang is a coral atoll with a population of approximately 2900. The majority of the population was visited with the exception of two villages which were on small detached islands.

Neither the means nor variances of these blood pressures showed the substantial rise which occurs in Western peoples with age. This was not due to different effects of arm circumference, because the relation between arm circumference and age is generally similar in both groups. Nor was there any evidence that it is due to any early selective mortality amongst those who might develop high blood pressure, or any other influence that might lower blood pressure in later years. Apparently these Gilbertese and Fijians had not yet been subject to some environmental influence which raised the blood pressure in later life.

It was interesting that the Fijians on Gau, who were incorporated into a comparison of Fijians and Indians by Lovell et al. (1960), had marginally less change in blood pressure with age than had the total Fijian population, which in the account of Lovell et al. included people in towns and on labour lines. Thus there is some suggestion that the population of Gau is not yet influenced by some factors operating on the main islands. At least in Maddocks' comparison the Gilbertese had lower mean pressures generally. There was slightly greater spread and skewness of the frequency distribution curves for the Fijians, as was seen in Lovell's study on the main island of Fiji for both Fijians and Indians. No data on salt intake of these populations were given.

Prior et al. (1968) studied two ethnically similar populations of Polynesians. One consisted of the entire adult population of the isolated coral atoll of PukaPuka in the Northern Cook Islands, where the people live under a subsistence-level economy with a diet composed chiefly of coconuts, taro and fish. The second group was 98% of a sample of adults who had lived under town conditions on Raratonga in the Southern Cook Islands for at least 10 years. They have a cash economy and ready access to a varied diet, which included salty tinned corned beef, tinned fish and other goods and salted coconut sauce. The PukaPukans had these items of diet occasionally, and did not add salt to foods in cooking. In Raratonga mean blood pressure increased with age, particularly in women, while in PukaPuka the rise in blood pressure with age was slight, and only seen in women. Blood pressure differences between the populations were greater than was accounted for by differences in height and weight, though the statistical model needed to derive this was complex. The data from dietary surveys and urine collections on a 24–48-h basis suggested a sodium intake in Raratonga about 50 mmol/day higher than in PukaPuka. However, the intake in PukaPuka was of the order of 50–70 mmol/day. No within-population correlation between blood pressure and 24-h urinary sodium output was demonstrated.

Murrill (1949) examined 251 pure Ponapean natives on a typical tropical island 1500 km south-east of Guam. No increase in blood pressure with age was noted. Mean systolic blood pressure of males aged from 20 to 59 years was 112 mmHg with diastolic pressure 76 mmHg. For 144 females from 18 to 59 years the mean systolic pressure was 110 mmHg and diastolic 76 mmHg. Murrill notes that the population was infested with hookworms but has no data as to whether this would have any influence on blood pressure.

Cruz-Coke et al. (1964) introduced their study of blood pressure on Easter Islanders by recording that Cruz-Coke and Covarrubias (1962) had shown in rural Chile that there was an age-dependent gradient of blood pressure, ranging from 0.037 mmHg per year in the poorest natives to 0.492 mmHg per year in the richest foreigners. On Easter Island there were two elements in the population: one group was termed 'Islanders' and the other 'Continentals'. Massive migration to the mainland of Chile was relatively recent and most of the Continentals had lived an average of only 3–5 years on the mainland and then returned. The authors went to some lengths to determine that the two populations examined were similar in terms of frequency of a number of genetic markers. The regression analysis of blood pressure on age showed that the blood pressure of the Continentals was significantly correlated with age, whereas in the Islanders it was independent of age. The immigrant leaves an isolated and peaceful life in the surroundings of the island for the stormy environment of the low-income group in a rising Latin-American country. As a result of this exposure there are, in keeping with an 'acculturated population', hypertensive subjects, whereas by contrast hypertension is absent in the Islanders. Unfortunately, no data on diet and sodium intake of the Islanders are available with this study.

Cruz-Coke and colleagues in 1973 studied two populations which were genetically homogeneous but in stongly different environments. This was in the Andean mountains in Chile at altitudes ranging from 300 to 4600 m. Hypertension was absent in the isolated primitive peoples of the highlands whereas it appeared in the more acculturated lowland populations. Though the sample was small—282 persons in 53 families—237 were related to one another. Hypertension in the lowlands appeared to be exclusively associated with some specific genotypes of the Rhesus and MN systems.

Bailey (1963) studied blood pressure in three Indonesian villages where there was general undernutrition. Rice was the staple foodstuff in one and cassava in two, protein intakes being higher in the former. Blood pressure levels were similar in each village. Systolic pressures (but not diastolic pressure) were low compared with other populations but rose with age. Thus low calorie and protein intakes did not prevent the rising systolic pressure with age.

Maddocks and Vines (1966) and Burns-Cox and Maclean (1970) both report that blood pressure is lower in tropical areas, associated with splenomegaly. Burns-Cox and Maclean studied a population which lived along the banks of the River Nenngiri in the highlands of Central West Malaysia—the Temiar-Senoi Orang Asli (Aborigines). These people live a life largely isolated from the rest of civilization. They are slim and eat a diet high in starch and low in salt. The observations were confined to 85 people over 15 years old studied in their houses along 65 km of river. No rise in blood pressure was apparent with age in the population, but those with splenomegaly had lower blood pressures than those without.

On the Australian mainland, Hicks and Matters (1933) examined Australian aboriginals in Central Australia who were living in their natural state by hunting and by grubbing for roots. Systolic and diastolic blood pressures were low. Nye (1937) examined 103 aboriginals at the Lockhart River Mission on Cape York Peninsula. The older members of the group had spent most of their lives in their primitive state. Hypertension appeared to be uncommon in these elderly people. He notes that the diet of the people in their native state had been almost entirely carnivorous. He states also that the people live in a state of communism in which their possessions, joys and sorrows are shared by the whole tribe and that there is none of the selfishness which causes so much of the stress and strain, both mental and physical, found in our civilization. Vandongen et al. (1962) report on a community living in Arnhem Land (Northern Australia) under semi-Europeanized conditions. The diet consumed is basically European: meat, flour, sugar and tea. There is no mention of salt. An indication of the extent to which the natives have been converted to European foods is the prevalence of dental caries. It was found that the blood pressure pattern of people seemed to parallel closely the typical European variance of blood pressure with age, although the aboriginal pattern was set at a generally lower level. It was, however, higher than the Central Australian aboriginals.

The study by Lot Page

An important study in this field has been made by Lot Page and colleagues (1974). Cardiovascular risk factors were analysed as part of a combined ethnographic, anthropometric and medical study of 1390 adult subjects in defined populations representing six Solomon Island societies (see Fig. 27-6). The six societies all had low levels of acculturation, and differed in habitat, way of life and exposure to Western civilization. Criteria for ranking the societies in respect of acculturation were developed, based on demographic changes with defined populations, secular increase in adult height, length and intensity of contact with Western influences, religious beliefs, education, availability of medical care, economy and diet. The six tribal groups were ranked by these criteria from most to least acculturated as follows: 1, Nasioi; 2, Nagovisi; 3, Lau; 4, Baegu; 5, Aita; 6, Kwaio. Three of the societies lived on the island of Bougainville, territory of Papua and New Guinea, and three on the island of Malaiti, British Solomon Islands Protectorate. Both islands, lying 5–7 degrees south of the Equator, are volcanic in origin and tropical in climate, with year-round tempera-

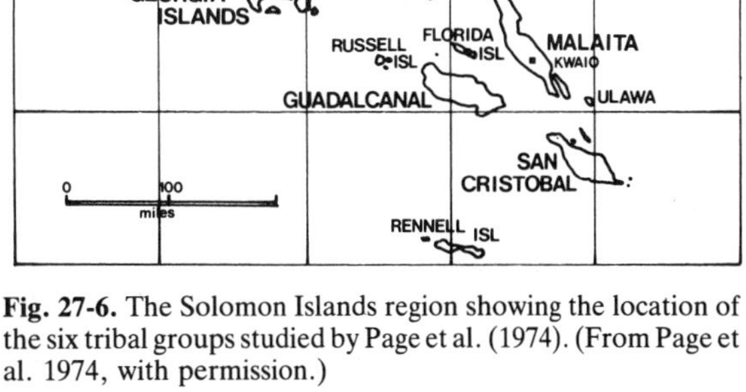

Fig. 27-6. The Solomon Islands region showing the location of the six tribal groups studied by Page et al. (1974). (From Page et al. 1974, with permission.)

tures averaging 20–30 °C and high annual rainfall. All six tribes are Melanesian and all were living in tribal villages in rural areas, where electricity and other Western conveniences were totally absent. Use of pipe tobacco was widespread amongst all age groups and both sexes in all six tribes. Alcohol intake was negligible. Physical health and nutrition were good in all six groups and clinical evidence of coronary heart disease and arteriosclerosis was absent.

The Nasioi people live in hills 15–25 km from the coastal town of Kieta in Eastern Bougainville. They had had continuous contact with European culture over 80 years. Their diet was sweet potato supplemented by purchased tinned meat, fish, rice and bread. Salt was used regularly in cooking. Pigs and chickens were raised. Bananas and other fruit were used. Two hundred and fifty-six Nasioi were studied.

The Nagovisi people lived on an inland coastal plain and had suffered severe privation during the Second World War. They had had continuous contact with Western culture since 1931 and a cooperative store supplied Western dietary items and other goods to them. Most households consumed some kind of European food every other day. Their economy, religion and diet were similar to the Nasioi. Four hundred and ninety-three persons were examined.

The Aita lived in small villages at an elevation of 1000–1200 m on the slopes of Bougainville's highest mountain. They were little known until 1964. With the influx of missionaries, cannibalism had been suppressed and most were nominally Christian. But the impact on local custom had been slight. Their staple diet was taro, supplemented with sweet potato, greens and fruit. They did not raise pigs or chickens. Rice and tinned fish were bought only occasionally in small quantities. Salt was not used in cooking. Four hundred and sixty-seven persons were examined.

The Kwaio lived in scattered hamlets of 10 to 20 persons in the rugged, hilly terrain of East Central Malaita. Contact with Europeans was slight and intermittent. They had retained traditional pagan religious beliefs and practices and Western education and medical care were virtually absent. They were swidden farmers (slash and burn). Their diet was 85% sweet potato supplemented by leafy vegetables, with grubs, insects and fresh-water prawns used as relishes. Pigs were eaten only on ceremonial occasions, mainly by men. Salt was used rarely during ceremonial occasions. Twenty-eight men were asked when they had last eaten salt, and this ranged from 1 week to 1 year previously, with a median of 18 weeks. European food was unavailable. Though young men typically worked on plantations elsewhere in the Solomon Islands for 2 or more years, they reverted to completely traditional ways on their return. Four hundred and forty-three persons were examined.

The Lau people lived on small artificial islands, ½–1 ha in area, in a large coastal lagoon in Malaita. The population density on these islands (400–500 persons per hectare) was very high, but there was unlimited space on the lagoon. The Lau were fishermen and obtained vegetables from gardens on the mainland and by trade with neighbouring Baegu. They retained their traditional beliefs and practices. Their diet was 15% fish, the rest mainly taro, sweet potato and greens, all cooked in copious amounts of sea water. Occasionally, pigs, shellfish and nuts supplied additional protein. Some rice, tinned fish and meat were purchased from trading ships. Four hundred and forty-two persons were examined.

The Baegu lived near the coast of north-eastern Malaita. They were predominantly swidden farmers of taro and sweet potato, beginning to change to settle down to agriculture and cash crop cultivation. Fish was obtained by trade with Lau. Canned fish and meat were occasional dietary items. Salt was seldom used in cooking. Four hundred and eighty-five persons were studied.

In all instances except the Kwaio, where there was an epidemic of influenza and only 78% of the population were studied, over 90% of the population were assessed. The degree of acculturation in the six societies on various criteria is shown in Table 27-4 (Page et al. 1974). In relation to dietary change, observations on habits over a period of 18 to 24 months were recorded by the cultural anthropologist of the group. Where trade stores were present, records were kept on the quantity of foodstuffs sold over a 6-month period, and the number of households served. To obtain data on utilization of salt, interviews based on recall were made. Random urine samples were collected to determine the sodium and potassium and creatinine content. However, tribal taboos prevented collection of more than a few urine samples in the less acculturated groups. Timed urine collections could not be obtained. Based on all data available, the authors' estimates of average daily salt intake for the six societies are as follows (see Table 27-4):

Nasioi and Nagovisi	50–130 mmol
Lau	150–230 mmol
Aita and Baegu	10–30 mmol
Kwaio	<20 mmol

The Lau, who had the highest intake, as noted above cooked their food in copious amounts of sea

Table 27-4. Indices of acculturation for six Solomon Island societies

Index	Nasioi	Nagovisi	Lau	Baegu	Aita	Kwaio
Length significant contact (yr)	85	30	50	25	6	25
Intensity of contact	+++	+++	++	+	++	+
Christianity	+++	+++	0	+	++	0
Medical care	++	++	+	+	+	0
Literacy/education	+++	+++	+	+	0	0
Western diet	+++	+++	++	+	+	0
Cash economy	+++	+++	++	+	+	0
Secular trend in height	Yes	Youngest males only	Youngest only	Youngest females only	No	No
Demographic change under 15/over 45	4.4	2.9	2.6	2.4	7.9	1.7
Habitat	Coastal hills near town	Inland plain, no town	Small islands, coastal lagoon	Inland hills	Mountains	Inland hills
Subsistence	Agriculture	Agriculture	Fishing	Swidden/agric.	Swidden	Swidden
Salt intake (mequiv/24 h)	50–130	50–130	150–230	10–30	10–30	<20
Acculturation order	1	2	3	4	5	6
Acculturation degree	+++	+++	++	+	+	0

From Page et al. (1974), with permission.

water. The three more acculturated societies showed higher cholesterol concentrations (Fig. 27-7), and uric acid levels were lower in these populations.

Figs. 27-8 and 27-9 show mean blood pressures of males and females in relation to age for all six societies. For both males and females the Lau had the highest systolic and diastolic pressures at almost all ages. An increase in systolic pressure with age was present in females in the three more acculturated societies but no significant age trend was present in the males. However, diastolic pressure fell with age among males in the three less acculturated societies and in the Nagovisi. Weight

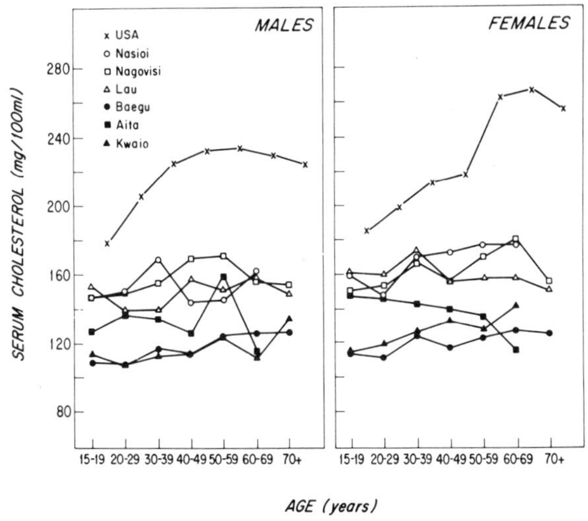

Fig. 27-7. Serum cholesterol levels by age for males and females in six Solomon Island societies. Values for a US population sample are included for comparison. (From Page et al. 1974, with permission.)

was negatively correlated with age in all groups studied, a relationship that was strong amongst females. Therefore the positive correlation of systolic blood pressure with age among the more acculturated females cannot be attributed to weight gain.

In noting that the Baegu, Aita and Kwaio populations, the three unacculturated populations, represent further examples of populations with low blood pressure, the authors state that in these groups nutrition, physique and physical fitness appear to be good. Malaria was present in the Baegu and to a small extent in the Kwaio, but was uncommon in the Aita. No significant haemoglobin changes were present, and haemoglobin values in these groups resembled those of the three more acculturated populations. Thus the data were against the hypothesis that the low blood pressure in Solomon Islanders was due to poor health. They also state that analysis of electrocardiograms showed striking absence of codable abnormalities in all groups, and a lower frequency of most abnormalities associated with coronary disease than in any population previously recorded.

Overall, the data provide further support for the hypothesis that the failure of blood pressure to rise with age may be biologically normal. With acculturation, a constellation of changes occurs. These include increase in serum cholesterol, reduction of serum uric acid, and an age-related rise in blood pressure, which expresses itself earliest and most notably in the female. Page et al. draw attention to Maddocks' (1967) report on three primitive groups of villagers in New Guinea, showing declining blood pressure with age. By

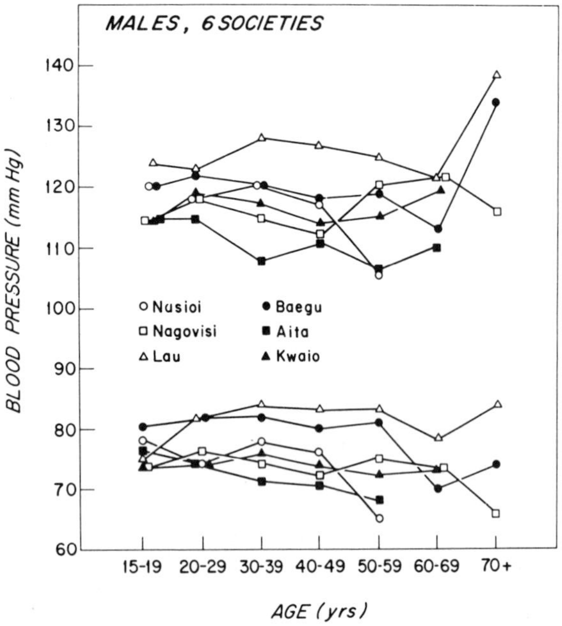

Fig. 27-8. Mean values for systolic (upper curves) and diastolic (lower curves) blood pressure by age in adult males of six Solomon Island societies. (From Page et al. 1974, with permission.)

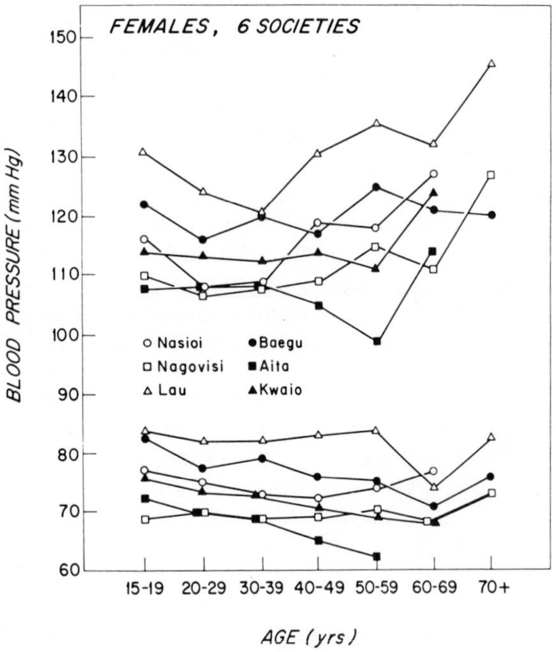

Fig. 27-9. Mean values for systolic (upper curves) and diastolic (lower curves) blood pressure by age in adult females of six Solomon Island societies. (From Page et al. 1974, with permission.)

contrast, in town dwellers near Port Moresby, New Guinea, blood pressure rose with age, with the steepest increase in systolic pressure in the female. Page et al. reflect that a process of acculturation is associated with biological changes which lead to the development of risk factors for cardiovascular disease. Thus it is important to identify the factors common to those societies in which these changes occur, and which are absent from those in which they do not.

For the Solomon Island populations, several factors commonly present in changing societies can be excluded. All of them lack motor vehicles and developed roads. Thus patterns of locomotion and physical activities remained unchanged. Similarly, telephones, electricity, industrial and household equipment, noise, pollution and smog were all absent. There had been some changes in tribal custom where Christianity replaced pagan belief, but social and family roles had remained essentially unchanged. Further, though entry into cash economy and wage-earning employment may have introduced stresses which are difficult to assess, these activities were confined to males, who had shown less change, at least in blood pressure, than females. There were crowded living conditions among the Lau, who lived on small, artificial islands, but even so, land was partitioned and it has been a traditional mode of life and was voluntary. Both men and women left the islands during the day to engage in fishing and gardening. Page et al. noticed that a similar pattern of voluntary crowding had been reported in the !Kung of Botswana, where blood pressure remains low throughout life (Draper 1973).

The authors suggest that the most constant change among the more acculturated Solomon Island populations that was not shared by the less acculturated, was the adoption of Western dietary items, especially salt, tinned meat, fish and rice. Their data did not suggest that this change had had any consistent effect on body weight or other indices of body fat, such as ponderal index and skinfold measurement. The Lau, the third in acculturation ranking, were the heaviest and then came the Aita, which were fifth in rank (Table 27-4). Furthermore, in all three societies which show rising blood pressure with age, weight decreased with age. Thus suspicion falls most heavily on salt intake, which was substantially greater in all the more acculturated groups. Of particular interest were the Lau, who had the highest salt intake as a result of cooking vegetables in sea water and who exhibited the highest systolic and diastolic blood pressures in both sexes in nearly all age groups of the six societies.

Simpson (1979) states that perusal of Figs. 27-8 and 27-9 in relation to variations in blood pressure at different ages between acculturated and unacculturated groups leaves him singularly unimpressed. On this point it might be agreed that Page's division between acculturated and unacculturated may be somewhat arbitrary, and Page et al. (1974) do say at the outset that all six societies were at a low level of acculturation. The picture overall could be seen as one of little or no change in blood pressure with age, but with a statistically significant increase in systolic pressure of females with age in the three more acculturated societies and a decrease in diastolic pressure of males with age in the three more primitive groups. The trends do not appear to be explained on a weight basis. The Lau are the group which were strikingly different, and they were the ones with a European level of salt intake. They were, however, the heaviest and had the largest arm circumference, but weight did decline with age.

Finally, the authors note that the overall low prevalence of codable electrocardiogram items is consistent with an almost total lack of coronary heart disease in these subjects. Even in the more acculturated groups which are beginning to exhibit biological changes, these effects have not been translated into evidence of coronary heart disease. The observed differences between the groups may represent the earliest antecedent of cardiovascular disease in transitional people, and the authors consider it will be of importance to follow these societies further. It would seem to be also of great interest in this study to know for how long the Lau had been following the practice of cooking in sea water, reported by Ivens (1929) in his book *Island Builders of the Pacific*.

Prior et al. (1974) reported on the blood pressures of the 914 persons aged 15 years and older in the Tokelau Islands (96% of the population). There was a very small increase in blood pressure with age. The population was of particular interest in that 56% of calorific intake was as fat, and of this 80% was derived from coconut oil which, unlike most vegetable oils, contains predominantly saturated fatty acids. Urinary sodium analysis showed 30–50 mmol/l, suggestive of a low sodium intake.

Africa

Kaminer and Lutz (1960) studied a small group of migratory Bushmen in the Kalahari region who lead a frugal life existing on game, melons, roots and other forms of wild vegetation. The blood pressures of both males and females were low and there was no evidence of any rise with age.

Shaper (1972), in discussing blood pressure and hypertension in tropical regions, notes that a survey on African men on the shores of Lake Victoria in the 1920s showed a population with moderately low mean blood pressure, which fell even lower in subjects over the age of 40 (Donnison 1929). Later studies in similar but not identical groups again showed low blood pressure patterns, but with no fall with age (Williams 1941). However, in 1960, community studies on the shores of Lake Victoria, with again similar but not identical groups, showed blood pressure patterns with age similar to those seen in economically advanced Western countries (Shaper and Saxton 1969). This historical information from East Africa was quoted as a possible indication that social and economic change may be associated with changes in blood pressure over the period of time, but it was stressed that the populations differ too much in their selection to be strictly comparable.

Better evidence that groups with low blood pressure may suffer change over a period of time came from studies of Samburu nomadic warriors of northern Kenya entering the army (Shaper et al. 1969). From a diet consisting of milk, meat and blood, these young men had changed to a diet made up of 53% carbohydrate, 20% protein and 27% fat. In their home districts the Samburu have very little access to salt and while they may obtain salt when their cattle are taken to salt licks, no salt is added to food and salt is not bought. Calculation of the sodium content of their diet of milk and meat suggests an intake of 100–150 mmol/day and loss in sweating could be considerable. Army diet contained 300 mmol/day (Shaper et al. 1969). They showed a significant rise in systolic blood pressure, not after their first 6 months in the army when their weight gain was considerable, but after an average of 2 years and when their weight gain was less than in the earlier period. After 6 years in the army, systolic blood pressure was still significantly higher than in age-matched controls. Shaper states that this phenomenon could possibly have something to do with the selection of who goes into the army and who does not. Therefore, it would be interesting to follow the influence on the raised systolic blood pressure when the subjects revert to their original way of life as nomads.

In discussing the societies with low blood pressure, Shaper notes that there is lean body build and a virtual absence of obesity. Where blood pressure rise is observed, there is usually, but not always (Page et al. 1974), an increase in fatness. It is known in developed countries that there is a strong association between body weight and the prevalence of hypertension. Obese persons are much more likely to have high blood pressure than those who are not overweight, the risk of hypertension developing in a previously normotensive being proportional to the degree of overweight. Shaper notes that this relationship is almost certainly not simple and direct. In the Framingham community it was found that hypertensives have an increased tendency to develop obesity, but also that obese normotensive subjects subsequently developed an excess of hypertension, but only after some time.

Truswell et al. (1972) described a study carried out on the !Kung bushmen in northern Botswana in collaboration with a Harvard group of anthropologists who had been living in close association with the bushmen. The bushmen live in camps of 20 or more people on the basis of an extended family group, each camp being associated with a waterhole. About 70% of the food by weight is vegetable, collected by the women; a variety of nuts, fruits, roots, bulbs and leaves are available varying with the season. The most important food is the nut of the Mongongo tree, because it is abundant, keeps well and is rich in protein and fat. Meat is provided by some of the men, who hunt with poisoned arrows. They occasionally obtain milk. They have very little salt; there are no deposits of salt in the area and the principal means of obtaining it is to walk 50 km to a settlement.

Blood pressures were measured in a total of 152 bushmen (79 men and 73 women) aged 15 to 83 years. In male bushmen systolic and diastolic pressures both showed a slight but definite downward trend with increasing age throughout adult life (Fig. 27-10). In the women the line joining decade means of systolic pressure rose by about 10 mmHg to a higher step in the second half of life. Diastolic pressures stayed the same until the two oldest age groups, in which there was a slight drop. The authors note that the arm circumferences were at the lower end of the range for Western adults. In both sexes they were a little

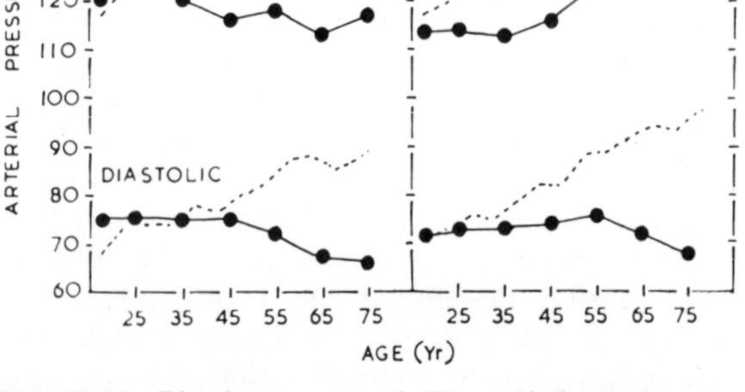

Fig. 27-10. Blood pressures of !Kung bushmen by age, compared with standard figures for a group from London. (From Truswell et al. 1972, with permission.)

higher between 30 and 49 years of age and a little lower in the very young and the very old. In relation to any changes in weight with age in the two sexes, it was found that all values for bushmen were low and there was little change with age except below the age of 20 and over the age of 69. An extremely interesting finding in relation to the sodium status of hunter-gatherer communities was the low sodium and chloride excretion in the urine, representing about 2 g/day of sodium chloride. Potassium excretion was 3–4 times higher than sodium.

In discussing the data and the question of whether these low blood pressures are due to absence of mental stress rather than another cause, the authors draw attention to the fact that isolated and peaceful communities have been studied and shown to have hypertension (Abrahams et al. 1960; Hawthorne et al. 1969). There was no evidence from physical examination of the bushmen which would suggest that any chronic disease contrived the low blood pressure; also correction for arm circumference did not explain the result.

Sever et al. (1980) studied tribal and urban Xhosa people in southern Africa. Blood pressures were high and rose significantly with age in the urban group whereas in the tribal group they were low and rose little with age. Indices of obesity including weight, skinfold thickness and ponderal index were significantly greater in urban dwellers and there were strong correlations with arterial pressure. Dietary sodium and urinary sodium/creatinine ratios were significantly higher in the urban people but there was no within-population relationship between either dietary sodium or urinary sodium/creatinine ratio and blood pressure. However, dietary sodium intake was relatively high in both communities: 162±69 mmol/day in the tribal and 240±66 mmol/day in the urban. Plasma noradrenaline and plasma renin activity were similar in the two groups.

Asia

China

Morse and Beh (1937) made observations on aboriginal groups living in the mountains in the provinces of Szechwan, Kweicheo and Yunnan in China. The peoples of the three groups lived at altitudes between 1000 and 3000 m, and in a fairly rigorous climatic environment. None lived in villages or cities, but were scattered about the countryside. They are agricultural and pastoral people and some are hunters. In all cases their staple diet was corn, not too abundant in amount; some ate some rice. Meat was not used as a general diet, but was almost entirely limited to special occasions. Milk, butter and cheese were not used, and vegetables were sparsely eaten. Thus they lived on a low protein and low fat diet. In approximately 400 people between the ages of 16 and 70 who were examined, blood pressure was uniformly low and there was no tendency towards an increase in blood pressure with age.

Iran

Page et al. (1978) have made a study of potential importance on a group of pastoral nomads, the Qash'qai tribe of southern Iran, who are claimed to be only slightly acculturated. They migrate semi-annually in response to the needs of their herds for favourable temperature, pasture and water. They are a lean active people who do not gain weight with age. The main dietary staples are bread, rice and dairy products, which results in a high sodium intake.

Analysis of the blood pressure of 268 individuals aged 15 years and older showed both systolic and diastolic pressure increasing rapidly with age in both sexes. Ponderal index was negatively correlated with blood pressure in males and not correlated in females. Changes in body mass did not account for trends in blood pressure. The systolic pressure was >140 mmHg in 18% of males and 17% of females of the entire population. Diastolic pressure was >90 mmHg in 17% of males and 23% of females. Figures for those 30 years old and above were as follows:

	>140	>90	>140/90	>160	>95	>160/95 (mmHg)
Males (%)	22	20	12	7	11	7
Females (%)	24	29	18	8	20	8

Sodium excretion in males averaged 186±106 and in females 141±73 mmol/24 h. The differences were said to be highly significant. Mean urinary Na/K ratio was 3–4 in both sexes. In the male sodium excretion was found to correlate with systolic blood pressure and in the female Na/K ratio correlated with systolic pressure. No significant correlations were found between sodium excretion and diastolic pressure. Thus this society could be an example of a traditional unacculturated society with a strong upward trend of blood pressure with age similar to that seen in an industrialized society but with the change unrelated to weight change. Habitual high sodium intake is a possible cause.

Thailand

Surveys made in Thailand by the US Interdepartmental Committee on Nutrition for National Defense showed average intake of Buddhist farmers was 9.1 g per day as derived from analysis of food composites. With military personnel, the average daily intake similarly derived was 18 g. In both instances a small portion only (1–2 g) was derived from actual table salt or salt added with cooking, the bulk coming from fish sauces etc. Much of that condiment was discarded as plate waste. Thus urinary analysis would be essential for any accuracy of appraisal of intake. There is a low incidence of hypertension in Thailand, the survey showing little rise in blood pressure until old age. Henry and Cassel (1969), however, cite the Thai data as an instance of no rise in blood pressure with age notwithstanding a high salt intake, and also suggest a similar situation with the citizens of Norelsk, a Russian industrial city in the Arctic, who eat a Western diet (Henry and Stevens 1977).

South and Central America

Kean (1944) made observations on the Cuna Indians who live on the San Blas archipelago, a tiny string of islands which stretches along the Atlantic coast from the northern part of Columbia to within 150 km of the Canal Zone. The Indians are short, stocky, dark people. Most of the islands are within a few kilometres of the mainland where many of the Indians have small plantations, consisting mostly of coconut trees which they visit daily. Fishing, and minor agricultural pursuits on the mainland, provide their food, which consists mostly of coconut milk, dried or boiled green bananas, and fish. Other items of diet are plantain, sugar cane, cocoa, yams, mangoes, oranges, papaya and, rarely, peccary, fowl or deer. They have had little contact with the rest of the Republic of Columbia and appear to be in very good health. In the 408 Indians examined, there was a virtual absence of hypertension and the average blood pressure of the entire group was 105 mmHg systolic and 69 mmHg diastolic. No rise with age occurred. No significant difference between males and females was encountered.

Lowenstein (1961) made a study of the blood pressures in two tribes of Brazilian Indians on the shores of two southern tributaries of the Amazon. The Mundurucu used to live in a savannah-like forest, but 45 years previously had moved and settled along the Cururu River near a Franciscan mission. Each family, consisting of a husband, wife and children, lived in an individual home. Seventy-seven members aged from 16 to 60 plus, representing all the adults, with a few living up-river, were examined. The second tribe, the Caraja, had always lived along the Araguaia

River. They had been in constant contact with civilization for some time, but were still pagans and practised their rites and dances. Eighty-nine over the same age group as the Munduruсu were examined. There was a significant rise in systolic and diastolic blood pressures with age in the Munduruсu men, whereas no such rise was found in the Caraja men. In the females a similar rise was seen amongst the Munduruсu (though not statistically significant) but the blood pressures of the Caraja females remained more or less constant with age. The two tribes differ in their way of life in that the Caraja are more primitive and live more or less as their ancestors did before they had more intimate contact with civilization. The Munduruсu on the other hand have accepted some of the culture of Catholic Brazil. This includes dietetic changes, such as the use of table salt instead of plant ashes, and the use of tobacco.

Oliver et al. (1975) have made a most important study on the Yanomamö Indians of northern Brazil and southern Venezuela. They are a relatively undisturbed tribe of inland tropical rain forest Indians. Approximately 12 000 to 15 000 Yanomamö are distributed among some 150 villages in an area of approximately 250 000 km^2 (100 000 square miles). Semi-permanent contacts of any of their villages with non-Indians date back only to the early 1950s. Any mission stations are, for the most part, located on navigable rivers and since the Yanomamö are not primarily a riverine people, even today there are probably villages not yet visited by non-Indians. The data suggest that they have been in a high degree of isolation for at least the past several thousand years. The material culture is extremely simple. The major staple of their diet is the cooking banana supplemented by irregular additions of game, fish, insects and wild vegetable foods. The tribe has had no access to sodium chloride except where the substance has been introduced by Caucasians. That is, it was not a trade item in the indigenous culture. The study of blood pressure was made on all available inhabitants of 16 independent villages. This was combined with an endeavour to get a reliable appraisal of sodium status and activity of the renin–angiotensin system.

In the 506 subjects in the study, as has been usually observed, both systolic and diastolic pressure increased in the first and second decade of life (Table 27-5). However, blood pressure did not continue to increase thereafter, but even seemed to decline slightly. The slope of the regression is negative in all four analyses, but significance is not achieved in any case. The authors note, but consider unlikely, the possibility that the absence of a rise in blood pressure is due to the effects of mortality rates being high throughout adult life. Twenty-four hour collections of urine were made, and in 26 instances were adequate in content to permit the assumption that they were full collections. The average 24-h excretion was 1 mmol of sodium and 152 mmol of potassium. These data were supported by the analysis of random on-the-spot urine samples; thus in 45 spot samples of adult males, the mean concentration of sodium was 2 mmol/l and of potassium 107 mmol/l, and for 21 spot samples on 21 adult females, many of whom were lactating and/or pregnant, the mean concentration of sodium was 1 mmol/l and of potassium 145 mmol/l. Simultaneous plasma sodium concentration obtained from randomly selected males averaged 140 mmol/l. Three controls on expedition members sampled at the time averaged 142 mmol/l. Table 27-6 shows the estimations of urinary aldosterone on tribal members on whom the 24-h collection was made, and also the plasma renin values determined at the same time. Controls on members of the expedition are recorded in the same table. In both instances the values in the Indians equalled or exceeded the range seen in normal Westerners after a short period on a low-sodium diet.

The authors remark that, notwithstanding the difficulties of field observation, it seems clear that the very low 24-h sodium excretion observed in most of the subjects—of the order of 1 mmol—must correspond to sustained sodium intake lower than any previously recorded for man. Several individuals were excreting only 2 mg of sodium per 24 h, which is a remarkable physiological performance. They note that the Yanomamö are seldom obese and rarely demonstrate weight gain with advancing age. They are physically a highly

Table 27-5. Blood pressures obtained in the Yanomamö Indians.

Age	No. of subjects	Systolic Mean	Systolic s.d.	Diastolic Mean	Diastolic s.d.
Males					
0–9	59	93.2	8.9	58.6	9.2
10–19	63	107.5	9.6	66.9	8.6
20–29	58	108.4	8.6	69.1	7.3
30–39	30	105.9	8.9	69.4	5.7
40–49	27	106.6	7.6	67.1	6.8
50+	7	100.0	8.2	63.7	8.1
Females					
0–9	60	95.7	12.0	61.6	8.0
10–19	72	104.9	9.7	64.5	10.8
20–29	62	99.8	10.0	62.6	6.6
30–39	32	99.5	10.5	62.9	6.3
40–49	19	97.6	11.4	62.2	16.8
50+	17	105.7	17.7	64.1	7.3

From Oliver et al. (1975), with permission.

Table 27-6. Urinary excretion of sodium, potassium, chloride and aldosterone compared with plasma renin activity and blood pressure in Yanomamö Indians and control subjects.

Subjects	Na (mmol/24 h)	K (mmol/24 h)	Cl (mmol/24 h)	Aldosterone excretion (μg/24 h)	Renin activity (ng ml^{-1} h^{-1})	Blood pressure (mmHg)[a]
Indians						
1622[b]	3.44	152.64	7.06	76.9	7.92	102/60
1631	0.53	234.36	6.05	155.4	3.93	98/72
1668	0.30	128.00	5.44	59.9	8.68	88/52
1671	0.33	150.00	6.38	74.3	19.97	104/60
1690	0.31	175.74	9.31	65.8	11.36	94/40
1691	0.38	114.03	7.56	49.8	7.96	94/60
1699	0.71	194.36	12.56	27.3	10.21	90/50
16104	0.90	390.72	15.18	164.9	7.98	98/50
16116	0.61	178.56	16.37	42.0	53.79	102/50
16119	0.42	194.65	29.58	62.9	8.91	122/60
16120	6.76	291.10	22.58	40.5	3.43	120/52
Mean ±s.d.	1.34 ±2.01	200.38 ±80.17	12.55 ±7.80	74.52 ±44.94	13.10 ±14.17	
Controls						
E.M.	208	75	241.9	3.8	2.52	120/80
H.S.	254	54	233	3.5	9.61	142/82
W.O.	193	52	201	1.0	1.46	122/78
A.B.	142[c]	44[c]	171.9[c]	—	6.19	120/82

From Oliver et al. (1975), with permission.
[a]Blood pressures obtained with subjects in the sitting position.
[b]Expedition identification number for individual subjects.
[c]Concentration in mequiv/l on spot sample only.

active people with an overtly aggressive life-style, matters to be taken into account before simplistically attributing a low blood pressure to the effects of absence of salt.

Noting that the high aldosterone level is probably related to the mechanisms of sodium balance in these people, Oliver et al. draw attention to the fact that chronic potassium loading has not effected suppression of plasma renin in these subjects. This finding is different from that recorded by Macfarlane, Skinner and members of the Howard Florey Institute in relation to the Chimbu in highland New Guinea, where aldosterone was high and renin normal. Oliver et al. (1975) remarked that it would appear that the stimulus of sodium deficiency has superseded any effect that potassium suppression might have in people on a chronic low sodium/high potassium intake. Thus the two stimuli, i.e. high potassium and high renin, would be acting to increase aldosterone secretion. The authors, noting the importance of aldosterone in the maintained equilibrium of these people, raise the question of whether it or any other corticosteroid might be involved in the mobilization of sodium from bone to maintain effective plasma volume during periods of potential crisis, such as haemorrhage, gastrointestinal electrolyte loss, pregnancy and lactation.

Oliver et al. (1981) report subsequent studies on pregnancy and lactation in the Yanomamö, and concurrently make a comparison with the Guaymi Indians of Panama. There are approximately 30 000 of these living in the *cordillera* of western Panama and eastern Costa Rica. Though they have had generations of contact with acculturated peoples, their relative inaccessibility has resulted in comparative isolation as documented genetically. Their diet consists primarily of rice, beans, yams and tropical fruit with meat sporadically. Salt is added to cooked vegetables and used to preserve meat. The data in the study confirmed the earlier report on the Yanomamö with regard to blood pressure and urinary sodium excretion. The Guaymi Indian males had a mean sodium excretion of 66.8±8.9 mmol/24 h. This reflected liberal use of salt, though output was less than that of expedition members.

Breast milk was assayed in the Yanomamö and the levels of sodium (5–9 mmol/l) and potassium (9–17 mmol/l) found not to deviate from those determined in Caucasian society with patients in normal sodium balance. The values derived in the comparison of metabolic status are shown in Tables 27-7 and 27-8. The outstanding finding was the exceedingly high aldosterone concentrations in the urine of the pregnant Yanomamö women.

Table 27–7. Physical characteristics and results of laboratory analyses of blood and urine samples in male Guaymi and Yanomamö Indians and in male expeditionary controls.

	Guaymi				Yanomamö					Yanomamö (spot samples)			
	Indians		Controls		Indians		Controls			Indians		Controls	
Variables	*n*	Mean ±s.d.	*n*	Mean ±s.d.	*n*	Mean ±s.d.	*n*	Mean ±s.d.	Variables	*n*	Mean ±s.d.	*n*	Mean ±s.d.
Age (yr)	4	36.0±5.6	2	39.5±14.8	6	29.5±8.2	3	38.3±11.2					
Weight (kg)	4	62.3±7.5	2	70.5±9.7	6	54.9±2.2	3	74.2±9.5					
Blood pressure (mmHg)	4	116.0±4.3/ 73.8±13.6	2	117.0±1.4/ 82.0±0.0	6	110.3±11.8/ 68.8±3.8	3	120.0±2.0/ 70.0±2.0					
Pulse (beats/min)			2	67.0±1.4	6	80.7±7.8	3	68.7±8.1					
Serum aldosterone (ng/dl)	4	10.3±6.6	2	13.4±13.7	6	147.5±142.1[a]	2	21.6±9.8					
Plasma renin activity ($ng\ ml^{-1}\ h^{-1}$)					6	9.7±6.1	3	3.6±1.9					
Urine													
Na (mmol/24 h)	4	66.8±8.9	2	148.6±129.3	6	1.0±0.9	3	165.4±67.7	Na (mmol/l)	6	0.5±0.2	3	99.0±59.6
K (mmol/24 h)	4	71.8±29.0	2	47.7±7.3	6	302.7±104.6	3	76.4±18.0	K (mmol/l)	6	226.0±68.8	3	72.0±32.2
Na/K	4	1.0±0.4	2	2.9±2.3	6	0.003±0.002	3	2.3±1.2	Na/K	6	0.002±0.001	3	1.4±0.7
Creatinine (g/24 h)	4	1.4±0.7	2	1.5±0.8	6	1.9±0.4	3	1.8±0.1					
Creatinine (mg/ml)	4	0.7±0.2	2	1.0±0.6	6	1.9±0.4	3	1.8±0.1	Creat (mg/ml)	6	1.3±0.2	3	1.3±0.5
Aldosterone (ng/ml)	4	5.7±3.1	2	15.7±18.4	6	72.5±17.2	3	11.3±2.5	Ald (ng/ml)	6	79.9±93.6[b]	3	9.9±8.0
Ald/Creat	4	7.7±1.9	2	12.5±11.3	6	39.2±6.3	3	6.1±1.0	Ald/Creat	6	65.9±80.7[c]	3	12.0±14.8[d]
Aldosterone (μg/24 h)	4	10,4±8.0	2	13.3±11.2	6	72.4±17.2	3	11.4±2.5					

From Oliver et al. (1981), with permission.
[a]This high standard deviation is due to one outlying value of 376.
[b]This high standard deviation is due to one outlying value of 266.
[c]This high standard deviation is due to one outlying value of 227.3.
[d]This high standard deviation is due to one outlying value of 29.1.

Table 27–8. Physical characteristics and results of laboratory analyses of blood and spot urine samples in female Guaymi and Yanomamö Indians.

	Guaymi				Yanomamö					
	Non-pregnant		Pregnant		Non-pregnant, non-lactating		Non-pregnant, lactating		Pregnant	
Variables	*n*	Mean ± s.d.	*n*	Mean ± s.d.	*n*	Mean ± s.d.	*n*	Mean ± s.d.	*n*	Mean ± s.d.
Age (yr)	9	25.0 ± 4.7	7	21.6 ± 3.1	16	23.3 ± 6.7	16	22.0 ± 5.3	4	22.5 ± 1.0
Weight (kg)	9	55.6 ± 7.6	7	58.1 ± 7.8	16	43.7 ± 4.2	16	44.0 ± 4.1	4	44.9 ± 5.3
Blood pressure (mmHg)	9	113.6 ± 7.9/ 74.9 ± 9.4	7	114.6 ± 7.1/ 69.7 ± 8.1	16	114.8 ± 12.0/ 71.0 ± 6.4	16	107.3 ± 7.8/ 67.0 ± 5.7	4	111.0 ± 11.8/ 72.0 ± 9.8
Pulse (beats/min)					16	85.1 ± 17.0	16	83.0 ± 11.7	4	98.0 ± 9.5
Serum aldosterone (ng/dl)	8	16.5 ± 6.3	5	85.5 ± 51.1	16	47.8 ± 59.3[a]	15	34.6 ± 25.0	4	155.9 ± 245.3[b]
Plasma renin activity ($ng\ ml^{-1}\ h^{-1}$)					14	6.2 ± 4.1	16	5.0 ± 2.6	4	25.6 ± 6.4
Urine										
Na (mmol/l)	9	119.0 ± 52.7	7	77.8 ± 30.0	16	1.4 ± 3.0	16	0.8 ± 0.7	4	0.7 ± 0.4
K (mmol/l)	9	138.0 ± 46.6	7	105.8 ± 47.1	16	215.6 ± 86.8	16	264.7 ± 71.4	4	262.0 ± 66.6
Na/K	9	1.0 ± 0.6	7	0.9 ± 0.8	16	0.007 ± 0.013	16	0.003 ± 0.002	4	0.003 ± 0.002
Creatinine (mg/ml)	9	0.9 ± 0.3	7	0.6 ± 0.3	16	0.8 ± 0.3	16	0.9 ± 0.4	4	1.3 ± 0.2
Aldosterone (ng/ml)	9	7.5 ± 5.5	7	92.1 ± 110.1[c]	16	38.8 ± 31.4	16	46.2 ± 62.4[d]	4	585.3 ± 48.7
Ald/Creat	9	8.3 ± 5.7	7	128.2 ± 97.2	16	49.7 ± 33.9	16	43.1 ± 36.7	4	459.4 ± 38.8

From Oliver et al. (1981), with permission.
[a]This high standard deviation is due to one outlying value of 256.
[b]This high standard deviation is due to one outlying value of 524.
[c]This high standard deviation is due to one outlying value of 330.
[d]This high standard deviation is due to one outlying value of 264.

Values were 585.3±48.7 ng/ml, which was far in excess of concentrations in pregnant Guaymi women (92.1±110.1 ng/ml) or in Yanomamö males (viz. 24-h collections, 72.5±17.2 ng/ml; on-the-spot samples, 79.9±93.6 ng/ml). The comparison was similar when viewed against simultaneous creatinine determinations. Plasma aldosterone showed a similar effect, and plasma renin was similarly greatly elevated in the pregnant Yanomamö, but the lactating non-pregnant females had values in the same range as the non-pregnant non-lactating females and the males.

Whereas there were evident difficulties in collection of samples under these conditions and particularly ones in relation to 24-h specimens of urine in females, the result seems clear-cut in terms of evidence of a massive response of the renin and aldosterone systems to the impact of pregnancy in a 'no-salt' culture. Though plasma and urine aldosterone concentrations were increased in the Guaymi Indians, the concentration of urinary aldosterone was significantly lower ($P<0.001$) than in pregnant Yanomamö. It was of considerable interest that the data did not show any distinction between hormonal levels of those with stress of prolonged lactation (3 months to 3 years) and those women with the stress imposed by life-long existence on a low sodium intake. Allowing up to 5 or more mmol/day of sodium lost in milk, and the fact that urinary sodium excretion is usually 1 mmol/day, it could be argued a considerable additional stress would result from lactation.

W.J. Oliver (personal communication) remarks that the lactating (and pregnant) women take extra food and probably one of the greatest sources of sodium is the cooking ashes which adhere to peeled plantains and other foods after roasting. This would increase with greater food intake. He found on analysis that the sodium content of samples of cooking ash varied from 1 mg/100 g to 271 mg/100 g, presumably reflecting differing capacities to accumulate sodium of the various forest species used for fire. Temporary mobilization of skeletal sodium might also occur. The same issue arises over the course of pregnancy, where 500–700 mmol of sodium accumulate in products of conception plus growth of uterus and breasts (Gray and Plentl 1954; MacGillivnay and Buchanan 1958)—a mean of about 2–3 mmol/day. Alterations in sodium appetite with changed food selection are an important possibility for acquiring this sodium, as discussed in Chapter 23.

It seems to me a matter of great interest whether such people, especially in conditions of pregnancy and lactation, would show an increased preference for weak or moderate-strength sodium solutions in a controlled taste experiment. It also would be of great value to determine the exact time course of inhibition of the extremely high aldosterone secretion rate in the face of rapid administration of salt and the contemporaneity, or otherwise, of changes in the renin–angiotensin system. Another issue is whether years of sustained high or very high levels of aldosterone (Oliver et al. 1981 remark that the levels in urine exceed by 14-fold those observed in patients with severe hypertension and adenoma) would discernibly modify the response characteristics of endocrine and renal systems as assayed after 6–12 months of salt intake characteristic of Western civilization. Also, in the face of such change, would the eventual incidence of hypertension, which seems a feasible outcome, be much greater than in a comparable age and otherwise matched group of Westerners? Or would the very low sodium intake over the anabolic period of life and the influence of this during differentiation of physiological systems ameliorate the impact of a Western level of salt intake when imposed for the first time in adult life?

Reverting to the pregnant Yanomamö, Oliver et al. (1981) have drawn attention to the complexity of physiological factors operating to maintain sodium homeostasis in such subjects. Factors promoting urinary sodium loss include an increase in glomerular filtration rate, reduction in renal vascular resistance, decrease in plasma oncotic pressure, increased plasma angiotensinase and a progressive increase in plasma progesterone which has natriuretic effects. The increase in plasma renin substrate and renin concentration presumably lead to the very high aldosterone which counteracts these effects; and, also, hepatic aldosterone extraction is reduced in pregnancy.

It is of particular interest to look at these metabolic studies along with the contemporaneous anthropological studies (Chagnon and Hames 1979). The medical studies showed the Yanomamö to be in good physical condition, well nourished, and to show no signs of protein deficiency. Chagnon and Hames, in challenging the hypothesis of a possible relationship between protein consumption on the one hand and warfare and demographic practices such as infanticide and male–female dominance relationships on the other, made a 13-month study of protein intake of a Yanomamö village of 35 individuals on the lower course of the Padamo River, a large affluent of the Upper Orinoco. Hunting in the deep forest was the most important source of game and a 216-day survey showed 2179 kg of game taken of which 1307 kg was edible. The 60-day survey of fish catches showed 182.4 kg of edible material taken. Thus there were 260 g of bulk edible meat per day per capita, which provided 52 g protein per day (ca 20% of edible meat). Allowing for the fact that the Yanomamö are small relative to Western people, the protein intake is well above basal requirements: approximately two and a half times the lowest estimate of basal need. The authors propose that the idea of Yanomamö warfare being determined by competition for locally scarce protein is invalidated not only by this fact, but also by the observation that the most enduring and extensive wars occur between villages that do not share overlapping or adjacent hunting territories. They recount specific instances of ambush and murder which occurred during this field study conducted by Dr Hames.

Some tribes studied by Dr Oliver (personal communication), where urinary sodium excretion was only 1 mmol/day, were also on a high tributary of the Upper Orinoco, and were of the same stock and reasonably close to the Yanomamö studied by Chagnon and Hames geographically and in life-style including dietary habits. If the US Departmentof Agriculture *Composition of Foods* is used as source, the sodium content of 60–70 mg/100 g of edible portion for beef and fish cited could result in a daily intake of 7–8 mmol of sodium by the Yanomamö ingesting 260 g of meat and fish daily. It is possible that the sodium content of tropical game and fresh-water fish may be somewhat lower. Allowing for the elements in turnover discussed above in relation to pregnancy and lactation, extremely low urinary sodium output of females would seem explicable. With the male, the considerable distances travelled in the high environmental temperatures while hunting, with associated sweating, would account also for sodium loss notwithstanding high blood aldosterone concentrations. W.V. Macfarlane's observations (personal communication) on highland New Guineans, who also had extremely low urinary sodium, indicate a sweat sodium content of 10–12 mmol/l. Again the low urinary sodium of the Yanomamö male is explicable.

Larrick (1980) found that the Waorani Indians of Ecuador who live in the Amazon rain forest have low sodium intake and low blood pressure ($103\pm12.8/65\pm10.7$ mmHg for men and $93.8\pm8.5/60.8\pm5.0$ mmHg for women) which does not increase with age.

Shattuck (1937) recorded blood pressures in Guatemala and found they were lower than those found in the USA. Saunders (1933) found that systolic blood pressure of the Mayas and the Mestizo in Yucatan were also lower than those recorded in the USA.

Glanville and Geerdink (1972) studied the blood pressure of Amerindians from Surinam. The population supported themselves largely by hunting and gathering and their nutritional status was representative of a group living under primitive conditions in a favourable environment. The observations were made at mission stations on the headwaters on a number of rivers. It was found that after a gradual rise in blood pressure during childhood and adolescence, mean pressures tended to reach a plateau during adult life. Systolic pressure in males showed a small but statistically significant decline with age, while in adult females it showed a small but significant increase. Diastolic pressure did not alter significantly with age in adults of either sex. The population was well nourished but there was no tendency towards obesity. Skinfold thickness tended to remain stable throughout adult life. The tribes concerned had a diet rich in protein from animals and fish, but relatively low in carbohydrate. There was no tobacco smoking. This population, then, was similar to the New Guineans in relation to blood pressure history with age but differed, as did other feral groups, in that protein intake was high and carbohydrate low. This speaks against any emphasis on the low-protein diet of New Guineans as the basis of the absence of a blood pressure rise with age.

Hoobler et al. (1963) found that 588 adult male rural Guatemalan Indians, who constituted the majority of the population sample, exhibited a low mean blood pressure: 113/73 mmHg at age 45. There was little tendency for the level to rise between the ages of 20 and 60 years. Two smaller sub-samples consisting of Indians of the same tribal origin but engaged in factory and managerial work in the midst of a Westernized culture showed higher blood pressures approaching the level of the small sample of Latin origin living in low socioeconomic circumstances. Abdominal adiposity and increasing arm circumference correlated with the higher blood pressure but did not entirely explain the highly significant differences between the populations. The authors concluded that acculturation was responsible for emphasizing or expressing the inborn and widely varying racial predisposition to develop elevated blood pressure with advancing age.

Caribbean region

Saunders and Bancroft (1942) made a study of the blood pressure of black and white populations in the Virgin Islands of the United States. The results showed that Virgin Island blacks have mean systolic and diastolic blood pressures which are higher than those of whites in either the Virgin Islands or the USA (the former being slightly higher). In both localities, after middle age females have a higher blood pressure than the males. A greater proportion of Virgin Island blacks have pressures which fall at the high and low ends of the range than do residents of the USA.

In commenting on the issue of the stress and strain of civilization affecting the black population, and therefore being conducive to hypertension, the authors remark that this may not be a logical explanation for high blood pressure in the Virgin Islands. On the contrary, although they are poor, the islanders are on the whole contented, happy and carefree. One must look for other factors. Saunders and Bancroft draw attention to the poor diet and incidence of vitamin deficiencies as possibly relevant, and also to a racial difference.

Humphries (1957) in reporting on hypertension in the Bahamas, recorded data on 462 persons predominantly of African stock. He found that 189 (41%) had systolic blood pressures of 150 mmHg or over. This included both sexes and all ages from 18 years upwards. In 71 (15%) systolic pressure was 180 mmHg or more. He draws attention to the fact that in the Bahamas the salinity of water drunk has always been high—except in Nassau, New Providence Island, during the 4 years preceding the survey—and analyses showed levels of up to 20 mmol/l sodium in well water. Filtration takes place from the sea and there is a certain amount of wind-blown salt. There are no rivers in the Bahamas. On the other islands, where the opportunities for refrigeration of foodstuffs are almost non-existent, pickled foods are consumed much more. On some, salt pork and salt fish are eaten in large quantities. Bahamians also eat much salt with their food and dislike being put off salt. Given that these conditions operated in the outer islands, it was interesting that there was some suggestion that the incidence of hypertension was greater in the outer regions. Numbers, however, were not very high in the comparison.

Johnson and Remington (1961) studied blood pressure levels in black and white residents of Nassau, and found a definite racial difference in blood pressure. The progression of blood pressure with age was almost linear in the blacks, whereas in the whites it showed a flattening or plateau effect from the time adult ages were reached until middle life, when mean blood pressure again began to rise. This was seen in the means of both systolic and diastolic pressure, but most noticeably in regard to systolic. The data on blacks appeared to resemble those on blacks in other parts of the Western Hemisphere. The authors suggested that it would be very interesting to have the data on the blood pressures in St Kitts in the Leeward Islands, since there might be a low salt intake there.

Schneckloth et al. (1962) published a detailed study of blood pressure in St Kitts. They note that the descendants of former slaves now form about 98% of the population, which is about 35 000. Based on a study of about 1575 villagers aged 20–29 years, blood pressures were found to be higher than those of white groups of like age in the Bahamas and in the USA; the means of blood pressures are higher in the Kittian women than in men at 35 years of age and over. However, their data show that salt intake was not low in this area. Based on casual specimens of urine and measurement of creatinine, it appeared that the mean output of sodium chloride was about 10 g (ca 200 mmol) per day. Analyses of food specimens representing the total daily intake of four families in three villages also suggested a sodium chloride intake of about 100 mmol/day.

North America

Boyle (1970) remarks that morbidity and mortality from cardiovascular disease are known to be higher in the south-eastern United States than in the nation as a whole. Further, the incidence of hypertensive disease is markedly higher in blacks than in whites. Physicians in the Charleston, South Carolina, area believe these patterns are accentuated in that biracial area. It is a sea coast community characterized by low population mobility. In 1960 the population was 216 000 of which 64% was white and 36% was Gullah black, a people closer to West African black than the general American black. Mortality statistics in the area showed that the peak death rate of Charleston black men occurred at age 52 years, 15 years younger than the white men, and for black women the death rate peaked at 57 years, 20 years younger than the whites. The Charleston blacks displayed markedly higher death rates from cerebrovascular and hypertensive disease than American blacks in general and 10–20 times those of the total US population.

A cross-sectional population base sample was drawn from the community based on the 1950 census. In the blacks systolic blood pressure was significantly elevated above that of the whites (Fig. 27-11). Further, the two sexes did not parallel each other, the systolic pressures of black women being significantly higher than those of the men and increasing more sharply with age. Diastolic pressures in the blacks showed a notable increase with age in the earlier years, before any appreciable rise in those of the whites. The whites

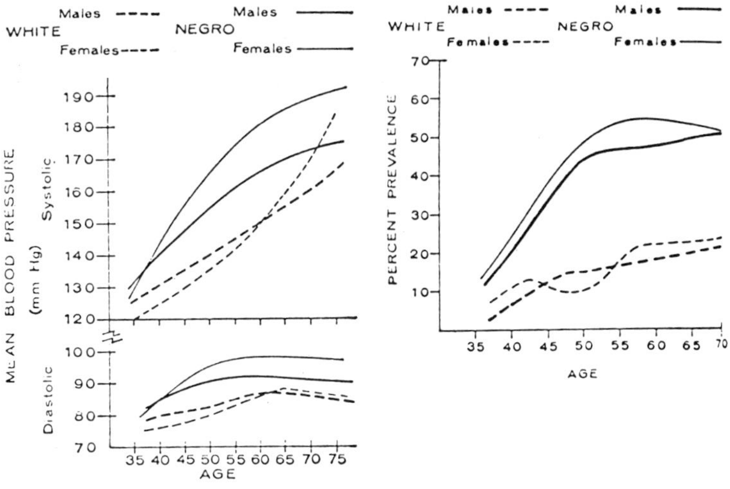

Fig. 27-11. Mean blood pressure by age in four race–sex groups (left) and percentage prevalence of definite hypertension by age in four race–sex groups (right). (From Boyle 1970, with permission.)

displayed a gradual rise in the prevalence of definite diastolic hypertension with increasing body weight, while in blacks there was no such clear correlation between body weight and hypertension.

Blood group studies confirmed earlier data suggesting that the Gullah black is genetically closer to the West Africans than are other American blacks. Using a photoelectric reflectant colorimeter, measurements were made of skin colour, and skin reflectance was found to be correlated very well with blood groups: the darker half of the population had a 30% incidence of blood group B, the same as the West African black, while the lighter half had 24% of the type B blood group, the difference being significant ($P<0.05$). The difference between the incidence of group B in the darker half of the Charleston blacks and in the average American black (21% type B in men) is significant at the $P<0.01$ level. A significant relationship was seen between age-adjusted blood pressure and skin colour in the blacks. In men, with each 5 units of decreasing skin reflectance (or darkening) the age-adjusted mean systolic blood pressure increased by 3.1 mmHg and the diastolic pressure by 1.3 mmHg. In women the adjusted systolic blood pressure increased by 1.2 mmHg and the diastolic pressure by 0.5 mmHg for each 5 units of decreasing skin reflectance.

Thomas (1927) made a study of the Greenland Eskimo. Apart from some Eskimos associated with small Danish settlements, the bulk of the population lived, after being weaned, on an exclusively carnivorous diet including fish. There was no edible vegetation. Their diet included meat of walrus, whale, seal, caribou, musk ox, arctic hare, polar bear, numerous sea-birds, geese, duck, etc., and fish, all eaten usually and preferably raw. Thomas notes that contrary to general opinion, the Eskimo eats relatively little fat or blubber. In 142 adults between the ages of 40 and 60, average blood pressure was 129 mmHg systolic and 76 mmHg diastolic. The highest pressure found was 170 mmHg systolic and 100 mmHg diastolic. There were eight others only whose systolic pressures were above 140 mmHg of which six were below 150 mmHg and the other two below 160 mmHg. The people at that time led a life of great physical activity, a large number of the men dying by violence, accidents, starvation or freezing before old age came, and being exposed to the severest extremes of cold. Their physical endurance was remarkable.

Thomas noted the difference between the Greenland and the Labrador Eskimo, who have had exposure to European food via trading of furs. With them, in contrast to the Greenland Eskimos, vitamin deficiencies in the form of scurvy and rickets were common. Overall, his survey suggested that though the Greenland Eskimo lives on an exclusively carnivorous diet, there was no increased tendency to vascular or renal disease, the latter being assessed on urinalysis of the population that was scrutinized for blood pressure.

Some evidence from a study of the Papago Indians in south-central Arizona is suggestive of a rapid increase in the incidence of hypertension. The data have deficiencies, but whereas 20–40 years ago the incidence of hypertension in Papago and Navajo Indians was thought to be 1%–2%, the result today is nearer 20%. Strotz and Shorr (1973) state that prior to the Second World War the Papago diet consisted mainly of meat derived from hunting and cattle-raising, and of flood-farmed beans and desert plants. Sea salt, which had to be hauled long distances from the Gulf of California, was the major salt source. The development of a cash economy with purchase of wheat flour, potatoes, rice and table salt for the first time has led to a diet of cooked carbohydrates, with salt and lard added liberally in cooking. Physical labour has also decreased.

Fodor et al. (1973) have examined four populations in Newfoundland comprising 1499 adult inhabitants. The prevalence of individuals with diastolic pressure over 100 mmHg in the age group over 50 was found to be three times higher in a coastal community than an inland community. Dietary survey showed the coastal inhabitants, who lived in relative isolation, to have higher mean sodium intake. Intake of salted fish, beef and pork was traditional. The incidence of hypertension in this coastal island community of Fogo was greater than in the USA, whereas in the inland community the rates were comparable.

Japan

Japan is of great importance in any discussion of the relation of salt and hypertension. The possibility that over-consumption of salt has influenced the incidence of hypertension and its sequelae is strong.

In Japan—in contrast to most Western countries—cerebrovascular lesions are a more common cause of death than heart disease. Japan has been called the apoplexy country. Watanabe (1953) produced figures showing that the death rate from apoplexy in Japan compared with the USA, Britain and West Germany was extraordinarily high: 4–8 and 3–6 times as high for people in their forties and fifties respectively. Large differences were noted in different parts of Japan. Highest incidence was in Akita prefecture where it is 2.3 times the average rate for people in their fifties in the rest of Japan. Rates in Iwate and Yamagata prefectures were also high whereas that in Fukui prefecture was the lowest at 0.66 the national average. There have been some doubts cast on the validity of these figures by the World Health Organization on the basis of certification and post-mortem data, and considerable efforts are being made to gain precise information in the course of combined hypertension and stroke-control programmes in Japanese communities (such as that reported by Hirota 1976).

Fig. 27-12 shows the distribution of death rate per hundred thousand males aged 30–59 years from cerebral haemorrhage in various regions of Japan. The salt intake of farmers in four regions has been added to the figure, which was prepared by Dahl (1960) and taken from data of Takahashi et al. (1957). Dahl (1960) quoted figures on examination of 5301 persons in Akita in northern

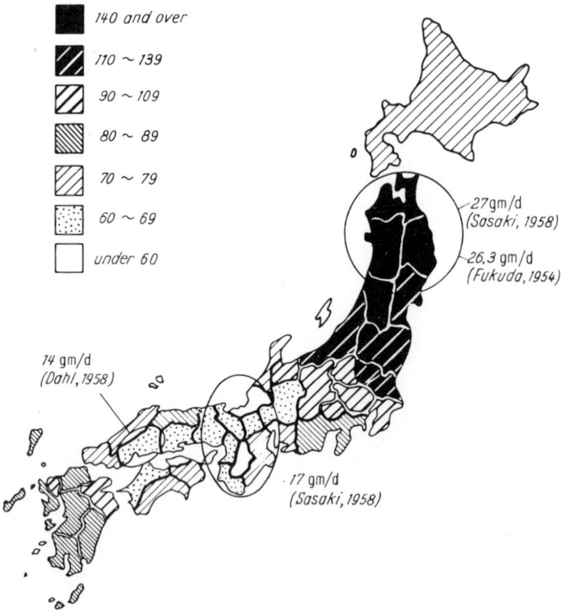

Fig. 27-12. Distribution of death rate (per 100 000 males aged 30–59 years) from cerebral haemorrhage in various regions of Japan. Average daily salt intake of farmers in four regions is also shown. (Figure prepared by Dahl 1960, using data from Takahashi et al. 1957. From Dahl 1960, with permission.)

Honshu of average age 45, showing an incidence of hypertension of 39%, compared with 21% in the region of Hiroshima in southern Japan. Kobayashi (1976) also states that the incidence of hypertension is much more frequent in Akita than, for example, in Hokkaido, and particularly in the young age groups.

The data of Takahashi et al. (1957) include comparison of age groups by decades from 20 to 80, and suggest a radical difference in the incidence of systolic pressure in excess of 150 mmHg and diastolic above 90 mmHg in Akita as compared with Kagochima in Kyushu. The possible importance of this finding lies in the very high salt intake in northern Honshu. Relevance of the data to the general problem is shown in Table 27-9.

Freis (1976) has noted the possibility that the age selection of subjects upon which Japanese figures are based could have contributed to the impression that hypertension was of particularly high incidence.

A most interesting study of the population of the villages of Japan has been made by Kondo (1962). His group made investigations in more than 840 villages over a period of 28 years, choosing those with a low frequency of migration. He remarks that by international standards the Japanese are short-lived. Only about 3% of the total population live to an age of over 70 years, this being about half

Table 27-9.

Group	Year	Sex	No.	Age (average)	Salt intake Average (g/day)	Range (g/day)	% hypertension[a]
Alaskan Eskimo	1958–1960	Both	20	38	4	1–10	0
Marshall Islanders (Pacific)	1958	Both	231	41	7	1.5–13	6.9
USA (Brookhaven)	1954–1956	Male	1124	36	10	4–24	8.6
Japan							
Hiroshima (S. Japan)	1958	Male	456	43	14	4–29	21
Akita (N. Japan)	1954	Both	5301	45	26	5–55	39

From Dahl (1960).
[a]Hypertension defined as blood pressures of 140/90 mmHg or over.

the number in any Western country. However, there are areas where longevity (over 70 years) is 8%–9% compared with others where it falls to as low as about 1%. Kondo states that in villages where rice is consumed in preference to other food articles, the percentage of old people is always low and cerebral haemorrhage is frequent even among the younger generations. The villages in the rice-producing districts of Akita, Aomori and Yamagata prefectures are good examples of such localities. Here an adult villager consumes as much as 930 g of rice daily, and the eating of large quantities of boiled rice always involves simultaneous consumption of large quantities of table salt as a condiment. In the villages where longevity is comparatively common, quantities of fish, meat or soya beans and vegetables are taken. Also, in villages where seaweeds are abundantly eaten, cerebral haemorrhage was very infrequent and there was a high percentage of old people. From personal discussion with Professor Kondo some years ago, it did not appear that any experimental data were available on the addition of seaweeds to the diet in areas of high rice consumption. He stated that coronary disease and nephrosclerosis were rare in the high salt areas.

There did not appear to be any data available on the incidence of eclampsia in pregnancy nor on whether, with patients with heart failure or cirrhosis, oedema was a common finding. However, the contrast was emphasized between islands such as Tachibana in Hiroshima prefecture where 12% of the female population were over the age of 70 and the diet included a high component of sweet potatoes and oysters, and Akita and Aomori where, in some villages, only 1% of the population were over the age of 70. The rice grown in the north-east, including Akita, is tasty and famous. There was not the same impetus to eat other foods, and the rice was eaten with condiments, pickles, cucumber, egg plant and with miso soups. This high salt intake with condiments by the Japanese farmers in Akita was not paralleled in the coastal area, where the consumption of salt was less as a consequence of considerable amounts of fish being eaten.

Dahl (1960) quotes Professor Fukuda's quantitative study on sodium intake in Akita made in 1954. From analysis of urinary sodium excretion it was found that of the 300 adults tested none was eating less than 5 g salt per day and only 10 were eating less than 10 g per day. In fact 250 of the 300 were consuming in excess of 20 g salt per day and the mean (Fig. 27-12) was over 25 g per day.

Sasaki (1962) and Takahashi et al. (1957) reported that the regional differences in mortality from apoplexy in Japan were parallel to regional differences in the level of blood pressure of the inhabitants. Further, the blood pressure measured during the cold season was higher than that measured during the warmer season. The blood pressure data for north-east Honshu are set out in Table 27-10, which may be compared with Fig. 27-2 from Sinnett and Whyte's data (1973a). A comparison was made of findings in Akita prefecture in the north-east and Okayama prefecture in the south-west with regard to salt intake and mortality. In Akita death rate from apoplexy was over 5000 per 100 000 population, while in Okayama it was a little over 3000 per 100 000 population. In Akita the death rate was high in the fourth and fifth decades.

Sasaki notes the results of investigations of the diet of nearly 6000 farming families all over Japan made in 1958 by the Ministry of Agriculture and Forestry. Possible correlation between the average intake of each nutritive element per day and mortality from apoplexy in middle age (30–59 years) was examined. A correlation at less than the 1% level of significance was found between mortality and salt intake, and also fat intake. No significant correlation was noted between mortal-

Table 27-10. Mean arterial pressure with standard deviation and frequency of persons with high blood pressure in the north-eastern district of Honshu, Japan.

		Males							Females					
		Systolic			Diastolic				Systolic			Diastolic		
Age group	*n*	*M*	s.d.	%	*M*	s.d.	%	*n*	*M*	s.d.	%	*M*	s.d.	%
Warmer season (from May to Oct.)														
12–14	485	118.4	13.4	2.9	58.2	13.3	0.8	451	123.0	13.3	2.4	65.4	12.7	0.9
15–19	610	131.4	14.2	10.2	74.3	13.5	13.3	1176	127.4	12.8	5.0	71.4	12.1	5.2
20–29	998	130.9	15.1	9.1	74.6	13.1	11.4	688	127.5	15.1	7.3	73.4	13.3	8.4
30–39	706	130.9	16.5	9.6	75.0	13.3	11.8	802	127.1	16.9	8.1	75.1	12.7	9.1
40–49	752	137.0	23.4	21.8	80.9	15.3	21.4	871	136.8	25.4	21.1	81.1	13.9	21.2
50–59	657	147.8	27.3	37.9	86.0	15.0	35.2	793	146.6	27.7	37.8	85.2	14.3	31.9
60–69	359	157.6	31.2	55.7	87.9	18.0	40.4	466	157.7	29.3	56.7	87.4	14.8	37.3
70–79	98	169.5	31.9	69.4	89.1	15.9	41.8	122	173.8	30.1	77.0	91.3	14.2	49.2
80–89	6	173.3			83.3			6	175.0			91.7		
Colder season (from Nov. to Apr.)														
6–11	287	108.8	13.0	0.3	63.8	11.5	1.4	299	109.6	15.0	0.0	62.8	11.9	2.0
12–14	294	122.1	14.9	3.7	68.0	13.1	5.1	283	122.7	15.9	5.3	66.5	11.8	1.8
15–19	72	139.0	13.4	19.4	67.5	13.3	5.6	114	133.0	13.2	8.8	71.1	13.0	5.3
20–29	167	142.1	15.0	28.7	72.7	14.3	10.8	308	127.3	12.2	8.4	71.5	12.4	6.8
30–39	204	141.9	19.0	28.4	76.2	14.6	12.7	350	133.3	19.1	11.7	76.4	13.9	10.9
40–49	283	147.6	27.0	35.0	83.9	15.1	28.6	476	140.9	24.4	27.1	81.6	13.7	23.5
50–59	309	159.8	28.0	58.9	89.9	14.0	48.5	440	157.1	30.0	52.5	87.8	14.1	39.1
60–69	203	171.1	30.3	74.4	90.7	15.5	48.8	275	168.5	29.8	69.5	90.4	13.9	49.8
70–79	66	185.0	35.2	81.8	92.9	17.8	59.1	98	177.1	29.6	89.8	95.2	14.8	62.2
80–89	6	196.7			95.0			13	185.0	33.7	84.6	84.2	12.7	30.8

From Sasaki (1962), with permission.
M, mean blood pressure in mmHg; s.d., standard deviation in mmHg; %, incidence of persons in whom systolic blood pressure was 150 mmHg or higher and diastolic pressure 90 mmHg or higher.

ity and intake of calories, protein, calcium, or vitamins A, B_1, B_2 or C. Sasaki notes that the Japanese take salt mainly in the form of soya sauce, seasonings or table salt, miso, pickles and other food. (Miso is made from soya beans, yeast and salt, the amount of salt varying from 4% to 15%.) An adult farmer in the north-eastern district takes 27 g/day of salt, while one in the Kinki district takes about 17 g. Tokamatsu is quoted as finding in farmers in Akita that the mean was 27 g/day on the basis of urinary excretion and 61 g/day as a maximum, whereas in Okayama prefecture in the south-west the mean was 15.4 g/day with 25.9 g as the maximum.

Sasaki (1962) states that the mortality from apoplexy differs between the neighbouring prefectures of Akita and Aomori. Table 27-11 records this difference although it is noteworthy that both prefectures are above the Japanese average. Sasaki records that the blood pressure of the inhabitants of Aomori prefecture is lower than that of those of Akita. Further, on the basis of a survey of 1110 middle-aged persons classified according to the number of apples they ate daily (Aomori produces 70% of Japan's apples) there was evidence that middle-aged people (40–59 years) eating three or more apples per day had lower blood pressures than those that did not. This effect of eating apples was attributed to the resulting increased potassium intake. It was claimed that eating six apples per day caused a fall in blood pressure in 18 of 38 middle-aged persons in Akita.

Kaneta et al. (1964) made a comprehensive study of diet in three villages in north-eastern Honshu: an inland farm village in the typical rice cultivating area, a village in the rice cultivating area near Sendai, and a village on a small island representative of rocky coast fishing villages. The villages had populations of over 3000 and the death

Table 27-11. Comparison between the death rate[a] from cerebral apoplexy in Akita and Aomori prefectures in 1955.

Prefecture	Male	Female	Total
Akita	277.9	162.6	218.6
Aomori	175.4	104.8	139.2
All Japan	116.8	82.8	103.5

From Sasaki (1962), with permission.
[a]Middle age (30–59 years) death rate per 100 000 population.

rate from cerebral haemorrhage in the previous 4 years for the fishing village was less than half that of the other two. The systolic blood pressure of the inhabitants in the fishing village was also significantly lower than in the two inland farm villages. The proportion of rice in the diet was greater in the two inland villages as was the consumption of miso and 'shoyu' as well as salted pickles, while in the fishing village consumption of seaweeds, barley, wheat, potatoes, beans and fish was greater with a correspondingly higher intake of protein, fat, calcium, phosphate, iron and several vitamins.

It is, perhaps, interesting to divert at this point and remark briefly on the acute effects of ingestion of large amounts of salt. Of course, the acute situation may be quite different from the physiological situation which prevails after weeks or years of action of a stimulus. Grant and Reischsmann (1946) and Lyons et al. (1944) examined the effects of ingesting 20–40 g/day for a few days on healthy young volunteers. This range, superimposed on a normal diet, involved intake of the order of that in the population in northern Honshu. After 2 days weight had increased, central venous pressure increased 50%–60%, and plasma, blood and extracellular fluid volume had increased 10%–20%. Blood pressure did not change but, whereas haematocrit fell, plasma protein did not, indicating mobilization from reserves. Thus this level of intake had fairly substantial immediate effects on the body fluids. McDonough and Wilhelmj (1954) gave a little less than 40 g salt per day for 23 days to a healthy 81-kg male. Body weight increased and there was some swelling of tissues coupled with gradual increase in systolic and diastolic pressure.

In Japan there may be some reservations about post-mortem certification of exact incidence of cerebrovascular accidents. And there are no comprehensive total population studies of incidence of hypertension in particular areas. Nonetheless the data provoke the question of whether the high salt intake has a damaging effect through causing hypertensive disease. The case might not be as strong, given some shortcomings of the data, if it were not set against the parallel and, in some cases more precise, information on the very low incidence of hypertension in unacculturated societies with low sodium intake. Some objections which have been raised to interpretation of the findings in northern Honshu in relation to salt intake and hypertension follow.

Takahashi (1962) has shown from the 1958 vital statistics data on deaths from cerebrovascular lesions and arteriosclerotic and degenerative heart disease in various age groups that the Japanese death rate from cerebrovascular lesions is very high (Takahashi 1962: WHO annual epidemiolo-

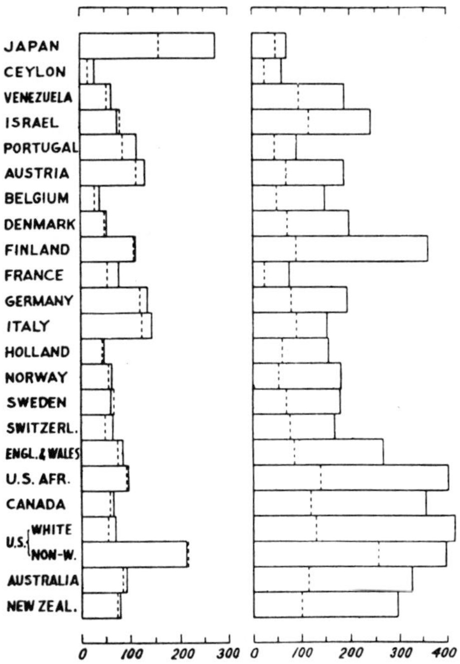

Fig. 27-13. Corrected death rate (per 100 000) for age group 30–69 from vascular lesions affecting the central nervous system (left) and arteriosclerotic and degenerative heart disease (right) for 22 countries. Male, continuous line; female, dotted line. (From Takahashi 1962, with permission.)

gical vital statistics 1961). His results are shown in Fig. 27-13. While remarking that the methods of certification and diagnostic criteria and coding on this issue in Japan may be imperfect, he states that there is other evidence on morbidity and mortality from cerebrovascular lesions in Japan. Nearly 800 public health centres, each of which serves a population of 100 000 to 200 000, have such data. For example, under the control of Akita public health centre there is a rice-growing rural district of 44 000 people. In 1959 it was found that 6% of the 50–54 year age group and 10% of the 55–59 year age group had been paralysed by stroke. In a single rice-field area there were 863 deaths from cerebrovascular lesions from 1955 to 1960 and all except one were diagnosed before death. He suggests it is uncommon in present-day Japan that cerebral haemorrhage or other cerebrovascular lesions are described as a cause on a death certificate without any diagnosis before death.

However, Professor Takahashi (1962, and also in personal discussion) has put emphasis on other factors being important as well as salt. The low calcium content of the Japanese diet and poor vitamin intake should be taken into account. Evidence of vitamin deficiency on various criteria came from the Japanese Nutritional Survey in 1959. The situation may have a parallel to that of the black people of the southern United States, who show the nearest mortality from cerebrovascular lesions to the Japanese nation. He also notes the relationship between calcium deficiency and cardiovascular disease, citing the data of Schroeder (1960) and Morris et al. (1961). The high carbohydrate content of the Japanese diet is paralleled in other countries—for example Sri Lanka and Yugoslavia—where the same instance of death from cerebral haemorrhage is not seen. My discussion with university colleagues in the Sendai area, and also with Dr Fukuda at Chiba drew attention to the fact of a very high level of physical exertion of the farmers in the Akita region and that this driving endeavour may in some way be related to the epidemiology referred to above.

Kobayashi (1957) proposed that there was a close correlation between the chemical composition of river water in Japan and the death rate from apoplexy. River water, as a result of weathering of rocks and soils of the river-bed, contains various kinds of inorganic matter. The rivers in Japan are characterized by a low calcium carbonate level, i.e. alkalinity, and in some districts of eastern Japan there are many rivers containing larger quantities of sulphate than carbonate. This is different from rivers generally in other parts of the world. In the Aomori and Akita prefectures the ratio of sulphate to carbonate in the river water is particularly high due to the volcanic sulphur-rich nature of the ground. Kobayashi's map of sulphate to carbonate ratios of Japanese rivers shows a close relation between those areas with a high ratio and those with the highest death rates from cerebrovascular accident. He records that the rivers of Thailand are carbonate-rich and that cases of apoplexy are scarce though rice is the staple food. This is also true in Okinawa and Formosa, whereas in Italy, a volcanic country, death rate from apoplexy is high. He thinks the role of calcium carbonate merits further enquiry.

In the discussion of Dahl's (1960) paper, Schroeder (1960) drew attention to the distribution of cadmium in the intestines of a fish eaten in Japan. It showed a cluster of high values in the north-eastern Honshu prefectures and low values in the west. Cadmium accumulates in the human kidney with age. So there could be factors other than salt which affect the death rate. There seems, however, to be little additional evidence available on this issue. Dr T. Fukuda (personal communication) has remarked on the fact that these are a very special type of fish and rather expensive, and probably not widely eaten.

Hatano (1974), in reviewing hypertension in Japan, reiterates its high incidence in the younger age groups in Akita prefecture, and the fact that this area has the highest death rate from stroke in Japan. He thinks this is not attributable to any inconsistency of death certification which has been considered. In Akita also the frequency of signs of associated organ damage, and incidence of stroke in middle age, were higher than in other regions.

Finally, it can be remarked that the Korean diet also appears to be very high in salt. The analysis of Cha and Suh (1970), which included analysis of additives such as soya bean paste, red pepper paste and Korean cabbage, indicated an average intake was 20 g/day of sodium chloride, but the range was wide. The Na/K ratio of diet was also high. Hypertension is reported as very common in the community (Hyun and Choi 1969; Hong and Suh 1972).

A high age-adjusted prevalence of hypertension has also been reported in Alexandrovka in the USSR where salt intake of 350 mmol/day is higher than in neighbouring villages, which have an intake of 160 mmol/day (Fatula 1977).

Sodium intake and blood pressure within Western populations

The attempts to relate incidence of hypertension to variations of salt intake within populations have produced contradictory results. There are large difficulties inherent in such studies and they may be more than methodological. Estimates of sodium intake based on salting of food may be inaccurate both because of subject unreliability and the fact that the main addition of salt may occur before the food reaches the table. Twenty-four hour urine collections give a closer approximation to intake. Dahl (1958a) studied one group of 70 subjects from whom 24-h specimens were collected for sodium analysis, and another group of 28 subjects from whom 24-h specimens were collected for from 6 to 38 days. These data all came from the Brookhaven National Laboratory community. The range of excretion of the two groups was from 60 to 400 mmol/day. The average and median levels of both groups were bunched closely at 180 mmol/day. However, it was noteworthy that day to day variation of those with multiple collections could be 2- to 4-fold.

In any event, if high sodium intake were to have an aetiological role in hypertension, it may be that intake in adult life is of minor significance relative to the implications of the level during infancy to adolescence. Adult data give little or no insight into this. Further, the development of hypertension could itself alter salt preference, giving some degree of aversion as suggested by animal studies.

Others have argued that hypertension may cause increased loss of sodium and, as a consequence, increased intake, though sodium restriction studies on patients with essential hypertension do not suggest any incapacity to conserve sodium. In relation to adolescent hypertensives, Kilcoyne (1977) reports that daily sodium intake, based on 24-h urinary excretion, was mainly in the 100–200 mmol range, and outputs of 300 mmol or more were noted. It may well be that, with food processing and the salting habits in Western diets, the low sodium intake seen in unacculturated societies is most uncommon, even in the young. Thus the determinant factor of whether hypertension ultimately develops may be genetic variation in the population in the face of sodium intake which is fairly uniformly high from an early age. In describing their analysis of hypertension in the young, Londe et al. (1977) note the high incidence of parental hypertension, and of obesity in the children affected.

Conclusions derived from surveys of sodium intake and hypertension in adults in industrialized Western communities are therefore open to criticism.

Ashe and Mosenthal (1937) measured the 24-h urinary excretion of chloride in 1000 ambulatory adults in New York City who were normal or had no disease other than hypertension. They found that (calculated as sodium chloride) the distribution was: (*a*) 4 g or less, 50 subjects; (*b*) 4–8 g, 416 subjects; (*c*) more than 8 g, 534 subjects. They found no differences in the amount of salt eaten by the 437 individuals with and the 563 without hypertension. They noted, however, that among females there were more 'low salt eaters' and among males more 'high salt eaters' (cf. Dahl 1958a).

Commencing in 1953, Dahl and Love questioned all members of the Brookhaven National Laboratory staff who reported to the staff clinic for annual physical examination. They were grouped according to salt intake and three groups were:

i) low intake: did not add and had never added salt to their food at table (135 subjects);
ii) average intake: added salt to their food if, after tasting, it was insufficiently salty (630 subjects);
iii) high intake: routinely added salt to foods customarily salted but did so before tasting them (581 subjects).

Dahl and Love (1954) note the limitations and deficiencies of such classification. They also consider that the group they studied was not a representative cross-section insofar as pre-employment policies would eliminate most applicants with severe diseases and so the population tested was probably healthier than the average. The age range of the subjects was 18 to 65 and males predominated.

The 105 hypertensives in the series were distributed as follows: low intake, 1; average intake, 43; high intake, 61. Dahl (1957) states that the probability of this distribution occurring by chance was found to be less than 1 in 1000 ($chi^2=16.1$, $P<0.001$).

Concurrent with this study, 24-h urine collections were made for between 6 and 38 days on 28 of the subjects. Groups of individuals who had been classified previously as having 'low' salt intakes showed significantly lower ($P<0.01$) average levels of urinary sodium excretion than individuals classified as having 'high' salt intakes. Individuals previously classified as having 'average' salt intakes did not differ statistically from the other two groups, being spread across them. When the 28 subjects in this study were reclassified into two groups according to the presence or absence of hypertension, it was found that the group with hypertension had significantly greater ($P<0.01$) sodium excretion than did those without hypertension.

In examining Dahl's figures, one noteworthy fact is that the mean 24-h urinary sodium excretion in the 'low salt' group was 155.9 ± 25.2 mmol and in the 'high salt' group 201.5 ± 30.5 mmol. The difference is not large and all subjects have, in fact, high sodium intake when considered against the data on the unacculturated and also transitional societies described above.

Dawber et al. (1967) did not find any correlation between hypertension and salt intake when the latter was assessed in the same way as by Dahl and Love. There was no evidence of blood pressure being related to 24-h urinary sodium output. Also there was little relation between a subject's statements on salt eating habits and his 24-h urinary output. Schlierf et al. (1980) found that in 383 men with normal blood pressure there was no relation between blood pressure and sodium intake as estimated from 24-h urine collections. They suggested there was probably high inter- and intra-individual variability in both sodium chloride intake and blood pressure. Liu et al. (1979) did not find any conclusive relation between blood pressure and salt intake in a relatively homogeneous population. There was large individual day to day variability in sodium excretion which highlights the problem of distinguishing one person from another.

Paul (1967) questioned 2000 men and found no correlation between salt eating habits and hypertension. He reported that frankly hypertensive individuals, given a choice of three test samples of soup with different salt concentrations, preferrred the two lower concentrations. Dahl (1972) in commenting on this draws the parallel with the genetically hypertension-prone rats studied by Wolf et al. which behaved in the same fashion when compared with a strain genetically resistant to hypertension (Wolf et al. 1965). This, however, tends to contradict the notion implicit in his Brookhaven study.

In one of the major comprehensive surveys in this field, Miall (1959) studied the general population of RhonddaFach, one of the Welsh coal-mining valleys. This was a follow-up study of the survey made initially on the same population by Miall and Oldham (1955). The influence of age on mean systolic and diastolic pressures in males and females is shown in Fig. 27-14. Amongst other interesting features, the study showed an important influence of an active reproductive life on the blood pressure of both women and men. Blood pressures were higher in childless women than women with a number of children. Similarly the blood pressure of men without children rose higher than that of those with families.

Miall also examined the influence of salt intake on blood pressures using the criterion suggested by Dahl and Love. He notes initially that some workers (for example Holley et al. 1951) found high serum sodium values in hypertensive subjects whereas others (Cottier et al. 1958) found no difference between hypertensives and normals. Miall found a sex difference in the response of the 953 people questioned: a smaller percentage of women than men were in the 'high salt' group and a larger proportion in the 'low salt' group, which is similar to the finding of Dahl and Love. They note that the difference in salt habits at the table may be due to loss of salt from sweating in men, or may simply indicate that women believe they have added sufficient salt in the kitchen. There was no suggestion of any systematic difference between the 'high salt' and 'low salt' groups of males in relation to the incidence of high blood pressure. However, among females there was a consistent difference between the two groups but in the opposite direction from that predicted by Dahl and Love for males: women who habitually added salt to their food showed lower systolic and diastolic pressures than those who did not. Not only had those women who habitually took extra salt lower blood pressures in 1958 than those not taking salt, but their pressures had increased less in the previous 4 years and except in old age they were lighter in weight. Separate analyses showed no correlation between parity and salt intake in women, nor between the energy expenditure of occupation and salt intake in men.

Miall determined the 24-h sodium output in urine for four groups of post-menopausal women in the sample. All women over age 50 with systolic pressures over 200 mmHg or under 150 mmHg were asked to provide 24-h urine specimens daily for a week. The completeness of the specimens was checked by measurement of urinary creatinine. Hypertensive women in this sample excreted less sodium than those with low pressure, whether in the high or the low salt intake group, but the

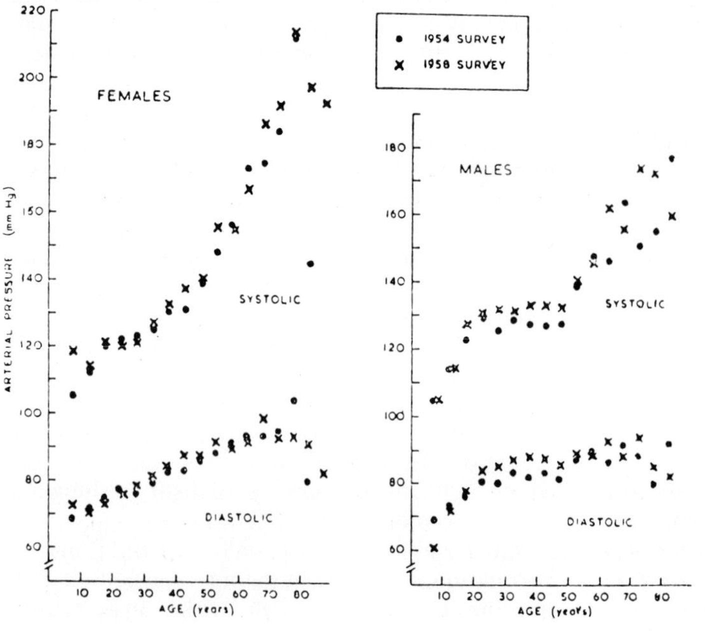

Fig. 27-14. Mean systolic and diastolic blood pressures by age in two surveys of the population of a Welsh mining valley. (From Miall 1959, with permission.)

differences were not significant at the 5% level. There was no difference in sodium output between hypertensive women who did or did not add extra salt to their food. Overall, the means of the four groups ranged from 122 mmol/24 h to 146 mmol/ 24 h. Thus the outputs were moderately high in all instances and statements about whether or not salt was added had virtually no influence on urinary sodium excretion levels actually found. Berglund (1980) draws attention to more recent Scandinavian studies suggesting a positive relation between salt intake and blood pressure in the normotensive part of the population but a significant negative relationship among the hypertensives. He notes that such studies have suggested by low-grade correlation coefficients that perhaps only about 10% of the variation in blood pressure could be attributed to variation in salt intake.

Phear (1958) investigated 20 male outpatients with hypertension (mean 219/127 mmHg, mean age 49.8 yrs) and 20 normotensives (mean age 56.3 yrs). The mean 24-h sodium excretion of the 20 men with hypertension was 156 mmol/day (s.d.=54); the mean excretion of the 20 with normal pressures was no different, being 154 mmol/day (s.d.=46).

Langford and Watson (1974) studied a population of 600 in Mississippi and were unable to show any relation between excretion of sodium, calcium and potassium, and blood pressure. Urinary Na/K of 97 black girls (4.26) was significantly higher than that of 101 white girls (2.98). In a detailed study on 108 black girls (aged 19–21 years), urinary excretion, overnight or per 24 h, did not predict the 6-day excretory pattern. Diastolic blood pressure was correlated with urinary Na/K ratio. Whereas a causal relation between sodium intake and occurrence of hypertension was not shown within this population, the authors proposed that in the salt-sensitive segment of the population blood pressure will be a direct function of sodium and an inverse function of potassium intake.

Simpson et al. (1978) found no evidence of a relation between blood pressure and sodium output or Na/K ratio in a population (Milton, New Zealand) where mean daily sodium excretion of men was 178 mmol and of women 140 mmol.

Hypertension and taste

The early work of the Harvard group showing that development of hypertension in rats by perinephric wrapping caused an aversion to salt was described in Chapter 22. The striking influence of endocrine disorders involving carbohydrate-active steroids—deficiency in Addison's disease and excess in Cushing's disease—upon chemoreceptor threshold and recognition of specific substances was also detailed (Chapter 10), and has been reviewed by Henkin (1975).

The evidence of the effects of these humoral disorders on taste and also olfactory physiology, as well as acceptance and rejection of mineral solutions, focuses attention on the question of any influence human hypertension might have on taste, olfaction, and acceptance and rejection behaviour. If humoral changes associated with hypertension caused any hedonic bias in salt intake it would be of importance to the matters considered in this chapter.

Animal studies

Wolf et al. (1965) studied the voluntary sodium intake in two strains of rats with opposite genetic susceptibility to experimental hypertension with salt intake (Dahl's S and R strains). The animals genetically predisposed to hypertension showed an aversion to sodium chloride even when tested when their blood pressure was only 10–20 mmHg above that of the control strain.

Catalanotto et al. (1972), by contrast, showed that when given free choice between sodium chloride and water, spontaneously hypertensive (SH) rats exhibited an increased preference for sodium chloride (0.15 M and 0.30 M) when compared with either Holtzman or Wistar normotensive controls. Also, the total fluid intake of the hypertensive animals was increased relative to the controls. The difference was substantial and sustained over experimental periods of 2 weeks. Further, McConnell and Henkin (1973) studied SH male rats and normotensive Wistar–Kyoto controls using a two-bottle preference test (0.3 M sodium chloride and water) for 16 weeks post weaning. Preference for the salt solution in the SH rats increased with age as blood pressure increased to the hypertensive range. Maximum levels of preference and hypertension were reached at 14 weeks of age and maintained thereafter. No changes in preference or blood pressure were observed in the normotensive controls during the same period. With SH rats in which preference for 0.3 M sodium chloride was measured for the first time at 14 weeks, and which had hypertension similar to that of the SH rats exposed to the salt solution from weaning, the increased preference for 0.3 M sodium chloride was the same. This behaviour was not associated with any increase in water intake. This increased preference for salt solutions of SH rats was seen with potassium salts as well as sodium but not with sour or bitter solutions of hydrochloric acid or quinine. It was only after the blood pressure of SH rats increased to levels that averaged 152 mmHg at 6 weeks of age that they showed a significantly increased preference for the salt solution compared with control rats. Thus it would be important to know the sodium excretion of the two groups before the onset of this increased preference in the SH group. But the situation was in marked contrast to the genetically salt-sensitive rats of Dahl.

Yamori et al. (1979) also found a large increased preference for 2% saline in stroke-prone and stroke-resistant SH rats. Stroke-resistant strains had less increase in blood pressure than stroke-prone strains and some impairment of renal salt handling was observed in the latter strains (Yamori et al. 1980). A low-protein diet aggravated the effects of high salt intake.

Fregly (1957) studied the influence of hypertension on the voluntary intake of 0.15 M sodium chloride by rats during pregnancy. Consistent with Richter's earlier data, mean intake of saline solution of control rats increased during the second week of pregnancy and reached a maximum during the third week. Intake remained high during week 4 at which time lactation was occurring. The sodium chloride intake of the pregnant hypertensive rats also increased during pregnancy and lactation; however, the absolute amount was less than in the controls so a relative salt aversion had developed despite an increase in sodium chloride intake. Implicit in this is the interesting fact that hypertension can modify the salt intake induced by the hormones of reproduction.

Forman and Falk (1979) have shown that sodium chloride solutions were more accepted and preferred by spontaneously hypertensive rats (SHR Okamoto strain) than by Wistar–Kyoto control rats. They showed, on the other hand, that rats with chronic renovascular hypertension produced by aortic ligation showed no difference from controls in sodium chloride acceptability. Thus a high blood pressure, of itself, appeared not to be the cause of increased salt acceptability in genetic hypertension.

Human studies

An early study of salt taste acuity was made at Johns Hopkins by Fallis et al. (1962) on a hypertensive and a control group, both groups mixed with respect to race and sex. The method of Richter and MacLean (1939) was used. The tests involved use of sugar and salt solutions, and distinction was on the basis of recognition of the solution as being different from distilled water, or actual identification of the specific taste. As regards the 'difference threshold' for salt, there was no difference between hypertensives and control subjects. However, the hypertensives did show a highly significant diminution in ability to identify saltiness as compared with their controls. In the case of sugar there was no difference in the threshold for difference or identification in relation to the two groups. The authors suggest that their data may indicate that hypertensives eat more salt because they cannot taste it as well as can normotensive individuals.

A study by Wotman et al. (1967) from Laragh's laboratory at Columbia University dealt with parotid and submaxillary secretions of hypertensive patients and also the salt taste threshold. Their subjects were derived from 60 ambulatory hypertensive patients (blood pressure above 140/90 mmHg). Over 90% had essential hypertension and more than half were receiving various drug treatments, including thiazide diuretics. Salivary flow and sodium concentration were lower in the hypertensives than in normal subjects. The authors also confirmed an elevated salt taste threshold as characteristic of patients with hypertension. Parenthetically, in relation to the reduced salivary flow, the effect of P113 ([Sar^1Ala^8]angiotensin II) would be interesting in the light of the report of McKinley et al. (1979) showing the rapid augmentation of salivary flow in the sodium-deficient sheep following administration of P113.

Henkin (1974) reported different results. Of 50 patients with essential hypertension, only four exhibited elevations of either detection or recognition thresholds for sodium chloride. Of 10 patients with primary aldosteronism, only one exhibited an elevation in taste threshold. There was no difference in salt taste acuity in the patients studied before operation when hypertensive or postoperatively after blood pressure had decreased significantly. In discussing these data and methodology, Henkin notes that if thresholds are measured only in the early morning hours the results will reflect higher values consistent with the circadian variation demonstrated for detection of sodium chloride. He also notes that ensuring an adequate time has elapsed after the last meal or usage of tobacco is an important aspect of achieving uniform test results. Otherwise a finding of higher than normal threshold can be made. Diurnal variation in taste acuity for salt was shown in relation to food intake by Irvin and Goetzl (1952). Free selection of meals was found to be preceded by a period of increasing and followed by one of decreasing acuity of salt taste. When meals were omitted the decrease in acuity of threshold for sodium chloride did not occur.

Jackson (1967) compared recognition threshold of salt in 80 people smoking at least 40 American cigarettes or the equivalent in cigars or pipe tobacco each day with that of 80 control non-smokers. Smokers required concentrations 12–14 times greater than required for recognition by 85% of the controls. It was stated that about four-fifths of the heavy smokers had a history of hypertension, but no data were cited. Krut et al. (1961) reported that smoking did not influence taste thresholds.

Though this issue of smoking, hypogeusia and salt taste appears unresolved, it is interesting to recall here some general observations made by Smith (1953). He notes that the physiological mechanisms by which salt influences flavours are obscure. Flavour is generally held to be a combination of gustatory and olfactory sensation. The four primary taste stimuli are therefore modified by what is simultaneously smelled, and the total of the stimuli blend to cause flavour. One prime effect of sodium chloride as a seasoning agent is to suppress 'rawness' and 'sourness' as the first step in the correction of undesirable tastes, and the flavour may be then controlled by further addition of sugar, pepper or other seasoning. Pepper and salt seasoning increases the total volume of flavour, suppressing some of the natural flavour without destroying identity, and accentuates desirable flavours. Monosodium glutamate, an adjunctive seasoning, also is capable of modifying rawness and other undesirable tastes. The possibility arises that an increase in salinity of foodstuffs influences odour during mastication. Since odours are more sharply perceived in substances with high vapour density, it may be suggested that salt tends to mobilize materials in the food which then escape and diffuse to the olfactory mucosa. If olfaction is augmented there may be a marked increase in flavour.

Schechter et al. (1974) investigated sodium chloride preference in 16 patients with essential hypertension and 26 normotensive volunteers. Each individual was admitted to hospital and placed on a constant diet containing 9 mmol of sodium per day for 4–7 days. The only source of fluids available to either patients or volunteers was the contents of two 2-litre flasks, one filled with distilled water and the other with a solution of 0.15 M sodium chloride. The subjects were unaware of which flask contained which fluid and they drank as little or as much of each as they desired. Patients with essential hypertension drank a significantly greater portion of their daily fluid as saline, the mean preference for saline being 2.5–2.8 times greater among the untreated hypertensive patients than the volunteers. The mean total fluid intake of the hypertensive patients was also significantly greater than that of the normal volunteers. Overall the daily mean sodium intake of the hypertensive patients was 4.8–7.3 times greater than that of the volunteers. There appeared to be no difference between the white and black patients in the results, but as Simpson (1979) points out the groups were poorly matched in this respect and also for age, although the authors propose data to suggest age did not influence the result. There were no data to determine whether the increased salt intake observed in hypertensives reflected an acquired preference which pre-dated any cardiovascular changes.

Swaye et al. (1972) compared 717 hypertensive and 819 normotensive subjects. They found a significant correlation between family history and hypertension, but no difference in the habits of salting food between the two groups.

Lauer et al. (1976) studied three groups of children aged 11–16 years on the basis of their mean blood pressure. One group was a random sample of 33 subjects from those children with blood pressures equal to or greater than the 95th percentile. The second group had pressures ranging from the 47.5th to 52.5th percentile and a third group of 33 subjects had pressures equal to or less than the 5th percentile. These values were 104±4 mmHg, 88±2 mmHg and 67±7 mmHg respectively. There were equal numbers of boys and girls in each group and no statistically significant sex-related differences were found in the blood pressure measurements between the two sexes. Salt threshold was determined by titrating solutions ranging from 1 to 60 mmol/l of sodium chloride on each subjects's tongue. Salt preference was tested by addition of salt by each subject to unsalted tomato juice and beef broth to individual taste. The samples were then analysed for sodium concentration.

There was no relationship between salt threshold and preference, nor did threshold or preference relate to blood pressure. Relative body weight was related to blood pressure, subjects in the highest pressure range being the most obese. The authors draw attention to Mayer's (1968) results from careful measurements of energy intake of obese children showing that they do not ingest more calories than their non-obese peers. The explanation lies in a difference in activity. For this reason they see it as difficult to implicate salt in the hypertension associated with childhood obesity if one follows the line of reasoning of Dahl that obese subjects eat more food and thereby ingest more salt.

Bernard et al. (1980) reported no difference between hypertensive and normal individuals in relation to salt intake. However, a subgroup of low-renin hypertensives had a significantly higher salt intake and gave higher hedonic ratings than matched normotensive controls. The question arises as to whether they were hypertensives with low renin because of hedonically determined high salt intake.

Contreras (1978) commenting on the discrepancies in the results with humans suggests that one obvious reason for confusion is the lack of procedural uniformity in experiments on taste acuity. Discrepancies could be attributed to the technique used in applying the stimulus. Area and duration of stimulation and temperature of solution are important variables affecting thresholds. Thus the drop technique used by Henkin probably does not stimulate the receptor surface to the same extent and degree as the sip method used by other investigators. Another criticism may be that some experiments used threshold techniques that are not controlled for adaptation. Water and solute tastes may be confused. Gustatory sensitivity may be influenced by medication effects. There is also the consideration that most useful information may not be in threshold determinations, which reflect only a small portion of the intensity domain. The evaluation of sensitivity to suprathreshold stimuli has not been made. Threshold measures cannot be used to predict the shape or slope of the stimulus concentration–sensitivity function since the slope is independent of the detection intercept.

Contreras goes on to note that Moskowitz and Abramson (1976) have used magnitude estimation techniques to quantify perceived saltiness of salted soups. The subjects were asked to select numbers from a limited scale to indicate gradations in perceived stimulus intensity. They found that hypertensive individuals tended to perceive suprathreshold salt stimuli as less intense than did controls. Using hedonic ratings, these same salt stimuli were also perceived as tasting more pleasant to hypertensive subjects than controls. Contreras raises the question of whether, with more comprehensive investigation along these lines, the evidence will reveal that hypertensive humans, like sodium-deficient rats, are less sensitive to the taste of salt.

Amelioration of hypertension by low salt intake

Ambard and Beaujard (1904) and Allen and Sherrill (1922) recommended restriction of sodium intake in the treatment of hypertension. Kempner (1944, 1948) set out his regime of the rice–fruit diet containing not more than 5 mmol/day of sodium and documented the changes in urine and plasma composition, weight loss, and the reduction in blood pressure produced in patients with hypertensive vascular disease. These changes were associated with a reduction in heart size and recession of retinopathy. The 24-h urinary sodium excretion and urinary Na/K ratio monitored on patients on the regime resembled those reported in unacculturated societies.

Grollman et al. (1945), based on the results of Grollman and Harrison (1945) showing a reduction in the blood pressure of hypertensive rats by rigid sodium restriction, advocated sodium restriction in the treatment of clinical hypertension, as did Volhard (1931). Dole et al. (1950) confirmed the benefit of the rice–fruit diet with a series of objective criteria in five of six cases of uncomplicated hypertension. With the diet, 3–5 months were required to reach a steady weight. Restriction of sodium but not of chloride was necessary for the antihypertensive effect. In a further analysis Dole et al. (1951) showed that reduction of sodium intake to 7 mmol/day caused significant reductions in average blood pressure. These reductions were due to limitation of dietary sodium, since during a preliminary equilibrating stage of 2–4 months, when 180 mmol of sodium chloride was given per day, no fall in blood pressure occurred. Thus the weight loss over this period, and the large variation in protein intake, were without effect on blood pressure. When sodium was withdrawn the patients lost 14%–20% of their exchangeable sodium during the first month of adaptation.

Dahl and Love (1954) in studying obese hypertensive individuals reported that weight reduction per se was not commonly associated with a fall in blood pressure, but a fall in blood pressure was correlated with curtailment of salt intake. Further it was suggested that the obese person, in eating more food, thereby also ate more salt and correspondingly had more hypertension.

Kempner's data were criticized by Ayman (1949) on the basis of lack of control and diversity of clinical material. Corcoran, Taylor and Page (1951) reported that it was necessary to restrict sodium to 20 mmol/day to achieve a response and that this was difficult if not impractical under outpatient conditions. Pickering (1980) considers the use of Kempner's diet quite impractical. Nye and Forrest (1951), noting the monotony of the diet, advocated a low-sodium diet based on tabulated values for vegetables and fruit as basic components. An interesting feature of this early paper was the attention drawn to familial incidence of susceptibility to lead poisoning, the authors raising the question of whether some degree of individual idiosyncrasy to sodium explains why hypertension is so frequently observed in several members of the same family. Renwick et al. (1955) showed that renal response and body weight changes in the acute phase of sodium withdrawal were similar in hypertensive and normotensive subjects.

The overall picture from a series of studies during 1945–55 appears to be that one-third of patients showed an excellent response to a low-sodium diet, one-third had a modest response and with the remainder there was little effect, though Dahl (1972) is of the view that even those whose blood pressure did not decline had a more benign clinical course on the low sodium intake. Murphy (1950) and Watkin et al. (1950) showed that the low-sodium diet caused a decrease in blood and extracellular volume. In relation to the influence of obesity, which may be complex, Kannel et al. (1967) in the Framingham study found with 5127 adult men and women that relative body weight, weight change under observation, and skinfold thickness were related to existing blood pressure, and to subsequent rate of development of hypertension. The risk of hypertension in normotensive individuals was proportional to the degree of overweight. Pickering (1968) states there is a definite correlation between blood pressure and body weight/height ratios. Stamler et al. (1978), in a survey of over a million people, showed the overweight section had a much higher percentage of hypertension than those of normal weight. Reisin et al. (1978) record that loss of weight without salt restriction lowers blood pressure in obese individuals. Amery et al. (1980) reviewed six intervention trials in hypertensive patients, and concluded that a decrease in body weight of 1 kg correlated with a decrease in blood pressure of 3 mmHg systolic and 2 mmHg diastolic for weight losses up to 10 kg. Tuck et al. (1981) studied 25 obese patients on a reducing diet which contained either 120 or 40 mmol per day of sodium. Plasma renin activity and plasma aldosterone declined with weight loss regardless of sodium intake. Mean arterial blood pressure fell significantly and equivalently in both groups. Loss of 20 kg in both groups resulted in 15–20 mmHg reduction in mean blood pressure. The authors consider reduction of plasma renin activity may contribute to decline in blood pressure.

Consideration of reduction of sodium in the diet brings us back to the question of whether hypertensives are found to excrete more sodium than normotensives within one population. Simpson et al. (1978) found the mean urinary sodium output of men and women referred to their hypertensive clinic was no greater than the mean values in the population of the town of Milton, New Zealand ($n=1206$) which they studied. Doyle et al. (1976) found that hypertensive subjects had a higher intake of sodium than normotensives.

Simpson (1979) is critical of much of the data which suggest a correlation between salt intake and hypertension. He notes that Morgan et al. (1975) state that the sodium output of normotensives ($n=50$, mean 111 mmol/day) is lower than that of mild hypertensives ($n=400$, mean 148 mmol/day); but while these groups are stated to be exactly comparable, no data are cited—not even the male to female ratio. The sodium output of 150 patients referred to Morgan et al.'s hypertension clinic (mean 185 mmol/day) is stated to be significantly different from that of the normotensives in the population survey, but, whereas the hypertensives are said to be from the same population, they are all male. Virtually all surveys show males have higher sodium excretion than females. Simpson notes that in a later study (Morgan et al. 1978) the sodium output of a more comparable group of relatively normotensive men and another group of mildly hypertensive men was found to be 191 mmol/day in both cases. In this study, sodium output was reduced from 191 to 156 mmol/day in a group of 31 men by sodium restriction, and a fall in blood pressure occurred relative to placebo treatment. Kawasaki et al. (1978) in a study on 18 hypertensive patients showed some patients responded to a reduction in salt intake by a fall in blood pressure.

Parijs et al. (1973) reported studies involving use of diuretics or low sodium diet or the two combined, from which a main conclusion was that reduction of sodium chloride intake from 10 to 5 g/day produced a fall in blood pressure of about 10/5 mmHg.

The data of Joossens et al. (1970) linking sodium intake and blood pressure are criticized by Simpson on the basis that they are not homogeneous but composed of three subgroups, one of which comprised all elderly people, who had a particularly high sodium/creatinine ratio in their urine. In this survey of 2027 people Simpson sees bias shown to sodium rather than body weight or age. Joossens (1979) contests this statement, saying there was a significant correlation between systolic blood pressure and salt intake independent of weight, height, pulse rate, urinary creatinine and age for males. Males were the larger group of this study in Lueven, The Netherlands, numbering 1314. The study included active and elderly people living in their own homes and they were part of a homogeneous group. The lower creatinine in the elderly was not surprising and is one of the reasons for including it as an independent variable. The important point from the Lueven study was that moderate reduction of salt intake from 12 to about 6 g/day produced a significant fall in blood pressure, independent of and additive to the influence of diuretics. There was, however, a weight loss.

In summing up his position on this issue of salt and hypertension, Simpson (1979) concludes that there is no good evidence within a population that hypertensive individuals habitually take more salt than the normotensives. He suggests that those who favour a relation of this type have the following fall-back positions, and he comments accordingly:

i) There is too much noise in the data. This assertion has some truth in it.
ii) There is too little variation in salt intake within a population. This is manifestly not true as sodium intake can vary from less than 70 to over 300 mmol/day.
iii) There is genetic variability in susceptibility to salt. This is true in rats and may be true in man, though it is surprising that it completely obscures the influence of salt eating habits.
iv) Once intake is over 60 mmol/day it does not matter how much more is taken (this was suggested by Freis 1976). This is a real possibility, and is highly relevant to (iii) above. Simpson remarks that it is, however, seemingly incompatible with the reductions in blood pressure claimed with minor reductions in salt intake such as halving the average, which would bring salt intake down to about 90–100 mmol/day in many countries.
v) The seeds of hypertension are sown in childhood and adolescence so that salt intake in later life will not affect blood pressure. Again this would mean that manipulation of salt intake in later life would be unprofitable.

Simpson suggests that apart from (ii) none of these explanations is impossible; they reflect the resistance of the salt–hypertension hypothesis, and how negative data can be explained away.

Overall, in his analysis, he appears to me to concede that the very high intake in north-eastern Honshu in Japan (350–450 mmol/day) has a relation to blood pressure and mortality from stroke. Though in his eyes there are some awkward facts in the studies of unacculturated societies and the causal connection between low blood pressure and low salt intake in terms of differing life-styles between communities, he does not appear to challenge the basic coincidence of low salt intake and the different pattern of blood pressure with age as reported in a considerable diversity of primitive communities. Thus the issue is the influence of salt intake in the range of 100–300 mmol/day on blood pressure in a Westernized society.

Joossens et al. (1979) have presented interesting data in relation to the overall decline in death rate from stroke and its association with decline in death rate from stomach cancer. This is seen between different countries. Thus plotting age-adjusted death rates from strokes against those for stomach cancer in 24 countries in 1974 gives a regression line with the equation: cerebrovascular accident=1.84+6.53×stomach cancer (r=0.81, P<0.002) (Fig. 27-15). Within the United States over time (Fig. 27-16), there is a similar regression line when annual death rates from cerebrovascular accident are plotted against those for stomach cancer from 1950 to 1974. Whereas the association could be spurious, the authors note the slope and intercept are of the same order of magnitude between different countries and within the USA over time. The probability of this occurring by chance is extremely low. The correlations derived by Joossens (1980) for England and Wales (Fig. 27-17), Japan (Fig. 27-18) and Switzerland (Fig. 27-19) show the same form. He contends that the lower stroke rate than that predicted from the stomach cancer rate for the years before 1955

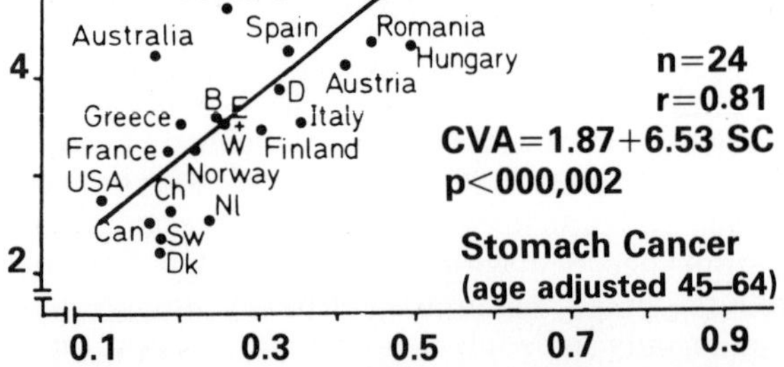

Fig. 27-15. Between-countries relationship between death rates (per thousand) from cerebrovascular accidents and stomach cancer in 1974. (From Joossens 1980, with permission.)

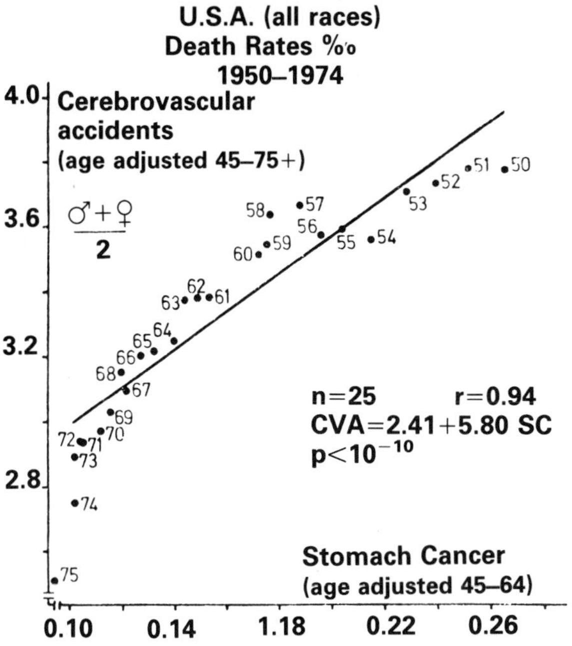

Fig. 27-16. Between-time-periods data on cerebrovascular accidents and stomach cancer averaged for both sexes. These data from the USA are applicable to both men and women. The year of observation is indicated next to each point. (From Joossens et al. 1980, with permission.)

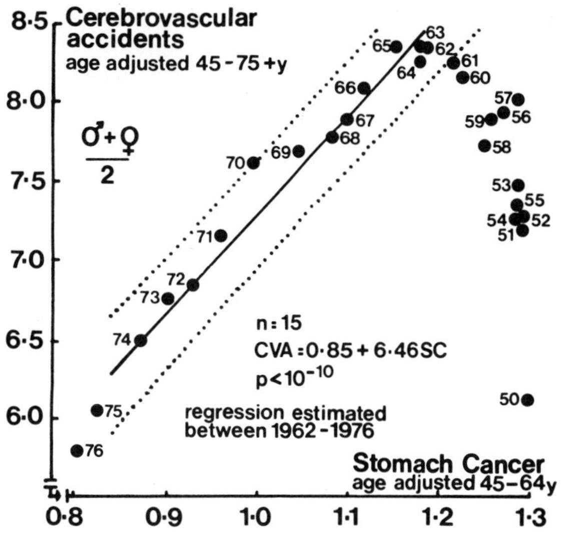

Fig. 27-18. Between-time-periods data relating cerebrovascular accidents and stomach cancer in Japan (1950–1976). (From Joossens et al. 1980, with permission.)

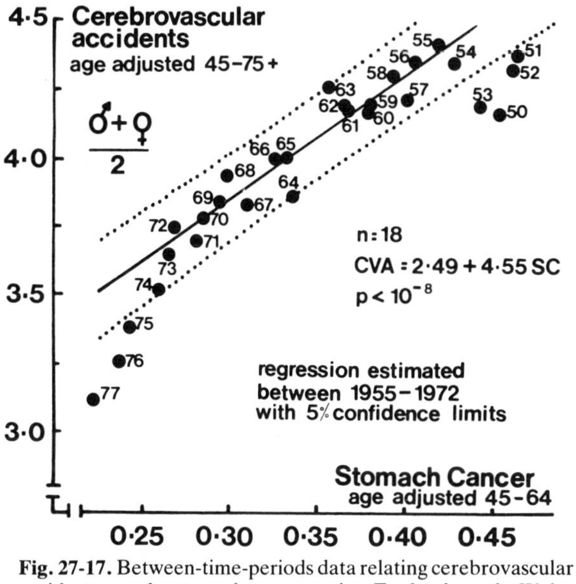

Fig. 27-17. Between-time-periods data relating cerebrovascular accidents and stomach cancer in England and Wales (1950–1977). (From Joossens et al. 1980, with permission.)

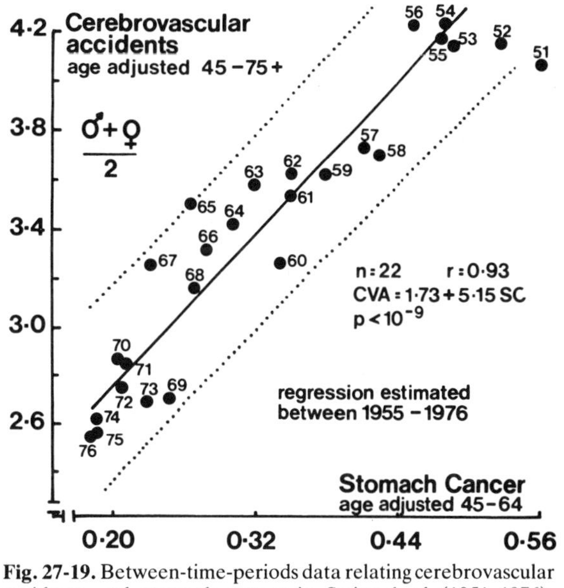

Fig. 27-19. Between-time-periods data relating cerebrovascular accidents and stomach cancer in Switzerland (1951–1976). (From Joossens et al. 1980, with permission.)

(1962 in Japan) is almost certainly due to underclassification of stroke. This was already documented in the USA in 1948 when the change from the fifth to the sixth revision produced a drastic increase in the classification of stroke that is evident after the linear part of the curves (ca 1955–68). From 1973 onwards in the UK, the USA and West Germany at least, the new deviation from linearity is most probably due to treatment of hypertension, which began increasing rapidly from 1968 onwards. The treatment would influence stroke but would be unlikely to influence stomach cancer.

In relation to the linear part of the curves, the

most important argument against a spurious origin of the stroke/stomach cancer relation is the similarity of the regression lines between time periods, between regions, and combined. Joossens suggests that an almost identical relation which holds under conditions so varied must be a relatively simple one. That is, each phenomenon separately can have a multifactorial causation, but the link between them must be either unique or at least predominant. A multifactorial link could not produce identical results under different conditions: the more factors involved in the link, the less chance there is of obtaining identical relations between and within countries. The linking factor, 'X', involves the stomach, as well as brain arteries, and may therefore be nutritional and should have the following properties:

i) It should raise blood pressure as this is the most important risk factor in stroke.
ii) It should tend to induce stomach cancer itself or facilitate the induction.
iii) It should be present in Western, Eastern European and Oriental diets. The concentration should be high in Japan (Fig. 27-18: note ordinate and abcissa scales compared with those in Figs. 27-17 and 27-19) and South Korea, medium in Eastern Europe and much lower in the West.
iv) It should be present in cereals, lard, sausages, miso soup, smoked and pickled foods, nearly absent in milk, and absent in fresh fruit and vegetables.
v) It should decrease with time, and in the USA this decline should begin around 1925.

Joossens states that the soil trace element theory of stomach cancer (Stocks and Davies 1964) might explain differences between regions and countries but not what is happening with time. This is also true in relation to blood groups and the greater incidence of cancer in blood group A, which cannot explain a decline quantitatively identical to the between-region relationship.

Joossens (1980) holds that factor X is salt. He cites well-documented material, also set out in this chapter, on epidemiology, animal experiments and amelioration of hypertension by restriction of salt, and notes the gradation in intake from Japan and South Korea with their extremely high levels through the intermediate stages of Eastern Europe to the lower level in the West, with the exception of Portugal (Miguel and Padua 1980) where intake is high. He records the evidence on measured decline in salt intake in Belgium, Japan and Switzerland over the past 20 years. Joossens et al. (1979) state that, based on 24-h urine specimens, average salt intake declined in Belgium from 15.3 g/day in 1966 to 8.2 g/day in 1978. He proposes the decrease, probable in all industrial countries, results from introduction of cooling techniques in the preservation of foods, which reduces intake of salted fish, meat, vegetables and cereals. The relation between salt intake and stroke clearly could be related to the influence of salt on blood pressure. The relation of salt and stomach cancer could involve the fact that reduction of food preservation by salting may involve a reduction, for example, of nitrosamines in diet as well as sodium chloride per se.

Joossens (1979) notes that salt by virtue of its influence on gastric osmotic pressure could damage stomach mucosa. Also, osmoreceptors at the entry of the duodenum could delay exit of a meal from the stomach for hours and contribute to atrophic gastritis which is, for example, a common condition in Japan. With diminished acidity in the stomach, high levels of nitrite may result and this can combine with food amines to produce nitrosamines leading to stomach cancer. Countries with documented high levels of salt intake like Japan, Portugal and Bulgaria and other Eastern European countries have the highest incidence of stroke and stomach cancer.

Cooling as a means of preserving food became popular in the US around 1925, in Western Europe after the Second World War, and in Eastern Europe after 1970. Joossens (1980) concludes by suggesting that, if his hypothesis be true, it follows that the most important contribution to public health in the field of non-infectious chronic disease over the last 30–40 years has been the mass introduction of refrigerators and deep-freezes.

Animal studies of induction of hypertension by high salt intake

Possibly the first animal experimental evidence of the role of salt in hypertension came from Grollman, Harrison and Williams (1940) who showed that various sterols were hypertensinogenic only if extra salt were eaten. By providing hypertonic saline as the sole source of fluid, Lenel et al. (1948) produced hypertension in the chicken. Sapirstein et al. (1950) showed that hypertension with cardiac hypertrophy could be produced in the rat by substituting 10% saline for drinking water.

A series of papers by Meneely and colleagues and Dahl and colleagues examined the influence of increasing amounts of salt in the diet of the rat on blood pressure (Meneely et al. 1953; Dahl 1960; Meneely and Dahl 1961). The experiments were conducted and designed as life span procedures.

Both male and female Sprague–Dawley rats were used. Distilled or demineralized water was freely available to the animals and the diets offered to the groups contained differing levels of sodium chloride. The authors regard three ranges as most revealing and realistically related to the human problem. The first range, the control level, was from 0.15% to 2%. The second range examined was 2.8%–5.6% which represented a moderate excess of sodium chloride. Finally 7%–9.8% sodium chloride in the diet represented a high level of excess salt.

Consistent and reproducible blood pressure changes were well established after 9 months of exposure to the diet (Fig. 27-20). Although mean blood pressure of each group was successively higher, and proportional to the sodium chloride content of the diet, there was a considerable scatter of individual values. Mean systolic blood pressure at control levels of salt was approximately 122 mmHg. At the lower level of excess salt feeding (2.8%) the mean blood pressure was 130 mmHg, but there were individual animals in this group that had blood pressures as high as 160 mmHg and others with blood pressures as low as 114 mmHg. Similarly at the high level of salt intake (9.8%) mean blood pressure for the whole group was 152 mmHg but there were rats with pressures as high as 205 mmHg and one rat had a pressure of

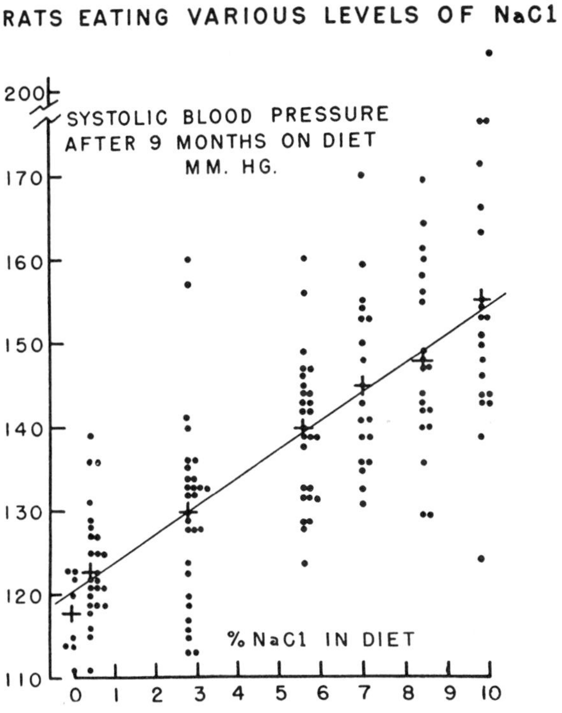

Fig. 27-20. Hypertensinogenic action of dietary sodium chloride in male rats. (After Ball and Meneely 1957, with permission.)

only 125 mmHg. Animals which were above or below the mean tended to remain in this relative position throughout life. These early experiments thus showed substantial differences in susceptibility to hypertension.

The most sensitive index of adverse effects of sodium chloride upon these rats was the survival time. No significant difference in survival was seen between control animals and those eating the lower levels of high salt (2.8%–5.6%) until the seventeenth month of the experiment. Thereafter the difference increased. The median duration of life for rats eating 5.6% sodium chloride was many months less than that of controls. With the rats eating 8.4% sodium chloride the median duration of survival was 8 months less than the controls. Meneely and Dahl (1961) suggest that the time at which the differences appeared would correspond roughly to age 51 in a human if 10 days for a rat is equivalent to a year for man.

In a subsequent experiment the authors showed that if these experiments were carried out on rats which were already 2 years of age, and which were placed on test diets containing 5.6% or 9.8% sodium chloride, the animals gradually became hypertensive but the effect was not as great as when the experiment was begun on young animals. If the procedure was carried out on young female rats, the level of blood pressure obtained was not as great as with male rats.

In a further series of experiments the procedure was modified by the addition of potassium chloride as well as increased amounts of sodium chloride. The protective effect of potassium chloride was dramatic. For example, when potassium chloride was added to the diet containing 5.6% sodium chloride there was no change in the hypertensinogenic action of the additional sodium chloride but there was a great improvement in survival, the median duration of life being increased by 7 months. At the very high levels of salt feeding the extra potassium did ameliorate the hypertensinogenic action of sodium chloride so that only moderate hypertension developed. Again the median duration of life was increased, by approximately 8 months. This role of potassium in ameliorating the hypertensinogenic action of sodium chloride has been discussed in detail in relation to physiological mechanisms by Meneely and Battarbee (1976a, b).

In commenting on the applicability of their finding to humans, the authors remarked that the levels of salt used may seem to be outside of the range of human consumption. Simpson (1979) makes the same point, remarking that a rat eats about 100 g food per kilogram of body weight per day and therefore the daily sodium intake in an experiment with an 8% sodium chloride diet is about 0.8% of body weight. For a 70-kg man this would represent 560 g of sodium chloride per day. However, Simpson also points out that there is a smaller turnover of food in a man, so that 8% sodium chloride in the dry diet would mean an intake of about 40 g or 680 mmol per day. In this regard Meneely and Dahl (1961) state that their levels were compared on a percentage of nutriment basis. Referring to Ball and Meneely (1957) they note that the average intake of sodium chloride for a normal adult is estimated at 7–15 g per day. A diet of 2500 kcal per day containing 500 g nutriment (without water and inert material) might therefore contain from 1.4% to 3% sodium chloride. Thus in the rat experiments the 2.8% sodium chloride approximately corresponded to the normal intake of 14 g sodium chloride daily in man; the control rat diet of 0.15%–2% to 0.75–10 g sodium chloride; and the moderate excess range (2.8%–5.6%) to 14–28 g per day. This last diet was frankly hypertensinogenic and life-shortening. Finally, the high level of excess salt (7%–9.8%) was equivalent to 35–49 g daily which would be a great excess and perhaps only reached in occasional individuals in northern Japan.

Battarbee et al. (1979) showed that Sprague–Dawley rats fed a moderately high sodium diet became hypertensive and that urinary excretion of catecholamines was moderately increased. Inclusion of a small potassium supplement in the diet ameliorated both hypertension and catecholamine excretion.

It is pertinent to cite some interesting new data on man at this point. Skrabal et al. (1981) studied 20 normotensive subjects (10 with a family history of hypertension) to determine whether moderate salt restriction and/or a high potassium diet had a beneficial effect on blood pressure regulation and prevention of hypertension.

High potassium intake (200 mmol/day) was contrived by inclusion in the diet of large amounts of fruit and vegetables, by use of a commercial salt substitute containing a mixture of potassium salts, mainly potassium chloride, and by taking daily a tablet of Kalinor (Nordmark-Werke) containing 40 mmol of potassium. In all subjects moderate reduction of sodium intake from 200 to 50 mmol/day over 2 weeks reduced the rise in blood pressure produced by various doses of noradrenaline. Of the 20 subjects it was found that 12 (8 with a family history of hypertension) had a fall in systolic or diastolic blood pressure of at least 5 mmHg with sodium restriction. There were no significant differences in plasma renin, aldosterone, antidiuretic hormone and catecholamine levels between responders (salt-sensitive subjects) and non-responders, but the salt-sensitive subjects had mean baseline diastolic pressure which was higher than that of salt-insensitive subjects by 13 mmHg. A high potassium intake reduced diastolic blood pressure by at least 5 mmHg in 10 of 20 subjects, of whom 7 of the 10 had a family history of hypertension and 9 responded to sodium restriction. The high potassium diet improved compliance with the low sodium regime, promoted sodium loss and increased sensitivity of the baroreceptor reflex which was postulated as indicative of how high potassium intake reduced blood pressure. In all subjects 2 weeks of the low sodium/high potassium diet reduced blood pressure rises caused by mental stress or noradrenaline infusion by 10 mmHg. Measurement in 5 subjects showed the low sodium/high potassium diet reduced sodium space by 3.91±0.34 litres (±s.e.m.; $P<0.005$) while weight loss was about 1 kg. Identification of salt-sensitive subjects with hypertension was also feasible in the study of Fujita et al. (1980).

Lemley-Stone et al. (1961) found that rats fed diets containing more than 2% sodium had increased sodium space, and increased exchangeable sodium and tissue sodium. Sodium space, exchangeable sodium and tissue sodium concentrations were not elevated in animals fed the same high levels of sodium chloride with potassium chloride added to maintain a less distorted Na/K ratio (i.e. about 1/1). Aoki et al. (1972), however, found no difference in blood pressure of spontaneously hypertensive rats of the Japanese strain when fed a 2.7% sodium chloride diet compared with a low sodium intake. But if 1% sodium chloride was also given as drinking fluid, the development of hypertension was accelerated and it was greater in degree. Simpson (1979) notes the New Zealand strain of genetically hypertensive rats is equally resistant to salt. Aoki et al. (1972) suggest that hypertension in spontaneously hypertensive rats is genetically determined to a major extent but may be altered in degree by salt intake.

Dahl's experiments with rats genetically prone to hypertension in the face of increased salt intake will be discussed further on. But it is relevant here, inrelation to the effect on rats of a human level of salt intake, to record the Dahl et al. (1970) experiment on the influence of consumption of commercial baby foods on blood pressure of these animals. Fifteen control animals of this strain which were litter-mates to the test animals ate special chow pellets containing 0.4% sodium chloride. The test group of 25 rats had access to commercially available baby foods which on analysis were shown to contain 70–220 mmol/kg of sodium. After 1 month the average blood pressure of the group eating the baby foods was higher than controls: 130.6 vs 119 mmHg. The difference increased thereafter and after 6 months averaged 190.2 vs 138.2 mmHg. Of the test rats, 12 died and 2 others became seriously ill after 8 months of study, whereas the animals maintained on the low sodium were all alive at 8 months.

While considering the data on rats, it is of great interest to recall a paper of Grollman and Grollman (1962) which does not seem to have received much attention. Various treatments to female rats were begun at specified times before, at the beginning of or during the course of pregnancy. After parturition, therapy was discontinued and the animal returned to its normal diet (commercial rat chow and tap water). The young were weaned at 20 days. The blood pressures of the offspring were determined in the unanaesthetized state at approximately monthly intervals, beginning at the age of 6 months for periods of up to 2 years, at which time their kidneys and hearts were weighed and examined microscopically at autopsy. As controls, untreated animals of the colony of the same strain as the experimental group were used.

In 10 litters involving 88 offspring, control mean blood pressure at 1 year of age was 114±7.4 mmHg. It was found that the provision of 1% sodium chloride as drinking water, or 0.1 mg desoxycorticosterone daily intramuscularly, or 5% ethanol in drinking water from the 7th day before pregnancy had no influence on the blood pressure of the young. However, the provision of a high salt intake (amount not specified) from 7 days before conception or from the 5th day of pregnancy onwards resulted in 34 offspring having a mean blood pressure of 150 mmHg. When provided from the 14th day of pregnancy in 12 offspring, the mean blood pressure 1 year later was 127±3.9 mmHg. A low-potassium diet begun at either 5 days before conception, at conception or at the 7th day of pregnancy also had a hypertensinogenic effect on the young. Administration of chlorthiazide during the course of pregnancy had this effect too, a fact which the Grollmans attributed to the drug producing potassium deficiency. Several corticosteroids, aldosterone, desoxycorticosterone, progesterone and cortisone also had this effect when given in high doses during the course of pregnancy. Post-mortem examination revealed hypertrophy of the left ventricle and histological examination of the kidneys revealed no morphologically evident lesion or any evidence of excretory renal insufficiency.

The authors pointed out that previous studies on the experimental induction of teratogenic disorders had dealt with structural malformations evident at birth. Their experiments demonstrated for the first time that prenatal measures may induce a disorder. They note that certain of the drugs had been used clinically as prophylactics against abortion and eclampsia and speculate that environmental factors affecting the foetus before birth might be relevant to instances of spontaneously occurring essential hypertension in the human. The embryo could be highly susceptible to environmental influences which exert little or no detrimental influence on the maternal organism.

Recently an important report has come from Contreras and colleagues (1981) who examined the effect of salt consumption by female rats during pregnancy on the salt preference of offspring in later life. Adult females consumed a diet containing either 0.12, 1 or 3% sodium chloride during pregnancy and their offspring were continued on the diet until day 30 of life. Thereafter all were maintained on a 1% sodium chloride diet and preference for salt was tested at 3 months of age by a two-bottle procedure. The male offspring with history of a high-salt diet showed a significantly stronger preference for saline than the other two groups, but this was not seen in the females. The explanation of the difference was not clear. When tested at 160 days under conditions where all rats had been on the same diet since day 30, blood pressures of the rats that were originally on the high-salt diet was significantly greater than those of the groups on lower sodium diets.

In the rabbit, Goldblatt (1969) did not find an increase in blood pressure with high salt intake.

Whitworth et al. (1979) studied the influence of sodium loading on normal sheep and sheep with reduced renal mass in the course of an investigation of hypertension generated by adrenocorticotrophic hormone. The sodium loading was high, being produced by adding 340 mmol sodium to the daily food and offering isotonic saline to drink in place of water. With this regime systolic, diastolic and mean arterial pressure increased. By 7 days mean arterial pressure was 13 mmHg above control. In sheep with reduced renal mass this rise was 26 mmHg after 7 days. The effect reversed on returning to normal sodium intake. It would be interesting to examine lower sodium chloride intake over much longer periods, but in this regard the data of Potter (1972) suggest that a sheep may be fairly resistant to the induction of hypertension by salt. Animals drinking 2% sodium chloride over 6 months to 4 years do not exhibit hypertension. In an unpublished study of Abraham, Baker and Denton (see Denton 1973) the pressor effect of angiotensin was examined in 3-year-old sheep which had been provided with drinking water of high salt content (270 mmol/l) since weaning. The data did not suggest that the pressor response of these animals differed from normal, nor were the control blood pressures remarkable in any way, though the question would warrant much more extensive quantitative investigation.

An important study on salt loading insofar as it involves a primate—the baboon (*Papio hamadryas*)—has been made by Cherchovich et al. (1976). Working at Sukhumi in the USSR they studied three groups of animals. Group one contained four young males born of mothers fed a high-salt diet during pregnancy and lactation and which themselves were fed a high-salt diet up till age 1 year. Group two contained eight males and eight females started on a high-salt diet at age 1.5 years. Ten members of this group were unilaterally nephrectomized at the start of the experiment. Group three contained four males and five females receiving a high-salt diet for 2.5 years. The animals in groups one and two did not do well and their health status determined the time of termination of the experiment. The baboons were given 1% normal saline to drink and food which contained 4% of sodium chloride (wet weight). This involved an intake of about 1.5 g of sodium per kilogram per 24 h. This is very high, and it is regrettable the study did not include some lower values.

In all experimental groups on a high-salt diet the values of mean blood pressure were increased due to an elevation of both systolic and diastolic pressures. This is taking into account the increase in blood pressure which occurred in normal control animals as age increased up to sexual maturity. In group one, mean blood pressure of control animals was 120/80 mmHg at 1 year whereas the salt-loaded animals had a mean blood pressure of 158/94 mmHg. The dependence of the magnitude of the reaction to salt on age that is observed with the rat was not as marked in the baboons. There was some evidence that the adult males were more susceptible to the hypertensive action of salt than the females. The animals exposed to the high-salt diet showed high glomerular filtration rates, as has also been reported in subtotally nephrectomized dogs (Langston et al. 1963) and intact rats (Coelho 1974). Diastolic blood pressure was elevated more in the unilaterally nephrectomized animals on high salt intake than those with kidneys intact. In the animals in which high salt intake was interrupted because of bad health after 1.5 years of exposure, the mean blood pressure returned to normal (from 119 ± 2.5 to 104 ± 2 mmHg: $n=4$) over 2 months, indicating that 'post-salt' hypertension had not yet developed.

Salt intake of infants and adolescents

There has been particular emphasis on the issue of salt intake of infants and adolescents for a number of reasons. The studies of Dahl on the induction of hypertension in rats by a high-salt diet showed that the young animals were the most susceptible. Furthermore, in strains with a genetic susceptibility to hypertension in the face of high salt intake, the feeding of commercially available baby foods induced hypertension and increased death rate of the animals compared with controls on a low-sodium chow (Dahl et al. 1970). The infant feeding practices of the early 1900s where the newborn were maintained solely on cow's milk to 1 year of age did not produce optimal growth or health, and deficiency diseases developed. Earlier introduction of solid foods led to great improvement (Guthrie 1968). Breast feeding provides adequate nutrition for a baby for the first 6–9 months of its life and early introduction of other foods is associated with many hazards, one of which is the danger of salt-loading. Instances such as described by Puyau and Hampton (1966) where a 15-month infant with coarctation of the aorta with chronic cardiac failure and oedema that was being fed homogenized milk and commercial baby food was found to have sodium intake of 50–60 mmol/day helped stimulate review of sodium intake contingent on these changed infant feeding practices.

Gamble et al. (1951) have estimated the sodium requirements in man as 76 mmol for each kilogram of body growth. A rapidly growing infant gaining 30 g/day would require 2.3 mmol/day. Allowing for this anabolic requirement, sweat losses and faecal and urine losses, daily sodium requirements are met generously by an intake of 8 mmol/day (ca 1 mmol/kg per day) (Fomon 1967). By contrast, Puyau and Hampton (1966) found in an evaluation of the sodium content of the diet of 1–14-month-old infants in the USA that intake was increased to 4.6 to 6.3 mmol/kg per day. These levels are in the range, in terms of sodium content per 1000 kcal, of the lower levels found by Meneely and Dahl (1961) to produce hypertension in rats. In terms of possible additional stresses on sodium balance, even in high ambient temperatures of 32–34 °C the sodium loss from skin is only 0.4–0.7 mmol/kg per day (Darrow et al. 1954), i.e. 4–7 mmol/day in a 1-year-old infant of ca 10 kg. Thus the levels in the diet are greatly in excess of metabolic requirement even allowing a liberal margin of safety in the instance of heat stress.

The findings of McCance and Widdowson (1957) indicate the kidney of the young may have limited capacity to handle the solute load imposed by a high-sodium diet. A study by Gamble et al. (1951) on effects of large loads of electrolytes in infants showed that large sodium retention produced by loads of sodium chloride caused extensive depression of potassium retention. The inverse effect of depression of sodium retention by large loads of potassium was observed. A large transfer of sodium into the intracellular compartment was found with administration of sodium bicarbonate, and, with potassium chloride load, sodium was extensively removed.

Given the contention of Dahl (1968a, b) and Mayer (1969) that the experience of thousands of generations has attested to the adequacy of the sodium of breast milk, the greatly increased sodium intake resulting from use of cow's milk and processed baby foods to which salt had been added came under critical review. For example in 1970 (Blair-West et al. 1970), following discussion of this issue, we managed to persuade manufacturers to reduce drastically the salt content of commercial baby food in Australia, and federal regulations on salt addition were introduced. Similar changes have taken place in the US and now most manufacturers add no salt. Previously many products approached isotonic saline in terms of sodium content of the fluid component, and it was asserted that salt was added primarily for the palate of the mother (Mayer 1969).

Another question has been whether the high sodium content of baby and infant food as it existed, and may still be contrived by maternal salting practices, habituates the young to a high level of salt acceptance before much elective behaviour is feasible.

When considering sodium intake in the young, it is important to recognize the large variation feasible as a result of the sodium content of water. The concentration in the municipal supply in the US may range from 0.4 to 1900 mg/l (White et al. 1967). In reconstituting a dried milk baby food, it is possible that infants in the high-sodium areas may receive an extra 1000 mg (44 mmol) of sodium per day from the water. Cooking food in such water which evaporates during the process may result in another sizeable addition to sodium intake. These considerations are highlighted in a report by Calabrese and Tuthill (1977). They examined 150 students from each of two schools in adjacent communities, in one of which the sodium content of the drinking water was 107 mg/l and in the other 8 mg/l. It was found that the students from the high-sodium community showed a blood pressure distribution curve characteristic of persons several years older. The difference in terms of sodium intake between the two communities was small (ca 4–8 mmol/day) when considered against other likely sources of sodium in food. Many other factors need to be considered—race, age, body weight, social circumstances, etc.—as Simpson (1979) pointed out.

In a follow-up study, Tuthill and Calabrese (1979) analysed these factors and found the two communities to be strikingly similar in sociodemographic characteristics known to be influential on blood pressure. The communities had similar size, income, education and ethnic composition in the 1970 census data. Students in the high-sodium community showed an upshift in both systolic and diastolic blood pressure in both sexes, although less pronounced in males (Fig. 27-21). Several major studies on US adolescents have reported no difference in systolic and diastolic pressure between the ages of 15 and 17 for females whereas over this time for males the average change in systolic pressure is 2 mmHg/year and in diastolic pressure about 1.4 mmHg/year. The upward shift of systolic and diastolic pressure in males from the high-sodium community is typical of a 2-year-older age group when compared with their low-sodium peer group. The highly significant 4–5 mmHg upward shift in systolic and diastolic blood pressure distributions for females in the high-sodium community represents approximately that of a 10-year-older age group compared with their low-sodium peer group. (The comparisons are based on the 1970–74 National Center for Health Statistics blood pressure data.)

The implication is that persons at the high end of the distribution would tend to become hypertensive some years earlier in the high-sodium community than the low-sodium community. The authors caution in relation to these striking data on the necessity of, amongst other things, a further check on the dietary data by urinary assay of sodium to confirm the overall differences, and this is planned with a younger group, as is a comprehensive survey on heavy metals in the drinking water. They quote two epidemiological studies from Rumania and Russia showing an upward shift in population blood pressure with high levels of sodium in the drinking water (Steinbach et al. 1974; Fatula 1977).

A recent study by Margetts et al. (1980) showed no difference in blood pressure of children in six towns of Western Australia where sodium concentration of water ranged between 1.5 and 9.7 mmol/l. Estimated sodium intake derived from local water as a percentage of total intake ranged from 0.9%–1.4% in the town with the lowest sodium concentration in the water to 7%–9% in the town with the highest. However, the estimates of total sodium intake indicated highest sodium intake occurred in boys in the town with the lowest

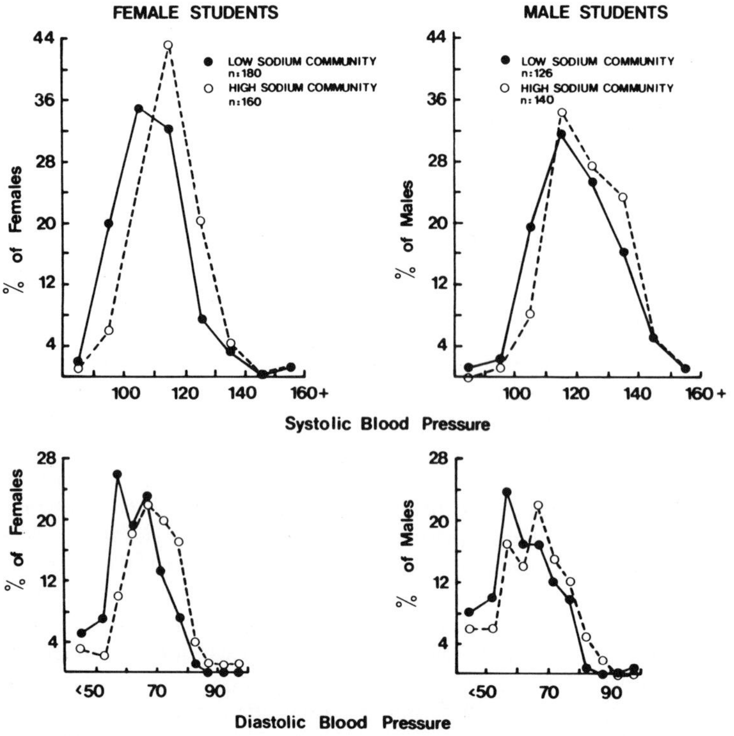

Fig. 27-21. Systolic and diastolic blood pressure distributions by sex for sophomore high-school students from two communities, in one of which sodium in drinking water was high (107 mg/l) and in the other low (8 mg/l), as of March 1977. (From Tuthill and Calabrese 1979, with permission.)

water sodium content, and little difference in urinary excretion of sodium between the two extremes.

These data highlight the importance of future studies. For example, in Perth, Western Australia, the sodium concentration in the water ranges from 1.6 to 5.2 mmol/l. In some country centres of Western Australia the concentration is higher: for example in Geraldton it ranges from 9.6 to 15.6 mmol/l. In Adelaide, South Australia, the average sodium content of the water is about 3.5 mmol/l and in many regions north of the capital the concentrations are higher, reaching 20–30 mmol/l in the Tod River reservoir supplying the Eyre Peninsula. Similarly in south-eastern Houston in Texas, USA, the concentration is about 10 mmol/l (Gonzales et al. 1979).

Apart from considerations of infant feeding, the data highlight the problems of management of cardiac, renal and liver patients where the therapeutic aim may be to restrict sodium intake to 500 mg/day. The American Heart Association proposed a maximum of 20 mg/l of sodium in drinking water as a standard to afford protection to high-risk segments of the population.

Staple foods such as cereals, bread, dairy products, meats, fish, puddings, have a moderate salt content but because they are eaten in large quantities may be a greater source of sodium than saltier foods such as olives, anchovies and salted snack foods.

The high sodium content of some processed foods, including those from fast-food outlets, which are becoming increasingly common, has been commented upon. An article in the *New York Times* (11 July 1979) republished analyses made by the Consumers Union that were reported in the March 1979 issue of *Consumer Reports.* Striking figures on sodium content included 1510 mg in one McDonald's Big Mac hamburger, 1152 mg in a Swanson's fried chicken dinner, 1186 mg in a Chef Boy-ar-dee Beefaroni, 925 mg in a cup of Del Monte green beans, 260 mg in 1 oz of Kellogg's Corn Flakes, 950 mg in a 10-oz serving of Campbell's tomato soup, 370 mg in a half cup of Kellogg's All Bran, 414 mg in one McDonald's apple pie, and 117 mg in 1 slice of Pepperidge Farm White Bread. An analysis of a wide variety of 'convenience' foods in Australia showed a large proportion with a very high sodium content of 150–600 mmol/kg (Dale 1979). An analysis in the UK (Bull and Buss 1980) showed also that a large proportion of sodium intake came from processed foods. Table 27-12 illustrates the effect of processing on the composition of food.

There is more than one reason for the high salt content of some of these products, as set out by Forsythe and Miller (1980). The salt is usually thought of as a flavour or flavour-enhancer in the foods and this is probably the sole reason in some cases. But it can play an essential role in food processing. Thus in bread, salt controls yeast

Table 27-12. Effect of food processing on sodium and potassium content.

Food item	Na content (mg/100 g)	K content (mg/100 g)	Na/K ratio
Corn			
Sweet corn	Trace	196	0.01
Canned corn	236	97	2.40
Corn flakes	1005	120	8.40
Potato			
Baked potato	4	503	0.01
Potato salad	528	319	1.70
Potato chips	1000	1130	0.90
Tomato			
Sliced tomatoes	3	244	0.01
Canned tomatoes	130	217	0.60
Tomato catsup	1042	363	2.90

From Altschul and Grommet (1980), with permission.

activity in the fermentation of dough, aiding in the prevention of objectionable bacterial action and 'wild yeast fermentation'; it has a strengthening effect on gluten enabling it to hold water and carbon dioxide and permitting an even expansion of the dough; and it improves crust colour.

In America 4.5 billion kg of processed meats are consumed each year, and they contain salt levels from 1.8% in fresh pork sausage to 10.6% in chip beef. Salt acts as a preservative in processed beef by lowering water activity and limiting microbial growth. It also solubilizes certain muscle proteins which aids gel structure and colour. One of the processes approved by the United States Department of Agriculture for destroying live trichinae in pork is treatment with 3.3% sodium chloride. Sodium salts as phosphates, nitrates, nitrites, caseinates etc. are also used in meat products. Salt plays an important functional role in the manufacture of butter, margarine and cheese products. It inhibits bacterial growth and prolongs shelf life. It plays a significant role in the control of ripening, flavour and development of texture in all cheeses. In other fermented foods such as sauerkraut, pickles and soya sauces, salt plays an essential role in control of microbiological flora and in allowing fermentation. Forsythe and Miller state that the elimination, or significant reduction, of salt usage would remove most of these processed foods from the market. Currently there are low-sodium products on the market including cheese, bread, cereals and canned vegetables. However, because they are low-volume products they cost more than the usual commercial products with high salt content.

The mechanism of hypertension induced by high salt intake

Animal experiments have given valuable insight into the manner in which high salt intake may initiate chronic high blood pressure. There are, however, many unresolved questions and it could not be assumed that the same sequence of events necessarily holds in man—though there is good evidence that major factors identified are involved (Freis 1976, 1979).

Early emphasis on the importance of extracellular fluid volume and plasma volume in renovascular hypertension came from the work of Ledingham (1953) and Floyer and Richardson (1961). Ledingham and Cohen (1963), summarizing the sequence of events occurring in renovascular hypertension, proposed (i) retention of salt and water, (ii) expansion of extracellular fluid and plasma volume, (iii) rise in central venous and right heart pressure, (iv) increase in cardiac output due to the Starling effect, (v) rise in blood pressure. This rise in blood pressure caused a diuresis which prevented further rise in extracellular fluid. After several weeks total peripheral resistance rose and cardiac output returned to normal. Thus a high cardiac output hypertension changed over time to a high peripheral resistance type which is characteristic of most chronic hypertensive states.

Ledingham's proposal was that protracted increase in cardiac output caused vasoconstriction by an autoregulatory process in the resistance vessels. When this happened cardiac output returned to normal because of increased afterload. Data involving induction of hypertension in conscious dogs by mineralocorticoid have indicated that sodium, apart from its effect on fluid volumes, also can interact with other factors to alter vascular resistance and hence arterial pressure (Onoyama et al. 1979).

The efficient autoregulating capacity of tissues is ratified in the instance of coarctation of the aorta where, despite large differences in blood pressure above and below, blood flows are similar in the arms and legs (Freis 1960).

Borst and Borst (1963), in the course of experiments on liquorice-induced sodium retention and hypertension, confirmed Ledingham's data in that venous filling pressure rose, followed by an increase in cardiac output and rise in blood pressure. The increased blood pressure permits the kidney to excrete the increased volume. In discussing why some individuals develop hypertension and others do not, they postulated that some had an unknown renal defect that required higher pressure to excrete the increased load of salt and water—a suggestion which Freis (1976) has remarked was prophetic.

Guyton and colleagues (1974) and Tobian (1974) have placed particular emphasis on the primary role played by the kidney in hypertension induced by increased load of salt. Guyton and colleagues expanded extracellular fluid (ECF) by salt- and water-loading dogs with a decreased renal mass. The haemodynamic sequence was the same as that recounted above. Expansion of ECF and a rise in blood pressure were associated with an increased urinary output. The thesis advanced was that the common factor in the development of any chronic rise in blood pressure is the requirement for the kidney to increase urine volume and sodium excretion to prevent a chronic rise in ECF volume. The level of blood pressure required will vary with individual functional capacity of the kidney to excrete an excess load of salt: the more impaired this capacity in the individual kidney, the steeper the slope of the blood pressure required. Experimental data on renal transplantation procedures in salt-sensitive and salt-resistant rats consistent with this view will be discussed in detail further on. The hypothesis advanced in the human has been that there is an inherited defect in the handling of sodium such that the kidney requires a higher than normal perfusion pressure to maintain ECF homeostasis in the presence of a high sodium intake (Guyton et al. 1974).

In the light of the transplantation experiments in rats showing that the phenotype response—blood pressure—was more influenced by the genotype of the renal graft than the genotype of the recipient (Dahl et al. 1972), there is thus the theoretical consideration that transplantation of a kidney from an individual who liberally ate salt lifelong with normotension to an individual with severe essential hypertension could cure or substantially ameliorate the condition, provided the structural events envisaged by Folkow et al. (1973) had not already produced irreversible changes!

The observations which were made in early studies of the benefit of the Kempner rice diet are consistent with this analysis. Murphy (1950) showed that with a diet of less than 8 mmol Na/day, serum sodium concentration did not fall, but plasma and extracellular volume fell ca 10% over 3 weeks and thus were accompanied by a fall in blood pressure. Watkin et al. (1950) made similar findings. Plasma volume was reduced 9% and ECF volume 15%. If 3 g sodium chloride were given daily, plasma volume was restored and blood pressure increased towards pre-treatment levels. One gram of salt per day increased plasma volume by a small amount but did not increase blood pressure. Thus there appeared to be a critical level of salt intake for the effect.

Freis (1979) has remarked that ageing may also diminish the capacity of the kidney to excrete effectively a salt load. In the overall picture there is clearly a diversity of factors. Excessive salt intake and body content may also directly influence resistance vessels increasing their sodium content and thickening the walls (Tobian and Binnion 1952). Increased intracellular sodium in muscle cells may activate ATPase, releasing free ionic calcium which activates contraction (Haddy 1974). Such processes contingent on sodium changes, could amplify other neurogenic and humoral influences, leading to structural changes which become progressive (Folkow et al. 1973). Onesti et al. (1975) have shown that volume hypertension in dialysed anephric subjects may be vasonconstriction hypertension. In instances where sodium content was purposefully raised in the subjects, 60% of those becoming hypertensive had the change without any preceding rise in cardiac output. In 20% of cases the change was due to rise in cardiac output with no evidence of any change in peripheral resistance. In 20% of cases both changes occurred. However, in some normotensive patients with end-stage renal disease raising body sodium did not cause hypertension.

Freis (1979) remarks that with ageing a number of homeostatic mechanisms associated with maintaining ECF homeostasis in the face of high salt intake may begin to fail. The kidney may then rely more on the blood pressure–urine volume relationship, and hypertension results. The curve relating renal arterial pressure and urine volume is much steeper in the isolated kidney than in the intact organism where all other homeostatic mechanisms can exert their effects. In this regard it is, however, salutary to note that, with increasing study, there is a growing evidence of a significant incidence of hypertension in adolescents (New et al. 1976; Loggie 1977; Londe et al. 1977; Kilcoyne 1977).

Freis (1976) summarizes his viewpoint by saying that whether due to renal deficiency in handling excessive ECF load, or to volume-induced enhanced responsiveness to other pressor influences, the difference in hypertension prevalence between acculturated and unacculturated societies appears to be due to the amount of dietary salt. It would seem wise on the basis of present knowledge, for individuals with a family history of essential hypertension to accustom themselves to a truly salt-free diet (less than 15 mmol Na/day) and to prevent their children from acquiring the habit of eating salted foods. An increased variety of unsalted foods would be available if there were increased public demand. It is possible, if not probable, that we already have the knowledge to prevent essential hypertension and its various complications.

This issue has become heavily politicized in the United States. For example, in testimony before the Senate Select Committee on Nutrition, Dickinson (1977) speaking for the Salt Institute, a trade association which represents the major interests of the world's salt manufacturers, made the point that setting an arbitrary low standard for salt intake would be like banning the use of penicillin because a very few people are allergic to it and it has caused death only in rare instances. This is a false analogy since the use of penicillin is for a major therapeutic benefit in the face of calculated risks, while high salt intake may be directly detrimental to a significant percentage of the population and of no particular benefit except in special therapeutic circumstances.

The Salt Institute in noting estimates that 21 million persons, less than 11% of the US population, may have high blood pressure and should be on a sodium-restricted diet, states that the source of this statement is not clear. They quote a letter from the American Heart Association to the effect that 'with advent of effective sodium-eliminating diuretics, the need for strongly restricted sodium diets has been sharply modified'. The AHA medical directors are quoted as adding that there continues to be a place for sodium-restricted diets in patients with congestive cardiac failure and uncontrollable hypertension. The Salt Institute says that when we consider only these two categories among all those who may have hypertension, the total number is only a fraction of the 21 million.

With regard to these arguments, it seems to me the statement of the AHA cited above could hardly be regarded as a scientific disclaimer of the relevance of low sodium intake. It would appear rather to endorse the central role of sodium in hypertension, while at the same time recognizing the great difficulties, including those contrived by the food processing industry, in reducing sodium in the diet. Thus, it welcomes the advent of the diuretics. Whether the advent of diuretics makes prudent moves on reduction of intake redundant is a moot point. As a preventive measure, is it better to have administration of diuretics to a wide segment of the population than to have public health measures which contrive a substantial reduction in overall salt intake? The argument that for 85%–90% of the population there may be no problem seems thin in the light of the morbidity and mortality for those remaining—ca 20 million in the US. Furthermore, in terms of policy, the State has to consider the cost to national resources of the impact on the health care system of the 10%–15% which do become hypertensive. The US Public Health Service at the end of the 1960s estimated the aggregate cost to the nation of all consequences of cardiovascular disease as $30.5 billion annually.

The Salt Institute summarizes its position by stating that there is definitely no need for a dietary goal that calls for the reduction of salt consumption. They add that there is no conclusive medical or scientific documentation that shows a detrimental effect of salt on the general population in amounts that are currently being consumed by the average American. In relation to the word 'conclusive' I suggest the appropriate position is the Scottish court verdict of 'not proven'. It is possible that there will never be conclusive evidence, for reasons also elaborated further on to the effect that accepting the opportunity still presenting to conduct such an experiment would be unethical. The current sales of salt within the USA are valued at about $0.5 billion, and ca 4% of the amount is for food processing and discretionary domestic use, but in value this represents near 20% of total sales, or $100 million (Dickinson 1980).

An overview, with some additional physiological considerations

Diet, sodium status and salt taste

During evolution, over the Miocene, Pliocene and Pleistocene—some 30 million years—the diet of hominoids has been predominantly herbivorous (see Chapter 4). For the latter 2–3 million years the facts are not certain but, in the main, the diet of hominids has probably been intermediate between that of herbivores and pongids on the one hand, and carnivores on the other. In a feral herbivorous people, sodium intake is of the order of 1–15 mmol/day, as ratified by direct measurement in a number of unacculturated societies. Further (Ball and Meneely, 1957; Meneely and Battarbee 1976a, b), this level of sodium intake is concomitant with a potassium intake twenty times higher. Indeed the Dutch anthropologists studying feral communities in the Western Highlands of New Guinea pointed to a K/Na ratio in diet as high as 200–400:1. In relation to the upper limit of sodium intake occurring naturally in foodstuffs, the Eskimos, an almost exclusively carnivorous people, are a prime example (Dahl 1958b). On the northern coast of Alaska their diet consisted principally of seal, walrus and whale. If one assumes a maximum intake of 6000 kcal/day in extreme environmental conditions, of which 25% is fat, 4500 kcal is supplied by lean meat; this is stated by Dahl to be equivalent to ca 2 kg of beef. This could involve a daily sodium intake up to 1.5 g or ca 65 mmol. The intake of the Maasai with a diet largely of milk, blood and meat gives a sodium intake of ca 80 mmol/day (Orr and Gilks 1931).

It is perhaps salutary to compare this upper limit with the estimates of the change in sodium intake a baby in a Western dietetic pattern may have over the first year of life. Human milk contains 0.5–0.6 mmol Na/100 kcal and breast-fed infants meeting their protein and calorie needs obtain 1 mmol Na/kg per day. Mayer (1969) points out this figure of 1 mmol/kg agrees well with calculated requirements of infants from birth to 1 year, with requirements for growth decreasing and faecal and insensible losses increasing during the period, leaving overall requirements constant. The change to formula and infant food increases the intake of the 1-year-old to ca 6 mmol/kg per day (Filer 1971). Many infants in this survey reviewed by Filer (1971) ingested more. Thus intake of infants on a body weight basis was approaching that recorded for the adult inhabitants of northern Honshu in Japan. Though water and protein requirements of the growing infant are also much higher on a weight basis than in the adult (Stewart 1967), the basic fact is that infant intake at 6–12 months has been 4–6 times greater than required, even in hot climates, where additional skin sodium losses in a 1-year-old might account for 4–7 mmol Na/day (Darrow et al. 1954).

Returning to these data of upper levels likely on a natural diet in the adult, it is noteworthy and consistent with Clark's (1970) viewpoint that intake of hunter-gatherer societies may be 60%–80% vegetarian, that the direct measurements on urine of the !Kung bushmen of Botswana indicated daily sodium intake of ca 30 mmol. The intake of many feral communities where meat plays little role in diet is much less.

The evidence of a substantially tropical mode of existence of man's progenitors also involves the problem of temperature control with sweating and salt loss. There are the additional stresses imposed by disease and the physiological demands of reproduction. These stresses could have been reduced when the habitat was of volcanic origin and thus there was a high salt content in soil, lakes and springs.

In humans on very low sodium intake, obligatory (or irreducible) daily sodium losses in urine, stools and skin range from 2 to 8 mmol (Dahl 1958b). Using data of Widdowson et al. (1951) giving the total body content of sodium of a 70-kg man as about 130 000 mg (2000 mg/kg of fat-free body weight), and assuming a birth weight of 3 kg and cessation of growth at 16 years, a daily requirement of 1 mmol Na/day can be estimated for growth (Dahl 1958b). It follows that the sodium intakes recorded for feral vegetarian societies are adequate for normal growth and physical vigour, as established clearly, for example, by Sinnett and Whyte (1973a). It is also clear that such a level of intake may be marginal in the face of any form of stress, including increased salt turnover with sweating due to temperature regulation in the tropics, and reproduction. The amount of sodium sequestrated in a 9-month human gestation could amount to 700 mmol, followed by the much larger provision of 6–9 mmol/day in lactation. In subtropical to tropical conditions, removed from sea coasts, deficiency could easily develop in the face of any additional stress, and many days to weeks of normal dietetic intake would be needed to repair sodium lost.

Overall it would seem likely that during primate evolution, selection pressures favouring emergence of behavioural mechanisms involving hedonic intake of salt, genesis of appetite in the face of deficit, and endocrinological mechanisms of sodium conservation would have operated at high intensity (Denton 1969, 1973; Blair-West et al. 1970; Gleibermann 1973). At least it is unlikely that the metabolic circumstances over this phase of mammalian evolution would have operated to favour regression or loss of those physiological mechanisms of sodium homeostasis which are demonstrably at a high degree of development in many herbivorous and omnivorous species of mammal and marsupial so far studied. The data on primates are limited but salt-lick-seeking behaviour of gorillas has been observed. More important, the spontaneous development of a salt-seasoning cultural pattern in a colony of Macaque monkeys, involving sea-water washing of potatoes frequently during consumption, has been beautifully recorded. This would appear to be strong evidence of a hedonic liking for salt in the primate. This innate liking and acceptance can be elaborated by cultural and ritualizing influences in humans. In the instances where cultural dietetic patterns of a community impose on all members a high sodium intake, such as in northern Honshu, there are no reports of numbers of inhabitants rejecting the pattern and dietetic practices. As is seen in urban Western communities with large consumption of processed foods, adaptation and acceptance of the taste milieu is rather general, though perhaps not universal.

It must be considered, however, that the

selection pressures on sodium homeostasis may have been reduced over the last million years or so of hominid evolution as a result of a substantial meat component episodically or continually in the diet as a result of hunting. Over some fifty thousand generations, as discussed earlier, some regression of both hedonic behaviour and appetitive response to deficiency might have occurred. Parenthetically it can be noted that changes in taste acuity in certain regards have been shown in human communities. Kalmus (1971) has reported that whereas 100% of the members of tribes of Amazonian Indians can detect phenylthiocarbamide, the ability to detect this bitter material has been lost in a significant proportion of Caucasians and also Asiatics. Richter (1956) suggests that in the course of the evolutionary process, man may be losing the ability to taste certain substances. Under feral conditions the detection of bitter may have high survival advantage for primates and hunter-gatherers in relation to the fact that 10% of plants contain toxic alkaloids and steroid-type compounds. The same pressure may not operate in agricultural and urbanized society. Richter has also, in discussing the influence of domestication, pointed to the reduction of adrenal size in the domesticated rat, including lipid content of the glomerulosa region, relative to the wild animal. The argument is that domestication has resulted in a reduction of cortical function, particularly in the capacity to conserve sodium. However, there is no evidence that these data are applicable to man.

Dahl's (1958b) study showed that in patients with good renal function, reduction of sodium intake to 2 mmol/day resulted in urinary outputs of 0.25–1.5 mmol/day. That is, urbanized Caucasians were no less effective in conserving sodium than the Yanomamö Indians with probably many thousand years of jungle existence on an exceedingly low sodium status. Further, though meat-eating in hunter-gatherer populations does increase sodium intake, there is some but not much capacity to store sodium in the body (increased extracellular fluid and in bone). The episodic impact of stresses involving sodium loss such as those instanced above, could cause sodium deficiency. That is, adoption of a hunter-gatherer mode of existence will ameliorate stress of this type, but not emancipate from it.

It is no phylogenetic caprice or accident that salt taste is one of the four primary modalities of taste in all vertebrates. The hedonic human liking or appetite for salt when it is available, and independently of any metabolic need, is an evolutionary legacy of behaviour of high survival value. It is part of the overall innate organization dedicated to salt ingestion along with the elements determining appetitive response to sodium deficiency and to the hormones of the reproductive process.

In this respect I would differ from Dahl's viewpoint (1960), reiterated by Meneely and Dahl (1961) and Freis (1976), which suggests that the appetite for salt is purely learned, though Dahl (1972) does draw a distinction between appetite for salt as seen in restaurants and most homes, and the hunger for salt in the gross deficit of Addison's disease. There is no doubt, of course, that learning plays a very big part through custom—particularly the traditional diet habits of the community in which the young are reared, and the influences to which individuals are later submitted. Further, learning or habituation may be predominant in the gradual augmentation of intake over years. Indeed even in a herbivorous creature such as the sheep, whose highly developed appetitive organization has been intensively investigated, it is obvious that learning mechanisms become richly superimposed on innately generated drives. This, of course, is characteristic of innate mechanisms in that most can be modified by learning.

But in the human case, the cultural influences are acting in the context of an innate propensity: the liking for the taste of salt and readiness to ingest it. It is presumably this taste which causes primitive peoples in Africa and New Guinea who have a diet with a high potassium content nonetheless to make wood-ash potassium salt in circumstances where they cannot get sodium salt, and where, with the very high potassium diet, there could hardly be any retroactive learned sense of benefit from ingestion. The lengths to which primitive peoples will go in terms of travel and dispute to obtain salt is consonant with J.B.S. Haldane's (1956) remark that in a hot country like India the desire for salt is imperious—a fact he suggests Gandhi understood better than the English Viceroy. Grasse (1956) has observed this hunger in the Pygmies of l'Oubangui.

It is the taste which attracts. This assertion is made notwithstanding the instances where salt has been offered to, and rejected by, primitive peoples who have not hitherto had access to it. The sudden exposure to the taste in a concentrated form easily soluble in saliva may cause rejection by virtue of the intensity of the stimulus. In experimental animals this is demonstrated by the preference–aversion curve. Gradual addition to food may win more ready acceptance. In this regard it might be noted that animals licking salt blocks and dissolving sodium into the saliva do not necessarily get a very high concentration in the mouth with each lick.

Intra-uterine and neonatal influences

If there ever were to be public health measures and advice in relation to ostensibly sensible restriction of salt intake, beyond the baby and infant food span, it would probably be more effective in the long term to recognize the propensity and the liking as physiological entities underlying the strength of the behaviour, rather than to see salt eating as an unfortunate habit. That is, it should not be seen any more than, for example, sweet taste as something acquired; both are elements which are fundamental in chemoreception and the central nervous organization of ingestive behaviour which has evolved. However, the ontogenesis of the two behaviours may be different. In the case of sweet taste, the facilitated intake by the newborn has been clearly established by Desor et al. (1977). They propose it to be innate, a sensory preference and an immediate hedonic response. Facilitation in their tests was unaffected by addition of salty or bitter tastes, but sour tastes suppressed the effect. With salt, the situation is more complex. Nowlis (1973, 1977) in studying the response to salt of newborns, using a specially constructed pressure-transducing nipple which monitored responses from the front and back of tongue, noted rejection behaviour invariably with quinine hydrochloride at 10^{-4} M. In some infants rejection also occurred with isotonic saline, whereas in others there was some mild ingestion of isotonic and hypotonic saline.

This situation is difficult to evaluate because from the twelfth week of gestation the human neonate swallows amniotic fluid (200–750 ml/day). In the sheep, sodium concentration of amniotic fluid ranges from 116 mmol/l at 100 days to 86 mmol/l at term (Mistretta and Bradley 1977). Any taste preference may be considerably delayed in its expression. Thus the introduction of glucose, or a bitter substance (Bitrex) into the amniotic fluid of sheep did not modify swallowing behaviour, though the amniotic fluid with Bitrex on withdrawal tasted bitter to the authors. They established in this outstanding series of experiments that in the last third of gestation the foetal taste receptors initiated chorda tympani responses—both on multiple- and single-fibre recording—similar to those in lamb and adult at least as far as salts were concerned. Sugars usually did not elicit a response. Bradley and Mistretta (1980) recorded from neurones of the tractus solitarius of neonate and foetal sheep. Foetal neurones from 84 days on responded to ammonium chloride and potassium chloride applied to anterior tongue but only after 114 days of gestation were responses obtained to sodium chloride and hydrochloric acid.

Mistretta and Bradley (1977) raise the possibility that intra-uterine taste experience is biologically significant and that prolonged exposure to chemicals at low concentration in the amniotic fluid contributes to neonatal rejection of chemicals at high concentration, or rejection of chemicals not present *in utero*.

In relation to any taste experience *in utero* being influential upon the neonate, in the case of sodium it would seem that the pattern of change of amniotic composition over the course of pregnancy is relatively fixed. Sodium deficiency has little influence on it, as we have confirmed in the course of extensive studies on the sheep foetus *in utero* in relation to ontogeny of steroid hormone secretion (Wintour et al. 1977). It is rather unlikely to be the genesis of any individual variation in post-natal preference. Thus, induction of sodium deficiency by parotid fistula in six pregnant sheep (fistula loss of 400–1200 mmol of sodium over 8–10 days) showed reduction of maternal plasma sodium from 147 ± 3 to 135 ± 5 mmol/l; amniotic sodium concentration decreased 5–10 mmol/l but foetal plasma sodium concentration was not changed significantly. When the ewes were killed at the end of the period and the foetal adrenals incubated, increased aldosterone production was not seen. Phillips and Sundaram (1966) produced very severe sodium depletion in pregnant ewes by drainage of all saliva from a unilateral parotid cannula. Maternal plasma sodium concentration fell 15 mmol/l, haematocrit rose from 40% to 54% and there was a 35% reduction in plasma volume, and by the end of the 6 days of drainage most animals had ceased eating. Sodium concentration of amniotic fluid fell by 14 mmol/l but no decrease in volume occurred. Foetal plasma sodium fell from 143 to 137 mmol/l. Allantoic fluid volume increased.

Human breast milk has a very much lower sodium concentration (5–10 mmol/l) than amniotic fluid but the fact that *in utero* the foetus swallows fluid containing 80–120 mmol/l of sodium through most of gestation, with central transmission of the stimuli, may have some relevance to the early acceptance by infants of cow's milk and processed food formulations giving sodium concentration close to the isotonic range. From the period before sodium was withdrawn from baby food, there does not appear to be evidence of widespread rejection of formulae with high sodium content. Indeed Stewart (1967) has suggested that this might validate the presence of salt in infant food at high levels, though Dahl (1968a, b) remarks that the infant has no discretionary power in such decisions, which are made by the mother.

Studies in taste preference in pre-school children (2–6 years) have shown a statistically significant preference for salted foods (Reynolds and Filler 1965). Children preferentially selected a beef stew with salt over one without salt. There is perhaps a similar implication in Guthrie's observation in 1968 that infants who begin to ingest solid foods during the first month after birth have consistently higher intakes of sodium than do infants who begin to take in solids at a later stage. Comprehensive study of the development of taste preference for salt in the very young would be of great interest.

Clear preferences with racial differences are seen in older children, as exemplified by the preference of black children for more concentrated salt and sugar solutions than their white counterparts (Desor et al. 1975). The authors emphasize the large individual variation for sweet preference observed in humans—analogous to the emphasis given in the opening section of this chapter to the individual variation in hedonic salt preference in animals. However, their study of preference for sweet and salt in 9–15-year-old monozygotic and dizygotic twins did not support the level of sweetness or saltiness preferred being under genetic control or for that matter under the influence of being raised in the same household.

The impact of technology and commerce

Against the background of a relative scarcity of salt over the course of primate and hominid evolution, a new situation has developed as a result of technology and commerce. Over a mere 100–200 years the people of Western and some Eastern societies now have abundant and cheap salt. In the case of the USA, for example, in 1968 the amount of table salt sold for discretionary use was 250 000 000 kg, or ca 3.4 g per capita/day. Total intake estimated from data on all aspects of food production and processing in another study suggested a total of 10–12 g/day (SCOGS—102-1978). The order of magnitude is broadly consistent with that of the clinical data on daily sodium excretion of adults on their accustomed diet.

The data from unacculturated societies, transitional peoples, and societies with very high salt intake, the animal experiments on hypertension with high sodium and low potassium intake, and the study of physiological mechanisms whereby the effects of high salt intake on extracellular volume may induce hypertension, integrate to raise a major question. Is a high maternal salt intake, followed by a high intake during infancy and childhood which then is sustained by cultural elaboration of hedonic preference, and amplified by involuntary intake as a result of the salting procedures of most food processing firms, of little or no relevance to the basic genesis of hypertension in modern societies? At any meeting involved in the analysis and dissection of the pathophysiology of human and animal experimental hypertension, the role of sodium is paramount in the discussion—because of its protean influence on the relevant parameters. This centres on (i) sodium balance, (ii) sodium distribution within the organism and the corresponding effect on volume of fluid compartments, (iii) the modulation of response characteristics of excitable tissues, and (iv) the effects on hormone secretion and action, involving in particular the renin–angiotensin system, antidiuretic hormone, catecholamines, and the steroid hormones. This is to name only four main contexts.

Is it feasible that dietary sodium loading over the early stages of differentiation and growth and during the elaboration of response mechanisms of a diversity of tissues may alter definitively, and possibly irreversibly, the characteristics of the adult organism?

Genetic susceptibility and environmental influences

Genetic susceptibility appears to be a factor in determining whether any environmental influence initiates hypertension. For example, high salt intake from infancy to middle age in modern affluent technological communities may only represent a permissive factor for the expression of other jointly sufficient and severally necessary factors, which in the long term contrive the hypertensive state. In other individuals with a genetic susceptibility, the high sodium and low potassium intake over decades may be the predominant factor.

If any major role in hypertension is to be validly attributed to change of diet, and specifically to a rise in sodium and fall in potassium intake, the appraisal of other factors is critical (Swales 1980). Lot Page (1976) has made a scholarly review of the issue which leads him to the view that, overall, there is persuasive evidence that excessive salt intake is highly important in induction of age-related increase in blood pressure in susceptible young individuals. However, he draws attention to other factors, especially weight change. Acculturation is usually associated with increasing weight, and it is difficult to separate this effect from other influences of changed diet and activity. Page's study (Page et al. 1974) is important in this regard.

In the six primitive Solomon Island societies he studied, weight declined with age in all groups, and there was no rise in blood pressure with males. Some rise in systolic pressure occurred in females of the three more acculturated societies. Maddocks and Rovin (1965) in their study of the Chimbu in the New Guinea Highlands found that there was a real fall in blood pressure with age, even when allowing for the fall in body build which occurred. They stated that some other factor must be operating.

Moreover, there are studies involving primitive groups in a transitional state moving to an acculturated pattern. Blood pressure increase with age is found to occur as is characteristic of the Western urban society. This would point clearly to the pattern in the unacculturated societies not being attributable to any genetic protection. The evidence points to some environmental factor or factors acting upon the population, though the variability of influence upon individuals may reflect genetic variation.

The contention that the reason that blood pressure does not rise with age in some communities is attributable to disease and its debilitating consequence gains some support in individual instances with, for example, correlation with splenomegaly. However, in some of the best-studied unacculturated societies where blood pressure is low, the people have a high level of physical performance and vigour (Shaper 1972).

Dahl has suggested that the blood pressure may rise with weight because the obese eat more, and in consequence have a higher sodium intake. On the other hand there does not seem any suggestion in the Japanese literature that the rural population of Akita which has high blood pressures is overweight. In personal discussion, Dr Fukuda has emphasized the intensity of work of the farmers, of the order of 12 h per day, and the general tenseness and strain. The high energy expenditure and high food intake, by virtue of dietetic custom, result in a very high salt intake.

The different impacts of environment on psyche and the resultant socially imposed nervous strain have been comprehensively analysed by Scotch (1960, 1963; Scotch and Geiger 1963) in a study of rural and urbanized Zulus. His finding of the different impact of family size on blood pressure of urbanized Zulu women from that of rural Zulu, and from that found by Miall and Oldham in the Welsh towns, points to increased stress of family responsibility on the urban Zulu women. This is due to the disruption of traditional patterns of support systems coupled with the problems of coping with a psychologically hostile environment. He found no evidence that consumption of a superior or inferior diet may have influenced the difference between the two groups, but regrettably there were no data on salt intake in the two groups.

It is perhaps apposite at this point to bring in some of the more recent views of Sir George Pickering on this issue of salt and hypertension (Pickering 1980). He has made a telling attack on the Dahl and Love (1954) Brookhaven study, proposing that the data show that the mean blood pressures of the three groups—high, average and low salt intake—were, in fact, the same. Hence salt had no effect on blood pressure. Their conclusions were derived, in his eyes, by drawing an arbitrary line between hypertension and normotension, and must imply, if there were more hypertensives in the high-salt group that there were also more hypotensives. The celebrated Dahl diagram on populations and percentage of hypertensives related to salt intake (Fig. 27-22) is similarly criticized (Pickering 1980) because of absence of information on the arbitrary line dividing hypertensives from normotensives and the evident way this could be biased if age were not strictly controlled—a fact not disclosed in Dahl's data.

However, these highly pertinent observations would probably not significantly alter the striking differences between blood pressure in the unacculturated societies and the inhabitants of northern Honshu, and the fact that there will be populations distributed in, at present, uncalculated relation along Dahl's 'imaginary' line joining the two extremes. Pickering (1980), remarking on the low blood pressure and lack of rise in pressure in middle age of people living in 'primitive conditions', and the rise which occurs when the life-style of Western societies is adopted, notes that change in salt intake is only one of many differences between the two life-styles. He goes on to say that having visited some of these unacculturated peoples he would certainly not consider salt intake as the primary reason for difference. He states: 'In my opinion the great contrast is between the certainty of behaviour in a society ruled by ritual and taboo and the uncertainty in Western societies in which life is a series of individual choices and decisions. Primitive peoples living in tribal societies differ in so many particulars that to select salt intake as the culprit is to me just bad science.'

In relation to the culprit Pickering appears to have arraigned, it is useful to turn immediately to the Yanomamö Indians, a low blood pressure society which has been more extensively studied metabolically than any other group by Oliver and colleagues (1975, 1981), and extensively studied anthropologically over 15 years by Napoleon Chagnon and colleagues. Chagnon (1977) in his

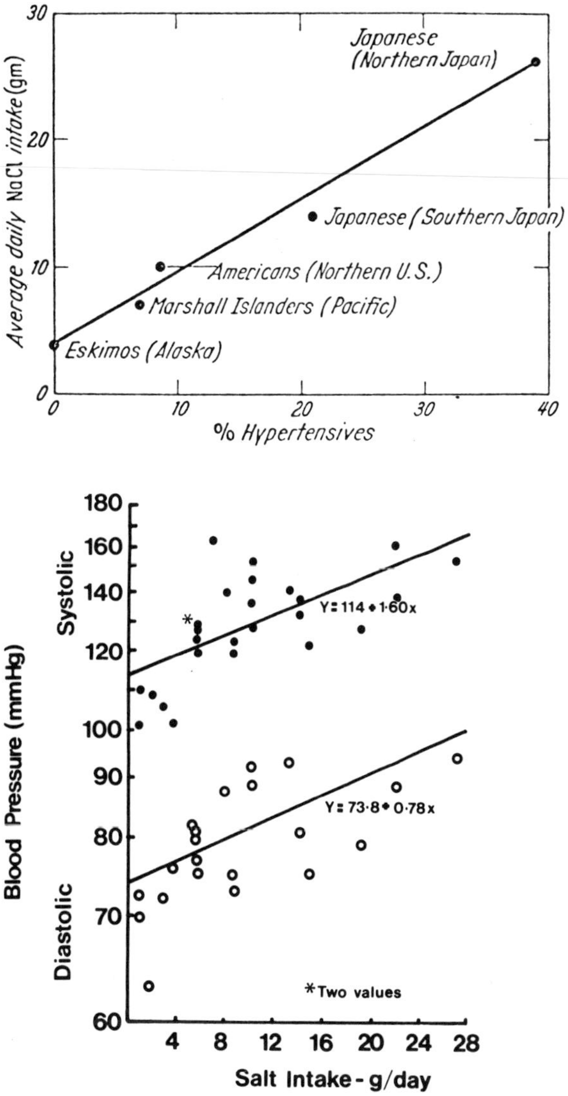

Fig. 27-22. Upper: Correlation of average daily salt (sodium chloride) intakes with prevalence of hypertension in different geographic areas and among different races. (From Dahl 1960, with permission.)

Lower: Gleibermann's (1973) analysis of correlation between blood pressure and salt intake in males in 27 populations drawn from the literature. The regression is plotted on a logarithmic scale. Systolic blood pressure (upper line), $P<0.0005$; diastolic pressure (lower line), $P<0.0005$. A similar correlation was seen in females. (After Gleibermann 1973; see also Froment et al. 1979.)

book *Yanomamö: The Fierce People* records in great detail the life of a people who engage in chronic warfare, and for whom much of life is charged with violence and tension. The Yanomamö goad each other within their own villages to the brink of an explosion. This hostility is projected on a larger scale in the negotiating of alliances between villages. Conflict is regulated through a series of graded escalations from duels, through club fighting, spear throwing, to raiding in a state of war, to the ultimate—the *nomohom*—massacre by treachery. Chagnon describes it as a politics of brinkmanship. War is only one form of violence in a graded series of aggressive activities. The culture calls forth aggressive behaviour but at the same time provides a regulated system in which expressions of violence may be controlled. A representative account by Chagnon (1977) illustrates this and also, as in many other of his descriptions, how the controls broke down.

> The village of Patanowa-teri split during the last month of my first field trip. One of the young men took the wife of another because she was allegedly being mistreated by him. This resulted in a brutal club fight that involved almost every man in the village. The fight escalated to jabbing with the sharpened ends of the clubs when the husband of the woman in question was speared by his rival and wounded. The headman of the village, a brother of Kaobawa, had been attempting to keep the fighting restricted to clubs. When the husband's rival speared his opponent, the headman went into a rage and speared him in turn, running his own sharpened club completely through the young man's body. He died when they tried to remove the weapon. The wife was then given back to her legitimate husband, who punished her by cutting both her ears off with his machete.
>
> The kinsmen of the dead man were then ordered to leave the village before there was further bloodshed. The aggrieved faction joined the Monou-teri and the Bisaasi-teri because these two groups were at war with their natal village, and they knew that they would have an opportunity to raid their own village to get revenge. The Monou-teri and the two Bisaasi-teri groups accepted these new arrivals—they were kinsmen and would actively prosecute the war against the Patanowa-teri. The hosts, of course, took several women from the refugees, the price a vulnerable group must pay for protection.

Chagnon makes clear that warfare between Yanomamö varies from region to region and may be extremely intense in some areas at particular times and almost non-existent in other areas. They do not, however, spend all or even a major fraction of their waking hours making war on neighbours or abusing their wives, and even the most warlike villages may have periods of relative peace during which time daily life is tranquil.

It is clear, though, that these people are frequently exercising options involving outcomes which vary from humiliation, through loss of property and women, to severe bodily injury and death. The idea that they are emancipated from mental stress, as compared with that arising from exercise of options in a Western community, by virtue of rituals which bestow certainty on behaviour, is hardly credible in the face of Chagnon's comprehensive studies unless we are to argue stress associated with attack, and fear of humiliation, of bodily injury and death does not access the hypertensinogenic neurohumoral mechanisms which are influenced by the chronic anxiety of Western life (the rent, the corporate peck order, etc.). Where the rice farmers of northern Honshu, working for long hours daily in the field, stand in relation to stress by virtue of exercise of options is an open question, but conceivably it is less than with the Yanomamö or Western urban man.

As Page (1976) states, there is universal agreement that susceptibility to hypertension has a genetic origin although the mechanism is still disputed. Platt (1963) has postulated inheritance through a single incompletely dominant autosomal gene whereas Pickering (1965, 1968), Kannel et al. (1967), Miall and Lovell (1967) and Ostfeld and Paul (1963) have favoured a polygenic inheritance as a graded character. Kass et al. (1975) and Zinner et al. (1975) have shown that variance of blood pressure within families was significantly less ($P<0.001$) than among children in the same age group studied. Heller et al. (1980) showed that first-degree relations of hypertensives had significantly higher blood pressure than their normotensive controls or relatives of controls ($P<0.001$). Pietinen et al. (1976) and Altschul and Grommet (1980) reported on the study of a small population and found a clear correlation between overnight sodium excretion and blood pressure, and weight and blood pressure, for the group that had a family history of hypertension and no correlation for the group that had no such family history.

Much evidence would favour the proposition that the genetic influence has a predisposing permissive nature, with development of hypertension depending on environmental factors. These factors affect populations throughout the world including those transitional or largely graduated from the unacculturated state. Racial differences would appear influential, as evidenced by the steeper gradients and higher prevalence of hypertension in the American blacks.

Animal experiments on genetic susceptibility

In terms of the main issue of this chapter, Dahl's experiments (Dahl 1972) have focused the question of whether genetic variability in susceptibility to hypertension may be due to vulnerability to the environmental factor of high salt intake. By selective inbreeding of rats over five to seven generations he was able to produce strains which were either highly susceptible or resistant to the effect of salt (Fig. 27-23). The level of potassium in the diet influenced this effect, high potassium

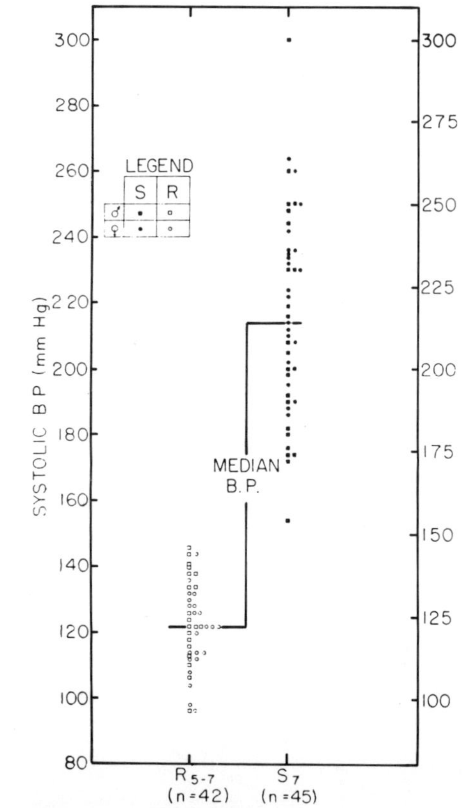

Fig. 27-23. Genetic influences on blood pressure responses to a high-salt diet in strains of rats sensitive (S) or resistant (R) to the effects of salt. S animals were tested after 1–2 months on salt, R animals after 10–12 months on salt. R_{5-7} and S_7 indicate the fifth to seventh generation inbred for resistance or sensitivity to hypertensinogenic effect of salt, respectively. (From Knudsen and Dahl 1966, with permission.)

ameliorating the action of sodium (Dahl et al. 1972). The susceptibility to salt was also paralleled by increased susceptibility to other hypertensinogenic stimuli: steroid hormones and renal artery constriction.

Dahl (1972) also crossed the two strains in order to examine further the genetic differences in vulnerability to sodium chloride in relation to blood pressure. He felt that these experiments demonstrated the difficulties in assessing the role of sodium chloride in genetically mixed populations, and hence in man (Fig. 27-24). Thus, in the R (resistant) strain, individual blood pressures were normal whereas they were raised in the S (sensitive) strain. This was not so with the various crosses. Thus among the BR (i.e. back-cross, $R=F_1 \times R$), whereas most animals had pressures that were normal or only slightly above those of the pure R strain, a few individuals developed markedly elevated pressures. The reciprocal was true for the BS (i.e. $F_1 \times S$) cross, in which most had high pressures but a few remained in the normal

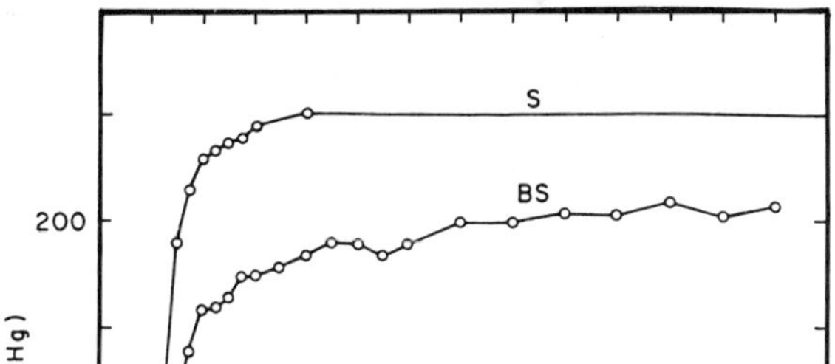

Fig. 27-24. Average blood pressures of different rat populations on a uniform high-salt (8% in chow) diet during a year of observation. The curves represent average cumulative systolic blood pressures for each cross. Cumulative blood pressures include the last blood pressure when the animal was in good health, and this value is carried forward whether an animal survived 6 weeks or 12 months on the regimen. No S rat survived beyond the sixteenth week. F_1=R×S; F_2=F_1×F_1; BS=F_1×S; BR=F_1×R. Although R, BR, F_1 and F_2 showed significant differences in blood pressure based on sex (males higher than females), the values have been pooled for the purposes of this graph which illustrates the definite genetic influence on the reaction to salt intake. (From Knudsen et al. 1970, with permission.)

range. Among the F_1 (R×S) and F_2 (F_1×F_1), the crosses in which average pressures were roughly mid-way between those of the parental R and S strains, marked hypertension in some and normal level in others were observed in members of both crosses.

Dahl suggests it is evident that despite foreknowledge of the average daily salt intake from weaning onwards as well as considerable information on the genetic background, predictions about the influence of sodium chloride on blood pressure might differ radically in these rat populations, depending on which individuals or groups were observed. This suggests how much more difficult it would be to make an assessment in a Western man in whom both lifetime salt intake and genetic background were unknown. This issue will be discussed at the end of this section.

The determination of the physiological basis of the difference between R and S strains has been an intriguing area of investigation. Dahl et al. (1972) showed that blood pressure was reduced by renal homografts from the R strain to the S with removal of the contralateral kidney, but raised with a transplant from the S strain to the R. The data indicated that the phenotype response—blood pressure—was more influenced by the genotype of the renal homograft than by the genotype of the recipient. This determinant role of the kidneys was elaborated by further transplantation experiments of Dahl et al. (1974) and by those of Bianchi et al. (1974) involving a strain of spontaneously hypertensive rats and a normal strain.

Tobian et al. (1977, 1978) have also shown the kidneys to be involved in this situation. They perfused the isolated kidneys of Dahl S and R rats with blood at different pressures. This was done on young rats before the S animals developed hypertension. At a normal inflow pressure the renal blood flow and glomerular filtration rate of S and R rats were not significantly different. At the levels tested the R kidneys excreted more sodium and water than those of the S strain did. Comparable outputs were obtained by increasing pressure to the S kidneys significantly higher than the R (Fig. 27-25). For example, at a normotensive inflow pressure of 130 mmHg the R kidneys excreted a mean of 146 μmol Na/min per 100 g of kidney, whereas at the same pressure the S kidneys averaged only 70 μmol. Thus at this level, the S kidneys from the hypertension-prone rats excreted

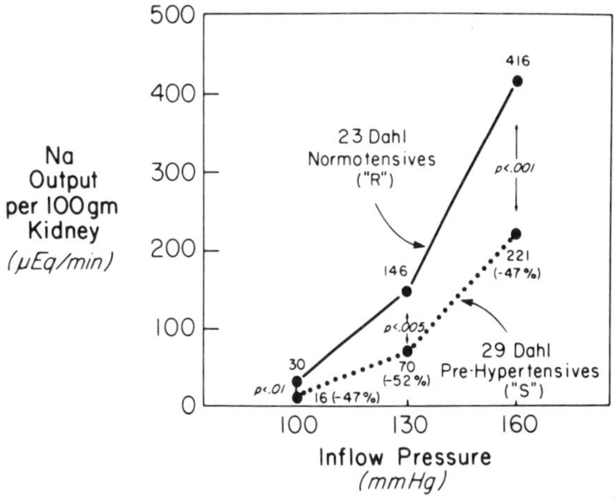

Fig. 27-25. Sodium excretion of the isolated kidneys from S and R rats at varying inflow pressures. The distance between the large black dot (the mean value) and the tip of the arrowhead showing the probability value represents the standard error of the mean (s.e.m.). (From Tobian et al. 1978, with permission.)

52% less sodium than comparable R kidneys ($P<0.005$). When perfused at 160 mmHg the S kidneys excreted about 50% more sodium per minute than the normotensive R kidneys perfused at 130 mmHg. That is, raising the blood pressure can completely overcome the natriuretic handicap. The explanation remains elusive, as Freis (1979) has remarked.

Tobian et al. (1979) have shown that if body sodium retention is prevented in the S rats by administration of a thiazide diuretic, there is little or no rise in blood pressure despite a very high 8% sodium chloride intake and the intrinsic limitation of natriuresis. Perfusion of the blood of S rats through normal 'neutral' kidneys at 125 mmHg results in only half the sodium excretion seen with blood from R rats. S rats appear to have a circulating antinatriuretic hormone or they lack a natriuretic hormone. Ikeda et al. (1978) have shown S rats have a much greater pressor response to introduction of hypertonic saline and angiotensin into the ventricles of the brain.

The experiments of Dahl and colleagues (1969) involving parabiosis have, like the outstanding experiments described above, emphasized the complexity of the phenomenon. When two rats of different strains were so united, hypertension could be induced by high sodium chloride intake in the normally resistant strain. The authors suggest that the rats normally prone to hypertension produce a pressor substance which can be transmitted by parabiosis; possibly this is a sodium-excreting hormone which has the capacity to produce hypertension when produced by a hypertension-prone individual. Rapp and Dahl (1972) in studies of steroidogenesis by the adrenals of the S and R strains showed a difference in 18-OH-desoxycorticosterone production, the output of the adrenals of the S strain being greater. However, they were able to account for only 16% of the blood pressure difference between the two strains on the basis of this. They concluded that the bulk of the blood pressure difference was accounted for by genes other than those controlling this element of steroidogenesis.

These intriguing data have been only cursorily summarized. To what extent they can be extrapolated to man is conjectural. The notion of inherited differences in functional capacity to handle chronic salt load, with higher blood pressures needed by some kidneys, could be a basis of the inherited tendency to hypertension in man. As Freis (1976) adds, an inherited defect in one or more of the ancillary feedback mechanisms for maintaining homeostasis of the extracellular fluid would equally explain the familial trend in hypertension. This, of course, embraces an increasingly wide field of humoral physiology, including steroid hormones and renin, the prostaglandins and kinins as well as central and peripheral neural regulation of the circulation (cf. Marx 1976).

If the inherited susceptibility to hypertension in man were related to sensitivity to salt, the formal possibilities include that it be an 'all or none' phenomenon with susceptible individuals showing it when chronic mean intake exceeded a certain threshold range. Simpson (1979) might be seen to accept a view along these lines in that he says in relation to northern Honshu that salt intake, blood pressure and mortality rate from stroke are highest in Japan, the salt intake is startlingly high (350–450 mmol/day or more), and it can be readily conceded that this sort of intake is undesirable. He adds that the question remains as to whether this fairly clear-cut conclusion that a very high sodium intake is bad should be extrapolated downwards to peoples with intakes of 170–200 mmol/day. Allowing that his assessment of the data on unacculturated peoples places more emphasis on undetermined factors than the virtually universal low salt intake, it might be inferred that he sees some threshold of effect operating in the range, for example, of 350 mmol/day.

Another possibility is that with increasing amounts of salt in the diet of the population, an increasing number of individuals exhibit the defect; this is consonant with the well-known Dahl diagram (Fig. 27-22). In this latter instance it might be argued that it is in those populations with the universal chronic high sodium intake that there may be the chance of showing some clear familial trends over three to four generations in incidence of hypertension or apparent resistance to it and its sequelae. In this respect, the rural communities of the Akita region might make a rewarding study.

Markers of genetic susceptibility

There is a substantial and growing body of evidence of disorder of intracellular electrolyte composition and sodium transport in essential hypertension. Tobian and Binnion (1952) were the first to describe an increase in the sodium content of renal arteries of hypertensive patients. Abnormalities of red cells were first demonstrated by Losse et al. (1960) and confirmed by von Gessler (1962), Wessels et al. (1967), Postnov et al. (1977) and Garay and Meyer (1979). There is general agreement that intracellular sodium concentration is raised and that demonstrable changes in sodium and potassium fluxes occur. Abnormalities have been demonstrated also in leucocytes by Edmondson et al. (1975) and Thomas et al. (1975).

Edmonson et al. (1975), in reporting a significant increase (mean 29%) in sodium and water content associated with depression (mean 30%) of the rate constant for active sodium efflux in the cells of hypertensives, remarked on the advantages for study of leucocytes over erythrocytes. They resemble body cells as a whole more closely in ultrastructure and metabolism, for the red cell lacks a nucleus and the capacity for aerobic respiration and protein synthesis. Also, sodium transport by leucocytes considerably exceeds that by erythrocytes. Changes in sodium content of lymphocytes in hypertensives have been shown also by Ambrosioni et al. (1979). Some groups have found the changes in sodium due to a change in permeability whereas others find a diminution in active transport. Wessels et al. (1967), Garay and Meyer (1979), Henningsen et al. (1979), Garay et al. (1980a, b) and Canessa et al. (1980) have found the same abnormality of transport in the normotensive relatives of essential hypertensive patients, whereas in several of the studies it was reported that the changes were not found in patients with secondary hypertension. Postnov et al. (1976) and de Mendonca et al. (1980) have reported changes in membrane permeability, and net Na^+ and K^+ fluxes respectively in the red cells of rats of various strains with spontaneous hypertension.

Postnov et al. (1977) in their report state that red cells of patients with hypertension differ from those of normotensive individuals by their higher passive permeability for sodium ions. Calcium depletion of the red cells by EDTA results in the removal of more calcium ions from the outer part of the red cell membrane in the case of hypertensive patients, suggesting altered calcium binding in the membrane. They suggest the findings may be a manifestation of a more widespread defect in cell membranes in essential hypertension.

Garay et al. (1980a, b, c) found that the net sodium/potassium flux ratios of all patients with secondary hypertension were the same as those of normotensive patients, whereas in patients with essential hypertension the ratios were always lower due to a high net sodium influx. Garay et al. (1980b) report there is a Na^+, K^+ co-transport system in normal erythrocytes which extrudes both internal Na^+ and K^+ and that this system is functionally deficient in erythrocytes of essential hypertensive patients and some of their descendants (Fig. 27-26). Thus the Na^+ and K^+ electrochemical gradients across cell membranes are believed to be maintained by the action of a Na^+, K^+ pump. In human erythrocytes this pump exchanges internal sodium for external potassium in approximately 1.5 ratio. In their studies they first loaded red cells with sodium. When sodium-loaded human erythrocytes are incubated in physiological conditions they recover their original low sodium/high potassium content. However, in normal healthy donors, the net Na^+ extrusion/K^+ influx ratio exceeded the 1.5 ratio predicted for Na^+,K^+ fluxes mediated by the sodium/potassium pump, whereas it was similar to this value in essential hypertensives and some of their descendants. The data suggest the net extrusion of a sodium load in an erythrocyte is accomplished by two different mechanisms—the Na^+,K^+ pump and Na^+,K^+ co-transport which operate against net Na^+ fluxes. Further that there is an inherited

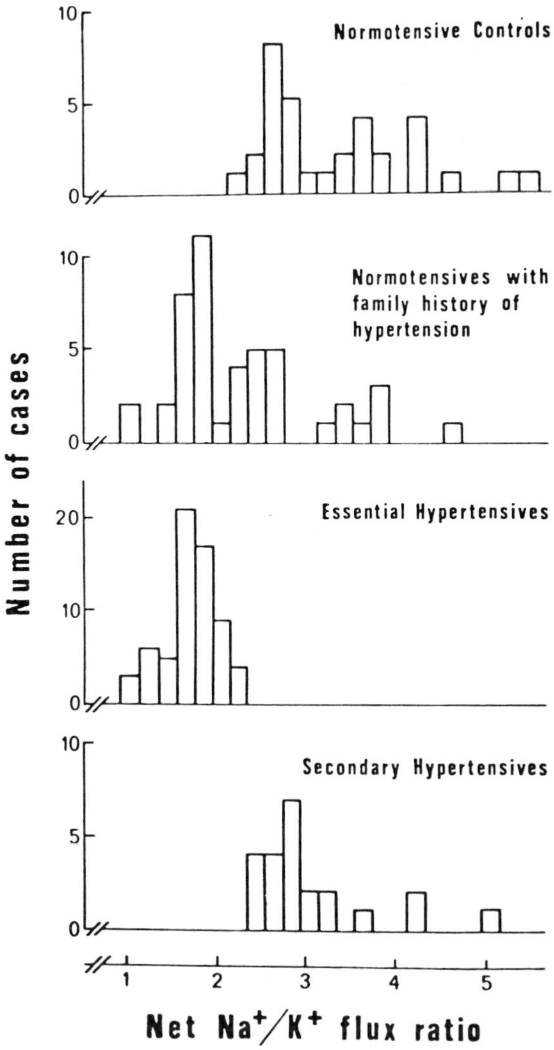

Fig. 27-26. Histograms of net sodium flux/net potassium flux for four populations: normotensive subjects with normotensive parents, young normotensive subjects with hypertensive parents, essential hypertensives, and secondary hypertensives with normotensive parents. (From Garay et al. 1980c, with permission.)

defect in the Na^+,K^+ co-transport system which may be genetically associated with essential hypertension. In avian red cells this system participates in regulation of cell volume and is under hormonal control.

Garay et al. (1980b) hypothesize that essential hypertension may result from an inefficient hormonal control of extrusion of a cell sodium load after excess sodium intake, due to functional disorder of the Na^+, K^+ co-transport system. Such a process in excitable cells of a high surface-to-volume ratio such as smooth muscle cells or catecholaminergic neurones may lead to a temporary or permanent increase in intracellular Na^+, producing critical changes capable of increasing blood pressure. The high Na^+,K^+ pump activity seen in normotensive offspring of hypertensives, and benign hypertensives, may represent a compensatory mechanism for extruding a cell sodium load and thus preventing severe hypertension in subjects with this genetic abnormality.

Meyer et al. (1981) studied inheritance of the red cell abnormality in subjects born to parents with essential hypertension. They found the deficiency occurred in 53% (52 out of 97) of those with one hypertensive parent and 73% (14 out of 19) of those with two hypertensive parents. Studies in 14 families over two to three generations showed erythrocyte cation abnormality in one or more members of each consecutive generation. They suggest that, in white people, abnormal erythrocyte cation transport is a biochemical disorder characteristic of essential hypertension, and is transmitted by a dominant and autosomal mode expressing a single abnormal gene.

Canessa et al. (1980) describe experiments showing that one of the pathways of sodium transport across the red cell membrane, Na^+/Li^+ countertransport, is faster in patients with essential hypertension than in control subjects. This transport system accepts only sodium or lithium and is not inhibited by ouabain. The maximum rate of Na^+/Li^+ countertransport was found to be 0.55 ± 0.02 (mean$\pm$s.e.m.) mmol/l of red cells per hour in 36 patients with essential hypertension and 0.24 ± 0.02 in 26 control subjects. First-degree relatives of eight patients with essential hypertension had rates of 0.54 ± 0.05 mmol. Patients with secondary hypertension had normal mean maximal rates.

In relation to the manner in which an abnormality of sodium and potassium transport, if also influencing arterial smooth muscle, might cause an alteration in vascular reactivity and hypertension, Blaustein (1977) has proposed that the tension of arteriolar smooth muscle is determined by intracellular calcium concentration, with an increase in concentration causing increased tension. As well as the Na^+,K^+ pump on the cell membrane there is a mechanism which moves calcium out of the cell in exchange for sodium. The activity of this sodium/calcium mechanism is influenced by intracellular sodium concentration so that a rise in concentration inhibits the exchange. Thus impairment of the main sodium/potassium pumping sufficient to cause a rise in intracellular sodium concentration in smooth muscle will depress sodium/calcium exchange, increase intracellular calcium and thus increase tension of muscle. The cells also become more sensitive to the action of vasoconstrictive agents (Overbeck et al. 1976).

de Wardener and MacGregor (1980) have proposed that the Dahl hypothesis that the rise in arterial pressure in salt-sensitive rats in due to a saluretic substance might be extended to essential hypertension in man. They suggest that the primary abnormality in essential hypertension in man is in the kidney, just as has been shown clearly by cross-transplantation kidney experiments in rats. The difficulty in sodium excretion causes increase in extracellular volume. This is continually corrected by increased concentration of a circulating substance which by inhibiting sodium reabsorption in the tubule increases sodium excretion. Thus the extracellular volume in hypertension is within normal limits but a sequence of events such as proposed by Blaustein (1977) and described above ensues, and increased tone of arterioles causes a rise in blood pressure. *According to this hypothesis the abnormality in sodium transport and increased intracellular sodium concentration are due to a circulating transport inhibitor and not to a hereditary defect of the cells themselves.*

Evidence for the latter viewpoint is that (i) plasma from a hypertensive man can alter the vascular reactivity of the rat into which it is injected (Michelakis et al. 1975), (ii) the sodium/potassium defects in red and white cells are not present in hypertensives treated with diuretics (von Gessler 1962; Thomas et al. 1975) whereas the direct action of diuretics on circulating cells, if any, should be to make the abnormalities worse, and (iii) the incubation of leucocytes from normotensive subjects in the plasma of patients with essential hypertension reduced the total sodium efflux rate constant and raised intracellular sodium to the same levels as those found in the leucocytes of patients with essential hypertension. This was due to a fall in the ouabain-sensitive component of the total sodium efflux rate constant (Poston et al. 1981).

Fitzgibbon et al. (1980) examined ^{22}Na efflux

from erythrocytes of normotensives and patients with mild to moderate hypertension. No difference was found when the cells were incubated in an artificial medium whereas when cells of both groups were incubated in their own plasma, the rate constant for efflux in the hypertensive subjects was decreased by an increased dietary intake of sodium. Woods et al. (1981) propose that the test system involving sodium-loaded potassium-depleted red cells used by Garay and colleagues might be criticized on the grounds of unphysiological conditions, and, accordingly, they have followed ^{86}Rb uptake by red cells, on the basis that rubidium uptake is handled by the sodium/potassium pump in the same way as potassium uptake. The activity of the ouabain-sensitive sodium/potassium pump was found using this method to be significantly greater in patients with untreated essential hypertension than in controls ($P<0.001$). The activity of the pump was also increased in a proportion of normotensive relatives of subjects with essential hypertension. They also made the interesting finding, in the light of other studies suggestive of slower excretion of a sodium load by normotensive black volunteers than whites, that rubidium uptake was significantly lower in normotensive black subjects than whites, the difference being in a ouabain-resistant pathway of cation transport.

The authors remark that it is difficult on present evidence to provide a unifying hypothesis to reconcile all data on disturbance of the various cation transport systems in essential hypertension. Nevertheless there is now impressive evidence that the abnormalities are not due to the raised blood pressure (since they are not seen in secondary hypertension) nor are artefacts of treatment. There are several ways abnormality of the ubiquitous Na,K-ATPase pump might affect sodium handling by the kidney or increase peripheral vascular resistance.

In relation to the notion (Laragh et al. 1979) that low renin hypertensives have a greater tendency to retain sodium and water, and that their renin does not respond normally to sodium restriction, Edmondson and MacGregor (1981) found the impairment of leucocyte sodium transport was significantly greater in eight patients whose plasma renin failed to rise to the normal range during a low-sodium diet than in 14 other patients whose renin system responded normally to sodium restriction. Forrester and Alleyne (1980) measured leucocyte electrolytes in pre-eclampsia and found sodium to be increased and sodium efflux depressed relative to normal primagravidae, and to the original pre-eclamptic subjects 6 months after delivery when blood pressure had returned to normal. They raise the question of whether the increased cell sodium affecting other tissues may underlie the hypertension of pre-eclampsia.

Clearly, in this field, great interest will attach to further incubation studies on leucocytes from normotensive individuals carried out in the serum of normotensive offspring of hypertensive patients, the red cells of the former showing the abnormalities described by Garay et al. (1980a, b, c), Meyer et al. (1981) or Canessa et al. (1980).

Natriuretic substances have been postulated by de Wardener and colleagues. One is proposed to have a molecular weight of less than 500 daltons and is short-acting (Clarkson et al. 1979). The other has a molecular weight possibly greater than 10 000 and is slow-acting (Clarkson et al. 1976). It is possible that such natriuretic substances are released when large natriuresis is caused by intracerebroventricular infusion of hypertonic sodium solutions, or that release of them is inhibited, associated with a large fall in sodium excretion, when sodium concentration in cerebrospinal fluid is reduced by intraventricular infusion of hypertonic 0.7 M mannitol-CSF in the water-deplete animal. These data have come from lateral and third ventricular infusion experiements on goats and sheep carried out in Stockholm and Melbourne (Andersson et al. 1969; McKinley et al. 1973; Leksell et al. 1981). The results, taken in association with the findings that lesions in the antero-ventral third ventricle involving the organum vasculosum of the lamina terminalis result in hypernatraemia and diminished ability to excrete a sodium load (Andersson et al. 1975; McKinley et al. 1980), are consistent with the idea of natriuretic hormone originating in the region or its release being controlled by a centre there.

A recent summary of the Lewis Dahl Symposium held at Brookhaven (Marx 1981) reported that de Wardener claimed to have identified a natriuretic substance in the hypothalamus which is an inhibitor of the sodium/potassium pump, and which is increased in the rat on a high sodium intake. Brody and Johnson, who have also shown that lesions of the anterior wall of the third ventricle reduce sodium excretion, have made collaborative experiments with Buckalew and Gruber whose work is cited immediately below. They have shown that such anterior wall lesions in animals with blood volume expansion prevent appearance of natriuretic hormone in blood whereas the non-lesioned animals have a high level of the hormone when volume-expanded.

Recently, Gruber et al. (1980) have reported the isolation of a substance in the plasma of dogs which competes with digoxin for two specific digoxin antibodies and is an inhibitor of Na,K-ATPase

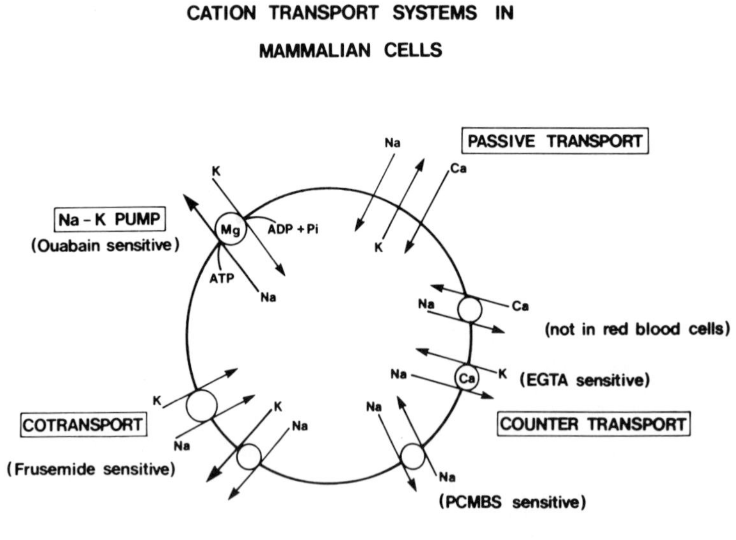

Fig. 27-27. Diagram of cation transport systems postulated to be involved in hypertension. These include the ouabain-sensitive sodium/potassium pump, a co-transport system and three countertransport systems. There is also passive transport of the sodium, potassium and calcium ions. (Kindly prepared by Dr Bruce Scoggins.)

activity. Increased amounts of the factor are found in the plasma of volume-expanded dogs, suggesting it is the putative natriuretic hormone. To identify this endogenous digitalis-like substance they used the same principle of approach as with the demonstration that biological substances which bind to the same receptors as drugs such as endogenous opioids and benzodiazepines may compete with antibodies specific to the drugs (Asano and Spector 1979).

A schematic outline of various membrane sodium and potassium transport systems, disorder of which might be involved in hypertension in man and animals, is shown in Fig. 27-27. Although there is no consensus as to which part of the system is altered in essential hypertension in man, it appears clear that membrane transport is different in normotensives and subjects with essential hypertension. Methodological differences may be responsible for the diversity of results which have been obtained so far. It is also not clear how these changes in membrane transport produce hypertension. A hypothesis based on suppression of the sodium/potassium pump has been proposed by Overbeck and colleagues (1976). Fig. 27-28 details the sequence of events that may be involved in hypertension associated with pump suppression.

Expression of genetic influence in the population

The increased prevalence of hypertension and steeper grades of blood pressure in the US and Caribbean blacks could be viewed in the light of the finding of Desor et al. (1975) on the much greater preference for concentrated salt solutions of the 9–15-year-old blacks over that of their white counterparts of the same age, and over either black or white adults. This may be a reflection, in some way, of overall dietetic habit resulting from stress or circumstances of a socioeconomic nature. However, other possibilities, admittedly quite speculative, should be considered.

The ancestors of the people of the United States reached its shores in many different ways. The survival advantage to tropical-dwelling humans and their ancestors of sodium appetite mechanisms has already been argued (Chapter 4). The passage from hominids of the Pleistocene in tropical/subtropical Africa to white settlement of America has been a circuitous one. Perhaps twenty to forty thousand generations of existence in Europe and Northern Europe occurred. It has been postulated, for example, that, of the many changes that occurred over this period, one involved a regression of skin pigmentation, since a white skin involved more efficient vitamin D production in the limited sunlight of Europe's cold north. This may have had profound implications for normal bone growth and therefore facility of labour and reproduction. The cold climate with absence of sweating may also have ameliorated stress on sodium homeostasis, and this may have been accompanied by regression of hedonic responses to sodium salts. On the other hand, the ancestors of black citizens of America reached the country two to four hundred years ago as a result of forcible transportation during the slave trade. Their progenitors may correspondingly represent twenty to forty thousand or more generations of existence in tropical to subtropical conditions largely in the interior of the continents, and with considerable stress on sodium homeostasis. Greater hedonic

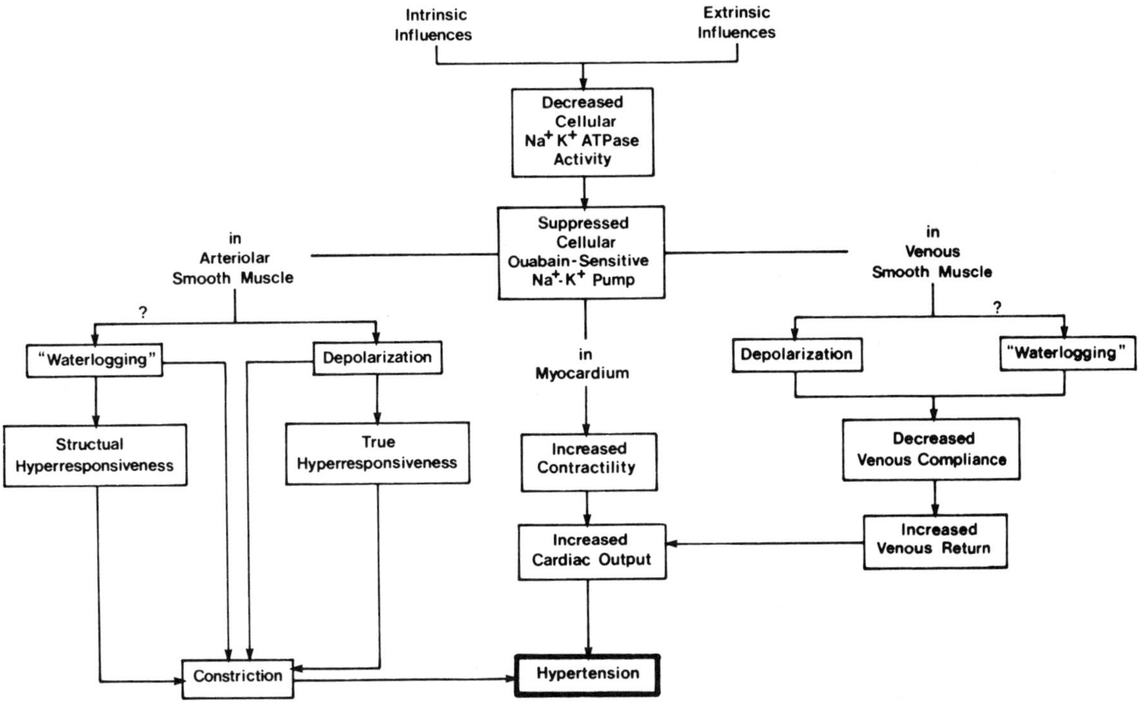

Fig. 27-28. Possible roles of depressed cellular ouabain-sensitive sodium/potassium pump in hypertension. (After Overbreck et al. 1976.)

liking of salt and less capacity to excrete it could result.

These conjectures at least point up the interest of studies on the hedonic salt preference of black children in central Africa to determine whether the same trend is seen as with those in the USA studied by Desor et al. (1975). It would also be valuable to analyse renal response to sodium load relative to that seen in children of Scandinavian or celtic descent. A positive result could, allowing the toll of hypertensive vascular disease, argue for some preventive measures on salt intake, analogous to those followed in Australia where fair-skinned people of celtic descent are advised rigidly to restrict skin exposure to sunlight in an effort to decrease the high incidence of skin cancer in people with this ethnic background. As discussed further on, the data of Luft et al. (1979) on sodium excretion by adult blacks are consistent with this proposition.

Much evidence suggests the primary defect in essential hypertension is within the kidney. This is in relation to its functional capacity to excrete effectively a moderate to high salt load over decades of life without entraining humoral sequelae which lead to a reset of the physiological regulatory systems with a higher blood pressure. There is little doubt that there are many of the population who can eat a high-salt diet for life and have no rise in blood pressure. Whereas, in unacculturated societies, blood pressure does not rise with age, on the above line of reasoning, as dietetic intake of salt increases in such a population in the period of transition to a Western way of life, individuals with genetically restricted functional capacity for excretion without blood pressure readjustment would be affected. This may be anything but a simple threshold. There may be polygenic determinants of a spread of functional capacity and hence level of susceptibility. Variation in hedonic salt intake, either aggravating the situation in some only marginally susceptible individuals, or ameliorating it in some highly susceptible individuals, is likely. Operation of the latter effect may be diminishing considerably in some segments of industrialized Western society, where processed foods and fast-food chain outlets ensure that a majority receive an obligatory high salt intake counteracting to a degree the progressively reduced intake envisaged by Joossens as following the introduction of refrigeration.

If there were any truth in a hypothesis which places the aetiology of essential hypertension primarily in the interaction between salt intake and a genetically determined functional capacity to excrete salt, experimental testing should reveal some support. The difficulties are great because of the time scale of the process. Comprehensive studies on physiological handling of a uniform salt load sustained over 1–4 weeks in relation to the time characteristics of excretion, changes in plasma and extracellular fluid volume, blood pressure, heart rate and endocrine secretions, could be revealing if it were feasible to do on a sufficiently large number in a community. This might involve the children of hypertensive parents at an age before there was any evidence of a blood pressure change, versus controls of the same age with exemplary family history.

Patients with essential hypertension have an accelerated natriuresis when given an intravenous saline load (Baldwin et al. 1958; Lowenstein et al. 1970) which is greatest in patients with low renin activity (Krakoff et al. 1970). The phenomenon can be seen in the normotensive sons of hypertensive fathers (Wiggins et al. 1978). Bianchi et al. (1978) found that normotensive children of hypertensive parents have a significantly raised renal plasma flow, and some tendency to lower renin activity which may suggest a slightly raised extracellular fluid volume.

Murray et al. (1978) studied the effect of increasing salt intake in eight normotensive males from 10 to 300, 800 and finally to 1500 mmol/day over a period of 16 days. Increase in salt intake in this acute fashion resulted in a progressive increase in weight, blood pressure, potassium excretion and creatinine clearance and a decrease in plasma renin activity and aldosterone concentration. The relation of systolic and diastolic blood pressure and sodium excretion was highly significant. But the increase after 3 days at 300 mmol/day was not significant. Cardiac output increased and calculated vascular resistance decreased. These data, and the suggestions the authors make on their being relevant to the conceptual framework integrating blood pressure and renal salt excretion, were an issue of debate in the recent Round Table on this subject of salt and hypertension at the Goteborg Symposium. Pickering (1980) emphasized the point that it was only after 800 mmol/day that any increase in blood pressure was seen. Tobian et al. (1979) remarked that there must be a lot of negative sodium/blood pressure correlations just because seven out of eight individuals may not be affected much by 170/200 mmol/day of sodium even if they might be affected by 800 mmol/day. However, apart from showing, perhaps, a quantitative relation between excretion of sodium load and blood pressure as postulated by Guyton and colleagues (1974), these data on acute increase of sodium load over but 9 days would seem to have little bearing on the issue of 30–40 years of exposure to 150–250 mmol/day of sodium versus 5–50 mmol/day in an unacculturated society.

Burstyn et al. (1980) in studies on normotensive volunteers found no change in blood pressure when sodium intake was trebled over 8 days, nor when potassium intake was doubled over 22 days in 21 volunteers. Sullivan et al. (1980) found six normal subjects placed on a 400 mmol/day sodium diet had a 16% increase in mean arterial pressure ($P<0.01$). Moderate sodium loading (200 mmol/day diet) caused a rise in blood pressure in only 4 of 27 normal subjects, whereas it increased in 14 of 19 borderline hypertensives. Their analysis indicated that many patients with borderline hypertension differed from normal subjects by displaying a blood pressure increase due to a disproportionate rise in cardiac index and an inadequate fall in total peripheral resistance in response to an acute increase in dietary sodium.

In a study from Luft's group (Grim et al. 1979), in order to examine potential mechanisms responsible for a greater incidence of hypertension in relatives of hypertensives than among relatives of normotensives, 43 normotensive first-degree relatives of known essential hypertensives and 43 age-, race- and sex-matched normal subjects with no family history of hypertension were subjected to sodium loading and depletion. The relatives had higher basal blood pressure and higher plasma renin activities before and after infusion of 2 litres of intravenous saline. The relatives excreted less ($P<0.05$) sodium on the day of the infusion. There was a correlation between age and blood pressure, and an inverse correlation between age and sodium excretion on the sodium-loading day, in relatives but not in the control subjects. There was also a correlation between both systolic and diastolic pressure on the sodium-loading day and the fractional excretion of sodium in the relatives but not in the control subjects. No differences were observed in sodium excretion with sodium depletion. The authors interpret these important data as suggestive that the tendency of blood pressure to increase with age is under genetic influence which may be mediated by the renal ability to excrete sodium. This in turn may be mediated by genetic influences on the renin system. The high salt intake of acculturated societies may make this chain of events manifest.

Luft and colleagues (1979) have also examined cardiovascular and humoral responses to extremes of sodium intake in black and white males with the aim of clarifying possible differences relevant to the increased prevalence of hypertension amongst blacks. Fourteen normotensive men were studied (seven black and seven white) and sodium intake was increased acutely in 3-day steps from a baseline of 10 mmol/day to 1500 mmol/day. Blood pressure and cardiac index increased, as in the study of Murray et al. (1978). Linear and quadratic regression analyses showed that blacks had higher blood pressures with sodium loading than whites (Fig. 27-29). An earlier report of the group showed blacks did not excrete an intravenous salt load as well as whites over a 24-h period, and there was a greater suppression of renin activity. The differences were not attributable to a difference in creatinine or *p*-aminohippuric acid (PAH) clearance. The authors also draw attention to the report of Grim et al. (1970) that no differences were found in sodium intake on a racial basis in a study in Georgia in the course of emphasizing the possibility of a renal difference in the susceptibility of blacks and whites to a hypertensinogenic effect of sodium.

Luft et al. (1979) also showed that whereas blacks and whites developed similar degrees of sodium retention after sodium loading, the potassium metabolism differed. Blacks accumulated more potassium when on a 10 mmol/day sodium intake and excreted less than the whites when sodium-loaded. In a follow-up study on three blacks and three whites where potassium losses associated with natriuresis were compensated for by increased intake, the blood pressure did not rise to the same degree, and natriuretic responses were greater. They note the effect may have been non-specific as chlorides other than potassium were not studied. The mechanism of effect was undetermined. Overall, Luft et al. (1979) see the data as indicative that blacks have an intrinsic reduction in the ability to excrete sodium compared with whites.

If the salt intake/genetic susceptibility interaction were the primary cause of high blood pressure it is easy to see physiologically, though extraordinarily difficult to prove, how other factors provisionally indicated might influence the process. At least it is feasible that psychogenic stress, chronic over years, may influence sympathetic nervous activity and, via that and other vectors,

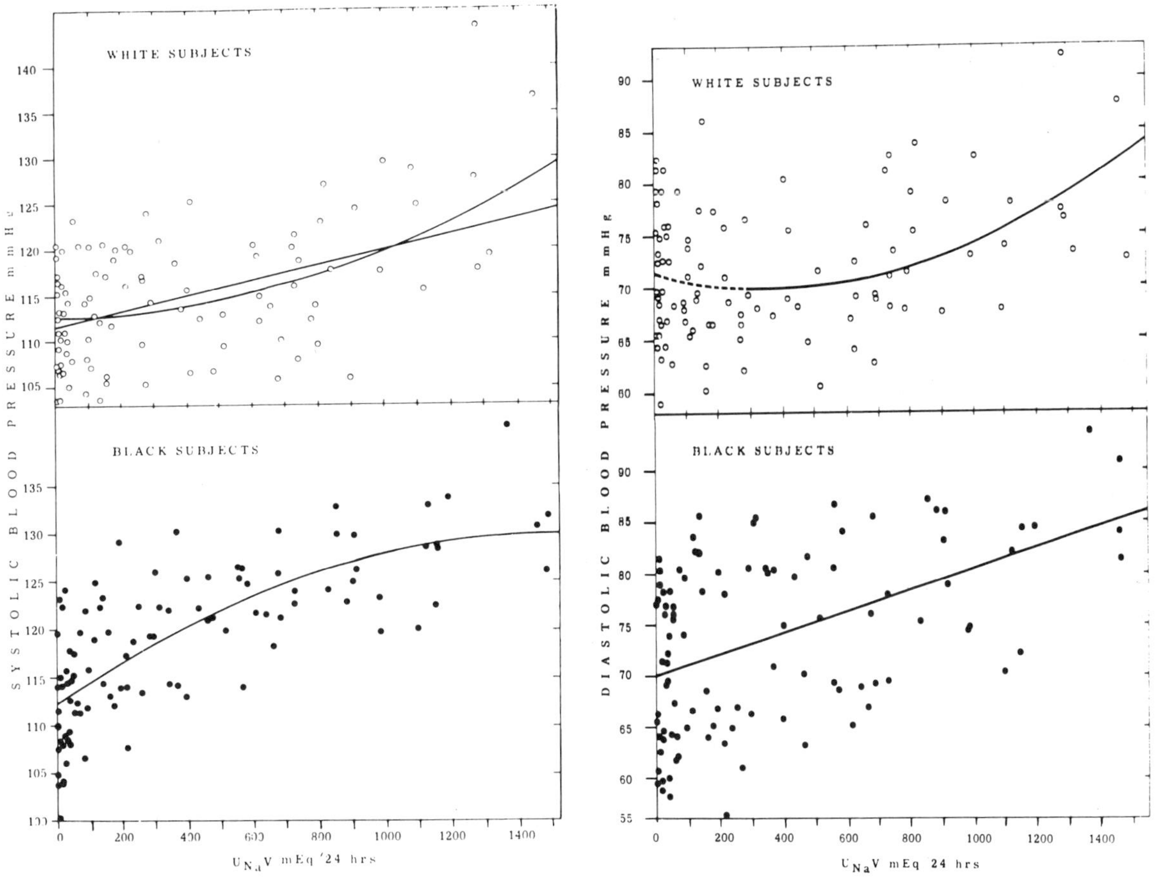

Fig. 27-29. Left: Relationship between urinary sodium excretion ($U_{Na}V$) and systolic pressure in whites and blacks. In whites the quadratic expression is convex downward; in blacks it is convex upwards ($P<0.005$).

Right: Relationship between urinary sodium excretion ($U_{Na}V$) and diastolic pressure in whites and blacks. The part of the parabola presented as a broken line is not representative of the actual data. In blacks the quadratic was no improvement over the linear expression. The linear regressions of whites and blacks were significantly different. (From Luft et al. 1979, with permission.)

hormone secretions. The integral effect over one or two decades of having continuously, or for many hours a day, increased levels of adrenal cortical hormone and noradrenaline could be the aggravation of any existing restricted capacity to eliminate salt—apart from any other hypertensinogenic action the hormones might have (Scoggins et al. 1978). There is some interest in the fact that with induction of an experimental neurosis in sheep, sustained high peripheral blood concentration of aldosterone was found over some weeks independently of an increase in ACTH (Blair-West et al. 1970).

In this regard, Henry and Cassel (1969) see psychological factors having a dominant role in essential hypertension, and obesity, salt and fat intake having a subsidiary influence. Both animal and human studies indicated that repeated arousal of the defence alarm response may be involved. They hold that in man such arousal occurs when previously socially sanctioned patterns of behaviour, especially those to which the organism has become adapted during critical early learning periods, can no longer be used to express normal behavioural urges. Thus difficulties of adaptation, as where there is status ambiguity, may result in years of repeated arousals of vascular autonomic and hormonal function due to the organism's perception of certain events as threatening. These, in turn, can lead to progressive and eventually irreversible disturbances such as renal hypertension, heart failure, or cerebrovascular disease.

Henry and Cassel (1969) reviewed a diversity of animal studies attesting to the influence of hypothalamic–reticular activation on blood pressure (Charvat, Dell and Folkow 1964). Pavlovian conditioning procedures in dogs have caused sustained high blood pressure (Gavlichek 1952; Pickering 1961a) and studies on monkeys and baboons suggest that induction of high blood pressure depends on interference with basic responses such as sexual or self-protective mechanisms (Miminoshvili 1960; Miashnikov 1962). High blood pressure has also been observed in chimpanzees undergoing aversive operant conditioning in preparation for space flight (Meehan et al. 1964).

The very interesting studies of Brod et al. (1959) have shown how harassment of a time stress can produce large physiological changes during conduct of mental arithmetic, the effects on hormonal and vegetative neural systems being typical of a defence alarm state. Thus there is elevation of blood pressure, cardiac output and heart rate. The blood flow through muscles is increased, but that through gastrointestinal tract and kidneys is greatly reduced. Henry and Cassel (1969) conclude that stimuli which lead to activation of defence alarm reactions are not necessarily associated with conscious fight or flight responses. The visceral responses may take place without any awareness. Within any social group there are systematic differences in the level of stimulation and hence of neuroendocrine arousal dependent on differences of response pattern and individual positions in the social hierarchy. Those in subordinate positions may have greater response, and the organism's response may depend on past history of stimulation and, in particular, upon early experience.

These considerations and those outlined in the discussion of the work of Scotch and Geiger (1963) also bring to focus the lack of data on the influence of ACTH and steroid secretion induced by it on salt appetite in man. The data in Chapters 18 and 23 show the large appetite which occurs in three animal species examined.

Conclusion

Overall, it can be recognized that civilization and technology, with greatly increased processing of foodstuffs in affluent societies, have introduced the issue of the subtle and aggregate effects of long-term exposure to excess of substances which are normal, indeed essential, components of diet. There are good grounds, but by no means a proven case, for suspecting excess salt intake, probably associated with reduced potassium intake, in the aetiology of hypertension in Western-type communities. If not a direct cause, perhaps the high salt intake throughout life is an important factor in the expression of other jointly sufficient and severally necessary factors which contrive the hypertensive state (Denton 1973, 1976). However, as already suggested, the proper assessment at this point would appear to be the Scottish court verdict: not proven.

One might wonder whether it ever can be proven, and whether all advice or measures at a public health level should be withheld until unequivocal proof is available. That is, whether assessment of the probabilities emerging in this chapter, and notwithstanding oversights therein, leaves the observer unable to say whether it makes any difference of consequence in a population whether daily salt intake life-long is 10–60 mmol/day as against 200–400 mmol/day. There can be little doubt that a large number of individuals can have an intake of 200–400 mmol/day of sodium yet have a low blood pressure in later life. Possibly the emergent data of a change in sodium/

potassium transport systems in red and white blood cells may provide a basis for identifying those at risk, by reflecting genetic susceptibility in terms of a premonitory rise in the plasma of a saluretic and hypertensinogenic agent. In the face of a malady affecting 15%–20% of the adult population in Western-type industrial societies and with major implications for morbidity, mortality and the costs to the health care system, many would feel the probabilities would make a preventive approach a good investment.

Of course, it would be possible to obtain convincing evidence on the relation between salt intake and hypertension for a population. In some of the few remaining unacculturated societies where ethnically similar people are also scattered and virtually isolated from one another (e.g. the Yanomamö or remote tribes in the western New Guinea Highlands), it might be possible to introduce salt in abundant quantity to some groups as the only change in life-style, and not do so for others, and allow a sufficient lapse of time to occur without other intrusions to observe the result. It could resolve the matter. As far as I see the probabilities I think it would be very unethical in practice as well as in principle, because blood pressure would rise in the exposed unacculturated communities.

References

Abrahams DG, Alele CA, Barnard BG (1960) The systemic blood pressure in a rural West African community. West Afr Med J 9: 45

Allen FM, Sherrill J (1922) Treatment of arterial hypertension. Journal of Metabolic Research 2: 429

Altschul AM, Grommet JK (1980) Sodium intake and sodium sensitivity. Nutr Rev 38: 393

Alwall N (1958) Forum Medicum (Istanbul) 11: 65

Ambard L, Beaujard E (1904) Causes de l'hypertension arterielle. Arch Gen Med Paris 1: 520

Ambrosioni E, Tantagni F, Montebugnoli L, Magnani B (1979) Intralymphocytic sodium in hypertensive patients: A significant correlation. Clin Sci 57: 325s

Amery A, Bulpitt C, Fagard R, Slassen J (1980) Does diet matter in hypertension? Eur Heart J

Andersson B, Dallman MF, Olsson K (1969) Evidence for a hypothalamic control of renal sodium excretion. Acta Physiol Scand 75: 496

Andersson B, Leksell LG, Lishajko F (1975) Perturbations in fluid balance induced by medially placed forebrain lesions. Brain Res 99: 261

Aoki K, Yamori Y, Ooshima A, Okamoto K (1972) Effects of high or low sodium intake in spontaneously hypertensive rats. Jpn Circ J 36: 539

Asano T, Spector S (1979) Identification of inosine and hypoxanthine as endogenous ligands for the brain benzodiazepine-binding sites. Proc Natl Acad Sci USA 76: 977

Ashe BI, Mosenthal HO (1937) Protein, salt and fluid consumption of a thousand residents of New York. JAMA 108: 1160

Ayman D (1949) Critique of reports of surgical and dietary therapy in hypertension. JAMA 141: 974

Bailey KV (1963) Blood pressure in undernourished Javanese. Br Med J ii: 775

Baldwin DS, Biggs AW, Goldring W, Hulet WH, Chasis H (1958) Exaggerated natriuresis in essential hypertension. Am J Med 24: 893

Ball OT, Meneely GR (1957) Observations on dietary sodium chloride. J Am Diet Assoc 33: 366

Bare JK (1949) The specific hunger for sodium chloride in normal and adrenalectomized white rats. J Comp Physiol Psychol 42: 242

Barnes R (1965) Comparisons of blood pressures and blood cholesterol levels of New Guineans and Australians. Med J Aust 1: 611

Battarbee HD, Funch DP, Dailey JW (1979) The effect of dietary sodium and potassium upon blood pressure and catecholamine excretion in the rat. Proc Soc Exp Biol Med 161: 32

Bays RP, Scrimshaw NS (1953) Facts and fallacies regarding the blood pressure of different regional and racial groups. Circulation 8: 655

Becker BJP (1946) Cardiovascular disease in the Bantu and coloured races of South Africa: Hypertensive heart disease. S Afr Med J 11: 107

Berglund G (1980) Should salt intake be cut down to prevent primary hypertension? Acta Med Scand 207: 241

Bernard RA, Doty RL, Engelman K, Weiss RA (1980) Taste and salt intake in human hypertension. In: Kare M, Bernard R, Fregly M (eds) Biological and behavioural aspects of salt intake. Academic Press, New York, p. 397

Bianchi G, Fox V, di Francesco GF, Giovannetti AM, Pagetti D (1974) Blood pressure changes produced by kidney cross-transplantation between spontaneously hypertensive rats (SHR) and normotensive rats (NR). Clin Sci Mol Med 47: 435

Bianchi G, Picotti GB, Bracchi G, Cusi D, Gatti M, Lupi GP, Ferrari P, Barlassina C, Colombo G, Gori D (1978) Familial hypertension and hormonal profile, renal haemodynamics and body fluids of young normotensive subjects. Clin Sci Mol Med 55: 367s

Bibile SW, Cullumbine H, Kirtisinghe C, Watson RS, Wikramanayake T (1949) Variation with age and sex of blood pressure and pulse rate, for Ceylonese subjects. Ceylon J Med Sci 6: 79

Blair-West JR, Coghlan JP, Denton DA, Funder JW, Nelson J, Scoggins BA, Wright RD (1970) Sodium homeostasis, salt appetite, and hypertension. Circ Res [Suppl 2] 26/27: 251

Blaustein MP (1977) Sodium ions, calcium ions, blood pressure regulation, and hypertension. A reassessment and a hypothesis. Am J Physiol 232: C165

Boe J, Humerfelt S, Wedervang F (1957) The blood pressure in a population; Blood pressure readings and height and weight determinations in the adult population of the city of Bergen. Acta Med Scand [Suppl 321] 157

Borst JGG, Borst EGA (1963) Hypertension explained by Starling's theory of circulatory homeostasis. Lancet I: 677

Boyle E (1970) Biological patterns in hypertension by race, sex, body weight and skin colour. JAMA 213: 1637

Bradley RM, Mistretta CM (1980) Developmental changes in neurophysiological taste responses from the medulla in sheep. Brain Res 191: 21

Brod J, Fencl V, Hejl Z, Jirka J (1959) Circulatory changes underlying blood pressure elevation during acute emotional stress (Mental Arithmetic) in normotensive and hypertensive subjects. Clin Sci 18: 269

Bull NL, Buss DH (1980) Contributions of foods to sodium intakes. Proc Nutr Soc 39: 30A

Burns-Cox CJ, Maclean JD (1970) Splenomegaly and blood pressure in an Orang Asli community in West Malaysia. Am Heart J 80: 718

Burstyn P, Hornall DEE, Watchorn C (1980) Sodium and potassium intake and blood pressure. Br Med J 281: 537

Calabrese EJ, Tuthill RW (1977) Elevated blood pressure and high sodium levels in the public drinking water: Preliminary results of a study of high school students. Arch Environ Health 32: 200

Canessa M, Adragna N, Solomon HS, Connolly TM, Tosteson DC (1980) Increased sodium–lithium countertransport in red cells of patients with essential hypertension. N Engl J Med 302: 772

Catalanotto F, Schechter PJ, Henkin RI (1972) Preference for NaCl in the spontaneously hypertensive rat. Life Sci 11: 557

Cha Kyung Ok, Suh Soon Kyu (1970) Study on sodium, chloride, potassium in Korean foods and well-water. Journal of Woo Sok University Medical College 7: 171

Chagnon NA (1977) Yanomamö: the fierce people, 2nd edn. Holt Rinehart & Winston, New York

Chagnon NA, Hames RB (1979) Protein deficiency and tribal warfare in Amazonia: New data. Science 203: 910

Charvat J, Dell P, Folkow B (1964) Mental factors and cardiovascular diseases. Cardiologia 44: 124

Cherchovich GM, Capek K, Jefremova Z, Pohlova I, Jelinek J (1976) High salt intake and blood pressure in lower primates (*Papio hamadryas*). J Appl Physiol 40: 601

Clark JD (1970) The prehistory of Africa. Praeger, New York

Clarkson EM, Raw SM, de Wardener HE (1976) Two natriuretic substances in extracts of urine from normal man when salt-depleted and salt-loaded. Kidney Int 10: 387

Clarkson EM, Raw SM, de Wardener HE (1979) Further observations on a low molecular weight natriuretic substance in the urine of normal man. Kidney Int 16: 710

Cobb S, Rose RM (1973) Hypertension, peptic ulcer and diabetes in air traffic controllers. JAMA 224: 489

Coelho JB (1974) Sodium metabolism and intrarenal distribution of nephron glomerular filtration rates in an unanesthetized rat. Proc Soc Exp Biol Med 146: 225

Comstock GW (1957) An epidemiologic study of blood pressure levels in a biracial community in the southern United States. American Journal of Hygiene 65: 271

Contreras RJ (1978) Salt taste and disease. Am J Clin Nutr 31: 1088

Contreras RJ, Kosten T, Berg AT (1981) The effect of prenatal and early postnatal sodium chloride on the saline preferences and blood pressure of adult rats. In: Abstracts of the 3rd Meeting of the Association of Chemoreceptor Sciences, Sarasota, Florida

Corcoran AC, Taylor RB, Page IH (1951) Controlled observations on the effect of low sodium dietotherapy in essential hypertension. Circulation 3: 1

Cottier PT, Weller JM, Hoobler SW (1958) Effect of an intravenous sodium chloride load on renal hemodynamics and electrolyte excretion in essential hypertension. Circulation 17: 750

Craig W (1918) Appetites and aversions as constituents of instincts. Biol Bull 2: 91

Cruz-Coke R, Covarrubias E (1962) Factors influencing blood pressure in a rural Chilean community. Lancet II: 1138

Cruz-Coke R, Etchevery R, Nagel R (1964) Influence of migration on blood pressure of Easter Islanders. Lancet I: 697

Cruz-Coke R, Donoso H, Barrera R (1973) Genetic ecology of hypertension. Clin Sci Mol Med 45: 55s

Dahl LK (1957) Evidence for an increased intake of sodium in hypertension based on urinary excretion of sodium. Proc Soc Exp Biol Med 94: 23

Dahl LK (1958a) Sodium intake of the American male: Implications on the aetiology of essential hypertension. Am J Clin Nutr 6: 1

Dahl LK (1958b) Salt intake and salt need. N Engl J Med 258: 1152

Dahl LK (1960) Possible role of salt intake in the development of essential hypertension. In: Bock KD, Cottier PT (eds) Essential hypertension. Springer, Berlin, p 53

Dahl LK (1968a) Letters to the Editor. Nutr Rev 26: 124

Dahl LK (1968b) Salt in processed baby foods. Am J Clin Nutr 21: 787

Dahl LK (1972) Salt and hypertension. Am J Clin Nutr 25: 231

Dahl LK, Love RA (1954) Evidence for relationship between sodium (chloride) intake in human essential hypertension. AMA Arch Intern Med 94: 525

Dahl LK, Knudsen D, Iwai J (1969) Humoural transmission of hypertension: Evidence from parabiosis. Circ Res [Suppl 1] 24/25: I–21

Dahl LK, Heine M, Litel G, Tassinari L (1970) Hypertension and death from consumption of processed baby foods in rats. Proc Soc Exp Biol Med 133: 1405

Dahl LK, Heine M, Thompson K (1972) Genetic influence of renal homografts on the blood pressure of rats from different strains. Proc Soc Exp Biol Med 140: 852

Dahl LK, Heine M, Thompson K (1974) Genetic influence of the kidneys on blood pressure. Evidence from chronic renal homografts in rats with opposite predispositions to hypertension. Circ Res 40: 94

Dale NE (1979) Sodium and potassium content of some Australian foods and beverages. Med J Aust 2: 354

Darrow DC, Cooke RE, Seagar WE (1954) Water and electrolyte metabolism in infants fed cow milk mixtures during heat stress. Pediatrics 14: 602

Dawber TR, Cannel WB, Cagan A, Donabedian RK, MacNamara PM, Pearson G (1967) Environmental factors in hypertension. In: Stamler J, Stamler R, Pulman TM (eds) The epidemiology of hypertension. Proceedings of an International Symposium. Grune & Stratton, New York, p 225

Denton DA (1969) Salt appetite. Nutrition Abstracts and Reviews 39: 1043

Denton DA (1973) Sodium and hypertension. In: Sambhi MP (ed) Mechanisms of hypertension. Proceedings of an international workshop conference in Los Angeles. Excerpta Medica, Amsterdam, p 46

Denton DA (1976) Hypertension: A malady of civilization? In: Sambhi MP (ed) Systemic effects of antihypertensive agents. Symposia Specialists, New York, p. 577

Denton DA, Nelson JF, Orchard E, Weller S (1969) The role of adrenocortical hormone secretion in salt appetite. In: Pfaffmann C (ed) Third International Symposium of Olfaction and Taste. Rockefeller University Press, New York, p 535

Desor JA, Greene LS, Maller O (1975) Preferences for sweet and salty in nine to fifteen year old and adult humans. Science 190: 686

Desor JA, Maller O, Greene LS (1977) Preference for sweet in humans: Infants, children and adults. In: Weiffenbach JM (ed) Taste and development, genesis of sweet preference. US Department of Health, Education and Welfare, Bethesda (DHEW publication no. [NIH] 77-1068), p 161

Dickinson WE (1977) No salt agreement: The Salt Institute's statement to the Senate Select Committee on Nutrition in opposition to dietary goals for the United States, specifically, a reduction of salt use. Snack Food Magazine, October

Dickinson WE (1980) Salt sources and markets. In: Kare M, Fregly MJ, Bernard RA (eds) Biological and behavioural aspects of salt intake. Academic Press, New York, p 49

Dodson PM (1980) Dietary fibre, sodium and blood pressure. Br Med J 280: 564

Dole VP, Dahl LK, Cotzias GC, Eder HA, Krebs ME (1950) Dietary treatment of hypertension. Clinical and metabolic studies of patients on the rice–fruit diet. J Clin Invest 29: 1189

Dole VP, Dahl LK, Cotzias GC, Dziewiatkowski DD, Harries

C (1951) Dietary treatment of hypertension. II. Sodium depletion as related to the therapeutic effect. J Clin Invest 30: 584
Donnison CP (1929) Blood pressure in the African native. Lancet I: 6
Doyle AE, Chua KG, Duffy S (1976) Urinary sodium, potassium and creatinine excretion in hypertensive and normotensive Australians. Med J Aust 2: 898
Draper P (1973) Crowding among hunter-gatherers: The !Kung bushmen. Science 182: 301
Dreyfuss F, Hamosh P, Adam YG, Kallner B (1961) Coronary heart disease and hypertension among Jews immigrated to Israel from the Atlas Mountain region of North Africa. Am Heart J 62: 470
Edmondson RPS, MacGregor GA (1981) Leucocyte cation transport in essential hypertension: Its relation to the renin–angiotensin system. Br Med J 282: 1267
Edmondson RP, Thomas RD, Hilton PJ, Jones N (1975) Abnormal leucocyte composition and sodium transport in essential hypertension. Lancet I: 1003
Edwards F, McKeown T, Whitfield AGW (1959) Arterial pressure in men over sixty. Clin Sci 18: 289
Epstein FH (1963) Epidemiological studies on the nature of high blood pressure. In: Proceedings of the 15th Annual Conference on the Kidney. Little Brown, Boston, p 263
Epstein FH, Eckoff RD (1967) The epidemiology of high blood pressure: Geographic distributions and aetiological factors. In: Stamler J, Stamler R, Pulman TN (eds) The epidemiology of hypertension. Grune & Stratton, New York, p 155
Fallis N, Lasagna L, Tetreau LT (1962) Gustatory thresholds in patients with hypertension. Nature 196: 74
Fatula MI (1977) Effect of water with a high sodium chloride content on the incidence of arterial hypertension and temporary invalidity. Gig Sanit 2: 7
Filer LJ (1971) Salt in infant foods. Nutr Rev 29: 27
Fitzgibbon WR, Morgan TO, Myers JB (1980) Erythrocyte ^{22}Na efflux and urinary sodium excretion in essential hypertension. Clin Sci 59: 195s
Floyer MS, Richardson PC (1961) Mechanism of arterial hypertension. Role of capacity and resistance vessels. Lancet I: 253
Fodor JG, Abit EC, Rusted IE (1973) Epidemiologic study of hypertension in Newfoundland. Can Med Assoc J 108: 365
Folkow B, Hallback M, Lundgren Y, Sivertson R, Weiss L (1973) Importance of adaptive changes in vascular design for establishment of primary hypertension studies in man and in spontaneously hypertensive rats. Circ Res 33[Suppl 1]: i–2
Fomon SJ (1967) Infant nutrition. Saunders, Philadelphia, p 141
Forman S, Falk JL (1979) NaCl solution ingestion in genetic (SHR) and aortic ligation hypertension. Physiol Behav 22: 371
Forrester TE, Alleyne GAO (1980) Leucocyte electrolytes and sodium efflux rate constants in the hypertension of pre-eclampsia. Clin Sci 69: 199s
Forsythe RH, Miller RA (1980) Salt in processed foods. In: Kare M, Bernard R, Fregly M (eds) Biological and behavioural aspects of salt intake. Academic Press, New York, p. 221
Fregly MJ (1957) The interaction of pregnancy and hypertension. Acta Physiol Pharmacol Neerl 5: 278
Freis ED (1960) Haemodynamics of hypertension. Physiol Rev 40: 27
Freis ED (1976) Salt, volume and prevention of hypertension. Circulation 53: 589
Freis ED (1979) Salt in hypertension and the effects of diuretics. Annu Rev Pharmacol Toxicol 19: 13
Froment A, Milon H, Gravier Ch (1979) Relation entre consommation sodée et hypertension artérielle. Contribution de l'épidémiologie géographique. Rev Epidém Santé Publ. 27: 437
Fujita T, Henry WL, Bartter FC, Lake CR, Delea CS (1980) Factors influencing blood pressure in salt-sensitive patients with hypertension. Am J Med 69: 334
Fukuda T (1954) Investigation on hypertension in farm villages in the Akita prefecture. J Chiba Med Soc 29: 490
Gamble JL, Wallace WM, Eliel L, Holliday MA, Cushman M, Appleton J, Hender A, Piotti J (1951) Effects of large loads of electrolytes. Pediatrics 7: 305
Garay RP, Meyer P (1979) A new test showing abnormal net Na and K fluxes in erythrocytes of essential hypertensive patients. Lancet I: 349
Garay RP, Dagher G, Pernollet M-G, Devynck M-A, Meyer P (1980a) Inherited defect in a Na^+,K^+-co-transport system in erythrocytes from essential hypertensive patients. Nature 284: 281
Garay RP, Elghozi J-L, Dagher G, Meyer P (1980b) Laboratory distinction between essential and secondary hypertension by measurement of erythrocyte cation fluxes. N Engl J Med 302: 769
Garay RP, Elghozi J-L, Dagher G, Meyer P (1980c) Abnormal erythrocyte cation fluxes in human hypertension. Adv Nephrol 10: 37
Gavlichek VA (1952) Changes of the blood pressure level during various functional states of the cortex in dogs. J Vyschey Nervnoy Dejatelmosty 2: 742
Geiger HG, Scotch NA (1963) The epidemiology of essential hypertension. A review with special attention to psychologic and sociocultural factors. I. Biologic mechanisms and descriptive epidemiology. J Chronic Dis 16: 1151
Gessler U von (1962) Intracellular and extracellular electrolyte changes in essential hypertension before and after therapy. Studies on erthyrocytes. Z Kreislaufforsch 51: 177
Glanville EV, Geerdink RA (1972) Blood pressure of Amerindians from Surinam. Am J Phys Anthropol 37: 251
Gleiberman L (1973) Blood pressure and dietary salt in human populations. Ecol Food Nutr 2: 143
Goldblatt H (1969) The effect of high salt intake on the blood pressure of rabbits. Lab Invest 21: 126
Goldring W, Chasis H (1944) Hypertension and hypertensive disease. The Commonwealth Fund, New York
Gonzales EA, Cech I, Smolensky MH (1979) Sodium in drinking water: Information for clinical application. South Med J 72: 753
Gover M (1948) Physical impairments of members of low-income farm families. VII. Variation of blood pressure and heart disease with age: Correlation of blood pressure with height and weight. Public Health Rep (Wash) 63: 1083
Grant H, Reischsmann F (1946) The effect of the ingestion of a large amount of sodium chloride on the arterial and venous pressure of normal subjects. Am Heart J 32: 704
Grasse PP (1956) Discussion. In: L'Instinct dans le comportement des animaux et de l'homme. Masson, Paris, p 629
Gray MJ, Plentl AA (1954) The variations of the sodium space and the total exchangeable sodium during pregnancy. J Clin Invest 33: 347
Greene L, Desor J, Maller O (1975) Heredity and experience: Their relative importance in the development of taste preference in man. J Comp Physiol Psychol 89: 279
Grim CE, McDonough JR, Dahl LK (1970) Dietary sodium, potassium and blood pressure: Racial differences in Evans County, Georgia. Circulation 42[Suppl III]: 85 (abstr)
Grim CE, Luft FC, Miller JZ, Brown PL, Gannon MA, Weinberger MH (1979) Effects of sodium loading and depletion in normotensive first-degree relatives of essential hypertensives. J Lab Clin Med 94: 764
Grollman A, Grollman EF (1962) The teratogenic induction of hypertension. J Clin Invest 41: 710
Grollman A, Harrison TJ (1945) Effect of rigid sodium restriction on blood pressure and survival of hypertensive rats. Proc Soc Exp Biol Med 60: 52

Grollman A, Harrison TR, Williams JR (1940) Effect of various sterol derivatives on blood pressure of rats. J Pharmacol Exp Ther 69: 149

Grollman A, Hudson TR, Mason MF, Baxter J, Crampton J, Reichsman F (1945) Sodium restriction in diet for hypertension. JAMA 129: 533

Gruber KA, Whitaker JM, Buckalew VM Jr (1980) Endogenous digitalis-like substance in plasma of volume expanded dogs. Nature 287: 743

Guthrie HA (1968) Infant feeding practices: A predisposing factor in hypertension. Am J Clin Nutr 21: 863

Guyton AC, Coleman PJ, Cowly AW, Manning RD Jr, Norman RA, Ferguson JD (1974) A systems analysis approach to understanding long range arterial blood pressure control and hypertension. Circ Res 35: 159

Haddy FJ (1974) Local control of vascular resistance as related to hypertension. Arch Intern Med 133: 916

Haldane JBS (1956) Les aspects physico-chimiques des instincts. In: L'Instinct dans le comportement des animaux et de l'homme. Masson, Paris, p 557

Hamilton M, Pickering GW, Roberts JAF, Sowry GSC (1954) The etiology of essential hypertension. The arterial pressure in the general population. Clin Sci 13: 11

Hatano S (1974) Hypertension in Japan: A review. In: Paul O (ed) Epidemiology and control of hypertension. Symposium Specialists, Miami, p 63

Hawthorne VM, Gillis CR, Lorimer AR, Calvert FR, Walker TJ (1969) Blood pressure in a Scottish island community. Br Med J iv: 651

Heller RF, Robinson N, Peart WS (1980) Value of blood pressure measurements in relatives of hypertensive patients. Lancet I: 1206

Henkin RI (1974) Salt taste in patients with essential hypertension and with hypertension due to primary hyperaldosteronism. J Chronic Dis 27: 235

Henkin RI (1975) The role of adrenal corticosteroids in sensory processes. In: Handbook of Physiology, Sect 7: Endocrinology, vol 6. American Physiological Society, Washington DC, p 209

Henningsen NC, Mattsson S, Nosslin B, Nelson D, Ohlsson O (1979) Abnormal whole-body and cellular (erythrocytes) turnover of $^{22}Na^+$ in normotensive relatives of probands with established essential hypertension. Clin Sci 57: 321s

Henry JP, Cassel JC (1969) Psychosocial factors in essential hypertension: Recent epidemiological and animal experimental evidence. Am J Epidemiol 90: 171

Henry JP, Stephens PM (1977) Stress, health and the social environment. A sociological approach to medicine. Springer, New York, p 208

Hicks CS, Matters RF (1933) Standard metabolism of the Australian aborigines. Aust J Exp Biol Med Sci 11: 177

Hipsley EH, Kirk NE (1965) Studies of dietary intake and expenditure of energy by New Guineans. South Pacific Commission, Noumea, New Caledonia (Technical paper no. 147)

Hirota Y (1976) A combined hypertension and stroke control programme in a Japanese community. In: Hatano S et al. (eds) Hypertension and stroke control in the community. WHO, Geneva, p 130

Hofman A, Valkenburg HA, Vaandrager GJ (1980) Increased blood pressure in schoolchildren related to high sodium levels in drinking water. J Epidemiol Community Health 34: 179

Holley HL, Elliot HC Jr, Holland CM Jr (1951) Serum sodium values in essential hypertension. Proc Soc Exp Biol Med 77: 561

Hong Myong Ho, Suh Soon Kyu (1972) Epidemiological and clinical studies of hypertension in Koreans. Journal of the Medical College of Korea University 9: 55

Hoobler SW, Tejada C, Guzman M, Pardo A (1963) Influence of nutrition and acculturation on the blood pressure levels and changes with age in the Highland Guatemalan Indian. Circulation 31 [Suppl 2]: 116

Hoshishima K, Yokoma S, Seto K (1962) Taste sensitivity in various strains of mice. Am J Physiol 202: 1200

Hughes I (1969) Some aspects of traditional trade in the New Guinea Central Highlands. In: Symposium on human adaptation and cultural change in Melanesia. Proceedings of the 41st ANZAAS Congress, Adelaide

Humphries SV (1957) A study of hypertension in the Bahamas. S Afr Med J 31: 694

Hunter JD (1962) Diet, body-build, blood pressure and serum cholesterol levels in coconut eating Polynesians. Fed Proc [2] 21: 36

Hyun Moo Sup, Choi Il Hoon (1969) Study on the blood pressure and sodium metabolism in long term prisoners. Journal of Woo Sok University Medical College 6: 589

Ikeda T, Tobian L, Iwai J, Gooseus P (1978) Central nervous system pressor responses in rats susceptible and resistant to sodium chloride hypertension. Clin Sci Mol Med 55: 225s

Interdepartmental Committee on Nutrition for National Defense (ICNND) (1958) Ethiopia. Nutrition Survey of the Armed Forces. US Government Printing Office, Washington

Interdepartmental Committee on Nutrition for National Defense (ICNND) (1959) Vietnam. Nutrition Survey of the Armed Forces, US Government Printing Office, Washington

Interdepartmental Committee on Nutrition for National Defense (ICNND) (1960a) Chile. Nutrition Survey of the Armed Forces. United States Government Printing Office, Washington

Interdepartmental Committee on Nutrition for National Defense (ICNND) (1960b) Columbia. Nutrition Survey of the Armed Forces. United States Government Printing Office, Washington

Interdepartmental Committee on Nutrition for National Defense (ICNND) (1960c) Thailand. Nutrition Survey of the Armed Forces. United States Government Printing Office, Washington

Irvin DL, Goetzl FR (1952) Diurnal variations in acuity of sense of taste for sodium chloride. Proc Soc Exp Biol Med 79: 115

Ivens WG (1929) Island builders of the Pacific. Lippincott, London

Ivinskis V, Kooptzoff O, Walsh RJ, Dunn D (1956) A medical and anthropological study of the Chimbu natives in the Central Highlands of New Guinea. Oceania 27: 143

Jackson JA (1967) Heavy smoking and sodium chloride hypogeusia. J Dent Res 46: 742

Janowitz HD, Grossman MI (1949) Hunger and appetite: Some definitions and concepts. Mt Sinai J Med (NY) 16: 231

Johnson BC, Remington RD (1961) A sampling study of blood pressure in white and Negro residents of Nassau, Bahamas. J Chronic Dis 13: 39

Johnson BC, Epstein FH, Kjelsberg MO (1965) Distributions and familial studies of blood pressure and serum cholesterol levels in a total community: Tecumseh, Michigan. J Chronic Dis 18: 147

Joossens JV (1979) Round table discussion of Robertson JIS, Wilhelmsen L, Morgan TO, Simpson FO: Salt intake and the pathogenesis and treatment of hypertension. Clin Sci 57: 453s

Joossens JV (1980) Stroke, stomach cancer and salt. In: Kertetoot H, Joossens JV (eds) Epidemiology of arterial blood pressure. Martinus Nijhoff Medical, The Hague

Joossens JV, Willems J, Claessens J, Claes J, Lissen SW (1970) Sodium and hypertension. In: Nutrition and cardiovascular diseases. Proceedings of the 7th International Meeting of Centrostudi lipidi alimentari-biologia E. Clinica della Nutrizione-fondazione Sasso, Rimini, p 91

Joossens JV, Kesteloop H, Amery A, (1979) Salt intake and

mortality from stroke. N Engl J Med 300: 1396
Josephson B (1971) Salt. Sartryck ur Nordisk Medicin 85: 3
Kagan A, Gordon T, Kannel WB, Dawber TR (1958) Blood pressure and its relation to coronary heart disease—the Framingham study. In: Hypertension, vol 7. Proceedings of the Council for High Blood Pressure Research. American Heart Association, Dallas, p 53
Kalmus H (1971) Genetics of taste. In: Beidler L (ed) Taste, Handbook of sensory physiology, vol IV. Springer, Berlin, p 165
Kaminer B, Lutz WPW (1960) Blood pressure in bushmen of the Kalahari Desert. Circulation 22: 289
Kaneta S, Ishiguro K, Kobayashi S, Takashi E (1964) An epidemiological study on nutrition and cerebrovascular lesions in Tohoku area of Japan. Tohoku J Exp Med 83: 398
Kannel WB, Brand N, Skinner JJ, Dawber PR, McNamara PM (1967) The relation of adiposity to blood pressure and development of hypertension: The Framingham study. Ann Intern Med 67: 48
Kannel WB, Schwartz MJ, McNamara PM (1969) Blood pressure and the risk of coronary heart disease: The Framingham study. Dis Chest 56: 43
Karvonen M, Keys A, Esko O, Flaminic F, Brozek J (1959) Cigarette smoking, serum cholesterol, blood pressure and body fatness: Observations in Finland. Lancet I: 492
Kass EJ, Zinner SH, Margolius HS, Lee YH, Rosner B, Donner A (1975) Familial aggregation of blood pressure and urinary kallikrein in early childhood. In: Paul O (ed) Epidemiology and control of hypertension. Stratton Intercontinental Medical Book Corp., New York, p 359
Kawasaki T, Delea CS, Bartter FC, Smith H (1978) The effect of high sodium and low sodium intakes on blood pressure and other related variables in human subjects with idiopathic hypertension. Am J Med 64: 193
Kean BH (1944) The blood pressure of the Cuna Indians. Am J Trop Med 24: 341
Kean BH (1951) Blood pressure studies on West Indians and Panamanians living on the isthmus of Panama. Arch Intern Med 68: 166
Kempner W (1944) Treatment of kidney disease and hypertensive vascular disease with rice diet: I. NC Med J 5: 125
Kempner W (1948) Treatment of hypertensive vascular disease with rice diet. Am J Med 4: 545
Kilcoyne MM (1977) Adolescent hypertension. In: New MI, Levine LS (eds) Juvenile hypertension. Raven Press, New York, p 25
Knudsen KD, Dahl LK (1966) Essential hypertension: Inborn error of sodium metabolism. Postgrad Med J 42: 148
Knudsen KD, Dahl LK, Thompson K, Iwai J, Heine M, Leitle G (1978) Effects of chronic salt ingestion: Inheritance of hypertension in rats. J Exp Med 132: 976
Kobayashi J (1957) On geographical relationship between the chemical nature of river water and death from apoplexy. Berichte des Ohara i. f. landwirtschaftlichte biologie Okayama Universität 2: 12
Kobayashi T (1976) Epidemiology of hypertension and stroke in Japan. In: Hatano S et al. (eds) Hypertension and stroke control in the community. WHO, Geneva, p 80
Kondo S (1962) Alimental habits and longevity. Jpn Heart J 3: 147
Krakoff LR, Goodwin FJ, Baer L, Torres M, Laragh JH (1970) The role of renin in the exaggerated natriuresis of hypertension. Circulation 42: 335
Krut LH, Perrin MJ, Bronte-Stewart B (1961) Taste perception in smokers and non-smokers. Br Med J i: 384
Langford HG, Watson RL (1974) Electrolytes and hypertension. In: Paul O (ed) Epidemiology and control of hypertension. Stratton Intercontinental Medical Book Corp., New York, p 119
Langston JB, Guyton AC, Douglas BH, Dawsett PE (1963) Effects of changes in salt intake on arterial pressure and renal function in partially nephrectomized dogs. Circ Res 12: 508
Laragh JH, Letcher RL, Pickering TG (1979) Renin profiling for diagnosis and treatment of hypertension. JAMA 241: 151
Larrick JW (1980) Letter to the editor. N Engl J Med 302: 1204
Lauer RM, Filer LJ, Reiter MA, Clark WR (1976) Blood pressure, salt preference, salt threshold, and relative weight. Am J Dis Child 130: 493
Ledingham JM (1953) Distribution of water sodium and potassium in heart and skeletal muscle in experimental renal hypertension in rats. Clin Sci 12: 337
Ledingham JM, Cohen RD (1963) The role of the heart in the pathogenesis of renal hypertension. Lancet II: 979
Lee RE, Schneider RF (1958) Hypertension and arteriosclerosis in executive and non-executive personnel. JAMA 167: 1447
Leksell LG, Congiu M, Denton DA, Fei DTW, McKinley MJ, Tarjan E, Weisinger RS (1981) Influence of mannitol-induced reduction in CSF Na on nervous and endocrine mechanisms involved in the control of fluid balance. Acta Physiol Scand 112: 33
Lemley-Stone J, Darby WJ, Meneely GR (1961) Effect of dietary sodium:potassium ratio on body content of sodium and potassium in rats. Am J Cardiol 8: 748
Lenel R, Katz LN, Rodbard S (1948) Arterial hypertension in chicken. Am J Physiol 152: 557
Lennard HL, Glock CY (1957) Studies in hypertension. VI. Differences in the distribution of hypertension in negroes and whites: An appraisal. J Chronic Dis 6: 186
Ling WK (1936) The blood pressure of the Chinese. Chin Med J [Engl] 50: 1773
Liu K, Cooper R, McKeever J, McKeever P, Byington R, Soltero I, Stamler R, Gosch F, Stevens E, Stamler J (1979) Assessment of the association between habitual salt intake and high blood pressure: Methodological problems. Am J Epidemiol 110: 219
Liu TY, Hung TP, Chen CM, Hau TC, Chen KP (1959) A study of normal and elevated blood pressure in a Chinese urban population in Taiwan (Formosa). Clin Sci 18: 301
Loggie JMH (1977) Prevalence of hypertension and distribution of causes. In: New MI, Levine LS (eds) Juvenile hypertension. Raven Press, New York, p 1
Londe S, Goldring D, Gollub SW, Hernandez A (1977) Blood pressure and hypertension in children: Studies, problems and perspectives. In: New MI, Levine LS (eds) Juvenile hypertension. Raven Press, New York, p 13
Lorenz KZ (1950) The comparative method in studying innate behaviour patterns. In: Physiological mechanisms in animal behaviour. Cambridge University Press, Cambridge, p 221 (Symposium of the Society for Experimental Biology 4)
Losse H, Wehmeyer H, Wessels F (1960) The water and electrolyte content of erythrocytes in arterial hypertension. Klin Wochenschr 38: 393
Lovell RRH, Maddocks I, Rogerson GW (1960) The casual arterial pressure of Fijians and Indians in Fiji. Aust Ann Med 9: 4
Lowenstein FW (1961) Blood pressure in relation to age and sex in the tropics and subtropics. A review of the literature and an investigation in two tribes of Brazil Indians. Lancet I: 389
Lowenstein J, Berenbaum ER, Chasis H, Baldwin SD (1970) Intrarenal pressure and exaggerated natriuresis in essential hypertension. Clin Sci 38: 359
Luft FC, Grim CE, Higgins JT, Weinberger MH (1977) Differences in response to sodium administration in normotensive white and black subjects. J Lab Clin Med 90: 555
Luft FC, Rankin LI, Bloch R, Weyman AE, Willis LR, Murray RH, Grim CE, Weinberger MH (1979) Cardiovascular and

humoral responses to extremes of sodium intake in normal black and white men. Circulation 60: 697

Lyons RH, Jacobson SD, Avery NL (1944) Increase in the plasma volume following the administration of sodium salts. Am J Med Sci 208: 148

Macfarlane WV, Skinner SL, Scoggins B (1970) Functional acculturation of Melanesians: Salts, hormones and blood pressure changes. In: Abstracts of the 42nd Congress of the Australian and New Zealand Association for the Advancement of Science

MacGillivnay I, Buchanan TJ (1958) Total exchangeable sodium and potassium in non-pregnant women and in normal and preeclamptic pregnancy. Lancet II: 1090

Maddocks I (1961) Possible absence of essential hypertension in two complete Pacific island populations. Lancet II: 396

Maddocks I (1967) The blood pressure of Melanesians. Med J Aust 1: 1123

Maddocks I, Rovin L (1965) A New Guinea population in which blood pressure appears to fall as age advances. Papua New Guinea Med J 8: 17

Maddocks I, Vines AP (1966) The influence of chronic infection on blood pressure in New Guinea males. Lancet II: 262

Margetts BM, Armstrong BK, Binns CW, Masarei JR, McCall MG (1980) Relationship between blood pressure and water sodium in rural school children. In: 5th Annual Conference of the Nutrition Society of Australia, Melbourne, p 161

Marvin HP, Smith ER (1942) Hypertensive cardiovascular disease in Panamanians and West Indians residing in Panama and the Canal Zone. Mil Surgeon 91: 529

Marx JL (1976) Hypertension: A complex disease with complex causes. Science 194: 821

Marx JL (1981) Natriuretic hormone linked to hypertension. Science 212: 1255

Mayer J (1968) Overweight: Causes, costs and controls. Prentice Hall Inc, New York, p 124

Mayer J (1969) Hypertension, salt intake, and the infant. Postgrad Med 45: 229

McCance RA, Widdowson EM (1957) Physiology of the newborn infant. Lancet II: 585

McConnell SD, Henkin RI (1972) Increased salt preference behaviour in spontaneously hypertensive rats (SHR). Clin Res 20: 387

McConnell SD, Henkin RI (1973) NaCl preference in spontaneously hypertensive rats: Age and blood pressure effects. Am J Physiol 225: 624

McDonough J, Wilhelmj CM (1954) The effect of excessive salt intake on human blood pressure. Dig Dis Sci 21: 180

McDonough JR, Garrison GE, Hames CG (1967) Blood pressure and hypertension among Negroes and whites in Evans County, Georgia. In: Stamler I, Stamler R, Pullman TN (eds) Epidemiology of hypertension: Proceedings of an International Symposium. Grune & Stratton, New York, p 167

McKinley MJ, Blaine EH, Denton DA (1973) Stimulation of renal sodium excretion by hypertonic stimuli to the third ventricle and effects of renal denervation on this response. Proc Aust Physiol Pharmacol Soc 4: 66

McKinley MJ, Denton DA, Hatzikostas S, Weisinger RS (1979) Effect of angiotensin II on parotid saliva secretion in conscious sheep. Am J Physiol 237: E56

McKinley MJ, Denton DA, Graham WF, Leksell LG, Mouw DR, Scoggins BA, Smith MH, Weisinger RS, Wright RD (1980) Lesions of the organ vasculosum of the lamina terminalis inhibit water drinking to hypertonicity in sheep. In: Abstracts of the Proceedings of the 7th International Conference on the Physiology of Food and Fluid Intake, Warsaw

Meehan JP, Fineg J, Mosely JD (1964) The effect of restraint and training on the arterial pressure of the immature chimpanzee. Fed Proc 23: 515

Meggitt MJ (1958) Salt manufacture and trading in the Western Highlands of New Guinea. Aust Museum Magazine 12: 309

Mendonca M de, Grichois M-L, Garay RP, Sassard J, Ben-Ishay D, Meyer P (1980) Abnormal net Na^+ and K^+ fluxes in erythrocytes of three varieties of genetically hypertensive rats. Proc Natl Acad Sci USA 77: 4283

Meneely GR, Battarbee HD (1976a) Sodium and potassium. Nutr Rev 34: 225

Meneely GR, Battarbee HD (1976b) High sodium–low potassium environment and hypertension. Am J Cardiol 38: 768

Meneely GR, Dahl LK (1961) Electrolytes in hypertension: The effects of sodium chloride. Med Clin North Am 45: 271

Meneely GR, Tucker RG, Darby WJ, Auerbach SH (1953) Chronic sodium chloride toxicity in albino rats. II. Occurrence of hypertension and syndrome of oedema in renal failure. J Exp Med 98: 71

Meyer P, Garay RP, Nazaret C, Dagher G, Bellet M, Broyer M, Feingold J (1981) Inheritance of abnormal erythrocyte cation transport in essential hypertension. Br Med J 282: 1114

Miall WE (1959) Follow-up study of arterial pressure in the population of a Welsh mining valley. Br Med J ii: 1205

Miall WE (1961) The epidemiology of essential hypertension. In: Proceedings of a symposium on pathogenesis of essential hypertension. State Medical Publishing House, Prague, p 89

Miall WE, Lovell HG (1967) Relation between change of blood pressure and age. Br Med J ii: 660

Miall WE, Oldham PD (1955) A study of arterial blood pressure and its inheritance in a sample of the general population. Clin Sci 14: 459

Miall WE, Oldham PD (1958) Factors influencing arterial blood pressure in the general population. Clin Sci 17: 409

Miall WE, Oldham PD (1963) The hereditary factor in arterial blood pressure. Br Med J i: 75

Miall WE, Kass EH, Ling J, Stuart KL (1962) Factors influencing arterial pressure in the general population in Jamaica. Br Med J ii: 497

Miashnikov AL (1962) Significance of disturbances of higher nervous activity in the pathogenesis of hypertensive disease. In: Cort JH, Fencl V, Hejl Z, Jirka J (eds) Symposium on the pathogenesis of essential hypertension. Pergamon Press/ Macmillan, New York, p 153

Michelakis AM, Mizukoshi H, Huang C, Nurakami K, Inagami T (1975) Further studies on the existence of a sensitizing factor to pressor agents in hypertension. J Clin Endocrinol Metab 41: 90

Miguel JP, Padua F de (1980) Epidemiology of arterial blood pressure in Portugal. In: Kerteloot A, Joossens JV (eds) Epidemiology of arterial blood pressure. Martinus Nijhoff Medical, The Hague

Miminoshvili DI (1960) Experimental neurosis in monkeys. In: Utkin IA (ed) Theoretical and practical problems of medicine in biology in experiments on monkeys. Pergamon Press, New York, p 53

Mistretta CM, Bradley RM (1977) Taste *in utero*: Theoretical considerations. In: Weiffenbach JM (ed) Taste and development. The genesis of sweet preference. US Department of Health, Education and Welfare, Bethesda (DHEW publication no. [NIH] 77-1068, p 51)

Morgan T, Carney S, Wilson M (1975) Interrelationship in humans between sodium intake and hypertension. Clin Exp Pharmacol Physiol [Suppl] 2: 127

Morgan T, Adam W, Gillies A, Wilson M, Morgan G, Carney S (1978) Hypertension treated by salt restriction. Lancet I: 227

Morris JN (1961) Opening remarks for session on epidemiology of essential hypertension. In: Proceedings of a symposium on the pathogenesis of essential hypertension. State Medical Publishing House, Prague, p 80

Morris JN, Crawford MD, Heady JA (1961) Hardness of local water supplies and mortality from cardiovascular disease. Lancet I: 860
Morse WR, Beh YT (1937) Blood pressure amongst aboriginal ethnic groups of Szechwan province, West China. Lancet I: 966
Moser M (1978) In: Freis ED (ed) The treatment of hypertension. University Park Press, Baltimore, p 96
Moser M, Harris M, Pugatch D, Ferber A, Gordon B (1962) Epidemiology of hypertension. II. Studies of blood pressure in Liberia. Am J Cardiol 10: 424
Moskowitz HR, Abramson R (1976) Altered taste preference in obese and renal patients. Paper presented to the Eastern Psychological Association meeting, New York, April
Moszkowski M (1911) Die Völkerstämme am Mamberano in Höllandisch-Neuguinea und auf den vorgelagerten Inseln. Zeitschrift für Ethnologie xliii: 315
Murphy RJF (1950) Effect of 'rice diet' on plasma volume and extracellular fluid space in hypertensive subjects. J Clin Invest 29: 912
Murray RH, Luft FC, Bloch R, Weyman AE (1978) Blood pressure responses to extremes of sodium intake in normal man. Proc Soc Exp Biol Med 159: 432
Murrill RI (1949) A blood pressure study of the natives of Ponape Island, Eastern Carolines. Hum Biol 21: 47
New MI, Baum CJ, Levine LS (1976) Nomograms relating aldosterone excretion to urinary sodium and potassium in the pediatric population: Their applications to the study of childhood hypertension. Am J Cardiol 37: 658
Nowlis GH (1973) Taste elicited tongue movements in human newborn infants: An approach to palatability. In: Bosma, JF (ed) Fourth symposium on oral sensation and perception. US Government Printing Office, Washington, p 292
Nowlis GH (1977) Discussion of Crook CK: Taste and the temporal organization of neonatal sucking. In: Weiffenbach JM (ed) Taste and development: The genesis of sweet preference. US Department of Health Education and Welfare, Bethesda (DHEW publication no. (NIH) 77-1068, p 159)
Nye LJJ (1937) Blood pressure in the Australian aboriginal with a consideration of possible ecological factors in hyperpiesia and its relation to civilization. Med J Aust 2: 1000
Nye LJJ, Forrest V (1951) Treatment of vascular hypertension by low-sodium diet. Med J Aust 2: 152
Oliver WJ, Cohen EL, Neel JB (1975) Blood pressure, sodium intake, and sodium related hormones in the Yanomamö: A 'no salt' culture. Circulation 52: 146
Oliver WJ, Neel JV, Grekin RJ, Cohen EL (1981) Hormonal adaptation to the stresses imposed upon sodium balance by pregnancy and lactation in the Yanomamö Indians, a culture without salt. Circulation 63: 110
Onesti G, Kun KE, Greco JA, del Guercio ET, Fernandez M, Swartz C (1975) Blood pressure regulation in end-stage renal disease and anephric man. Circ Res 36 [Suppl 1]: I–145
Onoyama K, Bravo EL, Tarazi RC (1979) Sodium, extracellular fluid volume, and cardiac output changes in the genesis of mineralocorticoid hypertension in the intact dog. Hypertension 1: 331
Oomen HAPC (1961a) The nutritional situation in western New Guinea. Trop Geogr Med 13: 321
Oomen HAPC (1961b) The Papuan child as a survivor. J Trop Paediatrics and Environmental Child Health 6: 103
Oomen HAPC (1967) Nitrogen compounds and electrolytes in the urine of New Guinean sweet potato eaters: A study of normal values. Trop Geogr Med 19: 31
Oomen HAPC, Spoon W, Heesterman JE, Ruinard J, Luyken R, Slump P (1961) The sweet potato as the staff of life of the Highland Papuan. Trop Geogr Med 13: 55
Orr JB, Gilks JL (1931) Studies of nutrition: The physique and health of two African tribes. HMSO, London (Medical Research Council, special report, series 155)
Ostfeld A, Paul O (1963) The inheritance of hypertension. Lancet I: 575
Overbeck HW, Pamnani MB, Akera T, Brody TM, Haddy FJ (1976) Depressed function of a ouabain-sensitive sodium–potassium pump in blood vessels from renal hypertensive dogs. Circ Res 38 [Suppl II]: II–48
Owens CJ, Newton CB (1978) Role of sodium intake in the antihypertensive effect of propranolol. South Med J 71: 43
Padmavati S, Gupta S (1959) Blood pressure studies in rural and urban groups in Delhi. Circulation 19: 395
Page LB (1976) Epidemiological evidence on the etiology of human hypertension and its possible prevention. Am Heart J 91: 527
Page LB, Damon A, Moellering RC Jr (1974) Antecedents of cardiovascular disease in six Solomon Island societies. Circulation 49: 1132
Page LB, Vandevert D, Nader K, Lubin N, Page JR (1978) Blood pressure, diet, and body form in traditional nomads of the Qash'gai Tribe, Southern Iran. Acta Cardiol (Brux) 33: 102
Parijs J, Joossens JV, Linden L van der, Verstreken G, Amery A (1973) Moderate sodium restriction and diuretics in treatment of hypertension. Am Heart J 85: 22
Parsons JA (1951) Sodium chloride: The salt of life. Discovery (Nov–Dec): 360
Paul O (1967) Discussion of paper by GR Meneely: The experimental epidemiology of sodium chloride toxicity in the rat. In: Stamler J, Stamler R, Pullman TN (eds) The epidemiology of hypertension. Grune & Stratton, New York, p 248
Paul O, Ostfeld AM (1965) Epidemiology of hypertension. Prog Cardiovasc Dis 8: 106
Perera GA (1961) General discussion. In: Proceedings of a symposium on pathogenesis of essential hypertension. State Medical Publishing House, Prague, p 56
Pfaffmann C (1960) The pleasures of sensation. Psychol Rev 67: 253
Phear DN (1958) Salt intake in hypertension. Br Med J ii: 1453
Phillips GD, Sundaram SK (1966) Sodium depletion of pregnant ewes and its effects on foetuses and foetal fluids. J Physiol (Lond) 184: 889
Pickering GW (1955) High blood pressure. Churchill, London
Pickering GW (1961a) The nature of essential hypertension. Churchill, London
Pickering GW (1961b) Concept of essential hypertension as a quantitative disease. In: Proceedings of a symposium on pathogenesis of essential hypertension. State Medical Publishing House, Prague, p 43
Pickering, GW (1965) Hyperpiesis: High blood pressure without evident cause: Essential hypertension. Br Med J ii: 959
Pickering GW (1968) High blood pressure. Churchill, London
Pickering GW (1980) Salt intake and essential hypertension. Cardiovascular Reviews and Reports 1: 1
Pickering GW, Roberts JAF, Sowry GSC (1954) The aetiology of essential hypertension. III. The effect of correcting for arm circumference on the growth rate of arterial pressure with age. Clin Sci 13: 267
Pietinen PI, Tindley TW, Clausen JD, Finnerty FA, Altschul AM (1976) Studies in community nutrition: Estimation of sodium output. Prev Med 5: 400
Platt R (1963) Heredity in hypertension. Lancet I: 899
Postnov YV, Orlov S, Gulak P, Shevchenko A (1976) Altered permeability of the erythrocyte membrane for sodium and potassium ions in spontaneously hypertensive rats. Pflügers Arch 365: 257
Postnov YV, Orlov SN, Shevchenko A, Adler AM (1977) Altered sodium permeability, calcium binding and Na–K-ATPase activity in the red blood cell membrane in

essential hypertension. Plfügers Arch 371: 263
Poston L, Swell RB, Wilkinson SP, Richardson PJ, Williams R, Clarkson EM, MacGregor GA, de Wardener HE (1981) Evidence for a circulating sodium transport inhibitor in essential hypertension. Br Med J 282: 847
Potter BJ (1972) The effect of prolonged salt intake on blood pressure in sheep. Aust J Exp Biol Med Sci 50: 387
Prior IAM, Evans JG, Harvey HPB, Davidson F, Lindsey M (1968) Sodium intake and blood pressure in two Polynesian populations. N Engl J Med 279: 515
Prior IAM, Stanhope JM, Evans JG, Salmond CE (1974) The Tokelau Island migrant study. Int J Epidemiol 3: 255
Puyau FA, Hampton LP (1966) Infant feeding practices. Am J Dis Child 111: 370
Rapp JP, Dahl LK (1972) Mendelian inheritance of 18- and 11-beta steroid hydroxylase activities in the adrenals of rats genetically susceptible or resistant to hypertension. Endocrinology 90: 1435
Reisin AR, Moden M, Silverburg DS, Eliahou HE, Moden B (1978) Effect of weight loss without salt restriction on the reduction of blood pressure in overweight hypertensive patients. N Engl J Med 298: 1
Renwick R, Robson JS, Stewart CP (1955) Observations upon the withdrawal of sodium chloride from the diet in hypertensive and normotensive individuals. J Clin Invest 34: 1037
Reynolds WA, Filler LJ (1965) Taste preference studies in preschool children. Motion picture film in colour and sound. University of Illinois
Richter CP (1956) Salt appetite in mammals: Its dependence on instinct and metabolism. In: L'Instinct dans le comportement des animaux et de l'homme. Masson, Paris, p 557
Richter CP, MacLean A (1939) Salt taste thresholds of humans. Am J Physiol 126:1
Rose G (1962) The distribution of mortality from hypertension within the United States. J Chronic Dis 15: 1017
Rose GA, Blackburn H (1969) Cardiovascular survey methods. WHO, Geneva
Sagan LA, Seigel DG (1958–1960) Adult health study: clinical and laboratory data. Atomic Bomb Casualty Commission. Japanese National Institute of Health, Nagasaki (Technical reports 12–63)
Sapirstein LA, Brandt WL, Drury DR (1958) Production of hypertension in rat by substituting hypertonic sodium chloride solution for drinking water. Proc Soc Exp Biol Med 73: 82
Sasaki N (1962) High blood pressure and the salt intake of the Japanese. Jpn Heart J 3: 313
Saunders GM (1933) Blood pressure in Yucatecans. Am J Med Sci 185: 843
Saunders GM, Bancroft H (1942) Blood pressure studies on negro and white men and women living in the Virgin Islands of the US. Am Heart J 23: 410
Schechter PJ, Horwitz D, Henkin RI (1974) Salt preference in patients with untreated and treated essential hypertension. Am J Med Sci 267: 320
Schlierf G, Arab L, Schelkenberg B, Oster P, Mordasini R, Schmidt-Gayk H, Vogel G (1980) Salt and hypertension: Data from the 'Heidelberg study'. Am J Clin Nutr 33: 872
Schneckloth RE, Corcoran AC, Stuart KL, Moore FE (1962) Arterial pressure and hypertensive heart disease in a West Indian negro population. First report of a survey in St Kitts, West Indies. Am Heart J 63: 607
Schroeder HA (1960) Relations between hardness of water and death rates from certain chronic and degenerative disease in the United States. J Chronic Dis 12: 586
Scoggins BA, Butkus A, Coghlan JP, Denton DA, Fan JSK, Humphery TJ, Whitworth JA (1978) ACTH-induced hypertension in sheep: A model for the study of the effect of steroid hormones on blood pressure. Circ Res 43/1: 176
Scotch N (1960) A preliminary report on the relation of sociocultural factors to hypertension among the Zulu. Ann NY Acad Sci 84: 1000
Scotch N (1963) Sociocultural factors in the epidemiology of Zulu hypertension. Am J Public Health 53: 1205
Scotch NA, Geiger HG (1963) The epidemiology of essential hypertension: A review with special attention to psychologic and sociocultural factors. II. Psychologic and sociocultural factors in aetiology. J Chronic Dis 16: 1183
Sever PS, Gordon D, Peart WS, Beighton P (1980) Blood pressure and its correlates in urban and tribal Africa. Lancet II: 60
Shaper AG (1964) Blood pressure studies in East Africa. Paper presented at the International Symposium on the Epidemiology of Hypertension, Chicago, p 3
Shaper AG (1972) Cardiovascular disease in the tropics. III. Blood pressure and hypertension. Br Med J iii: 805
Shaper AG, Saxton GA (1969) Blood pressure and body build in a rural community in Uganda. East Afr Med J 46: 228
Shaper AJ, Leonard PJ, Jones KW, Jones M (1969) Environmental effects on body build, blood pressure and blood chemistry of nomadic warriors serving in the army in Kenya. East Afr Med J 46: 282
Shattuck GC (1937) Possible significance of low blood pressure observed in Guatemalans and in Yucatecans. Am J Trop Med Hyg 17: 513
Short RV (1976) The evolution of human reproduction. Proc R Soc Lond [Biol] 195: 3
Short RV (1979) The future fertility of mankind. Roy Soc Encouragement of Arts J 127: 415
Simpson FO (1979) Salt and hypertension: A sceptical review of the evidence. Clin Sci 57: 463s
Simpson FO, Waal-Manning HG, Volli P, Phelan EL, Spears GFS (1978) Relationship of blood pressure to sodium excretion in a population survey. Clin Sci 55: 373s
Sinnett PF, Whyte HM (1973a) Epidemiological studies in a total highland population: Tukisenta, New Guinea. Cardiovascular disease and relevant clinical electrocardiographic, radiologic and biochemical findings. J Chronic Dis 26: 265
Sinnett PF, Whyte HM (1973b) Epidemiological studies in a Highland population in New Guinea: Environment, culture and health status. Human Ecology 1: 245
Skrabal F, Auböck J, Hortnagh H (1981) Low sodium/high potassium diet for prevention of hypertension: Probable mechanisms of action. Lancet II: 895
Smith JR (1953) Salt. Nutr Rev 11: 33
Stamler R, Stamler J, Riedlinger WF, Algera G, Roberts RH (1978) Weight and blood pressure: Findings in hypertension screening of 1 million Americans. JAMA 240: 1607
Stanhope JM (1968) Blood pressure of the Tinan-Aigram people, near Simbai, Madang district. Papua New Guinea Med J 11: 60
Steinbach M, Constantineau M, Harnagea P, Teodorina S, Georgecu M, Cretescu R, Manicatide M, Nicolaescu V, Suciu A, Vladescu C, Voiculescu M, Georgescu A (1974) On the aetiology of hypertension. Rev Roum Med Intern 12: 3
Stellar E, Hyman R, Samut S (1954) Gastric factors controlling water and salt solution drinking. J Comp Physiol Psychol 47: 220
Stewart RA (1967) Letter to the editor. Nutr Rev 25: 254
Stocks P, Davies RI (1964) Zinc and copper content of soils associated with incidence of cancer of the stomach and other organs. Br J Cancer 18: 14
Strotz CR, Shorr GI (1973) Hypertension in the Pago Indians. Circulation 48: 1299
Sullivan JM, Ratts TE, Taylor C, Kraus DH, Barton BR, Patrick DR, Reed SW (1980) Hemodynamic effects of dietary sodium in man. Hypertension 2: 506
Swales JD (1980) Dietary salt and hypertension. Lancet I: 1177

Swaye PS, Gifford RW, Berrettoni JN (1972) Dietary salt and essential hypertension. Am J Cardiol 29: 33

Switzer S (1963) Hypertension and ischemic heart disease in Hiroshima, Japan. Circulation 28: 368

Takahashi B (1962) An epidemiological approach to the relation between diet and cerebrovascular lesions and arteriosclerotic heart disease. Tohoku J Exp Med 77: 239

Takahashi E, Sasaki N, Takeda J, Ito H (1957) The geographic distribution of cerebral hemorrhage and hypertension in Japan. Hum Biol 29: 139

Thomas CB (1973) Genetic pattern of hypertension in man. In: Onesti G, Kim KE, Moyer JH (eds) Hypertension: Mechanisms and management. Grune & Stratton, New York, p 67

Thomas RD, Edmondson RPS, Hilton PJ, Jones NF (1975) Abnormal sodium transport in leucocytes from patients with essential hypertension and the effect of treatment. Clin Sci Mol Med 48: 169s

Thomas WA (1927) Health of a carnivorous race: A study of the Eskimo. JAMA 88: 1559

Tobian L (1974) Hypertension and the kidney. Arch Intern Med 133: 959

Tobian L, Binnion J (1952) Tissue cations and water in arterial hypertension. Circulation 5: 754

Tobian L, Lange J, Azar JT, Koop D, Coffee K (1977) Reduction of intrinsic natriuretic capacity in kidney of Dahl hypertension-prone rats. Circulation [Suppl III] 55/56: 240 (abstr)

Tobian L, Lange J, Azar S, Iwai J, Koop D, Coffee K, Johnson MA (1978) Reduction of natriuretic capacity and renin release in isolated blood perfused kidneys of Dahl hypertension-prone rats. Circ Res 43: 92

Tobian L, Lange J, Iwai J, Miller K, Johnson MA, Goossens P (1979) Prevention with thiazide of NaCl-induced hypertension in Dahl 'S' rats: Evidence for a Na-retaining humoral agent in 'S' rats. Hypertension 1: 316

Truswell AS, Kennelly BM, Hansen JDL, Lee RB (1972) Blood pressures of !Kung bushmen in Northern Botswana. Am Heart J 84: 5

Tuck ML, Sowers J, Dornfield L, Kledzik G, Maxwell M (1981) The effect of weight reduction on blood pressure, plasma renin activity, and plasma aldosterone levels in obese patients. New Engl J Med 304: 930

Tuthill RW, Calabrese EJ (1979) Elevated sodium levels in the public drinking water as a contributor to elevated blood pressure levels in the community. Arch Environ Health 34: 197

Vandongen R, Davivonges V, Abbie AA (1962) Aboriginal blood pressure at Beswick, South Western Arnhem Land, and correlation with physical dimensions. Med J Aust 49: 286

Volhard F (1931) Handbuch der inneren Medizin, vol VI. Springer, Berlin, p 1753

Wardener HE de, MacGregor GA (1980) Dahl's hypothesis that a saluretic substance may be responsible for a sustained rise in arterial pressure: Its possible role in essential hypertension. Kidney Int 18: 1

Watanabe S (1953) Japan: An apoplexy country. Weekly Asahi April 5, p 44

Watkin DM, Froeb HF, Hatch FT, Gutman AB (1950) Effects of diet in essential hypertension. Am J Med 9: 441

Wessels F, Junge-Hulsing G, Losse H (1967) Untersuchungen zur Natriumpermeabilität der Erythrozyten bei Hypertonikern und Normotonikern mit familiärer Hochdruckbelastung. Z Kreislaufforsch 56: 374

White JM, Wingo JG, Alligood LM, Cooper GR, Gutride J, Haydaker W, Benack TT, Dening JW, Taylor FB (1967) Sodium ion in drinking water. J Am Diet Assoc 51: 32

Whitworth JA, Coghlan JP, Denton DA, Hardy KJ, Scoggins BA (1979) Effect of sodium loading and ACTH on blood pressure of sheep with reduced renal mass. Cardiovasc Res 13: 9

Whyte HM (1958) Body fat and blood pressure of natives in New Guinea: Reflections on essential hypertension. Aust NZ J Med 7: 36

Widdowson EM, McCance RA, Spray CM (1951) Chemical composition of human body. Clin Sci 10: 113

Wiggins RC, Basar I, Slater JDH (1978) Effect of arterial pressure and inheritance on the sodium excretory capacity of normal young men. Clin Sci 54: 639

Williams AW (1941) The blood pressure of Africans. East Afr Med J 18: 109

Williams AW (1944) Hypertensive heart disease. IV. Heart disease in the native population of Uganda. East Afr Med J 21: 328

Wills PA (1958) Salt consumption by natives of the Territory of Papua New Guinea. Philippine Journal of Science 87: 169

Wilson JMG (1958) Arterial blood pressures in plantation workers in north-east India. Br J Prev Soc Med 12: 204

Wintour EM, Brown EH, Denton DA, Hardy KJ, McDougall JG, Robinson PM, Rowe EJ, Whipp GT (1977) In vitro and in vivo adrenal cortical steroid production by fetal sheep: Effect of angiotensin II, sodium deficiency and ACTH. In: Conti C (ed) Research on steroids VII. Elsevier, Amsterdam, p 475

Wolf G, Dahl LK, Miller NE (1965) Voluntary sodium intake in two strains of rats with opposite genetic susceptibility to experimental hypertension. Proc Soc Exp Biol Med 120: 301

Woods KL, Beevers DG, West M (1981) Familial abnormality of erythrocyte cation transport in essential hypertension. Br Med J 282: 1186

World Health Organization (1961) Annual Epidemiology Vital Statistics. WHO, Geneva

World Health Organization (1962) Calcium requirements: report of an FAO/WHO expert group. WHO, Geneva (Technical report series no. 230)

Wotman S, Mandel ID, Thompson RH, Laragh JH (1967) Salivary electrolytes and salt taste thresholds in hypertension. J Chronic Dis 20: 833

Yamori Y, Katsumi I, Horie R, Ooshima A, Ohtaka M, Nara Y (1979) Salt preference in stroke-prone and -resistant SHR. Jpn Heart J 20: 751

Yamori Y, Nara Y, Kihara M, Horie R, Ooshima A (1980) Sodium and other dietary factors in experimental and human hypertension: The Japanese experience. In: Laragh J (ed) Frontiers in hypertension research. Proceedings of a Conference at Cornell University School. Springer-Verlag, New York, p 46

Young PT (1944) Studies of food preference, appetite, and dietary habit. I. Running activity habit of the rat in relation to food preference. J Comp Physiol Psychol 37: 327

Young PT (1946) Studies of food preference, appetite and dietary habit. VI. Habit, palatability and the diet on factors regulating selection of food by the rat. J Comp Physiol Psychol 39: 139

Young PT (1977) Role of hedonic processes in the development of sweet taste preferences. In: Weifenbach JM (ed) Taste and development. US Department of Health, Education and Welfare, Bethesda (DHEW publication no. [NIH] 77-1068), p 399

Zinner SH, Martin LF, Sacks F, Rosner B, Kass EH (1975) A longitudinal study of blood pressure in children. Am J Epidemiol 100: 437

Subject Index